INTERNATIONAL ENCYCLOPEDIA OF ROBOTICS: APPLICATIONS AND AUTOMATION

VOLUME 1

EDITORIAL BOARD

INTERNATIONAL ENCYCLOPEDIA OF ROBOTICS: APPLICATIONS AND AUTOMATION

VOLUME 1

Richard C. Dorf, *Editor-in-Chief*

Shimon Y. Nof, *Consulting Editor*

A Wiley-Interscience Publication

John Wiley & Sons

New York / Chichester / Brisbane / Toronto / Singapore

Library of Congress Cataloging in Publication Data:

International encyclopedia of robotics. "A Wiley-Interscience publication."
1. Robotics—Dictionaries. I. Dorf, Richard C.
II. Nof, Shimon Y., 1946–
TJ210.4.I57 1988 629.8′92′0321 87–37264
ISBN 0–471–87868–5 (set)
ISBN 0–471–63512–4 (Vol. 1)

Printed in the United States of America

10 9 8 7 6 5 4 3 2 1

EDITORIAL STAFF

CONTRIBUTORS

James Acton, *Chula Vista, California,* Shipbuilding, robots in

Philip Adsit, *University of Florida, Gainesville, Florida,* Robots in agriculture

S. I. Ahmad, *Eastern Michigan University, Ypsilanti, Michigan,* Programming, task and motion planning

Shuhei Aida, *University of Electro-communications, Tokyo, Japan,* Robots in Japan

Mary K. Allen, *U.S.A.F., A.F.L.C., Wright Patterson AFB, Dayton, Ohio,* Machine loading/unloading

Joseph Alvite, *Mecanotron Corp., Roseville, Minnesota,* Interfacing, robots and sensors

Fred Aminzadeh, *Unocal Corporation, Brea, California,* Petroleum industry, robots in

Fred M. Amram, *University of Minnesota, Minneapolis, Minnesota,* Automation, human aspects

Joseph Anderson, *Tennessee Technological University, Cookeville, Tennessee,* Specifications

Roger Anderson, *Bryant College, Smithfied, Rhode Island,* Workers unions, and robots

Phillip Andrews, *Deloitte Haskins & Sells, Detroit, Michigan,* Integration, robots and CIM

James M. Apffel, *University of California, Davis, California,* Vision systems, industrial application

James M. Apple, *Systecon, Inc., Duluth, Georgia,* Material handling

Robert Ayres, *Carnegie-Mellon University, Pittsburgh, Pennsylvania,* Employment, impact

Alfred J. Baker, *New York City Police Department, New York, New York,* Police robots

Roman Baldur, *Ecole Polytechnic, Montréal, Québec, Canada,* Sensors, for safety

Bruce Ballard, *AT&T Bell Laboratories, Murray Hill, New Jersey,* Computational linguistics

Luc Baron, *Ecole Polytechnic, Montréal, Québec, Canada,* Sensors, for safety

Antal K. Bejczy, *Jet Propulsion Laboratory, Pasadena, California,* Software elements

Joseph Bellino, *General Electric, Plainville, Connecticut,* Safety, of operators

H. Bera, *South Bank Polytechnic, London, United Kingdom,* Sensors, special purpose

Philip L. Bereano, *University of Washington, Seattle, Washington,* Government policies

Edward Bernardon, *The Charles Stark Draper Laboratory, Inc., Cambridge, Massachusetts,* Apparel industry, robots in

John E. Biegel, *University of Central Florida, Orlando, Florida,* Learning and adaption

J. Browne, *University College, Galway, Galway City, Republic of Ireland,* Kitting

Timothy J. Bublick, *DeVilbiss Co., Toledo, Ohio,* Painting

William Burns, *Greenville Technical College, Greenville, South Carolina,* Maintenance and repair, robotic

Frank Busby, *Arlington, Virginia,* Underwater exploration, robots in

Perry Carter, *Brigham Young University, Provo, Utah,* Assembly, robotic, design for

John Cesarone, *Northwestern University, Evanston, Illinois,* Trajectories

Paul Chapman, *Micro Switch, Honeywell Corporation, Freeport, Illinois,* Sensors, evolution

J. Y. Chen, *Memorial Medical Center, Long Beach, California,* Surgery, robots in

Philip R. Chimes, *Pittsburgh, Pennsylvania,* Multiple robots systems, contol of

Tom Clareson, *College of Wooster, Wooster, Ohio,* Sci-Fi robot

John J. Craig, *Silma Inc., Los Altos, California,* Multicoordinated robotic devices

R. M. Crowder, *University of Southampton, Southampton, United Kingdom,* Sensors, touch, force, and torque measurement

Robert Crowder, *Ship Star Associates, Newark, Delaware,* Manufacturing automation protocol

John Cuadrado, *Octy, Inc., Fairfax Station, Virginia,* Vision systems, programming

Nicholas Dagalakis, *National Bureau of Standards, Gaithersburg, Maryland,* Testing

A. Davies, *University of Wales, Institute of Science and Technology, Cardiff, Wales, United Kingdom,* Market forecasts

Ray Davis, *Cheseborough Ponds, Inc., Clinton, Connecticut,* Packaging with robots

Rui J. P. Defigueiredo, *Rice University, Houston, Texas,* Space robots

Alan A. Desrochers, *Rensselaer Polytechnic Institute, Troy, New York,* Motion control

George Devol, *Devol Research Association, Wilton, Connecticut,* Robot systems, evolution

Robert Doornick, *International Robotics, New York, New York,* Show, home, and communication robots

Morris Driels, *Texas A&M University, College Station, Texas,* Research programs

John Dudley, *Seiko Instruments, U.S.A., Inc., Torrance, California,* Electronics industry, robots in

Marilyn Dulitzky, *Thomas J. Lipton, Inc., Englewood Cliffs, New Jersey,* Precision instruments industry, robots in

James Dunseth, *Vision Systems International, Yardley, Pennsylvania,* Vision systems, theory

Kornel F. Eman, *Northwestern University, Evanston, Illinois,* Trajectories

John Ettlie, *Industrial Technology Institute, Ann Arbor Michigan,* Management and robotics

H. R. Everett, *Naval Ocean Systems Center, San Diego, California,* Security and sentry robots

Daniele Fabrizi, *University of Pavia and the Italian Association of Robotics, International Federation of Robotics, Bergamo, Italy,* Appliance industry robots in

Charles H. Falkner, *University of Wisconsin-Madison, Madison, Wisconsin,* Leasing of robots

Francis Farrell, *Thomas J. Lipton, Inc., Engelwood Cliffs, New Jersey,* Precision instruments industry, robots in

Edward Fisher, *North Carolina State University, Raleigh, North Carolina,* Woodworking industry, robots in

Carl R. Flatau, *Telerobotics, Inc., Bohemia, New York,* Teleoperators, design of

James Fleck, *Edinburgh University, Edinburgh, Scotland, United Kingdom,* Organization and management

R. E. Floyd, *IBM, Boca Raton, Florida,* Technological forecasts

Philip Francis, *Motorola Inc., Shaumburg, Illinois,* Transportation industry, robots in

Andrew Frank, *University of California, Davis, California,* Walkers

Ernest Franke, *Southwest Research Institute, San Antonio, Texas,* Deriveters of aircraft wings, robotic

Hiroyasu Funakubo, *Shibaura Institute of Technology, Tokyo, Japan,* and *Ecole Polytechnique Fédérale de Lausanne, Switzerland,* Hospitals and nursing homes, robots in

Louis Galbiati, *State University of New York, Utica, New York,* Robot, revolution the

Scott Garlid, *University of Wisconsin-Madison, Madison, Wisconsin,* Leasing of robots

R. R. Gawronski, *University of West Florida, Pensacola, Florida,* Heuristics

Ludo F. Gelders, *Katholieke Universiteit Leuven, Heverlee, Belgium,* Reliability and maintenance

L. J. George, *University of Cincinnati, Cincinnati, Ohio,* Factory of the future—a case study

William Gevarter, *NASA, Mountain View, California,* Expert systems; Machine intelligence, its nature and evolution

Hassan Gomaa, *Wang Institute of Graduate Studies, Tyngsboro, Massachusetts,* Programming of multiple robot systems

James H. Graham, *University of Louisville, Louisville, Kentucky,* Controller architecture

Mikell Groover, *Lehigh University, Bethlehem, Pennsylvania,* Automation

William Gruver, *University of Kentucky, Lexington, Kentucky,* Programming

Richard Gustavson, *The Charles Stark Draper Laboratory, Inc., Cambridge, Massachusetts,* Economic justification, high level language

Daniel Hall, *Center for Occupational Research and Development, Waco, Texas,* Training of robotic personnel

P. A. Hancock, *University of Southern California, Los Angeles, California,* Sensors, integration

Dennis Harms, *Intelledex, Corvallis, Oregon,* Clean room applications

Roy Harrel, *University of Florida, Gainesville, Florida,* Robots in agriculture

Kevin Hartwig, *Holland, Michigan,* Sensors, new principles

Mitsuhiko Hasegawa, *Technological University of Nagaoka, Nagaoka, Japan,* Robots in Japan

Samad A. Hayati, *California Institute of Technology, Pasadena, California,* Calibration

Martin G. Helander, *State University of New York at Buffalo, Buffalo, New York,* Ergonomics, workplace design

David Hoska, *D&D Engineering, Inc., Shoreview, Minnesota,* Assembly robots

Vincent Howell, *Computer Consoles, Rochester, New York,* Integration of systems

T. C. Hsia, *University of California, Davis, California,* Servosystems, design of

Atlas J. Hsie, *SUNY College of Technology, Utica, New York,* Finishing

Kenichi Isoda, *Hitachi, Tokyo, Japan,* Inspection robot, advanced robotic inspection applications

A. Jain, *University of California, Davis, California,* Vision systems, industrial inspection

Anil K. Jain, *Michigan State University, East Lansing, Michigan,* Pattern recognition

Raymond A. Jarvis, *Monash University, Victoria, Australia,* Sensors, distance measurement

Mariann Jelinek, *Case Western Reserve University, Cleveland, Ohio,* Flexible manufacturing cells and systems

Lyle M. Jenkins, *NASA, Johnson Space Center, Houston, Texas,* Space robots

Jack Jeswiet, *Queens University, Kingston, Ontario, Canada,* Materials of robots

Wesley Johnston, *University of Southern California, Los Angeles, California,* Machine loading/unloading

E. Jonckheere, *Memorial Medical Center, Long Beach, California,* Surgery, robots in

M. Jones, *AT&T Bell Laboratories, Murray Hill, New Jersey,* Computational linguistics

Michael Kassler, *Michael Kassler and Associates, McMahons Point, Australia,* Mining, robots in

Stewart J. Key, *University of Western Australia, Nedlands, Western Australia,* Food processing, robots in

Steven H. Kim, *Massachusetts Institute of Technology, Cambridge, Massachusetts,* Information framework for robot design

Francis King, *Ford Motor Company, Dearborn, Michigan,* Vision systems for robotic guidance

Kerry E. Kirsch, *Kirsch Technologies, St. Clair, Michigan,* Inspection robot, applications in industry

Jerry Kirsch, *Kirsch Technologies, St. Clair, Michigan,* Inspection robot, applications in industry

Richard D. Klafter, *Temple University, Philadelphia, Pennsylvania,* Mobile robots, research and development

Charles Klein, *The Ohio State University, Columbus, Ohio,* Simulators, graphic

D. E. Koditschek, *Yale University, New Haven, Connecticut,* Robot control systems
Y. Koren, *University of Michigan, Ann Arbor, Michigan,* Numerical control
John G. Kreifeldt, *Tufts University, Medford, Massachussetts,* Ergonomics, human-robot interface
Andrew Kusiak, *University of Manitoba, Winnepeg, Canada,* Programming, off-line languages
Y. S. Kwoh, *Memorial Medical Center, Long Beach, California,* Surgery, robots in
Jack Lane, *Robotic Integrated Systems Engineering, Inc., Flint, Michigan,* Education, robotics
John Lamancusa, *Pennsylvania State University, University Park, Pennsylvania,* Sensors, ultrasonic
Stan Larsson, *ASEA AB, Vasteras, Sweden,* Foundry applications
J. F. Laszcz, *IBM, Atlanta, Georgia,* Product design
Kim Lau, *National Bureau of Standards, Gaithersburg, Maryland,* Testing
Jay Lee, *Robotics Vision Systems, Hauppauge, New York,* Tool changing
Kunwoo Lee, *Seoul National University, Seoul, Korea,* Terminology
S. H. Lee, *Northwestern University, Evanston, Illinois,* Trajectories
Jadran Lenarčič, *Institute Jožef Stefan, University of Edvard Kardelj, Ljubljana, Yugoslavia,* Kinematics
Martin D. Levine, *McGill University, Montréal, Québec, Canada,* Issues in robotics
H. Lewis, *University of Wales, Institute of Science and Technology, Cardiff, Wales, United Kingdom,* Market forecasts
Wen Lin, *University of California, Davis, California,* Speech systems
Z. Lin, *Concordia University, Montréal, Québec, Canada,* Trajectory planning
Elan Long, *CIMCORP, Aurora, Illinois,* Gantry robots
Pierre Lopez, *GARI/DGE/INSAT, Toulouse, France,* Control values from geometric model
James Lovett, *Center for Occupational Research and Development, Waco, Texas,* Training of robotics personnel
James Luckmeyer, *Southwest Research Institute, San Antonio, Texas,* Deriveters of aircraft wings, robotic
James T. Luxon, *GMI Engineering and Management Institute, Flint, Michigan,* Laser applications with robots
Anthony A. Maciejewski, *The Ohio State University, Columbus, Ohio,* Simulators, graphic
Oded Maimon, *Tel Aviv University, Tel Aviv, Israel,* Woodworking industry, robots in
Robert Malone, *Robert Malone and Associates, New York, New York,* Art, robots in; Literature, robots in; Movies, robots in
Alfred S. Malowany, *McGill University, Montréal, Québec, Canada,* Issues in robotics
E. H. Mamdani, *Queen Mary College, University of London, London, United Kingdom,* Control of robots using fuzzy reasoning
Abdol-Reza Mansouri, *McGill University, Montréal, Québec, Canada,* Issues in robotics
Danny McCoy, *B&D Sencorp Inc., Utica, New York,* Finishing
Kevin McDermott, *New Jersey Institute of Technology, Newark, New Jersey,* Original equipment manufacturers
B. C. McInnis, *University of Houston, Houston, Texas,* Control strategies
R. Meemarshi, *Tennessee Technological University, Cookeville, Tennessee,* Process planning
G. Menga, *Politecnici de Torino, Turin, Italy,* Quality control
John D. Meyer, *Tech Tran Consultants, Lake Geneva, Wisconsin,* Applications of robots
Alex Meystel, *Drexel University, Philadelphia, Pennsylvania,* Mobile robots
Lawrence T. Michaels, *Ernst & Whinney, Cleveland, Ohio,* Cost/benefit
Christian Michaud, *McGill University, Montréal, Québec, Canada,* Issues in robotics
David P. Miller, *Virginia Polytechnic Institute and State University, Blacksburg, Virginia,* Planning and problem solving
Anil Mital, *University of Cincinnati, Cincinnati, Ohio,* Desirability of robots; Factory of the future—a case study
Hiroyuki Miyamoto, *Tokyo Women's Medical College, Tokyo, Japan,* Prosthetics
Tom Moore, *Queens University, Kingston, Ontario, Canada,* Materials of robots
Philip Muilenberg, *La Jolla, California,* Undersea robots
Donald Myers, *National Bureau of Standards, Gaithersburg, Maryland,* Testing
Yasuo Nakagawa, *Hitachi, Tokyo, Japan,* Inspection robot, advanced robotic inspection applications
E. Thomas Napp, *Baxter Travenol, Round Lake, Illinois,* Medical equipment production, robots in
M. Narayanan, *Miami University, Oxford, Ohio,* Manufacturers of robots
Shimon Y. Nof, *Purdue University, West Lafayette, Indiana,* Ergonomics
John Nostrand, *New Canaan, Connecticut,* Service industry, robots in
P. O'Gorman, *University of Ulster at Jordanstown, Co. Anterim, Northern Ireland,* Kitting
Irv Oppenheim, *Carnegie-Mellon University, Pittsburgh, Pennsylvania,* Construction, robots in
Philip Ostwald, *Boulder, Colorado,* Economics, robotic manufacturing and products
William Palm, *University of Rhode Island, Kingston, Rhode Island,* Hands
Louis Panicali, *Tennessee Technical University, Cookeville, Tennessee,* Specifications
Yoh-Han Pao, *Case Western Reserve University, Cleveland, Ohio,* Flexible manufacturing cells and systems
H. M. Parsons, *Essex Corporation, Alexandria, Virginia,* Human factors
P. V. Patel, *Concordia University, Montréal, Québec, Canada,* Trajectory planning
David C. Penning, *DCP Associates, Palo Alto, California,* Unmanned factories
Noel Perrin, *Dartmouth College, Hanover, New Hampshire,* Human impacts
James Peyton, *Robotic Industries Association, Ann Arbor, Michigan,* Standards
Sandra Pfister, *Cincinnati, Ohio,* Personnel, robotics
David Pherson, *Battelle, Geneva, Switzerland,* Transducers
Lewis Pinson, *University of Colorado, Colorado Springs, Colorado,* Future applications
L. Pintelon, *Katholieke Universiteit, Leuven, Heverlee, Belgium,* Reliability and maintenance
Wallace Plumley, *Acworth, Georgia,* Aerospace industry, robots in
Gary Poock, *Monterey, California,* Programming by voice
Dejan Popovic, *Faculty of Electrical Engineering, Belgrade, Yugoslavia,* Handicapped, robots for
Alan Porter, *Georgia Institute of Technology, Atlanta, Georgia,* Futurism and robotics
Ronald D. Potter, *AMS Engineering Group, Norcross, Georgia,* End-of-arm tooling
James Pranger, *Robotic Industries Association, Ann Arbor, Michigan,* Standards
Hriday Prasad, *Ford Motor Co., Dearborn, Michigan,* Safety standards
Lute A. Quintrell, *Price Waterhouse, Cleveland, Ohio,* Cost/benefit
Mansouri Rahimi, *University of Southern California, Los Angeles, California,* Sensors, integration
H. Rahnejat, *Kingston Polytechnic, Kingston-upon-Thames, United Kingdom,* Sensors, special purpose
N. S. Rajaram, *University of Houston, Houston, Texas,* Automated guided vehicle systems
Carlos Ramirez, *Martin Marietta, New Orleans, Louisiana,* Safety of robot
Kenneth Ramsing, *University of Oregon, Eugene, Oregon,* Workers, unions, and robots
Paul G. Ranky, *The University of Michigan, Ann Arbor, Michigan,* Accuracy
Bahram Ravani, *University of California, Davis, California,* Teleoperator control using telepresence
Alan H. Redford, *University of Salford, Salford, United Kingdom,* Programmable assembly

I. S. Reed, *Memorial Medical Center, Long Beach, California,* Surgery, robots in

Peter A. Regla, *Elicon, La Habra, California,* Motion picture industry, robots in

P. P. L. Regtien, *Technicshe Hogeschool, Delft, The Netherlands,* Sensors, color sensing

Harry T. Roman, *Public Service Electric and Gas Company, Newark, New Jersey,* Electric power industry, robots in

Azriel Rosenfeld, *University of Maryland, College Park, Maryland,* Vision systems, history

Mark Rosheim, *Ross-Hime Designs, St. Paul, Minnesota,* Wrists

Fred Rossini, *Georgia Institute of Technology, Atlanta, Georgia,* Futurism and robotics

Zvi R. Roth, *Florida Atlantic University, Boca Raton, Florida,* Self-organizing and self-repair; Teaching a robot

A. Rovetta, *Politecnico de Milano, Milan, Italy,* Voice control of robots

Jerzy W. Rozenblit, *The University of Arizona, Tucson, Arizona,* Design and modeling concepts

Andrew P. Sage, *George Mason University, Fairfax, Virginia,* Cybernetics

Edward Sampson, *Cooper & Lybrand, Newport Beach, California,* Service industry, robots in

William A. Sanders, III, *U.S. Army Research Office, Research Triangle Park, North Carolina,* Vision systems, industrial application

Jorge Sanz, *IBM Research Laboratory, San Jose, California,* Vision systems, industrial inspection

George Saridis, *Rensselaer Polytechnic Institute, Troy, New York,* Adaptive control

F. Sassani, *University of British Columbia, Vancouver, British Columbia, Canada,* Research robots

Jerry W. Saveriano, *Saveriano & Associates, Carlsbad, California,* Pioneers in robotics

Susumu Sawano, *Tokico, Kawasaki-shi, Japan,* Sealing

E. M. Scharf, *Queen Mary College, University of London, London, United Kingdom,* Control of robots using fuzzy reasoning

Victor Scheinman, *Automatix, Inc., Billerica, Massachusetts,* Mechanical design of components; Transmission

George Schneider, Jr., *Lawrence Institute of Technology, Southfield, Michigan,* Interfacing robots with auxiliary equipment for education and training

Rolf Schraft, *Fraunhofer Institut fur Producktimstecnik und Automatsierung, Stuttgart, Federal Republic of Germany,* Robots in Western Europe

Mario Sciaky, *Sciaky S. A., Vitry-sur-Seine, France,* Welding robots

Warren P. Seering, *Cal Tech, Pasadena, California,* Mechanical design of components; Transmission

Mo Shahinpoor, *University of New Mexico, Albuquerque, New Mexico,* Dynamics

Yacov Shamash, *Washington State University, Pullman, Washington,* Teaching a robot

Loren Shaum, *SMI, Elkhart, Indiana,* Actuators

L. S. Shieh, *University of Houston, Houston, Texas,* Control strategies

Miroslaw J. Skbniewski, *Purdue University, West Lafayette, Indiana,* Construction, robots in

David G. Slaughter, *University of Florida, Gainesville, Florida,* Robots in agriculture

Bruce Smith, *ASEA, Troy, Michigan,* Welding robots

Sharon Smith, formerly, *International Personal Robot Congress, Conifer, Colorado,* Art, robotic

Steven E. Smith, *Triplex, Torrance, California,* Programmable controllers

Barry Soroka, *California State Polytechnic University, Pomona, California,* Programming, high level language

William Stallings, *Comp-Comm Consulting, London, United Kingdom,* Communications; Local area networks

Y. Stepanenko, *University of Waterloo, Waterloo, Ontario, Canada,* State-feedback robot control

Janet Strimaitis, *Zymark Corporation, Hopkington, Massachusetts,* Scientific laboratories, robots in

M. Sundaram, *Tennessee Technological University, Cookeville, Tennessee,* Process planning

Kazuo Tanie, *Mechanical Engineering Laboratory MITI, Tokyo, Japan,* Grippers

William R. Tanner, *Tanner Associates, Farmington Hills, Michigan,* Classification

T. J. Tarn, *Washington University, St. Louis, Missouri,* Software elements

G. E. Taylor, *University of Hull, Hull, United Kingdom,* Garment and shoe industry, robots in

P. M. Taylor, *University of Hull, Hull, United Kingdom,* Garment and shoe industry, robots in

William Teoh, *SPARTA, Huntsville, Alabama,* Teleoperators, research

Rajko Tomovic, *Faculty of Electrical Engineering, Belgrade, Yugoslavia,* Handicapped, robots for

James P. Trevelyan, *University of Western Australia, Nedlands, Western Australia,* Food processing, robots in

Robin Truman, *Plessey Network and Office Systems, Ltd., Beeston, Nottingham, United Kingdom,* Component assembly onto printed circuit boards

T. K. Truong, *Memorial Medical Center, Long Beach, California,* Surgery, robots in

Taizo Ueda, *Honda Foundation, Tokyo, Japan,* Robots in Japan

William Uhde, *UAS Automation Systems, Bristol, Connecticut,* Fabrication and machining applications

Noriyuki Utsemi, *Tokico, Kawasaki, Japan,* Sealing

A. Villa, *Politecnico de Torino, Turin, Italy,* Quality control

Donald A. Vincent, *Robotic Industries Associations, Ann Arbor, Michigan,* Associations, robotic

John Vranish, *Naval Surface Weapons Center, Dahlgreen, Virginia,* Deriveters of aircraft wings, robotic

Kenneth J. Waldron, *The Ohio State University, Columbus, Ohio,* Arm, design of

Charles Wampler, *General Motors Research Laboratories, Warren, Michigan,* Teleoperators, supervisory control

J. C. Wang, *Idaho State University, Pocatello, Idaho,* Control strategies

S. H. Wang, *University of California, Davis, California,* Flexible robots, control of

John Watteau, *University of California, Los Angeles, California,* Productivity

John G. Webster, *University of Wisconsin, Madison, Wisconsin,* Space robots, research; Teleoperator control using telepresence

John A. White, *Georgia Institute of Technology, Atlanta, Georgia,* Material handling

Daniel E. Whitney, *The Charles Stark Draper Laboratory, Cambridge, Massachusetts,* Part mating theory; Remote center compliance devices

Wilbert E. Wilhelm, *The Ohio State University, Columbus, Ohio,* Conveyor tracking

Theodore J. Williams, *Purdue University, West Lafayette, Indiana,* Economics, robot market and industry

Michael Wodzinski, *Selspot Systems (Selcom), Troy, Michigan,* Accuracy

R. F. Wolffenbuttel, *Technishe Hogeschool, Delft, The Netherlands,* Sensors, color sensing

Janet Worthington, *West Virginia Institute of Technology, Fayetteville, West Virginia,* Maintenance and repair, robotic

Chi-Huar Wu, *Northwestern University, Evanston, Illinois,* Compliance

Robert M. Wygant, *Western Michigan University, Kalamazoo, Michigan,* Ergonomics, robot selection

Yaming Yang, *Washington State University, Pullman, Washington,* Teaching a robot

R. Young, *Memorial Medical Center, Long Beach, California,* Surgery, robots in

Clara Yu, *Octy, Inc., Fairfax Station, Virginia,* Vision systems, programming

Bernard P. Zeigler, *The University of Arizona, Tuscon, Arizona,* Design and modeling concepts

M. Carl Ziemke, *University of Alabama, Huntsville, Alabama,* Teleoperators, research

Nello Zuech, *Vision Systems International, Yardley, Pennsylvania,* Vision systems, theory

FOREWORD

Looking Ahead

In 1939, when I was 19 years old, I began to write a series of science fiction stories about robots. At the time, the word *robot* had been in existence for only 18 years; Karel Capek's play, *R.U.R.,* in which the word had been coined, having been performed for the first time in Europe in 1921. The concept, however, that of machines that could perform tasks with the apparent "intelligence" of human beings, had been in existence for thousands of years.

Through all those years, however, robots in myth, legend, and literature had been designed only to point a moral. Generally, they were treated as examples of overweening pride on the part of the human designer; an effort to accomplish something that was reserved to God alone. And, inevitably, this overweening pride was overtaken by Nemesis (as it always is in morality tales), so that the designer was destroyed, usually by that which he had created.

I grew tired of these myriad-told tales, and decided I would tell of robots that were carefully designed to perform certain tasks, but with *safeguards built in;* robots that might conceivably be dangerous, as any machine might be, but no more so.

In telling these tales, I worked out, perforce, certain rules of conduct that guided the robots; rules that I dealt with in a more and more refined manner over the next 44 years (my most recent robot novel, *The Robots of Dawn,* was published in October 1983). These rules were first put into words in a story called "Runaround," which appeared in the March 1942, issue of *Astounding Science Fiction.*

In that issue, on page 100, one of my characters says, "Now, look, let's start with the three fundamental Rules of Robotics . . ." and he proceeds to recite them. (In later stories, I took to referring to them as "the Three Laws of Robotics" and other people generally say "Asimov's Three Laws of Robotics.")

I am carefully specific about this point because that line on that page in that story was, as far as I know, the very first time and place that the word *robotics* had ever appeared in print.

I did not deliberately make up the word. Since *physics* and most of its subdivisions routinely have the "-ics" suffix, I assumed that "robotics" was the proper scientific term for the systematic study of robots, of their construction, maintenance, and behavior, and that it was used as such. It was only decades later that I became aware of the fact that the word was in no dictionary, general or scientific, and that I had coined it.

Possibly every person has a chance at good fortune in his life, but there can't be very many people who have had the incredible luck to live to see their fantasies begin to turn into reality.

I think sadly, for instance, of a good friend of mine who did not. He was Willy Ley who, for all his adult life was wedded to rocketry and to the dream of reaching the moon; who in his early twenties helped found rocket research in Germany; who, year after year wrote popular books on the subject; who, in 1969, was preparing to witness the launch of the first rocket intended to land on the moon; and who then died six weeks before that launch took place.

Such a tragedy did not overtake me. I lived to see the transistor invented, and solid-state devices undergo rapid development until the microchip became a reality. I lived to see Joseph Engelberger (with his interest sparked by my stories, actually) found Unimation, Inc., and then keep it going, with determination and foresight, until it actually constructed and installed industrial robots and grew enormously profitable. His devices were not quite the humanoid robots of my stories, but in many respects they were far more sophisticated than anything I had ever been equipped to imagine. Nor is there any doubt that the development of robots more like mine, with the capacities to see and to talk, for instance, are very far off.

I lived to see my Three Laws of Robotics taken seriously and routinely referred to in articles on robotics, written by real roboticists, as in a couple of cases in this volume. I lived to see them referred to familiarly, even in the popular press, and identified with my name, so that I can see I have secured for myself (all unknowingly, I must admit) a secure footnote in the history of science.

I even lived to see myself regarded with a certain amount of esteem by legitmate people in the field of robotics, as a kind of grandfather of them all, even though, in actual fact, I am merely a chemist by training and a science-fiction writer by choice—and know virtually nothing about the nuts and bolts of robotics; or of computers, for that matter.

But even after I thought I had grown accustomed to all of

this, and had ceased marveling over this amazing turn of the wheel of fortune, and was certain that there was nothing left in this situation that had the capacity to surprise me, I found I was wrong. Let me explain . . .

In 1950 nine of my stories of robots were put together into a volume entitled *I, Robot* (the volume, as it happens, that was to inspire Mr. Engelberger).

On the page before the table of contents, there are inscribed, in lonely splendor *The Three Laws of Robotics:*

1. *A robot may not injure a human being, or, through inaction, allow a human being to come to harm.*
2. *A robot may obey the orders given it by human being except where such orders would conflict with the First Law.*
3. *A robot must protect its own existence as long as such protection does not conflict with First or Second Law.*

Never, until it actually happened, did I ever believe that I would really live to see robots, really live to see my three laws quoted everywhere.

Nor did it ever occur to me that I would live to see a vast three-volume *International Encyclopedia of Robotics* in which there would be enormous quantities of data and to which I would write the foreword (one that has already appeared in essence in the *Handbook of Industrial Robotics,* edited by Shimon Y. Nof).

It takes no great imagination to see that the *Encyclopedia* will increase in length and detail from edition to edition.

I see the world, and the human outposts on other worlds and in space, filled with cousin-intelligences of two entirely different types. I see silicon-intelligence (robots) that can manipulate numbers with incredible speed and precision and that can perform operations tirelessly and with perfect reproducibility; and I see carbon-intelligence (human beings) that can apply intuition, insight, and imagination to the solution of problems on the basis of what would seem insufficient data to a robot. I see the former building the foundations of a new, and unimaginably better society than any we have ever experienced; and I see the latter building the superstructure, with a creative fantasy we dare not picture now.

I see the two together advancing far more rapidly than either could alone. And though this, alas, I will not live to see, I am confident our children and grandchildren will, and that future editions of this *Encyclopedia* will detail the process.

Issac Asimov
New York, New York

PREFACE

Robotics and automation are critical ingredients in the world's efforts towards an improved standard of living for all. Automation, the automatic operation of processes, and robotics, which includes the manipulator, controller and associated devices, are all critical to effective operation of our plants, factories, and institutions. In this *Encyclopedia* we have taken both an encyclopedic and international view of this field of robotics. Thus, we include numerous articles written by international experts and have striven to include all the associated theoretical aspects of the field as well as most of the present and future applications of robots in the factory, office, and home.

The *International Encyclopedia of Robotics* defines the discipline and the practice of robotics by bringing together the core of knowledge and practice from the field and all closely related fields. The *Encyclopedia* is written primarily for the professional who seeks to understand and use robots and automation. The *Encyclopedia* has made significant contributions to the literature, not only because it brings many disciplines into one comprehensive reference, but also because it contains many articles that bring new or fresh insights.

The articles and the authors invited to write them were chosen with the cooperation of an editorial advisory board of distinguished authorities. The author of each article is a recognized research expert on the topic. Each article had a bibliography and extensive cross-references to other articles. The reader may start with almost any article and be led by cross-references to almost every other article in the *Encyclopedia*. There are more than 2000 tables and figures. Stressing readability, accuracy, and completeness of facts as well as overall usefulness of material, this great work brings you the result of years of labor and experience.

I became involved in the project to develop this *Encyclopedia* in the fall of 1984 when I was approached by Martin Grayson of John Wiley & Sons, Inc., and Professor Shimon Nof of Purdue University. Although I was warned by several people that this would involve a great effort, the opportunity to help create a definitive and comprehensive view of the field, authored by a wide variety of experts, each writing on his or her own area of expertise, and the promise of significant help from Wiley's Encyclopedia Department lured me onward. With the excellent assistance of Shimon Nof, the consulting editor, we put together an outstanding team of writers and reviewers. Michalina Bickford joined us as the managing editor and performed superbly.

Robotics is a relatively young field, and still has controversy about what it is and about what constitutes good and valuable research and application. Some researchers felt that an encyclopedia was premature. There was some controversy about the selection of articles. Nevertheless, I was extremely gratified with the number of people who were willing to take time from their already busy schedules to write and to review articles. Those involved constitute a significant percentage of all active practioners, from all the different companies and major research institutes and universities.

I am grateful to many people whose efforts have gone into making this *Encyclopedia:* Shimon Nof and Martin Grayson, who started it; the members of the editorial board, who defined it; Michalina Bickford, who managed it all; and the authors and reviewers, who created it.

Finally, my sincere appreciation goes to Joy, my wife who, as a humanist, has questioned and refined my views of the benefits and uses of robots, automation, and machines in the workplace and elsewhere.

RICHARD C. DORF
Davis, California

CONVERSION FACTORS, ABBREVIATIONS AND UNIT SYMBOLS

Selected SI Units (Adopted 1960)

Quantity	*Unit*	*Symbol*	*Acceptable equivalent*
BASE UNITS			
length	meter[†]	m	
mass[‡]	kilogram	kg	
time	second	s	
electric current	ampere	A	
thermodynamic temperature[§]	kelvin	K	
DERIVED UNITS AND OTHER ACCEPTABLE UNITS			
* absorbed dose	gray	Gy	J/kg
acceleration	meter per second squared	m/s^2	
* activity (of ionizing radiation source)	becquerel	Bq	l/s
area	square kilometer	km^2	
	square hectometer	hm^2	ha (hectare)
	square meter	m^2	
density, mass density	kilogram per cubic meter	kg/m^3	g/L; mg/cm^3
* electric potential, potential difference, electromotive force	volt	V	W/A
* electric resistance	ohm	Ω	V/A
* energy, work, quantity of heat	megajoule	MJ	
	kilojoule	kJ	
	joule	J	N·m
	electron volt[x]	eV[x]	
	kilowatt hour[x]	kW·h[x]	
* force	kilonewton	kN	
	newton	N	$kg{\cdot}m/s^2$
* frequency	megahertz	MHz	
	hertz	Hz	l/s
heat capacity, entropy	joule per kelvin	J/K	
heat capacity (specific), specific entropy	joule per kilogram kelvin	J/(kg·K)	
heat transfer coefficient	watt per square meter kelvin	$W/(m^2{\cdot}K)$	
linear density	kilogram per meter	kg/m	

magnetic field strength	ampere per meter	A/m	
moment of force, torque	newton meter	N·m	
momentum	kilogram meter per second	kg·m/s	
* power, heat flow rate,	kilowatt	kW	
radiant flux	watt	W	J/s
power density, heat flux density, irradiance	watt per square meter	W/m^2	
* pressure, stress	megapascal	MPa	
	kilopascal	kPa	
	pascal	Pa	
sound level	decibel	dB	
specific energy	joule per kilogram	J/kg	
specific volume	cubic meter per kilogram	m^3/kg	
surface tension	newton per meter	N/m	
thermal conductivity	watt per meter kelvin	W/(m·K)	
velocity	meter per second	m/s	
	kilometer per hour	km/h	
viscosity, dynamic	pascal second	Pa·s	
	millipascal second	mPa·s	
volume	cubic meter	m^3	
	cubic decimeter	dm^3	L(liter)
	cubic centimeter	cm^3	mL

* The asterisk denotes those units having special names and symbols.
† The spellings "metre" and "litre" are preferred by ASTM; however "er-" is used in the Encyclopedia.
‡ "Weight" is the commonly used term for "mass."
§ Wide use is made of "Celsius temperature" (t) defined by

$$t = T - T_0$$

where T is the thermodynamic temperature, expressed in kelvins, and $T_0 = 273.15$ by definition. A temperature interval may be expressed in degrees Celsius as well as in kelvins.
ˣ This non-SI unit is recognized by the CIPM as having to be retained because of practical importance or use in specialized fields.

In addition, there are 16 prefixes used to indicate order of magnitude, as follows:

Multiplication factor	*Prefix*	*Symbol*	*Note*
10^{18}	exa	E	
10^{15}	peta	P	
10^{12}	tera	T	
10^{9}	giga	G	
10^{6}	mega	M	
10^{3}	kilo	k	
10^{2}	hecto	h[a]	
10	deka	da[a]	
10^{-1}	deci	d[a]	
10^{-2}	centi	c[a]	
10^{-3}	milli	m	
10^{-6}	micro	μ	
10^{-9}	nano	n	
10^{-12}	pico	p	
10^{-15}	femto	f	
10^{-18}	atto	a	

[a] Although hecto, deka, deci, and centi are SI prefixes, their use should be avoided except for SI unit-multiples for area and volume and nontechnical use of centimeter, as for body and clothing measurement.

Conversion Factors to SI Units

To convert from	*To*	*Multiply by*
acre	square meter (m^2)	4.047×10^3
angstrom	meter (m)	$1.0 \times 10^{-10\dagger}$
atmosphere	pascal (Pa)	1.013×10^5
bar	pascal (Pa)	$1.0 \times 10^{5\dagger}$
barn	square meter (m^2)	$1.0 \times 10^{-28\dagger}$
barrel (42 U.S. liquid gallons)	cubic meter (m^3)	0.1590
Btu (thermochemical)	joule (J)	1.054×10^3
bushel	cubic meter (m^3)	3.524×10^{-2}
calorie (thermochemical)	joule (J)	$4.184^\dagger$
centipoise	pascal second (Pa·s)	$1.0 \times 10^{-3\dagger}$
cfm (cubic foot per minute)	cubic meter per second (m^3/s)	4.72×10^{-4}
cubic inch	cubic meter (m^3)	1.639×10^{-5}
cubic foot	cubic meter (m^3)	2.832×10^{-2}
cubic yard	cubic meter (m)	0.7646
dram (apothecaries')	kilogram (kg)	3.888×10^{-3}
dram (avoirdupois)	kilogram (kg)	1.772×10^{-3}
dram (U.S. fluid)	cubic meter (m^3)	3.697×10^{-6}
dyne	newton (N)	$1.0 \times 10^{-5\dagger}$
dyne/cm	newton per meter (N/m)	$1.0 \times 10^{-3\dagger}$
fluid ounce (U.S.)	cubic meter (m^3)	2.957×10^{-5}
foot	meter (m)	$0.3048^\dagger$
gallon (U.S. dry)	cubic meter (m^3)	4.405×10^{-3}
gallon (U.S. liquid)	cubic meter (m^3)	3.785×10^{-3}
gallon per minute (gpm)	cubic meter per second (m^3/s)	6.308×10^{-5}
	cubic meter per hour (m^3/h)	0.2271
grain	kilogram (kg)	6.480×10^{-5}
horsepower (550 ft·lbf/s)	watt (W)	7.457×10^2
inch	meter (m)	$2.54 \times 10^{-2\dagger}$
inch of mercury (32°F)	pascal (Pa)	3.386×10^3
inch of water (39.2°F)	pascal (Pa)	2.491×10^2
kilogram-force	newton (N)	9.807
kilowatt hour	megajoule (MJ)	$3.6^\dagger$
liter (for fluids only)	cubic meter (m^3)	$1.0 \times 10^{-3\dagger}$
micron	meter (m)	$1.0 \times 10^{-6\dagger}$
mil	meter (m)	$2.54 \times 10^{-5\dagger}$
mile (statute)	meter (m)	1.609×10^3
mile per hour	meter per second (m/s)	0.4470
millimeter of mercury (0°C)	pascal (Pa)	$1.333 \times 10^{2\dagger}$
ounce (avoirdupois)	kilogram (kg)	2.835×10^{-2}
ounce (troy)	kilogram (kg)	3.110×10^{-2}
ounce (U.S. fluid)	cubic meter (m^3)	2.957×10^{-5}
ounce-force	newton (N)	0.2780
peck (U.S.)	cubic meter (m^3)	8.810×10^{-3}
pennyweight	kilogram (kg)	1.555×10^{-3}
pint (U.S. dry)	cubic meter (m^3)	5.506×10^{-4}
pint (U.S. liquid)	cubic meter (m^3)	4.732×10^{-4}
poise (absolute viscosity)	pascal second (Pa·s)	$0.10^\dagger$
pound (avoirdupois)	kilogram (kg)	0.4536
pound (troy)	kilogram (kg)	0.3732
pound-force	newton (N)	4.448
pound-force per square inch (psi)	pascal (Pa)	6.895×10^3
quart (U.S. dry)	cubic meter (m^3)	1.101×10^{-3}
quart (U.S. liquid)	cubic meter (m^3)	9.464×10^{-4}
quintal	kilogram (kg)	$1.0 \times 10^{2\dagger}$

rad	gray (Gy)	$1.0 \times 10^{-2\dagger}$
square inch	square meter (m^2)	6.452×10^{-4}
square foot	square meter (m^2)	9.290×10^{-2}
square mile	square meter (m^2)	2.590×10^{6}
square yard	square meter (m^2)	0.8361
ton (long, 2240 pounds)	kilogram (kg)	1.016×10^{3}
ton (metric)	kilogram (kg)	$1.0 \times 10^{3\dagger}$
ton (short, 2000 pounds)	kilogram (kg)	9.072×10^{2}
torr	pascal (Pa)	1.333×10^{2}
yard	meter (m)	$0.9144^{\dagger}$

† Exact.

ABBREVIATIONS AND ACRONYMS

A	ampere
AACW	active adaptive compliance wrist
ac	alternating current (*noun*)
a-c	alternating current (*adjective*)
ACI	automatic component insertion
AFR	Air Force Regulation
AGV	automated guided vehicle
AGVS	automated guided vehicle system
AI	artificial intelligence
AMR	autonomous mobile robot
ANSI	American National Standards Institute
ASME	American Society of Mechanical Engineers
ASTM	American Society for Testing and Materials
ATC	automatic tool changes
ATE	automotive test equipment
avg	average
BCD	binary coded decimal
bpa	basic probability assignment
bps	bits per second
BWR	boiling water reactor
°C	degrees Celsius
ca	approximately (*circa*)
CAD	computer-aided design
CAE	computer-aided engineering
CAM	computer-aided manufacturing
CAPP	computer-aided production planning
CAT	computer-aided testing
CFR	Code of Federal Regulations
CIM	computer-integrated manufacturing
CL	control law
CMM	coordinate measuring machine
CMU	Carnegie Mellon University
CNC	computer numerical control
CPPP	computerized production process planning
CPU	central processing unit; control process unit
CRC	computer robot control
CRT	cathode ray tube
DAC	digital to analogue converter
dc	direct current (*noun*)
d-c	direct current (*adjective*)
DCF	discounted cash flow
DDM	direct drive motor
DMA	direct memory access
DNC	direct numerical control
DOD	Department of Defense (U.S.)
DOF	degree of freedom
DOT	Department of Transportation (U.S.)
DOM	design of maintenance
EEC	European Economic Community
eg	For example (*est gratia*)
EIA	Electronic Industries Association
EOD	explosive ordnance disposal
EPA	Environmental Protection Agency (U.S.)
est	estimated
ESU	emergency service unit
°F	degrees Fahrenheit
FAA	Federal Aviation Administration (U.S.)
FDM	frequency division multiplexing
FMC	flexible manufacturing cell
FMS	flexible manufacturing system
FOF	factory of the future
ft	foot
ft·lbf	foot-pound force (1.356 J)
FTAM	file, transfer, access, and management
g	gravitational acceleration
g	gram
gal	gallon (3.785 L in the U.S.)
GC	gas chromatography
GDB	global data base
gf	gram force (0.0098 N)
GNP	Gross National Product
GPS	general problem solver
gy	gray (10^{-2})
h	hour
hp	horsepower (746 W)
Hz	Hertz (cycles per second)
IC	integrated circuit
ICAO	International Civil Aviation Organization
IEEE	Institute of Electrical and Electronic Engineers
I/O	integrated circuit

IQR interquartile range
IRR internal rate of return
IRS Internal Revenue Service (U.S.)
ISO International Organization for Standardization
IWP intelligent work in process

J Joule (energy)
JIT Just-in-time
JPL Jet Propulsion Laboratory
K Kelvin
kgf kilogram-force (9.086 N)
kJ kilojoule
km kilometer
kPa kilopascal (0.145 psi)

L liter (volume)
LAN local area networks
lb pound (mass) (453.6 g)
lbf pound force (4.448 N)
LCD liquid crystal display
LED light-emitting diode
LTD long term debt
LTP local tracking problem
LVDT linear variable differential transformer

m meter
MAP manufacturing automation protocol
MARR minimum attractive rate of return
max maximum
mg milligram
MHS material handling system
MICAPP microcomputer-assisted process planning program
min minute; minimum
MLS master laboratory station
MMFS manipulating message format standard
MMS maintenance manipulator system
MP microprocessor
ms millisecond
MTBF mean time between failures
MTTR mean time to repair

N Newton
na not available
NBS National Bureau of Standards
NC numerically controlled
NRC Nuclear Regulatory Commission (U.S.)
NYPD New York City Police Department

OD outer diameter
OI operation interface
OLP off-line program
OSHA Occupational Safety and Health Administration (U.S.)
OTA Office of Technology Assessment (U.S.)

Pa Pascal (pressure)
PC programmable controller; programmable logic controller; personal computer
PCM pulse code modulation
psi pound (force) per square inch (6.893 kPa)
psig psi pressure gauge
PWA printed wire assembly
PWM pulse-width modulation
PWR pressure water reactor

qv which see (*quo vide*)

RAM random access memory
RCC remote center compliance
RCMC resolved motion rate control
R&D research and development
rf radio frequency (*noun*)
r-f radio frequency (*adjective*)
RGS remote guidance system
rh relative humidity
RIA Robotics Institute of America
RLL relay ladder language
RM remote mobile investigator
ROI return on investment
ROM read only memory
RPS robot programming system
RPV remotely piloted vehicle
RTM Robot time and motion

s second
SAT symmetric axis transform
SCA sensor-controlled automation
SCARA selective compliance-assembly robot arm
SIC Standard Industrial Classification
SMC surface mounted components
SMD surface mount devices
SME Society of Manufacturing Engineers
SOC self organizing control
SRBP synthetic resin bonded paper
SRI Stanford Research Institute
SUNY State University of New York
SUR speech understanding research
SUS speech understanding system

t metric ton
T temperature
TDL task description language
TDM time division multiplexing
TMI Three Mile Island
TMV technical maintenance vehicle

UIMS user interface management system
uv ultraviolet
UVS unmanned vehicle system

WIP work in progress
WM working memory
V volt
VLSI very large scale integration
vs versus

yr year

INTERNATIONAL ENCYCLOPEDIA OF ROBOTICS: APPLICATIONS AND AUTOMATION

VOLUME 1

A

ACCURACY

PAUL G. RANKY
The University of Michigan
Ann Arbor, Michigan

MICHAEL WODZINSKI
Selspot Systems (SELCOM)
Troy, Michigan

INTRODUCTION

Various types of performance test methods have been utilized since the inception of the industrial robot. Performance testing has many forms, from verifying the design goals by the manufacturer to the determination of the best robot for a particular application. Each type of testing is related to the various phases through which a robot model passes during its useful life. Three distinct phases of development usually occur during design:

Alpha Phase. This phase represents the fabrication and test of a prototype robot which was designed based on goals set forth by the design engineers in the functional and design specifications. Testing during this phase includes the verification of the design specifications and the determination of the system integrity. The final output of this phase is a proposed specification and a prototype robot.

Beta Phase. This phase consists of either preproduction or initial production units which are sent to specific customers on consignment. The beta sites are carefully selected to represent typical applications of the particular robot being evaluated. Both the robot manufacturers and the beta-site customers use this phase to determine whether certain features should be incorporated or eliminated. The output of this phase is the specification of the robot.

Production Phase. This phase consists of the actual robots produced in production quantities. Tests conducted during this phase are designed to determine whether or not the robot meets its specifications and how it performs in a particular application.

The tests performed in each of these phases may be similar or even identical, depending on the performance parameters to be measured. The type of information which can be obtained is often limited by the test equipment. This stems from the primary approach used currently to measure performance of industrial robots: measurement of robot position or path (1,2).

Many of the techniques used for the measurement of numerical-controlled (NC) machine tools were used to quantify robot performance, but it was quickly determined that performance parameters of robots were quite different, particularly in the case of non-Cartesian-type robots. One major advantage of machine tools is the ability to fix the *work* by means of rigid fixtures onto the table of the machine tool. The machine tool then has control over the *environment* of the process. The majority of current industrial robots, on the other hand, are spatial mechanisms which perform tasks on parts (or work) located *somewhere* in space. Since the robot designer does not know where these parts will be located by the robot users, the performance of the robot cannot be optimized for any particular area, or location. Whereas some machine-tool testing techniques are designed to measure the parts made and monitor the machine, robot testing techniques must determine the performance for all parts of the entire work envelope, regardless of the actual application, that can vary a lot. This places severe constraints on the test equipment, because not only is the position of the robot required, but this position must be of sufficicent accuracy to be meaningful. Most current industrial robots have work envelopes larger than 35.3 cubic ft (1 cubic meter). To be able to monitor a 35.3-cubic-ft (1-cubic-meter) volume at a resolution of an order of magnitude approximately 10 times greater than that rated for a typical robot currently exceeds the capabilities of most available test equipment.

MEASUREMENT PRINCIPLES

Most performance tests developed for industrial robots can be designed to measure either *Point-to-point* or *Path* behavior. *Point-to-point* tests measure parameters such as quasi-static repeatability, quasi-static accuracy, or dynamic stability. Path tests measure parameters such as kinematic repeatability or kinematic accuracy (1). Each type of test places different demands on the test equipment. The following general guidelines are recommended:

The measurement equipment should have a resolution of at least one order of magnitude greater than the value to be measured. For example, if a repeatability of 0.001 in. (0.0254 mm) is to be measured, then the test equipment should have a resolution of at least 0.0001 in. (0.00254 mm).

Noncontact gauging (eg, laser interferometers, proximity gauges, vision systems) is preferred over contact gauges (eg, LVDTs, dial indicators). Contact gauges have a tendency to *creep* over time due to the contact forces imparted from the robot during each measurement and they represent greater investment when interfaced to computers for real-time data evaluation purposes.

The robot-mounting platform, eg, mounting plate, should be rigid. Any shift in the mounting platform will appear as a shift in the robot. Usually this shift due to the dynamic forces is exacerbated when the robot is at full extension and comes full load.

Try to vary only one control parameter at a time for tests with multiple permutations. Performance parameters such as repeatability are dependent on parameters such as speed, settling time, robot temperature, payload, and extension. If many of these parameters are changed at the same time, it becomes difficult to determine which parameter is causing degradation in performance.

The test equipment should be time and temperature stable.

Many performance parameters are tested over prolonged periods of time. It is essential that the test equipment remain stable during the entire test period; thus, frequent test equipment recalibration may be necessary. One must be careful when testing and programming the robot, especially around singularity points, and all safety precautions must be followed. The test engineers should not be within the reach of the robot and between the robot and any other object when the robot is energized.

MEASUREMENT CONCEPTS

Often the performance parameters, such as repeatability, accuracy, and resolution, are confused.

Repeatability. Closeness of agreement of repeated position movements, under the same conditions, to the same location (3). (Note that "location" means a position (x,y,z) and an orientation vector (*O,A,T*).)

Accuracy. Degree to which the actual location corresponds to desired or commanded location.

Resolution. A measure of the smallest possible increment of change in the variable output of a device.

Resolution is determined by the position feedback devices, eg, optical encoders, internal to the robot mechanism. Resolution varies depending on robot position. On articulated robots it is usually best at the innermost points and worst at the outermost points. Both repeatability and accuracy are dependent on the resolution of the robot since the robot control must use the position information from the feedback devices to determine whether it is at the commanded taught point (in the case of static repeatability and static accuracy) or along the desired trajectory (in the case of kinematic repeatability and kinematic accuracy). Resolution is rarely specified by the robot manufacturer since it varies depending on the position of the robot and is usually of no consequence to the user. The resolution determines parameters such as repeatability and accuracy which are used to determine whether a robot will be able to perform a particular application.

Accuracy is a measure of how well a robot moves to the absolute position and orientation of the taught location everytime it is commanded to do so, whereas repeatability measures how well the robot returns to the taught location time after time. (Note again, "location" measuring a 3D position vector, x,y,z, pointing from the robot base coordinate system's origin to the tcp (*t*ool-*c*enter-*p*oint) and the orientation of the tool, *O,A,T,* as typically described by the so-called Eules angles.) Figure 1 shows graphically the difference between repeatability and accuracy as measured in one plane (4). Accuracy is a measure of the theoretical location or path, whereas repeatability is a measure of the relative placement of a location or path.

MEASUREMENT RESULTS

The results of each type of test have different forms. The most common form of data-analysis technique used in industry is process statistics. Control charts show the statistical behavior of performance parameters such as repeatability by determining the average, range, and standard deviation of a sample location (Fig. 2).

The following guidelines are generally used when evaluating process-control charts:

- A controlled process has both the average and range curves appear as random patterns between the control limits.
- The presence of one or more points beyond the control limit is evidence of instability.
- Runs (seven points in a row on one side of the average or intervals in a row that are consistently increasing or decreasing) are an indication of instability (5).

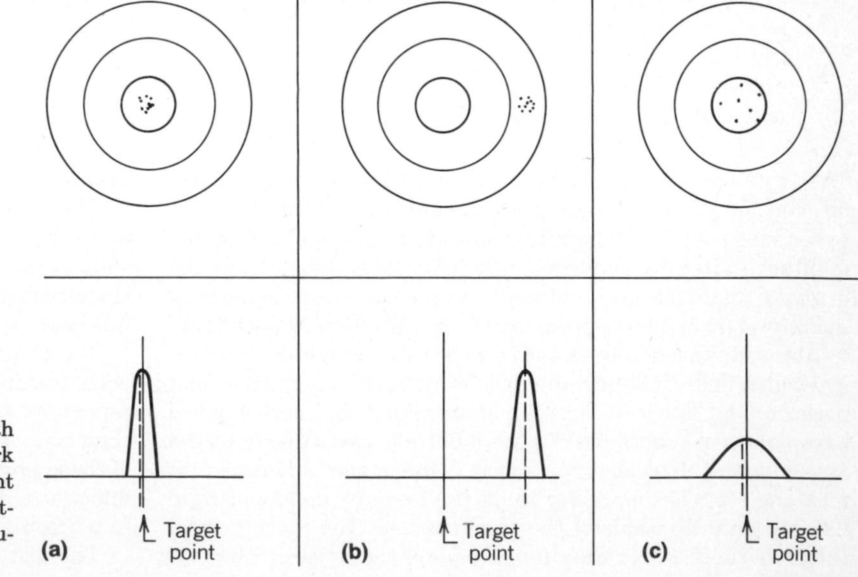

Figure 1. Repeatability versus accuracy. Note that both repeatability and accuracy are 3D vectors in the work area of the robot and that the graphs shown represent their worst values in one plane only (4). **(a)** High repeatability, high accuracy; **(b)** high repeatability, low accuracy; **(c)** low repeatability, high accuracy.

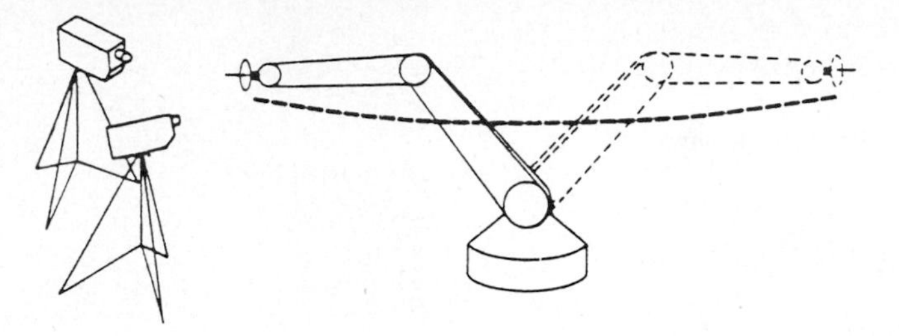

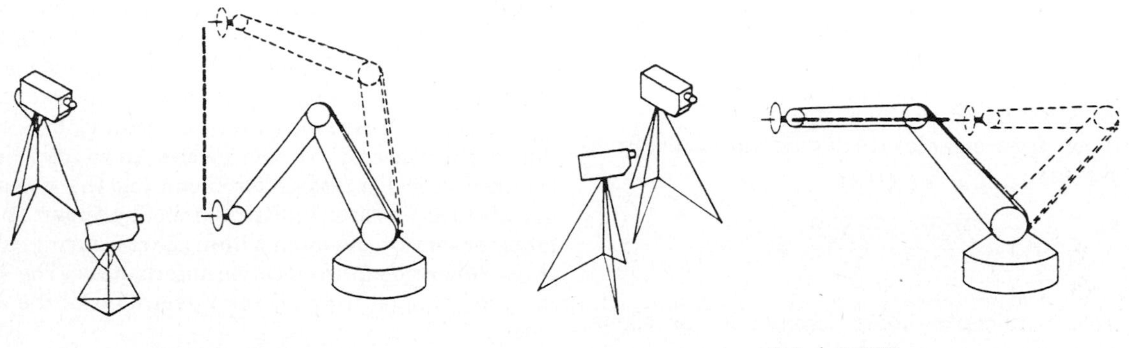

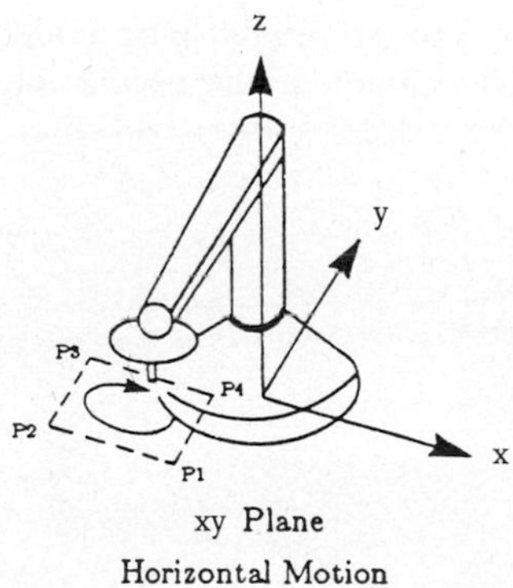

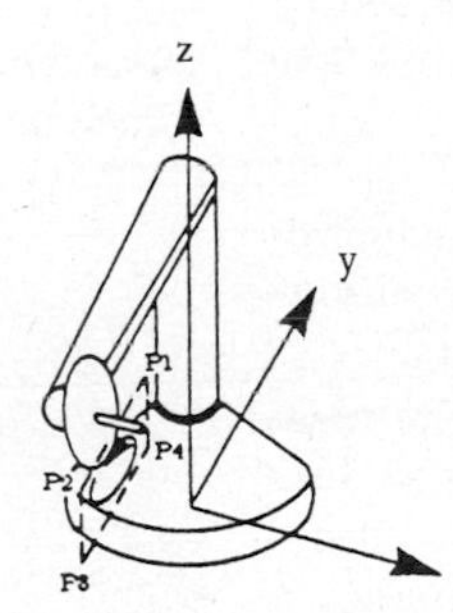

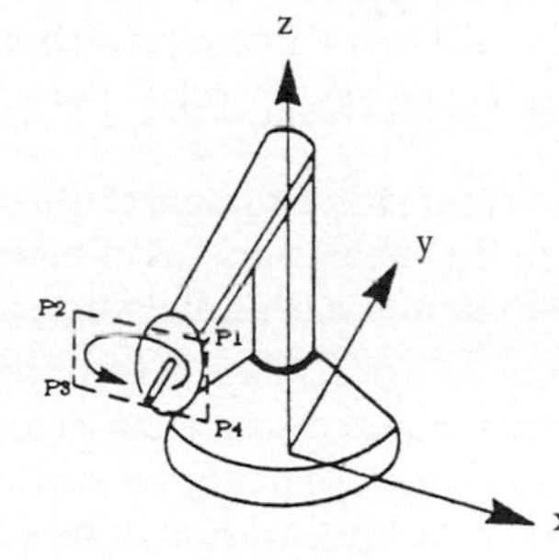

Figure 2. Ford RAACC repeatability test motions, and path-control test motions. Courtesy of Ford RAACC, Detroit, Mich.

The statistical formulas used for calculation of the average, range, and standard deviation are given below (6):

Average placement of the total samples ($\overline{\overline{X}}$):

$$\overline{\overline{X}} = \frac{\Sigma \overline{X}_i}{i}$$

where $\overline{X}_i$ = average of the individual subgroups

i = number of subgroups

Average range of the total samples (R):

$$\overline{R} = \frac{\Sigma R_i}{i}$$

where R_i = range of the individual subgroups

i = number of subgroups

Computed trial control limits:

Average placement

$$\text{upper limit} = \overline{\overline{X}} + A_2\overline{R}$$
$$\text{lower limit} = \overline{\overline{X}} - A_2\overline{R}$$

where A_2 is a constant based on the sample size
Range

$$\text{upper limit} = D_4\overline{R}$$

where D_4 is a constant based on the sample size
Note: For sample sizes below seven, the lower range limit is zero
Estimate of process standard deviation:

$$\sigma = \frac{\overline{R}}{D_2}$$

where D_2 is a constant based on the sample size
Standard deviation from samples (including variations from out of control points):

$$\sigma = \sqrt{\frac{N\Sigma X_i^2 - (\Sigma X_i)^2}{N(N-1)}}$$

Note: The resultant of the above formula is one sigma (1 σ). To determine the conventional ± 3 sigma, multiply the resultant of the above by ± 3

METHODS OF TESTING

Various robot test methods have been developed to determine the performance of industrial robots. The methods presented in this section illustrate basic principles that can be used to determine the different aspects of robot performance:

- The *"IPA method"* (of IPA, Stuttgart) illustrates a generalized approach to the measurement of both static-positioning errors and dynamic path-following parameters.
- The *"Ranky method"* describes a technique for the measurement of position and orientation errors obtained during static positioning and calculated statistically as three-dimensional error vectors, rather than scalar values.
- The *"Ford RAACC method"* describes techniques for the certification of industrial robots.
- The *"National Bureau of Standards (USA) method,"* using laser, and finally
- The *"Chesapeake Laser Systems method,"* using split laser beams in order to determine position errors.

The IPA Method

In order to measure the characteristics within the complete workspace of an industrial robot with sufficient accuracy, it is necessary to have exact coordination between the robot and the measuring equipment. The test bed must also be highly automated in order to reduce the amount of time required to determine the performance of the robot. With these requirements in mind, IPA constructed a test bed consisting of a measuring platform and a three-dimensional measuring device (Fig. 3) (7). The measuring device is movable along three sides of the measuring platform. This device allows for the measuring head to be positioned with an accuracy of 0.004 in. (0.01 mm) within the complete workspace. An adapter unit provides the mechanical interface for mounting the measuring cube quickly to any desired industrial robot's gripper. Three acceleration sensors are mounted within the measuring cube to obtain three-dimensional acceleration information. The velocity values are obtained through the integration of the acceleration values.

Figure 3. IPA measurement apparatus. Courtesy of IPA, Stuttgart, FRG.

Figure 4 shows the contactless three-dimensional measuring head and the measuring cube (fastened to the gripper of the industrial robot under test) (7). The measuring head can be rotated about any of its three axes. Since the measuring head is of cardanic construction, orientation of the path sensor can be obtained for any position in space without changing the intersection point of the sensor's axes in relation to the measuring device. The promixity gauges also allow for three-dimensional measurement of the oscillation of the gripper's

Figure 4. IPA measurement apparatus and test controller. Courtesy of IPA, Stuttgart, FRG.

end position. Potentiometers mounted at each revolute joint provide the angular position of the measuring head. This information combined with the position sensors on the measuring platform define the position of the measuring head for any point in time. Once the measuring head has been positioned, the proximity sensors provide three-dimensional displacement information of the robot under test. A computer is used to hold the measuring and the data-evaluation time to a minimum (Fig. 4) (7).

The original measuring apparatus was useful for the determination of point-to-point information, but the advent of continuous path-controlled robots required a different apparatus to be developed which could measure both path-accuracy and path-repeatability parameters (Fig. 5). A measuring head was developed which incorporated two promixity sensors. The measuring head was then attached to the gripper flange of the robot under test. The measuring head is held at a fixed distance from a calibrated measuring beam during programming. The distance from the measuring beam is recorded as the robot moves along the beam during programming. The robot is then run in automatic mode and a second path recorded. The test computer compares the two pathes and displays the results. According to the designers of this test bed, following a nonlinear, eg, circular, arc is not necessary, since a straight line, placed diagonal in space, requires the movement of all axes to maintain the orientation of the measuring head. This simplifies the data-acquisition process considerably.

Many robots displayed a strong temperature drift during initial start of the tests (7). To reduce the effect of the drift, the path tests were conducted after the robot was in a stabilized condition. For tests where the long-term effects were of interest, the temperature drift represented the largest problem to be overcome. In some cases temperature sensors, eg, thermocouples, were used to indicate the temperature stability of the robot. Tests of long duration were conducted in two phases. During the start-up phase the measurements were conducted over short cycles (approximately 3 min). After it was determined the robot had stabilized, the measurements were conducted over longer time intervals (approximately 10 min). Conducting the tests in this manner allowed the warm-up behavior of the robot to be adequately characterized.

The Ranky TEST-ROBOT Program

In general, manufacturing errors associated with each joint, arm, or tool together with the robot control system errors result in robot-hand position and orientation errors. The TEST-ROBOT computer program was developed to measure the robot-hand alignment errors and to characterize the orientation capability of industrial robots (8). Position repeatability and orientation information can be measured by the use of a test apparatus which can measure the position and the orientation of the robot end-effector and the related test gear. Figures 5 and 6 show the initial, ie, Mark I, test equipment which was used to verify the concept. (Although the test equipment uses dial indicators, noncontact displacement transducers are preferred and are currently available in the Mark II version.)

The true position of the robot hand, ie, tool, can be defined by a vector $\bar{r}$ pointing to any known point of the test component in the test-rig coordinate system (a point on the test cube is often used for simplicity) and the robot-hand orientation vector $\overline{N}$ defined by $\bar{n}_1$, $\bar{n}_2$, $\bar{n}_3$ independent unit normal vectors which define the orientation of the robot hand in the test-rig coordinate system. If a sufficient number or measurements are performed, then the evaluation of a series of such measurements

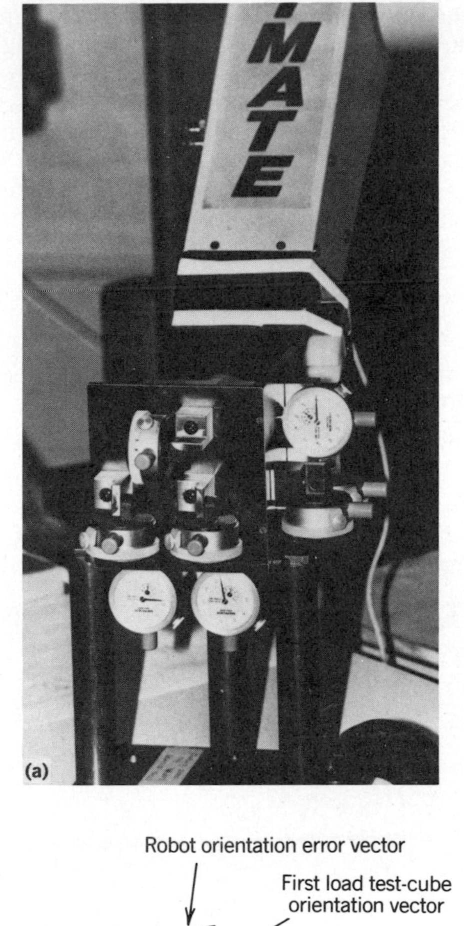

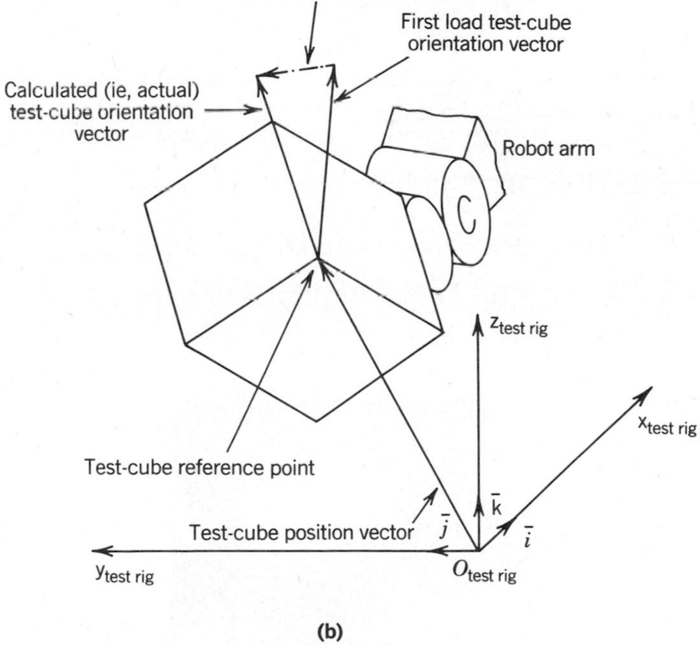

Figure 5. (a) Ranky Mark I version test apparatus, back view; **(b)** principle of measurement (8).

indicates the statistical average error of the position and orientation of the robot hand (8).

The core of the calculation method is based on the following mathematical model (Figs. 5 and 7). Consider Face 1 on the test cube. Having measured the indicated points in the test-rig coordinate system, vectors $\bar{p}_1$, $\bar{p}_2$, and $\bar{p}_3$ can be defined from points P_1, P_2, and P_3 on Face 1 of the test cube, respectively.

$$\bar{p}_1 = \begin{bmatrix} x_1 \\ y_1 \\ z_1 \end{bmatrix} \quad \bar{p}_2 = \begin{bmatrix} x_2 \\ y_2 \\ z_2 \end{bmatrix} \quad \bar{p}_3 = \begin{bmatrix} x_3 \\ y_3 \\ z_3 \end{bmatrix}$$

The orientation of Face 1 can be determined by calculating the normal (unit) vector ($\bar{n}_1$) to the face.

The $\overline{p_2p_1}$ and $\overline{p_2p_3}$ vectors can be defined as

$$\overline{p_2p_1} = \bar{p}_1 - \bar{p}_2 \quad \text{and} \quad \overline{p_2p_2} = \bar{p}_3 - \bar{p}_2$$

Both of these normal vectors lie in the plane of Face 1. A vector normal to Face 1 will be $\overline{p_2p_3} \times \overline{p_2p_1}$, since

$$\overline{p_2p_1} = (x_1 - x_2)\bar{\imath} + (y_1 - y_2)\bar{j} + (z_1 - z_2)\bar{k}$$

and

$$\overline{p_2p_3} = (x_3 - x_2)\bar{\imath} + (y_3 - y_2)\bar{j} + (z_3 - z_2)\bar{k}$$

then

$$\overline{p_2p_3} \times \overline{p_2p_1} = \begin{vmatrix} \bar{\imath} & \bar{j} & \bar{k} \\ x_3 - x_2 & y_3 - y_2 & z_3 - z_2 \\ x_1 - x_2 & y_1 - y_2 & z_1 - z_2 \end{vmatrix} =$$

$$\begin{aligned} = \; & [(y_3 - y_2)(z_1 - z_2) - (y_1 - y_2)(z_3 - z_2)]\bar{\imath} \\ & + [(x_1 - x_2)(z_3 - z_2) - (x_3 - x_2)(z_1 - z_2)]\bar{j} \\ & + [(x_3 - x_2)(y_1 - y_2) - (x_1 - x_2)(y_3 - y_2)]\bar{k} \end{aligned}$$

For simplicity, if

$$\begin{aligned} C_1 &= (y_3 - y_2)(z_1 - z_2) - (y_1 - y_2)(z_3 - z_2) \\ C_2 &= (x_1 - x_2)(z_3 - z_2) - (x_3 - x_2)(z_1 - z_2) \\ C_3 &= (x_3 - x_2)(y_1 - y_2) - (x_1 - x_2)(y_3 - y_2) \end{aligned}$$

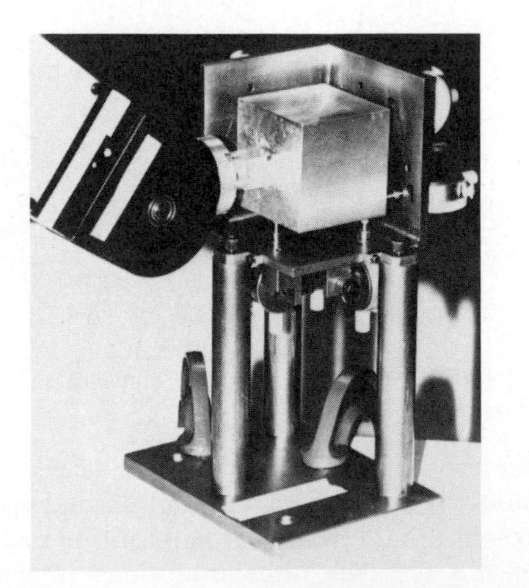

Figure 6. Ranky Mark I version test apparatus, side view.

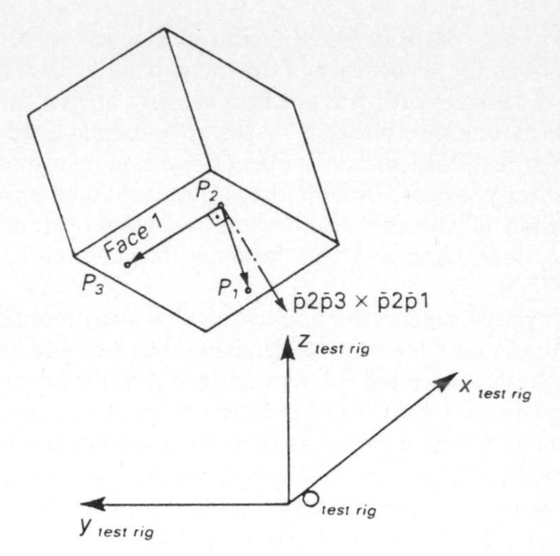

Figure 7. Ranky test method.

then

$$\overline{p_1p_3} \times \overline{p_2p_1} = C_{1\bar{\imath}} + C_2\bar{j} + C_3\bar{k}$$

where C_1, C_2, and C_3 contain only sensory feedback, ie, measured, data and are easily computed.

The unit normal to Face 1 is

$$\bar{n}_1 = \frac{C_{1\bar{\imath}} + C_2\bar{j} + C_3\bar{k}}{\sqrt{C_1^2 + C_2^2 + C_3^2}}$$

The other two unit vectors, $\bar{n}_2$ and $\bar{n}_3$, can be found independently on Face 2 and Face 3 after having measured points P_4, P_5, P_6, P_7, P_8, and P_9, respectively.

It is important to realize that by calculating and measuring any of the above three normal vectors, the robot test-rig and test-cube perpendicularity errors can also be detected. This redundant information serves as a self-calibrating and diagnostic feature, since the test cube is correctly manufactured and fully measured for perpendicularity before the robot-alignment error measurements are performed.

Having found $\bar{n}_1$, $\bar{n}_2$, and $\bar{n}_3$ normal unit vectors independently, the orientation vector of the robot hand can be defined:

$$\bar{n}_1 + \bar{n}_2 + \bar{n}_3$$

Assuming that $\bar{r}_{i+1}$ is the subsequent position of $\bar{r}_i$, generally the relative position error of the robot (ie, relative to the coordinate system of the test apparatus) can be calculated as

$$\overline{E}_i = \bar{r}_{i+1} - \bar{r}_i$$

The absolute position and orientation error calculation is based on the first test-cube load position and orientation. In other words, the first load (or set) position and orientation provides the base data for the calculations.

After a sufficient number of robot loading and unloading operations are performed, a statistical analysis of the position and orientation errors can be made. During the analysis, the measured as well as the calculated data can also be displayed and/or printed out by computer program. For user convenience and for experimentation, the input data table can be edited (partially or fully, ie, for each test-cube face and/or point) and the program can be rerun as many times as desired.

Several different industrial robots have been tested successfully with the described method and software. Figure 8 shows an example of some test data collected and the results of the robot-alignment error calculation program. The output shows the measured data as well as the calculated position and orientation error vectors together with their standard deviation and variance values. The results in this example are given in millimeters and show that the measured robot has reasonably large position errors and small orientation errors that the test-rig perpendicularity errors, ie, the measurement inaccuracies, are negligible.

Finally, a Mark-II version of the robot test device has been implemented recently by the first author and his students at the University of Michigan. The Mark-II version uses six noncontact sensors, a digital data logging card, and fast evaluation software running in an IBM PC/AT. This new device can be used both for off-line position/orientation error testing as well as for real-time recalibration purposes.

The Ford RAACC Testing Methods

The Ford Motor Company devised some tests after some discrepancies were found between the actual performance and the specified performance of the robots they had purchased. A group within the Robotics and Automation Applications Consulting Center (RAACC) became responsible for the certification of all robot models that were to be purchased by the Ford Motor Company. The major premise for the development of these tests was to use comparable test equipment and test procedures on each robot to enable comparisons from robot to robot. The information is then disseminated to each Ford plant to allow the application engineers to determine the *best* robot for the application at hand.

Following the RAACC method, robot under test is first put through a cycling test which consists of movement of all axes from one end of motion to the other at full rated speed and payload. The test is noninstrumented and run continuously for 40 hours with any duty cycle limitations strictly followed. The purpose of the test is to check for *infant mortality* problems, eg, undersized fuses, bolts not properly tightened, etc, and to get a feel for the integrity and reliability of the robot system.

The first test apparatus developed by Ford consisted of a proximity sensor system for the measurement of point-to-point behavior (Fig. 9). A test stand was constructed for placement of a test head (or *nest*—RAACC term) in various strategic locations within the work envelope of the robot under test. The test head consisted of three promixity sensors mounted orthogonally to each other. A *target* is mounted to the end of the robot and moved into the test head along a described path. The approach first taken by Ford was to measure the robot

```
************
Calculated results
************

Robot position error
*************

                              x             y             z

Standard deviation =       0.3722        0.1768        0.1129

Robot orientation error
***************

                              ī             j̄             k̄

Mean value          =      0.0005        0.0007       −0.0020
Variance            =      0.0000        0.0000        0.0000
Standard deviation  =      0.0020        0.0056        0.0026

Calculated, ie, actual, robot alignment
************************

Test-cube position vector    = [−1.3973 ī      5.3248 j̄      2.1436 k̄]

Test-cube orientation vector = [0.9795 ī       1.0308 j̄     −1.0001 k̄]

**>> Note: Both vectors are given in the test-rig coordinate system
           and relate to the predefined test-cube reference point

Test-cube face perpendicularity errors
*************************

Theoretical test-cube orientation = [0.0254 ī      − 0.0247 j̄     0.9992 k̄]

Measured test-cube orientation    = [−0.0292 ī       0.0202 j̄    −0.9994 k̄]
```

Figure 8. Partial computer output giving the measured input data and the statistically evaluated results based on three-dimensional robot position and orientation error analysis (8).

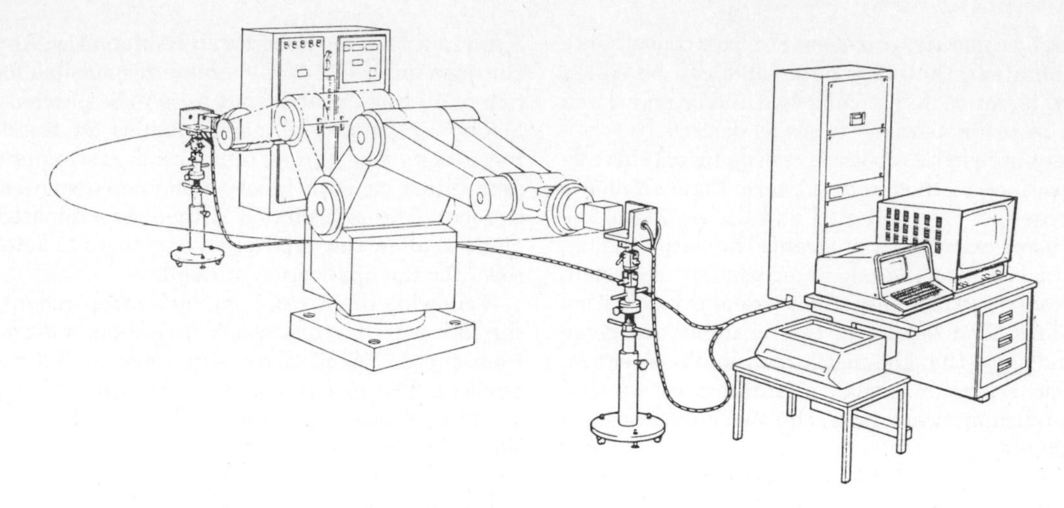

Figure 9. Ford RAACC repeatability apparatus. Courtesy of RAACC, Detroit, Mich.

under various conditions. The result was a series of tests consisting of three basic motions (sweep, in/out, up/down) and eight permutations on the robot condition (varying payload, reach, speed). One thousand continuous cycles were measured from a *cold start*. Then process statistics were used to determine the average placement, range, minimum reading, maximum reading, frequency distribution, and standard deviation (±3 sigma) of the data collected.

It became evident that there were too many permutations. Tests on each robot took over a month to perform, resulting in a lengthy backlog at the RAACC. A minimum requirement was adopted to test the robot under *worst case* conditions (ie, maximum reach, speed, and payload) for each of the three motions and to measure only the major axes for gross positioning behavior. The term *repeatable placement accuracy* (*RPA*) was coined (6):

> Repeatable placement accuracy is defined as the envelope of variance that the tool point was positioned after repeated cycles in relation to the original taught position. This should not be confused with "repeatability" which is the envelope of variance irrespective of original taught point (eg, the range of difference between the various locations of the tool point).

Each of the test motions were designed to measure either a particular axis or groups of axes to determine their contribution to RPA. The sweep motion was designed to limit motion to the base rotation axis on robots with rotary base axes or to the first linear axis of Cartesian-type robots. Through experimentation it has been determined that the base axes of robots which rotate about their base usually have the most detrimental effect on RPA. For this type of robot, the base rotation axis usually has the lowest resolution (when considering the full extension of the robot) and the highest frictional loading. The sweep test fixes all other axes and moves only the first axis. This motion would simulate many typical material-handling applications where the robot would pick up a part from one location (eg, a conveyor) and place the part at another location (eg, a pallet). The in/out motion is limited to the upper arm, forearm, and wrist axes (to maintain wrist posture) of articulated robots and to the second linear axis of Cartesian robots. This motion would simulate tool-loading applications. Similarly, the up/down test is limited to the same upper arm, forearm, and wrist axes of articulated robots and to the third linear axis of Cartesian robots. This motion would simulate palletizing and part insertion.

In addition to the capture of the end points, the path of the end-effector is measured as it approaches the measurement area. A strip-chart recorder is used to measure any undershoot or overshoot that may result and to determine when an output signal occurs in relation to the end-point position. The effect of warm-up on RPA is also evaluated to determine the drift tendencies of the robot. This allows the user to determine if exercise programs are required in the actual applications to enable the robot to maintain the required RPA.

The advent of the *Robot Check* system from Selspot Systems also allowed Ford to measure dynamically the path behavior of an industrial robot being considered for purchase. Process applications, eg, welding and sealant, require information on the ability of the robot to follow straight lines and arcs and negotiate corners. A test scheme was developed that consisted of a series of rectangles and circles positioned within three planes. Tracing each path at 10%, 50%, and 100% speed allows the path accuracy versus speed behavior to be characterized.

The path tests can be considered a form of *bench-mark* test. The goal is to determine the relative performance along a predefined path. All tests are conducted in a warm-start condition and at full payload. The rectangular and circular motions are placed in the center portion of the work envelope to allow the *best* performance available to be measured. One series of tests (ie, all three planes at all three speeds) is run using precision points (ie, acceleration/deceleration at each taught point) and another utilizing flyby intermediate points (ie, rough accuracy).

For each plane, graphs of the position of the three measured speeds are overlaid and measurements taken to determine any degradation with the increase in speed (Fig. 10). The 10% speed run is used as a *baseline* from which the 50% and 100% speed runs are measured. Graphs of the velocities are evaluated to determine the performance around the taught points (Fig. 11). Cycle times are also measured and the average velocity for the entire motion is determined.

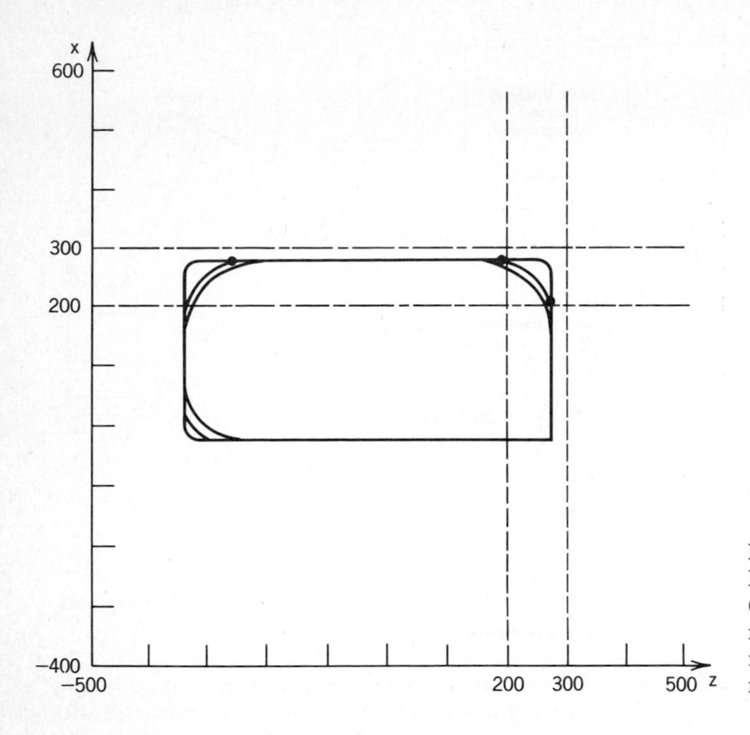

Figure 10. Typical path test position results using the RAACC method: ______, ROBARF. SXZ, robot A, rectangle (10% speed: xz plane: flyby); __.__, ROBARF. HXZ, robot A, rectangle (50% speed: xz plane: flyby); __-__, ROBARF. FXZ, robot A, rectangle (100% speed: xz plane: flyby). Position in millimeters.

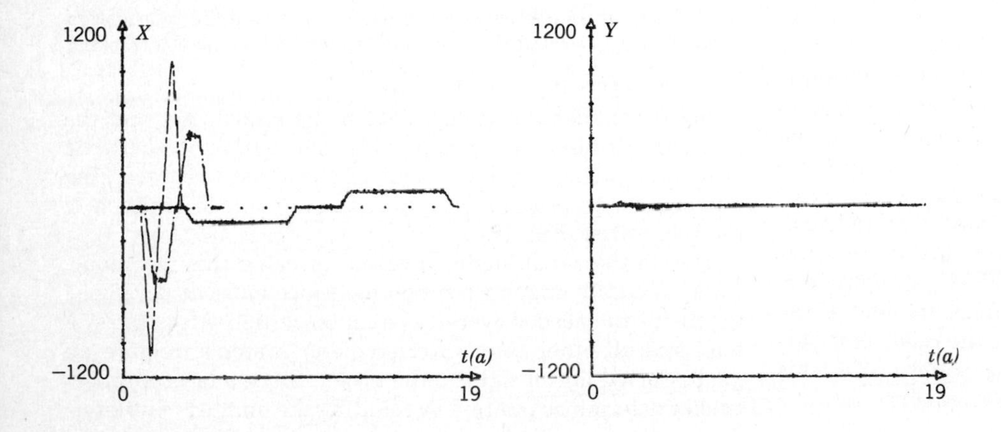

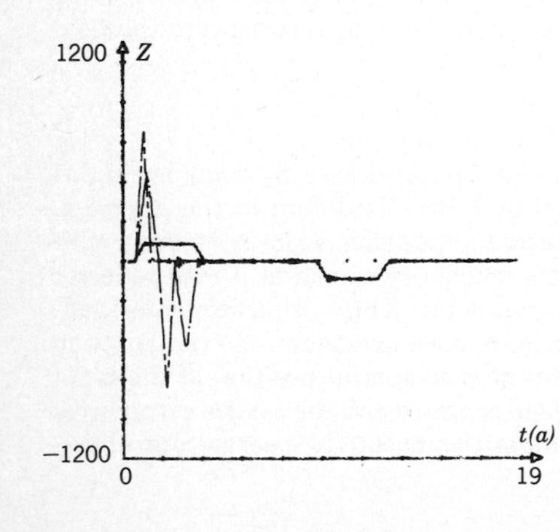

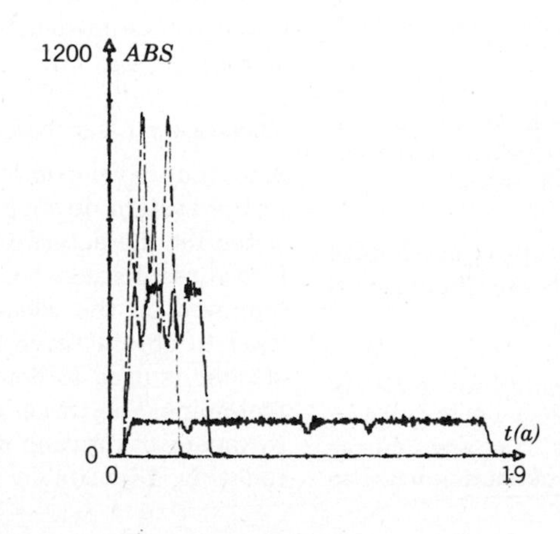

Figure 11. Typical velocity path test results using the RAACC method: ______, ROBARF. SXZ, robot A, rectangle (10% speed: xz plane: flyby); __.__, ROBARF. HXZ, robot A, rectangle (50% speed: xz plane: flyby); __-__, ROBARF. FXZ, robot A, rectangle (100% speed: xz plane: flyby). Velocity in millimeters per second.

The results from Ford certification tests provide the application engineer with valuable information regarding the integrity and actual performance capability of industrial robots. When properly designed, the tests can be used for the comparison and subsequent selection of a robot for a particular application.

FUTURE OUTLOOK

The approach taken for measuring performance by monitoring the spatial position of the robot places stringent demands on the measurement system. There are many types of equipment capable of measuring *static* behavior (eg, proximity sensors, LVDTs, vision systems), but there are few systems available with the required resolution or the measurement of *dynamic* behavior. By far the largest problem for measuring dynamic properties is the ability to measure the spatial position of the robot over a large volume (ie, greater than 1 cubic meter) with the required resolution (ie, better than 0.1 millimeter). Many robots have linear path accuracies in the range of 0.04 in. (1.0 mm). Using the *order-of-magnitude-better* rule requires the resolution of the test equipment to be in the range of 0.004 in. (0.1 mm).

Many types of technologies have been investigated to provide this resolution. For example, stereo vision systems have been used but their resolution is dependent on the field-of-view of the lenses. As the measurement volume increases, the resolution drops dramatically.

Probably the most promising technology is the use of laser interferometry. Most interferometers can measure a linear distance in the range of micro-inches. The use of interferometers for CNC machine tool and coordinate measuring machine (CMM) calibration is well-established. Most interferometer systems have automatic temperature and humidity compensation and their use is relatively straightforward. But traditionally, interferometers have been used for uniaxial measurements only, since light travels in a relatively straight line. The ability to provide three-dimensional measurement of the robot tool-point requires the laser to *track* the robot as it performs its motion. Much of the work being performed in this area is in the development of a tracking system (1).

National Bureau of Standards

A research group was formed in the National Bureau of Standards (NBS) for the measurement of industrial robot performance (9):

> The basic function of this group is to devise methodologies, instruments and standard test procedures for characterizing the accuracy, repeatability and dynamic performance of robots.

The system developed by NBS used the concept of one length and two angles for the determination of three-dimensional position (1):

> In this method the idea is to continuously control the angles of projection of a laser beam in order that the beam will impinge on the target mirror mounted to the robot wrist. The target mirror is also angularly controlled and returns the measuring beam to the source. The change in 'length' of the beam is measured by interferometry and this change in 'length,' when combined with the angular measurements, yields the position of the robot wrist in spherical coordinates. The angular orientation of the mirror on the robot wrist also can be used for information regarding the wrist orientation.
>
> Quadrature photodiodes are used to generate the information necessary to control the four angular servos (two on the tracking mirror and two on the target mirror).

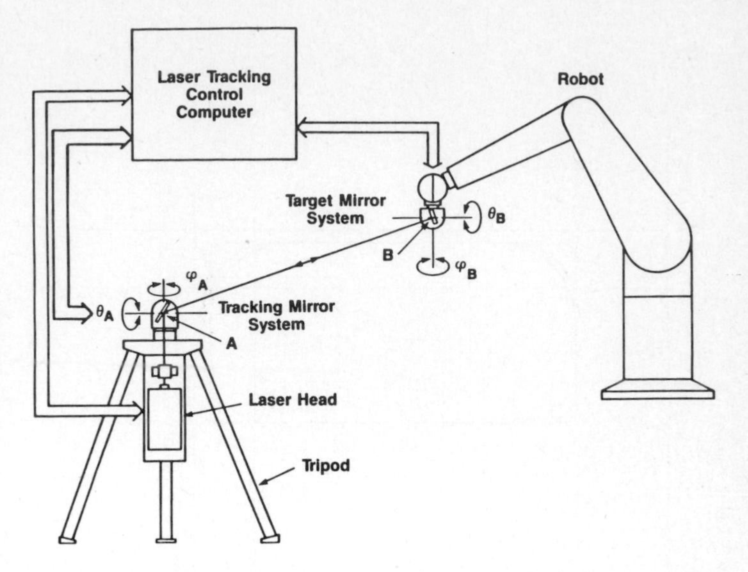

Figure 12. NBS Laser Metrology Systems's illustration of a conceptual 3D laser tracking system. Courtesy of NBS.

Using two tracking systems, one on the robot wrist and the other on the laser, can also provide information on the wrist orientation. The angular output of the robot-wrist tracking mirrors can be used to determine the angular orientation of the robot wrist (Fig. 12).

Due to the small angles involved in using this technique, a very accurate angular positioning device must be developed to provide the needed overall system resolution. Also, this system, and all other laser interferometer systems, require an unobstructed line of sight to the target. Since a laser interferometer determines position by counting the number of interference patterns as the target is moved along the laser beam, any obstruction in the path of the light will result in a loss of counts and an error in the measurement. (Vision systems have this problem to some extent, but most are able to recover and continue measuring when the light returns into the field of view.)

Chesapeake Laser Systems

A system developed by Chesapeake Laser Systems is similar to the system developed by NBS but differs in the approach taken for the determination of position. A laser beam is split into three distinct beams which are aimed at a retroreflector mounted on the robot tool-point. Three interferometers are used to obtain three length measurements. A triangulation scheme is used to determine the spatial position of the robot tool-point. The three laser beams track the targer retroflector by means of rotating mirrors connected to a servo-control system (Fig. 13) (10):

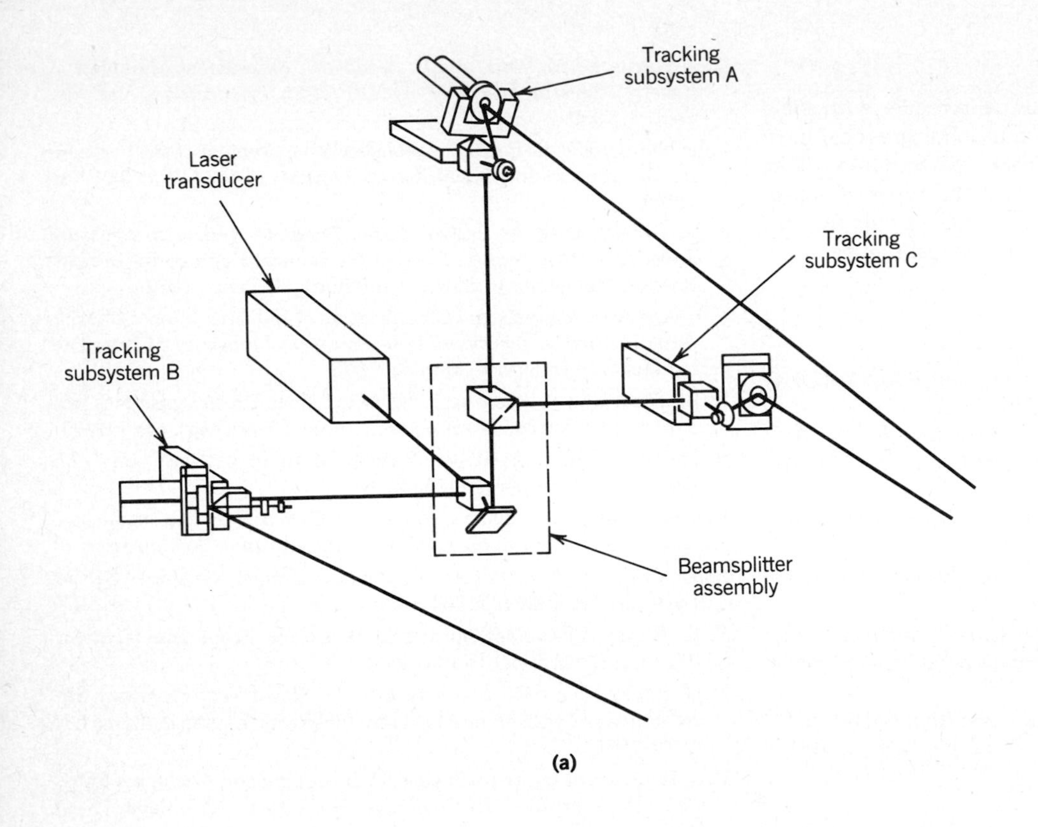

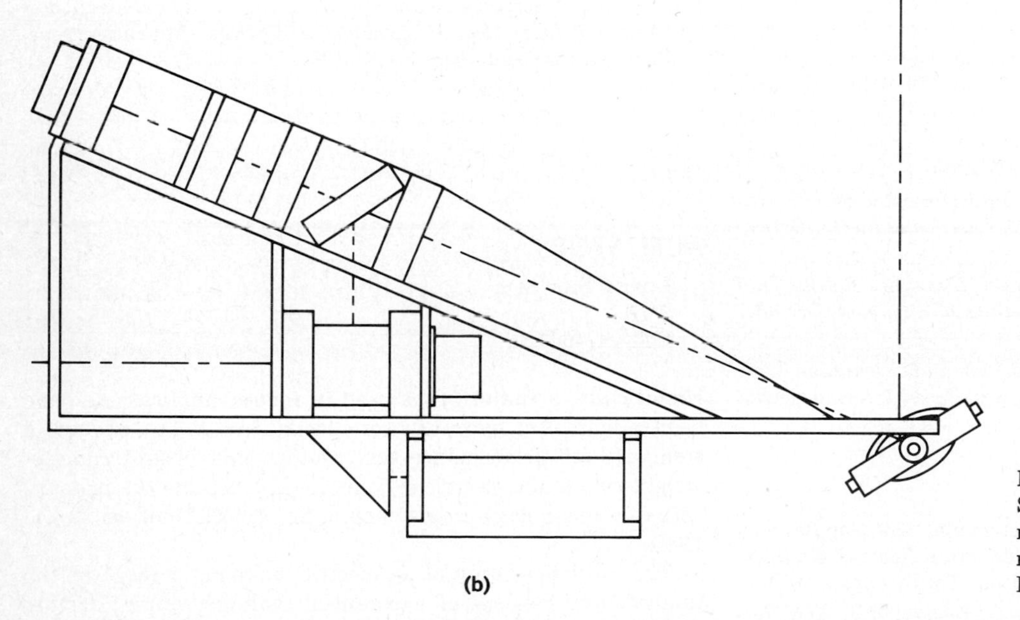

Figure 13. Chesapeake Laser Metrology Systems: **(a)** basic random path measurement system; **(b)** tracking interferometer module structure. Courtesy of Chesapeake Laser Metrology Systems.

It should be noted that knowledge of the angular orientation of the tracking reflectors in unnecessary, since the three cartesian coordinates of the target may be obtained solely from the three distance measurements and knowledge of the position of the tracking systems.

Wrist orientation information could be obtained by the addition of three more length measurements, ie, three more interferometers and associated equipment. All the laser beams (as much as six) must also be unobstructed in the Chesapeake system, otherwise the system will not be able to provide positional information. It should be emphasized, though, that a sufficient access must be available for just about every type of measurement system available. Vision systems and laser tracking systems differ only that they are usually located outside the work envelope and therefore require that the robot itself must not block the line of sight to the measurement equipment. In practice, this does not usually present any significant restrictions.

CONCLUSION

The purpose of this article is to acquaint the reader with some of the different equipment and testing methodologies that have been used for the measurement of robot performance. This discussion is not intended to describe all the types of equipment and measurement techniques currently available. See Refs. 9–12 for other sources of robot testing techniques.

BIBLIOGRAPHY

1. K. Lau and R. J. Hocken, "A Survey of Current Robot Metrology Methods," *Annals of the CIRP,* **33,** 2 (1984).
2. G. Arnold, *Continuous-Path Testing of Aerospace Robots, RIA/SME Technical Paper MS86-195,* Robot Institute of America, Ann Arbor, Mich., 1986.
3. Robot Institute of America, *RIA Robotics Glossary,* Robot Industries Association, Ann Arbor, Mich., 1984.
4. P. Albertson, "Verifying Robot Performance," *Robotics Today,* 33–36 (Oct. 1983).
5. Statistical Methods Office. *Continuing Process Control and Process Capability Improvement,* Operations Support Staffs, Ford Motor Co., Dearborn, Mich., Feb. 1984.
6. Robotics and Automation Applications Consulting Center, *Robot Assessment Program, Test Procedures, Test Equipment, Data Collection,* Engineering and Manufacturing Staff, Ford Motor Co., Dearborn Mich., Jan. 1986.
7. H. J. Warnecke, "Test Stand for Industrial Robots," *International Symposium on Industrial Robotics, 7th Proceedings,* Stuttgart, FRG, Oct. 19–21, 1977.
8. P. G. Ranky and C. Y. Ho, *Robot Modelling, Control, and Applications with Software,* IFS (Publications), Ltd., Bedford, UK, 1985.
9. K. Lau, R. Hocken, and W. Haight, *An Automatic Laser Tracking Interferometer System for Robot Metrology,* National Bureau of Standards, Gaithersburg, Md., Mar. 17, 1985.
10. L. B. Brown, "A Random-Path Laser Interferometer System," *International Congress on Applications of Lasers and Electro-Optics,* San Francisco, Nov. 11–14, 1985.
11. G. Arnold, "Continuous-Path Testing of Aerospace Robots," *Robotic Solutions in Aerospace Manufacturing Conference,* Orlando, Fla., Mar. 3–5, 1986.
12. P. G. Ranky and co-workers, *Robot Position and Orientation Error Measurement with a Non-contact Sensory Based Interchangeable Hand,* University of Michigan, Ann Arbor, Mich., 1986.

General References

H. J. Warnecke, R. D. Schraft, and G. Herrmann, "The Gap Between Required and Realized Properties of Industrial Robots," *4th International Symposium on Industrial Robots,* Tokyo, Japan, 1974.

H. J. Warnecke and R. D. Schraft, *Industrie-Roboter,* Krauskopf Verlag, Mainz, FRG, 1973.

G. Spur and B. H. Ames, "Acceptance Conditions for Industrial Robots," *3rd Conference on Industrial Robot Technology and 6th International Symposium on Industrial Robots,* Berlin, Mar. 24–26, 1976.

B. Scheffes, "Geometric Control and Calibration Method of an Industrial Robot," *12th International Symposium on Industrial Robots,* Renault, France, 1976.

F. Tanguy, "Assessment of Mechanical Performance on Industrial Robots," *6th International Conference on Industrial Robot Technology,* June 9–11, 1982.

H. J. Warnecke, M. Weir, and B. Brodbeer, "Assessment of Industrial Robots," *General Assembly of CIRP, 30th Proceedings,* Aug. 29–Sept. 4, 1980.

J. Lombard and J. C. Perrot, "Automatic Measurement of the Positioning Accuracy of Industrial Robots," *Annals of the CIRP,* **32** (Jan. 1983).

J. H. Gilby and G. A. Parker, *Laser Tracking System to Reassure Robot Arm Performance,* Dept. of Mechanical Engineering, Sensor Review, University of Surrey, Guildford, UK, Oct., 1982.

Y. Hasegawa, "Analysis and Classification of Industrial Robot Characteristics," *3rd International Symposium on Industrial Robots,* Verlag Moderne Industrie, Munich, 1973.

B. Brodbeer and R. Breitinger, "Entwurf eines Prüfprogrammes und Prüfstandes für Industrieroboter," *Industrie-Anzeiger* (Jan. 1976).

R. Breitinger and B. Brodbeer, "Prüfstand für Industrieroboter," *Jördesu und Lieben,* Nr. 15, 26 (1976).

VDI-Richtlinie 3254, Blatt 1, *Numerisch Gesteneste Werkzengmaschinen, Genauigreitsangaben, Begriffe und Statische Kenngrössen.*

Robot Check, A New Robotic Evaluation System, Selspine, Selspot Systems, Troy, Mich., 1986.

P. G. Ranky, "Test Method and Software for Robot Qualification," *The Industrial Robot* (June 1984).

P. G. Ranky, *The FMS Software Library,* Robot Test Program, IBM PC Software with manuals, Trent Polytechnic, Nottingham, UK, 1982–1985.

P. G. Ranky, "Integrated Software for Designing and Analyzing FMS," *3rd International FMS Conference,* Stuttgart, FRG, Sept. 11–13, 1984.

P. G. Ranky, *A Software Library for the Design and Optimization of Computer Integrated Manufacturing Systems,* Autofact/Europe, Basel, Switzerland, Sept. 25–27, 1984.

ACTUATORS

Loren Shaum
SMI
Elkhart, Indiana

Historically, actuator types used in robotic applications have been pneumatic, hydraulic, or electric. With the advent of smaller, high performance servomotors with large torque-to-weight ratios, the electric servomotor has become the primary actuator technology implemented for robotic motion in the 1980s.

The basic operation of an electric servomotor involves the fundamental concept of a potential (voltage source) driving current through a conductor. A magnetic field is generated, which can be controlled when current is flowing in a multiple conductor array (winding). Alternating polarized magnetic fields are generated when current is flowing in windings that are configured lengthwise over the circumference of a slotted cylinder (armature). Rotation results when these polarized fields are confronted by opposite polarity magnets mounted on a stationary frame (stator) enclosing the armature. The circulating current multiplied by the winding length results in a tangential force (to the winding). Torque at the motor shaft results from the multiplication of the armature radius

by this tangential force. Figure 1 provides an illustration of a simplified d-c motor.

The d-c motor is the predominant actuator utilized on articulated-arm robots today. However, brushless servomotors are now being introduced to robotic applications and will become the primary source of rotational actuation by 1990.

The remainder of this article discusses d-c servomotors and new a-c servomotors (brushless d-c) and actuators as now used in robotic applications.

DRIVE MECHANISM

To make the d-c or brushless servomotors operate in a particular mechanical configuration, various linkages are required. These linkages convert torque to a linear force or rotating motion which accommodates a particular robot configuration. Common linkages used in robotics include (*1*) gears, (*2*) rotating screws with associated nuts (ball bearing or Acme thread), (*3*) rubber belts (typically timing belts), (*4*) steel bands, and (*5*) harmonic drives. These linkages are used in robotic configurations where direct-drive actuation is not possible. (Direct drive implies that motor(s) are coupled directly to the load.) Early designs used gear and belt combinations. Designs are now more refined and use more efficient technology such as harmonic drives and small steel bands with high tensile strength. The largest application for direct-drive designs is in low cost, high speed assembly, and material-handling robots. Several designs that claim direct-drive actuation are currently available. However, some of these still retain belts or steel bands (even though linked by a 1:1 ratio) as connections between motor shafts and the moving member of the robot. All indirect-drive configurations have considerable torque losses due to inefficiencies in the linkages from the motor shaft to the load. Most indirect-drive arrangements require smaller motors. But if ratios in the linkages are required, the motors operate at substantially higher speeds, thus reducing life (brush life in d-c motors diminishes as speed increases).

Many designs incorporate harmonic drives to operate moving members that are arm-mounted. Waist motions on many robots use ring gears with a worm gear coupled to the motor shaft. Recent robot designs incorporate all actuators in the base of the assembly with a series of steel bands guided through the various robot linkages to achieve the motion at the appropriate robot joint. SCARA (simplified compliance assembly robot arm) makes use of harmonic drives for most joint actuations.

Servomotor technology commercially available in the early 1980s resulted in advances including higher speeds, lighter weights, and heavier payloads. These new technologies are now incorporated in many articulated-arm designs, including SCARA robots. These new technologies include (*1*) torque rings, (*2*) rare-earth d-c servomotors, (*3*) brushless servomotors, and (*4*) variable-reluctance stepping motors. The latter technology has recently been configured in arrangements that totally eliminate the conventional ring/worm gear combination for waist motions. These newer robotic designs (1984 and after) use variable-reluctance motors as direct-drive, high torque stepping motors. These motors are configured to fit the dimensions of the robot base. These so-called stepping torque rings result in good rotational speeds with reasonable repeatability. To obtain better repeatability, robot manufacturers incorporate encoder feedback to verify proper position (compensating for the open-loop stepping motor limitations).

As technology advances, these stepping rings will be replaced by high torque, brushless servomotors.

D-C SERVOMOTORS

With the advent of cost-efficient, rare-earth magnets, motor manufacturers can produce high torque motors commercially with very low weight. This large torque-to-weight ratio has established rare-earth d-c servomotors as the primary actuating means for articulated arms where motors are mounted on moving members. In some configurations, small rare-earth motors mated to harmonic drives provide best torque/speed/weight combinations for a given joint architecture. Recently, direct-drive actuation with rare-earth motors has been designed in. Many of the light weight, high speed assembly robots now use small rare-earth d-c servomotors in combination with a single belt and pulley combination. This allows the motor to be mounted more closely to the center of gravity, thus reducing the weight at the outer extremity of the robot arm.

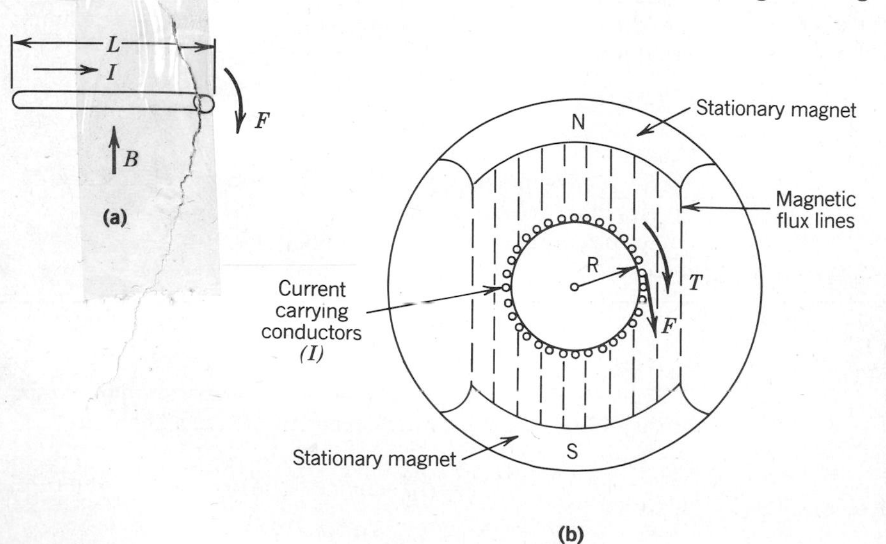

Figure 1. Principles of basic d-c motor operation. **(a)** Winding physics: F = force is perpendicular to conductor; B = magnetic-field strength (flux density); L = length of conductor; I = current through conductor. Force = BLI. **(b)** Basic configuration (cross section). d-c motor parameters for servo analysis: K_t = torque constant; K_v = voltage constant; J_a = motor armature inertia; T_r = rated torque (RMS); T_p = peak torque. Torque = $F \times R = BLI \times R$.

The operation of rare-earth d-c servomotors is the same as any other d-c servomotor. The difference lies in the flux density of the magnets. Substantially higher field strengths are available with rare-earth magnets without sacrificing motor size and weight. The result is smaller armatures with fewer windings. Consequently, high torques can be generated with less current, resulting in the additional savings of smaller power amplifiers.

In the future, the use of d-c servomotors in robotic design will be based on economic considerations. Economy results directly from the ability to produce higher torques with smaller motor frame sizes and lighter weight.

All d-c servomotors, whether standard magnet or rare-earth magnet, still require brushes to maintain current flow through the armature. Several industries have led a trend toward eliminating brushes on motors where contaminated environments or high speed operation is necessary. In many robot designs, motor speeds of 3000 rpm and greater are common because of large gear ratios in harmonic drives. By using rare-earth magnets in d-c servomotors, smaller gear reductions allow motors to run at slower speeds, typically 1000 rpm or less. These lower speeds have significantly enhanced brush life. Still, there is the stigma associated with brushes as a long-term maintenance item. Many large users of robotic equipment have implied that they prefer brushless motors as the technology becomes economically available.

Figure 2 shows typical d-c servomotors. Figure 3 is a cross-sectional view of a rare-earth servomotor.

A-C BRUSHLESS SERVOMOTORS

With the continuing pressure by end users to eliminate high wear items, an emotional trend is under way regarding the utilization of a-c brushless servo actuators in place of d-c servomotors.

In reality, an a-c brushless servomotor operates very much like a three-phase a-c synchronous-reluctance motor. The construction is similar except for the rotor. The rotor is constructed with magnets mounted on a small diameter shaft. The stator contains the three-phase windings. Because the conducting paths are stationary, no brushes are necessary to generate the rotating field. The stators are wound like conventional three-phase a-c motor stators, hence, the term a-c brushless servomotor.

A-c brushless motors cannot operate in a pure velocity-loop mode. They inherently require position feedback (encoder or resolver) to operate at any speed or position-control application (otherwise they can operate only at synchronous speed).

Figure 2. Typical d-c servomotor. Courtesy of Powertron Div., Contraves Goerz Corp.

Figure 3. Cross section of a rare-earth d-c servomotor. Courtesy of Powertron Div., Contraves Goerz Corp.

Because of the three-phase winding configuration, three amplifier power stages are needed for excitation. By controlling the energization sequence of each winding, rotation is established by virtue of opposing magnetic fields. The rotor follows because it is configured with alternating (and opposite) magnetic poles. Figure 4 is a brushless servomotor (with feedback). This is one example of many configurations now available. Typically, a-c brushless motors are smaller than d-c motors. This is primarily due to locating the magnets on the rotor. Since three-phase windings are used on the stator, there is less area necessary to create the same ampere-turns needed in a d-c motor.

Torque in the three-phase winding configuration is developed according to the actual angular position of the rotor. Torque can be determined by establishing phase A at 0°. Then the torque is determined by

$$T_a = K_t i_a \sin \theta$$

where T_a = torque developed from current flowing through the windings of phase A; K_t = torque constant of the motor (usually given in inch-pound per ampere); i_a = current flowing in phase A; and θ = electrical angular position.

If the phase current i_a is sinusoidal to provide self-commutating, then

$$i_a = i_p \sin \theta$$

where i_p = the maximum current that can be commanded to the amplifier driving the motor.

Phase A torque is now defined (by combining the two equations) as

$$T_a = K_t i_p \sin^2 \theta$$

The torque developed in phases B and C are similar except for the electrical angular displacement of each. Therefore

$$T_b = K_t I_p \sin^2 (\theta - 120°)$$

$$T_c = K_t i_p \sin^2 (\theta - 240°)$$

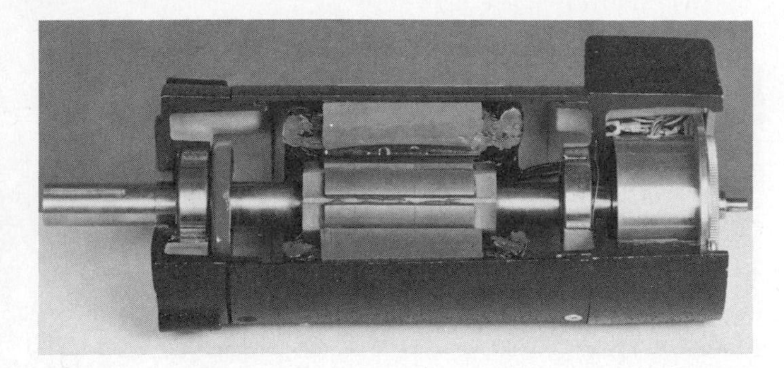

Figure 4. Cross section of a brushless servomotor. Courtesy of Powertron Div., Contraves Goerz Corp.

SERVO AMPLIFIERS

Established as a standard in the late 1970s, the pulse-width modulated (PWM) switching transistorized servo amplifier is the most common used for robotic actuations. Prior to this period, linear amplifiers and SCR controllers were the leading electronic components for d-c motor actuation. Today, practically all servo amplifiers utilize switching transistors. This technique incorporates the motor winding as part of the load in a full-wave power transistor bridge configuration.

The PWM design lends itself to small packages because of significant improvements in heat-dissipation characteristics (by switching rather than being "on" continuously). PWM technology also provides superior performance through current control (rather than voltage control, as in SCR controllers). These two obvious advantages have made SCR controllers and linear amplifiers virtually obsolete, particularly for robotic applications. There are still some large gantry mechanisms which might use large permanent magnet or field wound motors and SCR controllers. These configurations have substantial inertia and friction, which usually justifies the low bandwidth performance of SCR controllers. Articulated-arm robots typically require much higher performing motion-control systems. With this in mind, the best possible bandwidth (dynamic performance) between the servo amplifier and the servomotor is necessary. PWM technology is the key electronic element required to supply the bandwidth performance for a given application.

The new brushless servomotors use the same PWM switching technique except that there are three PWM power amplifiers. Each amplifier energizes one of the three-phase windings used in the stator.

All d-c and brushless servomotors have superior performance characteristics over variable-reluctance actuators because of peak torque capability in the amplifier. This, plus stabilized control loops (velocity and position), provides greater dynamic range and speed regulation.

Stabilization techniques are configured in many forms depending on the type of motion control utilized with the robot. Technology is available today which uses position-loop feedback without the requirement for the standard analogue, velocity feedback transducers (tachometers). These new techniques make use of microprocessors to calculate the velocity information from position feedback information. The latest designs involve digital implementation of all position and velocity information up to the point where digital information must be converted to analogue current commands. These are then amplified to sufficient ampere levels to drive the motors. At present, there are very few robots that provide full digital control to the power amplifiers. Later in the 1980s, this will become commonplace. Currently, there are too few people who sufficiently understand the motion-control algorithms associated with handling velocity and positional information in a microprocessor and having it converted appropriately to power-amplifier levels.

As amplifer stages become more digitized, they become simpler. The best design is full digital control loops with buffered power transistors necessary to create the appropriate PWM switching of current loads into d-c or brushless motors. Many suppliers make such products, but few robots incorporate this total digital technology.

Figure 5 is the fundamental servo-control loop for a conventional d-c servo amplifier. Figure 6 is a standard pulse-width modulated servo amplifier with axis-expansion capability.

Figure 7 shows a loop diagram for a conventional brushless servo amplifier. Figure 8 is a typical brushless servo-amplifier packaging configuration.

FEEDBACK DEVICES

To define a complete servo-actuator assembly as now used in robotic applications, it is necessary to include all the appropriate devices associated with providing information to the robot control system as it occurs at the motor shaft. This includes

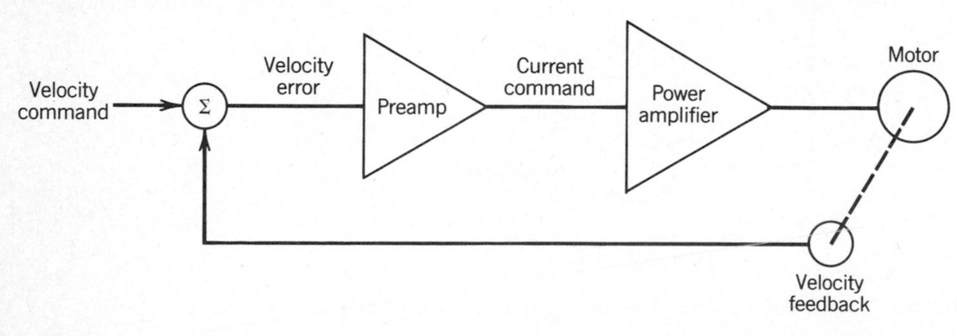

Figure 5. Conventional servo-amplifier loop.

Figure 6. Pulse-width modulated servo amplifier. Courtesy of Powertron Div., Contraves Goerz Corp.

velocity and position information. The most common actuators are currently comprised of the following components: (*1*) a d-c servomotor; (*2*) an analogue tachometer; and (*3*) an incremental encoder. Virtually all robots shipped in the early 1980s

Figure 8. Typical brushless servo-amplifier packaging. Courtesy of SMI.

use these elements in actuator assemblies that are the primary source for controlling motion.

With the advent of the microprocessor to handle velocity calculations, other transducers can be utilized on servo actuators to provide position and velocity information. Although resolvers have been available for some time, only in the last ten years have they been used for general position-control applications.

Many companies are now able to digitize resolver signals into appropriate position, velocity, and current information. With this capability, simplified current amplifiers can be used

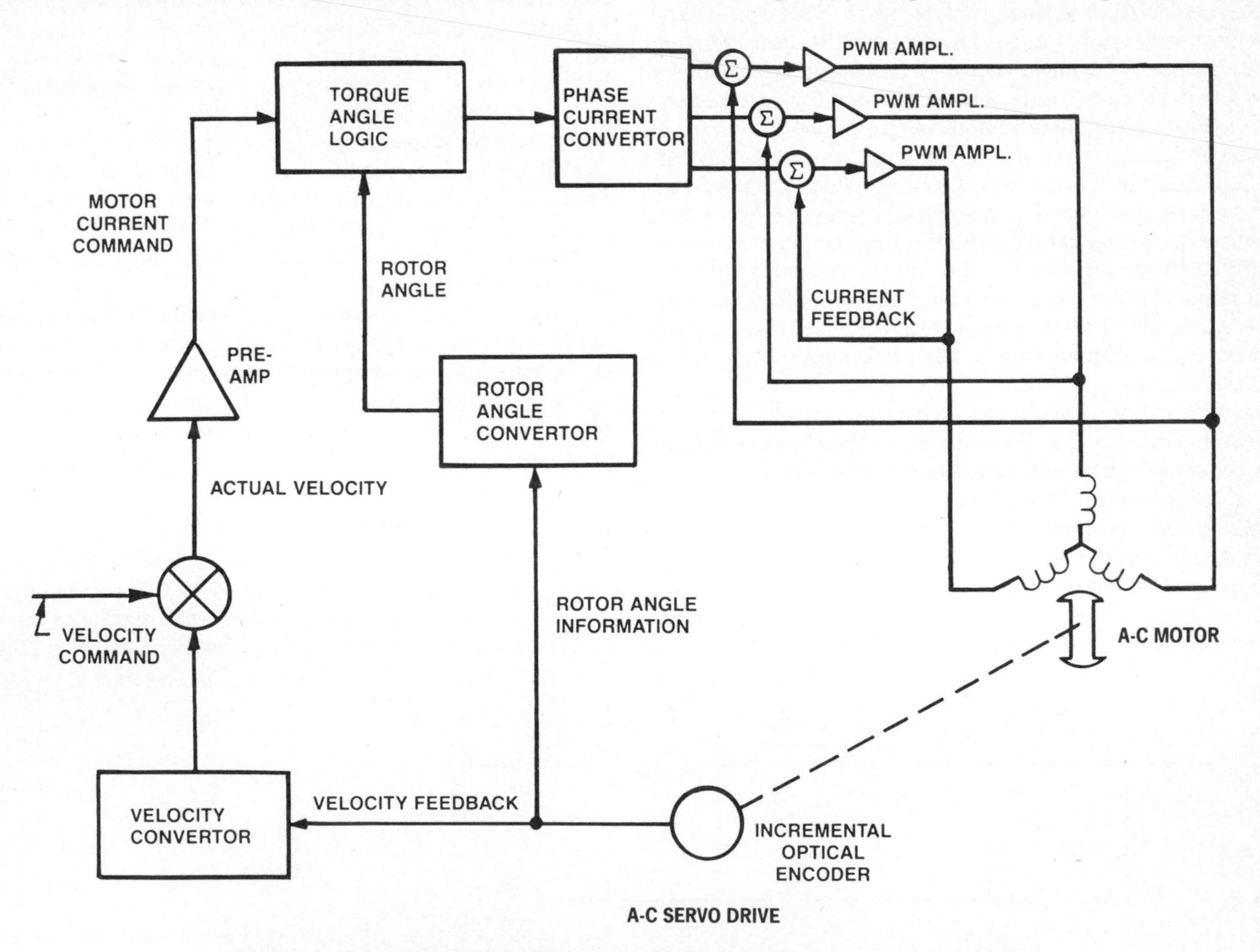

Figure 7. Conventional brushless servo-amplifer loop diagram. Courtesy of SMI.

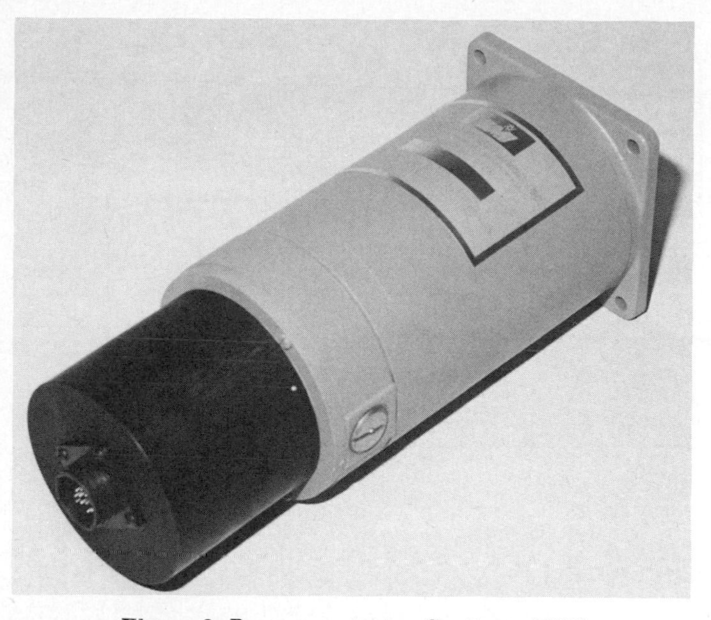

Figure 9. D-c servo actuator. Courtesy of SMI.

in conjunction with d-c or brushless motors and resolvers only (no other transducers are required).

The trend in the late 1980s will be toward this simplified amplifier/actuator concept.

Other position transducers utilized for special robot mechanisms include (*1*) linear encoders, (*2*) Inductosyn (trademark of the Farrand Corp.) scales, and (*3*) absolute encoders.

Linear encoders are used primarily for precision linear motions with short travel. Resolutions of 0.0001 in. are easily obtained; however, they are typically restricted to controlled-environment applications. Critical alignment and preventive maintenance procedures are necessary. Noise sensitivity is also of concern.

Absolute encoders are primarily used in military and aerospace applications as a means of providing absolute information. Absolute position information results from the ability to supply position feedback even if power is lost and restored. The larger the resolution, the larger the encoder. Cost also increases with size. The cost typically runs ten times that of a conventional rotary, incremental encoder.

Inductosyn slides are nothing more than a linear resolver. Because they are embedded in glass plate, they can provide higher repeatability than conventional rotary resolvers, which are wound on production machines. Inductosyns are typically used in machine-tool and other applications where controls are established for resolver-type feedback. However, Inductosyn slides provide resolutions of five times or better than conventional rotary resolvers. However, they are critical to align, require precision amplifiers to process signals, and must maintain constant air gap.

THE SERVO ACTUATOR

With this background on servo components, it is now appropriate to show the difference between a motor and an actuator. An actuator merely implies a motor with one or more transducers. Thus, a d-c servo actuator combines a d-c motor with one or more of the position and velocity transducers listed in this article.

Figure 9 shows a standard servo actuator consisting of a d-c motor, analogue tachometer, and rotary, incremental encoder (technology of the early 1980s). Figure 10 is an a-c brushless actuator consisting of an a-c brushless motor and resolver (the technology of the late 1980s).

THE COMPLETE SERVO LOOP

Since the actuator is the primary means of converting electrical energy (power) to mechanical power (horsepower), it is important to see how the total servo loop works in conjunction with the servo actuator. Figure 11 is an illustration of a servo loop with a servo actuator using tachometer and encoder feedback. The blocks are analogue from the summing junction of the tachometer and the velocity-command signal. The resultant of this summation combined with the compensation for the velocity loop is the current command. This is typical of all types of servo-control technology.

Figure 12 provides the same control loop except tachometers are omitted. Resolvers or encoders could be used, but typically resolvers are preferred today. The ability of robot manufacturers to incorporate the new control technology into their robot designs allows for a more efficient means of modeling particular robot geometries in microprocessors. These techniques will allow new robot controls to be more easily adapted to any type of robotic arm.

The control techniques shown in Figure 12 will be the key criteria in providing robotic capabilities for the high speed production requirements of the future.

Figure 10. A-c brushless actuator. Courtesy of CSR Div., Contraves Goerz Corp.

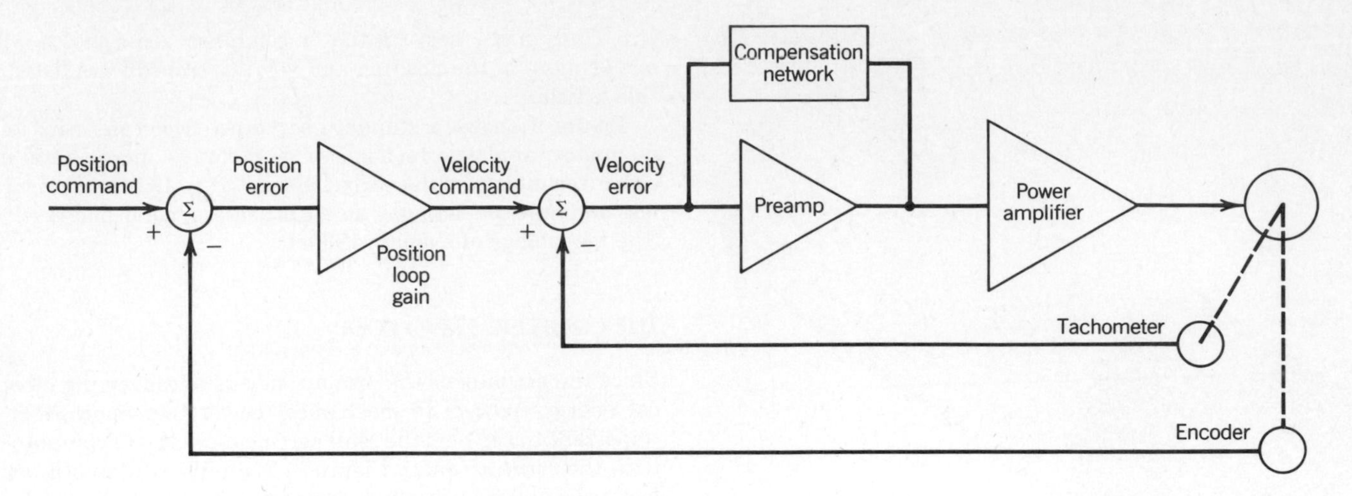

Figure 11. Servo loop with tachometer and encoder feedback (position loop). Courtesy of SMI.

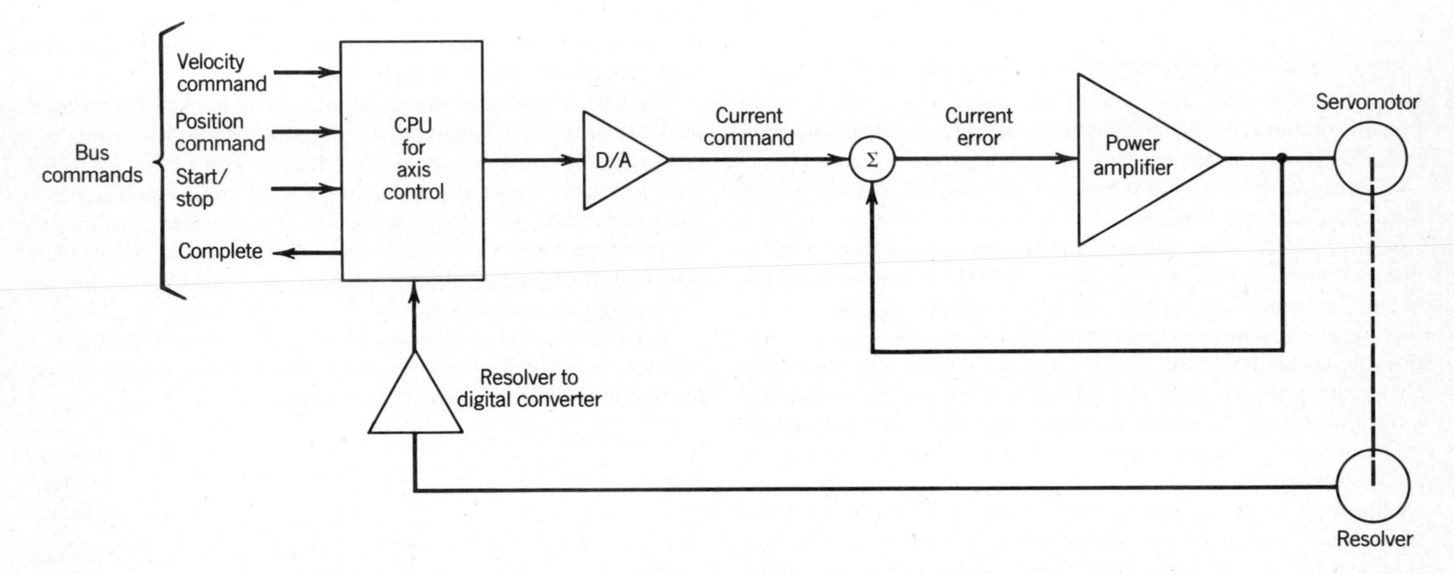

Figure 12. Servo loop with resolver feedback only (velocity loop in software). Courtesy of SMI.

ADAPTIVE CONTROL

George Saridis
Rensselaer Polytechnic Institute
Troy, New York

INTRODUCTION

In recent years, modern control techniques have been explored for possible controls of robot manipulators. This is mainly because robot mechanisms have been increasingly sophisticated, requiring higher precision of task execution and higher intelligence for task coordination while they operate in uncertain environments. Adaptive controls have been traditionally developed to drive systems with unknown or partially unknown dynamics operating in uncertain or unfamiliar environments (Fig. 1). Therefore, they provide a powerful tool for use of robotic control (see also Control; Expert Systems; Heuristics; Learning and Adaption).

Self-organizing control (SOC), a straightforward generalization of the above concepts, is defined as the discipline that treats dynamic systems with uncertainties operating in deterministic or stochastic environments. Self-organization is performed by the controller on-line to reduce the uncertainties pertaining to the systems and to attain desired output (1).

The name *self-organizing control,* created by Mesarovic (2) has been used to expand the idea of *adaptive control.* The word adaptive implies functions similar to its original "behavioral" meaning. The area of *learning control systems* has also been redefined to be compatible with definitions of self-organizing controls (3). The term learning systems is more inclusive, containing advanced decision-making, pattern recognition, and other functions resembling human behavior (4), and pro-

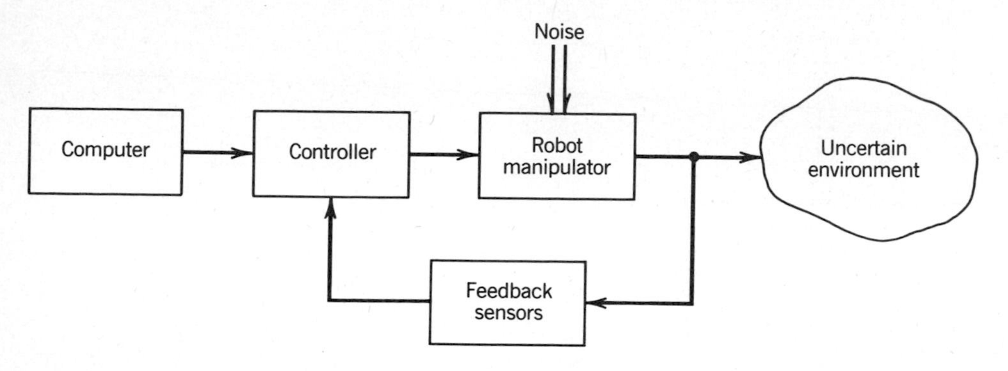

Figure 1. Computer control of robot manipulator.

vides a link for the next level of *intelligent control systems* suggested by Fu in Ref. 5 and defined by Saridis in Ref. 1.

DEFINITIONS

The area of *self-organizing control* (SOC) systems describes systems with characteristic features "beyond stochastic control systems." Common denominators of these features, namely, the system's uncertainties and accumulation of information, have been used in the definitions in order to provide room for all the researchers in the area. A duality has been retained in the name to preserve both the "technical" as well as the "behavioral" nature of the problems under consideration, and their common area of application is described to emphasize the unity of the area. The phrase self-organizing control process has been used to define the problems of technical origin, because of its technical meaning of providing on-line structural as well as parametric adjustment to the system. The word adaptive is of behavioral origin and has not been used as a name to avoid possible arguments regarding its interpretation. Instead, it has been used as an adjective to denote the type of adaptation. The concept of identification and the memory requirement, which for some researchers are the predominant features of the area, are implicitly contained in the terms accumulation of information and learning.

The following definitions are listed to describe a self-organizing control process and controller.

Definition 1

Self-organizing control process: A control process is called *self-organizing* if reduction of the *a priori* uncertainties pertaining to the effective control of the process is accomplished through information accrued from subsequent observations of the accessible inputs and outputs as the control process evolves.

Definition 2

Self-organizing controller: A controller designed for a self-organizing control process will be called *self-organizing* if it accomplishes *on-line* reduction of the *a priori* uncertainties pertaining to the effective control of the process as it evolves.

The word optimal has been replaced in the definitions by the word effective or improving, since in many problems strict optimality is not achievable. No specific mention is made regarding the deterministic or stochastic nature of the uncertainties so that both cases may be treated.

It was previously mentioned that self-organization is achieved either by reducing the uncertainties pertaining to the dynamics of the plant, where an explicit identification scheme is present, or by decreasing the uncertainties directly related to the improvement of the performance of the system. In the latter case, the information retrieved from the plant is used directly in the controller and the appropriate performance evaluator. To distinguish between the two categories of self-organizing controls, which represent two distinct design procedures, the following definitions are given where the word adaptive has been used to qualify the appropriate function.

Definition 3

Parameter-adaptive process: A self-organizing control process will be called *parameter-adaptive* if it is possible to reduce the *a priori* uncertainties of a parameter vector characterizing the process through subsequent observations of the inputs and outputs as the control process evolves.

Definition 4

Performance-adaptive process: A self-organizing control process will be called *performance-adaptive* if it is possible to reduce directly the uncertainties pertaining to the improvement of the performance of the process through subsequent observations of the inputs and the outputs as the control process evolves.

PARAMETER-ADAPTIVE CONTROL

The parameter-adaptive systems discussed in this section are represented by the block diagram of Figure 2. In such cases, the structure of the plant is assumed known *a priori* (in most cases linear), and the reducible uncertainties compose a vector θ of parameters and noise statistics. An identification procedure is used to "learn" sequentially the unknown vector $\theta^T \triangleq [\theta_1, \ldots, \theta_s]$, through its estimate $\hat{\theta}(k)$ at time k, or the probabilities associated with it

$$q_i(k) \triangleq \text{Prob}\,[\theta_i; k] \qquad k = 1, \ldots, s \quad \sum_{i=1}^{s} q_i(k) = 1 \tag{1}$$

Such identification algorithms are normally assumed to satisfy a Markov–Gauss-type relation

$$\hat{\theta}(k+1) = A_1(k)\hat{\theta}(k) + \xi(k) \tag{2}$$

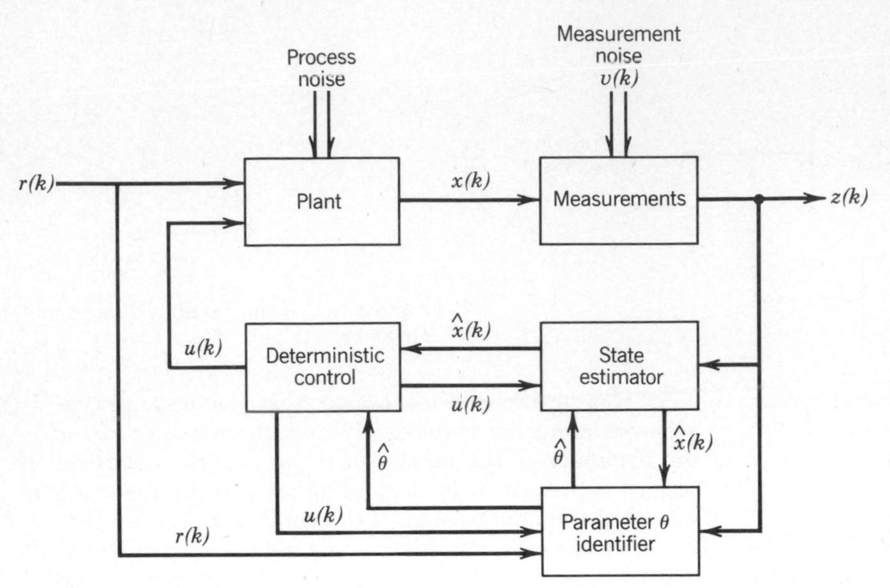

Figure 2. Block diagram of a general parameter-adaptive SOC system.

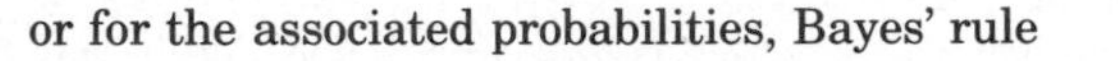

or for the associated probabilities, Bayes' rule

$$q_i(k+1) = \frac{p(z(k)/\theta_i)q_i(k)}{\sum_{i=1}^{s} p(z(k)/\theta_i)q_i(k)} \qquad i = 1, \ldots, s \tag{3}$$

In general, $A_1(k)$, a matrix, and $\xi(k)$, a noise vector of appropriate dimensions, are random variables depending on $Z^k = \{z(0), \ldots, Z(k)\}^T$ and $U^k = \{u(0), \ldots, u(k)\}^T$, the output and input past sequences, respectively. In certain instances, however, the system parameters are assumed to be random as a result of the interpretation of uncertainty and are defined to obey a Markov–Gauss relationship

$$\theta(k+1) = A_2(k)\theta(k) + \eta(k), \qquad \theta(0) = \theta_0 \tag{4}$$

where $A_2(k)$ is a known matrix, $A_2(k) = I$ for time invariant parameters, and $\eta(k)$ and θ_0 are white Gaussian random variables, with $\eta(k) = 0$ representing the fixed parameter case. In that case, the parameter identifier is defined by the appropriate estimator $\hat{\theta}$.

The information produced by the identifier is used to update a desired deterministic controller, which may or may not be asymptotically optimal, and a state estimator, possibly asymptotically optimal, when the states of the plant are not available for direct measurement. The various approximations to the optimal solution are obtained through relaxation of certain couplings among the various functions of the controller which constitute the dual control and, therefore, simplify its derivation.

Parameter-adaptive control for systems operating in a stochastic environment has been developed under various forms since about 1960. Bellman (6) was one of the first to present the basic ideas, and Mishkin and Braun (7) also discussed it along with the deterministic adaptive algorithms of the first stochastic equivalent. Some of the first parameter-adaptive SOCs were produced as an extension of the formulation by Kopp, Orford, and Cox's algorithms (8). Farison, Graham, and Shelton (9) used an open-loop feedback optimal (OLFO) approach as a parameter-adaptive SOC solution. Tse and Athans (10) extended this work. Lainiotis and his colleagues (11), Spang (12), and Bar-shalom and Sivan (13) have used the information-states approach for parameter identification, resulting also in an OLFO algorithm. Sworder (14) followed a similar procedure for parameter-adaptive SOC algorithms that effectively utilizes the information state in an approximation of the dual-control algorithm. Stein and Saridis (15) and Saridis and Dao (16) developed algorithms that minimize bounds on the optimal solution, producing also a measure of the approximation performed. Saridis and Lobbia (17) developed algorithms that utilize stochastic approximation algorithms for parameter identification in parallel with state estimations and deterministic control, after an idea suggested by Lee (8). Finally, Tse, Bar-shalom, and Meier (18) developed a parameter-adaptive SOC algorithm as a direct approximation of the dual-control formulation.

PERFORMANCE-ADAPTIVE CONTROL

Performance-adaptive methodologies have been developed to treat problems where dynamics are completely or partially unknown, operating in an unknown stochastic environment such as bioengineering or physiological systems, without limitations on the dimensionality of the problem. The simplicity of an algorithm that may be interpreted as decoupling subsystems and states, initial value-search techniques, and successive improvement schemes, which are obtained by relaxing the strict optimality conditions, are the factors that permit handling of large-scale systems. Instead, the ideas of asymptotic optimality or instantaneous improvement of a suitable performance criterion are applied and yield satisfactory results for problems of the complexity of self-organizing control systems. A word of caution is appropriate at this point: the performance-adaptive methods are developed to solve problems of higher degrees of complexity only. A typical performance-adaptive system is depicted in Figure 3.

In summary, performance-adaptive control systems represent the first step toward an intelligent decision-maker and

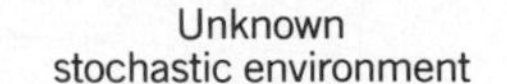

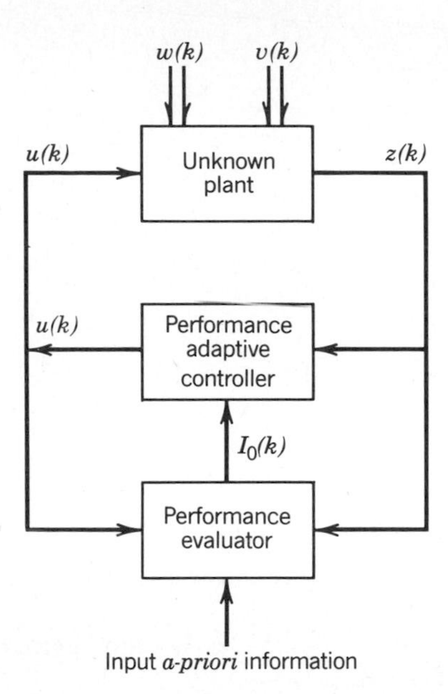

Figure 3. A typical performance-adaptive SOC system.

can handle the lower-level control function. On the other hand, they are on the top of the control hierarchy since they can treat advanced situations involving considerable uncertainties without much sacrifice of mathematical rigor (1).

Performance-adaptive systems evolved from various independent attempts to produce advanced decision-making devices for control purposes (2,4,19). One trend of thought was the natural extension of the idea of parameter-adaptive, which was popular in the 1950s. The work of Narendra and his colleague (20) is the most representative of this classical approach to generate a controller based on the performance evaluation of the system. A second trend of thought in performance-adaptive originated from the principles of learning systems and the applications of behavioral approaches such as reinforcement learning algorithms to the control problem. The work of Fu and his colleagues (21,22) has been used to implement a linear reinforcement algorithm and generate a performance-adaptive self-organizing controller. The idea in both cases is that the selection of a control action is based upon the improvement of a subgoal, eg, per-interval performance criterion, which is compatible with the overall performance of the system. The work of McLaren (23) with Fu is the most representative in this area. It was preceded by the important investigations of Fu and McMurty (21), and the work of Riordon (24) and Narendra and his colleague (20) followed and extended it. Mendel and McLaren (25) summarized the state of the art of the reinforcement techniques and learning automata algorithms.

A third school of thought has evolved through the application of stochastic approximation as a performance-adaptive self organizing technique. Tsypkin in the USSR (26) and Nikoliç and Fu (27) in the United States have proposed various formulations of the solution to the performance-adaptive problem, using stochastic approximation models. These methods represent a new point of view to a successful solution of the performance-adaptive SOC problem.

Bayesian estimation techniques being used for self-organizing control by Saridis and Dao (16) and Lainiotis and his coworkers (11) have referred to the parameter-adaptive algorithms mainly because of their parameter-searching nature.

The expanding subinterval algorithm, first developed by Saridis and Gilbert for attitude control of an orbiting satellite and then generalized by Saridis (3), represents another conceptually different approach to the performance-adaptive problem.

It produces a global asymptotically optimal solution to the control problem with completely unknown plant dynamics by searching for a minimum of a per-interval cost function defined over an expanding subinterval of time, thus relating the subgoal to the overall goal, a concept of behavioral methods. Structural considerations and stability investigation of the algorithm have been discussed (2).

The above algorithms represent the major approaches to the performance-adaptive problem.

Finally, the *model reference adaptive control algorithms* are described in detail by Landau (28). According to this method, the actual performance of the system is compared with the desired one generated by a "model" system. Control parameters are then upgraded to improve the instantaneous performance of the system. Figure 4 presents a block diagram of such a system. These algorithms complement another class of performance-adaptive controls called *self-tuning regulators* developed by Åström (29). In recent years, several adaptive control algorithms, based on linear multivariable control theory, pole placement, and other control methodologies, have been developed and are reported by Goodwin and co-workers (30) and Anderson and Ljung (31). However, these algorithms deal exclusively with linear systems operating mostly in deterministic environments, which discourages their application to nonlinear systems such as robot manipulators.

The above algorithms summarize the major approaches of performance-adaptive self-organizing control systems. However, they represent only the lowest stratum on the next generation of control systems, *intelligent controls*. Such systems are designed to execute anthropomorphic tasks with minimum interaction with a human operator, in unfamiliar structured or unstructured environments. The development of the design of such systems and their applications to robotics are discussed in Refs. 32 and 33.

APPLICATIONS OF ADAPTIVE CONTROL TO ROBOTICS

In all aspects of engineering, no new area is sufficiently justified unless it is demonstrated that there is enough demand from the applications point of view. Self-organizing controls are no exception to this rule. However, it must be understood that one should look for applications that are of the same level of sophistication as the SOCs. This is true for control of robot manipulators, dominated by nonlinear dynamics and operating in unfamiliar environments and carrying unspecified loads.

Under the above circumstances, an adaptive control solution to the manipulator control problem is highly desirable. Several attempts have been made in recent years to apply

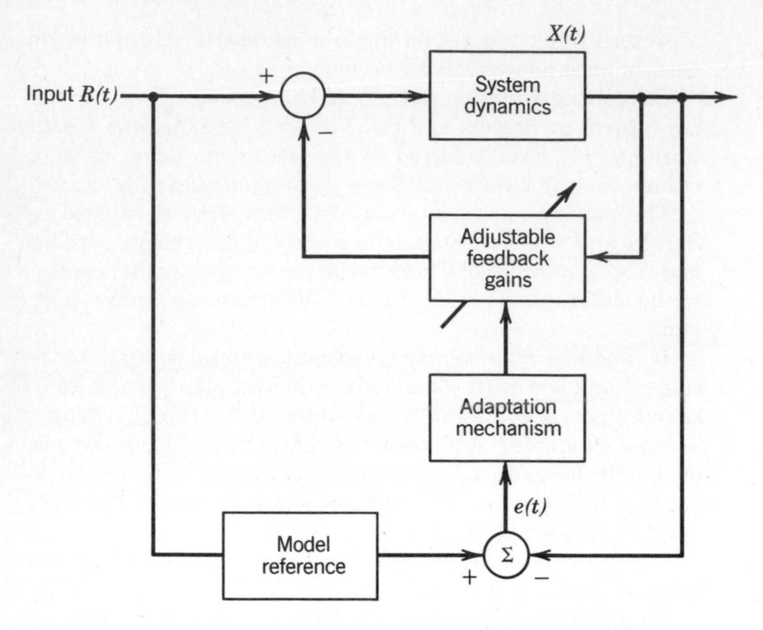

Figure 4. Model reference adaptive control system.

performance- or parameter-adaptive control algorithm for robot manipulation. Three of these are described below. In order to possess the real-time computational speed needed, all three algorithms require a linearization of the robot dynamics and therefore limit their effectiveness to rather small variations of motion about a nominal trajectory. It is hoped that in the future the more sophisticated performance-adaptive algorithms discussed in the previous sections will be tested on robotic controls without the requirement of model linearization for real-time operation.

MODEL REFERENCE ADAPTIVE CONTROL FOR ROBOTS

In 1981, Dubowski and DesForges published a paper that proposed the use of the model reference adaptive control approach to drive an n-degree of freedom robotic manipulator (34). The method assumes no interaction among the manipulator joints, and therefore the dynamics may be assumed decoupled with adjustable parameter $\alpha_i(t)$, $\beta_i(t)$:

$$\alpha_i(t)\ddot{x}_i + \beta_i(t)\dot{x}_i(t) + x(t) = \gamma_i(t) \qquad i = 1, \ldots, n \tag{5}$$

which may be forced to track a constant coefficient linear reference model

$$a_i\ddot{y}_i(t) + b_i\dot{y}_i(t) + y(t) = \gamma_i(t) \qquad i = 1, \ldots, n \tag{6}$$

In the above equations, $x^T(t) = [q_1(t), \ldots, q_n(t)]$ is the vector of actual joint coordinates, whereas $y^T(t) = [q_1^d(t) \ldots q_n^d(t)]$ is the model reference vector of desired joint coordinates; n is the degree of freedom of manipulation ($n = 6$ for most modern designs). Define the error vector

$$e_i(t) \triangleq y_i(t) - x_i(t) \qquad i = 1, \ldots, n \tag{7}$$

and the weighted error cost

$$J_i = \frac{1}{2}(k_{i2}\ddot{e}_i + k_{i1}\dot{e}_i + k_{i0}e_i)^2 \qquad i = 1, \ldots, n \tag{8}$$

where k_{ij} $i = 1, \ldots, n, j = 0, 1, 2$ are appropriately chosen positive coefficients.

Adaptation is obtained through adjustment of the parameters $\alpha_i(t)$ $\beta_i(t)$ $i = 1, \ldots, n$ to match the model's response. For this purpose, a steepest-descent algorithm is suggested:

$$\begin{aligned} \dot{\alpha}_i(t) &= (k_{i2}\ddot{e}_i + k_{i1}\dot{e}_i + k_{i0}e_i)(k_{i2}\ddot{u}_i + k_{i1}\dot{u}_i + k_{i0}u_i) \\ \dot{\beta}_i(t) &= (k_{i2}\ddot{e}_i + k_{i1}\dot{e}_i + k_{i0}e_i)(k_{i2}\ddot{w}_i + k_{i1}\dot{w}_i + k_{i0}w_i) \\ & \qquad i = 1, \ldots, n \end{aligned} \tag{9}$$

where $u_i(t)$ and $w_i(t)$ are the solutions of the following equation:

$$\begin{aligned} a_i\ddot{u}_i + b_i\dot{u}_i + u_i &= -\ddot{y}_i(t) \\ a_i\ddot{w}_i + b_i\dot{w}_i + w_i &= -\dot{y}_i(t) \qquad i = 1, \ldots, n \end{aligned} \tag{10}$$

and $y(t)$ is the time response of the model reference. The system is depicted in Figure 5.

ADAPTIVE ROBOTIC CONTROL WITH AN AUTOREGRESSIVE MODEL

In 1983, Koivo and Guo (35) published a paper on an adaptive linear controller for robot manipulators. They also assumed no interaction among the joint variables of an n-degree of freedom robot arm, thus decoupling the dynamics of the system. As in the previous case, such an approximation is valid when the joints of the robot move one at a time. The autoregressive model of such a linearized discrete-time system is given by

$$\begin{aligned} x_i(k) = \sum_{m=1}^{n} [a_i^m x_i(k-m) + b_i^m u_i(k-m)] \\ + a_i^0 e_i(k) \qquad i = 1, \ldots, n \end{aligned} \tag{11}$$

where $x(k)$ is the n-dimensional vector of joint coordinates, $u(k)$ the n-dimensional vector of input control torques at each joint, and $e_i(k)$ the modeling error assumed to be zero mean

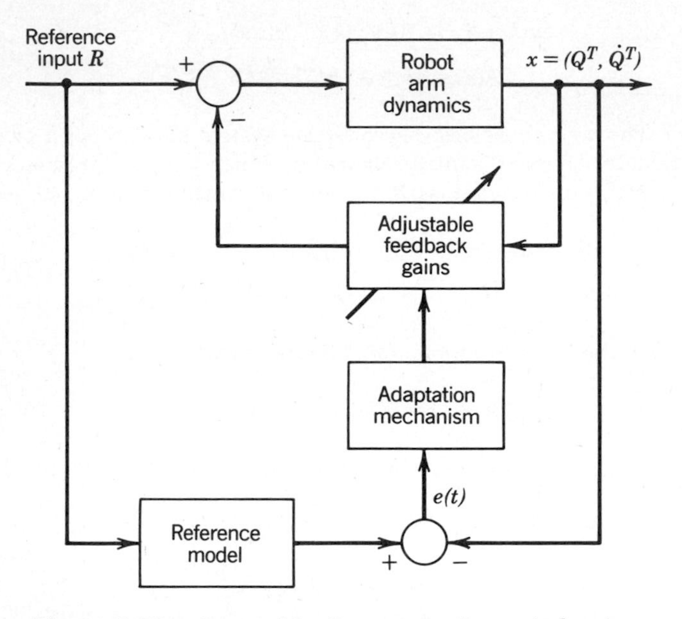

Figure 5. The robot model reference adaptive control system.

white Gaussian all at time $t = k\Delta T$. a_i^m, b_i^m, i, $m = 1, \ldots, n$ are appropriate but unknown model parameters:

$$\alpha_i = [a_i^1, \ldots, a_i^n, b_i^1, \ldots, b_1^n] \qquad i = 1, \ldots, n \quad (12)$$

Model identification is the first step in the development of this parameter-adaptive algorithm. For this purpose, one defines the accrued model error

$$E_i(x_i, N) = \frac{1}{N+1} \sum_{k=0}^{N} e_i^2(k) \qquad i = 1, \ldots, n \quad (13)$$

and the variable vector

$$y_i^T(k-1) = [1, x_i(k-1), \ldots, x_i(k-n), u_i(k-1), \ldots, u_i(k-n)] \quad (14)$$

By using a least-square fit algorithm, estimates of the unknown parameters to be used in the control part of the algorithm are obtained:

$$\hat{\alpha}_i(N) = \hat{\alpha}_i(N-1) + P_i(N)y_i(N-1)\,[x_i(N) - \hat{\alpha}(N-1)y_i(N-1)]$$

$$P_i(N) = \frac{1}{M_i}\left[\frac{P_i(N-1)y_i(N-1)y_i^T(N-1)P_i(N-1)}{M_i + y_i^T(N-1)P_i(N-1)y_i(N-1)}\right] \qquad i = 1, \ldots, n \quad (15)$$

where M_i is an appropriately defined "forgetting" factor $0 < M_i \leq 1$. The decoupled robot dynamics they may be modeled by is

$$x_i(k) = \hat{\alpha}_i y_i(k-1) + e_i(k) \qquad i = 1, \ldots, n \quad (16)$$

Optimal control is the second part of this parameter-adaptive control algorithm. Define a performance index per joint as the conditional expectation of the tracking error square and input effort square:

$$J_i^k(u) = E\{[x_i(k+2) - x_i^d(k+2)]^2 + \gamma_i u_i^2(k+1)/y_i(k)\} \qquad i = 1, \ldots, n \quad (17)$$

where $x_i^d(k)$ $i = 1, \ldots, n$ are the desired joint trajectories that the robot arm must track. Then an approximately optimal control may be obtained as a function of the estimated parameters (17):

$$u_i(k+1) = \frac{-\hat{b}_i^1(k)}{(\hat{b}_i^1(k))^2 + \gamma_i}\left[\hat{a}_i^0(k) + \hat{a}_i^1(k)(\hat{\alpha}_i^T y_i(k) + \sum_{m=2}^{n} \hat{a}_i^m(k)x_i(k+2-m) + \sum_{m=2}^{n} \hat{b}_i^m(k)\, u_i(k+2-m) - x_i^d(k+2)\right] \quad (18)$$

$$i = 1, \ldots, n$$

The procedure should be repeated for the length of the process. A block diagram illustrates the method in Figure 6. More recent work has extended the method to perform trajectory tracking in the Cartesian coordinates system (35).

THE PERTURBATION THEORY ADAPTIVE CONTROL METHOD

In 1983, G. Lee and B. Lee proposed another parameter-adaptive control algorithm which accounts for interactions of joint variables and considers the complete dynamic model for the n-degree of freedom manipulator obtained through the Euler–Lagrange formulation (33,36,37):

$$D(q)\ddot{q} + \dot{q}^T H(q)\dot{q} + G(q) = \tau \quad (19)$$

where q is the n-dimensional joint coordinate vector, $D(q)$ the $n \times n$ inertia matrix, $H(q)$ coriolis and centrifugal force $n \times n \times n$ tensor, and $G(q)$ the gravity n-dimensional vector; τ is an n-dimensional vector of joint control torques.

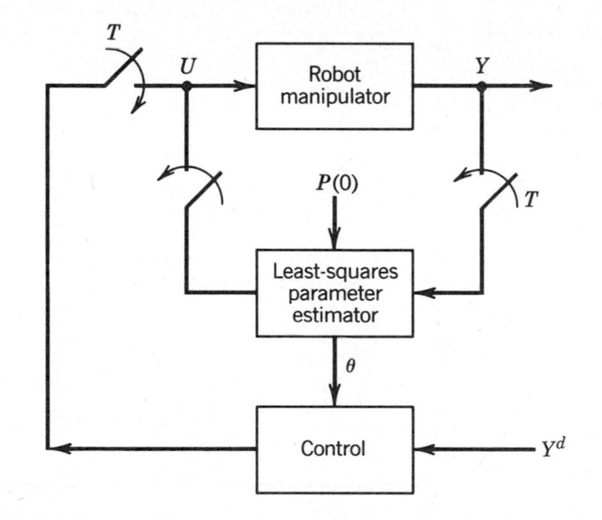

Figure 6. The adaptive robotic control system using an autoregressive model.

Performing the following transformation

$$x^T(t) \triangleq [q^T(t), \dot{q}^T(t)],$$
$$u^T(t) \triangleq [0,\tau^T]; \qquad D(q) > \tag{20}$$

one obtains the *state-variable* model

$$\dot{x}(t) = f(x(t),t) + b(x(t),t)u(t) \tag{21}$$

where $f(x,t)$ and $b(x,t)$ are appropriate nonlinear functions of the states. The control function $u(t)$ is sought to be a feedback control law of the form $u(x(t),t)$.

Set points on the desired trajectory are defined for tracking purposes, eg, $(q^d(k), \dot{q}^d(k), \ddot{q}^d(k))$, yielding state-variable set points $(x^d(k), \dot{x}^d(k))$ for $k = 0, 1, 2, \ldots, M$. Torques may be computed at the set points yielding $u^d(k)$.

Assuming good tracking, the state-variable equations (eq. 21) may be *linearized* about the desired trajectory $x^d(t)$ by Taylor series expansion and retaining of the linear terms. The resulting linear system is

$$\delta\dot{x}(t) = A'(t)\delta x(t) + B'(t)\delta u(t)$$

where

$$\delta x(t) \triangleq x(t) - x^d(t), \qquad \delta u(t) \triangleq u(t) - u_{\text{nom}}(t)$$
$$A'(t) = \frac{\partial f}{\partial x}(x^d(t)) \qquad B'(t) = b(x^d,t) \tag{22}$$

The system is then *discretized* at the set point $x^d(k)$, $k = 0, 1, 2, \ldots, M$.

$$\delta x(k+1) = A(k)\delta x(k) + B(k)\delta u(k)$$
$$A(k) = [I + A'(k)\Delta], \; B(k) = B'(k)\Delta \tag{23}$$

The now linearized discrete-time system may accept a parameter-adaptive control algorithm which uses a parameter identifier in parallel with an optimal controller (17). Define

$$\theta_{ik}^T \triangleq (a_{i1}(k), \ldots, a_{in}(k), b_{i1}(k), \ldots, b_{in}(k))$$
$$Y_k^T \triangleq (\delta x_1(k), \ldots, \delta x_n(k), \delta u_1(k), \ldots, \delta u_n(k)) \tag{24}$$

Then the state equations (eq. 23) are rewritten as

$$\delta x_i(k+1) = \theta_{ik}^T Y_k \qquad i = 1, \ldots, n \tag{25}$$

Using a least-square algorithm, one *identifies* the parameters

$$\hat{\theta}_{ik+1} = \hat{\theta}_{ik} - P_k T_k [Y_k^T P_k Y_k + M]^{-1} [Y_k^T \hat{\theta}_{ik} - \delta x_i(k+1)]$$
$$P_{k+1} = [P_k - P_k Y_k [Y_k^T P_k Y_k + M]^{-1} Y_k^T P_k] \frac{1}{M} \tag{26}$$

where M is a "forgetting" factor.

Every time the parameters $\hat{\theta}_{ik}$ are identified, yielding the matrices $\hat{A}(k)$, $\hat{B}(k)$, an approximately *optimal control* may be obtained to minimize

$$J(k) = \frac{1}{2}[\delta^T x(k+1)\, Q\, \delta x(k+1) + \delta^T u(k)\, R\, \delta u(k)] \tag{27}$$

where $Q = Q^T \geq 0$, $R = R^T \geq 0$. Then the approximately optimal control needed to drive the overall system is given by

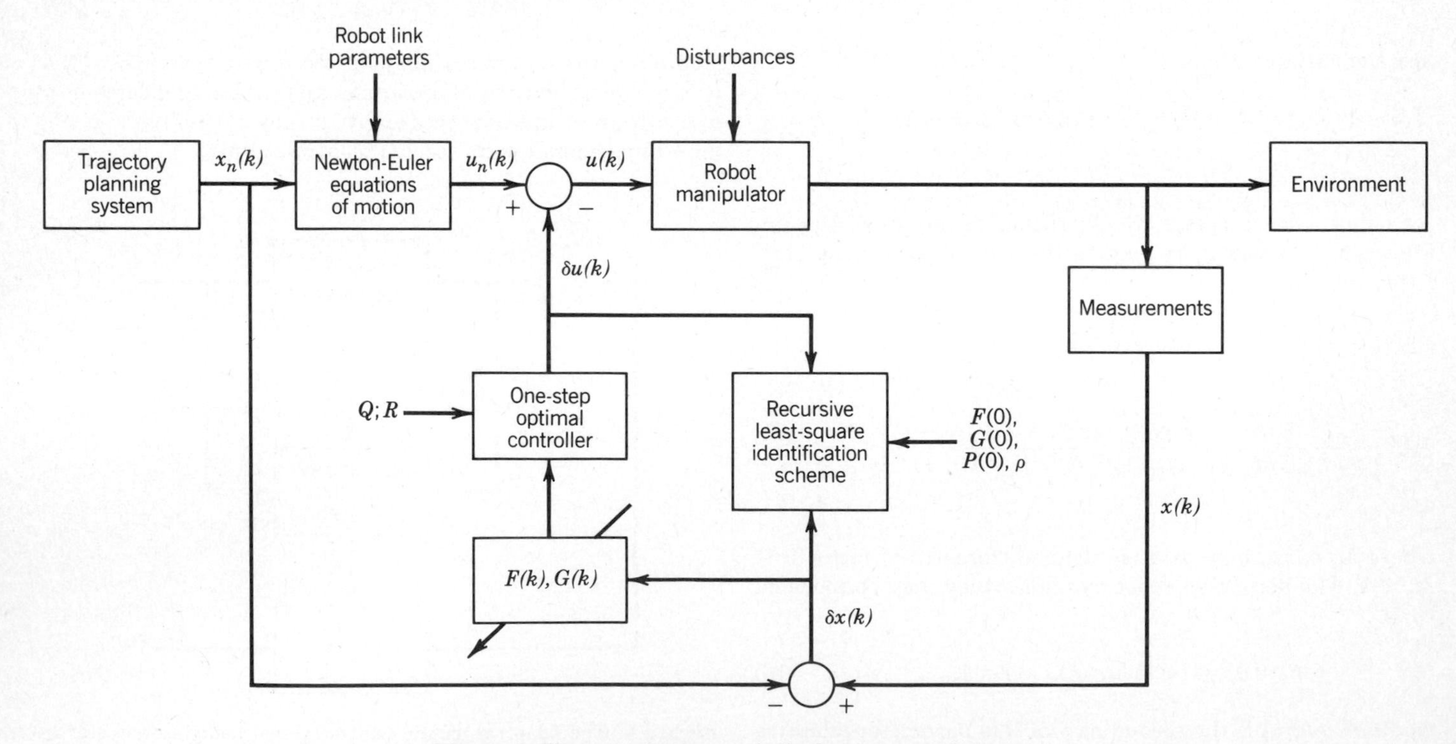

Figure 7. The perturbation method adaptive control system.

$$U^*(k) = U_{\text{nom}}(k) - [R + \hat{B}^T(k)\, Q\, \hat{B}(k)]^{-1}\, \hat{B}(k)\, Q\hat{A}(k)\delta x(k) \qquad (28)$$

The system is depicted in Figure 7. For properly selected set point with a reasonable discrete interval Δ, the controller may be implemented in real time. Satisfactory tracking was obtained both in simulation and in real time on a Unimation PUMA-600 robot arm.

CONCLUSIONS

An account of the most important parameter- and performance-adaptive control methods have been described and their applications to robot manipulators presented. Due to space considerations, many researchers in the area have inadvertently not been referenced. However, their work has helped to promote the field of adaptive control systems.

The robotic applications have just started to appear in the literature. The three methods described in this article are the most representative at the present time. They demonstrate feasible but somehow limiting application of adaptive controls to robotics. Further research aimed at utilizing simplified but nonlinear performance-adaptive control algorithms with reasonable real-time computational difficulties should prove valuable for the future of autonomous robot manipulators. However, the most challenging area of robotic research is the development of *intelligent controls* for machines minimally interacting with a human operator when they function in uncertain environments (1). Such controls designed with a sophisticated hierarchy of intelligence will still accommodate adaptive controls for the precise execution of desired tasks.

BIBLIOGRAPHY

1. G. N. Saridis, *Self-Organizing Control of Stochastic Systems,* Marcel Dekker, Inc., New York, 1977.
2. M. D. Mesarovic, "On Self-Organization Systems," in M. C. Yovits, G. T. Jacobi, and G. D. Goldstein, eds., *Self-Organizing Systems—1962,* Spartan Books, New York, 1962.
3. G. N. Saridis, "On A Class of Performance-Adaptive Self-Organizing Control Systems," in K. S. Fu, ed., *Pattern Recognition and Machine Learning,* Plenum Press, New York, 1971, pp. 204–220.
4. K. S. Fu, "Learning Control Systems," in J. T. Tou, ed., *Advances in Information Systems Science,* Plenum Press, New York, 1969.
5. K. S. Fu, "Learning Control Systems—Review and Outlook," *IEEE Transactions on Automatic Control,* **AC-15,** 210–221 (1970).
6. R. Bellman, *Adaptive Control Processes, A Guided Tour,* Princeton University Press, Princeton, N.J., 1961.
7. E. Mishkin and L. Braun, Jr., eds., *Adaptive Control Systems,* McGraw-Hill, Inc., New York, 1961.
8. R. C. K. Lee, *Optimal Identification, Estimation and Control,* The MIT Press, Cambridge, Mass., 1964.
9. J. B. Farison, R. E. Graham, and R. C. Shelton, "Identification and Control of Linear Discrete-Time Systems," *IEEE Transactions on Automatic Control,* **AC-12,** 438–442 (1967).
10. E. Tse and M. Athans, "Adaptive Stochastic Control for A Class of Linear Systems," *IEEE Transactions on Automatic Control,* **AC-15,** 38–51 (1972).
11. D. Lainiotis, J. G. Deshpande, and T. N. Upadhyay, "Optimal Adaptive Control: A Non-linear Separation Theorem," *International Journal on Control,* **15,** 877–888 (1972).
12. H. A. Spang, "Optimal Control of An Unknown Linear Plant Using Bayesian Estimation of the Error," *IEEE Transactions on Automatic Control,* **AC-10,** 80–83 (1965).
13. Y. Bar-shalom and R. Sivan, "On the Optimal Control of Discrete-Time Linear Systems with Random Parameters," *IEEE Transactions on Automatic Control,* **AC-14,** 3–8 (1964).
14. D. D. Sworder, *Optimal Adaptive Control Systems,* Academic Press, Inc., New York, 1966.
15. G. Stein and G. N. Saridis, "A Parameter Adaptive Control Technique," *Proceedings 4th IFAC,* Warsaw, as reported in *Automatica,* **5**(6) (1969).
16. G. N. Saridis and T. K. Dao, "A Learning Approach to The Parameter Adaptive Self-Organizing Control Problem," *Automatica,* **9,** 589–598 (1972).
17. G. N. Saridis and R. N. Lobbia, "Parameter Identification and Control of Linear Discrete Time Systems," *IEEE Transactions on Automatic Control,* **AC-17,** 52–60 (1972); "A Note on Parameter Identification and Control of Linear Discrete Time System," *IEEE Transactions on Automatic Control,* **AC-20** (1975).
18. E. Tse, Y. Bar-shalom, and L. Meier, "Wide-Sense Adaptive Dual Control of Stochastic Non-linear Systems," *IEEE Transactions on Automatic Control,* **AC-18,** 98–108 (1973).
19. J. Sklansky, "Learning Systems for Automatic Control," *IEEE Transactions on Automatic Control,* **AC-11,** 6–20 (1966).
20. K. S. Narendra and R. Wiswanathan, "A Two Level System of Stochastic Automata for Periodic Random Environments," *IEEE Transactions on Systems, Man and Cybernetics,* **SMC-2** (1972).
21. K. S. Fu and G. J. McMurtry, *A Study of Stochastic Automata as Models of Adaptive and Learning Controllers, Technical Report TR-EE 65-8.* Purdue University, West Lafayette, Ind., 1965.
22. M. D. Waltz and K. S. Fu, "A Heuristic Approach to Reinforcement Learning Control Systems," *IEEE Transactions on Automatic Control,* **AC-10,** 390–394 (1965).
23. R. W. McLaren, "Application of a Continuous-Valued Control Algorithm to the *On-Line* Global Optimization of Stochastic Control Systems," *Proceedings of National Electronics Conference,* **25,** 2–7 (1969).
24. J. S. Riordon, "An Adaptive Automation Controller for Discrete-Time Markov Processes," *Automatica-IFAC Journal,* **5,** 721–730 (1969).
25. J. M. Mendel and R. W. McLaren, "Reinforcement-Learning Control and Pattern Recognition Systems," in J. M. Mendel and K. S. Fu, eds., *Adaptive Learning and Pattern Recognition Systems: Theory and Applications,* Academic Press, Inc., New York, 1970, pp. 287–318.
26. Ya. Tsypkin, *Adaptation and Learning in Automatic Control* (translated from Russian by Z. J. Nikoliç), Academic Press, Inc., New York, 1971.
27. Z. J. Nikoliç and K. S. Fu, "An Algorithm for Learning Without External Supervision and Its Application to Learning Control Systems," *IEEE Transactions on Automatic Control,* **AC-11** (1966).
28. Y. D. Landau, *Adaptive Control,* Marcel Dekker, Inc., New York, 1979.
29. K. J. Åström, "On Self-Tuning Regulators," *Automatica,* **9,** 185–199 (1973).
30. G. C. Goodwin, P. J. Ramadge, and P. E. Caines, "Discrete-time Stochastic Adaptive Control," *IEEE Transactions on Automatic Control,* **AC-25,** 449–456 (1981).
31. B. D. O. Anderson and L. Ljung, "Special Issue on Adaptive Control," *Automatica,* **20,** 499–716 (1984).
32. G. N. Saridis, "Intelligent Robotic Control," *IEEE Transactions on Automatic Control,* **AC-28** (1983).

33. G. N. Saridis, ed., *Advances in Automation and Robotics,* Vol. 1, JAI Press, Inc., Greenwich, Conn., 1985.
34. S. Dubowski and T. DesForges, "The Application of Model Reference Adaptive Control to Robotic Manipulators," *Transactions of the ASME Journal of Dynamic Systems, Measurements, and Control,* **101,** 193–200 (1981).
35. A. J. Koivo and T. H. Guo, "Adaptive Linear Controller for Robotic Manipulators," *IEEE Transactions on Automatic Control,* **AC-28,** 162–171 (1983).
36. C. S. G. Lee and B. H. Lee, "Resolved Motion Adaptive Control for Mechanical Manipulators," *Proceedings of the Third Yale Workshop on Applications of Adaptive Systems Theory,* 1983, pp. 190–196.
37. C. S. G. Lee, M. J. Chung, and B. H. Lee, "An Approach of Adaptive Control for Robot Manipulators," *Journal of Robotic Systems* (1984).

General References

C. S. G. Lee and M. J. Chung, "An Adaptive Control Strategy for Computer-Based Manipulators," *Proceedings of the 21st Conference on Decision and Control,* Orlando, Fla., 1982, pp. 95–100.

G. J. McMurtry and K. S. Fu, "A Variable Structure Automation Used As A Multi-Modal Searching Technique," *IEEE Transactions on Automatic Control,* **AC-11,** 379–388 (1966).

K. Narendra and D. Streeter, "An Adaptive Procedure for Controlling Undefined Linear Processes," *IEEE Transactions* **PGAC-9** (1964).

AEROSPACE INDUSTRY, ROBOTS IN

WALLACE PLUMLEY
Acworth, Georgia

INTRODUCTION

Aerospace shares with other industries many of the common reasons for using robots—lower costs, improved output quality, and improved quality of life—but from a slightly different perspective and with a different set of constraints and problems. One of the best early studies of robotics in aerospace was done at Boeing Commercial Airplane Company and was reported in reference 1. The characteristics of large airframe manufacture include (1):

1. Very small production quantities.
2. A large quantity of active part numbers for both current and out-of-date models.
3. Very high tooling start-up costs and flowtime, with a high tool amortization rate per airplane produced.
4. High labor content, with manually operated and manipulated tools.
5. Minimal opportunity to design for mechanization.
6. Increasing and significant development of CAD/CAM techniques.

Many of the factors that make aerospace manufacturing unique are also important cost drivers supporting the implementation of robotics. The end goals for the industry's automation projects are easy to understand. The ability to allow a design engineer to sit at a computer terminal and create an electronic design for a part, and to then use that electronic design base to program robots and automated equipment to produce the part in an economical lot quantity of one without any hard tooling, is a very major goal. Improving the quality of the detail parts and the end item is another goal, as is improving the quality of life (a safer, more pleasant job) for the work force. The benefits for the Air Force of the implementation of robotics and automation in the airframe industry are: (*a*) reduced costs, (*b*) increased throughput and surge capability, (*c*) improved quality, and (*d*) manufacturing flexibility (2).

BATCH PRODUCTION

Aerospace production is typically based on small batch production—with the emphasis on small. The highest recent production rate for aircraft has been 20 ships per month on General Dynamic's F-16 fighter, and several programs have been at rates as low as 1 ship per month. These low production rates translate directly into small batches. An additional factor affecting batch size is the reluctance of aerospace engineers to compromise optimum part performance or design even slightly so that a part could be used in more than one location or application. This reluctance increases the number of unique parts and lowers the average batch size even more.

The batch size for the fabrication of detail parts varies widely among airframe companies and is determined on a part-by-part basis within these companies. The size of the individual batch is determined by the quantity of a particular part number needed per ship, the quantity needed per month, the setup cost versus run cost, the space required to store the completed parts, the possibility of damage in storage, and many other such factors. For large transport aircraft, which are usually built at very low production rates, the average batch size is about 15 parts, with a maximum batch size of perhaps 100 parts. For fighter aircraft, normally built in higher total quantities and higher production rates, the average batch size is 30–50 parts, with a maximum batch size of as many as a few hundred. Rockets and missiles span the entire range of quantities and rates, from very low to continuous production at high rates. Group technology appears to offer a partial answer by allowing the creation of part families, thus increasing the practical batch size, but this strategy has met with limited success in airframe manufacture. This lack of success seems to be primarily due to the difficulties involved in creating and maintaining a computer program that can manage the tremendous number of permutations and combinations resulting from the quantity of part numbers and the variety of production operations required. The recent lack of major new aircraft projects also means that most group technology projects have started with an existing parts base. Characterizing and grouping tens of thousands of existing designs, all of which are in production with completed plans and tooling, will not often lead to a reduction in cost of manufacture.

Most aerospace parts require high precision in some facet of manufacture, such as fastener patterns or contour. Such precision may be achieved by the careful application of skilled labor, or through the construction and use of dedicated or project tools. These tools may range in complexity from a simple drill guide to an assembly jig costing many thousands of dollars. While the number of project tools required per part varies greatly, at Lockheed-Georgia an average of 3.5 project tools is required to fabricate each part number (3). Each of these tools must be checked out of storage for each usage and then inspected for condition and engineering conformance. After the appropriate fabrication operations are completed and the parts are inspected and accepted, the project tools must again be carefully stored in a controlled location for the three to six months before the next batch of those parts will be made.

The intermittent production of small batches of parts and the low production rates combine to complicate part fabrication and assembly operations greatly. To produce a typical batch, the material must be prepared, and the project tools secured and checked. Fabrication operations at airframe manufacturers are broken down into individual functions, with a specific type of function being done in an organized area, usually called a cost center. Each batch of parts moves through a queue in the applicable cost centers, which are often well separated physically. As a result, fabrication of a small batch of parts may take 3–6 wk to accomplish 10 h of actual work. The fragmentary nature of the work, combined with the time span between repetitions of the same part number, means that there is virtually no chance that the production worker will remember any useful information from one cycle to the next.

REPEATABILITY

Another complication hindering the implementation of robotics in aerospace is the lack of repeatability in the detail parts. Since a contract may be for a very few airplanes or missiles, the planned production of a particular detail part may total fewer than ten parts. One obvious way of minimizing the final cost is essentially to hand-make the parts, which results in a great deal of hand fitting later in the assembly process. The final high-precision characteristic of aerospace is thus achieved by careful application of skilled labor for some parts, combined with very precise computer numerically controlled (CNC) machining and expensive project tooling for others. Many of the robotics projects in the airframe community have experienced difficulties because of this complication, when an automated system is installed in the middle of the production process so that the workcell is more precise than its input parts.

The prevailing wisdom of the 1970s seemed to indicate that robotics and small batch production were not compatible. "The more-or-less traditional approach to robot applications is on relatively short, repetitive cycles" (4). "Robots are most cost effective when utilized for applications where they can reduce the unit cost and operate in medium volume applications . . . if the production volume is very low annually, then manual labor will be the most cost effective" (5). Yet, later in the same paper the author says, "One can readily identify economic applications of flexible automated manufacturing that can be used in . . . aircraft manufacturing . . ." Given these intrinsic problems hampering robotic implementation in aerospace, the early lack of enthusiasm demonstrated by the airframe industry and the robot manufacturers can be easily understood. Robots were quietly appearing in support industries such as engine manufacturing, but efforts by the major aircraft producers were negligible. "The aerospace industry tried to use robots in the 1960s, but the robots of this era did not fit the job shop requirements, so these early models were rejected" (6). "Automated assembly operations have been employed outside the aerospace industry with resulting reductions in manhour requirements, increased quality and lower overall costs. These achievements have heretofore escaped the attention of manufacturing engineers. While this is understandable due to differences in mass production, single-purpose techniques and equipment versus the low production rates in the aircraft industry, it is no longer justified, based on identical advances in computer technology which have been adapted to the fabrication of aircraft parts" (7).

EXAMPLES

The fact that there now exist numerous implementations in airframe manufacture is at least partially a tribute to the determination and vision of the pioneers in the U.S. Air Force Wright Aeronautical Laboratories. The laboratories funded the development and demonstration of technically viable robotic systems and accelerated the transferral of the developed technology throughout the industry. The Air Force and the aerospace industry had led the original installation of numerically controlled (NC) machining during the 1950s in an attempt to lower the cost of the advanced weapons systems then in production. The "seed money" furnished by the Air Force had an effect far beyond what might have been expected and improved the productivity of much of the U.S. industry as NC equipment became commonplace. The Air Force hoped that a similar effort toward the development and implementation of robotics would have an equally drastic impact on costs and productivity in the 1970s, a sentiment shared by many in both the aerospace and robot industries.

One of the aerospace industry's earliest serious robotic development projects for use in airframe production began at the Fort Worth, Texas, plant operated by General Dynamics. "Material removal (drilling, routing, and milling) and riveting are by far the most common processes in aircraft production; therefore, the automation of these processes provides potential for the greatest production savings. The robotics effort at General Dynamics began in 1976 to reduce the cost involved in material removal" (8). This effort, supported by the Air Force, led to the airframe industry's first production implementation at the same plant. A jointed-arm hydraulic robot used interchangeable end effectors and project tooling to drill the composite (graphite/epoxy) vertical stabilizer fin on the F-16. Similar robotic workcells to drill the wing surfaces followed, and General Dynamics took the lead in terms of the number of robots used in production, a position it still holds.

At about the same time, Northrop Corporation in Hawthorne, California, supported by the Air Force, was developing a complex composite ply preparation system, which included a robot along with a computer-controlled cutting table, conveyor line, and vacuum-operated end effector. The system, while demonstrated extensively, saw only limited production use.

McDonnell Douglas in St. Louis, Missouri, had two projects underway for the Air Force. One project was intended to develop and demonstrate an automated riveting system including a robot, a conveyor system, and a vision system. The other project planned to develop a robot programming language similar to APT, the NC programming language developed under Air Force sponsorship in the 1950s that became a worldwide standard. These projects were technically successful, but neither resulted in a production implementation. McDonnell Douglas did develop and implement a successful robotic windshield frame/canopy drilling system at about the same time, using company funds. The language developed in the early project, called MCL (manufacturing control language), has since been the subject of follow-on projects aimed at extensions to the language and at production-oriented demonstrations (9).

Lockheed-Georgia Company installed a company-sponsored robotic riveting cell in production, utilizing a hydraulic jointed-arm robot (a Cincinnati Milacron T-3, which was virtually the aerospace industry's standard unit) and an automatic drill-riveter. Although this system was similar to that demonstrated by McDonnell Douglas, it was a simpler, independent development. Another company-sponsored system used a robot to rebuild NC milling machine cutters. A Cartesian robot, controlling a two-axis work positioner and a TIG (tungsten inert gas) welding system, automatically applied a cobalt alloy to worn or chipped cutter teeth. Figure 1 shows a welded cutter mounted in the work positioner. The portable robot control unit is seen behind the work positioner. Both of these robotic workcells were technically and economically successful and have remained in production.

Robotic painting systems were also implemented by airframers, including Boeing and Lockheed-Georgia, with a notable lack of economic success. The number and variety of detail parts to be painted, combined with a total inability to program the robots off-line, eliminated any hope for an acceptable payback in these applications. Subsequent specialized painting applications, including several missile and rocket body applications, have met with better results (10, 11).

Figure 1. A welded cutter mounted in the work position.

Figure 2. Drilling and routing cell in operation.

Robotic implementations within the aerospace industry have slowly but steadily increased over the last decade. Despite the fact—now obvious to everyone—that the industry will never be a high-volume user of robots, the robot vendors have developed a close relationship with those sophisticated users who are pushing the state of the art. The workcells and systems now in place in aerospace companies tend to be one of a kind. They also tend to be joint development projects, often with the Air Force, sometimes with other government agencies, robot vendors, or academic institutions.

Grumman Aerospace developed and installed a sophisticated drilling and routing cell that includes a track-mounted electric robot and four workstations. This cell cut production time by two-thirds on curved sheet metal parts and eliminated drilling and routing templates. Figure 2 shows the cell in operation. The operator is loading a part on one workstation while the robot routs at another of the four workstations. Development of this system was supported by a Naval Air Systems Command program (12).

Lockheed-Georgia Company developed and installed the MultiFunction Fabrication Robotic Workcell. This unit combines or eliminates ten separate production operations, including drilling, routing, sawing, and scribing. The travel and waiting times, setup times, and project tools were eliminated. The net result was a 75% reduction in the cost of production for a family of wing rib caps made from stretch-formed extrusion.

The only disappointment during the development of this workcell lay in the fact that although the robot could handle drilling and sawing without project tools, it was not capable of unguided routing to aircraft quality standards, but required guide templates. The elimination of these templates has been a long-term goal for the robotics engineers at Lockheed, and most other aerospace users, since the beginning of their efforts (10). The most effective method of achieving unguided routing and machining appeared to be through the use of a gantry-style robot, but there were difficulties with the units commercially available in 1984. Some of the robots were mechanically suitable, but could not be linked to a computer-aided design (CAD) database for off-line programming. A joint development effort with MTS Systems, a new entrant in the robotics field, appeared to offer the best chance of success. The completed LMT (light manufacturing tool) gantry robot was delivered in mid-1986, for final development and testing prior to production installation (11).

A design effort is also underway at Lockheed on a second (Mark II) MultiFunctional Workcell, using a newer design of hydraulic robot, which the engineers hope will allow unguided routing and machining.

The Extrusion Trim Center Robotic Workcell, delivered to Lockheed in late 1986, allows unguided routing by another method. An ASEA electric robot picks small (less than 28 in. (7 cm) long) pieces of rough-cut extrusion from a presentation conveyor, then manipulates them through a series of bench-mounted saws, shapers, drills, and sanders to make a family of wing rib braces.

A successful robotic implementation at General Dynamics' Fort Worth facility was a joint effort with the Air Force, under the Industrial Technology Modernization Project, and Cincinnati Milacron to develop an automated drilling method for F-16 canopy assemblies. The manual drilling operation was slow and difficult, requiring excessive operator effort. By using a track-mounted electric robot to operate automatic drills in fixtures at four workstations, an excellent payback was achieved, along with an improvement in output quality and quality of life for the operator (12).

An implementation unusual for aerospace was the result of a joint effort between the Vought Missiles and Advanced Programs Division of LTV Aerospace and Defense Company and CMW, Incorporated, of Clarksville, Arkansas. The most unusual feature of this system is that the planned production run is 400,000 small rockets for the Army's multiple launch rocket system (MLRS), each containing 644 bomblets. An automated assembly line, including numerous conveyors, workcells, robots, and vision systems, all contained within explosion barriers, assembles several thousand warheads per month at the rate of one every 2.8 min (13).

NASA and Rockwell's Rocketdyne Division have combined to develop a robotic welding system for the space shuttle main engine (SSME). Using gas tungsten arc (GTA) welding, the system will help make more than 3000 high-quality welds in a wide variety of superalloys, in material thicknesses in the range of 0.030–0.750 in. (0.08–1.9 cm) for each main engine. Complex sensors, including vision, will be required, as will elaborate fixturing. Future enhancements under development include off-line programming (14).

Sierracin Corporation, a manufacturer of aircraft canopies, and B. J. Systems, a systems house located in Goleta, California, developed a robotic system to remove scratches from in-process plastic transparencies. Manual scratch removal, essentially hand sanding and polishing, can take up to 60 h for a single scratch. A lead-through type of robot, taught a simple 2-min program that is repeated as many times as required, demonstrated acceptable scratch removal in 8 min (15).

Sierracin Corporation also has a joint effort underway with the Air Force Industrial Technology Modernization Project to develop robotic machining of the plastic transparencies. Cybotech is helping to test and develop a gantry robot for unguided machining of the canopies. The project includes tool changing and design of a universal holding fixture, along with off-line programming using a personal computer (16).

The continuing joint efforts of General Dynamics and the Air Force had succeeded in putting 13 robots on the production floor in Fort Worth by 1986. Applications include drilling the composite vertical fin, trimming of I&R (interchangeable and replaceable) aluminum panels, drilling the composite horizontal stabilizer and wings, deburring aircraft tubing, drilling canopy assemblies, and deburring machined parts. Excellent payback is reported, and development efforts will continue (17).

The Air Force has a major project currently underway aimed at automating airframe assembly. Numerous aerospace companies and universities are cooperating in this effort. Centers for Excellence in Aerospace Manufacturing Automation have been established at Stanford University and the University of Michigan by the Air Force Office of Scientific Research (AFOSR). These centers are intended to develop the technology base necessary for airframe automation, to train graduate students, and to become self-supporting. AFOSR is also funding work at SRI International in Menlo Park, California. The Defense Advanced Research Projects Agency (DARPA) is funding part of the work at Stanford and a project at the University of Utah.

DARPA and the Air Force Manufacturing Sciences Office are funding the Intelligent Task Automation (ITA) project. A team led by Honeywell is developing the technologies needed to assemble automatically a 17-part microswitch from randomly placed parts on a tray. Another team, led by Martin-Marietta, is developing the generic technologies necessary to inspect an F-15 fighter bulkhead automatically.

Two contractors, Grumman Aerospace Corporation and Rohr Industries, are participating in the Flexible Assembly Subsystems program, an important lead-in to a later Automated Assembly Center. These two companies are developing and demonstrating flexible fixturing, part location, and temporary and permanent part fastening. The approaches selected differ in concept and execution.

McDonnell Aircraft Company is continuing the enhancement of MCL, developed under an earlier contract. The enhancements are intended to improve the capabilities of MCL to interact with the factory, sensors, and databases. McDonnell is also utilizing MCL under a separate contract to program a gantry robot off-line to drill canopies and windshields. Grumman Aerospace is applying MCL to program the already-referenced electric robot mounted on tracks, utilizing four workstations to drill and rout aluminum skin panels without templates. Fairchild is implementing the assembly of a kit of sheet metal parts with MCL.

Robotic Vision Systems Incorporated has developed a 3-D

differential accuracy module, using a 3-D volumetric sensor to guide a robot end effector for the last 2 in. to its target.

This wide range of projects sponsored by AFOSR, DARPA, and the Air Force, was covered in detail in a paper presented at the Robots 9 Conference and was published in its proceedings. Space prohibits full coverage of this work here (18).

CONCLUSIONS

The total number of robots at work in the aerospace industry in 1986 is lower than 50, and this number will increase only slowly in the foreseeable future. These few machines have, however, exerted an influence far beyond what might be expected. The aerospace installations have extended the state of the art and have persuaded, allowed, and forced robot manufacturers to produce more precise, more capable systems. Through a combination of internal and joint development efforts, the aerospace industry has supported many manhours of skilled, well-financed, user-oriented application development. The results of these development programs have been, and will continue to be, diffused throughout the entire user community. Thus, aerospace efforts will result in more sophisticated, more cost-effective robotic installations for all users.

BIBLIOGRAPHY

1. R. Kuehn, Jr., "Requirements of Robotic Systems for Airframe Manufacture," presented at the 7th International Symposium on Industrial Robots, Oct. 19–20, Japan Industrial Robot Association, Tokyo, 1977.
2. T. J. Brandewie, "Air Force Robotics Research for Automated Sheet Metal Assembly," *Proceedings of the Robots 9 Conference,* Society of Manufacturing Engineers, Dearborn, Mich., 1985.
3. W. D. Dreyfoos and P. F. Stregevsky, "Robot Applications in Aerospace Manufacturing," in *Handbook of Industrial Robotics,* John Wiley & Sons, Inc., New York, 1985.
4. W. R. Tanner, "A User's Guide to Robot Applications," presented at the First North American Industrial Robot Conference, Oct. 26–28, 1976; in *Industrial Robots Volume 2: Applications,* Society of Manufacturing Engineers, 1981.
5. R. C. Dorf, "Robotics and the Automated Manufacturing Industry," *13th International Symposium on Industrial Robots and Robots 7 Conference Proceedings,* Society of Manufacturing Engineers, 1983.
6. T. Kusmierski, "Robot Applications in Aerospace Batch Manufacturing," Society of Manufacturing Engineers, Tech. Paper MS77–740, 1977.
7. W. D. Dreyfoos, "Robotics Research and Reality in Aerospace," Tech. Paper MS79–792, Society of Manufacturing Engineers, 1979.
8. D. M. Lambeth, "Robotic Fastener Installation in Aerospace Subassembly," *13th International Symposium on Industrial Robots and Robots 7 Conference Proceedings,* Society of Manufacturing Engineers, 1983.
9. B. O. Wood and M. A. Fugelso, "MCL, The Manufacturing Control Language," 13th International Symposium on Industrial Robots and Robots 7 Conference Proceedings, Society of Manufacturing Engineers, 1983.
10. M. C. Cowan, "Design Features of a Robotic Machining Workcell," *Proceedings of the Robots 10 Conference,* Society of Manufacturing Engineers, 1986.
11. M. Osheroff, K. King, T. Ngoc, and J. Just, "A Technology Forecast of Aerospace Robotics Applications," Society of Manufacturing Engineers, Tech. Paper MS85–197, 1985.
12. W. M. Harrison, "A New Automated Work-Cell for Manufacturing Aircraft Parts," Society of Manufacturing Engineers, Tech. Paper MS85–202, 1985.
13. J. L. Wilkerson, "Flexible Robotic System for F-16 Canopy Drilling," Society of Manufacturing Engineers, Tech. Paper MS85–201, 1985.
14. D. Leeds, "Flexible Assembly System for Rocket Warhead Manufacturing," Society of Manufacturing Engineers, Tech. Paper MS85–200, 1985.
15. T. A. Allison, R. F. Thompson, and C. S. Jones, "Advanced Robotic Welding for the Space Shuttle Main Engine," *Proceedings of the Robots 9 Conference,* Society of Manufacturing Engineers, 1985.
16. G. A. Grabits, "Polishing Aerospace Transparencies," Society of Manufacturing Engineers, Tech. Paper MS86–203, 1986.
17. G. A. Grabits and M. Murphy, "Robotic Drilling, Routing and Polishing of Aerospace Transparancies," presented at the Robotic Solutions in Aerospace Conference, Orlando, Fla., Mar. 3–5, 1986, unpublished.
18. L. McHale, "Robotic Applications in Aerospace at General Dynamics-Fort Worth," presented at the Robotic Solutions in Aerospace Conference, Orlando, Fla., Mar. 3–5, 1986, unpublished.

General References

G. K. Sweet, "Utilization of a Robot System to Apply Sprayable Ablative Material on Solid Rocket Booster Components for the NASA Space Program," Society of Manufacturing Engineers, Tech. Paper MS78–680, Dearborn, Mich., 1978.

P. A. Barone, "Robotic Paint Spraying at Fairchild Republic Company," *Proceedings of the Robots VI Conference,* Society of Manufacturing Engineers, 1982.

W. Marx, "Robotic Drilling and Routing for Complex Aerospace Parts," Society of Manufacturing Engineers, Tech. Paper MS84–223, 1984.

AGRICULTURE, ROBOTS IN.

See Robots in Agriculture.

APPAREL INDUSTRY, ROBOTS IN

EDWARD BERNARDON
The Charles Stark Draper Laboratory, Inc.
Cambridge, Massachusetts

INTRODUCTION

The apparel industry has grown to be the third largest consumer industry, with apparel shipments of over 51.6 billion dollars in 1982. Competition from foreign imports threatens to reduce the domestic apparel producers' market share. In 1975, imports made up 13.7% of the U.S. apparel market. Experts believe that by the late 1980s, the foreign share may increase to as high as 50% (1). Due to the lower wages paid to garment workers overseas, imported apparel can be sold at 20% less than U.S. goods, even after application of import duties. Thus, to maintain or increase domestic producers' share of the U.S. market, production costs must be reduced. The process of assembling pieces of fabric into a finished garment is an area that would seem to lend itself to cost-cutting through automation.

At present, a typical garment assembly room is characterized by rows of sewing machines and presses. These facilities are unchanged from operations of 40 years ago. Individual operators at each machine perform specialized tasks on pieces of cloth to produce garment subassemblies, which are eventually combined into a finished garment. Garment assembly consists of two basic operations: the first is cutting and second is sewing. Cutting is generally done on multilayers of material with a hand-held reciprocating knife cutter. Die cutting has been attempted by some but with little commercial success. Sewing or assembly of the cut pieces is carried out on isolated sewing machines, each of which has its own operator. Prepared or finished pieces are bundled, tied, and carted to the next sewing station. The consistent quality of the finished products is highly dependent on the operator's skill and dexterity.

The operations described above are labor-intensive and have been so for many years. Recent advances in computers and electronics have finally opened the door for major changes in the garment assembly process.

Cutting technology made a great leap forward in the 1970s when Gerber introduced a multilayer robotic cutter. The system consists of a variable-speed reciprocating knife mounted on an x–y positioner. The knife automatically adjusts for blade deflection when cutting multilayers of cloth. The high cost of the cutter has limited its use to larger apparel manufacturers. Using robots has virtually been nonexistent to this point except in long-term research projects. The high cost of robotic equipment and the inability of robots to handle limp cloth as dexterously as a human operator can be cited as the principal reasons for the lack of robots in this industry. The recent reduction in the price of robotic equipment has begun to open the door for potential use. Simple pick-and-place robots are sparingly used to load or remove a cloth workpiece from present-day sewing machines. The more complex operating of assembly, however, presents a greater challenge to robots.

Unlike the assembly of nonflexible metal or hard plastic parts, as are now beginning to be seen in automotive applications of robots, cloth assembly must address the problems caused by the inherent limpness of cloth. When transporting, sewing, or folding cloth, a system must be devised to maintain the orientation of a cloth workpiece to the robot. The various sections of the workpiece must also be kept in the proper orientation relative to each other. The tendency of woven and especially knit materials to stretch with little force makes such a task difficult.

The fact that garment sections are generally cut and prepared for assembly in multilayer stacks presents a challenge for automation. Thus, loading of a cloth workpiece requires a tool that is nimble enough to remove only the upper ply from a multilayer stack while not damaging the cloth. Folding of cloth also requires manipulation of multilayers when forming cuffs on pants, vents on sleeves, or aligning edges on different pieces of cloth.

Advanced research in automation of apparel technology is currently taking place in Japan, Europe, and the United States. Research in the United States is being conducted by the Textile Clothing Technology Corporation ([TC]^2).

Design of a machine that automates production of a variety of garment subassemblies while addressing some of the concerns stated above is presently under way at The Charles Stark Draper Laboratory, Inc. under supervision of TC^2 (2). This prototype system is a general-purpose, programmable work center that incorporates automated part recognition, material handling, sewing, and cloth manipulation, all operating under distributed microcomputer control. This article describes the design of the folding module and the steps taken to overcome the difficulties encountered in dealing with limp cloth.

FOLDING-MODULE DESCRIPTION

Sleeves for men's tailored suits were chosen for the folder-design study due to the variety and difficulty of operations during assembly. The task consists of a series of sewing and folding operations, as shown in Figure 1. The sequence begins with the sewing of the pre-cut inner and outer sleeve segments at the inseam (Fig. 1**a**). Next, the cuff vents are formed, first by opening the inner sleeve (Fig. 1**b**), then by folding and tacking of the vents (Fig. 1**c**). The sleeve is then prepared for closing by returning the inner sleeve to its original position (Fig. 1**d**). Finally, the sleeve is completed by matching the inner and outer sleeve elbow edges and sewing the elbow seam (Fig. 1**e**). All critical dimensions, such as vent depth and the alignment of inner and outer sleeve edges, must be held to within 1/16 in. (0.15 cm). This tolerance maintains a perfect "visual" alignment.

In order to automate the process described above, the sewing and folding machine, as shown in Figure 2, was designed. At the far right is a manual loading station. A hinged door, comprised of moving belts, lowers onto the loaded workpiece and slides it into the sewing module located in the center. Here all sewing on the workpiece is performed with aid from the vision system located over the loading area. Transport belts on the sewing module slide the sewn workpiece onto the vacuum folding table at the far left. Once the folder transport door is raised, the workpiece is held in position by the vacuum table. The vacuum table has a perforated reflective surface under which a series of shutters are located. The shutters are computer selected to accommodate various cloth styles and characteristics. All folding is accomplished by an IBM-7547 robot located near the folding table. A specialized cloth-handling end-effector is mounted on the robot (Fig. 3). The end-effector is comprised of 14 cloth pickers which have the ability to pick an upper ply of cloth from a multiply stack. The pickers are mounted on a flexible urethane spline which can be bent to match most "C"- or "S"-shaped cloth edges.

The spline shape is controlled by stepper motors located on the spline (Fig. 3). To determine the proper spline configuration and robot coordinates, a camera is located above the reflective folding table. Cloth-edge location data gathered by the camera is fed to an IBM PC. Information on the location of key workpiece points is calculated. The PC then takes the vision data and calculates the required spline shape and robot coordinates needed for folding. The manipulator folding mo-

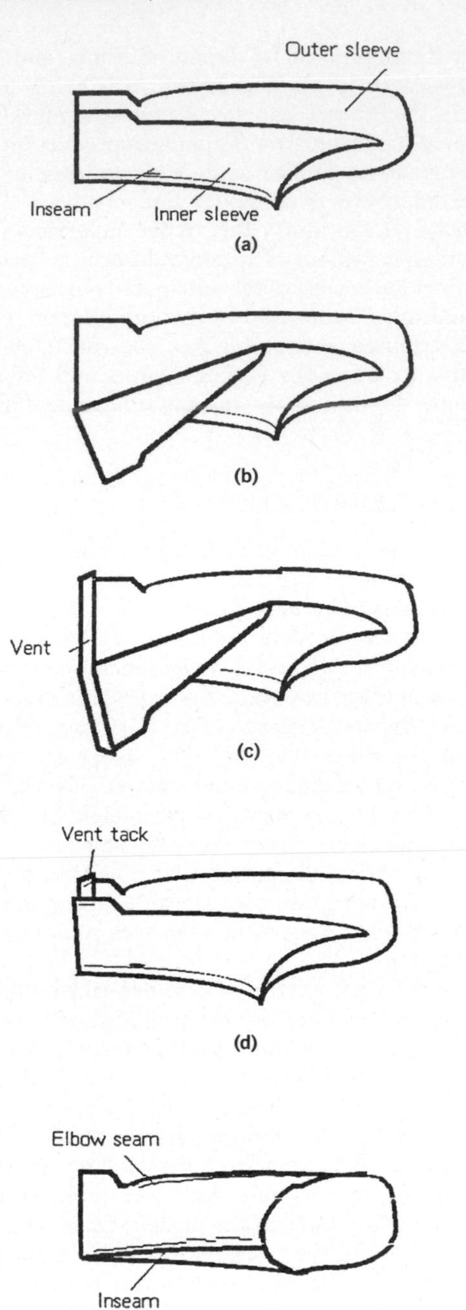

Figure 1. Sleeve-folding sequence.

tions are controlled by the IBM-7547 robot controller. When coordinates dependent on vision data are required by the robot, the IBM PC is signaled by the robot controller and the desired coordinates are downloaded. Once folding is complete, the PC signals the main sewing computer so that the workpiece can be ejected or slewed into the sewing machine for further sewing.

DEVELOPMENT OF A ROBOTIC CLOTH-FOLDING SEQUENCE

A men's tailored suit sleeve, as shown in Figure 1**a**, is composed of an inner and outer sleeve section. When assembling

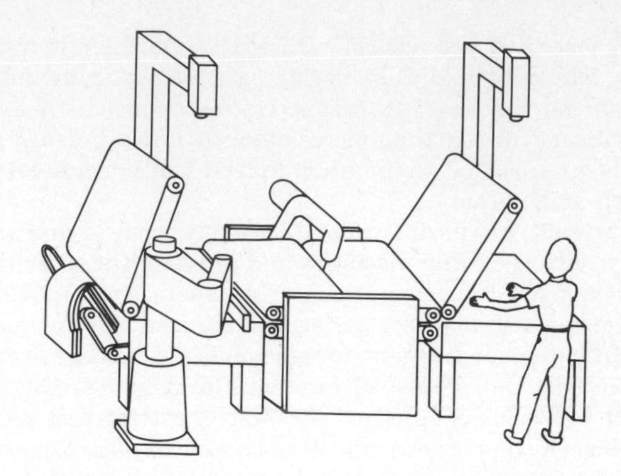

Figure 2. Automated sewing and folding machine.

a sleeve in the conventional manner, a process is devised which is well-suited to assembly by humans. Unfortunately, this process is not necessarily easily transferred to the automated folding system. The process must be altered in a way that allows automated assembly and maintains the final product tolerances and quality aspects. A significant difference between manual and automated folding is the manner in which the cloth is supported during the fold. In manual folding, the main cloth piece is held stationary during folding with a combination of the hands or fingers. End-effector technology has not yet reached the point where the dexterity and speed of the human hand can be achieved as is required for cloth folding. Therefore, with automated folding, the main cloth section is held in place by a vacuum. Folding is conducted on a vacuum surface under which a series of gates are located. The gates can be opened or closed to redirect and concentrate the vacuum in a specific region. With the workpiece properly held in place, the fold is made by grasping the desired section of cloth with the cloth-handling end-effector. The cloth is then placed at the desired folded position.

The manual folding process usually requires modification to be compatible with the vacuum table and folder. Modifications can be made to the workpiece, process, or both. Once altered, the process and workpiece are first tested to verify that a vacuum gate pattern exists that can hold the cloth during folding. During this initial testing, the cloth is held

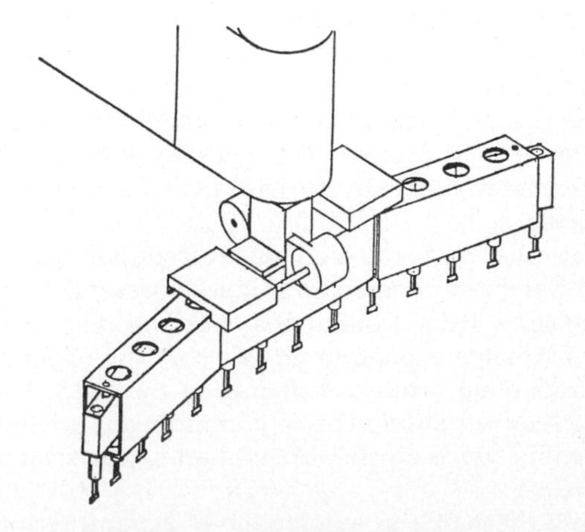

Figure 3. Cloth-handling end-effector.

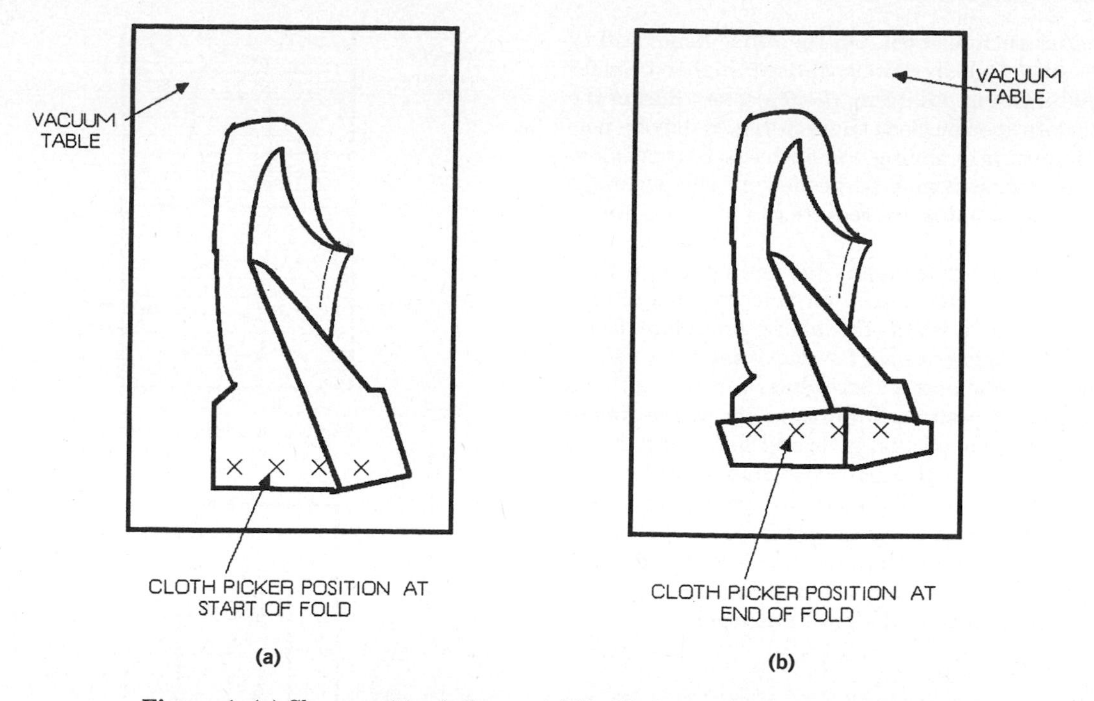

Figure 4. **(a)** Sleeve position before vent fold; **(b)** sleeve position after vent fold.

by the vacuum but moved by hand. A prospective vacuum gate pattern is selected and the desired cloth edge folded. The cloth-workpiece characteristics and position are then studied to determine if a proper fold has occurred. Proper folding is not achieved if during or after the fold any of the following occur:

1. The main cloth section is not held stationary during folding.
2. The cloth does not lie "flat" after the fold is completed.
3. The vacuum cannot hold the cloth in place after the fold is completed.

When any of the above occur, the vacuum pattern must be altered and folding reattempted. If the desired fold cannot be achieved with any vacuum pattern, then the workpiece or process must be further modified.

Once an acceptable fold is accomplished by hand, a similar fold motion must then be taught to the manipulator. Figure 4 shows a typical automated sleeve vent fold position before and after folding. In the initial position, the cloth is marked to indicate the robot pickup position. The manipulator is then put into the teach mode and moved to a coordinate that aligns the cloth pickers with the marks. This coordinate is then recorded for use in the folding sequence. Next, the desired cloth section is placed in the folded position by hand. The manipulator is then taught the final position by again aligning the cloth pickers to the appropriate marks.

With the fold start and end points known, a robot trajectory must be determined between the two points. Figure 5 shows the simplest trajectory from start (point 1) to finish (point 4). Here the manipulator is simply raised at cloth pickup, translated to the new position, and lowered. The height above the folding tables at points 2 and 3 must be high enough to avoid cloth bunching (Fig. 6) and low enough to not slide the main cloth section (Fig. 7). The height at which bunching or sliding occurs depends on cloth characteristics. Cloth with low coefficient of friction, such as polyester, will slip at lower heights

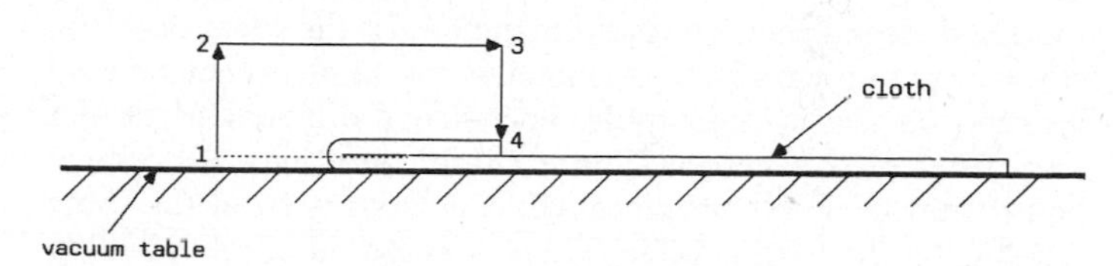

Figure 5. Simple robot fold trajectory: 1, pickup point; 2, same coordinate as pickup with raised z-axis; 3, translate to release point; 4, lower z-axis and release cloth.

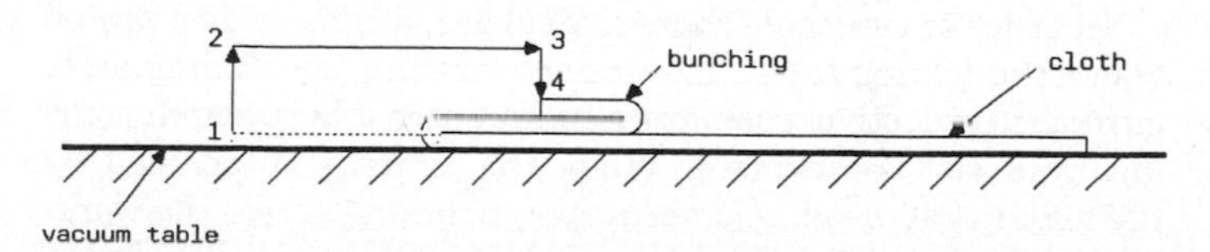

Figure 6. Cloth bunching caused by excessively low transfer height: 1, pickup point; 2, raise to translation height which is too low; 3, translation to release point causes bunching; 4, release cloth with improper fold.

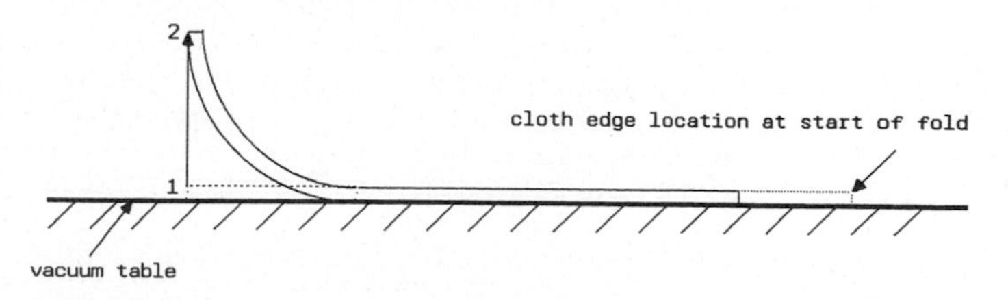

Figure 7. Main cloth section movement caused by excessively high transfer height: 1, pickup point; 2, raise to translation height which causes movement of main cloth section.

than nonsynthetics such as wool. On the other hand, stiff or thick materials such as heavy tweed require higher transfer heights to avoid bunching. Adapting the trajectory during the fold to accommodate specific cloth characteristics may be necessary to avoid bunching or sliding. For such cases, a trajectory such as shown in Figure 8 may be needed. In this example, four manipulator coordinates are required to assure a proper folding trajectory.

Once a desired trajectory is taught, the manipulator is then run through the fold at the maximum velocity that retains the desired fold characteristics. the above procedure is repeated for each of the folds needed for sleeve assembly. When completed, a folding sequence exists which can be run for a specific sleeve style and position. Unfortunately, sleeve dimensions vary even within a particular style and size. Dimensional variations result from the manual cloth-cutting techniques used by the apparel industry. Therefore, to assure proper folding despite these variations, a vision system is used to locate key points on the sleeve. Algorithms are then developed which determine robot coordinates from vision data. The following section gives examples of typical algorithms used.

VISION SYSTEM

The folding sequences taught to the robot are based on an assumed sleeve position and dimension. In the ideal case, the sleeve segments are always loaded at the same predetermined location on the loading table, and sleeve dimensions do not vary from piece to piece. Under these conditions, all critical points on the cloth workpiece will consistently be at the same location on the folding table. Thus, a single set of robot folding coordinates and spline shapes would be adequate to assure proper folding. Unfortunately, cloth dimensions vary well beyond acceptable cloth-edge misalignment, and exact loading cannot be assumed under factory operating conditions.

In order to overcome these difficulties, a camera is mounted above the folding table. To enhance viewing, the folding table surface is reflective, assuring a high contrast between the surface and cloth workpiece. When the camera is signaled by the folder that viewing is necessary, scanning across the workpiece (as shown in Fig. 9) is started. Each scan line has a light-to-dark and dark-to-light edge crossing, signifying the cloth-edge location.

The vision computer first filters stray points from the edge-crossing data and then finds key cloth contour changes or breakpoints on the workpiece. The filtering is done only in the "area of interest" to reduce calculation time. The folder determines the desired search area based on the fold being performed. For a vent fold (Figs. **1b–c**), the area of Figure 10 would be used; for a sleeve closing, the area of Figure 11 would be used. The severity of change in cloth profile that results in a breakpoint is determined by filter parameters (3).

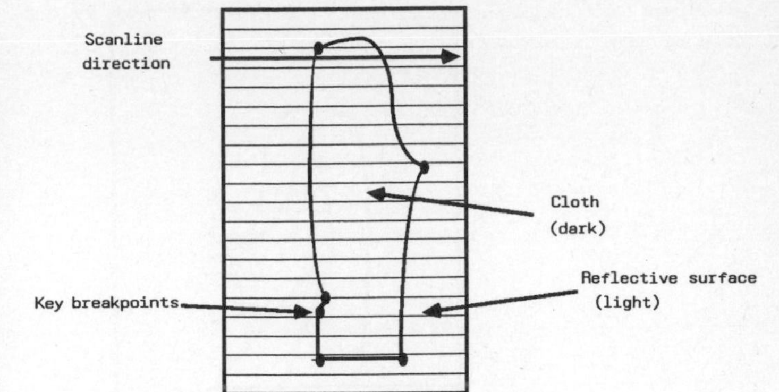

Figure 9. Camera scan of sleeve.

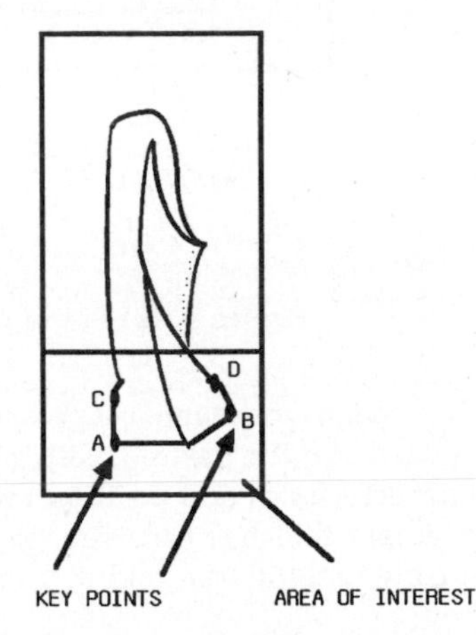

Figure 10. Area of interest during vent fold.

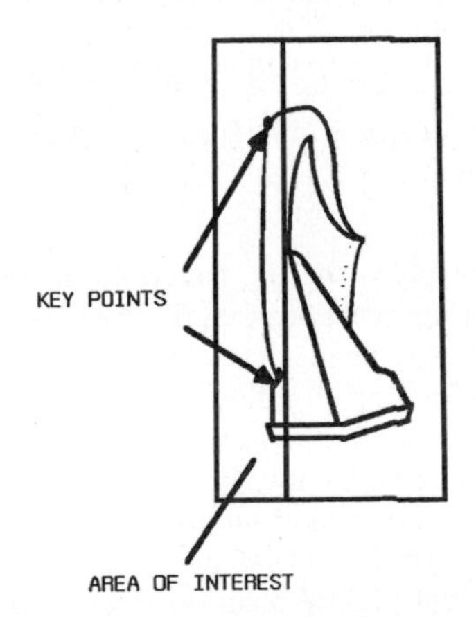

Figure 11. Area of interest during closing fold.

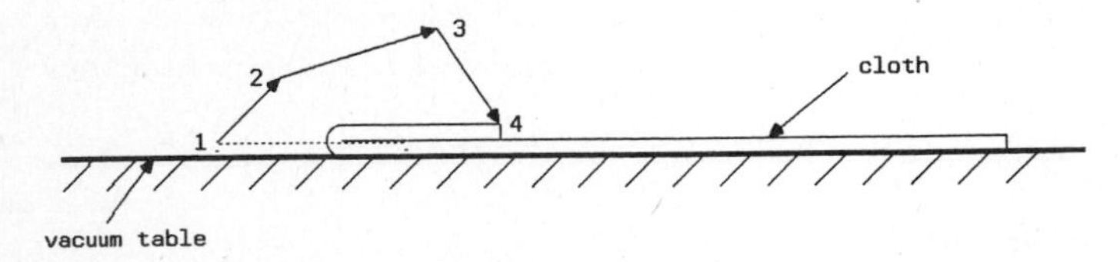

Figure 8. Modified folding trajectory to avoid cloth bunching or sliding.

In simple cloth folds, such as vent formation (Figs. **1b–c**), two breakpoints are needed for initial cloth pickup. Breakpoints A and B of Figure 10 are used to determine the actual sleeve orientation (X_{SA}, Y_{SA}, Θ_{SA}) at the start of the fold. This sleeve orientation differs, however, from the sleeve orientation of the manipulator teaching session (X_{ST}, Y_{ST}, Θ_{ST}). Therefore, the taught manipulator position (X_{MT}, Y_{MT}, Θ_{MT}) must be transformed to correspond to the actual sleeve orientation (X_{SA}, Y_{SA}, Θ_{SA}). The actual manipulator orientation for cloth pickup (X_{MA}, Y_{MA}, Θ_{MA}) is then found through equation 1 (4):

$$\begin{bmatrix} \cos(\Theta_{SA}-\Theta_{ST}) & \sin(\Theta_{SA}-\Theta_{ST}) & 0 \\ -\sin(\Theta_{SA}-\Theta_{ST}) & \cos(\Theta_{SA}-\Theta_{ST}) & 0 \\ 0 & 0 & 1 \end{bmatrix} \begin{bmatrix} X_{MT}-X_{ST} \\ Y_{MT}-Y_{ST} \\ \Theta_{MT}-\Theta_{ST} \end{bmatrix} + \begin{bmatrix} X_{SA} \\ Y_{SA} \\ \Theta_{SA} \end{bmatrix} = \begin{bmatrix} X_{MA} \\ Y_{MA} \\ \Theta_{MA} \end{bmatrix} \quad (1)$$

Other points along the fold trajectory are also found through equation 1. The sleeve breakpoints selected for robot coordinates transformation depend on which part of the trajectory is being considered. For a vent fold, points A and B are aligned relative to points C and D of Figure 10 during a fold. Points A and B are used for cloth pickup and initial trajectory transformations, whereas points C and D are used for transformation near the completion of the fold.

More complex folds involve not only changes in robot coordinates but also shaping of the cloth-picker spline. Spline shaping is required when cloth edges with curved contours must be moved and aligned. A typical example is the alignment of the inner and outer sleeve section before sewing of the elbow seam (Figs. **1d–e**). For this fold, information on the contour between the breakpoints of Figure 11 is required. This information is used to locate the robot end point, as well as to bend the cloth-picker spline. The spline must be placed so that the pickers align near the cloth edge and provide support. The contour is determined by applying a third-order polynomial least-square fit to the edge-crossing data which occurs between the breakpoints. The resulting equation coefficients are then passed to the folder. An example set of sleeve data points and the corresponding sleeve edge coefficients is shown in Figure 12.

Before the spline can be bent to match the edge contour, the contour must be transformed from the camera frame into the coordinate frame of the spline (Fig. 13). Brute-force transformation of the equation coefficients from the camera to spline coordinate frame would be time-consuming. Instead, four points are selected in the camera frame curve for transformation (Fig. 13). The transformation of these points into the spline coordinate frame is shown below (4):

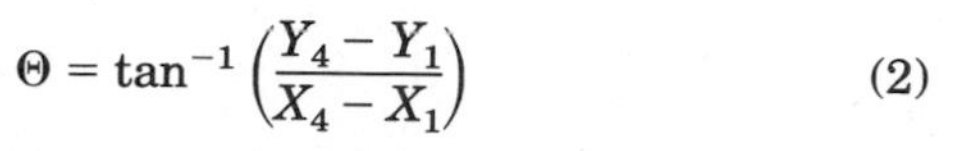

$$\Theta = \tan^{-1}\left(\frac{Y_4 - Y_1}{X_4 - X_1}\right) \quad (2)$$

$$\begin{bmatrix} \cos\Theta & \sin\Theta \\ -\sin\Theta & \cos\Theta \end{bmatrix} \begin{bmatrix} X_n - X_1 \\ Y_n - Y_1 \end{bmatrix} = \begin{bmatrix} XSP_n \\ YSP_n \end{bmatrix} \quad (3)$$

$X_1\, Y_1$ = breakpoint coordinate nearest camera origin
$X_n\, Y_n$ = table frame coordinate n
$XSP_n\, YSP_n$ = spline frame coordinate n

Points 1 and 4 are now located along the x axis of the spline coordinate frame. Since point 1 is also aligned with the y axis, the A_{SP} coefficient of equation 4 which defines the cloth contour becomes 0.

$$A_{SP} + B_{SP}\, XSP_n + C_{SP}\, XSP_n^2 + D_{SP}\, XSP_n^3 = YSP_n \quad (4)$$

Coefficients B_{SP}, C_{SP}, and D_{SP} are then found through equation 5:

$$\begin{bmatrix} XSP_2 & XSP_2^2 & XSP_2^3 \\ XSP_3 & XSP_3^2 & XSP_3^3 \\ XSP_4 & XSP_4^2 & XSP_4^3 \end{bmatrix} \begin{bmatrix} B_{SP} \\ C_{SP} \\ D_{SP} \end{bmatrix} = \begin{bmatrix} YSP_2 \\ YSP_3 \\ YSP_4 \end{bmatrix} \quad (5)$$

The coefficients found in equation 5 can now to be used to calculate cloth contour points in the spline coordinate frame. These points are needed to determine the proper spline shape.

The spline, which holds the cloth pickers, is a flexible member that can be modeled as a beam (Fig. 14). The spline is

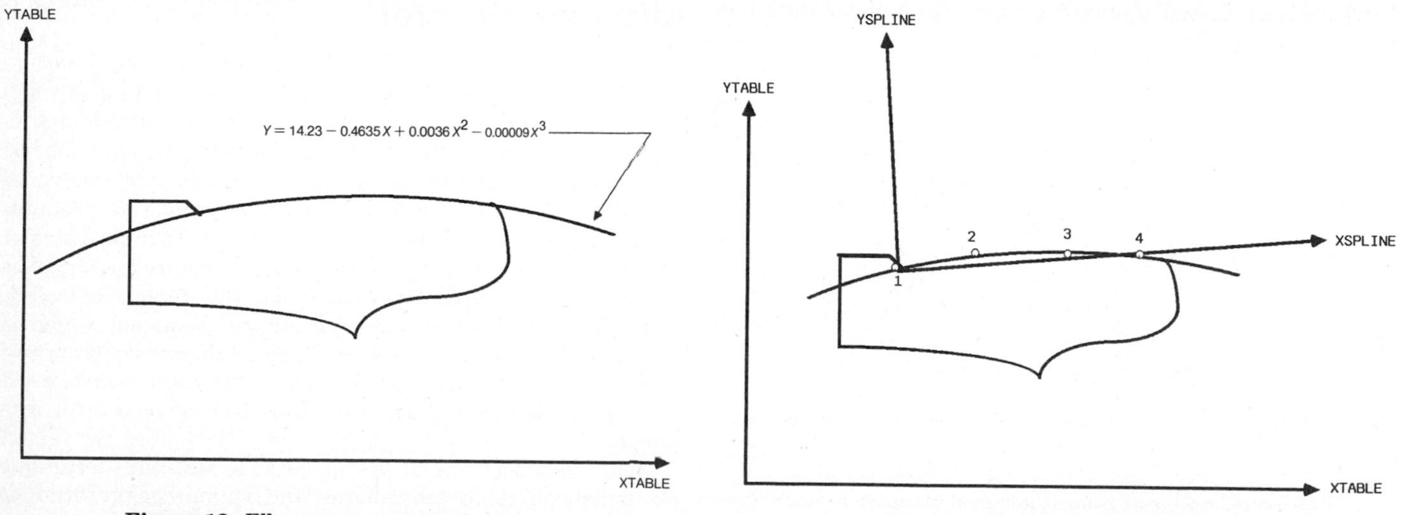

Figure 12. Elbow seam contour elevation.

Figure 13. Spline coordinate frame.

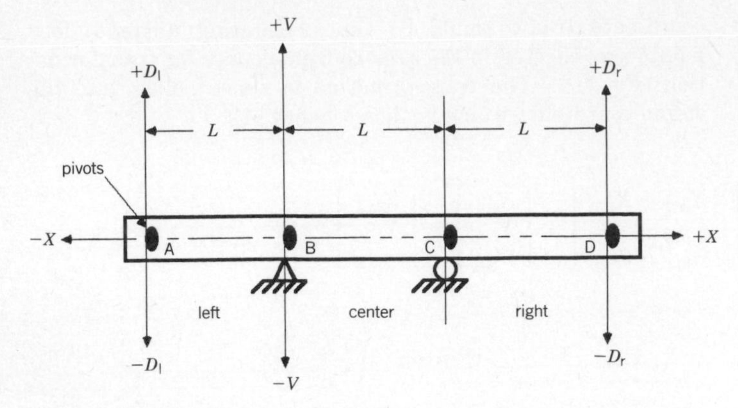

Figure 14. Spline represented as beam: L, pivot to pivot distance of 8 in. (20 cm); D_r and D_l, end pivot deflection which is controlled by spline stepper motors; A, B, C, and D, spline pivot points.

pivoted to a three-section support structure (Fig. 3). The inner pivots B and C are fixed relative to the robot end point. Outer pivots A and D rotate relative to the inner pivots to shape the spline. All pivots are separated a fixed distance L of 8 in. (20 cm). Based on beam theory, equations 6 and 7 are derived for deflection V along the left and right sections of the spline (5). Before spline deflection can be calculated, the position of the spline relative to the curve must be found. In order to orient the spline along a curve, the four points shown in Figure 15 are required. The center section of the spline will be placed such that the inner pivot points B and C are located at points 2 and 3, respectively. The robot end point is located halfway between these pivots (Fig. 16).

$$V_l = \left(\frac{4D_l - D_r}{15L^3}\right) X_l^3 + \left(\frac{4D_l - D_r}{5L^2}\right) X_l^2 - \left(\frac{7D_l + 2D_r}{15L}\right) X_l \quad (6)$$

$$V_r = \left(\frac{D_l - 4D_r}{15L^3}\right)(X_r + L)^3 + \left(\frac{8D_r - 2D_l}{5L^2}\right)(X_r + L)^2 + \left(\frac{11D_l - 29D_r}{15L}\right)(X_r + L) + \frac{3D_r - 2D_l}{5} \quad (7)$$

The spline must now be deformed such that the proper spline bends offsets, V_r and V_l, occur on the x axis at points 1 and

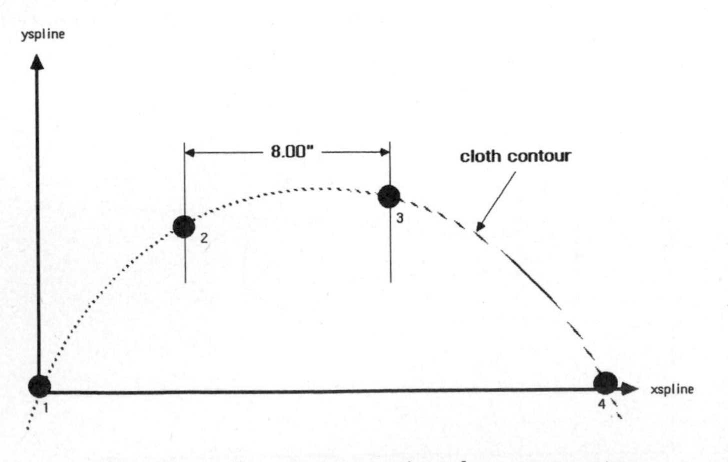

Figure 15. Spline-alignment points along curve contour.

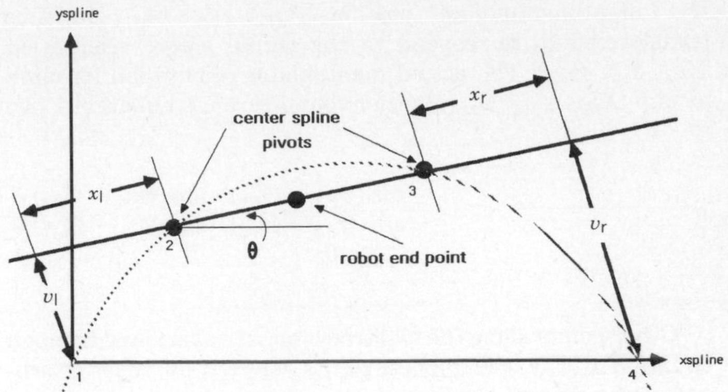

Figure 16. Geometric relationships required for spline bending.

4. For symmetric contours, a line through the inner spline pivots will be nearly parallel to the spline coordinates frame x axis. Based on this assumption, the small-angle approximation can be used in calculation of V_r and V_l. Figure 16 shows the geometric relationship that leads to equations 8–11. It must be noted that the contour curvature in Figure 16 is exaggerated to show spline orientation more clearly.

$$V_r = (X_4 - X_3) \sin \Theta + Y_2 \quad (8)$$

$$X_r = X_4 - X_3 - Y_3 \sin \Theta \quad (9)$$

$$V_l = Y_2 - X_2 \sin \Theta \quad (10)$$

$$X_l = X_2 + Y_2 \sin \Theta \quad (11)$$

The spline deflections V_r and V_l cannot be controlled directly. Instead, a deflection at pivots A and D must be found which results in the proper deflection at X_r and X_l. Substituting V_l, X_l into equation 6 and V_r, X_r into equation 7 yields two equations that can be solved for pivot deflections D_l and D_r. These deflections are then converted to stepper motor counts and downloaded to the stepper controller.

FOLDER CONTROL SYSTEM

The folder control system has the task of managing information from the vision system to determine the proper robot and end-effector movements. Figure 17 shows the folder subsystems. When a fold sequence is under way, all folder subsystems, except for the spline shape, are under direct control of the IBM-7547 robot. Most robot digital outputs drive solenoids which control air-activated devices, such as vacuum gates or the cloth pickers. When a spline bend is required, a signal is ent on a 7547 digital output channel to the stepper controller. This signal initiates the desired stepper sequence which is stored in the stepper controller. Since all digital input/output, as well as robot motion, is directed by the robot controller, it is programmed in AML/E, the IBM-7547 robot control language.

With incorporation of vision data, a real-time technique for update of robot coordinates and spline configuration is

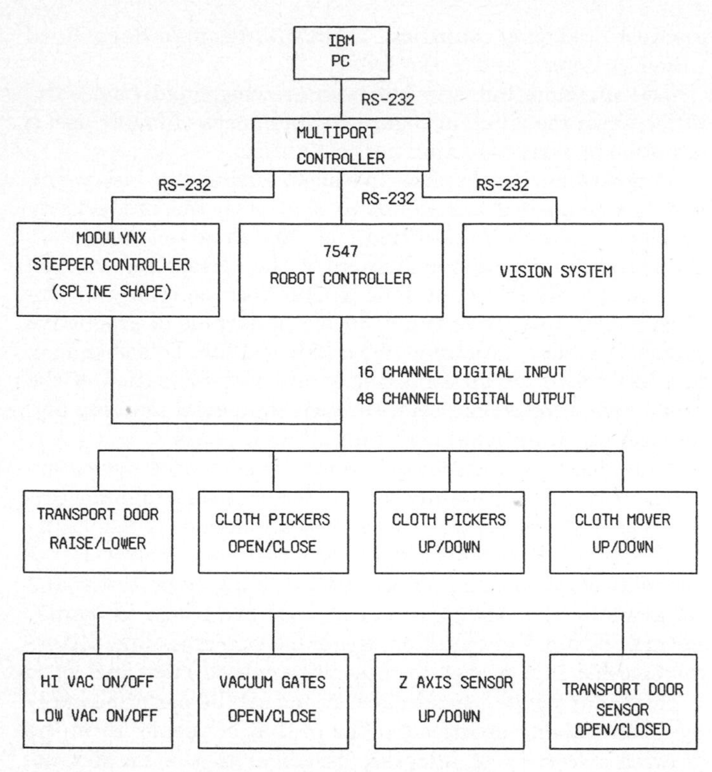

Figure 17. Folder subsystems.

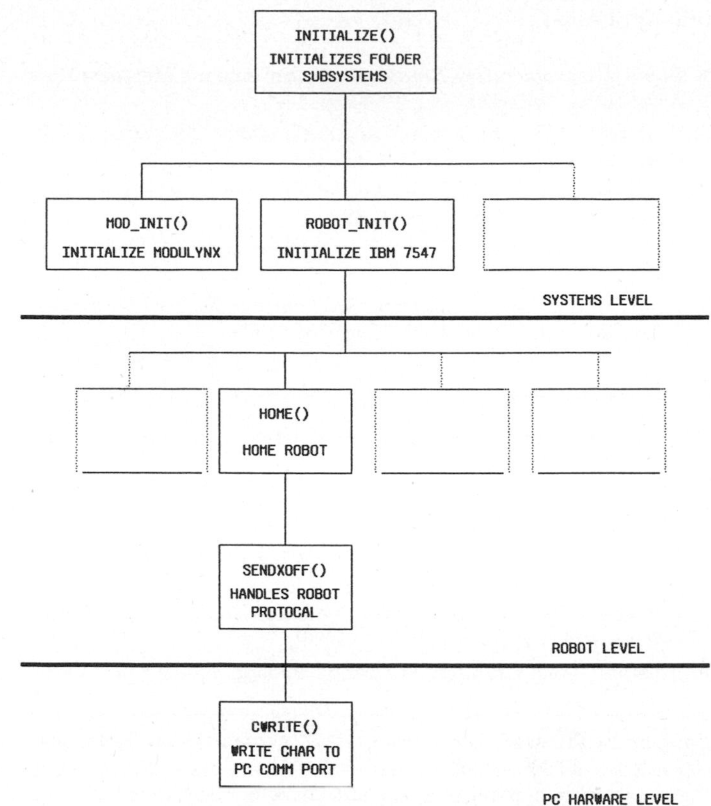

Figure 18. Sample high level C command functions.

needed. AML/E 4.0 robot control language allows real-time update of robot coordinates. A set of routines on the IBM PC use algorithms described earlier to take information from the vision system and then determine the desired robot coordinates. Additional routines handle the data format conversion and protocol between the PC and robot controller.

Even though the approach above is straightforward, the communication and data-management software can become tedious to modify. To alleviate this problem, folder control system libraries were developed in C on the IBM PC. C was chosen not only for its speed of execution but also for the ease with which multiple lower-order functions could be combined to form simple, higher-order functions. At the lowest order, a set of PC communication port functions were written in macro-assembler to ensure the utmost speed in communication. The next level of functions (written in C) utilized the comm-port functions to facilitate robot communication and folder system control. These functions do such things as download an altered robot coordinate or actuate robot teach box commands. At the highest level are the system functions. At the system level, functions from the communication, robot, and system libraries are combined to form the resulting folder control system command language. For example, to initialize the entire folder system, a single command function, initialize(), is used. Some of the functions that are combined to form initialize() are shown in Figure 18.

By combining the appropriate functions, folder controller programming is simplified. The example program shown below initializes the system and peforms a single vent and closing fold for a particular sleeve style and size. Arguments in these functions allow for a certain degree of flexibility.

```
/* This program initializes the folder control system and executes
   a right vent and right close fold                              */

main( )
{

/* initialize folder */
       initialize(comm_port);

/* execute vent fold */
       rvent(comm_port, robot_partition);

/* execute close fold */
       rclose(comm_port, robot_partition);
}
```

CONCLUSIONS

The apparel and textile industries present many opportunties for the application of automated systems. At the present time, systems are commercially available which automate apparel design, parts-cutting, as well as transfer of materials within the factory. The system described in this article goes one step further by demonstrating that a flexible and effective tool for the manufacture of suit sleeves can be devised with present technology. Future studies will extend this technology to other sections of the apparel industry.

BIBLIOGRAPHY

1. *Apparel Manufacturing Studies,* American Apparel Manufacturers Association, Arlington, Va., 1984, pp. 1–9.
2. S. Purdy, "TC²: The Vision," *Apparel Industry Magazine,* 40–44 (Feb. 1985).
3. A. Rosenfeld and A. Kak, *Digital Picture Processing,* 2nd ed., Academic Press, Inc., New York, 1982.
4. R. Paul, *Robot Manipulators,* The MIT Press, Cambridge, Mass., 1981.
5. Byars and Snyder, *Engineering Mechanics of Deformable Bodies,* Intext Education Publishers, New York, 1963.

General References

G. Berkstresser and K. Takeuchi, "Automation: A Fighting Chance," *Bobbin,* 50–66 (Mar. 1985).

S. H. Jones, "MITI: Progress Report," *Apparel Industry Magazine,* 46–49 (Feb. 1985).

S. Kry, "Apparel and the New Technologies," *Bobbin,* 180–185 (Oct. 1984).

J. Hamilton, "Robots in the Textile Industry: How Soon," *Industrial Fabrics Products Review,* 8 (Oct. 1984).

M. Gaetan, "Time and Manufacturing," *Bobbin,* 44–62, (July 1984).

Funding for this work is provided by the Textile Clothing Technology Corporation ([TC]²) which in turn is supported by 66 companies in the fiber textile apparel industry and the U.S. Department of Commerce (Grant No. 99-26-07176-10). Thanks also go to all those at the C. S. Draper Lab who contributed to this work. (The statements, findings, conclusions, recommendations, and other data in this report are solely those of the Draper Laboratory and do not necessarily reflect the views of the U.S. Department of Commerce.)

APPLIANCE INDUSTRY, ROBOTS IN

DANIELE FABRIZI
University of Pavia
and International Federation of Robotics
Bergamo, Italy

INTRODUCTION

The appliance industry is widespread in many countries and includes many products which differ in functions, dimensions, shape, structure, mechanical composition, appearance, and manufactured quantities.

It can be subdivided into the following classes: refrigerators, washing machines, cooking appliances, ventilation and suction appliances, water-heaters, house-cleaning appliances, heating appliances, electric hair-dryers, ironing appliances, radio and television sets, and others.

Roughly speaking, one can have an idea of the extent of the world production and division of some kinds of electric appliances from Table 1, where data relating to 1984 are reproduced.

In addition one can say that radio and television set production reached 200 million pieces, with the following list of the greatest producing countries: 1, USA; 2, People's Republic of China; 3, Japan; and 4, the USSR.

The appliance industry has been growing rapidly since the fifties, until reaching, in regard to appliances of larger use, a situation of overproduction in the eighties.

Competition has become too keen during the last years, and has compelled companies to deal with the productivity problem resolutely by resorting more and more to automation. Obviously the situation is different in the various sectors. The demand for electric household appliances, especially in Europe, shows a negative trend; in fact a surplus of productive capacity persists, estimated about 35% in 1985. This phenomenon has called for an adjustment and rationalization of the productive dimensions, even through industrial merging and integration, at national and international levels.

Until recently automation, where present, was almost exclusively "rigid" and mainly devoted to sheet-steel manufacturing, molding of plastics, spot welding, application of insulating materials, part manufacturing and assembly. However, the necessity of increasing productivity involving all processes and the availability of new electronic means have driven companies toward "flexible" automation, involving processes and sectors disregarded in the past. In assembly, where labor incidence is particularly high, certain operations can be performed only by means having abilities similar to that of people. Even the growing necessity of adapting products to the market demands, creating new styles and variations relating to the consumers' quickly mutable tastes, has weighed upon the introduction of "flexible" automation.

Products must meet the following requirements:

- Low costs, to cope with the high market competitiveness
- Quality and reliability
- Fast adaptation to the styles and requirements of the various countries

Many processes are common to the entire manufacturing sector, therefore, the same methods of automation are valid. Among those most easily adaptable to robotization are

- Painting, enameling, sandblasting
- Grinding
- Arc and resistance welding
- Loading/unloading of machine tools, presses, shearing machines, ovens
- Various material handlings
- Inspection, measure, checking
- Assembly of components (drums, tubs, electrical motors, various cells, compressors, switch-gears, printed-circuit boards, etc)
- Final assembly.

At present, many types of robots are available on the market and can be used in the above-mentioned processes.

Most work environments do not present special hygienic problems. Certain processes such as welding, painting, adhesive and sealing deposit, die-casting, etc, do require particular safety measures.

Table 1. Production of Some Electric Household Appliances, in Millions of Pieces, in 1984

Production Area	Refrigerators[a]	Washing Machines and Dryers[b]	Electric, Gas, Mixed, Cooking Appliances[c]
Western Europe	14	13	6
Eastern Europe	11	8	2
Other developed countries	13	18	14
Developing countries	9	8	2
Total	*47*	*47*	*24*

Position of countries in order of production.
[a] USA, USSR, Italy, Japan
[b] USA, People's Republic of China, Japan, Italy.
[c] Japan, USA, Italy, Federal Republic of Germany.

Until today the number of robots used in the appliance industry has remained markedly below that of robots used in motor-car industry, but the situation is rapidly changing. New products are often designed with an eye to the automation necessities.

POSSIBILITIES OF AUTOMATION

The various types of appliances that make up the category of the electric appliances, lend themselves considerably to the numerous applications of automation, widespread in the manufacturing industrial field at present (see also APPLICATIONS OF ROBOTS).

The dimensions of the appliances vary. Various processes, common to the whole industrial field, are adopted in part manufacturing and in assembly. In the smallest appliances (such as hair-dryers, ironing appliances, mixers), the use of plastic materials is widespread, and the output can sometimes reach such values as to justify types of "rigid" (nonflexible) automation, studied and fitted to that particular type of appliance.

The production of plastic parts can be automated easily by the use of presses furnished with microprocessor controllers. These controllers are able to control the production cycle automatically.

The pressing departments with high production (whose rate is limited only by the polymerization times of the materials) can function under the supervision of only one person, while another is assigned to the mold control and machine service.

The pieces, unloaded from the machines, are conveyed automatically, arranged in an orderly manner, and piled up in special containers, to be sent then to the following work places or to the final assembly. Cycle robots are sometimes used in operations of handling and orderly unloading from the machines.

All the components of the various types of electric appliances (motors, compressors, timers, ventilators, electron devices, clamps, switches, push-buttons, signal lamps, mechanic devices, hinges, locks) lend themselves to the various kinds of robotic automation. Robots are used less in the production of the single parts (where mechanic arms are sometimes resorted to), and more in the final assembly, where they play a fundamental part, alone, in a group, or in flexible lines. The first flexible lines able to assemble different kinds of devices are now available on the market.

The concept of flexibility (in contrast with that of rigidity) is valid both for *the single machine* (the typical example is the robot, able to process different products) and for *the whole production system,* able to conform itself to the variable market demands.

One of these lines is described below.

In the past small appliances, manufactured in large batches, were usually assembled automatically by means of monoscope machines (not flexible). Today, thanks to the flexibility of means, it is possible to extend this application widely.

Careful research has come to the conclusion that in the near future the greatest development of assembly automation will take place on objects composed of 20–25 parts (pieces) at most. Also many components of the electric, electronic, motor-car, electric appliance, and small mechanics industry, even manufactured in relatively limited batches (50,000–100,000 pieces a year), will be gathered to form quantities justifying the use of flexible automatic means from the economic point of view.

In order to comply with these requests, it was decided to study a modular automatic flexible assembly line, economically advantageous, for the assembly of different kinds of small devices or subgroups, with a maximum number of 20 components (Fig. 1).

The line is composed of 3 sections: assembly, testing, and packing. The testing and packing sections may not be included.

The assembly section is composed of a closed ring asynchronous motorized conveyor system, with pallet recirculation, articulated element chain, and blocking places (equipped with centering devices) on which all the other components of the system (robots, pick and place devices, feeders, stores) operate. The line can also be managed (upon simple modification) in a semi-automatic way, introducing one or more people for the manual assembly (asynchronous) of complex subgroups.

The assembly section is composed of several workstations (each used to realize different structure lines). The workstations are composed of the following modules:

- Handling and piece positioning module, with robot
- Handling and piece positioning module, with pick and place arm
- Modules for various kinds of processing: brazing, screwing, gluing, riveting, calking, etc

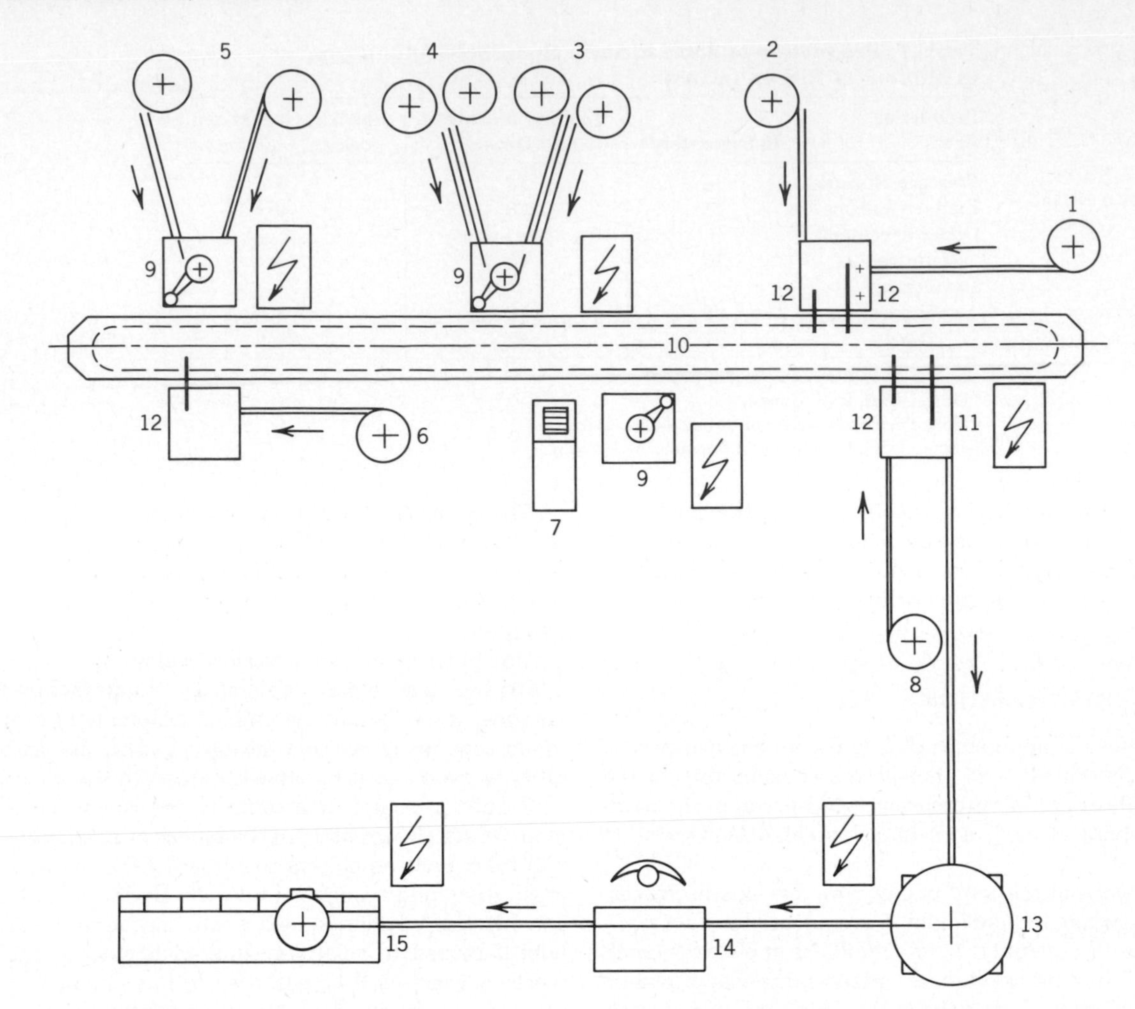

Figure 1. View of the flexible assembly line. 1–8, parts feeding; 9, robot; 10, conveyor; 11, programmable arm; 12, pick and place arm; 13, testing; 14, auxiliary processes and visual inspection; 15, packing.

- Feeding module with vibrator or other means
- Feeding module with automatic store
- Ordinary unloading module
- Unloading module with orderly disposition of the pieces

The line can work under the supervision of only one person. The high flexibility of the lines is achieved by means of the robots and other various devices. The robots are used for the handling of specific pieces (ie, not common to the family of appliances to assemble); in certain cases one must even resort to the automatic change of the grippers. After various analyses, a robot with a jointed-arm working on the horizontal plane was temporarily chosen.

The pneumatic pick and place arms are used for the handling of pieces common to the whole family and for which the same feeders can be used.

Feeding of the majority of pieces is made by means of vibrators, with conveyor belts, also functioning as buffer. For pieces that cannot be fed by the vibrator, the use of other kinds of feeders is provided, in particular a "tray-holder automatic store" from which every piece is picked up by a robot.

It is also possible to change the pallets automatically, in case of assembly of products having a totally different shape, by using the "pallet-holder automatic store" module.

All modules are managed by a distributed information system with the supervision of a personal computer, which controls the functioning of the whole system, classifies possible inefficiencies, and collects the statistical data of the production.

At the end of assembly, the products go to the testing section, and then to the packing section. The change of product requires only the feeding of new parts and the emptying of the feeders from the pieces of the previous product. In the worst cases the tooling time is about 30 min.

The assembly of some switches was experimented with using the line illustrated in Figure 1. In the final application it was equipped only with the assembly section and completed with the module for the orderly unloading of the pieces.

The assembly line can work at a rate of 12–15 pieces/minute and can handle 20–25 different pieces.

In the manufacture of television, radio, and tape-recorders, robots (the horizontal jointed type) are widely used in the assembly of electronic components on boards, especially when the quantities produced do not justify the use of rigid assembly machines.

The use of robots is also widespread in the handling of television kinescopes, both in the manufacturing and final assembly stages.

In regard to larger appliances, the problems of robotic automation become more complicated. Robots are used more and more in the manufacturing of sheet-metal cases for kitchens, refrigerators, dishwashers, and washing machines. In structure manufacturing, the robots are arranged in flexible automatized lines (ie, capable of operating on different frames). They carry out the handling of welding operations. The use of robots in the final assembly of these devices is much more difficult (see ASSEMBLY ROBOTS).

TRENDS

The present years have brought on a rapid evolution and reorganization in the production and business sectors. An intense robotization process is in progress in the electric appliance industry. This industry may be considered (after the automotive industry) among the most robotized.

The main needs that have promoted this rapid evolution may be summarized as follows:

- Reduction in labor cost
- Increase in productivity
- Quality improvement
- Improvement of the working environment
- Reduction in the development times of new products and quick adjustments to market demands
- Drastic reduction in the supply times of products
- Maximum flexibility of supply
- Reduction in stores and financial obligations
- Economical production of small batches

In industrial and university research laboratories, studies are in progress to improve the operative skills of robots especially for use in assembly.

Seven European countries have started, within the EUREKA program, a common research project for automatized assembly (FAMOS project). Their results will affect the electric appliance industry.

Robots will become more and more "skilled" thanks to the widespread use of sensors (especially in the field of touch and vision) and the use of techniques of artificial intelligence. They will be able to perform more and more complex operations that are limited to human operators at present. But, above all, the whole process of manufacturing will take place in a totally integrated system with continuous flow of information, orders, and materials.

The integration between the mechanic, electronic, and computer systems of flexible process, called CAD, CAM, CAE, CAPP, CAT, FMC, FMS, MHS, helps to form the global integrated system "CIM," which the "flexible, automatic factory" (or future factory) corresponds to, where all functions are totally integrated (see Glossary for explanation of abbreviations).

The push toward integration is taking place both horizontally (process, handling, check, assembly, packing) and vertically between machine shop (CNC, DNC), production management (CAM), and design and engineering (CAD, CAE).

APPLICATIONS

Manufacturing Line of Drum for Washing Machines

General Data. User: Candy Elettrodomestici SpA, Monza (MI), Italy; supplier: Camel Robot SpA, Senago (MI), Italy; number of robots: 10; product: drum for a washing machine; productive capacity (per shift): 2400 pieces; variations of the piece: 2; work shifts: possible; automatic cycle time: 12 s; investment payback period: 2 years.

Robot's Characteristics. HOO robot (Camel Robot) (Fig. 2); structure: cylindrical, modular; payload: 220 lb (100 kg); degrees of freedom: up to 6 (limited from time to time); axes: translation x (on the base) 59 in. (1500 mm), rotation 180°, translation z (elevation) 24 in. (600 mm), translation y (elongation) 39 in. (1000 mm); drive: hydraulic; grippers: these were designed expressly for any specific work executed in the various work places, and are equipped with electronic sensors and a certain degree of flexibility, to permit adjustment to the different variations; controller: microcomputer, PTP path, RAM and ROM memories, 4 kbyte memory capacity, off-line programming with teach pendant, number of steps: up to 1000.

Line Description. The line was studied for manual or automatic (total or partial) operation and can manufacture two different kinds of drum, at a rate of 300 pieces per hour. It is composed of 10 robots, operating machines, assembling and handling equipment; it is about 230 ft (70 m) long and contains six workstations.

The sheet-metal coil, unwound automatically, undergoes the operations of punching, shearing at desired length, and piling up in work place A. A robot picks up the slugs to be calendered. The calendered pieces are moved to station B, and by means of another robot, toward the fastening machine and then placed on the conveyor. In work place C, the superior rim and blade shaping is performed. In stations D and E, the shoulder fastening is carried out and in work place F the drum is completed (with the application of the starwheel

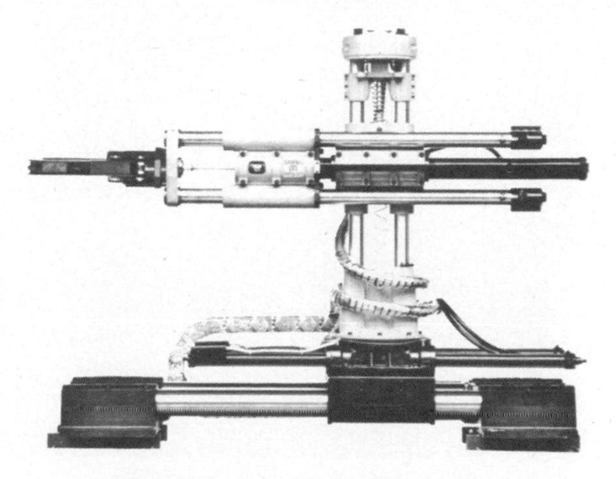

Figure 2. Camel Robot type HOO.

through riveting) and moved, on a continuous movement roller conveyor, toward the assembly area.

Work Place A: Band Calendering. The line (Fig. 3) begins with a sheet-metal coil (of the required width) which is unwound automatically. On the sheet-metal band the following operations are performed: shearing at the desired length; punching; hole flaring; and piling up in about 12 in. (300 mm) high packs.

The sheet metal pack is carried automatically to the gripping area of robot 1, where the following operations are performed: picking up of a sheared piece; positioning at one side of the calender; calendering (performed in different ways as regards the kinds of drum to manufacture: horizontal or vertical); programs permitting the desired combinations of horizontal and vertical bands are stored in the microcomputer's memory.

The end actuator of robot 1 is composed of 2 groups of vacuum suction cups (4 + 2), assembled on an elastic support, to stick perfectly to the band. Six vacuum suction cups are used for the handling of the horizontal band, four for the vertical one. The suction cups are placed so that they do not coincide with the holes of the two kinds of bands.

Work Place B: Band Folding and Band Fastening. The calendered bands are grasped by robot 2 and undergo the following operations: positioning on the folding machine; shifting, through robot 2, to the fastening machine; maintaining of position, with adequate overlapping of edges, and fastening; picking up, 90° rotation and placing on the conveyor for the horizontal bands, while the vertical bands are unloaded automatically (Fig. 4).

The end actuator of robot 2 consists of two groups, each with three grippers. The three grippers of the first group are designed to grasp the band at the outlet of the calender and to overlap the edges with the two superior grippers, in order to prepare them for the folding. The three grippers of the second group are designed to grasp the band, take it under the fastening machine, and place it, after a 90° rotation, on the conveyor, equipped with proper reference points.

Figure 4. Workstation B, band folding and fastening.

Work Place C: Rim and Blade Shaping. Robot 3 (Fig. 5) performs the following operations: picking up of the band from the conveyor; 110° rotation; and placing on the shaping machine.

Robot 4 (Fig. 5) performs the following operations: picking up of the band from the superior rim and blade shaping machine; 31° rotation; placing on the lower rim-shaping machine; placing of the band on the conveyor, in one of the two possible positions, according to the production needs.

The end actuators of robots 3 and 4 are equipped with grippers of the same structure. Robot 3 has three grippers, arranged at 120° and supported by a device that can rotate 110°. Robot 4 has six grippers: three are supported by a device that can rotate 30° and three by a device that can rotate 120°.

Work Place D: Shoulder Fastening. Robot 5 (Fig. 6) picks up the band from the conveyor and places it, oriented, on the stepping conveyor which has a template to receive the parts. Robot 6 (Fig. 6) picks up one shoulder at a time from the container, orients it, inserts it in the band present at the moment and commands the band-shoulder shifting toward the

Figure 3. Workstation A, band calendering.

Figure 5. Workstation C, rim and blade shaping.

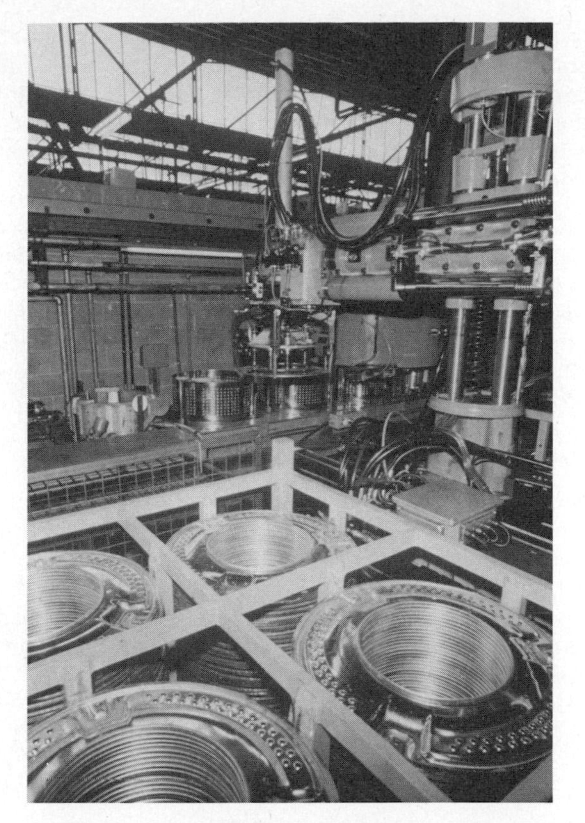

Figure 6. Workstation D, shoulder fastening.

fastening machine, that performs the operations. The end actuator of robot 5 is composed of three grippers arranged at 120°. The end actuator of robot 6 is composed of a 6-axis device (equipped with suction cups) fit for soft landing, suitable for 39 in. (1000 mm) vertical shifting, equipped with centering and orienting systems. The operations of picking the shoulders up are the following: descent into the box and into the selected pile; picking up of only one shoulder; centering of the shoulder with reference to the center of the suction cup (through four mechanical "fingers"); and shoulder rotation until identification (through optical sensors) of the correct position of orientation is reached.

Work Place E: Shoulder Fastening on Band. Robot 7 moves the drum from transfer device 1 to transfer device 2 (grasping it laterally) and rotates it 180°. The work performed by robot 8 is identical to that of robot 6, but it picks up the bottom plate instead of the shoulder.

Work Place F: Starwheel Riveting. Robot 9 picks up the drum from the conveyor and places it on the riveting machine. Robot 10 grasps the starwheel (preoriented) from the revolving table and settles it upon the drum in the desired position. The riveting machine sees to riveting the starwheel upon the drum. Robots 9 and 10 are equipped with traditional grippers.

The completed drum is grasped by hand, on a continuous movement roller conveyor, and moved toward the assembly area.

Manufacturing Line of Tub for Washing Machines

General Data. User: Candy Elettrodomestici SpA, Monza (MI), Italy; supplier: Camel Robot SpA, Senago (MI), Italy; number of robots: 8; product: tub for washing machine; productive capacity: 2400 pieces per 8-h shift; variations of the piece: 2; automatic cycle time: 12 s; investment payback period: 2 years.

Robot Characteristics. HOO robot, by Camel Robot (same as used for manufacture of the drum for washing machine).

Line Description. The line was studied for manual or automatic (total or partial) operation and can manufacture different kinds of tub at a rate of 300 pieces/hour. The line is composed of eight robots, operating machines, assembling and handling equipment; it is 131 ft (40 m) long and is composed of four workstations.

The sheet-metal slugs, obtained from the coil and unwound automatically, undergo the following oprations in work place A: calendering; folding; and fastening.

In work place B the bottom plate fastening takes place. In station C the flange and counter-flange assembly is performed. In the following station the bush coupling is applied to the counter-flange, then welding operations are performed. The completed piece is moved toward assembly.

Work Place A: Tacking. The line begins with a sheet-metal coil (of the required width) which is unwound automatically. After shearing at the desired length robot 1 performs the following operations: insertion into calender and calendering; unloading; positioning in the folding machine keeping its cylindrical position; folding, unloading, and positioning on the fastening machine, correctly overlapping the two sheet-metal edges; fastening and unloading on the stepping conveyor, after a 90° rotation.

Robot 2 performs the following operations: picking up from conveyor *a* and piece positioning on the welding machine; piece unloading (after welding) and placing on conveyor *a*. The end actuators of robots 1 and 2 were adapted to perform the complex operations and to give them the required degree of flexibility.

Robot 1 has a 6-gripper actuator, divided into two groups. The two superior grippers of the first group are assembled on slides, to permit the overlapping of the two band edges and its correct positioning on the folding machine. The second group's two superior grippers can pivot and with their rotatory motion they permit the positioning of the two edges, already bent, on the fastening machine. The third gripper is fixed and holds the inferior part of the band, keeping its cylindrical form. The end actuator of robot 2 is equipped with three grippers, set at 120°, able to rotate 90°, to place the band on the belt.

Work Place B: Bottom Plate Fastening. The bands, along the conveyor *a,* go through some processings after they have been positioned by means of a transfer device on a revolving table. At the same time robot 3 picks up the bottom plates (frontal or vertical according to the program) from conveyor *b* or *c* respectively and places them on the band which is on the revolving table. Bottom/band coupling occurs as follows: the bottom plate is laid on the band and simultaneously the gripper exercises a predetermined pressure on the bottom so as to complete the coupling of the two pieces. After this operation, the table rotates 180° and positions the pieces in the fastening work place. Then the piece is carried to the loading station where robot 3 moves it to line *d*, toward assembly. The end actuator of robot 3 is composed of 3 suction cups set at 120° and fixed to the support structure by means of an elastic support.

Work Place C: Flange, Counter-Flange Assembly. Robot 4 picks up the counter-flange from conveyor *e* (coming straight from the transfer lines of forging) and positions it, oriented, on the back flange, moved by conveyor *f*, this too coming straight from the transfer lines of forging.

The robot's gripper is electromagnetic and can rotate round its axis. The magnet is fixed to the support structure through an elastic support and is self-centering. While grasping, the sensors check the exact orientation of the piece and can stop the rotating device; the piece, grasped in the exact position, is positioned on the back flange.

Robot 5 picks up the pieces from conveyor *f*, positions them on the punching machine which makes the holes for the rivets. The grippers, self-centering, are equipped with four grasping "fingers" with balancing movement and were properly studied to guarantee maintenance of centering during movement.

Robot 6 unloads the pieces from the punching machine and places them in the special machine that performs the riveting. The grippers are similar to those of robot 5.

Work Place D: Bush Coupling. Robot 7 picks up from conveyor *g* the back flange/counter-flange group and positions it on the bush coupling machine; when the operation is over it unloads it and places it on conveyor *g*. Robot 8 picks up the piece from conveyor *g*, positions it in the welding machine and after the operation puts it back on conveyor *g*, to send it toward assembly.

Robotized Line for Spot Welding of a Kitchen Frame

General Data. User: IRE SpA Fabbrica Cucine, Cassinetta di Biandronno (VA), Italy; supplier: Bisiach & Carrù SpA, Venaria (TO), Italy; layout: Figure 7; number of robots: 4; product: kitchen frame (Fig. 8); variations of the piece: 6; productive capacity: 880 pieces per 7 h 20 min shift; manual cycle time: 6 min and 30 s; automatic cycle time: 30 s; investment payback period: 2.5 years.

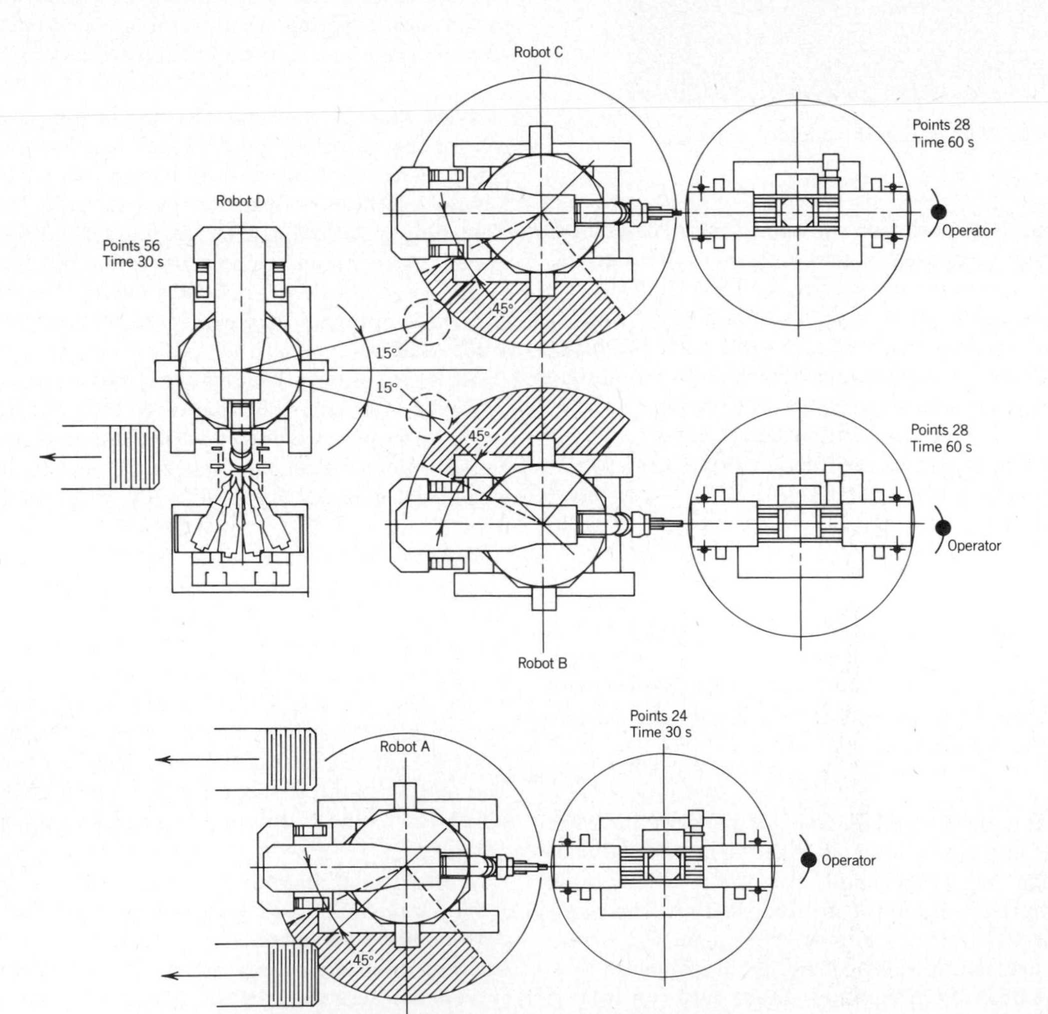

Figure 7. Robotized line for spot welding a kitchen frame.

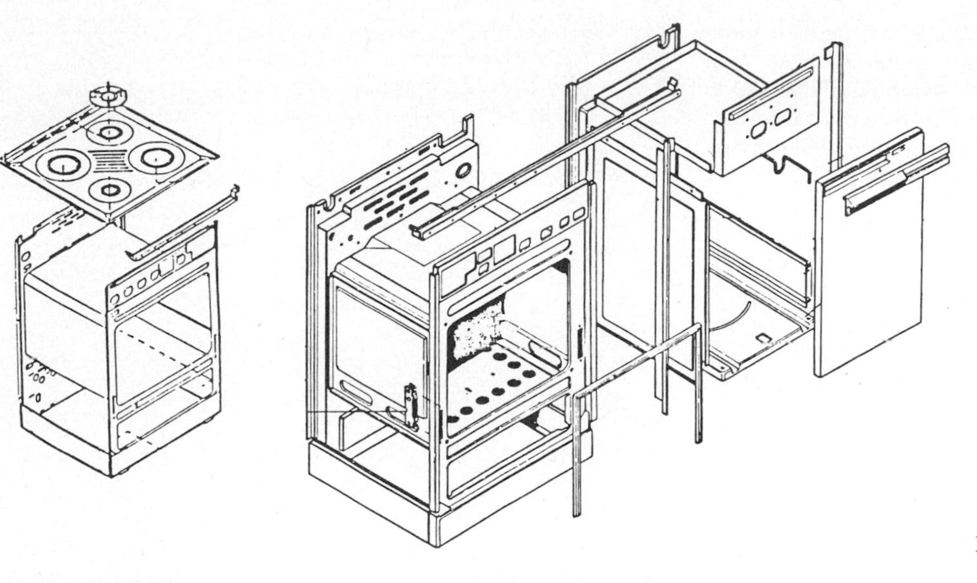

Figure 8. Kitchen frame.

Robot Characteristics

Type A. JKP6 robot, manufactured by Bisiach & Carrù (Fig. 9); structure: jointed-arm; payload: 287 lb (130 kg); degrees of freedom: 6; axes: 1st rotation: 300°; 2nd rotation 80°; and translation 59 in. (1500 mm); wrist: 1st rotation 480°; 2nd rotation 480°; 3rd rotation 480°; drive: d-c electric motor; grippers: multiple; controller: type Cobra 18, microcomputer, PTP path memory up to 96 K, programming with teach pendant.

Type B. JKP7 robot (Bisiach & Carrù); structure; jointed-arm; payload: 440 lb (220 kg); degrees of freedom: 7; axes: 1st rotation 300°; 2nd rotation 80°; translation 59 in. (1500 mm); wrist: same as type A; drive: d-c electric motors; grippers: each robot is equipped with two programmable grippers; controller: same as type A.

Type C. (same as type A).

Line Description. The line was studied for totally automatic processing (except certain operations of loading and unloading) and can produce different kinds of oven frames at 120 pieces/h. The line (Fig. 7) is composed of four robots (two of which are identical and operate in parallel to maintain the feed rate of the line) and the clamping equipment of the parts to weld. The line is co-ordinated by PLC. The piece undergoes, through three workstations, various operations of welding with total application of 38 + 56 points in about 30 seconds. Special multiple grippers are used.

The work places are arranged as follows.

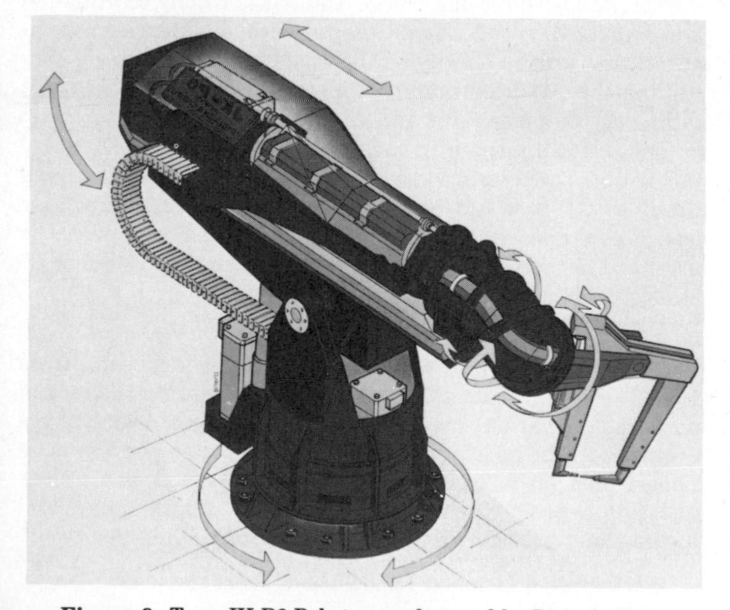

Figure 9. Type JK-P6 Robot manufactured by Bisiach-Carrù.

Work Place A: First Welding of the Frame. Robot A welds the oven frame with the help of a two position revolving table (180° rotation) by means of 3 grippers (with a transformer weighing ca 287 lb (130 kg) that work simultaneously. The operation regarding the application of 24 points and the unloading of the pieces on the chute is performed in 30 s. Loading is performed by hand (see Fig. 10).

Work Places B and C: Continuation of Welding. The piece is picked up from the two chutes of the preceding station,

Figure 10. Workstation A, first welding of the frame.

clamped on the respective revolving tables, acting as loading and working stations. Each robot uses a special head with two programmable grippers (7th axis) weighing about 352 lb (160 kg) (Fig. 11), so as to put two points simultaneously. These operations (performed in parallel) require 60 s per machine, with application of 28 to 32 points.

Work Place D: Completion of Welding. Robot D alternately grasps the pieces presented by robots B and C by means of a special gripper, and positions them under a fixed multiple spot-welder, equipped with eight grippers (Fig. 12). At the end of the process the robot unloads the piece upon a roller chute. The whole operation, with application of 38 to 56 points, is performed in 30 s (line frequency).

Robotized Flexible Cell for Preassembly of the Freezer Cell.

General Data. User: Merloni Elettrodomestici SpA, Fabriano (AN), Italy; various suppliers under co-ordination of Merloni Elettrodomestici; number of robots; 2; product: freezer cell (Fig. 13); variations of the piece: 4; productive capacity: 800 pieces per 7 h 30 min shift; manual cycle time: 50 s; automatic cycle time: 30 s; investment payback period: 3 years.

Robot's Characteristics. IRb60 robot, manufactured by ASEA; structure: jointed arm; payload: 132 lb (60 kg); degrees of freedom: 5: axes: 3 rotations; wrist rotation: 360°; wrist inclination: −75° +120°; maximum elongation: 6.5 ft (2 m); drive; d-c electrical motors with disk rotors, chopper feeding;

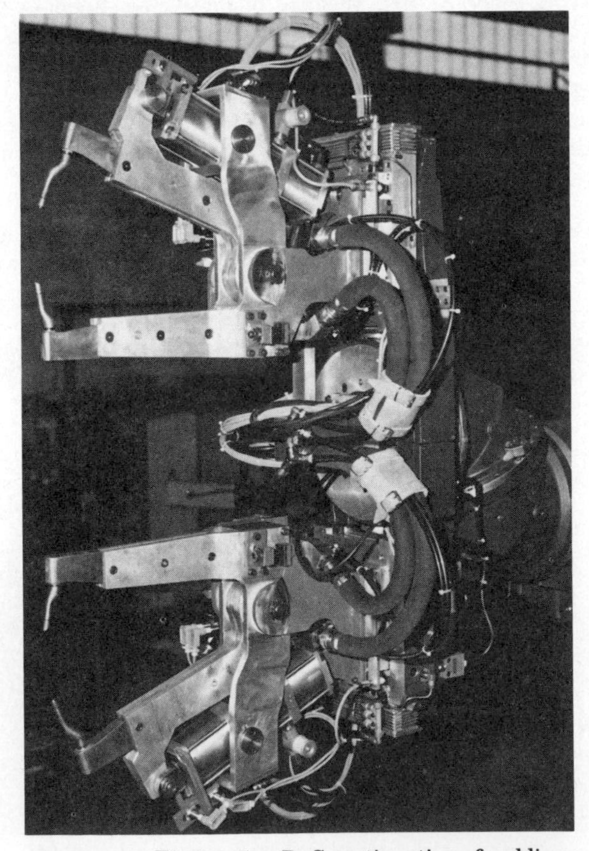

Figure 11. Workstation B, C continuation of welding.

Figure 12. Workstation D, completion of welding.

grippers: multiuse; controller: S2 type, PTP space-interpolated path, 22K capacity RAM memory, inter-active programming with control panel, 999 programs.

IRbG6 robot, manufactured by ASEA; structure: jointed-arm, suspended; payload: 13.2 lb (6 kg); degrees of freedom: 6; axes: 3 rotations; wrist rotation: 360°; drive: d-c electrical motors with disk rotors, chopper feeding; end actuator: extruder for hot sealing; controller: S2 type, PTP space-interpolated path, 22K capacity RAM memory, interactive programming with control panel, 999 programs.

Cell Description. The cell was studied for complete automatic functioning and can manufacture, at a rate of 120 pieces per hour, the whole family of "freezer cells" (Fig. 13), composed of 4 models at present. The cell includes two robots, a revolving table, three feeders, stores, motorized truck, envelope conformer, adhesive sealing pump. The cell is wholly coordinated and managed by a PLC, interfaced with a supervision computer, that sees to specifying the model to be assembled; controlling the produced quality; collecting the quality data and taking the due measures and diagnosis.

The cell is arranged in such a way that no supplementary fitting times are required (the equipment controls itself during the process). It would be even possible to manufacture only one piece economically. The part of "dedicated" equipment is arranged to receive new models even though the system operates on four models at present.

The work places are set out as follows.

Work Place A: Piece Feeding. IRb60 robot picks up pieces 1, 2, 3 (Fig. 13) from their respective automatic feeders and performs the various movements necessary for assembly. This begins with the picking up of piece 3 and placing on the positioner; then piece 2 is picked up and conformed in a proper tool; finally piece 1, which is assembled by the robot and held in the exact position until the subsequent sealing operation.

Work Place B: Sealing. IRb6G robot seals the joint zones of the three pieces.

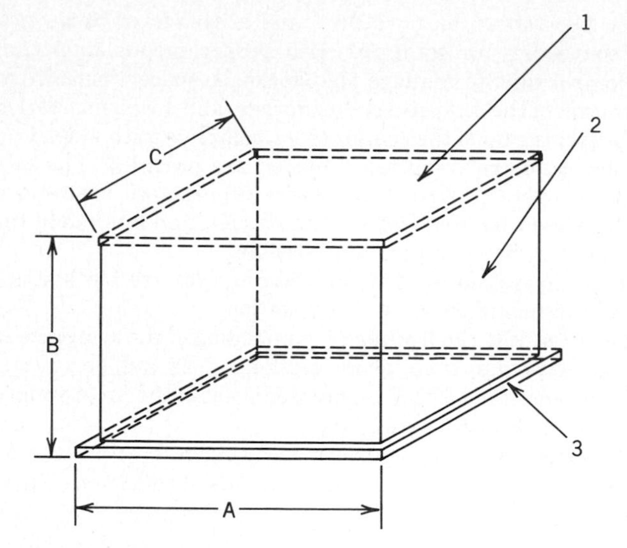

Figure 13. Freezer cell.

Type	1	2	3
A	427	550	550
B	415	430	430
C	242	300	351

Work Place A: Piece Unloading. IRb60 robot picks up the assembled and sealed piece and places it on truck 8, if available, otherwise in the store. Truck 8 carries the pieces to the following stations (manual at present but under automatization), where they are completed.

Handling of Kinescope Cell

General Data. User: Philips, Monza (MI), Italy; supplier: Camel Robot SpA, Senago (MI), Italy; layout: Figure 14; product: TV set kinescope; productive capacity: 1150 pieces per shift; variations of the piece: 6; manual cycle time: 30 s; automatic cycle time: 25 s; worked shifts: 2; investment payback period: 2.5 years.

Robot's Characteristics. EOO Camel Robot; structure: cylindrical, modular; payload: 220 lb (100 kg); degrees of freedom: 5; axes: rotation 270°, 23 in. translation Y (600 mm), translation Z (elongation) 39 in. (1000 mm), yaw 360°; parallel closing gripper; controller: microcomputer, PTP path, RAM and ROM memories, 4 kbyte memory capacity, off-line programming with teach pendant, number of steps up to 1000; gripper: with two axes, able to handle six kinds of kinescopes, grouped into two families (20 in.-22 in.-26 in. and 15 in.-17 in.-20 in.), equipped with parallel closing of the fingers and 360° wrist

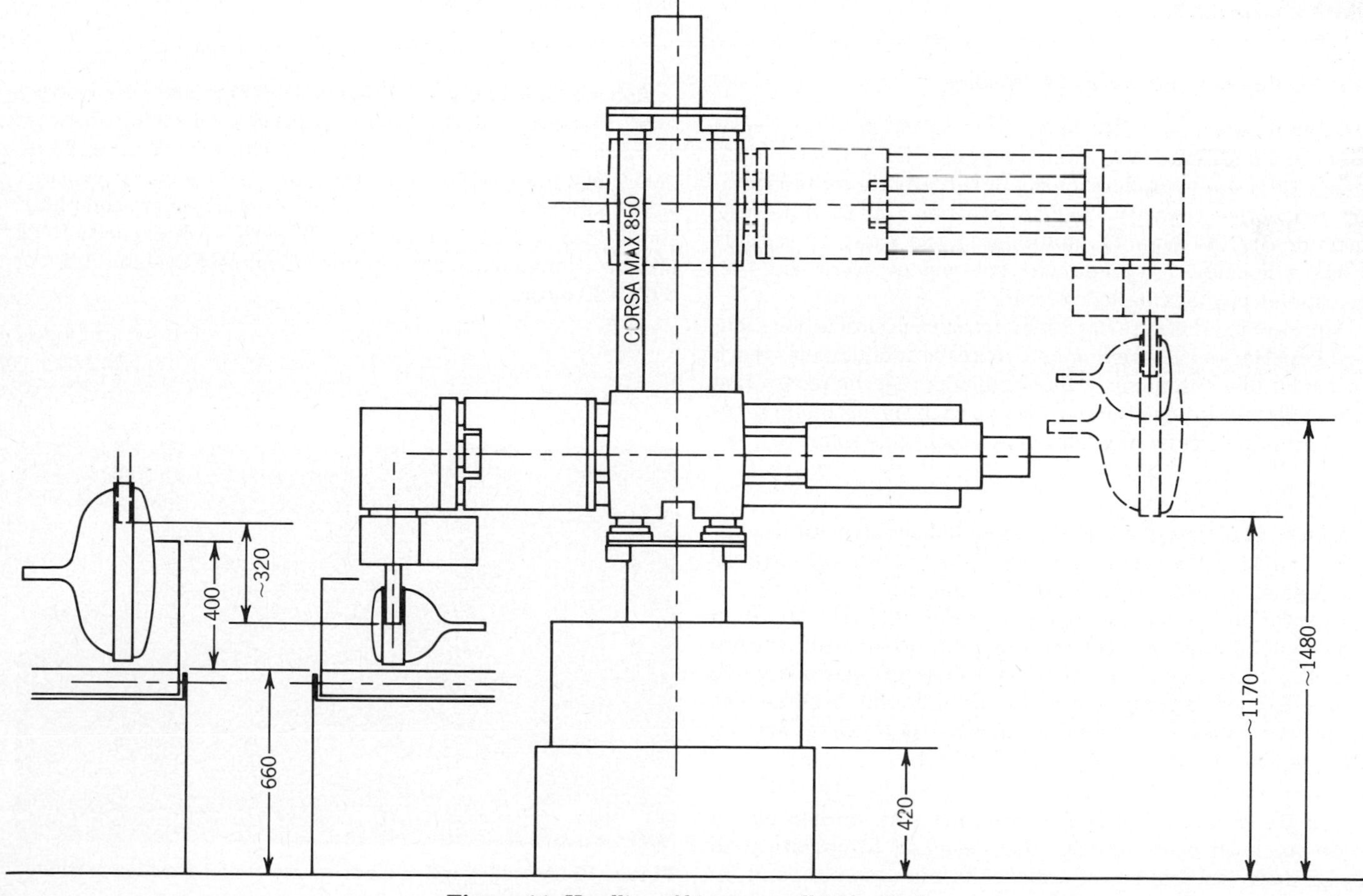

Figure 14. Handling of kinescope cell (side view).

rotation round the vertical axis (Z) of the group itself, servo-controlled and able to pick up kinescopes however oriented in regard to the horizontal plane.

Cell Description. The cell is located before the assembly shop of color TV sets with 15 in.-17 in.-20 in. and 20 in.-22 in.-26 in. screens. The kinescopes come in large boxes, on motorized and clutched rollers. A bar-code reader recognizes the type of kinescope in the box and automatically signals to the robot the kinds of kinescope to handle. The calling up of boxes to empty and the unloading of empty boxes are performed automatically by the robot's control unit. Changing of production schedules can take place any moment, according to the production necessities and is completely automatic, by means of a program selector. The robot can also take two types of kinescopes alternately, performing the complete emptying (according to the program) of the respective boxes, and controlling the respective conveyance systems. The assembly line is interfaced with the robot's microcomputer, which controls its store through sensors (max and min store).

Work Place A: Box Opening. The robot opens the box, which can then be moved to the next station.

Work Place B: Picking Up and Unloading of Kinescopes. The robot picks up the kinescopes and positions them on the pallets of the assembly lines, each able to receive three different kinds of kinescopes.

Handling Cell of Compressors for Welding

General Data. User: IRE SpA, Fabbrica compressori, Cassinetta di Biandronno (VA), Italy; supplier: Camel Robot SpA, Senago (MI), Italy; product: hermetic compressor for refrigerator; productive capacity: 2600 pieces per 7 h 20 min-shift; variations of the piece: 7; automatic cycle time: 10 seconds at 95% efficiency, 7 seconds with one welder; work shifts: 2; investment payback period: 3 years.

Aims are to eliminate hard work, reach an automation standard superior to the existing one, increase productivity, reach a considerable autonomy of the system as regards recognition of the different kinds of compressors, reach the optimum management of minimum and maximum stocks and improve quality.

Robot's Characteristics. EOI (Camel Robot); structure: cylindrical; payload (including gripper): 220 lb (100 kg); degrees of freedom: 4; axes: translation Y (elongation) 39 in. (1000 mm), rotation (on the base) 270°, translation Z (elevation) up to 33 in. (850 mm); gripper rotation: 90°; drive: electric; grippers: the robot is equipped with a special gripper adaptable to seven types of compressors having different heights. The gripper can rotate ±45° round the axis Z to perform the loading of the welder; controller: same as EOO robot.

Cell Description. The cell automation was worked out to eliminate hard work. It was also designed to operate with one or more welders out of order or in maintenance, and includes two robots, four welders, and a system of conveyors. The conveyors are equipped with proper means to permit the control unit to manage the stocks. A spacer, permitting separation of the compressors in storage, and a waiting station for the picking up of the compressors, complete with identification device of the compressor model, are provided. The cell, which can operate with seven types of compressors, is attended by one person for the system supervision, and has made the saving of two persons per shift possible.

The compressors arriving on the conveyor are blocked, selected, and moved to the waiting station.

Robot 1 sees to the loading and unloading of the compressor, first on S1 welder, then, after rotation, on S2 welder, to complete the operation. If the compressor is not of the programmed type, it goes on the loading line.

The same function is performed by robot 2 on S3 and S4 welders. The system meets the conditions of minimum, maximum, and emergency stocks.

Assembly Cell of Electrical Motors for Washing Machines

General Data. Supplier: DEA (Digital Electronic Automation SpA), 10024 Moncalieri (TO) Italy; layout: Figure 15; number of robots: 3; product: electrical motor (Fig. 16); variations of the piece: several; productive capacity: 144 motors per hour (80% efficiency); automatic cycle time: 20 s; and investment payback period: 3 years.

Robot's Characteristics. Pragma A 3000 robot (DEA); structure: cartesian, with three arms; payload: 22 lb (10 kg); degrees of freedom: up to 6; axes: X 52 in. (1320 mm), Y 19 + 23 in. (300 + 600 mm), Z 19 in. (300 mm); drive: d-c motors; grippers: with double and triple grip; controller: multi-micro-computer, simultaneous control of 6 axes, PTP path, 48 K capacity RAM memory, language programming, diagnosis system, 128 inputs, 120 outputs.

Figure 15. Assembly cell of electrical motors for washing machines.

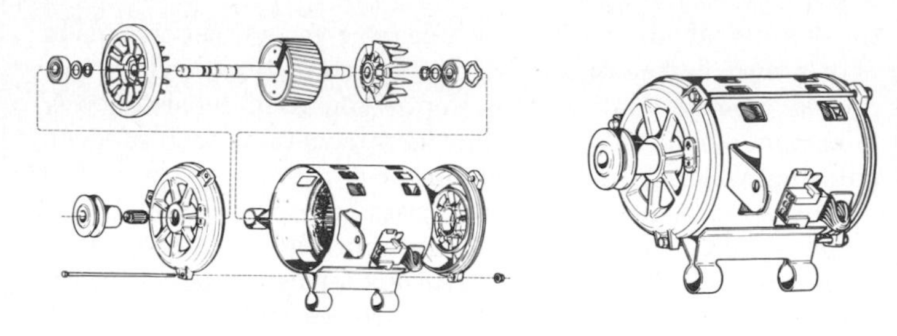

Figure 16. The assembled motors.

Figure 18. Compressor crankshaft assembly. Supplier, DEA 10024 Moncalieri (TO) Italy.

Cell Description. The cell was studied for completely automatic functioning and can produce, at a rate of 144 pieces per hour (80% efficiency), different kinds of induction motors for washing machines. Other models can be assembled upon limited intervention on the equipment. The line autonomy is 60 minutes. The cell is composed of three robots, operating units for particular operations (driving of fans and bearings, connection of seeger rings, simultaneous screwing of three tie-rods), and the electrical testing machine. In total there are 18 stations, besides those of loading and unloading. The outcome is the normal progression of the operations or the motor rejection.

The robots perform the whole assembly in the following order:

- the first robot, equipped with double gripper, handles the fans and motor
- the second robot, with triple gripper, picks up the washer and the bearings, and fits the complete rotor in the stator
- the third robot, with double gripper, positions the superior flange and connects the tie-rods.

The whole system is managed by the Pragma A 3000 robot's controller in an integrated method. The extensive use of force and presence sensors (upon the wrist, grippers, and equipment) makes the continuous control of the productive cycle possible. When an anomaly prevents the robot from going ahead (for example, a wrong connection of the seeger rings or rotor), the specific robot performs an alternative cycle, in order to overcome the trouble automatically. This adaptive behavior allows the system to achieve good reliability, which normally is over 80%. When the anomaly prevents the robot from going ahead, a precise signaling of the interested area is given, drastically reducing the search time of the failure.

OTHER APPLICATIONS

Figures 17–19 show other applications for the appliance industry.

Figure 17. Compressor valve assembly. Supplier, DEA, 10024 Moncalieri (TO) Italy.

Figure 19. Piston and connecting rod assembly. Supplier, DEA 10024 Moncalieri (TO) Italy.

Glossary

CAD = computer-aided design
CAE = computer-aided engineering
CAM = computer-aided manufacturing
CAPP = computer-aided production planning
CAT = computer-aided testing
CNC = computer numerical control
DNC = direct numerical control
FMC = flexible manufacturing cell
FMS = flexible manufacturing system
MHS = material handling system
NC = numerical control
PLC = programmable logic controller
RAM = random access memory
ROM = read only memory

The research and study in this article were conducted in some plants of Bassani Ticino and financed by the Istituto Mobiliare Italiano with funds made possible by a special law for research improvement.

APPLICATIONS OF ROBOTS

JOHN D. MEYER
Tech Trans Consultants, Inc.
Lake Geneva, Wisconsin

REASONS FOR USING ROBOTS

The first commercial application of an industrial robot took place in 1961, when a robot was installed to load and unload a die-casting machine. This was a particularly unpleasant task for human operators. In fact, many early robot applications were in areas where a high degree of hazard or discomfort to humans existed, such as in welding, painting, and foundry operations. Even though these early robots did not necessarily perform their tasks more economically than humans, the elimination of hazardous and unpleasant manual operations was sufficient justification for their use.

In recent years, robots have been used more in applications where they offer clear economic advantage over human workers. Although human labor rates have continued to escalate, the hourly operating and depreciation costs for robots have remained relatively constant. Thus, in many instances robots can perform tasks at considerably less cost than humans. Savings of 50–75% in direct labor costs is not uncommon.

Another closely related reason for using industrial robots is increased productivity. Robots are not only cheaper than manual labor, but frequently have higher rates of output. Some of this increased productivity is due to the robot's slightly faster work pace, but much is the result of the robot's ability to work almost continually, without lunch breaks and rest periods.

In addition to their economy and their ability to eliminate dangerous and unsocial tasks and increase productivity, robots are also used in many applications where repeatability is important. Although today's robots do not possess the judgmental capability, flexibility, or dexterity of humans, they do have the distinct advantage of being able to perform repetitive tasks with a high degree of consistency, which in turn leads to improved product quality. This improvement in consistency is important when justifying robots for applications such as spray painting, welding, and inspection.

These four benefits—reduced costs, improved productivity, better quality and elimination of unsocial and hazardous tasks—represent the primary reasons for using industrial robots in today's factories. In the future, an additional benefit, greater flexibility, is also expected to play a major role in robot justification. As flexible manufacturing systems and the totally automated factory become realities, the robot's ability to adapt to product-design changes and variations in product mix will become an increasingly important factor in their use.

ROBOT CAPABILITIES

In general, robots possess three important capabilities that make them useful in manufacturing operations:

1. Transport

One of the basic operations performed on an object as it passes through the manufacturing process is material handling or physical displacement (see MATERIAL HANDLING). The object is transported from one location to another to be stored, machined, assembled, or packaged. In these transport operations, the physical characteristics of the object remain unchanged.

The robot's ability to acquire an object, move it through space, and release it makes it an ideal candidate for transport operations. Simple material-handling tasks, such as part transfer from one conveyor to another, may only require one- or two-dimensional movements. These types of operations are often performed by nonservo robots. Other parts-handling operations may be more complicated and require varying degrees of manipulative capability in addition to transport capability. Examples of these more complex tasks include machine loading and unloading, palletizing, part sorting, and packaging. These operations are typically performed by servo-controlled, point-to-point robots.

2. Manipulation

In addition to material handling, another basic operation performed on an object as it is transformed from raw material to a finished product is processing, which generally requires some type of manipulation. That is, workpieces are inserted, oriented, or twisted in order to be in the proper position for machining, assembly, or some other operation. In many cases, it is the tool that is manipulated rather than the object being processed.

A robot's capability to manipulate both parts and tooling makes it very suitable for processing applications. Examples in this regard include robot-assisted machining, spot and arc welding, and spray painting. More complex operations, such as assembly, also rely on the robot's manipulation capabilities. In many cases, the manipulations required in these processing and assembly operations are quite involved and therefore either a continuous-path or point-to-point robot with a large data-storage capacity is required.

3. Sensing

In addition to transport and manipulation, a robot's ability to react to its environment by means of sensory feedback is also important, particularly in sophisticated applications such as assembly and inspection. These sensory inputs may come from a variety of sensor types, including proximity switches, force sensors, and machine vision systems.

State-of-the-art robots have relatively limited sensing capabilities. This is due primarily to the difficulty with which today's robots can be effectively interfaced with sensors and, to a lesser extent, to the availability of suitable low-cost sensing devices. As control capabilities continue to improve and sensor costs decline, the use of sensory feedback in robotics applications will grow dramatically.

In each application, one or more of the robot's capabilities of transport, manipulation, or sensing is employed. These capabilities, along with the robot's inherent reliability and endurance, make it ideal for many applications now performed manually, as well as some applications now performed by traditional automated means.

TYPES OF APPLICATIONS

By the end of 1983, there were approximately 8000 robots installed in the United States. These installations are usually grouped into the 7 application categories shown in Figure 1 (1). This figure also shows the major robot capabilities used in each application and the type of benefits obtained.

A brief description of each application category is contained in the following paragraphs (1–7).

1. Material Handling

In addition to tending die-casting machines, early robots were also used for other material-handling applications. These applications make use of the robot's basic capability to transport objects, with manipulative skills being of less importance. Typically, motion takes place in two or three dimensions, with the robot mounted either stationary on the floor or on slides or rails that enable it to move from one workstation to another. Occasionally, the robot may be mounted overhead. Robots used in purely material-handling operations are typically nonservo, or "pick-and-place," robots.

Examples of material-handling applications include transferring parts from one conveyor to another, transferring parts from a processing line to a conveyor, palletizing parts, and loading bins and fixtures for subsequent processing.

The primary benefits in using robots for material handling are to reduce direct labor costs and remove humans from tasks that may be hazardous, tedious, or exhausting. Also, the use of robots typically results in less damage to parts during handling, which is a major reason for using robots to move fragile objects. In many material-handling applications, however, other forms of automation may be more suitable if production volumes are large and no workpiece manipulation is required.

2. Machine Loading and Unloading

In addition to unloading die-casting machines, robots are also used extensively for other machine loading and unloading applications. Machine loading and unloading is generally considered to be a more complex robot application than simple material handling. Robots can be used to grasp a workpiece from a conveyor belt, lift it to a machine, orient it correctly, and then insert or place it on the machine. After processing, the robot unloads the workpiece and transfers it to another machine or conveyor. The greatest efficiency is usually achieved when a single robot is used to service several machines. Also, a single robot may also be used to perform other operations while the machines are performing their primary functions.

Other examples of machine loading and unloading applications include: loading and unloading of hot billets into forging presses; loading and unloading machine tools, such as lathes and machining centers; stamping press loading and unloading; and tending plastic injection-molding machines.

Although adverse temperatures or atmospheres can make robots advantageous for machine loading and unloading, the primary motivation for their use it to reduce direct labor costs. Overall productivity is also likely to increase because of the higher amount of time the robot can work when compared with humans. In machine loading and unloading, it is both the manipulative and transport capabilities that make use of robots feasible (see also MACHINE LOADING/UNLOADING).

3. Spraying

In spraying applications, the robot manipulates a spray gun which is used to apply some material, such as paint, stain, or plastic powder, to either stationary or moving parts. These coatings are applied to a wide variety of parts, including automotive body panels, appliances, and furniture. In cases where the part being sprayed is on a moving conveyor line, the robot's sequence of spraying motions is coordinated with the motion of the conveyor. Relatively new applications of spraying robots include the application of resin and chopped glass fiber to molds for producing glass-reinforced plastic parts and for spraying epoxy resin between layers of graphite broadgoods in the production of advanced composites.

The manipulative capability of the robot is of prime importance in spraying applications. A major advantage of their use is higher product quality through more uniform application of material. Further benefits are reduced costs by eliminating human labor, less waste coating material, and reduced exposure of humans to toxic materials (see also PAINTING).

4. Welding

The largest single application for robots at present is for spot welding automotive bodies. Spot welding is normally performed by a point-to-point servo robot holding a welding gun. Arc welding can also be performed by robots. However, seam tracking can be a problem in some arc-welding applications. A number of companies are developing noncontact seam trackers, which would greatly increase the use of robots for arc welding.

Robots are used in welding applications to reduce costs by

Manufacturing Operation	Sample Robot Applications
Material handling	Moving parts from warehouse to machines Depalletizing wheel spindles into conveyors Transporting explosive devices Packaging toaster ovens Stacking engine parts Transfer of auto parts from machine to overhead conveyor Transfer of turbine parts from one conveyor to another Loading transmission cases from roller conveyor to monorail Transfer of finished auto engines from assembly to hot test Processing of thermometers Bottle loading Transfer of glass from rack to cutting line Core handling Shell making
Machine loading/unloading	Loading auto parts for grinding Loading auto components into test machines Loading gears onto CNC lathes Orienting/loading transmission parts onto transfer machines Loading hot form presses Loading transmission ring gears onto vertical lathes Loading of electron-beam welder Loading cylinder heads onto transfer machines Loading a punch press Loading die-cast machine
Spray painting	Painting of aircraft parts on automated line Painting of truck bed Painting of underside of agricultural equipment Application of prime coat to truck cabs Application of thermal material to rockets Painting of appliance components
Welding	Spot welding of auto bodies Welding front-end loader buckets Arc welding hinge assemblies on agricultural equipment Braze alloying of aircraft seams Arc welding of tractor front weight supports Arc welding of auto axles
Machining	Drilling aluminum panels on aircraft Metal flash removal from castings Sanding missile wings
Assembly	Assembly of aircraft parts (used with auto-rivet equipment) Riveting small assemblies Drilling and fastening metal panels Assembling appliance switches Inserting and fastening screws
Other	Application of two-part urethane gasket to auto part Application of adhesive Induction hardening Inspecting dimensions on parts Inspection of hole diameter and wall thickness

Figure 1. Major categories of robot applications and rationale for use.

Application	Examples	Robot Capabilities Justifying Use: Transport	Manipulation	Sensing	Primary Benefits of Using Robots: Improved Product Quality	Increased Productivity	Reduced Costs	Elimination of Hazardous/ Unpleasant Work
Material handling	Parts handling Palletizing Transporting Heat treating	●					●	●
Machine loading	Die-cast machines Automatic presses NC milling machines Lathes	●	●			●	●	
Spraying	Spray painting Resin application		●		●		●	●
Welding	Spot welding Arc welding		●			●	●	●
Machining	Drilling Deburring Grinding Routing Cutting Forming		●	●		●	●	
Assembly	Mating parts Fastening		●	●		●	●	
Inspection	Position control Tolerance			●	●			

Figure 1. *(continued)*

eliminating human labor and to improve product quality through better welds. In addition, since arc welding is extremely hazardous, the use of robots can minimize human exposure to a harsh environment (see also WELDING ROBOTS).

5. Machining

In machining applications, the robot typically holds a powered spindle and performs drilling, grinding, routing, deburring, or other similar operations on the workpiece. In machining operations, the workpiece can be placed in a fixture by a human, another robot, or a second arm of the same robot performing the machining. In some operations, the robot moves the workpiece to a stationary powered spindle and tool, such as a buffing wheel.

Robot applications in machining are currently limited because of accuracy requirements, expensive tool designs, and lack of appropriate sensory feedback capabilities. Machining is likely to remain a somewhat limited application until both improved sensing capabilities and better positioning accuracy are achieved.

6. Assembly

One of the areas of greatest interest today is the development of effective, reasonably priced robots for assembly. Presently available robots can be used to a limited extent for simple assembly operations, such as mating two parts together. However, for more complex assembly operations, robots are subject to the same limitations as in machining operations, namely difficulties in achieving the required positioning accuracy and sensory feedback.

Examples of current robot assembly operations include the insertion of light bulbs into instrument panels, the assembly of typewriter-ribbon cartridges, the insertion of components into printed wiring boards, and the automated assembly of small electric motors.

More complex assembly tasks, however, typically cannot be performed by currently available robots. A number of companies are conducting research in sensory feedback, improved positional accuracy, and better programming languages that will permit more advance assembly applications in the future (see also ASSEMBLY ROBOTS).

7. Inspection

A small but rapidly growing number of applications are in the area of robot inspection. In these applications, robots are used in conjunction with sensors, such as a television camera, laser, or ultrasonic detector, to check part locations, identify defects or recognize parts for sorting. Such robots have been used to inspect valve-cover assemblies for automotive engines, sort metal castings, and inspect the dimensional accuracy of openings in automotive bodies.

As in assembly and machining operations, a high degree of accuracy and extensive sensory capabilities are required for inspection applications. These are expected to be among the high growth application areas as low-cost sensors and improved positioning accuracy evolve.

USER INDUSTRIES

The plants in which industrial robots are currently used are generally large, sophisticated operations. In addition to the automobile and foundry industries, other large and small equipment manufacturers are users of robots. Robots are found in both mass production and batch operations, and most companies that use robots also tend to use other advanced production tools such as computer numerical control, computer-aided design, and computer-aided testing.

A breakdown of current robot installations by application and industry is given in Table 1 (1). Not surprisingly, the industry with the most robots installed is the auto industry, with about 40% of the total U.S. robot population. Within the auto industry, welding is the most common robot application, with about 70% of all robots in that industry used for welding. Applications within the automotive industry should continue to dominate the market during the next decade. Robots are increasingly being used by automotive plants for assembly, part transfer, and machine loading in addition to traditional welding applications.

The foundry area also continues to be a major robot-using industry. Captive foundries, such as those operated by auto and heavy equipment companies, are principal users. In addition, independent foundries are increasing their use of robots. Robots were first used in foundries, and die-casting machine loading is still a significance application. Robots are also used to quench die-cast parts, rough trim parts, ladle molten metal, and produce investment castings.

In the electrical equipment and electronic industries, robots are used for material handling, assembly, and painting. For example, robots are used to lay wire for wiring harnesses for manufacturers of home appliances, computers, and other products requiring electrical wiring. Other electronics-industry applications for robots include assembly of microcircuits, insertion of components on printed circuit boards, visual inspection of loaded boards and substrates, switch and relay assembly, and placement of surface-mounted devices on printed circuit boards. In the material-handling area, robots are used to handle integrated circuit masks, palletize printed circuit-board components, and perform component burn-in handling. Finally, robots are being used for spraying in such applications as circuit-board coating or enclosure painting.

Heavy equipment manufacturers, such as construction or agricultural equipment companies, are using robots for applications similar to those in the automotive industry.

Similarly, aerospace firms, which currently use a relatively small number of robots, are likely to expand their usage significantly in the future. Aerospace companies appear to be more interested in the processing capabilities of robots than their part-handling capabilities. Robots are used for spraying paints and other surface coatings onto aircraft panel surfaces, for machining operations such as hole drilling, and for assembly. Aerospace companies are also using robots as components of fully integrated CAD/CAM work cells.

Robots are used in a number of other industries, such as plastics and other nonmetals. In the plastics industry, robots are used to automate injection-, extrusion-, and blow-molding processes (machine loading and unloading). Inexpensive robots are often used for removing parts from a mold. More advanced robots are used to unload 200-lb (9.1-kg) molds or to perform secondary operations on molds.

Many other applications in nonmetal manufacturing are also being performed by robots. In food, pharmaceutical, and cosmetics production, robots are used for material handling and packaging. In woodworking, robots are being installed to finish furniture (varnishing, painting, cutting, and assembly), for material transport, and for palletizing. Other potential ap-

Table 1. U.S. Robot Population by Application and Industry, 1982[a]

	Industy						
Application	Auto	Foundry	Nonmetals Light Mfg.	Electrical, Electronics	Heavy Equipment	Aerospace	*Total*
Welding	1[b]				1		2200 (35%)
Material handling		1	1	1			1550 (25%)
Machine loading	2	2			2		1250 (20%)
Spray painting, finishing	3		2	3	3	1	600 (10%)
Assembly				2		2	200 (3%)
Machine							100 (2%)
Other							300 (5%)
Total	2500 (40%)	1250 (20%)	1050 (17%)	700 (11%)	600 (10%)	100 (2%)	6200 (100%)

[a] Numbers show rank order of application by industry.
[b] 70–80% of robots in auto industry are used for welding.

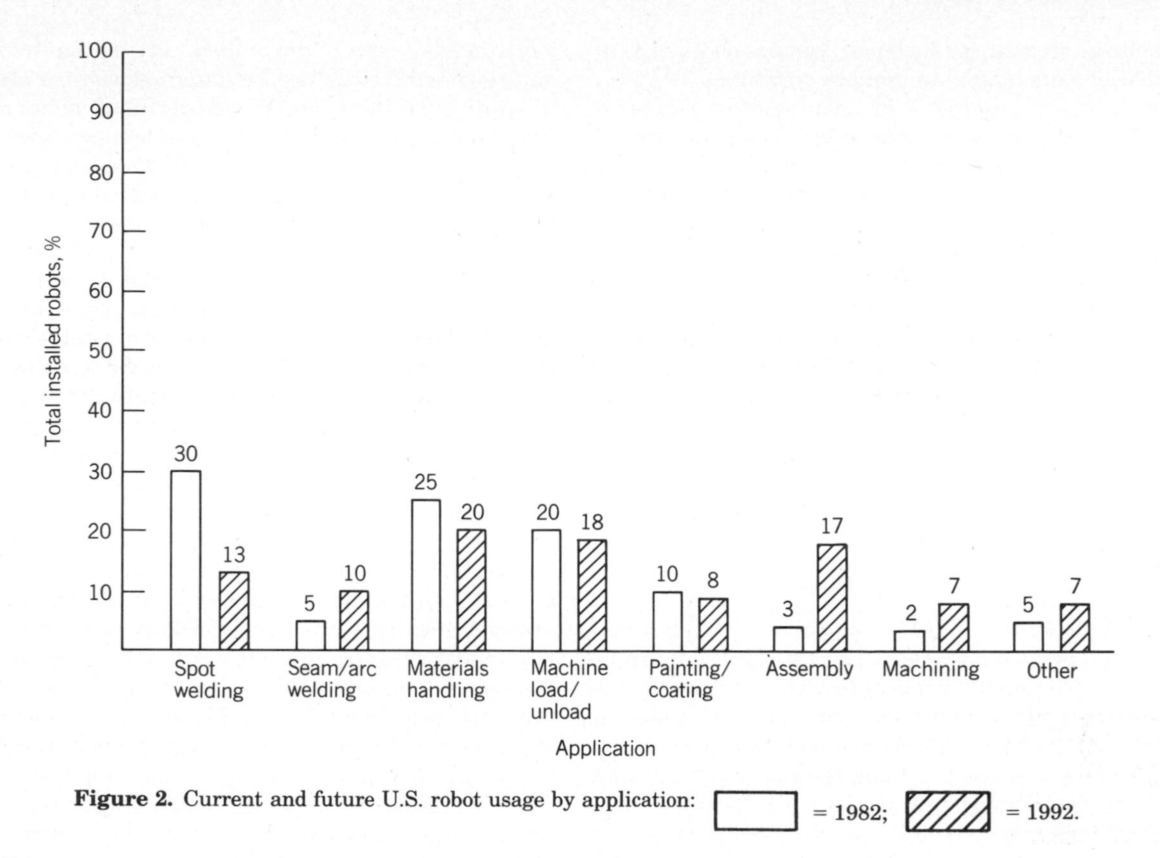

Figure 2. Current and future U.S. robot usage by application: ☐ = 1982; ▨ = 1992.

plications are under study for rubber processing (tire manufacturing), asbestos processing, glass loading and unloading where tactile sensing would be required, and even clothing manufacture. In many of these nonmetal processing applications, effective use of robots will require the incorporation of vision or tactile sensing capabilities.

Historically, robots were considered for applications in which they were used as independent pieces of automated equipment. However, an important new application of robots is in fully integrated manufacturing systems or cells. In such systems, raw materials flow into the work area, where operations such as machining, finishing, and inspection are performed. Support functions, including part design, tooling, programming, and flow scheduling, may also be integrated into the system to control the process. A key element of such a system is the integration of the robot with other CAD/CAM systems. The robot acts as the physical operator for the system and is responsible for moving workpieces from one stage to another.

Generally, these systems are best suited for advanced robots which have tactile or vision sensing capabilities. A vision system could be used to inspect incoming parts, guide the robot during operation, and inspect the finished parts before they move on to another station. The use of tactile feedback and universal grippers can assist in allowing the robot to place workpieces in various positions.

Use of these systems, which are primarily in the developmental stage, are expected to increase substantially in the future, with some observers forecasting that 20–30% of all robots purchased ten years from now will be interfaced with CAD/CAM systems. The next step will be the development of an integrated system that links several machining cells into a flexible computerized manufacturing system. In addition, many companies are investigating the integration of robots and other manufacturing tools, such as lasers.

Although robots are used in almost every industry and type of application, the majority of installations have been concentrated in relatively few plants and types of applications. For example, it is estimated that just ten plants contain nearly one-third of all robot installations and that the three categories of welding, material handling, and machine loading account for approximately 80% of all current applications. At the same time, it must be remembered that the market penetration of robots has been relatively limited in even the most common applications.

As robot technology begins to diffuse within industry, it will affect almost every manufacturer, from furniture producers to pharmaceutical firms. And as robot capabilities continue to improve, new applications will undoubtedly be uncovered.

FUTURE TRENDS

There is almost universal agreement that the use of industrial robots will increase dramatically during the next decade. Anticipated growth, in terms of annual sales increases, are on the order of 40%. At that rate, the U.S. installed base of industrial robots should reach 100,000 units somewhere in the early 1990s. If each robot displaces two workers, then about 200,000 jobs could be directly affected by robots. However, many new jobs would also be created, for such positions as robot programmers, troubleshooters, and maintenance personnel.

Along with this rapid growth in robot sales, a number of important technological and product developments are expected. These include the development of smaller and lighter-weight robots, as increase in payload capacity relative to the weight of the robot, and dramatically improved grippers. In the sensor area, significant improvements are expected in machine vision systems, tactile sensing, and low-cost force sensors. Major developments are also anticipated in robot control and programming capabilities, including the use of hierarchical control concepts and off-line programming. At the same time, robot prices are expected to decline as production rates increase.

These trends and anticipated developments can only help accelerate the use of robots for fabrication and processing applications. Improved capabilities, particularly in control and sensory technology, coupled with declining costs should make many presently difficult applications cost-effective realities in the future.

As robot sales continue to grow, some shift in applications is anticipated. Current and projected robot use by type of application is shown in Figure 2 (1). It is anticipated that traditional robot applications, such as spot welding, material handling, and painting, will decline somewhat in terms of their respective market shares, whereas other emerging applications, such as assembly and machining, will increase significantly. Since machining, assembly, and similar uses of robots generally require more sophisticated equipment and interfacing, their relative impact on future manufacturing operations will be even more prominent.

BIBLIOGRAPHY

1. *Industrial Robots: A Summary and Forecast,* Tech Tran Consultants, Inc., Lake Geneva, Wisc., 1983.
2. J. F. Engelberger, *Robotics in Practice,* AMACOM, New York, 1980.
3. V. D. Hunt, *Industrial Robotics Handbook,* Industrial Press, Inc., New York, 1983.
4. J. Hartley, *Robots at Work: A Practical Guide for Engineers and Managers,* IFS (Publications), Ltd., Bedford, UK, 1983.
5. W. R. Tanner, *Industrial Robots from A to Z: A Practical Guide to Successful Robot Applications,* The MGI Management Institute, Larchmont, N.Y., 1983.
6. W. R. Tanner, ed., *Industrial Robots,* Vols. 1 and 2, Society of Manufacturing Engineers, Dearborn, Mich., 1981.
7. H. J. Warnecke and R. D. Schraft, *Industrial Robots: Application Experience,* IFS (Publications), Ltd., Bedford, UK, 1982.

ARM, DESIGN OF

Kenneth J. Waldron
Ohio State University
Columbus, Ohio

INTRODUCTION

Design of a manipulation system is a complex task involving knowledge from many disciplinary areas. It is impossible to work on any aspect of the problem in isolation from the others since geometry, structure, actuators, sensors, controllers, coordination algorithms, and servo control all affect one another. For this reason, the design process becomes highly iterative, with later decisions requiring revision of earlier ones.

Although there are many sophisticated computer-aided engineering tools available that can and should be used in the design process, the initial problem facing the designer is that these tools cannot be applied until the system has reached a fairly high level of definition. Initially, of course, the system is completely undefined. The designer must, therefore, proceed on the basis of fundamental knowledge, and the use of relatively simple mechanical and informational models, to rough out the initial design. The following treatment will be primarily addressed to this conceptual stage of the design process. After it is completed, the full power of numerical analysis and simulation tools can be applied to further refine the design. Part of the skill of the conceptual designer lies in getting the dimensions and component specifications of the rough design close enough that the changes required after accurate analysis are evolutionary, rather than revolutionary, in nature.

A major objective in design is to size or select components and subsystems optimally within the design constraints. Consequently, a second theme of this presentation is the provision of bases for optimally sizing elements of the system.

Design of a robotic system is also intrinsically different from most machine design problems because robotic systems are designed for versatility. Their most important attribute is the capability of being easily reprogrammed to perform a wide variety of tasks. In contrast, most other machines are designed to perform a single task as well, and as economically, as possible. In a sense, in a robotic system, the design of a machine to perform a specific function is split between the design of the hardware and the design of the program, which is performed on site. The robot will not perform that function as well, or as fast, as a special-purpose machine would. However, design and manufacture of that special-purpose machine may be very expensive. By splitting the cost of system hardware design and manufacture among many different possible functions, each of which requires only the (relatively small) additional cost of programming and perhaps some fixturing, the economics may become attractive. An enhancement is the possibility of performing several different functions at the same workstation with the same machine, rather than with several specialized machines. It may be seen that the property of versatility should be the characteristic that drives the design of a manipulation system.

GEOMETRIC DESIGN

Here the kinematically important dimensions are determined to optimize geometric performance, as measured by working volume and orientation capability, within constraints that are largely imposed by the coordination algorithms used to drive the system. The kinematically important dimensions are the normal lengths of the links between successive joint axes, the angles between successive axes, and the offsets along axes between the common normals in successive members. These are described more precisely below.

The coordination algorithms constrain the geometry since

it is necessary that the equations on which they are based be simple to facilitate real-time computation. The forms of these equations are determined by the geometry. The coordination algorithms are also subject to the occurrence of mathematical singularities. These occurrences also are determined by the geometry.

These constraints drive the designer in the direction of highly simplified geometries so that the difficulties and costs are kept at a low level.

A Geometric Model

Before launching into the question of geometric optimization, it is necessary to develop a simple geometric description of a member of a manipulator. The parameters involved are shown in Figure 1. Here a_N, the *length* of member N, is defined as the length of the common normal between axis N and axis $N + 1$, and the *twist* α_N of member N is the angle between axis N and axis $N + 1$, measured positive clockwise about the common normal. The positive direction of the latter is from axis N to axis $N + 1$. r_N, the *offset* along axis N, is measured from the foot of the common normal of member $N - 1$ to that of member N. Finally, the *joint angle* about axis N is measured from the positive direction of the common normal of member $N - 1$ to that of member N. These four parameters are all that is required to define each member geometrically. For typical industrial robot geometries, they can be identified by inspection. Figure 2 shows the locations of the axes and common normals of a Stanford Arm. The corresponding values of the member parameters are shown in Table 1. The one additional entity needed to complete description of the geometry of the manipulator as a whole is the definition of a *reference point* fixed in the hand, or *end effector*. This point is needed to complete definition of the terminal member since it contains only one joint axis. The dimensions of this member are specified by adding an imaginary axis through the reference point parallel to the preceding joint axis. Thus, in the case of the Stanford Arm of Figure 2 and Table 1, axis 7 is imagined to pass through the reference point parallel to axis 6. Since axes 6 and 7 would then be coincident, the end effector is represented by a nonzero offset r_6 along joint axis 6. a_6 and α_6 are, of course, zero.

It is no accident that the simple link model above is almost universally applicable to industrial robots and teleoperation manipulators. In a serial chain mechanism, which includes the majority of industrial robots, each joint is actuated. The only joint types that match available actuators are rotary and sliding joints. Both are effectively accommodated in the above model. If joint N is a rotary joint, the offset r_N is constant and the joint angle θ_N is variable. Conversely, if the joint is

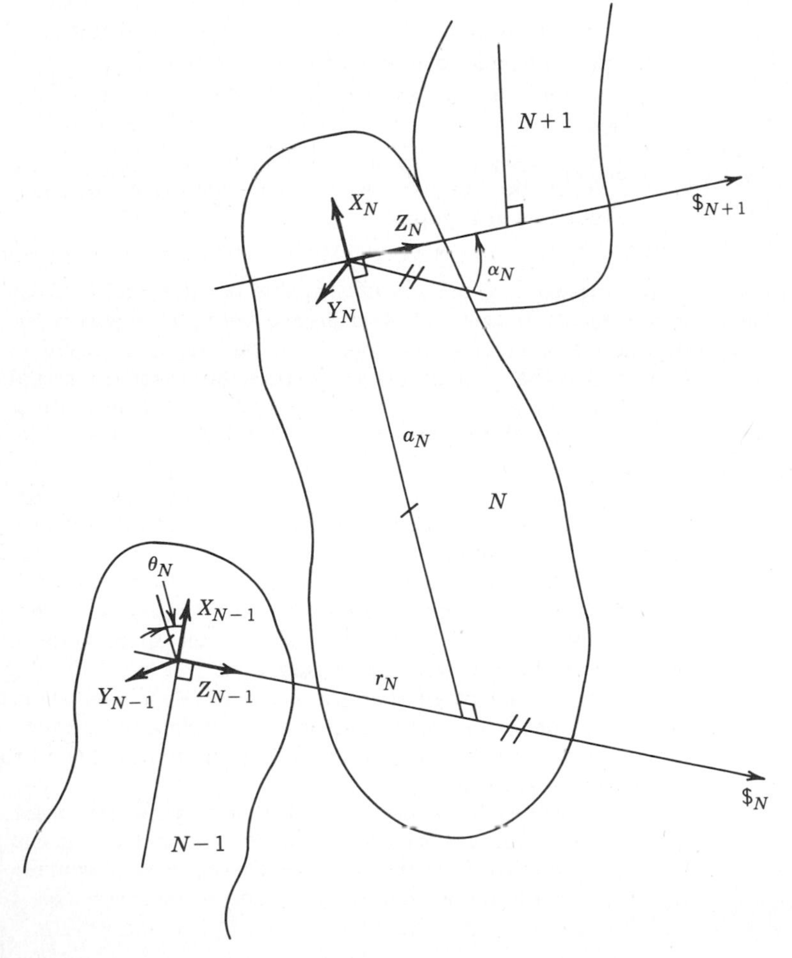

Figure 1. Geometric model of member N of a serial manipulator. $\$_N$ is the axis of joint N, and $\$_{N+1}$ is that of joint $N + 1$. The reference frame X_N, Y_N, Z_N is the reference frame affixed to member N. a_N is the link length, r_N is the offset on joint axis N, α_N is the twist angle between the two axes, and θ_N is the joint angle.

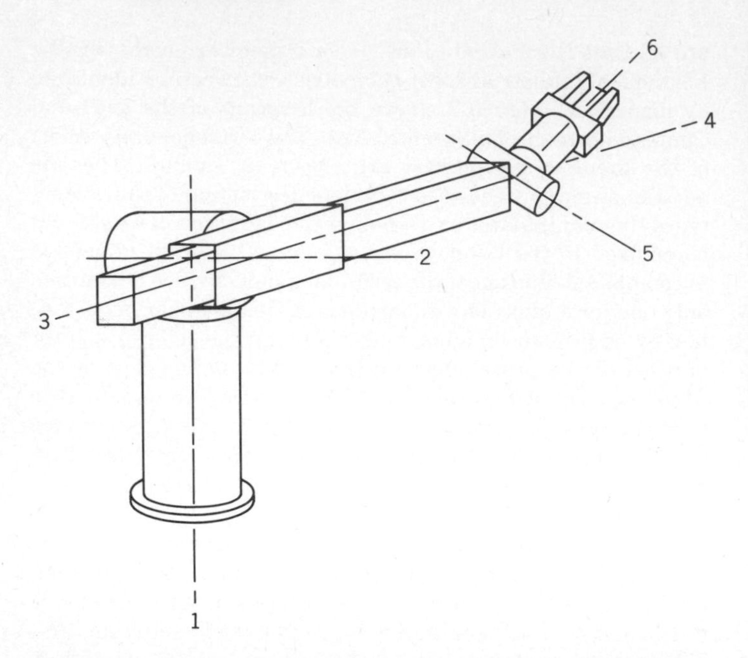

Figure 2. Location of axes and common normals for Stanford Arm. Joint 3 is a sliding joint. The remaining joints are revolutes. Axis 1: Vertical hinge. Axis 2: Horizontal hinge, intersects axis 1. Axis 3: Slide, offset from axis 1, intersects axis 2. Axis 4: Hinge, coaxial with axis 3. Axis 5: Hinge, intersects axes 3 and 4. Axis 6: Hinge, intersects axes 3–5. All successive intersecting axes are normal to each other.

a sliding joint, θ_N is constant and r_N is variable. In both cases, a_N and α_N are constant.

All industrial robot and teleoperator geometries are characterized by geometric specialties such as intersecting joints, parallel joints, successive joints normal to one another, and so on. There are a number of reasons for this, some of which were alluded to above. An obvious one is that the resulting simplified geometries have relatively tractable kinematic equations (1,2), leading to relatively simple coordination algorithms that can be run in real time on realistically sized control computers. This is one example of the interaction between geometry and control in manipulator design. Another is the fact that more general geometries lead to complex and unpredictable coordination singularities (3).

It is productive to think in terms of a *regional structure,* which is responsible for moving the reference point about in space, and an *orientation structure,* which is then responsible for moving the end effector into a desired orientation (4). Since production of significant lineal displacement by means of a rotary joint requires a long lever arm, the inboard joints are more effective for moving the reference point. Conversely, the ideal behavior of the orientation structure is rotation about an axis through the reference point without displacement. Thus, short lever arms to the reference point are favored, and the outboard joints are more effective for this purpose. Hence, the inner three joints and members of a six-degree-of-freedom manipulator usually comprise the regional structure, and the outer three joints and members comprise the orientation structure.

Table 1. Geometric Parameters of the Stanford Arm

Member	a_i	r_i	α_i	θ_i
1	0	0	90 deg	Variable
2	0	l	90 deg	Variable
3	0	Variable	0	0
4	0	0	90 deg	Variable
5	0	0	90 deg	Variable
6	0	m	0	Variable

The displacement of the reference point and the orientation of the end effector are not, in general, independent. Nevertheless, it is advantageous to make the coupling between lineal displacements and orientation movements weak. This is beneficial from both the geometric and the control points of view. Therefore, they are considered separately from the point of view of optimizing the geometry of a manipulator.

The Optimal Orientation Structure

The optimal capability of an orientation structure is the ability to rotate the end effector completely about any chosen axis through the reference point. That is, the orientation structure should possess maximal versatility in terms of orienting the end effector. It has been shown (5) that an orientation structure consisting of three rotary *wrist* joints can meet the above-stated performance optimum if, and only if, it satisfies the following conditions:

1. The three axes meet at a single concurrency point.
2. Successive axes are normal to one another.
3. The inner and outer joints are capable of complete rotation.
4. The middle joint is capable of at least 180 deg of rotation, covering the range $0 \geq \theta_5 \geq 180$ deg with the joint angle defined as in Figure 1.

In practice, it is usually not feasible to satisfy all of these requirements because of interference between members, or limitations of actuator motion. Nevertheless, it is useful to be aware of the optimal geometry since the closer the actual geometry approaches it, the closer the system will be to optimal orientation capability. Many industrial robots have three successively orthogonal rotary wrist joint axes that are concurrent. Thus, they satisfy conditions 1 and 2. The joint motion conditions 3 and 4 are harder to satisfy, and few, if any, systems meet them completely.

Sliding joints have not been mentioned because the function of the orientation structure is rotation, and sliding joints provide no capability for rotation. Thus, it is not appropriate to use them in the orientation structure.

It may also be noted that the geometry described above as optimal for the orientation structure is independent of the regional structure. Thus, it is applicable regardless of the regional structure geometry.

If the robot is to be used only with axi-symmetric tools, such as arc-welding electrodes, or drills, the sixth joint can be dispensed with. It is then optimal for the axis of symmetry of the tool to be normal to axis 5 and to intersect axes 4 and 5.

Optimization of the Regional Structure

The function of the regional structure is placement of the reference point in space and is quite different from that of the orientation structure. Consequently, the bases on which it should be optimized are quite different. The regional structure should provide maximal versatility in placement of the reference point. That is, the volume within which the reference point can be placed should be maximized. In any given function, it is likely that only a small portion of the working volume will be used. Nevertheless, the ability of the machine to be applied to many different functions is maximized by maximizing the working volume.

Since both sliding and rotary joints can be used in the regional structure, there are several different joint combinations that must be examined. First consider the case in which all joints are rotary joints. In order to simplify the discussion, only cases in which successive joint are parallel, or normal to one another, are considered. The results, however, have more general validity (5). An additional simplifying restriction is imposed. Each member has either the joint offset r_N or the length a_N equal to zero. Again, this is consistent with most industrial robot geometries. Actually, this constraint can be loosened to include slightly more general geometries, such as that of the PUMA, without invalidating the conclusions stated below (5).

The outer boundary of the workspace is determined by the arrangement of the two innermost joint axes. For a given *length* of the manipulator, defined as the maximum distance of the reference point from the first joint axis, the maximum volume contained by the outer boundary is achieved when the second axis intersects the first and is normal to it (5). In this case, the outer boundary is a sphere centered on the intersection of the first two joints, and with radius equal to the length of the manipulator. Any other location of the second axis gives a boundary that is a torus, which fits entirely within the sphere generated by the orthogonally intersecting axis arrangement.

The inner boundary of the workspace is determined by the location of the third axis. In a six-degree-of-freedom manipulator with optimal orientation structure geometry, it is impermissible for the third axis to be concurrent with the first two. If it were, the system would be geometrically singular in all positions and would only have five degrees of freedom of hand-positioning capability, rather than six. Thus, the length of member 2 must be nonzero, and within the restrictions imposed above, axis 3 must be either parallel or normal to axis 2. If axis 3 is located farther outboard than one-half of the length of the arm from the intersection of axes 1 and 2, there will be an inner boundary defining a void around the intersection point within which the reference point cannot be placed. If it is less than half the length of the arm from the intersection point, there will be an inner boundary within which location of the reference point is double-valued. Thus, geometrically, the optimum location of axis 3 is exactly one-half the length of the arm away from the intersection point (5). Mechanically, it may be better to place it a little farther out than this to diminish the potential for interference between the arm and its mounting structure.

Superficially, there is no geometric basis for preferring the case in which axes 2 and 3 are parallel over that in which they are normal to one another. At a deeper level, there is. If axes 2 and 3 are parallel, movements of the reference point in any plane containing axis 1 require no movement about axis 1. That is, there is a decoupling between movements in the plane containing axis 1 and movements normal to that plane. This is helpful from the point of view of coordination and also, in fact, from that of actuation energetics. No comparable decoupling occurs in the case in which axes 2 and 3 are normal to one another. Further, the case in which axes 2 and 3 are normal to one another leads to the generation of large torsional loads on member 2 and on joints 1–3. It is usually preferred to avoid this situation.

As a consequence of the above arguments, the optimal regional structure geometry satisfies the following conditions:

1. Axes 1 and 2 intersect and are normal to one another.
2. Axis 3 is parallel to axis 2 and is located halfway along the length of the arm.

In the above discussion, joint motion limits are ignored. The presence of limits on the motions of joints 1–3 in no way invalidates the above discussion. A limit on the motion of joint 1 reduces the optimal workspace volume from the interior of a sphere to that of a segment of a sphere. However, the possible volumes obtained by varying the position of joint 2 still lie completely within that spherical segment. Similar arguments apply in the cases of motion limits on joints 2 and 3. Optimally, joint 1 should be capable of complete rotation. Joint 3 should be capable of at least 180 deg in the range $0 \leq \theta_3 \leq 180$ deg. Joint 2 should be capable of 270 deg of rotation in the range $-180 \leq \theta_2 \leq 90$ deg (5). As was mentioned earlier, it is often impossible to meet these conditions.

It remains to relate the regional and orientation structures. Within the limits imposed above, two arrangements are possible (Fig. 3). In Figure **3a**, joint 4 intersects joint 3 and is normal to it. The resulting arrangement is usually referred to as having a *roll–pitch–roll* wrist. In Figure **3b**, joint 4 is parallel to joint 3, leading to a *pitch–yaw–roll* wrist. Both are widely represented in commercial industrial robots.

The case in which the regional structure includes sliding joints is qualitatively different from that in which it includes only rotary joints. If only rotary joints are used, the working volume is finite, regardless of the presence or absence of joint motion limits. This is not true if one or more sliding joints are used. In the absence of motion limits on the sliding joint, the workspace is infinite.

If all three joints in the regional structure are sliding joints, the optimal arrangement is for them to form a mutually orthogonal set. This configuration is quite common among industrial robots and includes the gantry arrangement. Combining this with an optimal orientation structure, an arrangement such as that in Figure 4 is obtained.

It is useful to have a figure of merit to quantify the effectiveness of a given structure in generating a workspace volume. A measure that has been used is the normalized volume index (NVI) (6). Here the actual working volume, including the effects of motion limits and so on, is compared to that of a sphere with radius equal to the *length sum* of the manipulator. The length sum used here is different from the length defined and

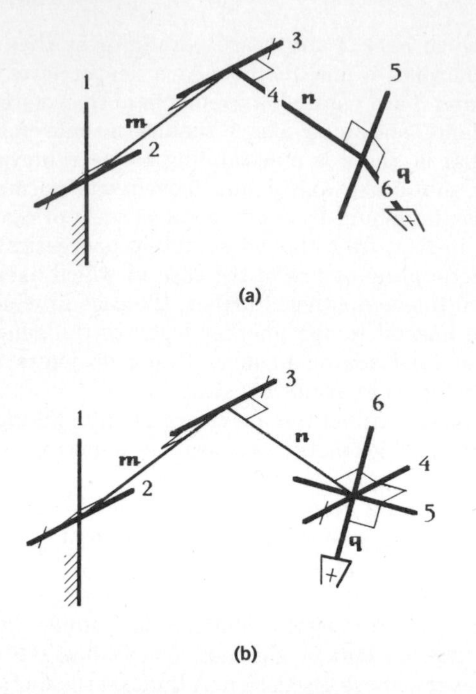

Figure 3. Optimal geometries for manipulators in which all joints are revolutes. The two alternatives differ only in the relationship of joints 3 and 4. **(a)** Arrangement has a *roll–pitch–roll* wrist. **(b)** Arrangement has a *pitch–yaw–roll* wrist.

used above. If it is represented by L, it is defined by the equation

$$L = \sum_{i=1}^{N} (a_i + r_i)$$

where N is the total number of joints in the manipulator. That is, it is the sum of the joint offsets and member lengths. In the case of a sliding joint, r_i in the equation must be interpreted as the length of travel between the motion limits of that joint, not as the variable position of the joint, which is the sense in which it has previously been used.

When making comparisons of regional structure geometries, it is convenient to neglect the portion of the working volume generated by rotation of the hand about the wrist. This volume is due to the length of the hand, and to any nonzero member lengths in the orientation structure. It is optimal to minimize these lengths since this diminishes the coupling between translation and orientation movements. For purposes of discussion of regional structure geometries, the contribution of the hand length and the orientation structure to the length sum is neglected.

A somewhat easier index to relate to, from the design point of view, is the ratio of the length sum to the cube root of the working volume. This index, which might be called the structural length index, Q_L, is not specialized to quasi-spherical geometries, as the NVI is, although it can be directly related to that index. It can be interpreted as the length sum the structure would have if the volume were scaled to unity. It gives a crude measure of the relative amounts of structure required by different geometries to generate comparable working volumes. The relationship between the two measures is $Q_L = (3/4\pi N_V)^{1/3}$ where N_V is the NVI.

For an ideal rotary joint regional structure, as exemplified by the geometries of Figure 3 without joint motion limits, $Q_L = (4\pi/3)^{-1/3} = 0.62$. Of course, in most real manipulator structures, this figure will be somewhat larger because of the effects of joint motion limits in reducing the volume. For a regional structure with orthogonal sliding joints, such as that in Figure 4, Q_L is minimized when the length of travel is the same on all three joints. The minimal value is $Q_L = 3.0$. As was mentioned above, this is a crude measure of the amount of structure needed to generate unit volume of workspace. The comparison of these numbers correctly reflects the fact that the sliding joint geometry of Figure 4 requires substantially more structure than the rotary joint geometries of Figure 3.

There are, of course, a number of possible arrangements in which the regional structure includes both rotary and sliding joints. Space does not permit an exhaustive discussion of all possible geometries. Two other geometries that have been widely used are shown in Figures 5 and 6.

The geometry shown in Figure 5 is the SCARA geometry. In this case, because axes 1–4 are parallel, the kinematic behavior and working volume are the same regardless of the rotary axis with which the sliding joint is placed coaxially. The sliding joint is commonly placed as axis 4, coaxial with axis 3, to minimize the inertia to which its actuator is subjected. Nevertheless, it should be regarded as part of the

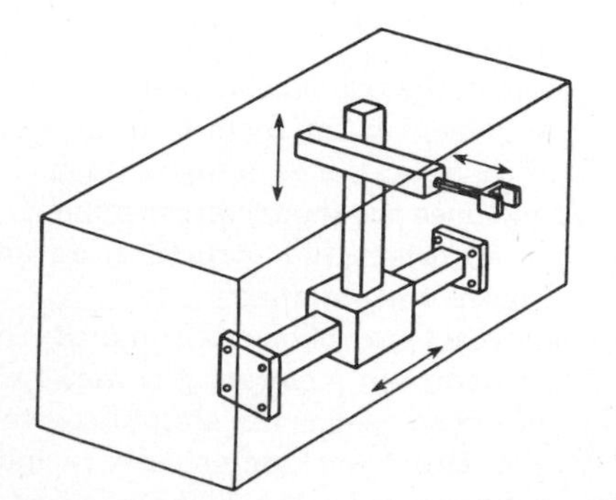

Figure 4. Orthogonal sliding regional structure. Gantry robots are of this type.

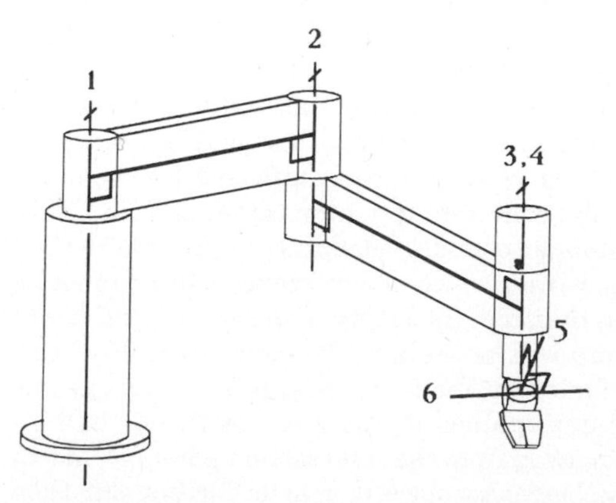

Figure 5. SCARA geometry. This geometry is suited to large horizontal and limited vertical displacements. It is popular in pick-and-place and assembly applications.

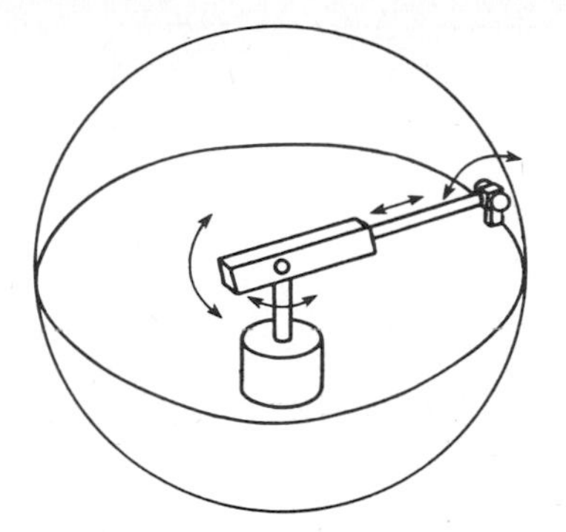

Figure 6. Telescoping geometry. A sliding joint replaces the "elbow" joint in the configurations of Figure 3.

regional structure. That is, in this case the regional structure consists of joints 1, 2, and 4, and the orientation structure consists of axes 3, 5, and 6. If joint motion limits are absent, the working volume of a SCARA structure is bounded by a right circular cylinder and two planes normal to the cylinder axis. If the maximum distance of the reference point from axis 1 is l and the length of travel of axis 4 is R_4, the radius of the cylinder is l and the distance between the truncating planes is R_4. The length sum is $l + R_4$. Thus,

$$Q_L = (l + R_4)/(l^2R_4)^{1/3}$$

Minimization of Q_L as a function of the ratio R_4/l gives $R_4 = l/2$ as optimal. The corresponding value of Q_L is 1.29. Once again, this figure is for an idealized geometry. A real SCARA structure with joint motion limits gives a somewhat higher figure.

Vertical and horizontal motions are completely decoupled by the SCARA geometry. This is advantageous from the points of view of both coordination and actuator energy consumption and dissipation. The geometry is well suited to operations requiring horizontal displacements that are large in comparison to the corresponding vertical displacements. This is typical of many pick-and-place and assembly-type operations.

The structure shown in Figure 6 has a sliding third joint in place of the elbow of those shown in Figure 3. It is found in industrial robots such as the UNIMATE 3000. The telescoping polar-coordinate-type configuration is advantageous from the point of view of producing relatively simple coordination equations. Here the structural length index depends only on the joint travel limits of joint 3, at least in the ideal case in which joint 1 can rotate completely and joint 2 can exceed the range $-90 \leq \theta_2 \leq 90$ deg. In this case, $Q_L = 0.62(1 - u_3/U_3)^{-1/3}$ where U_3 is the maximum value of r_3 and u_3 is its minimum value.

Other Measures of Geometric Performance

The dexterous or primary workspace (7,8) is a means of characterizing orientation performance. This is the portion of the workspace within which the hand can assume any specified orientation. Unfortunately, dexterous workspaces exist only for certain idealized geometries. Real industrial robots with joint motion limits almost never possess dexterous workspaces.

It is possible to define a measure somewhat analogous to the normalized volume index. This is the ratio of the area of the interception of the noncircular cone of possible orientations of a line fixed in the hand passing through the reference point and the concurrency point with a unit sphere centered on the concurrency point, to the area 2π, of that sphere (9). This measure is more suited to real geometries with joint motion limits.

The approach angle represents another means of characterizing orientation performance. It is much better suited to practical use. The portion of the workspace in which the hand can assume a given orientation is readily mapped (10). Figure 7 shows the portion of the workspace for a typical geometry that is reachable with the hand horizontal.

A central problem in coordination and control is the occurrence of *geometric singularities* (11). These occur whenever the system must generate trajectories defined in a fixed coordinate frame (relative to the earth). At a geometric singularity, the transformation that maps the velocity of the reference point, and the angular velocity of the hand, into joint rates breaks down. This transformation is characterized by the Jacobian matrix (12). It is not appropriate to discuss its character and derivation here (see DYNAMICS; KINEMATICS). However, for the present purpose it is necessary to note that it is a $6 \times N$ matrix whose elements are functions only of the joint position variables. Singular positions correspond to dependence of the linear system represented by this matrix. For a six-jointed manipulator ($N = 6$), geometrically singular positions may be identified as those in which $|\mathbf{J}| = 0$, where $\mathbf{J}$ is the Jacobian matrix.

It has been pointed out that operation in the neighborhood of singularities cause problems due to the ill-conditioning of the coordination algorithm. Thus, a requirement of the form $|\mathbf{J}| \geq \delta$, where δ is an arbitrary threshold value, is suggested. It is quite practical to use a condition such as this in real-time software for typical industrial robot geometries since $|\mathbf{J}|$ can be expanded to a rather simple expression (2). Another measure (13), as applicable to the case in which $N > 6$, has gained considerable acceptance. This is the manipulability index N_M, defined as $N_M = \sqrt{|JJ^T|}$. This index is too cumbersome for real-time application. For six-jointed manipulators, it reduces to $N_M = |\mathbf{J}|$. Figure 8 shows the portion of the workspace within which $N_M \geq 0.1$ for a typical geometry.

Parallel Geometries

So far, only serial chain geometries in which each member, except for the first and the last, is jointed to only two other members have been discussed. In current practice this is, by far, the most widely used type of structural geometry. However, an infinite variety of other geometries are possible, and at least a couple have proved to have practical advantages for some purposes.

The motion simulator shown in Figure 9 uses a six-degree-of-freedom parallel configuration called a Stewart platform. Six linear hydraulic actuators act in parallel between the base and end effector. It is very widely used in motion simulators for pilot training. Viewed as a manipulator structure, it offers great strength and stiffness, as compared to serial chain struc-

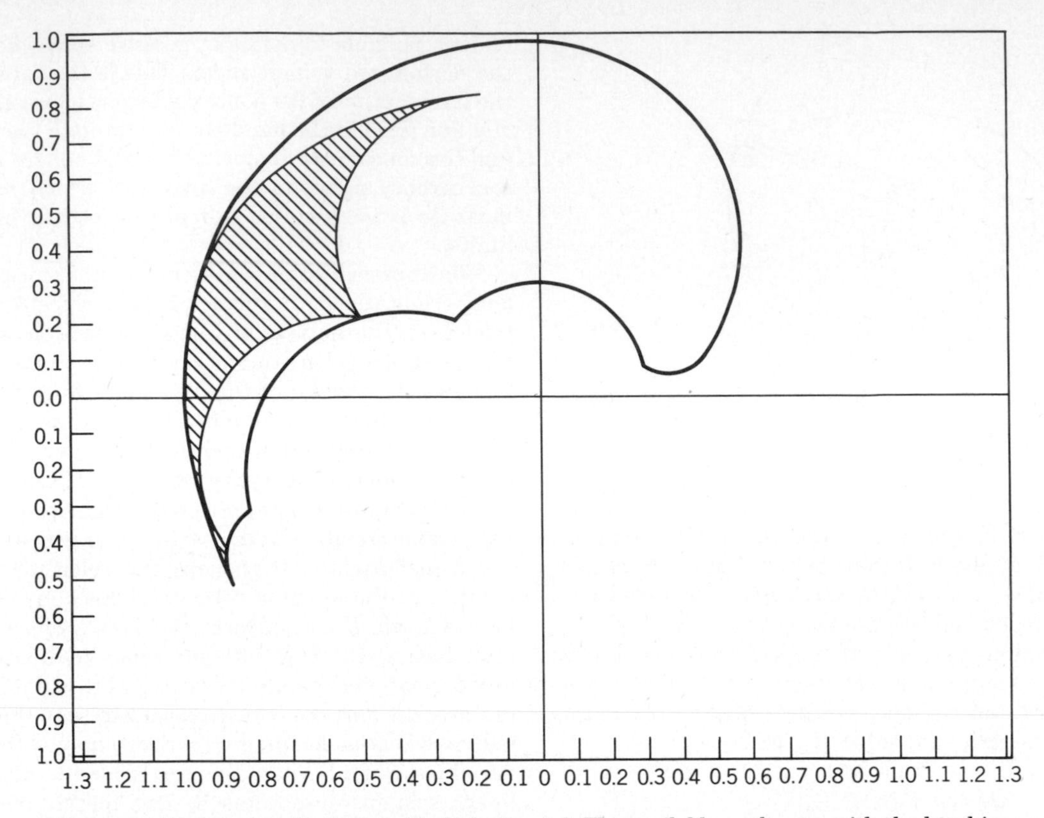

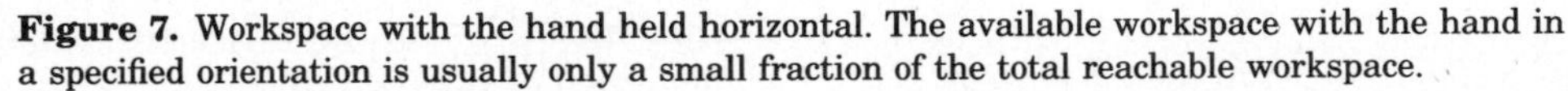

Figure 7. Workspace with the hand held horizontal. The available workspace with the hand in a specified orientation is usually only a small fraction of the total reachable workspace.

tures. The cost is relatively restricted motion capability. Its geometric capabilities are analyzed in reference 14.

In general, parallel structures offer the potential for greater stiffness at the cost of complexity and, possibly, reduced motion flexibility.

Another important example of a parallel geometry is the pantograph geometry, or rather geometries, since there are several variants. This has been employed as an industrial robot structure (15) and offers several important advantages. Vertical and horizontal motions can be completely decoupled. This is important both from the point of view of simplified control and for improvement of actuation efficiency. All regional structure actuators can be mounted inboard to improve the inertial properties of the system. For heavy duty systems, this type of structure lends itself to hydrostatic actuation with resulting great improvements in efficiency (16). Finally, the workspace closely approximates a cylindrical annulus with a rectangular generating curve, which lends itself very well to manufacturing applications. The geometric optimization of pantograph manipulators is treated in reference 17.

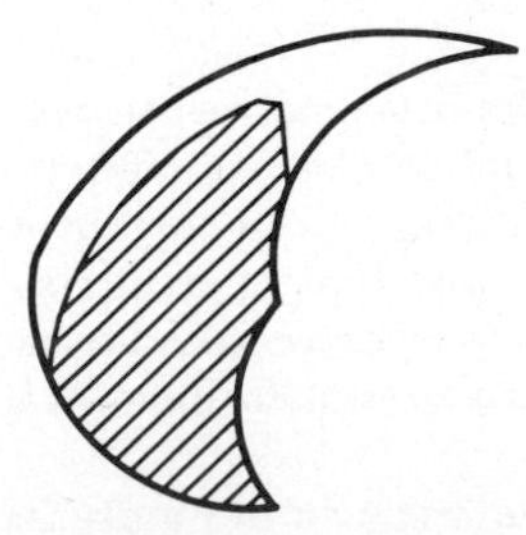

Figure 8. Workspace with manipulability index exceeding 0.1. The manipulator is a Cincinnati Milacron T3-746. Criteria of this type ensure avoidance of the neighborhoods of singular positions. Operation in the vicinity of a singular position usually degrades kinematic and dynamic performance.

Kinematically Redundant Geometries

It has long been recognized that six degrees of freedom is only the minimum requirement for a manipulator to have full, spatial positioning capability, and that larger numbers of degrees of freedom might offer improvements in performance. The human biological manipulation system not only has two arms, each with more than six degrees of freedom, but also has them mounted on a mobile platform. This makes possible a rich suite of strategies for making use of this highly redundant manipulation system. However, effective use of redundant degrees of freedom in an artificial manipulation system is not a simple problem.

Several systems have been successfully operated using one-, two-, or three-degree-of-freedom position adjustment systems mounted at the distal end of more or less conventional six-degree-of-freedom industrial robots (18,19), to give a total of seven to nine degrees of freedom. The key to all of these systems is direct sensing of the position of the workpiece relative to the end member of the manipulator. The position adjustment degrees of freedom are actuated at higher bandwidth and accuracy than the manipulator on which they are mounted, but are capable only of small motion amplitudes. The manipulator is programmed to follow a path approximat-

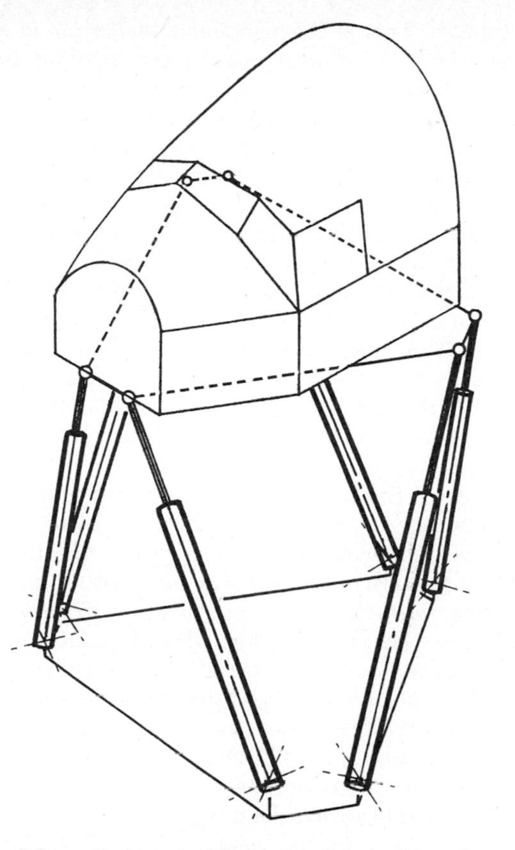

Figure 9. Schematic layout of flight simulator using a Stewart platform. The lower ends of the six linear actuators are connected to the base by passive universal joints. The upper ends are connected to the output member by spherical joints.

ing that necessary to perform a specified operation. This path is refined to higher precision by the position adjustment system using direct sensing of workpiece position.

The use of additional degrees of freedom in open-loop operation, as is typical of most current industrial robots, presents a very different problem. The most important issue here is determining the benefits that can be gained by the use of additional degrees of freedom since this determines the most effective locations for adding additional joint axes to the geometry.

The addition of one or more joints to an industrial robot might offer improvements in the following characteristics (20):

1. Workspace volume.
2. Geometric dexterity.
3. Ability to overcome geometric singularities.
4. Ability to use bracing strategies.
5. Reach inside capability.

It is necessary to examine each of these characteristics in detail to determine which offer the most significant improvements and the general locations of additional joints that might improve them.

As was demonstrated above, the maximum workspace volume is achieved when the two most proximal joints, that is, the shoulder joints (1 and 2), intersect orthogonally and when the elbow joint is exactly at the midpoint of the arm as measured from the hand reference point to the intersection of joints 1 and 2 and is normal to the line joining those points. If these conditions are met, the locations of the remaining joints make no difference. Thus, the effect of an additional joint on the workspace volume can only be to relieve the effects of joint motion limits. Although this may be useful in some situations, additional joints will not, in general, significantly improve workspace volume.

It has also been shown above that it is geometrically possible for the hand to assume any desired orientation, provided the three most distal axes (4–6) are successively orthogonal and are concurrent, and provided axes 4 and 6 are capable of complete rotation and axis 5 is capable of 180 deg of rotation, from $\theta_5 = 0$ to $\theta_5 = 180$ deg. Both of the configurations of Figure 3 satisfy these geometric requirements. However, in both cases it is mechanically difficult to meet the joint range of motion requirements. In the configuration of Figure 3**a**, it is possible to approach the requisite ranges of rotation of axes 5 and 6. However, the range achievable by joint 4 is usually on the order of 180 deg, far short of that needed to give complete orientation capability. In the configuration of Figure 3**b**, it is possible to approach, or exceed, complete rotation of joints 4 and 6. It is also possible to approach 180 deg of motion about joint 5. However, that motion be will be from −90 to +90 deg! rather than from 0 to 180 deg. This does not meet the requirement for full orientation capability, at least with respect to problems of axis reversal and geometric singularity. The addition of a seventh joint either at, or near, the wrist (taken to be the point of concurrence of axes 4–6) can certainly overcome the joint motion limitation problem (21) and permit full hand orientation capability.

Removal of geometric singularities should certainly be a primary objective of any redundant geometry (21). The singular positions of the configurations of Figure 3 are well understood (5,22). The requirement for an additional joint to remove these singularities is that its screw axis lie outside the 5th- (or lower) order screw system defined by the six original axes (23) in each of the singular positions. Another way of saying this is that its Plücker coordinates should be linearly independent of those of the original six joint axes in all singular positions.

The use of bracing strategies (24) when performing fine manipulation tasks has been, as yet, little explored, although it is a powerful means of raising the lowest structural natural frequency. Bracing is the kind of strategy used when a forearm rests against the desk while writing. One reason bracing has not been fully explored is that it requires a manipulator with more than six degrees of freedom, since maintaining single-point rolling and sliding contact with an object in the environment has the effect of reducing the number of degrees of freedom of the system by one. In order to be full effective, an additional axis should be located outboard of the bracing point, which, for the configurations of Figure 3 is most appropriately located on member 3. This is because the inevitably restrictive sizes of the edges or surfaces whose contact is used to produce the bracing action substantially reduce the effective ranges of motion of joints inboard of the bracing point.

The problem of reach inside capability is similar to that of bracing. If the manipulator must reach through a restricted

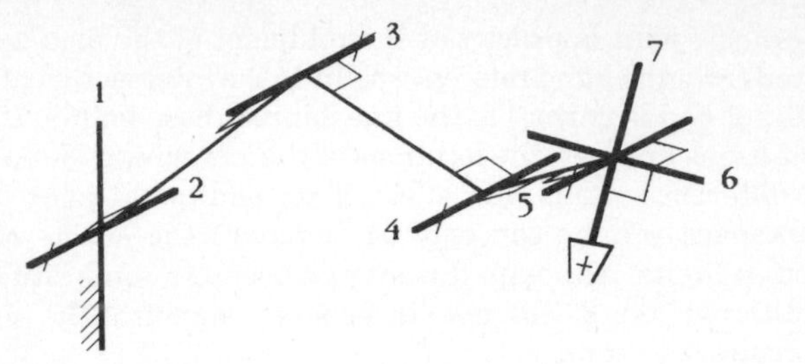

Figure 10. Suggested redundant configuration for improvement of manipulation performance and elimination of some singular positions.

opening, the ranges of motion of joints inboard of the opening are, effectively, restricted. Therefore, in order to provide increased capability for this type of operation, an additional joint should be located as distally as possible.

Figure 10 shows a configuration that has been suggested as having attractive properties from the point of view of *some* simplicity of the coordination equations, and the capability to improve dexterity, bracing capability, and operation when reaching through restricted portals. This configuration has a capability for removing singularities associated with the wrist and elbow joints (20). It is, however, subject to the shoulder-type singularity (21). The configuration of Figure 11 is superior in providing the capability of avoiding singularities, but lacks the capability for operation through restricted portals or in bracing modes.

DESIGN OF MANIPULATOR STRUCTURES

It is infeasible, in the space available, to treat the structural design of all the varied manipulator geometries that are in use. Consequently, in this section a simple model is used to optimize the selection of structural cross sections for "two-link" manipulators. As was demonstrated above, this geometry is optimal for rotary joint manipulators. A simple approach to the estimation of vibrational natural frequencies is also presented. Although the treatment is limited to this specific geometry, the principles and approach used are of general validity. Thus, this treatment is intended as a model that can be adapted to the design of structures for other geometries.

Most robotic manipulators are controlled in an *open-loop* manner with respect to hand position. For this type of manipulator, stiffness is the crucial structural requirement, rather than strength. This is because elastic deflection of the structure creates a position error at the hand. Without direct feedback of hand position there is no way of actively correcting this error. Systems in which *end-point* feedback is available present a totally different structural design problem. Teleoperator manipulators are a good example of this type of system. They are discussed later.

There are two different structural "stiffnesses" that are important. The first is *stiffness under static load.* Under static conditions, the joints can be assumed to be servoed to the commanded positions with a high degree of accuracy, regardless of load. Thus, the joints can be regarded as perfectly rigid, with all deflection taking place in the members. The second important "stiffness" is *dynamic stiffness,* that is stiffness with respect to structural vibrations. Here, much of the compliance is likely to be contributed by the joints, rather than the structural members. Static stiffness is important to absolute positioning accuracy. Dynamic stiffness is important because, if the natural frequencies of free structural vibrations are too low relative to servo bandwidth, the structural vibration modes will interact with the servo controllers to produce unstable, or oscillatory, behavior. The appropriate quantity for specifying dynamic stiffness is the lowest natural frequency.

Once again, the philosophy of this presentation is to provide simple models that can be used in the initial phase of design. Once the structure is fully defined, finite element and structural dynamic models should be used for more accurate analyses, on the basis of which the design can be refined.

The Manipulator Model

It has been shown above that the optimum geometry of the regional structure, for a manipulator with only revolute joints, has the first joint vertical, the second intersecting it orthogonally, and the third parallel to the second. The length of the second link is exactly half that of the entire manipulator. The optimal orientation structure has the three outboard joint axes concurrent, with successive axes meeting orthogonally.

Based on the above-stated geometric optimality conditions, the two-link structural model of the manipulator shown in Figure 12 will be used (25). The only joints considered are joints 1 and 2 at the "shoulder" and joint 3 at the "elbow." The three wrist joints are regarded as part of member B for the following reasons. The outboard end of member B contributes very little to static deflection both because the bending moments are low and because the cantilever length outboard

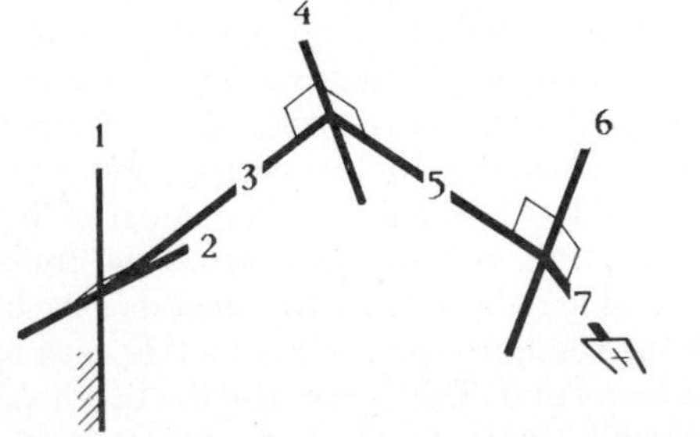

Figure 11. Suggested redundant configuration for elimination (with suitable coordination) of singular positions. This configuration does not improve flexibility of operation to the extent of that in Figure 10.

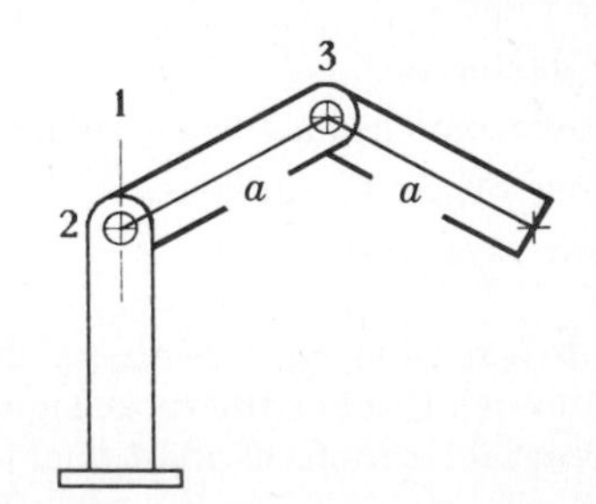

Figure 12. Two-link structural model of rotary joint manipulator. This permits reasonably accurate modeling of structural behavior for conceptual design purposes.

of a given cross section is small. Since the wrist joints are modeled as being rigid, when studying static deflection, little error will result from ignoring the wrist and changes in cross-section outboard of the wrist, thereby modeling member B as being of uniform cross section out to the hand reference point. Any weight due to the wrist actuators, over and above that which would be present if B were of uniform section similar to that at its inboard end, is lumped with the load and considered to be applied at the hand reference point. Similarly, the wrist joints contribute little to the lower-frequency vibration modes of the system since the inertia to which they are subject is relatively low. Thus, it will be assumed that the model of Figure 12 can be applied as a preliminary design model in both static and dynamic cases.

Computation of Static Deflection

Members A and B are assumed to have uniform cross sections. An objective, which will be addressed later, is to determine the optimal size ratio of section B to section A and the size of section A to keep static deflection under load within specified limits. It is first necessary to develop an equation for the vertical static deflection of the reference point.

The worst case for static deflection is the fully extended position shown in Figure 11. Bending deflections in all members are much larger than deflections in the longitudinal direction. Thus, deflection of the pylon due to the bending moment at the inboard end of member A produces a significant horizontal displacement of the hand reference point, but negligible vertical displacement. However, the angulation of the pylon at joint 2 is significant since, when multiplied by the length of the arm, it produces a significant vertical displacement of the reference point.

With the loading shown in Figure 13, with u representing the self-weight per unit of member A, v the self-weight per unit length of member B, U the weight of actuators at the elbow, and V the weight of actuators and load lumped at the hand, the bending moment at the interface of members A and B is

$$M_3 = aV + a^2v/2$$

where a is the length of each member. The shear force at the same section is

$$P_3 = U + V + av$$

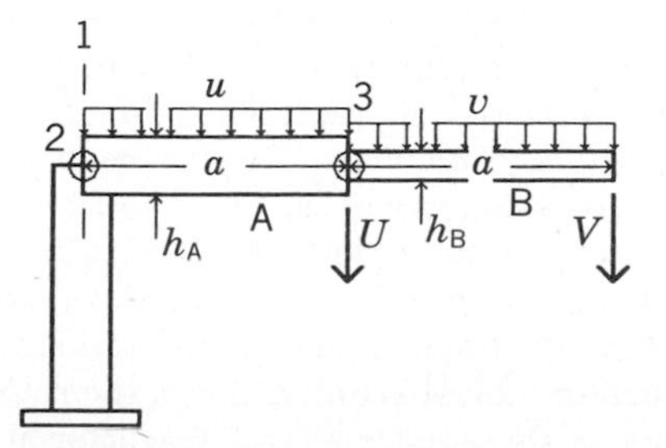

Figure 13. Loading pattern of structural model. The position shown is the worst case position for static deflection and free vibration.

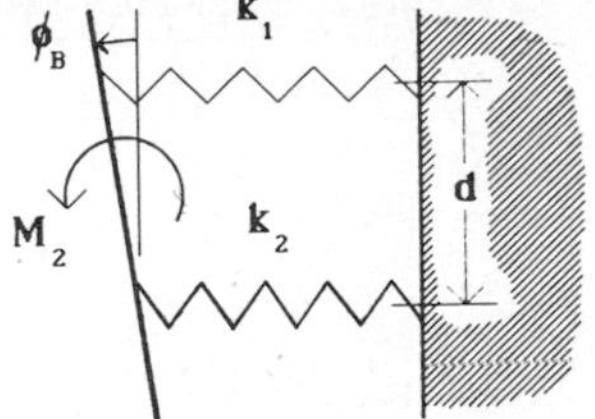

Figure 14. Bearing deflection model. The springs represent the radial compliances of two rolling element bearings, separated by distance d, supporting the vertical shaft of the first axis. M_2 is the bending moment at the shoulder, and ϕ_B is the resulting angular deflection due to bearing compliance.

At the shoulder, the bending moment is

$$M_2 = 2aV + aU + 3a^2v/2 + a^2u/2$$

Hence, using the Myosotis beam deflection formulas (26), the angular deflection of the pylon at this section is

$$\phi_P = \frac{M_2 h}{EI_P}$$

where I_p is the cross-sectional second moment of area of the pylon, assumed to be a uniform cantilever beam, and h is its height.

An additional significant angular deflection at this location is caused by deflection of the bearings of joint 1. This is usually the only location in this geometry at which bearing compliance must be considered. The geometry necessary for computation of the bearing deflection is shown in Figure 14. k_1 and k_2 are the radial stiffnesses of the upper and lower bearings, respectively, and d is the separation along the joint axis of the bearing planes. The resulting angular deflection is

$$\phi_B = \frac{M_2}{d^2}\left(\frac{1}{k_1} + \frac{1}{k_2}\right)$$

The relationship between compliance and size varies with bearing type and is a nonlinear function of load for the rolling element bearings usually employed in this location. Some data on stiffness characteristics of rolling element bearings can be found in reference 27. For a given type, radial stiffness data can usually be obtained from the manufacturer.

The total angular deflection at the shoulder is then

$$\phi_2 = \phi_P = \phi_B$$

The gradient of the beam at joint 3 due to deflection of member A is

$$\phi_3 = \frac{a^2}{6EI_A}(3U + 9V + au + 6av)$$

where I_A is the cross-sectional second moment of area of member A and E is Young's modulus. The cross-sectional moments and areas of beam sections commonly used in manipulators are given in Table 2.

Table 2. Section Parameters

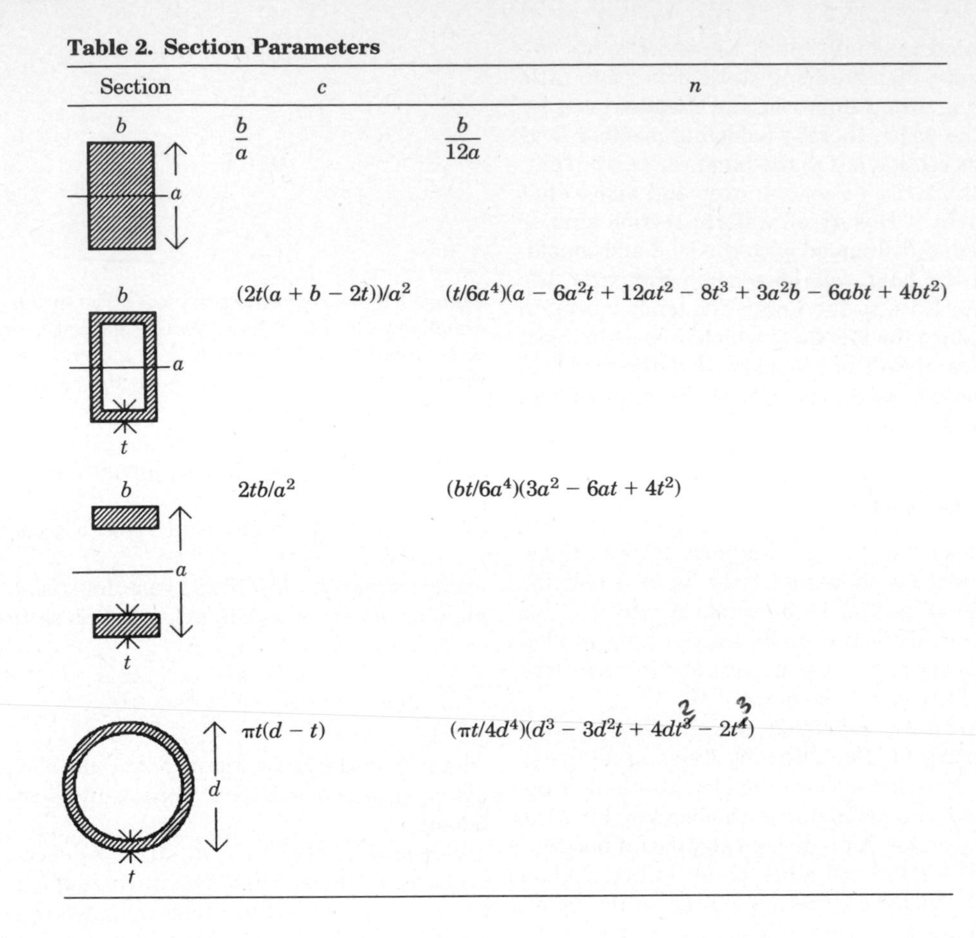

Section	c	n
Solid rectangle (b, a)	$\frac{b}{a}$	$\frac{b}{12a}$
Hollow rectangle (b, a, t)	$(2t(a + b - 2t))/a^2$	$(t/6a^4)(a - 6a^2t + 12at^2 - 8t^3 + 3a^2b - 6abt + 4bt^2)$
Two flanges (b, a, t)	$2tb/a^2$	$(bt/6a^4)(3a^2 - 6at + 4t^2)$
Tube (d, t)	$\pi t(d - t)$	$(\pi t/4d^4)(d^3 - 3d^2t + 4dt^3 - 2t^4)$

The vertical deflection at joint 3 is

$$\delta_3 = \frac{a^2}{24EI_A}(8U + 20V + 3au + 14av)$$

The vertical deflection of member B between sections 1 and 2 is

$$\delta = \frac{a^3}{24EI_B}(8V + 3av)$$

where I_B is the second moment of area of the cross section of B.

The total deflection at the outboard end is

$$y = \delta + \delta_2 + a\phi_3 + 2a\phi_2 \tag{1}$$

This equation can be used to compute the deflection given the structural cross sections and loads.

Optimal Selection of Cross Sections

As was mentioned above, the goal of this section is to determine the optimal size ratio of section B to section A to minimize deflection for a given total weight. For the present purpose, pylon deflections are not relevant. The pylon deflection term ($2a\phi_2$) is left out of equation 1, and y is now interpreted as the deflection due to bending of members A and B only.

Thus,

$$y = \delta + \delta_2 + a\phi_3 \tag{2}$$

Substitution in equation 2 for the deflections in terms of the loads followed by nondimensionalization gives (a detailed derivation is presented in reference 25)

$$Y = \frac{\gamma\eta}{24}(1 + \lambda^2)\left\{(20R + 56L)(1 + \lambda^2) + 7 + 38\lambda^2 + 8L\frac{(1 + \lambda^2)}{\lambda^4} + \frac{3}{\lambda^2}\right\} \tag{3}$$

where $Y = y/a$, $R = U/W$, $L = V/W$, and $W = W_A + W_B$ is the total structural weight. $W_A = au$ is the weight of member A, and $W_B = av$ is the weight of member B.

It has been assumed here that the cross sections of members A and B are geometrically similar. That is, if h_A and h_B are corresponding sectional characteristic dimensions, then $\lambda = h_B/h_A$ and $I_B/I_B = \lambda^4$, and $v/u = \lambda^2$. η is a geometric parameter depending on the geometry of the cross section. For a particular cross-section geometry, let $n = I_A/h_A^4$. Similarly, let $c = u/\rho g h_A^2$ where ρ is the density of the material (assumed to be the same in both members) and g is gravitational acceleration. If the structural material is uniform, c is the ratio of the cross-sectional area to the square of the characteristic dimension. n and c are tabulated, along with I, for typical cross sections in Table 2. Then η is defined by $\eta = c_2/n$.

γ is a nondimensional number depending on the material properties. It is defined by $\gamma = \rho^2 g^2 a^4/EW$ where E is the Young's modulus of the material, again assumed to be the same for both members.

Differentiation of equation 3 with respect to λ to find extremal values of Y gives, when $dY/d\lambda$ is equated to 0,

$$(40R + 112L + 76)\lambda^8 + (40R + 112L + 45)\lambda^6 - (16L + 3)\lambda^2 - 16L = 0 \quad (4)$$

This equation can be solved as a quartic in λ^2 using Ferrari's solution. The results are plotted in Figure 15, which gives λ as a function of L for different values of R. It can be seen from Figure 15 that R has little influence on the values of λ. That is, the weight of the elbow joint and associated actuators has little effect. Also, when the load is large relative to the structural weight, as would be typical of small manipulators, the optimum value of the sectional dimension ratio λ is about 0.7. When the load is small relative to the structural weight, the optimal λ drops to about 0.43. The majority of manipulators have optimal values of λ between 0.6 and 0.7.

Design Procedure

A design procedure for optimal choice of section depths to meet a specified deflection constraint can now be formulated. It is assumed that a, U, and V are known, the section geometry has been chosen, and the material properties E and ρg are known. It is necessary to include pylon deflection for this purpose (return to equation 1). It is suggested that a pylon cross section and material, bearings for joint 1, and the distance between those bearings should first be selected. In this way, the effective stiffness k_P of the pylon with the respect to rotation in the vertical plane can be calculated:

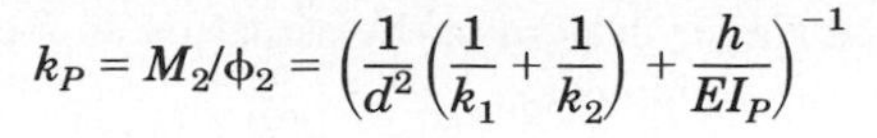

$$k_P = M_2/\phi_2 = \left(\frac{1}{d^2}\left(\frac{1}{k_1} + \frac{1}{k_2}\right) + \frac{h}{EI_P}\right)^{-1}$$

Equation 1 gives, when the deflections are substituted for in terms of the loads, and after substitution for section properties in terms of λ, a cubic equation in the total beam weight W:

$$AW^3 + BW^2 + CW + D = 0$$

where

$$A = \frac{a(3\lambda^2 + 1)}{k_P(1 + \lambda^2)}$$

$$B = \frac{a}{K_P}(4V + 2U) - Y$$

$$C = Q(1 + \lambda^2)(7 + 38\lambda^2 + 3/\lambda^2)$$

$$D = Q(1 + \lambda^2)^2 (20U + 56V + 8V/\lambda^4)$$

$$Q = \frac{a^4\rho^2 g^2 \gamma}{24E}$$

For a specified deflection, loads, and a chosen value of λ, the coefficients of this equation can be computed. Solution of the cubic gives the beam weight W, from which the sectional dimensions can be computed. The smallest positive root is the required solution. Since A, C, and D are always positive, and B must be negative (otherwise the deflection due to the pylon exceeds the specified deflection), there will, in general, be one negative and two positive roots, if all three roots are real. Note that if k_P becomes infinite, indicating that the pylon compliance can be ignored, the equation for W reduces to a quadratic that always has one positive and one negative root.

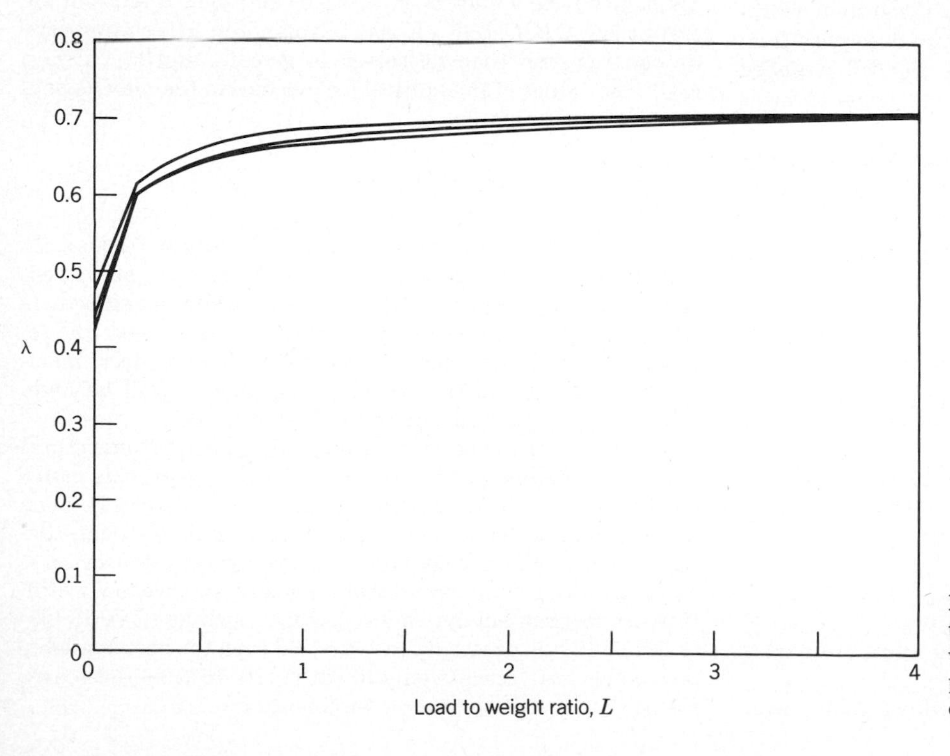

Figure 15. Plot of optimal structural cross section depth ratio for two-link structural model versus ratio of load to structural weight. This plot can be used to optimally select structural cross sections.

The procedure for designing the structure to meet a specified deflection is as follows:

1. Select the pylon cross section and material, the bearings for joint 1, and the distance between those bearings. Hence, compute the pylon stiffness k_P. Check that the deflection $y = 2a\phi_2$ resulting from the projected loads U and V is considerably less than the specified deflection.
2. Select a likely value of λ using Figure 13 and the expected range of L.
3. Specify the geometry of the cross section and compute η. Using the material properties, compute Q.
4. Compute the coefficients A, B, C, and D.
5. Solve the cubic equation. The smallest positive real root is W.
6. Use the calculated value of W and the specified values of U and V to compute R and L. Use Figure 15 to check the assumed value of λ. If necessary, use the computed value of λ and repeat steps 2 through 6 until the assumed and computed values of λ are consistent (no more than one iteration is usually required).
7. Use $h_A = (W/(a\rho gc(1 + \lambda^2)))^{1/2}$ to computer h_A and then $h_B = h_A$ to compute h_B. Hence, complete computation of the parameters defining each cross section.

Dynamic Deflection

The basic model shown in Figure 12 may also be used to study vibratory behavior. Here deflection of the members may be significant, but most often, the actuators contribute the largest compliances to the system. Therefore, for purposes of making a first estimate of the vibration modes and frequencies, the members are modeled as rigid with all compliance lumped at the joints.

Because of the geometry and symmetry of the system, vibratory modes in the plane of the manipulator are decoupled from movements out of that plane. For all cases, the full extended position shown in Figure 13 represents the worst case in terms of lowest natural frequencies.

One mode of vibration is a rigid body rotation of the whole arm about axis 1. The equation of motion of free vibration about axis 1 is then

$$J_1\ddot{\theta}_1 + k_1\theta_1 = 0$$

where

$$J_1 = \frac{Wa^2}{g}\left(\frac{1 + 7\lambda^2}{3(1 + \lambda^2)} + R + 4L\right)$$

and k_1 is the rotational stiffness of the actuator at joint 1. (Determination of actuator stiffnesses is discussed in a later subsection.) The natural frequency of vibration is

$$\omega_1 = \sqrt{k_1/J_1}$$

Turning now to the problem of in-plane vibration modes, the system model is shown in Figure 16.

The equations of motion for this model, which can be developed with about equal ease using either Newtonian or Lagrangian formulations, can be put in the form

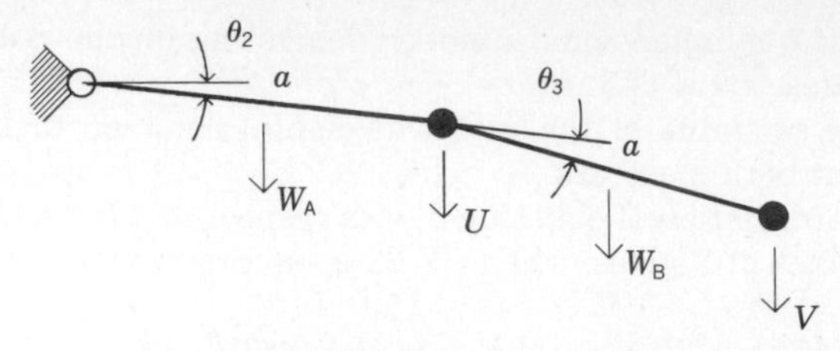

Figure 16. Model for estimating in-plane vibration natural frequencies.

$$A\frac{d^2\phi_2}{d\tau^2} + B\frac{d^2\phi_3}{d\tau^2} + K_2\phi_2 - K_3\phi_3 = 0$$

$$C\frac{d^2\phi_2}{d\tau^2} + D\frac{d^2\phi_3}{d\tau^2} + K_3\phi_3 = 0$$

where

$$A = \frac{2 + 9\lambda^2}{6(1 + \lambda^2)} + R + 2L, \quad B = \frac{\lambda^2}{2(1 + \lambda^2)} + L$$

$$C = \frac{5\lambda^2}{6(1 + \lambda^2)} + 2L, \quad D = \frac{\lambda^2}{3(1 + \lambda^2)} + L$$

and $\tau = t\sqrt{g/a}$, $K_2 = k_2/Wa$, $K_3 = k_3/Wa$. R, L, and λ are defined as in the preceding section.

The modal frequencies Ω_2 and Ω_3 are the positive real solutions of the equation

$$(AD - BC)\Omega^4 - (DK_2 + (A + C)K_3)\Omega^2 + K_2K_3 = 0$$

Thus, given the values of W, a, k_2, k_3 and λ, R, L as from the static case, A, B, C, D, K_2, K_3 can be computed. The characteristic equation can then be solved to give Ω_2 and Ω_3 (always real). The values of the natural frequencies in the dimensional time scale are given by

$$\omega_2 = \Omega_2\sqrt{g/a}, \quad \omega_3 = \Omega_3\sqrt{g/a}$$

The reason for estimating structural natural frequencies is that they should be substantially different to the servo bandwidth. Operation at values of the ratio of minimum structural natural frequency to servo cycle frequency of at least 3 is recommended (28). Because the model used here neglects member flexibility, the estimates of ω_1, ω_2, and ω_3 will be high. Therefore, it is appropriate to use higher values.

After the structure is fully defined, numerical structural dynamic modeling techniques should be used to provide better estimates of the lower natural frequencies. This can be done using standard finite element packages with elements designed to model joint compliances at the appropriate locations. An alternative, if the structural members are relatively stiff, is to use mechanism dynamics packages such as ADAMS (29) or IMP (30). It is necessary to warn that even the best available techniques are accurate only to within 10–15%, so these estimates should still be treated with caution.

Joint Stiffness

The procedure of the preceding subsection can only be used if the joint stiffnesses k_1, k_2, and k_3 are known. It is not difficult to estimate these stiffnesses for the more common types of actuator. Two actuator models will be discussed here: hydraulic actuators and electric servomotors with gear trains for speed reduction.

Consider the rotary hydraulic actuator shown in Figure 17. The volumes of the annular spaces on either side of the moving vane are, respectively,

$$V_1 = \frac{\theta_1}{2}(r_2^2 - r_1^2)w \quad \text{and} \quad V_2 = \frac{\theta_2}{2}(r_2^2 - r_1^2)w$$

where w is the axial width of the chamber. Since the bulk modulus K of the hydraulic fluid is defined by the equation

$$K = \frac{1}{V}\frac{dV}{dP}$$

where V is volume and P pressure, a small change in position $\delta\theta$ produces a torque

$$M = \frac{W(r_2^2 - r_1^2)}{2k}\left(\frac{1}{\theta_1} + \frac{1}{\theta_2}\right)\delta\theta$$

The actuator stiffness is $k = M/\delta\theta$ or

$$k = \frac{W(r_2^2 - r_1^2)}{2k}\left(\frac{1}{\theta_1} + \frac{1}{\theta_2}\right)$$

It may be seen from this that the stiffness is actually position-dependent. However, for the purpose of a first-order estimate of vibrational behavior, it may adequately be approximated by its minimum value, which is obtained when $\theta_1 = \theta_2$. In this case,

$$k = \frac{W^2(r_2^2 - r_1^2)^2}{kV}$$

where V is the total oil volume.

This model is only valid if the volumes of the lines between the control valve and the actuator can be neglected, and if those lines are sufficiently rigid that their compliance can be neglected in comparison to that of the hydraulic fluid. In general, actuator stiffness is inversely proportional to the active oil volume. Hydraulic linear actuators can be similarly modeled, except that there is then a nonlinear kinematic relationship between actuator displacement and joint position.

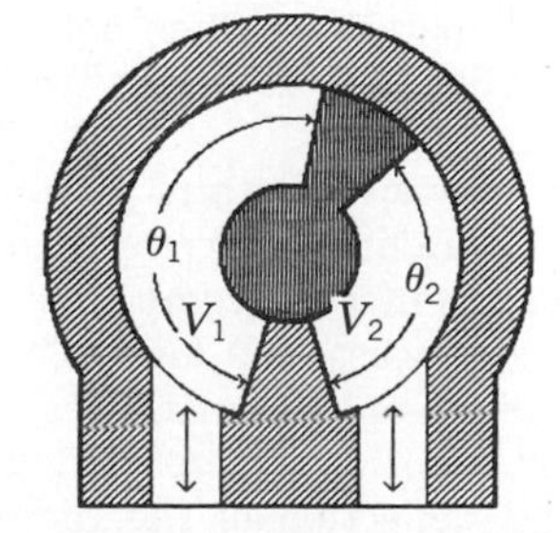

Figure 17. Rotary hydraulic actuator. Oil is pumped into either port and exhausted through the other. The rotating vane is attached to the output shaft and moves in response to the differential in oil pressure between the two chambers.

Turning now to the case of electric actuation, the low torque-to-weight ratio of an electric motor requires a speed-reducing transmission to produce adequate joint torque from a motor–transmission combination of reasonably light weight. The joint compliance can be regarded as being entirely due to the transmission. In fact, the final speed reduction is responsible for most of the compliance. The compliance can be computed recursively by using the equation

$$\frac{1}{k_N} = \frac{1}{k_N^*} + \frac{1}{R_N^2 k_{N-1}}$$

where k_N is the stiffness of the train as seen from the output shaft of the Nth speed reducer, R_N is the reduction ratio of the Nth reducer, defined as the ratio of input shaft speed to that of the output shaft, and k_N^* is the stiffness of the Nth reduction defined as that seen from the output shaft with the input shaft locked. It can be seen that the compliance of the reducer train inboard of reducer N is reduced by the inverse of the square of the reduction ratio of reducer N. Thus, if the final speed reducer has a high speed reduction ratio, compliance of the remainder of the train can be neglected.

For speed reducers such as harmonic drives, epicyclic gearheads, and so on, the manufacturer will usually provide compliance data, although bench testing of the chosen unit may be necessary to assure adequately accurate data. For purposes of estimating vibrational behavior, the stiffness of a gear pair with respect to rotation of the output gear is adequately approximated (31) by

$$k = C\,br^2$$

where b is the face width of the gears, r is the radius of the output gear, and $C = 2 \times 10^6$ psi ($C = 13{,}400$ MPa). Thus, the actuator stiffness for a joint with an electric servomotor and transmission is readily calculated.

Bearing Selection

The intermittent motion and frequent reversals, which are characteristic of joints in robotic systems in service, virtually mandate the use of rolling contact bearings. The loads seen by bearings in robotic systems vary from quite light to very severe. Horizontal rotary joints powered by rotary actuators see only the transverse shear load at that point in the arm, which is a comparatively light loading. However, some of the wrist joints, and certainly the first joint axis, if it is a vertical rotary joint, have much more severe loading conditions. Rotary joints actuated by linear actuators also carry much heavier loads since the joint torque is generated by a couple consisting of the actuator forces and a reaction at the bearings.

Sliding joints should also be arranged so that rolling elements can be used. Ball bushings are inexpensive and easy to mount. However, to prevent rotation about the bushing axis, it is necessary to arrange two sets in parallel. This arrangement becomes somewhat bulky. Roller assemblies, which are typically custom designed for the application, have been quite widely used in robotic systems.

Horizontal axis rotary joints see very little load and are very adequately supported by deep-groove radial ball bearings. Wrist joints also, which typically will have both radial and thrust loads, but for which the loads are relatively light, can be handled also with deep-groove radial ball bearings. The first axis represents a very different problem if, as is usual, it is a vertical rotary axis. Here heavy radial loads created by the bending moment of the arm and thrust loads generated by the weight of the arm, and of course also by inertial loading, must be carried. The bearings must also be designed with stiffness in view in this location. Tapered roller bearings provide a very good element for handling combinations of radial and thrust loads. Because the thrust load can be expected to be predominantly unidirectional, it is not necessary, in most cases, to use back-to-back tapered roller bearings, although it is certainly attractive to do so, particularly since this arrangement is readily preloaded to remove backlash and provide stiffness (Figure 18). In fact, all rolling element bearings in manipulators should be preloaded. Otherwise, their stiffness is very low. This is a result of the nonlinear load stiffness characteristic mentioned earlier.

If M is the bending moment in the manipulator at axis 1 and d is the distance between the axial planes of the tapered roller and cylindrical roller bearings, then each of them carries a radial load of M/d. If W is the vertical downward load from the weights and from any inertial loads on the system, then that is the thrust load to be carried by the tapered roller bearing. Of course, any inertial loads must also be included in M. For design purposes, it is necessary to estimate the worst case values of M and W.

Rolling element bearings have finite lives distributed statistically about a mean value that depends on the load carried and the total number of rotational cycles. Manufacturers' catalogues give the necessary equations to convert from a given set of service conditions to the standardized load ratings and lifetimes used in the bearing catalogue. However, in robotic service it is quite difficult to estimate the expected number of rotational cycles for a given service lifetime because of the undermined nature of the service to which an individual robot may be subjected. One way of doing this is to estimate a typical cycle time by assuming that a typical cycle consists of a specified number of moves, say, six to eight, and assuming that the moves are made either at maximum velocity or at some mean velocity, and then to divide the cycle time into the service lifetime expected for the bearings in hours and assume one revolution of each bearing per move. This is obviously a rough estimate, but should be adequate. It is likewise quite difficult to estimate loading on the bearings during each cycle. Worst case loads will typically be rare. A reasonable way to do this is to assume self-weight load only, assuming that the load is that exerted by the arm when fully extended as far as the inner joints are concerned.

Of course, since rolling element bearings do have finite life, and do require lubrication, it is necessary to make provision for both easy lubrication and reasonably easy removal and replacement of these bearings.

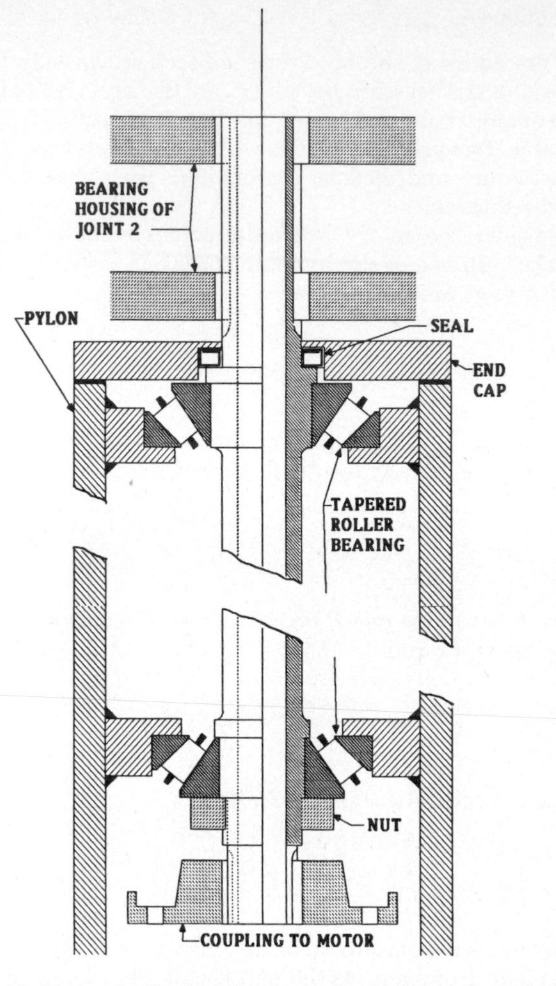

Figure 18. Tapered roller bearing arrangement for support of vertical first axis shaft. The bearings are preloaded against each other by the shaft nut to remove backlash. The shaft is hollow to increase longitudinal compliance for this purpose, but retains high stiffness in bending and torsion. Care must be taken when sizing elements to ensure that they can be assembled.

Composite Materials

Composite materials offer stiffness-to-weight ratios almost an order of magnitude greater than for metals (Table 3) (32). They are, therefore, very attractive for use in robot structures. However, their directional properties, and the necessity to minimize machining operations when using them make them considerably more difficult to deal with from the design point of view.

From the point of view of static deflection, the primary deformation mode is bending in the vertical plane. Therefore, the layup of a fiber-reinforced composite section should be designed primarily to resist this load. This means that the primary layup of the fibers will be at 0 deg to the longitudinal axis of the member. Another type of deflection that must be resisted is torsional deflection of the member due to offset loads in the hand or to angulation of a finite length hand at the wrist. Torsional deflection is resisted by a layup at plus or minus 45 deg to the longitudinal axis of the member. Figure

Table 3. Properties of Some Components of Composite Materials

Material	Density, lb/in.^{3} (g/cm^3)	Tensile Strength, ksi (MPa)	Tensile Modulus, $\times 10^6$ psi (MPa $\times 10^3$)	Specific Stiffness, $\times 10^7$ in. (m^2/s^2 $\times 10^6$)	Specific Strength, $\times 10^6$ in. (m^2/s^2 $\times 10^6$)
E Glass	0.092 (2.55)	500 (1430)	10.5 (30.1)	11.4 (11.8)	5.6 (0.56)
S Glass	0.090 (2.50)	650 (1870)	12.6 (36.2)	14.0 (14.5)	7.2 (0.75)
Graphite HS (high strength)	0.054 (1.50)	400 (1150)	40 (115)	74.2 (76.7)	7.4 (0.77)
Graphite HM (high modulus)	0.054 (1.50)	270 (770)	77 (220)	143 (147)	5.0 (0.51)
Boron	0.085 (2.36)	500 (1430)	55 (160)	64.7 (67.7)	4.7 (0.48)
Kevlar	0.052 (1.44)	525 (1510)	18 (51.7)	34.7 (35.9)	10.1 (1.05)
Epoxy	0.045 (1.25)	4–15 (12–43)	0.4–0.5 (1.15–1.45)	1–1.2 (1–1.2)	0.01–0.34 (0.001–0.035)
Steel (1040)	0.284 (7.83)	51 (170)	30 (83)	10.6 (10.6)	0.18 (0.02)
Aluminum (2024)	0.097 (2.70)	47 (120)	10 (30)	10.3 (11.1)	0.48 (0.04)

19 shows a design of a composite member suited to use in a manipulator. It is similar to designs discussed in reference 33. The zero-degree fiber layup as applied as a continuous belt around the outside of the member. Transverse shear loads are resisted by a high-density foam core. Load transfer to and from the member is via metal inserts. A thin, zero-degree ply may be laid on the lateral faces of the member to improve bending stiffness in the lateral direction, if dynamic compliance in that direction should prove to be of concern. The ±45-deg plies are applied alternately over the other plies. The cross section used for designing for stiffness in the vertical plane bending direction is simply the two cross sections of the zero-degree belt on the upper and lower surfaces of the member. The modulus of elasticity E is estimated from

$$E = E_f p + E_m(1 - p)$$

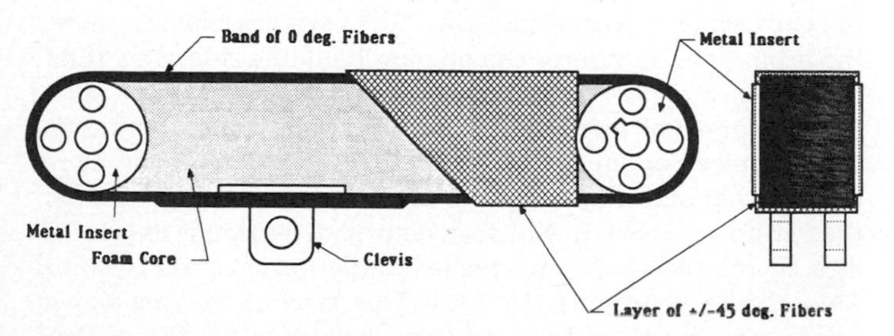

Figure 19. Schematic of wound-fiber composite member for manipulator. Loads are applied through metal inserts. Care must be taken to ensure integrity of the interfaces between metal and composite. These interfaces can be the sites of cracking due to differential thermal expansion during curing. They are also stress concentrators.

where E_f is the modulus of the fibers, E_m is that of the matrix, and p is the volume fraction of the fibers (32).

The volume fraction of the fibers would typically be about 70%. The most attractive reinforcing fiber materials are Kevlar and high-modulus graphite. The Young's modulus of high-modulus graphite is 77×10^6 psi (530,000) MPa and that of Kevlar is 18×10^6 psi (125,000 MPa). The modulus of the matrix varies considerably with composition, but for an epoxy would typically be about 44×10^4 psi (3000 MPa). Properties of typical materials are presented in Table 3. The design procedures used for optimizing the cross-sectional areas of the members and for designing the sections to meet a given deflection requirement described above can be applied by using this cross section and elastic modulus information.

The torsional deflection produced by a torsional moment M can be expressed by the equation (27)

$$\phi = \frac{Ml}{GI_P^*}$$

where l is the length of the beam, G is the shear modulus, and I_P^* is an equivalent sectional polar second moment of area, which is defined as follows:

$$I_P^* = \frac{A^2}{2\pi I_P}$$

Here A is the cross-sectional area and I_P is the actual sectional polar second moment of area. In the case of a cylindrical cross section or annular cross section, I_P^* becomes identical to I_P.

As was the case with in-plane bending stresses, the torsional stresses are considered as being resisted only by the

±45-deg fiber plies. Thus, the cross section that should be used in calculating A and I_p^* is the cross section of these plies alone.

The shear modulus G can be estimated from the Young's modulus calculated for the composite material as outlined above by using the relationship $G = E/2(1 + \nu)$ where E is Young's modulus and ν is the Poisson ratio of the material. Estimation of the Poisson ratio for a composite material is very difficult. However, since it varies only between 0 and 0.5, a reasonable estimate can be used for design purposes. If ν is equated to 0.5, which is the Poisson ratio given by an ideal incompressible material, a conservative estimate of G is obtained in the sense that it is lower than the actual value. Hence, the deflection estimated will be larger than the actual deflection, which is appropriate for design purposes.

The metal inserts needed to install bearings and to transfer loads to and from the member pose several problems. Since high-strength matrix materials must usually be cured at an elevated temperature, the difference in thermal expansion moduli between the metal insert and the composite material can lead to problems such as cracking within the composite material created by frozen thermal stresses. If this proves to be a difficulty with a particular design, it may be necessary to lay up the part with an insert, such as a Teflon plug, in place of the metal insert, until curing is complete. The dummy insert may then be removed, and the metal part cemented in. The recommended procedure is to perform boring and machining operations on the metal inserts after the part is fully cured and assembled, to ensure that dimensional tolerances can be accurately held. If actuation torques are applied to the member via the metal insert, it is necessary to design the interface between the insert and the composite material to have adequate strength to effect this load transfer. On a cylindrical interface surface, the shear stress τ is given by

$$\tau = \frac{M}{rA}$$

where M is the moment to be transferred, r is interface radius, and A is interface area. This stress should be within the allowable shear strength of the bonding material, adjusted by a suitable factor of safety to compensate for stress concentration effects. Likewise, when actuation loads are transferred to the member from a linear actuator by means of a clevis, such as that shown in Figure 19, the interface between the clevis and the composite plies should be designed to withstand the combination of normal and shear load, which is transferred. Once again, the critical strength is that of the bonding material, which is usually the matrix of the composite. In this case, it may be necessary to provide increased thickness of the composite material with plies equally distributed in all directions to provide properties in the composite near the interface that are approximately isotropic. In this way, stress concentration effects are diminished.

Composite members are somewhat less convenient to deal with than tubular metal members since they have a core consisting of a lightweight material such as structural foam. It is, therefore, impossible to run cables, hydraulic lines, or shafts through the interior of the members. Likewise, it is harder to mount some of the actuators inside members. However, it is possible to mold, or drill, relatively narrow passages into the core material to permit routing of cables through the member, without compromising strength. Composite materials can be expected to become increasingly important as manipulator performance limits are pushed higher.

Once again, the treatment provided here is directed toward conceptual design. Once a geometry has been defined, it is very important to analyze it using finite element modeling. Although several finite element packages accommodate orthotropic material properties and are, therefore, in principle suitable for analyzing composite components, some are very restricted in the types of element that can used. To model varying layup directions, it is necessary to calculate orthotropic material constants for each element individually. This can be done in a modeling package such as PATRAN G. The most versatile package for analyzing the resulting model is NASTRAN.

Structures for End-point Sensing Systems

If end-point feedback, providing a direct measure of the positional relationship between the hand and the workpiece, is available, a completely different design philosophy may be followed. In this case, deflection of the structure is not important since the relative positions of the end effector and the workpiece are directly measured. Therefore, static stiffness is no longer important. Design of such systems is done on a strength basis. As a consequence, the structures used in this type of system are much more compliant than those used when positioning is done in an open-loop manner. Likewise, actuation, power transmission, and speed reduction systems of relatively high compliance may be used. These include cable and band drives.

Teleoperation systems are the classic example of this class of manipulation structure. Here the operator's eye–hand coordination provides the end-point feedback needed to directly relate hand and workpiece position.

A number of autonomous end-point systems are now also in service. Typical examples are welding systems in which the seam is optically tracked (18). These systems are usually fitted with two-degree-of-freedom small-amplitude, high-bandwidth, end-point positioning systems permitting correction of the welding electrode path to follow the seam, as distinct from the preprogrammed path that approximates it. The welding arc may also be controlled by feedback from the optical system. Another example is the sheep-shearing robot developed by the University of West Australia (19). In systems such as these, the primary manipulator can be compliant and relatively inaccurate. Correction of errors is performed by means of the additional degrees of freedom and the end-point sensing system.

A simpler system configuration, from the mechanical point of view, that also allows design on the basis of strength rather than stiffness, is one in which structural deflection is measured directly relative to the undeflected configuration of the system. This can be done optically (31). This type of system allows active compensation for structural deflection by the control system.

OPTIMAL MASS DISTRIBUTION

The equations of motion, and other relationships needed in control system design, can be greatly simplified by proper distribution of weight among the members. The most important part of the manipulator, from this point of view, is the regional structure, and the optimal distribution of weight in this part of the manipulator has been exhaustively studied in reference 42. Conditions under which the inertia tensor appearing in the equations of motions is diagonalized, and also conditions under which it is invariant, were investigated. It is impossible to review here all of the different conditions pertaining to the various geometries that were found in that study. However, if it is assumed that the form of the regional structure for a rotary joint manipulator (Fig. 3) is optimal and that the centers of mass of members 2 and 3 lie in the plane through the intersection of axes 1 and 2 normal to joint 2, and the center of mass of axis 1 lies on that axis, then the optimal mass distribution is such that the center of mass of member 3 lies on joint axis 3, and the center of mass of the combined members 2 and 3 lies on joint axis 2. Also, the moment of inertia about the principal axis through the center of mass of member 3 parallel to the x axis of frame 3 is equal to that about the y axis of that member. If these conditions are satisfied, the inertia matrix is not diagonalized, but it is invariant, and in fact, it has only five nonzero elements. Further, the Coriolis centrifugal and gravitational terms disappear from the equations of motion. Obviously, the presence of an active wrist somewhat upsets this optimal mass distribution if it is used. However, if applied to lightly loaded high-speed systems, and making a suitable assumption about a lumped mass at the wrist, it is of value to attempt to meet the optimal mass distribution. The conditions on the locations of the centers of mass of members 2 and 3 can be met by counterweighting or, preferably, by appropriately placing the actuator masses. The most awkward condition to meet is the condition that the x and y axis moments of inertia of member 3 be equal, since one of these is a longitudinal axis moment of inertia and the other is a transverse axis moment of inertia. It is difficult to meet this condition without adding mass to the system, which is, of course, undesirable for a high-performance system.

ACTUATOR SELECTION

Three basic types of actuation system are in more or less widespread use in manipulation systems. These are electrical, hydraulic, and pneumatic systems (see ACTUATORS). The critical parameter, for robotic application of an actuator, is the ratio of actuation force to actuator weight for a linear actuator or the ratio of actuation torque to the product of weight and frame dimension for a rotary actuator. This is referred to as the force-to-weight ratio.

Electrical actuators have a rather poor force-to-weight ratio. In compensation, electric motors can be run at high speed to produce a respectable power-to-weight ratio. It is possible to reduce speed to that required at the joint by using a mechanical speed reducer so that the entire package has a good force-to-weight ratio. There is a cost associated with using speed reduction in terms of dynamic response, and that is discussed later. Electric actuation is clean, does not require bulky power supplies, is moderate in its maintenance requirements, and is flexible in application. Consequently, electric actuation is the system type of choice in the bulk of manipulation systems.

Hydraulic actuators offer the highest force-to-weight ratios available using current technology. Consequently, hydraulic actuators can be applied directly at the joints, without speed-reducing transmissions. Hydraulic actuation is the system of choice in heavy-duty manipulation systems with actuation requirements beyond the capabilities of electric systems.

Pneumatic actuation is inexpensive and is compatible with much light manufacturing, fixed automation technology. The power supply is convenient, requiring only a system of compressed air lines, which is usually already present. Unfortunately, pneumatic actuators have relatively low force-to-weight ratios and high compliance. This can be countered, to some extent, by using air motors operated at high speed, in combination with mechanical speed reducers. This is, essentially, the same solution used in the electric actuation case. However, at the present time, pneumatic actuation is limited to light-duty pick-and-place-type devices, using mostly simple on–off-type control elements.

Electric Actuation

The first task in actuator selection is the determination of the maximum torque and speed requirements seen by each actuator, or more effectively, the maximal torque–speed profile. For example, if joint 2 of the configuration shown in Figure 3, which is one of the most demanding actuation locations, is used, the maximum rate is encountered when the reference point is at minimum distance from axis 2 and is translated at maximum speed in a directional normal to that axis. The maximum static load at that joint occurs when the arm is fully extended in the horizontal direction under maximum load. However, the maximum dynamic loading may occur either at that position or when the reference point is at minimum distance from axis 2. This somewhat surprising effect is an artifact of the way robotic coordination software is usually set up. If, for example, a rate-control type algorithm is used, the reference point is translated at constant velocity for the majority of each movement. It is accelerated to that velocity in a small number of computation cycles at the beginning of the move and is decelerated in the same number of cycles at the end of the move. Thus, the angular acceleration of the joint is inversely proportional to the distance of the reference point from the joint axis. The moment of inertia of the arm about that axis decreases as the reference point is brought closer to it, but at a slower rate. In fact, it is characterized by two terms, one of which is constant and the other proportional to the square of the radial distance to the reference point. The constant term corresponds to the moment of inertia of the arm about axis 2 when it is fully folded at axis 3. The net result is that the inertia torque at axis 2 increases nonlinearly as the reference point approaches that axis. Most current manipulators are designed for reference point accelerations that are two to five times gravitational acceleration. For this range of accelerations, the inertia torque dominates the static load at this axis. Consequently, the worst case for actuator torque often occurs when the reference point is at minimal distance from the axis, and with the radius from axis 2 to the reference point horizontal. In this position, the reference point is accelerated to maximal velocity in the vertical upward direction in the specified number of computation cycles. The arm is assumed to be carrying a maximal load at the reference point.

Electric servomotors have torque–speed characteristics similar to those shown in Figure 20. At maximal excitation, the characteristic can be represented by the equation

$$\frac{\omega}{\omega_0} + \frac{T}{T_S} = 1$$

where ω_0 is no-load speed and T_S is stall torque. In terms of the output power,

$$P = T\omega = T_S\omega\left(1 - \frac{\omega}{\omega_0}\right)$$

The maximal power given by the system is then

$$P_m = T_S\omega_0/4$$

which occurs when $\omega = \omega_0/2$ and $T = T_S/2$.

In order to size the motor, it is necessary to recognize that the software is going to command a specified acceleration–time profile and to ensure that the actuator can follow that profile. If a constant acceleration from zero to the maximal reference point speed is assumed, the critical instant occurs at the moment at which the reference point reaches that speed. The maximum angular velocity at the joint occurs when the arm is at its innermost position. For the moment, it is assumed that this is also the worst case with respect to load torque. The angular velocity is then $V/r = \omega^*$ where V is the reference point speed and r is the minimum radius. The torque is $T^* = T_L + IV/(r\tau)$ where T_L is static load torque, including effects of both arm and load weight, I is the minimal moment of inertia about axis 2, and τ is the time allowed for acceleration. Given that a speed reducer will be used to match motor speed to joint speed, the important parameter is the operating point power $P^* = T^*\omega^*$. If η is the estimated transmission

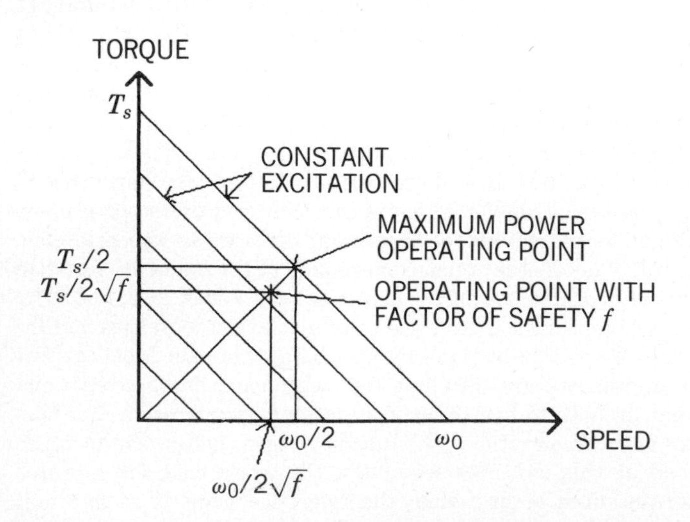

Figure 20. Idealized electric servomotor characteristics. Each line represents a constant level of excitation (command signal). T_s is stall torque and ω_0 is no-load speed. Output power varies parabolically with speed.

efficiency and f is a suitable factor of safety (say, 1.2–1.5), a motor can be selected by using the relationship

$$\omega_0 T_S = 4fP^*/\eta$$

with a table of manufacturer's data that will give no-load speed ω_0 and stall torque T_S for each motor size. This, of course, assures that the maximum power operating point of the motor is matched to the critical operating point of the joint. The required speed reducer ratio R is then determined from the relationship $R = (\omega/2\omega^*)\sqrt{\eta/f}$. The operating point is plotted in Figure 20.

This procedure is usually appropriate, even when the maximum joint torque condition occurs when the arm is fully extended. In this condition, the joint speed is relatively very low. Reference to Figure 20 indicates that, if joint speed is low, the torque can be significantly higher and still give an operating point within the maximal excitation characteristic. Thus, the maximum power condition still corresponds to the inner critical position.

It is often necessary to vary the above procedure, for a couple of reasons. First, it is usually desirable to avoid very large speed reduction ratios. The reflected inertia of the motor armature, as seen from the joint axis, increases as the square of speed reduction ratio. This, in turn, degrades dynamic response. Thus, it is often preferable to oversize the motor somewhat and operate lower on its speed range to reduce the required speed reduction. This may carry a further benefit in reducing electrical resistance losses. To reduce the critical operating speed to $k\omega_0$ where $0 < k \leq 0.5$, the factor of safety f is set equal to $\eta/4k^2$. The operating point torque is then kT_S, and the motor power at the operating point is $k^2\omega_0 T_S$ (Figure 20).

A second reason for varying the above procedure is heat dissipation. In manipulator service, motors must frequently operate in near-stall or stall conditions. Packaged motors intended for torque motor service do not usually present overheating problems in these conditions. However, there is a tendency to use high-technology, rare-earth motors supplied in component form. Motors of this type offer substantial performance benefits over conventional servomotors, and the component kit form allows the designer great flexibility in building the motor into a compact structure. An important feature is availability in brushless configurations since brush friction is a significant component of the friction torque seen at the joint. However, the full torque and speed capabilities of the motor may not be utilized without creating heat dissipation problems. In high-performance systems, peak loads are short-duration dynamic loads and can safely be allowed to approach the peak motor torque at the relevant speed. However, static loads must be kept below a level sustainable with acceptable temperature rise. Because the heat transfer characteristics of geometrically complex housings are difficult to predict, and because motor resistive losses are, to some extent, affected by the mode of speed control used, experimentation, or at least numerical simulation using finite element models, may be necessary to determine the appropriate motor size. The costs of this experimentation or simulation, in addition to the already high component costs of rare-earth motors, make it an expensive alternative. Nevertheless, because, as noted above, electric servo motor performance tends to be rather marginal for robotic service, rare-earth motors have become very popular in either packaged or component form.

Carried to its logical extreme, the idea of using oversized motors with low speed reduction ratios to improve dynamic response by avoiding large reflected inertia leads to the idea of direct drive, with no speed reducer. This has been an active area of recent research (36,37) and has led to commercial implementation in a SCARA geometry in which there are no heavily loaded rotary joints. The main problems with this approach are insufficient torque, even using the best available motor technology, and the large actuator compliance. This compliance is electromagnetic in origin. The motor field cannot hold the armature rigidly in place against a load torque. Electric actuator compliance is not usually noticed because of its reduction by the square of the speed reduction ratio when a speed reducer is used. There have been major recent advances in electric motor technology. It is quite possible that further advances will make direct electric drive viable for widespread application. For the moment, however, an intermediate approach using moderate speed reduction ratios seems most promising.

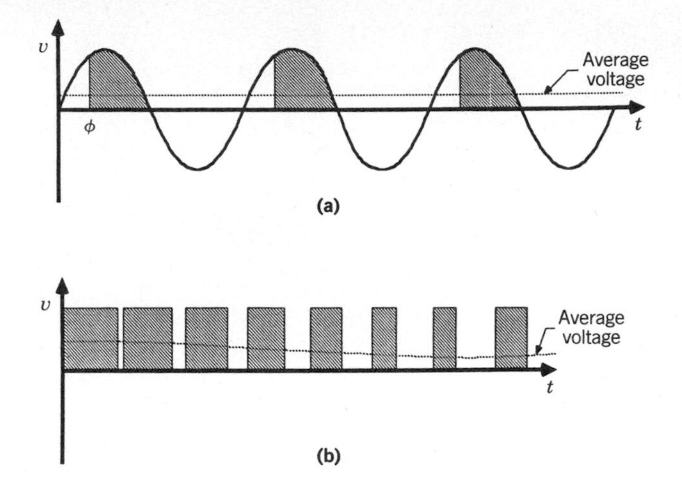

Figure 21. Principles of operation. **(a)** Phase control. **(b)** Pulsewidth modulation. Both operate to vary the average motor voltage.

Electric Controllers

In general, the electric power controllers used for manipulators fall into two classes. The first is the linear amplifier using power transistors. The second encompasses several kinds of time-controlled switching device.

In a linear amplifier, the back emf from the motor is, effectively, placed in series with a variable resistance. The supply voltage acts across this combination. The armature current drawn by the motor is proportional to the torque it is producing. The back emf is proportional to motor speed. Hence, the difference between the back emf and the supply voltage is applied across a resistive load. One result of this is that the power used by the actuator–controller combination is the product of the supply voltage and the armature current regardless of speed. That is, the net power used varies with torque, but not with speed. The difference between this power and the mechanical output power is resistively converted to heat. At low speed and high load, most of the heat is generated at the controller, but heat generation in the motor is also important because of the resistance of the armature coils. The latter is more serious since convective cooling, the most important means of cooling the armature, is ineffective at low speeds. Since, as previously remarked, manipulator actuators must operate for a significant portion of their operational time at near-stall, or stall, both the power cost and the accompanying heat generation problem associated with this arrangement are a concern.

The use of fast solid-state power switching devices, such as silicon control rectifiers, provides an alternative to the use of linear amplifiers. Phase control is a system in which an a-c waveform is fed to an electronic switch that can be turned on at a controlled phase interval after the positive-going zero crossing of the supply waveform. The switch stops conducting at the negative-going zero crossing. The result is a series of pulses of controlled width, as shown in Figure 21**a.** As a result, the *average* voltage supplied to the motor is varied. A full-bridged rectifier arrangement can be used to take advantage of both halves of the a-c waveform to supply a d-c servomotor. Alternatively, both halves can be phase controlled without inversion to supply an a-c servomotor.

Pulsewidth modulation (PWM) is a somewhat similar scheme applied to a d-c supply. PWM controllers provide relatively high dynamic response, but are only available for relatively low powers. In this case, solid-state switching is used to break the constant d-c supply voltage into a series of pulses of controlled width, as shown in Figure 21**b.** Again, the effect is to vary the average supply voltage to the motor.

These schemes eliminate the large resistive power dissipation inherent in the linear amplifier controllers and hence lead to lower system power demand. However, the pulse trains produced have a large content of high-frequency harmonics to which the motor cannot respond. These consequently contribute disproportionately to resistive losses. As a result, resistive dissipation in the motor, and the associated heat transfer problem, may be greater when using switching controllers, as compared to linear amplifier controllers.

Linear Actuation

In joints for which the required rotational range is on the order of 150 deg or less, linear actuation becomes possible, and even attractive. If electric actuation is used, the usual means of performing linear actuation is by means of a ball screw drive system. Extreme positions of the actuator should be arranged as shown in Figure 22**a.** These positions are those in which, for given joint torque and speed, the torque required of the motor is at a maximum. Figure 22**b** shows the position in which the motor torque is at a minimum and motor speed is at a maximum. At a given output power, that is, for a given product of joint torque and speed, of course the motor torque—speed product is also that power. This means that the maximum motor torque case corresponds to the minimum motor speed case. Of course, as was noted above, maximal load and speed conditions are a function of joint position. The worst case usually corresponds to one of the two positions in Figure 22**a.** Motor selection can be performed by power matching at

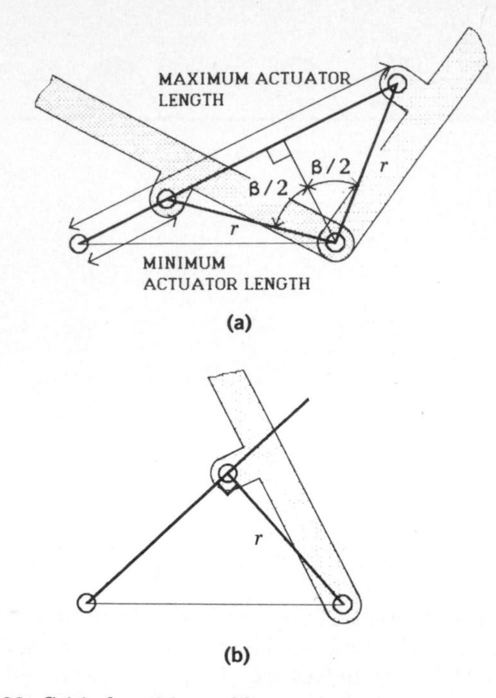

Figure 22. Critical positions of linear actuator operating on rotary joint. **(a)** These two positions are those of maximum actuator force and minimum actuator speed, for a given load and speed. **(b)** This position is that of minimum actuator force and maximum actuator speed.

that position. The required crank radius r for a given pitch p of the ball screw can then be determined since the effective speed reduction ratio is $p/(r \cos(\beta/2))$ where p is the pitch and β is the angular range of the joint.

Speed Reducers

It has been assumed in the foregoing that the motor can be run in its optimal speed range and that a speed-reducing transmission is used to reduce the motor speed to the desired joint speed. This, in fact, requires a quite large speed reduction ratio in most cases. This, in turn, implies a considerable loss of power in the speed reducer since, broadly speaking, mechanical losses in a speed reducer increase with reduction ratio. It is therefore very important to take account of the efficiency of the speed reducer when estimating the load torque to be seen by the motor.

Aside from good mechanical efficiency, the primary requirements for a speed reducer in robotic service are compactness and low compliance and backlash. Backlash is particularly troublesome since it causes nonlinearities in control and can lead to instability. Thus, robotic transmissions must usually be preloaded to eliminate backlash. Compliance in the speed reducer likewise detrimentally affects system dynamic response.

For rotary actuated joints, the most commonly used speed reduction systems are multiple-stage spur or helical gear trains, epicyclic gear trains, or harmonic drives. Conventional multistage gear trains are favored wherever the space is available to accommodate them. This is because they are simple, efficient, inexpensive, and comparatively stiff and strong. Epicyclic gear trains give a large speed reduction in a compact package, but have limited applicability in robotic service because, when packaged in gear heads, as is commonly done, the speed reducer may not be capable of handling the full torque that the motor can produce, and the compliance is likely to be high. Harmonic drives are quite commonly used since they give a large speed reduction in a vary compact package. They also are virtually backlash free. Harmonic drives are, however, quite compliant. They also have only modest mechanical efficiency, in the range of 70–80%.

As already pointed out, in a multistage gear train the final stage is far more important than the earlier stages in determining the overall compliance of the train. An equation was given permitting calculation of the compliance of a spur or helical gear train.

As was mentioned above, for linear actuation the usual element is a ball screw. This is a highly attractive speed-reducing system for robotic service since it is very stiff and is readily preloaded to eliminate backlash. At the same time, very high speed reductions are obtained in a very compact package. Thus, it is attractive to use ball screw linear actuation wherever the joint motion ranges are sufficiently restricted to permit it.

Hydraulic Actuation

A conventional hydraulic actuation system consists of a pressure-regulated hydraulic supply that provides essentially a constant pressure drop from the supply line to the reservoir return line, a hydraulic servovalve that controls the flow of hydraulic oil to the actuator, and a rotary or linear hydraulic actuator (38). This type of circuit is shown in schematic form in Figure 23. Hydraulic actuators have very high force-to-weight, or torque-to-weight, characteristics. It is, therefore, not necessary to use speed reducers. Consequently, hydraulic actuation systems can be relatively compact and lightweight for comparable load conditions.

Hydraulic actuators are sized for maximum load. Thus, if a rotary actuator is used (Figure 17), the actuator should be sized to carry the maximum joint torque divided by a suitable factor of safety. The maximum torque that can be developed by the actuator is the product of the supply pressure drop, the width of the actuator vane, and half the difference between the square of the outer casing radius and the inner casing radius. The required actuator torque determines the cross-sectional volume of the actuator and hence the flow rate required from supply to meet the velocity required. The power supply and the control valve must be sized to be capable of providing that flow rate without dropping the regulated supply pressure. The factor of safety used should be kept as low as possible since, if the actuator is oversized, the required supply flow is increased by the same amount, and because of the constant supply pressure drop, the power that must be produced at the supply is also increased. The output power remains the same, so the result is to increase the heat produced at the servo valve.

For linear actuation, the maximum actuator force for a desired joint torque corresponds to the positions shown in Figure 22**a.** Therefore, the actuator cross section must be sized to provide this force. The load that can be carried by the actuator

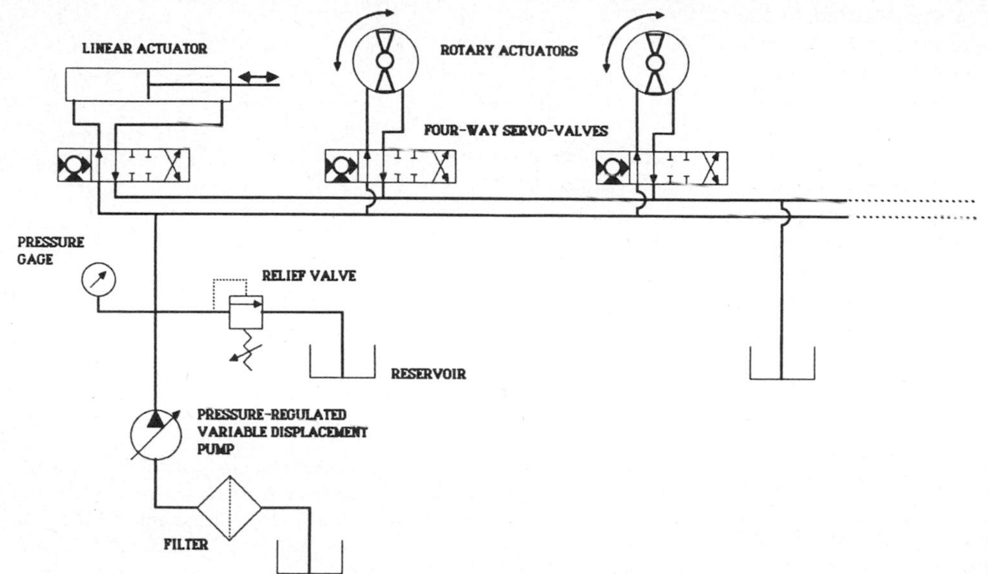

Figure 23. Schematic of typical hydraulic circuit for robotic actuation. A single pressure-regulated pump supplies all actuators. Control is via four-way, spool-type, electrohydraulic servo valves.

is the supply pressure drop times the effective cross-sectional area of the cylinder. If a one-sided cylinder goemetry is used, that is, if the piston rod extends out only one end of the cylinder rather than both ends, the cross section of the tube less the cross section of the piston rod should be used when calculating maximum load capacity. When the cylinder cross section has been determined to accommodate the loads in the positions shown in Figure 22**a**, the maximum actuator rate is determined in the position shown in Figure 22**b**. The cross section of the cylinder, and this joint rate, then determines the maximum flow rate that must be provided. When determining the maximum flow, the net cross section of the cylinder must be used, not the cross section reduced by the piston rod cross section.

When loads in one direction predominate, as is usual in manipulator shoulder and elbow joints, it is preferable to use the cylinder in compression since the larger area is available for expansion against the load. However, the ratio of barrel length to stroke should be relatively large in this arrangement to counter any tendency to column instability.

It has been noted in an earlier section that the compliance of the hydraulic actuator is determined by the active volume of the oil in the actuator. This means the volume on the actuator side of the control valve including the supply lines from the control valve to the actuator. A means of estimating this compliance was contained in that section.

Power Supplies

As has been mentioned, almost all hydraulic robots operate off a pressure-regulated supply (Figure 23). This is accomplished by means of a supply pump that has a servo regulator to attempt to maintain constant pressure at its outlet regardless of the output flow. Control of each actuator is accomplished by means of an hydraulic servo valve, which throttles the flow to control the actuator, and which is also capable of reversing the flow. The effect of this form of control is that the portion of the supply pressure drop that is not needed to support the load is dropped across the control valve. This can be very wasteful of energy in light-load, high-rate conditions in which almost the entire supply pressure differential is being dropped across the control valve. Thus, hydraulic systems of this type operate efficiently in low-rate, high-load conditions, but are wasteful of power in low-load, high-rate conditions.

Remote Actuation

Since actuators are relatively heavy, it is advantageous to place them as far inboard as possible. This is particularly true of the actuators for the first three joints, which tend to be relatively massive. In a manipulator with only rotary joints, such as in those configurations shown in Figure 3, actuators 1 and 2 present no problem since they are naturally mounted at the shoulder. Actuator 3 can also be placed at the shoulder with the use of a simple connecting rod arrangement, as shown in Figure 24. This is an elementary example of remote actuation.

Alternatively, the actuator for joint 3 may be mounted to

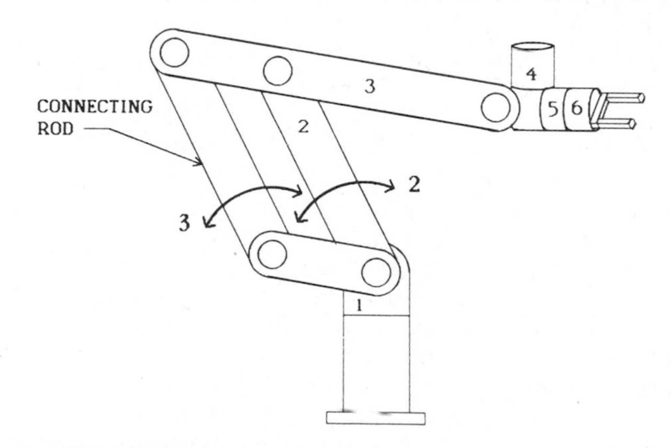

Figure 24. A simple example of remote actuation. Actuator 3 moves the connecting rod relative to the crank, which, along with member 2, rotates about the horizontal joint axis 2. Because of the parallelogram geometry, actuator 3 controls the angle of joint 3, even though it is not mounted to that joint.

the inboard end of member 2 with power transmitted to the joint via a shaft and bevel gear arrangement or by a roller chain (39). The gravitational unidirectional loading on joints 2 and 3 is often used to remove backlash in power transmission systems at those joints.

The wrist joints present a much more complex problem from the point of view of remote actuation. In a roll–pitch–roll wrist arrangement, the actuator for joint 4 can be placed inside member 3 and as far inboard as possible, and the joint can be driven by means of a shaft. Figure 25 shows an arrangement by which axis 4 in a pitch–yaw–roll-type wrist can be driven from an actuator mounted at the shoulder of the manipulator. Here the motion is transmitted to the joint by means of three roller chains, one mounted on the crank that drives the elbow joint as in Figure 24, another on the connecting rod, and the other in member 3. An important feature of this mechanism is the parallelogram formed by the elbow drive crank, the connecting rod, and members 2 and 3. This parallelogram automatically subtracts the elbow motion from the motion transmitted along the roller chains to joint 4. It is important to note that the configuration shown in Figures 24 and 25 is, despite its appearance, a simple serial chain manipulator configuration, not a parallel configuration. The elbow drive crank and connecting rod are added to the primary structure, which is the same as that of Figure 3**b**, solely for purposes of remote actuation.

Remote actuation of axes 5 and 6 is difficult because it is necessary, in order to gain any benefit, to mount the actuators at the inboard end of either member 2 or member 3. Thus, the motion must be transmitted past several independently moving joints. Mechanisms capable of doing this are complex and compliant. It is possible to configure a series of nested tubular shafts and nested bevel gear differentials at the joints that transmit motion from shoulder mounted actuators right to the wrist joints. However, this arrangement is extremely expensive and, because of the many gear contacts, is quite compliant. Another system that has been successfully used in situations in which actuator compliance is not a problem is a set of band drives (40). Here, speed reduction is done by means of block and tackle mechanisms. This technology has been applied very successfully in teleoperation systems.

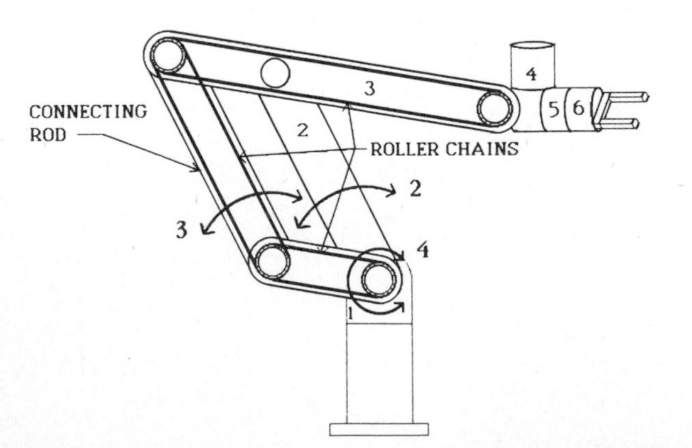

Figure 25. Remote actuation of joint 4. The actuator drives a sprocket mounted on axis 2 relative to member 1. Idler sprockets transmit the motion at the joints at either end of the connecting rod. The parallelogram geometry removes coupling with the motions of other joints.

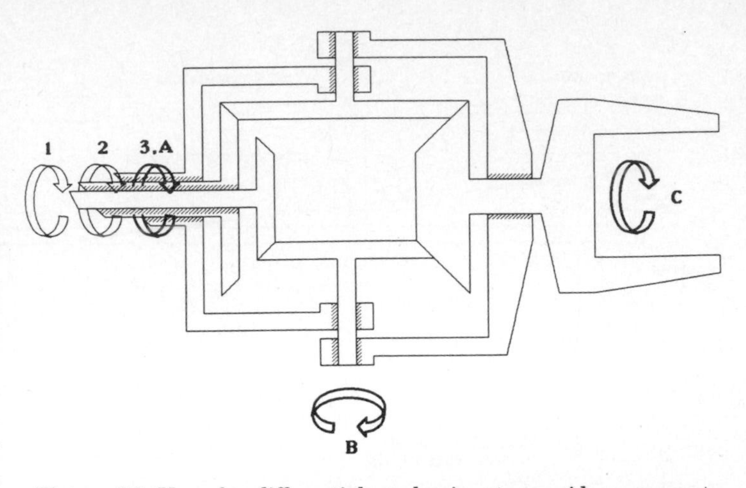

Figure 26. Use of a differential mechanism to provide a compact, coaxially driven wrist with three concurrent rotary axes. The mechanism is shown schematically with the trapezoidal shapes indicating bevel gears. The input rotations are indicated by 1, 2, and 3, and the output rotations by A, B, and C.

In remote actuation systems, the compliance of the elements transferring motion from the actuators to the joint is often a major problem. Shafts can usually be made quite stiff, but usually involve some form of gearing, which may introduce a compliance problem, particularly in the case of bevel gears or trains of differentials. Bands and cables, relatively speaking, are very compliant and should only be used when actuator compliance is not a problem. A roller chain is an alternative that is much stiffer, but that is also relatively bulky and presents much less flexibility in design (39). In this connection, it is useful to recall the observation made above that, in a series of speed reducers, if the final stage has a relatively large speed reduction ratio, the compliance of the inboard elements is usually negligible. Thus, if the major speed reducer is placed at the joint, compliance in the elements transmitting motion from a remote motor may not be important. A differential can be used to provide a very compact roll–pitch–roll wrist. This arrangement is shown in Figure 26. Drive difficulties lead to relatively restricted motion ranges and nonoptimal axis angles in most practical implementations.

SENSORS

In this section only internal, or proprioceptive, sensing is considered. This refers to the sensors that give information about joint positions and rates, and about actuator forces or torques, and which must be built into the arm structure (see also SENSORS, TOUCH, FORCE, AND TORQUE MEASUREMENT).

Position Sensors

The repeatability, that is, the ability to return to a prerecorded position within a specified error, of a robot is determined by the position sensors used on the joints. If L is the distance from a given joint axis to the reference point of a robot and e is the desired positioning error at the reference point, the angle that must be resolved by the position sensor at that joint is e/L. The ratio of 2π to that angle, expressed as a binary number and rounded up to the next larger power of 2, gives the number of bits per revolution that must be resolved by the position sensor at that joint. For repeatability within a fraction of a millimeter on a 1- or 2-m arm length, a resolution on the order of 14 bits/revolution is necessary at the first and second joints. At the wrist, it is usually possible to manage with about 12 bits.

Since an industrial robot must execute a very large number of cycles reliably with minimal maintenance, it is important to avoid position sensors that involve mechanical contacts. For this reason, potentiometers, although very useful in research prototype systems, are avoided in commercial use. The most often used position sensors are, consequently, resolvers and optical encoders (41). Neither of these devices involves any contact between the moving elements. Resolvers are electromagnetic devices in which a rotor coil carries a relatively high-frequency alternating excitatory voltage. Depending on the rotor angle, a voltage that has a sinusoidal variation, is excited in the stator coils. Resolvers are very compact and are quite inexpensive. They are available in a doughnut form, which is easy to build into a compact joint package. Resolvers are not capable of resolution on the order of 14 bits. The resolution can be improved by gearing up using spring-loaded instrument gears or a toothed belt system. To avoid the ambiguity resulting from multiturn use of the resolver, a two-resolver system, as shown in Figure 27 can be used. Resolvers produce an analogue output, which must be converted into digital form by means of an analogue-to-digital converter. Single-board computer configurations that have analogue-to-digital converters mounted directly on the computer board are available and are particularly convenient for this purpose.

Absolute encoders are optical devices in which a transparent disk has a photographically deposited pattern on it that either obscures or transmits a beam of light between an LED and a photo diode. As a result, a digital binary representation of each angular increment is generated and is transmitted to the computer over a parallel communication link. This can pose problems since single-board computes, such as are typically used in robots, have very limited numbers of parallel ports. Thus, it is necessary either to have a small processor servicing each joint or to use multiplexing, if optical absolute encoders are used.

Optical encoders are available with very high resolution capability. Fourteen-bit resolution is within the capabilities of moderately priced encoders. However, optical encoders of this resolution are quite bulky and can present packaging difficulties on a robot arm.

One thing of which the designer should be aware is that the binary code employed in optical encoders is not the regular binary code, but is Gray code. The reason for using this is that Gray code is set up so that only one bit is changed between any two adjacent subdivisions. In contrast, in regular binary code a substantial number of bits may change between two adjacent divisions. The problem is that any misalignment may lead to spurious signals when the encoder is close to the division boundary in the latter case. The Gray code removes the possibility of wildly incorrect codes being generated, since only one bit changes.

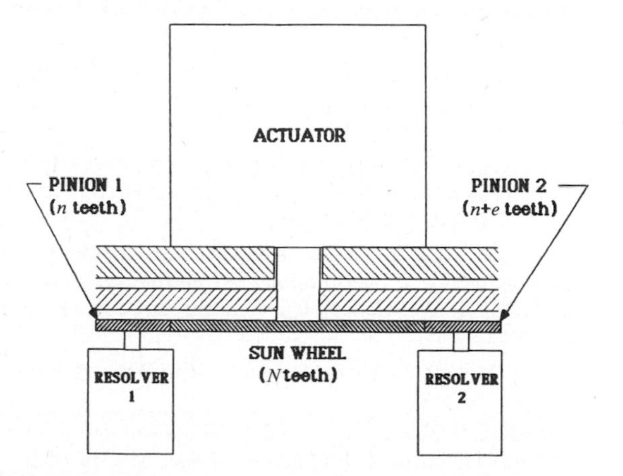

Figure 27. Arrangement of two geared-up resolvers to provide unique joint position data at high resolution. The tooth numbers of the two pinions differ by a small number e. The joint position may be obtained uniquely from the expression $\theta = n(n+1)(\phi_1 - \phi_2)/(Ne)$. The resulting resolution is improved by N/n over that of the individual resolvers.

Force Sensing

Force sensing is becoming increasingly important in industrial robots since it is becoming recognized that if the robot is to interact with fixed objects in the environment, force sensing is very necessary to avoid overload problems. Force sensing in a manipulator arm can be done by means of a six-axis load cell mounted in the wrist or in the base, or it may be done by means of sensors on the individual joints. If actuator forces are servoed, it is desirable to have force sensing in the joints. Conversely, if the control system is based solely on end-effector loads, a load cell in the wrist is often satisfactory. Force sensing in the joints is very easy to do in a hydraulic system since it requires only a differential pressure transducer connected across the actuator. Very compact differential pressure transducers with the necessary electronics packaged in the transducer are available. Direct actuator force sensing is somewhat more difficult in the case of electric actuators. Measurement of armature current is a method that has been employed, but it is a rather noisy measure. Direct sensing of torque in the joint shaft requires the introduction of a compliance in the shaft to generate sufficient deflection to be measurable by means of a strain gauge. This compliance is undesirable, and so this arrangement is not often used.

Six-axis load cells designed in a range of sizes for mounting in robot wrists or bases are available from several suppliers and are now in fairly wide commercial use. Most recent model industrial robots make provision for the use of feedback from devices such as this in their coordination software.

BIBLIOGRAPHY

1. D. L. Pieper, "The Kinematics of Manipulators Under Computer Control," Stanford Artificial Intelligence Laboratory, Stanford, Calif., Memo No. AI-72.
2. K. J. Waldron, S. L. Wang, and S. J. Bolin, "A Study of the Jacobian Matrix of Serial Manipulators," *Journal of Mechanisms, Transmissions, and Automation in Design* **107**(2), 230–238 (1985).
3. S. L. Wang and K. J. Waldron, "A Study of the Singular Configurations of Serial Manipulators," *Journal of Mechanisms, Transmissions, and Automation in Design,* in press.

4. H. C. Tsai and A. H. Soni, "Accessible Region and Synthesis of Robot Arms," *Journal of Mechanical Design* **103**(4), 803–811 (Oct. 1981).
5. R. Vijaykumar and K. J. Waldron, "Geometric Optimization of Manipulator Structures for Working Volume and Dexterity," *International Journal of Robotics Research* **5**(2), 91–103 (Summer 1986).
6. T. W. Lee and D. C. H. Yang, "On the Evaluation of Manipulator Workspace," *Journal of Mechanisms, Transmissions, and Automation in Design* **105,** 70–77 (Mar. 1983).
7. K. J. Waldron and A. Kumar, "The Dextrous Workspace," presented at the ASME Mechanisms Conference, Los Angeles, Calif., Sept. 20–Oct. 1, 1980, ASME Paper No. 80-DET-108.
8. K. C. Gupta and B. Roth, "Design Considerations for Manipulator Workspace," *Journal of Mechanical Design* **104,** 704–711 (October 1982).
9. Z. C. Lai and D. C. H. Yang, "A Study on the Singularity and Mobility of 6R Robots," presented at the ASME Mechanisms Conference, Cambridge, Mass., Oct. 7–10, 1984, ASME Paper No. 84-DET-220.
10. M. J. Tsai and K. J. Waldron, "Characterization of Manipulator Workspace Geometries," *Proceedings of the 15th International Symposium on Industrial Robots,* Tokyo, Japan, Sept. 11–13, 1985, pp. 887–894.
11. R. P. Paul and C. N. Stevenson, "Kinematics of Robot Wrists," *International Journal of Robotics Research* **2**(1), 31–38 (Spring 1983).
12. D. E. Whitney, "The Mathematics of Coordinated Control of Prosthetic Arms and Manipulators," *Journal of Dynamic Systems, Measurement, and Control* **94,** 303–309 (1972).
13. T. Yoshikawa, "Manipulability of Robotic Mechanisms," to appear in *International Journal of Robotics Research.*
14. E. F. Fichter, "Kinematics of a Parallel Connection Manipulator," presented at the ASME Mechanisms Conference, Cambridge, Mass., Oct. 7–10, 1984, ASME Paper No. 84-DET-45.
15. E. Nakano, T. Arai, K. Yamaba, S. Hashino, T. Ono, and S. Ozaki, "First Approach to the Development of the Patient Care Robot," *Proceedings of the 11th International Symposium on Industrial Robots,* Oct. 7–9, 1981, pp. 87–94.
16. K. J. Waldron, V. J. Vohnout, A. Pery, and R. B. McGhee, "Configuration Design of the Adaptive Suspension Vehicle," *International Journal of Robotics Research* **3**(2), (Summer 1984).
17. S. M. Song, J. K. Lee, and K. J. Waldron, "Motion Study of Two- and Three-Dimensional Pantograph Mechanisms," to appear in *Mechanism and Machine Theory.*
18. P. Marchal, J. Cornu, and J. M. Detriche, "Self Adaptive Arc-Welding Operation by Means of an Automatic Joint Following System," *Proceedings of Fourth CISM-IFToMM Symposium on Theory and Practice of Robots and Manipulators,* Zaburow, Poland, Sept. 8–12, 1981, pp. 464–474.
19. J. P. Trevelyan, P. D. Kovesi, and M. C. H. Ong, "Motion Control for a Sheep Shearing Robot," in M. Brady and R. Paul, eds., *Robotics Research: The First International Symposium,* M.I.T. Press, Cambridge, Mass., 1984, pp. 175–190.
20. K. J. Waldron and J. R. Reidy, "A Study of a Kinematically Redundant Manipulator Structure," *Proceedings of the 1985 IEEE International Conference on Robotics and Automation,* vol. 1, San Francisco, Calif., Apr. 7–10, 1986, pp. 1–8.
21. J. Hollerbach, "Optimum Kinematic Design for a Seven Degree of Freedom Manipulator," in H. Hanafusa and H. Inoue, eds., *Robotics Research: The Second International Symposium,* M.I.T. Press, Cambridge, Mass., 1985, pp. 215–222.
22. B. Roth, "Kinematic Design for Manipulation," presented at the NSF Workshop on Research Needed to Advance the State-of-Knowledge in Robotics, University of Rhode Island, Kingston, R.I., Apr. 15–17, 1980, pp. 110–118.
23. K. H. Hunt, *Kinematic Geometry of Mechanisms,* Cambridge University Press Cambridge, UK, 1978.
24. W. J. Book and V. Sangveraphunsiri, "Bracing Strategy for Robot Operation," A. Morecki, G. Bianchi, and K. Kedzior, *Theory and Practice of Robots and Manipulators,* Kogan Page, London, 1985, pp. 179–185.
25. K. J. Waldron, "Optimal Design of Manipulator Structures," *Proceedings of First National Seminar on Aerospace and Related Mechanisms,* Trivandrum, India, Aug. 22–23, pp. RB8.1–RB8.15.
26. J. P. Den Hartog, *Strength of Materials,* Dover, 1949.
27. F. Lhote, J.-M. Kauffmann, P. Andre, and J.-P. Taillard, *Robot Components and Systems: Robot Technology* Volume 4, Kogan Page, London, 1984.
28. W. J. Book and M. Majett, "Controller Design for Flexible, Distributed Parameter Mechanical Arms via Combined State Space and Frequency Domain Techniques," *Robotics Research and Advanced Applications,* ASME Book No. H00236, 1982, pp. 101–120.
29. N. Orlandea and T. Berenyi, "Dynamic Continuous Path Synthesis of Industrial Robots Using ADAMS Computer Program," *Journal of Mechanical Design* **103**(3), 602–607 (July 1981).
30. P. N. Sheth and J. J. Uicker, "IMP (Integrated Mechanisms Program), A Computer Aided Design Analysis System for Mechanisms and Linkages," *Journal of Engineering for Industry* **94,** 454–464 (1972).
31. D. B. Welbourne, "Fundamental Knowledge of Gear Noise—A Survey," *Proceedings of the Conference on Noise and Vibrations of Engines and Transmissions,* Institute of Mechanical Engineers Cranfield, UK, 1979, Paper No. C117/79.
32. D. R. Askeland, *The Science and Engineering of Materials,* Brooks/Cole, Monterey, Calif. 1984.
33. B. S. Thompson and C. K. Sung, "A Variational Formulation for the Dynamic Viscoelastic Finite Element Analysis of Robotic Manipulators Constructed from Composite Materials," *Journal of Mechanisms, Transmissions, and Automation in Design* **106,** 183–190 (June 1984).
34. T. G. Bartholet, "Compact, High Strength Robotic Manipulator Development," *Proceedings of Intelligent Systems for Army Logistics Support and Combat Engineering Operations Symposium,* Fort Belvoir, Va., Mar. 25–27, 1986, pp. 74–77.
35. C. R. Mischke, "Rolling Contact Bearings," in J. E. Shigley and C. R. Mischke, eds., *Standard Handbook of Machine Design,* McGraw-Hill, New York, 1986, Chapt. 27.
36. H. Asada and T. Kanade, "Design of Direct-Drive Mechanical Arms," *ASME Journal of Vibration, Acoustics, Stress and Reliability in Design* **105-3,** 312–316 (1983).
37. D. Schmitz, P. Khosla, and T. Kanade, "Development of CMU Direct-Drive Arm II," *Proceedings of the 15th International Symposium on Industrial Robots,* vol. 1, Tokyo, Japan, Sept. 11–13, 1985, pp. 471–478.
38. H. Merritt, *Hydraulic Control Systems,* John Wiley & Sons, New York, 1967.
39. K. M. Marshek, "Chain Drives," in J. E. Shigley and C. R. Mischke, eds., *Standard Handbook of Machine Design,* McGraw-Hill, New York, 1986, Chapt. 32.
40. J. Vertut, J. Charles, P. Coiffet, and M. Petit, "Advance of the New MA 23 Force Reflecting Manipulator System," in A. Morecki and K. Kedzior, eds., *Theory and Practice of Robots and Manipulators,* Elsevier, Amsterdam 1977, pp. 307–322.

41. E. O. Doebelin, *Measurement Systems,* McGraw-Hill, New York, 1983.
42. K. Youcef Toumi and H. Asada, "The Design of Open-Loop Manipulator Arms with Decoupled and Configuration-Invariant Inertia Tensors," *Proceedings of 1986 IEEE International Conference on Robotics and Automation,* vol. 3, San Francisco, Calif., Apr. 1–10, 1986, pp. 2018–2026.

ART, ROBOTIC

SHARON D. SMITH
formerly of International Personal Robot Congress
Conifer, Colorado

INTRODUCTION

Any discussion of robotic art must necessarily include a history of the robot. An examination of how, over the centuries, humans have turned their musings about human-made creatures that would serve and entertain them into the reality of the personal and industrial robots that exist today. This is important because, when limited technology prevented the creation of working robots, people created art of often astounding beauty to depict their belief in the possibility of automata. So a discussion of robotic art must include some discussion of the technological advances that paralleled the art of any particular period. Specifically, this discussion of robotic art must include at least a brief overview of the development of the computer since it was this device that enabled inventors to make the great leap from mechanical figures with simple, predictable actions to self-actuating ones that more closely simulate life. It marks the difference between clever toys and automata.

Though the word *robot* only dates from the 20th century, robots have really existed for hundreds of years, in dreams and legends, in religions and myths. Indeed, much of what is considered today to be historic robotic art is really the plans and designs (some very intricate and almost workable) for functioning robots that early inventors honestly hoped to bring to fruition. Or, as is seen in the very earliest historical depictions, what is now viewed as art, primitive cultures regarded as religious myths and relics.

But with the long-awaited arrival in the late 20th century of true, working robots—devices that really do dangerous or repetitive jobs in factories from Japan to the Federal Republic of Germany, early personal robots that may in coming decades help around the house—robotic art does not cease. Instead, the paths of art and industry merge, influencing one another and helping shape the future.

HISTORY: MYTHS, LEGENDS, AND EARLY MECHANICAL DEVICES

People have long been fascinated by stories of the creation of the human race, and from time to time throughout history have been said to have attempted to produce creatures of their own, sometimes to honor their own creator, sometimes out of rivalry. This is not a discussion of simple machines to assist humans in their work or clever gadgets to astound or entertain. Almost from the beginning, people have wanted to create self-actuating devices with which they could identify, but which would be at their command as they believed humans were at the behest of the gods or God.

Greek Mythology

The Greeks believed that Hephaestus (or Vulcan), the god of fire, fashioned a gigantic, animated statue out of bronze and gave it to King Minos to serve as a guardian for Crete. Talos could makes its body glow red-hot and, until the time of Jason and the Argonauts, is believed to have kept the island safe from invasion by crushing and burning strangers against its body.

Homer refers in the *Iliad* to mechanical gold females that served as assistants to Hephaestus. The fire god is also credited with constructing wheeled tripods that could roll themselves home unassisted.

Daedalus, a descendant of Hephaestus, was believed to have created self-powered, moving statues, one of which was a bronze soldier that helped in the fight against the Argonaut invasion of Crete.

Manually Operated Figures and Harnessed Power

The articulated dolls used by ancient religious leaders to make stories and myths more real to listeners can be classified today as the earliest physical representation of robotic art.

These robots, figures of clay or wood with movable limbs, mouths that opened and closed, and heads that turned, were typically moved by hand, and are the forebears of the puppets and mechanical toys of later centuries. From the Far East to Africa, these beautiful figures attest to the wish of early humans to imitate human movement and thought in creatures they made themselves.

Other examples of robotic ancestors that today would be viewed as early robotic art are the hydraulically powered figures that marked the hours by striking bells on ancient Egyptian water clocks, which appeared as early as 1500 BC. Around 100 AD, Hero of Alexander used fire to power a mechanism that controlled the doors of a temple. Syrian legend tells of a water-powered clock dating from 500 AD depicting Heracles using a club to strike the hours.

Judeo-Christian Tradition

Jewish legend early speaks of creatures called *golems,* from the Hebrew word meaning formless mass. The word even appears once in the Bible to allude to the human fetus.

The most famous golem tale concerns the 16th century Rabbi Judah Loew Ben Bezalel of Prague (1525–1609) who is said to have prevented a pogrom by invoking divine assistance to create an animated clay giant who saved the Jews of Prague.

Rabbis of earlier centuries were said to have had golems, both male and female, to serve and attend them. Both Rabbi Loew's golem and that of Rabbi Elijah Ben Judah of Chelm (1514–1583), however, were said to have become something more than mere servants. Legend holds that these creatures became menaces, a threat not only to their creators but to all humans (see ART, ROBOTS IN).

The early golem legends certainly presage the monster of Mary Shelley's *Frankenstein,* discussed later. These tales are also perhaps the first allusion to the potential dangers of human attempts to create automata, a practice viewed by many even to this day as an affront to God.

Legends of the Middle Ages

St. Thomas Aquinas, the early Christian theologian, is supposed to have destroyed an automaton that his teacher, Albertus Magnus (1193?–1280), spent 20 to 30 years building. The creature, denounced by St. Thomas as the work of the devil, was variously described as a beautiful, articulate female and as a mechanical servant that responded to a knock at the door and greeted visitors.

The English scientist and philosopher Roger Bacon (1214?–1294) is said to have invented a head that could speak. About the same time, two more examples of talking heads surfaced, in Spain. One was the creation of Don Enrique de Villena, known as a magician. The other was the mechanical tombstone of Don Alvara de Luna and his wife. The crypt, housed in the cathedral of Toledo, included figures of the couple that were said to kneel and pray during mass. Queen Isabella supposedly had the figures destroyed because they detracted from the cathedral's dignity.

The 15th-century inventor Regiomontanus is credited with creating an artificial eagle that flew in front of Emperor Maximillan I whenever the emperor entered the town of Nuremberg, and with the construction of an iron fly that could fly and then return to its inventor.

Clocks with mechanical figures to sound the hours began appearing in Europe around the 14th century. In 1354, the clocktower in Strasbourg, France, was graced with the world-renowned Strasbourg Cock. This mechanical bird could open and close its beak, extend its tongue, spread its wings, and crow. Around 1361, Nuremberg's Frauenkirche received a clock graced by a parade of puppets. A clock with two bronze giants to strike the hours was installed in Venice's Piazzi di San Marco sometime between 1497 and 1499, and continues to keep time even to this day.

By the end of the Middle Ages, almost every major town or city in Germany and Switzerland had a mechanical clock, many with beautifully rendered, intricate figures that acted out scenes from Bible stories and legends.

Leonardo da Vinci is known today both for his paintings and for his ingenious mechanical designs that were hundreds of years ahead of their time. In 1510 he succeeded in making a ferocious-looking mobile lion that displayed the French fleur-de-lis on its chest as a tribute to the artist's partron, the king of France.

The Roots of Computing

As early as 1642, a young man named Blaise Pascal had devised a mechanical calculator to help his father with his business accounts. Gottfried Willhelm Leibniz, in 1694, perfected a machine dubbed the "Stepped Reckoner" that performed addition, subtraction, multiplication, division, and the extraction of square roots.

Although not works of art, both of these inventions were essential in the development of the computer and deserve mention here. As mentioned earlier, without the computer, working robots would never have come into existence.

MECHANICS: 18TH AND EARLY 19TH CENTURY

From Watching-Making to Mechanical Dolls

The 18th century brought us many steps closer to having truly self-acting devices, powered by the purely mechanical interaction of springs and gears, and borrowing from the technology of clocks, watches, and even music box mechanisms. Although much of what had come before consisted largely of myths and wild legends, it is fortunate that many of the mechanical figures of the 18th century are still intact today, on display in museums.

Some of the best known inventors to use pure mechanics to create a semblance of life were Jacques de Vaucanson, the Jacquet-Droz family, and Henri Maillardet.

De Vaucanson (1709–1782) produced an automatic drummer, and a flutist that actually moved its fingers and played 12 different songs thanks to a bellows in its mouth.

But de Vaucanson's greatest work of art is probably his "Digesting Duck," so called because of its ability to digest (or at least, dissolve) what it had eaten. The gilded-copper fowl could also quack, swim, and flap its wings. Diagrams are all that remain of this work, but they are remarkable for their technical detail.

Pierre Jacquet-Droz (1721–1790) had already distinguished himself throughout Europe as a master Swiss clockmaker when he turned to the creation of life-size working dolls, having followed the work of de Vaucanson.

Droz *pere* is best known for his piece "The Scribe," an elegantly dressed young man seated at a writing desk. The figure can dip its quill pen in ink to write any text of up to forty letters. So intricate is its design that the automaton's eyes follow its hand as it moves across the page (see ART, ROBOTS IN).

Henri-Louis Jacquet-Droz followed in his father's footsteps. In fact, Henri-Louis is best known for "The Draughtsman," which physically resembles "The Scribe" but produces drawings instead of text. "The Draughtsman" has three interchangeable cams, enabling it to execute four sketches, including portraits of Louis XVI and his wife Marie Antoinette. Like de Vaucanson's flutist, "The Draughtsman" also has a bellows in its head, which allows it to blow dust off its sketch pad.

Droz *fils* was, in addition to being a robot maker, a composer. Henri-Louis's piece titled "The Musician" is the figure of a young woman seated at a keyboard instrument. This robot presses keys with all ten fingers, playing five simple tunes attributed to Henri-Louis. Adding even more humanity to his creation, Droz enabled the figure to move its eyes and head in time to the music and to bow at the conclusion of each piece.

The three Jacquet-Droz pieces made the rounds of Europe and were eventually acquired in 1906 by the Historical and

Archaeological Society of the Canton of Neuchâtel (Switzerland). Neuchâtel's Musée d'Art et d'Histoire, which painstakingly restored the pieces, has them on permanent display and activates them for public performances on the first Sunday of every month.

A Jacquet-Droz apprentice, Henri Maillardet, created a lifelike, creeping spider made of steel and a moving, hissing snake. These pieces have been lost. However, the Franklin Institute in Philadelphia owns what is perhaps Maillardet's most famous piece, "The Writing Child" or as it is sometimes called today "The Philadelphia Writer."

When "The Writing Child" was first acquired by the Institute, museum officials were not sure of its age or its origins. The device was damaged, by fire and by years of neglect, and clearly had not worked in years. But after they had restored its delicate mechanism, the figure wrote a short verse in French and signed it with words that solved the mystery of its origins: "Written by Maillardet's automaton."

Subsequent research determined that Maillardet created the piece sometime between 1805 and 1811, and that the figure kneeling at a writing desk had originally been dressed as a young boy in court. When the Institute acquired it, the figure was clothed as a French soldier. Today, "The Writing Child" wears a satin cap and dress. It can create seven pieces, including a short poem in English and a sketch of a 19th-century sailing vessel.

"The Writing Child" is a lovely, technically complex work of art that has been beautifully restored. Like "The Musician" and "The Draughtsman" by Henri-Louis Jacquet-Droz, it is also interesting as a work of robotic art that in turn creates art.

Jacquard's Loom

During the 4th century BC, Aristotle was one of the first visionaries to put into words a belief in automata. He wrote: "If every instrument could accomplish its own work, obeying or anticipating the will of others . . . if the shuttle could weave, and the pick touch the lyre, without a hand to guide them, chief workmen would not need servants nor masters slaves." (1)

It was not until 1804, however, that our instruments had reached this level of maturity. While the Jacquet-Droz family and Maillardet were creating devices intended to entertain, another European was inventing the first automatic machine for assisting people in their labors.

Joseph-Marie Jacquard is credited with inventing an industrial loom controlled by cards punched with holes. This innovation, which greatly speeded the production of cloth, was probably the single most important event of the entire Industrial Revolution because of its importance in the evolution of the computer. Jacquard's loom can also be regarded as a distant relative of the industrial robot.

Unlike Pascal's calculator and Leibniz's "Stepped Reckoner," however, Jacquard's loom enjoyed considerable commercial success, influencing the way cloth is made to this day. And, the principle on which it is based, the punched card, can still be found in use today in computer rooms throughout the world.

Shelley's Tale of *Frankenstein*

Writer Mary Wollstonecraft Shelley saw the Jacquet-Droz automata in 1816. The piece for which she is most famous, the frightening tale of *Frankenstein,* appeared just two years later. Some historians contend that there is a connection between the two events. It hardly matters today whether or not Shelley was so moved by these antimated figures depicting artificial life as to create one of literature's most lasting characters. What is important, especially in the context of this article, is the place of Dr. Frankenstein's monster in the history of robotic art (see also LITERATURE, ROBOTS IN).

Frankenstein is one of the earliest, and certainly today the most famous, *literary* examples of a human creation gone wrong, of a servant that threatens its master. No innocent figure making music or drawing pictures to delight and entertain. Instead, Dr. Frankenstein's creature is a golem straight from legend, but made from body parts of dead men instead of clay.

Much of the pathos and nuance has been lost since the tale's first appearance. Shelley's talent as a writer is rarely discussed when the work is reviewed today. And the underlying theme of the pain of rejection felt by all outcasts is largely ignored. But the work does stand as a moving, well-constructed warning of the dangers of hubris, that most human of emotions. A warning of what can happen when humans try to imitate God. A warning of the dangers of uncontrolled technology and irresponsible scientific experimentation.

It is a warning that spanned a century to take form in film and drama and other literary works.

ELECTRICITY: THE LATE 19TH CENTURY

Bell and Edison

When Mary Shelley wrote *Frankenstein,* scientists could only muse about the potentials of electricity, that energy source her fictional doctor used to spark life into his creature.

By the late 1800s, U.S. inventors such as Alexander Graham Bell and Thomas Edison finally discovered how to harness the power of electrical energy, and quickly presented the world with the most technologically advanced tools to date: the telephone and the light bulb. But this was also the time of the electric generator, the phonograph, and the first movie camera.

Edison even created a small robot by building a phonograph into a doll, which is said to have enjoyed much success. Evidently fascinated by the theme of artificial beings, Edison also produced the first Frankenstein film, in 1910, using the movie camera he invented. (See Robotic Art in Early Film and Drama section later in this article.)

Hollerith and IBM

Just 15 years before the dawn of our century, Herman Hollerith created a machine based on the punched cards of Jacquard's loom that, instead of making cloth, was used to help tabulate the 1880 U.S. census. Hollerith's invention reduced by two-thirds the time to complete the census. The machine was so successful that he co-founded the Computing Tabulating Recording Company in 1911 and the age of computers was launched.

Hollerith's company later became International Business Machines, known today as the industry-leading IBM.

ELECTRONICS: THE 20TH CENTURY

Robotic Art in Early Film and Drama

The birth of film coincides with a growing disillusionment with the Industrial Revolution. At first pleased with the help offered by machines, people soon began to see the disadvantages, and indeed, the dangers, of mechanization. Not surprisingly, many of our early films include robots, sometimes as character, sometimes as symbol. Almost always, they are viewed as a horror.

The first film version of *Frankenstein* was a 1910 film produced by Thomas Edison, directed by Searle Dawley, and starring Charles Ogle as the monster. Because the creature was mute, the tale was particularly well suited to silent film.

Since then, Frankenstein's monster has found its way onto the screen many times. Perhaps the most famous film depiction of Shelley's story, however, is the 1931 remake starring Boris Karloff. Remarkable makeup and Karloff's powerful injection of pathos into the character of the speechless creature make this James Whale film the standard by which all versions of this story have, to date, been measured.

In 1921, Karel Capek's play *R.U.R.* offered us the first literary version of mechanical man, created not from body parts like Frankenstein's monster, but of metal. Shelley's influence on the work can be clearly seen, however, in the dark view Capek takes of technology, which he believes is on the verge in real life of running amok and destroying its creators (see MOVIES, ROBOTS IN).

First produced in Prague and soon after in New York, *R.U.R.* is also significant in that it contributed the word *robot* to the English language. Indeed, today the word appears in some form in virtually all modern languages. Derived from the Czech word *robit* meaning to work, it also has roots in the Slav *robota,* which means menial or slave labor.

The play's name stands for Rossum's Universal Robots, a fictional manufacturer of mechanical creatures designed to replace human workers. Efficient but totally lacking in emotion, these robots are first thought to be an improvement on human beings since they do as they are told without question. But like Prague's 16th-century golems and Dr. Frankenstein's monster of the 19th century, these robots eventually turn on their masters. They destroy the human race, save for the one man spared so that he can continue to produce robots. The formula, unfortunately, has been lost in the destruction.

Despite the play's theme of utter hopelessness, it ends on a positive note when one robot couple discovers they have the capacity to care for one another. But Capek's depiction of the horror of mechanization is played over and over again in the coming decades, in film, fiction, and drama.

Science Fiction and Asimov's Three Laws

While Capek was creating a lasting impression on the stage, another form of literature in which artificial humans play a major role was grabbing our attention. Just after World War I, a whole new breed of fiction writer was born, creating a distinct genre of prose called *science fiction.*

Pulp magazines such as *Amazing Stories* and *Astounding Stories* fostered the growth of science fiction, and gave us short-story writers who wove often elaborate tales heavily laced with their ideas about the future of technology (see also LITERATURE, ROBOTS IN).

One of the most important writers to emerge from this time is Isaac Asimov. Perhaps the most prolific writer of our time and certainly one of our great thinkers, Dr. Asimov has produced more than 250 books on subjects ranging from the Bible and Shakespeare to math and robotics.

Probably best known as the "father of the modern robot story" (a title he bestowed on himself!), Asimov began writing robot stories in 1939 when he created Robbie, the robot nursemaid that is today almost as famous as its creator.

An avid science-fiction *reader,* Asimov did not set out to create a new literary style; he was simply trying to finance his way through Columbia University. Yet, in 1942 in a story titled "Runaround," Asimov coined a word that we would be hard-pressed to do without today—*robotics* to describe the study of robots. From this point, his place in the history of robotic art was assured.

It was also in "Runaround" that Asimov first listed his Three Laws of Robotics:

1. A robot may not injure a human being or, through inaction, allow a human being to come to harm.
2. A robot must obey the orders given by human beings, except where such orders would conflict with the first law.
3. A robot must protect its own existence, as long as such protection does not conflict with the first or second law.

Established by the human characters in Asimov's stories as a way to ensure that the robotic characters would remain harmless and totally under the control of people, these laws dictated design features built into the robot hardware, specifically into their highly advanced, so-called positronic brains.

Asimov says in the introduction to his first collection of robot stories, *The Complete Robot,* that these laws evolved from his belief that, until then, robot stories had generally fallen into two categories. Before Asimov began writing robot stories, they could generally be grouped into the class of Robot-As-Menace (potentially destructive and threatening to their creators) or Robot-As-Pathos (innocent victims of cruel humans). Robbie certainly falls into the second category: he is bought as a companion for a young girl whose mother takes a dislike to him and has him returned to the factory.

But with the writing of this story, Asimov began to see the possibility of a third portrayal and he began to depict robots as neither Menace nor Pathos.

"I began to think," says the author, "of robots as industrial products built by matter-of-fact engineers. They were built with safety features so they were not Menaces and they were fashioned for certain jobs so that no Pathos was necessarily involved." (2)

In some ways, Asimov viewed his fictional robots as nothing more than modern appliances, no more threatening than a washing machine or a vacuum cleaner.

And although Asimov never expected to influence the creation of real robots or to see his Three Laws go beyond his fiction, one can see his views depicted today in a variety of places. Engineers in the field of robotics and artificial intelligence even view them as an excellent guide for the creation of real robots.

Indeed, Joseph F. Engelberger, who founded Unimation, Inc., in the 1950s to commercially manufacture and market industrial robots for use in factories, once said he that became interested in robots in the 1940s after reading Asimov's stories.

Robotic Film Evolves

Most of the first films depict Robot—As—Menace. Even after Asimov began publishing his stories, this image persisted at the movies. It was not until the late 1950s that film-makers even began to hint at something beyond fear and loathing, a trend that culminated in the late 1970s with the positively loveable C3PO and R2D2 of George Lucas's *Star Wars* trilogy.

Some of the best known robot films include those listed below. See also Movies, robots in.

Frankenstein (1910): Thomas Edison's depiction of Mary Shelley's tale.

The Golem (1920): A German film that brings this Jewish tale to life.

Metropolis (1927): Fritz Lang's frightening vision of robot workers in the year 2000.

Frankenstein (1931): Boris Karloff's well-known and poignant portrayal of the monster, which sets the standards for the many Frankenstein films that have followed.

The Day the Earth Stood Still (1951): Starring Gort, a robot with the power to destroy the world.

Forbidden Planet (1956): Robbie the Robot makes his film debut.

2001: A Space Odyssey (1968) and *2010* (1984): In the first of this series, a computer named HAL runs amok, destroying astronaut companions on a long space flight. In the second, a computer called SAL tries to show us that they are not all bad.

Silent Running (1972): A peaceful film in which a botanist teaches robots to plant trees.

Westworld (1973): A robot gunfighter pursues human prey.

Sleeper (1973): Waking up in a hostile future world, director/star Woody Allen tries to escape disguised as a robot.

The *Star Wars* trilogy (beginning 1977): R2D2 and C3PO virtually steal the show from the films' human stars.

Demon Seed (1977): A human woman is forced by a robotic arm to do the bidding of a power-hungry computer.

Blade Runner: Good versus evil in this movie about killer robots pursued by a human bounty-hunter.

D.A.R.Y.L.: A government experiment turns sour as a human family becomes attached to a robot child.

Mainframes to Minicomputers

It was IBM engineers, working with researchers at Harvard, who created the first true computer in 1939. The Harvard Mark I filled a room and was controlled by punched paper tape (again, using Jacquard's theory); its memory was based on electromechanical relays. While it was so very much slower than today's computers, it was, thanks to its electronics, so very much faster than any calculating device that had come before it.

More significant, perhaps, than the Mark I was the ENIAC (Electronic Numerical Integrator and Calculator), which appeared just eight years later. Like its predecessor, ENIAC was a huge machine. But it was much faster, reducing the time to calculate some very complex, mathematical calculations from years to hours.

ENIAC was almost primitive by today's standards, being powered by thousands of vacuum tubes, which proved highly unreliable.

In the same year as ENIAC appeared, however, Bell Laboratories invented the transistor, leading the way to production of UNIVAC I (Universal Automatic Calculator), the world's first commercially produced computer. And again, the U.S. Census Bureau led the way by purchasing some of the first UNIVACs manufactured. General Electric also bought some 45 UNIVAC Is.

IBM soon became a leader in the manufacture and marketing of mainframe computers to business and government.

By the 1960s, the huge mainframe was supplanted by the more compact minicomputer. Electronics manufacturing had matured to the point where it was possible to fit hundreds of transistors on a silicon chip just a quarter-inch square. Manufacturers, such as Data General Corporation and Digital Equipment Corporation, could now provide the same computing power offered by IBM, but in a much smaller case and at a lower cost. Computers were no longer tools available to only the largest corporations. More and more smaller businesses could begin to exploit their power.

Microcomputers and Personal Robots

The next step in the evolution of the computer was Intel Corporation's introduction in 1974 of the *microprocessor.* This remarkable invention roughly doubled the processing capability of any chip before it and brought the cost of computing within range of private individuals.

In 1975, *Popular Electronics* Magazine published an article on a do-it-yourself computer kit available from a small electronics company in Albuquerque, New Mexico. Neither the company, MITS, nor its product, the Altair, exist any longer. But this was the true beginning of the personal computer.

By 1976, two young inventors named Steve Wozniak and Steve Jobs pushed us into the next stage of technological maturity by inventing the highly successful Apple personal computer. Following the tide, such companies as Atari, Commodore, Radio Shack, and the ever-present IBM began to market their own version of this high-tech phenomenon. Soon, families, particularly in the United States, were buying these machines both for recreational and educational purposes and as status symbols.

Once the personal computer (PC) was born, the personal *robot* was possible. Although industrial robots had made their appearance as early as the 1950s, it was not until the 1980s that great numbers of people were exposed to actual, three-dimensional automata that *began* to meet our expectations, to fulfill our ages-old fantasies about mechanical helpers. Existing technology was sufficiently reduced in size and cost, while at the same time its capability was increasing. Indeed, some early reviewers of the personal robots first marketed in the early 1980s called them "personal computers on wheels."

An oversimplification, this description does not take into account most people's expectations of what we have always dreamed robots could do. It is perfectly acceptable for a PC to sit on a desk, silently calculating spread sheets or balancing the check book, however, a robot must *move*. People also expect robots to communicate with something more than lines on a screen—a beep or a bell, though a voice, even an artificial one, is ideal. And, at minimum, a personal robot should have an arm or a gripper, enabling it to do something, to accomplish some physical task, such as fetching the evening paper.

Today's personal robots suffer from the same fate as early automata: our imaginations still exceed the abilities of available technology. It is becoming apparent that we will need advanced software based on something called *artificial intelligence* before robot helpers are possible. The hardware, too, still needs to be perfected. But some estimations predict that useful, working robots will begin taking their place in the home in as little as 10 years.

THE ROBOT EXHIBIT: HISTORY, FANTASY AND REALITY

In January 1984, The American Craft Museum in New York City opened the doors on "The Robot Exhibit: History, Fantasy and Reality."

The Robot Exhibit was described by American Craft Museum Director Paul J. Smith (3) as "the first comprehensive exhibition on the topic of robots by a museum in the United States. Presented as an overview of the subject—history, fantasy and reality—it traces the early beginnings of automation over a three-thousand year period to a wide variety of working robots in use today."

And as if to make quite clear the 20th-century marriage of Robotic Art and high technology, the exhibit's Guest Curator, Robert Malone, is not only an artist and writer, but a machine-tool designer as well.

A major portion of the exhibit was, quite naturally, dedicated to the image of the robot in art through the centuries. On display were a variety of interpretive robots and sculptures, toys ranging from the whimsical to the educational, handmade marvels from backyard inventors, and examples of both industrial and personal robots.

THE FIRST INTERNATIONAL PERSONAL ROBOT CONGRESS AND EXPOSITION

The First International Personal Robot Congress and Exposition took place in Albuquerque in April 1984. The site was chosen as a sort of tribute to the first personal computer company, MITS, which had marketed the do-it-yourself Altair kit from this New Mexico city in the 1970s.

IPRC '84 was the first organized gathering of all the different groups involved to that point in personal robots. Included were inventors who (typically in home workshops without benefit of commercial backing) had been creating robotic devices for decades, commercial manufacturers of personal robots and peripherals, students and educators interested in the educational value of personal robots, health professionals convinced that robots will someday aid the disabled and the elderly in caring for themselves in their own homes, and artists fascinated by the images evoked by robots and technology.

To clarify, personal robots are distinguished from industrial robots most notably in their size (personal robots are generally much smaller), in their cost (the few personal robots available today are less expensive, generally costing under $10,000, while their industrial counterparts cost hundreds of thousands of dollars), and in their uses (rather than being used in manufacturing, the prime function of today's personal robots is in education though similar devices may someday be used as household helpmates).

IPRC '84 consisted of a trade show with displays by the first commercial manufacturers of personal robots (RB Robot Corporation, Heath Company, and Androbot) and by individual inventors, dubbed Personal Robot Developers (PRDs). PRDs were given a highlighted place in the trade show. After all, it was these creative individuals who first believed in the possibility of personal robots and who, unknowingly, spawned a new industry. While limited by technology, as so many previous generations of automata have been, the PRDs' devices were remarkable in that they demonstrated the rudiments of how robots will, in the near future, assist humans on a personal level. A robotic lawn mower was one of the most popular PRD inventions on display.

In addition, IPRC '84 included seminars and workshops that explored such topics as The Business of Robots, The Future of Robots, Educational Robots, Robots in Health and Human Services, Hardware and Software Design, and Legal Issues.

(Clearly, however, in such areas as human services, the line between the personal and industrial blurs. Leading rehabilitation hospitals are today experimenting with industrial robot arms to do such tasks as assisting the disabled with feeding themselves.)

IPRC '84 participants also heard addresses by industry leaders such as Unimation founder Joseph F. Engelberger and Nolan Bushnell, founder of Atari and Androbot.

Among other firsts, IPRC '84 also marked the first time an international competition had been held soliciting artwork on the subject of robots and the robotics age. Cash prizes were awarded and the First Place entry was made into posters used to advertise the conference and trade show in schools, computer dealerships, and department stores worldwide.

Five judges with a range of backgrounds (an artist, a college art professor, a museum coordinator, and an administrator and a vice president of research and development for a personal robot manufacturer) viewed over 150 pieces to select the final 22 works that were displayed at IPRC '84.

The winning entry, by Randy Johnson of Chicago, was a sculpture of handblown glass and mixed media (electronic parts, tubing, springs, and steel enclosing a red digital clock face resembling a heart). Titled "O'Keeffe," the piece was a robotic rendition of George O'Keeffe's famous painting of a bovine skull dried and bleached by the desert sun.

According to artist Margaret Cooper, who organized the competition, served as a judge, and prepared the notes that accompanied the art displayed at IPRC '84 (4)—"Indeed, Johnson's sculpture looks both like a skull with horns and like a robot with arms."

"O'Keeffe" seems to embody the spirit of late 20th Century Robotic Art—it recalls an artistic predecessor, it makes use of the bits and pieces of computer equipment necessary to produce true working robots, and it is supremely clever at the same time.

The balance of the 21 pieces on display at IPRC '84 were paintings and water colors, and almost without exception, reflected a most positive view of robots and the mechanized world of the future that they represent. From the vivid, kinetic image of "Heart" to the kind and loving "Nanny Robot" with a human child cradled gently in its arms to the cool and peaceful "Good Morning" with a robot hand offering a cup of coffee and a flower to a human to the humorous vision of robots selling robots door-to-door in "Acme Robot Sales," there is no destruction here, no malevolence, no fear.

It is particularly fitting that author Isaac Asimov delivered the keynote address at IPRC '84, via satellite from a studio near his home in New York City. His Three Laws of Robotics seemed to be depicted throughout the art exhibited.

ART IN ACTION

Although the robots of our art and imagination have tended to resemble human beings, the first working robots were industrial robots, which bear only the slightest resemblance to the human race.

Still, people cannot seem to resist endowing even these machines with human traits, using them as artistic subjects to feed the fascination with automata. One example is the lovely short film "Ballet Robotique," which won a 1983 Oscar for best Live Action Short Film.

The piece depicts industrial robots in General Motors factories doing a variety of tasks. There is no dialogue or narration in the film. Instead, the robots dance and perform to classical music selections such as Bizet's "Carmen Suite" and Tsaikovsky's "1812 Overture." At times, it almost appears that these machines are conscious of the camera, making their movements even more dramatic.

"Each type of robot has its own personality," says Producer/Director Bob Rogers in an interview with *American Cinematographer* Magazine. (1) "One moves like a chicken. Another like a cat. One paint-spraying robot moves like seaweed, gracefully swaying and curling with the current."

With the help of special lighting and other technical aids, Rogers brings drama to the factory floor and adds life to his subjects.

For example, one paint-sprayer robot sports a pair of painted, glowing red eyes as it works in an atmosphere lit to resemble an underwater cavern. The result is that the device seems, on film, to come to life and to snoop along the car chassis looking for spots needing paint.

Although personal robots as an industry seem now to be in trouble (as of this writing, only a handful of companies are left from the dozen or so formed in the early 1980s to manufacture and sell personal robots), it is worth noting the connection between these creations and Robotic Art.

For instance, Topo and BOB from Androbot, Inc., share such an angular, interesting design, they seem almost to be modern art themselves. Others, like the RB5X designed and first marketed by RB Robot Corporation, look strangely like the offspring of such movie robots as R2D2.

Unfortunately, both of these companies is now defunct (or at least dormant) and their products might well survive only as kinetic sculptures of the 1980s. In talking with insiders, it appears that their products were just too far ahead of their time. This seems due, in large part, to the fact that the *idea* of working robots has been around, in one form or another, for so many centuries, people wanted them to immediately be able to do more than the technology would allow. People can travel to the moon and be kept alive with artificial hearts, but the dream of a self-actuated device to assist without harming remains beyond reach.

As soon as these creatures became a three-dimensional reality in the form of early personal robots, jumped off the movie screen and the page to become something touchable, consumers wanted them to be able to vacuum their homes, babysit the children, be companions to the grandparents, keep the family checkbook. Alas, that is years away.

While people wait for technology to catch up with their imaginations, the creation and enjoyment of Robotic Art will continue. This will help keep hold on to a dream dating back thousands of years.

BIBLIOGRAPHY

1. P. Berger, *The-State-of-the-Art Robot Catalog,* Dodd Mead & Co., New York, 1984.
2. I. Asimov, *The Complete Robot,* Doubleday & Co., Garden City, New York, 1982.
3. P. J. Smith, "Introduction," *The Robot Exhibit: History, Fantasy & Reality,* catalogue, American Craft Council and International Paper Company, New York, 1984.
4. M. Cooper, *Art Competition, International Robot Congress 1984,* Lakewood, Colorado, 1984.

General References

H. M. Geduld and R. Gottesman, *Robots Robots Robots,* Little Brown and Co., Boston, Mass., 1978.

N. B. Winkless III, *If I had a Robot . . . ,* dilithium Press, Beaverton, Ore., 1982.

R. Malone, "History of Robots," in *The Robot Exhibit: History, Fantasy & Reality,* American Craft Council and International Paper Co., New York, 1984.

T. W. Marrs, *The Personal Robot Book,* Tab Books, Inc., Blue Ridge Summit, Penn., 1985.

M. Higgins, *A Robot in Every Home,* Kensington Publishing Co., Oakland, Calif., 1985.

ART, ROBOTS IN

Robert Malone
New York, New York

INTRODUCTION

Early artists and craftspersons were probably as unaware of creating automatons or robots as they were unaware of creating art or technology. These terms and categories have been assigned to their work much later. It is best to be aware that a direct sense of the art may always have more validity than any critical view or analysis. The major role of the artist until the Renaissance was as craftsperson, as smith, as wood cutter, as reed weaver. Since the Renaissance a great self-consciousness on the part of the artist has led to self-expression, artist as philosopher, and artist as the soul of his or her society. It would be a mistake to think that even in the 18th century, artists and craftspersons sat down and told themselves, "Now I'm going to make an automaton." But, make a doll, a moving statue, a puppet, fulfill a commission, yes! (1)

Pre-automaton, Pre-robot Art

Technically speaking, the first "robot" art could not have occurred before the publication of Karel Čapek's play *R.U.R.* in Prague in 1921, where the word robot was born. However, the idea of artificial or machinelike creatures, human and otherwise, has a long and distinguished history that is as old and older than recorded history. Peoples in the most ancient and primitive cultures have had a predilection for making articulated figures, dolls, and puppets. With the gradual dawn of what we have come to refer to as civilization, they made moving statues, and objects that appear to have self-programmed motion. Robots, and robot art in this sense, are as old as art, and as old as technology.

There were early oral and written myths involving gods like Vulcan (Hephaestus) who created Talos, a bronze metal giant who could only be destroyed by removing a pin from his ankle, out of which ichor would run, presumably an early "battery" fluid. There was also Prometheus who, in bringing fire (knowledge) to mankind, is assigned the role of artificial life giver; and other myths of Chinese and Japanese origin, as well as later retreads which resurfaced as Pygmalion, Daedalus, the Golem, Frankenstein, etc. There are references in literature, usually copies of copies, and artist's reconstructions from later centuries, but there are no working automatons, or robots, from early centuries. The first original drawings are from after 1000 AD in Persia (2).

Automaton Art

The clock makers, smiths, and doll makers of Western Europe in the 14th to 19th centuries were the source of automatons of the greatest skill. They combined the arts of sculpture, dressmaking and costuming, wigs, and dolls with the craft of the metalworker, woodworker, clock maker, and scientific instrument maker. Their work falls into three broad categories: extensions of clockworks, extensions of music boxes, and free-standing and driven automatons. The automaton was principally created in the period between 1750 and 1850 and has as its root the same technological forces as the industrial revolution. Automatons were usually commissioned by royalty in the 18th century; however, they evolved into a mass-produced art during the mid-to-late 19th century, becoming just one more of the products flooding from machines and factories. Examples of the latter were in the form of talking dolls (such as Edison's) and the endless automatic cast-iron banks of the period.

Robot Art

The story of robots and art, after the coining of the word, seems to fall into two major categories: a period of investigation into the ideas and theories of machines and technology as they can contribute to art, and a later period of application of sophisticated technology toward artistic ends. The first period includes the cubists, dadaists, surrealists, and the organizing effects of the German Bauhaus. The second period occurred nearly 30 years later and concentrated on kinetic art, mobiles, cybernetic art, and the interweaving of electronics with aesthetic forms and feelings to produce what now can be called robot art.

EARLY HISTORY

Ancient peoples and their cultures used their surroundings, the human figures and the figures of animals, to create a simulacrum in native materials: straw, wood, stone, etc. This, along with pottery and weapons, is often all one has to go on to represent these peoples and their cultures. Some of these craftspersons went a step further in attempting to recreate reality. They imitated the arm and leg sockets by creating simple axles at the top and/or bottom of their figures' torsos; one such example is the Ibibio wooden sculpture from Nigeria (Fig. 1). Or they might hinge their version of a jaw (animal, human, or both) on a head. They added either string to pull, as in the North West Indian cultures, or levers to push, as

Figure 1. Ibibio sculpture in wood from Nigeria. Courtesy of The Metropolitan Museum of Art, The Michael J. Rockefeller Memorial Collection; gift of the Matthew T. Mellon Foundation.

in early Egyptian dolls. Out of these "experiments" came figures that were used for performance as puppets and marionettes, or shadow puppets. Other crafted efforts brought forth everything from children's dolls to shaman masks. Often these figures would be used with the culture's poetry to create theatre, or ceremony, like the Javanese Waygang, or the masks and ritual costumes of Sumatra, Bali, Liberia, Borneo, Japan, and Northwest Indian tribes such as the Bella Coola. The latter developed elaborate masks with moving jaws and secondary faces. These were less attempts to fool the eye into seeing an animal or godlike face than to give expression to the story of their common culture. They were a way of telling an animated story such as "Now I'm the moon, now I'm the sun." They could also tell a story using an analogy that referred to something deeply felt, for instance, their place in nature, as with the moon and the sun (2,3).

The next level of elaboration was probably an unconscious technique on the part of the artisans or "priests" responsible for creating the figures and masks. It was an effort toward verisimilitude, and may have been something of a scam as well. It mixed articulated heads or figures with hidden voice tubes, as in the jackal god Annubis from ancient Egypt (Fig. 2). This head had a speaking tube which presumably could be spoken through from behind the wall the Annubis was mounted on. Its jaws moved with the spoken word (primitive lip sync) This kind of statuary oracle became the very distant origin of those mechanical fortune-tellers so popular in the 19th century (4).

The Egyptian culture also created small doll models or statues of everyday life. These depicted simple activities such as baking, weaving, boating, and hunting. Some figures were articulated: for example, the baker had a movable arm to lift his grain-pounding stone (2).

Figure 3. Steam-powered Egyptian-style statue from the 1800s.

Figure 2. The head of the Egyptian god Annubis.

Greek and Roman

The Greeks were known to have the skill if not the interest to build machines. The archeological find that is now called "The Greek Computer" is a good example of this skill; of course, it was not a computer such as is known today, but it was a "programmable" star tracker. They also had the skill and the interest to sculpt, or cast, very large and elaborate figures which became the hallmark for Roman art and for many cultures up to today (5).

Hero of Alexandria, in the first century AD, took the science of the Greco–Roman world, and elements of this art and architecture, and added simple mechanical devices to create, at least on paper: automatic door closers for temples, moving statues (using steam power as in the 19th century reconstruction shown in Fig. 3), animal statutes that drank, a wind organ, and bird sculptures that sang. To design his concepts, Hero had to first understand (and therefore invent) pistons, siphons, uses of air pressure, etc. In fact, his inventions were probably not brought to physical reality until at least 1000 years later (6).

Persian

The science of the Greeks, and therefore Hero, was revived in Persia (present-day Iran) well before its rediscovery in Western Europe during the Renaissance. Such 13th-century sages as Al Jazaris presumably created automaton figures and environments which built on Hero's work and added the mysterious Persian touch, popularly recalled in the forms of flying horses and carpets, magical birds, and other wonders of the Arabian nights. His *A Book of the Knowledge of Mechanical Contrivances* consists of a series of fine Persian illustrations that contrast both the works in flattened perspective and the dramatic peacocks in his *Peacock Fountain.* His *Servant* is probably the first depiction of a "robot" servant with self-moving arms (Fig. 4). He created a hand-washing device, a servant, a system for lifting water, various continuous-action fountains, and other hydraulic devices. No evidence of actual automatons exists.

It was these same Persians that made astrolabes in the Greek manner; this might suggest that strongly defined automatons and robts are an outgrowth of cultures that have the ability to combine technology and art as a form of expression (2).

LATER HISTORY

Clocks and automatons can be intimately related, since the works of clocks became, in modified form, the functional works of automatons, glockenspiels, and automatonlike music boxes. Springs and weights were among the few reliable sources for nonhuman, nonanimal power. The means of delivering that power was through classical gears, wheels, cogs and cams, and lever and trip levers of various types.

Figure 4. Thirteenth-century drawing by Al Jazaris, *The Servant,* may be the first depiction of a robot. Courtesy of The Smithsonian Institution.

Figure 5. The oldest extant automaton is this cock from atop the cathedral clock tower in Strasbourg, France. The beak moves, the tongue thrusts out, the wings flap, and it crows the passing hours.

Oddly enough, the Chinese, whose invention of clockworks pre-dates the West, failed to go on to build automatons as we understand them in Western culture. There are odd and unsubstantiated literary references to moving statues, but no hard evidence. They had the capability, as witnessed by their observatories, Su Song's clock towers, water-driven mechanisms (clepsydras), and their ability to make large-scale castings (again in advance of the West) and large-scale ceramics. As far as is known, the Chinese did not "invent" either the scientific method as developed in Western Europe nor moving figures such as automatons (7).

The oldest existing automaton, one that still works after many rebuildings, is a piece of magnificent sculpture. It sits atop the clock tower of the cathedral in Strasbourg and is in the form of a cock (Fig. 5). First built in 1352, it incorporates several mechanisms: a beak mover, a wing flapper, a tongue thruster, and a means to crow, all of which are tied to the control of the clockwork's passing hours.

In the same clockwork tradition, the glockenspiel in Munich and the cathedral clock in the Marienkirche in Lubeck are good examples of sculpted automaton figures in a complex environmental setting that are set into the architecture of the building. The figures go through a programmed routine as per the instructions of the clock. The major influence behind these efforts could well be fantasy and fun, and the joy of a culture knowing for the first time the time of day. The builders used the top crafts of their day to produce such whimsical creations (2,6).

The Art of the Automaton

The craft of automatons reached its highest technical proficiency and artistic merit in the tradition of Jacques de Vaucanson (1709–1782), Pierre Jaquet-Droz (1721–1790), and son Henri-Louis Jaquet-Droz in Europe, which was at the time poised for a technological leap forward in an industrial revolution. The work of these artists was created most often by commission from royalty, the only ones who could afford their time-consuming art.

Vaucanson created three particularly famous automatons: *The Tabor Player, The Flute Player,* and *The Cannard.* The latter consisted of a series of weight- and spring-driven mechanisms that controlled wing flapping, eating and drinking, digestion, and defecation. It was considered a miracle in its time and left Vaucanson open to charges of necromancy. The questionable remains of the duck are in the Musée du conservatoire National des Arts et Métiers in Paris. *The Flute Player* was a man high atop a pedestal. It played several tunes most faithfully in a human manner.

Droz's *Scribe* (1770), and *Muscian* (1773) have all their automaton mechanisms built within the figures themselves. This was a feat never before accomplished, and a hard act to follow. Both figures are "programmable" to the extent that the changing of mechanical devices governing, or on the control drum, gives different performances. These figures can still be seen operating at Neuchâtel in Switzerland more than 200 years after their construction (Fig. 6) (6).

Baron Wolfgang von Kempelen's talking machine, and famous *Chess Player* of 1769 are objects of some controversy, since neither any longer exists, and the latter was exposed as a possible hoax. Regardless of its being honest or not, the *Chess Player* was unique for its time and can be considered the grandfather of chess and checker playing machines. If the descriptions and contemporary drawings can be trusted, it was an object of exotic interest. Talking machines date back to talking statues and heads. Vaucanson managed a crude mechanical voice box and von Kempelen's machine was known to have crude word-making capability. No further record exists of this artistry.

In 1760, Friedrich von Knauss built an automaton that made drawings. *The Scribe* could write up to 107 words and had a "keyboard" attached, by which new messages could be recorded. His drum "memory" was derived from music-box technology. The writing figure of von Knauss, unlike the figures of Droz and Vaucanson, is a cast-metal figure typical of contemporary sculpture, rather than a mannequin or doll (Fig. 7).

Figure 6. **(a)** Droz's automatons were the first to have their mechanisms completely self-contained. This is his *Scribe,* which can be seen in operation in Neuchâtel, Switzerland. **(b)** The mechanism.

Figure 7. Friedrich von Knauss's 1760 graphic automaton: more sculpture than mannequin.

The work of Vaucanson, Droz, and von Knauss was extended by the Maillardet family in the late 18th and early 19th century. They created figures considered magical in their time, including: *Le Grand Magicien,* which answered questions to it that were left in drawer; *The Draughtsman* (The Philadelphia Doll); and a hummingbird in a box. The latter was in the tradition of singing-bird music boxes made by the dozens of craftspersons in Europe during this period. These were high art and solid craft. *The Draughtsman* still exists in Philadelphia, and although not of the level of artistry of Droz, it is a very remarkable and still operable first-class automaton.

One of the last great automaton makers was Frenchman Robert Houdin (1805–1871), who built many trapeze artists, shooting figures, and circus performers. The style of these automatons led to more widely produced figures such as Edison's talking dolls and the great variety of automatic cast-iron banks. The late 19th century saw a mix of hand-produced automatons, principally in France and the United States, and the beginning of automatonlike toys in Germany, France, and the United States. The automaton, in an artistically diminished form had become a gift item for the middle and working classes.

The Machine Influences Art

Industry and technology built around steam power made the idea of independent and powerful machines a reality in the late 18th and 19th centuries. In factories, steam power revolutionized spinning, weaving, metalworking; on land and water, it revolutionized transportation with steam trains and boats. The applications seemed boundless. Self-acting machines with ball governors started the long progression toward servomechanisms and cybernetics. With calculating machines such as Charles Babbage's brass and mechanical Difference and Analytical Engines, the move was toward computers and robots. The power of this scientific and technological force was recognized by painters, such as England's Joseph Turner, as a positive force; others, such as poet/painter Willem Blake, saw a clockwork world as a threat. This ambivalence toward the effects of machines on art and life persisted. Lesser artisans resolved their feelings by decorating printing presses, stamping machines, and objects of mass production with scrollwork, artistic lettering, and in the Victorian era, with foliage or animal feet. This meant, at the least, that art and craft could influence the appearance of machines.

Paxton's Crystal Palace exhibition in London in 1850 can be considered a turning point in an awkward attempt at blending industrial and steam technology with art. Machine-made art reproductions stood alongside artfully decorated machines. The late 19th century saw the emergence of a new realism that acknowledged the world as changed from pastoral to urban, from agrarian to industrial. It included the down-to-earth views of Daumier, of the common man in common situations such as riding a steam train; and a direct realism in an American painter like Thomas Eakins and in a French painter such as Édouard Manet. This was a realism that recognized the scientific realities of the day and the formidable resources of the new technologies (8–10).

Literary Sources

The romance of the machine affected architecture, design, and the arts. An early example can be found in the writings and drawings of Leonardo da Vinci, whose ideas on mechanisms and biology stand as unique and far-reaching insights for his time. Da Vinci, unfortunately, did not share his ideas broadly during his lifetime, and his copious drawings and notes were hidden from inventors and scholars for centuries after his death. He not only anticipated most of the basic mechanical principles of today but also used nature's resources and structures to support his ideas for man-made mechanisms. For instance, his studies of birds' wings and bird flight gave him a nearly bionic concept in his designs of a wing for human flight.

This romance also ran deep within the literary culture of Europe and the Americas. Its major exponents were Edgar Allan Poe, Jules Verne, and H. G. Wells. Poe has been granted the title of early father of science fiction, and Verne and Wells' influence helped to integrate advanced machines into part of the common psyche. Poe's interest extended to actual automatons and included an exposé of von Kempelen's *Chess Player* (Fig. 8) (he thought a midget chess player entered the machine below the game board and secretly controlled the action). In the story *Meltonta Tauta,* Poe depicted a future world in a

Figure 8. Triumph of technology or elaborate hoax; the chess-playing machine of Baron Wolfgang von Kempelen (1769).

way that traditional science fiction would make familiar. Verne brought the machine's penetration of underwater and space landscapes into what readers felt was close to reality, and extended the concept of a machine-dominated future. Wells placed less stress on the machine, *per se,* but by implication built a world around miracle machines such as his time machine and the machine exo-skeleton of the invading Martians. The fruit of these efforts was a new vision of time, space, machine, and our relation to them. Olaf Stapelton and others (Arthur Clarke, Isaac Asimov, etc) have extended this kind of machine future vision into the science fiction of today's robots, time warps, laser weapons, and space travel (11–13).

New Art Visions

The new visions of Jules Verne were not alone in France. Artists found themselves preoccupied with new subject matter, new ways of seeing, and new ways of interpreting. Realism was joined and challenged by impressionism and pointillism. These powerful expressions gave way to Postimpressionism and the sculpture, paintings, and drawings of Pablo Picasso, Juan Gris, and George Braque. The word "cubism" was coined by Henri Matisse, an early practitioner of this kind of art; cubism became Braque's lifework and was Picasso's style for a time. Cubism turned the art of the turn of the century toward greater introspection and analysis; it principally used basic forms, geometric and otherwise, that made up the painter's visual reality. Art became an engineered vision and a step closer to things robotic. The result of cubism was a multiple vision of objects as perceived and a flattening of normal perspective. It may have been a response to the impact of physics' relativity; an extension of Cezanne's methods; or a move toward the kind of simplification that Matisse made a watchword throughout his career. Nevertheless, it happened, and to the viewer it meant the metamorphosis of faces into geometry, the creation of figures that seemed more clockwork than flesh; for the painter, sculptor, and printmaker, it meant a new vision. Startling visions were created by artists all over Europe; representative examples are Vasily Kandinsky's 1912 painting *Improvisation 28,* Jacques Lipchitz's stunning sculpture *Man with Mandolin* in 1917, and Constantin Brancusi's cleanly formed sculptural series of 1909–1916. These included abstract sculptural heads and figures that refined the fundamentals of Matisse and Picasso; the direction was toward simplicity, machine-style forms, and kinetic and robot art (14,15).

FUTURISM AND DADAISM

Futurism was the outgrowth of the literary mind of Emilio Filippo Tommaso Marinetti, who inflamed artists like Boccioni and Balla toward revolutionary action and acceptance of modern technology and its devotion to speed and dynamics. Its main thrust occurred before World War I.

In the hands of Umberto Boccioni sculptures such as his *Unique Forms of Continuity in Space* (1913), a product of a long line of sculptural experiments dating back to cubism, explored a new sense of motion, a machine sense of aesthetics. Boccioni drew his original forms from Matisse and Picasso, but he extended their vision into more motion and greater flow, closer to the airfoil forms of 20 years later. He expressed similar ideas in paintings such as *Dynamism of a Cyclist* (1913); the capturing of motion is similar to that in Giacomo Balla's famous painting *Dynamism of a Dog on a Leash* (1912). Both artists anticipated actual motion in art, a major ingredient in what would become cybernetic art and robot art. They were pushing the walls of static art to the limit (15).

The Dada Movement

As revolutionary as futurism was, another movement, eventually described as dada, was more revolutionary; it made a point of concentrating on the fantastic and making a benefit out of absurdity. First centered in Zurich, Switzerland, it included writers Hugo Ball and Richard Huelsenbeck, poet Tristan Tarza, and artists of the caliber of Jean (Hans) Arp who, in addition to his sculptures, created automatic drawings and addressed the question of machine forms in his painting and sculpture. Swiss dada also included Sophie Taeuber-Arp, who created *Dada Head* in 1920, and Raoul Hausmann, who built his robot-styled *Mechanical Head* in the same period.

Dada art spread as a movement to France. In 1917, Picabia painted *Amorous Procession,* a clockwork-style still life, and many other machine-parts paintings. Dada also spread to Germany; Kurt Schwitters explored new realms in collages and three-dimensional constructions, and Max Ernst developed endless variations on machine figures and heads in his painting (e.g., *Two Ambiguous Figures* (1919). Dada later came to New York through the visits of Picabia and Duchamp and the persistence of U.S. photographer Alfred Stieglitz (16).

Duchamp

Marcel Duchamp's work simultaneously defies category and creates categories. He invented the idea of mobile art work, thereby tremendously advancing the possibilities of robot art. His *Bicycle Wheel,* created in 1913, is composed of a simple bicycle wheel mounted upon a stool base. The piece shocked people but created interest in kinetic art. Duchamp was also responsible for the discovery of "found art," which has since become a regular art form, and a source for robot artists (see Bailey below). Duchamp's *Nude Descending a Staircase* series (nos. 1 and 2), which he painted in 1911, extended the vision of the cubists and the futurists into "caught motion" in painting. The second painting of the series (Fig. 9) shows greater emphasis on mechanical motion and is therefore more robotic than the first.

In his painted constructions *The Bride* (1912) and *The Bride Stripped Bare by Her Bachelors, Even* (*The Large Glass*) (1915–1923), Duchamp also anticipated later kinetic art and environmental art. In his own way, he opened up the horizons of art as widely as the cubists (16).

Figure 9. Duchamp: *Nude Descending a Staircase* (No. 2) (1912). Courtesy of Philadelphia Museum of Art, The Louise and Walter Arensberg Collection.

During these banner years of Duchamp, a movement called De Stijl flourished in northern Europe. Its principal artists were Jan Toorop, Mondrian, and Theo van Doesburg. Highly disciplined, this style enlarged on the geometrics of cubism; Mondrian took it nearly to its ultimate (*White on White*). De Stijl reinforced the notions of precision and machine simplicity and acted as yet another purifying force behind the modern movement. It has reemerged in contemporary art as Minimalism.

Constructivism was a movement that ran concurrently with De Stijl. Artists such as the Russian Vladimir Tatlin created the now-well-known *Monument to the Third International* in 1919–1920; Alexander Rodchenko and Naum Gabo helped articulate the movement in sculpture and constructed works from a Russian point of view that expressed the essence of the Postimpressionist zeal of France, particularly that of Matisse, Archipenko, and Picasso's sculpture (15–17).

R.U.R.

In the Vysehrad section of Prague, there is a grave in the cemetery of the Church of Saint Peter and Paul. Over the grave there is a bird feeder; it is the grave of Karel Čapek. The grave is no more than a good walk from the National Theatre in Prague, where on January 25, 1921, the play *R.U.R.* opened to a wildly enthusiastic audience. Čapek was a native of the same city, which is famous for Gold Maker's Lane (Zlata Ulicka), a tiny side street off the Hradcany, the most famous castle in Prague, and one noted as the model for another native's novel, Franz Kafka's *The Castle*. Gold Maker's Lane in earlier times housed the alchemists of Prague, who were among the greatest in Europe at their peak and were the future model for Dr. Frankenstein and all his imitators (Fig. 10) (16,18).

The Golem Background

Čapek and Kafka have as their distant "godfather" the famous Prague Rabbi Judah Low, reputed to be the creator of the artificial being the Golem, the Jewish revenge monster. Golem, after being built in the still-existing Alt-Neu Synagogue, operated through a "program" placed either in his mouth or on his brow; the "program," according to legend contained all the names of God. The Golem eventually went mad and had to be destroyed.

Rabbi Low's grave is still well-marked in the Old Jewish cemetery in Prague. It is interesting to note that Marvin Minsky (robot designer and AI specialist) claims to be a relative of Low, as do many other persons associated with robotics. As early as 1916, Danish and German film makers were exploring the Golem story for their experimental films (see MOVIES, ROBOTS IN), which were part of a wave of movies about mechanical people, monsters, and other imaginary creatures. It would be hard to believe that his own home-town spook did not, as well, influence Čapek (4,18,19).

Figure 10. Rossum's Universal Robots (*R.U.R.*) turn on their makers.

R.U.R. Robots and Reactions

R.U.R. described robots as artificially made creatures (made more as soup than as machine, however). The word robot stuck, and its association with creatures that could run amok also stuck, for a time, and on a worldwide scale. Čapek shared an enthusiasm for science and technology but also shared a fear of their consequences (18,19). On stage, his robot, basically a man dressed up in metalized suit, reflective, angular, and with stiff motions as interpreted by the actor, was a very powerful image. The metal men of Oz and fantasy illustration and early science stories such as those by A. Merritt and Edmond Hamilton had anticipated his machine men, but by worldwide staging of his work the look of his robots remained firm in the public mind. His idea shot round the world. One production of *R.U.R.* in Berlin in 1923 had the distinction of being designed by Frederick Kiesler, the sculptor and dadaist, who later went on to create new form vistas in sculpture 30 years later in the United States.

A year after *R.U.R.* played in New York City, another play of similar persuasion, and yet more powerful in its playing, opened at the Garrick Theatre on March 19, 1923. It was called *The Adding Machine* by Elmer Rice, and it continued the concept of machines as dulling, negative influences on individuals and cultures. Futurist Ruggero Vasari's *The Anguish of the Machines* (conceived in 1921 and published in 1925) worked this same theme. *R.U.R.* was, of course, unique, but the time was ripe for an expression of this point of view.

The reaction to *R.U.R.* was immediate and very specific. The robot became a metallic figure, as fully articulated as possible, with the ability to talk, move its limbs, smoke, and stand up. Radio technology supplemented by electric motors and gearing systems was able to create all these effects and the "Showbot" was born. Exhibition and homemade robots diffused Čapek's strongly negative image of the robot. People found show robots amusing, laughable, strangely intriguing, even awe inspiring (Fig. 11) (18).

THE BAUHAUS PERIOD

Making sense out of the machine and technology after the expert killing by machine, en masse, during World War I took some doing. The Bauhaus was one of the forces that took on this responsibility, and it was guided by a combination of artists, craftsmen, and architects that had never been assembled before. The Bauhaus in Weimar started in 1919, two years before the launching of *R.U.R.* It moved to Dessau in 1925, and after Nazi persecution, disbanded and then reassembled at various places in the United States (Black Mountain, IIT, Yale, Harvard).

Led by Walter Gropius, the Bauhaus attempted to reconcile the new technology with art and architecture with a depth that was lacking in its earlier counterpart movement in England under the guidance of William Morris. Fundamentals teachers such as Johannes Iten, Joseph Albers, and Moholy-Nagy brought craft and art skills to the Bauhaus which allowed a ready flow between art, craft, architecture, and technology, as well as modern design and a new machine-style theater. Bauhaus labs and studios matched new ideas with traditional skills and the needs of people and cultures which were contemporary. Works of note, such as the *Light Space Modulator* sculpture made in 1922–1930 by Moholy-Nagy, broke new ground in combining motorized motion with modern sculpture (16,20).

Oskar Schlemmer, who was both a sculptor and the innovative force in Bauhaus theater, created an Archipenko-style sculpture called *Abstract Figure* in 1921. It explored machine-form anatomy with the same rigor and distinction as the theater dances, masks, and experimental pieces he created between 1922 and 1927. The most robotlike of these pieces were *Triadic Ballet* (1922); *Figural Cabinet* (1923); *Trapenwitz* with its clowns (1926); *Space Dance* (1927); *Equilibristics,* which appeared as a definite forerunner of much later kinetic art and art by Calder (1927); and *Dance of Slats, Company of Masks,* and *Dance in Metal* (all 1927). Schlemmer produced within Bauhaus theater robot-style statements and they constitute, possibly, the strongest body of "robot" art by any one artist until the appearance of James Seawright's work, 30 to 50 years later.

Other robot "voices" in the Bauhaus included that of Rudolf Belling, who produced his piece *Head* in 1923, a strongly machinelike sculpture, influenced by Archipenko. Another was Bauhaus teacher Paul Klee, who in 1922 painted *The Twittering Machine,* which is well known for its simple automaton appearance.

A Negative View

Not all European artists looked upon the machine with joy or satisfaction. By far the more vocal of them viewed postwar machine technology and culture with hostility, derision, and scorn. In Germany and Austria, artists of the rank of George Groz and film maker Fritz Lang saw potential for evil—for example, in the machine's control over an impoverished mankind. They saw the machine as an easy tool for the oppressor. In Lang's 1926 classic film *Metropolis,* these fears were realized: science and technology went mad and the robot was at

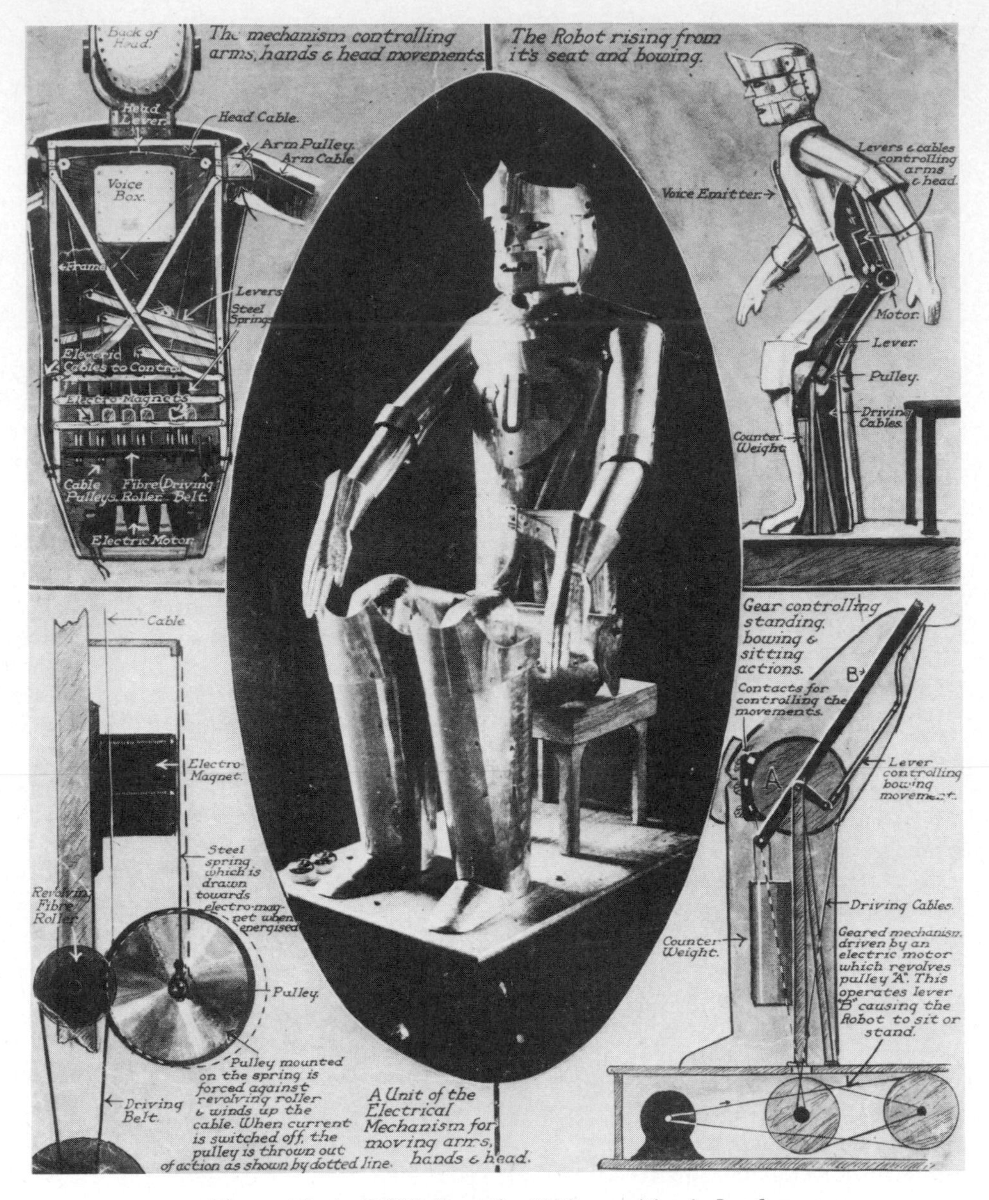

Figure 11. An R.U.R. from the 1928 exposition in London.

the center of the action. On a smaller scale, *Republican Automatons* (1920) (Fig. 12) cogently expressed Groz's feeling toward machines and their master.

MODERN MACHINE ART

The surrealist contribution to the art of robots is indirect and yet powerful. Artists such as Salvadore Dali and Max Ernst rarely created robots or machine men or creatures in any obvious way. Occasionally they did, however, as in Ernst's *The Elephant of Celebes* (1921), which depicts an automaton elephant with a mechanical end-effector, a rotating head, and a fully machinelike appearance (16,17).

The energy of cubism and surrealism emerged in the work of several artists in a way that emphasized the machinelike character of subject matter, technique, and type of media. The work of artists such as Léger, Man Ray (with photograms), de Chirico (with landscapes filled with machine-part mannequins), and later, Hans Arp and Calder (with his great range of work in mobiles and stabiles), was influenced by the pervasive mood of experimentation.

Mobiles

Calder and mobiles are nearly synonymous. Calder added motion to sculpture not as a gimmick or an additive, but as the essence of the work. By so doing, he enlarged the range of sculpture as a medium, advanced robot art, and provided impetus to the nascent robot artist. Calder's early circus sculptures, of wire and "tin," were a reexploration of simple automaton devices. By making his art more abstract, under the influence of this day, he was able to concentrate on pure motion effects. This is demonstrated in his motorized *The White Frame* (1934), which appears highly influenced by Duchamp, and later by

Figure 12. Groz: *Republican Automatons* (1920). Courtesy of The Museum of Modern Art, New York.

his *Lobster Trap and Fish Tail* (1939), which fully explores his own aesthetics in free-moving, hanging mobiles (15,16).

Science Fiction Illustration

The art of illustration may have reached its peak in the 19th century. A. Robida, for one, illustrated many books with robotlike creatures that had metal legs and axle joints. He produced fantasy illustration before science fiction as we know it existed. He worked alongside dozens of illustrators and graphic artists in creating an entire genre of illustration styles and techniques, including engraving and lithographic drawing (which, in their subtlety, became precursors of photography). Fantasy illustration in the late 19th century, as shown by boys' magazines, adventure magazines, and tall-tales magazines of the time, graduated into science fiction illustration with the proliferation and influence of science fiction pulps. In the 1930s, it was science fiction and fantasy material, along with romance magazines, that dominated the newsstands (Fig. 13).

To have science fiction illustration in the true sense, there had to be "science fiction," a term coined, according to legend, by Hugo Gernsback. Science fiction illustration abounded in his magazine, *Amazing Stories*.

The classic artist/illustrators of this genre were H. Wesso, Alex Raymond (Flash Gordon), Frank R. Paul (a greatly prolific illustrator), S. R. Drigin, and later, Kelly Freas, to name but a few. Robots, aliens and rocketships became the staple subjects for covers and dramatic inside illustrations. Stories of robots reached their culmination during this high point with the work of Isaac Asimov. Under the tutelage of editor John W. Campbell, the modern concept of the friendly robot was born. The friendly robot became a tradition that led to *Robby the Robot* of TV and film fame, and later to R2D2 and C3PO of *Star Wars* (21,22).

ART AND REAL ROBOTS

American scientist Vannevar Bush's Differential Analyzer, Englishman Alan Turing's Turing machine design, early experiments in servomechanisms in the late 1930s and early 1940s, and the later creation of electronic mice (Thomas Ross, 1938) and tortoises (W. Grey Walter, 1948) all were steps toward the electronic computer and to the more recent electronically controlled robot. Programmable machines constantly wavered between self-acting (and therefore robotlike) and self-thinking (computerlike) constructions. The reliability of subassemblies increased in the 1950s and 1960s, and the physical scale of "intelligent" machinery slowly diminished, ie, from tubes to transistors to microelectronics.

The artist was, most often, an onlooker to this progress in servomechanisms, robotics, microprocessors, and computers. Few artists could solder or follow circuits, and few engineers knew or even cared about art. However, mobiles, new visions, and early studies in kinetic art (23) were proof that art and technology were becoming more compatible.

Art and Technology Movement

In the period from 1950 to 1965 in the United States, there was a great surge of new art. Some ascribed it to the blossoming of the colleges and universities under the G.I. Bill, others to the financial and communications advantage of urban centers such as New York City, and still others to the influence of the beatniks. Influencing its thrust were such movements as Happenings (people such as Alan Kaprow, for instance) and an updated version of the kind of experimentation that occurred in Bauhaus theater. This artistic energy was joined by technical energy from such centers as MIT, Yale, and Stanford. Artist met engineer. Typically, U.S. art is influenced by many foreign individuals, and this movement was no exception; many of these artists eventually made their home in the United States.

One form this energy took was in a loosely formed movement that is best defined as Kinetics (16,23,24). One of the prime movers of kinetic art was Swiss-born sculptor Jean Tinguely; his work began, in the mid-1950's, with paintings of machines. By demonstrating his kinetic art in the garden of The Museum of Modern Art in New York in 1960, he communicated the Kinetic movement with great impact. He then produced *M.K. III* (1964), a kinetic sculpture in iron. Sculpture that moved by motors, gravity, and assistance from man proliferated. Another early exponent, also from foreign soil, Len Lye produced *Fountain II* in 1959 using steel rods attached to an electric motor for motion. This is the same Lye who had produced *The Birth of a Robot* film in 1935). The 1960s

Figure 13. Sci-fi magazines of the 1930s. Photograph by Ralph Gabriner.

were a hotbed for robotic art: sculptor Nicolas Schoffer's electrically driven sculptures, Nam June Paik's early work, and (English sculptor and collage maker) Eduardo Paolozzi's *St. Sebastian, No. 2.* Figure sculptor Edward Keinholz produced the amusing *The Friendly Grey Computer,* a motorized rocking chair with doll parts and a telephone speaking system that seemed to be laughing at the pretensions of robot art.

Art and technology as a movement was formally launched in the United States by Robert Rauschenberg and engineer Billy Kluver as EAT (Experiments in Art and Technology), Inc. In EAT, the artist found a marriage with the engineer; nevertheless, much of the work resulting from EAT's collaborative ventures was judged inconclusive by the art viewer. During the EAT period, artists began creating wildly imaginative light sculptures, large-scale wire sculptures, and massive environmental forms that were reminiscent of the work of dadaist sculptor and architect Frederick Kiesler. Some of the more notable artists included wire-forms artist Richard Lippold, light artists Julio Le Parc and Otto Peine, metal stabile artists Jesus Rafael Soto and Gunther Haese, and sculptural environmentalists Dan Falvin, Preston McClanahan, Chryssa, and Howard Jones.

To the art and technology movement should be added artists such as the following, whose work specifically addressed itself to robots. The French artist and designer François Dallegret created *The Machine* (1966), which is of monumental scale and makes use of light-sensitive tone production. Sculptor Ernest Trova produced the *Falling Man* series (1966), which influenced much of his later robot work. Richard Lindner painted a series of dramatic figures that combined a robot look and sadistic feeling. During this period, the tradition of the mobile was continued by Calder and sculptor George Rickey, who created motions and forms uniquely his own.

Cybernetics

As artists added sound to light and motion, a new art form that recalled the control theory of an earlier decade by Norbert Weiner was being developed: cybernetics. Artists were no longer content just to make things move or light up. They wanted to control either the program of the art or have the viewer create a pattern within the art piece through his or her interaction. Such was Takis's 1966 *Telemagnetic Musical,* an electromagnetic construction, and Nam June Paik's well-known 1965 piece, *Robot 456 with 20-Channel Radio Control + 10 Channel Data Recorder,* which acts like a robot but looks dismembered. In a similar spirit are Bruce Lacey's *Superman 2963* (1963) and *Boy, Oh Boy, am I Living!* (1964); both were motorized mannequin figures built within an iron-pipe framework. Artist Wen-Ying Tsa'i created the more abstract *Cybernetic Sculpture* (1970) out of rods on a motorized base, strobe lights, and a microphone (16,17,23,24). The 1960s and early 1970s saw a great deal of experimental cybernetic and kinetic art; however, the work of this period preceded the enormous enthusiasm for robots generated in the late 1970s and the advent of the inexpensive microprocessor and microcomputer (16,23,24).

CONTEMPORARY ROBOT ART

Robot art has diffused itself into the contemporary art scene. It maintains the tradition of kinetic art; it builds forms into working electronics; it extends sculpture, as well, into electronics as illusion (the appearance of electronics if not the reality); it can be integrated into various drawing and painting styles; and it can fall into the category of illustration photography. Robot art is less a movement than a pressure that is applied to contemporary art in general.

Sculpture

Dutch sculptor Paul von Hoeydonck may be best known as the first artist to have his art work placed on the moon (a small sculpted figure). He is probably the most prolific sculptor of robots and robot-related figures. Von Hoeydonck's specialty is not engineered working robots but bronzes and welded pieces that create the illusion of possessing electronic parts and of being working robot forms. His 1968 *CYB Head and Arm* (Fig. 14), and his *CYB Legs* are illustrative of this: their machine- and electronic-part construction suggests they will work if only switched on.

Ernest Trova has continued to create his smooth, polished metal sculptures which bridge the gap between classical sculpture and robotic forms. Trava's highly finished pieces are a stark contrast to the streamlined aluminum sculpture of Clayton Bailey, who may, to date, have made more robot forms than anyone. Originally a ceramic artist, Bailey creates robots from metal found in junk yards. His figures are often seven

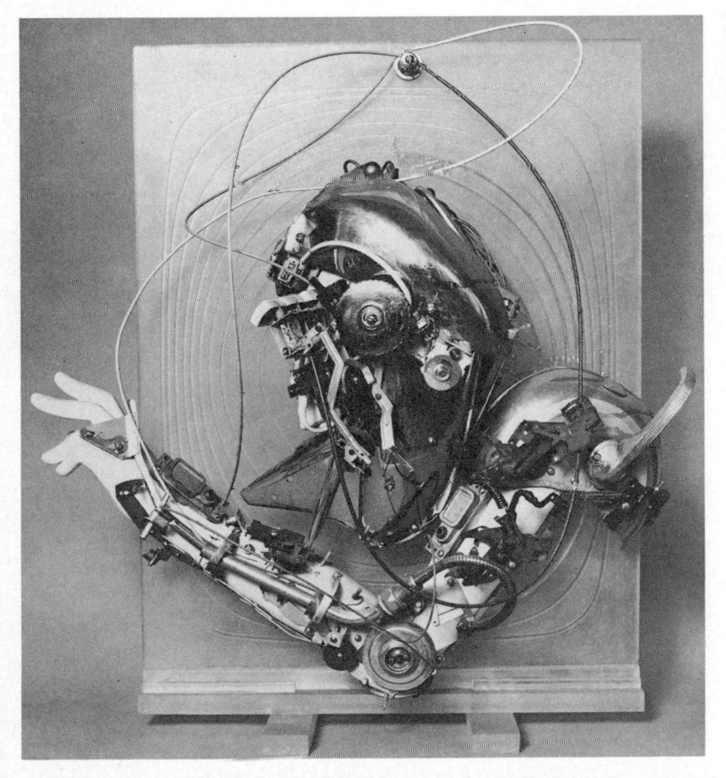

Figure 14. Von Hoeydonck: *CYB Head and Arm* (1968). Courtesy of von Hoeydonck.

Figure 15. The upper part of one of Clayton Bailey's sculptures.

feet or taller and frequently incorporate clocks and radios (usually old chrome types from cars). His predilection for art-deco forms leads him to forage in junk yards for such objects as airflow-style coffee pots, wringer-washer parts, and old toasters (Fig. 15) (23,25).

Toby Buonagurio's totemlike ceramic figures combine a lust for color, wild and exploratory forms, and a playful imagination. Her finishes include extremely vivid colors and silver, chrome, and gold. She is able to make her ceramics resemble a piece of acrylic, which is quite an accomplishment. Her inspiration seems to be drawn from two unlikely bedfellows: the Marquis de Sade and Walt Disney (25).

German sculptor Peter Vogel created cybernetic sculptures that are a lacework of wire and electronic parts. They play sound through light interruption or proximity and display a whimsical sense of humor. His *Going Up* and *Windy Noises* (both 1980) react to sounds through the use of photocells which change in tone or pitch depending on light intensity. Vogel has produced sculptures that physically dominate entire walls and change their form of response as the viewer walks along.

Marc van den Broek, a young sculptor from Wiesbaden, Germany, incorporates large hanging forms that appear bird- or planelike into his cybernetic sculpture. He calls them mutations from nature moving toward technology. His flying objects are made of soldered wire and stretched cloth and electronic parts. They share the world of Vogel, yet in form are more reminiscent of Calder mobiles, da Vinci's flying machine, or the Wright brothers' plane models. When approached, his sculptures, such as his 1983 *Manta* and *Solar 500,* respond through sound-activating sensors.

Photography and Robots

The influence of the early tradition of the dada photographers, such as Man Ray, on contemporary photography involving machines and robots would be hard to describe. As a machine-produced and processed art itself, photography has made use of surreal techniques and content in a very general way. However, there a few photographers who have carved a special niche for themselves by using the robot as a subject.

Les Krims is one such artist. His series of photos depicting student-made robots in an interior setting is titled *Uranium Robots* (1976). These art prints are unusual in that they are toned with uranium ore and are slightly radioactive. Shot in black and white, they seem less science fiction than pure photographic art.

Dan McCoy, whose work is most often seen in popular science magazines, photographs conventional industrial robots and personal robots in such a way that they become images of brilliant fantasy. Rather than using art to make robots, he uses robots to make art.

Illustration Plus

Robot illustration remains a staple of science fiction book covers and magazines. Rarely does it rise above the ordinary. However, there are exceptions among the painters: airbrush wizard Stanislaw Fernandez, John Schoenherr, designer and film illustrator Syd Mead, John Berkey, space artist Robert McCall, and English illustrators Jim Burns, Chris Foss (noted for his monster robots the size of a city), and Ian Miller. One illustrator who has dominated the mid-1980s with rather brutal images of cybernetic/bionic/robot figures and faces is Marshall Arisman. Although his work seems clearly influenced by English painter, Francis Bacon, Arisman has his own vision, which he applies to his biomechanical brutes.

Three-dimensional illustration has been produced by Walter Einsel, and two-dimensional illustration in the same vein by his wife, Naiad. They have created an entire menagerie of old-fashioned robots. Einsel's *Clock Man* (1981) has a moving arm and gongs that sound, and his *The Man Who Comes Apart* (1968) literally does: his hat tips, his nose folds down, he loses his neat collar, his chin drops, and his jacket opens to reveal a pulsating heart. His works are whimsical and beautifully crafted. Nick Aristovoulos, a sculptor who also produces three-dimensional illustration, most often works in acrylic and has produced stunning images of robot hearts, faces and figures (25).

Painting and Drawings

French designer/artist François Dallegret has continued since the 1960s to create drawings of robot-style machines and robotic figures and environments. These drawings have a wry sense of humor and drafting engineer's precision.

Swiss artist H. R. Giger's paintings and drawings may be best known through his creative association with the film *Alien.* Giger uses airbrush calligraphy to create a gothic vision of extra ordinary visual and emotion range. He has been creating biomechanical creatures with robotic overtones for nearly 20 years. Giger's figures seem like hybrids of well-oiled mechanisms and diabolical organic squid.

The collage, as practiced by Paolozzi, seems only too appropriate for robot art. The collage form is, after all, "art" in the Frankenstein style, since they are made out of arbitrary parts, assemblages of seemingly unconnected images. Sometimes these disjointed images can become connected as when they make the semblance of a robot figure or face. Robert Malone has created collages (Fig. 16), as has illustrator Jean-Claude Suares. Along with many others, they have extended the art of collage to include a contemporary view of artificial creatures and robots (21,26).

Robotics as Art

Throughout the history of robot art, the artist has drawn on his or her resources for its appearance and on science and electronics for its form, procedure, and method. The work of one artist, James Seawright, blends art and robotics in a way that does not compromise either. His art is a direct expression of the functional form that his figures, plants, or environments

Figure 16. A collage by Robert Malone.

Figure 17. The robot as currently portrayed by the toy industry. Photograph courtesy of Ralph Gabriner.

require; the robotics performs within the program, designed without added-on illusions or false promises.

Seawright developed his techniques in early pieces such as *Searcher* (1965–1966), *Watcher* (1965–1966) *Captive* (1966), and *Tower* (1967). Later he built upon his techniques to create room-filling viewer-response sculptures or environments. His recent work *House Plants* (1983) is computer-controlled and interacts with the viewer. It reacts to changes in light created by environmental or viewer change; the two "plants" respond with light displays and physical motions (23,25).

CONCLUSION

Art that calls for a complex human response, such as some kinetic art, environmental art, and the new robot art, may be the direction both art and our technology are taking. There are many types of art now calling for public response. Sometimes such art is quite approachable, but some works are not; for instance, it may be that the piece is too much a private vision of the artist. Robotic art, often more approachable and "user friendly" than other art forms (it may walk, talk, or relate to the viewer), could well become very popular, just as new electronic music has in the pop music industry (27). Certainly, the robot is currently a very popular image, as can be seen by toys on the market (Fig. 17).

The emergence of robotic art may also signal a closer association between the forces that make up our technological expression and the normal channels of culture. Culture may be getting closer to our technology and our technology may be becoming more human, more ergonomically aware. The fact that pacemakers, calculators, and wrist TVs can be worn exemplifies the first inclination. The trend toward "sculpted" car interiors (the instruments become a second skin,) space suits which are communications devices and life-support systems, and voice-responsive machines may be a case of the second inclination.

In the future, robotic art should become a more direct extension of new tools such as computer graphic systems, remote-controlled manipulators, and network machines which allow for the kind of team building associated with medeival cathedrals. Time-shared, space-shared robotic art could become the art of the future.

BIBLIOGRAPHY

1. H. Gardner, *Art Through the Ages,* Harcourt Brace Jovanovich, Inc, New York, 1976. As with Ref. 15, work is practically synonymous with an introduction to art.
2. A. Chapuis and E. Droz, *Les Automates,* Neuchâtel, Switzerland, 1947. One of many books by Chapuis, who was the foremost authority on automatons in the world, this is an exhaustively researched work with extensive background on clockworks, music boxes, and 16th–20th-century automatons.
3. D. Fraser, *Primitive Art,* Doubleday & Co., Inc., New York, 1962. Accurate commentary as far as it goes, but appears dated in its opinions and point of view.
4. J. Cohen, *Human Robots in Myth and Science,* G. Allen & Unwin, Ltd., London, 1966. Possibly the best book on the roots of robots in literature and mythology. A good companion volume would be McCorduck's *Machines Who Think.*
5. H. A. Groenwegen-Frankfort and B. Ashmole, *Art of the Ancient World,* Prentice-Hall, Inc., Englewood Cliffs, N.J., and Harry N. Abrams, Inc., New York, 1977. A fine text with magnificent illustrations which makes classical art a point of departure for most Western three-dimensional sculpture.
6. U. Eco and G. B. Zorzoli, *The Picture History of Inventions,* The Macmillan Co., New York, 1963. Contains an excellent section on Hero of Alexandria and another on 18th-century automatons. Generally a good treatment of inventions and their historical and technological settings.
7. J. Needham, *Science and Civilization in China,* Cambridge University Press, Cambridge, UK, 1959. The great comprehensive work on Chinese science and technology. See also Needham's *Science in Traditional China.*
8. L. Mumford, *Technics and Civilization,* Harcourt, Brace & World, New York, 1963. Pungent and penetrating, this is one of a dozen or more books by Mumford that traces the factual and conceptual history of machine culture.
9. L. Mumford, *The Myth of the Machine,* Vols. I and II, Harcourt, Brace & World, New York, 1967. An examination of the power and meaning behind machines within the historical context so familiar to Mumford.
10. S. Giedion, *Mechanization Takes Command,* Oxford University Press, Oxford, UK, 1948. The most comprehensive text, on the transition to our machine culture, it suffers slightly because it was written before the electronic revolution.

11. F. McConnell, *The Science Fiction of H. G. Wells,* Oxford University Press, Oxford, UK, 1981. An intelligent review of Wells' science fiction ideas, fantasy imagination, and influence.
12. I. F. Clarke, *The Pattern of Expectation,* Jonathan Cape, London, 1979. Covers the history of futurism from 1644 to 2001; the description of the period between the turn of the century and the aftermath of World War I is particularly good.
13. H. M. Geduld and R. Gottesman, eds., *Robots, Robots, Robots,* N.Y. Graphic Society, Boston, 1978. A series of fine essays on history, science fiction, and film.
14. *Machine Art, March 6 to April 30, 1934,* The Museum of Modern Art, New York, 1969. A comprehensive catalog review of machines in art and the effects of machines on art and art on machines during this period.
15. H. W. Janson, *History of Art,* Prentice-Hall, Inc., Englewood Cliffs, N.J., and Harry N. Abrams, Inc., New York, 1978. Another classic art history text which offers a different perspective than Gardner.
16. H. H. Arnason, *History of Modern Art,* Prentice-Hall, Inc., Englewood Cliffs, N.J., and Harry N. Abrams, Inc., New York, 1968. One of the best texts on modern art and contains individual biographies of most major artists.
17. S. Cheney, *A Primer of Modern Art,* Liveright Publishing Corp., New York, 1966. A good survey of the origins and practice of modern art, but slightly dated.
18. K. Čapek, *R.U.R.,* Washington Square Press, New York, 1969. Reading the play is better than reading about it. While not great theater, it has great imagination. This edition has fine historical material on Čapek, Prague, and the era, including excerpts from reviews from the 1920s.
19. T. Goldstein, *The Dawn of Modern Science,* Houghton Mifflin Co., Boston, 1980. A solid review of the science behind technology and its cultural interaction.
20. H. M. Wingler, *The Bauhaus,* Cambridge, Mass., 1978. The best text on the history and works of the Bauhaus, period by period and artist by artist.
21. R. Malone, *The Robot Book,* Harcourt Brace Jovanovich, Inc., New York, 1978. An contemporary survey of robots and their friends, heavily illustrated.
22. F. Antony, *One Hundred Years of Science-Fiction Illustration,* Pyramid, New York, 1975. A broad survey of the subject with many illustrations.
23. J. Burnham, *Beyond Modern Sculpture,* George Braziller, Inc., New York, 1973. A carefully documented survey of modern sculpture, particularly robotic sculpture, it provides a reasoned perspective on the aesthetics of sculpture from Burnhams' view.
24. D. Douglas, *Art and the Future,* Praeger Publishers, New York, 1973. A good review of the art and technology movement and the implications of art trends for the future.
25. R. Malone, *The Robot Exhibition* (Catalog), American Craft Museum, New York, 1984. Details the first major museum exhibition specifically devoted to robots; a good deal of attention is paid to the art of robots.
26. J. Reichardt, *Robots, Fact, Fiction, and Prediction,* Penguin Books, Harmondsworth, UK, 1978. A fine and well-researched survey of robots which gives an excellent chronology, it covers every area, from history to industry.
27. V. C. Ferkiss, *Technological Man, The Myth and the Reality,* Mentor/NAL, New York, 1969. An intelligent and penetrating study of the trend toward machine culture and automation.

General Reference

D. Kyle, *Science Fiction Ideas & Dreams,* Hamyln, London, 1977. Discusses science fiction as literature, art, and subculture.

ASSEMBLY, ROBOTIC, DESIGN FOR

Perry Carter
Brigham Young University
Provo, Utah

INTRODUCTION

Approximately 5% of all discrete products manufactured in the United States are assembled on special-purpose automatic machines. These machines are custom engineered and built for a specific product. They typically cost between $500,000 and $1 million. It is not unusual for the assembly machine to have a one-year lead time followed by several months of debugging. To justify such an investment, a product must have an expected market life of several years. The number of different product styles to be built on the machine must be small, and only limited product redesign can be expected. Also, the annual production volume to be run on the machine should be about 1 million assemblies per shift (1). With today's trends toward a variety of product styles, short market life cycles, and frequent engineering design changes, relatively few products meet the criteria justifying special-purpose assembly machines. As a result, most mechanical batch assembly is done by hand. This manual assembly typically accounts for between 30 and 50% of the total direct labor cost of discrete goods manufactured in the United States today (2,3).

Industrial robots offer the potential for economically automating a portion of this manual assembly. The principal advantage of robots over the special-purpose type of automation described above is that one robot can do the work of several special-purpose workheads, with their associated indexing or work transfer system, at a lower initial cost and with a shorter system development time. The robot system will have a longer cycle time than the special-purpose assembly system, but this may be acceptable where smaller batches are being considered and finished-goods inventories are being reduced (see Automation).

A second advantage of robots is their reprogrammability, or ability to change tasks as required to accommodate the product style variations, design changes, and new product introductions that typify modern batch manufacturing. This advantage is more difficult to exploit than the first because of the special-purpose nature (nonreprogrammability) of most of the ancillary equipment, such as feeders, orienting devices, pallets, fixtures, magazines, and so on, that is used with the robot (see Programmable assembly).

The rest of this article discusses the interdependences among the design of a product, the assembly equipment, the cycle time, and the cost of robotic assembly (see Assembly robots; Product design). The article is divided into five sections. The first section, **Choosing the Assembly System,** describes three possible robot assembly system configurations to illustrate how the choice of the system influences the effect that product design will have on the robotic assembly process.

The second and third sections, **Part Handling** and **Part Insertion,** discuss the relationships among part design and the ease of feeding and orienting the parts and the insertion or joining of parts to the assembly, respectively. It is particularly important to separate design for feeding from design for insertion because robotic assembly tends to offer a wider choice of attractive feeding techniques than does special-purpose assembly.

The fourth section, **Reducing the Number of Parts,** presents a useful technique for identifying parts that should be considered for elimination or combination with other parts. Reducing the number of parts in an assembly will usually have a greater effect on assembly costs than any other design modification and should be given correspondingly careful consideration by product designers.

The fifth section, **Design for Assembly Strategies,** presents two techniques for quantitatively evaluating product designs for robotic assembly, with simple example evaluations from each.

CHOOSING THE ROBOTIC ASSEMBLY SYSTEM

Before considering the design of parts and products for robotic assembly, the designer should determine the type of robot assembly system that will be used. Three typical system configurations will be presented here: the one-arm single-station system; the two-arm single-station system; and the multistation transfer-line system. Several product-design-dependent factors whose effect on assembly costs depends on the choice of system configuration are (*1*) the number of different grippers required, (*2*) the types of part feeders required, (*3*) parts that must be held in place while separate fasteners are added, and (*4*) the manual assembly operations required. The type of assembly system must be known beforehand to design a product for lowest-cost assembly with regard to these factors.

For the purpose of this article, robotic assembly refers to a system where one or more industrial robots, using suitable grippers or other tools, join several small, mechanical parts into a transportable unit or product. Typical robot assembly systems consist of one or more workstations, possibly connected by a work transfer device, with one or two robot arms working at each station. It is unlikely that more than two arms would be placed at any one station. The accumulation of idle time on the arms as they wait for the others to take their turns inserting parts would probably make a station with more than two arms uneconomical.

The elements of a basic *two-arm,* single-station assembly system are shown in Figure 1. The station includes two general-purpose robot arms, each of which has access to its own part feeders or magazines, a tool or gripper change station, and the common work fixture. One of the arms can also reach a conveyor or pallet for depositing completed assemblies. Surrounding the work fixture is a work envelope within which there may be only one arm at a time. The purpose of the envelope is to prevent collisions between the two arms. It represents the points where either arm can be stopped if necessary with no danger of interfering with the other arm. Protection is by computer control, which under certain circumstances may have to allow both arms to enter the envelope together. For example, a part may have to be held in place by one arm while another part is added by the other arm.

Each robot arm can reach its own station for gripper changes. To minimize the arm joint motions, the part presentation points preferably should be placed at the same height as the insertion point and at a common radius from the robot's base rotation axis. It may be assumed that, with the limita-

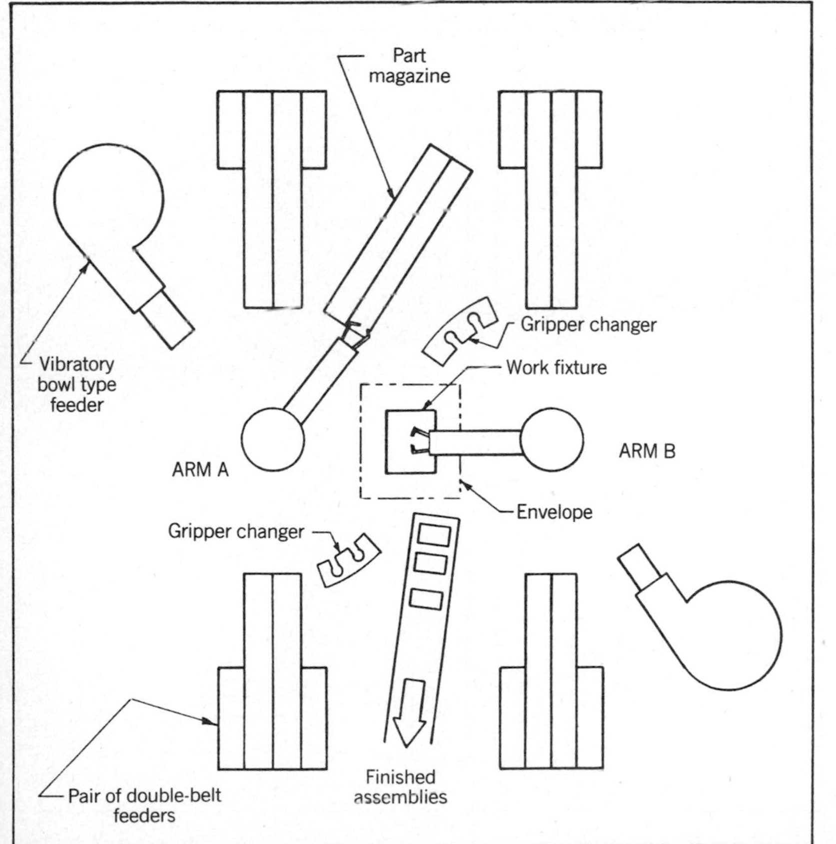

Figure 1. Two-arm assembly system.

tions of the robots' work envelope, no one feeder or magazine can be reached by both arms.

The basic *one-arm*, single-station assembly system is shown in Figure 2. At a one-arm station, any operation requiring the holding of one part while another is inserted will necessitate an additional special-purpose device of some kind, increasing the total capital investment in the system. Similarly, with only one gripper in use at a time, the need for gripper changes is greater at a one-arm station. The one-arm station will usually have the longest cycle time of any robot system, which means manually loaded part magazines and pallets may be economical.

A third type of robot assembly system consists of several one- or two-arm stations as described above, connected by a work conveyor as shown in Figure 3. This is called a transfer-type robot assembly system and offers the advantage of a variable cycle time by adjusting the number of operations performed, or parts assembled, at each workstation. Gripper changes can be virtually eliminated on such a system by allowing only parts that can be handled by the same gripper to be assembled at each station. With cycle times typically much shorter on transfer-type systems, higher-speed feeders can be used, and part design rules pertaining to such automatic feeding would apply.

The transfer-line system is of particular interest with parts requiring manual assembly. Most products include a small percentage of parts that cannot be handled automatically, such as parts having long wires attached, large, flexible gaskets, or flexible belts. These kinds of parts will continue to be assembled manually, although the rest of the assembly may be automated. For obvious safety reasons, such manual operations cannot be performed within the robot's work envelope. These operations require that the partial assembly be transferred away from the robot to a manual station and then possibly back to the robot for further work. On a single-station system this can be costly in terms of time and equipment, but on a transfer system it is not difficult to place a manual station on the conveyor line. Thus, when a transfer system is selected, parts requiring manual assembly may not increase product assembly cost as much as on a single-station system.

The choice of assembly system is largely economic and will depend mostly upon variables such as the production rate, the number of parts in the assembly, and the total volume to be produced. Figure 4 shows regions of economic application for various assembly methods where the equipment is assumed to operate two shifts per day and is amortized over a three-year period (4). The batch size represents the number of assemblies built during the three-year period. The limitation on the number of parts that can be assembled economically at a single station arises because of the increase in cycle time caused by each additional part, which causes a corresponding increase in the cost of using the peripheral equipment associated with individual parts.

This situation can be illustrated by a simple example. Consider a product consisting of 20 parts with an annual production of 800,000 assemblies. A proposed assembly system will work two shifts per day or 14.4 million (10^6) seconds per year. This means one assembly must be completed every 18 s. If 6 s is allowed as the average assembly time per part at a one-arm station, the cycle time for the 20-part assembly is 120 s. The number of such stations required to meet the production is 120/18 = 6.7, or 7. That means 7 robot arms and 140 part feeders are required.

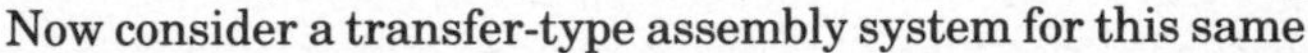

Now consider a transfer-type assembly system for this same

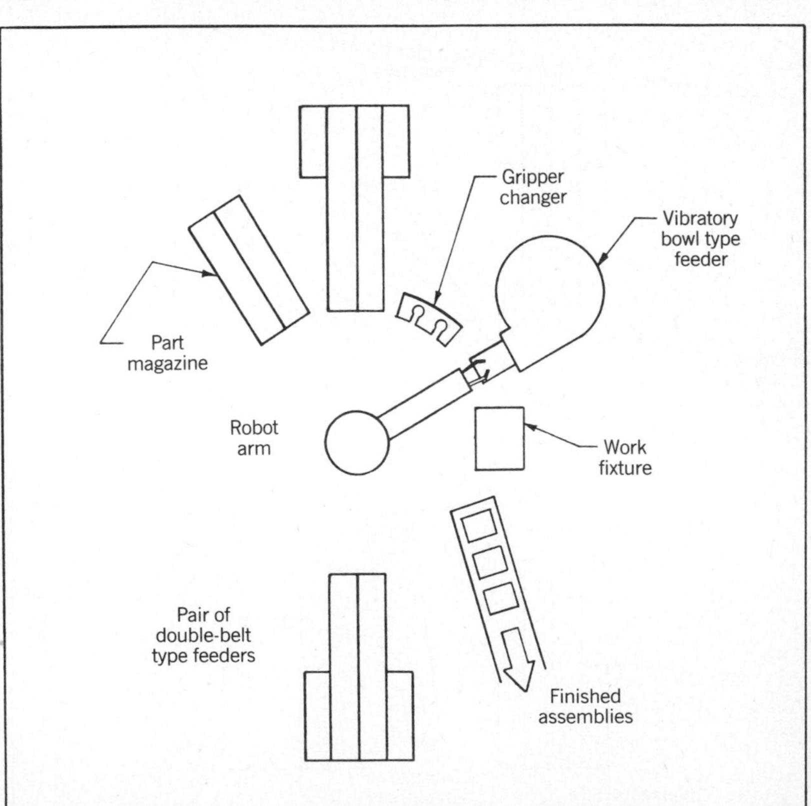

Figure 2. One-arm assembly system.

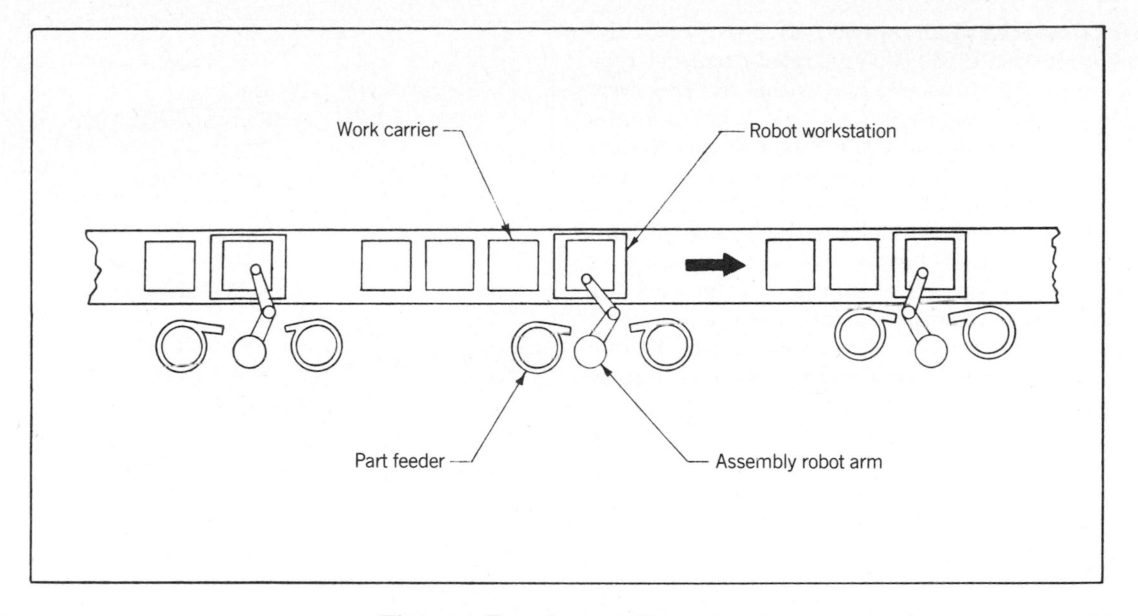

Figure 3. Transfer assembly system.

product. For a system cycle time of 18 s and a 6-s average assembly time for each part, each robot arm can assemble three parts. The number of stations on the transfer line will then be 20/3 = 6.66, or again, 7 stations, but with the transfer system only 20 feeders are required compared to the 140 feeders required for the 7 independent single-station systems. The cost of the pallet conveying equipment must, of course, be added to the total for the transfer system.

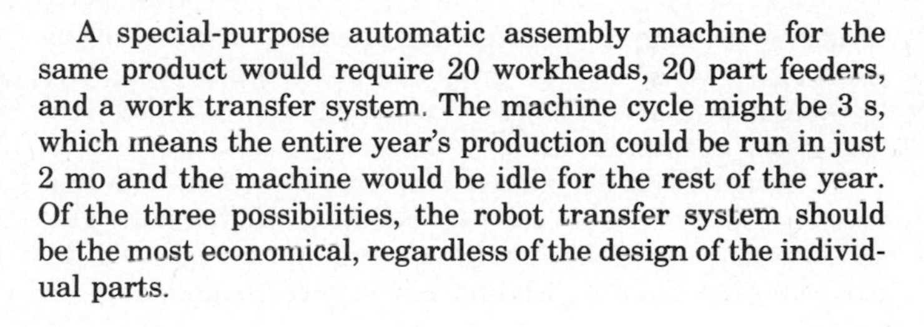

A special-purpose automatic assembly machine for the same product would require 20 workheads, 20 part feeders, and a work transfer system. The machine cycle might be 3 s, which means the entire year's production could be run in just 2 mo and the machine would be idle for the rest of the year. Of the three possibilities, the robot transfer system should be the most economical, regardless of the design of the individual parts.

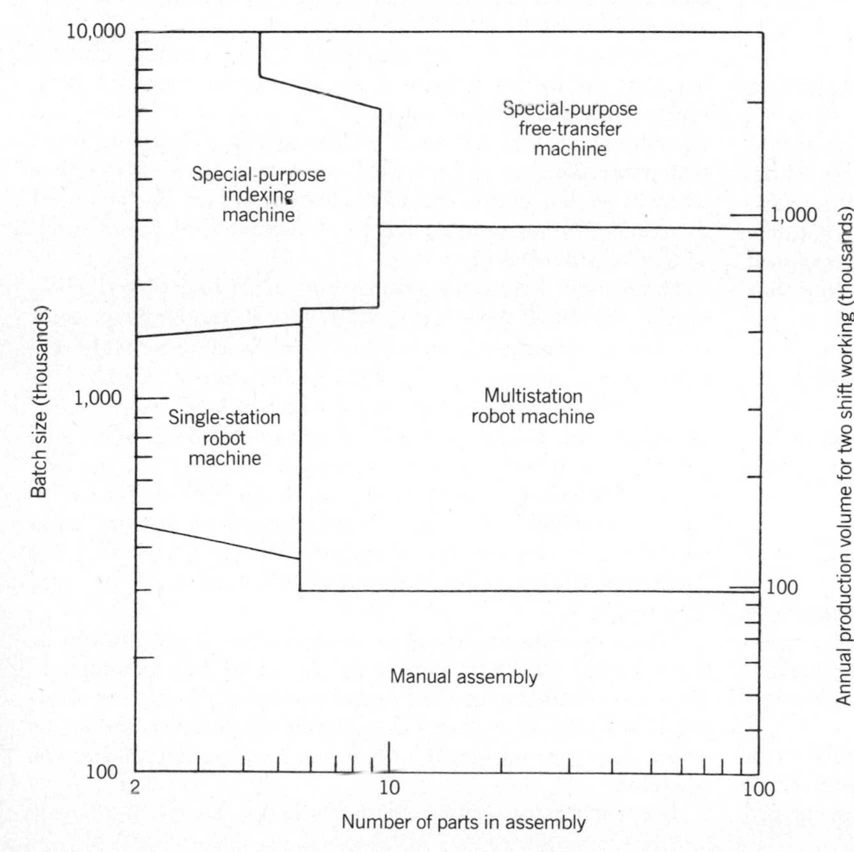

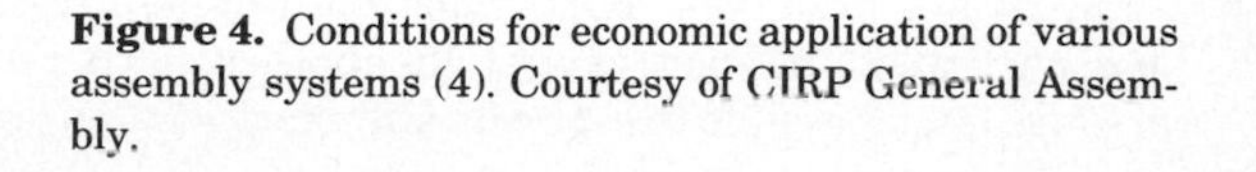

Figure 4. Conditions for economic application of various assembly systems (4). Courtesy of CIRP General Assembly.

If the product considered above had only 5 parts and the same required production of 800,000/yr, a robotic transfer-type system would required 2 stations to stay within the 18-s cycle time, assembling 3 parts at one station and 2 parts at the other. This system would require two robots, five part feeders and a conveyor. Alternatively, a two-arm single-station system could be used, where with two arms working together the assembly time per part would be reduced from 6 s to perhaps 4 s, giving a 20-s cycle time for the five-part assembly. At that rate, only 720,000 assemblies would be completed per year, necessitating some overtime operation of the system or perhaps some manual assembly. This single-station system would also require two robot arms and five feeders, but no conveyor, making it probably the better choice in this case.

This example has been simple, but illustrates the need to select first the most economic robot assembly system based on production requirements and the number of parts in the assembly. Product design can then be considered in light of the specific type of system to be built.

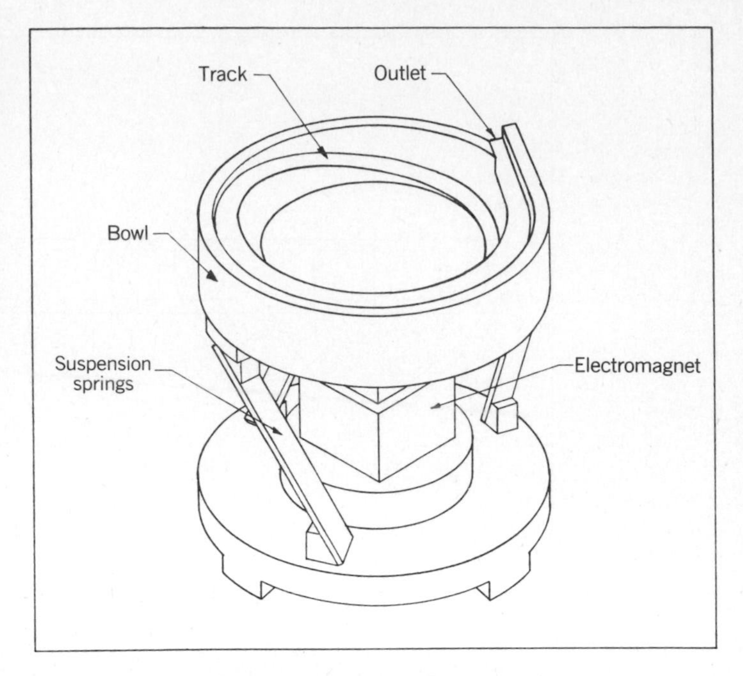

Figure 5. Vibratory bowl feeder.

PART HANDLING

The tasks involved in robotic assembly can be conveniently divided into two groups: (*1*) feeding, orienting, and presenting parts to the robot, and (*2*) gripping, inserting, securing, and inspecting the parts. These are referred to as "handling" and "insertion," respectively.

To design parts for efficient, low-cost handling, it is first necessary to determine how part handling will be done. This section discusses alternative part-handling methods and the particular constraints each imposes on part design.

Vibratory Bowls

The most versatile device for feeding and orienting small parts is the vibratory bowl. A typical bowl is pictured in Figure 5 and consists of a vibrating base, a stainless steel bowl with a spiral track along the inside and perhaps the outside walls, and tooling to orient the parts as they move along the track. Usually, a nonvibrating feed track is provided to hold a queue of oriented parts, with an escapement to allow the removal of one part at a time. Normally, the bowl and tooling, but not the vibratory base, must be custom made for feeding specific parts and cannot be reused. The cost of feeding parts with a vibratory bowl is proportional to the time between delivered parts. This results in low feeding costs per part on high-speed automatic assembly machines, but on robot assembly systems having much longer cycle times, vibratory bowl feeding may become uneconomical.

Design Rules

Part design guidelines for conventional vibratory feeding have been discussed in detail (1, 5–12). These rules generally apply to three aspects of bowl feeding: separating from bulk, conveying along a vibratory track, and orientation by mechanical tooling.

Separating from Bulk. Parts must not tangle. Projections on one part must not get caught in another part. Snap rings are almost impossible to separate automatically, as are springs with open coils on the ends or gaps between coils that are wider than the wire diameter. Parts must not nest. Parts that can nest as paper cups do are difficult to separate. Parts must not stick together. Parts coated with grease or oil may stick together in bulk, as will light parts having a static electrical charge. Parts must not be fragile or have delicate features that could be damaged by the continuous churning at the bottom of the bowl.

Conveying Along a Track. Parts must have a stable orientation that can be maintained as the part slides along the horizontal track and vertical bowl wall. Typically, parts push each other up the track and must not shingle or ride up on top of each other. The forces from pushing or being pushed by other parts must not cause the parts to reorient or be forced off the track. The parts must not be so flexible that they cannot be pushed up a track.

Orientation. Vibratory bowl tooling either rejects to the bottom of the bowl those parts that do not randomly present the desired orientation at the end of the track or actively reorients parts, improving the feeding efficiency of the bowl. If feeding efficiency is too low, more than one bowl may be used to supply the needed rate. To improve orienting efficiency, parts should be symmetrical. If possible, every stable resting aspect of the part should give the same orientation. For example, if a threaded stud has a longer thread on one end than on the other, two possible orientations exist. By making the thread lengths equal, presentation of either end gives the same orientation.

Where complete symmetry is impossible, as it usually is, the features causing asymmetry should be large enough to allow easy detection by mechanical devices. Parts having internal asymmetrical features that cannot be detected cannot be oriented in a bowl feeder unless some external indication is provided.

One constraint, critical for parts being fed for high-speed

automatic assembly machines, is that part quality be high, meaning that the proportion of parts that would cause a jam in a feeder or escapement must be very low. This rule may be somewhat relaxed for a robotic assembly system, again because of the longer cycle time. For example, assume that 1 part in every 100 causes a stoppage of 30 s on a special-purpose automatic assembly machine. If there are ten parts assembled every 3 s, one stoppage will occur every 30 s, and the machine will run only 50% of the time. Let parts of the same quality be fed at the aforementioned five-part robot assembly system with a 30-s cycle time. Now a 30-s stoppage occurs only every 10 min, resulting in less than 5% downtime. Clearly, part quality is less critical in designing for robotic assembly than for high-speed automatic assembly.

Another constraint is size. Vibratory bowls are seldom built for parts having a dimension greater than 150 mm. Typical robots used for assembly are not so limited, and most other part presentation methods allow larger parts.

Programmable Feeders

To build an economic assembly system that truly takes advantage of the reprogrammability of robots, the system should also have reprogrammable feeders. The robot's ability to change tasks between product style variations or redesigns is of little benefit if the part feeders cannot also be adapted easily. Several researchers have reported progress in the development of programmable feeders.

An adjustable tooling for the vibratory bowl feeder that allows a wide range of parts from the same family, say, headed parts, to be fed from the same bowl has been developed (13). Programming the bowl for a particular part is accomplished by manually fitting a sample part into the tooling and positioning and clamping the devices for the part dimensions.

A multitrack, linear vibratory feeder featuring interchangeable tracks has also been developed (14). The tracks are designed with computer assistance and machined on an NC milling machine using computer-generated tapes. The economic advantage of such a feeder is in both its division of cost over more than one part and its low part changeover cost. Rules for designing parts for programmable vibratory feeders such as these cannot be confidently stated until there has been more experience with their use. Generally, they will follow the pattern given for vibratory bowl feeders.

A type of programmable feeder with good potential for robotic assembly is the dual-opposed-belt feeder (15,16) shown in Figure 6. A variety of parts having at least one stable resting aspect can be fed and oriented on the same belt using simple mechanical blades having trips for reorienting. The key to the feeder's programmability is a microcomputer system that recognizes part orientation by a "signature" obtained as the part passes under a bank of photosensors above the belt. The part signature is programmed by passing the part under the sensor bank in each of its stable orientations. Recognizing which of the several possible orientations the part is in, the computer activates the necessary combination of trips to cause the part to rotate to the desired orientation as it moves along the belt. Upside-down parts are rejected to the return belt to be turned over and circulated again. For many parts, this type of feeder can be built for considerably less money than

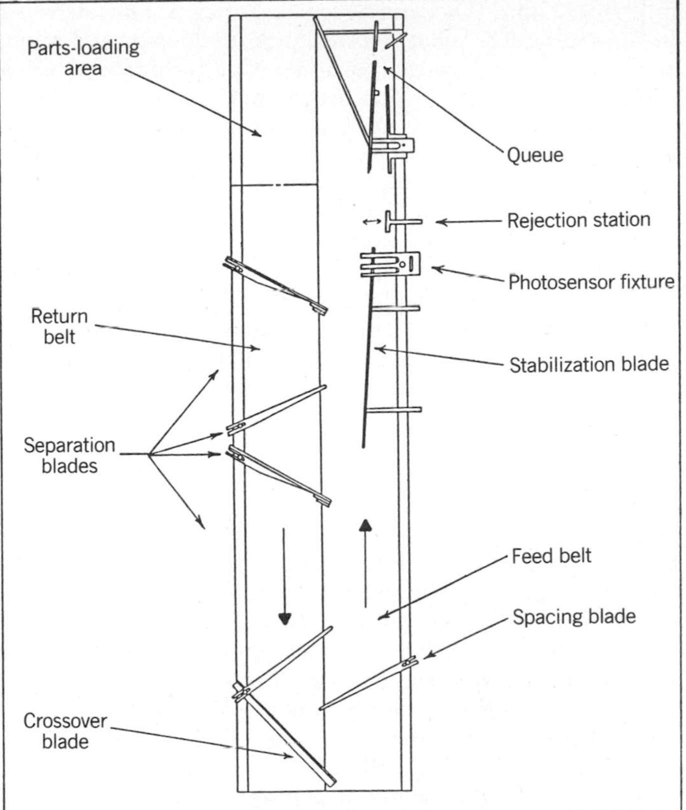

Figure 6. Dual-belt feeder (15). Courtesy of CIRP.

a vibratory bowl, and although not as versatile as the vibratory bowl in terms of the total variety of parts that can be fed and oriented, it offers the potential for retooling between parts when necessary.

For use with a belt-type feeder, parts should be designed to slide in a stable position along the side of a blade placed above the belt. Because the parts do not push one another but are carried by the belt, the design rules relating to pushing along a vibratory track do not apply. The belt feeder is more gentle on parts than a vibratory bowl, allowing more fragile parts to be fed. Features unidentifiable by the mechanical tooling of a vibratory bowl, such as paint spots or shallow, blind holes, may be recognized by optical sensors on a belt feeder and used for orienting. Optical sensors are more difficult to implement on vibratory feeders because of the motion of the parts. The restriction on part size imposed by vibratory bowls does not carry over to the belt feeder; however, rules regarding part tangling and nesting still apply.

An interesting part design consideration stemming from this work with inexpensive optical sensors is that the robot can be used to help orient certain parts. Rotational-type parts that require orientation about their axis of insertion, such as a gear with a keyway on its internal bore, may be extremely difficult to orient in any feeder using mechanical sensing. Such a part would be discouraged in any common list of design rules. However, if a small optical sensor is mounted in the tip of one of the robot gripper fingers, and provided the robot has a concentric roll motion about its wrist axis, the robot can position the gripper above the part and rotate it as shown in Figure 7 until the feature requiring orientation is detected by the sensor. The robot can then pick up the part and rotate it to the correct insertion position. This technique offers a little more flexibility in part design and reduces special-purpose feeder costs, but it carries a penalty of increasing the overall cycle time for assembly. Its most economic application would be with small batches, where the cost of special-purpose orienting equipment may be difficult to justify.

A final type of programmable feeding to be considered involves the application of machine vision. The concept of using machine vision to distinguish between a fixed number of possible part orientations (17) has been applied at Hitachi, Ltd., to the feeding of components for assembly of a portable tape recorder (18). In the Hitachi system, parts are fed one at a time from a multibowl vibratory feeder to a stage mounted under a camera. The parts are designed so that they will assume only a limited number of stable orientations, all of which have been taught to the computerized vision system. The actual image of the part is compared with the stored images until a match is found, and with the part's orientation thus identified, the computer sends out position data to the robot, allowing it to pick up the part properly for insertion. This system is less restrictive of part design than those discussed previously. Basically, the parts must be easily separable from bulk and have few stable orientations, which can be distinguished visually by features either in the silhouette or on visible surfaces.

Also providing flexibility of part design is the practice of using machine vision to allow the robot to pick up parts directly from a randomly loaded box or basket, much the way a human operator would. This technique is called "bin picking" and has been a subject of research at the University of Rhode Island (19,20). Parts to be picked from bins must be very easily separable and have clearly distinguishable orientations. Internal features or subtle asymmetry would be difficult to detect. The high cost and comparatively slow speed of current bin picking systems make the method difficult to justify economically.

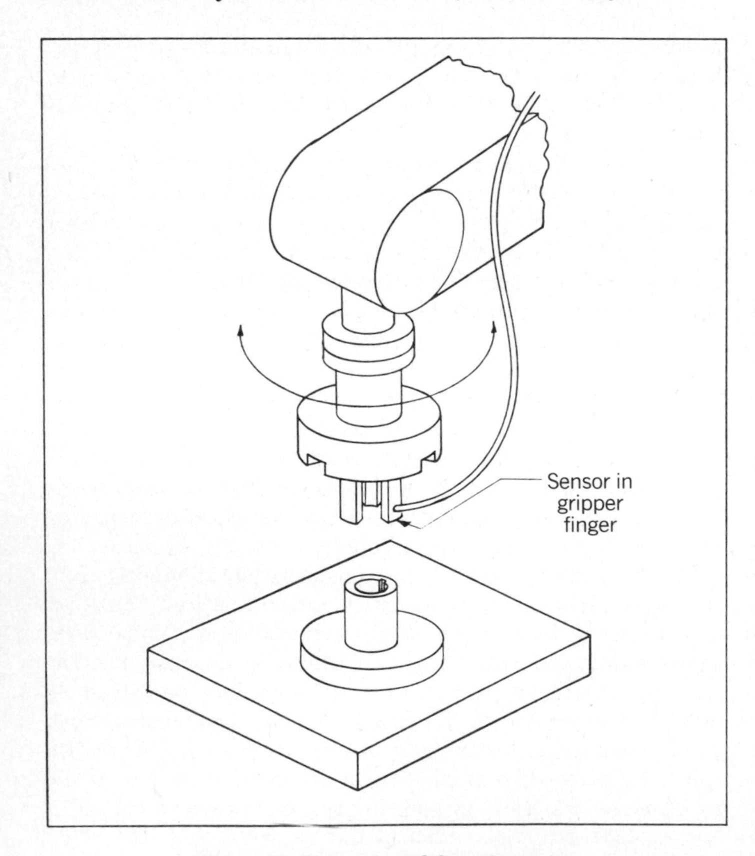

Figure 7. Gripper used for orientation.

Magazines

The nature of some parts makes them unsuitable for most types of automatic feeding. Circlips, snap rings, and open-ended coil springs often fall into this category, as do certain flexible parts such as wires, gaskets, belts, O-rings, felt seals, and rubber protective boots. Many of these kinds of parts can be assembled by the robot if suitable grippers are provided, but the separation, orientation, and presentation of the parts to the robot must be done manually. Manually loaded magazines may be used in these situations to keep the worker away from the robot workstation. Magazines may also be the most economical means of part presentation when batch sizes are small or cycle times are long, making it difficult to justify special-purpose equipment.

Part design rules applicable to magazine feeding are the same as for manual feeding and orienting. Magazines for use with special-purpose assembly machines usually must present all of the parts at the same place in the same orientation for the workhead. Robot assembly, however, offers the interesting advantage that pallets can be used to present parts in an array.

A related technique applicable to robot assembly is to place all the parts for one complete assembly on a special pallet referred to as a kit. The robot then removes the parts in turn from their oriented positions in the kit to build the assembly. Part design for kitting or palletizing again follows the guidelines for manual part handling, as the preparation of these kits will probably be done manually. It is important to see that the cost of preparing the kit does not approach the cost of assembling the product manually in the first place.

A final consideration in designing parts for feeding and orienting is to design the entire part manufacturing sequence such that part orientation is never lost. Most parts start out in some known orientation, but many have to be degreased, deburred, tested, or inspected between manufacturing and assembly. If the parts were placed directly in pallets or cartons that would preserve them in some stable orientation after the final handling, the overall cost of part presentation could be reduced. Damage occurring to parts handled in bulk could be greatly reduced if fixed-position packaging were always used (21).

The closer the production operation is to the assembly area, the easier it is to maintain part orientation. One approach is to manufacture parts right at the point of assembly. Spring winding and light sheet metal stamping are examples of processes that may be successfully performed at an assembly station.

This section has discussed several possible methods for presenting parts to a robot for assembly and how the constraints on product design can vary according to the handling method chosen. The choice of handling method will depend upon variables such as the product's market life and production rate and the individual part's suitability for automatic handling. The best method of making this equipment selection is through a quantitative comparison of the costs of using alternative part-presenting devices for robotic assembly. Vibratory bowl feeders, for example, are generally economical when the total production volume exceeds about 250,000 assemblies and the system cycle time is less than 60 s (22).

With the handling method and equipment thus chosen, parts can be designed to suit the particular requirements of the method, thereby minimizing the cost of feeding and orienting those parts.

INSERTION

As defined earlier, insertion includes those assembly activities involving the robot, such as part gripping, insertion, securing, and inspection.

Gripping

Other than reducing the total number of parts, which is discussed later, perhaps no other aspect of product design has as great a potential for cost avoidance as designing for ease of gripping. The cost of changing grippers at a robot assembly station is related to the increase in assembly cycle time. Assume, for example, that the time required for a robot to exchange grippers is about the same as the time to pick up and insert one part. If one part of a five-part assembly requires a different gripper, two gripper exchanges will be necessary at a one-arm robot station, and if the time to assemble one part or make one gripper exchange is 3 s, the gripper change will increase the cycle time from 15 to 21 s. If the robot and other general-purpose equipment are charged to the assembly at a rate of, for example, \$0.014/s, the gripper change increases the cost of assembling 500,000 units by \$42,000. Costs of this magnitude provide strong motivation for minimizing gripper changes, which should, logically, begin in the product design stage.

The first important consideration in designing to reduce gripper changes is the number of parts to be assembled by one arm. On a multirobot transfer system, each arm can have a different gripper, and the number of parts per gripper depends on the number of stations. Such a system allows quite a wide range of gripping requirements. At a two-arm robot assembly station, it may be possible to divide the parts into two groups, based on gripper requirements for insertion by the two different arms. All of the parts at a one-arm robot station, however, should be compatible with a single gripper if possible.

Efforts are being made to develop more versatile assembly grippers that can accommodate a larger range of part shapes and features. Designs exist that have three mechanical fingers that give two prehensile modes, one with the fingers equally spaced for picking up cylindrical shapes along the major axis and another with two fingers rotated together, giving a parallel jaw motion for grasping parts on two parallel flat surfaces (23,24). Another approach to versatility is to have two or more sets of fingers mounted together on one gripper, using either the robot's wrist pitch, or roll axis to change between them. Applications have also been developed where gripper changes are eliminated by combining parts into a loose subassembly in the gripper as they are picked up and then inserting the group of parts into the main assembly (25,26).

In those cases where gripper changes cannot be avoided, the increased cycle time may be divided over several parts by having the robot build more than one assembly at a time. This requires multiple fixtures with some means of indexing them in and out of the robot workspace (27).

The number of gripper changes occurring at a robotic assembly station depends on several factors, including the number of parts assembled by one arm, the versatility of the robot gripper, and the gripping requirements of the parts. All of these have to be considered in minimizing gripper changing.

Insertion Rules

Apart from the issue of gripping uniformity, guidelines for the design of parts for special-purpose automatic insertion usually apply equally well to parts intended for robot insertion (see references 4, 5, 8, 10–12, 28, 29) and refer generally to maintaining vertical insertion, providing means for alignment of mating parts, and reducing the number of fasteners in the assembly.

Insertion Direction. In designing parts for assembly by robots, deviations from vertical, straight-line assembly motions usually carry the penalty of increased robot cost. A practical specification for an assembly robot is that it have four degrees of freedom, as with the SCARA-type (30) arm, offering X, Y, and Z positioning with rotation about the Z (vertical) axis for part orientation, as described earlier. The SCARA configuration is particularly suited for vertical, straight-line insertions, as it is designed to be rigid vertically and to have compliance horizontally for easy alignment of parts having chamfers. Parts not meeting this vertical straight-line criterion usually require a robot having more than four degrees of freedom, with a correspondingly higher cost. Most insertion motions can be described in terms of cylindrical coordinates originating from the robot base centerline, as shown in Figure 8. Here Θ represents the base rotation angle, R represents motion extending radially from the robot centerline, and Z represents vertical motion parallel to the robot centerline. Within this coordinate system, nonvertical assembly motion in any R–Z plane requires an additional degree of freedom. Similarly, nonradial assembly motion in any R–Θ plane requires two additional degrees of freedom. Non-straight-line insertions are feasible with five- or six- degree-of-freedom robots, but increase both the cycle time and the programming time (31). Generally, any parts not assembled with a vertical, straight-line insertion motion increase the capital cost of the system.

Alignment. Assembly robots typically offer excellent repeatability. Many are capable of returning to previously taught positions within 0.1 mm. Although the robot may have this consistency, many production parts do not and therefore should have generous alignment aids, such as chamfers, provided in their design.

A natural property of any arm composed of several movable links is compliance. This can be an advantage for the insertion

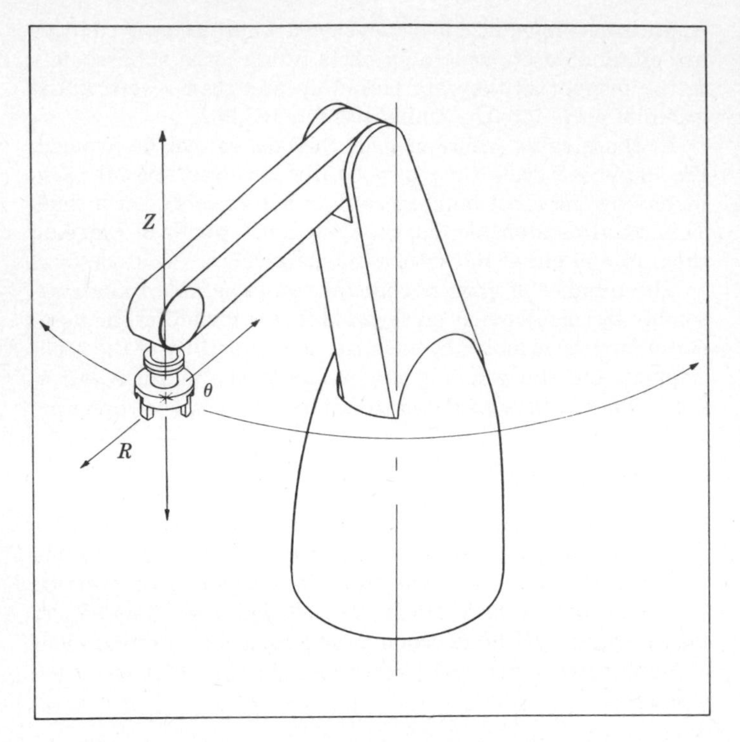

Figure 8. Cylindrical coordinate system for robotic assembly.

of close-fitting parts if chamfers are provided. In some cases, parts may require angular as well as lateral compliance. Special tools adaptable to the robot wrist, called remote center compliant devices (32), are available and provide such compliance. In any case, alignment by compliance requires that parts have chamfers to produce the necessary corrective forces. A related subject of research deals with the use of force sensing to insert parts without chamfers (33). This capability would relax the chamfer requirement on parts for robotic assembly, but at present would also carry a large cycle time penalty.

Related to designing for ease of part alignment is the need to provide access around the insertion point for the robot fingers and gripper, which are often bulky. To enable gripper sensing of insertion completion, sufficient clearance should be provided to allow the gripper to insert parts to the bottom, or limit, of their insertion depths.

Fasteners. Reducing the use of fasteners is critical to design for assembly, as the cost of their handling and insertion may account for a significant proportion of the total assembly cost. In many cases, designing parts that snap together using interlocking tabs or detents provides the necessary fastening at a much lower cost. Savings come from eliminating the cost of the fastener itself; eliminating the equipment cost for feeding, orienting, and securing the fastener, including automatic nut driving, screw driving, or riveting costs; and eliminating the portion of cycle time used for inserting and securing the fastener. To snap fit parts, most robots should be able to exert an insertion force several times greater than their rated end-of-arm load capacity since the rated capacity takes into account the large inertia forces generated by rapid acceleration and deceleration of the load. In one study, an ASEA IRb-6 robot having a 13-lb (5.9-kg) load capacity was able to exert a vertical insertion force of 40 lb (18 kg) without reaching its overload limit or reducing its speed (31).

Part Holding

A concern in the design of products for robotic assembly deals with parts needing to be held in place while another securing part is added to the assembly. For a one-arm station, there is no choice but to provide some additional device either to hold the loose part or to insert the securing part. One robot arm cannot usually do both. The additional device will be special-purpose, which may significantly increase the cost of assembling small batches.

At a two-arm station, it is possible to use one arm to hold the part while the other inserts the securing part. No additional capital equipment is necessary, but the assembly cycle time is increased. The analysis is similar to that of the gripper changing cost, where an increase in initial investment is offset by a reduction in the cycle time. If the cycle time is short and the batch large, a special-purpose holding device may be in order. If the cycle time is long and the batch small, it is probably more economical to use one robot arm to hold such a part. In either case, assembly cost can be reduced by avoiding such part designs.

Unloading. A situation similar to the part holding problem arises for the unloading of completed assemblies. Either the robot can pick up the assembly or some special-purpose unloading device can be used. For a one-arm station, there is a choice between the additional capital investment for an unloading device and the cost of increased cycle time for robot unloading.

The situation at a two-arm station is more complicated because the insertion order of the two arms affects the time available for unloading. The arm that inserts the first part should never unload the finished assembly because the other arm would have to wait through both the unloading of the assembly and the picking up of the first part. If the other arm unloads, the picking up of the first part can occur simultaneously with the unloading, and the total system waiting time is reduced. The overall cycle time is still increased, but not generally enough to justify the cost of an automatic unloading device. From a product design standpoint, if the robot is to unload, it should be able to grip the completed assembly without having to change grippers.

These discussions have emphasized that an important concern of the designer of products intended for robotic assembly is the significant effect on costs of any activity that increases the system cycle time. Manipulation of a partial assembly, for example, may demand little added capital, but can add significantly to the total assembly cost by increasing the cycle time. The next section discusses perhaps the most effective means of reducing assembly costs at the product design stage: that of minimizing the number of parts in the assembly.

REDUCING THE NUMBER OF PARTS

Regardless of the method of assembly, whether manual, robotic, or special-purpose automatic, the cost of assembly is usually proportional to the number of parts in the product.

Eliminating one part from a robotic assembly system eliminates the requirement for feeding, orienting, and gripping that part. It also reduces the total assembly cycle time, which has a large effect on the unit assembly cost. Two systematic approaches for reducing the number of parts in an assembly are value analysis and the method developed by Boothroyd and Dewhurst.

The value analyst asks questions about the definition, function, and cost of each part. The costs of possible alternative ways of performing the defined function are then estimated, attempting to determine the least-cost design that meets the product's functional and aesthetic requirements (34). The method emphasizes minimizing the cost of acceptably performing a function, but not always with regard to the cost of assembly. In most cases, the number of parts in the product will be reduced.

Another method, proposed by Boothroyd and Dewhurst, asks three questions of each part in an assembly. If the answer to any question is a yes, the part must separate. If the answers to all three questions are no, then the part should be considered a candidate for elimination or combination with other parts. The questions are

1. Will the part have to move relative to all other parts that have already been assembled?
2. Must the part be of a different material than, or be isolated from, all other parts already assembled? Only fundamental reasons concerned with material properties such as electrical or thermal conductivity are acceptable.
3. Must the part be separate from all other parts already assembled, to allow necessary assembly or disassembly of other parts? (35)

As mentioned earlier, screws, rivets, nuts, washers, bolts, or other fasteners should always be considered for elimination. The use of these questions is demonstrated in the next section with a sample design analysis.

DESIGN FOR ASSEMBLY STRATEGIES

Some manufacturing companies, seeking to standardize their product designs with respect to assemblability, have adopted quantitative systematic evaluation methods for use by their design engineers. Such systems call attention to the need for assembly-conscious design and tend to foster better communication between design and manufacturing functions. These design analysis methods have generally been developed for special-purpose automatic assembly, but similar systems based on designing for robotic assembly have also been reported. Two systems allowing quantitative evaluation of designs for robotic assembly of small mechanical products are reviewed here: one from Production Engineering Laboratory, NTH-SINTEF, Norway, and the other from the University of Rhode Island.

SINTEF System

The SINTEF method is intended to evaluate the suitability of an entire product for assembly by robots. Each part in the assembly is first scored on 14 different criteria, with discrete scores of 0, 1, 2, 4, or 8 points being given for each criterion. A 0 indicates good potential for automation, and an 8 indicates the need for manual assembly. Table 1 presents the 14 criteria and the allowable responses to each. The assembly as a whole is then evaluated on five more criteria, given in Table 2, and receives a score of 0, 1, 2, 4, or 8 for each. If on any of the criteria the part or product receives a score of 4 or 8, redesign should be considered. The scores of all the parts are entered on a work sheet, and the average of all the part scores is added to the five product criteria scores to give an overall product rating (36).

The result of this analysis is a numerical rating useful for comparing the robotic assembly potential of several different products.

University of Rhode Island System

The system proposed by Boothroyd and Dewhurst for design for robotic assembly estimates the cost of assembly for each part. The following example (37) illustrates the method of analysis. This example estimates the cost of robotic insertion only. The part-handling costs require separate data and work sheets.

The assembly considered is a small pneumatic piston shown in Figure 9. The parts are numbered in the reverse order of assembly, so entry on the work sheet (Table 3) starts with the highest-numbered part (7), the main block, which is first inserted into the work fixture. This is followed by insertion of the piston, the piston stop, and spring into the bore of the main block. The cover is then placed in position and held down against the spring force while the screws are inserted and secured.

One row of the work sheet is filled out for each of these parts in turn. The completed work sheet for assembly of the pneumatic piston is shown in Table 3. The correct entry for each column is

Column 1. Enter the I.D. number of the part.

Column 2. Enter the number of times the insertion operation is repeated consecutively.

Columns 3–7. Determine the appropriate two-digit insertion code from Table 4. The first digit is the row number of the classification system, and the second digit is the column number. Determine the relative robot cost, AR; the additional relative cost, AG, of grippers or special tools; the system time for the operation, TP; and the time penalty for any necessary tool or gripper change, TG, from the appropriate block in the classification system. The relative cost figures are dimensionless values indicating the cost of specifically indicated equipment relative to the cost of basic or standard equipment. Enter the code in column 3 and the four items of data in columns 4–7, respectively.

Column 8. Determine the total system time for the operation by multiplying the numbers in column 2 and column 6 and then adding the number in column 7.

Column 9. Enter the number in column 9 to determine the theoretical minimum number of parts. A 1 is entered for each part that satisfies any of three criteria for separate parts.

After completing the work sheet, the total assembly cost

Table 1. Criteria and Points Given to Parts[a,b]

Criteria Number	Part Points 0	1	2	4	8
P1 Weight	$0.1\ g < G \leq 2\ kg^c$	$0.01\ g \leq G \leq 0.01\ g$ $2\ kg < G \leq 6\ kg$	$G < 0.1\ g$ $6\ kg < G$		
P2 Size	$5\ mm < L \leq 0.5\ m$	$2\ mm \leq L \leq 5\ mm$ $0.5\ m \leq L \leq 2\ m$	$L < 2\ mm$ $L > 2\ m$		
P3 Delivery	Oriented		In bulk		Single-packed
P4 Separation	Easy in mechanized way			Possible in mechanized way	Tangle together
P5 Orientation	Easy in mechanized way			Possible in mechanized way	Impossible to handle automatically
P6 Flexibility	Inelastic			Possible to handle automation	Impossible to handle automatically
P7 Fragility	Not fragile		Cannot fall over 200 mm	Cannot fall over 50 mm	
P8 Quality	$F < 0.1\%$			$0.1\% \leq F \leq 1.5\%$	$F > 1.5\%$
P9 Securing	Self-securing		Screw, nail, fast gluing	Weld solder	
P10 Insertion	Linear			Linear + rotations	Coaxing
P11 Combined Movement	Two parts			Three parts	Four or more parts
P12 Tolerance	$T < 0.5$ mm		$0.1\ mm \leq T \leq 0.5\ mm$ or $T > 0.5$ mm and H/D, L/B, or H/B > 10	$T < 0.1$ mm or $0.1\ mm \leq T \leq 0.5\ mm$ and H/D, L/B, or L/B > 10	$T < 0.1$ mm and H/D, L/B, or L/B > 10
P13 Control	Unnecessary			Necessary	
P14 Force	$K > 20$ N		$20\ N \leq K \leq 60\ N$	$K > 60$ N	

[a] Reference 36. Courtesy of T. R. Langmoen, Norwegian Institute of Technology.
[b] G = Weight of part
L = Longest side of enclosing rectangle
F = Proportion of parts causing stoppage in an assembly system
T = Linear tolerance between two parts
H = Inserted length
D = Diameter of a circular insertion hole
L,B = Dimension of a rectangular insertion hole
K = Insertion force
[c] To convert g to oz, divide by 28.345; kg to lb, divide by 0.4536; mm to in., divide by 25.4; m to ft, divide by 0.304; N to dyne, divide by 1×10^{-5}.

Table 2. Criteria and Points Given to the Whole Product[a]

Criteria Number	Product Points[b] 0	1	2	4	8
P15 Weight	$0.1\ g < G \leq 2\ kg$	$0.01\ g \leq G \leq 0.1\ g$ $2\ kg < G \leq 6\ kg$	$G < 0.01\ g$ $G > 6\ kg$		
P16 Size	$5\ mm < L \leq 0.5\ m$	$2\ mm \leq L \leq 5\ mm$ $0.5\ m \leq L \leq 2\ m$	$L < 2\ mm$ $L > 2\ m$		
P17 Number of Parts	$n < 20$		$20 \leq n \leq 40$	$n > 40$	
P18 Basic Component	Yes	No			
P19 Assembly Directions	1, 2, or 3		4 or 5	6	

[a] Reference 36. Courtesy of T. R. Langmoen, Norwegian Institute of Technology.
[b] See Table 1, footnote [c] for conversion factors.

and the efficiency of the design are obtained with this procedure (39):

1. Add the number in column 8 to obtain the estimated total assembly time, TA.
2. Add the numbers in colum 5 to determine the total relative cost of additional grippers and robot tools, AC.
3. Assign the relative robot cost, AR_{max}, as the largest of the numbers in column 4.
4. Add the numbers in column 9 to obtain the theoretical minimum number of parts, NM.
5. Determine the total cost per assembly, CT, from the equation

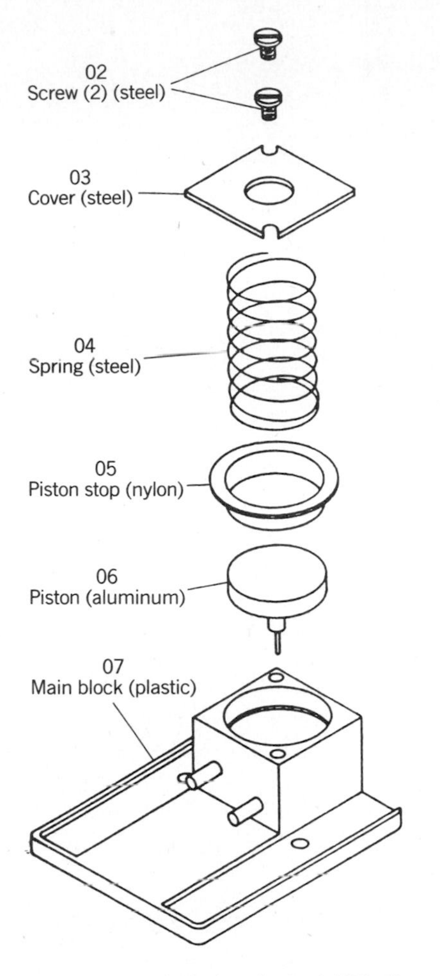

Figure 9. Pneumatic piston (existing design) (37). Courtesy of CIRP and *Machine Design*.

$$CT = (NR * CF + CA * AR) * CR * TA + 100 * ((CA * AC) + DF)/TB$$

where

AG additional relative cost for special tools and grippers
AR relative robot cost
NR number of different parts which can be fed and oriented in general-purpose programmable feeders
CA cost of standard assembly robot with two arms, each having four degrees of freedom and equipped with versatile grippers (in thousands of dollars)
CF cost of each programmable feeder (in thousands of dollars)
CR cost per second for using nondedicated equipment of unit value $1000
DF total cost of all dedicated parts presentation devices and work fixtures used with the assembly system (in thousands of dollars)
TA total assembly time (in seconds)
TB total batch size to be assembled (in thousands of assemblies)

For the present example, the work sheet gives AR = 1, AC = 0.1, TA = 24, and an assumed total batch size TB = 300. Reasonable values for the other parameters are

NR = 5 (all parts but the spring can be fed on programmable feeders)
CF = 5
CA = 150
CR = 0.003 cents
DF = 5 (for automatic handling of the coil springs)

This gives a total cost per assembly of

$$CT = 19.27 \text{ cents}$$

6. The theoretical minimum number of parts, NM, is four, which has been obtained by adding the figures in column 9. the "ideal" design for robot assembly has this minimum number of parts, where each part can be assembled in the basic system time of 2 s and none of the parts requires special grippers, tools, or part feeders. The cost of assembly of this ideal design is thus

$$CI = (NI * CF + CA) * CR * 2 * NM$$

where NI is the number of different parts (NI = NM). For the present example, this gives

$$CI = (4 * 5 + 150) * 0.003 * 2 * 4 = 4.08 \text{ cents}$$

7. Determine the design efficiency, DE, by simply comparing the actual and ideal designs. Thus,

$$DE = (CI/CT) * 100$$

For the present example,

$$DE = (4.08/19.27) * 100 = 21\%$$

Most of the reasons for this low efficiency rating are evident from the completed work sheet. First, the assembly contains seven parts, compared to a theoretical minimum of only four. Of these seven parts, four would present problems in robotic assembly. The piston requires a special gripper since it cannot be gripped on the outer diameter and then inserted into the bore. The screws necessitate a tool change, and the cover is difficult to align and needs to be held down. Finally, the open-ended springs would require a special part presentation device.

CONCLUSION

Flexible robotic assembly is still in the developmental stages. Some successful systems have been built, but they are most often applied to a single product and use a large proportion of special-purpose equipment. This article has emphasized the need to first select the particular robot assembly system configuration based on overall production parameters and then

Table 3. Design for Robot Assembly Work Sheet[a]

1	2	3	4	5	6	7	8	9	Name of Assembly
Part I.D. Number	Number of Times Operation Is Carried Out Consecutively	Robot Insertion Code	Relative Robot Cost AR	Additional Relative Cost, AG	System Operation Time	Time Penalty for Gripper or Tool Changes	Total System Time	Figures for Estimating Theoretical Minimum Number of Parts	Pneumatic Piston
7	1	00	1	0	2	0	2	1	Main block
6	1	20	1	0.05	2	4	6	1	Piston
5	1	00	1	0	2	0	2	1	Piston stop
4	1	00	1	0	2	0	2	1	Spring
3	1	02	1	0	4	0	4	0	Cover
2	2	42	1	0.05	2	4	8	0	Screw
			1 (AR_{max})	0.1 (AC)			24 (TA)	4 (NM)	300,000 (Total Batch Size, TB)

[a] References 37 and 38. Courtesy of CIRP and *Machine Design*.

Table 4. Portion of Draft Classification System for Robot Insertion[a]

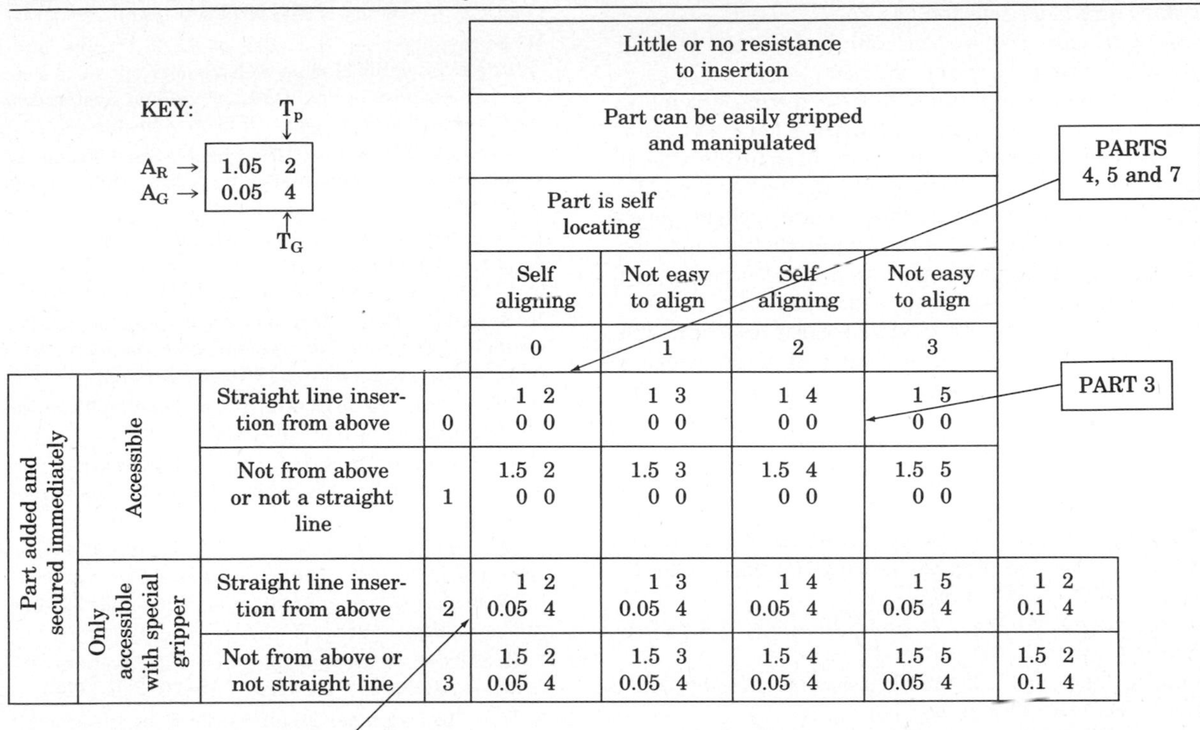

KEY: T_P (upper right), T_G (lower right), A_R (upper left), A_G (lower left): 1.05 2 / 0.05 4

				Little or no resistance to insertion				
				Part can be easily gripped and manipulated				
				Part is self locating				
				Self aligning	Not easy to align	Self aligning	Not easy to align	
				0	1	2	3	
Part added and secured immediately	Accessible	Straight line insertion from above	0	1 2 0 0	1 3 0 0	1 4 0 0	1 5 0 0	
		Not from above or not a straight line	1	1.5 2 0 0	1.5 3 0 0	1.5 4 0 0	1.5 5 0 0	
	Only accessible with special gripper	Straight line insertion from above	2	1 2 0.05 4	1 3 0.05 4	1 4 0.05 4	1 5 0.05 4	1 2 0.1 4
		Not from above or not straight line	3	1.5 2 0.05 4	1.5 3 0.05 4	1.5 4 0.05 4	1.5 5 0.05 4	1.5 2 0.1 4

PARTS 4, 5 and 7

PART 3

PART 6

				Little or no resistance to insertion			
				Part can be easily gripped and manipulated			
				Snap/push fit spire nuts etc		Screw, nut etc	
				Self aligning	Not easy to align	Self aligning	Not easy to align
				0	1	2	3
Part added and secured immediately	Accessible	Straight line insertion from above	4	1 2 0 0	1 3 0 0	1 2 0.05 4	1 3 0.05 4
		Not from above or not straight line	5	1.5 2 0 0	1.5 3 0 0	1.5 2 0.05 4	1.5 3 0.05 4
	Only accessible with special gripper	Straight line insertion from above	6	1 2 0.05 4	1 3 0.05 4		
		Not from above or not straight line	7	1.5 2 0.05 4	1.5 3 0.05 4		

PART 2

		0	1
Separate assembly operation where no part is added	8	1 2 0 0	1 2 0.05 4
		Snap/push fit	Screw tighten
		Easy with standard tool or gripper	

[a] Reference 38. Courtesy of CIRP and *Machine Design*.

complete the detailed product design for compatibility with the particular equipment selected.

Constraints on part design stem mainly from the method of feeding and orienting the parts, the need to minimize gripper changes and other nonproductive time during assembly, and keeping insertion tasks compatible with low-cost robots and equipment. Still, the most direct means of reducing assembly costs is to design products having as few parts as possible.

Adoption of a systematic design-for-assembly scheme gives designers a rational method of evaluating product designs for ease of assembly by calling attention, in a structured way, to the capabilities of common assembly equipment.

Product design guidelines for flexible robotic assembly will need frequent updating in the near future as improvements and developments continue in general-purpose grippers, part feeders, and robots.

BIBLIOGRAPHY

1. G. Boothroyd and P. Dewhurst, *Design for Assembly,* Department of Industrial and Manufacturing Engineering, University of Rhode Island, Kingston, R.I., 1983, p. 11.
2. J. L. Nevins and D. E. Whitney, "Assembly Research," *Automatica* **16,** 595 (1980).
3. G. Boothroyd, C. Poli, and L. E. Murch, *Automatic Assembly,* Marcel Dekker, New York, 1983, p. 1.
4. G. Boothroyd, "Economics of General-Purpose Assembly Robots," *paper presented at CIRP General Assembly,* Madison, Wis., Aug. 1984, fig. 12.
5. J. R. Bailey, "Product Design for Robotic Assembly," *Proceedings of the 13th International Symposium on Industrial Robots,* Chicago, Ill., 1983, pp. 11-44–11-57.
6. B. Lotter, "Using the ABC Analysis in Design for Assembly," *Assembly Automation,* 80–86 (May 1984).
7. R. Achterberg, *Product Design for Automatic Assembly,* Tech. Paper AD74-435, Society of Manufacturing Engineers, Dearborn, Mich. 1974.
8. M. M. Andreasen, S. Kahler, and T. Lund, "Design for Assembly—An Integrated Approach," *Assembly Automation,* 141–145, (Aug. 1982).
9. W. Eversheim and W. Muller, "Assembly Oriented Design," *Proceedings of the 3rd International Conference on Assembly Automation,* Stuttgart, FRG, 1982, pp. 177–189.
10. R. D. Schraft, "Assembly-Oriented Design-Condition for Successful Automation," *Proceedings of the 3rd International Conference on Assembly Automation,* Stuttgart, FRG, 1982, pp. 155–164.
11. J. F. Laszcz, "Product Design for Robotic and Automatic Assembly," *Proceedings of the Robots 8 Conference,* Detroit, Mich., 1984, pp. 6-1–6-22.
12. H. J. Warnecke and J. Walther, "Automatic Assembly—State-of-the-Art," *Proceedings of the 3rd International Conference on Assembly Automation,* Stuttgart, FRG, 1982, pp. 1–14.
13. J. L. Goodrich and G. P. Maul, "Programmable Parts Feeders," *IE,* 28–33 (May 1983).
14. A. K. Redford, E. K. Lo, and P. J. Killeen, "Parts Feeder for a Multi-arm Assembly Robot," *paper presented at the 15th CIRP International Seminar on Manufacturing Systems—Assembly Automation,* University of Massachusetts, Amherst, Mass., 1983.
15. D. C. Zenger and P. Dewhurst, "Automatic Handling of Parts for Robot Assembly," *Annals of the CIRP,* **33**(1) (1984).
16. D. Pherson, G. Boothroyd, and P. Dewhurst, "Programmable Feeder for Non-rotational Parts," *paper presented at the 15th CIRP International Seminar on Manufacturing Systems—Assembly Automation,* University of Massachusetts, Amherst, Mass., 1983.
17. W. B. Heginbotham, D. F. Barnes, D. R. Purdue, and D. J. Law, "Flexible Assembly Module with Vision Controlled Placement Device," *Proceedings of the 11th International Symposium on Industrial Robots,* Tokyo, Japan, 1981, pp. 479–488.
18. T. Suzuki and M. Kohno, "The Flexible Parts Feeder Which Helps a Robot Assemble Automatically," *Assembly Automation,* 86–92 (Feb. 1981).
19. R. B. Kelley, J. Birk, and A. Henrique, "A Robot System Which Acquires Cylindrical Workpieces from Bins," *IEEE Transactions on Systems, Man and Cybernetics,* **SMC-12** (2) (Mar.–Apr. 1982).
20. R. B. Kelley, "Heuristic Vision Algorithms for Bin-Picking," *Proceedings of the 14th International Symposium on Industrial Robots,* Gothenburg, Sweden, 1984, pp. 599–609.
21. B. Richardson, "Keep Control of Component Orientation," *Assembly Automation,* 30–32 (Feb. 1985).
22. G. Boothroyd, P. Dewhurst, and C. C. Lennartz, "Part Presentation Costs in Robot Assembly," *Assembly Automation,* 138–146 (Aug. 1985).
23. J. Motherway, "Versatile Grippers for Robot Assembly," *paper presented at the 15th CIRP International Seminar on Manufacturing Systems—Assembly Automation,* University of Massachusetts, Amherst, Mass., 1983.
24. F. R. Skinner, "A Decision and Cost Effectiveness Study of Robot Grippers," *Robotics Engineering,* 16–19 (Feb. 1986).
25. K. Hall, "Intra-gripper Technique for Robotic Assembly," *Proceedings of the Robots 8 Conference,* Detroit, Mich., 1984, pp. 17-22–17-30.
26. R. G. Abraham and co-workers, *Final Report—Programmable Assembly Research Technology Transfer to Industry,* Westinghouse, Oct. 1977.
27. A. S. Kondoleon, "Application of a Technological–Economic Model of Assembly Techniques to Programmable Assembly Machine Configuration," Thesis, Department of Mechanical Engineering, Massachusetts Institute of Technology, Cambridge, Mass., 1976.
28. J. A. Behuniak, "Product Design—The First Step in Assembly Automation," *paper presented at the 15th CIRP International Seminar on Manufacturing Systems—Assembly Automation,* University of Massachusetts, Amherst, Mass., 1983.
29. P. B. Scott, "Robotic Assembly: Design, Analysis and Economic Evaluation," *Proceedings of the 13th International Symposium on Industrial Robots,* Chicago, Ill., 1983, pp. 5-12–5-29.
30. H. Makino and N. Furuya, "SCARA Robot and Its Family," *Proceedings of the 3rd International Conference on Assembly Automation,* Stuttgart, FRG, 1982, pp. 433–444.
31. Y. S. Liu, "Robot Assembly Motion Times," Thesis, Technology Department, Brigham Young University, Provo, Utah, 1985.
32. D. E. Whitney, "Discrete Parts Assembly Automation—An Overview," *Transactions of the ASME, Journal of Dynamic Systems Measurement, and Control* **101,** 8–15 (Mar. 1979).
33. D. S. Seltzer, "Tactile Sensory Feedback for Difficult Robot Tasks," *Proceedings of the Robots 6 Conference,* Detroit, Mich., 1982, pp. 467–478.
34. L. D. Miles, *Techniques of Value Analysis and Engineering,* 2nd ed., McGraw-Hill, New York, 1972.
35. G. Boothroyd and P. Dewhurst, *Design for Assembly,* Department of Industrial and Manufacturing Engineering, University of Rhode Island, Kingston, R.I., 1983, p. 23.
36. T. R. Langmoen, "Assembly with Robots," Dissertation, University of Trondheim, Norwegian Institute of Technology, Trondheim, Norway, 1984.
37. G. Boothroyd, P. Dewhurst, and A. Redford, "Assessment of the

Suitability of Product Designs for Robot Assembly," *paper presented at the 15th CIRP International Seminar on Manufacturing Systems—Assembly Automation,* University of Massachusetts, Amherst, Mass., 1983.

38. G. Boothroyd, P. Dewhurst, and A. Redford, "Design for Assembly: Robots," *Machine Design* **56** (4), 23 (Feb. 1984).
39. G. Boothroyd and P. Dewhurst, *Design for Robot Assembly,* Department of Industrial and Manufacturing Engineering, University of Rhode Island, Kingston, R.I., 1985.

General References

K. G. Swift and A. H. Redford, "Assembly Classification as an Aid to Design and Planning for Mechanical Assembly of Small Products," *Engineering* **212** (12), 1081–1083 (Dec. 1977).

Hitachi, Ltd., "Quantitative Evaluation of Assemblability by Design Drawing," *Nikkei Mechanical,* 54–55 (Oct. 2, 1978).

R. Kelley and co-workers, "A Robot System Which Feeds Workpieces Directly from Bins into Machines," *Proceedings of the 9th International Symposium on Industrial Robots,* Washington, D.C., 1979, pp. 339–355.

G. Boothroyd and P. Dewhurst, "Parts Presentation and Feeding for Robot Assembly," *Annals of the CIRP* **31,** 377–381 (Jan. 1982).

J. L. Nevins and D. E. Whitney, *Programmable Assembly System Research and Its Application—A Status Report,* Tech. Paper MS82-125, Society of Manufacturing Engineers, Dearborn, Mich., 1982.

D. T. Pham, "On Designing Components for Automatic Assembly," *Proceedings of the 3rd International Conference on Assembly Automation,* Stuttgart, FRG, 1982, pp. 205–214.

T. Sakata, "An Experimental Bin-Picking Robot System," *Proceedings of the 3rd International Conference on Assembly Automation,* Stuttgart, FRG, 1982, pp. 615–619.

R. D. Zimmerman, *Updated Art—Parts Feeding,* Tech. Paper AD82-156, Society of Manufacturing Engineers, Dearborn, Mich., 1982.

M. M. Andreasen, S. Kahler, and T. Lund, *Design for Assembly,* IFS Publications, United Kingdom, and Springer-Verlag, Berlin, 1983.

G. Boothroyd, C. Poli, and L. E. Murch, *Automatic Assembly,* Marcel Dekker, New York, 1983.

F. L. Bracken, "Design of Data Processing Equipment for Automatic Assembly, *paper presented at the 15th CIRP International Seminar on Manufacturing Systems—Assembly Automation,* University of Massahcusetts, Amherst, Mass., 1983.

W. J. Macska, "G.E. Has 'Designs' on Assembly," *Assembly Engineering,* 16–18 (June 1984).

H. J. Warnecke, V. Schmidt-Streier, and B. Graf, "Flexible Magazine Systems for Production," *Proceedings of the 14th International Symposium on Industrial Robots,* Gothenburg, Sweden, 1984, pp. 529–540.

A. Wright, "Designing for Robotic Assembly," *Design News,* 82–91 (Mar. 1985).

ASSEMBLY ROBOTS

DAVID HOSKA
D & D Engineering, Inc.
Shoreview, Minnesota

INTRODUCTION

Of all robotic applications, assembly is one of the most difficult to implement. Before an industrial robot can be successfully interfaced with a flexible assembly system, the entire manufacturing process must be geared for automation. The requirements of even a simple flexible assembly system include the following:

- Product should adhere to design-for-assembly principles.
- Part shape must facilitate ease of orientation, presentation, and alignment.
- Product must flow from workstation to workstation in an orderly and trackable fashion.
- All parts must arrive at the correct location at the right time and in the proper order.
- Parts to be assembled must be of uniform quality and without defects.
- Assembly system must have effective error sensing and recovery.

These requirements make assembly applications both expensive and difficult to implement. The average cost of interfacing a robot with an assembly workstation is about $130,000. Only about 40% of this cost ($52,000) is used to purchase the robot. Another 35% ($46,000) is needed for such robot accessories as grippers, remote center compliance devices (qv), and feedback transducers. The remaining 25% is required for workstation installation. This implementation cost ($32,000) is higher than that of any other type of robotic application (1).

The relatively high accessory and installation costs for assembly applications is made evident if the cost breakdown of a typical robotic assembly workstation is compared with those of other types of robotic application (Fig. 1).

The cost of a robotic assembly system does not end with the workstation. Most applications also require the purchase and installation of part-presentation and part-orientation devices, specialized sensing systems, and product-inspection equipment. However, despite the high cost and the many problems associated with implementing flexible assembly systems, such results as increased productivity, reduced per-product assembly costs, and improved product quality often adequately justify the use of robots in medium-volume or batch-run applications. Robotic assembly systems are generally cost-effective when production rates are between 100,000 and 1,000,000 products per year. They are more expensive to implement than manual insertion methods but are usually less expensive than dedicated equipment. Also, they add a degree of flexibility not found in hard automation.

Currently, robotic assembly systems are being used for a variety of applications (2), including:

- The insertion of standard and nonstandard components into printed circuit boards.

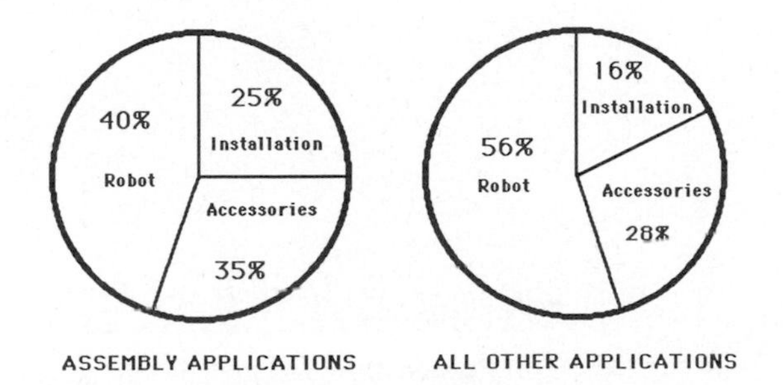

Figure 1. Comparison of average robot costs.

Figure 2. Robotic assembly station.

- The manufacture of wire harnesses.
- The assembly of disk drives in clean-room applications.
- The manufacture of automotive subassemblies.
- The manufacture of household appliances.
- The assembly of electrical components (switches and motors) and electronic equipment (videotape recorders, typewriters, word processors, printers, and computers) (Fig. 2).

This article concentrates on the assembly robot. However, the true role of the robot within an assembly system must not be overestimated. It is but one tool, albeit an important one, among many tools and devices that work together to perform assembly tasks. Some robots are specially designed to perform assembly tasks. These include but are not limited to the Westinghouse (Unimate) PUMA (programmable universal manipulator for assembly) and any of many models of horizontal jointed-arm robots referred to as SCARA (selectively compliant assembly robot arm) (Fig. 3).

Even these robots, however, are not capable of all assembly tasks. Some are more suited than others for assembly applications. Actually, many different types of robot have characteristics that enable them to perform a range of assembly tasks. To determine the robot specifications required for an assembly application, it is necessary to know the characteristics of various assembly operations and understand how the assembly process affects robot requirements.

CHARACTERISTICS OF ASSEMBLY OPERATIONS

The requisite specifications of a robot used for an assembly operation depend on the characteristics of both individual as-

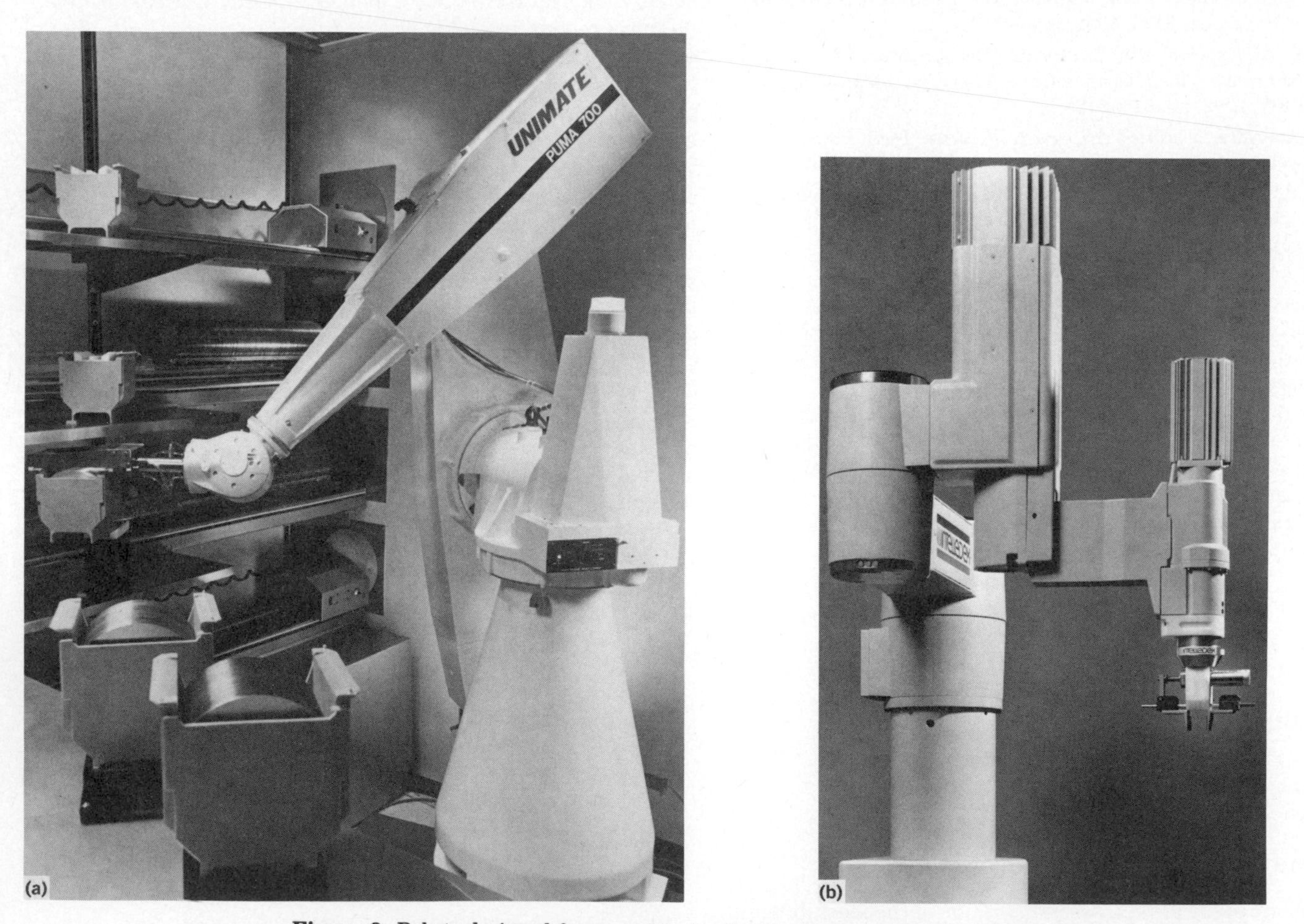

Figure 3. Robots designed for assembly. (a) Westinghouse PUMA, (b) SCARA robot.

sembly tasks and the assembly process in its entirety. Factors that must be considered include (*1*) product design features, (*2*) assembly tasks, (*3*) assembly processes, (*4*) insertion sequences, (*5*) assembly-system configurations, and (*6*) effective error recovery.

Product Design Features

Design-for-assembly (DFA) principles cover a wide range of part and product features. Several of them affect the characteristics required of the robot that performs the part-pickup and insertion tasks. Table 1 identifies several DFA practices and describes how each affects robot-specification requirements (3–7).

A product's design features directly affect the number and difficulty level of required assembly tasks. The basic goals of DFA principles are to minimize the number of parts, the number of required manipulator motions, and the orientation requirements of assembly tasks (8). A properly designed product can eliminate many of the implementation problems currently associated with assembly applications. Robots should not be relied on to compensate for and solve the assembly problems inherent in a poorly designed product.

Assembly Tasks

The specific tasks that a robot must perform during an assembly operation depend, of course, on the characteristics of the product. Most assembly tasks, however, can be grouped into one of several categories: initial calibration, product registration, part pickup, part insertion, fastening, and liquid dispensing (9).

Table 1. Effect of DFA Practices on Robot Specifications

DFA Practices	Robot Specifications
Number of product parts should be kept to a minimum. Parts should be incorporated into subassemblies, which can then be put together to produce a product. Each subassembly should contain no more than 10 to 12 parts.	Each part that must be presented, picked up, and inserted either adds complexity to an individual workstation or increases the number of stations required. If a robot must manipulate many insertion tools or move within a complex workstation, it must have a sophisticated control system and a manipulator with maximum orientation capabilities.
All parts should be inserted using straight-line motion from an above position.	Insertion from an above position allows gravity to aid the insertion process. Parts that are not inserted from above or with straight-line motion increase the orientation requirements of the manipulator and may also necessitate the use of specialized end-of-arm tooling.
Assemblies and subassemblies should be designed so that insertion and fastening tools can easily reach the desired location and perform the required task.	The greater the difficulty of part insertion and securement, the greater the number of required manipulator and wrist axes and the more complex the required insertion tooling.
Whenever possible, all parts inserted into the same assembly or subassembly should be able to be inserted and secured with a single tool.	A large number of insertion tools usually necessitates the use of either quick-change or multiple-device tooling. Complex tooling affects both the payload-capacity and the manipulator-flexibility requirements of the robot.
Parts should be compliant and self-aligning. Parts that fit into depressions should be chamfered, tapered, or have radii on mating edges.	The more difficult it is to align parts, the greater the repeatability requirements of the robot. Alignment problems can also require the use of either remote center compliance devices, which increase required payload capacity, or sensing systems, which increase control-system requirements.
Parts should fit together only one way and not require a secondary operation for alignment. If possible, they should be inserted and secured with one-step, single-handed adjustment procedures.	Each time a secondary operation is added to an insertion or securement task, the robot's manipulator-flexibility and tooling requirements are increased.
If product-strength requirements permit and the product does not need to be disassembled, parts should be designed to be pushed or snapped rather than screwed together.	If a part is snapped together or inserted with an interference fit, the assembly base, the part, and the manipulator linkages must all be rigid enough to withstand required insertion forces.
The use of coil springs, keys and keyways, and retaining rings that require special insertion tools should be avoided.	Such devices as coil springs, keys and keyways, and retaining rings require specialized tooling and increase the orientation and flexibility requirements of the robot manipulator.

Initial Calibration. The robot must have precise information about the relative locations of all workstation equipment. Therefore, the robot's control system must have the sophistication required for the programming of calibration routines.

Product Registration. The robot must know the precise location and orientation of both presented parts and part-insertion positions. If possible, part and product bases should be positively located with guides, latches, or other devices. If presentation equipment cannot be used to locate parts mechanically, the robot's control system must interface with sensing equipment and use off-line program routines to direct manipulator and wrist motion. Product registration is especially difficult in such applications as DIP (dual in-line package) insertion.

Part Pickup. The task of accurately grasping a part, without damaging or dislocating it, is often difficult, especially if the part is small and delicate. The robot's control system must precisely regulate manipulator and wrist motion to enable the tooling to approach part pickup and insertion locations at exact z heights and x–y orientations. Tooling must often be designed specifically for the assembly task.

Some applications also require the tooling to be fitted with transducers (contact, proximity, range, or force) or sensing systems (vision or tactile sensing). In such cases, the robot's control system must be capable of interfacing with the sensors.

Part Insertion. Such operations as tab-in-slot, peg-in-hole, and washer- or sleeve-over-peg usually require clearance fits. Insertion tasks, therefore, consist of the positioning and alignment of each part. However, if insertion requires an interference or snap-together fit, the robot must also forcibly seat each part. Because tight fits require parts to be closely toleranced, the parts must be prechecked for size accuracy before insertion. Tight fits also require the robot to have a high repeatability rating and may require the use of alignment aids.

Fastening. Parts that mate with a clearance fit must be secured, either through the insertion of another part or through a fastening operation. Suggested operations include screwdriving and crimping. The robot either manipulates the fastening tool or positions the product so that dedicated fastening equipment can perform the required operation. The use of such complex fastening methods as nutrunning, riveting, welding, soldering, or the application of adhesives should be avoided because they increase manipulator-flexibility and tooling requirements.

Liquid Dispensing. Liquids that must be accurately dispensed include acids, lubricants, sealants, adhesives, and such finishing materials as paints and varnishes. Like fastening operations, dispensing applications require the robot either to manipulate the dispensing tool or to interface the product with dispensing equipment.

Of the six assembly tasks described above, precise registration and locating of parts is usually the most critical. Currently, this task can be facilitated by several alignment methods:

- Vision sensing identifies part or insertion locations by interpreting part features (Fig. 4).

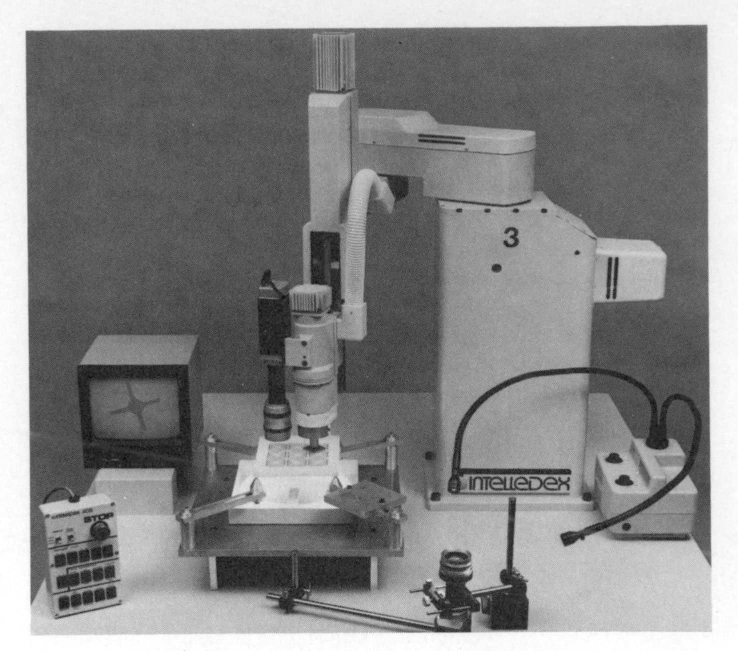

Figure 4. Vision sensing of part location.

- Tactile sensing systems interpret either a contact or pressure pattern and thereby provide information about the location of parts or insertion positions.
- Vibration insertion processes use mechanical devices to vibrate the product base until the insertion position aligns with a robot-held part (10).
- Search-and-find programs direct the robot manipulator to move from point to point within a preset pattern until the correct insertion position is located.
- Remote center compliance devices counteract parallel and angular misalignment and thus help seat slightly dislocated or misaligned parts (Fig. 5).

Assembly Processes

Along with identifying required assembly tasks, it is also necessary to determine which assembly process is most appropriate for the product. Depending on the design of the product, the assembly process can be additive, multiple-insertion, or a combination of the two. Additive assembly requires a series of parts to be added to a base in a specific sequence. This process is used when product design requires parts to be layered. Multiple-insertion involves the insertion of a series of

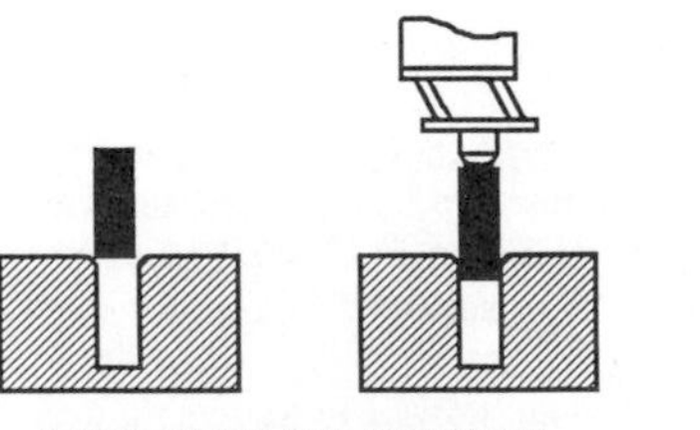

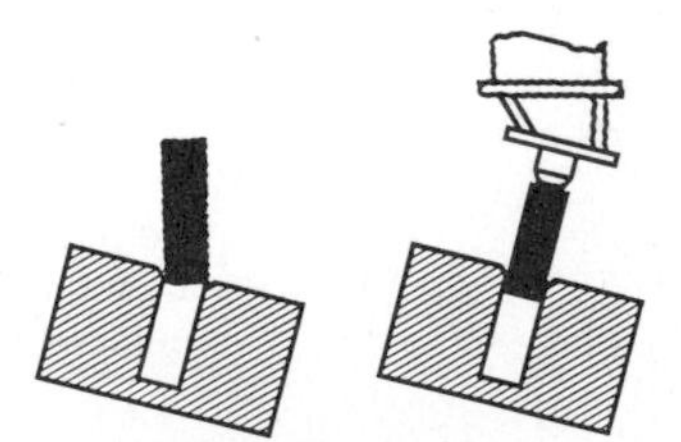

Figure 5. Effect of remote center compliance.

parts in different locations on the product base. Parts are not layered (Fig. 6) (see also ASSEMBLY, ROBOTIC, DESIGN FOR).

The characteristics of the assembly processes affect the qualities required of the robot selected to perform the insertion tasks. If parts are layered in an additive assembly process, required tooling orientations are usually minimal and there are fewer problems with clearance than if a multiple-assembly process is used. With additive assembly, parts are usually inserted easily with straight-line vertical motion. With the multiple-insertion process, however, clearances are tighter, and the manipulator and wrist need the flexibility required to move tooling around or between previously inserted parts.

One disadvantage of the additive-insertion method is that it often requires additional equipment at the workstation. Because a missing component or the insertion of a faulty component immediately produces a defective product, each part should be inspected after insertion. Therefore, the workstation must contain test equipment and the robot must be able to interface the product with the inspection equipment. Also, the robot's control system must be capable of interfacing with a master controller so that parts and product defects can be tracked.

Actually, most assembly operations use a combination of additive and multiple-insertion processes. Of the two processes, the multiple-insertion method is usually the more efficient. It reduces the problems associated with defective or missing parts and adds flexibility to the assembly sequence.

Insertion Sequence

The type of assembly process or combination of processes directly affects the insertion sequence. Before a sequence is devised, a planning chart should be developed. This should contain information on each task, including task goal, task difficulty, basic orientation and motion requirements, tooling or fixturing specifications, inspection requirements, approximate cycle times, and other factors that may affect the decision as to which insertion sequence is best for the assembly operation.

Often, if the assembly method is additive, the sequence options are limited. Determining an efficient insertion sequence is more important if multiple-insertion assembly can be used. The insertion sequence should minimize the number of times a part must be flipped or reoriented during the assembly process. It should also minimize tool-orientation requirements, maximize clearance between parts, minimize the number of required insertion tools, and allow for verification of part and product quality at the end of the assembly process.

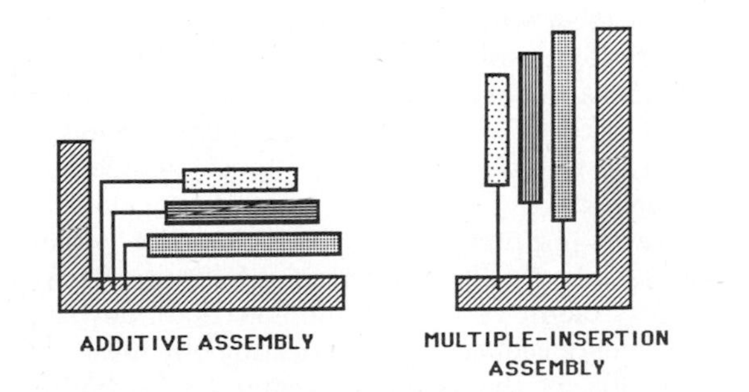

Figure 6. Comparison of additive and multiple-insertion assembly.

It is often useful to devise parts trees for several alternative assembly sequences. Each parts tree should describe the relationship of parts to one another and indicate such features as subassemblies, fixtures, and required reorientations and alignments (11) (Fig. 7).

Assembly System Configuration

After an efficient insertion sequence is devised, the assembly system must be designed. This requires an evaluation of various types of sensors and ancillary equipment that can be used to perform required assembly tasks. Several factors must be considered, including part size and weight, shapes and sizes of ancillary workstation equipment, device-location constraints, device-interaction requirements, and manipulator motion and orientation requirements. The assembly system must also meet the economic-justification standards of the company. Therefore, such plant operations data as loaded labor rate, man hours per product, and required rate of return must be considered. Finally, several important production questions must be addressed:

- Do production levels stay constant or fluctuate throughout the year?
- Must the assembly system accommodate variations in product style?
- How long will the product stay in production before major changes are required?

Only after all factors are considered can the physical and cost parameters of various types of assembly system be evaluated.

One major decision that must be made is whether the assembly operation should be performed by a serial limited-task or a parallel multiple-task system (12).

Serial Limited-task Assembly Systems. The workstations of a serial assembly system are linear. Each workstation is designed to complete one or possibly two tasks in the assembly process. The product is held in the required orientation by dunnage or fixtured pallets and is usually transferred from one workstation to another by a conveyor system. At each station, the dunnage or pallet is either locked in position on the conveyor or removed from the conveyor and precisely located. A limited number of operations, often just one, are then performed on the product. In this type of system, robot stations may or may not be integrated with either dedicated automation or manual operator stations (Fig. 8).

In limited-task systems, the operations performed at each workstation are relatively simple. Therefore, robots can generally perform their required task(s) with few degrees of freedom, limited decision-making capabilities, and a low level programming language. Each workstation contains the ancillary equipment required for the completion of only one or two tasks. Thus, tooling requirements are usually minimal.

In some serial assembly systems, tasks related to the assembly process, such as applying adhesives or finishing materials, are robotically performed to a product or subassembly as its fixtured base passes by the station on a continuously

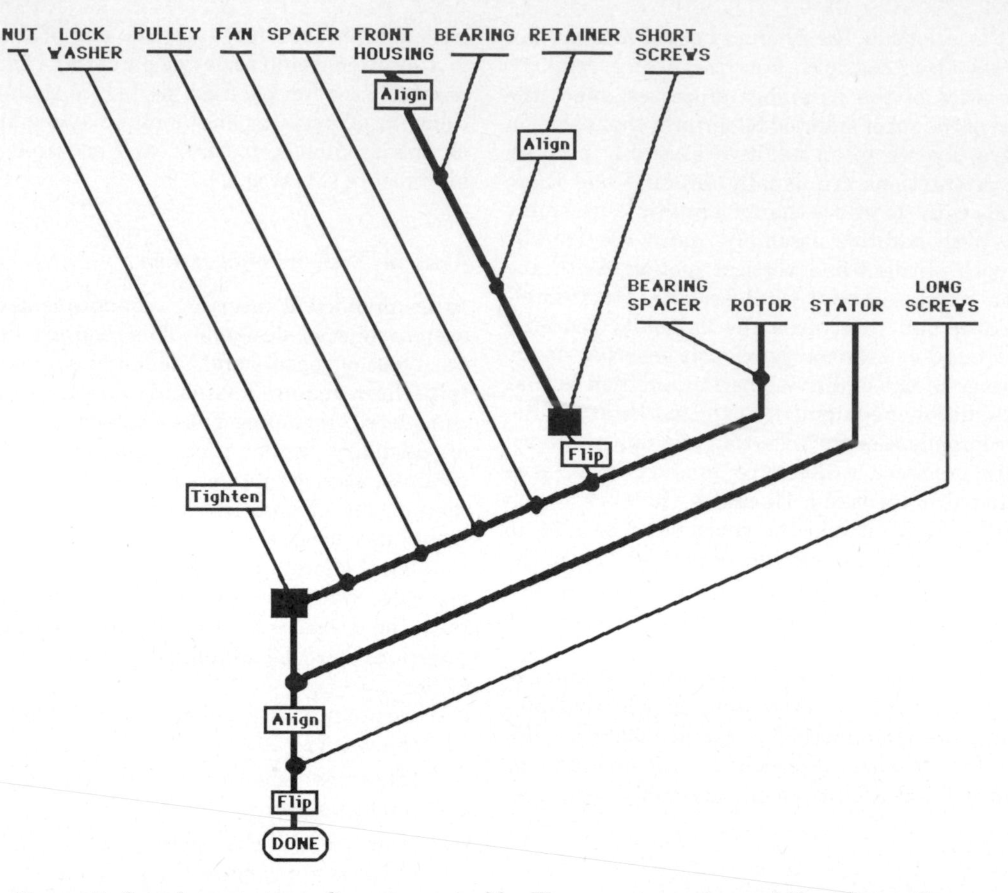

Figure 7. Sample parts tree: alternator assembly. There are two subassemblies (■), two flips, and three fixtures.

moving conveyor. Although this method reduces part-in-process time, it requires the use of robots with complex control systems. The process of synchronizing robot operation to a moving line, referred to as line tracking, is extremely difficult when high material placement accuracies are required. Also, this type of system forces each product application operation to be completed in a strictly controlled and limited amount of time.

Serial assembly systems are often used when consistent volumes of products are required. A major disadvantage for use, however, is that each workstation can affect the operation of the entire system. Therefore, all stations must be interfaced with a master controller. Also, the production level is set by the process rate of the slowest, least efficient workstation. A failure at a single station can shut down the entire line.

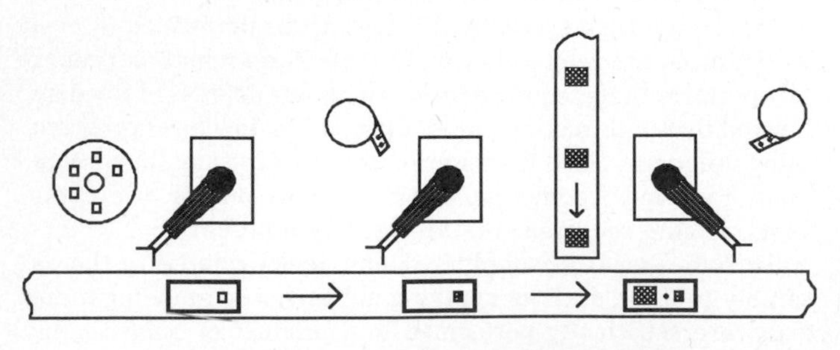

Figure 8. Serial limited task assembly system.

Parallel Multiple-task Assembly Systems. Parallel assembly systems consist of one or more multiple-task workstations, each of which is designed to enable either a single robot or two robots working together to complete all tasks required to assemble a product or subassembly. Each workstation, therefore, contains all the ancillary equipment needed for completion of every assembly operation (Fig. 9).

Multiple-task workstations are specially suited to batch manufacturing, where small numbers of differing products or differing styles of the same product must be assembled (Fig. 10). Parallel-line systems are more flexible than serial-line systems because production volume can be varied readily by removing or adding stations as required. Also, if one station is operating at reduced volume or has a major failure, the rest of the system is not affected. A multiple-task workstation requires a robot with a large work envelope, five or more degrees of freedom, and complex decision-making abilities. Often, these robots also have high level programming languages. The diversity of tasks that must be completed usually requires the robot to be fitted with either quick-change or multiple-device tooling.

The tasks of properly designing a product, identifying required assembly tasks, determining the most effective assembly process (additive vs multiple-insertion), devising an efficient assembly sequence, and selecting the most appropriate configuration for the assembly system (serial limited-task or parallel multiple-task) should be addressed before industrial

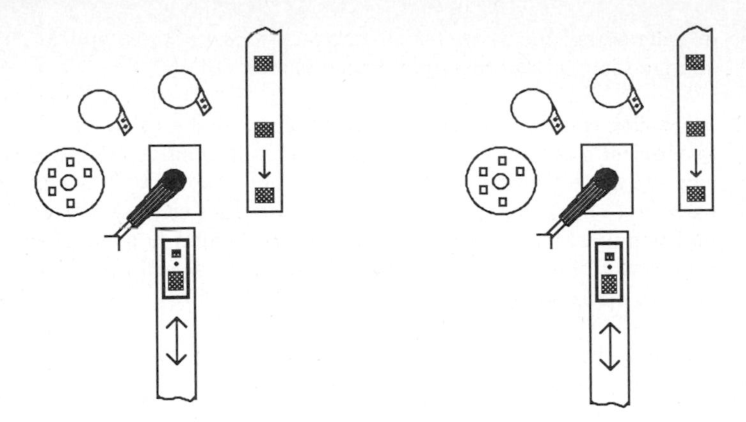

Figure 9. Parallel multiple-task assembly system.

robots—the station operators—are selected. Only after the assembly system is designed and all characteristics and requirements of the assembly process are known can an informed decision be made concerning required robot specifications.

Error Recovery

Automated assembly systems require equipment to be carefully and routinely maintained, parts to be quality tested and 100% within required specifications, and all insertions to be verified. Each phase of the assembly process must therefore be closely monitored for error occurrence. Assembly robots must be able to respond to errors and take corrective action in real time. Either the robot control system or a workstation controller must have the ability to monitor station activities. The memory capacity of the robot must be large enough to store program subroutines that enable the robot to correct or recover from problems that could occur in the workcell (13).

ROBOT SPECIFICATIONS

Analysis of the types of robots currently performing assembly tasks leads to the identification of several common characteristics. The type of robot typically used in assembly applications is electrically operated, servo-driven, and has at least four, but preferably five or six, degrees of freedom. Manipulator repeatability ratings range from 0.01 in. (2.5 mm) in large robots with heavy payload capacities to 0.0005 in. (0.01 mm) for small robots having light payload capacities.

Whichever robot is selected for an assembly application, the manipulator must coordinate all linear and/or rotary motion produced by its joints. The robot's ability to coordinate joint motion and thereby position end-of-arm tooling at exact locations within an assembly workstation depends on several characteristics. They include the manipulator's coordinate system, tooling characteristics, wrist axes, payload capacity, repeatability, and path-control and hierarchical control capabilities.

Manipulator Coordinate Systems

Any of five coordinate systems can enable a robot's manipulator arm to move within the three major axes (x, y, and z), thereby providing three degrees of freedom: base travel, vertical elevation, and horizontal extension. These systems are rectangular, cylindrical, spherical, revolute, and horizontal jointed arm (SCARA). They differ in the combinations of linear and rotary motion each uses for the major axes of motion, the joints that control each degree of freedom, and the shape of the work envelope produced.

Theoretically, manipulators used to perform assembly tasks can have any coordinate system. However, robots currently used to assemble products usually have a revolute or SCARA coordinate system. A growing number of rectangular- and cy-

Figure 10. Multiple task workstation.

lindrical-coordinate robots are also being successfully used for assembly operations.

Table 2 identifies some of the characteristics of the four manipulator coordinate systems of robots used for assembly operations.

Each coordinate system has capabilities and limitations that affect that robot's ability to perform various assembly tasks. The coordinate system must provide the manipulator with the flexibility required to enable the wrist to position and orient the tooling at all required pickup and insertion locations.

Rectangular-coordinate robots have manipulators whose motion is easy to control. Consequently, they are able to perform precision tasks rapidly and with a high degree of repeatability. Their motion flexibility, however, is limited because the manipulator cannot easily reach around or over other objects. The work envelope of rectangular-coordinate robots is also small because the manipulator is limited to performing tasks on one side of its base.

Manipulator motion in cylindrical-coordinate robots is also easy to control. Because their base joint is rotary, the manipulator can perform tasks on more than one side of its base. However, like rectangular-coordinate robots, motion flexibility is limited because the robot cannot easily reach around or over other objects.

Both rectangular- and cylindrical-coordinate robots can be used successfully for assembly tasks in which parts can be grasped and inserted with vertical straight-line motion. They are suited for use in limited-task workstations but cannot always provide the positioning flexibility required for multiple-insertion stations. However, if suspended from a gantry system, the ability of these robots to reach various workstation positions can be increased.

The three rotary joints of a revolute-coordinate robot enable the manipulator to reach around, over, and down into other objects. These robots have large work envelopes and are well-suited for such applications as multiple-insertion workstations that require complex manipulator motions and tooling orientations.

SCARA robots also have large work envelopes and have some motion flexibility to enable the manipulator to reach around other objects. However, their work envelopes have limited vertical height, which makes them unsuited for some multiple-task applications that require the tooling to reach locations at a wide range of z heights. Also, some models do not have a servo-z axis. Therefore, they are limited to performing insertion tasks at a single level. Without a servo-z axis, SCARA robots can only be used in limited-task workstations where they are required to perform a single insertion task.

The coordinate system, which determines motion characteristics within the three major axes, is only one of the factors affecting a robot's ability to perform various assembly tasks. Equally important are the end-of-arm tooling and the robot wrist.

Tooling Characteristics

If a robot is required to perform more than one task, it may need more than one piece of tooling. Any of three methods can provide some degree of tooling flexibility: tooling installed at each end of the arm (see also END-OF-ARM TOOLING), quick-change tooling, and multiple-device tooling (14).

Tooling Installed at Each End of the Arm. If the robot has a cylindrical-coordinate system and the manipulator forearm can swing 180° during program execution, use of tooling on each end of the forearm can increase productivity and keep additional costs to a minimum. If such a system is used, the robot's work envelope should be large enough to perform tasks on both sides of its base.

Quick-change Tooling. Quick-change tooling consists of a tool holder into which any of several tools can be easily inserted. Tools are stored in a rack within the workstation and the robot is programmed to change tools when necessary. Such tooling increases required control-system capabilities and payload capacity. It also affects work-envelope and motion-flexibility requirements.

Multiple-device Tooling. Multiple-device tooling consists of several gripping devices mounted on a single carrier. It is especially suitable for use in multiple-insertion workstations because it enables the manipulator to first pick up several parts from various part-presentation devices and then move to a fixtured base to insert each part into the assembly. It also allows for faster cycle times than can be achieved by a quick-change tooling system. The use of multiple-device tooling affects robotic applications in several ways. The weight of the tooling, along with the weight of several simultaneously handled parts, can appreciably increase the end-of-arm payload and thus require the use of a larger robot than would be necessary if simple tooling were used. Multiple-device tooling also requires more maneuvering room than other tooling types. The need to manipulate complex tooling within a workcell that contains such obstacles as part-presentation and fastening devices, guards, and inspection equipment may create clearance problems. This can result in the need for a revolute-coordinate or SCARA robot capable of moving around, over, and into objects. Also, the additional space required for multiple-device tooling may affect the shape and size requirement of the work envelope.

Wrist Axes

Few tasks can be performed with the limited motion provided by the three degrees of freedom of the robot manipulator. Therefore, robots are fitted with wrists that provide up to three additional degrees of freedom: roll, pitch, and yaw. Not all robot manufacturers provide a three-axis wrist. Often only one axis—pitch or roll—is available. The ability of four-axis robots to orient and align parts is limited. Although some vertical straight-line motions can be performed with four axes, most assembly tasks require five or even all six degrees of freedom.

The number of wrist axes has a direct effect on the robot's ability to orient parts. The greater the number of wrist axes, the greater the orientation flexibility of the robot. If possible, workstations should be designed to minimize the articulation and orientation requirements of the manipulator. This helps minimize both programming difficulties and task cycle times.

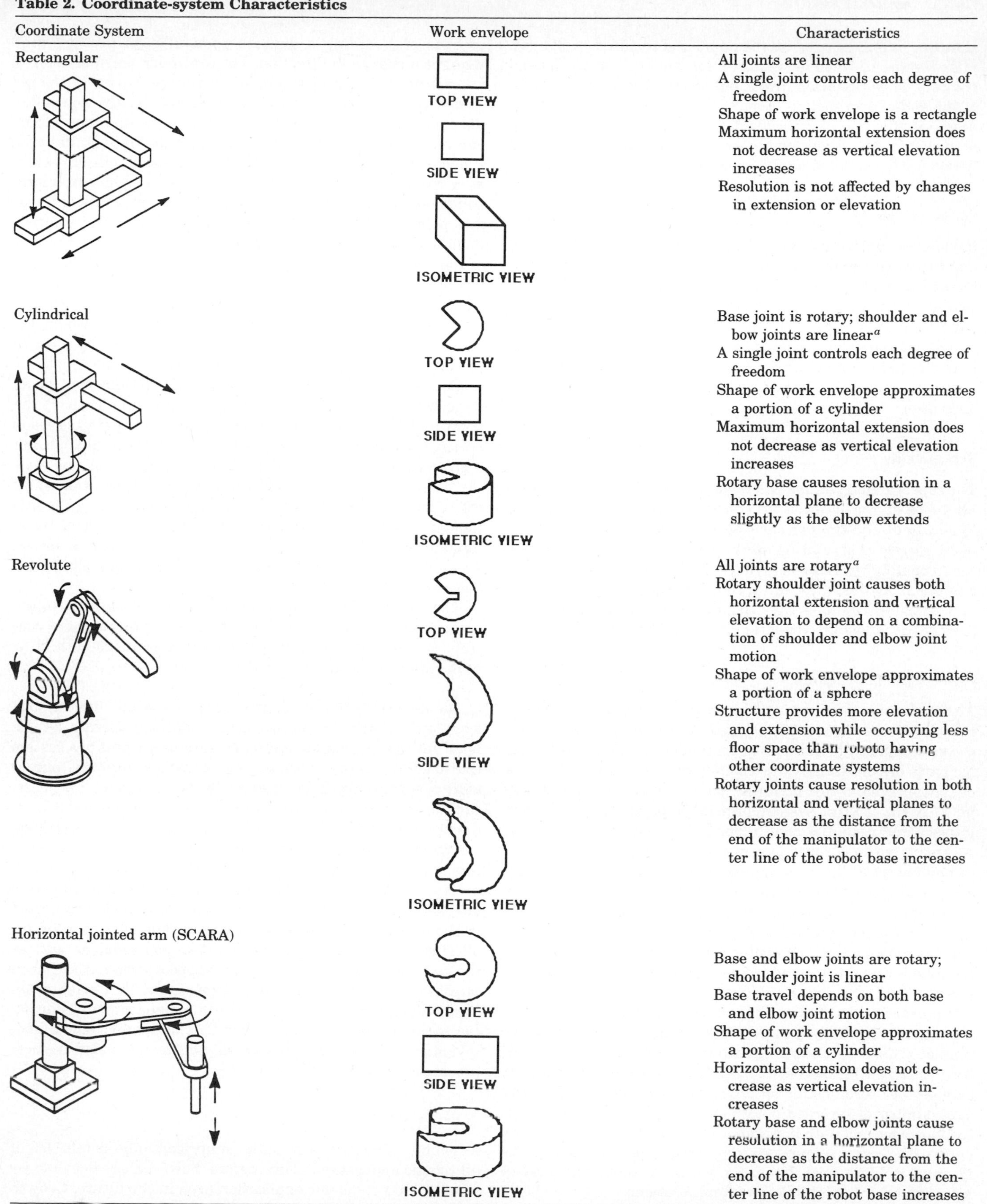

Table 2. Coordinate-system Characteristics

Coordinate System	Work envelope	Characteristics
Rectangular	TOP VIEW SIDE VIEW ISOMETRIC VIEW	All joints are linear A single joint controls each degree of freedom Shape of work envelope is a rectangle Maximum horizontal extension does not decrease as vertical elevation increases Resolution is not affected by changes in extension or elevation
Cylindrical	TOP VIEW SIDE VIEW ISOMETRIC VIEW	Base joint is rotary; shoulder and elbow joints are linear[a] A single joint controls each degree of freedom Shape of work envelope approximates a portion of a cylinder Maximum horizontal extension does not decrease as vertical elevation increases Rotary base causes resolution in a horizontal plane to decrease slightly as the elbow extends
Revolute	TOP VIEW SIDE VIEW ISOMETRIC VIEW	All joints are rotary[a] Rotary shoulder joint causes both horizontal extension and vertical elevation to depend on a combination of shoulder and elbow joint motion Shape of work envelope approximates a portion of a sphere Structure provides more elevation and extension while occupying less floor space than robots having other coordinate systems Rotary joints cause resolution in both horizontal and vertical planes to decrease as the distance from the end of the manipulator to the center line of the robot base increases
Horizontal jointed arm (SCARA)	TOP VIEW SIDE VIEW ISOMETRIC VIEW	Base and elbow joints are rotary; shoulder joint is linear Base travel depends on both base and elbow joint motion Shape of work envelope approximates a portion of a cylinder Horizontal extension does not decrease as vertical elevation increases Rotary base and elbow joints cause resolution in a horizontal plane to decrease as the distance from the end of the manipulator to the center line of the robot base increases

[a] Some robots that have three rotary manipulator joints have restricted link motion that results in a cylindrically shaped work envelope. Even though these robots are sometimes referred to as cylindrical, their operational characteristics more closely resemble those of a revolute-coordinate than a cylindrical-coordinate robot.

However, it may not reduce the number of required wrist axes. The degrees of freedom required for an application depend not only on workstation layout but also on the shape and size of the end-of-arm tooling. Even if the workstation is properly designed, the use of multiple-device or quick-change tooling may require the manipulator wrist to have yaw, a degree of freedom not available on five-axis robots.

Payload Capacity

Payload is affected by both the weight of parts being handled and the weight of the tooling. If the robot is fitted with multiple-device tooling, payload must include both the tooling weight and the combined weight of all parts the tooling must handle simultaneously. The manipulator's required speed and horizontal extension must also be considered. As either manipulator speed or the distance from the robot base to the load increases, a robot's ability to handle its rated payload safely decreases.

Repeatability

The repeatability ratings of a robot must enable the manipulator to meet the location and insertion requirements of the application. The smaller the clearance between parts and the more closely that mating parts are toleranced, the greater the repeatability requirements of the application. Although alignment methods can aid a robot to reach required pickup and insertion locations, they should not be used to compensate for vastly inadequate repeatability capabilities.

Path-control Capabilities

Although some pneumatic pick-and-place mechanisms are fitted with mechanical stops and used to perform assembly tasks, the vast majority of the reprogrammable robots used to assemble parts are electrically driven and servo-controlled. The type of control system can be either point-to-point or continuous-path. If the control system is point-to-point, it usually has interpolated-path and continuous-path playback capabilities.

Among various robot manufacturers, there is a difference in opinion as to the meaning of the terms point-to-point and continuous-path. This confusion stems from the development of advanced control systems without corresponding development of industry-accepted definitions. Although the terms continuous-path and point-to-point refer to the method by which the robot's control system receives and processes manipulator-position data, they are often used to describe robot types.

Continuous-path robots have control systems that require a programming method called walkthrough. As either the manipulator itself or a teaching arm is manually moved through the desired path, the control system records about 60 position changes per second. During program playback, therefore, the vast number of recorded points enables the manipulator arm to follow an exact path in a smooth continuous motion. Continuous-path control is found mainly in robots designed for painting and welding operations, for which manipulator path is more important than initial or final locations.

On the other hand, point-to-point robots are programmed by means of the leadthrough (teach pendent) method, which consists of recording a discrete point at each position that requires a change in direction. The controller's memory stores information concerning the required joint positions for the recorded program points only. In early robots, the controller did not coordinate joint motion as the manipulator moved from one recorded point to another. As a result, the manipulator arm did not follow a controlled path between points. However, as a result of advancements in computer technology, many point-to-point control systems now offer two playback options: interpolated-path and continuous-path control. The interpolated-path option is a playback feature that enables the end of the manipulator arm to move in a specified path between recorded points. This option carefully regulates the speed of all joint actuators so that the path of the end-of-arm tooling is controlled, usually following a straight line between recorded points. Interpolated-path motion can also be used to produce arc motion between recorded points.

Continuous-path control is a playback feature that deals more with manipulator speed than manipulator path. If the point-to-point control system has a continuous-path playback option, the computer directs the manipulator arm to approach a given point. However, before the manipulator links decelerate as the end of a manipulator arm approaches the point, the controller sends a new set of commands directing the manipulator to move toward the next point. As a result, the manipulator arm does not stop at each point. Rather, it moves smoothly and continuously through an entire series of recorded points.

The confusion concerning robot control capabilities exists because some robot manufacturers refer to point-to-point control systems that have interpolated- or continuous-path playback options as continuous-path systems.

Hierarchical Control Capabilities. No serial limited-task workstations and few multiple-task stations operate as isolated entities. Most assembly workstations are part of a system that contains many stations, each of which performs one or more of several required assembly tasks. Therefore, it is necessary for workstations to be interfaced with a master controller (facility control computer) that coordinates system activities, interfaces workstations with part-transfer devices, and monitors and evaluates production data (15).

The decision as to how to achieve the needed communication affects the control-system capabilities required of robots used in the assembly system. A workstation controller coordinates all station activities. Each workstation controller as well as the control systems for ancillary support systems are then interfaced with the facility control computer. This master controller should then be interfaced with a management information system computer, which handles production scheduling, inventory control, maintenance management, and statistical process and quality control (Fig. 11).

CONCLUSION

Assembly applications in 1985 comprised only about 11% of all robotic applications. Forecasters, however, predict that assembly will be a major application area in the future. Growth rates are envisioned to be from 23 to 32% per year. By 1995, assembly systems are expected to make up about 30% of all

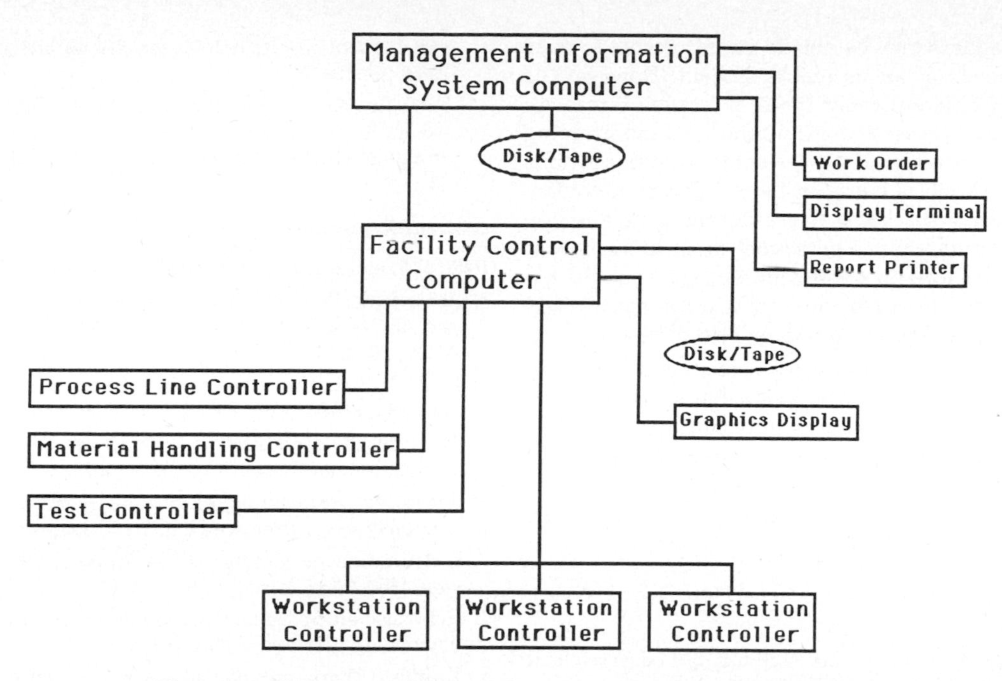

Figure 11. Computer architecture for management information system.

robotic applications (2). Highest usage will be in the automotive, light-manufacturing, and electrical/electronic industries.

Robotic assembly will probably not come into common use until technological advances occur in positioning accuracy and sensing capabilities. Robots must also become less costly, more repeatable, and faster. Areas that require improvement include vision sensing, tactile sensing, ease and standardization of programming, manipulator speed, positioning capabilities, and manipulator repeatability.

No easy methods and few shortcuts exist for implementing a robotic assembly system. Selection of the robot is actually the easiest and one of the last of the many tasks that must be performed to get an assembly system fully operational. Proper preparation for the flexible assembly system is essential. Required tasks, which can be done simultaneously or in systematic stages, include the following:

Adapt Factory Layout to Provide for Efficient Product Flow. Tasks include the relocation of equipment and the implementation of part-transfer and -presentation devices. If possible, a major goal of this stage should be to captivate and maintain proper orientation of parts and products throughout the entire assembly process. Each time part or product orientation is lost, another task (subsequent reorientation) is added to the assembly sequence.

Design or Redesign the Product in Accordance With DFA Principles. Few assembly systems can overcome the problems associated with a product that is not designed for ease of assembly. Those that do are complex, expensive, and often difficult to keep running at maximum efficiency. No new product should be designed without assuming that its assemblage will eventually be automatic. Proper product design facilitates even manual assembly methods. Thus, benefits can be realized even if production specifications never warrant the use of robotic assembly systems.

Determine Part Quality and Tolerance Requirements. An important step in preparing for automated assembly is to assess the current quality of parts received from various vendors. Often parts are not consistently within tolerance and have a higher percentage of defects than is acceptable for automated systems. New part-acceptance specifications may need to be developed and vendors must be worked with closely to reduce the percentage of defective and out-of-tolerance parts.

Determine Assembly Method, Sequence, and System Configuration and Then Develop Workstations That are Manually Operated but Which Contain the Ancillary Equipment That Would be Required if the System Were Robotically Operated. Manually operated workstations, in which the operator interfaces with presentation devices and inspection equipment, are easily justified because of the resulting increase in productivity. More importantly, however, they enable plant engineers to troubleshoot the mechanics and flow scheme of an assembly system before it becomes fully automated.

Concept and Build the End-of-arm Tooling and Then Purchase and Install Robots in One or More Workstations. An entire assembly station does not need to be automated at one time. If is often wise to install robots in only one or two stations and work out final implementation problems before an entire line of robots is purchased. The factors that must be addressed when selecting a robot have already been discussed.

Additional considerations include the following:

- A robot's most important characteristic is its ability to perform a wide range of tasks. Therefore, even if a robot is selected to perform assembly operations in a specific workstation, it should have enough sophistication to perform different tasks in another workstation.

- Many robot vendors may be able to supply robots to meet the specific needs of an assembly system. However, as the number of different robot types or brands increases, the problems and costs of programming and maintaining the robots also increase. If an assembly system contains five robots and each is supplied by a different manufacturer, engineers must know five different programming methods, and maintenance personnel must know how to maintain and troubleshoot five different systems. Such a situation can lead to confusion and is often more costly than if robots are limited to one or two different types.
- All robots selected for an assembly system should be able to be interfaced with a system controller to receive and send both commands and status information. Even if an assembly system consists of one multitask station operated by a single robot, the robot's control system should accommodate eventual integration with a master computer system.

BIBLIOGRAPHY

1. *Industrial Robots: A Summary and Forecast,* 3rd ed., Tech Tran Consultants, Inc., Lake Geneva, Wisc., 1986.
2. D. N. Smith and R. C. Wilson, *Industrial Robots: A Delphi Forecast of Markets and Technology,* Society of Manufacturing Engineers, Dearborn, Mich., 1982.
3. M. M. Andreasen and co-workers, *Design for Assembly,* IFS (Publications), Ltd., Bedford, UK, and Springer-Verlag, Berlin, 1983.
4. D. R. Hoska, "Practical Approach to Robotic Assembly," *Robots 9 Conference Proceedings,* Vol. 1, Society of Manufacturing Engineers, Dearborn, Mich., 1985, pp. 7/21–7/41.
5. G. Boothroyd and A. H. Redford, *Mechanized Assembly,* McGraw-Hill, Inc., New York, 1968.
6. G. Boothroyd and P. Dewhurst, "Design for Assembly: Automatic Assembly," *Machine Design,* 87–92 (Jan. 26, 1984).
7. J. R. Bailey, "Product Design for Robotic Assembly," *Robots 7 Conference Proceedings,* Vol. 1, Society of Manufacturing Engineers, Dearborn, Mich., 1983, pp. 11/45–11/54.
8. G. Boothroyd, C. Poli, and L. Murch, *Automatic Assembly,* Marcel Dekker, Inc., New York, 1982.
9. K. Shea, "Computer Integrated Flexible Robotic Assembly," *Robots 10 Conference Proceedings,* Society of Manufacturing Engineers, Dearborn, Mich., 1986, pp. 7/27–7/41.
10. B. D. Hoffman and co-workers, "Vibratory Insertion Process: A New Approach to Non-Standard Component Insertion," *Robots 8 Conference Proceedings,* Vol. 1, Society of Manufacturing Engineers, Dearborn, Mich., 1984, pp. 8/1–8/9.
11. R. E. Gustavson and co-workers, "Assembly System Design Methodology and Case Study," *Robots 8 Conference Proceedings,* Vol. 1, Society of Manufacturing Engineers, Dearborn, Mich., 1984, pp. 6/64–6/65.
12. K. D. Soldner and K. L. Spitznagel, "A Multi-Task Robotic Workcell for Electronic Assembly," *Robots 8 Conference Proceedings,* Vol. 1, Society of Manufacturing Engineers, Dearborn, Mich., 1984, pp. 8/45–8/55.
13. D. D. Wintch and C. F. Sermons, "Development of a Robotic Assembly Line: Achieving Success," *Robots 9 Conference Proceedings,* Society of Manufacturing Engineers, Vol. 1, Dearborn, Mich., 1985, pp. 9/83–9/101.
14. D. R. Hoska, "Effect of End-of-Arm Tooling Design on Robot Selection," *Proceedings of Seminar on Robotic End Effectors: Design and Applications,* Robotics International of SME, Dearborn, Mich., 1984, pp. 12–15.
15. H. N. Graiser, "Automated Assembly Facility for Consumer Electronics Products," *Robots 8 Conference Proceedings,* Vol. 1, Society of Manufacturing Engineers, Dearborn, Mich., 1984, pp. 6/67–6/83.

General References

Refs. 3, 5, and 8 are good general references.

R. Achterberg, *Product Design for Automatic Assembly, Technical Paper AD74–435,* Society of Manufacturing Engineers, Dearborn, Mich., 1974.

N. H. Cook, "Computer-Managed Part Manufacture," *Scientific American,* **232,** 86–93 (Feb. 1975).

G. G. Dodd and L. Rossol, eds., *Computer Vision and Sensor-Based Robots,* Plenum Press, New York, 1979.

M. P. Groover, *Automation, Production Systems, and Computer-Aided Manufacturing,* Prentice-Hall, Inc., Englewood Cliffs, N.J., 1980.

J. J. Harvey and co-workers, *Design Guide for Assembly and Automation,* IBM Corp., Armonk, N.Y., 1982.

E. Kafrissen and M. Stephans, *Industrial Robots and Robotics,* Reston Publishing Co., Reston, Va., 1984.

L. D. Miles, *Techniques of Value Analysis and Engineering,* 2nd ed., McGraw-Hill, Inc., New York, 1972.

D. Nitzan and C. A. Rosen, "Programmable Industrial Automation," *IEEE Transactions on Computers,* **C-25,** 1259–1270 (Dec. 1976).

R. C. Smith, *Design of a Modular Programmable Assembly System,* SRI International, Menlo Park, Calif., 1982.

Automated Assembly, 2nd ed., Society of Manufacturing Engineers, Dearborn, Mich., 1986.

Robots 7 Conference Proceedings, Vol. 1, Society of Manufacturing Engineers, Dearborn, Mich., 1984.

Robots 8 Conference Proceedings, Vol. 1, Society of Manufacturing Engineers, Dearborn, Mich., 1984.

Robots 9 Conference Proceedings, Vol. 1, Society of Manufacturing Engineers, Dearborn, Mich., 1984.

Robots 10 Conference Proceedings, Society of Manufacturing Engineers, Dearborn, Mich., 1984.

"Grouping Parts into Families for Automatic Assembly," *Assembly Automation,* 22 (Feb. 1982).

ASSOCIATIONS, ROBOTIC

Donald A. Vincent
Robotic Industries Association
Ann Arbor, Michigan

The robotics industry in the United States and Canada is served by three associations—two trade associations and a professional society. From time to time, other associations publish articles or put on programs that overlap with those of the organizations discussed below.

NATIONAL SERVICE ROBOT ASSOCIATION

900 VICTORS WAY
P.O. BOX 3724
ANN ARBOR, MICHIGAN 48106
(313) 994-6088
JEFFREY A. BURNSTEIN, MANAGING DIRECTOR

Established

Established in 1984, as a trade association to promote the use of service robots for education, entertainment, health care, and household chores. Managed by the Robotic Industries Association (RIA), but has its own separate Board of Directors. (See separate RIA item.)

Membership

Service robot manufacturers and developers. NSRA membership is also open to individuals.

Activities

Sponsors International Service Robot Congress & Exposition, for manufacturers, developers, component suppliers, educators, software writers, and hobbyists. Also sponsors a variety of other promotional events and provides speakers for government, industry, and academia.

ROBOTIC INDUSTRIES ASSOCIATION (RIA)

900 VICTORS WAY ANN ARBOR, MICH. 48106
313-994-6088
DONALD A. VINCENT, EXECUTIVE VICE PRESIDENT

Established

Established in 1974, as the Robot Institute of America, to promote the use and development of industrial robotic technology. Name was changed to present designation in 1984.

Membership

As a trade association, membership is entirely corporate. Total membership, as of June 1987, exceeds 300. Composed of industrial robot manufacturers, distributors, users, accessory suppliers, systems integrators and other consulting/service organizations, universities and other research organizations, and government units. Primarily U.S. in makeup, but also includes Canadian organizations and U.S. affiliates of European and Japanese robot manufacturers.

Activities

Industry Statistics. Started gathering domestic industry statistics in 1983, with confidentiality of proprietary data maintained by using an independent accounting firm to collect and tabulate the data. Now published quarterly, covering sales, shipments, and backlogs.

Standards Development. RIA has been named by the American National Standards Institute (ANSI) and the International Standards Organization (ISO) as the U.S. Technical Advisory Group Administrator for "industrial robot" standards. As such, RIA is responsible for working with all groups in the United States involved in formulating industrial robot standards. Seven standards-development groups have been established within RIA: safety, terminology, electrical/human/mechanical interfaces, information and communications, and performance.

Expositions. RIA sponsors the world's largest industrial robot exposition. Exhibitors and attendees from virtually all industrial nations are drawn to the exhibits and concurrent technical conference. (The conference is sponsored by Robotics International of the Society of Manufacturing Engineers (RI/SME). (See separate RI/SME item.) RIA also sponsors/cosponsors robot exhibits held elsewhere in the United States and overseas.

Government Relations. Recognizing the impact that government trade, investment, and tax policies and regulations have on the rate that industry adopts robots, RIA articulates its members' needs and recommendations to Congress, federal officials, and the press. RIA is now recognized as the "voice" of U.S. robotic technology.

Education. Strong user demand for practical trade-related robotic information led to workshops and seminars on topics such as robotic assembly, welding, vision, controls, and safety; flexible manufacturing systems, cost justification, etc. At first restricted to member groups, the workshops and seminars are now open to nonmembers. RIA also cosponsors other workshops on standards and robot-related technology.

To reach those people unable to attend the seminars/workshops, RIA has produced a film, *Commitment to Robotics,* and publishes/distributes trade literature.

Individuals making significant contributions to robotics are recognized through RIA's annual Joseph F. Engelberger Awards, the industry's most prestigious honor, named after the "father" of industrial robotics.

ROBOTICS INTERNATIONAL OF THE SOCIETY OF MANUFACTURING ENGINEERS (RI/SME)

P.O. BOX 930, DEARBORN, MICH. 48121
(313) 271-1500
GENE KORTE, ASSOCIATION MANAGER

Established

Established in 1980, as an individual member association, within SME, dedicated to the dissemination of technical applications, research, and managerial information to professionals working in industrial robotics. RI/SME is the world's largest society serving individuals in the field of robotics.

Membership

Over 10,000 individuals, meeting regularly in over 40 local chapter groups in the United States, Canada, and overseas in Israel, Europe, and Australia. Members are typically engineers or scientists, college or university educators, consultants, corporate technical/marketing directors or executives, civilian and military government personnel, and specialists or executives from insurance, labor, and private training organizations.

Nationally, reflecting the diversity of member interests and needs, RI is organized into 10 topical divisions: Research & Development, Assembly Systems, Welding, Aerospace, Material Handling Systems, Casting and Forging, Military Systems, Human Factors and Safety, Education and Training and Remote Systems.

Activities

Sponsors/cosponsors conferences, workshops, and seminars. The RI conference held concurrently with RIA's annual exposition is the preeminent annual technical conference in industrial robotics. (See separate RIA item.) Topical seminars and "hands-on" workshops are held throughout the year at various locations in the United States and Canada, the more popular ones being repeated at additional sites.

Workshops/seminars being held in the latter half of 1987 are representative:

Automated clean room processes robotics
Applying sensors in robot applications.
Developing end-of-arm tooling for industrial robots.
Robots in clean room applications.
Automated electronic assembly.
Robotic education and training.
Integrating lasers with robots.
Developing robot work cells.
Applying robots in materials handling systems.
Resistance welding with robots.

In addition to the bound technical conference/seminar papers, RI publishes a bimonthly journal, *Robotics Today,* the leading U.S. industrial robotics journal; RI also produces educational films and videotapes for distribution to business firms and academia.

AUTOMATED GUIDED VEHICLE SYSTEMS

N. S. Rajaram
University of Houston
Houston, Texas

INTRODUCTION

Automated guided vehicle systems (AGVS) are as the term indicates a system of driverless, usually identical vehicles guided by a control mechanism consisting of some form of guided path embedded in the floor, by remote radio control or by a combination of these and other signaling mechanisms. The term AGVS refers to the entire system of vehicles, guiding mechanisms, traffic-control devices as well as the central and on-board computers. AGVS are finding applications in a wide range of material-handling tasks including places like warehouses, factory floors, docks, and even the space program, for loading, unloading, and transporting materials and parts without a human operator on board (see also MATERIAL HANDLING). An intriguing development is the evolution of AGVS and mobile robots to the point of soon becoming virtually indistinguishable. The example system discussed later is a manifestation of this merging of robotics and AGVS technologies. Even systems such as the so-called robot submarines and the autonomous land vehicle being developed by the U.S. Department of Defense can be regarded as descendents of AGVS.

Although AGVS resemble robots in some respects, particularly in their increasing use of microprocessors for control and communication, their historical origins are quite different. A computer-controlled vehicle developed in the 1960s by Stanford University under contract to the National Aeronautics and Space Administration to investigate the use of remotely controlled rovers for the exploration of the lunar surface, has strong claims to be regarded the true ancestor of AGVS. Also, unlike a robot which functions for extended periods of time without human intervention, a vehicle in an AGVS is essentially a remotely controlled device. This distinction though is likely to continue to erode with the development of autonomous land vehicles, and, as carts in the AGVS acquire more on-board processing capabilities in the form of sensors, computers, and possibly in the not too distant future, vision and artificial intelligence.

Clearly, the concepts and the technology underlying AGVS would not have gained such widespread following in such a relatively short time if they did not confer many benefits upon its users. Among the advantages possessed by AGVS the following are worthy of note.

1. Ease of modification. A significant advantage of AGVS over "hard" transportation systems such as fixed conveyors and railed vehicles is the relative ease with which the guidepaths of AGVS can be modified. The route of an AGVS can be altered relatively easily by modifying the guidepaths of the carts and changing the traffic control systems and software. Modifying the layout of hard transportation systems on the other hand is significantly more expensive and time consuming. This edge in flexibility can only increase as AGVS acquire greater on-board intelligence through the integration of more sophisticated hardware and software.
2. Maintenance and reduced downtime. The components of an AGVS such as carts can be maintained without having to shut the whole system down. This enables the design of AGVS with redundancies so that a maintenance schedule requires no downtime of the whole system, at least in principle. Generally speaking, the major maintenance item in an AGVS is the recharging of the cart batteries needed approximately every eight hours. The other maintenance items include those of motors, controllers, traffic-control switches and the associated computer hardware and software. Again, with increased computerization of systems, it is becoming possible to include on-board diagnostics and monitoring, making a fair amount of maintenance automatic.
3. Load variability. In flexible manufacturing systems and similarly flexible warehouses, the load attributes of materials to be handled and transported can often change with or without any changes in the routing and other characteristics. This can be accommodated in an AGVS by substituting vehicles with the appropriate load capabilities for existing ones without the need for extensive modifications to other parts of the system.
4. Economy. There are estimates (1) that attribute as much as 30–40% of the average manufacturing cost of products to the cost of material flow. Although this includes the

costs of distribution to customers and of shipping raw materials to the manufacturing plant, the cost of material flow within the plant can still be significant. Because of their ease of maintenance and flexibility, AGVS can offer significant cost advantages also.

In addition, AGVS offer a few other advantages such as safety, real-time monitoring and control, and the relative ease with which they can be made to function with minimal damage to the environment. Nevertheless, the major advantage that they offer over hard transportation systems is flexibility in the transportation network and load bearing capabilities. This flexibility also applies in designing plant layouts. Thus, as the manufacturing industry increasingly moves towards flexible systems, it can reasonably be expected that the scope of AGVS will continue to expand.

It should be recognized that an AGVS is a system as well as a subsystem. It is composed of vehicles, traffic-control devices, on-board and central computers, and other components. At the same time, it is also a subsystem in a manufacturing or warehousing facility. This article discusses different aspects of an AGVS viewing it primarily at the system level.

TECHNICAL DESCRIPTION

At the system level, an AGVS can be regarded as being composed of two subsystems: transportation and information. The transportation system is composed of the following three components: the vehicles, the traffic network, and the load-handling systems. The information system consists of the vehicle control, the network (traffic) control, and the control for load-handling systems. The information system is necessarily *distributed,* with the control and communication functions being shared among computers, transceivers, and other necessary electronic systems both at a central location as well as on-board the vehicles and on-site various points in the network. The load-handling systems serve as the physical interface between the vehicles and the material-handling stations making up the network. Figure 1 gives a schematic representation of an AGVS system. (see Operation and Example Application for details)

Driverless transport systems consist basically of guided vehicles and a traffic system. Because of the extreme variability in both the type of materials to be transported and distances to be traversed, transportation systems exhibit considerable diversity. Distances of travel can vary from a few hundred feet to as much as several miles. These vehicles can be individual carts or trucks, tractors, tractor trailers, and transporters of other kinds. The term truck is used in a generic sense and does not denote any specific kind of transporter. Several commonly used types of transporters are examined next. The description and the classification given here essentially follows that in Ref. 2.

Vehicle Systems

Vehicle systems can roughly be classified into tractor trailers, pallet trucks, skid tractors, and the largest category type, special-purpose transporters.

Tractor trailers that consist of a truck hauling a train are especially suited for transporting materials in situations where there are relatively few destinations and a large material flow. Docks, large warehouses, and shipyards might be regarded as typical facilities where such trailers are used. Trucks used in such situations have the further advantage that they can have a diesel or gasoline engine in addition the usual electric motor, thereby reducing the drain on the battery. Trucks used for such applications typically have a single front wheel for steering control and the drive may either be front-wheel, rear-wheels or both. At this time, trailer loads

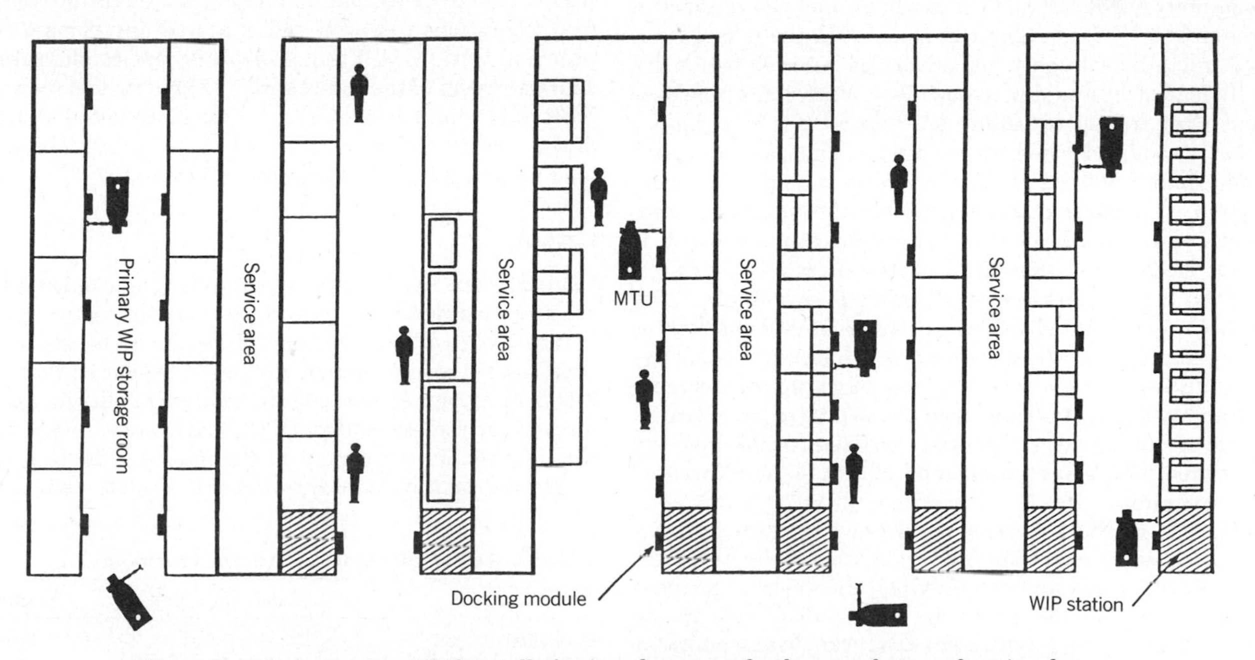

Figure 1. Schematic of an AGVS installation in a clean room for the manufacture of semiconductors. WIP = work-in-process. Courtesy of The FMS Corporation.

in the range of ca 440 to 44,000 lb (2,000 to 20,000 kg) are fairly typical, with drive capacities from about 134 f. (1 kW to 7 kW) and with battery capacities ranging from 300 to 1,000 Ah (10.8×10^5 to 36.0×10^5 C). The maximum speed is of the order of 3.2 ft (1 m) per second, with a lowering of the speed necessary when negotiating bends and turns. In the case of some trains it is not uncommon to have a driver seated truck, as for instance at airports. The truck brakes are typically electromagnetic disk units. Because of desirable stability properties and the accompanying simplicity of design, three-wheeled trucks with a swiveling bolster steering mechanism are probably the single most widely used, especially inside factories. Other steering systems used are differential steering, axle-pivot steering, and all-wheel steering.

Pallet trucks are individually steered vehicles equipped with inductive steering for transporting individual pallets, though units capable of transporting multiple pallets are not unknown. They can usually transport payloads in the range of ca 3300 to 4400 lb (1,500 to 2,000 kg), with a power-drive capacity ranging from 6.7 to 134 hp (0.5–1.0 kW), and with battery capacities from 720×10^3 to 180×104 (200 to 500 Ah). Pallet trucks can be used either individually or as trailers in trains hauled by tractors. As with tractor trains, their speed is typically in the range of 3 ft (1 meter)/sec with some reduction in speed during turning and cornering. As just noted, pallet trucks are inductively steered, the working of which can briefly be described as follows. If the truck deviates from the guide path, a voltage proportional to the magnitude of the deviation is induced, which is used to correct the path of the truck. Thus, there is no physical contact between truck motor and the guide wire embedded in the floor. Pallet trucks may be designed and installed in AGVS to move either unidirectionally in a closed traverse loop or to have a reversing mechanism. Further, this reversing may be automatic or remotely controlled. Pallet trucks are especially useful where material flow density is small to medium, and the number of destinations or load handling locations is relatively large.

Skid tractors are highly maneuverable vehicles which are typically loaded from a device such as a platform, a conveyor or some other transport/loading unit placed at a level higher than that of the truck load bearing surface. Thus, a skid tractor is loaded from a storage shelf or a conveyor, and never from the ground. Its performance characteristics are similar to those of pallet trucks except for the steering mechanism which is based on either the differential or the all-wheel principle. These have the advantage over other kinds of steering of enabling both forward and backward travel as well as tighter cornering ability. All-wheel steering is preferred particularly for large trucks in view of increased maneuverability including the possibility of diagonal and even transverse mode of travel. Skid tractors are among the most versatile trucks and can be integrated with workstations and materials handling and processing centers.

In addition to these three general types of transport systems, AGVS use an enormous variety of others which at present defy classification and can only be called special purpose or hybrid trucks. These trucks are obtained by integrating different features of different kinds of trucks discussed so far as well by incorporating different types load handling and other systems mounted on their chasses, depending upon the type of applications. Such special purpose trucks should be considered the rule rather than the exception in the majority of AGVS installations today. An example of such a special-purpose transport system for wafer handling and clean-room applications in the semiconductor industry is discussed later.

Load-Handling Systems

The load-handling system serves as the physical interface between the transport vehicle and the loading/unloading points in the traffic network of an AGVS. As can be expected, there is considerable variability in the different systems and subsystems used for load handling at locations in the network as well as on the trucks themselves. These load-handling systems may or may not include humans in the loop. It is also possible that the same network might have different types of load handling systems at different locations, some with humans in the loop and others not. An airport is an example where such a hybrid approach is often to be found.

Load-handling systems can consist of conveyors of different types, fork lifts mounted on the station or the truck itself, fork-lift trucks, telescopic forks, lifting jacks, and many other types of devices. Invariably the needs of the materials to be handled dictate the type of load-handling mechanism to be used. The entire system including the type of transportation system and the controls is affected by the choice of the load-handling mechanism chosen. Two basic decisions have to be made in choosing any load-handling strategy. First, whether the load is to be handled at the floor level, or at a level determined by the storage and plant requirements. Second, whether only one of the two systems—the truck and the load-handling station, should be active during handling or both; and if only one, which one. Once this essentially strategic decision is made, it is necessary to decide upon the numerical mix of trucks and load-handling stations, as well as their locations which have to be estimated based upon quantitative methods from operations research and expected and desired throughputs and workloads. Frequently, different configurations and characteristics of the combination of layouts and material flow have to be simulated. See Ref. 2 for more detailed discussion of the characteristics of different types of active and passive load-handling devices and their interfaces.

Control

Control of an AGVS entails performing the following: Truck control; load-handling control; and traffic (network) control.

These control functions have to be shared between on-board processors and the central computer, which further requires integrating some form of communication facility for data transfer and for remote control of the various entities making up the information component of the total AGVS. Table 1 below gives a schematic representation of the relationships between

Table 1. AGVS Information System Taxonomy

Truck Control	Load Handling	Traffic Control
On-board		On-board
Central		Central
On-board and central		

the various control, data transfer, and communication functions.

Most of the functions for truck control are managed by on-board controllers assisted by a central computing facility via the communication channels. The basic functions involved in truck control are steering control, route control, drive or actuator control, and the load-handling control. As just noted, any or all of these functions may be carried out in conjunction with the central computer. The initiation of any operation, in particular that of travel has to be done through a command input facility such as a keyboard which may be located on-board or at a remote facility. Although it is technically feasible to have all control functions on-board using microprocessor-based control, operational and planning considerations dictate that at least some of the higher level control functions be performed from a remote facility. Typically, steering control and control of all or part of the load handling are done by on-board controllers. In a highly automated system in which the material flow has attained a steady state, the actuation functions, namely, starting, acceleration, and stopping may also be entirely on-board. Nevertheless, even the most highly automated transporters can have their functions manually overridden from the central facility, or from one or more remote locations.

Clearly, in a system which depends as heavily on data transfer and remote control as an AGVS, different types of devices for handling information are needed. Only the inductive steering component when used is not part of the information network. These devices consist of active and passive components for transmitting and receiving signals. These are installed on-board each truck, on floor installations, and at the central facility. In addition, there may also be hand-carried units in use by supervisory staff and for emergency control. The switching commands between the truck and the floor installation necessary for automatic operation are typically transmitted by 24 V electromagnets which control contacts in the guide path laid in the floor. In addition, coils needed for the inductive magnetic field in the steering is located on the truck body. These typically perform the actuation-control functions. The general commands for conveying the coded locations and carrying out specific transportation operations are input using a keyboard with a console. Infrared or radio frequency transmitters and receivers can be located on-board to transmit and receive signals to and from the central computer if such control is used. Many of these topics are best understood when studied with the aid of a specific example. The next section gives examples.

OPERATION AND EXAMPLE APPLICATION

The AGV system whose operation is to be discussed in this section is the one known as Model 4020 Automated Material Handling and Transport System developed by FMS Corporation as part of its computer-integrated manufacturing (CIM) system for clean-room operations that arise widely in semiconductor manufacturing. This system is designed to transport wafer cassettes, cassette boxes, lead-frame magazines, or other material containers between workstations and process equipment. Its singular feature is a transport vehicle that is a hybrid between a truck and a robot. This and other features make the FMS Model 4020 something of a bridge between conventional AGV systems in widespread use and trends increasingly likely in future systems, and for that reason, chosen here for the purpose of illustration.

As just noted, the transporter is a truck with a robotic arm which under the control of a central minicomputer distributes material containers to or from workstations and work-in-process (WIP) inventory stations. Production-control software on the minicomputer provides the framework to monitor material flow to manage WIP inventory, thus, providing a degree of integration in the transition from production management to control. A typical Model 4020 AGVS consists of a minimum of one transporter, ten docking modules and operator interfaces, a central control computer, at least two intelligent WIP stations, and the associated software previously alluded to. Figure 2 gives an illustration of the mobile transporter. Figure 1 displays an implementation of the AGV system. In implementing such a system, initially a single transporter would be used to transfer materials between intelligent WIP stations along the periphery of the manufacturing areas. At locations with automated processing stations, additional transporters would be added to the system to distribute materials, possibly automating material handling at all of the workstations, Operator assistance would be used only when needed. A brief description of the system components and their functions follows.

Mobile Transport Unit

The mobile transport unit (MTU) is an automated guided vehicle specially designed for transporting and handling of wafer cassettes, boxes and other containers from one workstation to another in clean-room environments used in the production of very large scale integrated (VLSI) circuits. Transfer between the MTU and production workstations is performed by the on-board robotic arm (Fig. 3). This arm is equipped with customized end-effectors to meet specific applications. Further, a combined sensing and interlocking mechanism prevents the operation of the robot arm unless it is docked at a workstation. Also, the safety mechanism does not allow the robot arm to operate if it senses the presence of anyone in its proximity. The MTU robot arm for the workstation interface can be programmed via a teach pendant.

An unusual feature of this AGV system is the guidance

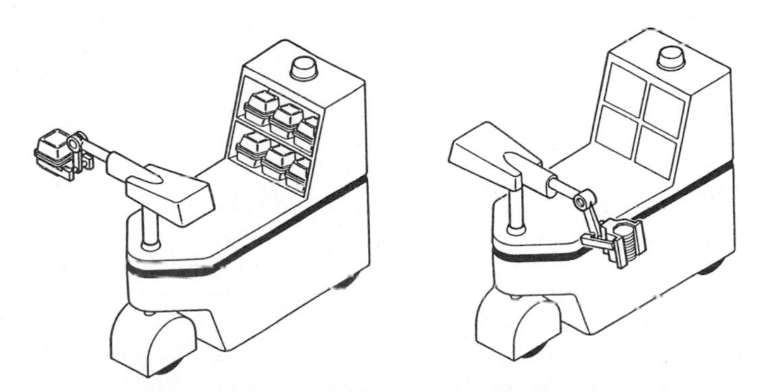

Figure 2. Mobile transport units (MTU) with robot arms and customized end-effectors. Courtesy of FMS Corporation.

and control mechanism which uses no fixed path guides such as stripes or wires commonly used with most systems. The MTU uses a sophisticated navigational system composed of an optically encoded dead-reckoning system for distance monitoring, a sonar proximity sensor for obstacle detection and collision avoidance, and a gyroscopic system for position determination. These systems enable the MTU to navigate accurately in a manufacturing area and position itself within a few inches of the zero reference point at each workstation location. Two-way communication is provided by an infrared-optical broadcast medium between MTU and the central computer. Travel positioning and workstation status information can be up- downloaded by this communication channel, which can also be used for passing instructions for redirecting the path of particular MTUs when needs arise. The MTU carries on board a main computer as well as several satellite processors which perform dedicated functions including vehicle management and diagnostics, communication with the central computer, housekeeping functions such as monitoring battery power, collision avoidance, and measuring course accuracy. Power is provided by sealed batteries which get charged each time the MTU docks with a workstation, thereby enabling the MTU to operate 24 hours a day if necessary. In addition, the MTU gives forth an audible alarm whenever it encounters an obstacle and comes to a stop if the obstacle does not move away from its path.

Docking Modules

The physical interface between the MTU and each workstation is accomplished through the docking modules. These modules, one active mounted on board the MTU, and the other passive mounted on the workstation, provide the precise alignment and recalibration necessary between the position of the MTU and the workstation. In addition, they also perform the secondary function of charging the MTU batteries. During docking, the MTU aligns itself beside a workstation without making actual physical contact. The on-board docking module optically scans the passive docking module on the workstation. Three angular measurements are taken which are used to calculate a lateral position offset with respect to the workstation. This data is used to recalibrate the MTU position with respect to the manufacturing area as well as for providing the information necessary for repositioning the robot arm to load or unload a container from a workstation. In effect the docking modules provide the communication between the fixed coordinate frame of the stationary workstation and the varying coordinate frame of the MTU.

Operator Interface

The operator interface (OI) is a small display terminal and keypad mounted on each workstation as shown in Figure 3. It is used by manufacturing personnel to communicate with the central computer, and thus, used only when there is need for a change in the production run. The OI is hardwired to the central computer and provides a data link between the workstation and the central computer. It can be used to enter instructions, verify status and data such as process sequences and/or wafer lot locations, form new lots, or provide or access

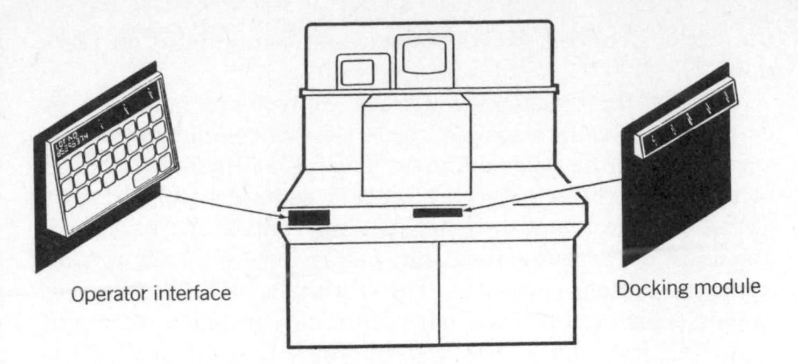

Figure 3. Production workstation with enhanced view of the operator interface and docking module. Courtesy of the FMS Corporation.

any other relevant information. Lot-tracking integrity is maintained through the OI. There is also facility for a bar code reader to allow automatic lot tracking.

Activities of the cassette elevator (from which cassettes are picked by the MTU robot arm) and of the production work are monitored by sensors to ensure operating integrity and cassette security. The latter function serves as a precautionary measure, so that cassettes are not removed from designated places by accident or design, a point apparently of some considerable concern to semiconductor manufacturers. In addition, any change in the status due to breakdowns or outages can be communicated to the central computer via the operator interface.

Central Control Computer

The central computer controls the entire manufacturing system and not only the AGV system, thus integrating the manufacturing operations and material flow. It supervises all MTU guidance and location monitoring functions by directing instructions to the MTUs. In addition, it also serves as the interface between workstations and the MTUs when information is conveyed through the OI. A single central computer used in the Model 4020 system can control five MTUs and fifty workstations. The central computer gathers real-time data about product lot location, status of the production stations, and CIM systems fault analysis. In addition, it enables new machines to be brought on-line, initiates new process sequences, and aids in the formation of new lots and modification of existing ones. Communication between the central computer and the MTUs is via the infrared broadcast link. The central computer is a minicomputer running under the UNIX operating system, an increasingly popular choice, especially for distributed systems and networks.

Intelligent Work in Process Station

Intelligent work in process (IWIP) stations provide local storage and inventory control in the manufacturing system. These are placed in the system according to the production plan and can store up to sixteen wafer cassettes or boxes which can be expanded if necessary. It interfaces to the MTUs via the docking modules as previously explained. Each IWIP station is hardwired to the central computer through its operator interface. A sensor system maintains lot location control and security.

The AGV system just described (FMS Model 4020) represents perhaps the upper limit of the commercial AGVS technology. Its high level automation has been possible because of its design as part of a computerized manufacturing system for a specific industry, namely, semiconductor manufacturing, an industry already characterized by a very high degree of automation.

APPLICATIONS AND STANDARDS

Since their arrival on the commercial scene, growth in the scope of applications of AGVS has been quite impressive, and there are indications that this growth might if anything be on the increase. In fact, it is quite possible that the applications realized so far represent a small fraction of their true potential. The situation is further complicated by the fact that AGVS are acquiring capabilities more commonly associated with robots, capabilities such as manipulation and vision. In addition, ongoing research by U.S. government agencies such as the Defense Advanced Research Projects Agency (DARPA) and the National Aeronautics and Space Administration (NASA) has the specific goal of developing vehicles with artificial intelligence, thereby providing such vehicles with the capability of operating autonomously for extended time periods.

Applications

It can therefore be seen that both the technology and the applications base of AGVS are at this time still in an extremely fluid state. Nonetheless, most AGVS applications can be classified into the following categories.

1. Manufacturing applications. In manufacturing, AGVS are installed primarily for controlling material flow, especially in situations where flexibility in the layout is deemed essential. Increasingly, they are also being used in the actual setup of processing stations as in the example discussed in the previous section.
2. Warehousing applications. These include applications arising in a wide range of facilities such as docks, airports, industrial plants or any other places where materials have to be stored and retrieved.
3. Transportation applications. These refer to places such as airports, freight stations, shipyards and other similar facilities where no manufacturing is done but a great deal of material handling and transporting take place.
4. Hostile-environment applications. Under this category, one can include remotely controlled vehicles such as the mars rover, and robot submarines used in undersea exploration. It has to be noted that the potential of AGVS for hazardous work in hostile environments has not been exploited to the extent that one would have hoped. Applications such as nuclear-material handling, mining, and work in polar regions which seem to be natural candidates for AGVS technology, for both economic and humanitarian reasons seem to have been almost entirely unaffected by it.
5. Military applications. Although it is possible to conceive of myriads of applications in the military, the fact that the vehicle has to operate almost entirely in an unstructured environment relegates the larger portion of such possibilities to the speculative realm. This remains so until AGVS acquire substantially greater degree of intelligence through the integration of sensors of various kinds, and more importantly, artificial intelligence. Technological challenges to the development of such advanced systems are certainly formidable. Nevertheless, it is an active area of research in the United States, and no doubt in other countries as well. It is possible that some vehicles capable of surveillance might be among the first such autonomous systems to appear in the next several years.

Standards

With the AGVS technology still in a state of evolution flux, there seems to be little possibility of any comprehensive standards for the foreseeable future. The areas in which one can expect to see some standards emerging are in the arena of manufacturing applications, particularly in data management and communication. For instance, with the manufacturing industry increasingly moving towards a factory communication standard in the GM sponsored Manufacturing Automation Protocol (MAP), it is likely that AGV systems will sooner or later be required to be compatible with it. Similarly, the AT&T UNIX operating system might become at least the *de facto* standard for the management of various on-board and on-site processors and centralized computers. It is possible that the d-c line voltage used for the drives in transporters might be standardized at 24 V. In any event, the initiative for any such standardization must come from the manufacturing sector. Other sectors of AGVS users have not yet attained the same level of sophistication and integration.

FUTURE TRENDS

Given the range of potential applications of AGVS, it is difficult to be precise about future developments particularly in light of the fact that important technological advances are likely to come from military and other research which do not have a direct economic motive. Yet, it can be stated with some confidence that the evolution of AGVS will be driven by integration of new capabilities and the goal of greater autonomy. With this trend in place, it is only natural to expect AGVS to increasingly supplant "hard" material-flow technologies based on fixed rails and conveyors. If in addition, vehicles in AGVS also take on the capabilities of robots, they would make extremely attractive alternatives to conventional robots on fixed bases, especially in plants aspiring to be flexible manufacturing systems. In order for this to happen, vehicle systems in AGVS will have to acquire significantly greater sensory and decision making capabilities than is the norm today. The FMS Model 4020 system described in this article can be regarded as a prototype in that direction. In summary, it can with some confidence be said that AGVS systems of the future will embody an ever increasing degree of integration of transportation, material handling, and possibly material processing functions, ie, integrate material flow and robotics.

BIBLIOGRAPHY

1. H. B. Maynard, *Handbook of Modern Manufacturing Management,* McGraw-Hill Publishing, New York, 1980.
2. T. Muller, *Automated Guided Vehicles,* IFS Publications, Bedford, UK, 1983.

General References

Ref. 1 is a good reference on manufacturing. It contains discussions of cost and problems of material flow in the larger context of manufacturing management.

Ref. 2 is one of the few books containing all aspects of AGVS. Its examples and bibliography are largely limited to European sources, particularly the FRG.

AGVS at Work, IFS Publications, Bedford, UK, 1986. Summary of experience with AGVS in North America.

P. Ranky, *The Design and Operation of FMS,* IFS Publications, Bedford, UK, 1983. Places AGVS and robots in the broader context of flexible manufacturing systems.

Proceedings of the International Conference on Automated Guided Vehicles and *Proceedings of the International Conference on Automated Material Handling,* IFS Publications, Bedford, UK.

H. Nasr, "Getting to Grips with Material Handling Problems," *Robot System User* (Oct. 1985).

AUTOMATION

Mikell P. Groover
Lehigh University
Bethlehem, Pennsylvania

WHAT IS AUTOMATION?

In its modern usage, automation can be defined as a technology that uses programmed commands to operate a given process, combined with feedback of information to determine that the commands have been properly executed. Automation is often used for processes that were previously operated by humans. When automated, the process can operate without human assistance or interference. In fact, most automated systems are capable of performing their functions with greater accuracy and precision, and in less time, than humans are able to do.

The process in an automated system is one that requires power to actuate or drive it from one physical condition to another. The physical condition can be defined in terms of mechanical, electrical, or chemical states. For example, there are many manufacturing processes in which the shape of the product is produced by transforming it from one mechanical state to a more desirable state. Other products are made by changing their chemical or electrical properties. In each case, power in some form is required to accomplish the process. Accordingly, one of the conditions that must be satisfied in order for a system to be classified as an automated system is that the controlled process uses power (energy) which results in a change in physical state.

The technology of automation has become strongly associated with and dependent on computer technology. Today, computers provide the principal means for programming and controlling an automated system. As computer technology has become more and more sophisticated, the automated processes which depend on it have become more sophisticated. Modern automated systems are able to control physical processes with accuracies measured in millionths of an inch, detect and identify problems related to their operations, make decisions, report their own performance, and interact with humans if that becomes necessary. As a result of the application of computers in automated systems, the technology of automation has reached a level where automated systems can perform functions that humans are not capable of performing.

The terms "automated system" and "computer system" are often used interchangeably. A computer system is sometimes referred to as an automated system, and vice versa. Although the two terms are closely associated, as suggested above, it is appropriate to recognize a principal distinction between them. The distinction is that an automated system causes some physical action or process to occur whereas a computer system results in the generation of data, information, and/or calculations. A computer system can be utilized as a component in an automated system to store programs of processing commands, perform control calculations, make decisions, etc, but these various functions are then converted into actions by other components of the automated system. Computers are also used in applications that have little or no association with automation. These applications include data processing and engineering analysis. Similarly, automated systems can be implemented without digital computers. Mechanical or electrical devices can be used to define and store the control programs for the automated system. The relationship between computer systems and automated systems might be conceptualized by means of the Venn diagram in Figure 1. The two categories overlap, but they also can exist on their own as well.

Robotics is a technology closely associated with automation. Industrial robotics can be defined as a particular field of automation in which the automated machine (ie, the robot) is designed to substitute for human labor. To do this, robots possess certain human-like characteristics. Today, the most common human-like characteristic is a mechanical manipulator that is patterned somewhat after the human arm and wrist. The robot's manipulator can be programmed to move through a series of positions to perform some useful task, such as loading

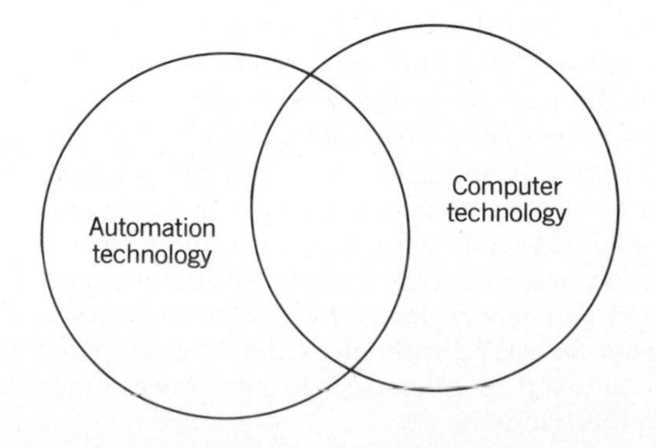

Figure 1. Venn diagram showing relationship between automated systems and computer systems.

and unloading a machine tool, spray painting a metal part, or spot welding an automobile car body. The motion sequence will be repeated until the robot is reprogrammed to accomplish some alternative task. In the future, industrial robots will have other human-like characteristics in addition to the manipulator. These characteristics might include: two arms instead of one, vision and other advanced sensors, greater intelligence to perform more complicated tasks, and the ability to move around the factory.

Industrial robots are related to computer technology. Indeed, robotics has been described as a combination of machine tool technology and computer science. The reason is that virtually any robot designed today uses a computer (either a microcomputer or programmable logic controller) as its controller or "brain." The controller stores the programs that define the tasks performed by the robot. The computer is also an important component in the feedback control system used to correctly position the manipulator in the workspace.

The relationship between robotics, automation, and computer technology can be pictured as shown in Figure 2. Robotics is a subset of automation technology. Since robots rely on the computer, a portion of robotics lies within the scope of computer technology.

Automation is a technology that has been applied widely to a variety of fields including household appliances, control of automobile engines, automatic bank teller machines, industrial manufacturing processes, and robotics. The discussion in this article of the encyclopedia focuses on automation as it relates to manufacturing and robotics.

HISTORICAL DEVELOPMENT OF AUTOMATION

The term automation is attributed to a Del Harder, who was an engineering manager at the Ford Motor Company in the 1940s. The word was coined to describe the increased use of mechanized devices and automatic controls in the production lines at Ford's plants. The automobile industry has made significant contributions to the development of automation technology. However, the origins of automation precede the automobile industry by centuries.

A term often associated and sometimes confused with automation is mechanization. Mechanization refers to the use of powered devices to drive machines. It involves the substitution of mechanical power (eg, wind, running water, steam, and other forms of energy) for human or animal power. Mechanization has its origins in the Industrial Revolution during which the steam engine and other important machines were invented to magnify the skills, strength, and power of humans. Strictly speaking, mechanization still requires humans to control or guide the machine in performing its function. Automation, on the other hand, has the capability for self-regulation. An automated machine is able to operate by itself without the need for human control or guidance.

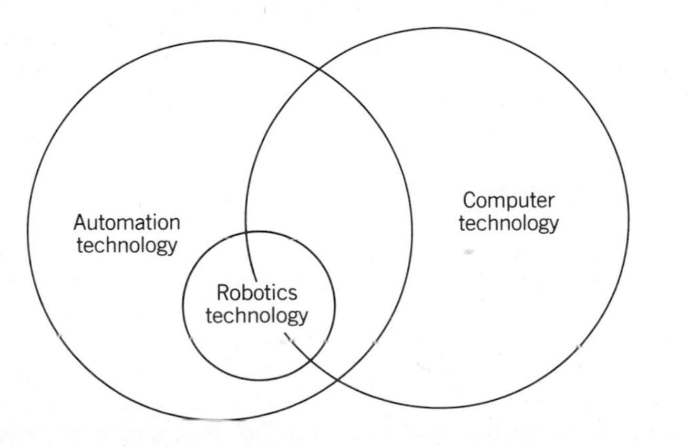

Figure 2. Venn diagram showing relationship of robots to automated systems and computer systems.

Table 1 provides a summary of many of the major historical milestones that are related to the development of mechanization and automation. The table begins with the initial efforts

Table 1. Significant Historical Developments in Automation and Robotics

1765	Development of James Watt's steam engine. The Watt steam engine is often used to mark the beginning of the Industrial Revolution in Great Britain. During the industrial revolution, great strides were made in the development of mechanization, a predecessor technology to automation.
1775	J. Wilkinson's boring machine, an important development in machine-tool technology. The boring machine contributed to the success of Watt's steam engine by improving the precision and accuracy of the cylinder hole.
1800	Eli Whitney's concept of interchangeable parts manufacturing, demonstrated in the production of muskets. Whitney's development is often considered as the beginning of mass production.
1800	H. Maudslay's screw-cutting lathe.
1801	J. M. Jacquard's loom, a programmable weaving machine controlled by punched cards. The Jacquard loom was used for making complicated weaving patterns. This was a significant development in automation because it demonstrated the feasibility of a programmable production machine.
1860	Brown & Sharpe automatic screw machine.
1866	J. R. Brown's precision grinding machine.
1887	Tolbert Lanson's Monotype machine, a type-setting machine controlled by punched tape.
1913	Henry Ford's mechanized assembly machine introduced for automobile production.
1924	Transfer machine developed for automobile plant in England.
1946	ENIAC, the first general-purpose electronic computer developed by J. W. Mauchly and J. P. Eckert.
1952	Numerical control (NC) developed at Massachusetts Institute of Technology for control of machine-tool axes.
1954	C. W. Kenward develops first industrial robot design in Great Britain. British patent issued in 1957.
1954	George Devol develops "programmed article transfer," considered to be the first industrial robot design in the United States. Patent issued in 1961.
1958	Kearney & Trecker's Machining Center, a highly automated NC machine tool with 4-axis control, tool changer, and pallet shuttle.
1958	Development of LISP (for List Processor) programming language for artificial intelligence (AI), destined in the future to be an important technique in automated systems. Although research in AI precedes 1958, the development of LISP represents a convenient date to use for the start of this branch of computer science.

Table 1. (*continued*)

1959	First experiments in machine vision occurred during the late 1950s. We are dating it 1959 although some of the work precedes this date.
1960	First Unimate robot introduced based on Devol's designs. Unimate installed in 1961 for tending die-casting machines.
1961	First computer-type programming language for NC developed, called APT (Automatically Programmed Tooling).
1963	Ivan Sutherland's "Sketchpad" project, the first demonstration of images being created and manipulated on a CRT by computer graphics. This might be used to identify the beginning of CAD/CAM, although the first commercially available CAD/CAM systems did not appear until the late 1960s.
1973	First computer-type robot programming languages developed at Stanford Research Institute. First commercially available language for programming robots was VAL, released in 1979 by Unimation, Inc.
1981	First direct-drive robot manipulator developed at Carnegie-Mellon University. Direct drive means that the joint motors are located at their respective joints. The conventional design is to locate the motors in the base of the robot using transmission linkages to drive the joints.

by James Watt in 1765 to design and build a practical steam engine. His design was an improvement of an earlier version called the Neucomen engine, developed around 1710. The Watt steam engine is generally used to date the start of the Industrial Revolution. However, there were many other developments that preceded the Industrial Revolution, and which set the stage for subsequent achievements in mechanization and automation.

The first of these earlier developments that might be mentioned were the basic tools and implements used by humans, some of which can be traced back to prehistoric times. The evolution of the basic implements progressed to include the wheel, the lever, the pulley, and other mechanical devices that extended the power of human muscle. The next milestone in the historical evolution of mechanization involved the generation of power readily available from nature. This power was in the form of windmills, waterwheels, and animal power (eg, horses, oxen). Man had discovered a means of driving machines that did not require his own muscles to operate.

The next developments in our history of automation are generally associated with the Industrial Revolution. The Industrial Revolution began around 1765 and continued for the next 100 years or so. Its impact is still very much being felt today. Some of the important inventions and developments made during the period are listed in Table 2. What makes the Industrial Revolution important in the present discussion of automation is that it resulted in the evolution of the factory system of production and in the development of steam power, the first great man-made source of power. Until the Industrial Revolution, the production of goods was accomplished largely in cottages by individual workers using manual craft methods. Many of the mechanical inventions of the Industrial Revolution required large sums of money to build or purchase. The machines also needed human workers to operate. Both of these factors resulted in the collection of workers into mills or factories. As suggested by the entries in Table 2, many of the early factories were concerned with textile production. Later, steelmaking and the fabrication of mechanical machinery became oriented to factory methods. The rise of the factory demanded new techniques of production, which in turn led to mechanization and, subsequently, automation.

Table 2. Inventions and Developments Associated with the Industrial Revolution

1765	James Watt's steam engine. This invention was to be the source of power to drive the factories in England and the rest of the world for more than 100 years.
1767	Spinning jenny, invented by J. Hargreaves between 1764 and 1767 to spin cotton into yarn.
1769	Water frame, by R. Arkwright, another machine to spin cotton.
1775	Start of significant developments in machine-tool technology, such as Wilkinson's boring mill (1775) and Maudsley's screw-cutting lathe (1800).
1778	S. Crompton's mule, a spinning machine that combined the best features of the spinning jenny and the water frame.
1785	E. Cartwright patented the power loom, a machine for weaving yarn into cloth. Although this machine was not particularly successful, it demonstrated the potential for mechanizing the weaving process.
1785	Watt's steam engine first used to operate a cotton mill.
1793	Eli Whitney's cotton gin, a machine for separating the cotton seeds from the fibers used to make thread.
1801	Jacquard's programmable loom.
1807	First voyage of the first successful steamboat, the Clermont, by inventor Robert Fulton.
1825	Invention of the locomotive by George Stevenson, which led to the development of the railroad transportation industry.
1855	H. Bessemer developed a process of steel-making called the Bessemer converter for producing steel in large quantities.
1867	Open-hearth steel-making process developed and patented by M. Martin.

The steam engine represented the first important source of mechanical power that people could generate themselves. The significance of this invention was that the energy required to drive machinery was no longer dependent on available wind, water, or animal power. Also, the amount of energy possible from the steam engine far exceeded that which could conveniently be collected and transmitted from natural sources at that time in history. Steam engines could be used to provide the power needs of entire factories or to drive locomotives and steamboats. As a milestone in the history of mechanization, the Watt steam engine stands as a most significant achievement.

The concept of feedback control was demonstrated by the flying-ball governor used to regulate the amount of steam fed to a steam engine's cylinder(s). Until the invention of this device by James Watt, the steam input had to be regulated by a person who operated a valve leading to the cylinder. The flying-ball governor consists of a weighted ball on the end of a hinged lever. The lever is mechanically coupled to the output shaft of the steam engine in such a way that as the rotational speed of the shaft is increased, the ball is forced to move outward by centrifugal force. This motion, through a mechanical connection to the steam intake valve, is used to control the steam to the cylinder. As the rotational speed of the engine speeds up, the valve opening is reduced, thereby slowing the engine. As the engine slows down, the reduced centrifugal force causes the ball to retract, thereby increasing the valve opening to allow more steam to enter. The flying-ball governor is an elegant example of feedback control, in which, the measured output of the system is used to increase or decrease the input. Today, feedback control remains an important principle in automation.

It was during the Industrial Revolution that substantial improvements were made in machine-tool technology. When Watt began work on his steam engine, one of the difficulties he faced was that the bore on the engine cylinder could not be machined with sufficient accuracy to prevent significant losses in steam pressure, thereby reducing the efficiency of the engine. Around 1775, an Englishman, John Wilkinson, invented a boring machine that enabled the cylinder holes on steam engines to be bored with a very high degree of accuracy and roundness. This boring machine contributed greatly to the success of the steam engine.

Other advances in machine tool technology made in England during the period included the development of the screw-cutting lathe (around 1800), the planer (during the early 1800s), and the shaper (around 1836). The first precision grinding machine was designed around 1866 by J. R. Brown of the Brown & Sharpe Company in the United States.

Another major advance in machine-tool technology came much later and was concerned with the automatic control of a machine. In the late 1940s, a U.S. industrialist, John Parsons, developed a means of defining the positions of a machine tool table using punched cards containing coordinate axis data. Parsons demonstrated the concept to the United States Air Force which later sponsored a research and development effort at the Massachusetts Institute of Technology. This effort led to the development of the first numerical control (NC) machine tool at MIT in 1952.

Computer technology and related advances in electronics have contributed to the development of automation. Although Charles Babbage is sometimes credited with inventing the computer (circa 1834), his was a mechanical machine that suffered from the difficulties of operating a complex mechanical device. The first electronic general purpose digital computer was designed and built by John Mauchly and J. P. Eckert in 1946. The machine, called ENIAC, was developed to compute ballistics trajectory tables for the U.S. Army. It was a very large machine, consisting of many cabinets full of vacuum tubes, and occupied an entire room. The trend in computer technology since that time has been toward smaller and smaller units that are more powerful yet less costly than their predecessors. Today, computers have become an integral part of automation technology. Many automated systems, especially systems that are highly complex, are controlled by computers. Sophisticated control programs that operate the automated equipment can be stored in the computer's memory, and the computer's CPU (central processing unit) is able to calculate control strategies and make operating decisions with much greater speed and accuracy than humans can accomplish.

The digital computer has spawned developments in many other fields which have influenced and continue to influence the development of automation. These fields include control theory, numerical control, off-line programming of numerical control machines, simulation, artificial intelligence, machine vision, and other sensor techniques.

BUILDING BLOCKS OF AUTOMATION

The definition of automation and the subsequent discussion of its historical development permit the identification of four basic attributes found in automated systems. These attributes might be termed the building blocks of automation. They are

1. Power to perform some action or process.
2. Programmed commands to define the action or process performed.
3. Feedback controls.
4. Decision-making.

Although examples of automation can be cited in which the four building blocks are not all readily apparent, most modern automated systems used in production today exhibit these four features in some way.

Power to Perform some Action or Process

A factory production system is designed to perform some useful activity, such as removing metal in a machining operation, or loading materials into a storage bin, or spray painting a car body panel. In each of these activities, power is required in order to perform the operation. Today, electrical power is the most common and versatile form because it can be generated from so many alternative sources (eg, coal, fuel oil, atomic energy, etc) and because it can be converted into so many other energy forms to accomplish work (eg, mechanical, hydraulic, and pneumatic). Electrical energy can even be stored by means of long-life batteries in applications where transmission of electricity is not feasible.

Whether the production system is automated or not, power is required to perform the activity. The activities performed by automated production systems can usually be classified into two types: (*1*) processing and (*2*) transfer and/or positioning. The processing activities are those associated with a given manufacturing operation. The manufacturing process transforms a starting workpart into a more desirable and valuable form. Examples of important manufacturing processes include metal machining, bending of sheet metal, forging, plastic molding, etc. These activities are usually performed by production machines such as lathes, presses, forge hammers, and injection-molding machines. In all of these machines, power is needed to perform the process.

The second type of activity performed by an automated system is transfer and/or positioning. This refers to the movement of the part before and after each manufacturing process and between manufacturing operations. The term transfer is typically used to denote the movement of the part between operations, and positioning refers to the procedure of accurately locating the part prior to a given processing operation. Material handling is the name commonly used to describe this collection of activities that occur in manufacturing.

Programmed Commands to Define the Activity Performed

The activities that are performed by an automated system are contained in a program. The program specifies the step by step actions that are carried out by the system. It defines what actions are to be accomplished, the sequence in which they must be performed, and when to perform them. The actions defined by the program are often performed in a cyclical manner. In many production operations, this cycle of actions constitutes what is called the work cycle. For every work cycle, a unit of product is completed in the process.

The program of commands can be relatively simple, or relatively complex and sophisticated. In perhaps an extreme example of a simple program, consider a thermostat in a house heating system. It consists of a single command: to maintain the temperature in the room at a specified level (or, more accurately, within a given temperature range). Depending on the rate at which heat is lost from the room to the outside environment, the heating system cycles between a condition in which the thermostat turns on and turns off the heating unit. The length of the operating cycle is determined by the rate of heat loss. In other simple programs typical of factory operations, a limited number of well-defined actions are performed continuously. Every work cycle is repeated with little or no variation from one cycle to the next. Examples of these kinds of programs include moving parts from one location to another (called pick-and-place operations), machine loading activities, and simple pressworking operations.

In more complicated programs, the number of separate actions performed by the automated system may be large, and the actions themselves may be quite complex. Sometimes the program may require communication of signals between different pieces of equipment in the automated system. These signals are referred to as interlocks and they are used to ensure coordination of the actions of the various components of the system. The series of actions that constitute the work cycle must be performed in the right order. Any given action performed by one component of the system must be completed before another component begins its portion of the cycle. For example, in a robot machine-loading application, the production machine must not begin its processing operation until the robot has loaded the workpart into the fixture and the holding device of the fixture is closed on the part.

Another feature in an automated system that makes the program of commands complex is that the program may contain decision-making routines. In these cases, one or more decisions must be made during the work cycle in response to variations in the operating environment. This is discussed later as a separate automation building block.

The programmed commands are contained in the controller unit of the automated system in any of several different forms. Today, computer memory is a very common means of storing the program. This is the typical storage media in computer control and programmable logic controllers. The commands are digitally encoded in computer memory and played back to execute the programmed work cycle. Other forms of program storage related to the computer include magnetic tape, magnetic disks and diskettes, and punched paper tape (traditional form used in numerical control). All of the preceding forms of program storage media can be read into and from computer memory with the appropriate peripheral devices (ie, tape reader, disk drive, tape reader and punch). A significant advantage of computer memory and associated forms of storage is that the program can be readily changed. If an error in the program is discovered, or improvements in the program must be made, or the program itself must be changed to accommodate a new product, these changes can be made with relative ease. In alternative forms of storage, the programmed commands are not as readily altered. These other storage media include mechanical and electrical devices such as cams, timer drums, mechanical stops, electronic sequencers, and electrical pegboards.

Feedback Controls

Feedback control is used in an automated system to determine that the programmed commands are correctly executed. As illustrated in Figure 3, a feedback control system consists of five basic elements: (*1*) input, (*2*) output, (*3*) process being controlled, (*4*) sensing elements, and (*5*) controller and actuating elements. The input to the system represents the desired set point value of the output. In an automated system, the input at a given moment corresponds to a desired control action that is to be taken in the work cycle. This action is associated with a particular command in the program. For example, one of the commands in a robot program may require the robot to position the end of its manipulator at a specified point in the work space. A feedback control system is used to determine that the point has been reached. The output in the feedback control system is the variable that is being measured relative

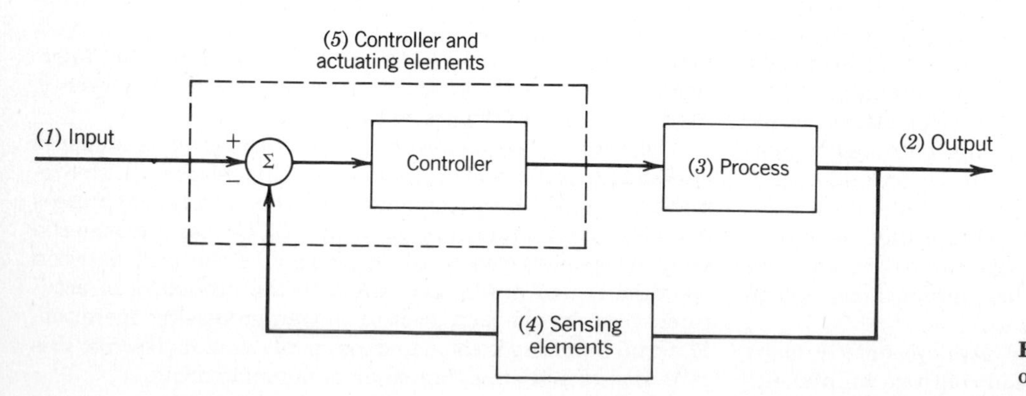

Figure 3. Configuration and components of a feedback control system.

to the input. In our robot positioning example, the output is the actual position of the end of the manipulator. The relationship between the input and output is determined by the process being controlled, the sensing elements, and the controller and actuating elements. For an industrial robot, the process is most appropriately represented by the manipulator configuration and its individual mechanical joints. The sensing elements are the devices in the feedback loop used to measure the value of the output. For a robot, a separate sensor is typically used for each joint in the manipulator. Optical encoders are commonly used for this purpose in robotics.

The controller and actuating elements are the components of the control system that cause changes to occur in the process that in turn determine the output values. The operation of the controller generally involves a comparison of the input value and the measured output value followed by some control computation whose objective is to determine how the difference between them can be reduced. The action resulting from this computation is applied to the process by the actuator. For an industrial robot, the controller and actuating elements consist of the computer control unit that calculates the values of the joints required to achieve the point in the workspace, the drive motors which power the joints, and the transmission hardware used to connect the motors to the joints.

Feedback control systems are often referred to as closed loop control systems. Not all control systems are closed loop. A second type of control that is used in many automated systems is called an open loop system. In this type of control, pictured in Figure 4, there is no feedback loop to verify that the input command has been executed. The system is considered sufficiently reliable that this verification is deemed unnecessary. For example, a programmed command to actuate an electrical switch may not require a feedback signal to confirm that the switch has been thrown. Because open loop systems require fewer components than closed loop systems (ie, the feedback loop is eliminated), they are generally less expensive. On the other hand, if a high level of accuracy and dependability is required in the control of the output, closed loop control is preferred.

Decision-making

Some automated systems are programmed with the ability to make decisions during operation. The decision-making capability is in the form of logic subroutines in the control program. By linking these subroutines in the control program to sensors that monitor the process and its environment, the system can respond to changes and irregularities in a logical manner. In effect, the control system makes decisions. There are a number of reasons for endowing an automated system with the capacity to make decisions. These reasons include: process optimization; interaction with humans; error detection and recovery; and safety.

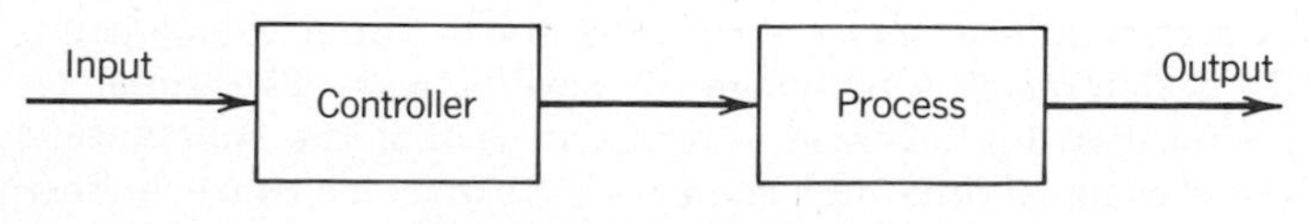

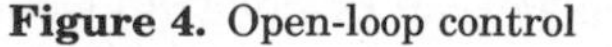
Figure 4. Open-loop control

Process Optimization. Process optimization involves controlling the system in such a manner as to optimize some economic or quality objective for the process. For example, the objective may be to maximize the yield of good product or to minimize the cost per unit of product. In many process optimization problems, the system controller must make decisions about how to best achieve the defined performance objective. In these cases, the controller uses sensory data collected from the process, and decides according to a preprogrammed algorithm the appropriate modifications to make in the process to optimize it.

A strategy called adaptive control is often used for these types of control problems. In adaptive control, the system is assumed to be subjected to unpredictable external variations that affect its operation and performance. For a manufacturing process, these unpredictable external variations may consist of the day-to-day variations in raw materials, wearing of critical tools or equipment components, changes in air temperature or humidity (that might influence performance for some processes), etc. If the internal control parameters of the system are fixed, as in a feedback control system of the type described above, it would be unable to cope with these variations. Therefore, the system is provided with the capability to monitor its own performance and to modify its internal control mechanism to seek an optimal performance level.

To accomplish adaptive control of a process, the control program performs three functions (the so-called functions of adaptive control): identification, decision, and modification. All three functions must be present in some form or another for adaptation to occur. The identification function determines the current performance of the system or process. This performance is usually defined in terms of a performance objective (eg, maximize yield, minimize cost). The current performance is evaluated by using sensor measurements collected on-line during operation and calculating the corresponding value of the performance objective. The decision function in adaptive control is concerned with the problem of deciding how the control mechanism should be changed to improve process performance, thus driving the system toward an optimal state. The decision may be to adjust some internal parameter(s) of the controller, or to make changes in the inputs to the process, or some similar decision. Third, the modification function involves the implementation of the decision. It is concerned with making the physical changes indicated by the decision function. The sequence and structure of the three functions of adaptive control are illustrated in Figure 5.

There are various other strategies that can be employed in process optimization, many of which requiring decisions to be made during the operation as in adaptive control. New strategies currently being researched will make use of artificial intelligence to figure out the most appropriate decisions to achieve optimal performance, rather than relying on preprogrammed logic contained in the controller.

Interaction with Humans. Another reason for automated systems to make decisions is to interact with humans. The system must be capable of making decisions and taking different actions depending on the instructions and inputs entered by humans. An automatic bank teller machine is a common example of this capacity to interact with humans. The customer enters a variety of instructions in response to prompts by the

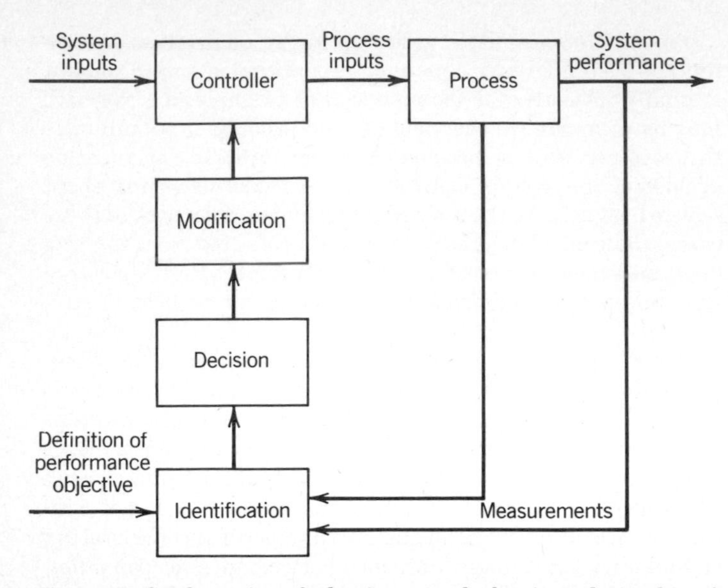

Figure 5. Configuration of adaptive control, showing relationship of its three functions.

machine, and the machine must make decisions about whether or not a valid identification number has been entered, amounts of cash to be dispensed, changes in balance, etc. As far as the customer is concerned the teller machine is exhibiting a modest level of intelligence in its ability to respond to assorted requests.

Error Detection and Recovery. Automated systems have traditionally been programmed to call a human worker when a malfunction arose during the normal operating cycle. Today, computer controlled systems are being programmed to identify these malfunctions using sensors, determine their cause, and execute routines to recover from the difficulty. This capability is referred to as error detection and recovery.

A decision-making capability is required in order to implement an error detection and recovery strategy in automated systems. The procedure is divided into two steps: detection and recovery. Error detection is concerned with the use and interpretation of sensor data to determine when an error or malfunction has occurred during system operation. This is accomplished, for example, by using certain sensors to identify one type of malfunction and other sensors to identify other malfunctions. Once the error has been identified, a recovery maneuver is executed by the controller to restore the system back to normal operation. The recovery maneuver must usually be designed specifically for the type of error or malfunction that has been identified. Possible recovery procedures include: stop the current cycle and remove the cause of the malfunction, adjust the process during the current work cycle, or abandon the current work cycle and start a new cycle.

Safety. Safety procedures implemented by the automated system are a special case of error detection and recovery. A decision-making capability is often required by the automated system in many safety hazard situations. When the system detects by means of its sensors that an unsafe condition has occurred, it must respond to the problem by taking appropriate action. This action may be as simple as stopping the process and alerting a human operator of the safety hazard. Or it may involve a more sophisticated procedure designed to remove the identified safety hazard and restore the system to regular operation.

TYPES OF PRODUCTION AUTOMATION

As the term is applied in production, automation can be defined as a technology that involves the application of mechanical, electronic, and computer-based systems to operate, control, and manage manufacturing systems. Automated production systems can generally be classified into three basic types:

1. Fixed automation.
2. Programmable automation.
3. Flexible automation.

These three types correlate to a large extent with the kind of production that is accomplished. Fixed automation is limited to large volumes of product being made where the variations in product are limited. Programmable automation is usually applied to low and medium volumes of production and is equipped to deal with relatively large variations in product configuration. Flexible automation is a relatively new form of automation and has thus far been applied in the mid-volume production range. Figure 6 illustrates the generally accepted application characteristics for the three types.

Fixed Automation

In fixed automation, the sequence of the production process is fixed by the configuration of the equipment. This type of automation is also sometimes referred to as "hard" automation to emphasize the hardened configuration of the production machinery. The individual processing (or assembly) operations performed in a fixed automated system are generally uncomplicated. However, the coordination and control required to integrate multiple operations into a single system is what makes fixed automation complicated. Some of the features associated with fixed automation are

Used for continuous production of identical or nearly identical parts (or products) in high volumes.
High production rates.
Specialized equipment designed to perform a specific sequence of processing operations with a high level of efficiency.
Relatively inflexible to changes in product design.

Much of the development work in fixed automation has been accomplished in the automobile industry. Mechanized assembly lines were first used around 1913 to move the partially assembled cars past manually operated workstations. Mechanized transfer lines for machining engine and transmission components date back to around 1924. In the automobile industry, the products and their components are generally made in high volumes, and for many of the internal components the design remains unchanged for several years. The economic justification (qv) of fixed automation is that the substantial investment of time and money which must be made in the specialized equipment can be divided among a large number

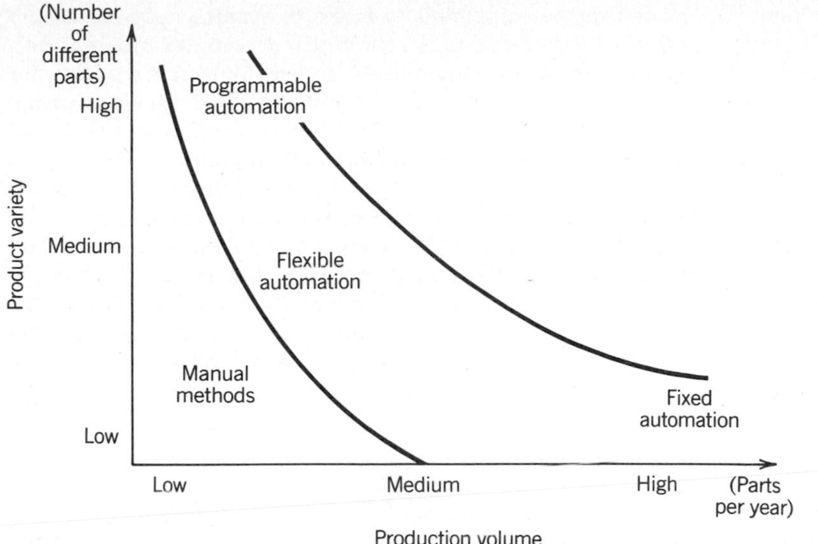

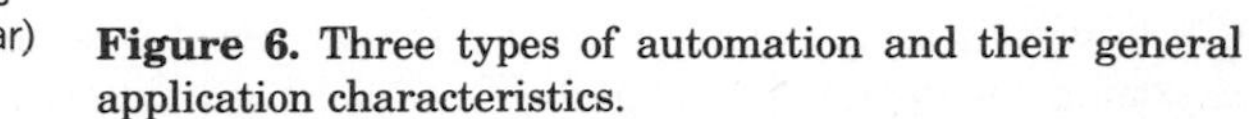

Figure 6. Three types of automation and their general application characteristics.

of units. This division of the equipment cost over many units and the low work cycle time associated with fixed automation result in a low unit cost of the product relative to alternative methods of production. The economic characteristic of operation is compared with the other two types of automation in Figure 7.

Programmable Automation

As its name suggests, programmable automation is represented by a production system that can be programmed and reprogrammed with relative convenience. The capability to be programmed means that the sequence of processing operations can be changed to accommodate variations in product style. Programmable automated systems are therefore used in industries that make varieties of products in low to medium volumes. The majority of products made throughout the world fall into this category.

The features that characterize programmable automation can be summarized as follows:

Used for low-to-medium volume products. The products are usually made in batches.

Lower production rates than fixed automation.

Equipment designed to accommodate variety of product configurations.

Readily reprogrammed to change over from one product configuration to another.

The typical operation of a programmable automated system involves the production of parts (or products) in batches. The batch size can be anywhere from several parts to several thousand parts. Prior to running a given batch the system controller must be reprogrammed, and the physical setup of the equipment must be changed over. Reprogramming involves entering the commands that correspond to the particular part design being made. The physical setup includes the tools, workholding fixtures, arrangement of the worktable, and any machine settings that must be adjusted. In general, the setup must be adapted to the configuration of the particular workpart being made. The reprogramming and physical changeover of the production system take time. Therefore, systems of the programmable automation type generally operate with periods of time during which no production occurs (this is when the setup and reprogramming are accomplished), followed by the produc-

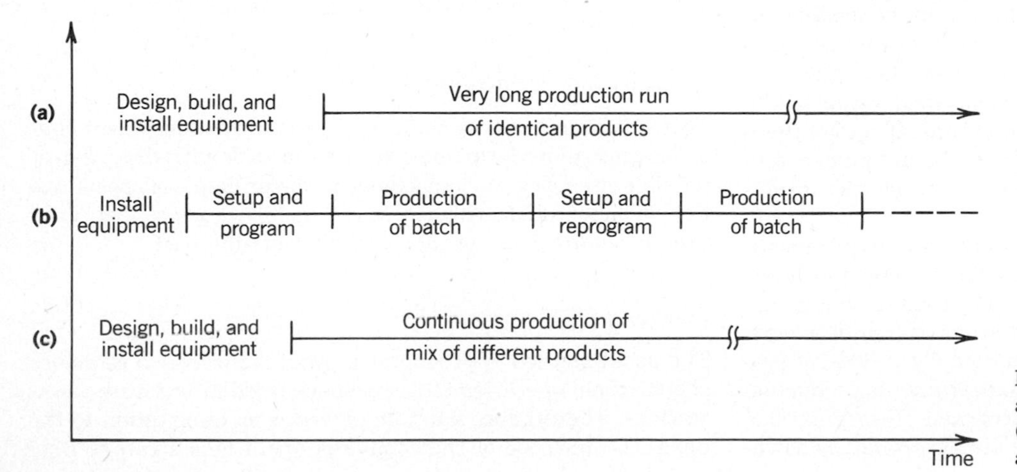

Figure 7. Comparison of operating characteristics of three types of automation: **(a)** fixed automation, **(b)** programmable automation, **(c)** flexible automation.

tion run period (when the batch of parts is actually produced). This characteristic operation is portrayed in Figure 7. Because of the costs of setup and reprogramming for each batch, the economics of programmable automation require a batch size that is sufficient to minimize these costs. Nevertheless, the nature of batch production is such that programmable automation usually represents an attractive method of production compared to traditional manual methods.

Programmable automation can be traced back to the Jacquard loom developed in France around 1801. More recent advances include numerical control first demonstrated in 1952, and industrial robots first developed around 1961.

Flexible Automation

The trouble with programmable automation, the way it has been defined above, is that there are interruptions in the flow of production as a result of the programming and setup requirement between each batch. Flexible automation represents a means of addressing this deficiency. Flexible automation is an extension of programmable automation because the production system must be programmed for each different part made. However, there are certain differences in operation that distinguish flexible automation from programmable automation.

A flexible automated system is capable of producing a mix of different products with virtually no time lost for physical changeovers from one product to the next. The system is able to produce the various products without the downtime for setup and reprogramming that is characteristic of programmable automation. As a consequence, the system can efficiently make the products in varying combinations and schedules to meet changing demand requirements without the need to operate in a batch production mode. At the present state of the technology, the product variety that can be accommodated in flexible automation is not as great as in programmable automation. However, flexible automated production systems are a relatively recent innovation, and their capabilities are likely to expand in the future. The first significant example of flexible automation was the flexible manufacturing system (FMS), introduced in the late 1960s. These flexible manufacturing systems consisted of a group of machining stations (NC machine tools) connected together by a material handling system, all operating under computer control.

The two technological requirements that make flexible automation possible are (*1*) the capacity to reprogram the equipment for different products with no lost production time, and (*2*) the capability to change over the physical setup of the equipment, again with no lost production time. If both of these requirements can be accomplished, then the automated production system can manufacture the various products continuously rather than in batches.

Off-line programming is what makes the first requirement, reprogramming without downtime, possible. Instead of interrupting the operation of the production system to change the program, the programs for new parts are prepared at a separate site, generally using computer-assisted methods of programming. This permits the system to continue its production of parts under previously prepared programs. New programs prepared at the remote computer site are electronically downloaded to the equipment to make new parts introduced onto the production system. In the future, computer systems with graphics modeling capabilities (called CAD/CAM systems, for computer-aided design and computer-aided manufacturing) will be used to automatically prepare part programs based on part geometry data entered by the product designer.

The second requirement for flexible automation, changing the physical setup with no lost production time, is generally accomplished on an FMS by means of pallet fixtures that are loaded off-line with new parts while the system is engaged in the production of the previous parts. The material handling system, operating under computer control, transfers the pallet fixture from the loading station to the machine tool scheduled to make the particular part. The machine tool has a variety of different cutting tools available in a tool storage drum for use on the part. Some parts require machining by more than one machine tool in the system. The material-handling system is programmed to make the corresponding transfers. When the machining sequence is completed, the part is transferred to an unloading station. The loading and unloading stations are manned by human operators and represent the interface between the FMS and the other systems in the factory.

The important features of flexible automation, as illustrated by an FMS, include

- Considerable preplanning required to identify the parts (or products) that will be made on the system and to specify the most appropriate equipment to make these parts.
- Continuous production of variable mixtures of parts (or products)
- Medium production rates, typically between those of fixed automation and programmable automation
- Flexibility to deal with a limited range of variations in product design.

The operating characteristics of flexible automation, compared to the other two types, are portrayed in Figure 7.

AUTOMATION STRATEGIES

The design of an automated production system involves the implementation of various principles or strategies whose purpose is to improve process productivity and product quality. Ten of these basic strategies are listed here:

Specialization of Operations

This principle involves the use of production equipment that is designed to perform one operation or task with the greatest possible efficiency. It is analogous to the principle of specialization of labor (also called division of labor) which has been used for many years to increase labor productivity.

Combined Operations

The usual method of producing a product involves a sequence of individual operations that are performed at separate workstations. The product must be moved from one station to the next. The purpose of the combined operations strategy is to combine several of the individual operations in sequence at a single workstation, thereby saving handling time and expense.

Simultaneous Operations

This principle extends the previous strategy by not only combining the processing steps at a single workstation, but also performing the operations at the same time rather than in sequence.

Integration of Operations

The term "integration of operations" refers to a production system in which the individual workstations, each performing a particular processing step on the product, are physically connected together by means of a handling mechanism that transfers the parts from station to station. This type of integration represents an alternative to the second strategy (combined operations), and usually results in higher production rates since several products can be processed simultaneously (one product at each workstation).

Increased Flexibility

The objective of this principle is to design the production system so that it can be used for more than a single product. This permits the equipment to be used for a variety of products, making the system adaptable to changes in demand patterns and production schedules. Programmable automation and flexible automation make use of this strategy.

Improved Material Handling and Storage

This strategy involves a number of principles in the applied field of material handling. These principles include minimizing distances that materials must be moved, and sometimes including temporary storage zones between workstations to act as buffers against irregularities in production rates.

On-line Inspection

The traditional approach to inspection in industry is to perform the inspection process after processing has been completed. This principle involves the use of automatic on-line inspection in which the product is inspected during processing so that the information can be used to correct the process immediately. This means that processing errors and variations can be corrected either on the current unit of product or on the following product.

Process Control and Optimization

This principle involves the use of a variety of control schemes and optimization methods to operate the individual process more efficiently. A well-developed theory of process control exists for this purpose.

Plant Operations Control

This strategy involves the overall factory operations rather than the individual processes. It is concerned with optimum production scheduling algorithms, shop floor control, material-handling system management (if applicable), and other approaches that can be applied in the individual factory.

Computer Integrated Manufacturing

This strategy extends the previous strategy to the corporate level. It is concerned with integration of the design function with manufacturing through the use of CAD/CAM, corporate management information systems and data bases, computer networking, and other techniques to improve corporate operations.

EXAMPLES OF AUTOMATED PRODUCTION SYSTEMS

In this section, some of the important examples of automated production systems are described, including industrial robots. In each case, the features of the system are described in light of the previous discussion.

Early Machine Tools

The evolution of mechanization and automation in the manufacturing industries is based largely on developments in machine tool technology. Some of these developments are listed in Table 1. One early example of machine tool automation was the automatic screw machine, an invention credited to Brown & Sharpe around 1860. Although early versions of the automated screw machine do not possess all of the four automation building blocks described above, it nevertheless stands as a remarkable achievement for its time.

The automatic screw machine (another machine that is very similar is called an automatic bar machine) is a highly mechanized lathe used for repetitive machining of small turned parts. An important early application involved the manufacture of screws and similar threaded hardware items; hence, the name screw machine. The program for the machine tool is contained in a series of mechanical stops and cams that regulate the cycle of machining operations. The stop settings and the cams must be designed specifically for the part that is made on the machine. Because of the time and expense involved in preparing the program (setting the stops and making the cams), automatic screw machines are usually employed for medium-to-high production jobs. Automatic screw machines are technically classified as belonging to the programmable automation class, although they are often employed as fixed automated systems for very high production.

Some automatic screw machines have multiple spindles (eg, a six-spindle automatic bar machine). These machines are capable of performing six machining operations simultaneously on six different workbars. This mode of operation increases the production rate and reduces the unit cost of the product.

Transfer Line

A transfer line is an automated production system consisting of a series of automatic workstations with a parts handling mechanism for transferring parts from one station to the next. It is another example of automation whose roots are in the machine tool industry. The processing or assembly operations performed on a transfer line are done progressively, each station contributing work to the part as it moves through the sequence. A raw workpart begins at one end of the line and is processed through each workstation until it is completed at the final station. The transfer line illustrates several of the automation strategies, including specialization of operations and integration of operations. The significant advantage of the transfer line is that many parts (the number is theoretically equal to the number of workstations on the line) are being processed simultaneously at any moment, each at a different station. An example of an automated transfer line is illustrated in Figure 8.

The term transfer line derives from the fact that the configuration of the system is a production line and that the parts are transferred between workstations. The generic form of the transfer line involves a straight-line flow of work. Other transfer line work flows include L-shapes, U-shapes, and loops. The same principle of progressive processing of the work at stations can also be accomplished using a circular configuration. This type of system is called a dial-indexing machine because the work is transferred from station to station using a rotating circular table called a dial.

One of the problems in the operation of transfer lines, especially lines with many workstations, is downtime. When one workstation on the line malfunctions or breaks down, the upstream and downstream stations are affected. Parts at upstream stations cannot be transferred to the broken-down station, and the broken station is not producing parts to supply the downstream stations. The entire line must be temporarily stopped while the broken-down station is repaired. This causes a high proportion of nonproductive time on the production line. To alleviate this problem, transfer lines are sometimes designed with parts storage buffers located between certain workstations along the line. The purpose of these buffers is to separate the line into groups of workstations that can operate somewhat independently of each other.

In addition to storage buffers, transfer lines can also be designed with manually operated workstations, and automatic inspection stations. The manual stations are included for operations that are difficult to automate. Assembly operations that require adjustments or calibration procedures are examples of these kinds of operations. Automatic inspection stations are included to monitor certain quality characteristics of the products made on the line. These inspections are generally performed on a 100% basis in which every part is inspected, rather than on a statistical sampling basis in which only a small portion of the parts are checked, usually by human inspectors.

Transfer systems (transfer lines and dial indexing machines) are examples of fixed automation. The parts or products made on these systems are almost always made in large quantities at high production rates, and there is no real deviation in product design. When changes are made in the product design, either a substantial changeover of the production line is required, or a new production line must be built thus rendering the previous system obsolete.

There is no need to alter the programmed commands which regulate and coordinate the basic operation of a transfer system. The same work cycle is repeated at each workstation for every part made without variation. Before computer control was used in production systems, the program for a transfer line was contained in the form of control circuits consisting of electromechanical relays. Since there was little or no need to change the programs (except to work out the initial bugs in the logic that might be present), these hard-wired controls were sufficient for early transfer lines. As the need for more sophisticated controls increased (eg, process optimization, error detection, decision-making for repair diagnosis), relay controls were replaced by programmable logic controllers and similar controls. These computer-based systems permit the programming of more complex logic and allow for easier program debugging.

Numerical Control

Numerical control (NC) is a form of programmable automation in which the processing equipment is controlled by a program consisting of numbers, letters, and other symbols. As described previously, NC had its beginnings in the early 1950s in machine tool applications. In the traditional operation of an NC machine, the program is coded on a punched tape (one inch wide with eight holes across, each hole representing a binary

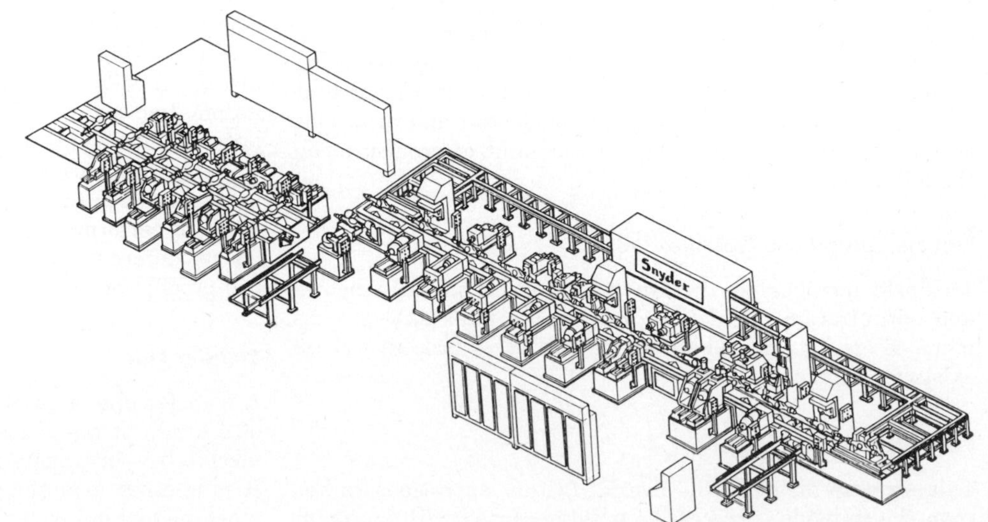

Figure 8. Example of a transfer line. Courtesy of Snyder Corp.

digit). The program is called a "part program" because it refers to the workpart being machined. The tape is read into a tape reader which is a component of the machine tool control unit. The tape determines the sequence of operation of the NC machine required to process the particular workpart. Today, other forms of storage media, such as disks and magnetic tape, are gradually replacing punched tape for storing NC part programs. In addition, modern numerical control systems are implemented using computers as their control units. This has given rise to the name computer numerical control, or CNC, to describe these computer controlled NC systems.

The principal applications of numerical control are still in the machine-tool industry, and include processes such as milling, turning, drilling, and related metal machining operations. In addition to machining, other applications include assembly, drafting, and dimensional inspection. In all of these applications, the common operating feature is that the relative position of a tool (or other processing element) is controlled relative to the workpart (or other object) being processed.

The control of the tool in numerical control is achieved by means of an axis-coordinate system as illustrated in Figure 9. The coordinate system consists of the three principal axes, x, y, and z. Additional axes can be defined as rotational axes about the three linear axes. These rotational axes are, respectively, the a, b, and c axes. To use these axes during programming, the part programmer directs the motion of the tool by commanding it to move to defined points or along defined paths in the coordinate system. The APT language (Automatically Programmed Tooling) is one of the important NC-part programming languages. The following APT statement illustrates the use of the x, y, and z axes for one form of motion control in NC part programming: GOTO/2,4,1. This would direct the tool to move to the point in the axis system defined by $x = 2$, $y = 4$, and $z = 1$. This statement is a point-to-point motion command in numerical control, in which the tool is directed to a particular point in the machine tool workspace and the programmer is not particularly concerned about the path that is taken to get to the point. The other type of motion command in NC programming is contouring. This involves a continuous path motion in which the tool is directed along a defined path in the workspace. An example of a contouring command in NC is GOFWD/PL2,TO,PL3.

This command would move the tool forward (GOFWD = go forward) along a plane defined in the workspace of the machine tool called PL2. The forward motion would continue until the tool came in contact with plane PL3, another plane in the workspace. By defining the various points, lines, and planes that specify the geometry of the part to be machined, and directing the cutting tool with motion commands similar to the above, the program can be prepared to provide the required machining sequence for the workpart.

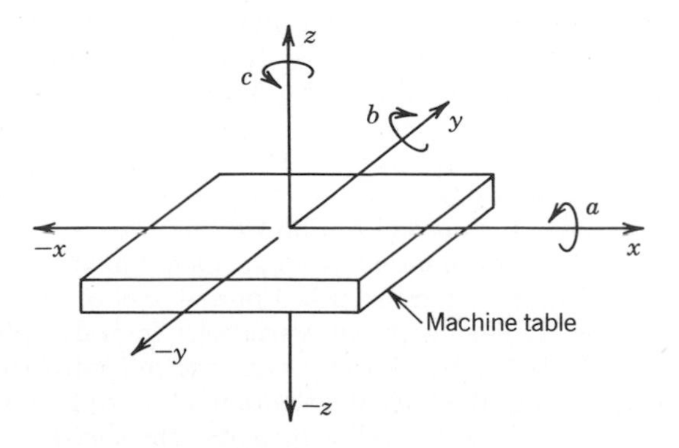

Figure 9. Machine-tool coordinate system in NC.

The operation of an NC system is one which exhibits three of the basic building blocks of automation: power to drive the machine tool in the cutting process, programming (the NC part program), and feedback control. The NC machine tool uses closed loop feedback control to determine that the programmed coordinate positions are achieved by the tool relative to the work table. Some NC machining systems also possess a decision-making capability in the form of adaptive control. In adaptive control machining, the machine tool controller monitors the performance of the machining process and makes decisions about changes in feed rates and cutting speeds. For example, adaptive control systems can distinguish whether the tool is cutting in metal or moving through air, and decide the appropriate feed rate in either case.

NC Machining Centers

A highly automated version of NC machine tool is the machining center. It is a multifunction machine first introduced in the late 1950s by Kearney & Trecker Corporation. The modern machining center can perform under numerical control a variety of different machining processes such as milling, drilling, boring, and related operations. Until the machining center was developed, these processes were generally accomplished on separate machines (eg, milling machine, drill press, and boring machine). In addition, the machining center features several innovations designed to reduce the machining cycle time. Among these innovations are

Automatic tool-changer, which exchanges cutting tools between the spindle chuck and a tool storage unit on the machine under NC control. The automatic tool-changer reduces the time normally required to change tools.

Workpart reorientation, accomplished by means of a worktable whose rotational position can be controlled through the NC part program. This permits several sides of the part to be positioned for machining.

Two-position worktable, so that the operator can be unloading the previous part and loading the next part onto one table position, while the other table is presenting the current part to the spindle for machining. This means that the machine tool does not stand idle while parts are being loaded and unloaded.

In today's advanced machining centers, the concept of two-position worktables has been extended to multiple-position pallet shuttles. This type of shuttle system, consisting of perhaps a dozen or more part positions, allows the machine tool to be loaded in advance for several hours of processing during which the machining center operates in an untended mode.

Each of the automatic features of the machining center help to increase the efficiency and utilization of the production machine.

Automated Guided Vehicle Systems

An automated guided vehicle system (AGVS) is a material handling system rather than a production system. It represents a relatively sophisticated and highly automated form of material-handling system. An AGVS uses independently operating vehicles that are guided along defined pathways and powered by on-board batteries. The pathways are determined by guidewires in the floor or paint strips on the surface of the floor. The guidewire method uses wires imbedded slightly below the floor surface which emit a magnetic field that can be sensed and tracked by the vehicles. The paint strip method uses a painted pathway (about one inch wide) that reflects light. Each vehicle has optical sensors that are capable of seeing the paint strip so that the vehicle control unit can follow the defined pathway. The batteries commonly used on-board the guided vehicles are rechargeable and possess long cycle lives. A typical AGV battery will last eight to 16 hours between recharging.

The feature that makes automated guided vehicles somewhat unique among material handling systems is that they can be programmed to make pickups and deliveries to particular destination points. This achieves a high degree of flexibility in their applications. Rather than continuously deliver large quantities of materials along fixed routes as other mechanized handling systems do, the vehicles in an AGVS can move along their own routes and make individual deliveries to different drop-off locations in the system.

Flexible Manufacturing Systems

A flexible manufacturing system (FMS) is a highly sophisticated example of automation. An FMS is an integrated production system consisting of a group of processing stations (usually NC machine tools) connected together by an automated material-handling system, all operating under computer control. As its name suggests, an FMS is a form of flexible automation. It is capable of processing a variety of part configurations although there are limits to the amount of variation that an FMS can accommodate. It can also tolerate variations in the mix of parts made on the system, so that if requirements change over time, the production schedule can be adjusted accordingly. In the ideal operation of an FMS, these changes in production can be made without the need to shut the system down for reprogramming or changeovers in the physical setup.

The most common applications of flexible manufacturing systems have thus far been in the machining area. There are several hundred machining type FMS installations in operation throughout the world at the time of this writing (1986). Additional applications are in other metalworking processes, inspection, and assembly. In nearly all of these applications, flexible manufacturing systems have been used to satisfy medium volume production requirements, as illustrated in Figure 6. It is anticipated that future flexible systems will have the capability to extend the range of economic production both in the direction of higher as well as lower production volumes.

As suggested by the definition above, the three components of an FMS are the processing workstations, the material-handling system, and the computer controls. For machining-type FMSs, the processing stations are predominantly CNC machine tools. Parts are delivered to the machines using automated guided vehicles, conveyors, or other forms of material-handling devices. Industrial robots are used in some systems to handle parts between workstations. The overall control of the system is accomplished by the computer control unit. The NC part programs are transmitted electronically to the individual machine tools, and status information is collected on parts made, cutting tool status, and machine downtime, to monitor overall production performance of the system. The physical interface between the FMS and the outside world is usually accomplished by means of human workers who load raw workparts onto the system and unload finished parts from it.

Figure 10 shows a flexible manufacturing system with eight CNC machine tools. The system was installed by Cincinnati Milacron for a major aerospace plant. The system is designed to produce approximately 600 different machined parts under NC control. Because aerospace is typically a low volume production situation, this FMS is designed to economically produce these parts in relatively low quantities for an FMS. Two loading and unloading stations are shown at the upper left of the drawing. Human workers fixture the parts onto special pallets that are transported to the various workstations in the system. An automated guided vehicle system is utilized as the material-handling system to deliver workparts between the load/unload stations and the workstations. The FMS also has several inspection stations where machined parts are checked for dimensional accuracy. The computer control room for the system is shown to the right of the drawing.

Industrial Robots

Strictly speaking, an industrial robot is an example of programmable automation. As defined previously, an industrial robot is a programmable machine that possesses certain humanlike features. Although the robot itself is an example of programmable automation, robots are often used in production systems that represent either fixed automation or an approximation of flexible automation. An illustration of these two cases is in robotic spot-welding lines used to assemble car bodies in the automobile industry. Some of these assembly lines weld the same body styles in a dedicated, continuous fashion. There is absolutely no variation in the product made on these kinds of production line. Accordingly, although the robots are capable of being reprogrammed, the applications involve the repetition of the same motion patterns for each car body that moves long the line. In effect, the robots are being used as components in a fixed automated production system.

Other robot welding lines are designed to perform welding cycles on body styles that are not the same. The difference might be between sedans and station wagons of the same basic model, or the models themselves might be different. In these applications, the robots must use different motion and spot welding cycles to deal with the variations in body style. A sensor (usually an optical sensor) is utilized to identify which body style is next in the line, and the central control computer or programmable logic controller indicates the corresponding program for the robot to use. To a significant degree, such a

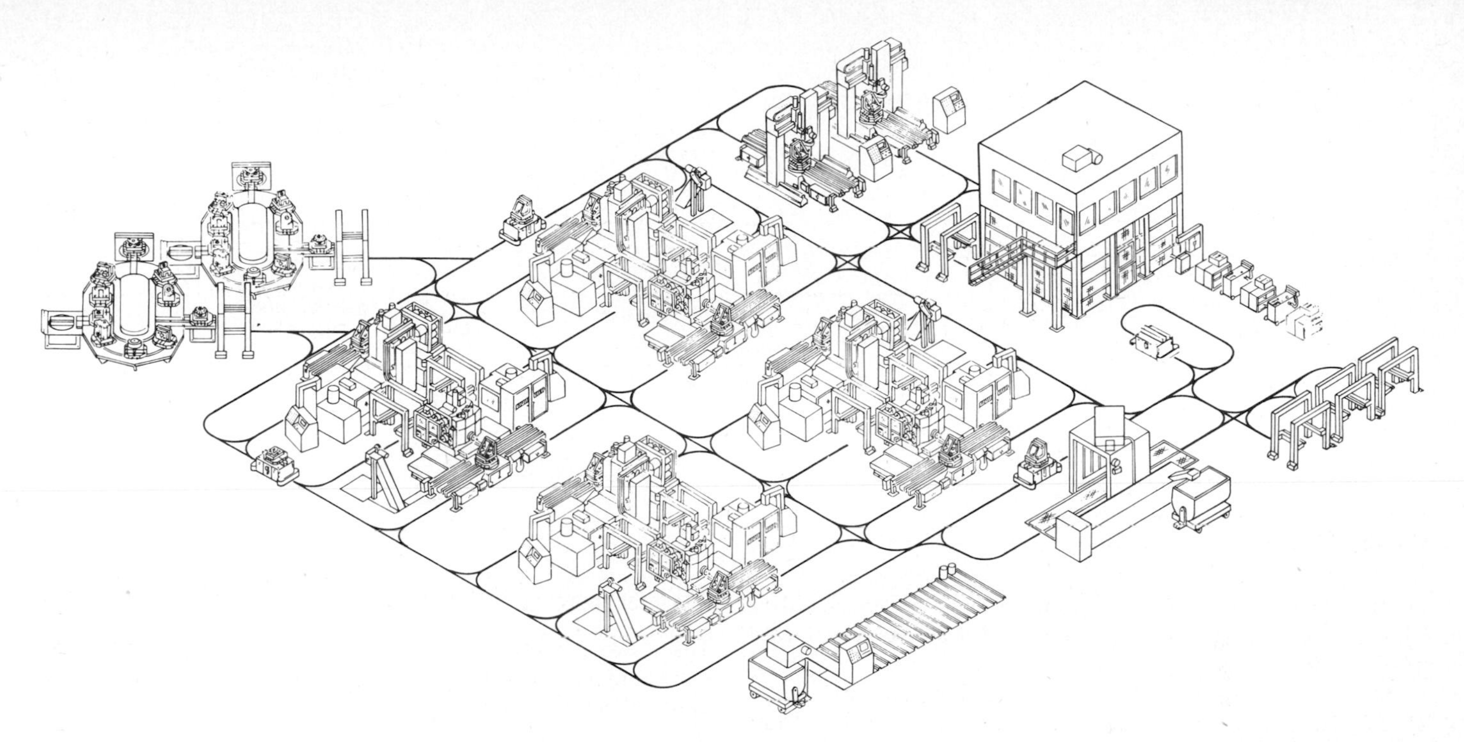

Figure 10. Flexible manufacturing system Courtesy of Cincinnati Milacron.

production line is flexible in the sense that the mixture of body styles can vary from day to day, corresponding to changes in demand for the different types of cars. This type of automated line does not completely satisfy our definition of flexible automation because the line cannot be reprogrammed for a new body style, which has not been produced before on the line, without shutting it down to introduce the new program.

COMPUTER INTEGRATED MANUFACTURING

The examples of automation considered above are all ones in which physical actions or processes are performed. Materials are processed, or moved, or both. Although the preceding examples do not cover all of the possibilities, the physical activities that take place in a typical factory can be classified into four categories: (*1*) manufacturing processes; (*2*) assembly operations; (*3*) material handling and storage; and (*4*) inspection. A common feature of these activities is that they all come in physical contact with the product in one way or another. Automation technology can be applied to each of the four activities to create smooth-running systems that operate with little or no human intervention. These automated systems are sometimes referred to as "islands of automation."

In addition to the physical activities in manufacturing, there are other functions that must be performed which do not come in physical contact with the product. Instead, these other functions deal with the data, information, and knowledge that are applied to operate the factory. They are information processing activities that are performed in support of the physical activities. A useful conceptual model of the relationship between the physical activities and the information processing functions in the factory is illustrated in Figure 11. The physical manufacturing activities are performed in the "pipe" of the model which is supposed to represent the actual factory. Raw materials flow in one end of the pipe and finished products flow out the opposite end. While in the pipe, the materials are processed, assembled, handled, and inspected. It is inside the pipe that automation, as it is defined in this article, is applied. Surrounding the pipe is the cycle of information processing activities. This cycle of activities can be divided into four functional areas: (*1*) business functions; (*2*) product design; (*3*) manufacturing planning; and (*4*) manufacturing control.

The business functions deal with sales and marketing, order entry, customer billing, and other related activities that interface the factory with the outside world. These business activities represent the beginning of the information processing cycle because the order to produce enters the firm as part of this function. This area also represents the end of the cycle because customer billing for the products made in the factory occurs here.

Product design is concerned with all of the activities that support the specification of the product for manufacturing. It includes product development, engineering analysis, prototype fabrication, engineering drafting, preparation of bills of material, product approval by top management, and any other activities associated with the product design function in a company. The output of product design feeds into manufacturing planning.

Manufacturing planning logically follows product design, although some of the planning functions must be accomplished concurrently with design. Manufacturing planning is concerned with preparing the plans needed to produce the product. These planning activities include process planning (preparation of the route sheets), master production scheduling, make-or-buy decisions, material requirements planning

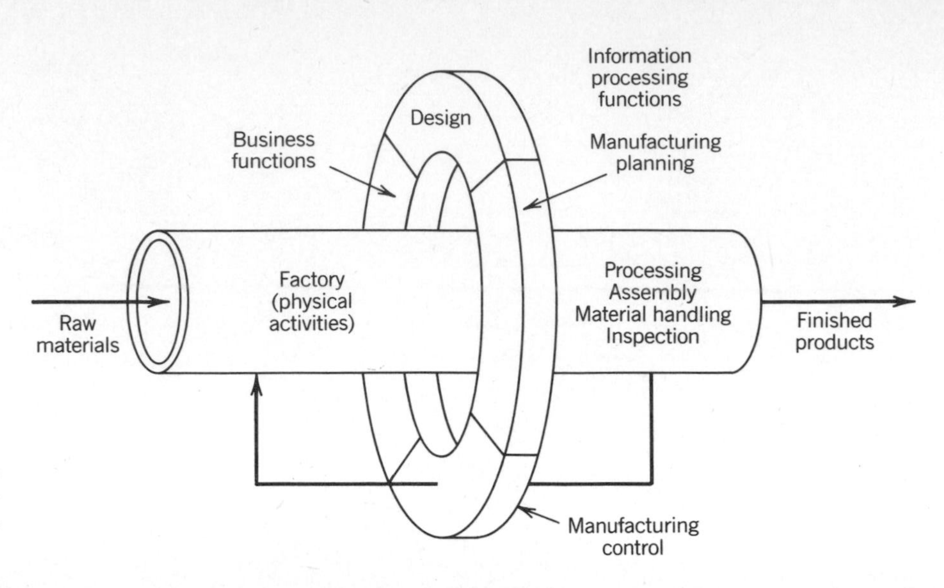

Figure 11. "Pipe and ring" model of the factory, separating the physical activities that can be automated from the information processing activities that support manufacturing.

(MRP), capacity planning, and preparation of purchase orders. The output of manufacturing planning flows into manufacturing control.

Manufacturing control is concerned with managing and controlling the production operations in accordance with the manufacturing plans that have been prepared. It is the function in the information processing cycle which intersects with the physical activities in the factory. Typical control activities included in this function are shop floor control, inventory control, and quality control.

Substantial opportunities exist to integrate the four functions in the information processing cycle into one interconnected computer system. Computer-integrated manufacturing is the term used to denote this type of computer system in the firm. A fully implemented CIM system in a company would possess the following features:

It integrates all of the functions in the organization that are related to manufacturing.

It connects these various functions by means of a communications network.

It uses a common data base or a set of common data bases that are accessible by all of the functions.

It provides the information and communications framework to permit the "islands of automation" in the factory to be physically interconnected.

The factory system is driven by data, information, and knowledge.

The technologies and systems normally associated with CIM include computerized order entry, CAD/CAM, computer-aided quality control, computerized MRP, manufacturing resource planning (MRP II), and computerized shop floor control systems (eg, automatic identification systems using bar codes and other techniques). In effect, computer-integrated manufacturing is concerned with the automation of the information processing cycle, just as automation technology is concerned with the physical activities in the factory.

REASONS FOR AUTOMATING

There are a number of economic and technical reasons for installing an automated production system. These reasons for automating include the following:

1. **Higher productivity and lower cost.** Automation reduces the labor content of the production operation and usually increases the rate at which parts are made. Both of these factors tend to increase labor productivity and reduce the cost per unit produced.
2. **Increased safety.** By removing human operators from active participation in potentially dangerous production jobs, safety is improved.
3. **Improved uniformity and quality of product.** Automated production tasks are typically performed with a higher degree of precision and consistency than tasks accomplished by manual labor. This precision and consistency usually mean greater uniformity and quality of product.
4. **Reduced work-in-process and manufacturing lead time.** Work-in-process inventory represents unfinished product. The producer has invested money in raw materials, labor, and equipment to make the product but cannot receive payment until it has been completed. Automation tends to reduce the amount of unfinished product held in the factory, and it also reduces the time required to complete the associated production processes.
5. **Better corporate image.** Finally, by automating its factories and maintaining a technological advantage over its competitors, a company tends to improve its corporate image. This translates into higher employee morale and better customer acceptance of its products.

BIBLIOGRAPHY

General References

P. P. Bose, "Basic of AGV Systems," Special Report 784, *American Machinist and Automated Manufacturing,* 105–122 (March 1986).

M. P. Groover and J. C. Wiginton, "CIM and the Flexible Automated Factory of the Future," *Industrial Engineering,* 74–85 (Jan. 1986).

M. P. Groover, M. Weiss, R. N. Nagel, and N. G. Odrey, *Industrial Robotics: Technology, Programming, and Applications,* McGraw-Hill Book Company, New York, 1986.

M. P. Groover, *Automation, Production Systems, and Computer Integrated Manufacturing,* Prentice-Hall, Inc., Englewood Cliffs, N.J., 1987.

J. Jablonowski, "What's New in Machining Centers," Special Report 763, *American Machinist,* 95–114, (Feb. 1984).

J. Zygmont, "Flexible Manufacturing Systems—Curing the Cure-all," *High Technology,* 22–27 (Oct. 1986).

AUTOMATION, HUMAN ASPECTS

FRED M. AMRAM
University of Minnesota
Minneapolis, Minnesota

BACKGROUND

> Let us attempt to sift out two main themes which have interested social observers, and to suggest the variety of views which can be discovered among these observers. These two themes are: *What does the machine "do" to the [individual] who attends it?* and *What does the progressive addition of machinery "do" to the institutions of society which contain it?* (1)

The point at which technology and the humanities intersect is the concern of the "philosophy of technology" (2). That intersection has also been of concern to sociologists, economists, and government officials. Recently, scientists who work at the frontiers of technology have become more concerned about the impact of their discoveries on individuals and institutions. "Astroscientist" Carl Sagan, for example, argues eloquently that humanity's survival depends to a considerable degree on maximizing the cooperation between human and machine intelligence (3).

Automation as well as bioengineering are technologies that are dramatically impacting on human affairs. Automation's impact has a history and a body of quantitative research (albeit limited) and has been explored in fiction. It can therefore be examined from diverse perspectives.

Automation is pervasive in its impact on individuals and institutions—as are all machines. The grand piano and Chopin are as much a product of the first industrial revolution as are the automobile and pollution. The "total" impact of automation cannot be assessed or reported. Scholars examine the frontiers of knowledge in limited segments and break through at specific points. This exploration examines a few themes, especially work and household settings, which have been of interest to modern researchers and which give some perspective to future research. Special focus is given to robotics—a theme which has historically been of interest to individuals and groups in many cultures. If the past can be a guide, then now, more than ever, robots will have an impact on psychology as well as on myth.

To provide some perspective on a discussion of the human aspects of automation, this preview suggests that humans historically have tried to breathe life into their artifacts. The temptation to anthropomorphize robots is especially great and is part of the relationship between humans and their robotic machines. Second, robotics is examined as a subset in the field of automation, albeit an important and especially interesting subset. A third exploration acknowledges that automation is not restricted to the world of work and notes that the impact of automation, especially robotics, on the whole human experience is worthy of examination. Finally, it is important to understand that an occasional robot in the factory, office or home does not have the same impact on individuals and on institutions as in settings where automation is pervasive.

Anthropomorphism

> A human being will always suppose that, the more human a robot is, the more advanced, complicated, and intelligent he will be (4).

Asimov's *The Robots Of Dawn* points to a most human tendency—to anthropomorphize artifacts. A boat is named after a sweetheart, a tool is blamed for faulty workmanship, and an automobile is described as "temperamental." Indeed, researchers report that workers name robots and endow them with human qualities (5). A U.S. patent granted in 1883 replaced human workers in an almost bizarre vision of an "automation shoe factory." Interestingly, simple machines were made to appear as "androides" of both men and women (see Fig. 1).

Once artifacts (especially robots) are perceived as "alive" they become part of our ethical system, and it becomes easy to attribute a value structure to them. Sinful robots are portrayed in a movie advertisement:

On Saturn 3
the ultimate
man-made robot
possesses
everything . . .
Violence, Evil, Lust.

Helpful robots are portrayed in the cartoon, which shows a young boy who commands three little humanoid toys, "Clean up my room, Squad Six!"

A vast collection of myths and movies documents the perception that "the anthropomorphic notion of a machine which can possibly out-think man is not easy to assimilate or live with."

> The machine in general and the computer in particular seem to have touched some raw nerve, some irrational and deep-rooted fear—a fear reflected in legends of man-made robots who turn on their creators. (6)

The implications are many. If artifacts are perceived as having life, if robots are perceived as humanoid, the relationship between user and tool is affected. It is difficult to objectively diagnose an automobile's failure if the vehicle is perceived as "cranky" or even "capricious." Furthermore, the "android syndrome" may inhibit creativity. By restricting perceptions of robots to the human form, one limits the view of the machine's potential. Robots, as machines, can move in more directions than humans, can lift more weight, have greater endurance, and have potentials not yet imagined.

On the other hand, designers of personal robots can be guided by the understanding that a robot can be perceived as a cute friend (R2D2).

Robots as Part of Automation

> Recent advancements . . . constitute the first stages of what coming generations will look upon as a second industrial revolution. (7)

Mechanization marked the beginning and growth of the industrial "revolution," and automation marks its end. Diebold speaks of a second industrial revolution (7), whereas Ayres

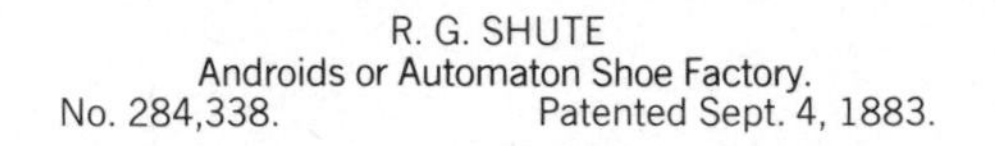

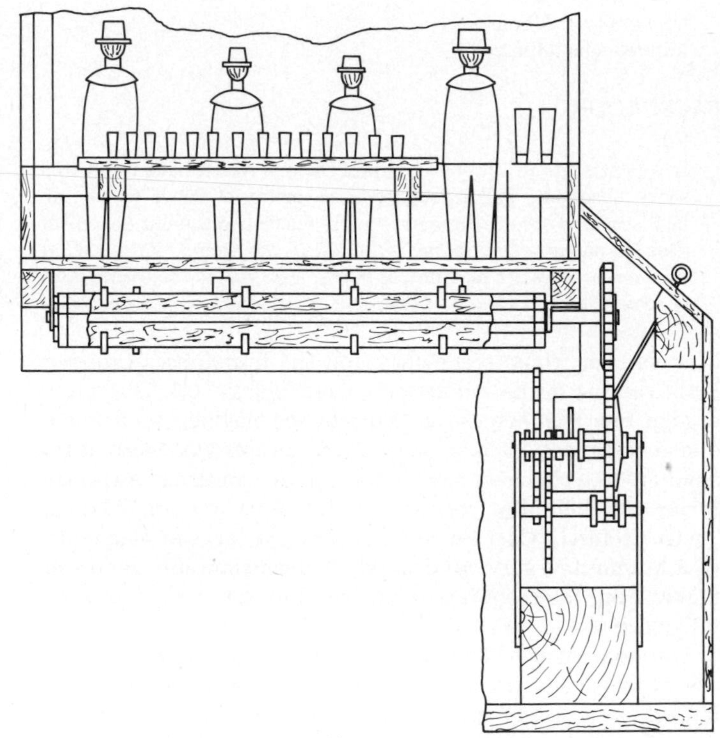

(a)

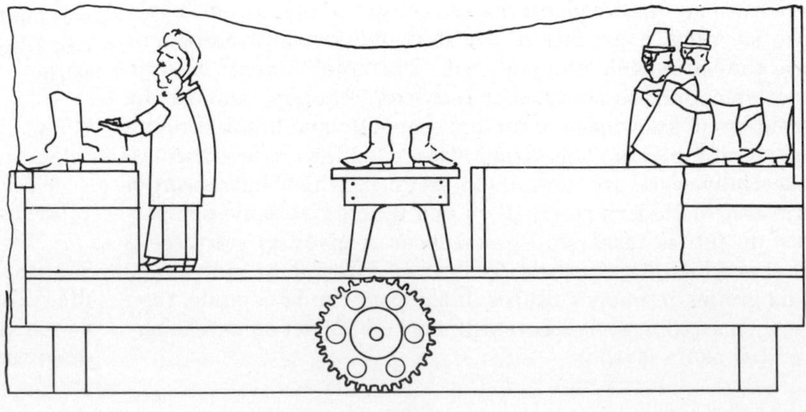

(b)

Figure 1. R. G. Shute's "Androides or Automation Shoe Factory." (U.S. Pat. 284,338, Sept. 4, 1883). **(a)** Underside view of the toothed shaft and levers for operating the androides. **(b)** Rear elevation. **(c)** Transverse section of the two views of androides and their operative mechanisms.

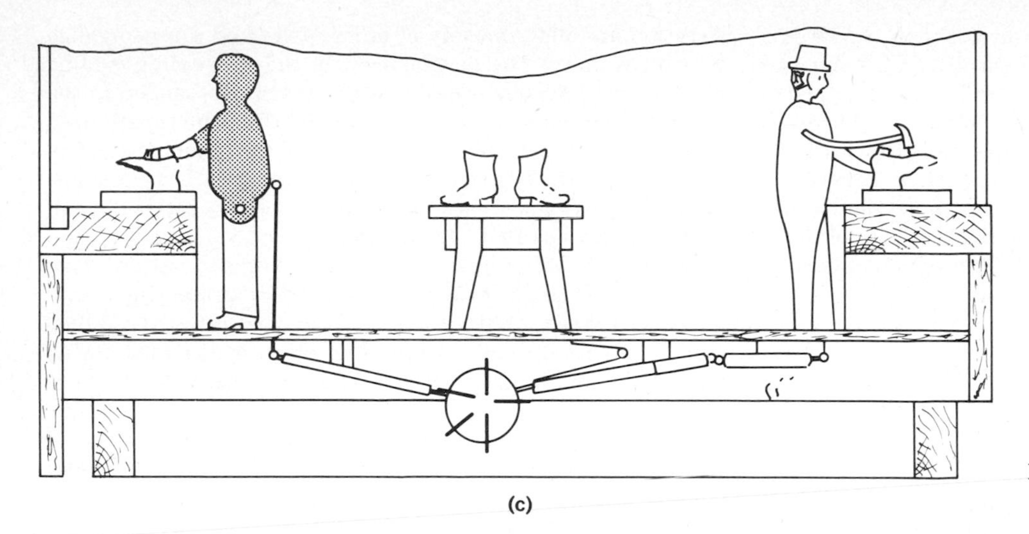

(c)

Figure 1 *(continued)*

asserts that the United States has entered a third industrial revolution (8). The precise historical markers of technological change are less important than the notation that automation is not simply an extension of mechanization, but rather "a process which substitutes programmed machine-controlled operations for human manipulations. It is the fruit, so to speak, of cybernetics and computers" (9).

"Automation, based on the principle of feedback, institutes a new type of technology that controls its own operations" (10). Although feedback and its impact on supervision will be discussed later, it must be noted here that as automation, with its feedback capabilities, increases, control moves from workers (tool users) to machines. A brief historical examination, adapted from reference 11 will point out the implication for machine users.

Craft production systems include workers with significant skills who produce and/or repair all, or a substantial part, of their product. *Mechanized production systems* include workers who operate special-purpose machines and who make only a small and isolated contribution to the total product. *Automated production systems* include workers who monitor an integrated production system. Although these latter workers have a larger scope of the product's development than those who work in mechanized production systems, they may not participate in the actual production.

How do robots fit into this model? "An intelligent robot should be able to think, sense, and effect" (12). Ideal automation, utilizing even rudimentary artificial intelligence, is similarly expected to have sensing, memory, decision-making, and implementation functions. This seeing, hearing, touching, thinking, working model of automation becomes an accessible public image of a sophisticated robot and a seductive picture of a humanoid. Although robots are simply a part of the entire automation process, lay observers, because of the peculiarities of language and mythology, often use the terms "automation" and "robotics" interchangeably. Note the anthropomorphizing in this statement about automation by an expert in advanced manufacturing technologies: "Thinking is primarily a brain function. Sensing (seeing and touching) and effecting (moving and manipulating) are primarily body functions" (12).

Automation at Work and Elsewhere

Most discussions of automation focus primarily on the world of work. It is, however, also useful to regard automation outside the factory and office. The robotic traffic officer or baby sitter is also a worker. The home security robot or entertainment robot may have more sociological implications than the factory robot. It is easy to imagine a society with more "personal" robots than industrial robots, and their impact on the home and on personal relationships could be enormous.

An advertisement proposes, "Consider the kitchen that thinks." The automated kitchen will certainly influence work patterns as well as leisure aspirations in the twenty-first century, as much as the washing machine, dryer, refrigerator, and dishwasher have affected sex roles in the twentieth century.

One Robot or Many

> "On the planet there are fifty robots to each human being on the average. . . . Most . . . are on our farms, in our mines, in our factories, in space. If anything, we suffer from a shortage of robots, particularly of household robots. Most Aurorans make do with two or three such robots, some with only one. . . ."
>
> "How many human beings have no household robots at all?"
>
> "None at all. That would not be in the public interest. If a human being, for any reason, could not afford a robot, he or she would be granted one which would be maintained, if necessary, at public expense." (13)

An occasional household robot is interesting. If most people have one "appliance" (such as an automobile), the "appliance" quickly modifies the culture by becoming a part of it. If there are "fifty robots to each human being," as suggested in the fiction cited, they become central to society. An examination of the human implications of robotics must distinguish between the effect of the occasional robot and the effect of "many."

Similarly, the impact of one or two robots in a large production setting may be significant, but it is not the same as that of the "factory of the future" (FOF) built around total computer-integrated production. The difference is not one of degree; it is a difference in kind. The first simply involves a special tool in a traditional environment. The FOF involves computer-integrated design, inventory, traffic, assembly, quality control, packaging, and perhaps, shipping. The former is controlled primarily by humans. The latter requires little human interference other than that of support personnel. An exploration of human facts in manufacturing must distinguish between the impact of an occasional robot in a traditional setting and that of a computer-integrated production facility.

Work

> Bowed by the weight of centuries he leans
> Upon his hoe and gazes on the ground,
> The emptiness of ages in his face,
> And on his back the burden of the world.
> Who made him dead to rapture and despair,
> A thing that grieves not and that never hopes,
> Stolid and stunned, a brother to the ox? (14)

In the early days of agriculture, a farmer pushed a plow. Farmer and plow were one. The farmer provided both brain and brawn. Then horsepower (or ox power) replaced human muscle power. The human driver could build a "talking," affectionate relationship with a "beast of burden." Tasks and relationships changed again as the farmer became the tractor's mechanic. But still, while steering the machine, the farmer provided brain power. Now plans exist for agricultural robots that plow the land, sow the seed, cover the furrow, water the plant, and pick the crop, unaided by humans.

Tools change, and such changes affect the user and the culture. Flexible automation—especially robotics—is a tool change that will have a major sociological effect in the factory, in the office, and in the professions.

WORK

Factory, Office, and . . .

> Automation may reduce the distinction between factory and office, between hand and brain work. . . . Blue-collar operators in the new clean, office-like factories will be responsible employees, much like the white-collar employees in the new, more mechanized, factory-like offices. (15)

Most commentary on automation pictures an automated factory. It is noteworthy, however, that factories have ever-enlarging offices which, like the "office industries" (eg, insurance companies), experience automation as surely as every other work setting. The office begins to simulate the factory.

> The new office is rationalized: machines are used, employees become machine attendants; the work, as in the factory, is . . . standardized for interchangeable, quickly replaceable clerks; it is specialized to the point of automatization. (16)

An early nineteenth-century secretary, probably male, had substantial skills, such as bookkeeping, arithmetic, and office managing. Such work was analogous to that of craft workers in production settings. More recently, the secretary who took dictation and did a variety of other tasks had a broad range of responsibility. The development of the typewriter, adding machine, and diverse other business machines has led to the mechanization of office work. Now each worker has specialized tasks and is responsible for only a small portion of the entire work process (the typing pool is a clear example). Data processing machinery makes the office worker, in many cases, a monitor (see Robots as Part of Automation).

Work satisfaction analysts are beginning to gather some evidence to assess if the new electronic data processing equipment is upgrading office work, with opportunities for skilled machine operators, or downgrading office work, with boring monitoring and input functions. Thus far, the judgments are mixed (4).

There are reports of major cuts in employment directly attributable to office automation. "One corporation cut its office staff by 45 percent (120 people) [and] reduced office space requirements by 20 percent . . ." (17).

> Considering that the technology is still in its infancy and that 50 million Americans are currently employed in office jobs, the potential social impact of office automation is likely to be far more pervasive than the potential impact of factory automation. (17)

Automation, of course, has moved beyond factories and offices. "I envision a time when a laboratory that currently has five chemists doing routine jobs will one day have four robots and one chemist to supervise the progress" (18).

Automation in agriculture is documented by the University of California's development of a tomato harvesting machine and the breeding of a tomato suitable for automatic harvesting. The predicted consequence is the replacement of two-thirds of tomato harvest jobs and the elimination of four-fifths of California tomato farmers.

Robotics Today reports, "Scientists are extending the boundaries of robotics" as they go to work with humans in many settings. Included are agriculture (transporting pepper plants and orange harvesting), medicine (brain surgery), and construction (excavation), and as one reads in daily newspapers, they are in the sea, in the sky, and under the ground (19). In all these areas, automation will have an impact on workers, or will displace them.

Transition Stress

> Organizational structure lends predictability and stability to human communication, and thus facilitates the accomplishment of administrative tasks. (20)

Individuals, as well as organizations, crave stability. Change, although sometimes exciting, even stimulating, when not predicted or predictable, is disconcerting. Uncontrolled change is frequently harmful. *Cybernetics* makes the analogy to physiological phenomena.

> The conditions under which life, especially healthy life, can continue in the higher animals are quite narrow. A variation of one-half degree centigrade in the body temperature is generally a sign of illness. (21)

Each introduction of a new component (eg, new machinery) affects other components. Clearly, the introduction of robots into the work setting affects the least predictable, least controllable, and most vulnerable component—the human worker. The disruptive effects of introducing flexible automation can lead to "transition stress."

The problems facing workers in settings where robots are being introduced gradually are significantly different from the problems associated with employment in a setting where robots are well integrated. Researchers often confuse the stress of transition with stress associated with human–robot interaction. The stress which grows out of a response to change, any change, must be treated as transition insecurity rather than a response to a particular kind of environment.

Experienced workers know their jobs. When the task is suddenly done differently, a worker may feel that "my" job has been changed, and transition insecurity sets in. It may be best to transfer a different worker to the newly automated job—a worker who does not come with rigid habits and who expects to learn a new task. A new employee in an altered setting sees a modified job as a new job and comes with new expectations. Similarly, the old worker coming to a different task comes with a relatively open mind and also expects to learn a new job.

Excessive transferring can, however, cause dissatisfaction. Japan, whose plants are more automated than those in Europe or the United States, reports that "the strains produced by large-scale automation are beginning to affect the style and substance of Japanese labor–management relations . . . [the workers'] lot is to endure the frequent job reassignments management imposes on them. Some workers, complaining of psychological fatigue, have even quit. . . . The psychological impact is far greater than outside observers imagine" (22). Clearly, workers in transition need emotional support, recognition that they are in transition, and praise when they adapt well.

It is important to understand that a single robot in a traditional setting may create transition stress. That situation, however, is in no way the same as a totally automated work setting as perceived in the FOF. The latter may lead to "technostress," which will be discussed later (see HUMAN IMPACTS).

Supervision and Control

> Automation . . . is brought to full fruition only through a thorough exploitation of its three major elements, communication, computation, and control—the three "Cs". (23)

Cybernetics, subtitled "Control and Communication in the Animal and the Machine," adds "feedback" to the "three Cs." Simply, an information source sends a message that causes something to happen. Measurement of the information transmittal and its effect is "feedback." The entire process could become the basis for a definition of supervision, perhaps supervision by a central computer.

> The rapid advance of computing technology has vastly augmented the capabilities of process control systems. Modern distributed control systems can monitor thousands of process variables and alarms, delivering a constant stream of information to the control room. With such systems, a small crew of operators can regulate the operation of immensely complex industrial processes.

In the FOF, a central supervisory computer will send instructions to inventory and to diverse workcells to control the behavior of the automatic machinery. Decisions by the computer will be made on the basis of feedback provided by local computers concerning the progress and quality of the product and about the status of each machine.

Initial decision making and planning will, under any circumstances, be done by humans. After that, one sees the potential of computer/supervisor control over the on-line workers, be they human or robot.

> No longer free to plan his work, much less to modify the plan to which he is subordinated, the individual is to a great extent managed and manipulated in his work. (25)

Telephone operators have been expressing anger that computers monitor the quantity and quality of work, and report the results to human bosses.

> For the telephone industry . . . the computer speeds work that employees are expected to do and allows it to be carefully monitored. *In effect, the computer becomes the foreman.* (26)

The potential of "smart," interactive machines acting as work partners, if not supervisors, is great. The implications for supervision theory can be interesting. For example, supervisors currently have, on average, a control span of six people. Will robots be counted as supervisees? Will a supervisor have, on average, a control span of four humans and two robots or five robots and one human? Will robot supervisors control perhaps three humans and three robots? It is interesting to speculate on a modification of the "Automation Shoe Factory," illustrated earlier, which would include several humanoid bosses.

More imminent is the possibility of reduced supervision ratios. Historically, more direct supervision has been available to those in intellectually or psychologically demanding positions. The new tools are certainly more complex than precomputer machines, and one can anticipate smaller work teams or even a modified "buddy system." Documentation for this assertion is already available. Several studies report that workers report closer supervision from all levels: a reflection of "both a decrease in the number of workers per foreman and an increase in the amount of time spent by each foreman in direct supervision of the line" (27).

Power and Self-Esteem

How will machine-controlled systems affect the self-esteem of employees? What impact will loss of control have on worker satisfaction? Alienation stems from a sense of powerlessness, almost literally, of being "out of control."

> A person is powerless when he is an object controlled and manipulated by other persons or by an impersonal system (such as technology) and when he cannot . . . modify his domination. (28)

Certainly, the telephone operators, cited previously, feel out of control. Robots, with their current limitations, are not very tolerant of ambiguity or disorder. Consequently, workers sometimes complain of becoming handmaidens to machines, of running their errands and fixing their environment. A relevant observation is that the first industrial revolution involved the transfer of physical skills from human workers to machines. "The second industrial revolution, now in its infancy, involves the transfer of intelligence from man to machine" (29). Self-esteem is clearly threatened when robots are stronger, faster, smarter, and sometimes, in charge.

There are some who take a different view:

> It does not say in the manner of the old type of machine, "You must move a handle or else I can't go on." Automation is the exact opposite of mechanization. The man in charge extends his faculties but remains himself. He does not become a slave. He stays in the center of things and becomes the *real* master. (30)

Another assessment of "perceived powerlessness at work" (Table 1) suggests:

> It is clear from the percentages that few craftsmen (19 percent) experience lack of freedom and control in their work. At the other extreme, 94 percent of the assemblers feel a sense of powerlessness. For monitors, less than half (42 percent) are on the high end of the powerlessness scale. (31)

Table 1. Perceived Powerlessness in Work by Phases in the Human–Machine Relationship (31)

Perceived Powerlessness in Work	Phases in the Human–Machine Relationship		
	Craft ($N = 113$)	Mechanized ($N = 93$)	Automated ($N = 85$)
Above median	19%	84%	42%
Below median	81	6	58
Total	100	100	100

$X^3 = 114.67$, $df = 2$, $P < 0.001$, $\bar{C} = 0.77$

This statement has important qualifications. Especially important is the note that, in the case of office workers, "lack of freedom and control on the job showed a tendency to *increase* as the level of technology advanced, up to and including computer operators."

Although not all the votes have been counted, enough theory and research have been published to cause managers and human resource specialists to be alert to a potential decrease in worker self-esteem and a corresponding increase in worker alienation, especially in office settings.

Worker Isolation

> Workers reported: (*1*) much closer and more constant attention required by the new jobs, (*2*) even fewer jobs where work was paced by the operator, (*3*) greater distance between work stations . . . , and (*4*) fewer jobs involving teamwork. These changes affected patterns of social relations on the job. Many workers were "virtually isolated socially." (32)

For most employees, work is largely a social activity. One returns home to report that Betty returned from her honeymoon, the boss was in a foul mood again, or the company softball team won four in a row. Rarely does one report production-related information, such as the completion of 18 letters on the word processor or the assembly of 80 carburetors. On-the-job pranks, teasing, and gossip provide much of the pleasure of work. When socialization needs are not met, alienation sets in, with its potential for lack of product interest, or worse, the potential for product sabotage.

Men at Work reports that, in the industrial setting, increased automation decreases opportunities for co-workers to interact.

> In the older mills, men . . . [formed] small work teams whose members could and did interact readily with each other. Members . . . in the new plant were so isolated physically that they had no opportunity to interact with one another. . . . (33)

On the factory floor, the introduction of robots leads to work-cells. One worker may attend one or more cells. Automation does away with the shoulder-to-shoulder work that provided ample social contact. Even in extremely noisy settings where conversation was difficult, visual contact was constant. It may be that the absence of visual contact is even more isolating than the absence of verbal contact (34).

Some workers report that the speed and complexity of the new machinery requires greater concentration, thus forbidding distractions and leading to even greater isolation. Office workers who spend much of their workday entering data into computer files (eg, IRS employees) report that the required concentration leads to isolation, boredom, tension, and consequent errors.

Isolation has effects beyond precluding some of the important social aspects of work. Productive communication also diminishes with physical isolation. "Two people who are on the same floor, if separated by more than 25 yards, will rarely have any significant communication" (35).

Recommendations must depend on the work setting. Purposeful distractions, such as coffee breaks, may not be cost-effective unless they reduce error in especially boring and/or tension provoking work. Access to radio may reduce isolation, although radio is not an interactive medium. More novel are computer links to other workstations, which increase the potential for both social and productive interaction.

As robots become more "responsive" they may provide the social interaction needed by humans. Scientists are already working on intelligent robots with whom humans will be better able to share information. The same systems that help improve productive communication between workers and robots can enhance social communication. It is noteworthy that most workers anthropomorphize their robots, and many expect the camaraderie of R2D2 and similar friendly robots.

Another solution to the worker isolation problem is to enhance the relationship between on-site production workers and visiting maintenance workers. Cooperation between production workers and technical support personnel is necessary to keep automatic machinery in top working order and to catch potential breakdowns as early as possible. To encourage task cooperation, management can encourage social cooperation. The maintenance technician or engineer might be perceived as a good doctor who is both technically competent and friendly. Both robot and on-site worker will be well treated. Or the roving technician may be perceived as a "visitor" who diminishes tedium and provides social contact. In earlier days, the visiting repairman, salesman, and postal worker served similarly to reduce the isolation of the homebound housewife (34). Perhaps the social aspect of the maintenance worker's job should be explicitly stated to prevent a potential problem.

> If many robots were being placed on-line and the support personnel had great demands on their time, then we might find some conflict between support personnel and the operators. (36)

Similarly, there are some critical labor relations issues that could lead to competition rather than cooperation.

> These issues include jurisdictional lines between operators, maintenance personnel, and programmers; labor grades; pay scales; training. . . . (37)

Worker isolation is as great a danger in the office as it is in the factory. Post-World-War-II offices introduced machines which minimized socializing. Highly efficient typewriters led to the "quantity-oriented" typing pool. Dictating machines ended the era of "taking dictation." Modern telephones with extensions and switching capabilities diminished dependence on secretaries and helped to isolate the boss.

Modern automated office machinery includes word processors that enable an increasing number of bosses to do their own typing. Spreadsheets, databases, and other filing systems permit unassisted information retrieval. In many cases, the office worker does little more than "input data" (analogously to the former filing clerk) at a terminal for most of the workday.

Indeed, the terminal is at least as isolating as the typewriter was, with the same potential for social separation experienced by factory workers. Office workers are desperate for more social stimulation. Reports from the IRS office, and elsewhere, are that some computer terminal workers, in spite of contrary regulations, are wearing earphones connected to radios—a difficult situation for supervisors.

Communication Relationships with Robots

> . . . we tend to think of *things* not *relations,* of *items* but not *contexts*. . . . (38)

Humans build incredibly close relationships with the elements of their environment (plants, animals, boats, cars) and even anthropomorphize their inanimate "friends" and "enemies." If one accepts that humans need relationships with each other, that automation may lead to isolation, and that humans can build relationships with their tools, one can understand that humans can, and will, build relationships with robots. Indeed, robots, more than other tools, are "smart," with capabilities of "sensing" and interactive "behavior," even mobility.

To examine the nature and quality of human–robot relationships, it is appropriate to use a modified version of interpersonal communication theory. For example, robots inspire competitive feelings.

> By our second visit, workers seemed resigned to the fact that a robot would always be able to outproduce a human worker. . . . Although objectively the operators controlled the robot and the two milling machines, they subjectively felt that there was competition between them and the machine. (39)

Competitive communication relationships foster low self-esteem and may lead to alienation, which could be acted out as sabotage. (It is noteworthy that competition with robots is more likely in situations where a new robot has been introduced than in situations where new workers enter a situation in which standards have not yet been established.)

How does one build closer, cooperative relationships between humans and robots? Interpersonal communication theory suggests that a beginning step is to open communication channels that are clear and understood.

Communication researchers, without exception, agree that at least two-thirds of human communication is nonverbal. Key elements are the nuances of gesture, touch, distance, tone of voice, dress, and odor—elements which robots have not yet "learned." Although tone of voice is a skill rapidly being mastered by computers, robots rely mostly on buzzers, lights, gauges, and occasionally, spoken and written language when communicating with humans. Of these options, our culture most values spoken language as a communicator of friendship.

> Once we have established linguistic communication with the machine, we will have opened up a whole new vista of responses to the most intricate processes of thought and action. To capitalize on these advancements in man–machine relations will require that the computer be thought of in humanistic terms. (40)

Voice processing technology continues to improve and has already made it possible for industrial robots to call aloud when they need the help of a human operator. At Stanford University, robots are being "taught" to respond to the spoken commands of quadriplegics. Brian McGrattan has created a new computer language that is designed to control robot movement.

> The single command Move, for example, replaces the sophisticated electronic input that tells the arm where to go and what to do. . . . Using the language, McGrattan can program a specific lab function, store it on disk, and recall it any time it is needed. . . . (18)

A truly "human-friendly" language would be natural language rather than the awkward program languages now in use. Great strides have been made in the development of "talking" machines, but they will never be perfect! Every human knows that natural language is context bound and inherently ambiguous. Human–machine communication will probably be no better than human interpersonal communication, although robots are less likely to distort purposefully or to participate in unconscious defensive communication disruptions.

The important point is that natural language, however imperfect, can help humans build relationships with machines—essential when one perceives the debilitating effect of isolation at work. Furthermore, humans resent having to learn the machine's language, not only because it is difficult and "foreign" but also because it gives "power" to the machine. Clearly, there is greater comfort in one's own territory speaking one's own language.

Technostress

> The men saw the jobs in the new mill as radically different. Jobs were "physically easier," but "mentally harder," demanding alertness, continuous "watching," and thought. Responsibility was seen as much greater, and, initially, there was anxiety over the serious consequences of a wrong move. The men were jumpy and uncertain. . . . (41)

Earlier comments pointed out that stress brought on by change (transition stress) is not the same as the specific stress symptoms caused by a specific stimulus. The physical and psychological symptoms brought on by extended exposure to extremely bright light are a specific response to a specific stimulus. Technostress refers to the specific stresses caused by technological devices to which modern society is exposed.

Physical stresses are addressed by the science of ergonomics. Among the many issues addressed by ergonomics is the impact of long hours of sitting at a computer terminal, not only on the back but also on the eyes. Ergonomics can, for example, measure the physical consequences of diverse seating options as well as effects of various screen lighting and coloring configurations.

More difficult to address is the psychological stress brought on by the components of automation. The stress of competing with machines that are faster than human workers is a theme already documented. Similarly, the stress of isolation from other humans has been explored in some detail, with the recommendation that managers integrate isolated workers into an appropriate communication network and provide them with a sense of belonging.

The introduction of automation may alter activities as well as interactions (42). In mechanized systems, workers are special-purpose machine operators, generally physically active and occasionally very active. In automated systems, as previously stated, workers are monitors. The task of watching is probably more stressful than most tasks requiring lifting and placing. Psychologists have therefore long encouraged exercise as a tension reliever. In cases where stress-provoking monitoring is the only or primary work activity, assignment of multiple tasks and the opportunity to walk from location to location seems advisable to alleviate tension. Interestingly, workers who have been relieved of heavy work by automation often report that their new task of monitoring machinery produces boredom, a cause of substantial stress. Boredom, however, is not a new problem to human resource specialists, and the relevant literature is filled with prescriptions.

Occasionally, workers are confronted with tasks more difficult than those for which they are technically or psychologically prepared. They report that relating to a complex and, incidentally, expensive piece of machinery requires greater concentration and generates a sense of pressure. Responsibility for a $500,000 automated workcell is, after all, different from using a $30 wrench. Rewards for the added responsibility (such as increased pay) are appropriate, but only compensate for the stress, without reducing it. More relevant are stress reducers such as frequent periods of rest and relaxation, opportunities for diversion (eg, social communication), and occasional job rotation.

When robots are perceived as unsafe, their presence may lead to stress. Rather than treat this fear exclusively as a psychological problem, it seems that every possible safety feature should be built into such devices so that their safety sells itself and they then become the stress reliever. Asimov's "Three Laws of Robotics" have served to assure the safety of robots in his many stories (43). A real-life counterpart would be encouraging.

Finally, stress has been used here in a negative sense, to mean *distress*. Some workers, in jobs where flexible automation has been introduced, report the kind of positive stress associated with adventure. They welcome the break in tedium. Further inquiry into this area is warranted and may provide clues to introducing automated systems with minimum distress to workers.

Training Versus Socialization

> First we make our tools and then our tools remake us.
> —Winston Churchill

Child development experts understand that preschool and early childhood education serve primarily to socialize children into the culture. A major goal of university education (although unstated) is to teach behavior styles so that graduates in the working world act out role behavior appropriate to their selected careers and work settings. One learns how a doctor or teacher behaves. When studying the impact of changing technology on humans, one must focus clearly on the need for resocialization. How to do the job is no more important than how to behave in the new setting.

Employers, labor unions, and educational institutions have made a cliché of retraining. Teaching technical skills is the easiest part of helping workers adapt to automation. Human resource departments in schools and corporations are beginning to value the need for personal retraining to enhance worker adaptation to new jobs.

After assessing the potential consequences of automation—stress, lowered self-esteem, isolation—one sees why workers' socio-emotional needs are taking on as much importance as information and skill needs.

Changing Employment Patterns

> Ideally, raw materials come into the back door, and the completed product goes out the front door without being touched by human hands. (44)

If this dream becomes reality, one may ask "Where have all the workers gone?" Most economists project that, in the long range, work demands will be redistributed, and the labor market will adapt to increasing automation. Nevertheless, there are others who predict that, at least in the short run, worker displacement will be at crisis levels. The human impact of automation will be visible in employment patterns. The picture is probably not as gloomy as described in a flier calling for a rally to protest automation:

If robots and computers were invented for society's benefit, then why are people standing in unemployment lines while somewhere in the metropolis their chairs are being occupied by microchips?

A futurist predicts that computer technology will provide improved communication systems (eg, telecommunications), reducing the need for the U.S. Postal Service. One is not told, however, the fiscal or emotional cost of retraining 750,000 postal workers. Different jobs will appear in new information distribution systems, in the new leisure industries, and in human services. But some workers will not be retrained; others are not retrainable.

Ford Motor Company plans to have more than 7000 robots in use during the last decade of the twentieth century. The President of Ford was unable to predict how many workers would be replaced, but promised that Ford would try to use attrition rather than layoffs to lessen the impact. Attrition will lessen the impact on this generation of workers, but not on the next generation when they look for jobs.

Predictions for Europe are no more hopeful than for the United States. *Research Institute* notes that "unemployment remains a persistent European Community structural problem. It's still too high, was 11% in 1984, could climb to 11.5% in 1985" (45).

Some potentially negative indicators include:

1. Plans to raise the social security eligible retirement age and to reduce benefits for early retirees. This means more older workers will remain in the labor market and will continue to compete for a decreasing number of jobs.
2. Legislation to tax a portion of social security payments, thus providing an added disincentive to retire.
3. Proposals to decrease student loans and grants to low- and middle-income students. A potential effect is to put more young people in the labor pool instead of delaying their full-time employment while they attend colleges and universities.

Some potentially positive indicators include:

1. Statistics that demonstrate that fewer young people are entering the labor market, a reflection of declining birth rates in the United States and Europe.
2. Worker pension plans are becoming a widespread employee benefit and provide security for those choosing early retirement.
3. Longer vacations and other forms of distributing work opportunities are becoming commonplace. Voluntary job sharing is a sophisticated device permitting two persons who choose part-time employment to share a single job, often to the advantage of both the employees and the employer. Enforced job sharing during periods of economic distress and/or in distressed industries has been discussed as a possible human resource policy (46).
4. Training and retraining opportunities not only provide reemployment opportunities for many, but also keep some people out of the labor market for significant periods. Experts state that "⅓ of the work force must be retrained every 10 years or so for the foreseeable future" (47).

For the older person, the only solution is an adequate pension or severance pay allowance, which would give him a minimum income for his needs, or the adoption, as is already present in many collective bargaining contracts, of an *attrition principle* which guarantees each worker *his* job and allows the employer to reduce his work force only because of death or retirement. For the younger person, the answer, in general, lies in a satisfactory degree of education which allows him or her to change not only jobs but occupations when necessary in order to attain the flexibility of skills in a complex economy. These are prosaic answers, but necessary ones. (48)

A forecast of permanent mass unemployment would almost certainly be incorrect. Shifting employment patterns will, no doubt, lead to new patterns of work distribution. As automation reduces the demand for work, the most startling effect on the individual and the culture will be the reduced workweek. Except for those whose talents are in especially high demand, the biblical workweek will be reduced. "Six days shall you labor . . ." is becoming passé. To spread the work among large numbers of eligible workers, each individual will spend fewer hours on the job. The impact on the culture will be as enormous as it has been in the past several hundred years of reductions in the workweek. A futurist bible may call for two days of rest . . . or even three!

Historically, a six-day workweek, with minimal vacation periods and limited sick-leave opportunities, was the norm. Modern labor contracts are written around a 5-day workweek with as many as 13 paid holidays. Allowances for illness are liberal, including pregnancy, maternity, and even paternity leaves. It is noteworthy that the concept of parenting leave could only become acceptable in a society where the demand for human labor is reduced.

Paid vacation days simply increase vacation opportunities. Congressional action to establish Martin Luther King's birthday as a national holiday was in keeping with a trend toward fewer workdays. Aside from its merits of honoring an important American, the holiday makes a significant impression on labor statistics. If 100 million workers (a realistic number) take one day off from work, they leave a gap of 100 million workdays—the equivalent of 400,000 work years (calculated at 250 workdays per person per year) (49).

On one level, these statistics indicate a potentially major impact on unemployment figures. On another level, one sees a clear trend in how individuals spend their time. Work becomes a less important part of life and eventually becomes devalued, and leisure becomes an ever-increasing focus of society's attention.

BEYOND WORK

In the sweat of thy face shalt thou eat bread.
—Genesis

Karl Marx saw meaning in productive labor as the "first necessity in life." His concern was with the quality of the worker's experience while earning a living. As work becomes a less central part of life, "meaning" may come from activities beyond the worksite. And if work in the automated setting were to become less satisfying, as some predict, there would be all the more reason to look elsewhere.

As practice, craftsmanship has largely been trivialized into "hobbies," part of leisure, not of work. (50)

Finding Personal Worth

Genesis, Marx, and Mills provide a perspective from which to look at life beyond work in a society where automation is increasing, at work and in the home.

In spite of inexpensive, high-quality, mass-produced vegetables, gardening is ever more popular—with a blind insistence that home-grown tastes best. Strenuous weekend wood splitting brings aching muscles and satisfaction in a household with central heating. Home carpentry, antique restoration, and wilderness camping are a few modern efforts to regain a sense of control, a sense of "can do," a sense of accomplishment. As automation frees (or bores) workers, they strive to find a sense of personal worth.

> If the work . . . people do is not connected with its resultant product, and if there is no intrinsic connection between work and the rest of their life, then they must accept their work as meaningless in itself, perform it with more or less disgruntlement, and seek meanings elsewhere. (51)

Crafts and farming gave individuals visible documentation of accomplishment. Western value systems have for centuries cherished work (paradoxically, also perceiving it as punishment for the sins of Adam and Eve). Now a time of transition forces individuals, as well as cultures, to reexamine human purpose.

What little sense of accomplishment there was in washday has vanished with the automatic washing machine, dryer, and permanent-press clothing. The sewing machine still requires a human operator with substantial talent, but the vacuum cleaner, automatic oven, and blender have taken some of the self out of housework, along with the drudgery. One sign of changing values in response to automation is a radical modification in sex roles with women's active pursuit of self in education and at work (16).

If automation controls housework and production, one must turn elsewhere for signs of personal worth. Some predict a "TV-and-beer" society. That, however, is not likely because of humanity's need to create, to be productive. More likely is a continuing growth of adult education. "Life-long learning" will become more than a cliché. An expansion of volunteer work will continue as individuals reach out to each other to avoid isolation (52).

Robots for the Physically Disabled

> One of the most advanced upper-limb prostheses for patients without arm muscles is the computerized artificial arm . . . [which] enables the patient to comb his hair, eat, scratch his back . . . and . . . to accomplish every essential human task. (53)

Robot-as-worker is a traditional view hitherto often discussed. A somewhat newer, but growing, field with interesting human consequences is robot-as-helper. Tokyo University has produced a computerized machine that opens bottles, pours drinks, and carries them to disabled and bedridden patients. The hospital aide robot has potential as a courier and messenger. Its strength can be put to use in moving patients, and with natural language ability, it would provide entertainment and company for the lonely and bedridden.

At Stanford University, robots are being developed to ease the life of quadriplegics. Interactive voice-actuated programs turn robot arms into aides that can lift, pour, feed, and perform a variety of other useful tasks. A wheelchair-confined person can take the robot arm to another room on its mobile cart. The economic advantages of being able to turn the pages of a book or drink a beer, without the help of a human attendant, are a bonus tacked onto the real advantages of independence, pride, and power.

NASA has developed a voice-controlled wheelchair that, in combination with the voice-controlled arm, gives paralyzed individuals the ability to turn doorknobs, pick up objects, and perform other self-care tasks.

Sensitive to the control and self-esteem issues raised earlier, researchers are careful not to make the robots too smart or too independent. They want to be sure that the human clients are in absolute control and are in no way controlled by the machines. At Stanford, each improvement is customer-tested for mechanical effectiveness as well as emotional impact (54).

Robots in the Home

> When children imagine a domestic robot, they think of a companion in their home; when adults imagine a domestic robot, they think of a servant. (55)

The potential of robots in the home takes many forms: from the already available mobile, home security system to the fantasy android helpers and companions. The family, a fragile institution, will be affected by the intrusion of automation—positively and negatively.

Two charming stories by Isaac Asimov set the stage. In "Robbie," a mother becomes jealous of the close relationship between her small daughter and the baby sitter/playmate robot. In "Satisfaction Guaranteed," a housewife first becomes envious of the superb household skills of a new robotic butler and then develops a temporary romantic relationship with it (him?) (56). The stories, although fantasy, provide social commentary (43).

Will robots have feelings? Scientists claim that machines cannot be designed to have emotions. Nevertheless, the term "warmware" has been invented to describe purposeful "irrational" or "appealing" behavior.

The Asimov stories and robots with "warmware" become less far-fetched with the imminence of a real robot pet. An important manufacturer is tooling up to produce a mobile pet that is sensitive to light and sound and that has temperament. The custom microprocessor allows the robot "to act merry or mopish, peppy or pooped." One prediction, perhaps exaggerated, is that the pet "will replace biological animals as . . . companions (57). Yet Arthur C. Clarke projects:

> Soon these electronic animals will have memories and vocabularies superior to those of any living creature. . . . [They] would be a

kind of robot companion, matching the intelligence of the growing child. . . . (58)

Some futurists see a long-range future where householders are quite comfortable with, even dependent on, robotic baby sitters, playmates, pets, helpers, and so on. Perhaps they will not take the precise form currently described. Yet ever more "intelligent" household machinery is certain.

Dependence on machinery is not new in Western culture. It is difficult to imagine a household without a telephone. Relationships, personal and business, are built around it. Television has invaded the family, and many individuals would experience withdrawal symptoms if it were removed. Similarly, one could develop dependence on a generally useful and truly "human-friendly" personal robot. An entire household of helpful and/or entertaining robots could become a habit just as much as the conveniences that are now taken for granted, which did not exist in the time of this generation's parents and grandparents.

Attractive as the household friends and helpers are, many people still express resistance. Many reveal a sense of uneasiness with the prospect of smart robots that work, even live, in the home. Surveys at the University of Minnesota indicate that many humans fear lack of control and loss of self-esteem. It is noteworthy that these same issues turn up when examining automation in the world of work (see Supervision and Control, Power and Self-Esteem) (59).

The fear of losing control takes many forms. An important one is "I can't fix it." Early tools were simpler. A jiggle or a replacement part usually solved the problem. The old tools, mostly mechanical, were understandable. The microwave oven, with its timers and sensors, is beyond most people's comprehension. Indeed, the microwave seems dangerous. "Will it contaminate me or my environment if it malfunctions?" Similarly, a robot (especially if anthropomorphized) seems frightening to the uninitiated because it may accidentally, or even purposefully, strike out—"and I can't fix it or control it."

Basic to the loss of control is the inability to communicate with the smart machine. "I need to learn its language" and "I need to learn its rules" are statements that point to the feared loss of self-esteem, as are, "It's smarter than I am" and "Can I live up to its expectations?" Psychologists are familiar with similar feelings of inadequacy in interpersonal relationships, where they also interfere with positive cooperation.

Adults were asked to speculate about the advantages and disadvantages of an ideal household robot. The reservations outweighed the virtues and included:

- Would it do the job the way I want it done?
- Might it break something?
- Can I trust it while I'm gone?
- What if I'm counting on it and it breaks down?
- Can I afford the maintenance (energy/food)?
- Could I be replaced?

From a psychiatric orientation, these reservations parallel those adults express about having children, relating to children, and trusting children to do work. To some, robots in the home may be perceived as clever (and threatening) children.

The Future

> The human future depends on our ability to combine the knowledge of science with the wisdom of wilderness.
>
> —Charles Lindbergh

Technology has social consequences—slow, sometimes imperceptible in short time frames, but nevertheless enormous when seen from the perspective of history. Tractors, for example, provided another step in the gradual destruction of the family farm as surely as they made available cheaper food and greater health. As larger, more sophisticated, and more expensive farm machinery became available, small farmers could no longer afford the ever-increasing capital investments. Furthermore, the larger, more efficient machinery no longer made sense for relatively small plots of land. Hence, the corporate farm emerged, with its access to the money and land that made capital-intensive rather than labor-intensive farming possible. Similarly, automobiles brought social and cultural consequences such as suburbia and Big Macs.

Ultimately, there are two pervasive consequences of automation. The first deals with the human–machine relationship. There is significant evidence, from ancient mythology to the modern factory, that humans cannot resist creating robots in their own image. This need to anthropomorphize machinery will continue to lead designers to develop androids (Figure 1). The visual form of these robots may have a large impact on individual and cultural psychology. The relationship with intelligent machines, especially household robots, may be more intimate than with former mechanical tools and may parallel interpersonal relationships.

The second pervasive consequence of automation is the decrease in human work involvement, on the job as well as at home. Although this trend will be gradual, perhaps continuing in small spurts, it is clear. The potential for civilization can be phenomenal. A look into history may help to create a vision of the future.

The flourishing of culture in ancient Athens was made possible by a slave class that was essentially invisible. The slaves were not referred to and did not exist. They apparently were looked on as machines, and to this day, there is not much information regarding the size of the slave population, although there were probably enormous numbers of them. Perhaps citizens of the twenty-first century will, like the citizens of ancient Athens, devote themselves to the pursuit of beauty and truth, supported by the work of a new slave class—the robots.

BIBLIOGRAPHY

1. R. L. Heilbroner, "The Historic Debate," in John T. Dunlop, ed., *Automation and Technological Change,* Prentice-Hall, Inc., Englewood Cliffs, N.J., 1962, p. 17.
2. L. Hickman and A. al-Hibri, *Technology and Human Affairs,* C. V. Mosby Co., St. Louis, Mo., 1981, p. v.
3. C. Sagan, "In Praise of Robots," in H. M. Geduld and R. Gottesman, *Robots Robots Robots,* New York Graphic Society, Boston, Mass., 1978, p. 165.
4. I. Asimov, *The Robots of Dawn,* Ballantine Books, New York, 1983, p. 391.

5. L. Argote, P. S. Goodman, and D. Schkade, "The Human Side of Robotics: How Workers React to a Robot," *Sloan Management Review* **31–41,** 33 (Spring 1983).
6. F. R. Bahr, "The Man–Machine Confrontation," in L. Hickman and A. al-Hibri, *Technology and Human Affairs,* C. V. Mosby Co., St. Louis, Mo., 1981, p. 103.
7. J. Diebold, *Automation: The Advent of the Automatic Factory,* D. Van Nostrand Co., Inc., New York, 1952, p. 2.
8. R. U. Ayres, *The Next Industrial Revolution,* Ballinger Publishing Co., Cambridge, Mass., 1984, p. 99.
9. D. Bell, Preface, in L. Bagrit, *The Age of Automation,* The New American Library, New York, 1965, p. xvii.
10. J. M. Shepard, *Automation and Alienation,* M.I.T. Press, Cambridge, Mass., 1971, p. 138.
11. *Ibid.,* p. 7.
12. R. K. Miller, "Artificial Intelligence: A New Tool for Manufacturing," *Manufacturing Engineering,* 61 (Apr. 1985).
13. Ref. 4, p. 100.
14. E. Markham, "The Man With The Hoe," *A Pocket Book of Verse,* Washington Square Press, New York, 1940, p. 283.
15. Ref. 10, p. 116.
16. C. W. Mills, *White Collar,* Oxford University Press, New York, 1953, p. 209.
17. Ref. 8, p. 155.
18. D. L. Cooke, "New Knowledge," *Syracuse University Magazine,* 19 (Apr. 1985).
19. R. Schreiber, "Robots Unlimited: Reaching Beyond the Factory," *Robotics Today,* 43 (Dec. 1984).
20. E. M. Rogers and R. Agarwala-Rogers, *Communication in Organizations,* Collier Macmillan, London, 1976, p. 6.
21. N. Wiener, *Cybernetics,* M.I.T. Press, Cambridge, Mass., 1961, p. 114.
22. I. Saga, "Japan's Robots Produce Problems for Workers," *The Wall Street Journal,* 24 (Feb. 28, 1983).
23. L. Bagrit, *The Age of Automation,* The New American Library, New York, 1965, p. 36.
24. Ref. 12, p. 60.
25. Ref. 16, p. 226.
26. W. Serrin, "Technology Takes Toll on Operators," *Minneapolis Tribune,* 10D (Nov. 27, 1983).
27. F. C. Mann, "Psychological and Organizational Impacts," in J. T. Dunlop, ed., *Automation and Technological Change,* Prentice-Hall, Inc., Englewood Cliffs, N.J., 1962, p. 47.
28. R. Blauner, *Alienation and Freedom,* University of Chicago Press, Chicago, Ill., 1964, p. 32.
29. "Robots Join the Labor Force," *Business Week,* 73 (June 9, 1980).
30. Ref. 23, p. 40.
31. Ref. 10, p. 24.
32. Ref. 7, p. 47.
33. W. F. Whyte, *Men at Work,* Richard D. Irwin, Inc., Homewood, Ill., 1961, p. 219.
34. F. M. Amram, "Worker Isolation and Productivity" *Robotics Today,* 22 (Feb. 1984).
35. Ref. 20, p. 102.
36. Ref. 5, p. 32.
37. J. Howard, "Focus on the Human Factors in Applying Robotic Systems," *Robotics Today,* 34 (Dec. 1982).
38. Ref. 9, p. xix.
39. Ref. 5, p. 37.
40. Ref. 6, p. 107.
41. Ref. 27, p. 45.
42. R. U. Ayres and S. M. Miller, *Robotics Applications & Social Implications,* Ballinger Publishing Co., Cambridge, Mass., 1983, p. 95.
43. L. D. Allen, *Science Fiction: An Introduction,* Cliffs Notes, Lincoln, Neb., 1973, pp. 31–35.
44. D. Hoska, "Meeting The Challenge Of The Steel-Collared Worker," *Encounters,* 16 (Mar./Apr. 1983).
45. *Research Institute Recommendations,* Jan. 25, 1985, p. 1.
46. Ref. 8, p. 207.
47. *Research Institute Recommendations,* Apr. 15, 1983, p. 3.
48. Ref. 9, p. xxiii.
49. F. M. Amram, "Spread-the-Work Demands," *Robotics Today,* 29 (June 1984).
50. Ref. 16, p. 224.
51. *Ibid.,* p. 228.
52. F. M. Amram, "The Human Touch," *Robotics Today,* 9 (Spring 1981).
53. J. Reichardt, *Robots: Fact, Fiction, and Prediction,* Penguin Books Ltd., Middlesex, England, 1978, p. 123.
54. F. M. Amram, "The Human Touch," *Robotics Today,* 14 (Aug. 1982).
55. Ref. 53, p. 112.
56. I. Asimov, "Robbie" and "Satisfaction Guaranteed," *The Complete Robot,* Doubleday & Company, Inc., New York, 1982.
57. "Rover's A Robot," *Robotics Age,* 34 (May 1985).
58. A. C. Clarke, "Robots in the Nursery," in H. M. Geduld and R. Gottesman, *Robots Robots Robots,* New York Graphic Society, Boston, Mass., 1978, p. 163.
59. F. M. Amram, "The Psychology of Robot/Human Interface," *Robotics Today,* 16 (Dec. 1983).

B

BIN PICKING. See KITTING.

C

CALIBRATION

Samad A. Hayati
California Institute of Technology Jet Propulsion Laboratory
Pasadena, California

INTRODUCTION

The subject of calibration is related to the notions of precision and accuracy. In robotics, the word "calibration" refers to procedures which either initialize the robot's configuration at start-up or determine the robot's geometric parameters to enhance the positioning accuracy.

At the present time, calibration techniques in robotics may be categorized into two groups. The goal of the first group is to find the exact configuration of the robot at power-up time and initialize the controller software accordingly. In industrial applications, this is the common definition for robot calibration. The second group encompasses various techniques developed to increase the absolute positioning accuracy of robots for those applications where the robot end effector must be positioned very accurately without first being "taught" those positions. An accurate model for the robot's kinematics is necessary to achieve this goal. These two techniques are described further in the remainder of this article.

POWER-UP CALIBRATION

Robots with good accuracy are equipped with servo-controlled motors. To control a motor, a sensor must measure the precise position of the motor shaft angle in real time. In servo motors, the position information together with the angular velocity is used in a feedback control system to zero out any discrepancy between the desired and actual positions. However, the position information from either an encoder or a potentiometer cannot be used to control the robot unless the readings reflect the true position of the robot. The goal of the calibration is, therefore, to initialize the joint measurements to their "true" values. The calibration procedure consists of placing the robot in a prespecified position and reading the joint angle measurements.

Most robots are equipped with incremental encoders for feedback control. Incremental encoders can provide very-high-resolution angular information that ranges from 200 to 5000 divisions per motor shaft turn. For example, the encoders used in the Unimation PUMA robots provide about 1000 divisions per motor turn. One turn of a motor corresponds to a fraction of the actual joint motion due to the gear train that exists between the motor shaft and the actual link of the robot. This means that the resolution is even higher for the link angle measurements. The incremental encoders must be initialized, however, because their information is not absolute. It is for this reason that those robots equipped with incremental encoders must first be calibrated.

One calibration method is the use of a potentiometer for each joint in addition to an optical encoder. Potentiometers are connected to the robot controller via an analogue-to-digital converter. They are used to obtain absolute reference points for the joint measurements. At start-up time, the robot can be in any configuration. The potentiometers provide a rough estimate of the robot's joint measurements with two to three degrees of accuracy. This information can be used to initialize the optical encoders as described below. Each optical encoder circle has one zero index. The robot is placed in a known position (for example, upright), and the encoder count for a joint is set to a number such as 100,000 octal. Then the joint is moved until the first zero index is reached. Call this encoder count value A. The joint is moved again until the zero index line is encountered again. Let us call this number B, and the difference between A and B is DELTA. With this information, the accurate position of the joint can be found if the number of times the encoder has crossed its zero index line while traveling from the initial position to the present is known. For example, the encoder count at the nth crossing is

$$\text{encoder-count} = A + (n - 1) * \text{DELTA}$$

The variable n is unknown, however, if the robot is turned on when in an unknown position. Since the potentiometers can provide a rough measure of the absolute position of the joint, one can determine n by servoing the joint until it reaches a zero crossing. Since the potentiometer accuracy is better than the angle which corresponds to DELTA, one then can determine n without ambiguity from

$$n = \text{round}\ ((\text{pot-angle} - A)/\text{DELTA}) + 1$$

where pot-angle is the potentiometer reading and round is a function that gives the nearest integer to its argument. After repeating the above procedure for all the joints, the robot is said to have been calibrated. The potentiometers are used only in this initialization process.

GEOMETRIC LINK PARAMETER CALIBRATION

In this article the term "position" refers to the distance and orientation of an object relative to a reference coordinate frame. Position inaccuracies are caused by many factors, such as link parameter errors, clearances in the mechanisms' connections, wear, thermal effects, flexibility of the links and the gear train, gear backlash, encoder resolution errors, errors associated with relating the theoretical robot base coordinate frame to the world coordinate frame, and control errors. Some of these errors are time varying. For example, the position error caused by the flexibility of the robot is a function of the robot configuration, its load, and various dynamic effects. Although in principle it might be possible to control the robot to compensate for these effects (eg, see references 1 and 2) practical techniques are not yet developed. On the other hand, part of the positioning error is due to some biases that could be compensated. For example, the link twist error (3) will result in a substantial position error at the end effector whose magnitude is equal to the error in the angle multiplied by an equivalent link length.

The accuracy problem of mechanisms has been recognized and studied extensively. (For examples, see references 4–7.) These works are concerned with the accuracy problems associated with planar mechanisms. In the recent past, several researchers have devoted their attention to similar studies for the accuracy problem of robot manipulators (8–21).

Background

Link geometry parameters are those quantities that are used in robot kinematic equations together with joint state variables to establish relations between joint space and Cartesian space. Joint space specifies robot configuration in terms of joint angles for revolute or rotational joints and joint distances for prismatic or sliding joints. Cartesian space characterizes the robot's position, usually the position of its end effector, in a Cartesian coordinate system. The joint states are related to the position of the end effector through forward and inverse kinematics (qv). Differential relationships are expressed in terms of the Jacobian or inverse Jacobian matrices. This means that using a Jacobian matrix one can relate small joint angle variations to the corresponding Cartesian position changes of the end effector. In order to describe mathematically the robot's end effector position in terms of the joint states (joint angles if the robot has only revolute joints), each link of the robot is assigned a coordinate frame. These coordinate frames are attached to the links and consequently move with them as the robot moves about. It is possible to relate each coordinate frame to the next one by using homogeneous transformations.

An homogeneous transformation is a 4×4 matrix that uniquely identifies a coordinate frame relative to another coordinate frame. The top left 3×3 portion of this matrix designates the orientation of the frame relative to a reference frame. The fourth column is a 4×1 vector whose first three elements are the x, y, and z coordinates of the origin of the coordinate frame. The forward kinematics equation is obtained by multiplying the transformation matrices for all the links, link 1 through the last link.

The elements of the homogeneous transformation for each link are functions of the geometric parameters of the link as well as the joint state. Link parameter assignment is not unique. One method to determine the link parameters was introduced (22) and later expanded upon (1). This method is widely used in the analysis, design, and control of serial link manipulators. According to this technique, each revolute link is completely characterized by four parameters, and each prismatic link is defined by three parameters.

If all of these parameters are known perfectly, then the robot end effector can be placed in any position with very high precision as long as the other error sources that were mentioned earlier are not significant. Since the link parameters are not known precisely, their error causes inaccuracies in the positioning of the end effector.

Calibration Technique

The goal of a geometric calibration is to estimate the link parameters so that the kinematic equations more accurately represent the actual robot geometric model. The calibration task cannot be undertaken by simple procedures since the mapping of these errors to the end effector frame is itself a function of the robot's position.

To obtain an analytic tool for the geometric calibration, a model is required to relate the end-effector position errors to those of the link parameters. A differential relation can be obtained to map a position error vector in one frame to any other frame (1). Assuming that the link parameter errors can be modeled as small variations of the nominal parameters, one can map all the error sources to the end-effector frame (for details, see references 18 and 19). Assuming that accurate end-effector measurements are available, one can relate any discrepancy between the computed and measured positions to the link parameter errors. Since any measurement will introduce random errors of its own, these should also be included in the analysis. Following this procedure, a linear observation equation can be obtained to relate the bias errors to the actual measurements of the end-effector positions.

The calibration procedure consists of placing the end-effector in various positions, computing the difference between the actual end-effector position and the theoretical one, and then feeding this information to a computer program to solve for the link parameter errors.

Inverse Kinematics

To utilize the results obtained from the above geometric calibration technique, one must incorporate the improved knowledge of the link errors in the forward and inverse kinematic equations of the calibrated robot. The modification for forward kinematic equations is very simple. All one has to do is update the numerical values for the link parameters in the software. The modification of the inverse kinematics, unfortunately, is not so simple. It has been shown (23) that only "simple" robots have closed-form inverse kinematic solutions. One condition for a six-link robot to be simple is that the joint axes of the last three links intersect at a single point. Most industrial and research robots have been designed to possess this property so that their inverse kinematics can be computed in closed form. This means that certain link parameters are designed to be zero. However, due to manufacturing tolerances, the actual link parameters of a robot differ from the ideal ones by small amounts. Thus, the postcalibration model of the robot, which more accurately represents the physical system, is that of a "nonsimple" one. To solve the inverse kinematic equations for these robots, one can build on the solution obtained for their ideal model. The inverse kinematic equations are solved by first finding the closed-form solution for the ideal model and then computing small variations to be added to the joint angles by utilizing the Jacobian of the postcalibration model (24).

CONCLUSION

The subject of geometric calibration is a new area in robotics. Although many researchers have developed calibration techniques to improve the accuracy of robots, as of today there is no method which is universally accepted. The need for geometric calibration will increase as robots become more autonomous and require higher positioning accuracy.

BIBLIOGRAPHY

1. A. Zalucky and D. E. Hardt, "Active Control of Robot Structure Deflections," presented at the Winter Annual Meeting of the American Society of Mechanical Engineers, Phoenix, Ariz., Nov. 14–19, 1982, pp. 83–100.
2. W. J. Books and M. Majette, "Controller Design for Flexible Distributed Parameter Mechanical Arms Via Combined State Space and Frequency Domain Techniques," *Journal of Dynamic Systems, Measurement, and Control* (Dec. 1983).
3. R. Paul, *Robot Manipulators: Mathematics, Programming, and Control,* The M.I.T. Press, Cambridge, Mass., 1982.
4. R. E. Garrett and J. Hall, "Effect of Tolerance and Clearance in Linkage Design," *Journal of Engineering for Industry ASME Transactions Series B* **91,**(1), 198–202 (Feb. 1969).
5. S. A. Kolkatkar and K. S. Yajnik, "The Effect of Play in Joints of Function Generating Mechanisms," *Journal of Mechanisms,* **5,** 521–532 (1970).
6. S. G. Dhande and J. Chakraborty, "Analysis and Synthesis of Mechanical Error in Linkages," *Journal of Engineering for Industry Transactions of ASME* **95,** 672–676 (Aug. 1973).
7. S. Dubowsky, J. Maatuk, and N. D. Perrira, "A Parameter Identification Study of Kinematic Errors in Planar Mechanisms," *Journal of Engineering for Industry Transactions of ASME,* **79** (May 1975).
8. J. Chen, C. B. Wang, and J. C. S. Yang, "Robot Positioning Accuracy Improvement Through Kinematic Parameter Identification," *Proceedings of the 3rd Canadian CAD/CAM and Robotics Conference,* Toronto, Canada, June 1984.
9. L. P. Foulley and R. B. Kelley, "Improving the Precision of a Robot," *Proceedings of the First IEEE International Conference on Robotics,* Atlanta, Ga., Mar. 13–15, 1984, pp. 62–67.
10. A. Mukerjee and D. H. Ballard, "Self-Calibration in Robot Manipulators," *Proceedings of the 2nd IEEE International Conference on Robotics and Automation,* St. Louis, Mo., Mar. 25–28, 1985, pp. 1050–1057.
11. G. J. Agin, "Calibration and Use of a Light Stripe Range Sensor Mounted on the Hand of a Robot," *Proceedings of the 2nd IEEE International Conference on Robotics and Automation,* St. Louis, Mo., Mar. 25–28, 1985, pp. 680–685.
12. S. Ahmad, "Second Order Nonlinear Kinematic Effects, and Their Compensation," *Proceedings of the 2nd IEEE International Conference on Robotics and Automation,* St. Louis, Mo., Mar. 25–28, 1985, pp. 307–314.
13. D. E. Whitney, C. A. Lozinski, and J. M. Rourfe, "Industrial Robot Calibration Method and results," *Proceedings of the ASME Computers in Engineering Conference,* Las Vegas, Nev., Aug. 1984, pp. 92–100.
14. C. H. Wu, "The Kinematic Error Model for the Design of Robot Manipulators," *Proceedings of the 1983 American Control Conference,* San Francisco, Calif., June 1983.
15. A. Dainis and M. Juberts, "Accurate Remote Measurement of Robot Trajectory Motions," *Proceedings of the 2nd IEEE International Conference on Robotics and Automation,* St. Louis, Mo., Mar. 25–28, 1985, pp. 92–99.
16. B. W. Mooring, "The Effect of Joint Axis Misalignment on Robot Positioning Accuracy," *Proceedings of the ASME Computers in Engineering Conference,* Chicago, Ill., Aug. 7–11, 1983, pp. 151–155.
17. K. J. Waldron, "Positioning Accuracy of Manipulators," *Proceedings of N.S.F.,* University of Florida, Gainesville, Fla., Feb. 1978, pp. 111–141.
18. C. H. Wu and C. C. Lee, "Estimation of the Accuracy of a Robot Manipulator," *IEEE Transactions on Automatic Control* **AC-30,** 304–306 (Mar. 1985).
19. S. Hayati, "Robot Arm Geometric Link Parameter Estimation," *Proceedings of the 22nd IEEE Conference on Decision and Control,* San Antonio, Texas, Dec. 1983, pp. 1477–1483.
20. S. Hayati and M. Mirmirani, "Improving the Absolute Positioning Accuracy of Robot Manipulators," *Journal of Robotic Systems,* **II** (IV), (Dec. 1985).
21. N. G. Dagalakis and D. R. Myers, "Adjustment of Robot Joint Gear Backlash Using the Robot Joint Test Excitation Technique," *The International Journal of Robotics Research* **4**(2), (Summer 1985).
22. J. Denavit and R. S. Hartenberg, "A Kinematic Notation for Lower-Pair Mechanisms Based on Matrices," *ASME Journal of Applied Mechanics,* 215–221 (June 1955).
23. D. L. Pieper, "The Kinematics of Manipulators Under Computer Control," Ph.D. dissertation, Department of Computer Science, Stanford University, Stanford, Calif., 1968.
24. S. Hayati and G. Roston, "On Inverse Kinematic Solutions for Near-simple Robot Manipulators and its Application to Robot Calibration," *Proceedings of the International Symposium on Robot Manipulators: Modeling, Control and Education,* Albuquerque, N.M., Nov. 12–14, 1986, pp. 41–50.

CLASSIFICATION

William R. Tanner
Tanner Associates
Farmington Hills, Michigan

Several schemes are used in the classification of industrial robots, including *principle of operation, operating mode, configuration,* and *application.* Most of these are descriptive of some physical or behavioral characteristic of the device, such as geometry or operation. The classification of industrial robots serves a useful function for comparison of like characteristics and is especially helpful in the selection of a particular robot for a specific application. Computerized robot directories, which the prospective robot user will find helpful in the selection process, make use of classifications in both the specification and search processes.

No U.S. or international standards have yet been established for classifying robots. However, the classifications described in the following are commonly used throughout U.S. industry, by robot suppliers and robot users alike. The schemes are not mutually exclusive; some classifications, in fact, imply others; for example, "continuous-path" robots are all "servo-controlled." Also, no single, all-encompassing classification method exists; more than one classification is usually needed to describe a robot totally. In the development of a robot description based on classification, the order or hierarchy of the process usually begins with principle of operation, followed by operating mode and configuration. Application may also be specified as a final descriptor.

PRINCIPLE OF OPERATION

Robots are most broadly classified as fixed stop (nonservo) or servo-controlled. These terms refer to the methods of controlling the robot's positions in its work envelope, that is, the interactions between the controller and the robot arm. The principles of operation are best explained by describing the sequence of operation of each type of robot.

Fixed Stop

Fixed Stop Robot. This type is also known as nonservo robot or open-loop robot. The robot has stop-point control, but no trajectory control. Each of its axes has a fixed mechanical limit at each end of its stroke and can stop only at one or the other of these limits (1).

A typical operating sequence for a fixed stop robot would be

1. At program start, the controller initiates output signals to control valves and/or external devices.
2. The control valves open, allowing the flow of air or oil to the actuators, and the manipulator's axes begin to move.
3. The valves remain open and the axes of the manipulator continue to move until they are restrained by physical contact with fixed stops.
4. Limit switches or other sensors signal the end of movement to the controller, which then commands the control valves to close.
5. The controller then indexes to the next step and initiates new output signals. These may again be to control valves for actuators and/or to external devices.
6. The process is repeated until the entire sequence of steps has been executed.

The operation of an hydraulically actuated fixed stop robot is shown schematically in Figure 1.

The significant features of a fixed stop robot are

- The manipulator's various axes move until the limits of travel (fixed stops) are reached. Thus, there are usually only two positions for each axis to assume. These fixed stops are adjustable to suit the particular application.
- Often, only one axis will move at a time, even though movement of more than one is required for end-of-arm positioning. Simultaneous motions of two or more axes are not coordinated; that is, all axes may not reach their limits of travel at the same time, and the end-of-arm trajectory is not guided during movement.
- The controller provides the capability for many motions in a program, but only to the fixed stops of each axis.
- It is feasible to activate intermediate stops on some axes to provide more than two positions; however, there is a finite limit to the number of such stops that can be installed.
- Deceleration at the approach to the stops may be provided by the actuators or by external shock absorbers.
- Drives are usually either pneumatic or hydraulic.
- This class of robots is usually restricted to the performance of single programs; however, the programmed sequence can be conditionally modified (truncated) through appropriate external sensors.
- Relatively high speed is possible.
- Repeatability to within 0.010 in. (0.25 mm) or better can be achieved.
- These robots are relatively low in cost, and simple to operate, program, and maintain.
- These robots have limited flexibility in terms of program capacity and positioning capability.

Servo-controlled

Servo-controlled Robot. This type is driven by servomechanisms, that is, motors whose driving signal is a function of

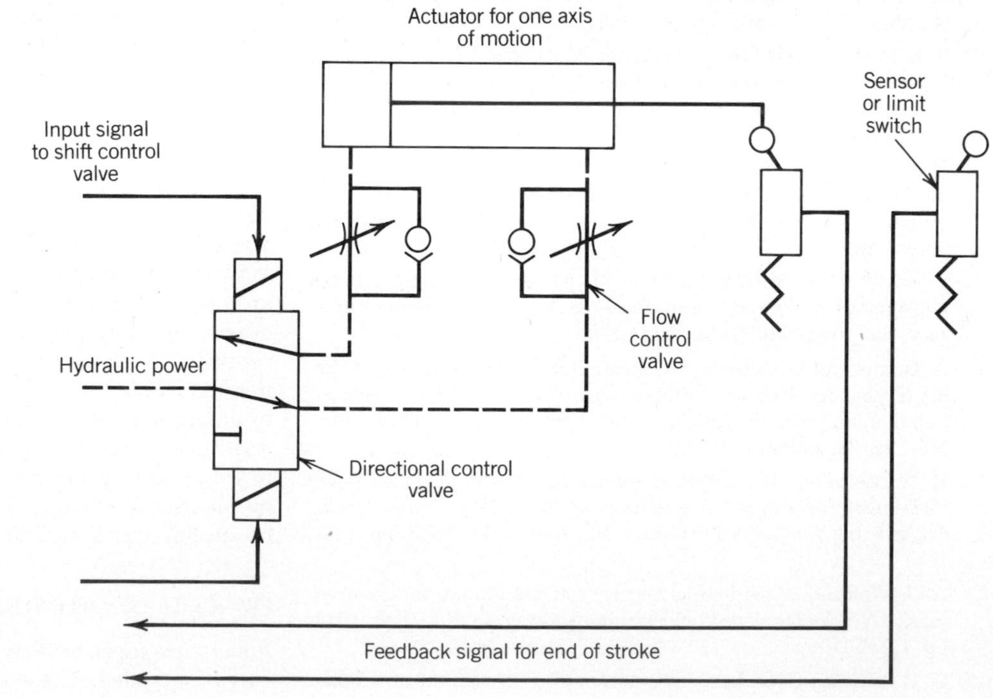

Figure 1. Typical fixed stop hydraulic robot control schematic.

the difference between command position and/or rate and measured actual position and/or rate. Such a robot is capable of stopping at, or moving through, a practically unlimited number of points in executing a programmed sequence (1).

A typical operating sequence for a servo-controlled robot would be

1. At program start, the controller addresses the memory location of the first programmed position and also determines the actual position of the manipulator's various axes as measured by the position feedback system.
2. These two sets of data are compared and their differences (commonly called error signals) are amplified and transmitted as "command signals" to servo valves or servo amplifiers for the actuator of each axis.
3. As the actuators move the manipulator's axes, feedback devices such as encoders, potentiometers, resolvers, and tachometers send position (and in some cases, velocity) data back to the controller. These "feedback signals" are compared with the desired position data and new error signals are generated, amplified, and sent as command signals to the servo valves or servo amplifiers.
4. This process continues until the error signals are effectively reduced to zero, whereupon the energy flow to the actuators is stopped and the axes come to rest at the desired position. In some designs, brakes are applied to certain axes to restrain motion and resist external forces.
5. The controller then addresses the next memory location and responds appropriately to the data stored there. This may be another positioning sequence for the manipulator and/or a signal to an external device.
6. The process is repeated sequentially until the entire set of data (program) has been executed.

The operation of an hydraulically actuated servo-controlled robot is shown schematically in Figure 2.

The significant features of a servo-controlled robot are

- The manipulator's various axes can be commanded to move to and stop anywhere within their limits of travel, rather than only at the extremes.
- Simultaneous motion of more than one axis is common; however, the movements may not be coordinated so that all axes reach the desired end-of-arm position at the same time.
- It is possible to control the trajectory, velocity, acceleration, and deceleration of the various axes as they move between programmed points.
- Generally, the controller's memory capacity is large enough to store many more commands than a fixed stop (nonservo) robot.

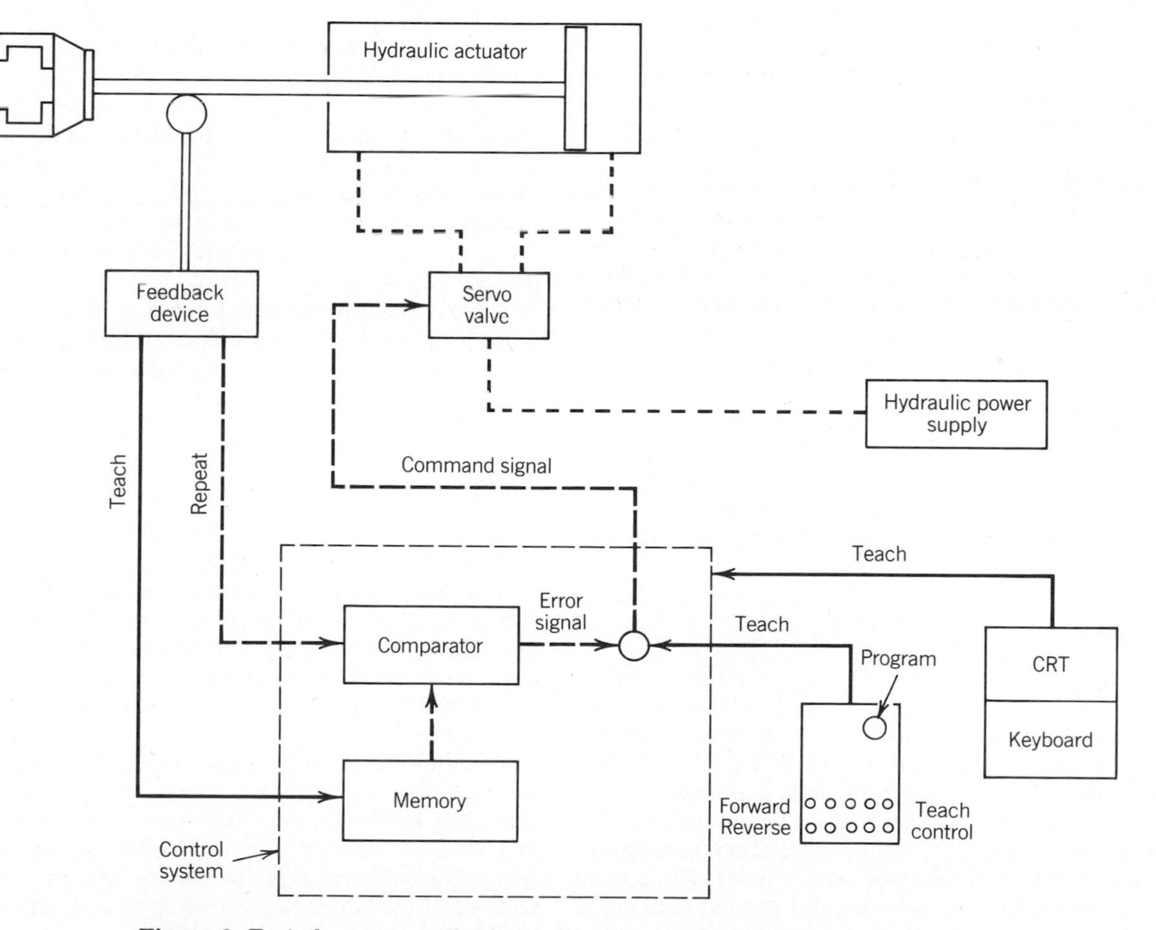

Figure 2. Typical servo-controlled hydraulic point-to-point robot control schematic.

- Both continuous-path and point-to-point motions are possible.
- Drives are usually hydraulic or electric and use state-of-the-art servo control technology.
- The axes of the manipulator can be programmed to any position within the limits of their travel, for maximum utility.
- Control systems may permit storage and execution of more than one program, with random selection of programs from memory in response to external signals.
- Subroutining and branching capabilities may permit the robot to take alternative actions within a program when commanded externally or internally.
- An end-of-arm positioning repeatability of 0.060 in. (1.5 mm) or better can be achieved.
- These robots are more expensive and more complex to operate, program, and maintain than fixed stop robots.

OPERATING MODE

In addition to describing or classifying them on the basis of their principle of operation, robots may also be described in terms of their operating mode. There are two modes of operation for industrial robots: point-to-point and continuous-path.

Point-to-point

Point-to-point Control. This is a robot motion control in which the robot can be programmed by a user to move from one position to the next. The intermediate paths between these points cannot be specified (1).

Point-to-point System. This is the robot movement in which the robot moves to a numerically defined position, stops, and performs an operation, moves to another numerically defined position and stops, and so on. The path and velocity while traveling from one point to the next generally have no significance (1).

In the point-to-point mode, a robot performs useful work only when all axes are stopped, at desired positions in its work envelope. Most parts-handling and some tool-handling operations are performed in this mode. The robot's trajectory (path between successive points) is usually not programmed; it is an internal function of the controller. All fixed stop robots and some servo-controlled robots operate in the point-to-point mode.

Continuous-path

Continuous-path Control. This is a type of robot control in which the robot moves according to a replay of closely spaced points programmed on a constant time base during teaching. The points are first recorded as the robot is guided along a desired path, and the position of each axis is recorded by the control unit on a constant time basis by scanning axis encoders during the robot motion. The replay algorithm attempts to duplicate that motion. Alternatively, a continuous-path control can be accomplished by interpolation of a desired path curve between a few taught points (1).

Continuous-path System. This is a type of robot movement in which the tool performs the task while the axes of motion are moving. All axes of motion may move simultaneously, each at a different velocity, in order to trace a required path or trajectory (1).

In the continuous-path mode, the robot performs useful work while the arm is moving. The arm's trajectory may be programmed directly (lead-through) or the controller may calculate the path between successive points. A few parts-handling and some tool-handling operations are performed in this mode. Continuous-path operations can be performed only by servo-controlled robots.

CONFIGURATION

A third classification scheme describes the geometric or mechanical arrangements of the robot arms. Mechanical arrangements of robot arms are widely varied, but most fall into one of six configurations. Three common robot configurations are described in terms of their coordinate systems. These are rectangular or Cartesian, cylindrical, and spherical or polar. Another three configurations are named according to appearance or function. These are anthropomorphic or jointed-arm, SCARA, and gantry.

Rectangular

Rectangular Robot. This is also known as a rectangular coordinate robot, Cartesian coordinate robot, Cartesian robot, rectilinear coordinate robot, or rectilinear robot. It is a robot that moves in straight lines up and down, and in and out. The degrees of freedom of the manipulator arm are defined primarily by a Cartesian coordinate axis system consisting of three intersecting perpendicular straight lines with their origin at the intersection. This robot may lack control logic for coordinated joint motion (1).

A rectangular or Cartesian robot is configured with a horizontal arm assembled to a vertical lift axis, which is mounted on a linear traverse base, thus producing an X–Y–Z axes robot, as shown in Figure 3.

Cylindrical

Cylindrical Robot. This is also known as a cylindrical coordinate robot or columnar robot. It is a robot, built around a column, that moves according to a cylindrical coordinate system in which the position of any point is defined in terms of an angular dimension, a radial dimension, and a height from a reference plane. The outline of a cylinder is formed by the work envelope. Motions usually include rotation and arm extension (1).

A cylindrical robot is configured with a horizontal arm assembled to a vertical axis, which is mounted on a rotating base. The horizontal arm can move in and out and move up and down on the vertical axis. In addition, this arm assembly can rotate left and right about the vertical axis; thus, the motions of the three major axes form a portion of a cylinder, as shown in Figure 4.

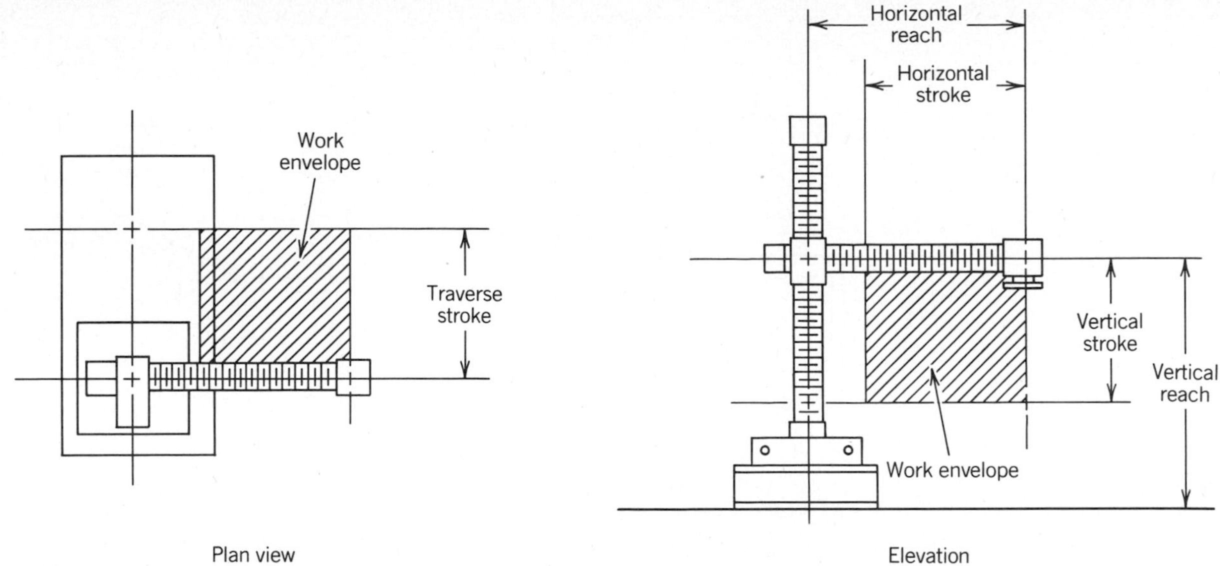

Figure 3. Rectangular robot configuration.

Spherical

Spherical Robot. This is also known as a spherical coordinate robot or a polar robot. It is a robot operating in a spherical work envelope, or a robot arm capable of moving with rotation, arm inclination, and arm extension. It can be a cylindrical coordinate robot with the addition of pitch (1).

A spherical or polar robot configuration consists of an arm which moves in and out in a reach stroke, but utilizes a pivoting vertical motion instead of a straight-line vertical stroke. In addition, the arm can rotate left and right about a vertical pivot axis. Thus, the motions of the major axes form a portion of a sphere as a work envelope, as shown in Figure 5.

Anthropomorphic

Anthropomorphic Robot. This is also known as a jointed-arm robot. It is a robot with all rotary joints and motions similar to those of a person's arm (1).

An anthropomorphic or jointed-arm robot contains rotary joints (called shoulder and elbow) that are mounted on a rotating base to provide the three major axes of motion, as shown in Figure 6.

SCARA

Selective Compliance Robotic Arm for Assembly (SCARA). This is a horizontal-revolute configuration robot designed at

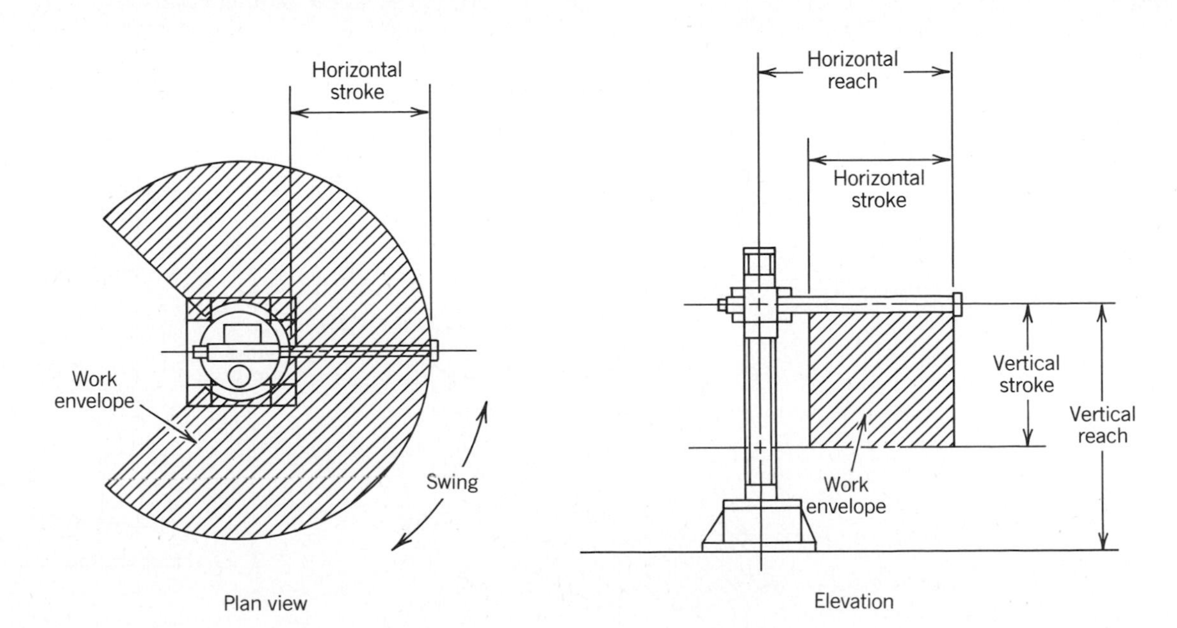

Figure 4. Cylindrical robot configuration.

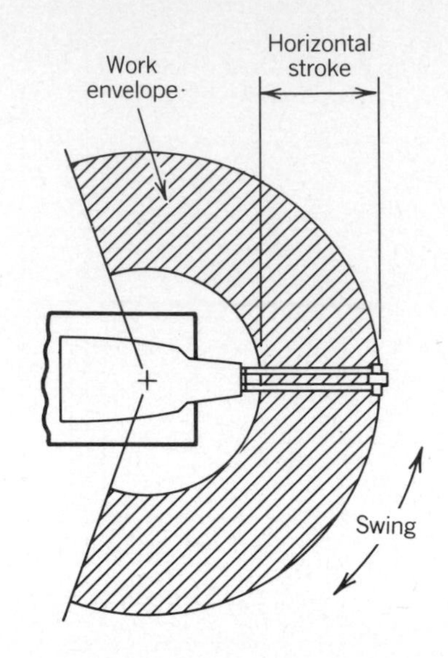

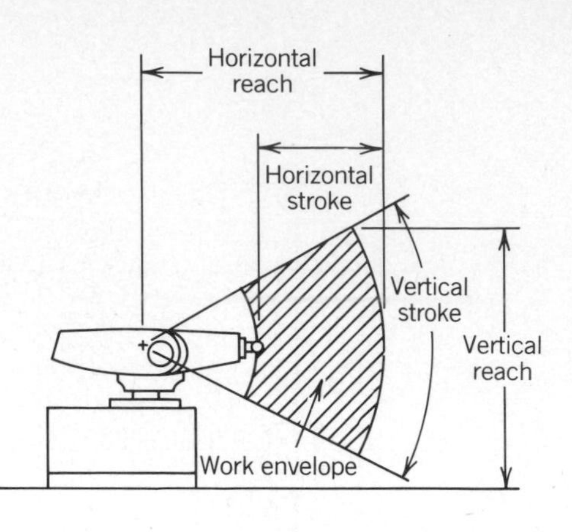

Figure 5. Spherical robot configuration. Plan view

Japan's Yamamachi University. The tabletop-size arm sweeps across a fixtured area and is especially suited for small-parts insertion tasks (1).

A SCARA robot is basically an anthropomorphic or jointed-arm structure operating in a horizontal plane, as shown in Figure 7. It typically has two or three horizontal servo-controlled joints (shoulder, elbow, and sometimes, wrist) and one vertical nonservo or servo-controlled axis. By removing power and "relaxing" one or more of the horizontal joints, the robot structure becomes compliant to external forces, but at the same time its position is not affected by gravity. In this condition, close-tolerance insertion tasks may be accomplished, with the arm moved by external forces to compensate for minor errors in position.

Gantry

Gantry Robot. This is an overhead-mounted rectilinear robot with a minimum of three degrees of freedom and normally not exceeding six. Bench-mounted assembly robots that have a gantry design are not normally included in this definition. A gantry robot can move along its x and y axes traveling over relatively greater distances than a pedestal-mounted robot at high traverse speeds while still providing a high degree of accuracy for positioning. Features of a gantry robot include large work envelopes, heavy payloads, mobile overhead mounting, and the capability and flexibility to operate over the work area of several pedestal-mounted robots (1).

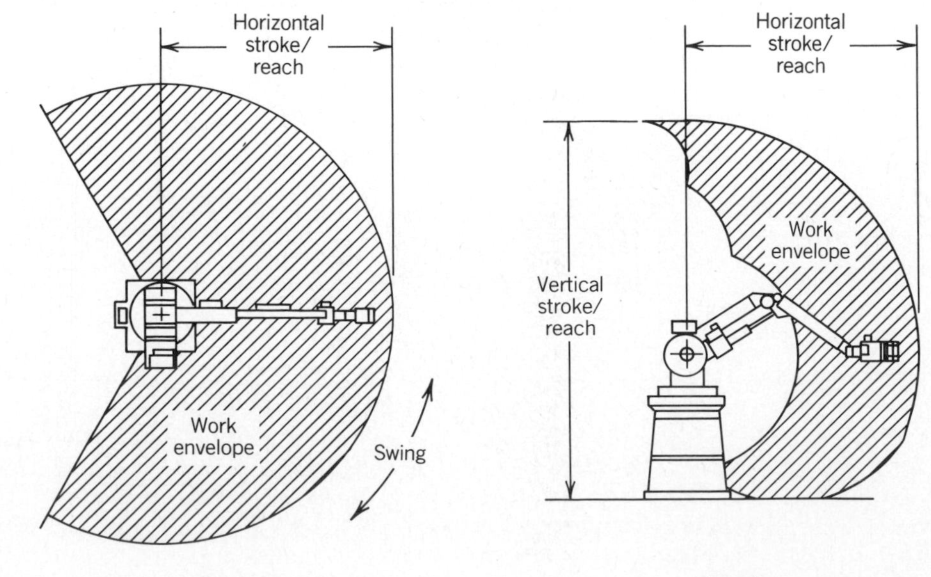

Figure 6. Anthropomorphic robot configuration.

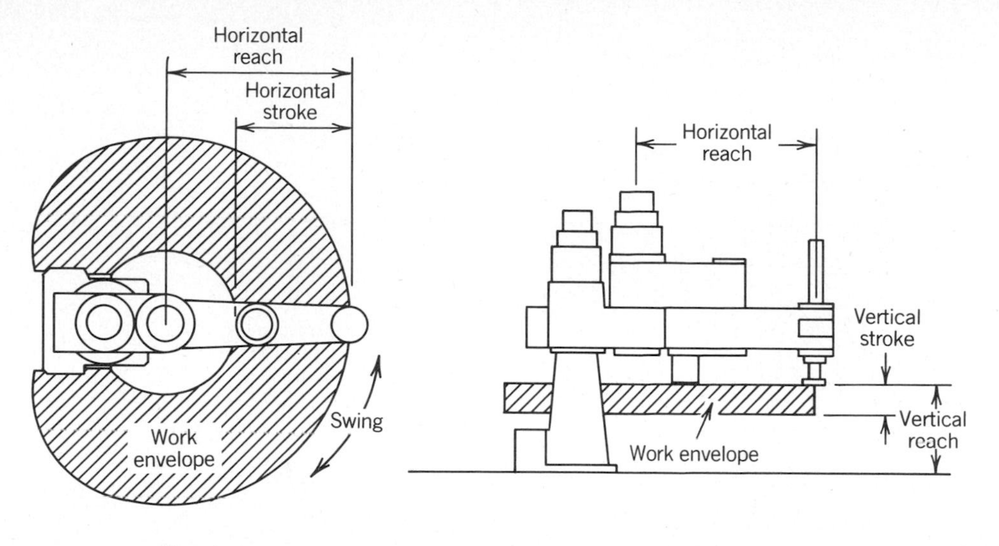

Figure 7. SCARA robot configuration.

Gantry Robot Coordinate System. The *x, y,* and *z* axes of a gantry robot consist of the following components:

x Axis. Runway. This is the longitudinal axis, normally the passive side rails of the superstructure of the gantry robot.

y Axis. Bridge. This is the transverse axis, an active member of the robot riding on the runway rails and supporting the carriage of the gantry robot.

z Axis. Telescoping Tubes or Masts. This is the vertical axis, supported by the carriage (1).

The gantry robot is not always a unique geometric or coordinate system; it may be one of the robot configurations previously described mounted in an elevated position above the workplace. Although any of the robot arms may be used in this way, the most common gantry robots are rectangular, as shown in Figure 8.

APPLICATION

Another classification that is sometimes used is the intended application of the robot. Most industrial robots are of such a design as to be usable in a variety of applications; conversely, most robot applications can be performed by a variety of robots. A few applications, however, have unique characteristics and/or requirements, for which robots have been specifically designed. These include painting, assembly, clean room operations, and education/research.

Painting Robots

Painting robots are all servo-controlled and continuous-path (see also PAINTING). Manipulator structures are designed for low weight, low inertia, and high rigidity to be able to replicate complex trajectories at speeds of up to 6½ ft (2 m) per second. The most common configuration for painting robots is anthropomorphic. Because much of the robotic painting uses solvent-based materials, these robots are generally designed and certified as intrinsically safe for operation in an explosive environment. The requirements of high-speed movement and intrinsic safety are usually (but not always) satisfied by using hydraulic actuator systems. The robot controller and hydraulic power supply are separate from the manipulator so that they can be located outside the hazardous (explosive) environment. Lead-through programming is usually used for teaching painting robots.

Assembly Robots

Assembly operations involve robots placing parts together in a preprogrammed sequence (see ASSEMBLY ROBOTS). Although assembly includes mechanical parts of all descriptions, most robotic assembly at this time is inserting electronic components into (or placing components onto) printed circuit boards in the manufacturing of electronic products. This generally requires a robot with a high level of repeatability. High speed and smooth acceleration/deceleration are also desirable attributes of assembly robots. With increasing frequency, assembly robots are interfaced to external devices such as vision, force, range, or tactile sensors. Most assembly robots are small, have payload capacities less than 11 lb (5 kg) and are electrically driven. Rectangular (bench- or overhead-mounted), cylindrical, or SCARA configurations are most common, and assembly robots are usually point-to-point and servo-controlled. Microprocessor- or computer-based controllers are used, and off-line programming or computer-aided design (CAD) systems are generally used for teaching.

Clean Room Robots

In the manufacture of semiconductor devices, most of the processing is done in a "clean" environment where airborne particulates are closely controlled (see CLEAN ROOM APPLICATIONS). Human workers (even when suitably clothed, masked, etc) are a major source of contaminants, and the operations performed by humans are primarily the movement of objects into and out of automatic processing devices; thus, robots are increasingly being introduced into these environments. The ordinary industrial robot, however, is also a source of contamination and a special class of robots, known as clean room robots, has been developed for these applications. Clean room robots are usually electrically driven assembly robots that may have

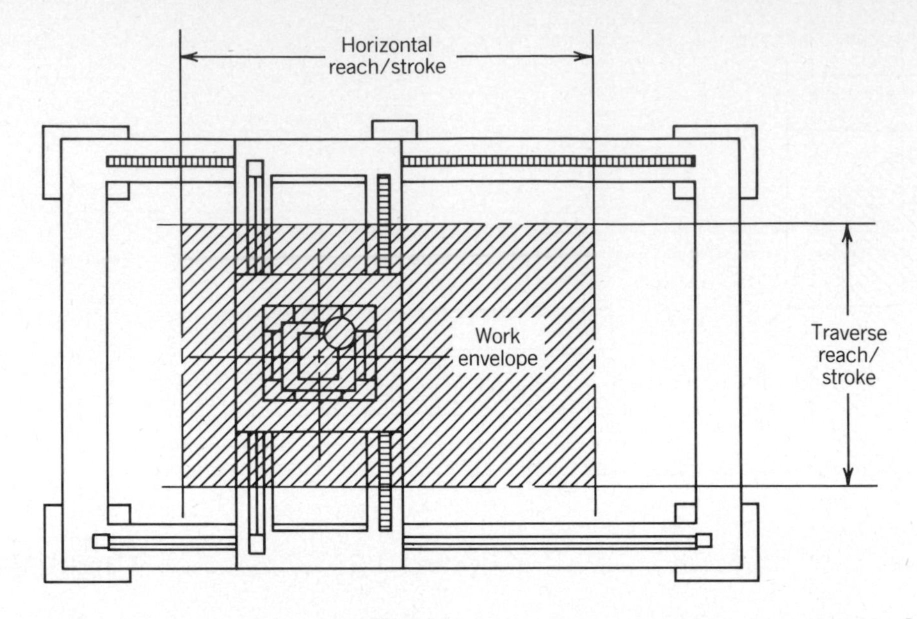

Plan view

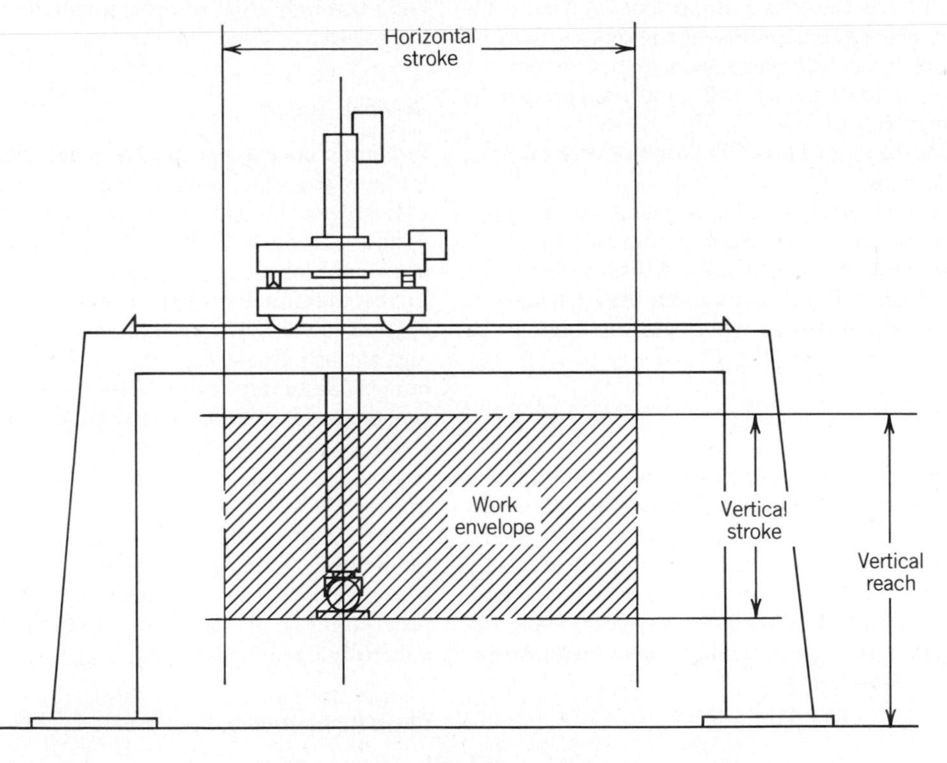

Elevation

Figure 8. Gantry robot configuration.

special drives and lubrication systems and are often shrouded in flexible, noncontaminating envelopes. Air may be extracted from the envelope to further prevent release of particles into the atmosphere. Clean room robots are mostly servo-controlled, point-to-point, and microprocessor-controlled. Rectangular, cylindrical, and jointed-arm configurations are most common.

Education/Research Robots

Robots used in education and research are generally low in cost, are small in size and weight, and have limited performance capabilities. They may be placed on desk or table tops for the purposes of illustrating robot theory, application, and programming. Stepping motors and cable/pulley systems are often used for drives, and the anthropomorphic configuration is most common. Structures usually are not robust, and speed and repeatability may be low as a result. These robots may not have a dedicated controller; often only a computer interface is provided. Programming is usually by means of a high-level language, through the computer/controller.

OTHER CLASSIFICATIONS

The Robotic Industries Association (RIA) conducts an annual census of the world robot population and publishes a *Worldwide Robotics Survey and Directory*. In it, the robot population is broken down by type into three groups:

1. Reprogrammable, servo-controlled continuous-path.
2. Reprogrammable, servo-controlled point-to-point.
3. Reprogrammable, nonservo-controlled point-to-point.

In addition, they report population by applications, which include: spot welding, arc welding, painting/coating, finishing, assembly, loading/unloading, material handling, casting, and others (see WELDING ROBOTS; PAINTING; FINISHING; ASSEMBLY ROBOTS; MACHINE LOADING/UNLOADING; MATERIAL HANDLING).

PROPOSED STANDARDS

Although no official standards for classification of industrial robots exist worldwide, several organizations are presently developing draft standards for classification, and local standards have been developed in Japan and elsewhere.

In the United States, the American Society for Testing and Materials (ASTM) is drafting a *Robot Classification Guide*, F-28.03, which classifies or categorizes industrial robots by (2):

Primary Power Sources. Pneumatic, hydraulic, electric, or any combination.

Types of Motion Control. Servo or nonservo, point-to-point or continuous-path.

Payload. Very light, light, medium, heavy, very heavy.

Programming Method. Set points, memory devices, keyboard or on-line terminal, off-line, download from host computer or CAD system.

Mechanical Configuration. Cylindrical, rectilinear, spherical, articulated.

Degrees of Freedom. Three translational (X,Y,Z), three rotational.

Application. Spray painting, arc welding, spot welding, assembly, educational, general-purpose, tool manipulation.

The R15.07 Terminology Subcommittee of the RIA R15 Committee on Robotic Standards has produced and published an "RIA Robotics Glossary," which includes definitions of many of the classification terms commonly used and may, in the future, develop a draft standard for robot classification. The RIA is also the Administrator of the U.S. Technical Advisory Group (USTAG) for the International Standards Organization's Industrial Robot Subcommittee (ISO/TC184/SC2) and is involved in the development of international standards for industrial robots.

Two proposals on industrial robot classification have been submitted to the International Standards Organization ISO/TC184/SC2 WG committee for consideration. A French proposal would classify industrial robots by the following criteria (3):

- Servo-controlled axes.
- Type of programming.
- Control characteristics, including type of control and control schemes.
- Mechanical structure.
- Main power source.
- Number of axes.
- Scale of the robot.
- Mobility or not.

A Japanese proposal would classify robots according to (4):

- Programming.
- Control.
- Main power source.
- Mechanical configuration.
- Field of application.
- Size/load.
- Mobility.
- Environment.

In Japan, industrial robots are classified, per Japanese Industrial Standard JIS-B-O 134-1979, by two schemes (5):

- Input of Information and Teaching Method
  - Type A. Manual manipulator.
  - Type B. Fixed-sequence robot.
  - Type C. Variable-sequence robot.
  - Type D. Playback robot.
  - Type E. Numerically controlled robot.
  - Type F. Intelligent robot.
- Motion Geometry
  - Cylindrical coordinate.
  - Polar coordinate.
  - Cartesian coordinate.
  - Articulated.

GLOSSARY

Most of the following robotics term definitions are taken from the *RIA Robotics Glossary*, published in 1984 by the Robot Institute of America (now Robotic Industries Association), 900 Victors Way, P.O. Box 3724, Ann Arbor, Mich. 48105 and are included with their permission.

Actuator. A motor or transducer that converts electrical, hydraulic, or pneumatic energy to effect motion of the robot.

Analogue. An expression of values which can vary continuously; eg, translation, rotation, voltage, or resistance. (Contrasted with **Digital.**)

Anthropomorphic. An adjective with the literal meaning "of human shape." An anthropomorphic robot is one that performs tasks with motions similar to those of a human.

Cartesian. A coordinate system whose axes or dimensions are three intersecting perpendicular straight lines and whose origin is the intersection (also described as rectilinear). A robot whose manipulator arm degrees of freedom are defined primarily by Cartesian coordinates.

Continuous-path. A control scheme whereby the inputs or commands specify every point along a desired path of motion.

Controller. An information processing device whose inputs are both desired and measured position, velocity, or other pertinent variables in a process and whose outputs are drive signals to a controlling motor or actuator.

Cylindrical. A coordinate system that defines the position of any point in terms of an angular dimension, a radial dimension, and a height from a reference plane. These three dimensions specify a point on a cylinder. A robot whose manipulator arm degrees of freedom are defined primarily by cylindrical coordinates.

Degrees of Freedom. One of a limited number of ways in which a point or a body may move or in which a dynamic system may change, each way being expressed by an independent variable and all required to be specified if the physical state of the body or system is to be completely defined.

Digital. Control involving digital logic devices that may or may not be complete digital computers.

Encoder. A rotary feedback device that transmits a specific code for each position. A device that transmits a fixed number of pulses for each revolution.

End Effector. An actuator, gripper, or mechanical device attached to the wrist of a manipulator by which objects can be grasped or otherwise acted upon.

Feedback. The signal or data sent to the control system from a controlled machine or process to denote its response to the command signal.

Fixed Stop. A robot with stop-point control, but no trajectory control. That is, each of its axes has a fixed limit at each end of its stroke and cannot stop except at one or the other of these limits. Such a robot with N degrees of freedom can therefore stop at no more than $2N$ locations (where location includes position and orientation). Some controllers do offer the capability of program selection of one of several mechanical stops to be used. Also called a nonservo robot.

Hydraulic Motor. An actuator consisting of interconnected valves and pistons or vanes that converts high-pressure hydraulic or pneumatic fluid into mechanical shaft translation or rotation.

Lead-through. Programming or teaching by physically guiding the robot through the desired actions. The speed of the robot is increased when programming is complete.

Manipulator. A mechanism, usually consisting of a series of segments, jointed or sliding relative to one another, for the purpose of grasping and moving objects, usually in several degrees of freedom.

Passive Accommodation. Compliant behavior of a robot's end point in response to forces exerted on it. No sensors, controls, or actuators are involved.

Pitch. The angular rotation of a moving body about an axis perpendicular to its direction of motion and in the same plane as its top side.

Point-to-point. A control scheme whereby the inputs or commands specify only a limited number of points along a desired path of motion. The control system determines the intervening path segments.

Potentiometer. An analogue feedback device based upon tapping the voltage at a point along a continuous electrical resistive element.

Power Supply. In general, a device that converts a-c line voltage into one or more d-c voltages. In hydraulic robots, an electrically driven mechanical pump that generates a flow of fluid under pressure.

Program. *(1) n.* A sequence of instructions to be executed by the computer or robot controller to control a machine or process. *(2) v.* To teach a robot system a specific set of movements and instructions to accomplish a task.

Record–playback Robot. A manipulator for which the critical points along desired trajectories are stored in sequence by recording the actual values of the joint position encoders of the robot as it is moved under operator control. To perform the task, these points are played back to the robot servo system.

Remote Center Compliance (RCC). A compliant device used to interface a robot or other mechanical workhead to its tool or working medium. The RCC allows a gripped part to rotate about its tip or to translate without rotating when pushed laterally at its tip. The RCC thus provides general lateral and rotational "float" and greatly eases robot or other mechanical assembly in the presence of errors in parts, jigs, pallets, and robots. It is especially useful in performing very close tolerance or interference insertions.

Repeatability. Closeness of agreement of repeated position movements, under the same conditions, to the same location.

Resolver. A transducer which converts rotary or linear mechanical position into an analogue electrical signal by means of the interaction of electromagnetic fields between the movable and the stationary parts of the transducer.

Roll. The angular displacement of a moving body about the principle axis of its motion.

Servo-controlled. A robot driven by servomechanisms, ie, motors whose driving signal is a function of the difference between commanded position and/or rate and measured actual position and/or rate. Such a robot is capable of stopping at or moving through a practically unlimited number of points in executing a programmed trajectory.

Settling Time. The time for a damped oscillatory response to decay to within some given limit.

Spherical. A coordinate system, two of whose dimensions are angles, the third being a linear distance from the point of origin. These three coordinates specify a point on a sphere. A robot whose manipulator arm degrees of freedom are defined primarily by spherical coordinates.

Tachometer. A rotational velocity sensor.

Teach. (1) To move a robot to or through a series of points to be stored for the robot to perform its intended task. (2) (See **Program.**)

Work Envelope. The set of points representing the maximum extent or reach of the robot hand or working tool in all directions.

Wrist. A set of rotary joints between the arm and end effector that allow the end effector to be oriented to the workpiece.

Yaw. The angular displacement of a moving body about an axis that is perpendicular to the line of motion and to the top side of the body.

BIBLIOGRAPHY

1. J. Jablonowsi and J. W. Posey, "Robotics Terminology," in S. Y. Nof, ed., *Handbook of Industrial Robotics,* John Wiley & Sons, Inc., New York, 1985, pp. 1271–1303.
2. *Robot Classification Guide,* American Society for Testing and Materials (ASTM), ASTM Draft Standard FXXX-XX (Jurisdiction F28.03), Oct. 21, 1985.
3. *Industrial Manipulating Robots Classification,* International Standards Organization, French Proposal ISO/TC184/SC2 WG1 N31, July 1985.

4. *Manipulating Industrial Robots Classification,* International Standards Organization, Japanese Proposal ISO/TC184/SC2 WG1 N46, Nov. 1985.
5. K. Yonemoto, "Industrial Robots: Their Development Process, Current Status of Their Popularization, and Their Future Status," Japan External Trade Organization, Tokyo, Japan, 1982.

CLEAN ROOM APPLICATIONS

DENNIS HARMS
Intelledex
Corvallis, Oregon

Intelligent assembly robots are finding their way into the cleanest manufacturing environments yet achieved by the semiconductor industry. In fact, these specially designed robots can reduce particulate contamination by orders of magnitude over that existing a few years ago. In addition to providing a degree of cleanliness not possible using human workers, the robots improve process control, increase process flexibility, and result in a more cost-effective product.

Some of the newer intelligent robots are targeted specifically at the semiconductor industry. Communications protocol, sensory ability, and powerful control languages capable of advanced decision making help ease the task of integrating these systems into current integrated circuit (IC) manufacturing systems (see COMPONENTS INSERTION).

This article addresses the need for robots capable of performing in super clean rooms, design requirements of these robots, and existing applications where clean room robots are working for the semiconductor industry (see ELECTRONICS INDUSTRY, ROBOTS IN).

THE NEED FOR AUTOMATION

Clean room manufacturing is opening its doors to the newest breed of intelligent assembly robot. This is because robots are addressing problems the semiconductor industry needs to overcome in order to continue advancements in IC manufacturing. The problems discussed here relate to cleanliness, process control, and process flexibility.

Cleanliness

The demand for cleaner environments for producing ICs is increasing as the sophistication of techniques for manufacturing increases. The semiconductor industry desires to move beyond class 100 clean rooms, and definitions for class 10 and even class 1 environments are currently being standardized. The de facto class 10 standard stipulates that there be no more than 10 particles of 0.5 μm or larger per cubic foot, whereas the de facto class 1 standard allows no more than 1 particle of 0.3 $\mu m/ft^3$.

Although these definitions are emerging, the ability of current technology to actually measure a class 1 standard is questionable. In fact, a new technology questions the validity of air or fluid *volume* as an appropriate measure of cleanliness for microelectronic manufacturing. Instead, it proposes that the measurement of particles per area of water *surface* is a more accurate indicator of the percentage of acceptable products that a particular manufacturing environment is capable of producing. This is because the number of particles on the water surface would be a better indicator of the number of particles likely to settle on the surface of a wafer. Instruments that make it possible to accomplish this measurement routinely in the work environment are now available.

The need for increasingly cleaner manufacturing rooms is due in large part to the increasing geometries of wafers and the decreasing geometries of the ICs produced on those wafers. The industry is moving beyond very-large-scale integration and acquiring ultra-large-scale integration capabilities. Unless there is a corresponding increase in the cleanliness of the manufacturing environment, the decreased geometry of a chip will cause more failures because of particulate contamination. A random particulate of 0.1 μm size will cause a failure in a chip of 1-μm geometries, but it might not affect a chip of 4-μm geometries.

People are the major source of contamination in a clean room today. Even a person fully outfitted in the cumbersome, bulky clean room outfit generates the majority of particulates in the clean room. This problem has resulted in the need for and the emergence of sophisticated robots capable of working in super clean rooms. These computer-based robots are capable of a high degree of process control while handling material without generating particulates.

Process Control

The ability to accurately control all aspects of the manufacturing process provides the manufacturer with an accurate estimate of yield. This results in the correct number of products being produced by the desired time with the least amount of waste. Since the value of an 8-in. (200 mm) processed wafer can be as high as $10,000, the competitive nature of the chip market does not allow a manufacturer an excess of unacceptable products because of the inconsistencies in the manufacturing process. For example, the process may require a cassette to be dipped for 30 s the first time an acid bath is used and 33 s the second time the bath is used because of chemical aging. An error of a few seconds in the acid bath can result in a cassette full of wafers that are unacceptable. The use of a robot in the process allows the manufacturer to sense the level of acidity in a tank and adjust the time accordingly. An example of a robot in a clean room is shown in Figure 1.

The robot also provides a more accurate means of holding onto the large wafers. An 8-in. wafer grasped with a vacuum pick is more securely held by a robot with an unwavering path between source and destination. This helps reduce the amount of breakage due to the inherent unsteadiness of the human arm.

The robot is also capable of reading the bar code on a cassette and verifying that the proper batch of cassettes has been delivered. If the recipe specifies that a certain number of cassettes of type A are to be processed, the robot will ensure the materials are correct before starting the process.

Better process control also results in greater throughput and a higher yield. A system that requires no coffee breaks

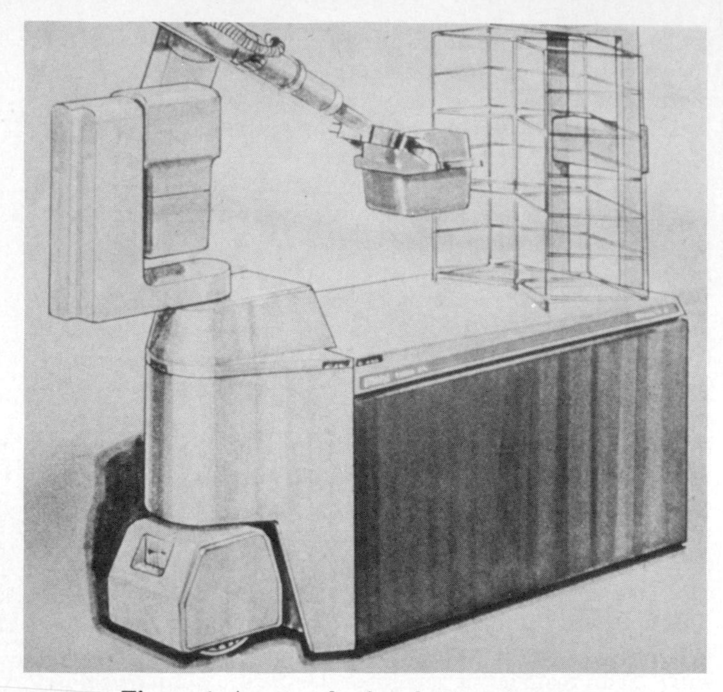

Figure 1. An example of a robot in a clean room.

and can work an extra shift at no extra expense results in higher productivity and increased cost-effectiveness.

A robotic system also allows a quicker, more accurate method of tracking inventory. For example, information on the status of a particular run of wafers is immediately available through a terminal or personal computer attached to the system.

Process Flexibility

Computer-based robots perform tasks accurately and consistently. If a flexible manufacturing process requires slightly different processing for different runs, a robot has the ability to remember accurately which process is currently being used. If the current process calls for the cassette to be dipped for exactly 27.5 s in solution A, instead of 32.5 s, as was the case in the previous process, the robot will not make a mistake due to fatigue or a misunderstanding.

A computerized robotic system accessed through a terminal gives an operator the flexibility to program quickly a desired change in the process on-line. If a priority lot has been identified due to market conditions, then the operator can simply input the appropriate instructions and the robot will make the necessary adjustments in the processing queues. Similarly, if mixing lots on a single run is desired, with variable processes required for different lots, a robot has no difficulty keeping track of which lot is undergoing a specific process at any given time. An example is intermixing runs of CMOS and NMOS memory chips on the same line.

The previous arguments all support the need for automation in the clean room. Greater cleanliness, process control, and process flexibility are most easily accomplished by automating the workcell with robotic and other automation equipment designed to facilitate these tasks.

DESIGN OF CLEAN ROOM ROBOTS

Clean room robots have specific design requirements. They include super cleanliness, special communication abilities, a reliable product that is easy to repair, programming languages capable of advanced decision making, and sensory capabilities that provide intelligent feedback about the ongoing process.

Cleanliness

There are several critical areas in the design of a clean room robot that significantly affect cleanliness. The major sources of contaminants from a robot handling wafers and cassettes are the joints nearest the end effector or manipulator. Rotary joints tend to produce less contaminants than sliding joints because of the greater effectiveness of seals used to entrap particles. These seals can be conventional O-rings or lipseals, or labyrinth seals with an accompanying lubricant. An example of an ultra-clean labyrinth seal is a ferro–fluid seal; this is a combination of a magnetic washer that fits around the motor shaft, and a ferro–fluid held in place magnetically, occupying the space between the shaft and the washer. This seal serves a dual purpose: lubrication and entrapment of any possible particulates inside the arm.

A source of molecular contamination is material sublimation onto and from the devices actually gripping the wafers or cassettes. Certain types of lubricants as well as construction materials sublimate at an accelerated rate under elevated temperatures. A lubricant with a low vapor pressure tends to sublimate the least in any environment and should therefore be selected for use. The specific environment determines which combination of plastics and metals is most suitable for construction materials. The metal may be coated with an epoxy or plated with nickel to prevent sublimation.

The design of the gripper mechanism itself and the process used in handling wafers and cassettes affect particulate generation. Wafer grippers designed by the robot manufacturer Intelledex contain cables within tubes that connect to the foot, which actually touches the wafer. These cables wind and unwind to cause the gripper to grasp a wafer. Individual O-ring seals are employed to ensure that particle generation does not occur due to the action of the cables. A metal bellows surrounds the tube and extends the foot radially to grasp a wafer. It is glued into place with epoxy resin, thereby sealing particulates that may be generated from the tube movement.

A cassette gripper, also designed by Intelledex, uses ball-bearing slides to drive the jaws of the gripper laterally. The slides are enclosed on three sides to prevent any particles generated from being blown into the wafer area by the laminar flow. Rotary movement is accomplished with a jaw rotate shaft protruding from a housing through a lubricated O-ring, which prevents particle escapement. Coil cord cabling enters through rubber grommets, which serve as particle seals, and all electrical wires are contained within the arm and end effector to eliminate particle generation.

The manufacturing process can reduce possible contamination if no part of the robot or end effector is allowed to pass directly over the wafers and the wafers are kept in a vertical orientation as much as possible. The grip used on the cassette or wafer must be sure and accurate because two surfaces accidentally brushing together can generate particles in the air.

There are a couple of design changes in the wafers and cassettes that could facilitate the handling of each by robots. One particular cassette is designed with insufficient clearance between the end wall and the last wafer. This makes it difficult for a gripper to pick up that wafer. This problem could be eliminated by cutting the end wall as low as possible, moving the slot pattern towards the bar end, or eliminating the ribs on the end walls of the Teflon cassette.

Wafers could be designed with a slot rather than the typical flat area. The slot could be used for location information more readily than the flat and could also help reduce the tendency of the wafer to roll toward the flat when the cassette is being moved.

The motion of the robot itself can produce particulates during the pickup and placement of parts. A robot capable of very smooth motion does not vibrate the wafers within the cassette as it is moving them. Wafers vibrating within the cassette produce particulates by scraping against the cassette walls. Smooth motion also ensures a higher degree of accuracy when placing the wafer into or removing it out of the cassette or fixture used in processing. An example of a robot gripping a cassette is shown in Figure 2.

Communications

A cost-effective clean room robot must be able to integrate easily with other clean room equipment in order to achieve a successful application. The robot must communicate down to tanks, rinser/dryers, and other clean room automation, and up to the host computer or a terminal used to communicate to the system.

The networking required is greatly simplified through use of a communications standard. The standard used by the semiconductor industry is known as SECS-II and should be incorporated in a robot designed for clean room applications.

Ease of Repair and Reliability

A clean room does not benefit from a robotic system that suffers from chronic problems. Repair of any system generates particles from disassembly of the parts as well as the extra traffic and equipment necessary to remove them. A system that is designed to be modular can facilitate the process necessary to bring the system back on-line. Once it has been determined which part of the system has failed, a replacement can be installed and the defective part removed and repaired at the robot manufacturer's facility rather than the clean room.

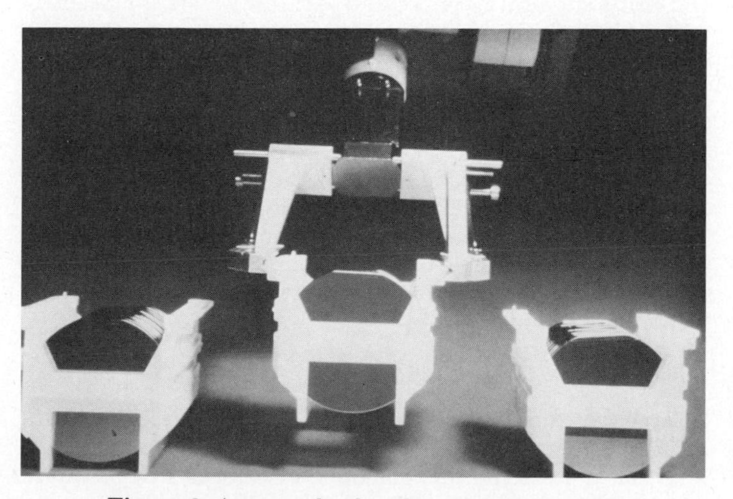

Figure 2. An example of a robot gripping a cassette.

If the arm of the robot needs repair, the process of replacing it with a temporary arm can be tedious. Since the hardware involved cannot be exactly duplicated in the production process, the dimensions from one joint to another are not exact. This can require many hours of reteaching the points of a program to the replacement robot in order to achieve the accuracy needed to resume production. Intelledex approached this problem by providing a built-in calibration routine. Calibration enables accuracy by eliminating the variances caused by mechanical inaccuracies in the manufacturing process. The calibration routine requires the robot to touch a few accurately placed points from several different approach angles. The error that the robot makes is measured and then used in the calculation of calibration constants. These constants are used by the software in performing the calculations that accurately position the robot, thus allowing the robot to compensate for the mechanical variances.

Another time-saving software feature is an orientation routine. If replacement of the robot or the worktable is necessary due to repair, then the exact orientation of the robot to the worktable has been altered. A robot with no ability to orient itself will again require hours of point teaching to resume production. Intelledex developed a software orientation routine that eliminates this need by teaching the robot three points that define a plane and its origin in world coordinates. These points in effect precisely locate the workpiece with respect to the worktable. They are again stored in memory and referenced as part of the process of movement to a particular point on the workpiece.

Another important design consideration is the approach taken in case of a power failure. There are three possible problems that could result from an unexpected loss of power: the arm could crash, the end effector could release the workpiece, or the robot could be unable to resume the process where it left off.

Robots that use servomotors need a mechanism to freeze their position in the event of a power failure. This generally consists of a braking mechanism that automatically stops the arm when loss of power occurs. Robots that use stepper motors freeze when power is cut off because electricity is necessary to allow any movement.

The end effector must hold onto the cassette or wafer if it is in the middle of a move when the power fails. A spring mechanism on the gripper that requires power to make it release the part can be used.

A good system is able to maintain its position when the power goes off and continue where it left off when the power resumes. This "fail-safe" method requires no operator intervention to reset the equipment and material manually back to the start of the process and reduces particulate contamination caused by such an intervention.

Programming, Control

Creation and maintenance of programs to perform a particular production task are greatly simplified with a high-level programming language. A high-level language provides the sophisticated controls and structures necessary to implement desired movement under varying conditions. It has the ability to branch and make decisions based on information supplied by the sensors. An example of this ability is the following code, written in ROBOT BASIC, a high-level language developed at Intelledex. It instructs a robot to pick up a cassette full of wafers and check a sensor to determine if it was successful.

```
100 MOVE Q
200 TCLOSE
300 IF SINP(5) = GOT THEN GOTO 600
400 GOSUB 1000
500 GOTO 100
600 SMOOTHNESS .1
```

Line 100 moves the end effector over the location of a cassette, and line 200 closes the tool (gripper) on it. Line 300 checks the serial input port to which the Hall switch sensor is interfaced in order to verify cassette presence. If the cassette is present in the gripper, the program will branch to line 600. Line 400 branches the program to a subroutine where a correction can be made to the location coordinates. The program returns to line 500 after the subroutine and is sent to line 100 to attempt the pickup again. Line 600 sets the smoothness of a straight-line move for the delicate handling of the cassette.

Although it may be possible to accomplish the task with a machine language, much more time is necessary initially to write and then effect any changes the program may need in the future. This is primarily due to the lack of common-sense labels that are available in a high-level language and the sheer bulk of a machine language, which generally has five commands to accomplish what a high-level language can do in a single command.

The orientation and calibration routines mentioned earlier can also ease the task of programming, particularly for applications that require several robots to perform the same task. Without these two routines, each robot must be taught all of the points after they have been positioned on the line. With orientation and calibration, the same program can be used by all the robots, with only a short time required for each robot to calibrate and learn the proper orientation.

A simulator that allows programming off-line and testing and debugging the code on a personal computer can save a considerable amount of time and reduce the need to tie up the production line for extensive on-line programming when a new program is being initiated.

Sensors and Intelligence

The sensory ability of a robotic system can determine the success and feasibility of a particular application. Sensors vary from those measuring gripper force and position to integrated vision systems. Robotic systems with sensory intelligence necessitate a control language sophisticated enough to allow for flexibility and decision making. The language must analyze the sensory information and send instructions to react accordingly. Again, this requires a high-level language with control structures capable of making decisions based on data supplied by sensory equipment.

The end effector must be highly accurate in order to grasp a wafer or cassette without producing particulates. Generally, this accuracy is obtained through the use of a potentiometer, which provides feedback to the motor controlling gripper movement. A force sensor can also be involved in this process to detect a preprogrammed level of pressure exerted on a part.

Vision can be used to "see" whether an object is present or to verify proper positioning of an end effector or part. Intelledex developed a high-level language known as VISION BASIC, which allows programs to include special vision commands with the robot commands. An example of code that searches for a cassette on a tray is as follows:

```
100 POSITION 1
110 MWAIT
120 DELAY 2
130 VFIND "CASSETTE", Q, M
140 IF M(0) = 1 GOTO 130
150 GOSUB 500
```

Line 100 instructs the robot arm to move to the best camera viewing position, and line 110 stops the program until the arm has had a chance to get into position. Line 120 delays the process to allow movement vibrations to dissipate. Line 130 instructs the vision system to locate the cassette at a visual depth Q and store its exact location in M. Line 140 tests to see if the cassette was located, and line 150 sends the program to a subroutine starting at line 500.

These sensory capabilities help incorporate "intelligence" into the robotic system or, in other words, the ability to make decisions based on information received from sensors.

CURRENT APPLICATIONS

The following applications illustrate how robots are currently being used in clean rooms.

Loading Plasma Etch Station

A robot controls a plasma etch workstation. The robot has a dual-action end effector that contains two sets of jaws to hold two wafers opposite each other.

During a typical cycle, a cassette of raw wafers is presented to the robot, which grasps a single wafer and turns to a hexagonal column containing four processed wafers on each of the six sides. The end effector removes a processed wafer and immediately rotates 180 deg to place an unprocessed wafer in the now-empty slot. The processed wafer is placed in a cassette slot, and the cycle continues until the column is filled with unprocessed wafers ready to be etched.

The use of the dual-action end effector improves throughput because the robot is able to exchange immediately a raw wafer for a processed one each time it is at the cassette or the column.

Automated Wet Chemical Processing

A large wet processing system at Flurocarbon controlled by a robot consists of ten separate baths, two rinser/dryers, and a special process control computer (PCC). The PCC is interfaced to the robot via an RS-232-C serial port and communicates to the tanks through an auxiliary I/O module containing optically isolated switches.

The robot is initially given the recipe, which specifies into which baths the cassettes are to be dipped and for how long. It monitors the solution in the tanks and can issue a command to drain and refill a tank when the solution has aged beyond acceptable limits. It then signals the PCC to start the process, which may require filling a particular bath, bringing others up to temperature, or turning on recirculating pumps. The robot then sends an activation signal, which starts the beginning of a cycle.

The PCC monitors the processes occurring in the baths and rinser/dryers and alerts the robot to any problems. If a problem occurs, the robot issues a stop command, shutting down the process, and takes appropriate action.

The robot also controls the amount of time the wafers are in the baths and signals the PCC at the end of a normal processing. The robot then removes the cassette and places it into the rinser/dryer. Following the drying, the cassette is removed and the same process continues until another recipe is programmed.

At National Semiconductor, a similar system uses a robot that is mounted on an overhead track. This system comprises a chemical process cleaner, a megasonic cleaner, and two rinser/dryers. Since this system is smaller than the previous example, a process control computer is not necessary to handle the I/O. Instead, the robot monitors and controls all aspects of the process. Communication between the robot and the tanks and rinser/dryers, such as signals to open and close doors or turn on the process chambers, occurs directly through the auxiliary I/O module.

Wafer Metallization System

CVC uses a system with a robot to load disks onto a rotostrate, load and unload the rotostrate into the metalization chamber, and place the processed disks back into the cassette.

The end effector is dual-function, allowing it to load the disks onto the rotostrate and pick up the rotostrate to place it into the tank.

In a typical cycle, the robot removes the processed rotostrate and sets it down on a worktable. It then places another rotostrate into the tank to process while it unloads processed disks into a cassette and loads raw disks onto the now-empty rotostrate.

The wafers are handled on the back side with a vacuum tip. When it releases the wafer, air is exhausted back into a pneumatic enclosure. The jaws that handle the rotostrate are made of a special black nylon to prevent particle emission.

A fiber-optic sensor in the end effector detects the presence of a wafer in the rotostrate, and a vacuum-level sensor detects if the wafer has been picked up. A limit switch is used to detect rotostrate presence.

Hard Disk Manufacturer

A manufacturer of hard disk media uses a table-mounted robot to handle the delicate disks in a testing area. The robot controls the workplace, which contains three testers, communicating over RS-232-C serial ports. It places disks on the testers, polls the results, and sorts the disks into the appropriate bins. An example of a robot holding a disk is shown in Figure 3.

The robot is capable of servicing three testers at one time, providing maximum efficiency. Communications are routed to the proper tester through a special "black box" switching device.

The robot initially uses an optical sensor to verify the presence of a disk in a cassette, then picks it up and places it on an available tester, which checks the disk's magnetic quality. Accuracy in this task is extremely important because a single microscopic scratch can destroy one of the costly disks.

The robot queries the testers to determine which is available and analyzes the results of the completed tests. The tested disks are sorted into one of several bins, depending on their quality according to the tests.

Disk-Drive Assembly Station

In this application, two robots and a vision system are used to assemble Winchester hard disk drives. An operator places trays with the disk drives in front of the workstation. A vision system locates the position of the spindle and screw holes and passes the information to a robot. One robot places the disks and spacers, followed by a tip cap, and then drives the screws into the head assembly. The other robot installs the read/write heads between the disks.

The robots receive all communications from parts feeders and other clean room automation through auxiliary I/O. Sensors for the system are chosen to maintain the stringent cleanliness required for a class 100 clean room environment.

Vision is used to provide spindle location information without resorting to a contact method that might generate particles. Ventilation fans are used to clear away possible dirty air from parts-feeding mechanisms.

The screwdriver in the gripper has a vacuum that sucks away dirt as the screw is driven or picked up. The end effector that picks up the disks contains Delrin spools on the ends of wire shafts that contact the sides of the disks, which emit virtually no particulates. The wires are compliant so that when

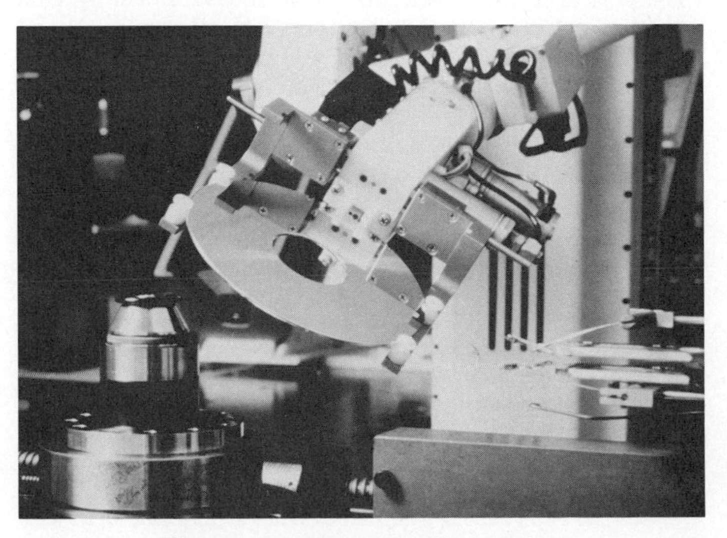

Figure 3. A robot arm holding a disk.

the disk is lowered over the spindle with less than 0.0015 in. (6×10^{-3} cm) clearance, it will self-center without excessive rubbing. Tests conducted on particle count, when the robot was in the act of inserting a disk, revealed a count of zero particles.

Proximity sensors (which use a magnetic field) are used to confirm parts presence in the feeders and in the end effector used to pick up the disks. A vacuum level sensor is used to confirm the presence of the screw in the gripper. These sensors are all noncontact, non-particle-emitting devices.

Mobile Robot Vehicle for Clean Room Fab

This system, developed by FMS, consists of three basic components: a robot mounted on a mobile robot vehicle for transport, an intelligent WIP station for material buffering and inventory monitoring, and a computer system to provide information gathering and system control. At each station, an operator interface provides communication to the central control computer and, if available, the MIS system host computer.

Material tracking and direction can be performed through the MIS system or from a supervisory terminal. A menu-driven inquiry system provides material lot status, transfer transaction listing, and equipment status. Other system functions provide lot history files, the date and time of transfer, and the position of robotic vehicles within the fab area.

The mobile robot vehicle (Fig. 1) provides material transport and transfer functions for workstations or machines. Both the vehicle path and the robot's transfer movements are taught with a teach pendant. Vehicle control is provided by the control computer through an infrared communication link. The vehicle's narrow dimensions minimize air turbulence as it travels through the clean room aisles.

Navigational control is provided through three guidance systems: a gyroscope for direction, an optical encoder for distance, and sonar for obstacle detection and collision avoidance.

The vehicle is capable of transferring materials between stations, and a simple change of end effector permits handling any type of material container.

Docking at a workstation is accomplished with a module consisting of a mirror and an optical measurement system. The exact vehicle position is determined by triangulation, and the vehicle recalibrates its position at each docking.

Implementation into existing clean rooms requires connection of operator interfaces to the central computer, mounting of docking modules and mirrors on the workstation, and driving in the vehicle. If WIP stations are required, they are easily incorporated adjacent to workstations or at the aisle's ends.

CLEANER AUTOMATION FOR PRECISE MANUFACTURING

The need for cleaner standards in today's IC manufacturing environment has been presented. Ultra-large-scale integration demands the presence of fewer contaminants that potentially can destroy valuable products. Manufacturers need more precise control over the process and the ability to perceive which changes will produce a superior product.

Sophisticated robots that can provide a manufacturer with the necessary automation to remain competitive now exist. Robots designed to work in super clean rooms permit a much higher standard of cleanliness than is possible with human beings. Advanced sensory abilities provide the robot with the information necessary to make decisions affecting the quality of the product, and high-level languages allow the manufacturer to respond to this information. Accurate records that permit the manufacturer immediate access to information concerning inventory and current production are readily available.

Many applications that make use of the latest robotic technology to aid in the manufacture of ICs are currently on-line. These facilities operate under stringent standards of cleanliness and successfully produce large quantities of superior products. Meanwhile, research continues to produce even cleaner automation capable of working under the stricter standards that will undoubtedly be necessary for future manufacturing.

COMMUNICATIONS

William Stallings
Comp-Comm Consulting
London, United Kingdom

A key aspect of the use of robots is communications. It is a rare circumstance in which a robot or robot-like device remains operational for an extended period of time without some form of communication with other devices.

The term communications has two aspects:

- The reliable exchange of bits of data between two devices.
- The integration of two or more devices so that they can successfully cooperate in the exchange of information as well as control, alarm, and status signals.

The first aspect is generally referred to as data communications, whereas the second is generally referred to as communications architecture (1) (see also CONTROLLER ARCHITECTURE). Both aspects are examined in this article.

DATA COMMUNICATIONS

Analogue and Digital Transmission

In any sort of robotics environment, it is likely that both analogue and digital data will be used. *Analogue data* take on continuous values on some interval. For example, voice and video are continuously varying patterns of intensity. Most data collected by sensors, such as temperature and pressure, are continuous-valued. *Digital data* take on discrete values; examples are text and integers.

In a communications system, data are propagated from one point to another by means of electric signals. An *analogue signal* is a continuously varying electromagnetic wave that may be transmitted over a variety of media, depending on frequency; examples are wire media, such as twisted-pair and coaxial cable, fiber optic cable, and atmosphere or space propagation. A *digital signal* is a sequence of voltage pulses that may be transmitted over a wire medium; for example, a constant positive voltage level may represent binary 1 and a constant negative voltage level may represent binary 0. The principal advantages of digital signaling are that it is generally cheaper than analogue signaling and is less susceptible to noise interference. The principal disadvantage is that digital signals suffer more from attenuation than do analogue signals.

Data Encoding Techniques

Both analogue and digital data can be represented, and hence propagated, by either analogue or digital signals. In a mixed environment, it often makes sense to convert analogue signals to digital, or vice versa, so that transmission facilities can be effectively shared. Digital data can be represented by analogue signals by use of a *modem* (modulator/demodulator). The modem converts a series of binary (two-valued) voltage pulses into an analogue signal by modulating a *carrier frequency*. The resulting signal occupies a certain spectrum of frequency centered about the carrier and may be propagated across a medium suitable for that carrier. The most common modems represent digital data in the voice spectrum and hence allow those data to be propagated over ordinary voice-grade telephone lines. At the other end of the line, the modem demodulates the signal to recover the original data. Various modulation techniques are discussed below.

In an operation very similar to that performed by a modem, analogue data can be represented by digital signals. The device that performs this function for voice data is a *codec* (coder/decoder). In essence, the codec takes an analogue signal that directly represents the voice data and approximates that signal by a bit stream. At the other end of a line, the bit stream is used to reconstruct the analogue data.

Digital Data, Analogue Signals

The basis for analogue signaling is a continuous constant-frequency signal known as the *carrier signal*. Digital data are encoded by modulating one of the three characteristics of the carrier: amplitude, frequency, or phase, or some combination of these. Figure 1 illustrates the three basic forms of modulation of analogue signals for digital data:

- Amplitude-shift keying (ASK).
- Frequency-shift keying (FSK).
- Phase-shift keying (PSK).

In all these cases, the resulting signal contains a range of frequencies on both sides of the carrier frequency. That range is referred to as the *bandwidth* of the signal.

In ASK, the two binary values are represented by two different amplitudes of the carrier frequency. In some cases, one of the amplitudes is zero; that is, one binary digit is represented by the presence, at constant amplitude, of the carrier, and the other by the absence of the carrier. ASK is susceptible to sudden gain changes and is a rather inefficient modulation technique. On voice-grade lines, it is typically used only up to 1200 bits/s.

In FSK, the two binary values are represented by two different frequencies near the carrier frequency. This scheme is less susceptible to error than ASK. On voice-grade lines, it is typically used up to 1200 bits/s. It is also commonly used for high-frequency (3–30 MHz) radio transmission. It can also be used at even higher frequencies on local networks that use coaxial cable.

In PSK, the phase of the carrier signal is shifted to represent data. Figure 1 is an example of a two-phase system. In this system, a 0 is represented by sending a signal burst of the same phase as the previous signal burst sent. A 1 is represented by sending a signal burst of phase opposite to the previous one. PSK can use more than two phase shifts. A four-phase system would encode two bits with each signal burst. The PSK technique is more noise resistant and efficient than FSK; on a voice-grade line, rates up to 9600 bits/s are achieved.

Finally, the techniques discussed above may be combined. A common combination is PSK and ASK, where some or all of the phase shifts may occur at one of two amplitudes.

Analogue Data, Digital Signals

The most common example of the use of digital signals to encode analogue data is *pulse code modulation* (PCM). PCM is based on the sampling theorem, which states:

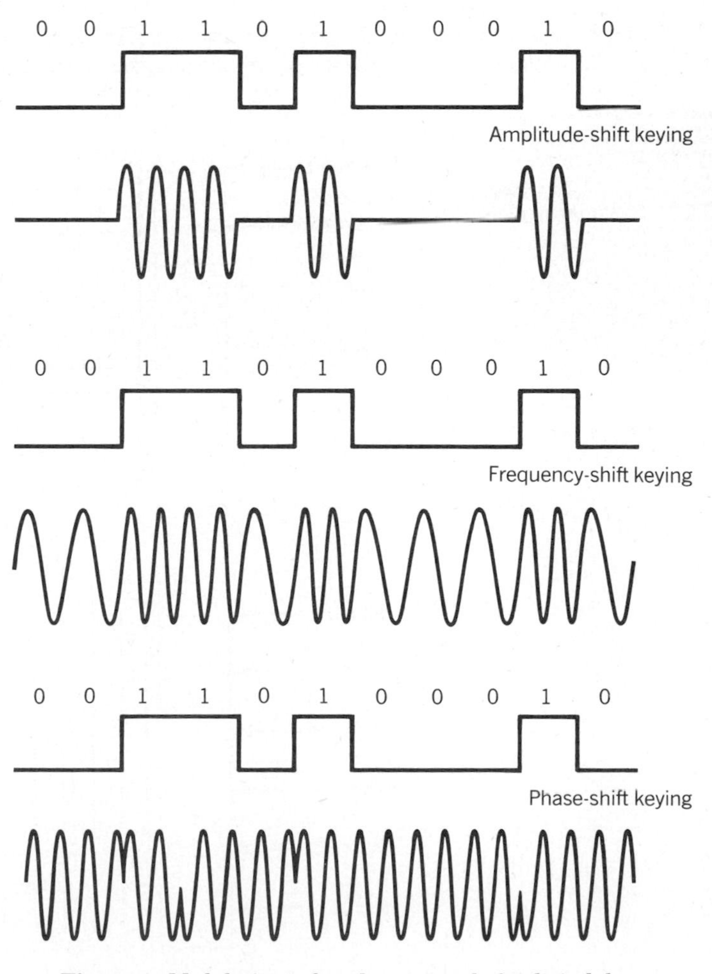

Figure 1. Modulations of analogue signals for digital data.

> If a signal $f(t)$ is sampled at regular intervals of time and at a rate higher than twice the highest significant signal frequency, then the samples contain all the information of the original signal. The function $f(t)$ may be reconstructed from these samples by the use of a low-pass filter. (2)

If voice data are limited to frequencies below 4000 Hz, a conservative procedure for intelligibility, then 8000 samples/s is sufficient to completely characterize the voice signal. Note, however, that these are analogue samples. To convert to digital, each of these analogue samples must be assigned a binary code. Figure 2 shows an example in which each sample is approximated by being "quantized" into one of 16 different levels. Each sample can then be represented by four bits. Of course, it is now impossible to recover the original signal exactly. By using a seven-bit sample, which allows 128 quantizing levels, the quality of the recovered voice signal is comparable to that achieved via analogue transmission. Note that this implies that a data rate of 8000 samples/s × 7 bits/sample = 56 kbits/s is needed for a single voice signal.

PCM can, of course, be used for other than voice signals. For example, a color TV signal has a useful bandwidth of 4.6 MHz, and reasonable quality can be achieved with ten-bit samples, for a data rate of 92 Mbits/s.

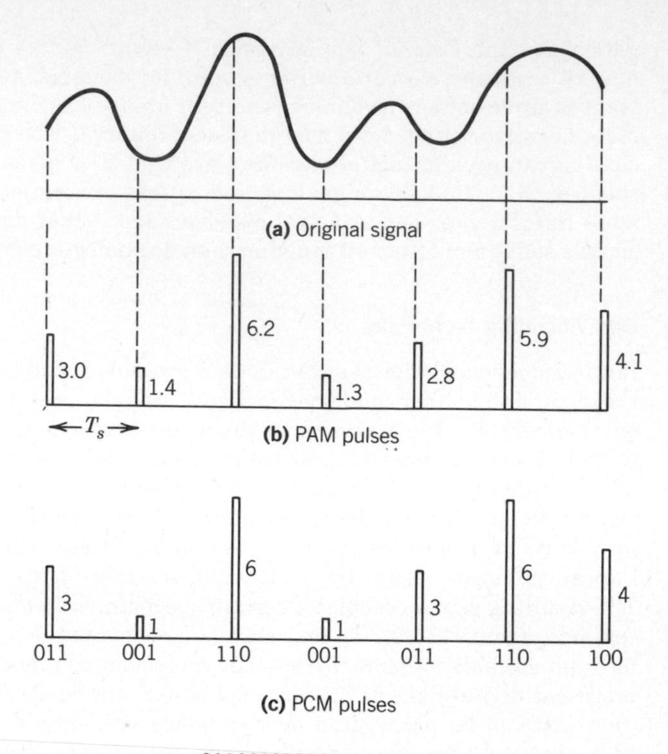

Figure 2. Pulse code modulation.

Multiplexing

In both local and long-haul communications, it is almost always the case that the capacity of the transmission medium

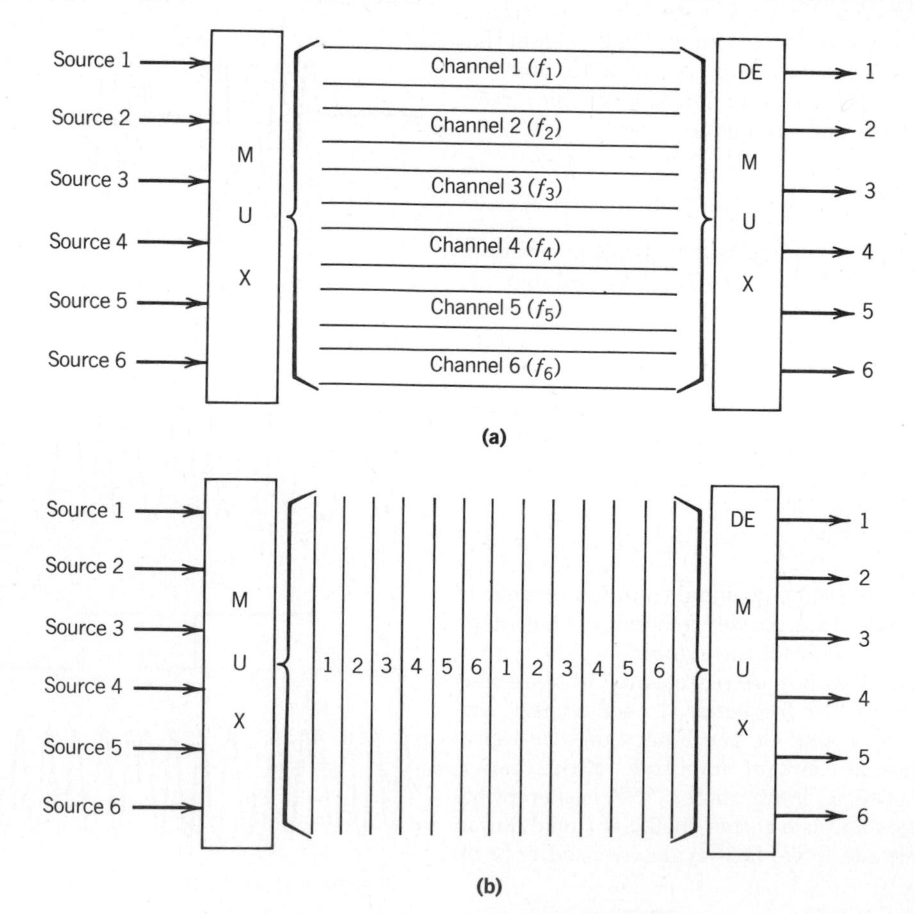

Figure 3. Multiplexing. **(a)** FDM. **(b)** TDM.

exceeds that required for the transmission of a single signal. To make cost-effective use of the transmission system, it is desirable to use the medium efficiently by having it carry multiple signals simultaneously. This is referred to as *multiplexing,* and two techniques are in common use: frequency-division multiplexing (FDM) and time-division multiplexing (TDM).

FDM takes advantage of the fact that the useful bandwidth of the medium exceeds the required bandwidth of a given signal. A number of signals can be carried simultaneously if each signal is modulated onto a different carrier frequency and the carrier frequencies are sufficiently separated that the bandwidths of the signals do not overlap. A general case of FDM is shown in Figure 3**a.** Six signal sources are fed into a multiplexer (MUX), which modulates each signal onto a different frequency ($f_1, \cdots, f_6$). Each signal requires a certain bandwidth centered around its carrier frequency, referred to as a *channel.* To prevent interference, the channels are separated by guardbands, which are unused portions of the spectrum.

TDM takes advantage of the fact that the achievable bit rate (sometimes, unfortunately, called bandwidth) of the medium exceeds the required data rate of a digital signal. Multiple digital signals can be carried on a single transmission path by interleaving portions of each signal in time. The interleaving can be at the bit level or in blocks of bytes or larger quantities. For example, the multiplexer in Figure 3**b** has six inputs that might each be, say, 9.6 kbits/s. A single line with a capacity of 57.6 kbits/s could accommodate all six sources. Analogously to FDM, the sequence of time slots dedicated to a particular source is called a *channel.* One cycle of time slots (one per source) is called a *frame.*

The TDM scheme depicted in Figure 3**b** is also known as *synchronous TDM,* referring to the fact that time slots are preassigned and fixed. Hence, the timing of transmission from the various sources is synchronized. In contrast, asynchronous TDM allows time on the medium to be allocated dynamically.

Asynchronous and Synchronous Transmission

A fundamental requirement of digital data communication (analogue or digital signal) is that the receiver know the starting time and duration of each bit that it receives.

The earliest and simplest scheme for meeting this requirement is asynchronous transmission. In this scheme, data are transmitted one character (of 5–8 bits) at a time. Each character is preceded by a start code and followed by a stop code (Figure 4**a**). The *start code* has the encoding for 0 and a duration of one bit time; in other words, the start code is one bit with a value of 0. The *stop code* has a value of 1, and a minimum duration, depending on the system, of from one to two bit times. When there are no data to send, the transmitter sends a continuous stop code. The receiver identifies the beginning of a new character by the transition from 1 to 0. The receiver must have a fairly accurate idea of the duration of each bit in order to recover all the bits of the character. However, a small amount of drift (eg, 1% per bit) will not matter since the receiver resynchronizes with each stop code. This means of communication is simple and cheap, but requires an overhead of two to three bits per character. This technique is referred to as *asynchronous* because characters are sent independently from each other. Thus, characters may be sent at a nonuniform rate.

A more efficient means of communication is synchronous transmission (Figure 4**b**). In this mode, blocks of characters or bits are transmitted without start and stop codes, and the exact departure or arrival time of each bit is predictable. To prevent timing drift between transmitter and receiver, their clocks must somehow be synchronized. One possibility is to provide a separate clock line between transmitter and receiver. Otherwise, the clocking information must be embedded in the data signal. With synchronous transmission, there is another level of synchronization required, to allow the receiver to determine the beginning and end of a block of data. To achieve this, each block begins with a *preamble* bit pattern and ends with *postamble* bit pattern. The data plus preamble and postamble are called a *frame.*

COMMUNICATIONS ARCHITECTURE

When computers, terminals, and/or other data processing devices exchange data, the scope of concern is broad. Consider, for example, the transfer of a file between two computers. There must be a data path between the two computers, either directly or via a communication network, but more is needed. Typical tasks to be performed are

- The source system must either activate the direct data communication path or inform the communication network of the identity of the desired destination system.

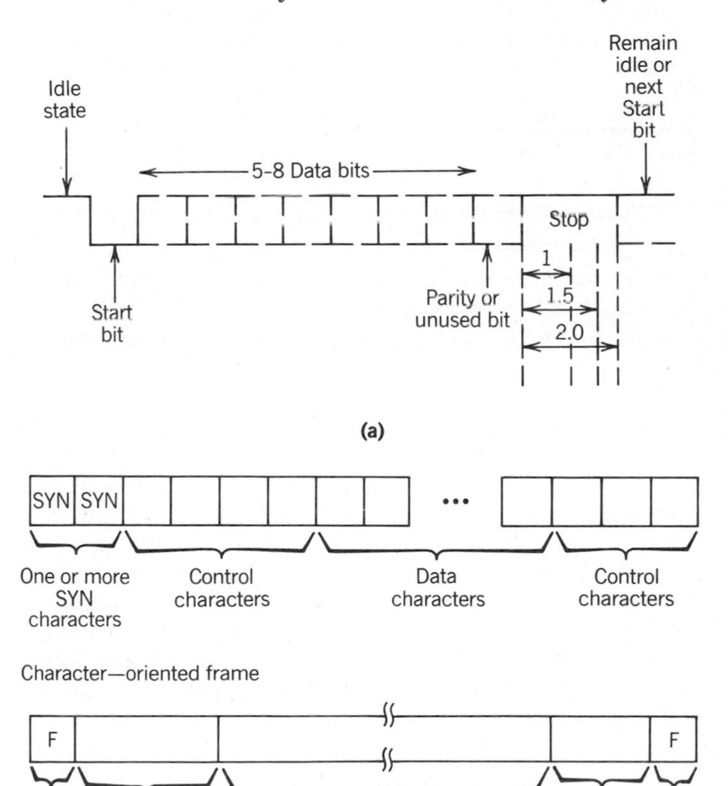

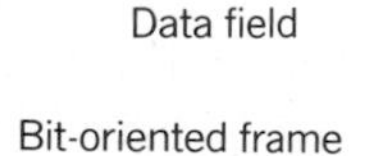

(b)

Figure 4. **(a)** Asynchronous transmission. **(b)** Synchronous transmission.

- The source system must ascertain that the destination system is prepared to receive data.
- The file transfer application on the source system must ascertain that the file management program on the destination system is prepared to accept and store the file.
- If the file formats used on the two systems are incompatible, one system or the other must perform a format translation function.

It is clear that there must be a high degree of cooperation between the two computer systems. The exchange of information between computers for the purpose of cooperative action is generally referred to as *computer communications*. Since a similar level of cooperation is required between a user at a terminal and a computer, the term is often used when some of the communicating entities are terminals or other communicating devices.

In discussing computer communications and computer networks, two concepts are paramount: protocols and computer-communications architecture.

Protocols

For two entities to communicate successfully, they must "speak the same language." What is communicated, how it is communicated, and when it is communicated must conform to some set of conventions mutually acceptable to the entities involved. The set of conventions is referred to as a *protocol,* which may be defined as a set of rules governing the exchange of data between two entities (3). The key elements of a protocol are

- *Syntax,* which includes such things as data format, coding, and signal levels.
- *Semantics,* which includes control information for coordination and error handling.
- *Timing,* which includes speed matching and sequencing.

A protocol may be either standard or nonstandard. A nonstandard protocol is one built for a specific communications situation or, at most, a particular model of computer. Thus, if K different kinds of information sources have to communicate with L types of information receivers, $K \times L$ different protocols are needed without standards and a total of $2 \times K \times L$ implementations are required. If all systems shared a common protocol, only $K + L$ implementations would be needed. The increasing use of distributed processing and the decreasing inclination of customers to remain locked into a single vendor dictate that all vendors implement protocols that conform to an agreed-upon standard.

The Need for a Standardized Architecture

When work is done that involves more than one computer, additional elements must be added to the system: the hardware and software to support the communication between or among the systems. Communications hardware is reasonably standard and generally presents few problems. However, when communication among heterogeneous (different vendors, different models of the same vendor) machines is desired, the software development effort can be a nightmare. Different vendors use different data formats and data exchange conventions. Even within one vendor's product line, different model computers may communicate in unique ways.

As the use of computer communications and computer networking proliferates, a one-at-a-time special-purpose approach to communications software development is too costly to be acceptable. The only alternative is for computer vendors to adopt and implement a common set of conventions. For this to happen, a set of international, or at least national, standards must be promulgated by appropriate organizations. Such standards have two effects:

- Vendors feel encouraged to implement the standards because, due to their wide usage, their products would be less marketable without them.
- Customers are in a position to require that the standards be implemented by any vendor wishing to propose equipment to them.

However, the task of communication in a truly cooperative way between applications on different computers is too complex to be handled as a unit. The problem must be decomposed into manageable parts. Hence, before one can develop standards, there should be a structure or *architecture* that defines the communications tasks.

This line of reasoning led the International Organization for Standardization (ISO) to establish in 1977 a subcommittee to develop such an architecture. The result was the *Open Systems Interconnection* (OSI) reference model (4), which is a framework for defining standards for linking heterogeneous computers. The OSI model provides the basis for connecting "open" systems for distributed applications processing. The term "open" denotes the ability of any two systems conforming to the reference model and the associated standards to connect.

Concepts

A widely accepted structuring technique, and the one chosen by ISO, is *layering*. The communications functions are partitioned into a vertical set of layers. Each layer performs a related subset of the functions required to communicate with another system. It relies on the next-lower layer to perform more primitive functions and to conceal the details of those functions. It provides services to the next-higher layer. Ideally, the layers should be defined so that changes in one layer do not require changes in the other layers. Thus, one problem has been decomposed into a number of more manageable subproblems. The resulting OSI reference model has seven layers, which are listed with a brief definition in Table 1.

Figure 5 illustrates the OSI model. Each system contains the seven layers. Communication is between applications in the systems, labeled AP X and AP Y in the figure. If AP X wishes to send a message to AP Y, it invokes the application layer (layer 7). Layer 7 establishes a peer relationship with layer 7 of the target machine, using a layer-7 protocol. This protocol requires services from layer 6, so the two layer-6 entites use a protocol of their own, and so on down to the physical layer, which actually passes the bits through a transmission

Table 1. The OSI Layers

1. Physical	Concerned with transmission of unstructured bit stream over physical medium; deals with the mechanical, electrical, functional, and procedural characteristics to access to physical medium.
2. Data link	Provides for the reliable transfer of information across the physical link; sends blocks of data (frames) with the necessary synchronization, error control, and flow control.
3. Network	Provides upper layers with independence from the data transmission and switching technologies used to connect systems; responsible for establishing, maintaining, and terminating connections.
4. Transport	Provides reliable, transparent transfer of data between end points; provides end-to-end error recovery and flow control.
5. Session	Provides the control structure for communication between applications; establishes, manages, and terminates connections (sessions) between cooperating applications.
6. Presentation	Provides independence to the application processes from differences in data representation (syntax).
7. Application	Provides access to the OSI environment for users and also provides distributed information services.

medium. Note that there is no direct communication between peer layers except at the physical layer. That is, above the physical layer, each protocol entity sends data *down* to the next-lower layer in order to get the data *across* to its peer entity. Even at the physical layer, the OSI model does not stipulate that two systems be directly connected. For example, a packet-switched or circuit-switched network may be used to provide the communications link. This point should become clearer below, where the network layer is discussed.

The attractiveness of the OSI approach is that it promises to solve the heterogeneous computer communications problem. Two systems, no matter how different, can communicate effectively if they have the following in common:

- They implement the same set of communications functions.
- These functions are organized into the same set of layers. Peer layers must provide the same functions, but note that it is not necessary that they provide them in the same way.
- Peer layers must share a common protocol.

To assure the above, standards are needed. Standards must define the functions and services to be provided by a layer (but not how it is to be done—that may differ from system to system). Standards must also define the protocols between peer layers (each protocol must be identical for the two peer layers). The OSI model, by defining a seven-layer architecture, provides a framework for defining these standards.

Figure 6 illustrates the OSI principles in operation. First, consider the most common way in which protocols are realized. When application X has a message to send to application Y, it transfers those data to an application entity in the application layer. A *header* is appended to the data that contain the required information for the peer-layer-7 protocol (encapsulation). The original data, plus the header, are now passed as a unit to layer 6. The presentation entity treats the whole unit as data and appends its own header (a second encapsulation). This process continues down through layer 2, which generally adds both a header and a trailer (eg, HDLC). This layer-2 unit, called a *frame,* is then passed by the physical layer onto the transmission medium. When the frame is received by the target system, the reverse process occurs. As the data ascend, each layer strips off the outermost header, acts on the protocol information contained therein, and passes the remainder up to the next layer.

Layers of the OSI Model

Physical Layer. Most digital data processing devices have limited data transmission capability. Typically, they generate

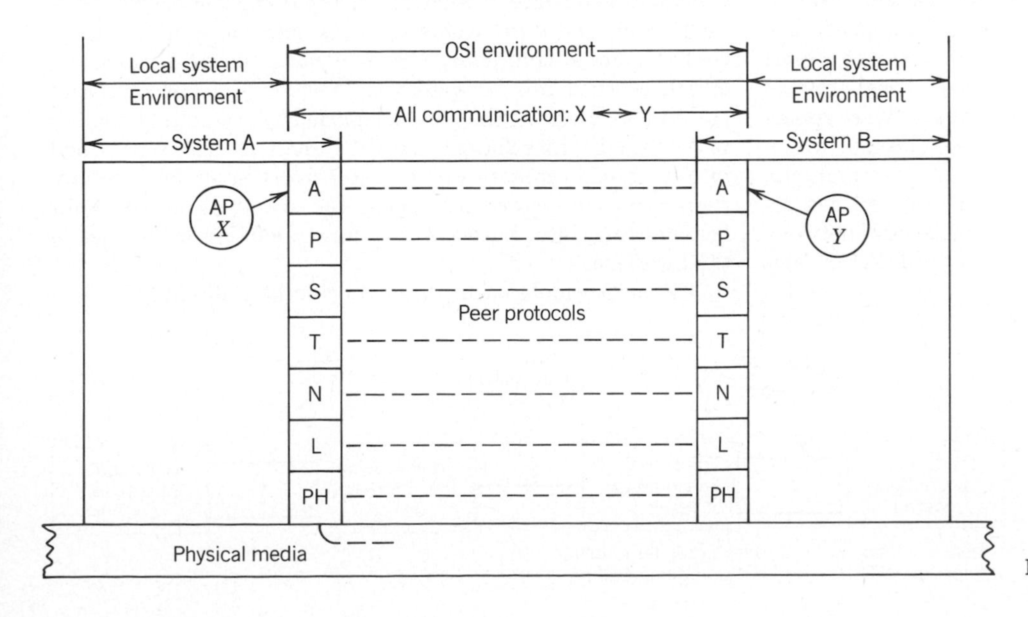

Figure 5. The OSI environment.

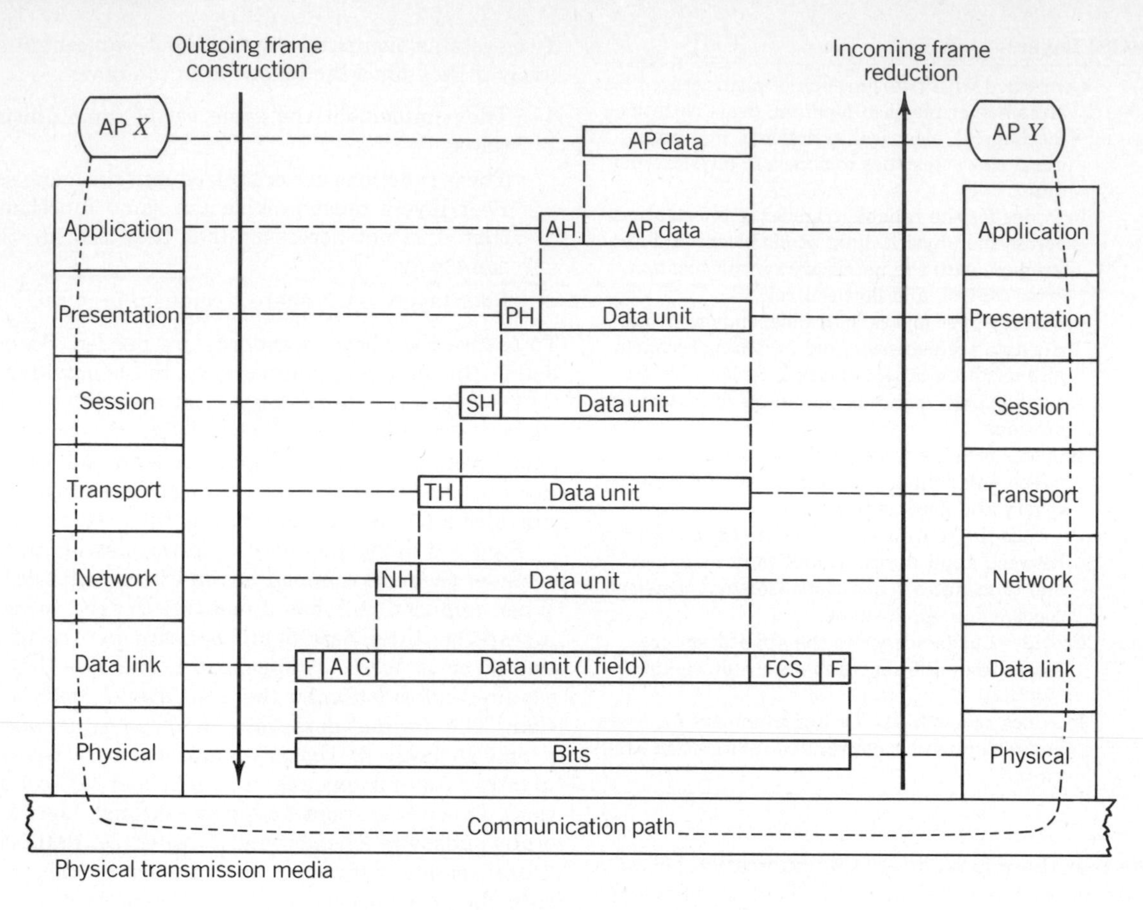

Figure 6. OSI operation.

only simple digital signals using two voltage levels to encode binary data. The distance across which they can transmit data is generally limited. Consequently, it is rare for such a device to attach directly to a long-distance transmission medium. The more common situation is depicted in Figure 7. The devices under discussion, which include terminals and computers, are generically referred to as *data terminal equipment* (DTE). DTE makes use of the transmission system through the mediation of *data circuit-terminating equipment* (DCE). An example of the latter is a modem used to connect digital devices to voice-grade lines. Most, but not all, physical protocol standards employ the model depicted in Figure 7. More generally, a physical protocol refers to the interface through which a device transmits and receives data signals. For example, in the context of local networks, a physical protocol defines the interface between an attached device and the local network transmission medium; there is no model of DTE/DCE employed.

The interface has four important characteristics (5,6): mechanical, electrical, functional, and procedural.

The *mechanical* characteristics pertain to the actual physical connection of the DTE and DCE. Typically, the signal and control leads are bundled into a cable with a terminator plug, male or female, at each end. The DTE and DCE must each present a plug of opposite gender at one end of the cable, effecting the physical connection. The *electrical* characteristics have to do with the voltage levels and timing of voltage changes. These characteristics determine the data rates and distances that can be achieved. *Functional* characteristics specify the functions that are performed by assigning meaning to the various interchange circuits. Functions can be classified into the broad categories of data, control, timing, and ground. *Procedural* characteristics specify the sequence of events for transmitting data, based on the functional characteristics of the interface.

The most widely used physical protocol is RS-232-C.

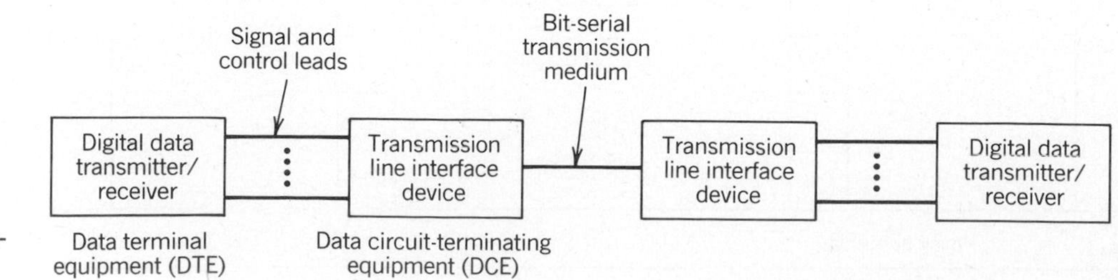

Figure 7. Generic interface to transmission medium.

Data Link Layer. A physical interface or protocol provides only a raw bit stream service, which is subject to error. A data link protocol is used to manage the communication between two connected devices and to transform an unreliable transmission path into a reliable one. The key elements of a data link protocol are (7,8):

Frame Synchronization. Data are sent in blocks called frames. The beginning and end of each frame must be clearly identifiable.

Use of a Variety of Line Configurations. Devices may be connected by a point-to-point link or a multipoint link. Examples of the latter are (1) a multidrop line that connects multiple terminals to a single computer port and (2) a local network.

Flow Control. The sending station must not send frames at a rate faster than the receiving station can absorb them.

Error Control. The bit errors introduced by the transmission system must be corrected.

Addressing. On a multipoint line, the identity of the two stations involved in a transmission must be known.

Control and Data on Same Link. It is usually not desirable to have a separate communications path for control signals. Accordingly, the receiver must be able to distinguish control information from the data being transmitted.

With the use of a data link protocol, the next-higher layer may assume virtually error-free transmission over the link. However, if communication is between two systems that are connected via a network, the connection comprises a number of data links in tandem, each functioning independently. Thus, the higher layers are not relieved of an error control responsibility.

The most widely used data link standards are HDLC and two very similar standards, LAP-B and ADCCP.

Network Layer. The basic service of the *network layer* is to provide for the transparent transfer of data between transport entities. It relieves the transport layer of the need to know anything about the underlying data transmission and switching technologies used to connect systems. The network service is responsible for establishing, maintaining, and terminating connections across the intervening communications facility.

There are a large number of intervening communications facilities that might be managed by the network layer. The simplest case—a direct link between stations—may not require a network layer at all since a data link layer can perform the necessary management. The most common use of the network layer is handling the interface with a communications network. The network protocol in a device attached to the network must give the network sufficient information to switch and route data to another station.

X.25 is a widely accepted layer-3 standard that is used to communicate across most public data networks, such as Telenet and Tymnet. It can also be used to interface to a local network.

Transport Layer. The purpose of the transport layer is to provide a reliable mechanism for the exchange of data between processes in different systems (9). It ensures that data units are delivered error free, in sequence, with no losses or duplications. The transport layer may also be concerned with optimizing the use of network services and providing a requested quality of service to session entities. For example, the session entity might specify acceptable error rates, maximum delay, priority, and security. In effect, the transport layer serves as the user's liaison with the communications facility.

The size and complexity of a transport protocol depend on the type of service it can get from layer 3. For a reliable layer 3, a minimal layer 4 is required. If layer 3 is unreliable, the layer-4 protocol should include extensive error detection and recovery. Accordingly, ISO has defined five classes of transport protocol, each oriented toward a different underlying network service.

Session Layer. The *session layer* provides the mechanism for controlling the dialogue between applications (10). At a minimum, the session layer provides a means for two application processes to establish and use a connection, called a *session*. In addition, it may provide the following services:

Dialogue Type. This can be two-way simultaneous, two-way alternate, or one-way.

Recovery. The session layer can provide a checkpoint mechanism, so that if a failure of some sort occurs between checkpoints, the session entity can retransmit all data since the last checkpoint.

Presentation Layer. The *presentation layer* is concerned with the syntax of the data exchanged between application entities. Its purpose is to resolve differences in format and data representation. The presentation layer defines the syntax used between application entities and provides for the selection and subsequent modification of the representation to be used.

Examples of presentation protocols are teletext and videotex, encryption, and virtual terminal protocol. A virtual terminal protocol converts between specific terminal characteristics and a generic or virtual model used by application programs.

Application Layer. The *application layer* provides a means for application processes to access the OSI environment. This layer contains management functions and generally useful mechanisms to support distributed applications. Examples of protocols at this level are virtual file protocol and job transfer and manipulation protocol.

Status of Standards

In a few short years, the OSI model has achieved nearly universal acceptance. It provides not only a framework for developing standards, but also the terms of reference for discussing communications system design. Standards for the lowest three layers of the OSI model are well defined and widely implemented. ISO has recently issued standards for layers 4 and 5 and is continuing work on layers 6 and 7. These long-awaited standards are welcome news for the customer and users. Customers can now begin the migration from proprietary communication architectures to the OSI model and ISO standards.

MANUFACTURING AUTOMATION PROTOCOL

The increasing use of automation in the manufacturing process, including computers, programmable controllers, and robots, has meant an increasing dependence on communications.

These devices must interact and be controllable to be effective. Data communications and networking have become the key to improving productivity in the factory (11) [see also MANUFACTURING AUTOMATION PROTOCOL (MAP)].

Increasing the use of communications means, of course, increasing communication costs. But the observed costs are higher than what might be expected. General Motors (GM) reports (12) that recent appropriation requests for plant floor computer equipment have allocated as much as 50% of their total to communication and networking costs. The reason for this is that today's plants utilize programmable equipment from a wide range of manufacturers, and each requires its own proprietary protocol and interface. Typically, a company's data processing department may purchase equipment from one manufacturer, while plant engineering uses another, and manufacturing buys from a third. Table 2 is a representative example (13).

In response to the problem, in 1980 GM set up a task force to define a manufacturing automation protocol (MAP). The objective of MAP is to define a local network and associated communications architecture for terminals, computing resources, programmable devices, and robots within a plant or complex (14). It sets standards for procurement by GM units and provides a specification for use by vendors who want to build networking products for factory use that are acceptable to GM. The GM strategy has three parts:

1. For cases in which international standards exist, select those alternatives and options that best suit GM's needs.
2. For standards currently under development, participate in the standards-making process to represent GM's requirements.
3. In those cases where no appropriate standard exists or is under development, develop and promulgate a GM standard.

It is turning out that, at least in this case, what is good for GM is good for the country. The response to MAP has been impressive. A MAP user group has been set up under the sponsorship of the Society of Manufacturing Engineers. Over 200 companies, all customers for automation-related data processing and communication products, are members of the user group and have endorsed MAP. There has also been activity on the vendor side. In November 1985, GM sponsored a multivendor MAP demonstration at the Autofact Factory Automation trade show in Detroit, Michigan. Equipment from over 30 vendors, including semiconducter, microcomputer, minicomputer, and mainframe vendors, was networked together. All vendors demonstrated at least part of the MAP suite of protocols. This broad acceptance, by both customers and vendors, means that in the fields of factory automation and robotics, MAP has become the standard specification for communications (15). The remainder of this article describes the elements of MAP, which can be conveniently grouped as follows: communications network, midlevel protocols, and application-oriented protocols.

Table 2. Devices Typically Found in an Automotive Assembly Plant

Device	Quantity	Primary Manufacturers
Programmable controls	1200	A-B, Gould, Square D
Robots and automation devices	600	GMF, Cincinnati Milacron
Computers	200	IBM, DEC, Motorola
Total	*2000*	

Communications Network

GM's specification of a communication network is driven by the sophisticated communications strategy that has evolved to meet its requirements. These requirements reflect those that are obtained in other factory and robotics environments. Among the key areas are the following:

- Work force involvement has proven to be a valuable tool for GM's quality and cost-improvement effort. In an attempt to provide facts about the state of the business, employees are told GM's competitive position in relation to quality and costs. This information is communicated by video setups at numerous locations in the plant complex.
- An indirect effect on manufacturing costs has been an escalating cost of utilities. To try to control this area, GM measures usage of water, gas, pressurized air, steam, electricity, and other resources—often by means of computers and programmable controllers.
- GM is investigating, and in some cases implementing, asynchronous machining and assembly systems that are much more flexible than the traditional systems of the past. To facilitate flexibility, the communication requirements increase an order of magnitude.
- To protect its large investment in facilities, GM uses closed-circuit TV surveillance and computerized monitoring systems to warn of fires or other dangers.
- Accounting, personnel, material and inventory control, warranty, and other systems use large mainframe computers with remote terminals located throughout the manufacturing facility.
- The nature of process-control and factory environments dictates that communications be extremely reliable and that the maximum time required to transmit critical control signals and alarms be bounded and known.

To interconnect all of the equipment in a facility, a local network is needed (see LOCAL AREA NETWORKS). The requirements listed above dictate the following characteristics of the local network: high capacity, the ability to handle a variety of data traffic, a large geographic extent, high reliability, and the ability to specify and control transmission delays.

These characteristics are best provided by the use of a broad-band local network with token-bus access control. These terms are briefly described below.

The broad-band transmission medium is standard coaxial cable of the kind used in cable television (CATV) distribution systems. The topology of this type of local network is a branching tree; there is a main trunk, and subsidiary or branching cables can be attached to the trunk and to other branches by signal-splitting taps. All of the devices are attached to the network by taps onto the trunk or one of the branches. A transmission from any one device propagates throughout the medium and may be received by any other device. To achieve communication, data are transmitted in packets. Each packet contains data plus the destination address. The capacity of the broad-band cable is divided into channels, just as in a CATV system. Each device is tuned to one of the available channels. Channels may be utilized for digital data, video, or voice transmission.

The rationale for selecting broad-band cable as the standard communications medium is as follows:

- Broad-band allows multiple logical networks to exist simultaneously (different channels) on the same medium. Therefore, wiring installation and future wiring modifications are minimized.
- Broad-band not only supports high-speed data requirements, but also handles voice and video requirements.
- Broad-band components were developed for outdoor use and are designed to be rugged and noise-resistant. Thus, they are suitable for harsh factory environments.
- Broad-band cable is capable of handling large volumes of data over relatively long distances, and is thus suitable for a factory or factory complex.

Because the local network is a shared-access medium, some protocol is needed for regulating access and assuring that only one device transmits at a time. The protocol specified by MAP is token bus. All of the devices attached to the medium and tuned to the same channel are ordered in a logical, round-robin ring. At any time, one device is in possession of the "token" and may transmit. When it is done transmitting, that device then passes the token to the next device in logical order, and so on. Token bus has two characteristics that are desirable in a process-control or other real-time environment:

- A priority mechanism may be used to give priority to time-critical functions and/or functions with high throughput requirements.
- By setting an upper limit on the time that each device can hold the token, a fixed upper bound on the maximum transmission delay for each device is set. This is important for certain control or alarm signals that must get through within a specified time.

On top of the token-bus specification, MAP specifies the use of a data link control protocol suitable for local networks, known as logical link control (LLC). LLC handles the addressing required for a shared medium. MAP specifies the simplest version of LLC, which includes no error control or flow control. This is most efficient, since these functions are performed by the transport layer.

The communications network specifications for MAP are all part of the IEEE 802 standard for local networks and correspond to layers 1 and 2 of the ISO standard. At the present time, MAP does not require a layer-3 protocol since the simple local network architecture does not require the routing and address functions normally performed by layer 3.

Midlevel Protocols

MAP specifies the use of ISO transport and session protocol standards as layers 4 and 5 of the architecture. In the case of the transport protocol, Class 4 transport is specified. As was mentioned, ISO has defined a family of five transport protocols (Class 0 through Class 4). Class 4 is the most complex because it assumes the least about the underlying network. Thus, regardless of the nature of the communications facility, transport Class 4 assures reliable, sequenced data delivery. The combination of ISO session and Class 4 transport provides a data exchange facility that is both reliable and easy (for application programs) to use.

Application-Oriented Protocols

The current version of MAP specifies two application-oriented protocols: file transfer, access, and management (FTAM) and the manufacturing message-format standard (MMFS). These are likely to be the two application-oriented protocols with broadest applicability.

FTAM is a recently approved ISO standard that specifies the services and protocols for moving files between heterogeneous systems. The initial protocol features mostly file transfer capabilities, which allow entire binary or text files to be transferred between systems. File access features are currently under development by ISO and will be adopted for MAP. The difference is that file transfer necessitates the transfer of the entire file, whereas file access allows users to access individual records from remote machines.

MMFS was developed by GM and its vendors to provide necessary services and messaging to robots and programmable devices. It has been reissued, with minor modifications, by the Electronic Industries Association (EIA) as draft standard 1393A. When the final standard becomes available, it will be adopted as part of MAP. MMFS specifies common message formats, commands, and responses for communication between automation devices, regardless of the manufacturer. STOP, START, UPLOAD, DOWNLOAD, and other commands and responses will be the same for all devices on the network. Standardization efforts include:

- Common messaging that spans different types of devices. For example, robots will support common messages with programmable controllers.
- Standard methods of raw data representation, such as floating point and integer.
- Standard error messages and status queries.
- Powerful high-level commands transferring large quantities of data.

Standardization of application messaging will provide the network user with an unprecedented benefit: the ability to achieve standard application-software systems. Software packages would be independent of the automation device addressed. This would allow the use of common software in a variety of facilities and manufacturing processes.

Summary

Communication and networking have become essential components of today's factory. Although automation remains the key to increased productivity, protocol standardization is an essential element of that strategy. GM seeks to solve that problem with MAP. Because MAP is based on international standards and has widespread support, this solution is available to other companies of more modest size.

BIBLIOGRAPHY

1. W. Stallings, *Data and Computer Communications* 2nd ed., Macmillan, New York, 1988.
2. International Telephone and Telegraph Corp., *Reference Data for Radio Engineers,* Howard W. Sams, Indianapolis, Ind., 1975.
3. A. Tanenbaum, "Network Protocols," *ACM Computing Surveys* (Dec. 1981).
4. *Basic Reference Model for Open Systems Interconnection,* International Organization for Standardization, IS7498, New York, 1983.
5. H. Bertine, "Physical Level Protocols," *IEEE Transactions on Communications* (Apr. 1980).
6. F. McClelland, "Services and Protocols of the Physical Layer," *Proceedings of the IEEE* (Dec. 1983).
7. D. Carlson, "Bit-Oriented Data Link Control Procedures," *IEEE Transactions on Communications* (Apr. 1980).
8. J. Conard, "Services and Protocols of the Data Link Layer," *Proceedings of the IEEE* (Dec. 1983).
9. W. Stallings, "A Primer: Understanding Transport Protocols," *Data Communications* (Nov. 1984).
10. W. Emmons and A. Chandler, "OSI Session Layer Services and Protocols," *Proceedings of the IEEE* (Dec. 1983).
11. S. McGarry, "Networking Has a Job to Do in the Factory," *Data Communications* (Feb. 1985).
12. M. Hall, "Factory Networks," *Micro Communications* (Feb. 1985).
13. K. Hughes, "Factory Communications Becoming Standard," *Systems and Software* (May 1985).
14. General Motors, *Manufacturing Automation Protocol (MAP) Protocol Specification, Version 2.1,* General Motors Technical Center, Warren, Mich., Mar. 31, 1985.
15. E. Keller, "GM-MAP Support, New Products Continue Growing," *Systems and Software* (Mar. 1985).

General References

J. Martin, *Telecommunications and the Computer,* Prentice-Hall, Englewood Cliffs, N.J., 1976. A quite readable and broad discussion of data communications.

R. Freeman, *Telecommunication Transmission Handbook,* John Wiley & Sons, Inc., New York, 1981. A detailed coverage of data transmission topics.

J. Martin, *Computer Networks and Distributed Processing,* Prentice-Hall, Englewood Cliffs, N.J., 1981. An informal and very readable discussion of communications architecture and the OSI model.

A. Tanenbaum, *Computer Networks,* Prentice-Hall, Englewood Cliffs, N.J., 1981. A discussion of protocols using the OSI model. Averages about one chapter per layer.

J. McNamara, *Technical Aspects of Data Communications,* Digital Press, Bedford, Mass., 1982. A coverage of data communications with a computer hardware and electrical engineering orientation. It is quite approachable, even for those with little background in these areas.

W. Stallings, *Data and Computer Communications* 2nd ed., Macmillan, New York, 1988. A textbook that covers both data communications and communications architecture.

W. Stallings, *Handbook of Computer-Communications Standards, Volume I: The Open Systems Interconnector (OSI) Model and OSI-Related Standards,* Macmillan, New York, 1987. A tutorial and reference book that describes the OSI model and provides detailed descriptions of standards at all seven layers.

COMPLIANCE

Chi-haur Wu
Northwestern University
Evanston, Illinois

INTRODUCTION

When industrial robots were introduced, they were universal transfer machines and could be programmed to perform many different tasks, removing the need for a product-dependent economic justification. After the integration of computer technology, robot flexibility and productivity increased. With the advent of computer control, tasks such as seam welding, routing, and drilling appear to be suitable for automation using industrial robots. In general, an industrial robot is a position-controlled mechanical device, which means that it should have the ability to perform a required motion by positioning and orienting its end-effector. This motion is parameterized by time and called a trajectory. However, when robot motion is constrained by the task and requires that the trajectory be modified continuously by contact forces or tactile stimuli in the environment, during motions such as opening a door, turning a crank, inserting a peg into a hole, assembling parts into a product, and so on, the control of position only is not enough to accomplish the task; in addition, the robot needs the control and monitoring of the exerted forces.

Through these basic differences in motion behavior and controller design, there are actually two types of motion, unconstrained and constrained, in robot maneuvers. Unconstrained motion means that the robot is moving in its workspace without contact with the environment, and normally this is the only motion industrial robots are allowed. The second type of motion occurs when the robot is required to be continuously in contact with the environment, and its movement is constrained. The control of this constrained motion requires a controller that combines both position control and force control. This type of robot control is called compliance control and is extremely important for robot applications in industrial automation, especially for those tasks that require constrained motion. Although now, due to its design complexity, almost no commercial robots have this compliance capability, its importance is well recognized.

In general, through the differences in motion behavior, robot motion control can be grouped into three categories: pure position control, pure force control, and compliance control.

Pure Position Control

This type of motion control is designed to handle unconstrained motion. Every industrial robot is provided with pure position control, which is the control of positions and orientations of the robot's end-effector in some predetermined coordinate system, so that the robot can perform an unconstrained task. Normally, there are two types of motion trajectory planning: joint path planning (1) and Cartesian path planning (2,3). A number of position control schemes have been proposed to control the robot motion at either joint space or Cartesian space. For example, resolved motion rate control (RMRC) controls a desired Cartesian path velocity (4,5); current torque control (1), which is also called the computed torque method (6), is performed at joint space; resolved acceleration control is a Cartesian path control and is servoed at Cartesian space (7); resolved motion force control is also a Cartesian path control performed at Cartesian space (8); and so on.

Since a robot's configuration changes continuously throughout the motion, the performance of a classical control system with fixed gains will deviate when fast motion is required. To minimize position servo errors caused by those changing dynamics, a feedforward compensation of robot dynamics is necessary. Robot dynamics can be calculated through either Lagrangian formulation (9–12) or Newton–Euler formulation (13).

Realizing the significant effect of robot dynamics, in recent years a utilization of the adaptive control concept to eliminate the nonlinearity of a robot's dynamics has also been applied to position control. Different approaches, such as model referenced adaptive control (14–17), self-tuning control (18), decoupling control (19,20), resolved motion force control (8), and so on, have been proposed.

Although many sophisticated position control schemes have been achieved in the research environment, industrial robots are still designed with a PID controller for its simplicity and ease of implementation. The motion trajectory can be planned at either Cartesian space or joint space, but the PID controller is designed at joint space.

Pure Force Control

The main purpose of designing a force control is to control accurately the desired forces and moments that should occur between the end of a robot and its environment so that the robot can comply with the constrained motion. To design a good force control, proper force sensors are needed. The advantages and disadvantages of different types of sensors used for a robot force control have been described (21). Based on the sensor's placement in relation to a robot, the existing force-sensing devices have been classified into three categories: force pedestal, wrist and finger, and joint torque (22).

Force pedestal sensors are mounted on the support platform with which the object is to be handled (23). This kind of sensor is product dependent and so is difficult to justify economically; however, it has high structural stiffness and good sensitivity. Wrist and finger sensors (21,22,24) are mounted on or close to a robot hand. This type of sensor has been used most frequently in designing a force control because it has the advantages of sensitivity and reducing the uncertainty that might be caused by the large weight and inertia of a robot in measuring the forces and moments occurring at its hand. However, due to the low bandwidth of the manipulator structure, a force servo system using this type of sensor will result in low servo bandwidth and poor response time (25). The third type of sensor is the joint torque sensor, which is mounted on the joints of a robot to measure joint torques (26,27); however, joint torque can also be estimated by utilizing the relation between joint velocity and joint motor current (1). This sensing technique has been used successfully for many years in master–slave robot arms to sense the developed joint torques directly at the slave arm (28,29). Joint torque sensors were also employed (30) to detect the interaction forces produced when two manipulators were used to lift and move simultaneously a large box. Due to the heavy weight and inertia of a robot manipulator, joint torque sensing has the disadvantage of inferring the forces and moments at the robot hand. However, it has the following advantages. (1) It detects not only forces and moments applied at the robot hand, but also those applied at other points on the manipulator. (2) It can withstand large impulsive forces and moments applied at the robot hand because it is protected by the inertia of the robot's links and the stiffness of the robot's structure. (3) A torque servo system with a joint torque sensor can have high bandwidth and fast response (26,27).

The first demonstration of force feedback control of a robot manipulator was made (31) based on the idea of RMRC (4). Then a single-axis hydraulic manipulator with a sensor on its end-effector was built and its performance analyzed (32). Based on the concept of coordinated continuous axis motion (5) and RMRC a formal representation of vector force feedback control called accommodation force control was presented (25), in which a force feedback matrix was utilized to convert a sensed force vector from a wrist sensor into a responsed velocity vector, which was used as feedback velocity in Cartesian compliant coordinates. The method used an inverse Jacobian matrix in the servo loop, which established the connection between sensed forces and commanded joint velocities. Some other researchers (1,33–35) presented an open-loop force control that utilized the relation between joint velocity and joint motor current to estimate joint torques, in which the effect of friction could also be minimized. From the above results, the method was enhanced and a closed-loop force control using a wrist sensor was described (36). Instead of using wrist sensor, a joint sensor was designed to mount on a joint actuator, and then a closed-loop joint torque servo system with minimized friction effect was also developed (26,27,37). Another kind of Cartesian force control is active stiffness control (38–40), in which the user is allowed to specify the stiffness of the robot hand in all six degrees of freedom, and by giving the nominal position of the hand, the control of position error signals is equivalent to a force control. A comparison of stiffness control and accommodation control is given in reference 41. A hybrid position/force control was also developed to control robot forces and positions at the same time (42). Another study on the effect of wrist sensor stiffness on the performance of a robot force control was also given (43).

In general, the control of force can be either at Cartesian space or at equivalent joint space, as long as the constrained forces can be satisfied. Another important design consideration is how to cope with the position control system so that the compliance can be controlled. Therefore, in order to increase the speed of robot operations in performing compliance, a fast, reliable force control system using a proper sensing technique is crucial.

Compliance

The third type of robot control is compliance. This occurs when the robot manipulator's position and orientation are constrained by task geometry, and in order to accomplish the task, the robot control has to have the ability to react to contact forces and moments or stimuli during the motion. Although all the tasks that require compliant motion are easily performed by humans, they are generally beyond the capabilities of current industrial robots. This deficiency is one of the major problems limiting the range of robot applications. Therefore, the control of compliance is one of the major issues in improving robot capability and versatility.

The key to controlling compliance is the method of integrating force sensing into a manipulator's programming and control system. Currently, there exist two approaches to providing compliance: passive and active. Passive compliance is a purely mechanical device and can be designed with springs and other elastic members that change to follow a path. A robot equipped with this special tool can perform some special tasks requiring sensory feedback; however, this type of compliance is somewhat task dependent. An example of this is the famous remote center compliance (RCC) device developed at the Charles Stark Draper Laboratory (44–46). This device is mounted on the robot hand and used for insertion tasks. The second approach is active compliance, which requires control motion to be generated in response to contact forces and moments during the motion. In order to provide active compliance, the robot control has to combine the control of both position and force. It generally requires that a robot be equipped with general-purpose force sensors, touch sensors, or some special force-sensing techniques for force control (47,48). The advantages of this approach are that it is not task dependent and can be programmed to perform any task and that the robot can also use its sensors to monitor the operation. In general, a robot equipped with sensors and compliance control has improved capability and versatility and is thus conceptually attractive. In reality, aside from the difficult aspects of the realization of sensor devices and the complicated process necessary to extract the control information from sensor inputs, the design of such a compliance control system is still under research.

Many methods to provide active compliance have been proposed. However, in general they can be categorized into three basic approaches. All the force control techniques were developed to satisfy at least one of these approaches. The first approach is based on the concept of selecting free joints for partially constrained motion and then servoing these free joints with forces and other joints with positions (1,22,26,27,35, 36,49). Because the compliance is selected at joint space, this approach will have to compensate the position errors caused by force-servoed joints through software design. The second approach is based on the concept of a hybrid controller that controls both positions and contact forces generated at the robot hand to provide compliance (42,50,51). The third approach is based on the concept of force feedback strategy. This approach was initiated by Whitney's accommodation force control (25,41), in which the desired compliant motion was velocity and the sensed forces were converted into an accommodated velocity feedback vector through a force feedback matrix. The control method uses RMRC (resolved motion rate control) to convert the net velocity errors into the desired robot joint velocities and control the robot. Instead of velocity modification, a new method called active stiffness control (38–40) has also been proposed, in which users are permitted to specify the stiffness of the robot hand in all six degrees of freedom so that, by giving the nominal position of the hand, control of the error signals in position will provide simultaneous position and force control. Each of the three aforementioned approaches has its own advantages and disadvantages. Through identifying the disadvantages, a combination of the three approaches has also been proposed (52).

The development of the aforementioned approaches was from the point of view of controlling compliance. However, the success of implementing these approaches and providing compliance relies heavily upon a fundamental analysis of the basic physics of the compliant task, which means knowing how to model precisely the robot's work environment. In other words, the method of partitioning the constraints of compliant motion, based on the analysis of task geometry, into position and force constaints, and satisfying simultaneously the required position and force of the compliant task during the execution of robot motion, is imperative to the success of a robot compliance control. The results of this analysis will generate a set of motion control strategies (53) to supervise the control of compliance and to satisfy the task contraints. The generation of this control strategy is so important because the control strategy becomes the intelligent part of an automated compliance control.

The importance of robot compliance has long been recognized, and some fundamental theories and approaches have been achieved. Although the implementation of robot compliance control is still well behind the applications in industry, it will definitely affect the success of future industrial automation. A summary of past achievements is given in this article so that the reader can understand what robot compliance is and how to find the related references.

THEORETICAL BASIS

Compliant motion occurs when the position and orientation of a robot manipulator are constrained by a task, such as inserting a round peg in a hole. In this task, the peg's motion is constrained by the hole. The peg can only rotate about its own axis and translate up or down along the axis. If the robot that holds the peg has only a position control system, then the motion of the peg and the force exerted between the peg and hole will be so unpredictable that the peg may just jam the hole and cause the robot to fail such a simple task. To accomplish the task, those unpredictable movements have to be controlled. The only way to control them is to give the task the required constraints, which are on the force freedom in addition to the position freedom. Therefore, a combination of position and force control is necessary for this simple task.

In addition to designing a compliance control system to provide compliance, there is another important issue, which is how to generate compliant motion strategies according to the task geometry so that the compliance can provide proper position freedom and force freedom and thus agree with the task constraints. Without these strategies, the robot cannot properly control the compliant motion and accomplish the constrained task. Mason (53) was the first person to present a formal theory that partitions a constrained task into position and force constraints so that a set of compliant motion control strategies can be generated for compliance control. These strategies will then supervise compliant motion control and command the robot to perform the task. As a result, this motion control strategy theory, together with the active compliance approach and the passive compliance approach, forms the theoretical basis for current robot compliance, which is briefly described in this section.

Compliant Motion Control Strategy

To generate a compliant motion control strategy, the distinction between position control and force control must be identified. Pure position control and pure force control are dual concepts, which means that pure force control has no positional freedom at all and pure position control is the natural result of applications that involve no force freedom. Therefore, compliant motion control should be a hybrid mode of control that gives both positional and force freedom. Based on this idea, Mason introduced a configuration space called C-surface that allows only partial positional freedom; in other words, freedom of position occurs along C-surface tangents, whereas freedom of force occurs along C-surface normals. In this case, a hybrid mode of control will give control of force along the C-surface normal and control of position along the C-surface tangent at the compliant coordinates, which is a coordinate frame at the robot hand, where the occurred compliance needs to be controlled.

To derive the motion control strategy, Mason introduced three terminologies: natural constraints, goal trajectory, and artificial constraints. A robot manipulator was modeled as an ideal effector, representing an object with no physical dimensions and with attributes of position and force only, in an ideal domain. The natural constraints, representing the task configuration in the real world, are a set of linear equations relating the components of the ideal effector's force and velocity. The goal trajectory that is consistent with the natural constraints gives the desired ideal effector's positions and forces, which are a function of time. Artificial constraints are a set of control strategies for the ideal effector and are represented as linear equations on the effector's force and velocity. In this ideal domain, the control strategy synthesis problem is to find a set of artificial constraints so that the desired goal trajectory, given the natural constraints, will be reproduced. From these artificial constraints, a set of compliant motion control strategies can then be generated so that the freedom of position and force of the task motion required at the compliant coordinates can be maintained. By following these control strategies, a robot manipulator equipped with active compliance can then accomplish the constrained task.

In Mason's ideal domain, the ideal effector's position, velocity, and force are represented as points in vector space $\mathbf{R}^6$:

$$\boldsymbol{p} = [p_x\, p_y\, p_z\, q_x\, q_y\, q_z]^t,$$

$$\boldsymbol{y} = [v_x\, v_y\, v_z\, w_x\, w_y\, w_z]^t,$$

and

$$\boldsymbol{f} = [f_x\, f_y\, f_z\, g_x\, g_y\, g_z]^t$$

where the first three elements of $\boldsymbol{p}$, $\boldsymbol{y}$, and $\boldsymbol{f}$ are translational components and the last three are rotational components. As for the task configuration, it is represented by a smooth hypersurface consisting of the possible positions of the ideal effector. This hypersurface is called the ideal C-surface and is assumed to be connected and smooth, that is, having a unique tangent space at each point. Its tangent space is a vector flat with the same dimension as the ideal C-surface. The dimension of the ideal C-surface may vary from 0 to 6. An ideal C-surface of dimension 0 is a point, corresponding to the absence of any freedom of motion; on the other hand, an ideal C-surface of dimension 6 contains all the points of $\mathbf{R}^6$ near the effector's position, representing complete freedom of motion.

Since the ideal effector position lies in the ideal C-surface at all times, the ideal effector velocity will lie in the vector subspace parallel to the tangent space. This constraint can be expressed as a set of linear equations on the velocity coordinates, representing the constraints of position. Similarly, if the tangential forces are neglected, the effector's force is restricted to be orthogonal to the tangent space. This constraint can be expressed as a set of linear equations on the ideal effector's force coordinates, representing the constraints of force. These linear equations on the ideal effector velocity and force are the natural constraints. For example, consider the ideal C-surface for compliant coordinates located at the center of a peg in a hole as shown in Figure 1. The peg is able to translate along and rotate about the z axis, but it cannot translate along or rotate about the other axes, so that the motion is constrained at those axes in the ideal case. Similarly, force along and torque about the z axis should be constrained to be free and thus are zero if frictions are zero in the ideal domain, whereas forces and torques for the other axes are unconstrained. Therefore, the natural constraints for this task configuration are

$$v_x = 0 \quad w_x = 0$$

$$v_y = 0 \quad w_y = 0$$

$$f_z = 0 \quad g_z = 0$$

The above equations are homogeneous linear equations and represent the ideal constraints of the task; thus, they restrict the effector's velocity and force to vector subspaces $\mathbf{S_v}$ and $\mathbf{S_f}$, respectively. Where $\mathbf{S_v}$ is the v_z–w_z plane, representing the freedom of position, and $\mathbf{S_f}$ is spanned by the f_x, f_y, g_x,

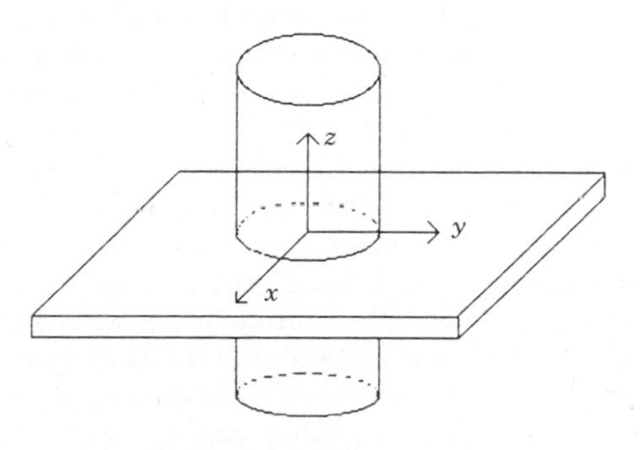

Figure 1. Peg in a hole.

and g_y axes, representing the freedom of force. In addition, these two subspaces are orthogonal complements.

The purpose of finding the natural constraints is to determine the artificial constraints, which represent the freedom of position and force of the task plus the desired goal trajectories. Since the natural constraints are the constraints of the task, a natural choice for the task to have free motion is a set of constraints that are orthogonal to the derived natural constraints; therefore, this set of constraints is chosen as the artificial constraints. So, the artificial constraints have the same form as the natural constraints, except that the equations need not be homogeneous because of the addition of goal trajectories. Since the degrees of freedom are divided into position and force for the task, the artificial constraints also consist of two parts. The artificial velocity constraints can be derived based on the condition that the components along $\mathbf{S_v}$ are equal to the goal trajectory velocities. As for the artificial force constraints, they are the condition that the components along $\mathbf{S_f}$ be equal to the goal trajectory forces or some values that satisfy considerations of task stability. In the example of Figure 1, the artificial constraints can be derived as

$$f_x = 0 \qquad g_x = 0$$
$$f_y = 0 \qquad g_y = 0$$
$$v_z = k_1(t) \quad w_z = k_2(t)$$

if $k_1(t)$ and $k_2(t)$ are the goal trajectory velocities and all the force trajectories are equal to zero.

As in these artificial constraints, the compliant motion control strategy consists of both position control and force control, to satisfy, respectively, the artificial velocity constraints and the artificial force constraints, so that an active compliance control can be applied. In the above example, the control strategies are force servoing along x and y axes, torque servoing around x and y axes, position servoing along the z axis, and orientation servoing around the z axis. By following these strategies, the implementation of the compliance will accomplish the task.

Active Compliance

Applications of a robot manipulator in assembly tasks require constraints in position and orientation of the manipulator. For example, inserting a peg into a hole, the position and orientation of the manipulator are required to comply with the constraints relative to and normal to the hole axis of insertion; otherwise, the task will be failed. The relative position and orientation between the peg carried by the manipulator and the hole can be sensed by the forces and moments occurring at the insertion interface. These forces and moments occur when the joint torques generated from the actuators of the manipulator are in excess of those necessary for providing the required motion. Hence, if the joint torques can be properly controlled to comply with the contact forces and moments and at the same time the task motion can be maintained, the robot can achieve the constrained task. This kind of compliant motion provided by a robot is called active compliance. Based on this concept, three fundamental approaches have been proposed: free-joints selection, hybrid position/force control, and force strategy control.

Free-joints Selection Method. The free-joints selection algorithm was initially proposed in reference 1, and was then formalized in references 22 and 27. The concept of this algorithm is to select joints to match the required degrees of force freedom of the task and then to perform force servo on the selected joints while position servoing the other joints. At the end, the position errors caused by the compliant motion will be approximately compensated through the changes in the coordinate transformation. The force control of this method is based on joint torque control (1,27,36).

Joints Selecting Algorithm. Since the motor friction has a significant effect on providing compliance, the joints selected to be force servoed should have the most freedom to minimize this effect. Based on this concept, the results of predecessors have been formalized, and a selecting method including the effect of friction has been proposed (27). The algorithm of this method is to pick joints with the most work caused by the compliant forces in the direction of compliant motion. The procedure is as follows:

1. Calculate the equivalent compliant joint torques $\boldsymbol{T}$ corresponding to the desired compliant forces $\boldsymbol{F}$ at the compliant Cartesian coordinates through the Jacobian transpose as follows:

$$\boldsymbol{T} = [\mathbf{J}]^{\mathbf{t}} \boldsymbol{F} \tag{1}$$

where $\boldsymbol{T}$ is an $N \times 1$ vector for a robot manipulator with N joints, $[\mathbf{J}]$ is the Jacobian matrix, and the superscript $\mathbf{t}$ represents the matrix transpose.

2. Calculate joints' position changes in the directions of forces. For example, if a force is required to comply in the direction of the x axis, then the direction of free motion is also in the direction of the x axis. In order to represent this free motion, a vector with a unit in the x direction, $\boldsymbol{u} = [1\ 0\ 0\ 0\ 0\ 0]^t$, is formed. Then the position changes in the joints due to this motion are calculated as

$$\mathbf{d}\boldsymbol{q} = [\mathbf{J}]^{-1} \boldsymbol{u} \tag{2}$$

where $[\mathbf{J}]^{-1}$ is the Jacobian inverse (when $N = 6$) or pseudoinverse (when N is not equal to 6).

3. Calculate the absolute work caused by the compliant forces in the direction of free motion for each joint by

$$dW_i = |T_i\, dq_i|, \qquad \text{for } i = 1, \cdots, N$$

4. Calculate a weighting factor $\boldsymbol{P}$ based on the static frictions $\boldsymbol{f}$ of robot joints by

$$P_i = \max\{f_j, 1 \le j \le N\}/f_i, \qquad \text{for } i = 1, \cdots, N$$

Therefore, a joint with less friction will have more weights than other joints, and this joint will have more freedom.

5. Calculate the selecting factor for each joint by

$$S_i = dW_i P_i, \qquad \text{for } i = 1, \cdots, N$$

Select the number of joints with the largest selecting factors to match the number of components in $\boldsymbol{F}$, and those joints will be the force-servoed joints.

This selecting method includes not only the freedom of motion, but also the effect of friction. Therefore, the selected joints will have the most freedom and the least frictional effect to provide the required compliant forces.

Compensation of Joint Coupling Errors Resulting from Compliant Motion. Once the force-servoed joints have been operated for a brief time, their motion is also coupled into other noncompliant axes so that the positions of the joints being position servoed must be altered to correct for modifications in the net motion of the manipulator along those directions for which no force is to be applied. These coupling errors can be compensated by a method developed in references 22 and 36. This method is briefly described in the following.

The trajectory of the manipulator at each servo sampling time during the compliant motion can be represented as the following transformation equation (12):

$$^{\mathbf{base}}\mathbf{T}_{\mathbf{0}} * {}^{\mathbf{0}}\mathbf{T}_{\mathbf{N}} * {}^{\mathbf{N}}\mathbf{T}_{\mathbf{tool}} = {}^{\mathbf{base}}\mathbf{T}_{\mathbf{com}} * {}^{\mathbf{com}}\mathbf{T}_{\mathbf{tool}} \tag{3}$$

where ${}^{\mathbf{b}}\mathbf{T}_{\mathbf{a}}$ is a homogeneous transformation between the two coordinates **a** and **b; base, tool,** and **com** represent the base, tool, and compliant motion coordinates, respectively. At each servo sampling time, the preplanned joint positions can be solved through ${}^{\mathbf{0}}\mathbf{T}_{\mathbf{N}}$ using a joint solution program (54). Comparing these desired positions with the actual joint positions sensed from joint encoders, the errors in joint positions, $\mathbf{d}q$, due to the compliant motion can be obtained. Therefore, the equivalent Cartesian errors ${}^{\mathbf{N}}\mathbf{d}x = [dx\ dy\ dz\ \delta x\ \delta y\ \delta z]^t$ at joint N coordinates of the manipulator can be calculated through multiplying a Jacobian matrix as in the following equation:

$$^{\mathbf{N}}\mathbf{d}x = [\mathbf{J}]\ \mathbf{d}q. \tag{4}$$

These Cartesian errors can be used to form a differential change transformation $\boldsymbol{\delta}\mathbf{T}_{\mathbf{N}}$ at ${}^{\mathbf{0}}\mathbf{T}_{\mathbf{N}}$, and this error transformation has the following form:

$$\boldsymbol{\delta}\mathbf{T}_{\mathbf{N}} = \begin{bmatrix} 1 & -\delta z & \delta y & dx \\ \delta z & 1 & -\delta x & dy \\ -\delta y & \delta x & 1 & dz \\ 0 & 0 & 0 & 1 \end{bmatrix} \tag{5}$$

As the result of this error changes, there is an equivalent differential change transformation $\boldsymbol{\delta}\mathbf{T}_{\mathbf{com}}$ at the compliant coordinates ${}^{\mathbf{base}}\mathbf{T}_{\mathbf{com}}$. Hence, in order to compensate these coupling errors, the trajectory expressed in equation 3 has to be modified as

$$^{\mathbf{base}}\mathbf{T}_{\mathbf{0}} * {}^{\mathbf{0}}\mathbf{T}_{\mathbf{N}} * \boldsymbol{\delta}\mathbf{T}_{\mathbf{N}} * {}^{\mathbf{N}}\mathbf{T}_{\mathbf{tool}} = {}^{\mathbf{base}}\mathbf{T}_{\mathbf{com}} * \boldsymbol{\delta}\mathbf{T}_{\mathbf{com}} * {}^{\mathbf{com}}\mathbf{T}_{\mathbf{tool}} \tag{6}$$

The only unknown transformation, $\boldsymbol{\delta}\mathbf{T}_{\mathbf{com}}$, in the above equation can be solved as

$$\boldsymbol{\delta}\mathbf{T}_{\mathbf{com}} = ({}^{\mathbf{N}}\mathbf{T}_{\mathbf{tool}} * {}^{\mathbf{com}}\mathbf{T}^{-1}{}_{\mathbf{tool}})^{-1} * \boldsymbol{\delta}\mathbf{T}_{\mathbf{N}} * ({}^{\mathbf{N}}\mathbf{T}_{\mathbf{tool}} * {}^{\mathbf{com}}\mathbf{T}^{-1}{}_{\mathbf{tool}}) \tag{7}$$

and $\boldsymbol{\delta}\mathbf{T}_{\mathbf{com}}$ has the same form as in equation 5 containing six elements.

To compensate the coupling errors from the compliant motion, but without overcompensating, those elements in $\boldsymbol{\delta}\mathbf{T}_{\mathbf{com}}$ that do not correspond to the compliant axes will be set to zero, representing noncompliance. The remaining nonzero elements along directions of force application will be used to modify the compliant coordinate frame for the next servo sampling time by the following equation:

$$^{\mathbf{base}}\mathbf{T}_{\mathbf{com}} = {}^{\mathbf{base}}\mathbf{T}_{\mathbf{com}} * \boldsymbol{\delta}\mathbf{T}_{\mathbf{com}} \tag{8}$$

After this modification, ${}^{\mathbf{base}}\mathbf{T}_{\mathbf{com}}$ will be renormalized to maintain orthogonality. In this manner, the compliant coordinate system moves around in response to the compliances, and at the same time, the noncompliant axes are unaffected.

Implementation. The implementation of this approach is as follows:

1. Calculate the equivalent compliant joint torques $\boldsymbol{T}$ corresponding to the desired compliant force $\boldsymbol{F}$ using equation 1.
2. Select the force-servoed joints to match the degrees of force freedom of the compliant motion through the joints selecting algorithm.
3. Perform force control on the force-servoed joints to provide the desired joint torques calculated from equation 1, while position servoing the other joints.
4. Compensate the joint coupling errors of the compliant motion through changes in the coordinate transformation.
5. The calculation of joint torques and the selection of free joints should be evaluated at a rate commensurate with the rate of change of the manipulator's configuration in order to provide a continuous match.

The force control scheme used in this approach is based on a joint torque servo system which can be an open-loop control utilizing the relationship between the joint velocities and the motor currents (1) or a closed-loop joint torque control through designed joint sensors (27). The detailed design can be found in the references.

Hybrid Position/Force Control. The second fundamental type of active compliance control is called hybrid position/force control (42). This approach is based on a theory of compliant motion (53) described in the earlier section. Once the natural constraints are used to partition the degrees of freedom of the compliant coordinates at the robot hand into two orthogonal subsets of position constraints and force constraints, and the artificial constraints are specified with the desired position and force trajectories, then the compliant motion control strategies have already been derived and all that remains is to control the manipulator. Therefore, this approach is simply the design of a low-level control to implement those control strategies, which will regulate the motion and force at the compliant coordinates about their desired trajectories.

The design of this low-level controller is based on the concept of the following control law:

$$T = \mathbf{G_f(s)}\, \mathbf{S}\, {}^c\boldsymbol{F}_e + \mathbf{G_p(s)}\, (\mathbf{I} - \mathbf{S})\, {}^c\mathbf{X}_e \qquad (9)$$

In the above equation, T is an $N \times 1$ vector representing applied joint torques for a robot with N joints. ${}^c\boldsymbol{F}_e$ and ${}^c\mathbf{X}_e$ are two $M \times 1$ vectors representing the force errors and position errors, respectively, at the compliant coordinate system, which has M degrees of freedom, where M is less than or equal to 6, depending on the compliant constraints at compliant coordinates. S is a binary $M \times M$ diagonal matrix with all the off-diagonal terms equal to zero. This matrix is called the compliance selection matrix, in which the jth element, S_{jj}, of the diagonal terms represents the selection of force servo or position servo for the jth axis of the compliant coordinates where $j = 1, \cdots, M$. The jth axis is under force control if S_{jj} is equal to 1, and it is under position control if S_{jj} is zero. $\mathbf{I}$ is an $M \times M$ identity matrix. $\mathbf{G_f(s)}$ and $\mathbf{G_p(s)}$ are two $N \times M$ matrix functions representing force and position compensation functions, respectively.

The compliance selection matrix $\mathbf{S}$ and its complement $\mathbf{I{-}S}$ are used to separate the directions in the compliant coordinates in which force and position are under control. These directions form the constraints that allow the motion and force trajectories to perform the task. However, $\mathbf{S}$ may be a function of the position of the manipulator and must be recalculated for each position of the arm when the motion constraints of the robot hand are due to contact at a point between the manipulator and the environment other than the hand (51).

Figure 2 illustrates the above concept of a hybrid position/force control system. A kinematic transformation function

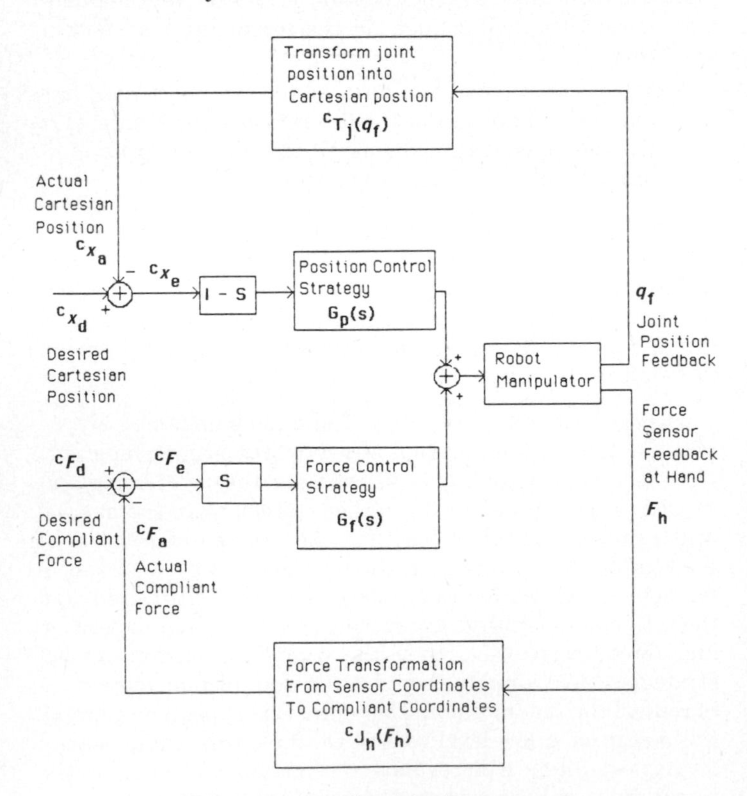

Figure 2. Hybrid position/force control.

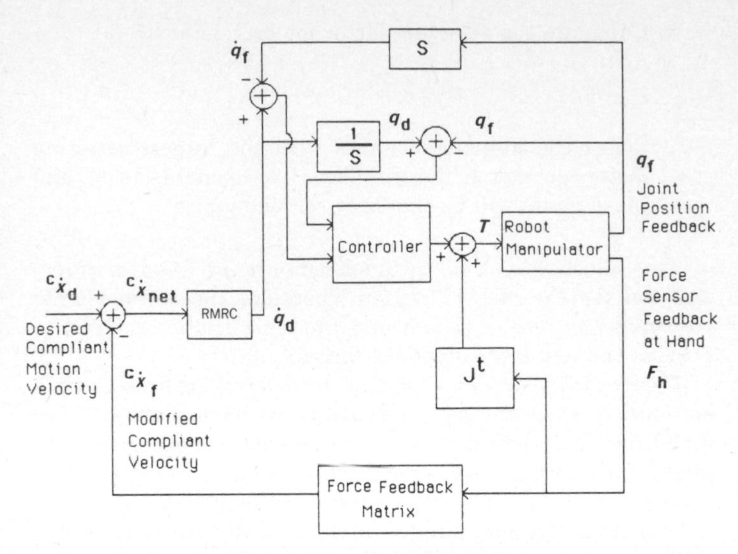

Figure 3. Accommodation control.

${}^c\mathbf{T}_j(\boldsymbol{q}_f)$ transforms the sensed actual joint positions $\boldsymbol{q}_f$ into actual Cartesian position and orientation ${}^c\mathbf{X}_a$ at the compliant coordinates. Another Jacobian transformation ${}^c\mathbf{J}_h$ between the force sensor's coordinates and the compliant coordinates transforms the sensed force $\boldsymbol{F}_h$ at the force sensor into the actual force feedback ${}^c\boldsymbol{F}_a$ at the compliant coordinates. ${}^c\boldsymbol{X}_d$ and ${}^c\boldsymbol{F}_d$ represent desired position and force trajectories at the compliant coordinates.

Although each degree of freedom in the compliant coordinates is controlled by only one loop following the control strategies generated from the artificial constraints, each robot joint is controlled cooperatively by both sets of position and force loops. This is the central idea of hybrid control. This idea has been expanded (51) to control a robot with all types of constraints due to contact with the environment.

Force Strategy Control. This type of compliance based on the force feedback strategy, which means a strategy to transform the contact forces and moments between the robot and the environment into coordinate motion feedback. This idea was initiated by Whitney's accommodation control (25), which is also called damping control (41). Figure 3 depicts the control scheme. The approach uses a force feedback matrix to convert the contact forces and moments measured at a force sensor into accommodated Cartesian velocity feedback commands at the compliant coordinates. By comparison with the desired Cartesian velocity commands at the compliant coordinates, RMRC converts the net modified Cartesian velocity errors into the desired robot joint velocity commands. These commands then go through a controller to drive the robot to contact the environment and build up the contact forces and moments at the force sensor. This control creates small motions to null selected axes of forces or moments at the compliant coordinates while maintaining motions on the other axes. One disadvantage of this method is the use of an inverse Jacobian matrix in RMRC. Some other force feedback strategies have also been discussed (41).

Instead of velocity modification, a new method using position modification, called active stiffness control, was proposed by Salisbury (38). This approach allows users to specify the stiffness of the robot hand in all six degrees of freedom so that, by giving the nominal position of the hand, control of the error signals in position provides simultaneous position and force control. The idea of this approach is based on the following spring stiffness equation:

$$\boldsymbol{F} = [\mathbf{K}]\boldsymbol{X} \tag{10}$$

where $\boldsymbol{F}$, $\boldsymbol{X}$, and $[\mathbf{K}]$ represent the six forces and moments, the six Cartesian displacements between the desired and actual position commands, and a stiffness matrix at any defined compliant coordinates fixed in the hand, respectively. Using the Jacobian relation in equations 1 and 4, a stiffness equation between the joint torques and joint displacements can be derived as

$$\boldsymbol{T} = [\mathbf{J}]^t[\mathbf{K}][\mathbf{J}]\, \mathbf{d}\boldsymbol{q} \tag{11}$$

and a joint stiffness matrix was defined by Salisbury as

$$[\mathbf{K}_q] = [\mathbf{J}]^t[\mathbf{K}][\mathbf{J}] \tag{12}$$

Since the Jacobian matrix can be calculated for any coordinates fixed in the hand, any point relative to the hand can be chosen as the compliant coordinates, and this point will be called the stiffness center.

By modifying the positional and orientational errors at this stiffness center and letting the user specify the stiffness $[\mathbf{K}]$ at this center, the contact forces and moments occurring at this point can be controlled through equation 10 to meet different physical constraints of an assembly task. For example, by specifying a low stiffness in one direction, a low contact force would be ensured in that direction; conversely, a high stiffness would make the hand follow closely the desired position command. Therefore, the control of the desired position and stiffness at the selected compliant coordinates allows simultaneous position and force control.

The implementation of this stiffness control was made at joint coordinates using equation 11, to simplify the control computation. It has the advantage of requiring only the calculation of the Jacobian transpose and not its inverse as in accommodation control. The joint stiffness matrix in equation 12 is nondiagonal and highly coupled among joints, which means that the position errors at one joint will affect the commanded torques in all the other joints. Consequently, due to this coupling effect, the robot is able to control the compliant motion at the stiffness center. Figure 4 illustrates the concept of this control. More detailed design information can be found in reference 38. An in-depth frequency-domain analysis of this stiffness control can be found in references 39 and 40.

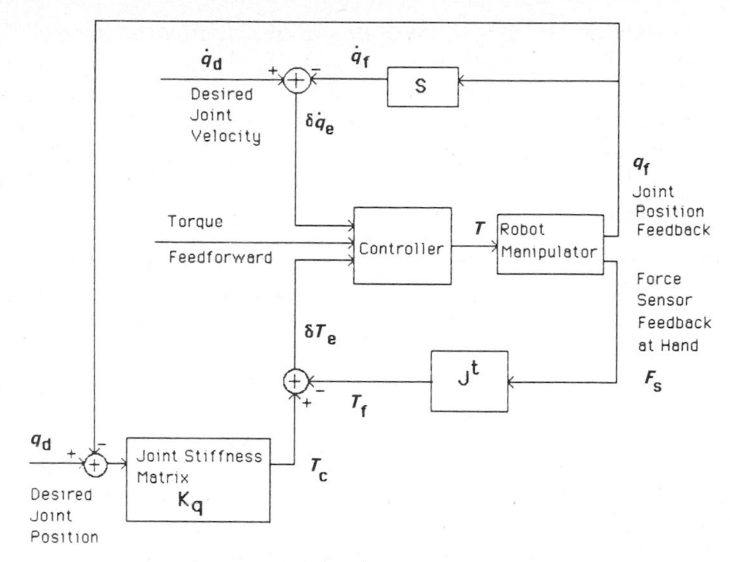

Figure 4. Active stiffness control.

Passive Compliance

This kind of compliance means that the force strategy of the task at the compliant coordinates will be executed automatically by a mechanical device mounted at the robot's wrist, without any sensing, computation, or robot motion command modifications. An example of this kind of device is RCC, which was developed at the Charles Clark Draper Laboratory (44–46). The development of this device was the confluence of the theoretical and experimental results of sensing force vector analysis, geometric and friction analysis, and wedging and jamming analysis of part mating at the Draper Laboratory. The goal was to create a passive compliant wrist that could execute the fine motions required for closed-clearance insertions without the use of active sensors and controls. During the same period, another fine-motion device, called Hi-Ti-Hand, that could also achieve closed-clearance insertions was developed by Hitachi, Ltd., Japan (55). The above devices were developed based on the results of thorough studies of insertion tasks; therefore, some comparable studies of other assembly tasks might lead to the development of other useful passive compliance devices.

The design of RCC involves constructing two types of linkage. The first type allows the peg to rotate about its tip if it is angularly misaligned with the hole. Then the other type of linkage, cascaded with the first type, allows the peg to move from side to side to correct lateral errors. As a result, these linkages can respond to either rotational or translational motions at the tip of the peg. The detailed diagrams of the device can be found in references 44–46. Currently, RCC devices are being used everyday in industry, and similar commercial versions are also available.

To increase the flexibility through realizing the advantages and limitations of such a passive device as RCC, an active adaptive compliant wrist (AACW) (56,57) that consists of active servos was also developed to provide all the potential benefits of active force feedback. Through another mechanical modification (58,59), the developed RCC wrist can automatically adjust for parts of different lengths and weights; in addition, the designed sensors in the wrist can also be used both to readjust the wrist and to provide the force information for the active compliance control.

The results of these developments show that a passive device designed with excellent passive properties and sensing capabilities is good enough to accomplish many tasks with no active control; however, a passive device with active servos can provide more flexibility and versatility. It indicates that there are advantages to controlling a compliant wrist instead of controlling a compliant robot for the fine compliant motion. Following the same principle, major research is also devoted to the design and control of compliant robotic hands with sensors (60,61).

FUTURE OUTLOOK

The importance of robot manipulators in increasing quality and productivity in manufacturing has long been recognized. To increase the range of robot applications, the control of compliance is a key element in enhancing robot versatility and capability, especially in parts assembly applications. From the results of past research and development, a fundamental theory of compliant motion was developed and several approaches to providing the compliance were also proposed. In coping with these ideas, a number of force sensors and force controls were also developed to handle the compliant constraints.

Although some important findings in understanding and providing compliance have been made, some principal problems remain to be solved, and more work is needed in this area. Beginning with the fundamental theory, Mason's work gave a guideline to the synthesis of a compliant task; however, it is not easy to generate the compliant motion control strategy for any compliant task without the help of an expert. A more useful goal is an assembly expert system that can synthesize the task and automatically generate the compliant motion control strategy for the constrained task. By reaching this goal, a link to the design of an intelligent manufacturing system can be made.

Three basic approaches to providing active compliance have been proposed, but all three have deficiencies. For example, in the free-joints selection method, the selection of compliant motion is in joint space and thus causes joint motion coupling errors; therefore, a modification of the coordinate transformation through software design is necessary. To maintain the requirement of continuous motion, the whole process becomes very complicated. Using current microprocessors, the drawback to hybrid position/force control and accommodation force control is computational, due to the involvement of the Jacobian inverse and the direct control in Cartesian space. Active stiffness control is very attractive, but the method fails as the manipulator approaches any kinematic singularity. Many researchers have identified these drawbacks and proposed modified approaches. For example, one modified approach that combines the basic concepts of the above three has also been proposed (52); in it, the desired compliance is specified in Cartesian space and the control is accomplished in joint space. Some other researchers (62,63) also included the dynamics into the design of compliance control and modified the basic approaches. Since the control of compliance is such a delicate operation for a robot to perform, more efficiency and accuracy considerations are needed in the design of the system. This leaves enough room for more modified or new basic approaches in the future.

All the developed approaches to providing compliance were successfully demonstrated through experiments, but the basic type of servo is a design of a simple PID controller to make the system simple and workable. Therefore, some principal problems related to the control stability issue in the control of compliance still remain. The problems include: how the sensitivity and force resolving time of the sensor will affect the control; how to design a more advanced control law and robust analysis of the system; how the sampling time will affect the stability of the discrete system; what is the effect of robot stiffness and natural frequency on servo stability and how to control the compliance of a flexible arm; what is the effect of dynamics and kinematics on the system; how to plan motion trajectory through software implementation so that the compliance system will have minimized errors; and so on. Many researchers will be heading in these directions and trying to solve these important problems; for example, a thorough stability analysis of active stiffness control was done recently (39,40).

Another important element of compliance control is sensors, and the kind of device that should be used. Each kind of sensor has its own advantages; for example, wrist and finger sensors are the best for measuring contact forces, but joint torque sensors are excellent for controlling applied joint torques needed by the compliant motion and minimizing the motor friction effect. Therefore, a combination of different sensors may be the best choice for future implementation. In addition, a passive compliance device can also be used as a sensor; for example, the instrumented remote center compliance (IRCC) device (64) developed at the Draper Laboratory can provide active sensory feedback along with the passive mechanical linkages. Since a passive compliance device designed for a certain task is the best tool in providing compliance for that task, a combination of active and passive compliance may be the best compliance control for the future (56–61).

Up to now, the developed approaches to robot compliance have provided the basic knowledge in this area. However, compliance is still well behind vision in both theory and level of application in industry, especially in understanding compliant motion in assembly tasks and control stability issues. More effort is needed to solve these basic problems. Until it is made, a fully automated CAD/CAM assembly system can be viewed as: the CAD system designs the parts and analyzes the motion for mating parts, and the assembly expert system synthesizes the compliant motion and generates the control strategy for the mating parts; then the CAM system machines the parts, and the robot follows the compliant control strategies and planned trajectories to assemble the parts. This picture of the system will be a part of a future intelligent manufacturing system.

BIBLIOGRAPHY

1. R. Paul, "Modeling, Trajectory Calculation and Servoing of A Computer Controlled Arm," Stanford University Artificial Intelligence Laboratory, Stanford, Calif., AIM-177, 1972.
2. R. Paul, "Manipulator Cartesian Path Control," *IEEE Transactions on Systems, Man and Cybernetics* **SMC-9,** 702–711 (Nov. 1979).
3. R. H. Taylor, "Planning and Execution of Straight Line Manipulator Trajectories," *IBM Journal of Research and Development* **23**(4), 424–436 (1979).
4. D. E. Whitney, "Resolved Motion Rate Control of Manipulators and Human Prostheses," *IEEE Transactions on Man–Machine Systems* **23**(2), 47–53 (1969).

5. D. E. Whitney, "The Mathematic of Coordinated Control of Prostheses Arms and Manipulators," *Transactions of ASME, Journal of Dynamic Systems, Measurement, and Control,* 303–309 (Dec. 1972).
6. B. R. Markiewicz, "Analysis of the Computed Torque Drive Method and Comparison with Conventional Position Servo for a Computer-controlled Manipulator," Jet Propulsion Laboratory, Pasadena, Calif., Memo 33–601, 1973.
7. J. Y. S. Luh, M. W. Walker, and R. Paul, "Resolved-Acceleration Control of Mechanical Manipulators," *IEEE Transactions on Automatic Control* **AC-25**(3), 468–474 (June 1980).
8. C. H. Wu and R. Paul, "Resolved Motion Force Control of Robot Manipulators," *IEEE Transactions on Systems, Man and Cybernetics* **SMC-12**(3), 266–275 (1982).
9. A. K. Bejczy, "Robot Arm Dynamics and Control," Jet Propulsion Laboratory, Pasadena, Calif., Tech. Memo 33–669, 1974.
10. A. K. Bejczy, "Dynamic Models and Control Equations for Manipulators," Jet Propulsion Laboratory, Pasadena, Calif., Tech. Memo 715–19, 1979.
11. J. M. Hollerbach, "A Recursive Formulation of Lagrangian Manipulator Dynamics," *IEEE Transactions on System, Man and Cybernetics* **SMC-10**(11), 730–736 (Nov. 1980).
12. R. Paul, *Robot Manipulators: Mathematics, Programming, and Control,* M.I.T. Press, Cambridge, Mass., 1981.
13. J. Y. S. Luh, M. W. Walker, and R. Paul, "On-line Computational Scheme for Robot Manipulators," *Transactions of ASME, Journal of Dynamic Systems, Measurement, and Control* **102,** 69–76 (June 1980).
14. S. Dubowsky and D. T. DesForges, "The Application of Model Referenced Adaptive Control to Robotic Manipulators," *Transactions of ASME, Journal of Dynamic Systems, Measurement, and Control* **101**(3), 193–200 (Sept. 1979).
15. I. D. Landau, "Model Reference Adaptive Controllers and Stochastic Self-tuning Regulators—A Unified Approach," *Transactions of ASME, Journal of Dynamic Systems, Measurement, and Control* **103,** 404–416 (Dec. 1981).
16. A. Balestrino, G. De Maria, and L. Sciavicco, "Adaptive Control of Manipulators in the Task Oriented Space," *Proceedings of the 13th ISIR,* Chicago, Ill., Apr. 1983.
17. C. S. G. Lee and M. J. Chung, "An Adaptive Control Strategy for Mechanical Manipulators," *IEEE Transactions on Automatic Control* **AC-29**(9), 837–840 (Sept. 1984).
18. A. J. Koivo and T. Guo, "Adaptive Linear Controller for Robotic Manipulators," *IEEE Transactions on Automatic Control* **AC-28**(2), 162–171 (Feb. 1983).
19. M. Vukobratovic and D. Stokic, "On Engineering Concept of Dynamic Control of Manipulators," *Transactions of ASME, Journal of Dynamic Systems, Measurement, and Control* **102,** 108–118 (June 1981).
20. M. Vukobratovic and N. Kircanski, "Decoupled Control of Robots Via Asymptotic Regulators," *IEEE Transactions on Automatic Control* **AC-28**(10), 978–981 (Oct. 1983).
21. S. S. M. Wang and P. M. Will, "Sensors for Computer Controlled Mechanical Assembly," *The Industrial Robot* (Mar. 1978).
22. B. E. Shimano, "The Kinematic Design and Force Control of Computer Controlled Manipulators," Stanford University Artificial Intelligence Laboratory, Stanford, Calif., AIM-313, 1978.
23. J. L. Nevins and co-workers, "Second Report on Exploratory Research in Industrial Modular Assembly," Charles Stark Draper Laboratory, Cambridge, Mass., 1974.
24. V. D. Scheinman, "Design of a Computer Controlled Manipulator," Stanford University Artificial Intelligence Laboratory, Stanford, Calif., AIM-92, 1969.
25. D. E. Whitney, "Force Feedback Control of Manipulator Fine Motions," *Transactions of ASME, Journal of Dynamic Systems, Measurement, and Control,* 91–97 (June 1977).
26. C. H. Wu and R. Paul, "Manipulator Compliance Based on Joint Torque Control," *Proceedings of the 19th IEEE Conference on Decision and Control,* Albuquerque, N.M., Dec. 1980, pp. 88–94.
27. C. H. Wu, "Compliance Control of a Robot Manipulator Based on Joint Torque Servo," *The International Journal of Robotics Research* **4**(3), 55–71 (Fall 1985).
28. R. C. Goertz, "Fundamentals of General Purpose Remote Manipulators," *Nucleonics* **10**(11), 36–42 (1952).
29. R. C. Goertz, "Manipulators Developed at ANL," *Proceedings of the 12th Conference on Remote System Technology,* 1964, pp. 117–136.
30. E. Nakano and co-workers, "Cooperational Control of the Anthropomorphous Manipulator 'MELARM,'" *Proceedings of the 4th International Symposium on Industrial Robots, Society of Manufacturing Engineers,* 1974, pp. 251–260.
31. R. C. Groome, "Force Feedback Steering of a Teleoperator System," Master's Thesis, Department of Aeronautics and Astronautics, Massachusetts Institute of Technology, Cambridge, Mass., 1972.
32. M. A. Jilani, "Force Feedback Hydraulic Servo for Advanced Automation Machine," Master's Thesis, Department of Mechanical Engineering, Massachusetts Institute of Technology, Cambridge, Mass., 1974.
33. H. Inoue, "Computer Controlled Bilateral Manipulator," *Bulletin of the JSME 14* **69,** 199–207 (1971).
34. D. Silver, "The Little Robot System," M.I.T. Artificial Intelligence Laboratory, Cambridge, Mass., AIM-273, 1973.
35. K. Takase, H. Inoue, K. Sato, and S. Hagiwara, "The Design of an Articulated Manipulator with Torque Control Ability," *Proceedings of the 4th International Symposium on Industrial Robots,* Tokyo, Japan, Nov. 1974, pp. 261–270.
36. R. Paul and B. Shimano, "Compliance and Control," *Proceedings of the IEEE Joint Automatic Control Conference,* San Francisco, Calif., 1976, pp. 694–699.
37. J. Y. S. Luh, W. Fisher, and R. Paul, "Joint Torque Control by a Direct Feedback for Industrial Robots," *IEEE Transactions on Automatic Control* **AC-28**(2), 153–161 (Feb. 1983).
38. J. K. Salisbury, "Active Stiffness Control of a Manipulator in Cartesian Coordinates," *Proceedings of the 19th IEEE Conference on Decision and Control,* Albuquerque, N.M., Dec. 1980, pp. 95–100.
39. H. Kazerooni, T. B. Sheridan, and P. K. Houpt, "Robust Compliant Motion for Manipulators, Part I: The Fundamental Concepts of Compliant Motion," *IEEE Journal of Robotics and Automation* **RA-2**(2), 83–92 (June 1986).
40. H. Kazerooni, P. K. Houpt, and T. B. Sheridan, "Robust Compliant Motion for Manipulators, Part II: Design Method," *IEEE Journal of Robotics and Automation* **RA-2**(2), 93–105 (June 1986).
41. D. E. Whitney, "Historical Perspective and State of the Art in Robot Force Control," *Proceedings of the 1985 IEEE International Conference on Robotics and Automation,* St. Louis, Mo., Mar. 1985, pp. 262–268.
42. M. H. Raibert and J. J. Craig, "Hybrid Position/Force Control of Manipulators," *Transactions of ASME, Journal of Dynamic Systems, Measurement, and Control* **102,** 126–133 (June 1981).
43. R. K. Roberts, R. P. Paul, and B. M. Hillberry, "The Effect of Wrist Force Sensor Stiffness on the Control of Robot Manipulators," *Proceedings of the 1985 IEEE International Conference on*

Robotics and Automation, St. Louis, Mo., Mar. 1985, pp. 269–274.

44. P. C. Watson, "A Multidimensional System Analysis of the Assembly Process as Performed by a Manipulator," Charles Stark Draper Laboratory, Cambridge, Mass., P-364, Aug. 1976.
45. J. L. Nevins and D. E. Whitney, "Computer Controlled Assembly," *Scientific American* **238**(2), 62–74 (1978).
46. J. L. Nevins and D. E. Whitney, "Assembly Research," *The Industrial Robot,* 27–43 (Mar. 1980).
47. J. W. Hill and A. J. Sword, "Manipulation Based on Sensor-directed Control: An Integrated End Effector and Touch Sensing System," paper presented at the 17th Annual Human Factors Society Convention, Washington, D.C., 1973.
48. R. Bolles and R. Paul, "The Use of Sensory Feedback in a Programmable Assembly System," Stanford University Artificial Intelligence Laboratory, Stanford, Calif., AIM-220, 1973.
49. C. C. Geschke, "A System for Programming and Controlling Sensored-Based Robot Manipulators," Ph.D. Dissertation, University of Illinois, Urbana, Ill., Dec. 1978.
50. K. G. Shin and C. P. Lee, "Compliant Control of Robotic Manipulators with Resolved Acceleration," *Proceedings of the 24th IEEE Conference on Decision and Control,* Ft. Lauderdale, Fla., Dec. 1985, pp. 350–357.
51. H. West and H. Asada, "A Method for the Design of Hybrid Position/Force Controllers for Manipulators Constrained by Contact with the Environment," *Proceedings of the 1985 IEEE International Conference on Robotics and Automation,* St. Louis, Mo., Mar. 1985, pp. 251–259.
52. H. Zhang and R. P. Paul, "Hybrid Control of Robot Manipulators," *Proceedings of the 1985 IEEE International Conference on Robotics and Automation,* St. Louis, Mo., Mar. 1985, pp. 602–607.
53. M. T. Mason, "Compliance and Force Control for Computer Controlled Manipulators," *IEEE Transactions on Systems, Man and Cybernetics* **SMC-11**(6), 418–432 (June 1981).
54. R. Paul, B. Shimano, and G. E. Mayer, "Kinematic Control Equations for Simple Manipulators," *IEEE Transactions on Systems, Man and Cybernetics* **SMC-11**(6), 449–455 (June 1981).
55. T. Goto, T. Inoyama, and K. Takeyasu, "Precise Insertion Operation by Tactile Controlled Robot "Hi-Ti-Hand" Expert 2," *Proceedings of the Fourth International Symposium on Industrial Robots,* Tokyo, Japan, Nov. 1974, pp. 209–218.
56. H. Van Brussel and J. Simons, "The Adaptable Compliance Concept and Its Use for Automatic Assembly by Active Force Feedback Accommodation," *Proceedings of the 9th International Symposium on Industrial Robots,* Washington, D.C., 1979, pp. 167–181.
57. H. Van Brussel, H. Thielemans, and J. Simons, "Further Development of the Active Adaptive Compliant Wrist (AACW) for Robot Assembly," *Proceedings of the 11th International Symposium on Industrial Robots,* Tokyo, Japan, Oct. 1981, pp. 377–384.
58. M. R. Cutkosky and P. K. Wright, "Position Sensing Wrists for Industrial Manipulators," *Proceedings of the 12th International Symposium on Industrial Robots,* Paris, France, June 1982.
59. M. R. Cutkosky and P. K. Wright, "Active Control of a Compliant Wrist in Manufacturing Tasks," *Transactions of ASME, Journal of Engineering for Industry* **108,** 36–43 (Feb. 1986).
60. J. K. Salisbury and J. J. Craig, "Articulated Hands: Force Control and Kinematic Issues," *International Journal of Robotics Research* **1**(1), 4–17 (Spring 1982).
61. S. Jacobsen and co-workers, "The Utah/MIT Hand, Work in Progress," *International Journal of Robotics Research* **3**(4), 21–50 (Winter 1984).
62. N. Hogan, "Impedance Control of Industrial Robots," *Robotics and Computer Integrated Manufacturing* **1**(1) (1984).
63. O. Khatib, "The Operational Space Formulation in Robot Manipulator Control," *Proceedings of the 15th International Symposium on Industrial Robots,* Tokyo, Japan, Sept. 1985, pp. 165–172.
64. T. L. DeFazio, D. S. Seltzer, and D. E. Whitney, "The Instrumented Remote Center Compliance," *Industrial Robots* **11**(4), 418–432 (Dec. 1984).

COMPONENT ASSEMBLY ONTO PRINTED CIRCUIT BOARDS

Robin Truman
Plessey Network and Office Systems, Ltd.
Beeston, Nottingham, United Kingdom

INTRODUCTION

Modern manufacture with its competitive demands requires that the most mechanized means be used in the pursuit of excellence. Electronics manufacture is one of the most emergent and changing areas within manufacturing. It is natural, therefore, that the methods used in the assembly of electronic boards, the heart of any electronic system, have occupied the minds of equipment suppliers and electronics manufacturers; this is especially the case in recent years, as evidenced by the growth in electronics and robot exhibitions and conferences. Such efforts have resulted in the availability of a wide range of highly specialized and capable dedicated board assembly equipment to cater for the many facets of board interconnection technology. From this array of mechanized assembly M/Cs the robot has emerged as a means to fulfill tasks that cannot be justified or accomplished with as much flexibility and effect by other methods.

PRINTED BOARD TECHNOLOGY

Printed Wiring Assemblies

An appreciation of printed circuit board (PCB) terminology, design, and manufacture (1) is essential to understanding component assembly problems and requirements. PCBs, sometimes called bare boards (Fig. 1), are normally produced from either copper-coated epoxy glass laminate or synthetic resin bonded paper (SRBP) and are assembled with a wide variety of components to form printed wiring assemblies (PWAs) (2) (Fig. 2).

Printed wiring assemblies have been used in vast quantities for many years by the electronics industry, and their use is growing and spreading progressively to a wide range of other industries. This, combined with changing board interconnection technology, has induced suppliers of manufacturing equipment and components and circuit board manufacturers to seek improved manufacturing methods.

Modern electronics manufacture uses computer-aided design (CAD) (Fig. 3) to produce the original circuit design and board drilling instructions, the circuit design then being used to produce photographic art-work tools for board printing. Af-

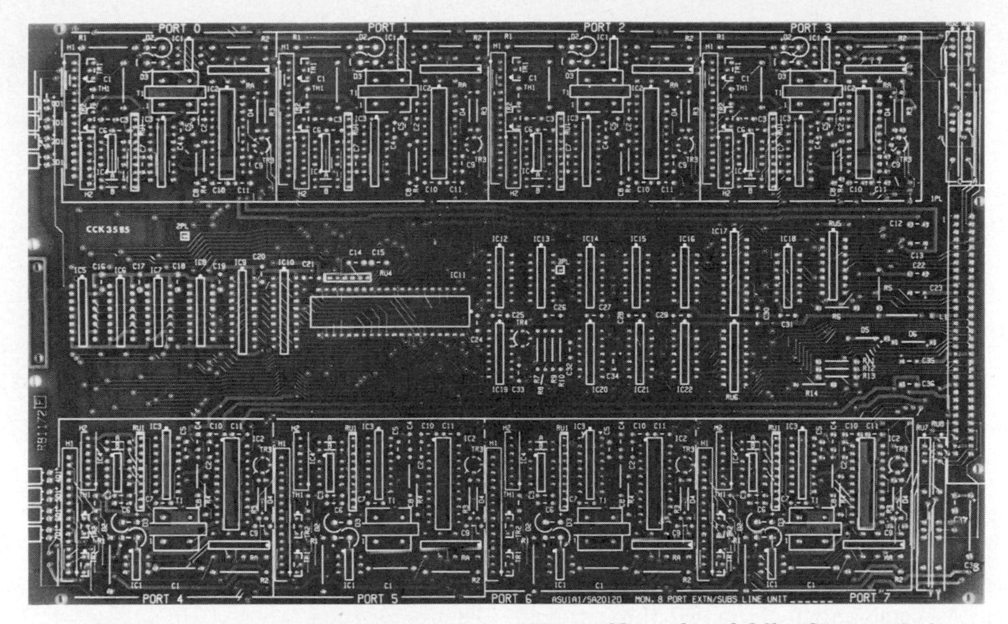

Figure 1. Printed circuit board. Courtesy of Plessey Network and Office Systems, Ltd.

ter drilling and printing, the boards are chemically etched to form the circuit pattern.

Circuit boards are either single- or double-sided (ie, with a circuit pattern on either one or both sides), and during board manufacture, to align the art work on opposite sides, pins are used to locate in tooling holes made through the board and art-work film. Tooling holes provide a reference for precise art-work alignment during subsequent component assembly. After the circuit pattern has been etched on the board, solder resist is applied to the board surface to ensure that during soldering (after component insertion) the solder is confined to component termination pads. Double-sided boards have plated-through-holes (PTH) (Fig. 4) to assure the integrity of joints after components have been fitted and soldered.

Circuit Technology

Traditionally, plated-through-hole boards and leaded components have been used to make circuit connections, but interconnection technology is moving increasingly toward surface-mounted components (SMCs) (3,4). Surface mount offers several advantages over leaded components, including improved performance, reduction in board size and, in some cases, reduced cost. Currently, it is estimated that approximately 150,000 components are available in standard form and 15,000 in chip form for surface mounting. Total transition from through-hole (insertion) to surface mount is therefore dependent on component manufacturers producing a total range of components in chip form, and will take many years. In the

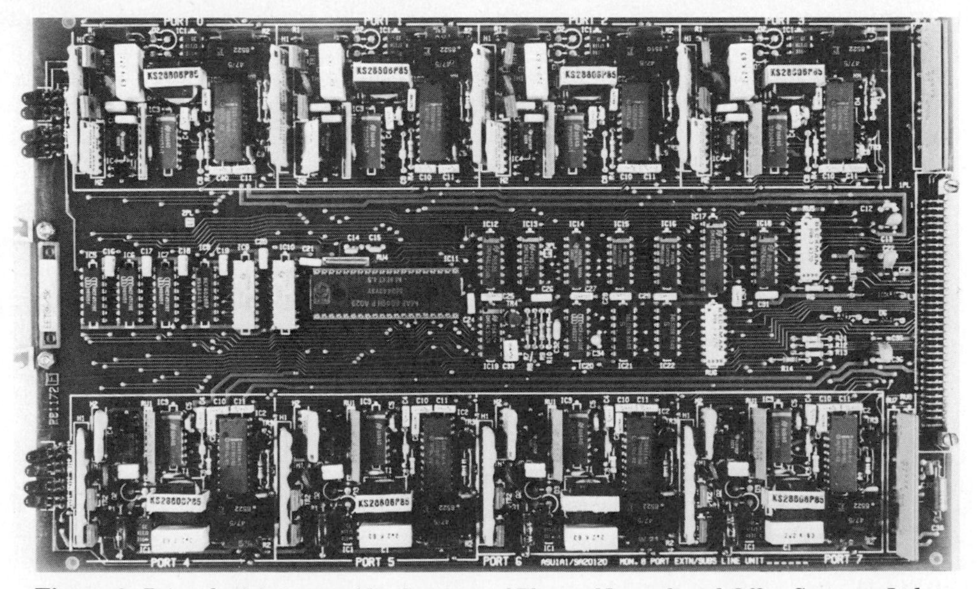

Figure 2. Printed wiring assembly. Courtesy of Plessey Network and Office Systems, Ltd.

Figure 3. Computer-aided circuit design. Courtesy of Plessey Network and Office Systems, Ltd.

interim, boards will be either through-hole, surface mount, or "mixed technology," a combination of the two (through-hole and surface mount on the same board).

Surface-mounted components can be assembled on either one or both sides of the board, and when used with leaded (inserted) components, the assembly sequence is influenced accordingly. SMCs are secured to board pads using solder paste, adhesive, or both and are subsequently connected by either wave soldering, when adhesive only is used, or solder reflow (vapor phase, infrared etc), when solder paste is used. On mixed-technology boards with SMC on both sides and where it is possible to clinch retain inserted components, a common approach is to assemble and solder the SMCs on the leaded component side first, followed by the leaded components and lastly, the SMC on the reverse side. The whole assembly is then wave-soldered, the SMCs on the underside passing through a double solder wave to overcome shadowing and bridging.

With this assembly sequence, care must be exercised to clear the surface-mounted components during subsequent leaded component insertion. Assembly of inserted components first may cause clinched components to fall off during surface-mounted assembly. Simple clinching may not be practical when component leads are stiff or so configured that damage to the component may result. In such cases, components may be retained by slight distortion of one or more leads or by the application of adhesive to the underside of the component before insertion.

Component Placement

Inserted Components. A wide variety of leaded components can be placed into boards using dedicated automatic component insertion (ACI) machines (5), some of the earliest of which were for the insertion of axial components (Fig. 5). This latter M/C is used in conjunction with a component sequencing machine (Fig. 6). More recently, extremely capable and highly automated machines have been built (Fig. 7) for the insertion of axial and radial leaded components (Fig. 8).

Surface-mounted Components. Surface-mounted components require specialized equipment (Fig. 9) to apply adhesive and place the components on the board.

Most components for surface-mounted application are carried on 8-mm tape, but larger (flat-pack) components are magazine-fed. Parts are picked and placed, usually by a vacuum head with centralizing jaws.

Dedicated machines for surface-mounted, axial, radial, and dual in-line pack (DIP) placement can successfully assemble the majority of components to a board, but up to 40% of components cannot be assembled by these means. Such components are designated "odd form" and are increasingly being addressed for insertion by robot suppliers.

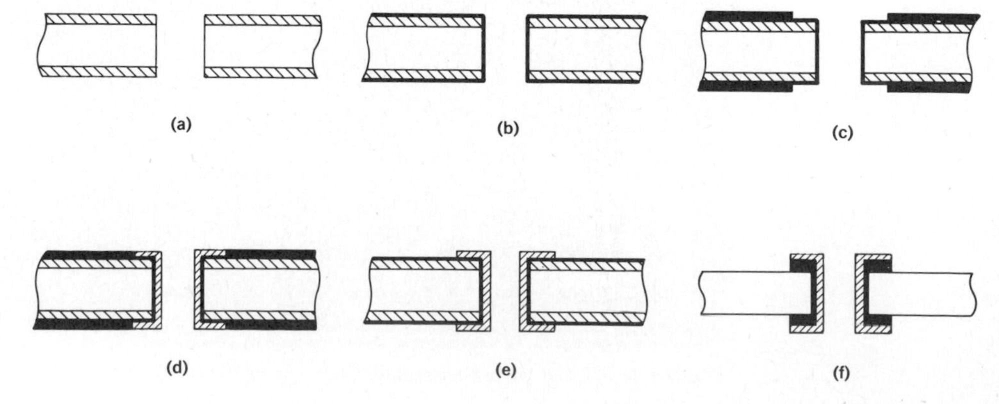

Figure 4. Stages in making plated-through-hole printed circuit boards: (**a**) copper-clad laminate, holes drilled; (**b**) electroless copper-coated; (**c**) plating resist applied; (**d**) electroplated copper plus tin–lead; (**e**) plating resist removed; (**f**) etched.

Figure 5. Axial component insertion machine. Courtesy of Plessey Network and Office Systems, Ltd.

THE PLACE OF ROBOTS IN COMPONENT INSERTION

Extent of Installations

Although robots have been commercially available for over a decade, it is only in recent years that they have been developed for component insertion, possibly resultant from coincident need and robotic development.

Accordingly, the installed base of operational integrated robotic component insertion lines is not high, with most progress in Japan and the United States, but there is considerable interest in Europe, where individual and dual robot cells and integrated lines are being increasingly applied to electronic assembly. At the current rate of technological development, integrated lines of ACI and robotic component insertion machines will increase rapidly in use and application and will become one of the greatest growth areas for robots over the next decade.

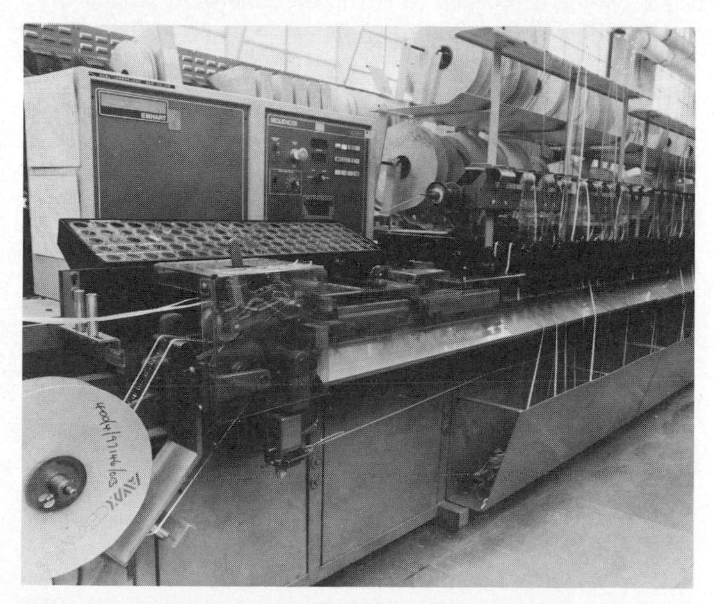

Figure 6. Component sequencing M/C. Courtesy of Plessey Network and Office Systems, Ltd.

Figure 7. Axial and radial component insertion M/C. Courtesy of Plessey Network and Office Systems, Ltd.

Reasons for Robot Use

Because dedicated ACI machines generally have higher placement rates (3000 parts per hour and upward) than currently available robots, whenever component types and quantities permit, they are normally preferred by PWA manufacturers. However, robots, through their inherent ability for ready reconfiguration to meet a wide variety of component and board

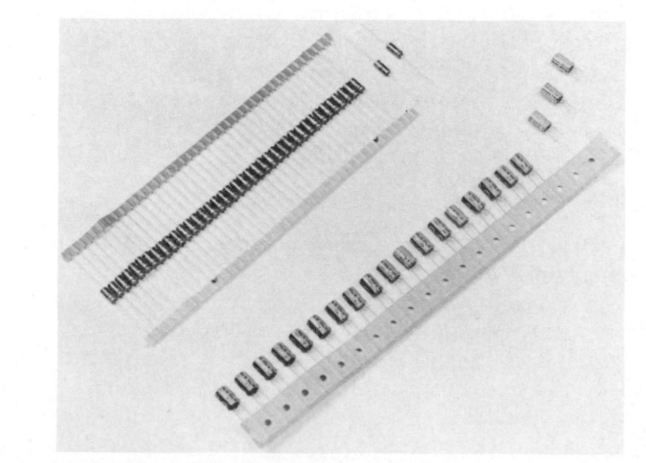

Figure 8. Axial and radial components on tape reels. Courtesy of Plessey Network and Office Systems, Ltd.

Figure 9. Surface-mounted component placement M/C. Courtesy of Plessey Network and Office Systems, Ltd.

assembly requirements, are more flexible than ACI, and fully robotic insertion systems are used by some electronic manufacturers. When used with ACI, robots can populate on average 95% of board components, leaving the remaining 5% for assembly by hand or other means. Component form, creating feeding and gripping problems, most frequently inhibits robotic insertion.

SYSTEM SPECIFICATION AND OPERATION

The Problem of Time Scale

The principal difficulty in specifying robot lines, as with most capital projects, is that of time scale from inception, through justification, authorization, build, and commissioning. This cannot normally be achieved in less than 15 months and can approach 2 years before a line achieves the planned level of performance.

With such timetables, a high level of assurance is required in the life of the product for which the equipment is to be provided. Herein lies one of the principal advantages of robots, that of their inherent flexibility, enabling the machines to be reconfigured to accept and place different components as design changes or different board requirements occur during the equipment procurement period.

Simulation Techniques

Much of the difficulty the project engineer has in specifying equipment lines exactly is caused by not truly knowing how the proposed equipment will perform under different board volumes and configurations and under different board-throughput conditions. This uncertainty can be greatly reduced by the use of computer simulation techniques, which permit the modeling of lines using a color graphic display and the posing of "what if" situations to the model. Using these techniques, the effect on throughput times and inventory with different cell configurations and board requirements can be examined. A number of proprietary systems are available for production-line simulation.

Component Identification for Robotic Insertion

At the earliest stage of project evaluation, known board and component placement requirements need to be tabulated, with as much future prediction as possible. It is convenient to mount components that are required to be addressed for possible robotic insertion, when identified, on a display card (Fig. 10) with double-sided adhesive tape, together with identity, quantities, form in which supplied (eg, trays, tape reels, loose, etc), and lead cutting, forming, and clinching requirements.

This information can be photographed to assist discussion with potential robot suppliers. A typical board, before and after population by the components to be addressed, and samples of components from each supplier to be used are essential to robot supplier understanding of precise needs. This enables the robot supplier to assess the practicality of component insertion and to prepare a proposal, cost, and delivery.

In considering practicality, the equipment supplier seeks ease of solution and normally wishes not to tackle components that are difficult to feed or grip, as the cost of mechanization may prove prohibitive compared with manual methods.

Board-design Considerations

It is imperative that designers of circuits, and especially board layout engineers, design with manufacturing and test methods in mind. Problems of component insertion are considerably eased by obeying design rules on such vital aspects as component spacing, orientation, and placement relationship with other components. The choice and nature of components (ie, whether they are axial, radial, surface-mounted, DIP, or odd form), as has been shown, significantly influence manufacturing technique and costs. Board-conveying problems are simplified if boards are designed with one dimension standard so that fixed-width conveyors and board-feeding magazines can be used. For the above reasons, board designers and industrial engineers are increasingly working more closely, often using common CAD workstations, and thus improving mutual understanding (6).

If boards cannot be designed to standard dimension, there are four principal solutions to the conveying problem:

- The boards can be designed in frets of multiple boards up to a standard width, with break-out perforations (Fig. 11), or on thinner boards with scoring for individual separation after component insertion and soldering (and perhaps test). This is a good and commended solution.
- The board can be located and transported in a standard width carrier. This system presents problems of carrier cost and recycling but simplifies board location.
- A computer-controlled variable-width conveyor can be used. An expensive solution, this requires that boards of one size are cleared from the line before the next is processed. It also presents problems in matching board magazines, feeding and off-loading the line, with the conveyor.

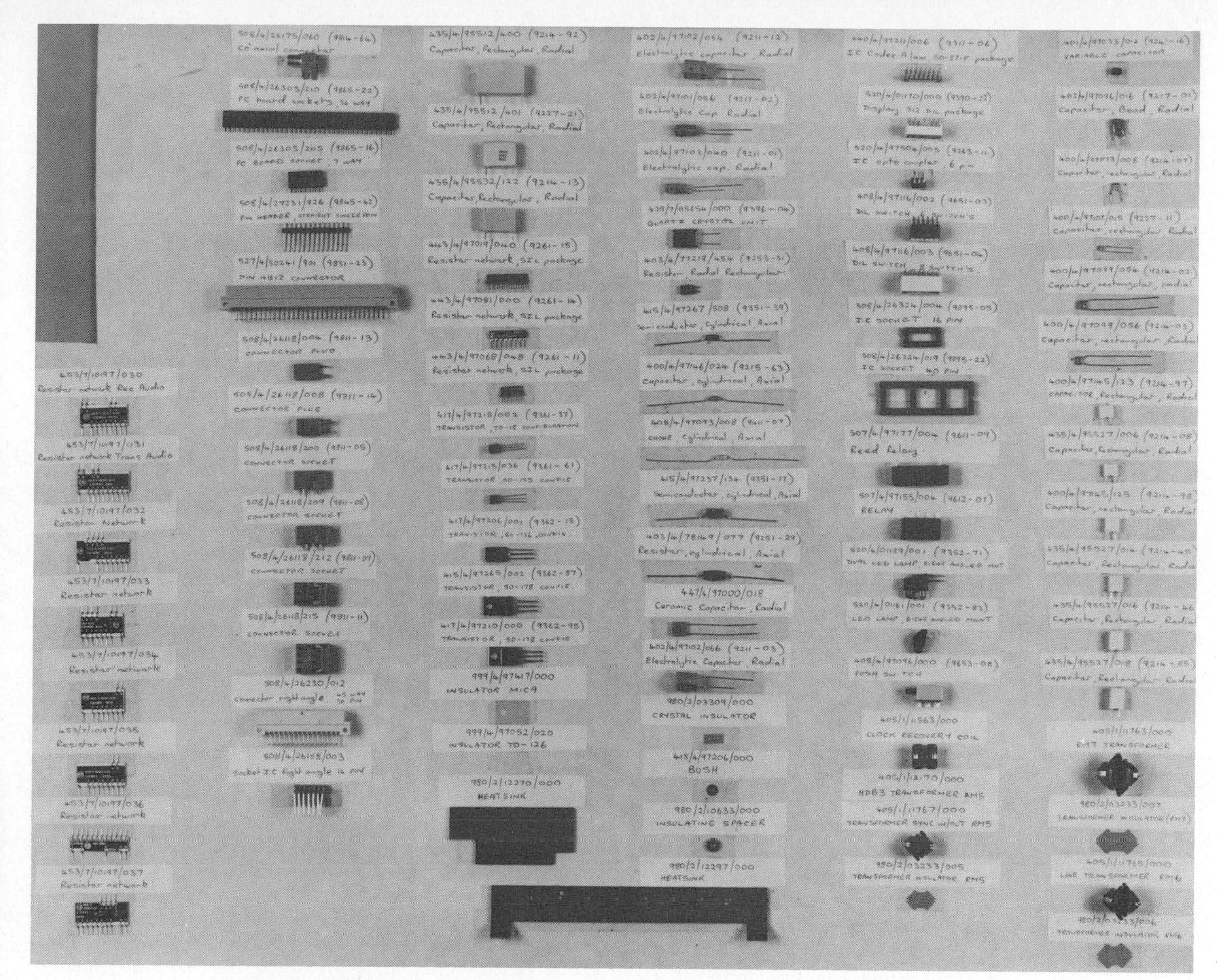

Figure 10. Components displayed for consideration for robotic insertion. Courtesy of Plessey Network and Office Systems, Ltd.

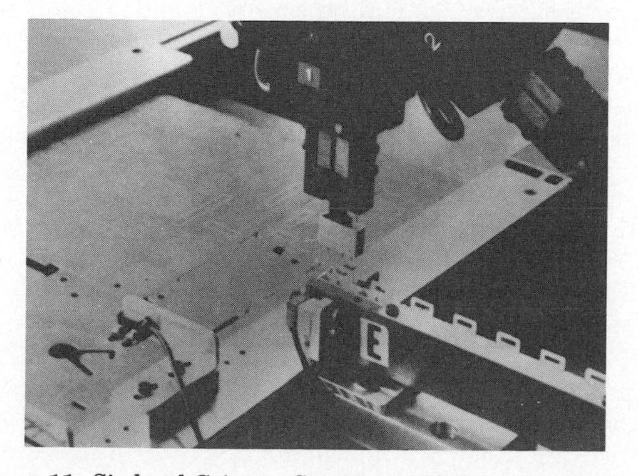

Figure 11. Six-head Gripper. Courtesy of Phillips, Croyden, UK.

- With careful and modular design, conveyor width can be manually adjusted to accommodate different board widths.

Line Configuration and Build

Depending on the variety, quantities, and nature of boards to be handled, different solutions to component insertion are offered (7) (Fig. 12). Small quantity, high variety boards can be assembled, using a gantry or SCARA robot cell, to feed and off-load boards and to place components from feeders into the board, which is located in a fixed position on the machine table (8). Alternatively, boards can be fed manually or by conveyor. Ancillary lead preforming and qualifying operations can be accommodated within the work envelope.

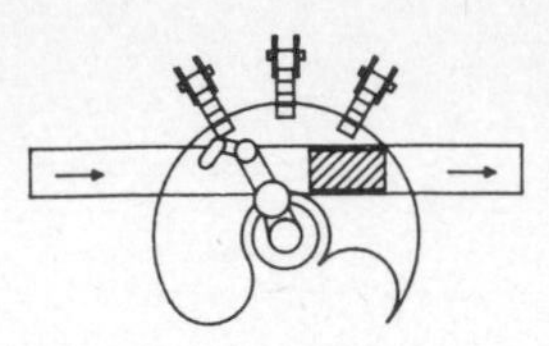

SINGLE CELL - HIGH VARIETY

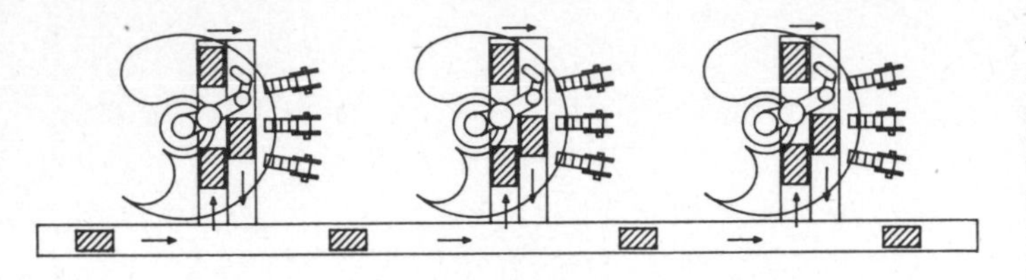

MULTIPLE CELL
HIGH VARIETY, HIGH/LOW VOLUME

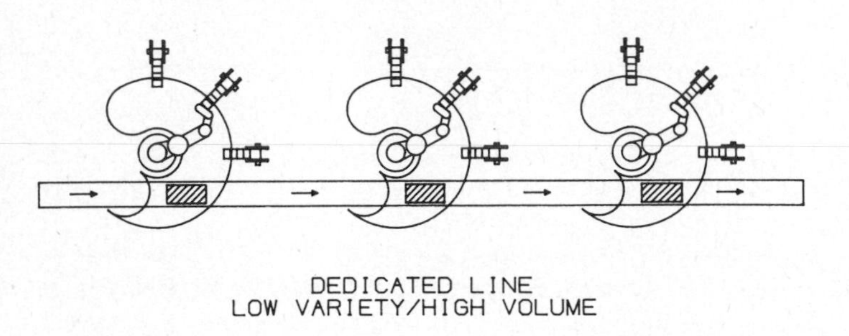

DEDICATED LINE
LOW VARIETY/HIGH VOLUME

Figure 12. Robotic component insertion line configurations to suit different board requirements. Courtesy of Plessey Network and Office Systems, Ltd.

High volume, moderate-variety work is best handled by a number of robots integrated into lines, linked by board-carrying conveyors. Such systems fall into two categories: those addressing volume production of a single board type and those dealing with a range of board types and sizes. The latter configuration presents the greatest difficulty, but both must feed a wide range of different components, in the former case to a family of the same board type.

High variety, low volume is best addressed by a series of robotic cells arranged as conveyorized spurs to a main feed to permit buffers of boards to be held at each cell and for cells to be bypassed to meet different board build configurations.

Higher volume, lower-variety requirements are met by a less flexible approach with a single conveyor passing through dedicated robot stations. The need for line balancing (9) on this system is paramount because of the absence of buffering; hence, the cycle time for the line will be that of the slowest station.

This is by no means an exhaustive list; line configurations (10) are only constrained by the imagination of the line designer, prompted by individual need. In the future, line designs will be influenced by the specific requirements of the PWA manufacturer and by developments in component supply packaging, robots, feeders, end effectors, conveyors, and vision systems.

Equipment users have three choices when considering the design and construction of robotic component insertion systems:

1. To design and build the proposed system "in house."
2. To have the robot supplier design and build the system.
3. To have the system designed and built by a systems integrator.

Advantages and disadvantages exist with all three options and the choice is influenced by the experience and capability of the user and by commercial and contractual considerations. In-house construction of systems has much to commend it and should be seriously considered if sufficient skill and confidence exists within the user organization. Advantages include the following:

- Cash flow is better, as costs are spread and incurred as the work progresses.
- Experience is gained in house, rather than financing the experience of suppliers.
- The system can be grown steadily with each aspect of implementation being properly digested before the next is tackled.
- Confidence with and knowledge of the system are gained by industrial and maintenance engineers, overcoming much of the apprehension and problems associated with the "big-bang" installation of major systems into an organization.
- Costs will probably be less than by other methods.

Disadvantages are that skills gained by integrators are denied the user and timetables may be protracted, but both of these possibilities can be overcome with sufficient energy, skill, and tenacity. Robot suppliers can normally be relied upon to make the installation of their equipment a success.

Having the robot supplier build the system is the next-best choice, but the supplier may not always wish to undertake the task and may not, in the eyes of the user, have the requisite skills. If the robot supplier fulfills the foregoing criteria, there are many reasons to follow this route, the principal one being avoidance of ambiguity over contractual responsibility. The robot supplier in all probability needs support from other equipment suppliers to fulfill the contract and is highly motivated to ensure that performance criteria, deadlines, and costs are met.

The use of system integrators to provide robotic component insertion lines is growing and is the route taken by the majority of users. The reasons are

- It removes responsibility for the physical building of the system from the user.
- It is contractually convenient.
- The integrator is often able to refer to previous installations and the user can frequently see systems under construction at the integrator's establishment, encouraging confidence in capability. As the technology is still new, there is a "comfort factor" in working with an organization with a proven record.

Disadvantages are that arguments over the division of responsibility can develop, causing frustration and delay. Costs can be high, as margins need to be made by robot suppliers, peripheral equipment suppliers, and integrators. The latter, despite previous experience, is often faced with unforeseen, "first-time" dilemmas and needs to construct costs to cover contingencies. The user must choose the integrator carefully, as many are small and emergent, with little previous implementation experience in robotic component insertion.

Whichever route is taken, error recovery must be built into the system to cover all conceivable failure contingencies with facilities to provide for new ones as they are encountered (9). In designing the line, the aim is to achieve a balanced load on each robot cell; this is calculated from quoted placement times for each component. The number of robots is determined by the throughput of boards required and possibly by inventory considerations.

Modularity in system design permits extension by being able to change robots, grippers, and component and board feeders as board requirements change. It also allows conveyor extension and adaption.

The SCARA Robot

Much has been written (11,12) on the SCARA (selective compliance assembly robot arm) robot (Fig. 13); suffice it to say that by definition this robot was designed for assembly tasks and most major suppliers of robots have a SCARA design within their range. The positional repeatability of better than ± 0.002 in. (± 0.05 mm) and its ability to comply at the final point of operation make it ideal for precise assembly operations as encountered in electronic component insertion. SCARA robots are now commonly specified for such tasks, but robots of other configurations are also used. Since compensation can be made for positional inaccuracies, it is repeatability, ie, the ability to repeat a program sequence accurately, that is often most important. The exception is where off-line programming to a defined position is carried out (perhaps driven from CAD-derived data).

Figure 13. SCARA robot. Courtesy of Evershed Robotics, Ltd. (UK).

The SCARA robot can insert one component about every 3.5 seconds, although faster insertion speeds are claimed by some suppliers. If component integrity (soundness, polarity, etc) is required to be confirmed automatically within the insertion cycle, insertion time is extended to approximately 5 seconds, thus making the use of this practice economically questionable. Ultimately, the success of automated assembly depends on consistent component and board quality.

Gripper (End-effector) Design

Grippers for component insertion developed from early mechanical and pneumatic devices; current designs incorporate force sensing, permitting a range of components to be grasped by the same grippers (13) (see also GRIPPERS). Tactile sensing (14) has wide application in component insertion and is often used to sense whether component leads are entering the holes in the board into which placement is being attempted. It is common for a search routine (9) to be made when the initial insertion attempt fails; should placement not be made, the faulty component is discarded, a new one selected, and the sequence repeated.

There are two basic options in gripping electronic components: to grip either on the body or on the leads (Fig. 14). Gripping on the body affords simpler gripper design, but care must be taken not to crush sensitive components. Unless a system of force sensing is used, a wide range of grippers may be required to cover all the parts to be handled. It can be arranged for the robot to change its own grippers, but this inevitably adds to cycle time. The close proximity of other components on the board may prevent gripping on the leads (through inadequate space for the gripper), in which case components must be gripped on the body.

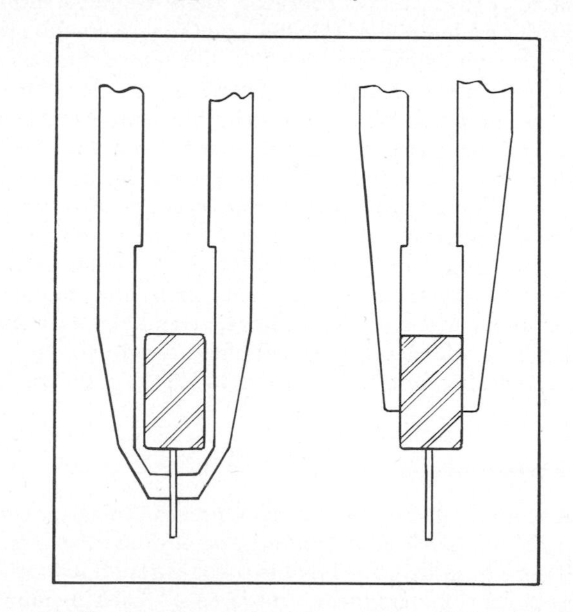

Figure 14. Component gripping alternatives. Courtesy of Plessey Network and Office Systems, Ltd.

The major disadvantage of gripping on the body of components results from the lack of a precise relationship between body and leads. This problem may be overcome by having the robot head or gripper release, align, and reclamp the component, referencing the leads to the board coordinates by using a "V" or conical location qualifying fixture, before finally placing the component into the board. In effect, the component is allowed to float, to reference the leads via the fixture, which is in a precise position relative to the board datum. Gripping on the leads is a more positive component locating system and enables a range of components to be gripped with the same gripper using a "swan-neck" gripper design. With both gripper systems, lead cropping and preforming is usually carried out before component placement in the board. Preforming creates a positive location feature on the leads to create standoff from the board and to prevent component movement during subsequent board handling and soldering. After placement, the component leads may be clinched from below to secure them into the board. Clinching can either be passive or active, ie, using either a static or moving clinch plate below the board.

Force-sensing grippers are expected to be developed in the future, but the problem of the relationship of the leads of a component with the body will continue, aggravated by dimensional differences on like components obtained from different suppliers.

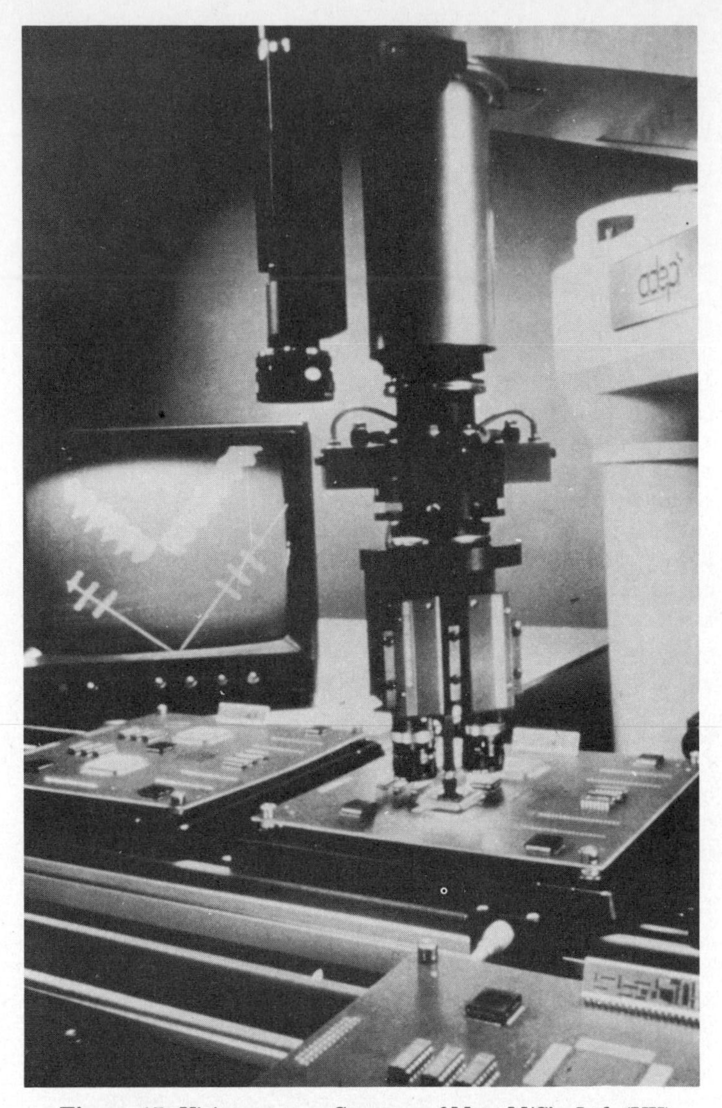

Figure 15. Vision system. Courtesy of Meta M/C's, Ltd. (UK).

Vision Systems

Vision systems (Fig. 15), like with other orientation systems, can be said to be a means of creating order out of disorder in component and board feeding (15,16) (see also Vision systems, implementation).

Vision systems are developing rapidly, but their use as offered, for the general solution of component insertion problems, has to be questioned, as the cost of a vision system currently approximates that of a high level, intelligent, programmable SCARA robot. However, there is much international work and interest in this field, and improved and lower-cost systems can be expected to solve difficult orientation and location problems. Commonly, a vision system works by a TV camera scanning the work, referencing the image received to programmed instructions, and directing the robot with which it is associated accordingly. Vision systems have developed to a stage where they can identify and locate randomly placed and partially overlapping parts traveling on a moving conveyor, signaling a robot to pick up the parts sequentially. As they are currently expensive solutions for normal component insertion applications, initial emphasis should be on working with suppliers and component manufacturers to supply parts in a form suitable for auto insertion, ideally in tape reels.

Robot Programming

Two basic methods are used in robot programming for component insertion: teach programming and off-line programming.

In the first method, the robot is taken through its sequence of operation by the engineer, who uses a "teach pendant" to fix each point and activity in the sequence. Once taught, the program can be "replayed" and the robot will follow the sequence, any slight positional error resulting from robot-arm inertia being readily corrected. Where the sequence of operation is complex and the robot must interact with a range of other equipment, such as conveyors, readers, sensors, and other machines, it is usual to program "off line." Individual manufacturers of high level, programmable, intelligent robots have developed their own programming languages, and program training is usually included in the equipment purchase price. The programming engineer, having set out the programming sequence required to perform the task, programs the robot using its own language. Programming must take account of the system's peripheral equipment, which is linked with the robot controller by its input/output (I/O) ports. A RS 232 port, or comparable access, is required for off-line programming and interfacing with other computer-based equipment.

After programming, commissioning lead times can be shortened by physically simulating the ultimate layout using wooden boards, stands, etc, to represent conveyor and other locations. This simulation can be performed away from the ultimate location and permits programming and robot cycles to be tested during the delivery period of peripheral equipment. When the system is finally constructed, minor positional discrepancies due to off-line programming are corrected.

Robot Simulation

Graphic computer simulation of integrated robot lines has been mentioned, but the simulation of actual robot cells is a different field of work, affording excellent prospects for assisting robot selection and configuration (17). A number of notable systems are in continuing development, using CAD workstations to simulate robot operation dynamically on an animated graphic display (Fig. 16). These systems have a range of popular robot types within their software and permit the ready insertion of other types, which, when tested in the proposed environment, aid robot choice and installation. The interaction of robots with each other and with peripheral equipment (typically component feeders, conveyors, board feeders, etc) can be simulated and the effects of overlapping work envelopes on multiple robot installations can be assessed. Once simulated, the actual robot program can be established from the CAD source and the various elements of the simulation system employed to complete the robotic programming aspects of the installation.

Justification

Arguments for robotic insertion are principally made on direct labor savings over manual insertion, especially if volumes are such as to require three-shift working (18–20), but interest is growing in the savings in inventory (21) achievable through reduced work in progress from automated printed wiring assembly lines. Boards produced within one month of order give a 12-times inventory turn, generally held to be an acceptable stockholding position, but using fully automatic assembly lines, lead times of 4 days are possible, thereby creating dramatic inventory savings.

To achieve such results, equipment utilization would probably be low, as sufficient insertion stations need to be provided to prevent delays when processing small quantities and large varieties of boards. Some trade-off is therefore necessary between inventory savings and capital equipment costs, and the importance of each to the specific business situation must be considered when drawing up the specification for the line.

Benefits from improved quality from automatic component insertion should result, but, in the early stages of mechanization implementation, a deterioration in quality is possible. The reasons for this are difficult to generalize, but component supply quality and early teething problems with equipment and parts feeding are common factors. Despite the occasional servo failure, robots have a track record of reliability; it is peripheral equipment that is usually more troublesome.

Mechanization in itself will not improve quality, but proper attention to board and component quality and care in equipment design and specification will result in improved consistency in the finished assembly. Close cooperation is required

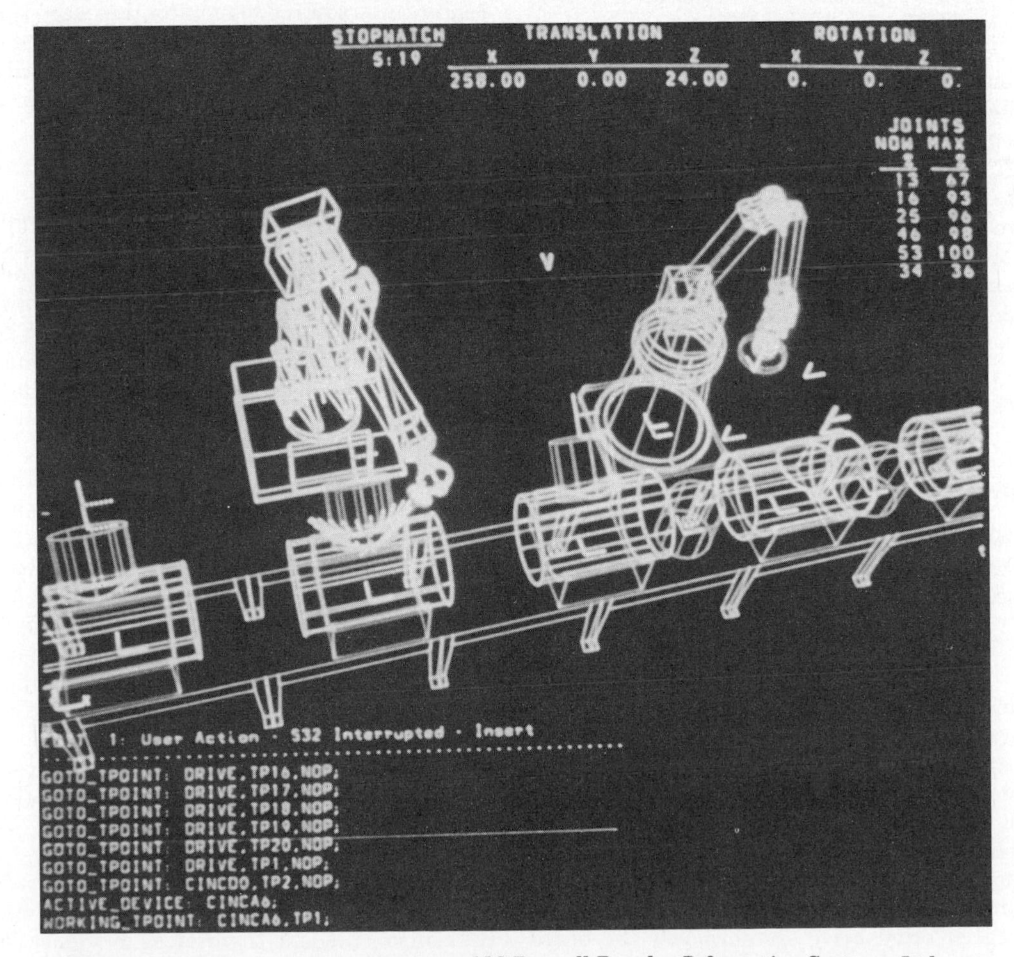

Figure 16. Robot simulation. Courtesy of McDonnell Douglas Information Systems, Ltd.

among board and circuit designers, component and equipment suppliers, and industrial engineers specifying and commissioning robot lines to build quality into the mechanized process. If static sensitive devices are involved, static protection must be incorporated in the line design.

System Operation

Very few systems operate trouble-free immediately upon installation, but many potential problems can be avoided by engineers spending time with the builder during the latter stages of construction and agreeing on appropriate actions in particularly troublesome areas. Just before equipment delivery, training of staff directly responsible for line operation is usually offered by the supplier.

Operation and maintenance manuals normally form part of the contractual agreement for system supply. Installation and commissioning responsibility should contractually be vested in the supplier, although in-house staff may be used on the actual installation. Commissioning is the stage at which the system is operated for the first time in its manufacturing environment. Under the best circumstances, this crucial activity can take three to six months (or longer). Common problem areas are in the quality of boards and components and in peripheral equipment. Physical dimensions of circuit boards, particularly hole locations, are obvious areas for attention, but not so evident is the need for the board to be flat within reasonably close limits. Components from different suppliers are a frequent cause of difficulty due to different body to lead dimensions and packaging variations. Robots are capable of many millions of operations before mechanical wear and tear begin to cause failure (usually servomotors). Bowl feeders will often be troublesome until bowl tooling flights become progressively refined.

Two types of staff are required for operation: skilled engineers capable of mechanical and electrical maintenance and line-feeding operatives to keep component magazines charged and to clear jams and minor stoppages. On a six/eight robot line, two line feeders and one skilled engineer are required on each shift that the line is operated. It is important that several people are equally knowledgeable about the line, particularly with regard to start-up, stop, and clearance of minor faults.

PERIPHERAL EQUIPMENT

Because conveyorized board lines offer more scope and have wider application than fixed board location systems, this will be the focus in discussing peripheral equipment.

Conveyors

A wide range of proprietary conveyor systems are available to system builders, and already fixed-width and variable-width conveyors have been discussed. In this article, a detailed discussion of conveyor types would be inappropriate, but the best advice is to keep it sound but simple. The aim is to move the board smoothly from workcell to workcell. Boards may be conveyed with or without carriers, from station to station by belt, chain, or merely on a bearing surface. In the latter instance, a center shaft, drag, or other drive system below the board provides the necessary propulsion.

In the simplest configuration, each robot has a balanced work load and the board moves from station to station through the line until completed. A simple development of this principle is where a conveyor is used to transport boards through workcells but with the ability to bypass certain cells if required. Complex variants of this latter arrangement are possible to cater for a wide variety of boards and components, with "sidings" and "buffers" to bypass stations and compensate for imbalance between cells. Final location of the board is normally by shot pins in the board or carrier to give a positional accuracy of ±0.002 in. (±0.05 mm) or better. Adjuncts to conveyors are board stack feeders and unloaders for which proprietary equipment is available, but catering for different board widths is an added complication. Board magazines are commonly used to feed from and into, and to transport boards between processes. More mechanized systems use automatic guided vehicles to transport magazines to and from the line (10).

Readers and Sensors

In order to monitor progress through the system, a means of automatically identifying individual boards is used, the most common of which is bar coding (22). Such a system allows the robotic component insertion line to be integrated with factory material planning and quality systems.

The identification code may be incorporated onto the board or carrier, or both, and is read as the board enters the line. Sensors are required to monitor the board through the various stages of the system and to initiate the activities involved in the handling and assembly sequence. A variety of sensor types and makes are available to fulfill this task.

Component Feeders

A wide range of component feeders are used in robotic component insertion, the aim being to bring the components to be fed into an ordered state for precise and consistent presentation to the robot grippers. The degree of difficulty in feeding and gripping usually determines whether automated component insertion is possible. Even though components are difficult to handle, investment in seeking a solution may be justified if quantities and manual assembly costs are high. Specialist companies have developed and should be sought to solve difficult feeding and gripping problems.

Vision systems, already mentioned as a possible solution to complex parts location and orientation, have not yet found significant application in circuit-board assembly, but this situation is expected to change as improved lower-cost systems become available. Components supplied on tape reels are the most readily handled, as feeding systems are well-proven. Components so supplied include capacitors, resistors, diodes, etc. DIPS are usually supplied in plastic tubes (sticks) and fed from chute feeders.

Components supplied packed in trays can be handled, but such loose initial location usually necessitates precision qualification via intermediate tooling.

Vibratory bowl feeders, used extensively over recent decades to serve a wide range of mechanized equipment, find equal application in robotic component insertion, wherever the components lend themselves to orientation, can withstand bowl handling, and cannot otherwise be brought into order.

Bowl-feeder tooling is highly specialized and not normally undertaken by the user company. Custom-tooled bowl feeders are relatively expensive but it is advisable to buy the best possible, since poor bowl feeders seriously impair total line efficiency. Before contemplating bowl feeding, improved packaging should be examined. At some stage in their manufacture, the components must be in an ordered state and the aim should be to retain this order at the supplier, even if this adds marginally to component cost.

Successful mechanization depends on sound initial design, consistent and ordered component supply, and carefully conceived and well-executed systems.

CASE STUDIES

Case studies of actual installations are vital to a full comprehension of the state of the art in robotic component insertion technology and the following details describe working lines from some of the leading robot suppliers and electronics manufacturers.

Fukaya Television Line, Toshiba

The Toshiba Corporation's Fukaya Works, among the world's largest plants manufacturing color TV receivers, makes extensive use of factory automation in a wide range of manufacturing activities to reduce costs and maintain competitiveness (23). Accordingly, there has been considerable emphasis on establishing the most advanced methods of manufacturing television circuit boards, considered to be the most important aspect of the total mechanization program. Between 1975 and 1980, Toshiba installed 37 ACI machines at a cost of $6 million as the first phase of a total mechanization plan. In 1982–1983, a further $3 million was invested in the second phase, including robotic insertion of odd-form components. The line, claimed to be the first in Japan, was completed in April 1983 (Fig. 17). The television board contains approximately 500 components comprised of 70% regularly shaped items (resistors, diodes, etc) and 30% irregularly shaped items (condenser, transformer, rheostat connector, etc).

Odd-form components are inserted using 9 Toshiba TSR-700H robots, which together with the ACI machines, assemble 95% of all components. Unique features of the Toshiba line are that each individual robot can insert, cut, and clinch several different components, verifying each for polarity, etc, as it is inserted.

The line uses a fixed-width conveyor to feed frets of 4 or 5 different sized boards, dependent on TV model, the boards being retained in frets through wave solder and test. The line is highly flexible and employs 6-head pneumatic robot grippers (Fig. 11) on each machine to reduce equipment cost and component placement time. In all, 34 different components can be handled and a balanced load to each robot is achieved by careful selection of parts to be assembled at each station (Fig. 18). After each insertion, leads are cut and clinched using a x–y coordinate plate positioned below the board and programmed by the robot controller. Shot pins are used for precise board location and the following accuracies are achieved:

- Robot positioning accuracy ±0.002 in. (±0.05 mm)
- Conveyor positioning accuracy ±0.002 in. (±0.05 mm)
- PCB hole positioning accuracy ±0.004 in. (±0.1 mm)
- Component lead positioning accuracy ±0.002 in. (±0.05 mm)

The above tolerances are claimed to give 99.9% assurance of insertion (checked by a conductivity test through component lead wires). Labor savings of 70% are claimed on a cost comparison between robotic and manual insertion, with equipment depreciation over 5.5 years and annual increased labor costs

Figure 17. Robotic component insertion line for the manufacture of televisions. Courtesy of Toshiba Corp., Fukaya, Japan.

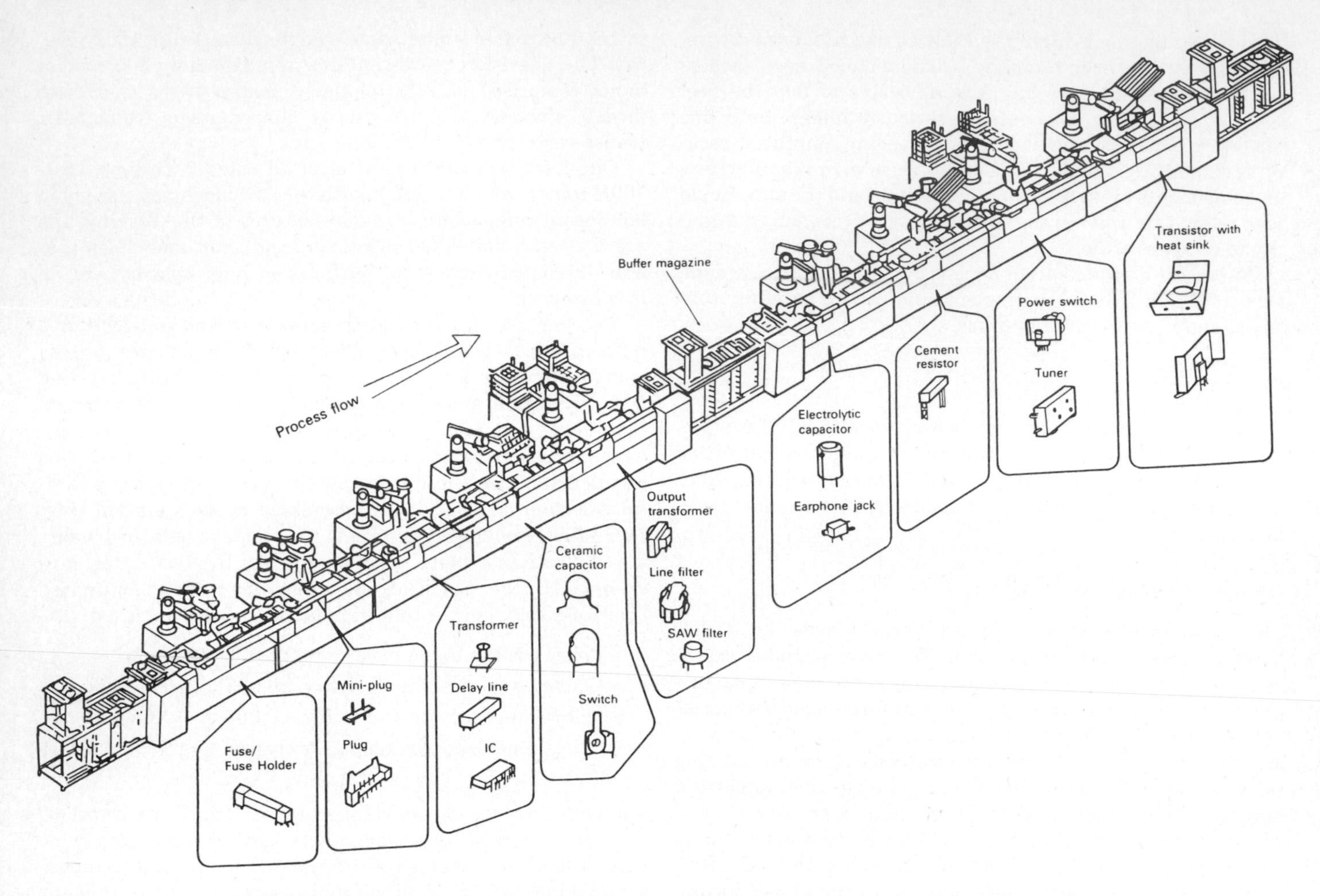

Figure 18. Schematic of Toshiba robotic component insertion line for television board manufacture. Courtesy of Toshiba Corp., Fukaya, Japan.

of 6%. Toshiba emphasizes that packaging of bought-out components had to be improved to achieve this result. Components are received as follows:

- Trays—heavy parts such as tuners, transformers, and power switch
- Sticks—ICs and delay lines
- Tape reels—ceramic capacitors and switch
- Loose for chute magazines—cement resistor and transistor/heat sink
- Loose for bowl feed—fuse holders, plugs, and jacks

Each cell in the line can be controlled independently, simplifying line changing. The system is linked to the business computer (Fig. 19) for associativity with production schedules and for management reports. Linking board programming to computer-aided design has simplified operator instruction, reducing setup time per robot and x–y table to 0.5 hours. The robots and x–y tables were programmed using ALT (assembly language for Toshiba robots). The operator's work station control panel facilitates start-up, shut down, rectification of faults, and removal of faulty components. The total system is operated from a main console, which houses the Westinghouse controller, the conveyor control, line start, safety controls, and associated power supplies. Error recovery is provided at each robot station to cover the following conditions so that corrective action can be taken: no component available; low component supply; failure to insert (force and height detection); component rejection; programmable number of insert re-tries; inability to locate PCB; performance statistics. Additionally, line load and unload is equipped with error detection.

The line is designed to be extensible to include error logging and recovery; board tracking by bar-code readers; additional feeder for other boards; and quality control and data-collection reporting.

Plessey Telecommunications and IBM

The Plessey Company is a leading multinational electronics manufacturer, and telecommunications form a substantial part of its operations. At the Plessey plant in Liverpool, UK, large numbers of circuit boards are required for the manufacture of an advanced digital main exchange system designated System X. In October 1985, a robotic component insertion line (Fig. 20) was installed for the assembly of the odd-form components on one of the high volume line cards (designated Mark III SLU). The line was justified on reduced manual assembly content and improved quality and inventory control. Payback

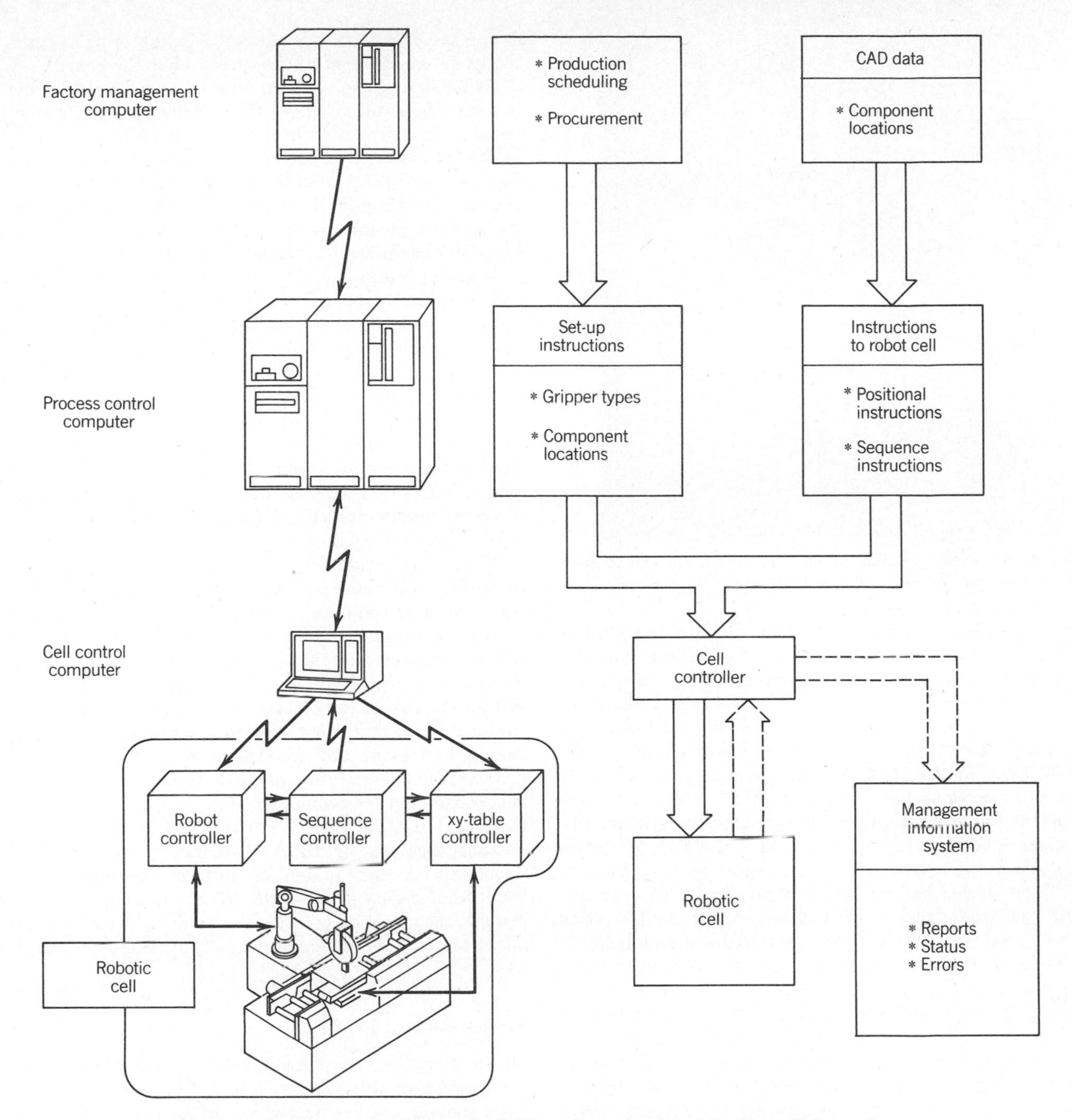

Figure 19. Computer control hierarchy, TV line. Courtesy of Toshiba Corp., Fukaya, Japan.

was calculated to be within two years and the line was designed with built-in flexibility to meet future board requirements. The line is comprised of 6 IBM 7545 SCARA robots, a line conveyor, and control equipment and was built for Plessey by V.S. Engineering, a UK integrator.

The MkIII SLU contains 250 components, approximately 150 of which are inserted by ACI machines.

Figure 21 shows the chart prepared by Plessey of the components to be inserted by the robot line, with quantities, packaging, preforming, and clinching information. Ninety-six components of 8 varieties are inserted, the capability of the line being 210,000 boards per year on three shifts, 5 days per week (230 days per year).

The sequence of operation is

Module 1: magazine unload stage
Module 2: thyristor insertion
Module 3: thermistor insertion
Module 4: relay and fuse insertion
Module 5: resistor insertion
Module 6: hybrid (long) insertion
Module 7: hybrid (short) insertion
Module 8: magazine load stage

The function of the magazine unload station is to accept magazines loaded with 10 component boards. Each magazine is

Figure 20. Robotic component insertion line for the manufacture of System X exchange boards. Courtesy of Plessey Major Systems, Ltd.

registered in position and indexed vertically until boards have been fed sequentially to the line. Once emptied, the magazine returns to its original position, is indexed clear, and a new magazine is indexed in. The magazine permits linking the robotic line with ACI equipment. The robotic modules (2, 3, 4, 5, 6, and 7) are similar in concept and each equipped with an IBM 7545 SCARA robot with a gripper mounted on the z axis to suit the components being handled.

Each module has an independent pneumatic system, vacuum generator, electrical system, and a positioning system to lift the board against datum faces, located by pins in datum holes. Each station has sensors to check for board presence. Module 2 (thyristor insertion) is arranged with double-acting grippers capable of placing two components simultaneously into the board. Two components are fed into nests at the final insertion spacing. The gripper tooling is then lowered, locating in each hole in the thyristor heat sink and preforming the component legs. Cutters in the nest then chop the legs to length and the gripper lifts the two components out of the nest for insertion. Components are fed from stick magazines, situated in matched pairs; 16 of these components are inserted into each board.

Module 3 inserts 16 thermistors per board from a tape reel. A single gripper holds the components while leads are cut to length and then transfers and inserts the components into the board. After insertion, an anvil under the board clinches the leads.

At module 4, relays and fuse links are fed from magazine tubes and presented to the gripper with a datum reference from the component legs. Since the relay legs as supplied, are vulnerable to being out of position, built-in straightening blocks are incorporated in the relay nest. The gripper is arranged to overtravel as it descends to pick up the relay, forcing the component legs into conical holes to align them prior to insertion. The robot picks up either two fuses or two relays as the build sequence demands. Gripping is by vacuum and the leads are clinched by the anvil under the board.

At module 5, two resistors are dispensed by tape-reel feed to a cutting position where the resistors are separated from the tape and presented to a preform position. The first form in the legs is created by the overtravel of the gripper head. Separate air cylinders in the preform position create the second and third bends; the gripper lifts the two resistors out of the preform position with its outer fingers, maintaining the leg positions during insertion to the board.

Modules 6 and 7 insert two types of hybrid circuit (long and short) to the board. The hybrid components are fed from a double pallet of trays, each component being nested at a suitable pitch for gripping. The robot selects a component, places it in a qualifying fixture, referencing the body to the leads, and regrips and inserts the component.

The board-transfer system only has to accommodate one width of board. A characteristic of the board-feed conveyor is its smooth acceleration and deceleration to prevent the dislodging of inserted components. The line is unloaded at module 8, where boards are placed into magazine racks (a reversal of the process at module 1).

The control system is based upon a MTE Westinghouse controller, IBM robot controllers, and a supervisory IBM PC, which permits communication with other computer systems.

Each of the six workstations has an operator panel. Personnel protection is afforded by fixed screening, with provision for line component feeding without stoppage or guard removal. Within the cost of the system, provision was made for a trial run at the builder's plant, delivery, installation, and commissioning at Plessey, and training of mechanical and electrical personnel in the operation and maintenance of the line. Vision was considered for component verification but discounted in the initial installation.

The system was built in conformance with British Standard Specifications and guidelines and in accordance with the Health and Safety at Work Act. When running automatically, transfer from station to station is achieved in 90 seconds. All mechanical movements are monitored using noncontact sensors.

Hewlett-Packard and Seiko

At the Portable Computer Division of the Hewlett-Packard Company, Corvallis, Oregon, 5 Seiko RT 3000 robot cells (Fig. 22) are being used to assemble surface-mounted components to circuit boards for hand-held calculators (24). Before assembly, the boards are screen-printed with solder paste in areas where component leads or end terminations are subsequently soldered using vapor-phase soldering. The robots place chip resistors, capacitors, inductors, 28-pin SOIC packages, and 3 sizes of quad pack, the latter having leads with a center-line distance of 0.031 in., calling for high precision in placement. Maximum permissible misalignment of leads with board pads is ±0.006 in.

Cyanoacrylate adhesive is used to bond the larger components to the printed circuit board, whereas tiny chip components are bonded in place with an epoxy, dispensed by the robot to the PCB, prior to component assembly.

Because the circuit traces on PCBs are relatively imprecise,

Summary Chart—Inserted Parts

Parts No.	Quantity	Name	Shape	Packing	Preforming	Lead cut/clinch	Insertion Shape	Drawing	Diameter, mm	
									Lead	Hole
1	24	Relay		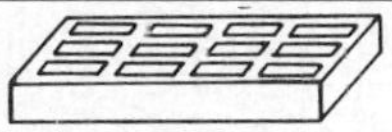 100 pcs/tray	None	Yes		Fig. 01 ICKB0017 AAN	0.65 × 0.3	1.2–1.35
2	8	Thyristor (Transistor)		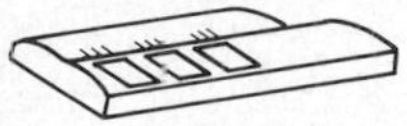 45 pcs/stick	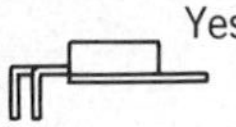 Yes	"	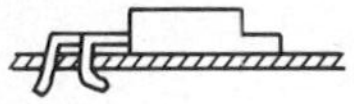	Fig. 02 IOGB0049 AAT	0.45 × 0.3	1.2–1.35
3	8	"	"	"	"	"	"	Fig. 02 ICGB0049 ABL	0.45 × 0.3	1.2–1.35
4	16	Thermistor	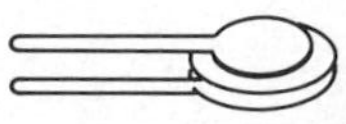	 Tape	None	"		Fig. 03 ICRE0010 AAL	0.6	0.95–1.1
5	8	Hybrid IC A	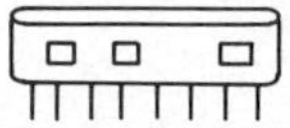 SIP	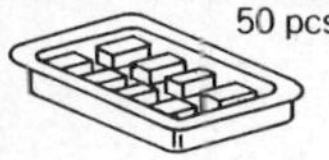 50 pcs/Tray	"	"		Fig. 04 ICDB0027 AAP	0.8	0.95–1.1
6A	8	Hybrid IC B	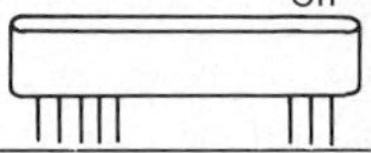 SIP	"	"	"	"	Fig. 05 ICDB00228 ABJ	0.8	0.95–1.1
6B	(8)	Hybrid IC C	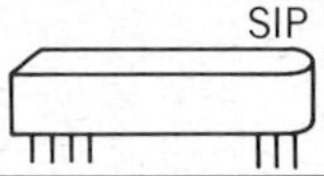 SIP	"	"	"	"	Not fixed	0.8	0.95–1.1
7	8	Fuse		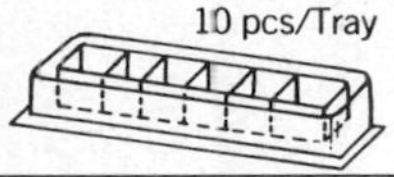 10 pcs/Tray	"	(Yes)		Fig. 06 ICFA00014 AAK	0.9 (diagonal)	1.2–1.35
8	16	Resistor	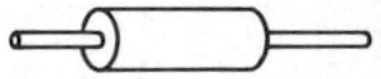	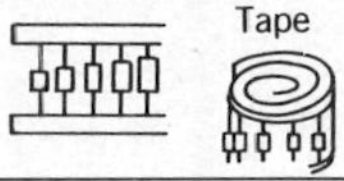 Tape	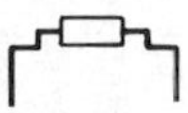 Yes	Yes	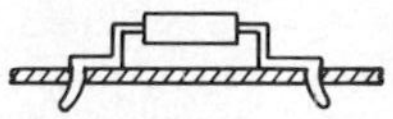	Fig. 07 ICRD0003 ALQ	0.88	1.2–1.35

Figure 21. Components inserted on Systems X robotic component insertion line. Courtesy of Plessey Major Systems, Ltd. (UK).

Figure 22. Robotic surface-mounted component assembly line. Courtesy of Hewlett-Packard Co., Corvallis, Oregon.

Hewlett-Packard has developed a system of target traces (of known position relative to placement pads) that is designed onto all surface-mounted boards. Target traces are located by a Skan-a-matic retroreflective photo detector mounted on the robot end effector, referencing the trace to the board pads.

Quad packs are gripped on the body of the component, which gives control of the device leads to ±0.007 in., insufficiently accurate for successful placement. This is overcome using a V-nest qualifying fixture to align the leads precisely with the gripper. After release, centering, and regrip, the quad pack is taken to the cyanoacrylate dispenser, where adhesive is applied to the underside. The robot next pushes the component onto its board location until a force transducer on the end effector signals that the component has been placed. The robot holds for 2 seconds to allow adhesive cure before continuing with its next placement.

Chip resistors and capacitors are presented to the robot on 8-mm tape; placement accuracies of these components being less critical, recentering is unnecessary. The SOIC is dispensed from a custom feeder which precisely locates the component in a V-grooved nest, permitting direct placement.

Problems arose through the inherent inaccuracy in robot positioning (0.003 in.) relative to programming within its work envelope. Since robot positional repeatability was acceptable, to eliminate this error a compensating program was written for each board position, which remained valid until retooling or reprogramming occurred.

The cells have been proven over many millions of placements. Cycle times are 6 seconds for board target search, 4 seconds for placement of chip resistor/capacitor, 0.75 seconds for epoxy dispensing, and 5.5–8 seconds for place and bond quad pack. The process has permitted the building of otherwise difficult to assemble boards and has improved quality. Through the manufacture of these boards, Hewlett-Packard has reduced board size and cost and achieved improved circuit performance.

Thorn EMI–Ferguson and Hirata

Ferguson, a leading UK manufacturer of color televisions, produces approximately 1 million units annually, and this involves the assembly of 750 million printed circuit board components at their factories in the south of the United Kingdom. It is natural that with the competitiveness of the industry and with such volumes, manufacturing lines are highly mechanized. A major requirement is for main chassis boards measuring 375 × 262 mm, for the Ferguson range of TV receivers. These are produced on a conveyorized robotic component insertion line using 4 Hirata AR 450 SCARA robots.

Thorn EMI–Ferguson is a vertically integrated company producing their own single-sided boards from copper-plated paper phenolic board. Boards are designed on Racal CAD equipment, and at the same time that they are screen-printed, tooling holes are produced with reference to the circuit art work. The tooling holes are used to locate the boards during power-press piercing of through-holes and for conveyor shot-pin location during subsequent component insertion. To maintain order throughout the manufacturing process, boards are held in stack magazines containing 20 boards during each stage of assembly. Universal, TDK, and National Panasonic ACI machines are used for the major part of the board population with axial and radial lead components. The boards contain few ICs and no surface-mounted components. All of the components are mounted on one side of the board.

The robotic concept, which is of modular construction, was originally supplied to Ferguson by Hirata, the robot manufacturer that assumed total responsibility for the engineering and construction of the system. The system has since been extensively engineered within Thorn EMI–Ferguson. Teach programming is used to program the robots. Boards are transported by friction on twin chains driven by low power independent electric motors situated below the conveyor. To achieve accurate positioning for component insertion, piloted and shouldered shot pins are located from below in the board tooling holes. The action of shot-pin insertion lifts the boards, trapping them under side rails. Because of their size and the weight of components, boards are required to be supported during assembly by a center rail situated between the conveyor chains.

Although of fixed width, the modular design of the conveyor allows it to be readily adapted to take different sized boards.

The assembly sequence through the line is as follows:

1st Robot station: insertion of 4 Toko coils

2nd Robot station: insertion of 5 Toko coils

3rd Robot station: insertion of 4 potentiometers

4th Robot station: insertion of 2 Toko coils and 2 potentiometers

Ferguson closely specified all aspects of the mechanization through their own in-house industrial engineering unit, which has the capability for design and precision manufacture within its own tool-room facility; the component feeders for the line were developed by this unit. Components are supplied from the manufacturer in sticks for magazine feeding through the Ferguson-designed unit.

The relationship between leads and body on the Toko coils

are sufficiently precise to allow gripping on the component body. The leads are tensioned outward and "sprung" in during insertion to retain the component in the board. Below the board, an x–y plate containing anvils moves synchronously with insertion to clinch the component leads. The potentiometers have less body to lead consistency and hence are gripped on the leads during insertion. A natural standoff from the board surface is created by the shape of the potentiometer lead as supplied. Total throughput is 4500 boards per day (single extended shift).

Northern Telecom and IBM

One of the largest robotic component insertion systems in the world is being used by Northern Telecom in Raleigh, North Carolina. The system, developed by Dynamac, a division of American Robot, uses 40 IBM 7535 SCARA robots to insert odd-form components in PWAs for digital switches (25).

Robotic assembly was chosen as the manufacturing solution, since the components could not be inserted by ACI equipment and a system that would be expandable and flexible to changes in design and production requirement was needed.

Northern Telecom's robotic component insertion installation comprises three lines: two employ 12 robots and complete one board assembly with up to 18 components every 10 seconds; the third line has 16 robots and produces one board every 8 seconds. All three lines, having built-in flexibility, currently produce the same assembly. Each line has 12 assembly stations and boards are transported on purpose-designed pallets between stations by a Bosch conveyor. Boards are barcoded and pallets digitized for tracking through the system. An Allen Bradley PLC-3 controller directs the data received from the bar-code reader and assembly stations. The total system is managed by a HP 1000 host computer.

Details of the robotic insertion stations are as follows. Station 1 inserts a radial capacitor from tape feeders (two are provided). Station 2 is identical to station 1. Station 3 feeds box capacitors via a plastic stick feeder to a two-axis transfer device which qualifies the lead positions relative to the component body. The SCARA robot is fitted with compliant grippers, allowing the component to be gripped on its body while maintaining the relationship with the leads, established by the qualifying fixture. After insertion, component leads are crimped by a fixture below the board. Stations 4, 5, and 6 assemble an enclosed relay using a similar technique to that employed at station 3 except that vacuum pickup is used. Station 7 inserts an edge connector; the component is chute-fed to a pick and place unit which uses mechanical grippers to transfer it before insertion to a lead/body qualifying fixture. Stations 8 and 9 feed an IC socket and DIP, again using chute feeders and inserting via a lead qualifying fixture. Station 10 feeds an RM 10 transformer, which presents exceptional difficulty through its design and the indeterminate relationship of the transformer tags with the body. This is overcome by a combination of qualifying fixture and insertion search routines. Stations 11 and 12 insert ceramic resistor networks using a vacuum pickup; a lead qualifying fixture is used and the leads are crimped after insertion.

After assembly, the boards are automatically lifted off the carrier and fed into the feed conveyor of a wave-soldering machine. The empty carriers are recirculated to the loading station. A repair loop is incorporated in the line to accommodate boards having missing components. Information on misfed components is related to the board-carrier coding and displayed on a visual display unit at the manual rework station. Malfunction of any robotic station is automatically handled by diversion of the boards to manual assembly stations.

The first line was built by Dynamac in 11 months, with commissioning requiring a further 6 months. The system is claimed to run at 98% efficiency.

THE FUTURE

There is no doubt that the future of robots in electronic assembly is secure. Their inherent flexibility and precision, particularly when associated with high resolution vision systems, make robots capable of fulfilling tasks not readily accomplished by dedicated hard automation. It must be said, however, that dedicated equipment manufacturers are increasingly addressing the insertion of odd-form components. Such machines will no doubt have a place in circuit-board assembly, but it is difficult to see how they can match the flexibility of the robot. Integrated lines of ACI, robots, soldering, automatic test equipment (ATE), and rework stations, currently rare, will grow in application, probably with great concentration on board-throughput times and "just in time" (JIT) inventory, rather than on machine utilization.

The design and proving of robot systems off line using advanced CAD simulation aids will gain popularity as such systems' capabilities improve and costs reduce.

Computer integrated manufacture (CIM) (26), as yet only conceptual, will gradually become a reality as automated cells are linked through integrating protocols (such as MAP and IGES) currently under development. Within this giant technological leap, robotic lines will be embraced to link with business computers, to tie in automatic component insertion systems with materials requirements planning (MRP), inventory, and financial control.

Robotic component insertion (also ACI and ATE) will derive their control programs from computerized circuit design and layout. These changes will be progressive over the next decade, spurred by major manufacturers with the purchasing power to influence component suppliers and with sufficient financial and competitive motivation to push out the technological frontiers.

BIBLIOGRAPHY

1. *BS.6221, Part 3, 1982, Guide for the Design and Use of Printed Wiring Boards,* British Standards Institution, London.
2. *BS 6221, Part 20, 1984, Guide for the Assembly of Printed Wiring Boards,* British Standards Institution, London. A highly recommended definitive reference document detailing the requirements for the assembly of printed wiring boards.
3. J. P. McCarthy, "Design for Automated Manufacture of Surface Mounted Electronics Assemblies," *Proceedings of the 6th International Conference on Assembly Automation,* IFS (Publications), Ltd., Bedford, UK, May 1985. Very knowledgeable reading.
4. H. Nakahara, "SMT Expands Options in Japan." *Electronic Packaging and Production* (Jan. 1986).

5. H. W. Markstein, "Automatic Component Insertion/Placement Systems Ready for the Automated Factory," *Electronic Packaging and Production* (Aug. 1984).
6. J. Hardie, "Seven Deadly Sins of P.C.B. Designers," *Electronic Production* (Jan. 1986).
7. J. S. Nuthall, "FMS for PCB Assembly," in Ref. 3.
8. B. Rooks, "Robot Assembly Gives Flexibility in Computer Manufacture," *The Industrial Robot* (Sept. 1985).
9. G. Worthington, "The Design and Performance of Robotic Systems," *Technical Horizons,* Issue No. 4, IBM Corp., Portsmouth, UK, Autumn 1985. A very comprehensive paper.
10. P. Fenner, "Automatic Factory Approach to Electronics Assembly." in Ref. 3.
11. H. Makino and N. Furuya, "Selective Compliance Assembly Robot Arm," *Proceedings of the 1st International Conference on Assembly Automation,* IFS (Publications), Ltd., Bedford, UK, Mar. 1980.
12. H. Makino and N. Furuya, "Scara Robot and its Family," *3rd International Conference on Assembly Automation,* May 1982.
13. P. K. Wright and M. R. Cutsosky, "Design of Grippers," in S. Y. Nof, ed., *Handbook of Industrial Robotics,* John Wiley & Sons, Inc., New York, 1985.
14. M. Ogorek, "Tactile Sensors," *Manufacturing Engineering* (Feb. 1985).
15. *Proceedings of the 4th International Conference on Robot Vision and Sensory Controls* IFS (Publications), Ltd., Bedford, UK, Oct. 1984.
16. W. B. Heginbotham and E. P. Tuite Dalton, "An Assessment of Proprietary Vision Systems in Relation to Practical Production Engineering Applications," *Proceedings of the 2nd European Conference on Automated Manufacturing,* IFS (Publications), Ltd., Bedford, UK, 1983.
17. Y. Young and R. Marshall, "Factory Simulation, Putting Robots in their Place," *CADCAM International* (Feb. 1986).
18. C. H. Mangin and S. D'Agostino, "Printed Circuit Board Assembly Picks Up on Automation," *Electronics* (Jan. 12, 1984).
19. P. J. Hughes, "Making Automated Assembly Pay the Non-technical Factors," in Ref. 3.
20. A. J. Pendlebury, "Costs and Benefits of Assembly Automation," in Ref. 3.
21. C. H. Mangin, "Organising Flexible Electronic Assembly," *Assembly Engineering* (Nov. 1984).
22. "Bar Codes Speed Progress Chasing," *Electronics Manufacture and Test* (June 1985).
23. N. Tanaka, T. Nogouchi, and M. Mibuka, "DNC Robots and Mini Computers make Flexible Assembly Line for Colour TV," in Ref. 3.
24. J. M. Altendorf, *Precision Surface Mount Assembly Using Robots,* Hewlett-Packard, Corvallis, Ore.
25. "Phone Company Connects with Switch to Robots," *Robotics World* (Mar. 1986).
26. D. D. S. Bowles, "Robotics," *Autofact 1985 Conference Proceedings,* Computers and Automation Systems Association of SME, Society of Manufacturing Engineers, Dearborn, Mich., Nov. 1985.

General References

Automatic Component Insertion

J. Page Walton, "Developments in PCB Component Insertion Equipment," *Electronic Production* (Jan. 1986).

Surface Mount

A. C. Snow, "Some Practical Observations on the Processes Involved in Surface Mounted Technology," *ICT's 10th Annual Symposium on Circuit Technology,* June 1984; also published in *Circuit World,* **II**(2) (1985).

R. Pound, "S.M.T is Poised for Penetration of European Manufacturing," *Electronic Packaging and Production* (Jan. 1986).

K. Barnet, "Selectively Plated PCB's for 'Surface Mount'," *Electronic Production* (Jan. 1986).

"Automation Thrives as Companies Strive for SMT Capability," *Electronic Packaging and Production* (May 1985).

Robotic Component Insertion

M. Decollibus, "Robotic Application to Odd Form Component Assembly in Printed Circuit Boards," *EMTAS '84 Conference,* Society of Manufacturing Engineers, Dearborn, Mich.

W. H. Kintner, Jr., "Robots for Electronics Assembly," *Robots 9 Conference Proceedings,* Society of Manufacturing Engineers, Dearborn, Mich., June 1985.

W. H. Schwartz, "Robots in PC Board Assembly," *Assembly Engineering* (Aug. 1985).

Grippers

B. J. Schroer, W. Teoh, and R. D. Stiles, "Parameters Affecting the Robotic Insertion of Nonstandard Electronic Components," *Robots 9 Conference Proceedings,* Society of Manufacturing Engineers, Dearborn, Mich., June 1985.

Robot Grippers, IFS (Publications), Ltd., Bedford, UK, April 1986.

K. Tanie, "Design of Robot Hands," in S. Y. Nof, ed., *Handbook of Industrial Robotics,* John Wiley & Sons, Inc., New York, 1985.

Tactile Sensors

K. A. Morris, "Tactile Sensing Aids Assembly," *Robotics World* (Nov. 1985).

Programming

M. P. Deisenroth, "Robot Teaching," in S. Y. Nof, ed., *Handbook of Industrial Robotics,* John Wiley & Sons, Inc., New York, 1985.

T. F. Young and co-workers, "Off-line Programming of Robots," in S. Y. Nof, ed., *Handbook of Industrial Robotics,* John Wiley & Sons, Inc., New York, 1985.

Vision Systems

T. Kothari, "The Role of Vision Systems in Robotic Applications," *Robots 8 Conference Proceedings,* Vol. 2, Society of Manufacturing Engineers, Dearborn, Mich., 1984.

G. J. Agin, "Vision Systems," in S. Y. Nof, ed., *Handbook of Industrial Robotics,* John Wiley & Sons, Inc., New York, 1985.

R. Keeler, "Vision Guides Machines in Automated Manufacturing," *Electronic Packaging and Production* (May 1985),

U. Rembold and C. Blume, "Interfacing a Vision Systems with a Robot," in S. Y. Nof, ed., *Handbook of Industrial Robotics,* John Wiley & Sons, Inc., New York, 1985.

Case Studies

C. Carberry, "Robots Ease Bottleneck in PCB Production," *Production Engineer* (Mar. 1986).

Robot Reliability

G. E. Munson, "Industrial Robots: Reliability, Maintenance & Safety," in S. Y. Nof, ed., *Handbook of Industrial Robotics,* John Wiley & Sons, Inc., New York, 1985.

CIM

J. Davis and T. Yates, "Too Early for the Factory of the Future," *New Scientist* (Jan. 1986).

S. L. McGarry, "Just in Time and Computer Integrated Manufacturing: Friends or Foes? *Autofact 1985 Conference Proceedings,* Computers and Automation Systems Association of SME, Society of Manufacturing Engineers, Dearborn, Mich., Nov. 1985.

COMPONENTS, MECHANICAL, DESIGN OF.

See MECHANICAL DESIGN OF COMPONENTS.

COMPUTATIONAL LINGUISTICS

B. Ballard and M. Jones
AT&T Bell Laboratories
Murray Hill, New Jersey

INTRODUCTION

Research in computational linguistics (CL) is concerned with the application of a computational paradigm to the scientific study of human language and the engineering of systems that process or analyze written or spoken language. The term *natural-language processing* (NLP) is also frequently used, especially with regard to the engineering side of the discipline. As an historical note, the term *computational linguistics* included the study of formal languages and artificial computer languages (e.g., ALGOL), as well as natural languages, until the middle 1960s, but this entry concerns CL as it is presently conceived.

Theoretical issues in CL concern syntax, semantics, discourse, language generation, language acquisition, and other areas, whereas areas for applied work in CL have included automatic programming, computer-aided instruction, database interface, machine translation, office automation, speech understanding, and other areas. Historically, much CL research has been done by researchers whose language interests overlap with interests in such related disciplines as AI, cognitive science, computer science and engineering, information science, linguistics, philosophy, psychology, and the speech sciences. The middle 1970s, however, witnessed an increase in hybrid efforts, so that present efforts in CL typically draw from and contribute to work in one or more of these cognate areas.

This entry serves primarily as an overview of the primary topics in CL. It begins with a historical introduction to the field, followed by brief remarks on some of the more important theoretical problems, and concludes with pointers to the literature. Since space has permitted only a general statement of the *goals* of a theory or implementation, with occasional *examples* of either I/O behavior or internal representation formalisms, conclusions cannot be drawn from this entry alone concerning the capabilities of the work to be described. More detailed information is available in the separate entries related to the topics considered here.

Early Work (1950–1965)

Most CL work prior to 1960 concerned *machine translation,* as defined below, but the advent of *transformational grammar* and the emergence of paradigms for *information retrieval* also played an important role in the formation of a CL community. Following is a discussion of the essential work in these three areas.

Machine Translation. Many of the first attempts at using computers to process natural language concerned the problem of translating from one natural-language text into another. Although actual computer programs seeking to solve this task were not written until the early 1950s, the idea of mechanical translation can be traced to conversations as early as 1946 between Warren Weaver and A. D. Booth. The initial impetus came in 1949, when Weaver wrote and privately circulated a paper titled "Translation" (1). This paper, along with a detailed account of initial work in machine translation, can be found in Locke and Booth (2).

Most early work on machine translation, also known as automatic translation, mechanical translation, or simply MT, was conducted in the United States and the USSR, where the political and military interests in natural-language translation were especially strong. There were also two British projects and some work done in Italy, Israel, and elsewhere. Typically, efforts at machine translation, which predated the important work in linguistics and computer science on syntax, grammars, and languages, were based on word-by-word translation schemes. In particular, no attempt was made to "parse" sentences (i.e., determine their syntactic structure) and, at least as significantly, no attempt was made to actually "understand" the material to be translated. A characterization of the basic approach of word-for-word processing can be found in Ref. 3.

As an example of what had been achieved by about 1960, the first sentence of a 1956 Russian article yielded the output "'razviti' electronics (allowed permitted) (considerably significantly considerable significant important) to (perfect improve) (method way) 'fiz' (measurement metering sounding dimension) (speed velocity rate ration) (light luminosity shine luminous)," where parentheses indicate uncertainty on the part of the system and where *razviti* and *fiz* were unknown and thus untranslated (*fiz* derives from a proper name). From this output, a human posteditor produced "Development of electronics permitted considerably to improve method Fizeau of measurement of speed of light," which may be compared with the fully human translation "The development of electronics has brought about a considerable improvement of Fizeau's method of measuring the velocity of light." This example is discussed in detail in Oettinger (3).

Concerning the distinction between fully automated as opposed to machine-assisted human translation, even Bar-Hillel, an outspoken detractor of much MT work, observed that "word-by-word Russian-to-English translation of scientific texts, if pushed to its limits, is known to enable an English reader who knows the respective field to understand, in general, at least the gist of the original text, though of course with an effort that is considerably larger than that required for reading a regular high quality translation" (4). Nevertheless, researchers and government funding agencies continued to anticipate systems that would provide "fully automatic high-quality translation" (FAHQT). It was with respect to this more ambitious goal that the Automatic Language Processing Advisory Committee (ALPAC) was formed in April of 1964 "to advise the Department of Defense, the Central Intelligence Agency, and the National Science Foundation on research and development in the general field of mechanical translation of foreign languages." In essence, the committee found that "there has been no machine translation of general scientific text, and none is in immediate prospect" (5). They further observed that, in some cases, "the postedited translation took slightly longer to do and was more expensive than conventional human translation" and also noted that "unedited machine output from scientific text is decipherable for the most part, but it is sometimes misleading and sometimes wrong (as is postedited output to a lesser extent)."

Although the ALPAC committee had presumably intended its report to effect "useful changes in the support of research," their findings resulted in the virtual elimination of federal funding for work in MT. As a consequence, very little work was done, and few papers published, for roughly a decade. Since the middle 1970s, however, a number of projects have been spawned or reactivated. The entry on machine translation provides technical details and also discusses more recent work in the area.

Transformational Grammar. In 1957 an event occurred that not only revolutionized the world of linguistics but left a lasting impression on philosophy, psychology, and other areas. That event was the publication of a short monograph by Noam Chomsky entitled *Syntactic Structures* (6) that explored the implications of automata theory for natural language. In it, Chomsky first argued that the sentences of a natural language cannot be meaningfully generated by a finite-state machine or by any context-free grammar, or at least that "any grammar that can be constructed . . . will be extremely complex, ad hoc, and 'unrevealing'" (7). He then proposed a theory of what he called *transformational grammar* (TG) and began to work out its details.

At the most abstract level the theory of TG involves specifying a set of "kernel" sentences of a language; an assortment of "transformations," such as verb tensing and passive voice; and an ordering in which transformations are to be carried out. For example, to avoid producing a sentence such as "John are liked by the students," the passive transformation must apply to the kernel sentence "The students liked John" before the rule for subject–verb agreement. The entry on transformational grammar provides details of the theory.

With the publication of *Syntactic Structures,* Chomsky had argued for, if not established, the efficacy of a transformational component, but he recognized that TG would have to be "formulated properly in . . . terms that must be developed in a full-scale theory of transformations." As a suggestive first step, his appendix provided a sample grammar for a very small subset of English that included 12 content words and fairly elaborate auxiliary verb structures. The period from 1957 to 1965 was one of intense activity by Chomsky and several students, culminating in 1965 with the publication of *Aspects of the Theory of Syntax* (8) and its far-reaching theory of *deep structure,* which relates to an internal sentence-independent representation of (the meaning of) the sentence.

Although TG has had an uneven impact on CL, centered mostly around matters of syntax, its influence on early work in CL is evidenced through bibliographic references and, more substantively, by concepts and borrowed terminology that appeared in the CL literature of the 1960s. In the long term the hypothesis of TG most significant for work in CL is that an understanding of the syntax, or structure, of natural-language sentences can be arrived at on a solely grammatical basis, without considering the real-world properties (e.g., meanings) of the terms being discussed. This notion, sometimes known as the "autonomy of syntax," continues to provide a useful, if regrettable, division in categorizing current work in CL, as the debate continues as to what interactions are desirable, or necessary, between the structural (syntactic) and interpretive (semantic, pragmatic) components of a theory or implementation.

Information Retrieval. It is fairly well known that the emergence of the modern digital computer occurred during the 1940s and that the problems first solved by these computers were numerical in nature and often military in origin. During the 1950s computers were increasingly called upon to provide access to large volumes of nonnumeric data for such purposes as database retrieval and on-line bibliographic search.

Most systems of the 1950s and early 1960s that provided for English "inputs" were directed toward bibliographic search and other library services, and these efforts coalesced into a field that became known as "information retrieval" (IR), which is "concerned with the structure, analysis, organization, storage, searching, and retrieval of information" and has grown to include procedures for "dictionary construction and dictionary look-up, statistical and syntactic language analysis methods, information search and matching procedures, automatic information dissemination systems, and methods for user interaction with the mechanized systems" (9). Although little association remained between CL and IR by the middle 1960s, early work in IR did overlap that being done by the early workers in CL. The evolution of work in IR is chronicled in Refs. 9–12.

Broadening Interests (1960–1970)

In contrast to the 1950s, during which time CL researchers concentrated primarily on machine translation, the 1960s witnessed the application of CL techniques to database retrieval, problem solving, and other areas. For the most part, these early NL systems provided quite limited forms of interaction and were often based on techniques specifically tailored for a single domain of discourse. Nevertheless, the work represented interesting and important, if tenuous, first steps at seeking computational solutions to problems of human language processing. In addition, Raphael notes that these programs "contain the seeds, or at least surfaced the issues, that led to many of today's major computer science concepts: semantic net representations, data abstraction, pattern matching, object oriented programming, syntax-driven natural language analysis, logic programming, and so on" (13).

One important aspect of CL implementations of the 1960s, largely without counterpart in CL work of the 1950s, was that the "processing" to be done required programs to *understand* their inputs to some nontrivial degree. For example, although Bobrow recognized that "we are far from writing a program that can understand all, or even a very large segment, of English" (14), he claimed that "a computer *understands* a subset of English if it accepts input sentences which are members of this subset and answers questions based on information contained in the input" (15). This issue was not without controversy, however, as suggested by Giuliano's complaint that an "arbitrary heuristic procedure . . . which is used in several computer programmed systems" does not "become a principle through its use" (16). To this argument, Simmons (17) responded that "theory often lags far behind model building and sometimes derives therefrom" and further maintained that the early systems represented "truly scientific approaches to the study of language" (18).

The following discussion seeks to convey a sense of the problems addressed by NL applications in the 1960s. They are grouped in terms of question-answering, problem-solving, consultation, and miscellaneous systems.

Question-Answering Systems. One of the first fully implemented data retrieval systems was BASEBALL (qv), "a computer program that answers questions posed in ordinary English about data in its store" (19). This system was designed to interact with a primitive database, stored as attribute-value pairs, that contained information about the month, day, place, teams, and scores for American League baseball games. An example input is "What teams won 10 games in July?"

Another early program was SAD SAM, designed to "parse sentences written in Basic English and make inferences about kinship relations" (20). This system comprised two modules, one for parsing (the syntactic appraiser and diagramer, SAD) and one for semantic analysis (the semantic analyzing machine, SAM). The basic operation of the semantics module involved searching a previously constructed parse tree for words denoting kinship relationships in order to construct a family tree, which was stored as a linked structure.

The SIR system had the goal of "developing a computer [program] . . . having certain cognitive abilities and exhibiting some humanlike conversational behavior" (21). The system was similar to SAD SAM in allowing a user to input new information, then ask questions about it. However, SIR emphasized relations such as set-subset, part-whole, and ownership, as suggested by the following: Every boy is a person. A finger is part of a hand. Each person has two hands. John is a boy. Every hand has 5 fingers. How many fingers does John have?"

The DEACON system, which was designed to answer questions about "a simulated Army environment" (22), represents an important precursor of the database frontends of the 1970s. Its internal "ring"-like data structures could be dynamically updated, thus enabling users to supply new information ("The 425th will leave Ft. Lewis at 21950!") as well as ask questions ("Is the 638th scheduled to arrive at Ft. Lewis before the 425th leaves Ft. Lewis?"). In reflecting upon their experiences with DEACON, the authors noted that "perhaps the most significant new feature needed is the ability to define vocabulary terms in English, using previously defined terms" (23). This realization led directly to the REL system and its successors.

The REL system (Rapidly Extensible Language) represented the logical continuation to the work with DEACON, and its primary goals were "to facilitate the implementation and subsequent user extension and modification of highly idiosyncratic language/data base packages" (24). An example customization is

def:power coefficient:high speed memory size/add time

From a theoretical standpoint, REL was based on the notion that an English language subset could be treated as a formal language "when the subject matter which it talks about is limited to material whose interrelationships are specifiable in a limited number of precisely structured categories" (25). The first sizable application of REL was to an anthropological database at Caltech of over 100,000 items. As indicated below, work on the REL project continued well into the 1970s, until the system, now quite advanced over its early prototypes, was renamed ASK.

Another early database interface, CONVERSE, was designed as an "on-line system for describing, updating, and interrogating data bases of diverse content and structure through the use of ordinary English sentences" (26). It was intended to strike "a reasonable compromise between the difficulties of allowing completely free use of ordinary English and the restrictions inherent in existing artificial languages for data base description and querying" (26). An example input is "Which Pan Am flights that are economy class depart for O'Hare from the city of Los Angeles?" In addition to question-answering capabilities, the system included facilities for English-like data definitions and English-like means of populating the database.

Problem Solving. The STUDENT (qv) system was designed as "a computer program that could communicate with people in a natural language within some restricted problem domain" (14). It sought to solve high-school-level algebra word problems stated in what the author considered a "comfortable but restricted subset of English" by constructing an appropriate set of linear equations to be solved. As an example of STUDENT's capabilities, a sample problem is: The price of a radio is $69.70. If this price is 15% less than the marked price, find the marked price. Having been given equations such as "distance equals speed times time" and by employing a number of clever pattern-matching techniques (e.g., "years older than" is converted to "plus"), STUDENT could answer some very complex problems. Its designer suggested that the system "could be made to understand most of the algebra story problems that appear in first-year high-school textbooks" but noted that "the problems it cannot handle are those having excessive verbiage or implied information about the world not expressible in a single sentence" (27).

In building upon Bobrow's work with STUDENT, Charniak set out to develop a program to solve calculus word problems (28). This led to the implementation of the CARPS system, which was restricted to freshman-level problems related to rates, as Bobrow had in fact suggested in his paper. An example input processed by CARPS follows.

Water is flowing into a conical filter at the rate of 15.0 cubic inches per second. If the radius of the base of the filter is 5.0 inches and the altitude is 10.0 inches, find the rate at which the water level is rising when the volume is 100.0 cubic inches. (29).

In reflecting on his work, Charniak noted that, despite its advances over STUDENT, "a very powerful calculus word problem solver will require a good deal of 'common sense' knowledge" (30). In advancing this claim, Charniak foreshadowed an interest on the part of the AI community in the 1970s in story understanding.

Consultation. One of the most widely publicized AI programs of the 1960s, the ELIZA (qv) system, was designed "for the study of natural language communication between man and machine" (31). In essence, ELIZA was intended "to respond roughly as would certain psychotherapists (Rogerians)" since "the psychiatric interview is one of the few examples of categorized dyadic natural language communication in which one of the participating pair is free to assume the poise of knowing almost nothing of the real world" (31). An excerpt from a "typical" conversation with ELIZA follows.

"I need some help, that much seems certain."

WHAT WOULD IT MEAN TO YOU IF YOU GOT SOME HELP

"Perhaps I could learn to get along with my mother."

TELL ME MORE ABOUT YOUR FAMILY

"My mother takes care of me."

WHO ELSE IN YOUR FAMILY TAKES CARE OF YOU

"My father."

YOUR FATHER (31)

As can be inferred after considering a dozen or so responses from the system, ELIZA sought to match its current input against one of a known set of patterns. It then selected one of possibly several transformations associated with that pattern. Actually, patterns were associated with a keyword, and the algorithm considered various priorities in choosing among candidate matches. The idea of maintaining a "script" of data separate from the algorithms of the program itself was not without precedent, but ELIZA carried this out more fully than had previous systems.

In addition to its technical contributions and the excitement it caused, the system convinced at least its designer that "the whole issue of the credibility (to humans) of machine output demands investigation" (31). This thought led Weizenbaum to his widely publicized social criticisms of AI research (32). An interesting and also famous follow-up of ELIZA, in which the program played the role of the patient rather than the analyst, is reported in Colby et al. (33).

Miscellaneous. Within the tradition of information retrieval established in the 1950s, but with greater attention to syntax and other linguistic issues, Protosynthex sought to accept natural English questions and search a large text to discover the most acceptable sentence, paragraph, or article as an answer (17). The system was applied to portions of Compton's Encyclopedia, and an example of a question posed to the system is "what animals live longer than men?" The project continued for several years and evolved into "a general purpose language processor . . . based on a psychological model of cognitive structure that is grounded in linguistic and logical theory" (34).

A few systems were designed to produce English output, as described by Simmons (17). One system, NAMER, was designed to generate natural-language sentences from line drawings displayed on a matrix (35). It produced sentences such as "the dog is beside and to the right of the boy." Another system, the Picture Language Machine (36), would be given a picture and a sentence as input and, after translating both the picture and the English statement into a common intermediate logical language, would determine whether the statement about the picture was true. An example input is "all circles are black circles."

Formalism Developments (1965–1970)

In addition to system-building activities, a number of formalisms were developed during the 1960s, especially in the latter half of the decade, relating to linguistic, psychological, and other aspects of natural languages. Based on experiences with previous attempts to construct natural-language-processing systems, and upon developments in linguistics and various areas of AI, these formalisms provided more sophisticated ways of representing the results of a partial or complete analysis of inputs to an NL system. A few of the more important of these formalisms are summarized here, namely, augmented transition networks (see Grammar, augmented transition networks), case grammar (qv), conceptual dependency (qv), procedural semantics (qv), and semantic networks (qv). Further details appear in the individual entries.

By extending the expressiveness of the transition network models described by Thorne et al. (37) and Bobrow and Fraser (38), which were themselves based on the basic finite-state machine model stemming from work in formal language theory, Woods developed an *augmented transition network* (ATN) model for the syntactic analysis of natural-language sentences (39). One of the primary advantages of the ATN model over its predecessors rested in its "hold-register" facility, which allowed information to be passed around in a parse tree under construction. This enabled the handling of deeply nested structures and other syntactic complexities. The hold-register facility derives, at least in spirit, from the desire to construct the "deep structure" corresponding to a sentence under analysis, a concept deriving from work in transformational grammar.

The theory of *case grammar,* as proposed by Fillmore (40), expands on the view that "the sentence in its basic structure consists of a verb and one or more noun phrases, each associated with the verb in a particular case relationship." For instance, Fillmore observes that understanding the sentence "The hammer broke the window" involves recognizing that the noun *hammer* acts differently from *John* in John broke the window." Specifically, it is an *instrument* ("the inanimate force or object causally involved in the action") rather than an *agent* ("instigator of the action identified by the verb"). Fillmore's original theory included these and six additional case roles. One important aspect of case grammar theory is its distinction between "surface" roles (e.g., subject) and "deep" cases (e.g., agent or instrument). Bruce (41) provides a survey of ways in which the notion of case grammar was taken up by computationalists in the 1970s.

Having adopted a view that language-processing systems should not produce a syntactic analysis of an input divorced from its meaning, Schank proposed a *conceptual dependency* (CD) model of language and exhibited its operation in the context of an implemented parsing system (42). Deriving loosely from ideas to be found in Hays (43), Kay (44), and Lamb (45), CD is based on a small number of "conceptual categories," including picture producers (PPs), PP assisters (PAs), actions or abstract nouns (ACTs), and ACT assisters (AAs). Developments in the original theory, including more sophisticated conceptual categories such as mental information transfer (MTRANS) and ingestion (INGEST), are outlined in Schank (46). In addition to its central role in the development of the MARGIE system, discussed below, CD contributed to philosophical discussions concerning the role of "primitives" in theories of meaning.

In seeking to develop "a uniform framework for performing the semantic interpretation of English sentences" (47), Woods devised a framework that he termed "procedural semantics" that acted as an intermediate representation between a language analyzer, e.g., a question-answering (qv) system, and a back-end database retrieval component. In essence, the idea behind procedural semantics is to define, given a particular database, a collection of "semantic primitives" that comprise a set of predicates, functions, and commands. This strategy was first demonstrated in the context of a hypothetical question-answering system for an airlines reservation system and was soon to be used in building the LUNAR system, as described below.

Motivated by work in linguistics and psychology and attempting to formulate "a reasonable view of how semantic information is organized within a person's memory" (48), Quillian proposed a memory model that has come to be known as a *semantic network*. Although precursors of semantic networks are to be found in the use of property lists by designers of early NL systems, Quillian provided a theoretical and more formal treatment. In essence, a semantic network consists of a set of "nodes," typically representing objects or concepts, and various "arcs" connecting them that are typically labeled to indicate a relation between nodes. Quillian's initial use of his network structures involved their role in making inferences and finding analogies. Semantic networks have been important not only because of the many systems that incorporate them but also in their contribution to the development in the middle 1970s of various theories of knowledge representation. The evolution of semantic network structures, together with a discussion of applications based on them, is traced by Simmons (49), Findler (50), and Sowa (51).

A Turning Point (ca. 1970)

In the aftermath of disappointing results from work in machine translation (qv) in particular and the difficulty of constructing sophisticated natural-language-processing systems in general, two natural-language projects in the early 1970s captured a degree of attention that served to boost the confidence of AI researchers regarding the prospects for broadly based, well grounded NL systems. These projects, which are discussed in turn, were the SHRDLU (qv) system of Winograd (52) and the LUNAR (qv) system described in Woods et al (53) and Woods (54).

SHRDLU. Winograd's SHRDLU system provided a natural-language interface (qv) to a simulated robot arm in a domain of blocks on a table. The system could handle imperatives such as "Pick up a big red block," questions such as "What does the box contain?" and declaratives such as "The blue pyramid is mine." Since SHRDLU maintained information about its actions, it could also be asked questions such as "Why did you pick up the green pyramid?" to which the system might respond "to clean off the red cube."

The primary design principle of SHRDLU was that syntax, semantics, and reasoning about the blocks world should be combined in understanding natural-language input. The main coordinator of the system was a module (effectively a parser) consisting of a few large programs written in a special programming language called PROGRAMMAR, which was embedded in LISP. These programs corresponded to the basic structures of English (clauses, noun groups, prepositional groups, etc.) and embodied a version of the systemic grammar theory of Halliday (55). A semantics module that was similarly organized coordinated with the parser and made calls to a reasoning system programmed in MICROPLANNER (qv), a theorem-proving (qv) language. *Procedural representations* for most of the knowledge in the system gave SHRDLU a considerable amount of flexibility to integrate semantic and pragmatic tests, to apply heuristic procedures for anaphora resolution, etc. The success of the procedural representations sparked the procedural–declarative controversy (56), which led to the identification of important knowledge representation issues.

In the final analysis, many have agreed with Wilks (57) that SHRDLU's power seems to derive in large measure from the constraints of its small, closed domain and that the techniques would fail to scale up to larger domains. Furthermore, the grammatical coverage of SHRDLU was spotty in the sense that "although a large number of syntactic constructions occur at least once in sample sentences appearing in the published dialog, our attempts to combine them into different sentences (involving no new words or concepts) produced few sentences that Winograd felt the system could successfully process" (58). Nevertheless, SHRDLU was an impressive demonstration system that rekindled the hope of truly "natural" language-understanding systems and touched upon many still unsolved research topics.

LUNAR. The task of LUNAR, a system deriving from the work discussed above on procedural semantics, was to provide lunar geologists with a natural-language interface to the Apollo moon rock database. The system had three main components. The first phase formed a syntactic parse using an elaborate ATN grammar and a dictionary of 3500 words. The parser created a deep-structure representation, which was then passed to a rule-driven semantic interpreter. The antecedent of a semantic rule specified a tree fragment to be matched against the deep-structure representation plus semantic conditions on the matched nodes. The right side of a semantic rule was a procedural template for the final, retrieval component. For example, the sentence

What is the average concentration of aluminum in high alkali rocks?

was translated as

```
(FOR THE X13 /
  (SEQL (AVERAGE X14 /
    (SSUNION X15 / (SEQ TYPEAS):T;
      (DATALINE (WHQFILE X15) X15
        (NPR*X16 / (QUOTE OVERALL))
        (NPR*X17 / (QUOTE AL203)))):T)):T;
  (PRINTOUT X13)).
```

The database was a flat file containing 13,000 entries. Run time performance of the system was acceptable; the sentence above was parsed in just under 5 s. In an informal demonstration of the system at the Second Annual Lunar Science Conference held in Houston in January of 1971, 78% of the 111 requests were handled without error. After correcting minor dictionary coding errors, this rate was improved to 90%.

In discussing the coverage of the system, Woods considered the syntactic coverage to be "very competent" but noted that "if a [lunar geologist] really sat down to use the system to do some research he would quickly find himself wanting to say things which are beyond the ability of the current system" (54). In summary, the LUNAR system demonstrated that a sizable, important database problem could be handled using the techniques of ATNs and procedural semantics.

A Variety of Application Areas (1970–1984)

Following the technical advances of the 1960s and in the wake of the rather dramatic work of Winograd and Woods, the period from the early 1970s witnessed a variety of applied natural-language projects. Application areas include database interface, computer-aided instruction (qv), office automation (qv), automatic programming (qv), and the processing of scientific text. These areas are discussed in turn.

Database Interface. Typical NL database interfaces operate by translating English or other natural-language inputs into a formal database query language to be run against an existing relational or other database management system. For a number of reasons, this formed the most frequent application area for applied NL work in the 1970s. First, the growing presence of database systems in business and industry resulted in a rapid increase in the number of potential computer users, many of whom preferred not to have to learn a "formal" computer language. Second, the idea of database query followed logically from the question-answering mode of many NL systems of the previous decade. Third, the NL system designer could, by starting with an existing set of data and by assuming an implemented back-end retrieval module, avoid the need to address low-level representation issues.

An early attempt at providing natural-language access to a relational database is the RENDEZVOUS system, which emphasized human factors and concepts deriving from the database world, without less attention to techniques developed in AI and CL. The primary design goal of the system was "to accept queries stated in *any* English, grammatical or not, rejecting only those that are clearly outside the domain of discourse supported by the data base at hand" (59). To accomplish this, RENDEZVOUS temporarily ignored portions of the input it could not recognize and, to compensate, engaged the user in "clarification dialog" to refine its understanding. A representative initial input to RENDEZVOUS is "I want to find certain projects. Pipes were sent to them in Feb. 1975." To help ensure reliable processing, RENDEZVOUS provided a paraphrase of its current understanding of the user's request. In situations where a phrase was ambiguous to the system, a request for clarification was generated. Although RENDEZVOUS tended to overburden the prospective user with too many seemingly pointless questions (e.g., it thought that "37" in "part 37" might be a quantity on order rather than a part number), it addressed human-factors issues that were taken up by PLANES (qv) and other systems.

A sharp contrast to the previous system in terms of linguistic sophistication is found in TQA, formerly REQUEST (60), whose syntactic processing is based on the principles of transformational grammar. This decision was made "in an attempt to deal with the complexity and diversity that are characteristic of even restricted subsets of natural language" (60). In essence, the parsing involves applying transformations in *reverse* to reconstruct the deep structure associated with a question to be answered. Details concerning this process, and further motivation for it, are given in Petrick (61). Initial applications included Fortune 500 data and a database on White Plains land usage. During the latter field study, a set of "operating statistics" was collected (62).

A system intermediate between the previous two in terms of linguistic sophistication is LADDER (63), whose syntactic processing was based on a "semantic grammar" designed around the object types of the domain at hand, such as ship and port, rather than linguistically motivated lexical and syntactic categories, such as noun and verb (see Semantic grammar). The system provided an ability "for naive users to input English statements at run time that extend and personalize the language accepted by the system" (64). The specific set of "tools" seeking to "facilitate the rapid construction of natural language interfaces" (64) was called LIFER (qv). The system contained facilities for the user to add synonyms and define paraphrases; mechanisms to handle ellipsis (incomplete inputs) were also provided. For example, after asking "What is the salary of Johnson?" the user could type "position and date hired" and the system would answer the question "What are the position and date hired of Johnson?" The development of LADDER helped to reassure the CL community that "genuinely useful natural language interfaces can be created and that the creation process takes considerably less effort than might be expected" (64). Hershman et al. (65) describe an experiment in which LADDER was used for a simulated Navy search-and-rescue operation.

Other interesting and important NL database systems were constructed, but space permits only brief descriptions. The TORUS system (66) represented an early attempt at formulating an "integrated methodology" for designing NL database front ends. It employed a semantic network representation, and the prototype was developed around a database of graduate student files. As described in Thompson and Thompson (67), work continued on the REL system mentioned earlier and eventually gave rise to the desk-top systems POL and ASK (68,69). In addition to manifesting refinements and extensions over earlier work, ASK was extended to allow for French as well as English inputs. Results of an experimental study with the REL-ASK family appear in Thompson (70). EUFID (71,72) was also designed to be database independent and, like LADDER, provided a "personal synonym" facility designed to be "forgiving of spelling and grammar errors" (71).

Several systems seeking to address the issue of cooperative response were designed. For instance, the construction of CO-OP (73) was based on the belief that "NL systems must be designed to respond appropriately when they can detect a misconception on the part of the user." For example, the probable presumption of a user who asks, "how many students got a grade of F in CIS 500 in spring 1977" is that the course was in fact given at the time in question. If this were not the case, CO-OP would so inform the user, rather than simply give the literal but misleading answer "nobody." The PLANES system (74) was based on the notion that an effective NL system "must be able to help guide and train the user to frame requests in a form that the system can understand." According to the designers, the work derives in spirit from Codd's work (59) with RENDEZVOUS. For instance, PLANES incorporated novel techniques of "concept case frames" for generating dialogue to flesh out an incomplete understanding of the user's request and "context registers" for handling pronouns and other anaphora. JETS, a successor to PLANES, responded to some "interesting questions about the conceptual completeness of question-answering systems" (75) that arose during experiments with the earlier system (76).

A number of high-quality European systems were developed, each manifesting some interest in domain independence. The USL system (77) was designed in Heidelberg as a domain-independent, German language database front-end. It incorporated a "revised version" of a parser built by Martin Kay, and "the method of [semantic] interpretation used in the REL system . . . was taken as a point of departure" (78). Additional grammars were constructed to enable USL to answer questions posed in English, Dutch, and Spanish as well as German. A user study with the USL system was reported by Krause (79). PHLIQA was built in Eindhoven to answer English questions about information stored in a CODASYL database (80). It operated on hypothetical data concerning computers installed at European establishments and was intentionally structured to include "some features that cause difficulties . . . and which are also found in 'real' databases" (81). Attention was paid to isolating the parts of the system that depended on the chosen data base, and an effort was made "to derive the parts dependent on a data base in a systematic way from the structure of the data base and its subject domain" (81). HAM-ANS (82) was developed in Hamburg as "a robust and cooperative system" to enable natural-language access in German to database and other software services.

The ROBOT system (83,84), rather than emphasizing linguistically complex (e.g., deeply nested) English structures, represented a database front-end with concern for report generation facilities, as suggested by the representative input "Print a report of direct commission, net loss incurred, and change in INBR for region B, sorted by net loss." The system was interesting in its use of the database as a world model in which to carry out disambiguations, but its primary importance is that it led in the late 1970s to a product for mainframe database query, INTELLECT (qv) and the concomitant founding of Artificial Intelligence Corporation.

More recent systems for database and information retrieval, which represent an important direction but for which space does not exist and for which a historical perspective is not yet possible, are mentioned below in the section on current trends. Also, there has been a small amount of attention to providing natural-language facilities for database *update,* as opposed to query. Examples of this work are found in Salveter (85) and Davidson and Kaplan (86).

Computer-Aided Instruction. The first attempt at incorporating AI and CL techniques into an integrated system for computer-aided instruction (CAI) was the SCHOLAR (qv) program (87). By representing the information to be learned in *semantic network* structures, this system was designed to be independent of the actual "lesson" at hand, which for the prototype consisted of information about South American geography. In particular, "no specific pieces of text, questions, with their predicted answers, errors, or anticipated branching form part of this data structure" (87). An example interaction with SCHOLAR follows.

> THE LANGUAGE IN BRAZIL IS FRENCH. RIGHT OR WRONG?
>
> "Wrong"
>
> YOU ARE CORRECT. THE LANGUAGE IN BRAZIL IS PORTUGESE.

As indicated, the system included a nontrivial English-generation component. More significant, however, was the provision for "mixed-initiative" dialogues, where the user could ask questions of the system at any time rather than supply an answer for a pending question. For example, if the system had just asked "What is Brazil," the student might delay answering the question by asking the system to "Tell me something about Peru." In addition, the system allowed its user to switch to a question-answering mode, thus acting as a question answerer for the data at hand. Historically, however, SCHOLAR was most important by establishing the feasibility of mixed-initiative NL systems.

In continuing the work begun with the initial SCHOLAR prototype, Collins et al. (88) extended the system's basic matching mechanisms to enable it to operate in situations in which requested information is missing. The resulting system would sometimes make inferences and sometimes inform the questioner that the requested information was not available. In the latter case, related or partial information might be given. As an example, the system would respond to the question "Is Bolivia a capital?" by answering "No, Bolivia is a country, not a capital." In formulating this response, SCHOLAR has used its knowledge that country and city are mutually exclusive object types and only a city can be a capital.

Another extension to the work of SCHOLAR is represented by the SOPHIE (qv) system, which sought to encourage "a wider range of student initiatives" by presenting "a 'reactive' environment in which the student learns by trying out his ideas rather than by instruction" (89). To explore this idea, SOPHIE was designed as an electronics lab troubleshooting system that would "mimic the roles of a human teacher" (89). To use the system, a student would receive a schematic diagram of a power supply and be told the control settings of a simulated faulty circuit. An example interaction follows.

"What is the output?"

THE OUTPUT VOLTAGE IS 11.7 VOLTS.

"What is it in a working circuit?"

IN A WORKING CIRCUIT THE OUTPUT VOLTAGE IS 19.9 VOLTS.

"What is the voltage between R8 and Q1?"

THE VOLTAGE AT N1 IS 47.6 VOLTS.

When the student finally settled on a diagnosis of the problem and recommended a course of action, such as "replace Q6," SOPHIE would seize the initiative and begin asking questions of the student. In addition, SOPHIE could alter the parameters of its simulated circuit, thus providing "what-if" capabilities. Despite the fact that it was carefully designed for a troubleshooting application by the use of a "semantic" grammar and thus lacked the domain independence of SCHOLAR, SOPHIE contained a variety of non-NL capabilities that themselves proved to be interesting and important.

Subsequent to the efforts reported above, Weischedel et al. (90) constructed a system to aid students in learning first-year German. The designers were interested, among other things, in enabling computers to deal with ungrammatical sentences, and in their chosen setting, it was mandatory for the system to respond meaningfully to inputs that were linguistically flawed as well as those that were factually incorrect. An example of such a response follows (the system's question translates as "Where did Miss Moreau learn German?").

WO HAT FRAULEIN MOREAU DEUTSCH GELERNT?

"Sie hat es gelernt in der Schule."

ERROR: PAST PARTICIPLE MUST BE AT END OF CLAUSE.

A CORRECT ANSWER WOULD HAVE BEEN:

SIE HAT DEUTSCH IN DER SCHULE GELERNT.

In addition to detecting incorrect grammar in the context of an otherwise acceptable response, the system was able to recognize when an input was incorrect or, more subtly, correct but not fully responsive to the question. As suggested above, the tutoring program dealt with reading comprehension, and the prototype was applied to several "lessons," each consisting of a paragraph. Concerning generality, the designers pointed out that "the texts that appear in foreign language textbooks very rapidly surpass the ability of artificial intelligence systems (90). They also observed that "there does not seem to be any way to tune the system to particular types of errors," which means that an instructor would have to construct each lesson by hand, unlike for the semantic network approach adopted for SCHOLAR (which carefully avoided storing textual information). Perhaps the most significant outcome of the project was to demonstrate ways in which "ill-formedness" can extend to morphological, semantic, and pragmatic problems as well as syntactic ones.

The ILIAD system (91) was conceived as a way of helping instruct people having a language-delaying handicap (e.g., deafness) or who are learning English as a second language. It included a powerful English generator based on the transformational grammar model.

Office Automation. The SCHED system was based on techniques and formalisms developed for the automatic programming system NLPQ described below and presented an initial study of "the feasibility of developing systems which accomplish typical office tasks by means of human-like communication with the user" (92). Although the long-range goal of SCHED was to provide an on-line system to review and update one's own desk calendar and those of fellow office workers, the implemented system was restricted to information pertaining to a single user. An example input for SCHED is

Schedule a meeting, Wed, my office, 2 to 2:30, with my manager and his manager, about 'a demo'.

to which the system would respond by stating in English its understanding of the input. Subject to user verification, the system would issue an appropriate command to a resident calendar management system. In situations where a user input failed to supply all necessary information, SCHED was able to ask for specific information, thus providing for mixed-initiative conversations reminiscent of the previously mentioned work in CAI.

The GUS system, similar in spirit to SCHED, though quite different in its methods, was "intended to engage a sympathetic and highly cooperative human in an English dialog, directed towards a specific goal within a very restricted domain of discourse" (93). In particular, GUS played the role of a travel agent able to assist a user in making a round trip from a city in California. Although its implementation was apparently less robust than SCHED, its designers suggested that "the system is interesting because of the phenomena of natural dialog that it attempts to model and because of its principles of program organization" (93).

The VIPS system, which seeks to "allow a user to display objects of interest on a computer terminal and manipulate them via typed or spoken English imperative sentences" (94), is unusual in that it incorporates a hardware voice recognizer into an NLP. Initial applications have been to the numerical domain of its predecessor (the automatic programming system NLC) and to text editing, where objects may be referenced either in English or by use of a touch-sensitive display screen.

ARGOT represents a "long-term" research project seeking to "partake in an extended English dialogue on some reasonably well specified range of topics" (95). The initial task domain for ARGOT was that of a computer-center operator.

The UC system is designed as "an intelligent natural-language interface that allows naive users to communicate with the UNIX(TM) operating system in ordinary English" (96). It answers questions such as "How do I print the file fetch.1 on the line printer?"

Finally, some research has applied CL techniques to the analysis, as opposed to processing, of written texts. One such system is the Writer's Workbench (97), which, upon scanning the draft of a document, flags words, phrases, and sentence types that have been felt to negatively affect the readability of the text. For example, the previous sentence could be improved by eliminating the passive voice ("have been felt") and unsplitting the infinitive ("to negatively affect"). Another system, CRITIQUE, formerly EPISTLE (qv) (98), similarly performs stylistic analyses but also diagnoses several classes of grammatical errors. For example,

Your statement of deficiencies *have* not been completed.

(should be *has*)

Automatic Programming. The term *automatic programming* (qv) refers to the effort to relieve a computer user from the details needed to get the machine to perform a desired task. In certain ways attempts at a natural-language solution to this problem are similar to the work on problem-solving systems (STUDENT and CARPS) described earlier, although no specific techniques from these systems appear to have been employed. A survey of early NL work in automatic programming appears in Heidorn (99).

One early system, described by Gelb (100), was designed to answer simple probability questions such as "What is the probability of getting two or more heads or exactly three tails when four coins are tossed once?"

Another early system is NLPQ (101), which produced simulation programs in GPSS for a class of queueing problems. An example of an initial specification of a user's problem follows: When a vehicle arrives at a station, it leaves there immediately if the length of the line at a pump in the station is not less than 2. Seventy-five percent of the vehicles are cars and a fourth are trucks. There is just one pump. A simulation run time of 8 h is desired. At this point the system would take control and begin asking questions to clarify and extend what it had been told. For example, it would begin by asking "How often do the vehicles arrive at the station," and the user might answer "The arrivals of vehicles are normally distributed with a mean of 8 minutes." Having acquired all requisite information, NLPQ would first produce an English statement of its understanding of the entire problem and then a complete program for the task.

As reported in Green (102), the PSI system included modules for English input and output, along with many other components, in a large-scale automatic programming system. An excerpt from a session with the system follows.

"[The program] first inputs a concept. Then it repeatedly accepts an input scene from the user, classifies it based on whether or not it fits the concept, and outputs this classification to the user."

WHAT IS THE STRUCTURE OF THE CONCEPT?

"A concept is a collection of relations. Each relation is just a string."

WHAT IS THE EXIT TEST OF THE LOOP . . .

Due to the complexities involved in program synthesis, PSI was tested only on problems related to sorting, graph reachability, and determining prime number.

Unlike the preceding work, the NLC system (103) chose to regard an English input as a statement in a very-high-level program. The system acted as an interpreter, providing an interactive matrix-domain programming environment with visual feedback as each English command was processed. An example input follows.

"Subtract six from the next to last positive number in the row that was multiplied by 5."

In addition to simple imperatives, NLC provided for loops ("repeat"), conditionals ("if . . ."), and procedure definitions ("define a way to . . ."). An experiment study of programming with the system is described in Biermann et al. (94), and an application of the system for college sophomore-level linear algebra instruction is discussed in Geist et al. (104).

Scientific Text Processing. Based on many years of work developing a comprehensive grammar for English (105), a group of researchers constructed a system intended "to allow the health care worker to create [a] medical report in the most natural way—in medical English, using whatever syntax is appropriate to the information (106,107). After gathering reports in English and converting them to a textual database form, the system could be interrogated as though it were a conventional database system, again using English for inputs. Some examples of the types of inputs gathered in a clinical setting are the following:

X-rays taken 3-22-65 reveal no evidence of metastatic disease.

Chest X-ray on 8-12-69 showed no metastatic disease.

3-2-65 chest film shows clouding along left thorax and pleural thickening.

and an example question is "Did every patient have a chest X-ray in 1975?" A distinctive feature of the system is its method of creating so-called information format structures, which are similar to structured database records but capture the information initially supplied in textual form. Defining an information format for a particular application involves, first, isolating *word classes* by syntactic properties, e.g., the verbs *reveal* and *show* are alike in taking *X-ray* as a subject, although *X-ray* and *film* are alike in taking *show* as a verb, and, second, defining the *columns* of the table from the word classes so that any input sentence will have a paraphrase like that shown above (108). The system, which is unusual in containing aspects of both database and information retrieval, was subsequently adapted to the domain of navy messages, as described in Marsh and Friedman (109).

Current Trends

Domain-Independent Implementations. Several domain-independent database systems have already been mentioned, but the intensity of effort at enhancing the transportability of systems for this and other applications areas should be noted. In particular, a number of projects are seeking either to allow users themselves to carry out a customization or to have the system adapt itself automatically to a user or a domain of discourse. Representative examples of this work include Haas and Hendrix (110), Hendrix and Lewis (111), Mark (112), Thompson and Thompson (68,69), Wilczynski (113), Warren and Pereira (114), Bates (115), Ginsparg (116), Grosz (117),

Ballard et al. (118), and Grishman et al. (119). Also, several papers deriving from a recent workshop on transportability (120) have appeared, including Damerau (121), Hafner and Godden (122), Marsh and Friedman (109), Slocum and Justus (123), and Thompson and Thompson (124).

The Reemergence of Machine Translation. As indicated earlier, the ALPAC report of 1966 nearly eliminated U.S. government funding of projects in machine translation (MT). Naturally, this caused a marked decline in the amount of work being done in the area and in the number of papers published. Nevertheless, due in part to progress in AI and other areas of CL, a gradual resurgence of interest in MT has occurred over the past decade, and the field gives evidence of becoming well populated once again. A bibliography of about 450 publications since 1973 related to MT can be found in Slocum (125), along with summary papers on several of the major full-scale translation systems in existence. Papers from a recent conference on MT (126) are also available.

The Commercialization of NLP. As indicated above, Harris's database front-end, ROBOT, became the proprietary software of Artificial Intelligence Corporation in the late 1970s. Under the name INTELLECT, this system was for several years virtually the only natural-language product on the market. In the early 1980s, however, several well-known NL researchers, including Gary Hendrix and Roger Schank, formed or became associated with start-up ventures. In recent years products from these and other companies have been appearing for database and other applications [e.g., a developing expert system interface is discussed in Lehnert and Schwartz (127)]. More recently researchers at Carnegie-Mellon University and other academic institutions have formed companies; Texas Instruments has produced a menu-based natural-language-like interface; at least one project at BBN Laboratories is slated for commercial release; and other corporate flirtations are occurring. In addition to database query, machine translation systems are also being sold, and prospects exist for additional application areas. All of these activities, as well as an overview of ongoing research into the theoretical and applied side of CL, are reported in Johnson (128).

Theoretical Issues

Having thus far conducted a chronological review of projects within CL that relate more or less directly to specific applications, this section provides a brief overview of some of the major theoretical topics associated with CL. The discussion relates specific systems to the theoretical issues, but it primarily emphasizes theoretical techniques and formalisms that have contributed to the classification of research in CL. The topics are parsing and grammatical formalisms, semantics, discourse understanding, text generation, cognitive modeling, language acquisition, and speech understanding. Further information is available in separate articles.

Parsing and Grammatical Formalisms. Parsing (qv) issues have been of central interest in CL since its inception, when CL included the study of formal languages and programming languages, as well as natural languages. As the term is used here, *parsing* refers to the process of assigning structural descriptions to an input string. Classically, parsers have used various forms of phrase structure grammars and have assigned phrase structure markers to produce derivation trees. Parsers with access to semantic and pragmatic knowledge, however, may build semantic descriptions directly without explicitly creating derivation trees.

Direction of Analysis. Parsing strategies are often classified as top-down (or goal-driven) if they begin with the start symbol and backward chain from the consequents of rules to their antecedents. Recursive-descent parsers, the PROLOG execution procedure for definite-clause grammars (DCGs) (129), and the usual execution procedure for augmented transition network (ATN) grammars (39) all use top-down approaches (see Processing, bottom-up and top-down). Bottom-up (or data-driven) techniques proceed in a forward direction from the terminal symbols (words) in the grammar toward the start symbol. Left-corner (including shift-reduce) parsers (130), word-based parsers (131) and its descendants (132), chart parsers (44,133–136), and deterministic parsers (137,138) are primarily bottom-up.

Parsers can also be classified according to how they analyze the input string: from left to right, from right to left, or from arbitrary positions in the middle outward. The left-to-right ordering is simple and natural and lends itself to easy bookkeeping. It is also of theoretical interest for parsers that attempt to model aspects of cognitive processes, such as attention focusing, that are dependent on temporal ordering. Middle-out, bottom-up parsers have been used particularly in speech systems (139,140), where the parser can use its analysis in regions of greatest certainty to help in noisy or unintelligible regions, which would cause trouble for a rigid left-to-right parser.

Parsing techniques bear a close relationship to grammatical formalisms, although a particular grammar or class of grammars can sometimes be parsed in a variety of ways. ATN grammars, for example, have been parsed by both top-down/left–right and bottom-up/middle–out methods. Another technique for matching grammar and parser is to preprocess a grammar into an equivalent grammar suitable for a particular parsing method.

Search and Nondeterminism. Controlling the search effort for a parse and handling nondeterminism are major problems for parsers. Many actual parsers use a blend of top-down and bottom-up techniques. As a simple example of this, almost every recursive descent (top-down) parser uses some kind of (bottom-up) scanner to identify the tokens in the input. Part-of-speech classifications in a lexicon are a form of bottom-up information. Another method for improving top-down parsing techniques is the use of a precomputed left-branching reachability matrix that can be used to decide whether the next input symbol can appear in the leftmost branch of a derivation tree headed by a particular nonterminal. Word expert parsing (qv) (132) uses an idiosyncratic combination of top-down expectations and bottom-up processing.

The three principal methods for dealing with nondeterminism are backtracking, parallelism, and transforming the grammar so that a deterministic algorithm (perhaps using bounded look-ahead) can efficiently parse it. Backtracking parsers pursue one alternative at a choice point and return to select another alternative on failure of the first one. Forcing failure after a successful parse can cause the backtracking parser to find additional parses. The standard execution procedure of PROLOG provides such a facility directly for logic grammars such as DCGs, which can be represented as PROLOG programs. Backtracking techniques are especially popular and natural for context-free grammar formalisms. Context-free grammars, which have a single consequent nonterminal, lend themselves to backward-chaining execution methods that work nicely in conjunction with backtracking.

Parallel parsing methods keep track of multiple derivations at each point in the processing. The derivations can be developed concurrently using sequential algorithms and machines or truly in parallel if multiple processing resources are available. Interest in parallel approaches has increased as parallel hardware is becoming available; it has also been buoyed by the resurgence of connectionist and neural network research (141–143). Chart parsing (qv) algorithms use a particularly efficient way of recording which derivations have already been found to cover substrings of the input string. By splitting the computation at choice points, backtracking methods can also be parallelized.

Another alternative, explored by Marcus (137), is factoring the rules in such a way that limited look-ahead is sufficient to resolve most of the nondeterminism. Although not all of English, for example, can be treated deterministically in this manner, the parser interestingly fails on many of the same "garden path" sentences that cause people trouble. Cases of lexical and structural ambiguity that the parser cannot resolve are left for other modules.

Word-based parsing systems generally attempt to incorporate enough knowledge to determine a unique interpretation. When ambiguity cannot be resolved without look-ahead, two possibilities present themselves. One technique is to let a later constituent complete the interpretation; for example, verbs can be responsible for assigning the role of the subject noun phrase. Another solution is to spawn demons that check for the appearance of disambiguating items. For example, sense-specific demons might check for the presence of particular particles to detect multiword verbs.

Grammatical Formalisms. Many different grammatical formalisms have been used by natural-language-processing systems. One of the earliest systems, the Harvard Syntactic Analyzer (144), recognized context-free grammars. Transformational grammar (TG) theories had a direct influence on the Mitre (145) and Petrick (146) parsers and an indirect influence on many others. The UCLA grammar combined TG with case grammar theory (147). In simplest terms, transformational grammars specify a set of (usually context-free) base phrase structure rules, a set of structure mapping rules, and various conditions, filters, or principles that generated structures must satisfy. Since TG is stated as a generative theory, parsers must try to guess which transformations must have been applied by effectively inverting the rules. This has proved to be quite difficult in practice.

One of the most comprehensive computer grammars of English has been developed at the Linguistic String Project (105). The grammar consists of a set of 180 BNF phrase structure rules, 180 restriction rules that check feature conditions, string-transformation rules, and ellipsis rules. Additional sublanguage categorizations are added to the lexicon together with domain-specific restriction rules to increase parsing efficiency.

ATN grammars (39) have been very influential on computational approaches to language processing. They augment recursive transition-network grammars, which recognize context-free languages, with actions and tests that give them the recognition power of Turing machines. With suitable self-restraint, however, one can produce disciplined, well-structured grammars in many different grammar-writing styles. The LUNAR grammar (53) was a quite detailed, large grammar (see Grammar, ATN). Other augmented phrase structure formalisms include the DIAMOND grammars at SRI [e.g., DIAGRAM (148)] and the APSG (augmented phrase structure grammar) formalism used in the CRITIQUE system developed at IBM (149). The systemic grammar theory of Halliday (55) has been incorporated in many NLP systems, notably in Winograd's SHRDLU system (52) and in the large NIGEL grammar (150).

More recent work in linguistics has revived interest in nontransformational theories of phrase structure grammars, particularly context-free grammars. These theories hold that notational augmentations to phrase structure grammars can express such difficult, "transformational" phenomena as movement nontransformationally and, furthermore, in most cases, that there exist equivalent context-free grammars. From a parsing perspective it will be most useful if the augmentations can be processed on the fly with little overhead above that required for context-free parsing. The augmentations include metarules, complex features, and principles of feature instantiation (151). The major theoretical frameworks include generalized phrase structure grammar (GPSG) (152), tree adjoining grammar (TAG) (153), head grammar (HG) (154), lexical functional grammar (LFG) (155), and functional unification grammar (FUG) (156).

Providing for Ungrammaticality. It has been observed that "while signficant progress has been made in the processing of correct text, a practical system must be able to handle many forms of ill-formedness gracefully" (157). When the ill-formedness in question is syntactic in origin and when the expected deviations can be grouped into a manageable number of classes, it is possible to prepare for "errors" by explicitly including extra rules in the system grammar so that a predictably deviant input is in fact treated as though it was grammatical. Due to the possible ambiguities that this practice introduces and to confront general situations where either the full range of errors cannot be predicted or the intended meaning cannot be recovered, more sophisticated mechanisms are called for.

Attacks on the problem of ungrammaticality are represented by work described by Weischedel and Black (158), Hayes and Mouradian (159), Kwasny and Sondheimer (160), Jensen et al (161), Weischedel and Sondheimer (162), Granger (163), and Fink and Biermann (164). It is also worth noting that, for some applications, grammatical errors are part of the problem being addressed rather than a regrettable accident. Examples include the German language CAI system and the text-critiquing systems mentioned earlier. In addition, most AI work in speech understanding is fundamentally concerned with the rampant and perhaps inherent uncertainties associated with speech-recognition devices. These uncertainties actually make error conditions the rule rather than the exception.

Semantics. Semantics concerns the study of meaning. In the context of CL, this most often relates to problems of finding and representing the meaning of natural-language expressions. The previous discussion has already touched on several approaches to semantics, including conceptual dependency, procedural semantics, and semantic networks. Some others are preference semantics and other decompositional systems, Montague semantics, and situation semantics. Details on each of these can be found in the entry on semantics.

Among the more significant questions to be asked of an approach to semantics, at least insofar as its relevance to CL is concerned, are what sorts of *noncompositionality* does the system involve and what role, if any, is played by *primitives*. In essence, the idea behind compositional semantics is to determine the meaning of an entire unit under analysis (phrase, sentence, text) in a systematic (ideally simple) way from the meanings of its parts. This approach has obvious advantages in terms of being tractable for incorporation into an automated scheme for language understanding. The idea behind primitives (qv) (in its strong sense) is to determine a finite set of terms that by themselves can express the meaning of any word and, by implication, the meaning of any utterance. Of the semantic schemes discussed earlier, conceptual dependency (qv) adheres to this goal, where procedural semantics (qv) does not. Many interesting debates on these and other issues of semantics have occurred, as discussed by Jackendoff (165).

In building entire NL systems, many designers have attempted to separate syntax from semantics by performing syntactic analysis first and then converting the resulting structure (produced by the parser) to a meaning representation (see Natural-language understanding). In other cases, however, the two processes have been much more tightly integrated. Whereas problems in CL related to syntax have largely involved issues also addressed by the field of conventional (as opposed to computational) linguistics, problems of semantics have typically concerned work in philosophy. In the context of AI, important work related to natural-language semantics is to be found in the area of *knowledge representation* (qv).

Discourse Understanding. Discourse understanding includes natural-language-processing phenomena that span individual sentences in multisentence texts or dialogue. The work in discourse understanding acknowledges that the syntactic and semantic representations of sentences in discourse contexts relate both explicitly (e.g., by clue words such as *now, but, anyway*) and implicitly (e.g., by world knowledge) to the representations of other sentences in the discourse.

As an example of how an amazing amount of complexity can enter into even simple interchanges, consider the following brief dialogue:

Q: Can you tell me where John is?

A: Oh, he was hungry for one of Joe's pizzas. He'll be back soon.

The petitioner's use of a yes–no question is an example of an *indirect speech act*. In an indirect speech act one illocutionary act is performed indirectly by way of performing another (166–168). The yes–no question is interpreted as a form of politeness instead of a more direct utterance such as "Where is John?" Grice (169) noticed that conversational participants follow cooperative principles that he subcategorized as quantity (be informative), quality (be truthful), relation (be relevant), and manner (be brief). The response in the example above meets Gricean notions of appropriateness, but it, too, is indirect in communicating both where John is and why he is there. To infer John's location from his state of hunger and desire requires a plan and goal analysis from pragmatic (extralinguistic) knowledge. The final part of the answer responds to an inferred petitioner's goal of being copresent with John by suggesting that he will be back soon.

In cooperative conversations Grice noted that speakers caused listeners to make certain inferences, which he termed *conversational implicatures*. Hirschberg (170) has studied a class of implicatures called *scalar implicatures*. In the sentence "some people left early," for example, the hearer may reasonably conclude that "not all people left early." A cooperative response occasionally requires that faulty presuppositions in the question be corrected. For a database query such as "How many juniors failed CS 200?" an answer of "none" is misleading if there were no juniors enrolled. The CO-OP system performed this type of presupposition checking (73). Another type of cooperative response involves informing the user of discontinuities (171). In a flight reservation database one might want to know of any flights leaving before noon. It might be helpful to suggest one at 12:05 P.M. or one the next or previous day if none are otherwise available.

A natural idea for discouse understanding was to extend some of the concepts of grammars and schemas from sentence parsing to discourse. Conversation-related work includes the Susie Software system (172,173) and discourse ATN grammars (174). In story understanding Rumelhart (175) and Coreira (176) developed the idea of story grammars. Many of the language-understanding systems of Schank and his students use knowledge structures [such as scripts, plans, memory organization packets (MOPs), and thematic abstraction units (TAUs)] to guide discourse-understanding processes. Dialog-games (177) were an attempt to use a goal-centered theory for dialogues. Litman (178) has also integrated work on planning and discourse.

Focus is an important technical notion in discourse work that relates to the shifts in attention during comprehension. Focus influences many aspects of language understanding, including choice of topic, syntactic ordering, and anaphoric reference. Grosz (179) did pioneering work on *global focus,* i.e., how attention shifts over a set of discourse utterances. *Immediate focus* represents how attention shifts over two consecutive sentences. Sidner (180) used focus to disambiguate definite anaphora by tracking three things: the immediate focus of the sentence, a potential focus list created from discourse entities in the sentence, and the past immediate foci in a focus stack.

The resolution of anaphora is an important problem within discourse understanding. Early techniques principally used a simple history list of discourse entities combined with a heuristic method for selecting them (often a variation of the most recently encountered entity satisfying the reference). The simple techniques are inadequate, largely because they fail to account for focus effects and because discourse referents do not have to be explicitly mentioned (e.g., the referent of *he* in the sentence "I got stopped yesterday for speeding, but he didn't give me a ticket"). Besides Sidner's technique described above, there are several other notable approaches (see Ref. 181 for a more detailed account). Other methods include concept activatedness (182), task-oriented dialogue techniques (179), logical representations (183), and discourse cohesion (184,185).

Text Generation. Text generation is the process of translating internal representations into surface forms. The forms of internal representations have included deep structure, semantic networks, conceptual dependency graphs, and deduction trees. The *strategic* component of a generation system chooses what to say—the message to be conveyed including any propositional attitudes. The *tactical* component determines how to say it.

The earliest systems generated sentences at random to test grammars (186,187). Later AI efforts used generation techniques as a part of paraphrase systems, which parsed input strings into meaning representations and then generated back out into surface representations. Klein (188) used dependency grammars that generated a semantic dependency tree and a standard phrase structure derivation tree. Dependency trees from multiple sentences were related by nominal coreference links. A generation grammar matched portions of the dependency trees. Simmons and Slocum (189) produced sentences from a semantic network using an ATN modified for generation. Eventually, a parser was added to fully automate the paraphrase process (49). Similarly, Heidorn (190) reported an algorithm based on an augmented phrase structure grammar for producing English noun phrases to identify nodes in a semantic network. Goldman (191) used a discrimination net for conceptual dependency graphs. The net tested the primitive action types and roles to select an appropriate surface verb. This generator was later used as a part of the MARGIE system (192).

The generation technique in SHRDLU (52) is an example of the template-based approach that has predominated in generation techniques. The program used several types of patterned responses including completely canned phrases such as "ok," parameterized phrases such as "sorry, I don't know the word ________," and more complex parameterizations that involved the substitution of determiners, discourse phrases, and dictionary definitions. Small programs were responsible for formatting the descriptions of objects and events. For example, the definition for the event PUTON was

```
(APPEND (VBFIX (QUOTE PUT)) OBJ1
(QUOTE (ON)) OBJ2)
```

A heuristic pronominal substitution mechanism improved the quality of the responses, allowing for the generation of noun phrases as complex as "the large green one that supports the pyramid."

Although most generation systems of the 1970s used techniques similar to SHRDLU, two different, important generation programs appeared in 1974. Davey's PROTEUS program (193) described tic-tac-toe games. The program had a rich understanding of the tactics of the game and could provide natural summarizations at an appropriate, high level. An example is "I threatened you by taking the middle of the edge opposite that and adjacent to the one I had just taken but you blocked it and threatened me." The ERMA program (194) embodied a cognitive model of human generation that mimicked the real-time false starts and patching of utterances. The model was developed by studying transcripts of psychoanalysis sessions to determine a patient's reasoning patterns. As an example, the program generated "you know for some reason I just thought about the bill and payment" as a gentle way of beginning to argue that "you shouldn't give me a bill."

Interest in generation work has revived in the 1980s with a number of new research projects. Mann et al. (195) provide a survey of text generation projects. Some of the major projects are as follows. The transformational grammar generation system of Bates and Ingria (91), a very syntactically powerful generaor, was used in a CAI application. McDonald's generator, MUMBLE (196), models spoken language and concentrates on the fluency and coverage of the tactical component. The KDS system (197) used a "fragment-and-compose" paradigm in which the knowledge structure is divided into small propositional units, which are then composed into large textual units. Mann and Mattheissen (150) used a systemic grammar (NIGEL) for the tactical component in a text generation system. In describing a system for generating stock market reports, Kukich (198) proposed a "knowledge-intensive" approach to sentence generation; similarly specialized techniques form the basis of the generator for the previously mentioned UC project (199). The KAMP system (200) views generation as a planning problem of proving what to say. In her TEXT system, McKeown (201) adapted ideas of text schemas and focus from discourse-understanding research to the task of answer generation in a natural-language database system.

Cognitive Modeling. In the late 1960s at Stanford, Roger Schank, while working on a parser for an automated psychiatrist project with Kenneth Colby, developed a meaning representation known as conceptual dependency (CD). Having been exposed to machine translation as a graduate student, Schank was convinced that more of the underlying meaning of sentences needed to be represented. In particular, certain inferences were included in the CD graphs. The basic scheme was centered on approximately a dozen primitive action concepts. The translation of "X hit Y," for example, was approximately "X propelled some Z from X to Y which resulted in the state of Y and Z being in physical contact." The first fairly complete system, MARGIE, included a parser (conceptual analyzer), an inferencer, and a text generation system (192).

In an interesting early retrospective of the CD paradigm, Schank offered this perspective on the situation that he faced in the late 1960s (202):

Thus, my point was that Chomsky was wrong in claiming that we should not be attempting to build a point by point model of a speaker-hearer. Such a model was precisely what I felt should be tackled. Linguists viewed this as performance and thus uninteresting. I took my case to psychologists and found them equally uninterested. Psychologists interested in language were mostly psycholinguists, and psycholinguists for the most part bought the assumptions of transformational grammar (although it seemed very odd to me that given the competence/performance distinction, psychologists should be on the side of competence).

Schank's emphasis on semantic representations was supported by others [notably the work on preference semantics by Wilks (203)] but has been slow to make a large impact on practical systems. Perhaps the slow acceptance was a result of methodology [the "free form speculation approach to theory building" (202)], of general attitudes inherited from linguistic theory, of the emphasis in CD systems on the I/O behavior of programs instead of formal computational models, and of the difficulty in discovering and representing conceptual knowledge structures.

One problem that plagued the inferencer in MARGIE was how to control the potential inferences that could be made. Later CD-based systems made inferences organized from knowledge sources such as scripts (204,205), plans and goals (206), beliefs (207), episodic memory (qv) (208), and thematic abstraction units (209). Scripts provide prepackaged causal and temporal links for stereotypical situations. For less structured situations the links are created dynamically by a plan and goal analysis. Inferences are also affected by one's beliefs (e.g., conservative/liberal political beliefs) and memory of past events. Although many of the ideas of schematic inference and planning are being incorporated in recent work, the difficulty of identifying and integrating a wide range of semantic and pragmatic representations remains a difficult problem for AI and CL.

Language Acquisition. Computational language acquisition (qv) research subdivides in much the same way that AI research generally does. Some researchers attempt to automate the acquisition of linguistic expertise by any efficacious method; other work is explicitly aimed at cognitive modeling and tries to be faithful to the psycholinguistic data on language acquisition. Most of the language-learning systems are concerned primarily with learning syntactic rules.

New computational approaches to language acquisition have generally followed developments in linguistics or natural-language-processing techniques. The ZBIE system (210) learned foreign language rules from input pairs consisting of a semantic representation and a surface string. For example, the representation (be (on table hat)) was paired with the sentence "The table is on the hat." To the extent that the appropriate syntactic structure of a sentence bears a particular relationship to the semantic structure, the semantic representation can guide in the induction of syntactic rules. Anderson's graph deformation condition (211) is a statement of this principle. Klein's AUTOLING program (212) derived a transformational grammar in cooperation with a linguist informant. The derived grammars contained context-free phrase structure rules and transformations. Harris (213) produced a language-learning system for a simulated robot. The system performed *lexicalization,* the process of mapping words to concepts, and the induction of a Chomsky normal-form grammar. Berwick (214) investigated learning transformational grammar rules of the sort embodied in a Marcus parser.

Reeker (215) explicitly modeled a child's acquisition of language with a problem-solving theory. The grammar was represented by context-free syntactic rules paired with a semantic representation modeled after conceptual dependency notation. The system received as input an "adult sentence" and its meaning. A heuristic reduction process formed a reduced sentence, which was then compared against a "child sentence" produced from the meaning by the child's current grammar. If a difference in the derived sentences was obtained, the grammar was adjusted. The AMBER system (216) similarly compares input sentences to internally generated sentences to identify discrepancies. The CHILD system (217) receives an adult sentence and a conceptual dependency representation of visual input. The model builds lexical definitions similar to those of other word-based parsers.

The psychologist John R. Anderson has made many contributions to language acquisition research. His LAS system (211) accepted sentence–scene description pairs and learned an ATN grammar that was used for both recognition and generation. The scene descriptions were encoded in the HAM associative network representation (218). Following this work, he developed a series of cognitive models and learning theories based on a hybrid architecture, called ACT (adaptive control of thought). An elaborate version of the model, ACT* (219), uses a production system to control spreading activation processes in a semantic network. Anderson has studied the learning of production rules for language generation, which is viewed as a problem-solving activity in ACT*.

Speech Understanding. The problem of understanding spoken natural language involves virtually all of the issues discussed above as well as others of its own (see Speech understanding).

Further Reading

In addition to the many references already cited and the discussions and references in related articles, Feigenbaum and Feldman (220) and Minsky (221) contain descriptions of, and Simmons (17,34) discusses, early work in natural-language processing; Rustin (222) and Zampolli (223) consider the status of several question-answering systems of the early to middle 1970s; Kaplan (224) contains brief summaries of several dozen projects underway in the early 1980s; and the brief articles in Johnson and Bachenko (225) give prospects for work in several areas of CL. Tennant (173) provides a fairly broad introduction to natural-language processing and contains technical details and historical remarks, as do the articles in Barr and Feigenbaum (226) and Lehnert and Ringle (227). Grishman (228) provides a general introduction to technical problems in the field; matters of parsing and grammatical formalisms are discussed in King (229), Winograd (230), Sparck Jones and Wilks (231), and Dowty et al. (232); an interesting discussion of cognitive approaches to semantics is Jackendoff (165); Brady and Berwick (233) contains papers on discourse. Schank and Riesbeck (234) and Simmons (235) present the actual mechanisms by which specific processors have been constructed. Harris (236) has written a recent textbook on natural-language processing (see Natural-language understanding).

Many articles have appeared in conference proceedings, including the annual meeting of the ACL, the biennial International Conference on Computational Linguistics (COLING), conferences sponsored by the American Association for Artificial Intelligence (AAAI), the biennial International Joint Conference on AI (IJCAI), a Conference on Applied Natural Language Processing, and two conferences on Theoretical Issues in Natural Language Processing. A primary journal is *Computational Linguistics* (formerly the *American Journal of Computational Linguistics*), and other important journals include *Artificial Intelligence*, the *Canadian Journal of Artificial Intelligence*, and *Cognitive Science*.

BIBLIOGRAPHY

1. W. Weaver, W. Locke and A. Booth (eds.), in *Machine Translation of Languages,* MIT Press, Cambridge, MA, pp. 15–23, 1955.
2. W. Locke and A. Booth (eds.), *Machine Translation of Languages,* MIT Press, Cambridge, MA, 1955.
3. A. Oettinger, *Automatic Language Translation,* Harvard University Press, Cambridge, MA, 1960.
4. Y. Bar-Hillel, The Present Status of Automatic Translation of Languages, in F. Alt (ed.), *Advances in Computers,* Vol. 1, Academic Press, New York, pp. 102–103, 1960.
5. National Research Council, *Language and Machines: Computers in Translation and Linguistics,* Report by the Automated Language Processing Advisory Committee (ALPAC), National Academy of Sciences, Washington, DC, p. 19, 1966.
6. N. Chomsky, *Syntactic Structures,* Mouton, The Hague, 1957.
7. Reference 6, p. 34.
8. N. Chomsky, *Aspects of the Theory of Syntax,* MIT Press, Cambridge, MA, 1965.
9. G. Salton, *Automatic Information Organization and Retrieval,* McGraw-Hill, New York, 1968.
10. J. Becker and R. Hayes, *Information Storage and Retrieval Tools, Elements, Theories,* Wiley, New York, 1963.
11. D. Hays (ed.), *Readings in Automatic Language Processing,* American Elsevier, New York, 1966.
12. K. Sparck Jones and M. Kay, *Linguistics and Information Science,* Academic Press, London, 1973.
13. B. Raphael, Hewlett-Packard, personal communication, July 1983.
14. D. Bobrow, Natural Language Input for a Computer Problem-Solving System, in M. Miusky (ed.), *Semantic Information Processing,* MIT Press, Cambridge, MA, pp. 133–215, 1968.
15. Reference 14, p. 146.
16. V. Giuliano, "Comments on the article by Simmons," *CACM* **8**(1) p. 69, (1965).
17. R. Simmons, "Answering English questions by computer: A survey," *CACM* **8**(1), 53 (1965).
18. Reference 17, p. 70.
19. B. Green, A. Wolf, C. Chomsky, and K. Laughery, BASEBALL: An Automatic Question Answerer, in E. Feigenbaum and J. Feldman *Computers and Thought,* McGraw-Hill, New York, 1963.
20. R. Lindsay, Inferential Memory as the Basis of Machines which Understand Natural Language, in E. Feigenbaum and J. Feldman (eds.), *Computers and Thought,* McGraw-Hill, New York, p. 221, 1963.
21. B. Raphael, SIR, a Computer Program for Semantic Information Retrieval, in M. Minsky (ed.), *Semantic Information Processing,* MIT Press, Cambridge, MA, p. 33, 1968.
22. J. Craig, S. Berezner, C. Homer, and C. Longyear, DEACON: Direct English Access and Control. *AFIPS 1966 Fall Joint Computer Conference,* p. 366.
23. Reference 22, p. 376.
24. F. Thompson, P. Lockemann, B. Dostert, and R. Deverill, REL: A Rapidly Extensible Language System, *ACM National Conference,* p. 400, 1969.
25. Reference 24, p. 404.
26. C. Kellogg, A Natural Language Compiler for On-line Data Management, *AFIPS 1968 Fall Joint Computer Conference,* pp. 473–492.
27. Reference 14, p. 204.
28. E. Charniak, Computer Solution of Calculus Word Problems, *Proceedings of the First International Joint Conference on Artificial Intelligence,* Washington, DC, pp. 303–316, 1969.
29. Reference 28, p. 305.
30. Reference 28, p. 309.
31. J. Weizenbaum, "ELIZA: A computer program for the study of natural language communication between man and machine," *CACM* **9**(1) 36–45 (1966).
32. J. Weizenbaum, *Computer Power and Human Reason,* W. H. Freeman, San Francisco, CA, 1976.
33. K. Colby, S. Weber, and F. Hilf, "Artificial paranoia," *Artif. Intell.* **2,** 1–25 (1971).
34. R. Simmons, "Natural language question answering systems: 1969," *CACM* **13**(1) 15–30 (1970).
35. R. Simmons and D. Londe, NAMER: A Pattern Recognition System for Generating Sentences about Relationships between Line Drawings, Report TM-1798, System Development Corp., Santa Monica, CA, 1964.
36. R. Kirsch, Computer Interpretation of English Text and Picture Patterns, *IEEE Trans. Electron. Comput.,* **13,** 363–376 (1964).
37. J. Thorne, P. Bratley, and H. Dewar, The Syntactic Analysis of English by Machine, in D. Mitchie (ed.), *Machine Intelligence,* Vol. 3, American Elsevier, New York, pp. 281–299, 1968.
38. D. Bobrow and B. Fraser, An Augmented State Transition Network Analysis Procedure. *Proceedings of the First International Joint Conference on Artificial Intelligence,* Washington, DC, pp. 557–567, 1969.
39. W. Woods, "Transition network grammars for natural language analysis," *CACM* **13,** 591–606 (October 1970).
40. C. Fillmore, The Case for Case, in E. Bach and R. Harms (eds.), *Universals in Linguistic Theory,* Holt, Rinehart and Winston, New York, pp. 1–90, 1968.
41. B. Bruce, "Case systems for natural language," *Artif. Intell.* **6,** 327–360 (1975).
42. R. Schank and L. Tesler, A Conceptual Parser for Natural Language, *International Joint Conference on Artificial Intelligence,* pp. 569–578, 1969.
43. D. Hays, "Dependency theory: A formalism and some observations," *Language* **40,** 511–524 (1964).
44. M. Kay, Experiments with a Powerful Parser, *Proceedings of the Second International Conference on Computational Linguistics,* Grenoble, August 1967.
45. S. Lamb, "The semantic approach to structural semantics," *Am. Anthropol.* (1964).
46. R. Schank, "Conceptual dependency: A theory of natural language understanding," *Cog. Psychol.* **3,** 552–631 (1972).
47. W. Woods, Procedural Semantics for a Question-Answering System, *AFIPS 1968 Fall Joint Computer Conference,* pp. 457–471.
48. M. Quillian, Semantic Memory, in M. Minsky (ed.), *Semantic Information Processing,* MIT Press, Cambridge, MA, pp. 216–270, 1968.
49. R. Simmons, Semantic Networks: Their Computation and Use for Understanding English Sentences, in R. Schank and K. Colby

(eds.), *Computer Models of Thought and Language,* W. H. Freeman, San Francisco, CA, pp. 63–113, 1973.

50. N. Findler (ed.), *Associative Networks: Representation and Use of Knowledge in Computers,* Academic Press, New York, 1979.
51. J. Sowa, *Conceptual Structures: Information Processing in Mind and Machine,* Addison-Wesley, Reading, MA, 1984.
52. T. Winograd, *Understanding Natural Language,* Academic Press, New York, 1972.
53. W. Woods, R. Kaplan, and B. Nash-Webber, *The Lunar Sciences Natural Language Information System: Final Report,* Report 2378, Bolt Beranek and Newman, Cambridge, MA, 1972.
54. W. Woods, Lunar Rocks in English: Explorations in Natural Language Question Answering, in A. Zampolli (ed.), *Linguistic Structures Processing,* North-Holland, Amsterdam, pp. 521–569, 1977.
55. M. Halliday, "Categories of the theory of grammar," *Word* **17,** 241–292 (1961).
56. T. Winograd, Frame Representations and the Declarative-Procedural Controversy, in D. Bobrow and A. Collins (eds.), *Representation and Understanding,* Academic Press, New York, pp. 185–210, 1975.
57. Y. Wilks, Natural Language Understanding Programs Within the A.I. Paradigm: A Survey and Some Comparisons, in A. Zampolli (ed.), *Linguistic Structures Processing,* North-Holland, Amsterdam, pp. 341–398, 1977.
58. S. Petrick, On Natural-Language Based Computer Systems, in A. Zampolli (ed.), *Linguistic Structures Processing,* North-Holland, Amsterdam, pp. 313–340, 1975. Also appears in *IBM J. Res. Dev.* **20**(4), 314–325 (1976).
59. E. Codd, R. Arnold, J. Cadiou, C. Chang, and N. Roussopoulos, Seven Steps to RENDEZVOUS with the Casual User, in J. Kimbie and K. Koffeman (eds.), *Data Base Management,* North-Holland, pp. 179-200, 1974.
60. W. Plath, "REQUEST: A natural language question-answering system," *IBM J. Res. Dev.* **20**(4), 326–335 (1976).
61. S. Petrick, Transformational Analysis, in R. Rustin (ed.), *Natural Language Processing,* Algorithmics, New York, pp. 27–41, 1973.
62. F. Damerau, "Operating Statistics for the Transformational Question Answering System," *Am. J. Computat. Ling.* **7**(1), 30–44 (1981).
63. Hendrix, G. E. Sacerdoti, D. Sagalowicz, and J. Slocum, "Developing a natural language interface to complex data," *ACM Trans. Database Sys.* **3**(2), 105–147 (1978).
64. Hendrix, G. Human engineering for applied natural language processing. *Proc. of the Fifth Int. J. Conf. on Artificial Intelligence,* Cambridge, MA, 1977, pp. 183–191.
65. R. Hershman, R. Kelley, and H. Miller, User performance with a natural language query system for command control. Tech. Report TR 79-7, Navy Personnel Research and Development Center, San Diego, Ca., 1979.
66. J. Mylopoulos, A. Borgida, P. Cohen, Roussopoulos, J. Tsotsos, and H. Wong, TORUS: A Natural Language Understanding System for Data Management, *Proc. of the Fourth IJCAI,* Tbilisi, Georgia, pp. 414–421, 1975.
67. F. Thompson and B. Thompson, Practical Natural Language Processing: The REL System as Prototype, in M. Rubinoff and M. Yovits (eds.), *Advances in Computers,* Vol. 3., Academic Press, New York, pp. 109–168, 1975.
68. F. Thompson and B. Thompson, Shifting to a Higher Gear in a Natural Language System, *National Computer Conference,* pp. 657–662, 1981.
69. B. Thompson and F. Thompson, Introducing ASK, a Simple Knowledgeable System, *Conference on Applied Natural Language Processing,* Santa Monica, CA, pp. 17–24, 1983.
70. B. Thompson, Linguistic Analysis of Natural Language Communication with Computers, *Proceedings of the Eighth International Conference on Computational Linguistics,* Tokyo, pp. 190–201, 1980.
71. M. Templeton, EUFID: A Friendly and Flexible Frontend for Data Management Systems, *Proceedings of the Seventeenth Annual Meeting of the ACL,* pp. 91–93, 1979.
72. M. Templeton and J. Burger, Proglems in Natural-Language Interface to DBMS with Examples from EUFID, *Conference on Applied Natural Language Processing,* Santa Monica, CA, pp. 3–16, 1983.
73. S. Kaplan, Indirect Responses to Loaded Questions, *Theoretical Issues in Natural Language Processing,* Vol. 2, pp. 202–209, 1978.
74. D. Waltz, "An English language question answering system for a large relational database," *CACM* **21**(7), 526–539 (1978).
75. T. Finin, B. Goodman, and H. Tennant, JETS: Achieving Completeness through Coverage and Closure, *Proceedings of the Sixth International Joint Conference on Artificial Intelligence,* Tokyo, Japan, pp. 275–281, 1979.
76. H. Tennant, Experience with the Evaluation of Natural Language Question Answerers, *Proceedings of the Sixth International Joint Conference on Artificial Intelligence,* Tokyo, Japan, pp. 874–876, 1979.
77. H. Lehmann, "Interpretation of natural language in an information system," *IBM J. Res. Dev.* **22**(5), 560–571 (1978).
78. Reference 77, p. 560.
79. J. Krause, Results of User Study with the User Specialty Language System and Consequences for the Architecture of Natural Language Interfaces, Technical Report 79.04.003, IBM Heidleberg Scientific Center, 1979.
80. W. Bronnenberg, S. Landsbergen, R. Scha, W. Schoenmakers, and E. van Utteren, "PHLIQA-1, a question-answering system for data-base consultation in natural English," *Philips Tech. Rev.* **38** 229–239, 269–284 (1978–1979).
81. Reference 80, p. 230.
82. W. Hoeppner, T. Christaller, H. Marburger, K. Morik, B. Nebel, M. O'Leary, and W. Wahlster, "Beyond domain-independence," *Proceedings of the Eighth Int. J. Conf. on AI,* Karlsruhe, FRG, pp. 588–594, 1983.
83. L. Harris, "User-oriented data base query with the Robot natural language system," *Int. J. Man–Mach. Stud.* **9,** 697–713 (1977).
84. L. Harris, "The ROBOT system: natural language processing applied to data base query," *ACM Natl. Conf.* 165–172 (1978).
85. S. Salveter, Natural Language Database Updates, *Proceedings of the Nineteenth Annual Meeting of the ACL,* Unversity of Toronto, pp. 67–73, 1982.
86. J. Davidson and S. Kaplan, "Natural language access to data bases: Interpreting update requests," *Am. J. Computat. Ling.* **9**(2), 57–68 (1983).
87. J. Carbonell, "AI in CAI: An artificial intelligence approach to computer-assisted instruction," *IEEE Trans. Man–Mach. Sys.* **11,** 190–202 (1970).
88. A. Collins, E. Warnock, N. Aiello, and R. Miller, Reasoning from Incomplete Knowledge, in D. Bobrow and A. Collins (eds.), *Representation and Understanding,* Academic Press, New York, pp. 383–415, 1975.
89. J. Brown and R. Burton, Multiple Representations of Knowledge for Tutorial Reasoning, in D. Bobrow and A. Collins (eds.), *Rep-*

resentation and Understanding, Academic Press, New York, pp. 312–313, 1975.

90. R. Weischedel, W. Voge, and M. James, "An artificial intelligence approach to language instruction," *Artif. Intell.* **10,** 225–240 (1978).
91. M. Bates and R. Ingria, Controlled Transformational Sentence Generation, *Proceedngs of the Nineteenth Annual Meeting of the ACL,* Stanford University, pp. 153–158, 1981.
92. G. Heidorn, Natural Language Dialogue for Managing an Online Calendar, *Proceedings of the Annual Meeting of the ACM,* Washington, DC, pp. 45–52, 1978.
93. D. Bobrow, R. Kaplan, M. Kay, D. Norman, H. Thompson, and T. Winograd, "GUS: A frame-driven dialog system," *Artif. Intell.* **8**(2), 155–173 (1977).
94. A. Biermann, B. Ballard, and A. Sigmon, "An experimental study of natural language programming," *Int. J. Man–Mach. Stud.* **18**(1), 71–87 (1983).
95. J. Allen, A. Frisch, and D. Litman, ARGOT: The Rochester Dialogue System, *Proceedings of the Second National Conference on Artificial Intelligence,* Carnegie-Mellon University and University of Pittsburgh, Pittsburgh, PA, pp. 66–70, 1982.
96. R. Wilensky, Talking to UNIX in English: An Overview of UC, *Proceedings of the Second Annual Conference on Artificial Intelligence,* Pittsburgh, PA, pp. 103–105, 1982.
97. N. MacDonald, L. Frase, P. Gingrich, and S. Keenan, "The Writer's Workbench: Computer aids for text analysis," *IEEE Trans. Commun.* **30,** 105–110 (January 1982).
98. G. Heidorn, K. Jensen, L. Miller, R. Byrd and M. Chodorow, "The EPISTLE text-critiquing system," *IBM Sys. J.* **21**(3), 305–326 (1982).
99. G. Heidorn, "Automatic programming through natural language dialogue: A survey," *IBM J. Res. Dev.* **20**(4), 302–313 (1976).
100. J. P. Gelb, Experience with a Natural Language Problem-Solving System, *Proceedings of the Second International Joint Conference on Artificial Intelligence,* London, pp. 455–462, 1971.
101. G. Heidorn, Natural Language Inputs to a Simulation Programming System, Ph.D. Dissertation, Technical Report NPS-55HD72101A, Naval Postgraduate School, Monterey, CA, 1972.
102. C. Green, A Summary of the PSI Program Synthesis System, *Proceedings of the Fifth International Joint Conference on Artificial Intelligence,* Cambridge, MA, pp. 380–381, 1977.
103. A. Biermann and B. Ballard, "Toward natural language computation," *Am. J. Computat. Ling.* **6**(2), 71–86 (1980).
104. R. Geist, D. Kraines, and P. Fink, Natural Language Computation in a Linear Algebra Course, *Proceedings of the National Educational Computer Conference,* pp. 203–208, 1982.
105. N. Sager, *Natural Language Information Processing: A Computer Grammar of English and Its Applications,* Addison-Wesley, Reading, MA, 1981.
106. L. Hirschman, R. Grishman, and N. Sager, From Text to Structured Information: Automatic Processing of Medical Reports, *Proceedings of the AFIPS National Computer Conference* pp. 267–275, 1976.
107. R. Grishman and L. Hirschman, "Question answering from natural language medical data bases," *Artif. Intell.* **7,** 25–43 (1978).
108. N. Sager, Natural Language Information Formatting: The Automatic Conversion of Texts to a Structured Data Base, in M. Yovits (ed.), *Advances in Computers,* Vol. 17, Academic Press, New York, pp. 89–162, 1978.
109. E. Marsh and C. Friedman, "Transporting the linguistic string project system from a medical to a Navy domain," *ACM Trans. Ofc. Inform. Sys.* **3**(2), 121–140 (1985).
110. N. Haas and G. Hendrix, An Approach to Acquiring and Applying Knowledge, *Proceedings of the First National Conference on Artificial Intelligence,* Stanford University, Stanford, CA, pp. 235–239, 1980.
111. G. Hendrix and W. Lewis, Transportable Natural-Language Interfaces to Databases, *Proceedngs of the Nineteenth Annual Meeting of the ACL,* Stanford University, pp. 159–165, 1981.
112. W. Mark, Representation and Inference in the Consul System, *Proceedings of the Seventh International Joint Conference on Artificial Intelligence,* Vancouver, BC, pp. 375–381, 1981.
113. D. Wilczynski, Knowledge Acquisition in the Consul System, *Proceedings of the Seventh International Joint Conference on Artificial Intelligence,* Vancouver, BC, pp. 135–140, 1981.
114. D. Warren and F. Pereira, "An efficient easily adaptable system for interpreting natural language queries," *Am. J. Computat. Ling.* **8**(3–4), 110–122 (1982).
115. M. Bates, Information Retrieval Using a Transportable Natural Language Interface, *Proceedings of the International ACM SIGIR Conference,* Bethesda, MD, pp. 81–86, 1983.
116. J. Ginsparg, A Robust Portable Natural Language Data Base Interface, *Proceedings of the Conference on Applied Natural Language Processing,* Santa Monica, CA, pp. 25–30 (1983).
117. B. Grosz, TEAM: A Transportable Natural Language Interface System, *Proceedings of the on Applied Natural Language Processing,* Santa Monica, CA, pp. 39–45, 1983.
118. B. Ballard, J. Lusth, and N. Tinkham, "LDC-1: A transportable, knowledge-based natural language processor for office environments," *ACM Trans. Ofc. Inf. Sys.* **2**(1), 1–25 (1984).
119. R. Grishman, N. Nhan, E. Marsh, and L. Hirschman, Automated Determination of Sublanguage Syntactic Usage, *Proceedings of the International Conference on Computational Linguistics,* Stanford, pp. 96–98, July 1984.
120. B. Ballard (ed.), "Special issue on transportable natural language processing," *ACM Trans. Ofc. Inf. Sys.* **3**(2) 104–230 (1985).
121. F. Damerau, "Problems and some solutions in customization of natural language database front ends," *ACM Trans. Ofc. Inf. Sys* **3**(2), 165–184 (1985).
122. C. Hafner and K. Godden, "Portability of syntax and semantics in DATALOG," *ACM Trans. Ofc. Inf. Sys.* **3**(2), 141–164 (1985).
123. J. Slocum and C. Justus, "Transportability to other languages," *ACM Trans. Ofc. Inf. Sys.* **3**(2), 204–230 (1985).
124. B. Thompson and F. Thompson, "ASK is transportable in half a dozen ways," *ACM Trans. Ofc. Inf. Sys.* **3**(2), 185–203 (1985).
125. J. Slocum (ed.), "Special issues on machine translation," *Computat. Ling.* **11**(2–4) (1985).
126. S. Nirenburg, (ed.), *Machine Translation: Theoretical and Methodological Issues,* Cambridge University Press, New York, 1987.
127. W. Lehnert and S. Schwartz, EXPLORER: A Natural Language Processing System for Oil Exploration, *Proceedings of the Conference on Applied Natural Language Processing,* Santa Monica, CA, pp. 69–72, 1983.
128. T. Johnson, *Natural Language Computing: the Commercial Applications,* Ovum, London, 1985.
129. F. Pereira and D. H. D. Warren, "Definite clause grammars for language analysis: A survey of the formalism and a comparison with Augmented Transition Networks," *Artif. Intell.* **13,** 231–278 (1980).
130. D. Chester, "A parsing algorithm that extends phrases," *Am. J. Computat. Ling.* **6**(2), 87–96 (1980).
131. C. Riesbeck, Conceptual Analysis, in R. Schank (ed.), *Conceptual Information Processing,* North-Holland, Amsterdam, pp. 83–156, 1975.

132. S. Small, Parsing and Comprehending with Word Experts (A Theory and Its Realization), in W. Lehnert and M. Ringle (eds.), *Strategies for Natural Language Processing,* Lawrence Erlbaum, Hillsdale, NJ, 1982.
133. D. Younger, "Recognition and parsing of context-free languages in time n^3," *Inf. Ctr.,* **10,** 129–208 (1967).
134. J. Earley, "An efficient context-free parsing algorithm," *CACM* **13**(2), 94–102 (February 1970).
135. R. Kaplan, *A General Syntactic Processor,* Algorithmics, New York, 1973.
136. W. Ruzzo, S. Graham, and M. Harrison, "An improved context-free recognizer," *ACM Trans. Program. Lang. Sys.* **3,** 415–562 (July 1980).
137. M. Marcus, *A Theory of Syntactic Recognition for Natural Language,* MIT Press, Cambridge, MA, 1980.
138. R. Milne, "Resolving Lexical ambiguity in a deterministic parser," *Computat. Ling.* **12**(1), 1–12 (1986).
139. W. Woods, "Optimal Search Strategies for Speech Understanding Control," *Artificial Intelligence* **18,** 295–326 (1982).
140. L. Erman, F. Hayes-Roth, V. Lesser and D. Reddy, "The Hearsay-II speech understanding system," *Computing Surveys* **12,** 213–253 (1980).
141. G. Cottrell, A Model of Lexical Access of Ambiguous Words, *Proceedings of the Fourth Conference of the AAAI,* Austin, TX, pp. 61–67, August 1984.
142. M. Jones and A. Driscoll, Movement in Active Production Networks, *Proceedings of the Twenty-Third Annual Meeting of the Association for Computational Linguistics,* pp. 161–166, July 1985.
143. D. Waltz and J. Pollack, "Massively parallel parsing," *Cog. Sci.* **9**(1), 51–74 (1985).
144. S. Kuno, "The predictive analyzer and a path elimination technique," *CACM* **8** 453–462 (1965).
145. A. Zwicky, J. Friedman, B. Hall, and D. Walker, The MITRE Syntactic Analysis Procedure for Transformational Grammars, *IFIPS Proceedings Fall Joint Computer Conference,* Spartan, Washington, DC, pp. 317–326, 1965.
146. S. Petrick, A Recognition Procedure for Transformational Grammars, Ph.D. Dissertation, MIT, Cambridge, MA, 1965.
147. R. Stockwell, P. Schachter, and B. Partee, *The Major Syntactic Structures of English,* Holt, Rinehart and Winston, New York, 1973.
148. J. Robinson, "DIAGRAM: A grammar for dialogues," *CACM* **25**(1), 27–47 (January 1982).
149. G. Heidorn, Augmented Phrase Structure Grammars, in B. Webber and R. Schank (eds.), *Theoretical Issues in Natural Language Processing,* Cambridge, MA, pp. 1–5, 1975.
150. W. Mann and C. Mattheissen, Nigel: a Systemic Grammar for Text Generation, in Freedle (ed.), *Systemic Perspectives on Discourse: Selected Theoretical Papers of the Ninth International Systemic Workshop,* Ablex, Norwood, NJ, 1985.
151. M. Kay, Functional Grammar, *Proceedings of the Fifth Annual Meeting of the Berkeley Linguistic Society,* pp. 142–158, 1979.
152. G. Gazdar, Phrase Structure Grammar, in P. Jacobson and G. Pullum (eds.), *The Nature of Syntactic Representation,* D. Reidel, Dordrecht, pp. 131–186, 1982.
153. A. Joshi, How Much Context-Sensitivity is Required to Provide Reasonable Structural Descriptions: Tree Adjoining Grammars, in D. Dowty, L. Karttunen, and A. Zwicky (eds.), *Natural Language Processing: Psycholinguistic, Computational and Theoretical Properties,* Cambridge University Press, New York, 1984.
154. E. Proudian and C. Pollard, Parsing Head-Driven Phrase Structure Grammar, *Proceedngs of the Twenty-Third Annual Meeting of the Association for Computational Linguistics,* pp. 8–12, July 1985.
155. J. Bresnan and R. Kaplan, Lexical-Functional Grammar: A Formal System for Grammatical Representation, in J. Bresnan (ed.), *The Mental Representation of Grammatical Relations,* MIT Press, Cambridge, MA, 1982.
156. M. Kay, Functional Unification Grammar: A Formalism for Machine Translation, *Proceedings of Coling 84,* Menlo Park, pp. 75–78, 1984.
157. J. Allen (ed.), "Special issue on ill-formed input," *Am. J. Computat. Ling.* **9**(3–4), 123–196 1983.
158. R. Weischedel and J. Black, "Responding intelligently to unparasable inputs," *Am J. Computat. Ling.* **6**(2), 97–109 (1980).
159. P. Hayes and G. Mouradian, "Flexible parsing," *Am. J. Computat. Ling.* **7**(4), 232–242 (1981).
160. S. Kwasny and N. Sondheimer, "Relaxation techniques for parsing ill-formed input," *Am. J. Computat. Ling.* **7**(2), 99–108 (1981).
161. K. Jensen, G. Heidorn, L. Miller, and Y. Ravin, "Parse fitting and prose fixing: Getting a hold on ill-formedness," *Am. J. Computat. Ling.* **9**(3–4), 147–160 (1983).
162. R. Weischedel and N. Sondheimer, "Meta-rules as a basis for processing ill-formed output," *Am. J. Computat. Ling.* **9**(3–4), 161–177 (1983).
163. R. Granger, "the NOMAD system: Expectation-based detection and correction of errors during understanding of syntactically and semantically ill-formed text," *Am. J. Computat. Ling.* **9**(3–4), 188–196 (1983).
164. P. Fink and A. Biermann, "Correction of ill-formed input using history-based expectation with applications to speech understanding," *Computat. Ling.* **12**(1), 13–36 (1986).
165. R. Jackendoff, *Semantics and Cognition,* MIT Press, Cambridge, MA, 1983.
166. J. Searle, Indirect Speech Acts, in P. Morgan and J. Cole (eds.), *Syntax and Semantics,* Vol. 3, *Speech Acts,* Academic Press, New York, pp. 59–82, 1975.
167. P. Cohen and C. Perrault, "Elements of a plan-based theory of speech acts," *Cog. Sci* **3,** 177–212 (1979).
168. J. Allen and C. Perrault, "Analyzing intention in utterances," *Artif. Intell.* **15**(3), 143–178 (1980).
169. H. Grice, Logic and Conversation, in P. Morgan and J. Cole (eds.), *Syntax and Semantics,* Vol. 3, *Speech Acts,* Academic Press, New York, pp. 41–58, 1975.
170. J. Hirschberg, "Toward a Redefinition of Yes/No Questions," *Proc. of Tenth Int. Conf. on Computational Linguistics,* Stanford, CA, pp. 48–51, 1984.
171. L. Siklossy, Question-Asking Question-Answering, Department of Computer Science Report TR-71, University of Texas, Austin, TX, 1977.
172. G. Brown, *A Framework for Processing Dialog,* MIT Press, Cambridge, MA, June 1977.
173. H. Tennant, *Natural Language Processing,* Petrocelli, New York, 1981.
174. R. Reichman, "Extended person-machine interface," *Artif. Intell.* **22**(2), 157–218 (March 1984).
175. D. Rummelhart, Notes on a Schema for Stories, in D. Bobrow and A. Collins (eds.), *Representation and Understanding,* Academic, New York, 1975.
176. A. Coreira, "Computing story trees," *Am. J. Computat. Ling.* **6**(3–4), 135–149 (1980).

177. J. Levin and J. Moore, "Dialog-games: Metacommunications structures for natural language interaction," *Cog. Sci.* **1**(4), 395–420 (1977).
178. D. Litman and J. Allen, "A plan-based recognition model for subdialogues in conversation," *Cog. Sci.* **11** (1987).
179. B. Grosz, Focusing and Description in Natural Language Dialogues, in *Elements of Discourse Understanding,* Cambridge University Press, pp. 84–105, 1981.
180. C. Sidner, "Focusing for interpretation of pronouns," *Am. J. Computat. Ling.* **7**(4), 217–231 (1981).
181. G. Hirst, "Discourse-oriented anaphora resolution in natural language understanding: A review," *Am. J. Computat. Ling.* **7,** 2, pp. 85–98.
182. R. Kantor, The Management and Comprehension of Discourse Connection by Pronouns in English, Ph.D. Dissertation, Ohio State University, 1977.
183. B. Webber, *A formal Approach to Discourse Anaphora,* Garland, New York, 1978.
184. J. Hobbs, "Coherence and coreference," *Cog. Sci.* **3**(1), 67–90 (1979).
185. A. Lockman, Contextual Reference Resolution, Ph.D. Dissertation, Columbia University, May 1978.
186. V. Yngve, Random Generation of English Sentences, *Proceedings of the International Conference on Machine Translation of Languages and Applied Language Analysis,* National Physical Laboratory, Symposium No. 13, Her Majesty's Stationery Office, London, pp. 66–80, 1962.
187. J. Friedman, "Directed random generation of sentences," *CACM* **12**(1), 40–46 (1969).
188. S. Klein, "Automatic paraphrasing in essay format," *Mechan. Transl.* **8**(3), 68–83 (1965).
189. R. Simmons and J. Slocum, "Generating English discourse from semantic networks," *CACM* **15**(10), 891–905 (1972).
190. G. Heidorn, Generating Noun Phrases to Identify Noun Phrases in a Semantic Network, *Proceedings of the Fifth International Joint Conference on Artificial Intelligence,* Cambridge, Mass., p. 143, 1977.
191. N. Goldman, Conceptual Generation. in R. Schank, *Conceptual Information Processing,* North-Holland, Amsterdam, pp. 289–371 1975.
192. R. Schank, *Conceptual Information Processing,* with contributions from N. Goldman, C. Rieger, and C. Riesbeck, Vol. 3 of *Fundamental Studies in Computer Science,* North-Holland, Amsterdam, 1975.
193. A. Davey, *Discourse Production,* Edinburgh University Press, Edinburgh, 1979.
194. J. Clippinger, "Speaking with many tongues: Some problems in modeling speakers of actual discourse," TINLAP-1, 68–73 (1975).
195. W. Mann, M. Bates, B. Grosz, D. McDonald, K. McKeown, and W. Swartout, "Text Generation: The state of the art and literature," *JACL* **8,** 2 (1982).
196. D. McDonald, Natural Language Generation as a Computational Problem: An Introduction, in M. Brady and R. Berwick (eds.), *Computational Models of Discourse,* MIT Press, Cambridge, MA, pp. 209–265, 1983.
197. W. Mann and J. Moore, "Computer generation of multiparagraph English text," *Am J. Computat. Ling.* **7,** 17–29 (1981).
198. K. Kukich, Design of a Knowledge-Based Report Generator, *Proceedings of the Twentieth Annual Meeting of the ACL,* Cambridge, MA, pp. 145–150, 1983.
199. P. Jacobs, "PHRED: a generator for natural language interfaces," *Computat. Ling.* **11**(4), 219–242 (1985).
200. D. Appelt, *Planning English Sentences,* Cambridge University Press, New York, 1985.
201. K. McKeown, *Text Generation,* Cambridge University Press, New York, 1985.
202. R. Schank, Inference in the Conceptual Dependency Paradigm: A Personal History, Yale University, Department of Computer Science, Research Report 141, September 1978.
203. Y. Wilks, "A preferential, pattern-seeking semantics for natural language inference," *Artif. Intell.* **6,** 53–74 (1975).
204. R. Schank and R. Abelson, *Scripts, Plans, Goals, and Understanding,* Lawrence Erlbaum, Hillsdale, NJ 1977.
205. R. Cullingford, Script Application: Computer Understanding of Newspaper Stories, Research Report 116, Yale University, Department of Computer Science, 1978.
206. R. Wilensky, *Planning and Understanding,* Addison-Wesley, Readng, MA, 1983.
207. J. Carbonell, "POLITICS: Automated ideological reasoning," *Cog. Sci.* **2,** 27–51 (1978).
208. J. Kolodner, *Retrieval and Organizational Strategies in Conceptual Memory: A Computer Model,* Lawrence Erlbaum, Hillsdale, NJ, 1984.
209. M. Dyer, *In-Depth Understanding,* MIT Press, Cambridge, MA, 1983.
210. L. Siklossy, "A language-learning heuristic program," *Cog. Psychol.* **2,** 479–495 (1971).
211. J. Anderson, "Induction of augmented transition networks," *Cog. Sci.* **1**(2), 125–157 (April 1977).
212. S. Klein, Automatic Inference of Semantic Deep Structure Rules in Generative Semantic Grammars, Technical Report 180, Computer Science Department, University of Wisconsin, Madison, May 1973.
213. L. Harris, "A system for primitive natural language acquisition," *Int. J. Man–Mach. Stud.* **9,** 153–206 (1977).
214. R. Berwick, Computational Analogues of Constraints on Grammars: A Model of Syntax Acquisition, *Proceedings of the Eighteenth Annual Meeting of the ACL,* Philadelphia, PA, pp. 49–54, 1980.
215. L. Reeker, "A problem solving theory of syntax acquisition," *J. Struct. Learn.* **2,** 1–10 (1971).
216. P. Langley, "Language acquisition through error recovery," *Cog. Brain Theor.* **5,** 211–255 (1982).
217. M. Selfridge, Inference and Learning in a Computer Model of the Development of Language Comprehension in a Young Child, in W. Lehnert and M. Ringle (eds.), *Strategies for Natural Language Processing,* Lawrence Erlbaum, Hillsdale, NJ, pp. 299–326, 1982.
218. J. Anderson and G. Bower, *Human Associative Memory,* Winston and Sons, Washington, DC, 1973.
219. J. Anderson, *The Architecture of Cognition,* Harvard University Press, Cambridge, MA, 1983.
220. E. Feigenbaum and J. Feldman (eds.), *Computers and Thought,* McGraw-Hill, New York, 1963.
221. M. Minsky (ed.), *Semantic Information Processing,* MIT Press, Cambridge, MA, 1968.
222. R. Rustin (ed.), *Natural Language Processing,* Algorithmics, New York, 1973.
223. A. Zampolli (ed.), *Linguistic Structures Processing,* North-Holland, Amsterdam, 1975.
224. S. Kaplan (ed.), "Special section on natural language processing" *SIGART Newslett.* **79,** 42–108 (1982).
225. C. Johnson and J. Bachenko (eds.), "Applied computational lin-

guistics in perspective," *Am. J. Computat. Ling.* 8(2), 55–84 (1982).

226. A. Barr and E. Feigenbaum (eds.), *The Handbook of Artificial Intelligence,* Vol. 1, William Kaufmann, Los Altos, CA, 1981.
227. W. Lehnert and W. Ringle (eds.), *Strategies for Natural Language Processing,* Lawrence Earlbaum, Hillsdale, NJ, 1982.
228. R. Grishman, *An Introduction to Computational Lingistics,* Cambridge University Press, New York, 1986.
229. M. King (ed.), *Parsing Natural Language,* Academic Press, London, 1983.
230. T. Winograd, *Language as a Cognitive Process,* Vol. 1, *Syntax,* Addison-Wesley, Reading, MA, 1983.
231. K. Sparck Jones and Y. Wilks, *Automatic Natural Language Parsing,* Ellis Horwood, Chichester, UK, 1985.
232. D. Dowty, L. Karttunen, and A. Zwicky (eds.) *Natural Language Parsing,* Cambridge University Press, Cambridge, UK, 1985.
233. M. Brady and R. Berwick, *Computational Models of Discourse,* MIT Press, Cambridge, MA, 1983.
234. R. Schank and C. Riesbeck, *Inside Computer Understanding,* Lawrence Erlbaum, Hillside, NJ, 1981.
235. R. Simmons, *Computations from the English,* Prentice-Hall, Englewood Cliffs, NJ, 1984.
236. M. Harris, *Introduction to Natural Language Processing,* Reston Publ. Co., Reston, VA, 1985.

Reprinted from the S. C. Shapiro, ed, *Encyclopedia of Artificial Intelligence, 1987.*

CONSTRUCTION, ROBOTS IN

Irving J. Oppenheim
Carnegie-Mellon University
Pittsburgh, Pennslyvania

Miroslaw J. Skibniewski
Purdue University
West Lafayette, Indiana

INTRODUCTION

There is widespread interest (1,2) in applying robotics to construction, to bring productivity gains to this large but diffuse industry and to extend construction to environments inaccessible to humans. Conventional factory robots are of limited applicability because the construction environment is not permanently structured or maintained. Construction robots, therefore, confront the challenges of task complexity, robot mobility, obstacle avoidance, domain recognition, large force demands, and more. Prototype models and demonstrations are nonetheless evident for a number of construction applications. Some early examples are motivated by the merit of removing workers from hazardous environments, but there exist other examples where work quality and economic efficiency justify the robotization. Moreover, the research issues are recognized and progress is being reported; in most cases the research impetus is shared with other application areas as well.

The various construction robots of the future will predominantly be intelligent in being sensor based with extensive processing or with machine intelligence. In this article, the current steps toward that future state are outlined. However, this article also presents less advanced robots (including playback robot examples) where they have found application in the construction domain. In this way, a comprehensive survey of current construction robotics is provided, along with examples indicative of future developments.

CATEGORIZATION OF CONSTRUCTION OPERATIONS

There are a number of different ways to categorize the construction industry. For instance, *building* construction includes commercial, industrial, and residential, and *heavy* construction includes roads, bridges, and dams. However, construction applications share certain basic operations. For example, the basic operations in building construction have been described as follows (3):

Element Placement Operations

1. *Building:* This consists of placing repetitive basic structural elements such as bricks and concrete blocks to obtain a rigid structure or part. At present, it usually involves the use of a binding agent such as mortar or adhesives and is work of a repetitive character, requiring relatively high accuracy and consistency.
2. *Positioning:* This involves placement of (typically) large, heavy components in their service locations. It is presently performed by several laborers using building cranes and requires flexibility of movement and reasonably high accuracy on the part of the laborers as well as supporting machinery.
3. *Connecting:* This is the set of operations needed to achieve joint action between different parts of the structure. It often requires special tools and high work accuracy on the part of laborers.
4. *Inlaying:* This is a type of *building* process (*1*), but is instead applied to existing structural surfaces. It involves placement of small elements attached to each other on a structural base for the purpose of obtaining a continuous surface.
5. *Sealing:* This is the application of a sealant to the joint edges of structural elements to obtain an uninterrupted and isolating surface.

Surface Treatment Operations

1. *Finishing:* This is a mechanical treatment of raw structural surfaces to obtain surface quality or utility. It is a repetitive and often hazardous task requiring protective equipment, continuous control, and high accuracy.
2. *Coating:* This describes the spreading or spraying of a liquid substance on the structural surface. It is a repetitive and hazardous task requiring protective clothes, high control, and accuracy.
3. *Covering:* This describes the placing of sheets of material over existing surfaces. It requires high manipulation accuracy.

Filling Operations

1. *Concreting:* This consists of pouring the concrete mix into previously prepared formwork to create structural volume. It requires strength and endurance on the part of laborers.
2. *Excavating:* This is the act in which the site is brought to a controlled geometry from which construction proceeds.
3. *Backfilling:* This describes replacing the empty space between foundation walls and the ground with soil. It requires transferring large volumes of soil with mechanical pushers and backhoes.

In addition to the above operations, there are other elementary activities necessary to perform a successful construction project. They include, but are not limited to, inspection, testing, and operation control.

SUITABILITY OF THE EXISTING ROBOTIC TECHNOLOGY

Construction operations are generally unique, and commercially available robotic systems are at the present time largely unsuited for such work. The reasons for this are quite complex. They include the need for sturdiness and roughness of equipment at construction sites, which is very different from most manufacturing environments. However, there are numerous other technical problems specific to the nature of an *ill-structured* construction environment, which are largely unsolved at present. Therefore challenges facing construction robotics are greater than those facing robotics in most manufacturing applications. The research problems include:

- *Robot Mobility:* Mobile-based equipment is essential for most on-site construction applications. Mobility requires sophisticated navigational capabilities involving obstacle avoidance, surprise sensors and surprise-handling algorithms, robot vision systems, new control systems and data processing units, and so on.
- *Robot Sensing:* Construction robots will have to use sensors for vision, pattern recognition, and proximity sensing in order to perform in an unstructured environment.
- *Construction Robotic Grippers:* Further development of robotic grippers is needed for broader potential use in construction operations. Emphasis should be put on developing new types of grippers particularly suitable for specific operations.
- *Control Systems:* Available control systems have significant limitations on their ability to modify robot behavior in response to sensed conditions. Also, response time to these conditions is not yet satisfactory to perform most work tasks. Computational capabilities will have to be significantly expanded to handle large amounts of sensory data and to process them in an acceptable amount of time.
- *Robot Accuracy:* In construction work, the design accuracy of robots is likely to be affected by intensive wear. Measures must be taken to assure proper positioning accuracy of a robot for each specific task, possibly by self-calibration procedures. This calls for greater use of *servo control* computations than presently employed in manufacturing robots.
- *Weight of Hardware:* Most existing industrial robotics hardware structures are relatively massive and unwieldy; at maximum they can lift and handle objects representing only about 10% of their own weight. To avoid overloading structures under construction, this proportion must be altered to levels more typical of construction equipment.
- *Hardware Stability Problems:* Most objects to be handled by construction robots are heavier than their counterparts in manufacturing, and the reach of any construction robot arm will be greater than that of a manufacturing robot. Therefore, considerably greater robot flexibility must be anticipated, with possible stability implications.
- *External Factors:* External factors such as weather conditions, extreme temperatures, dust, and excessive vibrations affect most construction environments. Influences referred to in cybernetics as "noises" can significantly affect the level of responsiveness of robot sensors and dampen the precision of manipulator performance. In most of the development efforts, designers of robotic sensors and manipulators must always take these constraints into account, again demanding special control mechanisms for such new applications.

Present robot technology nonetheless offers some capabilities that can be employed in construction applications. Spray robot technology is well developed, and a number of early construction applications have originated with that function. Similarly, certain sensing functions are presently reliable; an example would be a single-channel touch sensor. Again, there exist construction applications that are satisfied with this limited but very accessible technology. Another obvious example is the fundamental capability of a robot to perform repetitive motion, whether programmed algorithmically or taught. This capability suits particular construction applications at the present time and is being exploited where appropriate.

EXISTING PROTOTYPES AND OPERATING MODELS

Although construction robots are in general not commercially marketed, there have been significant attempts to robotize a number of narrow applications, some of which appear to be technically and potentially economically feasible. These attempts have so far covered virtually every major area of construction operations, such as:

- *Surface Finishing:* There are a shotcrete robot by Kajima, a fireproofing spray robot by Shimizu, a slab finishing robot by Kajima, and a wall climbing robot by Nordmed Shipyards.
- *Tunneling:* Robotic-type controls have been introduced in drilling and in shield driving by Kajima.
- *Excavation:* There have been a robotic excavator (REX) demonstration by Carnegie-Mellon and DRAVO, automatic grading control, and a diaphragm wall excavating robot by Kajima.
- *Structural Element Placement:* A reinforcement-placing robot has been developed by Kajima.

- *Construction Inspection:* A core-boring robot and magnetic sensing of concrete reinforcement have been developed by Carnegie-Mellon.

A number of technical and corporate publications describe existing field examples (4–6). These and other examples are now discussed in some detail.

Examples of Robots for Surface Finishing

Shotcrete Robot. In the new Austrian tunneling method, shotcrete application takes as much as 30% of the total time; improving the efficiency of this one task can bring about significant benefits. Normally, a skilled operator is needed to regulate the amount of concrete to be sprayed and the quality of hardening agent to be added, both of which depend on the consistency of the concrete. Kajima Corporation has developed and implemented a computer-controlled applicator (Figure 1) by which high-quality shotcrete placement can be achieved. The special features of this system are the following:

- The concrete is fed and jetted by compressed air.
- The accelerator, a dry powder, is mixed into the concrete at a point approximately 2 m before the mouth of the nozzle.
- The rate of shotcrete application is in the range of 4–6 m^3/h and varies with the consistency of the concrete.

The required air volume and pressure vary with the consistency of the concrete; the appropriate rate of shotcrete application is controlled by computer. As a result, the work can be performed without the presence of an engineer familiar with the characteristics of concrete and applicator equipment.

Three employed types of automated shotcrete nozzle manipulation include remote control, semiautomatic remote control, and robot playback. The equipment described here can be classified as semiautomatic remote control, and the playback type robot was developed by Ohbayashi-Gumi, Ltd. and Kobe Steel, Ltd. The first unit of this type is now in use on the work site in Japan.

Figure 1. Kajima shotcrete robot (4). Courtesy of *IABSE Proceedings*.

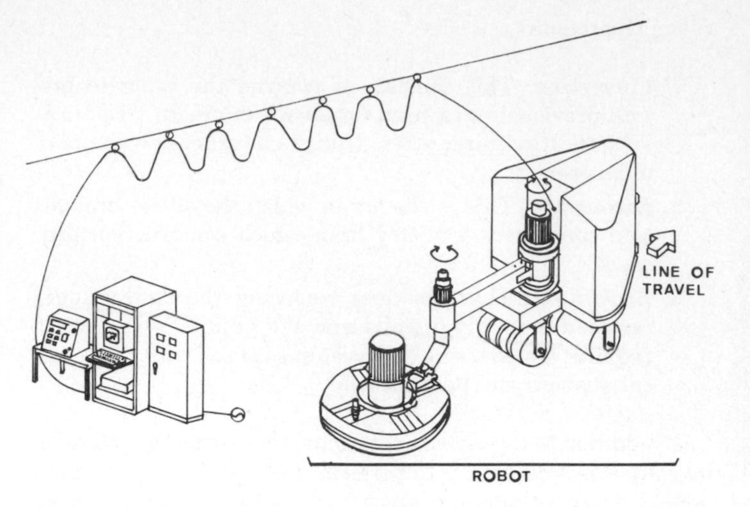

Figure 2. Kajima slab finishing robot. Courtesy of M. Saito, N. Tanaka, K. Arai, and K. Banno, Mechanical Engineering Development Department, Kajima Corporation.

Slab Finishing Robot. Finishing the rough surface of a cast-in-place concrete slab after pouring usually requires laborious human hand work, often performed at night and in adverse weather. The robot designed for this task by Kajima Corporation (Figure 2) is mounted on a computer-controlled mobile platform and equipped with mechanical trowels that produce a smooth, flat surface. By means of a gyrocompass and a linear distance sensor, the machine navigates itself and automatically corrects any deviation from its prescheduled path. It is controlled by a Z80 8-bit microprocessor and is connected by an optical fiber transmission system to a host computer, enabling monitoring of robot position by graphic display.

The entire system consists of a main unit with a mobile platform, a horizontal articulated arm with a rotary trowel, a host computer, a console, and a power supply unit. The operator inputs the course, the starting position coordinates, the number of arm swings, the angle of the trowel and number of trowel rotations, and the degrees of the turns at each corner. Once the robot starts operation, no further instructions are necessary. As the robot advances, it pulls the trowel arm, which swings back and forth. The robot features a gyrocompass to keep the robot from tilting, a rotary encoder to determine distance traveled, and sensors to detect obstacles. This mobile floor finishing robot is able to work to within 1 m of walls. It is designed to replace at least six skilled workers.

Fireproofing Spray Robot. Shimizu Company has developed two robot systems for spraying fireproofing material onto structural steel. The first version, the SSR-1 (Figure 3), was built to use the same materials as in conventional fireproofing, to work sequentially and continuously with human help, to travel and position itself, and to have sufficient safety functions for the protection of human workers and of building components. For the spraying function itself, the KTR-3000 (Kobelco-Trallfa spray robot) was initially employed because of

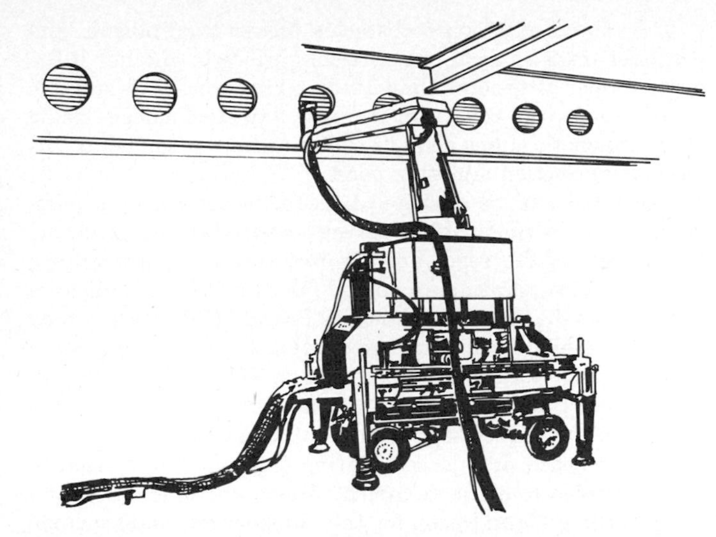

Figure 3. Shimizu fireproofing spray robot (4). Courtesy of *IABSE Proceedings*.

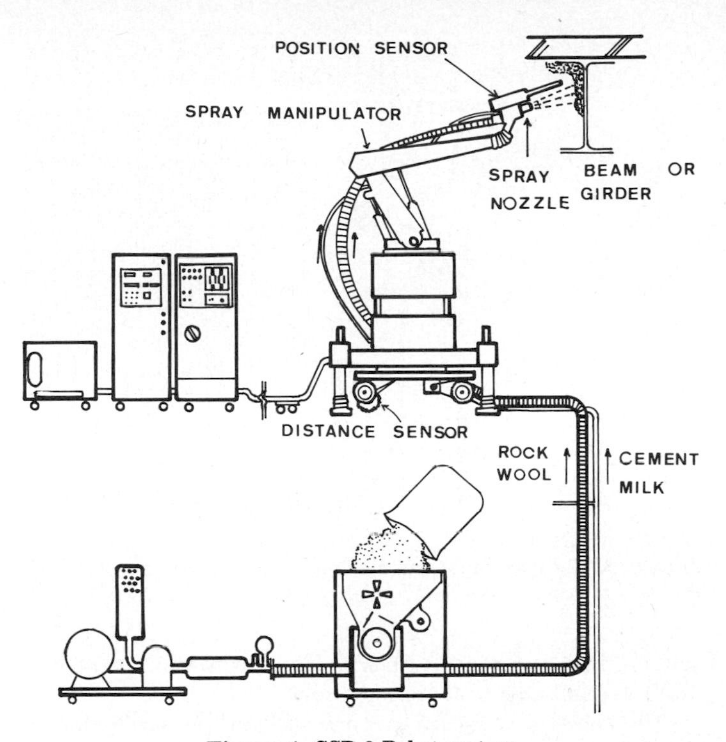

Figure 4. SSR-2 Robot system.

its large memory module capacity, its spray nozzle weight capacity (greater than 3 kg, satisfactory for the fireproofing work), and its use of continuous path (CP) control.

The manipulator consists of four modules: the base, a vertical arm, a horizontal arm, and a wrist. It has six degrees of freedom of motion, which are operated by a playback control system consisting of an electrohydraulic servo control. The height and width of the operating area are approximately 2 m × 3 m. Electric wires, hydraulic hoses, and material-handling hoses are mounted on supports, set at 1 m intervals, which can move smoothly across the floor. The manipulator is mounted on a mobile platform weighing 220 kg, which has four outriggers for stable positioning when spraying. The manipulator is controlled by an independent controller, and the mobile platform follows a wire path and has a sequential controller that controls both the platform and the manipulator. The manipulator control equipment computer robot control (CRC) system has CP/PTP teaching and CP playback control functions.

The work efficiency was evaluated by measuring the spray time per specific area. As a result, it was found that the processing speed was almost twice as fast as that of the conventional manual method. However, new additional tasks were involved: placement of the path wire and hoisting and initial positioning of the robot system. The specific gravity of the rock wool sprayed (a major determinant of the work quality) is nearly the same as that achieved by a human worker. The biggest problem is the thickness dispersion, which is caused by the inability to supply the material uniformly for spraying.

A second robot system, the SSR-2 (Figure 4), was developed to improve some of the job-site functions of the first prototype. The new features included the introduction of a new positioning system defined in relation to the overhead beams being sprayed, self-traveling and tracking of the robot, eliminating the path wire for guiding the robot, and improving the feeder for supplying the rock wool more uniformly.

The SSR-2 manipulator itself is fundamentally similar to that of the SSR-1, and the mobile platform has additional sensors for step counting and obstacle detection. The control system of the SSR-2 has a traveling device controller for path control and for position calculation, based on a 16-bit system (TMS-9995, 16 kbyte ROM, 14 Kbyte RAM). The positioning system of the mobile platform, free of any path wire, is the main improved feature of the SSR-2. As slight elevation deviations exist due to floor unevenness, the SSR-2 must adjust its position through sensor information. The SSR-2 measures its position by touch, by gauging the distance to the web and the bottom of the beam flange above.

From an economic viewpoint, the SSR-2 can spray faster than a human worker, but requires time for transportation and setup. The SSR-2 takes about 22 min for one work unit, whereas a human worker takes about 51 min. The SSR-2 does not require much personnel power for the spraying preparation, only some 2.08 workdays compared to 11.5 for the SSR-1. This shortening of preparation time contributes considerably to the improvement of robot system economic efficiency. As the positional precision of the robot and supply of the rock wool feeder were improved, the irregular dispersion of the sprayed thickness decreased and became nearly equal to that applied by a human worker.

Wall Climbing Robot. Nordmed Shipyards of Dunkerque, France, developed the RM3 robot (Figure 5) for marine applications, including video inspections of ship hulls, γ-ray inspections of structural welds, and high-pressure washing, deburring, painting, shotblasting, and barnacle removal (7). The RM3 weighs 206 lb (93kg) and has three legs, one arm, and two bodies. Magnetic cups on its hydraulic actuated legs allow the RM3 to ascend a vertical steel plate, such as a ship's hull, at a speed of 8.2 ft/min (150 m/h). RM3 has a cleaning rate of 53,800 ft^2/d (5000 m^2/d) and a 320-ft (98-m) range. Nordmed

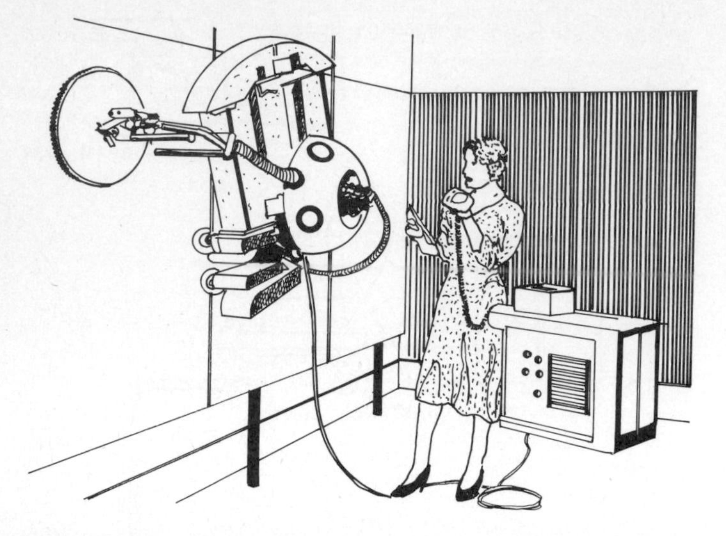

Figure 5. Nordmed Shipyards climbing robot (7). Courtesy of *Robotix*.

entered into a joint venture with Renault to use a version of RM3 to paint chemical storage tanks.

The robot is designed to work without any scaffolding in the following environments:

- Visual inspection for prefabricated blocks.
- X-ray, γ-ray, ultrasonic examination of prefabricated block, and on-board grinding and burr removing.
- Wire brushing before painting.
- High-pressure painting.
- Recycled shot blasting on shell painting joints.

The robot cannot pass obstacles higher than 50 mm and it cannot transfer from one vertical surface to another if the angle formed is greater than 10 deg. However, with suitable modifications, the robot can traverse any steel surface using electromagnetic adhesion pads or flat or curved nonmetal surfaces using suction adhesion pads.

The robot can be operated by remote control or preprogrammed to perform its work automatically. The electrical components of the robot are powered through a low-voltage cable also incorporating an optical fiber link. The intelligence system uses the Texas Instruments Pascal MPP programming language, a high-level language developed for real-time applications using systems designed around TMS 9900 family microprocessors. It is multitask, and its core gives a processing speed performance close to that of an assembler.

An aerospace firm is considering the use of a version of the RM3 robot to paint its aircraft shells and to do its γ- and X-ray testing. Modification for this application would include using vacuum cups to fasten the robot to the aircraft skin. Other potential uses include cleaning or washing the sides of buildings, applying ground coatings, brushing or spraying in radioactive environments inside nuclear power plants, carrying and handling objects in radioactive environments, milling, machining, cutting, and welding in various areas of the construction industry.

Examples of Robots for Tunneling

Five-Boom Drilling Robot. Robot drilling machines of both playback type and numerical control type have been developed and implemented. Kajima Corporation has adopted the playback system and implemented a machine with up to five booms (Figure 6). This fully automated excavating machine is a major

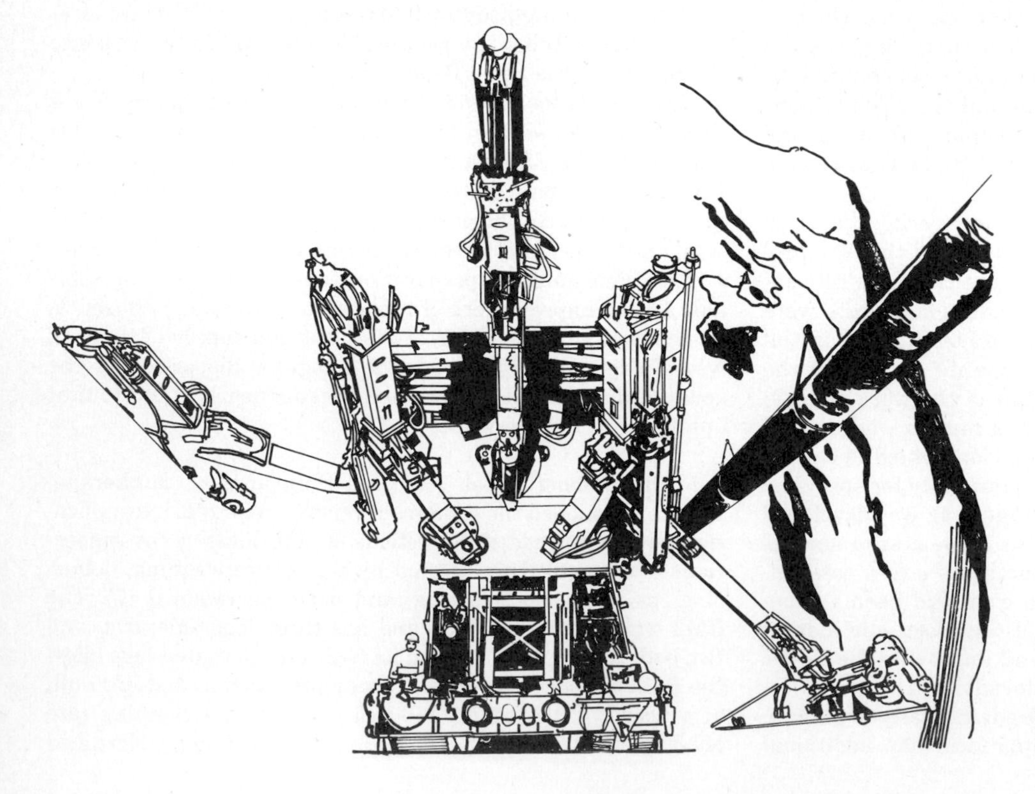

Figure 6. Kajima tunnel drilling robot (4). Courtesy of *IABSE Proceedings*.

contribution to semiautomated technology in tunneling works and makes it possible to execute a series of drilling, blasting, mucking, and shotcreting operations simultaneously on both the upper and lower halves of a tunnel bore. By setting this machine in the basic position for the face to be drilled and starting the automatic drilling device, the machine automatically drills the face in accordance with a previously memorized drilling pattern.

Adopting an automatic drilling machine has the following advantages over previous methods:

- Skilled drillers are not required.
- One person can operate more than one machine.
- Drilling can continue even during the operator's rest period.
- A correct drilling pattern and depth of holes can be secured.
- The drilling time is shortened.
- Workers are liberated from the environment.

Shield Driving. Shield driving is employed in the construction of most tunnels in Japanese urban areas. In shield driving, there would be value in automation and robotization of operations at each step: driving control, removal of excavated material, shield attitude–position control, backfill grouting, segment erection, and handling of materials. Shield equipment manufacturers and general contractors are all performing technical development aimed at final objectives of automating and robotizing all steps of operations. Kajima has developed a system for driving control and attitude–position control. In the operation of slurry shields (being one type of mechanical excavation shield), driving is accompanied by the monitoring of data, including the pressure within the face chamber, the revolving cutter torque, the volume of excavated material, and the properties of the slurry. These data are measured separately in conventional tunneling, and the development of corrective measures arises through the judgment of experienced engineers and skilled operators. However, the relationships between the various data items are not necessarily clear, and much reliance is placed on intuition. With the Kajima system, a determination is made by gathering and analyzing data in real time. With this system, it is possible to attain and maintain a stable driving condition by statistically analyzing the various data obtained in the initial stage of driving and repeatedly feeding these back into the shield operation.

Attitude control is also of great importance in shield driving. The general practice has been to survey line and grade by hand at 5–10 m intervals and correct the direction of the shield in accordance with analysis of survey data. However, the shield would go off-line, and the construction period and cost would be adversely affected by major directional corrections and by weakening of the surrounding ground resulting in ground settlement. In the robotic system, it is possible to monitor continuously the deviation of the shield from the planned line by direction angle, lateral distance, vertical distance, and pitch angle. Because the driving jacks can be controlled from the amount of deviation, the attitude and position of the shield can be controlled in real time; at the same time, the survey operation is eliminated and further major labor saving becomes possible. At the present time, jack operation based on the measured data has not been automated, but technical development is in progress to link these in the near future.

Examples of Robots for Excavation

REX. Carnegie-Mellon University has developed a robotic excavator (REX) to unearth buried utility piping by mapping an excavation site, planning the digging operations, and controlling excavation equipment. Explosive gases are sometimes ignited accidentally during blind digging of gas utilities, and REX reduces the human injuries and property losses attributed to such explosives; it also has the potential to decrease costs and increase productivity for utility excavation. The REX currently uses a sensor-built surface model to plan its digging action. REX interprets sonar data to build accurate surface and object depth maps to model the excavation site. Based on the surface topography and the presence and location of target pipes, appropriate trajectories are generated and executed. The benign end tooling developed for REX is a supersonic air-jet cutter; this air-jet cutter can dislodge material without the direct contact encountered with bucket excavation. Results to date have been promising (8), and an unmanned, benign excavation in a simulated laboratory excavation has been demonstrated. The research is currently implementing a distributed control architecture for increased speed and efficiency.

Ultradeep Diaphragm Wall Excavator. In the construction of in-ground LNG storage tanks of 100,000 kL or larger capacity in soft reclaimed coastal land, an ultradeep diaphragm wall is constructed. This wall typically descends to an impermeable layer to surround the tank and prevent the inflow of groundwater. In order to make this diaphragm wall effective in shutting out the groundwater, it is important to secure precision in vertical excavation by the diaphragm wall excavator. In an attempt to solve this problem, an automatic excavation system was developed by Kajima Corporation. By controlling the attitude of the machine during excavation, and by controlling the load in accordance with the physical properties of the soil being excavated, it has been possible to secure an excavation precision of over 1/1000.

Automatic Grading Control. A limited but significant example of robotics is now widely used in excavation operations in the San Francisco bay area (9). Excavation operations such as grading (scraping) and trenching (for subsurface drains and utility lines) are a major part of site preparation. A critical element in such work is the control of the invert elevation to which the excavation is performed. Conventionally, this has been done manually using levels or string lines, creating a tedious and discontinuous operation.

An automation mechanism was developed that is presented here as an intelligent robotic example. It can be defined as intelligent because it senses, thinks, and acts. It is a single-channel control that automatically scrapes or grades to the specified invert elevation. It consists of a laser level plane, a sensor mounted on the excavator blade, and a microprocessor that servos blade position to the specified invert elevation. The operator is left to drive the machine, and blade control is handled automatically and smoothly. In addition to decreasing labor demand, higher machine speeds are possible, as is improved efficiency from the continuity of the operation.

This example differs from most of the others in that it is a single-channel robot, but one which nonetheless contributes highly to productivity. Its simplicity and robustness have made it an example of robot technology that has entered the marketplace. It is supported not by advanced corporate technology, but instead by local application skills using modern products such as laser levels and microprocessors.

Examples of Robots for Assembly

Reinforcement Placing Robot. A robotic adaption for reinforcement placing was developed by Kajima Corporation (Figure 7). It carries up to 20 reinforcement bars, automatically placing them in floor slabs and walls according to a variety of preselected patterns. On many construction projects, rebars for such applications often have diameters in the range of 33 mm, are 8 m long, and weigh more than 100 kg, requiring considerable labor to position. According to the company, this robot has achieved 40–50% savings in labor and 10% savings in time on a number of projects (such as nuclear power plants) requiring heavily reinforced foundations.

Examples of Robots for Inspection

Robotized Core Boring. Carnegie-Mellon University developed a rover in use in the radioactively contaminated areas at the Three Mile Island (TMI) nuclear power plant in Pennsylvania. As a recent robotic tooling (Figure 8), a device was developed to recover drilled concrete core samples. In this application, they are recovered to provide samples of contamination and its characterization with depth from the surface. However, the broader problems of concrete coring and of concrete or rock drilling are pertinent to construction robotics. The tooling employs robotized drilling modules. In the present application, the vehicle positioning remains a teleoperated function with televised feedback. The application demonstrates the pertinence of robotic technology at one level in the system, teleoperation at others, and pure mechanical design at others (10).

Figure 7. Kajima reinforcing bar placement robot (4). Courtesy of *IABSE Proceedings*.

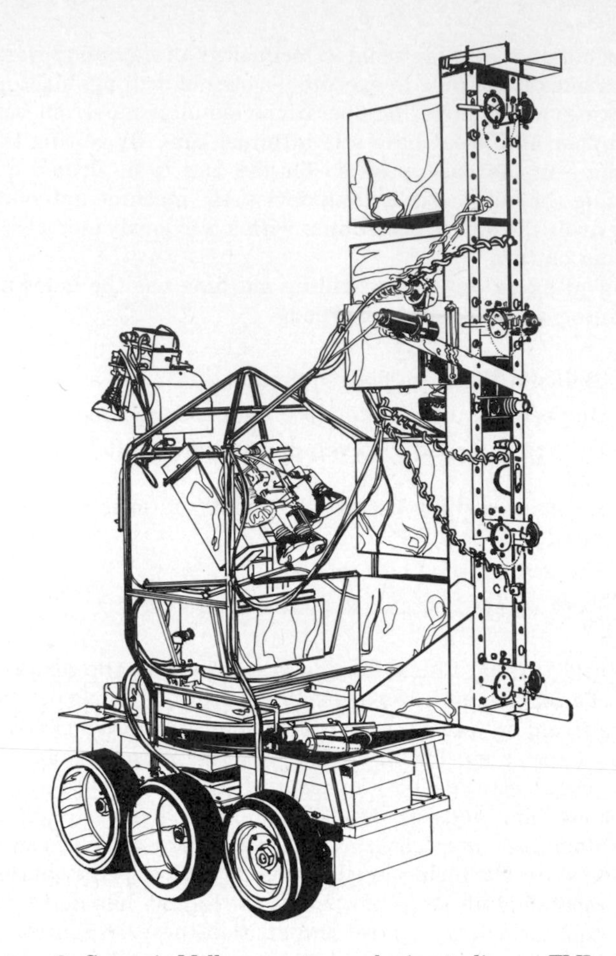

Figure 8. Carnegie-Mellon remote core-boring tooling on TMI rover. Courtesy of Construction Robotics Laboratory, Carnegie-Mellon University.

Magnetic Sensing with a Robot. Carnegie-Mellon University developed an intelligent magnetic sensor to automatically scan, size, and map embedded steel such as reinforcing bars in concrete and pipes in the ground. The mapping of embedded concrete reinforcing is useful to validate as-built structures, and the mapping of buried pipes is important in the excavation of utility lines. In the prototype, a robotic scanner moves a magnetic sensor in a rectangular area, and the output is digitized to form magnetic images (11). Low-level processing extracts features from the data; a postprocess compares data features to a template library of ferrous object patterns and predicted anomalies that result from an idealized ferrous source. Future efforts are to integrate an expert system to guide deductions, verify solutions, and control strategy.

Tile Inspection Robot. Kajima Corporation has developed a tile inspection robot (Figure 9). The traditional method consists of manually tapping each individual tile with a hammer and judging its adhesion by the sound produced. The robot inspects wall tiles automatically; a microprocessor-based system in the ground console analyzes the sound that a robotic tapping head produces. The adherence strength and location of each tile are automatically recorded.

Figure 9. Kajima tile inspection robot (4). Courtesy of *IABSE Proceedings.*

SYSTEMS OF THE FUTURE

The Near Future

Research and development continues for other construction applications for which no prototype device can be cited. However, such studies are likely to reach some level of field demonstration in the near future. Therefore, two representative applications are presented.

Building Construction and Prefabrication. Building construction often draws upon prefabricated components or assemblies. Producers, builders, and researchers have begun the study of systems to robotize the construction process or key portions thereof. One example is in concrete block placement, for which laboratory small-scale demonstrations have been performed at Carnegie-Mellon University and elsewhere. In those studies, a robot was able to build a sample block wall to a design database containing door and window openings. Studies included examples with random block size. The robot would measure the block and then process a task plan based upon that information, working at all times toward the design geometry. It is important also to recognize that when robotization proceeds, the block will evolve from the present one (constrained to be handled by humans) to blocks of larger size. This will lead to changes in block characteristics, such as the use of reinforcement or mechanical interlock, which cannot be accommodated at present. Therefore, a new material type will be put into service, with the promise of greater efficiency.

Prefabricated building panels or modules are an advanced technology in many countries. The advantages of mechanization and automation are recognized by the proponents of that technology, and their interest logically extends to automation of tasks after the component leaves the factory. Studies of robot technology have been started, and some level of demonstration should be expected in the near future.

Earth Structures. The building of dams and embankments represents the forming of an earth structure to an intended large-scale geometry. The elemental acts of deposition or removal are repeated in a long and largely repetitive pattern. Robotization of these elemental tasks will permit continuous execution of such an accretion process. Moreover, the robot devices will operate under the control of a central computer environment combining the design information, the monitored field information, the task management, and the project database. This type of application shares the characteristics of surface mining, and it is likely that developments will propel both of these major application areas together.

The Far Future

The far future will feature exploitation of domains such as the subsurface and outer space. The act of exploiting such domains will require some ordering of that environment or some type of construction. It is noteworthy that conventional construction exposes robotics to the challenges of an uncontrolled environment, and that those challenges reappear in this long-term perspective on the role of robots.

An important element of the future is the expanded role to be filled by the computer. An integrated computer environment (12) is envisioned to support a project through numerous stages including:

- Conceptual design.
- Detailed design.
- Fabrication.
- Materials handling and site management.
- Erection and construction.
- Construction management.
- Operation.
- Maintenance and repair.
- Decommissioning.

Such an integrated computer environment is an ultimate objective for a more perfect engineering of constructed facilities. The robot system is needed to make such a model complete. Note that some portion of this total scope is reflected in the earth structure application cited earlier.

RESEARCH FRONTIERS

Research problems were identified earlier when addressing the suitability of existing robot technology. It is clear that construction robotics will advance with new results in many areas including:

- Robot mobility.
- Vision and pattern recognition.
- Navigation and positioning.
- Sensing and sensor-based control.
- Dynamics and control.
- Obstacle detection and avoidance.

- Hierarchical control and planning.
- Knowledge-based expert systems.

In some cases, researchers in construction robotics are major contributors to these more general research areas, and in others the lead role originates elsewhere. In any event, the reader is referred to the appropriate articles on those and many other topics.

There are research frontiers in construction robotics, important to other applications, which may not yet be widely known. One is the problem of domain modeling, constituting the development of a computer model of an environment in which robots move and work, interacting with domain objects both physically and functionally. Researchers (13) are developing a domain model for robotic construction and maintenance in facilities such as power plants. An object-oriented programming language is employed, and all entries are treated as objects in that sense. Another example, originating in construction robotics but pertinent to many applications within and without robotics, is the representation and manipulation of geometric information in knowledge-based expert systems. Construction robotics is an application drawing heavily on expert systems and at the same time dealing with physical objects. Previous attempts to operate on geometric information have been tedious and incomplete. Research proceeds at Carnegie-Mellon University for a more fundamental geometric modeling system, one designed to support expert system applications. Similarly, work proceeds on the application of expert systems to the control of heavy equipment such as mining machines; this is another example of research directions common to construction and other application areas. A summary statement of research frontiers raised by construction robotics is generated by the broad, complex system nature of the problem. An autonomous robot system will feature decision capabilities at the reflexive, tactical, and strategic levels. It will be cognitive in spatial terms and in force terms. It will engage issues of perception, representation, abstraction, and modeling. Although the present examples of construction robots may be limited in number and scope, autonomous construction as a research area is a crucible for some of the most far-reaching problems in robotics.

BIBLIOGRAPHY

1. *First International Conference on Robotics in Construction,* Department of Civil Engineering, Carnegie-Mellon University, Pittsburgh, Pa., 1984.
2. *Second International Conference on Robotics in Construction,* Department of Civil Engineering, Carnegie-Mellon University, Pittsburgh, Pa., 1985.
3. A. Warszawski, "Application of Robotics to Building Construction," Carnegie-Mellon University, Pittsburgh, Pa., Tech. Rep., 1984.
4. Y. Sagawa and Y. Nakahara, "Robots for the Japanese Construction Industry," *IABSE Proceedings* **7,** (P-86/85), (May 1985).
5. T. Yoshida and T. Ueno, "Development of a Spray Robot for Fireproof Treatment," *Shimizu Technical Research Bulletin* (4), (Mar. 1985).
6. S. Suzuki, "Construction Robotics in Japan," presented at the *Third International Conference on Tall Buildings,* Chicago, Ill., 1986.
7. "Robot Climbs the Walls," Editorial, *Robotix News,* (1, 6 July/Aug. 1985).
8. W. Whittaker and co-workers, "First Results in Automated Pipe Excavation," presented at the *Second International Conference on Robotics in Construction,* Carnegie-Mellon University, Pittsburgh, Pa., June 1985.
9. B. Paulson, "Automation and Robotics for Construction," *ASCE Journal of Construction Engineering and Management* **111** (3), 190–205, (Sept. 1985).
10. W. Whittaker and co-workers, "Three Remote Systems for TMI-2 Basement Recovery," presented at the *ANS Meeting on Remote Technology at TMI,* San Francisco, Calif., 1986.
11. B. Motazed, *"Inference of Ferrous Cylinders from Magnetic Fields,"* Ph.D. dissertation, Carnegie-Mellon University, Pittsburgh, Pa., Dec. 1984.
12. S. Fenves and D. Rehak, "Role of Expert Systems in Construction Robotics," presented at the *First International Conference on Robotics in Construction,* Carnegie-Mellon University, Pittsburgh, Pa., June 1984.
13. W. Keirouz, D. Rehak, and I. Oppenheim, "Object-Oriented Domain Modeling of Constructed Facilities for Robotic Operations," presented at the *First International Conference on Applications of Artificial Intelligence to Engineering Problems,* Southhampton University U.K., Apr. 1986.

General References

Y. Hasegawa, "Robotization of Reinforced Concrete Building Construction," presented at the *II International Symposium on Industrial Robots,* Tokyo, Japan, 1981.

J. S. Albus, A. J. Barbera, and M. L. Fitzgerald, "Programming a Hierarchical Robot Control System," *Proceedings of the 12th International Symposium on Industrial Robots,* Paris, France, 1982.

Y. Hasegawa, "Robotization of Building Construction Part 1," *Robot* (33), (1982).

Y. Hasegawa, "Robotization of Building Construction Part 2," *Robot* (34), (1982).

T. Pavlidis, *Algorithms for Graphics and Image Processing,* Computer Science Press, Rockville, Md., 1982.

T. Furukawa and S. Kikukawa, "Introduction of Robots to the Construction Work Site," *Robot* (38), (1983).

K. Goto and co-workers, "Self-Climbing Inspection Machine for External Wall," *Robot* (38), (1983).

Y. Hasegawa, "Robotization of Construction Work," *Robot* (38), (1983).

S. Matsubara, "Floor Cleaning Robot," *Robot* (38), (1983).

P. Sjolund and M. Donath, "Robot Task Planning: Programming Using Interactive Computer Graphics," presented at the Robots 7 Conference, Chicago, Ill., 1983.

D. Sriram and co-workers, "Potential for Robotics and Artificial Intelligence in Tunnel Boring Machines," Robotics Institute, Civil Engineering and Construction Robotics Laboratories, Carnegie-Mellon University, Pittsburgh, Pa., Tech. Rep., Jan. 1983.

J. Crowley, "Dynamic World Modeling for an Intelligent Mobile Robot Using an Ultra-Sonic Ranging Device," Robotics Institute, Carnegie-Mellon University, Pittsburgh, Pa., Tech. Rep. 84–27, 1984.

B. Yamada, "Development of Robots for General Construction and Related Problems," presented at the *Research Conference,* Material and Construction Committee, Architectural Institute of Japan, Oct. 1984.

G. Agin and co-workers, "Three Dimensional Sensing and Interpretation," Robotics Institute, Carnegie-Mellon University, Pittsburgh, Pa., Tech. Rep. 85–1, 1985.

M. Skibniewski and C. Hendrickson, "Evaluation Method for Robotics Implementation: Application to Concrete Form Cleaning," presented at the *Second International Conference on Robotics in Construction,* Carnegie-Mellon University, Pittsburgh, Pa., June 1985.

M. L. Maher, I. J. Oppenheim, and D. R. Rehak, "Computer Developments for Mining Robotics," presented at the *Second International Conference on Robotics in Construction,* Carnegie-Mellon University, Pittsburgh, Pa., June 1985.

W. L. Whittaker, J. Crowley, I. J. Oppenheim, and co-workers, "Mine Mapping by a Robot with Acoustic Sensors," presented at the *Second International Conference on Robotics in Construction,* Carnegie-Mellon University, Pittsburgh, Pa., June 1985.

M. P. Rosick, "Obstacle Avoidance for a Mobile Manipulator," Master's thesis, Carnegie-Mellon University, Pittsburgh, Pa., Sept. 1985.

A. Warszawski and D. Sangrey, "Robotics in Building Construction," *ASCE Journal of Construction Engineering and Management* **111** (3), (Sept. 1985).

CONTROL.

See ROBOT CONTROL SYSTEMS.

CONTROLLER ARCHITECTURE

JAMES H. GRAHAM
University of Louisville
Louisville, Kentucky

INTRODUCTION

This article presents an overview of digital control architectures that are appropriate for control of robotic manipulators and robotics-related equipment. Despite the relative youth of the robotics field, a large number of control algorithms and control strategies have been proposed for control of robot manipulators and related equipment (1–12). It is beyond the scope of this article to describe or summarize these control alternatives, except as they affect controller architectures. References 1–7 provide details of some specific control algorithms. References 8–12 contain overviews of various control methods (see also CONTROL; ADAPTIVE CONTROL). Obviously, the more recently reported work represents recent research results, which generally have not yet been implemented in commercial equipment.

Most of the control algorithms described in the cited literature can be efficiently implemented on the conventional computer architectures discussed in this article. The application of specialized computers for robot and robot systems control is a current research topic and is discussed in the final section of the article.

SIMPLE CONTROL ARCHITECTURE

Most commercial controllers for industrial manipulators rely on an analogue servo system for control of the prime movers (actuators) for the individual joints. The simplest type of control architecture consists of a system for loading a set of joint trajectory points into the analogue servos. The digital controller commands the start of motion and checks for stopping conditions. This is shown schematically in Figure 1.

The controller for this simple architecture can easily be implemented by a single-card microcomputer shown in Figure 2. These systems typically consist of a microprocessor unit and its support chips (clock generator, bus controller, etc), memory chips, and peripheral input/output (I/O) chips. Most single-board computers provide both parallel and serial I/O channels. The serial interface typically operates with an asynchronous protocol in the range of 300–9600 bits/s with an Electronic Industries Association (EIA) standard RS-232 electrical interface. This type of interface is very good for communicating with a video display terminal or with other remote equipment that can tolerate low to medium data rates.

For higher data rate communication, particularly within the controller, the parallel peripheral interface chips can be used. By transferring 8–24 parallel bits at one time, much greater data rates can be achieved. Line lengths must be kept to a minimum, and special shielding and grounding techniques should be observed (13). Most parallel peripheral chips provide a number of control lines in addition to the data lines, such as READY, STROBE, and so on. These signals can be used to establish handshaking protocols between the sending and receiving devices. For information about digital logic components and microprocessors, see references 13–15. For use of microprocessors in control applications, see references 16 and 17.

One final note on the simple controller concerns the case where stepping motor actuators are used in the joints instead of hydraulics or direct current (d-c) electrical motors. Stepping motors are inherently digital devices, rotating clockwise or counterclockwise one unit with each input pulse (18). Thus, the analogue control servoboards of Figure 1 can be replaced with simpler digital-based interface circuitry. Several integrated stepping motor chips are available for use in such applications. The microcomputer controller's function is to keep track of the present position and end point position. The controlling joint movement simplifies sending the correct number of pulses to the appropriate joint stepping motors. This architecture has been implemented in a number of inexpensive, cable-controlled manipulators for handling light loads at moderate speeds.

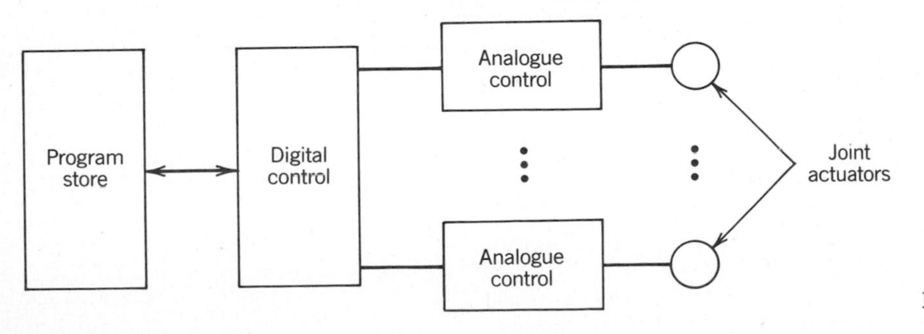

Figure 1. Simple controller architecture.

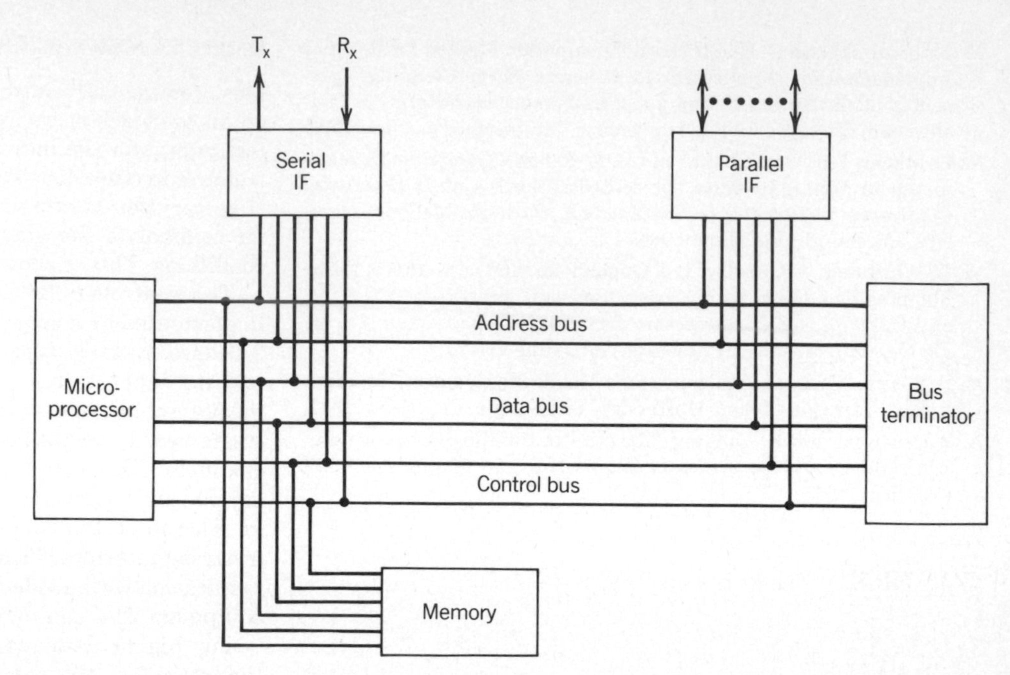

Figure 2. A simple microcomputer architecture.

AN EXTENDED CONTROLLER ARCHITECTURE

The simple controller architecture of the preceding section is too limited to be used for most industrial manipulators. In addition to the basic joint angle control function previously described, the robot controller must also serve as an interface to human operators for purposes of maintenance, setup, programming, debugging, and so on; it must, in most cases, have facilities for long-term, nonvolatile, program storage; it may have responsibility for interfacing with other industrial equipment, either through a general network interface or through dedicated control lines; and it may have responsibilities for monitoring internal and external conditions related to proper and safe operation of the robot, such as load ranges, movement limits, temperatures, and so on. Figure 3 shows an expanded block diagram of the robot controller incorporating these extended functions.

Human Interaction

The operator interface allows the robot controller to interchange information with a human operator. The interface may be very simple or very elaborate; it may be built into the controller chassis, or it may be a separate unit; and it may or may not allow the robot to be moved in a restricted motion regime (ie, one joint at a time).

Regardless of the physical configuration of the operator interface, its prime functions are

1. To convey information about the internal state of the robot to the operator.
2. To allow the operator to specify control instructions to the robot.

Relevant internal information may include joint positions, ranges, conditions (OPEN/CLOSED for end effector), current

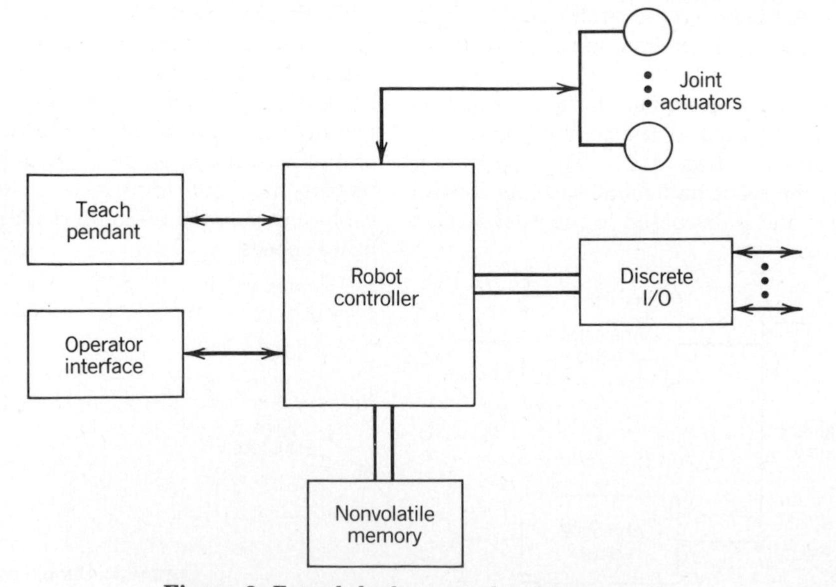

Figure 3. Extended robot control architecture.

operating mode, current program, current program step, and any abnormal conditions.

Control instructions may include STOP/START, go to home or calibration position, enter program (teach) mode, leave program mode, enter automatic mode, store program, and retrieve program. Many robot controllers provide a manual control or teach pendant as a controller feature.

Teach Pendant

Despite advances in off-line trajectory planning, much work remains to be done before robots can be programmed by off-line integrated CAD/CAM. Also, there are certain jobs, such as spray painting, that are not easily described analytically. For these reasons, the teach pendant will remain an important component of the human–robot interface for many industrial robots. The teach pendant allows the operator to move individual joints or to make movements in a selected coordinate system (Cartesian, tool, etc). When a desired position is achieved, it can be recorded by pushing a button on the pendant.

The architectural consequences of incorporating a teach pendant are not severe. Basically, additional parallel control lines are required for the pendant, and additional storage must be allocated to store the teach mode programs and data. Since the robot is typically placed in a special teach mode condition, the control computer may supervise teaching operations. Some controller architectures allow special hardware to interface with the pendant.

Program Storage

In the extended architecture, facilities are provided by software and hardware to record program information for future retrieval and use. Since the random access memory of the control computer is generally semiconductor memory chips, they will for the most part lose their contents if power is interrupted or withdrawn. Therefore, a provision must be made to save the robot programs and associated data permanently. This may be accomplished by one or more of the following magnetic storage media: magnetic bubble memory, magnetic core, floppy disks, hard disks, magnetic tape. (It is noted that floppy disks are not acceptable for many industrial environments because of particulate contamination.)

Architecturally, program storage requires an additional I/O channel in the controller architecture. There are several commercial semiconductor chips that implement control of magnetic media (13). In most cases, these chips effect a direct memory access (DMA) into the volatile semiconductor memory to efficiently transfer the stored data.

External Interface

This section deals with the interface between the robot and other external equipment. There are two common forms of external interface provided by commercial controllers: discrete (usually binary) inputs and outputs and a local computer area network (LAN) interface.

Discrete I/O can be provided by the type of parallel I/O circuits discussed in the section on the simple controller. There would typically be an additional buffer level of electronic or photooptic relays to isolate the controller electronics from the external world.

LAN interfaces are more complicated and would typically require a separate processor or specialized chip set to handle the encoding and decoding of the LAN protocols.

ARCHITECTURAL ALTERNATIVES

Some of the internal configurations that are used to achieve the functional performance previously discussed are presented here. The basic partition is between a single-processor, centralized architecture and a multiple-processor, distributed architecture. Within the latter category are several subclasses such as joint-parallel architectures, function-oriented architectures, and hierarchical architectures (19–22). This section will briefly address the issues related to the centralized versus distributed architectures and then will examine the characteristics of the major architectural classes and subclasses.

Centralized Control Architecture

The simple controller architecture shown in Figure 1 is a good example of the single-processor, centralized architecture. All computation is done, and all decisions are made, by the single central processor unit (CPU). The advantages of this architecture are its simplicity and resulting low cost. The disadvantage is the relative lack of desirable control features.

Some of the extended controller features may also be included in a centralized architecture by careful design. As mentioned previously, many of the special control features are handled in separate control regimes, and thus do not interact simultaneously with each other, as with, for example, teaching and playback. For those functions that are simultaneous, interrupt handlers can be implemented (15). As a practical matter, however, most commercial controllers use multiple processors to implement extended controller functions.

Joint-Parallel Architecture

One convenient multiple-processor configuration is to allocate a processing element for each of the manipulator joints that are to be controlled. A separate, more powerful processor can be utilized to coordinate the individual joint control processors and to handle external interfaces. This approach was adopted for at least one well-known mid-sized, electrically actuated robot series.

The advantage of this approach is that additional computing power is made available to augment the centralized controller during joint control intervals. This is typically the most time-critical phase of controller operation and is one of the most computationally demanding. The disadvantages that relate to this increased complexity of architecture are more circuitry, increased interunit communication, and more potential failure modes.

Overall, this architecture is adequate, if not optimal, for the task of robot control. It derives from similar architectures used in machine control and is thus, in some senses, a proven architecture. This architecture is shown in Figure 4.

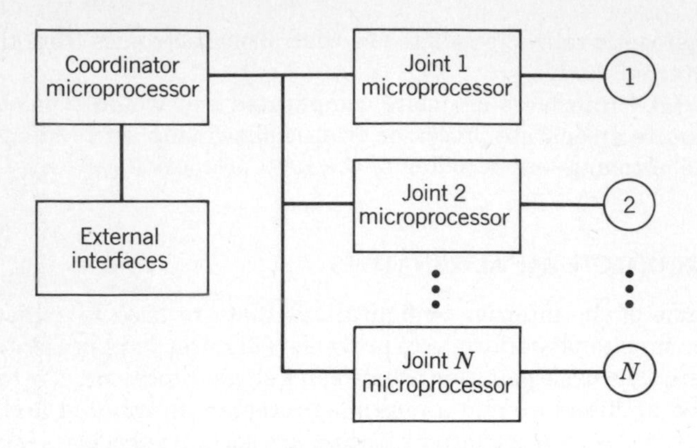

Figure 4. Joint-parallel control architecture.

Functional Parallel Architecture

An alternative to the joint-parallel is the multiprocessor architecture, which divides controller tasks among processing elements on a functional basis. Thus, there may be one processor (or more) allocated to joint control, a processor allocated to the operator interface, a processor for external equipment interface, and a processor to manage the nonvolatile (permanent) storage. This functional distribution architecture is shown in Figure 5. Advantages include the ability to provide more features with low-cost components. Disadvantages include the increased circuitry complexity.

Hierarchical Multiprocessor Architecture

An hierarchical multiprocessor architecture has been suggested by several research studies as being suitable for robot control (23–27). Hierarchical systems have been employed successfully in other control applications, notably process control (28). The central idea is that control authority is decomposed into a series of levels of increasing sophistication but decreasing precision. Hierarchical control is attractive because it decomposes the problem into a set of manageable subproblems that can be controlled with simple microprocessor-based controllers and permits multilevel sensory interaction. The disadvantage of hierarchical control is the increased complexity of circuitry and communications. One form of this architecture is shown in Figure 6.

FUTURE ARCHITECTURES

There are several discernible trends for future architectural development of robot controllers. One is the inevitable performance improvements in both speed and features in microcomputers, which are fueled by progress in integrated electronics. This makes possible the incorporation of extended features with simple controller architectures, reducing the number of processors required. This, in turn, reduces parts, circuitry, and other hardware costs, but tends to increase the amount of software development and related costs.

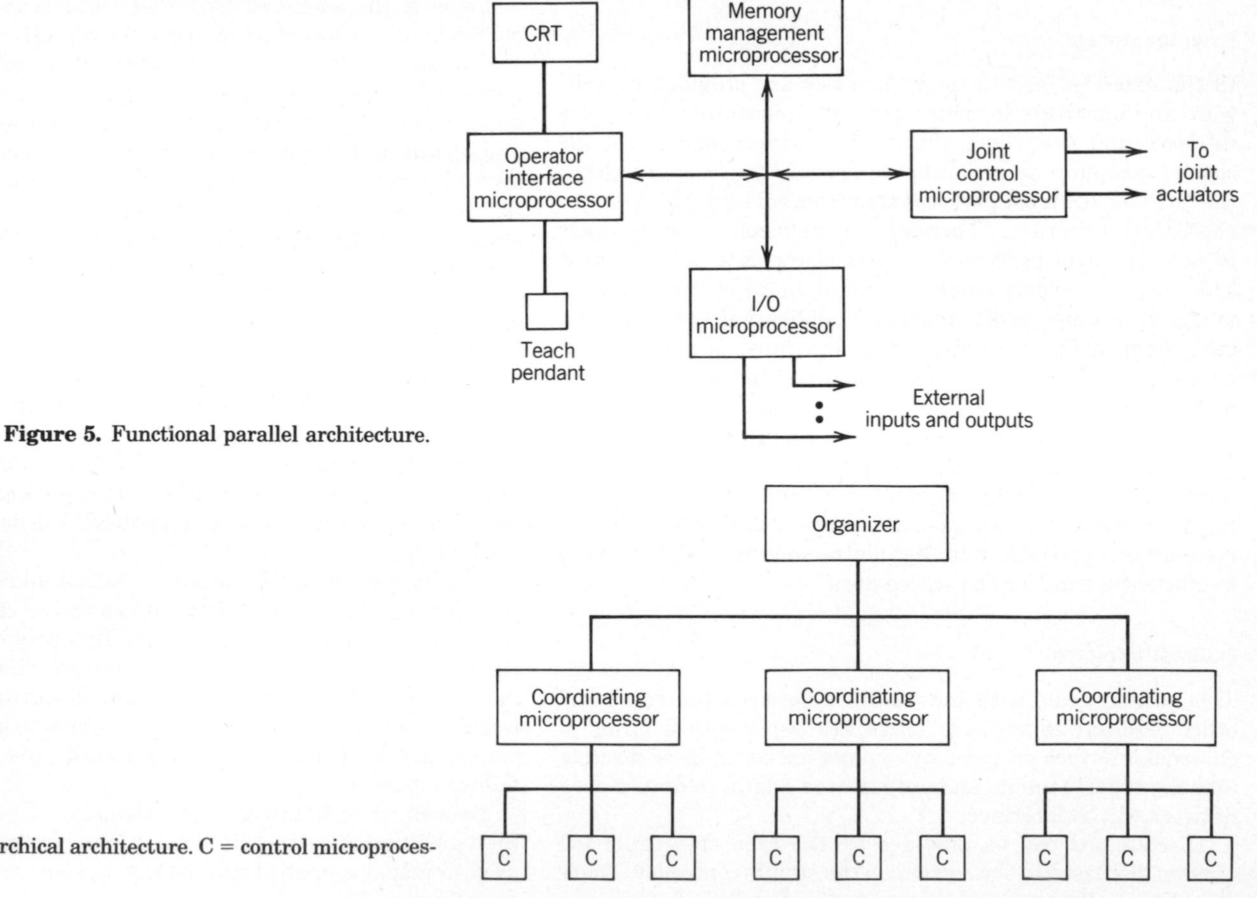

Figure 5. Functional parallel architecture.

Figure 6. Hierarchical architecture. C = control microprocessor.

A second likely development will be increased emphasis on reliability. A functional breakdown of the controller will at least impair, and more likely, disable the robot. Downtime of any piece of equipment reduces productivity and raises expenses, especially if the robot is a critical component of some larger manufacturing cell. In addition, controller failure can create possible safety hazards to human operators and maintenance personnel. Current reliability approaches involving parts testing, burn-in and redundant parts, and circuits are expensive. It is hoped that new chip architectures with built-in redundancy will help solve this problem.

A final development regarding future architectures is the investigation of specialized integrated architectures, which have become feasible due to customized integrated circuit design tools. Although this technology is just developing and most of the work reported to date is still in the early research stage (29–31), it appears to hold significant promise for enhanced controller performance.

BIBLIOGRAPHY

1. R. P. Paul, "Modeling, Trajectory Calculation, and Serving of a Computer Controlled Arm," Stanford Artificial Intelligence Laboratory, Stanford, Calif., AIM-177, Nov. 1972.
2. A. K. Bejczy, "Robot Arm Dynamics and Control," Jet Propulsion Laboratory, Pasadena, Calif., Tech. Rep. 33–669, Feb. 1974.
3. D. E. Whitney, "Resolved Motion Rate Control of Manipulators and Human Prostheses," *IEEE Transactions on Man–Machine Systems* **MMS-10** (2), 47–53 (June 1969).
4. R. P. Paul, "Manipulator Cartesian Path Control," *IEEE Transactions on Systems, Man and Cybernetics,* **SMC-9,** 702–711 (Nov. 1979).
5. J. Y. S. Luh, M. W. Walker, and R. P. Paul, "Resolved-Acceleration Control of Mechanical Manipulators," *IEEE Transactions on Automatic Control* **AC-25,** 468–474 (June 1980).
6. S. Dubowsky and D. T. Des Forges, "The Application of Model Reference Adaptive Control to Robotic Manipulators," *Transactions of ASME, Journal of Dynamic Systems, Measurements, and Control* **101,** 193–200 (Sept. 1979).
7. K. D. Young, "Controller Design for a Manipulator Using the Theory of Variable Structure Systems," *IEEE Transactions on Systems, Man and Cybernetics* **SMC-8** (2) 101–109 (Feb. 1978).
8. J. Y. S. Luh, "An Anatomy of Industrial Robots and Their Controls," *IEEE Transactions on Automatic Control* **AC-28** (2), 133–153 (Feb. 1983).
9. M. Brady, J. M. Hollerbach, T. L. Johnson, T. Lozano-Perez, and M. T. Mason, eds., *Robot Motion: Planning and Control,* M.I.T. Press, Cambridge, Mass., 1982.
10. C. S. G. Lee, R. C. Gonzalez, and K. S. Fu, eds., *Tutorial on Robotics,* IEEE Computer Society Press, Silver Spring, Md., 1983.
11. G. N. Saridis, *Advances in Automation and Robotics: Theory and Applications,* JAI Press, Greenwich, Conn., 1984.
12. W. E. Snyder, *Industrial Robots: Computer Interfacing and Control,* Prentice-Hall, Englewood Cliffs, N.J., 1985.
13. H. S. Stone, *Microcomputer Interfacing,* Addison-Wesley, Reading, Mass., 1982.
14. F. J. Hill and G. R. Peterson, *Digital Logic and Microprocessors,* John Wiley & Sons, Inc., New York, 1984.
15. J. F. Wakerly, *Microcomputer Architecture and Programming,* John Wiley & Sons, Inc., New York, 1981.
16. D. A. Cassell, *Microcomputers and Modern Control Engineering,* Reston Publishing Co., Reston, Va., 1983.
17. C. L. Phillips and H. T. Nagle, *Digital Control System Analysis and Design,* Prentice-Hall, Englewood Cliffs, N.J., 1984.
18. B. Kuo, *Incremental Motion Control,* SLR Publishing, Champaign, Ill., 1979.
19. R. H. Lathrop, "Parallelism in Manipulator Dynamics," Artificial Intelligence Laboratory, Massachusetts Institute of Technology, Cambridge, Mass., Tech. Rep. #754, Dec. 1983.
20. C. A. Klein and W. Wahawism, "Use of a Multiprocessor for Control of a Robotic System," *International Journal of Robotics Research* **1** (2) 45–49 (Summer 1982).
21. J. H. Graham and G. N. Saridis, "Two Level, Multimicroprocessor Digital Controller," *Journal of Automatic Control Theory and Applications* **8** (1) (1980).
22. J. H. Graham and J. W. Burnett, "Microcomputer Control of a Robotic Arm in a Distributed Computer Environment," *Proceedings of the American Control Conference,* Washington, D.C., June 1982, pp. 1184–1185.
23. J. S. Albus, *Brains, Behavior, and Robotics,* BYTE Books, Peterborough, N.H., 1981.
24. A. J. Barbera, M. L. Fitzgerald, and J. S. Albus, "Concepts for a Real-Time Sensory-Interactive Control System Architecture," *Proceedings of the 14th Southeastern Symposium on System Theory,* Apr. 1982, pp. 121–126.
25. J. S. Albus and co-workers, "A Control System for an Automated Manufacturing Research Facility," *Proceedings of the Robots 8 Conference, Society of Manufacturing Engineers,* Detroit, Mich., 1984.
26. J. H. Graham and G. N. Saridis, "Linguistic Methods for Hierarchically Intelligent Control," School of Electrical Engineering, Purdue University, West Lafayette, Ind., Tech. Rep. EE 80–34, Oct. 1980.
27. J. H. Graham and G. N. Saridis, "Linguistic Decision Structures for Hierarchical Systems," *IEEE Transactions on Systems, Man and Cybernetics* **SMC-12** (3), 325–333 (May 1982).
28. M. D. Mesarovic, K. Macko, and Y. Takahara, *Theory of Hierarchical Multilevel Systems,* Academic Press, New York, 1970.
29. D. E. Orin and co-workers, "Pipeline/Parallel Algorithms for the Jacobian and Inverse Dynamics Computation," *Proceedings of the IEEE International Conference on Robotics and Automation,* St. Louis, Mo., Mar. 1985, pp. 785–789.
30. R. Nigam and C. S. G. Lee, "A Multiprocessor-Based Controller for the Control of Mechanical Manipulators," *Proceedings of the IEEE International Conference on Robotics and Automation,* St. Louis, Mo., Mar. 1985, pp. 815–821.
31. F. Ozguner and M. L. Kao, "A Reconfigurable Multiprocessor Architecture for Reliable Control of Robotic Systems," *Proceedings of the IEEE International Conference on Robotics and Automation,* St. Louis, Mo., Mar. 1985, pp. 802–806.

General Reference

J. Graham, ed., *Specialized Computer Architectures for Robotics and Automation,* Gordon & Breach Publishers, New York, 1986.

CONTROL OF ROBOTS USING FUZZY REASONING

E. M. SCHARF
E. H. MAMDANI
Queen Mary College, University of London
London, United Kingdom

INTRODUCTION

Human operators are able to control very complex industrial processes, such as blast-furnaces and cement kilns, which are themselves not amenable to the detailed mathematical description required by the application of conventional control theory. In these cases, the operators can often formulate their experience in linguistic terms (1–4), as is evidenced by the existence of rule books for the operation of cement kilns.

Zadeh (5) introduced fuzzy logic to provide a mathematical formulism for linguistic statements and reasoning. Mamdani and co-workers (1, 6–8) in the 1970s then demonstrated the relevance of these ideas to industrial control engineering. The core idea was to model the actions of the operator or controller without attempting to find a mathematical model of the process: for conventional control systems, a detailed model of the process is required in order to design the controller.

Subsequently, there have been numerous applications of fuzzy dynamic controllers; mainly from Japan and Europe. Large-scale industrial examples include cement kilns (2, 3, 6) water purification (9) and treatment (10), a heat exchanger (11), and a sinter plant (12). Smaller scale applications comprise the pioneering steam-engine controller (13), marine autopilots (14), as well as control of a diesel engine (15), train speed (16), car speed (17) and satellite positioning (18).

The experience and ability of athletes to coordinate complicated high-speed limb movements is one of many examples which defy accurate knowledge elicitation. This calls for the use of a learning mechanism whereby the system can form and modify the rules according to which it acts. Work on this was first reported in Ref. 8 and was subsequently found to be relevant to robot control (19) (see also ADAPTIVE CONTROL; CONTROL; LEARNING AND ADAPTION).

THEORETICAL CONSIDERATIONS

Crisp Subsets

Items which have unique descriptions (eg, length) can be said to belong to *crisp subsets*. Thus, in the example (Fig. 1) an item, by virtue of its measurement, can be regarded as short, medium, or long. This can be described by crisp subsets which map each measurement into either 0 or 1. The formal description is thus

$$\chi_A(u): u \rightarrow (0,1)\ u \in U$$

In this case, U is the universe of discourse and u is a particular measurement. A is the property (eg, short, medium, or long).

One may then define the following operations: union AUB, intersection $A \cap B$ and complement $\bar{A}$. There are a number of properties: (1) $AUA = A$, $A \cap A = A$; (2) $AUB = BUA$; (3) $AU(BUC) = (AUB)UC$; (4) $\overline{AU(B \cap C)} = (AUB) \cap (AUC)$; (5) $\overline{(AUB)} = A \cap B$ and (6) $A \cap \bar{A} = 0$, $AU\bar{A} = 1$.

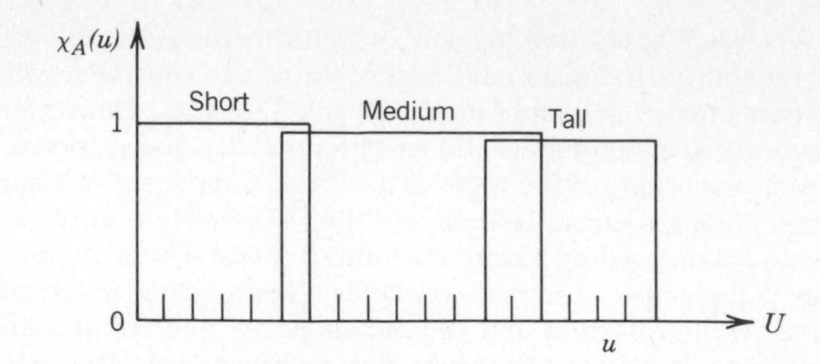

Figure 1. Example of crisp subsets (6). Courtesy of Academic Press.

Fuzzy Subsets

In Zadeh's generalization (5), each element u of the universe of discourse U may belong to the subset "A" to a degree varying between 0 and 1. This can be formalized as:

$$\mu_A(u): u \rightarrow [0,1]\ u \in U$$

and Figure 2 gives an example.

All the properties of crisp subsets apply to fuzzy subsets except number 6 where "=" now becomes "≠".

There are three additional properties for fuzzy subsets:

1. Union: $\mu_{A \cup B}(u) = \max(\mu_A(u), \mu_B(u))$
2. Intersection: $\mu_{A \cap B}(u) = \min(\mu_A(u), \mu_B(u))$
3. Complement: $\mu_{\bar{A}}(u) = 1 - \mu_A(u)$

Mathematical Representation of Logical Connectives

Linguistic rules are generally of the form

IF A AND B THEN C

where A and B are the antecedents and C the consequent. A and B could be "long" and "short" and would thus have the same universe of discourse U of length as illustrated in Figure 3. They could have two different universes of discourse, thus A and B could be "short" and "very young" with universes of discourse of "length" and "age." This could be represented by the 3-D situation of Figure 4; the intersection of the shapes formed by the projection of each of the fuzzy subsets can thus represent "short AND very young"; this uses the intersection property (number 2) above.

For computer implementation, the universes of discourse are discrete and this enables the following, more formal, statements to be made.

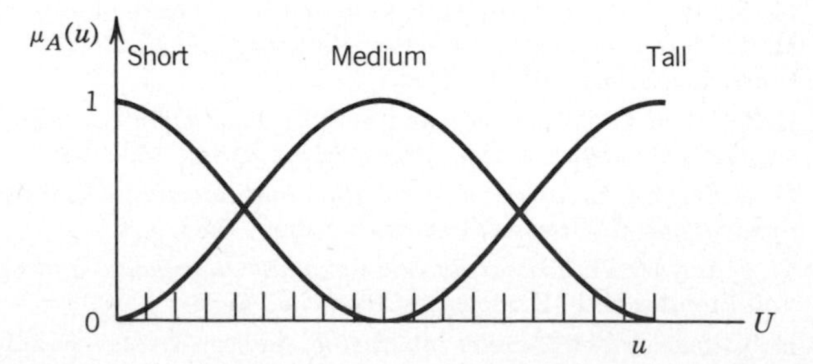

Figure 2. Example of fuzzy subsets (6). Courtesy of Academic Press.

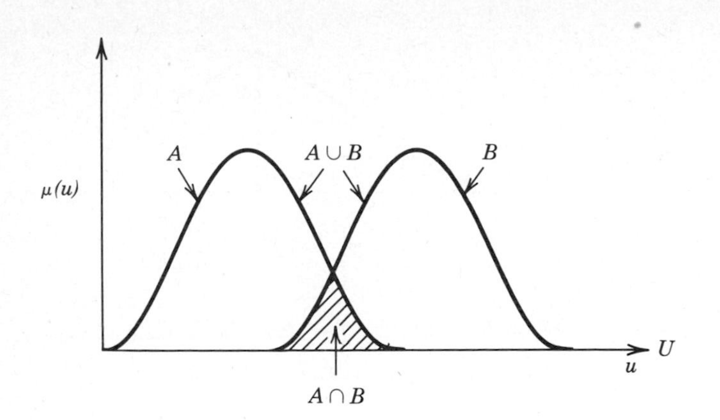

Figure 3. Geometrical interpretation of logical connectives "AND" and "OR" for fuzzy subsets on the same universe of discourse.

The nth rule of a set of rules is

$$(\text{RULE})_n = \text{IF } [(A_t \text{ is } A_n) \text{ AND } (B_t \text{ is } B_n)] \text{ THEN } [C_t \text{ is } C_n]$$

A_t, B_t and C_t could be error, change of error and change in output at a given sampling time t. A_n, B_n and C_n are linguistic terms such as "Positive Small" or "Negative Very Big" which are names of fuzzy subsets on the universes of discourse U, V, and W of A, B, and C. For computer implementation, U, V, and W are sequences of integers, $1 \ldots I, 1 \ldots J, 1 \ldots K$; in a practical case, these could be 1 to 13 or −6 to +6. A_n, B_n and C_n are associated with membership vectors representing fuzzy subsets $\mathbf{A}_n$, $\mathbf{B}_n$ and $\mathbf{C}_n$ on U, V, and W.

Thus $\mathbf{A}_n = \{\mu_n(u_i), \ldots \mu_n(u_i), \ldots \mu_n(u_r)\}$ and for $\mathbf{B}_n$ or $\mathbf{C}_n$, u is replaced by v or w, I by J or K.

Thus, if $I = 6$, $J = 7$, and the following (almost triangular) fuzzy subsets are defined:

$\mathbf{A}_n$ = positive medium (PM) = {0.0,0.0,0.3,0.7,1.0,0.7} and

$\mathbf{B}_n$ = negative small (NS) = {0.3,0.7,1.0,0.7,0.3,0.0,0.0}.

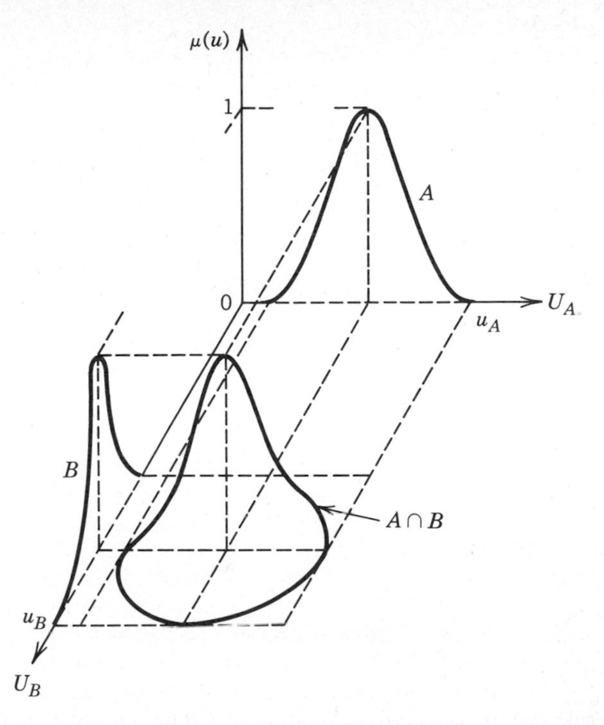

Figure 4. Geometrical interpretation of logical connective "AND" for fuzzy subsets on different universes of discourse.

The condition is $(\text{COND})_n = (A \text{ is PM}) \text{ AND } (B \text{ is NS})$.

This can be represented geometrically as shown in Figure 5 as well as by the *relation matrix* of Figure 6, noting that

1. $x_{ij} = \min(\mu_n(u_i), \mu_n(v_j))$ and consequently,
2. it can be seen that the "min" function models AND.

The elements of a fuzzy subset are usually assumed to have values in the range 0 to 1, where 1 indicates the maximum *degree of fulfillment* (DOF) and 0 no DOF. Hence,

$$0 \leqslant \mu_n \leqslant 1.$$

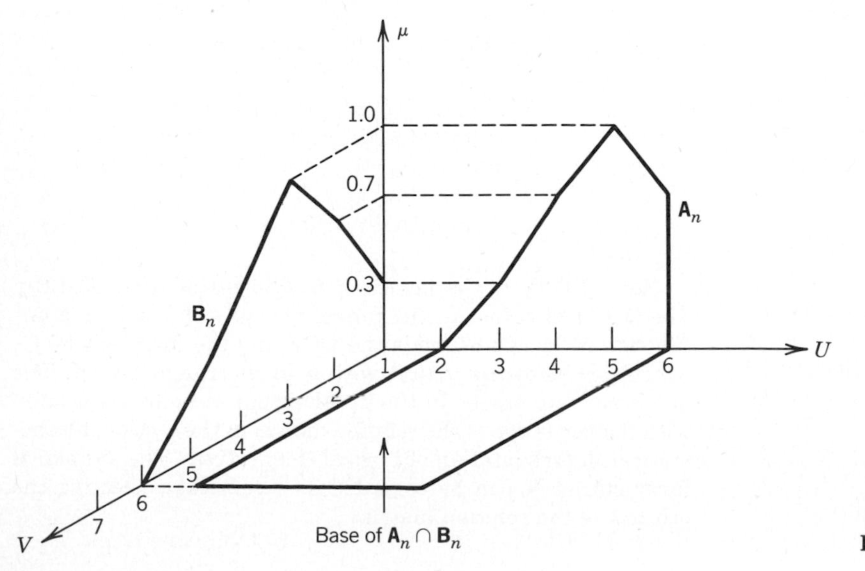

Figure 5. Geometrical interpretation of $\mathbf{A}_t \cap \mathbf{B}_t$.

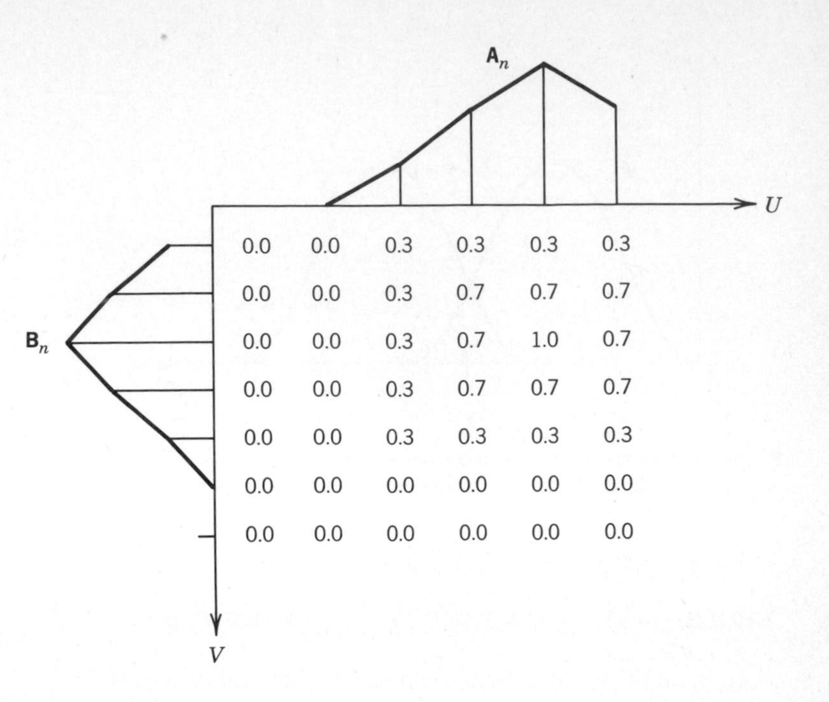

Figure 6. Relation matrix for $\mathbf{A}_t \cap \mathbf{B}_t$.

The THEN connective can be modeled by a number of functions "f" and research (1) has found "min" to work best here as well. Consider a rule

$$(\text{RULE})_n = \text{IF } (A \text{ is PM}) \text{ THEN } (C \text{ is NS}).$$

This could be represented by the relation matrix of Figure 6 with

1. "COND" replaced by "RULE" and
2. $x_{ik} = f(\mu_n(u_i), \mu_n(w_k))$.

The numerical example would also be furnished by Figure 5, using "min" for "f".

The $(\text{RULE})_n$ with two antecedents A and B will have three universes of discourse U, V and W and it can be represented by a three dimensional matrix with each element given by

$$x_{ijk} = \min(\mu_n(w_k), \min(\mu_n(u_i), \mu_n(v_j)))$$

which, from the properties of fuzzy sets, can be rewritten as

$$x_{ijk} = \min(\mu_n(w_k), \mu_n(u_i), \mu_n(v_j)) .$$

Decision Making

This is the process of finding the value of the consequent C_t resulting from one (A_t) or more (A_t, B_t, . .) specific inputs. This is done using the *compositional rule of inference* (CRI), which is (5) the term for *modens ponens* when applied to fuzzy logic: "o" is used to denote the CRI.

For just one input A_t with a fuzzy subset

$$\mathbf{A}_t = \{\mu_t(u_1), \ldots, \mu_t(u_i), \ldots, \mu_t(u_r)\}$$

one can write

$$\mathbf{C}_t = \mathbf{A}_t \text{ o } (\text{RULE})_n = \{x_1, \ldots x_k, \ldots x_k\}$$

where $x_k = \max_i (\min(\mu_t(u_i), \mu_n(u_i, w_k))$ and $\max_i$ means "maximum over all i". (ie, $i = 1$ to I).

For control engineering applications, the input has a single value resulting from the measurement process. In this case, A_t becomes

$$\mathbf{A}_t = \{0, 0, 0, \ldots .1, \ldots 0, 0, 0\}$$

or

$$\mathbf{A}_t = \{0 \text{ for all } u_i \text{ such that } i \neq q\}$$
$$\{1 \text{ for } u_i \text{ for which } i = q\}$$

Hence, x_k reduces to

$$\begin{aligned} x_k &= \max_{i=q}(\min(\mu_t(u_i), \mu_n(u_i, w_k)) \\ &= (\min(1, \mu_n(u_q, w_k)) \\ &= \mu_n(u_q, w_k) \\ &= \min(\mu_n(u_q), \mu_n(w_k)) \\ &= \min((\text{DOF})_n, \mu_n(w_k)). \end{aligned}$$

Here $(\text{DOF})_n$ is the *degree of fulfillment* of rule $(\text{RULE})_n$. The DOF has values in the range, typically of 0 to 1. A "min" function is formed between the DOF and the fuzzy subset C_n which has elements $\mu_n(w_k)$ with k in the range 1 to K. The above last two stages in the development of x_k are associated with the formation of the relation matrix in the section, Mathematical Representation of Logical Connectives. The simplified fuzzy subset $\mathbf{A}_t$ can be regarded as a template selecting the qth row of the relation matrix.

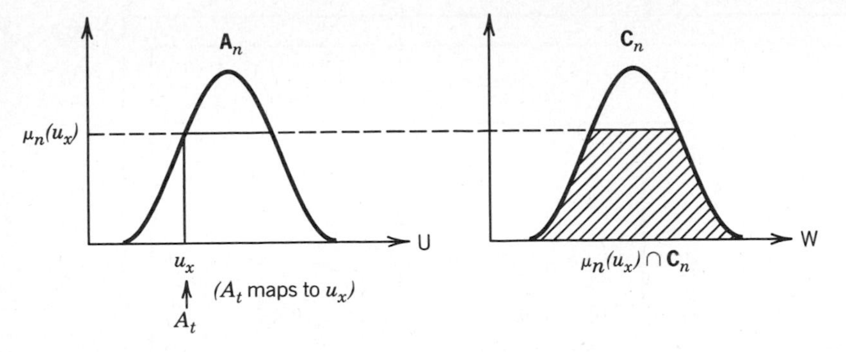

Figure 7. Drawing an inference from a rule—the "THEN" connective.

Figure 7 shows how the inference mechanism produces an output fuzzy subset on the universe of discourse W, given

1. a unique input A_t
2. the rule $(\text{RULE})_n$ = IF (A is A_n) THEN (C is C_n)
3. fuzzy subsets $\mathbf{A}_n$ and $\mathbf{C}_n$

This simplification of Zadeh's full theory (5) considerably reduces the computational effort as explained in greater detail in Ref. 6.

This can easily be extended to rules with more than one antecedent. Thus with two antecedents with universes of discourse U and V, the degree of fulfillment for rule $(\text{RULE})_n$ becomes

$$(\text{DOF})_n = \min(\mu_n(u_q), \mu(v_r))$$

where the unique, measured, inputs are now u_q and v_r.

Combining Rules

Rules are combined to form an *output action set* w which is a fuzzy subset on the output universe of discourse W. The connective can be regarded as ELSE or ALSO which is implemented by the "max" function as Figure 8 shows.

The output action set may be defuzzified by using a suitable heuristic (see The Nonlearning Controller).

ALGORITHMIC IMPLEMENTATION OF DYNAMIC FUZZY CONTROL

Overview

The fuzzy controller algorithm may be considered in two stages. The nonlearning algorithm embodies the original ideas of Mamdani and other workers (2–4, 6, 7). It uses a fixed set of rules; these and the inputs to the algorithm are processed by an inference mechanism to determine a suitable output; this is similar to the conventional idea of an expert system. Subsequent studies (1, 8) introduced an additional learning mechanism, whereby new rules could be created and old ones updated in response to new experience. This modified algorithm is termed the *self-organizing controller* (SOC) and is summarized in Figure 9.

The Nonlearning Controller

This comprises three major blocks: (1) input, (2) controller or inference mechanism, (3) output.

For control applications the input section usually uses the error between setpoint and process output to give both the error and rate of change of error. These two quantities are mapped to (typically) one of 13 discrete levels using gains GE and GCE as scaling factors to yield the discrete quantities E and CE.

The controller uses E and CE to index the rule base in which the numbers associated with the rules are positions on the output universe of discourse W. The precise way the rules are interpreted depends on the fuzzy subsets which are assumed to be associated with them. For control applications with short time constants (eg, robotics) the almost triangular shape

$$\{\ldots, 0, 0.3, 0.7, 1.0, 0.7, 0.3, 0, \ldots\}$$

is usually adequate and gives a fast processing time. This means that rules one index distant ($E\pm1,C\pm1$) from the main

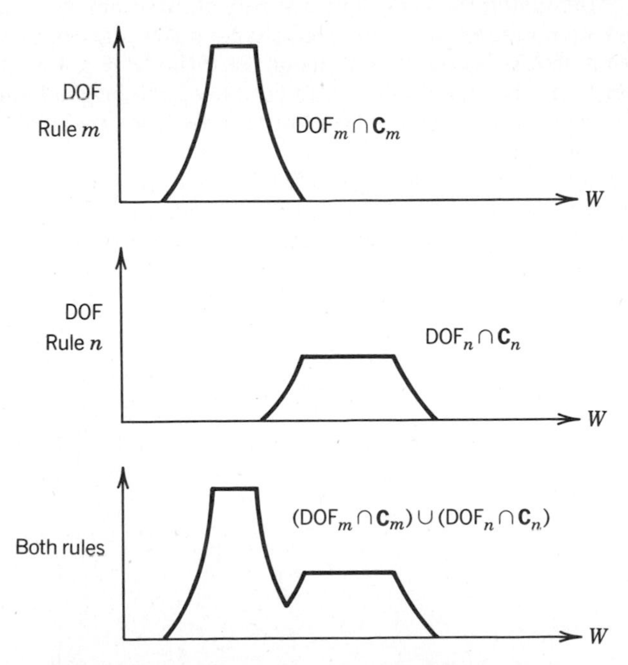

Figure 8. Combining rules—the "ELSE" connective.

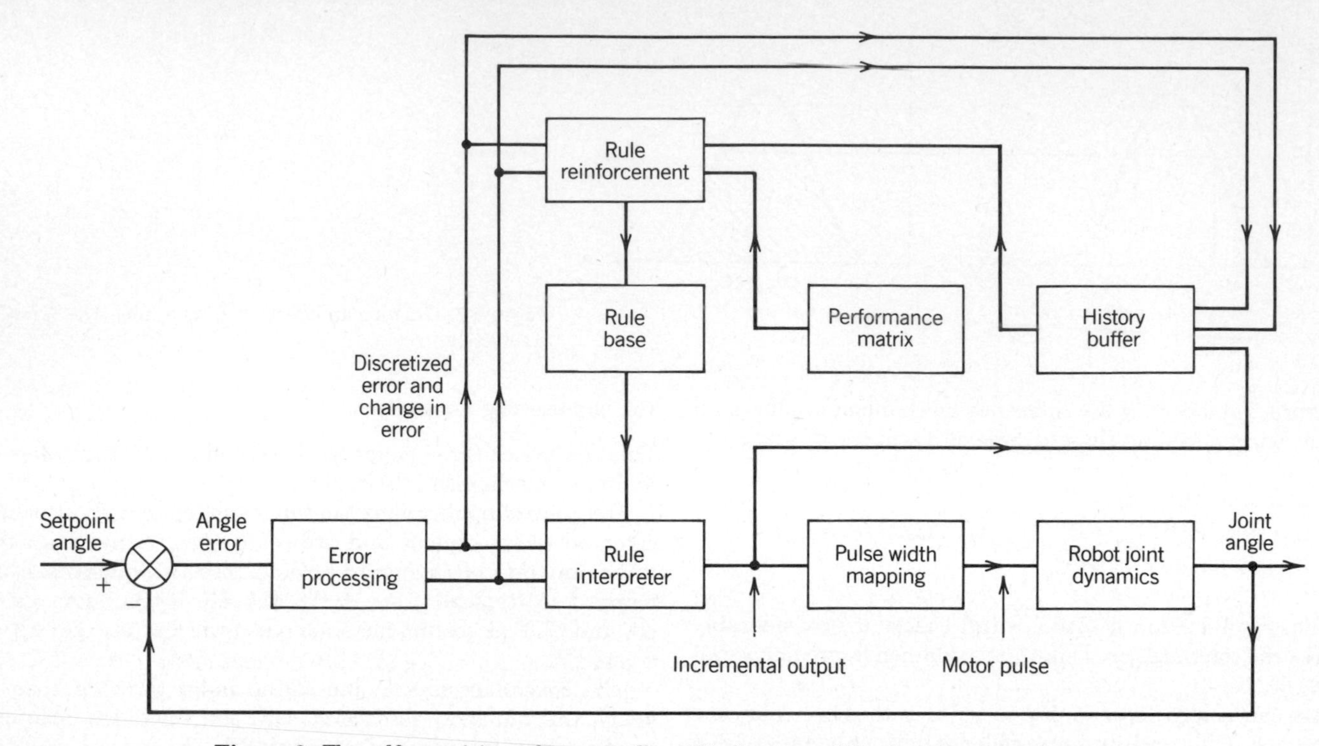

Figure 9. The self-organizing robot controller (20). Courtesy of North Holland Publishing Co.

rule have a degree of fulfillment (DOF) of 0.7 on the output action set (output fuzzy subset) being formed on the output universe of discourse W. For rules two indexes distant ($E\pm2, CE\pm2$), the DOF is 0.3. In this, way a shape (eg, Figure 8) with a maximum of 4 levels may be obtained. Figure 10 shows a typical rule base generated using such a shape. Other, almost Gaussian, shapes have been used for control applications less demanding of time (3, 4).

When using the compositional rule of inference, the "min" operation results in a truncated shape of the fuzzy subset as a contribution to the output action set if the DOF is less than unity. Another possibility (2, 6) is to keep the original shape and width of the fuzzy subset but to scale down its height by a factor equal to the DOF.

The defuzzification of the controller output (the output action set) can be achieved by the mean of maxima (MM) method (21, 22) or by the center of gravity (CG) method (3, 23). The MM method takes no account of rules triggered below the extreme levels and is thus considered as not being truly fuzzy. The CG method rarely considers control actions at the extremes of the action range. The MM method produces a better transient response, the CG method a better steady state response (24).

The resultant, unique quantity, is often multiplied by a scaling factor (GO) and is then, in many controllers, fed into an accumulator (integrator) whose output is, in turn, scaled and fed to the process being controlled.

The Learning Capability

The above system assumes that knowledge of how to control the system is already available. This may not always be so and thus calls for a method which can guide the system to the desired state of zero error and zero rate of change of error.

E \ *CE*	−6	−5	−4	−3	−2	−1	0	+1	+2	+3	+4	+5	+6
−6	.	.	.	.	.	.	−6	.	.	.	.	.	.
−5	.	.	.	.	.	−6	.	.	.	.	.	.	.
−4	.	.	.	.	.	.	.	.	.	.	.	.	.
−3	.	.	.	.	.	.	−3	.	.	.	.	.	.
−2	.	.	.	.	.	.	−5	+1	.	.	.	.	.
−1	.	.	.	+1	+1	−1	−2	+2	.	.	.	.	.
0	.	.	+2	+4	+3	.	0	+5	+5	.	.	.	.
+1	.	.	.	.	+5	.	.	+6	.	.	.	.	.
+2	.	.	.	.	.	+6	.	.	.	.	.	.	.
+3	.	.	.	.	.	+6	+6	.	.	.	.	.	.
+4	.	.	.	.	.	.	.	.	.	.	.	.	.
+5	.	.	.	.	.	.	.	.	.	.	.	.	.
+6	.	.	.	.	.	.	.	.	.	.	.	.	.

Figure 10. Example of a rule base (20). Courtesy of North Holland Publishing Co.

		CE												
		−6	−5	−4	−3	−2	−1	0	+1	+2	+3	+4	+5	+6
	−6	0	0	−1	−2	−3	−4	−6	−6	−6	−6	−6	−6	−6
	−5	0	0	0	−1	−2	−3	−4	−4	−5	−5	−6	−6	−6
	−4	0	0	0	0	−1	−2	−3	−3	−4	−5	−5	−6	−6
	−3	+1	0	0	0	0	−1	−2	−2	−3	−4	−5	−5	−6
	−2	+2	+1	0	0	0	0	−1	−1	−2	−3	−4	−5	−6
	−1	+3	+2	+1	0	0	0	−1	−1	−1	−2	−3	−4	−5
E	0	+4	+3	+2	+1	+1	0	0	0	−1	−1	−2	−3	−4
	+1	+5	+4	+3	+2	+1	+1	+1	0	0	0	−1	−2	−3
	+2	+6	+5	+4	+3	+2	+1	+1	0	0	0	0	−1	−2
	+3	+6	+5	+5	+4	+3	+2	+2	+1	0	0	0	0	−1
	+4	+6	+6	+5	+5	+4	+3	+3	+2	+1	0	0	0	0
	+5	+6	+6	+6	+5	+5	+4	+4	+3	+2	+1	0	0	0
	+6	+6	+6	+6	+6	+6	+6	+6	+4	+3	+2	+1	0	0

Figure 11. Yamazaki's performance index table (20). Courtesy of North Holland Publishing Co.

This can be done by using a performance index table or matrix which is itself derived from linguistic rules of the form

"IF (E is X AND CE is Y) THEN Z"

where Z is to be read as "the process output should really be different by P units" and the quantity P can be positive or negative. Two popular performance matrices are those found in Refs. 8 and 23; the latter is shown in Figure 11 and was generated using the triangular fuzzy subset described above.

A FIFO buffer is used to contain triplets (E_i,CE_i,CO_i), corresponding to the system state i samples in the past, to a depth of 8 or 10 samples; CO_i is the incremental controller output fed to the integrator-accumulator.

A simple incremental model is assumed for the SISO process, the model consisting of a gain, usually unity, and a time delay of m samples.

Thus, the performance index table yields a desired reinforcement P on basis of the present inputs (E,CE). The value of the rule at (E_m,CE_m) in the rule matrix/table is then changed to (CO_m+P).

A modification to this learning process considered the mean of three past controller outputs about the past sample m (23).

The Multi-Input-Multi-Output (MIMO) Form of SOC

In the MIMO case there is one rule matrix/table for each controller output (7). For an I-input MIMO system, the nth rule for the kth incremental controller output would be

$$
\begin{aligned}
(\text{RULE})_{nk} = \text{IF } & (E_1 \text{ is } E_{n1}) \text{ AND } (CE_1 \text{ is } CE_{n1}) \text{ AND} \\
& (E_2 \text{ is } E_{n2}) \text{ AND } (CE_2 \text{ is } CE_{n2}) \text{ AND} \\
& \ldots \\
& (E_1 \text{ is } E_{nI}) \text{ AND } (CE_1 \text{ is } CE_{nI}) \\
\text{THEN } & (CO_k \text{ is } CO_{nk}).
\end{aligned}
$$

Dimensional considerations preclude the use, for the rule base (matrix or table), of the array data structure often used in the SISO case. Rules can be stored and searched, either in a sequential fashion (25), or in threaded linked list form (26).

Learning can be performed by a MIMO controller, (22), by using the performance table of the SISO controller to relate a given controller input pair (E_i,CE_i) to the desired change P_i in the ith process output. A simple, incremental, (Jacobian) model of the process is used to take account of cross-coupling and time delay effects of the process. This model can then be used to relate each P_i to changes in the appropriate rules, one in each rule base, in the manner described for the SISO system. It should be stressed that the model does not demand a detailed knowledge of the process being controlled and the values of the parameters of the model are not critical to the successful operation of the controller.

Application to the Control of Robot Dynamics

The first major industrial application of fuzzy control was to cement kilns (1–4) and used the nonlearning algorithm. The controller applied to robot dynamics (19, 24, 26–28) uses the *Self-Organizing* Fuzzy algorithm to enable the robot to adapt to changes in loading and robot configuration. There are additional differences. (*1*) The algorithm acts in the direct forward path of each robot actuator control loop, the cement kiln controller merely supplying time-varying setpoints to the PID (three-term) controllers of the kiln actuators. (*2*) The sampling times for the robot need to be of the order of 20 ms compared to several seconds for the cement kiln. (*3*) The integration effect of the robot motor makes the output integrator of the SOC algorithm unnecessary.

Setpoint tracking and step input experiments have been performed (24, 26, 27) using a revolute joint robot with a reach of 3.2 ft (1 m) and a lifting capability of 3.3 lb (1.5 kg). The joints were driven as separate decoupled systems with one SOC per joint. The fuzzy controller compared very favorably with conventional PID controllers. Figure 12 and Table 1 show a load-cycling experiment. The robot moved in a vertical plane, the lower (shoulder) movement being driven by a SOC or PID algorithm and the upper (arm) movement being driven by a separate (NE544) controller chip as shown in Figure 12k. The tip load was 2.3 lb (1.035 kg) and the sampling period 20 ms. The responses (a–h) for the shoulder movement are shown close to the final setpoint, the vertical scale being 18 units per degree. It can be seen how the response improved with learning (eg. Figure 12 a and b). Figure 12 c and f show the offset that appeared when learning was inhibited. Figure 12 i and j for the PID controller with constant gains show how the dynamics of the robot arm system changed with added load.

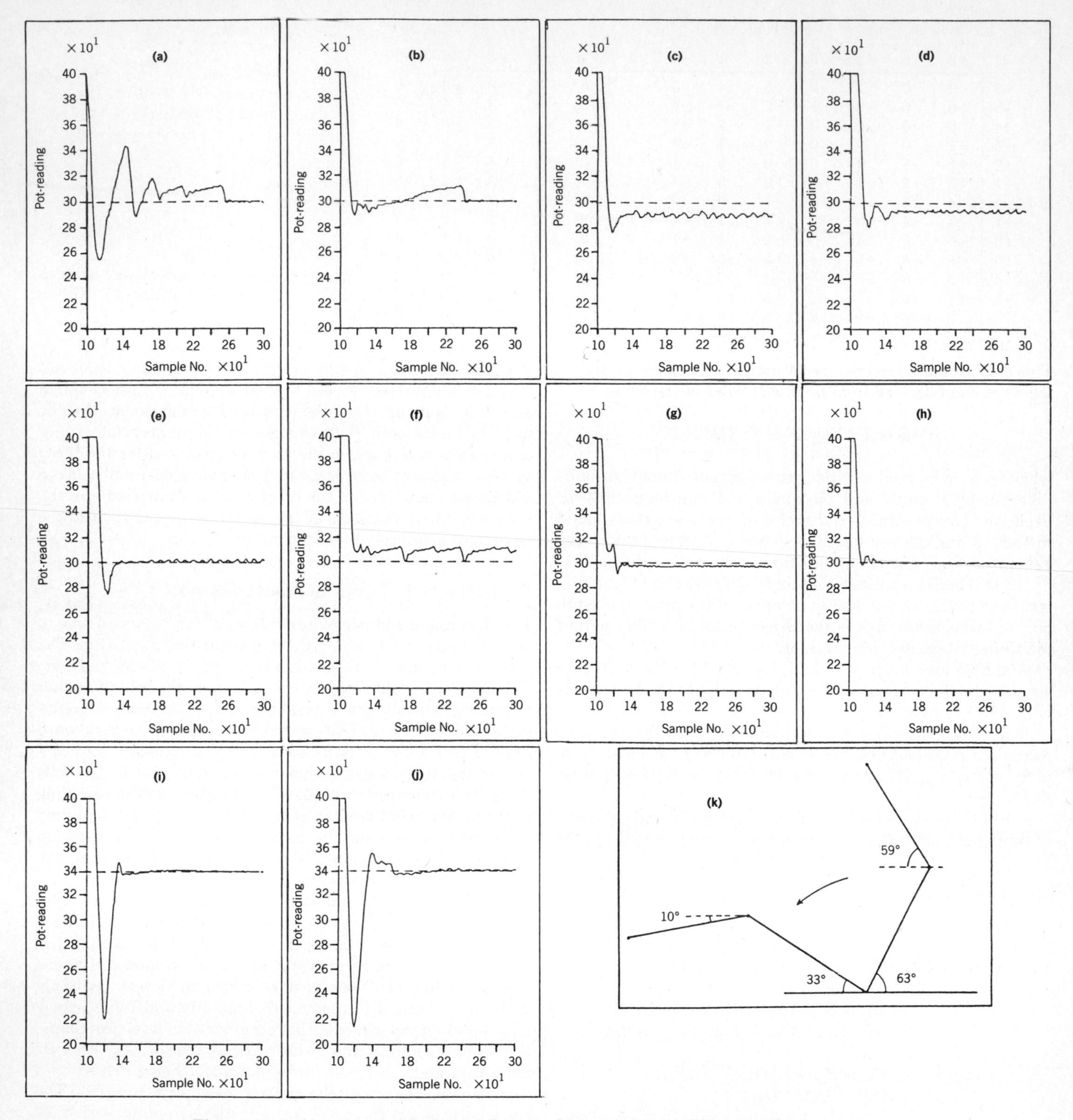

Figure 12. Load cycling experiments: a-h, using SOC; i, j, using a PID controller; k, range of angular movements (set point inputs) (20). Courtesy of North Holland Publishing Co.

Enhancements to the Fuzzy Controller

SOC has many parameters which could effect its performance and have been the subject of numerous investigations.

Ref. 29 indicates that the shape of the fuzzy sets do not seem to be very important, since SOC will tend to correct for deficiencies in these. However, it was shown (30) that (*1*) the output accuracy of the system can be improved by considering the input variables as continuous as opposed to discrete, (*2*) the inclusion of an error integral term improves convergence

Table 1. Details for Load Cycling Tests

Figure 12, pt	Controller	Learn	Load
a	fuzzy	y	0
b	fuzzy	y	0
c	fuzzy	n	y
d	fuzzy	y	y
e	fuzzy	y	y
f	fuzzy	n	o
g	fuzzy	y	o
h	fuzzy	y	o
i	PID		y
j	PID		y

and (3) that widening the dynamic range of control will result in a quicker process response.

Methods have been investigated for choosing controller parameters to improve system performance (30–32). The stability of the fuzzy controller has been investigated in Refs. 33 and 34, the latter using describing functions.

Constraints on the formation of rules to prevent "rule threshing," the continued change in neighboring rules near the controller setpoint with no change in controller performance, has been studied in Ref. 25. The performance of the robot SOC has been improved by keeping rule formation away from the corners of the rule base (26).

The question of introducing a finer action near the setpoint has been investigated in Ref. 35 who have proposed using a separate rule base close to the setpoint. Scharf and Mandic (26), for their robot controller, have achieved a fine action by making the output gain a function of the controller input error. The possibility of automatic parameter adjustment was considered (30). For robot control, the possibility of using SOC (1) with force feedback (eg, by measuring actuator armature current) and (2) as a supervisory PID controller has also been considered (26). The effect of controller input noise has been investigated (36, 37).

The speed of execution of the fuzzy controller is especially important in robot control applications where loop sampling times may have to be lower than 10 milliseconds. It has been shown (28) that a SISO SOC system occupies only about 2kbytes of machine code and has a response time not exceeding 2.5 ms when an 8 MHz 68000 microprocessor is used. A p-MOS implementation of a nonlearning fuzzy controller with a response time of 670 ns is described in Refs. 38 and 39; this has 7 quantized input levels, and uses MIN-MAX operations and the center of gravity method for defuzzification of the output.

Variants of the inference mechanism, the defuzzification process, the performance index matrices and the learning process have been discussed above (see The Nonlearning Controller and The Learning Capability).

HIGHER LEVEL CONTROL OF ROBOTS AND MANUFACTURING PROCESSES

Introduction

Robotics and manufacturing applications of fuzzy logic have often been directed to control above the level of the intimate control loop. These applications are in the areas of robot dynamics, robot planning, natural language understanding and modeling of the human operator.

Control of Robot Dynamics

Saridis (40, 41) has proposed a robot dynamic controller (Fig. 13) with a structure different to that studied by Mamdani and co-workers (6, 26).

At the lowest level a mathematical self organizing controller (SOC) works on the error between angular setpoint and actual output for the movement of each link in such a way as to minimize a function of the squares of error and actuator drives. The feedback control for the ith joint movement is given by

$$u_i(z) = K_i f_L(z_i) + C_i f_N(\mathbf{z})$$

where z_i is the output from the ith joint, $\mathbf{z}$ is the vector of all such outputs, f_L is a linear function and f_N is a nonlinear function absorbing the dynamic cross-coupling effects from the other joints. K and C are constants.

The intermediate control level is the coordinator which forms the outer control loop and takes account of the dynamic cross coupling between the links. A fuzzy learning automaton was developed (42) that takes as inputs the joint outputs and their rate of change (ie, z, $\dot{z}$ or Θ, $\dot{\Theta}$); in accordance with a minimum energy criterion, the appropriate updates to K and C are evaluated for each joint.

The highest level is the organizer which forms the link between the desired angular setpoints and linguistic statements or, in the case of a prosthetic device, the electromagnetic signals emanating from voluntary movement in the patient's

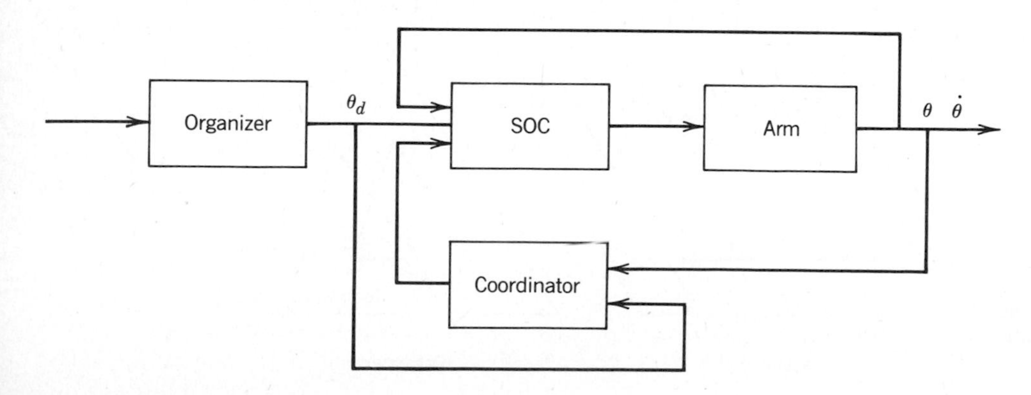

Figure 13. Saridis' controller.

remaining musculature. For this level, Saridis proposes the use of fuzzy grammars (43) for the generation of command strings for the organization of the arm's movement (Fig. 13).

Simulation studies (41) suggest that the number of learning iterations can be as high as 200 as opposed to 3 or 4 associated with the controller studied in Refs. 6 and 27 and discussed in the section, Application to the Control of Robot Dynamics.

Fuzzy Programs and Robot Planning

The use of a program of fuzzy instructions has been reported in Ref. 44. This is a development from earlier work (45). The fuzzy program is obeyed by a three legged "inchworm robot" moving in "real space." Reference is made (Fig. 14) by the interpreter of the fuzzy program to a "map" data structure that contains the robot's position and can give information on the robot's immediate surroundings.

There are four fuzzy instructions: (1) GO ABOUT *n* STEPS; (2) GO TO *x*; (3) GO ABOUT *n* STEPS TO *x*; (4) TURN TO THE RIGHT/LEFT.

There are two supplementary instructions which give initial values to the system: (5) HEAD EAST/WEST/NORTH/SOUTH; (6) START FROM *x*.

A fuzzy instruction contains several nonfuzzy machine instructions, each associated with a particular degree of fulfillment. When a fuzzy instruction is interpreted, the machine instruction with the highest degree of fulfillment (DOF) is selected first ("Max Method"). Machine instructions with a DOF below a certain threshold are not chosen as the interpretation of the fuzzy instruction. Backtracking to the previous instruction occurs if the present fuzzy instruction yields no executable machine instruction. A program can fail if, at any point, no executable machine instruction is found.

The authors substantiate the functioning of their programs with the aid of several simulations and forsee the use of their ideas in any situation where a robot is commanded from a distance (eg, space exploration).

Ref. 46 gives some initial ideas on robot planning. A subset of English to be mapped by a semantic interpreter into an "intermediate representation language" (IRL) which is in turn "compiled" into QLISP is envisioned. In its simplest form, an IRL consists of a sequence of fuzzy vectors, each vector containing a fuzzy length and direction. H is a hint, a selection of fuzzy vectors, expressed in the IRL. The compiler tests the hint on a maze M, and, as a check on robustness, the physical perturbations $M_1, M_2, \ldots$ of the maze M. The compiler keeps a track of failures and can readjust failure levels; backtracking is also possible. Such a compiler should do significantly better than the depth and breadth first searches which are part of the standard toolkit for artificial intelligence.

Natural Language Understanding

Saridis (40) has proposed the use of fuzzy grammars for his "organizer" which acts as a linguistic front-end to his robot controller.

Sondheimer's (47) study recognizes that the designer of systems for the vocal control of mechanical devices faces a number of difficulties in allowing for references to the position, orientation and direction of motion of objects and actions in space.

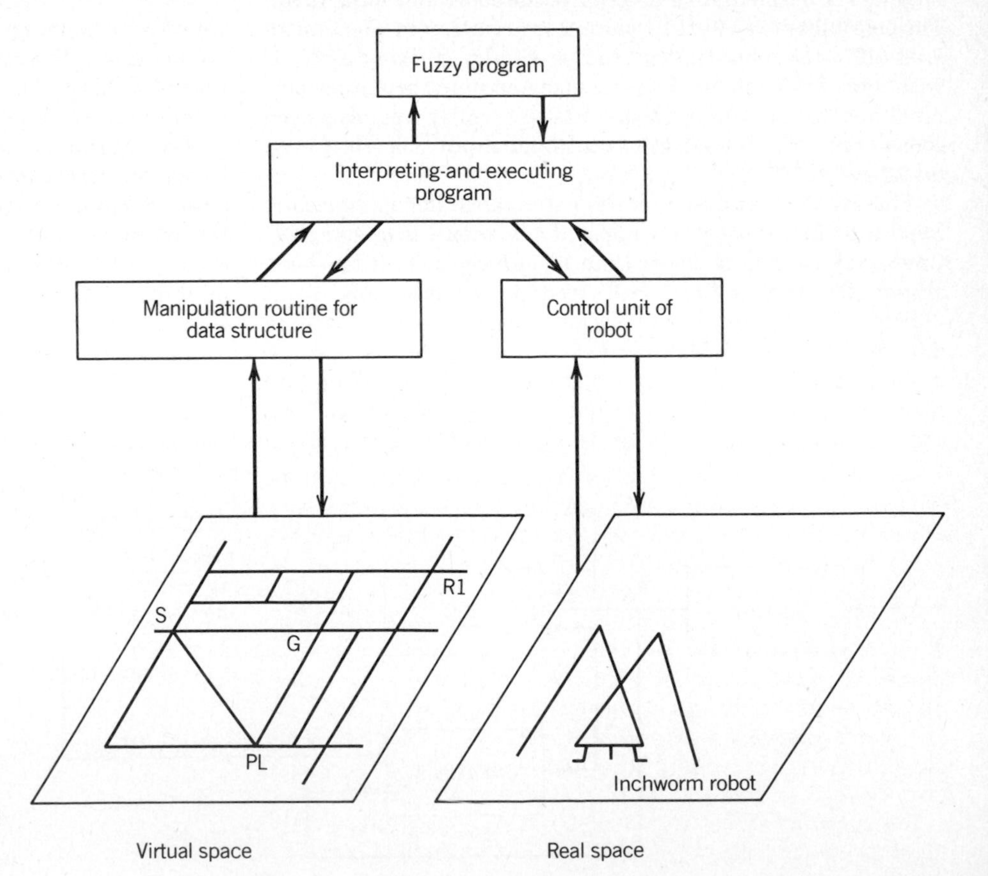

Figure 14. Using a fuzzy program for robot guidance (44). Courtesy of Hemisphere Publishing Corporation, New York.

The conclusion is that the best way of allowing for the ambiguous interpretation of spatial references is to restrict the syntactic and semantic structure of spatial references even though trade-offs between naturalness and expressiveness would have to be introduced. Ref. 47 recognizes the value of linguistic symbols.

"PRUF," which is a meaning representation language for natural languages has been introduced (48). The PRUF representation of the English object description "The red fish which is quite small" becomes

```
(OBJECT (category    = FISH)
        (n(color)    = RED)
        (n(size)     = QUITE SMALL)
        (determiner = definite))
```

"n" is a fuzzy descriptor; thus n(color) is characterized by a possibility distribution over perceived colours of fish and the n(size) descriptor contains the linguistic modifier "quite" which modifies the possibility distribution associated with the fuzzy linguistic value "small."

PRUF is used in the implementation of a communication model between two people about visual patterns (49). The AI-Language "L-Fuzzy" is used (50); this is a dialect of the LISP based language FUZZY and can directly represent fuzzy sets with linguistic modifiers. Fuzzy sets are interpreted as possibility distributions and the linguistic modifiers are represented by procedures which modify these distributions.

Experiments have been conducted with a human operator giving to a robot ambiguous instructions in terms of fuzzy membership and linguistic vagueness and concerned with distance, direction, catch, and place (51). The robot also has a knowledge base which relates requirements of speed, position and gripper pressure to the handling of objects such as paper cups, blocks, cylinders and spheres which are part of the robot's world. The robot is able to recognize the instructions and select the movements appropriate to the objects it is dealing with. Work is now in hand to include vision processing in this system.

Fuzzy Model of the Human Operator

The modeling of a human operator's response to fault conditions in the turning process are described in Ref. 52; the system is a nonlearning controller. Each linguistic control strategy is an implication to recover the cutting status in an undesirable situation in the generation of a chip (workpiece), (eg, undesirable chip flow, chatter, and tool fracture). The fuzzy implications are written: if P then Q; if P then R.

The antecedent P is the status in chip generation (ie, chip length); this can be: snarled, excessively/very/normally long, all right, or normally/excessively short. The consequent Q is the required change in feed rate and can be positive very_big/big/medium/small, zero or negative small/medium. The auxiliary adjustment R represents the change in cutting speed and can assume values negative big/medium/small/very_small or positive small/medium.

The authors present results (52) but do not provide a definite evaluation of them; however, their rules are derived from linguistic statements and they give a discussion of this method of modeling the human operator.

A nonlearning fuzzy controller is used in the study of parking a car (53). Fuzzy rules are derived, not from linguistic statements, but directly from a human's actions in reversing a model battery powered car (15.7 in. long and 7.7 lb in weight) into a garage. About 500 samples are taken of the human's performance; these samples relate the 2-D coordinates x, y and the car's angle Θ to f, b and s which are the required angles of the front wheels moving forwards and backwards and the speed which is bounded between $\pm$ 10 units.

Fuzzy rules are of the form:

$$R^i\text{: } IF\ x_1 \text{ is } A_1^i,\ x_2 \text{ is } A_2^i, \ldots\ x_n \text{ is } A_n^i, \\ \text{THEN } u^i = p_0^i + p_1^i x_1 + \cdots + p^i x_n$$

where x_j are the input variables, u^i is the output from the ith implication, A_j^i are fuzzy variables and p_j^i are the coefficients of a linear equation.

There are, respectively, 18, 16, and 6 rules for forward and backward steering and speed regulation. The authors have reported some experimental success.

Success has been reported in modeling the behavior of a human operator controlling a servo system with a saturation nonlinearity (54). The main set of rules is of the form.

$$\text{IF } e \text{ is } A \text{ AND } ce \text{ is } B \text{ THEN } u \text{ is } c$$

where e and ce are error and its rate of change, the error being between the output of the controlled system and the operator's own input. "u" is the change in input to the controlled system, via a third order smoothing model of the neuromuscular system. An additional set of rules relates separately $|e|$ and $|ce|$ to the sampling period inherent in the neuro-muscular model. These rules are derived from measurements of the operator's pursuit tracking behavior of a step input to a nonlinear servosystem.

The authors confirmed that the operator's control strategy could be described by a set of linguistic control rules and they also established the grade which each control rule contributes to the generation of the output variable "u." They then refined the model of the human operator and verified this experimentally.

Conclusions

Fuzzy logic provides a mechanism for the algorithmic implementation of linguistic statements; it is thus a means of dealing with systems which do not yield readily to accurate characterization. Areas of application include robot dynamics and motion planning, natural language understanding, and modeling of the actions of the machine-tool operator. Fuzzy reasoning has already produced some interesting possibilities when coping with these systems and the range of applications is growing steadily.

BIBLIOGRAPHY

1. E. H. Mamdani, J. J. Ostergaard, and E. Lembessis, "Use of Fuzzy Logic for Implementing Rule-Based Control of Industrial Processes," in H.-J. Zimmermann, L. A. Zadeh, and B. R. Gaines,

eds., *TIMS/Studies in the Management Sciences 20,* Elsevier Science Publishers B. V., North Holland, Amsterdam, 1984, pp. 429–445.

2. I. G. Umbers and P. J. King, "An Analysis of Human Decision Making in Cement Kiln Control and the Implications for Automation," *Int. J. Man-Machine Studies* **12,** 11–23 (1980).
3. I. P. Holmblad and J. J. Ostergaard, "Fuzzy Logic Control: Operator Experience Applied in Automatic Process Control," *FLS Review (F. L. Smidth & Co., Valby, Copenhagen)* **45,** 11–16 (1981).
4. P. M. Larsen, "Industrial Applications of Fuzzy Logic Control," *Int. J. Man-Machine Studies* **12,** 3–10 (1980).
5. L. A. Zadeh, "Outline of a New Approach to the Analysis of Complex Systems and Decision Processes," *IEEE Trans. SMC-3* **1,** 28–44 (1973).
6. E. H. Mamdani, "Process Control Using Fuzzy Logic," in *Designing for Human-Computer Communication,* Academic Press, London, 1983, pp. 311–336.
7. S. Assilian and E. H. Mamdani, "An Experiment in Liguistic Synthesis with a Fuzzy Logic Controller," *Int. J. Man-Mach. Stud.* **7,** 1–13 (1974).
8. T. J. Procyk and E. H. Mamdani, "A Linguistic Self-Organising Process Controller," *Automatica* **15,** 15–30 (1979).
9. O. Yagishita, O. Itoh, and M. Sugeno, "Application of Fuzzy Reasoning to the Water Purification Process," in M. Sugeno, ed., *Industrial Applications of Fuzzy Control,* North Holland, Amsterdam, 1986, pp. 19–40.
10. R. M. Tong, M. B. Beck, and A. Latten, "Fuzzy Control of the Activated Sludge Wastewater Treatment Process," *Automatica* **16,** 659–701 (1980).
11. N. K. Sinha and J. D. Wright, "Application of Fuzzy Control to a Heat Exchanger," in *Special Interest Discussion Section on Fuzzy Automata and Decision Processes, 6th IFAC World Congress,* Boston, Mass., Aug. 1975.
12. D. A. Rutherford and G. A. Carter, "A Heuristic Adaptive Controller for a Sinter Plant," *Proceedings of the 2nd IFAC Symposium in Mining, Mineral and Metal Processing,* Johannesburg, 1976.
13. S. Assilian, *"Artificial Intelligence in the Control of Real Dynamic Systems,"* PhD Thesis, Queen Mary College, University of London, 1974.
14. J. van Amerongen, H. R. van Nauta Lemke, and J. C. T. van der Veen, "An Autopilot for Ships Designed with Fuzzy Sets," in H. R. van Nauta Lemke, ed., *Digital Computer Applications to Process Control,* IFAC and North Holland, Amsterdam, 1977, pp. 479–87.
15. Y. Murayama, T. Terano, S. Masui, and N. Akiyama, "Optimising Control of a Diesel Engine," in Ref. 9, pp. 63–72.
16. S. Yasanobu and S. Miyamoto, "Automatic Train Operation by Predictive Fuzzy Control," in Ref. 9, pp. 1–18.
17. S. Murakami and M. Maeda, "Automobile Speed Control System Using a Fuzzy Logic Controller," in Ref. 9, pp. 105–24.
18. S. Daley and K. F. Gill, "A Design Study of a Self-Organising Fuzzy Controller," *Proc. Instn. Mech. Engrs,* **200** (C1) 59–69 (Jan. 1986).
19. E. M. Scharf, N. J. Mandic, E. H. Mamdani, "A Self-Organising Algorithm for the Control of a Robot Arm," *International Journal of Robotics & Automation,* Vol. 1, No. 1, 1986, pp. 33–41.
20. Ref. 9, pp. 44, 48, 51, 52.
21. T. R. Andersen and S. B. Nielsen, *"An Efficient Single Output Fuzzy Control Algorithm for Adaptive Applications,"* Servo Laboratory, Technical University of Denmark, 1982.
22. T. J. Procyk, *"A Self-Organising Controller for Dynamic Processes,"* PhD Thesis, Queen Mary College, University of London, 1977.
23. T. Yamazaki and E. H. Mamdani, "On the Performance of a Rule-Based Self-Organising Controller," *IEE Conference on the Applications of Adaptive and Multivariable Control,* Hull, UK, July 19–21, 1982, pp. 50–55.
24. E. M. Scharf and N. J. Mandic, *Development of Learning Algorithms for Applications in Control of Robot Arms,* SERC Report GR/B 7940.0, April 1984.
25. E. Lembessis, *"Dynamic Learning Behaviour of a Rule-Based Self-Organising Controller,"* PhD Thesis, Queen Mary College, University of London, Sept. 1984.
26. N. J. Mandic, E. M. Scharf and E. H. Mamdani, "The Practical Application of a Heuristic Fuzzy Rule-Based Controller to the Dynamic Control of a Robot Arm," *IEE Proceedings-D, Part D,* **132** (4), 190–203 (July 1985).
27. E. M. Scharf and N. J. Mandic, "The Application of a Fuzzy Controller to the Control of a Multi-Degree of Freedom Robot Arm," in Ref. 9, pp. 41–61.
28. E. M. Scharf, E. H. Mamdani, and M. Krishnamurthy, *The Application of a Fuzzy Rule-Based Self-Organising Controller to a Demanding Industrial Robotics Requirement,* SERC Report GR/C 9136.9, March 1986.
29. K. Sugiyama, *Comparison of Various Fuzzy Logic Controller Algorithms,* Internal Report, Department of Electrical and Electronic Engineering, Queen Mary College, University of London, 1984.
30. K. Sugiyama, *"Analysis and Synthesis of the Rule-based Self-Organising Controller,"* PhD Thesis, Queen Mary College, University of London, June 1985.
31. M. Braae and D. A. Rutherford, "Selection of Parameters for a Fuzzy Logic Controller," *Fuzzy Sets and Systems,* Vol 2, North Holland, 1979, pp. 185–199.
32. S. Daley and K. F. Gill, "A Design Study of a Self-Organising Fuzzy Controller," *Proc. Instn. Mech. Engrs,* **200** (C1), 59–69 (Jan. 1986).
33. R. Tong, "Some Properties of Fuzzy Feedback Systems," *Memo UCB/ERL M78,* University of California, Berkeley, Calif., 1978.
34. W. J. M. Kickert and E. H. Mamdani, "Analysis of a Fuzzy Feedback Controller," *Fuzzy Sets and Systems,* Vol. 1, 1977.
35. T. Yamazaki and M. Sugeno, "A Microprocessor Based Fuzzy Controller for Industrial Purposes," in Ref. 9, pp. 231–240.
36. M. J. Rickwood, "The Self-Organising Controller and its Performance when Controlling Noisy Systems," *Final Year Project,* Department of Electrical and Electronic Engineering, Queen Mary College, University of London, May 1983.
37. R. Tanscheit and E. M. Scharf, "Experiments with the Use of a Rule-Based Self-Organising Controller for Robotics Applications," *Proceedings of the First IFSA Congress,* Palma, Mallorca, July 1985.
38. T. Yamakawa and H. Kabuo, "Synthesis of Membership Function Circuits in Current Mode and its Implementation in the p-MOS Technology," *Proc. 2nd Fuzzy System Symposium,* Japan, June 16–18, 1986, pp. 115–121.
39. T. Yamakawa, "High-Speed Fuzzy Controller Hardware System," *Proc. 2nd Fuzzy System Symposium,* Japan, June 16–18, 1986, pp. 122–130.
40. G. N. Saridis, "Self-Organising Control and Applications to Trainable Manipulators and Learning Prostheses," *Proc. 6th IFAC World Congress,* Boston, Mass. Pt I, August 1975, pp. 50.4/1–9.
41. G. N. Saridis and H. E. Stephanou, "Fuzzy Decision-Making in Prosthetic Devices," in M. M. Gupta and G. N. Saridis eds., *Fuzzy Decision Making in Prothetic Devices,* Elsevier, New York, 1977, pp. 387–402.
42. W. G. Wee and K. S. Fu, "A Formulation of Fuzzy Automata and

its Application as a Model of Learning Systems," *IEEE Transactions on Systems Science and Cybernetics,* SSC-5, 3, July 1969, pp. 215–223.
43. E. T. Lee and L. A. Zadeh, "Note on Fuzzy Languages," *Infor. Sci.* **1** (6) 421–434 (Oct. 1969).
44. M. Uragami, M. Mizumoto and K. Tanaka, "Fuzzy Robot Controls," *Journal of Cybernetics* **6,** 39–64 (1976).
45. K. Tanaka and M. Mizumoto, "Fuzzy Programs and their Execution," in L. A. Zadeh and co-eds., *Fuzzy Sets and their Application to Cognitive and Decision Processes,* Academic Press, New York, 1975, pp. 41–76.
46. J. A. Goguen, "On Fuzzy Robot Robot Planning," in Ref. 45, pp. 429–447.
47. N. K. Sondheimer, "Spatial Reference and Natural-Language Machine Control," *Int. J. Man-Mach. Stud.* **8,** 329–336 (1976).
48. L. A. Zadeh, "PRUF - A Meaning Representation Language for Natural Languages," *Int. J. Man-Machine Studies* **10,** 395–460 (1978).
49. C. Freksa, "Communication about Visual Patterns by Means of Fuzzy Characterizations," *Proceedings of the XXIInd International Congress of Psychology,* Leipzig, GDR, July 1980.
50. C. Freksa, *L-FUZZY - an AI Language with Linguistic Modification of Patterns,* ERL Memo M80/10, University of California at Berkeley, 1980.
51. K. Hirota, Y. Arai, and W. Pedrycz, "Robot Control Based on Membership and Vagueness," in M. M. Gupta and co-eds., *Approximate Reasoning in Expert Systems,* North Holland, Amsterdam, 1985, pp. 621–635.
52. Y. Sakai and K. Ohkusa, "A Fuzzy Controller in Turning Process Automation," in Ref. 9, pp. 139–152.
53. M. Sugeno and K. Murakami, "An Experimental Study of Fuzzy Parking Control Using a Model Car," in Ref. 9, pp. 125–138.
54. K. Matsushima and H. Sugiyama, "Human Operator's Fuzzy Model in Man-Machine System with a Nonlinear Controlled Object," in Ref. 9, pp. 175–186.

CONTROL STRATEGIES

J. C. Wang
Idaho State University
Pocatello, Idaho

B. C. McInnis
L. S. Shieh
University of Houston
Houston, Texas

INTRODUCTION

The purpose of manipulator-path control is to follow a desired trajectory by adjusting the torques or forces applied by the actuators to remove any deviations from the desired path. In current industrial practice, control-system design is based upon the use of an independent servomechanism for each joint, ie, the feedback controllers dedicated to each joint are designed to move the manipulator joints independently. These controllers do not attempt to model the detailed dynamics of the interacting manipulator joints in determining actuator signals. Instead, the coupling effects, friction and other nonlinearities are viewed as disturbances that the feedback control system must reject. This kind of independent single-input-single-output (SISO) servo design has the advantage of simplicity and hence enhances the robustness of the system in the face of unmodeled dynamics. On the other hand, since a robotic manipulator is intrinsically a multivariable coupled nonlinear system, in order to guarantee acceptable worst-case performance, such independent SISO servo design inevitably results in sacrifices of performance, ie, reduction in operating speed, excessive damping, and errors in dynamic positioning (1,2). These issues are considered briefly in this article.

To improve the performance of manipulator-path control, a variety of approaches have been proposed and promising preliminary results have been obtained. However, these results have been based primarily on theoretical or simulation studies. The concepts of these methods are presented below.

An intuitive approach is to cancel the disturbances caused by gravity and the interactions between joints by feed-forward compensation. This approach, which is called the computed torque technique, has received a great deal of attention in the literature. When this method is extended to the use of workspace variables, it is called the resolve-acceleration control method.

The commonly used perturbation method (dynamic linearization) for a nonlinear control-system design has been proposed for manipulator control. In this approach, the overall nonlinear-system model is split into a nominal model and a perturbation model. The nominal input can be precomputed using the nominal model with the nominal trajectory preplanned. The linear-state feedback-control based on either optimal control or pole placement is then derived from the perturbation model to compute the corrective input on-line. The overall input is then composed of the nominal input and the input computed using feedback from the joints.

Although recent developments in adaptive-control theory may also be applied to manipulator control, the value of adaptive control in robotics is questionable (see also Adaptive control). Both self-tuning control and model-reference adaptive control have been proposed, but may not be of practical value because of the necessity of evaluating parameter estimation algorithms within a short sampling interval and oscillations occuring before the adaptive control algorithms converge.

All the above-mentioned approaches along with others such as variable structure control, optimal control, minimum-time and minimum-time-fuel control are discussed in this article. For simplicity this discussion of control strategies is based upon continuous-time models and continuous-time controllers. For the implementation of these techniques in computer control a discrete-time model and discrete-time design methods are required.

SERVO DESIGN FOR A SINGLE LINK

A robotic manipulator is a positioning device in that each of its joints is driven hydraulically, pneumatically, or electrically within a feedback loop. This section illustrates conventional servo design of a single joint driven by an electrical actuator, eg, a permanent magnet d-c motor, through a gear chain. The schematic diagram for the electrical driven system is shown in Figure 1. (**See** Nomenclature for explanation of symbols.)

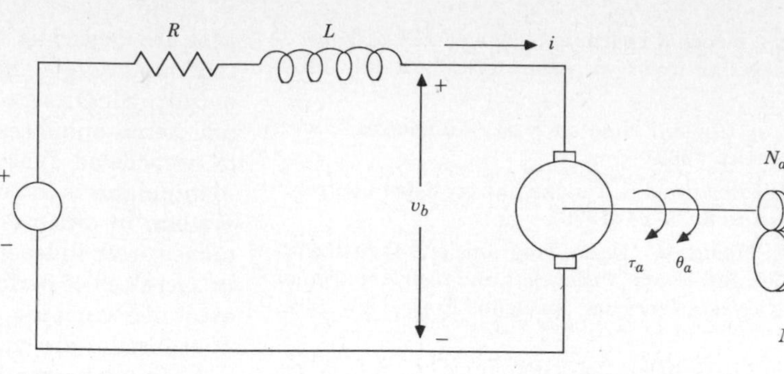

Figure 1. Schematic diagram for an electrical drive system.

When the motor armature is rotating, an emf voltage v_b, which is proportional to the product of flux and speed, is induced, ie

$$v_b(t) = K_b\dot{\theta}_a(t) \tag{1}$$

where K_b is the back emf constant. The voltage equation of the armature circuit is

$$v_a(t) - v_b(t) = Ldi(t)/dt + Ri(t) \tag{2}$$

Suppose the d-c motor is operated in its linear range, the torque delivered by the motor is

$$\tau_a(t) = K_t i(t) \tag{3}$$

where K_t is the torque constant.

The torque τ_a is applied to drive the actuator and the load. The dynamic equation for the actuator is (see Fig. 2)

$$\tau_a - \tau_l' = J_a\ddot{\theta}_a + B_a\dot{\theta}_a \tag{4}$$

where

$$\tau_l' = r_a F \tag{5}$$

The dynamic equation for the load is (see Fig. 3)

$$\tau_l = J_l\ddot{\theta}_l + B_l\dot{\theta}_l \tag{6}$$

where

$$\tau_l = r_l F \tag{7}$$

If the assumption is that backlash is negligible, the following relations are obtained.

$$r_l/r_a = n_g \tag{8}$$

Thus

$$\tau_l' = (1/n_g)\tau_l \tag{9}$$

and

$$\theta_l = (1/n_g)\theta_a \tag{10}$$

$$\dot{\theta}_l = (1/n_g)\dot{\theta}_a \tag{11}$$

$$\ddot{\theta}_l = (1/n_g)\ddot{\theta}_a \tag{12}$$

Then, by combining equations 4 and 6 and using equations 9–12

$$J\ddot{\theta}_a + B\dot{\theta}_a = \tau_a \tag{13}$$

where

$$J = J_a + (1/n_g)^2 J_l \tag{14}$$

$$B = B_a + (1/n_g)^2 B_l \tag{15}$$

The equation for the overall drive system can be obtained by combining equations 1–3. The armature inductance L is small and can usually be neglected, ie, $L \simeq 0$. Then

$$(RJ/K_t)\ddot{\theta}_a + (RB/K_t + K_b)\dot{\theta}_a = v_a \tag{16}$$

and its Laplace transformation form

$$\frac{\Theta_a(s)}{V_a(s)} = \frac{K_t}{RJs^2 + (RB + K_tK_b)s} \tag{17}$$

When backlash is assumed to be negligible, ie, $\theta_a = n_g\theta_l$, the task of controlling the angular displacement of the load shaft $\theta_l(t)$ to follow the desired angular displacement $\theta_l^*(t)$ is equivalent to controlling the angular displacement of the actuator shaft $\theta_a(t)$ to follow the desired angular displacement $\theta_a^*(t)$ where $\theta_a^*(t) = n_g\theta_l^*(t)$. For a feedback controller the error signal

$$e(t) = \theta_a^*(t) - \theta_a(t) \tag{18}$$

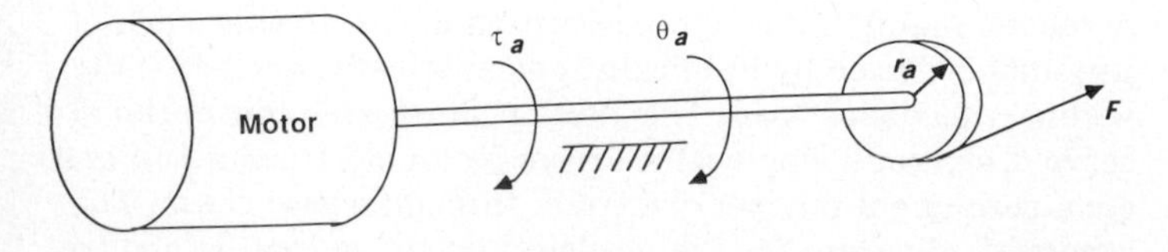

Figure 2. Schematic representation of the actuator-gear assembly.

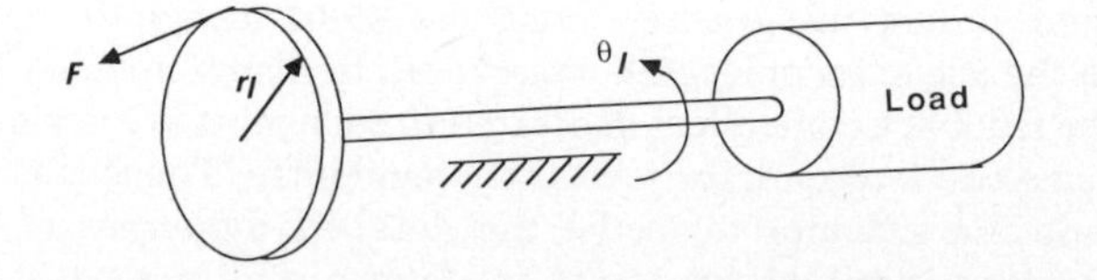

Figure 3. Schematic representation of the gear-load assembly.

is used and for proportional control the applied voltage is

$$v(t) = K_p[\theta_a^*(t) - \theta_a(t)] \tag{19}$$

where K_p is the proportional gain. Also, in order to inject some damping into the system, a negative feedback signal

$$v_f(t) = -K_v\dot{\theta}_a(t) \tag{20}$$

of the actuator shaft velocity is added to supplement the effect of back emf. The block diagram of the resulting controller is shown in Figure 4 where the inductance of the armature winding is neglected, ie, $L \simeq 0$. The closed-loop system becomes

$$\frac{\Theta_a(s)}{\Theta_a^*(s)} = \frac{K_pK_t}{RJs_2 + [RB + (K_b + K_v)K_t]s + K_pK_t} \tag{21}$$

$$= \frac{\omega_n^2}{s^2 + 2\zeta\omega_n s + \omega_n^2} \tag{22}$$

where

$$\zeta = \frac{RB + (K_b + K_v)K_t}{2(RJK_pK_t)^{1/2}} \tag{23}$$

$$\omega_n = (K_pK_t/RJ)^{1/2} \tag{24}$$

Thus, by specifying the desirable damping ratio ζ and the undamped natural frequency ω_n, the proportional gain K_p and the velocity feedback gain K_v can be determined. For a robotic manipulator, it has been suggested that a critically damped system, ie, $\zeta = 1$, should be designed (1). The value of ω_n can be chosen by specifying the desirable settling time $t_s = 4/(\zeta\omega_n)$. Consideration of structrual resonance provides an upper bound. For a conservative design with a safety factor of 200%, ω_n should be no more than one-half of the structural resonant frequency (1,3).

EQUATIONS OF MOTION OF MANIPULATORS AND SISO SERVO DESIGN FOR MULTIPLE LINKS

By applying either the Newton-Euler or Lagrange's equations, dynamical equations of a robotic manipulator of n joints can be obtained and written in vector-matrix notation as (4)

$$\mathbf{D(q)\ddot{q}} + \mathbf{h(q, \dot{q})} + \mathbf{g(q)} = \boldsymbol{\tau}(t) \tag{24}$$

where $\boldsymbol{\tau}$ is an $n \times 1$ force or torque vector applied to links; $\mathbf{q}, \mathbf{\dot{q}}, \mathbf{\ddot{q}}$ are $n \times 1$ joint position, velocity and acceleration vectors; $\mathbf{D(q)}$ is an $n \times n$ generalized mass matrix; $\mathbf{h(q, \dot{q})}$ is an $n \times 1$ vector containing Coriolis and centrifugal terms; and $\mathbf{g(q)}$ is an $n \times 1$ vector containing the gravity terms.

Note that the $n \times 1$ vector $\mathbf{h(q, \dot{q})}$ can be expressed in the form

$$\mathbf{h(q, \dot{q})} = \begin{bmatrix} \mathbf{\dot{q}}^t\mathbf{C}_1\mathbf{(q)\dot{q}} \\ \cdot \\ \cdot \\ \mathbf{\dot{q}}^t\mathbf{C}_n\mathbf{(q)\dot{q}} \end{bmatrix} \tag{25}$$

where each $\mathbf{C}_i\mathbf{(q)}$ is an $n \times n$ matrix. When the effects of viscous friction is taken into account, an additional term $\mathbf{V\dot{q}}$ can be added in the equation 24 as

$$\mathbf{D(q)\ddot{q}} + \mathbf{V\dot{q}} + \mathbf{h(q, \dot{q})} + \mathbf{g(q)} = \boldsymbol{\tau}(t) \tag{26}$$

where $\mathbf{V}$ is an $n \times n$ diagonal viscous friction matrix.

To include the dynamics of the actuators, we combine the equations 1–3 and write it in vector form. Assuming $L \simeq 0$,

$$\mathbf{v}_a = \mathbf{K}_b\boldsymbol{\dot{\theta}}_a + \mathbf{RK}_t^{-1}\boldsymbol{\tau}_a \tag{27}$$

where $\mathbf{K}_b$, $\mathbf{K}_t$ and $\mathbf{R}$ are $n \times n$ constant diagonal matrices. Also equation 4 is written in vector form

$$\boldsymbol{\tau}_a - \boldsymbol{\tau}_l' = \mathbf{J}_a\boldsymbol{\ddot{\theta}}_a + \mathbf{B}_a\boldsymbol{\dot{\theta}}_a \tag{28}$$

where $\mathbf{J}_a$ and $\mathbf{B}_a$ are $n \times n$ constant diagonal matrices.

For simplicity assume that all the joints are revolute, ie, $\mathbf{q}$ is the angular displacement vector of driven links. Thus, equations 9–12 can be written in a vector form as

$$\boldsymbol{\tau}_l' = \mathbf{N}_g^{-1}\boldsymbol{\tau} \tag{29}$$

$$\boldsymbol{\theta}_a = \mathbf{N}_g\mathbf{q} \tag{30}$$

$$\boldsymbol{\dot{\theta}}_a = \mathbf{N}_g\mathbf{\dot{q}} \tag{31}$$

$$\boldsymbol{\ddot{\theta}}_a = \mathbf{N}_g\mathbf{\ddot{q}} \tag{32}$$

where $\mathbf{N}_g$ is an $n \times n$ constant diagonal matrix of which each diagonal element is the gear ratio of the corresponding joint.

Combining equations 26–28 and using equations 29–32, then

$$\mathbf{D'(q)\ddot{q}} + \mathbf{V'\dot{q}} + \mathbf{h'(q, \dot{q})} + \mathbf{g'(q)} = \mathbf{v}_a(t) \tag{33}$$

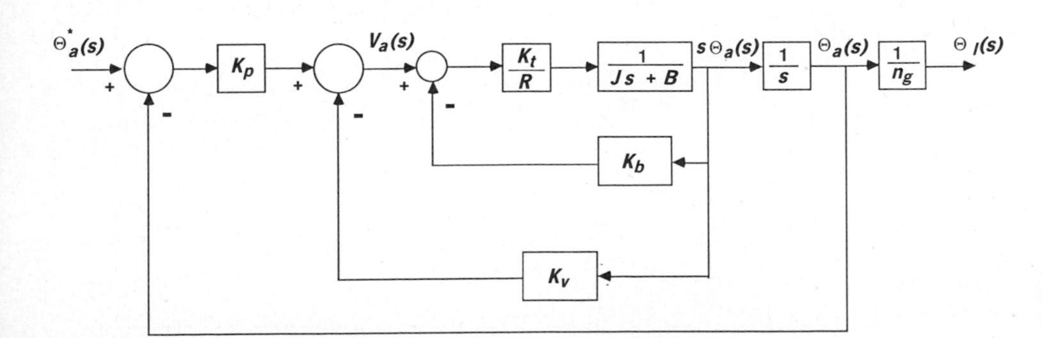

Figure 4. Block diagram of the positioning controller.

where

$$\mathbf{D}' = \mathbf{R}\mathbf{K}_t^{-1}(\mathbf{N}_g\mathbf{J}_a + \mathbf{N}_g^{-1}\mathbf{D}) \quad (34)$$

$$\mathbf{V}' = \mathbf{N}_g\mathbf{K}_b + \mathbf{R}\mathbf{K}_t^{-1}(\mathbf{N}_g\mathbf{B}_a + \mathbf{N}_g^{-1}\mathbf{V}) \quad (35)$$

$$\mathbf{h}' = \mathbf{R}\mathbf{K}_t^{-1}\mathbf{N}_g^{-1}\mathbf{h} \quad (36)$$

$$\mathbf{g}' = \mathbf{R}\mathbf{K}_t^{-1}\mathbf{N}_g^{-1}\mathbf{g} \quad (37)$$

For the discussion of SISO servo design for multiple links, equation 33 for each joint i is written as

$$D'_{ii}(\mathbf{q})\ddot{q}_i + \sum_{j=1, j\neq i}^{n} D'_{ij}(\mathbf{q})\ddot{q}_j + V'_{ii}\dot{q}_i + h'_i(\mathbf{q}, \dot{\mathbf{q}}) + g'_i(\mathbf{q}) = v_{ai} \quad (38)$$

The term

$$\sum_{j=1, j\neq i}^{n} D'_{ij}(\mathbf{q})\ddot{q}_j + h'_i(\mathbf{q}, \dot{\mathbf{q}}) + g'_i(\mathbf{q}) \equiv w_i(\mathbf{q}, \dot{\mathbf{q}}) \quad (39)$$

can be considered as a disturbance caused by the coupling interaction of the links and by the gravity effect, and equation 38 can be rewritten as

$$D'_{ii}\ddot{q}_i + V'_{ii}\dot{q}_i + w_i = v_{ai} \quad (40)$$

Note that the effective moment of inertia D'_{ii} is a function of the configuration of joints, and w_i is a function of joint positions and joint velocities. When the SISO servo design discussed in the previous section is applied for a robotic manipulator of multiple links, the capability of the manipulator is inevitably restricted because of the coupling interaction.

Because of the disturbance w_i, there exists a steady-state position error. The position error may be reduced by increasing the proportional gain K_p. However, for the consideration of structural resonance the undamped natural frequency ω_n must be appropriately upper bounded which in turn imposes an upper bound on K_p (see eq. 24).

One way to reduce the disturbance w_i is to lower the speed of the manipulator. It can be seen clearly by noting that the component h'_i of the disturbance is

$$h'_i = \frac{R}{K_t n_g} h_i \quad (41)$$

and equation 25 h_i is a quadratic form of the speed of the manipulator. Thus, in order to achieve good performance the independent SISO servo design approach inevitably limits the working speed of the manipulator.

Though ideally it is desirable to design a critically damped system, this may not be feasible because the value of the effective moment of inertia D'_{ii} depends on the configuration of joints and is time varying. Thus, a compromise is made so that an underdamped system design is avoided. This restriction ($\zeta \geq 1$) imposes a lower bound on the velocity feedback gain K_v. From equation 23 (D'_{ii} can be considered as J) it is seen that for the design of the fixed K_v, the maximum value of D'_{ii} should be used to determine the lower bound of K_v. Therefore, a severely overdamped system results.

THE INVERSE METHODS

As discussed in the previous section, when the conventional independent SISO servo design is used, it has been necessary to reduce manipulator velocities, to use excessive damping, and to sacrifice precision in dynamic positioning. Since the difficulties in designing a control system for a robotic manipulator stem from the elimination of the disturbances caused by the interactions between joints and from the intrinsic nonlinear and time-varying characteristics of the system, an intuitive approach is to decouple the system and to apply feedforward compensation for the disturbances. This inverse method was investigated (5) and called the computed-torque technique (4,6).

A more general approach has also been proposed (7) in which nonlinear state-feedback is applied to achieve decoupled control. This method reduces to the inverse method in the special case when the state coordinates correspond to joint coordinates. The computed torque method is discussed in this

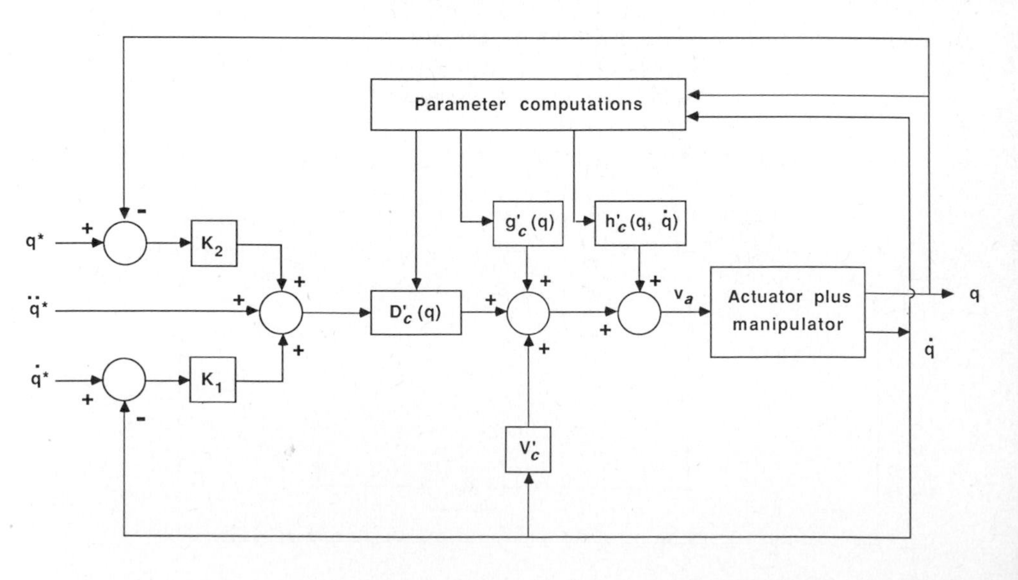

Figure 5. Block diagram of the computed-torque control system.

section along with a variation called resolved-acceleration control (8).

The basic idea of the computed torque technique is to use a feedforward compensation for the terms $\mathbf{V}'\dot{\mathbf{q}}$, $\mathbf{h}'$ and $\mathbf{g}'$ in equation 33 and a PD-feedback control. The control input is of the form

$$\mathbf{v}_a = \mathbf{D}_c'(\mathbf{q})\{\ddot{\mathbf{q}}^* + \mathbf{K}_1[\dot{\mathbf{q}}^* - \dot{\mathbf{q}}] + \mathbf{K}_2[\mathbf{q}^* - \mathbf{q}]\} + \mathbf{V}_c'\dot{\mathbf{q}} + \mathbf{h}_c'(\mathbf{q}, \dot{\mathbf{q}}) + \mathbf{g}_c'(\mathbf{q}) \quad (42)$$

where $\mathbf{K}_1$ and $\mathbf{K}_2$ are constant gain matrices, and where $\mathbf{D}_c'(\mathbf{q})$, $\mathbf{V}_c'$, $\mathbf{h}_c'(\mathbf{q}, \dot{\mathbf{q}})$ and $\mathbf{g}_c'(\mathbf{q})$ are the calculated counterparts of $\mathbf{D}'(\mathbf{q})$, $\mathbf{V}'$, $\mathbf{h}'(\mathbf{q}, \dot{\mathbf{q}})$ **and** $\mathbf{g}'(\mathbf{q})$, respectively. The block diagram of the control scheme is shown in Figure 5.

This technique is an effective method of manipulator control provided $\mathbf{q}$ converges to $\mathbf{q}^*$. The convergence criterion may be interpreted as follows. Suppose the computed parameters are exactly equal to their actual counterparts, ie,

$$\mathbf{D}_c'(\mathbf{q}) = \mathbf{D}'(\mathbf{q}) \quad (43)$$

$$\mathbf{V}_c' = \mathbf{V}' \quad (44)$$

$$\mathbf{h}_c'(\mathbf{q}, \dot{\mathbf{q}}) = \mathbf{h}'(\mathbf{q}, \dot{\mathbf{q}}) \quad (45)$$

$$\mathbf{g}_c'(\mathbf{q}) = \mathbf{g}'(\mathbf{q}) \quad (46)$$

Then substitute the control law (eq. 42) into equation 33,

$$\mathbf{D}'(\mathbf{q})\{\ddot{\mathbf{q}}^* - \ddot{\mathbf{q}} + \mathbf{K}_1[\dot{\mathbf{q}}^* - \dot{\mathbf{q}}] + \mathbf{K}_2[\mathbf{q}^* - \mathbf{q}]\} = \mathbf{0} \quad (47)$$

Let $\mathbf{e}_q = \mathbf{q}^* - \mathbf{q}$ be the joint position error. Since $\mathbf{D}'(\mathbf{q})$ is always nonsingular, equation 47 reduces to the following error equation

$$\ddot{\mathbf{e}}_q + \mathbf{K}_1\dot{\mathbf{e}}_q + \mathbf{K}_2\mathbf{e}_q = \mathbf{0} \quad (48)$$

Therefore, the gains $\mathbf{K}_1$ and $\mathbf{K}_2$ can be chosen so that the desired dynamic characteristics of the position error be achieved and error $\mathbf{e}_q = \mathbf{q}^* - \mathbf{q}$ approaches zero asymptotically.

However, the large amount of computations for evaluating the system parameters $\mathbf{D}'(\mathbf{q})$, $\mathbf{h}'(\mathbf{q}, \dot{\mathbf{q}})$ and $\mathbf{g}'(\mathbf{q})$ deters the usage of the computed-torque technique in real-time control. Several approaches to improve the computational efficiency have been reported. For example, realizing that the velocity-dependent term $\mathbf{h}(\mathbf{q}, \dot{\mathbf{q}})$ is a quadratic form of the manipulator's speed $\dot{q}$ (see eq. 25), Ref. 9 proposes using a look-up table organized by only position variables for the system parameters. Thus the required computation can be drastically reduced. Another approach is to use the computationally efficient set of forward and backward recursive equations (10,11). Ref. 12 proposes implementing the computed-torque method replacing $\ddot{q}_i(t)$ for each joint i in the forward and backward equations that arise in the Newton-Euler derivation of equation 24 by the expression

$$\ddot{q}_i^*(t) + k_1[\dot{q}_i^*(t) - \dot{q}_i(t)] + k_2[q_i^*(t) - q_i(t)] \qquad i = 1, \ldots, n \quad (49)$$

In the computed-torque technique, all the feedback control is done and the desirable dynamic stability property is achieved at the joint level. Since a robotic manipulator is basically a positioning device, it may be more desirable to have the manipulator controlled in workspace coordinates, especially in performing tasks which utilize force-torque sensing and/or position sensing of the end effector.

In order to use workspace coordinates, the techniques of resolved motion-rate control and resolved acceleration control have been developed (8,13,14).

Let $\mathbf{s}$ be the vector

$$\mathbf{s} = \begin{bmatrix} \mathbf{p} \\ \boldsymbol{\phi} \end{bmatrix} \quad (50)$$

where $\mathbf{p}$ and $\boldsymbol{\phi}$ are the position and orientation vectors, respectively, of the hand with respect to the base coordinates.

The vector $\mathbf{s}$ is related to the joint position $\mathbf{q}$ by a usually complex form

$$\mathbf{s} = \mathbf{f}(\mathbf{q}) \quad (51)$$

Differentiating equation 51 with respect to time,

$$\dot{\mathbf{s}} = \mathbf{J}(\mathbf{q})\dot{\mathbf{q}} \quad (52)$$

where $\mathbf{J}(\mathbf{q})$ is the Jacobian of $\mathbf{f}$ with respect to $\mathbf{q}$.

If the inverse Jacobian matrix $\mathbf{J}^{-1}(\mathbf{q})$ exists at $\mathbf{q}(t)$, then the joint velocities $\dot{\mathbf{q}}(t)$ of the manipulator can be computed from the hand velocities $\dot{\mathbf{s}}(t)$ using equation 52.

$$\dot{\mathbf{q}} = \mathbf{J}^{-1}(\mathbf{q})\dot{\mathbf{s}} \quad (53)$$

Given the desired hand velocities $\dot{\mathbf{s}}$, the resolved-motion-rate control uses equation 53 to compute the joint velocities $\dot{\mathbf{q}}$ and to indicate the rates at which the joint motors must be maintained in order to achieve a steady hand motion along the desired Cartesian path.

Further differentiating equation 52 results in

$$\ddot{\mathbf{s}} = \mathbf{J}\ddot{\mathbf{q}} + \dot{\mathbf{J}}\dot{\mathbf{q}} \quad (54)$$

Solving for $\ddot{\mathbf{q}}$,

$$\ddot{\mathbf{q}} = \mathbf{J}^{-1}[\ddot{\mathbf{s}} - \dot{\mathbf{J}}\dot{\mathbf{q}}] \quad (55)$$

To obtain the equation for the resolved-acceleration control law in a form similar to that given for the computed torque method in equation 42, replace $\ddot{\mathbf{q}}$ in the dynamic equations of motion of the manipulator with

$$\mathbf{J}_c^{-1}[\ddot{\mathbf{s}}^* + \mathbf{K}_1(\dot{\mathbf{s}}^* - \dot{\mathbf{s}}) + \mathbf{K}_2(\mathbf{s}^* - \mathbf{s}) - \dot{\mathbf{J}}_c\dot{\mathbf{q}}] \quad (56)$$

where $\mathbf{J}_c$ is the computed counterpart of $\mathbf{J}$ and

$$\mathbf{s}^* = \begin{bmatrix} \mathbf{p}^* \\ \boldsymbol{\phi}^* \end{bmatrix} \quad (57)$$

where $\mathbf{p}^*$ and $\boldsymbol{\phi}^*$ are the desired position and orientation vectors, respectively, of the hand with respect to the base coordinates.

The control law expressed in workspace variables is of the form

$$\mathbf{v}_a = \mathbf{D}_c'\{\mathbf{J}_c^{-1}[\ddot{\mathbf{s}}^* + \mathbf{K}_1(\dot{\mathbf{s}}^* - \dot{\mathbf{s}})] + \mathbf{K}_2(\mathbf{s}^* - \mathbf{s}) - \dot{\mathbf{J}}_c\dot{\mathbf{q}}]\} + \mathbf{V}_c'\dot{\mathbf{q}} + \mathbf{h}_c' + \mathbf{g}_c' \quad (58)$$

The convergence criterion can be interpreted by assuming all the computed parameters be exactly equal to their actual counterparts, ie, in addition to equations 43–46,

$$\mathbf{J}_c = \mathbf{J} \quad (59)$$

$$\dot{\mathbf{J}}_c = \dot{\mathbf{J}} \quad (60)$$

Substitute the control law (eq. 58) into the dynamic equation 33 and use equation 55 then,

$$\mathbf{D}'\mathbf{J}^{-1}[(\ddot{\mathbf{s}}^* - \ddot{\mathbf{s}}) + \mathbf{K}_1(\dot{\mathbf{s}}^* - \dot{\mathbf{s}}) + \mathbf{K}_2(\mathbf{s}^* - \mathbf{s})] = \mathbf{0} \quad (61)$$

Thus, by choosing appropriate gains $\mathbf{K}_1$ and $\mathbf{K}_2$, desirable dynamic stability at the hand level can be achieved. The advantage of the resolved-acceleration control over the computed torque technique is that the desired dynamic characteristics of the position error can be achieved in workspace coordinates instead of joint coordinates (compare equation 61 with 47). However, the advantage of decoupling with respect to end-effector coordinates is obtained at the price of the computation of $\dot{\mathbf{J}}(\mathbf{q})$ and $\mathbf{J}^{-1}(\mathbf{q})$.

POLE PLACEMENT DESIGN WITH COMPENSATION OF THE GRAVITY TERMS

The computed-torque technique discussed earlier can be considered as decoupled pole placement control with cancellation of the velocity dependent and gravity terms, and its performance depends largely upon the accuracy of the feedforward compensation. The compensation of the gravity term $\mathbf{g(q)}$ is justified since it only involves the position variables and also can be determined experimentally by measuring the torque required to maintain constant manipulator joint positions. The pole placement design with gravity term compensated has been proposed for the control design of simple locomotion systems (15) and for multivariable PD-control of robotic manipulators (16) in which the velocity dependent terms are ignored. However, the compensation of the velocity dependent terms is more error-prone because of its more complicated computation and the additional error introduced by the velocity measurement.

In an effort to reduce dependence upon feedforward compensation, scheme has been proposed (17) in which velocity dependent terms are retained and used in the computation of the feedback control.

The basic idea is as follows. For simplicity, the dynamics of the actuator is ignored; however, the result is appliable when it is added in the dynamic equation as can be seen from that the dynamic equation 33 of the overall system including the actuator has the similar form as equation 26.

Note that the Coriolis and centrifugal term $\mathbf{h}(\mathbf{q}, \dot{\mathbf{q}})$ is a quadratic vector form of $\dot{\mathbf{q}}$ (see eq. 25) and can be also expressed as

$$\mathbf{h}(\mathbf{q}, \dot{\mathbf{q}}) = \mathbf{E}(\mathbf{q}, \dot{\mathbf{q}})\dot{\mathbf{q}} \quad (62)$$

where $\mathbf{E}(\mathbf{q}, \dot{\mathbf{q}})$ is an $n \times n$ matrix.

The dynamical equation 26 of the manipulator can be rewritten as

$$\mathbf{D(q)}\ddot{\mathbf{q}} + [\mathbf{V} + \mathbf{E}(\mathbf{q}, \dot{\mathbf{q}})]\dot{\mathbf{q}} = \boldsymbol{\tau} - \mathbf{g(q)} \quad (63)$$

or

$$\ddot{\mathbf{q}} = -\mathbf{D}^{-1}(\mathbf{q})[\mathbf{V} + \mathbf{E}(\mathbf{q}, \dot{\mathbf{q}})]\dot{\mathbf{q}} + \mathbf{D}^{-1}(\mathbf{q})[\boldsymbol{\tau} - \mathbf{g(q)}] \quad (64)$$

Define a $2n$-dimensional state vector

$$\mathbf{x}^t(t) \equiv [\mathbf{q}^t(t), \dot{\mathbf{q}}^t(t)] \quad (65)$$

and an $n \times 1$ compensated input vector

$$\mathbf{u}(t) \equiv \boldsymbol{\tau}(t) - \mathbf{g(q)} \quad (66)$$

Equation 64 can be expressed in a linear time-varying state variable representation as

$$\dot{\mathbf{x}}(t) = \mathbf{A}(t)\mathbf{x}(t) + \mathbf{B}(t)\mathbf{u}(t) \quad (67)$$

where

$$\mathbf{A}(t) = \begin{bmatrix} \mathbf{0} & \mathbf{I} \\ \mathbf{0} & -\mathbf{D}^{-1}(\mathbf{q})[\mathbf{V} + \mathbf{E}(\mathbf{q}, \dot{\mathbf{q}})] \end{bmatrix} \quad (68)$$

$$\mathbf{B}(t) = \begin{bmatrix} \mathbf{0} \\ \mathbf{D}^{-1}(\mathbf{q}) \end{bmatrix} \quad (69)$$

Therefore, the linear time-varying state-feedback control law is given by

$$\mathbf{u}(t) = -\mathbf{K}(t)[\mathbf{x}(t) - \mathbf{x}^*(t)] \quad (70)$$

where

$$\mathbf{K}(t) = [\mathbf{K}_p(t), \mathbf{K}_v(t)] \quad (71)$$

is the $2n \times n$ feedback gain matrix and

$$\mathbf{x}^*(t) = \begin{bmatrix} \mathbf{q}^*(t) \\ \dot{\mathbf{q}}^*(t) \end{bmatrix} \quad (72)$$

is composed of the desired path $\mathbf{q}^*(t)$ and the desired velocity $\dot{\mathbf{q}}^*(t)$.

The feedback gain $\mathbf{K}(t)$ can be obtained by pole-placement design method as follows.

Substituting the control law (eq. 70) into the system equation 67,

$$\dot{\mathbf{x}}(t) = [\mathbf{A} - \mathbf{BK}]\mathbf{x}(t) + \mathbf{BKx}^*(t) \quad (73)$$

Given $n \times n$ diagonal matrices $\boldsymbol{\Lambda}_1$ and $\boldsymbol{\Lambda}_2$ with the desired poles as entries on the diagonal, it is noted that the desired poles are the eigenvalues of the canonical matrix

$$\boldsymbol{\Lambda} = \begin{bmatrix} \mathbf{0} & \mathbf{I} \\ -\boldsymbol{\Lambda}_1\boldsymbol{\Lambda}_2 & \boldsymbol{\Lambda}_1 + \boldsymbol{\Lambda}_2 \end{bmatrix} \quad (74)$$

Thus, equate

$$\mathbf{A} - \mathbf{BK} = \boldsymbol{\Lambda} \tag{75}$$

ie,

$$\begin{bmatrix} \mathbf{0} & \mathbf{I} \\ -\mathbf{D}^{-1}\mathbf{K}_p & -\mathbf{D}^{-1}(\mathbf{V} + \mathbf{E} + \mathbf{K}_v) \end{bmatrix} = \begin{bmatrix} \mathbf{0} & \mathbf{I} \\ -\boldsymbol{\Lambda}_1\boldsymbol{\Lambda}_2 & \boldsymbol{\Lambda}_1 + \boldsymbol{\Lambda}_2 \end{bmatrix} \tag{76}$$

Therefore

$$\mathbf{D}^{-1}\mathbf{K}_p = \boldsymbol{\Lambda}_1\boldsymbol{\Lambda}_2 \tag{77}$$

$$-\mathbf{D}^{-1}(\mathbf{V} + \mathbf{E} + \mathbf{K}_v) = \boldsymbol{\Lambda}_1 + \boldsymbol{\Lambda}_2 \tag{78}$$

and

$$\mathbf{K}_p = \mathbf{D}\boldsymbol{\Lambda}_1\boldsymbol{\Lambda}_2 \tag{79}$$

$$\mathbf{K}_v = -\mathbf{D}(\boldsymbol{\Lambda}_1 + \boldsymbol{\Lambda}_2) - \mathbf{V} - \mathbf{E} \tag{80}$$

The overall input torque becomes

$$\boldsymbol{\tau}(t) = -\mathbf{K}(t)[\mathbf{x}(t) - \mathbf{x}^*(t)] + \mathbf{g}(\mathbf{q}) \tag{81}$$

The overall control scheme is shown in Figure 6. There are two things to be noted. First, for the implementation of this approach in computer control a discrete-time model must be used to develop the control law (this also applied to other approaches); however, the canonical structure of the matrix **A**, (eq. 68), is destroyed in the discrete-time model. A method (18) for finding the required similarity transformation for transforming the system equation into block companion form can be used. Finally the determination of the discrete-time version of the state feedback gain can also be reduced to explicit expressions similar to equations 79 and 80. Second, assuming that **D** and **E** are constant (eg, by piecewise approximation), the system equation 63 is of type 1 with the input $\mathbf{u} = \boldsymbol{\tau} - \mathbf{g}$. Thus if the computed system parameters are exactly the same as the true ones, the elimination of the steady-state position errors will be assured by the state feedback control, but not for the position errors when the set point is a ramp input (ie, the manipulator is programmed to move at constant velocity). In order to eliminate the ramp-input position errors and disturbances (eg, caused by modeling errors and by piecewise approximation), integral control may be applied. This can be done by extending the state variable equation to include the integral of the position error and by deriving the state feedback control law based on the extended state variable system. Both aspects mentioned above have been employed (17).

However, for multiple-input-multiple-output (MIMO) systems, assigning only closed-loop poles may lead to the nonuniqueness of feedback gains, ie, different feedback gains may result in the same set of closed-loop poles, but different sets of associated eigenvectors. Thus, the nonuniqueness of selecting feedback gains provides extra freedom for matching the closed-loop structural requirements. For example, state-feedback decomposition of MIMO systems by block-pole placement has been proposed in which by assigning the latent roots and the associated latent vectors of the characteristic matrix polynomial, not only satisfactory dynamic performance but also parallel decomposition of the closed-loop system can be achieved (19).

LINEAR STATE FEEDBACK CONTROL BASED ON THE PERTURBATION EQUATIONS

If the path is known in advance, the perturbation method (dynamic linearization) for nonlinear control system design may be applied. This is based on the observation that, when the desired trajectory $\mathbf{s}^*(t)$ in the task-space (world space) of the hand (gripper) of a manipulator is preplanned, the corresponding trajectory $\mathbf{q}^*(t)$ in the joint-space can also be precomputed using the plant inverse dynamics and so can the nominal applied torque $\boldsymbol{\tau}^*(t)$ using the dynamic equation along the nominal joint-space trajectory $\mathbf{q}^*(t)$. Thus, a linear state-feedback control law based on the perturbed dynamic equation can be used to compute the corrective torque. The applied torque is then composed of the feedforward nominal torque and the feedback corrective torque.

To illustrate the idea, rewrite the dynamic equation 26 of the manipulator as

$$\mathbf{D}(\mathbf{q})\ddot{\mathbf{q}} + \mathbf{n}(\mathbf{q}, \dot{\mathbf{q}}) = \boldsymbol{\tau} \tag{82}$$

where

$$\mathbf{n}(\mathbf{q}, \dot{\mathbf{q}}) \equiv \mathbf{V}\dot{\mathbf{q}} + \mathbf{h}(\mathbf{q}, \dot{\mathbf{q}}) + \mathbf{g}(\mathbf{q}) \tag{83}$$

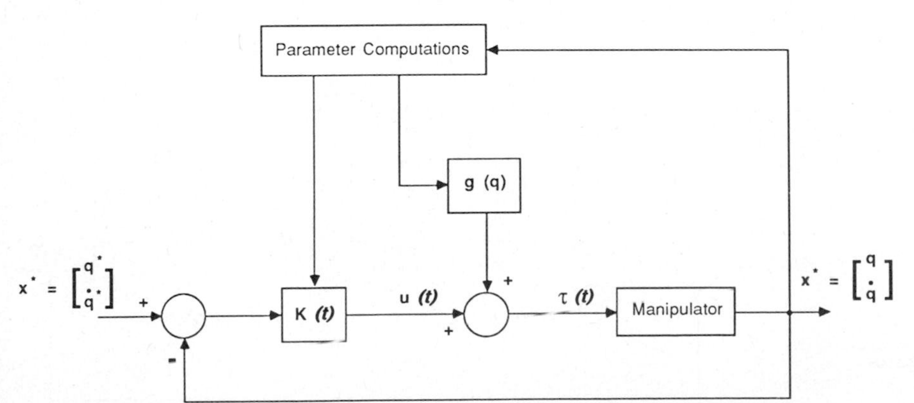

Figure 6. Block diagram of pole placement design with gravity compensation.

Let the $2n$-dimensional state vector for the system be denoted by

$$\mathbf{x}^t(t) \equiv [\mathbf{q}^t(t), \dot{\mathbf{q}}^t(t)] \equiv [q_1, \ldots, q_n, \dot{q}_1, \ldots, \dot{q}_n] \equiv [x_1, \ldots, x_{2n}] \quad (84)$$

Equation 82 can be expressed in state space form as

$$\dot{\mathbf{x}}(t) = \mathbf{f}(\mathbf{x}(t), \boldsymbol{\tau}(t)) \quad (85)$$

or

$$\begin{bmatrix} \dot{x}_1(t) \\ \cdot \\ \cdot \\ \dot{x}_n(t) \\ \dot{x}_{n+1}(t) \\ \cdot \\ \cdot \\ \dot{x}_{2n}(t) \end{bmatrix} = \begin{bmatrix} x_{n+1}(t) \\ \cdot \\ \cdot \\ x_{2n}(t) \\ a_1(\mathbf{x}) + \mathbf{b}_1(\mathbf{x})\boldsymbol{\tau}(t) \\ \cdot \\ \cdot \\ a_n(\mathbf{x}) + \mathbf{b}_n(\mathbf{x})\boldsymbol{\tau}(t) \end{bmatrix} \quad (86)$$

where $a_i(\mathbf{x})$ is the ith element of the $n \times 1$ vector $-\mathbf{D}^{-1}(\mathbf{q})\mathbf{n}(\mathbf{q}, \dot{\mathbf{q}})$ and $\mathbf{b}_i(\mathbf{x})$ is the ith row of the $n \times n$ matrix $-\mathbf{D}^{-1}(\mathbf{q})$. From a Taylor Series expansion it can be shown that the dynamic equation (85) can be approximated by a nominal equation of motion

$$\dot{\mathbf{x}}^*(t) = \mathbf{f}(\mathbf{x}^*(t), \boldsymbol{\tau}^*(t)) \quad (87)$$

plus a time-varying linearized perturbation equation

$$\delta\dot{\mathbf{x}}(t) = \mathbf{A}(t)\delta\mathbf{x}(t) + \mathbf{B}(t)\delta\boldsymbol{\tau}(t) + \mathbf{d}(t) \quad (88)$$

where

$$\delta\mathbf{x}(t) = \mathbf{x}(t) - \mathbf{x}^*(t) \quad (89)$$

$$\delta\boldsymbol{\tau}(t) = \boldsymbol{\tau}(t) - \boldsymbol{\tau}^*(t) \quad (90)$$

$$\mathbf{A}(t) = \frac{\partial \mathbf{f}}{\partial \mathbf{x}}\Big|_* \quad (91)$$

$$\mathbf{B}(t) = \frac{\partial \mathbf{f}}{\partial \boldsymbol{\tau}}\Big|_* \quad (92)$$

and $\mathbf{d}(t)$ accounts for the modeling error and other disturbances.

Note that $\mathbf{A}(t)$ and $\mathbf{B}(t)$ are the $2n \times n$ Jacobian matrices of $\mathbf{f}(\mathbf{x}(t), \boldsymbol{\tau}(t))$ evaluated at $(\mathbf{x}^*(t), \boldsymbol{\tau}^*(t))$.

Having derived the linearized perturbation model (eq. 88), the linear state-feedback control law is introduced as

$$\delta\boldsymbol{\tau}(t) = -\mathbf{K}(t)\delta\mathbf{x}(t) \quad (93)$$

to compute the corrective torque $\delta\boldsymbol{\tau}(t)$ to obtain the overall input torque, $\boldsymbol{\tau}(t)$, ie,

$$\boldsymbol{\tau}(t) = \boldsymbol{\tau}^*(t) + \delta\boldsymbol{\tau}(t) \quad (94)$$

The overall control scheme is shown in Figure 7.

There are several design methods available for choosing the appropriate feedback gain **K.** For example, pole placement design as discussed earlier can be used.

Another approach (20,21) is to use asymptotic regulator theory. That is, the feedback gain **K** is computed by minimizing the following performance index

$$\mathbf{J} = \int_0^\infty (\delta\mathbf{x}^t\mathbf{Q}\delta\mathbf{x} + \rho\delta\boldsymbol{\tau}^t\mathbf{R}\delta\boldsymbol{\tau})\mathbf{dt} \quad (95)$$

where **Q** and **R** are constant symmetric semi-postive definite and positive definite matrices, respectively, and ρ is a nonnegative scalar.

With the assumptions that in equation 88, $\mathbf{d}(t) = 0$ and **A** and **B** are constant (using a piecewise approximation for small-time intervals) the asymptotic solution of **K** is readily available in the optimal control literature (22). That is,

$$\mathbf{K} = \mathbf{R}^{-1}\mathbf{B}^t\mathbf{P}/\rho \quad (96)$$

where **P** is the solution of the algebraic Riccati equation

$$\mathbf{PA} + \mathbf{A}^t\mathbf{P} + \mathbf{Q} - \mathbf{PBR}^{-1}\mathbf{B}^t\mathbf{P}/\rho = 0 \quad (97)$$

In the work of Ref. 21, the selection of **Q** and **R** is based on the design method of implicit model following (23). A p-th order model for the disturbance $\mathbf{d}(t)$ is also incorporated in the extended state space equation and it results in the so-called integral multivariable control of order p.

The advantage of liner state-feedback control based on the linearized perturbation equation is the elimination of the enormous amount of on-line computation because, if the desired

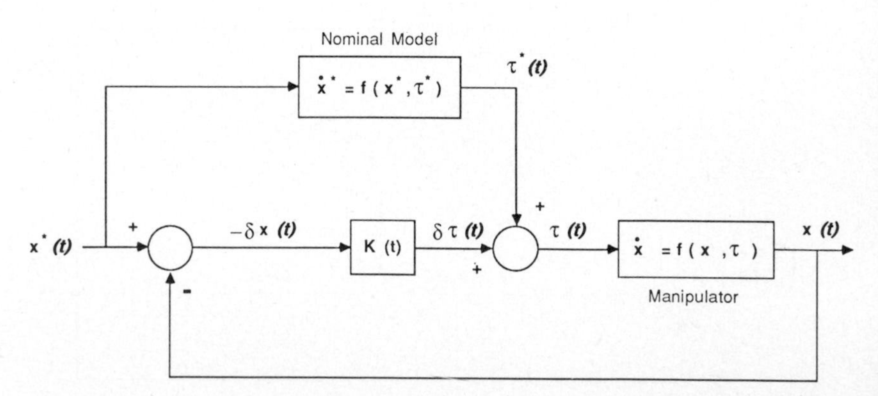

Figure 7. The block diagram of state feedback control based on the perturbation model.

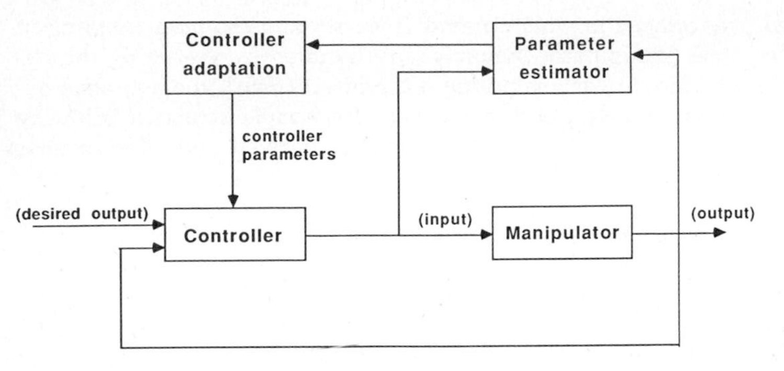

Figure 8. Self tuning control of a manipulator.

trajectory is preplanned, the matrices **A** and **B** in equation 88 can then be precomputed and so can the feedback gain **K.**

ADAPTIVE-CONTROL METHODS

All the previously discussed design methods for manipulator control are based on the assumption that an accurate system model is available, i.e. all the system parameters can be predetermined. However, modeling errors such as the unmeasurable friction and the uncertainty in the system parameters may render the system model inaccurate. Therefore, instead of using the known system model for the control design, the adaptive control methods, eg, self-tuning control (24,25) and model reference control (26), may be applied in which the system parameters are assumed unknown and a suitable adaptation law is used to adjust the controller parameters on-line based on the input/output information of the controlled plant (manipulator).

In self-tuning control a design procedure is first chosen which can be used when the plant parameters are known and this is applied to the unknown plant using recursively estimated values of these parameters. Its block diagram is shown in Figure 8.

For example, a self-tuning pole placement design has been proposed (27), in which an independent SISO control loop is used for each joint by assuming that the interactions between joints are neglected and the gravity term is feedforward compensated. In other work (28,29), the control design is based on the minimization of the weighted output prediction error in joint-space and workspace, respectively.

If the desired trajectory is preplanned, then the dynamic equation of the manipulator can be split into a nominal model and a perturbation model as mentioned earlier. Thus, the self-tuning control method can also be applied to compute the corrective input for the unknown perturbation model while the nominal input is computed using the known nominal model for feedforward compensation. This has been reported in Ref. 30 for joint-space tracking control and in Ref. 31 for workspace tracking control.

Another adaptive control method, the direct Model Reference Adaptive Control (MRAC) or the adaptive model following control (AMFC), has also been proposed for the control design of robotic manipulators. The block diagram of MRAC is shown in Figure 9.

The purpose of MRAC is to design an adaptive controller that drives the manipulator to follow the desired outputs (manipulator positions and velocities) generated by the reference model as closely as possible.

For the consideration of the overall system stability, the adaptive model reference controller should be designed using either Lyapunov's Direct Method or Popov's Hyperstability Theory in the general case. For example, the work in Refs. 32 and 33 is based on the hyperstability approach and the Lyapunov's direct method has been adopted in Ref. 34.

Most investigators have assumed a system model, of which all the parameters are unknown, to design either MRAC or self-tuning control. Thus, because of the large amount of computation involved in the parameter estimation, the control performance relies heavily on the initial parameters selected; in addition, theoretically there is a problem of model mismatching. However, a robotic manipulator is actually a partially known system, ie, almost all the physical quantities can be measured or precomputed accurately and the most uncertain one involves friction torques or forces. Therefore, instead of using adaptive control for the overall control system design, adaptive friction compensation (35) may be more suitable. In

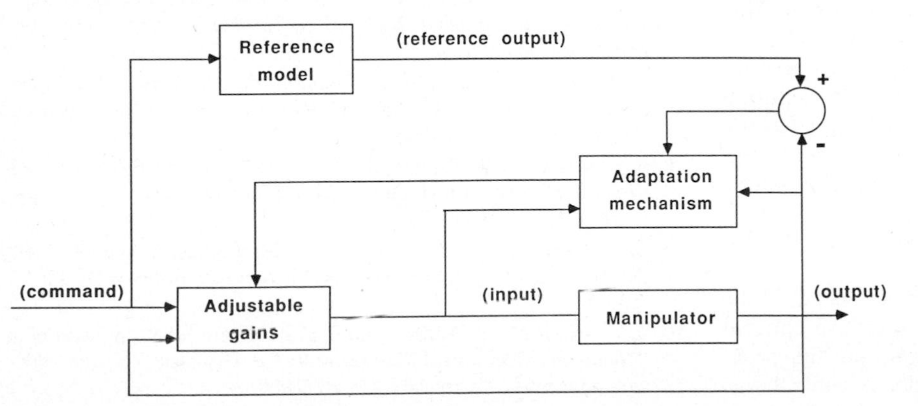

Figure 9. Model reference adaptive control of a manipulator.

this method the parameters of a friction model are estimated on-line to compute the friction torques or forces which are to be feedforward compensated in combination with the feedback control design based on the friction-free model of known parameters.

OTHER APPROACHES

Besides these control methods discussed in the previous sections, other methods have also been proposed. For example, the theory of variable structure systems (VSS, ie, a class of system with discontinuous feedback control) has been applied to design a variable structure control for multijoint manipulators (36). The main feature of VSS is that the so-called sliding mode occurs on a switching surface. While in sliding mode, the system remains insensitive to parameter variations and disturbances. Hence this insensitivity property of VSS enables the elimination of interactions among the various joints of the manipulator.

An approximate theory of optimal control has been developed (37) to design a suboptimal controller for manipulator trajectory-tracking problem. However, it was found that the approximate optimal controller was very sensitive to parameter variations as well as structural inaccuracies of the model. An improvement has been suggested (38) to use a hybrid (two-level) controller in which the first level is a digital controller designed by the approximate theory of optimal control and the second level is an analogue compensator based on additional feedbacks (joint torques or accelerations) which are closed around each joint to form minor loops imbedded in the first level controller.

The trajectory-tracking problem of manipulator control in the presence of input contstraints has been considered (39) by using a so-called Optimal Decision Strategy (ODS). The ODS is a pointwise optimal control law which is derived by minimizing the deviation between the vector of joint accelerations and a desired joint acceleration vector, subject to the input constraints. The design of the optimal control law is reduced to the solution of a quadratic programming problem and the solution gives an on-line feedback control scheme.

There are many applications that do not require robotic manipulators to strictly follow a prescribed path. Collision with obstacles can be avoided by specifying a few appropriate intermediate points in the workspace for the manipulator to pass through. For example, for most manufacturing tasks, it is desirable to move the manipulators at their highest speed to minimize the task cycle time between two locations. This prompted the investigation of the time-optimal control problem for mechanical manipulators (40). Because of the nonlinearity of the equations of motion, only a numerical solution is available for this problem and it is optimal only for the specific initial and final conditions. Therefore, as an alternative to the numerical solution, the suboptimal feedback control was obtained by approximating the nonlinear system by a linear system, and analytically, an optimal control for the linear system was found (40).

In addition to the consideration of operating speed, if the use of energy has been taken into account, then the optimal solution of minimum time-fuel problem is required. This problem was investigated and a suboptimal feedback control was developed in which the nonlinearity and the joint coupling in the manipulator dynamics are handled by averaging the dynamics at each sampling interval (41). With the averaged dynamics, a feedback controller of a simple structure allowing for on-line implementation can be derived and offers a near minimum time-fuel solution.

Nomenclature

R = armature-winding resistance, ohms
L = armature-winding inductance, henrys
i = armature-winding current, amperes
v_a = applied armature voltages, volts
v_b = back emf, volts
θ_a = angular displacement of the actuator shaft, radians
θ_l = angular displacement of the load shaft
N_a = number of teeth of the pinion on the actuator side
N_l = number of teeth of the gear on the load side
r_a = pitch radius of the pinion, m or in.
r_l = pitch radius of the gear
J_a = moment of inertia of the actuator, kg-m^2 or oz-in.-s^2
J_l = moment of inertia of the load
B_a = viscous-friction coefficient of the actuator, kg-m-s/rad or oz-in.-s
B_l = viscous-friction coefficient of the load
τ_a = torque delivered by the motor, N-m or oz-in.
F = force transmitted at the contacting point of the mating gears, N or oz
τ_l = torque delivered to the load ($= r_lF$)
τ_l' = torque applied to the actuator by the link ($= r_aF$)
n_g = gear ratio ($= N_l/N_a$)

BIBLIOGRAPHY

1. R. P. Paul, *Robot Manipulators: Mathematics, Programming, and Control,* MIT Press, Cambridge, Mass., 1981.
2. C. S. G. Lee, "Robot Arm Kinematics, Dynamics and Control," *Computer* **15,** 62–80 (1982).
3. J. Y. S. Luh, "Conventional Controller Design for Industrial Robots—A Tutorial," *IEEE Trans. Systems, Man and Cybernetics,* **SMC-13**(6), 298–316 (1983).
4. A. K. Bejczy, *Robot Arm Dynamics and Control,* Tech. Memo 33-669, Jet Propulsion Laboratory, Pasadena, Calif., Feb. 1974.
5. R. P. Paul, *Modelling, Trajectory Calculation, and Servoing of a Computer Controlled Arm,* AIM 177, Stanford University, Artificial Intelligence Laboratory, Nov. 1972.
6. B. R. Markiewicz, *Analysis of the Computer Torque Drive Method and Comparison with Conventional Position Servo for a Computer-Controlled Manipulator,* Tech. Memo 33-601, Jet Propulsion Laboratory, March 1973.
7. E. Freund, "Fast Nonlinear Control with Arbitrary Pole-Placement for Industrial Robots and Manipulators," *Int. J. Robotics Research* **1**(1), 65–78 (1982).
8. J. Y. S. Luh, M. W. Walker, and R. P. Paul, "Resolved Acceleration Control of Mechanical Manipulators," *IEEE Trans. Automatic Control* **25**(3), 468–474 (1980).
9. M. H. Raibert and B. K. P. Horn, "Manipulator Control Using the Configuration Space Method," *Industrial Robot* **5**(2), 69–73 (1978).
10. J. Y. S. Luh, M. W. Walker, and R. P. Paul, "On-line Computational Scheme for Mechanical Manipulators," *J. Dynamic Systems, Measurement and Control* **102,** 69–76 (1980).

11. B. C. McInnis and C. K. Liu, "Vectorial Methods for the Kinematics and Dynamics of Robotic Manipulators," *IEEE J. Robotics and Automation* **RA-2**(4) (1986).
12. R. Nigam, and C. S. G. Lee, "A Multiprocessor-Based Control for the Control of Mechanical Manipulators," *IEEE J. Robotics and Automation* **RA-1**(4), 173–182 (1985).
13. D. E. Whitney, "Resolved Motion Rate Control of Manipulators and Human Prostheses," *IEEE Trans. Man-Machine Systems,* **MMS-10**(2), 47–53 (1969).
14. D. E. Whitney, "The Mathematics of Coordinated Control of Prostheses and Manipulators," *J. Dynamic Systems, Measurement and Control* **94,** 303–309 (1972).
15. H. Hemami and P. C. Camana, "Nonlinear Feedback in Simple Locomotion Systems," *IEEE Trans. Automatic Control* **21**(6) 855–860 (1976).
16. A. Peltomaa and A. J. Koivo, "Multivariable PD-Control of Robotic Manipulators," *Proc. 5th IASTED Int. Symp. Robotics and Automation,* New Orleans, La., 1984, pp. 82–86.
17. R. J. Norcross, J. C. Wang, B. C. McInnis, and L. S. Shieh, "Pole Placement Methods for Multivariable Control of Robotic Manipulators," *J. Dynamic Systems, Measurement, and Control,* **108,** 340–345 (1986).
18. L. S. Shieh and Y. T. Tsay, "Transformation of a class of Multivariable Control Systems to Block Companion Forms," *IEEE Trans. Automatic Control* **27**(1), 199–203 (1982).
19. L. S. Shieh, Y. T. Tsay, and R. E. Yates, "State-Feedback Decomposition of Multivariable Systems via Block-Pole Placement," *IEEE Trans. Automatic Control* **28**(8), 850–852 (1983).
20. M. Vukobratovic and N. Kircanski. "Decoupled Control of Robots via Aysmptotic Regulators," *IEEE Trans. Automatic Control* **28**(10) 978–981 (1983).
21. S. Desa, and B. Roth, "Synthesis of Control System for Manipulators Using Multivariable Robust Servomechanism Theory," *Int. J. Robotics Research* **4**(3), 18–34 (1985).
22. B. D. O. Anderson and J. P. Moore, *Linear Optimal Control,* Prentice-Hall, Inc., Englewood Cliffs, N.J., 1971.
23. E. Kreindler and D. Rothschild, "Model-Following in Linear-Quadratic Optimization," *AIAA J.* **14**(7), 835–842 (1976).
24. K. J. Astrom, U. Borisson, L. Ljung, and B. Wittenmark, "Theory and Application of Self-Tuning Regulators," *Automatica* **13**(5), 457–476 (1977).
25. G. C. Goodwin and K. S. Sin, *Adaptive Filtering, Prediction and Control,* Prentice-Hall, Inc., Englewood Cliffs, N.J., 1984.
26. Y. D. Landau, *Adaptive Control—The Model Reference Approach,* Marcel Dekker, Inc., New York, 1979.
27. R. G. Walters and M. M. Bayoumi, "Application of a Self-Tuning Pole-Placement Regulator to an Industrial Manipulator," *Proc. 21st IEEE Conf. Deci. Contr.,* Orlando, Fla., 1982, pp. 323–329.
28. A. J. Koivo and T. H. Guo, "Adaptive Linear Controller for Robotic Manipulators," *IEEE Trans. Automatic Control* **28**(2), 162–171 (1983).
29. A. J. Koivo, "Self-Tuning Manipulator Control in Cartesian Base Coordinate System," *J. Dynamic Systems, Measurement and Control* **107,** 316–323 (1985).
30. C. S. G. Lee and M. J. Chung, "Adaptive Perturbation Control with Feedforward Compensation for Robot Manipulators," *Simulation* **44**(3), 127–136 (1985).
31. C. S. G. Lee and B. H. Lee, "Resolved Motion Adaptive Control for Mechanical Manipulators," *J. Dynamic Systems, Measurement and Control* **106,** 134–142 (1984).
32. B. K. Kim and K. G. Shin, "An Adaptive Model Following Control of Industrial Manipulators," *IEEE Trans. Aerospace and Electronic Systems,* **AES-19**(6), 805–813 (1983).
33. S. Nicosia and P. Tomei, "Model Reference Adaptive Control Algorithms for Industrial Robots," *Automatica* **20**(5), 635–644 (1984).
34. K. Y. Lim and M. Eslami, "Adaptive Controller Designs for Robot Manipulator System Using Lyapunov Direct Method," *IEEE Trans. Automatic Control* **30**(12), 1229–1233 (1985).
35. C. Canudas, K. J. Astrom, and K. Braum, "Adaptive Friction Compensation in DC Motor Drives," *IEEE Int. Conf. Robotics and Automation,* San Francisco, Calif., April, 1986, pp. 1556–1561.
36. K. K. D. Young, "Controller Design for a Manipulator Using Theory of Variable Structure Systems," *IEEE Trans. Systems, Man and Cybernetics,* **SMC-8**(2), 101–109 (1978).
37. G. N. Saridis and C. S. G. Lee, "An Approximation Theory of Optimal Control for Trainable Manipulators," *IEEE Trans. Systems, Man and Cybernetics,* **SMC-9**(3), 152–159 (1979).
38. G. L. Luo and G. N. Saridis, "Robust Compensation for a Robotic Manipulator," *IEEE Trans. Automatic Control* **29b,** 564–567 (1984).
39. M. W. Spong, J. S. Thorp, and J. M. Kleinwaks, "The Control of Robot Manipulators with Bounded Inputs," *IEEE Trans. Automatic Control* **31**(6), 483–490 (1986).
40. M. E. Kahn and B. Roth, "The Near-Minimum-Time Control of Open-Loop Articulated Kinematic Chains," *J. Dynamic Systems, Measurement and Control* **93,** 164–172 (1971).
41. B. K. Kim and K. G. Shin, "Suboptimal Control of Industrial Manipulators with a Weighted Minimum Time-Fuel Criterion," *IEEE Trans. Automatic Control* **30**(1), 1–10 (1985).

This work was supported in part by the U.S. Army Research Office, under contract DAAL-03-87-K-0001, the U.S. Army Missile R&D Command, under contract DAAH01-85-C-A111, and NASA-Johnson Space Center, under contract NAG9-211.

CONTROL VALUES FROM GEOMETRIC MODEL

PIERRE LOPEZ
GARI / DGE / INSAT
Toulouse, France

INTRODUCTION

In training task programming (through active (1) and passive robot (2) or active compliance (3) the servo-control set values **v** driving joint motions (in other words the actuator control signals **u** according to the selected definitions) are obtained directly from the experimental configuration (in training controls the coordinates are transformed by the mechanical articulated system (MAS) itself).

In computer task programming

- Whatever the nature of the machine-human communication (a conversational work-mode with a key-board/ screen system of graphic or other control (4,5); a word-based conversational work-mode with a key-board / screen combination (6) etc),
- whatever the speed range (position (7), kinematic (8), dynamic (9) or strain control (10)),

- whatever the control law (analogical or digital PID controller (11), adaptive controller (12) etc),
- whatever the actuators (electrical, hydraulic or pneumatic (13)),

the control unit is used to transform operational coordinates (**x**) into generalized ones (**q**) so that the joint motions determining the new effector location (position and orientation) are carried out. This transformation implies the use of the homogeneous matrix T_o^n, locating the effector-bound coordinate frame R_n (*n*th moving link) in relation to the base-linked coordinate frame R_o. In fact this matrix, characteristics of MAS represents the coordinate transformer used by constructors or the geometric model of the theoretician roboticists. With respect to homogeneous formalism (transformation matrices and punctual transformations) this article successively deals with

1. the considerable part played by geometry in MAS controls in various speed ranges.
2. the coordinate transformation in different modelings and at several speeds.
3. the taking into account of actuators (modeling and calculation of motor control signals $\mathbf{u}^*$).
4. the taking into account of the servo-control (modeling, control law, calculation of set values $\mathbf{v}^*$).

With every control and task context it is shown that it becomes more and more interesting for the base robot configuration to include useful software modules such as the coordinate transformer, the Jacobian matrix, the actuator model and the control law. The roboticist has to integrate appropriate modules (for task programming, decision and exteroceptive data processing) with those applicable in elaborate control laws (adaptive, hierarchical, fuzzy, etc) and those required in dynamic controls (providing these controls are compatible with real time constraints) (see also ADAPTIVE CONTROL; CONTROL; CONTROLLER ARCHITECTURE; CONTROL STRATEGIES).

IMPORTANCE OF GEOMETRY IN THE CONTROL OF THE MAS AT VARIOUS SPEED RANGES

An open-loop kinematic chain robot manipulator (*n* links moving around the base, *n* revolute (R) or prismatic (P) joints, n_a actuators with $n_a \leq n$) can be considered as a multivariable dynamic system; the actuator control signals **u** are among the inputs; the situation **x** of the effector-bound coordinate frame, the generalized coordinates **q**, speeds $\dot{\mathbf{q}}$ and accelerations $\ddot{\mathbf{q}}$ are among the outputs (see Fig. 1).

The task imposes the law $\mathbf{x}(t)$ to the effector situation. As operational (for $\mathbf{x}(t)$), generalized (for $\mathbf{q}(t)$), actuator control (for $\mathbf{u}(t)$) and set value (for $\mathbf{v}(t)$) spaces are distinct the general tracking problem (GTP) may be broken down into $(n + 1)$ problems (14) (Given uncoupled joint variables and an actuator for every joint axis)

$$\text{GTP} = \text{GCP} + \sum_{i=1}^{n} (\text{LTP})_i \tag{1}$$

GCP means general coordination problem: when $\mathbf{x}^*$ is imposed, find $\mathbf{q}^*$ to have $\mathbf{x}^* = \text{F}(\mathbf{q}^*)$ where F is the direct geometric

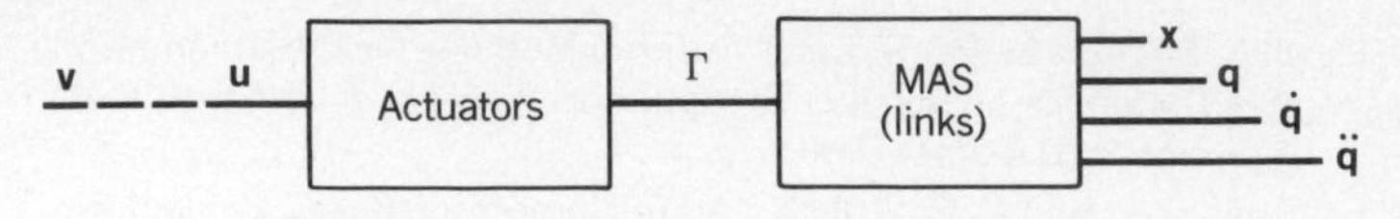

Figure 1. Actuator-MAS input and output.

model. LTP means local tracking problem: when $q_i^*(t)$ is imposed find $u_i^*(t)$ or $v_i^*(t)$ such that $q_i(t)$ follows $q_i^*(t)$.

The motion to be performed by end effector $\mathbf{x}^*(t)$ must often take the following constraints into account: initial and final situations have a null-speed; the intermediary situation $\mathbf{x}^*(k)$ at the moment *k* has a set speed $\dot{\mathbf{x}}^*(k)$. To the triplet $(k, \mathbf{x}_k^*, \dot{\mathbf{x}}_k^*)$ corresponds the triplet $(k, \mathbf{u}_k^*, \mathbf{v}_k^*)$ as elaborated by the system control unit in various contexts: open or closed-loop on the environment with or without compliance or artificial intelligence in the decisional part, etc. Given a key-board/screen-aided machine-human communication for the case of an environmental open-loop robot performing a task and a control unit with the five following modules: F as the geometric model ($\mathbf{x}^* = \text{F}(\mathbf{q}^*)$); J as the differential mold ($d\dot{\mathbf{x}}^* = \text{J}\, d\mathbf{q}^*$); $\boldsymbol{\Gamma}^*(\mathbf{q}^*, \dot{\mathbf{q}}^*, \ddot{\mathbf{q}}^*, \boldsymbol{\psi})$ as the dynamic model ($\boldsymbol{\psi}$ bringing the inertial parameters together); TIAA as the taking into account of actuators ($\mathbf{u}^* = \text{TIAA}(\boldsymbol{\Gamma}^*)$); and CL as control law ($\mathbf{v}^* = \text{CL}(\mathbf{u}^*)$).

The first three modules are associated with the geometry of the MAS; the latter two with the actuator type and the choice of the control law (PID), etc); the notation * shows the expected result calculated after the computer task programming in the context of the set machine-human communication. Figures 2 **a**, **b**, and **c** represent the module implementation in the case of low (**a**), average (**b**) and high (**c**) speeds (**u** can be mistaken for **v** in the case of servo-control programming; the TIAA can be a simple proportionality depending on the actuator type). Thus the geometric model is used in the three following cases: alone in Fig. 2**a**), together with J in Fig. 2**b**), or with J and $\boldsymbol{\Gamma}(\boldsymbol{\psi})$ in Fig. 2**c**). The TIAA and CL modules are also used in these three cases. Thus, in robot control through external computer it is important for the constructor to provide (or the roboticist to set up) the modules F, J, $\boldsymbol{\Gamma}(\boldsymbol{\psi})$, TIAA and CL. In simple cases (electrical stepper motor, digital PID) the modules TIAA and CL are "transparent" with respect to their relative elementary nature and robustness (15). On the contrary, in the case described in Fig. 2**c**) the CL, TIAA and of course $\boldsymbol{\Gamma}(\boldsymbol{\psi})$ modules become significant (14,16) for direct current electrical motors with sophisticated control laws.

COORDINATE TRANSFORMATION IN VARIOUS MODELS AND SPEEDS

Consider the general case with speeds that can be high, unbalanced gravity, non-negligible centrifugal and Coriolis forces, heavy loads to manipulate; suppose that nonlinear phenomena (slacks, elasticities, distortions) that cannot easily be modeled are negligible. In the course of time the evolution of the MAS configuration is then directed by the dynamic equations which can be written more simply (14), in a clear way related to the generalized forces Γ_i applied to joints (17) and in a linear way related to the dynamic parameters ψ_j (16):

$$\boldsymbol{\Gamma}_{n\times 1}(\mathbf{q}, \dot{\mathbf{q}}, \ddot{\mathbf{q}}, \boldsymbol{\psi}) = D_{n\times m}(\mathbf{q}, \dot{\mathbf{q}}, \ddot{\mathbf{q}})\, \boldsymbol{\psi}_{m\times 1} \tag{2}$$

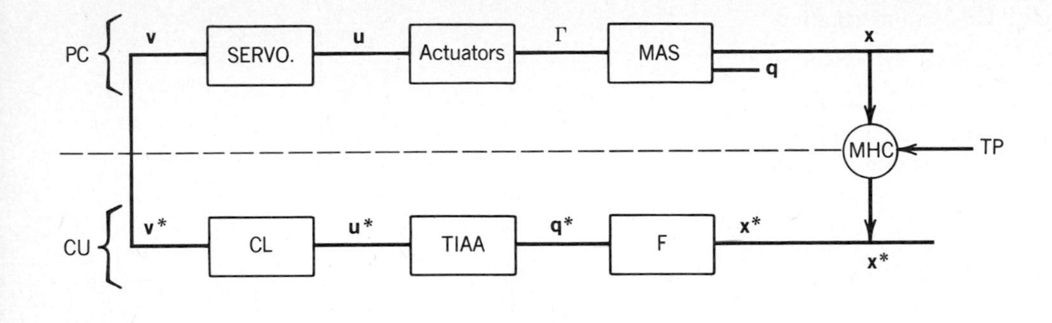

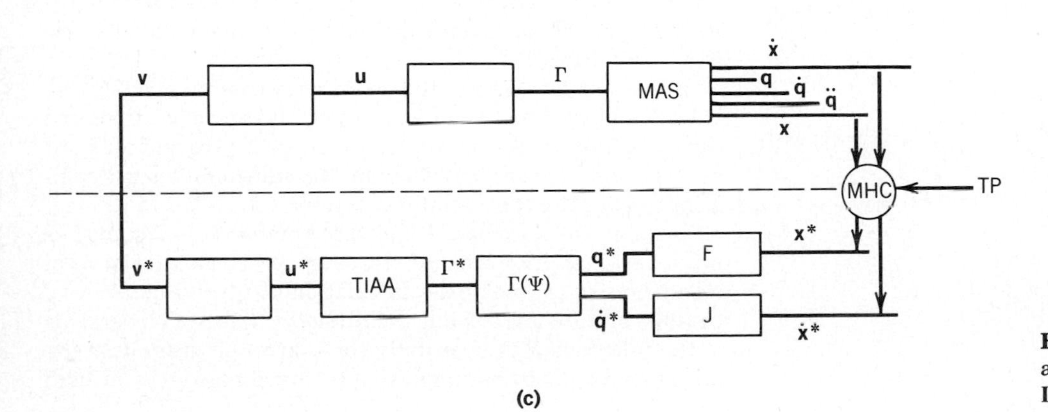

Figure 2. Geometric model used **(a)** alone; **(b)** together with J; **(c)** with J and **Γ(ψ).**

The vector ψ represents the inertial parameters of the different MAS components including the base and the effector (with the manipulated load). The parameters characterize the MAS dynamics and appear in the inertia matrices of this kind.

$$J_i = \begin{vmatrix} I_i & (XY)_i & (XZ)_i & X_i \\ & J_i & (YZ)_i & Y_i \\ & & K_i & Z_i \\ & & & M_i \end{vmatrix} \tag{3}$$

where $X_i = M_i\, \bar{x}_i$; $(XY)_i = \int xydm$; $J_i = \frac{1}{2}(Jy_i + Jz_i - Jx_i)$; J_i is connected with the link C_i, it is symmetrical and defined in the coordinate frame R_i which is bound to the link C_i; *a priori* ten parameters characterize the dynamics of C_i; M_i is the mass of C_i; $\overline{x}_i$ is the abscissa of C_i center of gravity; Jx_i is the inertia moment with respect to the x_i axis; and $(XY)_i$ is the inertia product from C_i into R_i.

At average or low speed dynamic equations can be simplified (with $\dot{\mathbf{q}}$ and $\ddot{\mathbf{q}}$ terms becoming negligible) these equations can be replaced by the balanced equations describing only the situation $\mathbf{x}$ (position and orientation) of the coordinate frame C_n with respect to the coordinate frame C_o. In any case the parameters defining this situation intervene. These parameters can be found in the following homogeneous transformation matrix T_o^n (14,18):

$$T_o^n = \begin{vmatrix} M_o^n & P_o^n \\ 0\ 0\ 0 & 1 \end{vmatrix} \tag{4}$$

where M_o^n is a matrix representing in the chosen configuration (spherical, direction cosines, Euler and Bryant angles, etc) the orientation of R_n with respect to R_o; in the same representation P_o^n gives the coordinates x_{no}, y_{no} and z_{no} of the origin

O_n from R_n in the coordinate frame R_o. This transformation matrix is the product $\prod_{i=1}^{i=n} T_{i-1}^{i}$ of the matrices T_{i-1}^{i} having the same structure as T_o^n. In the homogeneous transformation formalism the matrix T_{i-1}^{i} also describes the punctual transformation ${}^{i-1}A_i$ making from R_{i-1} the coordinate frame R_i. The following important results ensure (1,4):

$$\left.(\vec{v})_{R_{i-1}} = T_{i-1}^{i}\,(\vec{\mathbf{v}})_{R_i}\right\}\text{for the same vector } \vec{\mathbf{v}}$$

$$\left.(\overrightarrow{O_{i-1}M})_{R_{i-1}} = T_{i-1}^{i}\,(\overrightarrow{O_i\ M})_{R_i}\right\}\text{for the point } M \text{ itself}$$

$$\left.\begin{array}{l}(\vec{v}')_{R_{i-1}} = {}^{i-1}A_i\,(\vec{v})_{R_{i-1}}\\ (\overrightarrow{O_{i-1}M'})_{R_{i-1}} = {}^{i-1}A_i\,(\overrightarrow{O_{i-1}M})_{R_{i-1}}\end{array}\right\}\text{in the case of the punctual transformation}$$

(${}^{i-1}A_i \equiv T_{i-1}^{i}$; Moreover it must be inferred:

$${}^{i-1}A_i\,(\overrightarrow{O_{i-1}M})_{R_{i1}} = T_{i-1}^{i}\,(\overrightarrow{O_i\ M'})_{R_i}$$

that is to say that a point M and its mapping M' have the same coordinates respectively in R_{i-1} and R_i).

The Denavit and Hartenbert formalism gives to T_{i-1}^{i} (or ${}^{i-1}A_i$) a form permitting the data processing iteration and describing in an optimal way the MAS joints in every architectural configuration (19–21). The four parameters of DH (α_i, a_i, θ_i and r_i) are defined as follows:

$$T_{i-1}^{i} = \begin{vmatrix} C\theta_i & -S\theta_i\cdot C\alpha_i & +S\theta_i\cdot S\alpha_i & a_i\cdot C\theta_i\\ S\theta_i & +C\theta_i\cdot C\alpha_i & -C\theta_i\cdot S\alpha_i & a_i\cdot S\theta_i\\ 0 & S\alpha_i & C\alpha_i & r_i\\ 0 & 0 & 0 & 1\end{vmatrix} \tag{6}$$

$$\left.\begin{array}{l}\alpha_i = (\overrightarrow{z_{i-1}}, \vec{z_i}) \text{ round } \vec{x_i}\\ a_i \text{ along } \vec{x_i}\\ \theta_i = (\overrightarrow{x_{i-1}}, \vec{x_i}) \text{ round } \overrightarrow{z_{i-1}}\\ r_i \text{ along } \overrightarrow{z_{i-1}}\end{array}\right\}\text{(see Fig. 3)}$$

Two rules must be respected when setting up of the coordinate frames which are necessary to go iteratively (according to DH meaning) from R_o to R_n (21):

First rule: z_{i-1} must be the common perpendicular to x_{i-1} and x_i.
Second rule: x_i must be the common perpendicular to z_{i-1} and z_i.

These two rules completed by the introduction of any x_o (often in the symmetry plane of the robot base with z_o as vertical axis) and a dummy coordinate frame R_{n+1} (defining α_n, θ_n and a_n) are sufficient for the roboticist to describe every architectural configuration of the joints in a robot whether it is anthropomorphic or not (21): going from R_{i-1} (z_{i-1}) the roboticist may have to define coordinate frames in order to obtain R_i (z_i) (using the two rules mentioned above). The case of the RPPRRR mechanical articulated system is taken in Figure 4

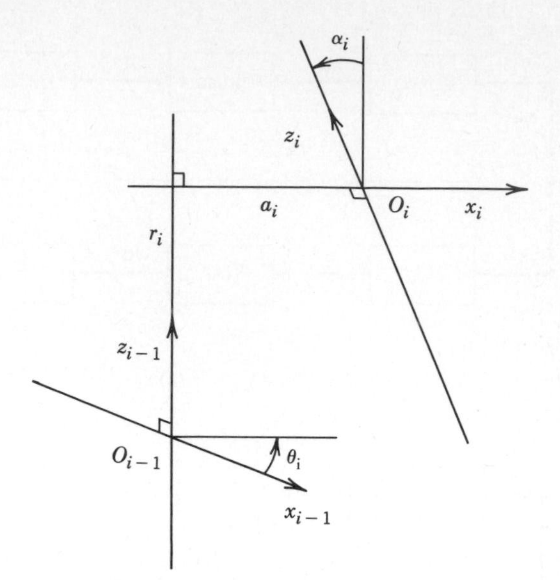

Figure 3. Definition of Denavit-Hartenberg parameters.

as an example (eg, TH8 from Renault); as the origins 04 and 05 are distinct, the arrangement in z_4 and z_5 naturally leads to defining the coordinate frame R_4' in order to respect the two rules used to determine x_5 and z_4 (when setting up the successive coordinate frames drawing up the DH parameter table). Without R_4' the two former rules cannot be complied with as the DH parameter definitions are not respected (see Fig. 5 and Table 1).

In fact, at low speeds the coordinate transformer T_o^n provides the MAS model and allows the $\mathbf{q}^*$ vector to be calculated corresponding to the $\mathbf{x}^*$ vector set in task programming. At average and high speeds, knowing the value of $\mathbf{q}^*$ is not sufficient to calculate the actuator control $\mathbf{u}^*$; the other MAS models such as the Jacobian J and the dynamic $\Gamma = D\psi$ models must be taken into account. In every practical case several algorithms (4,18) make the calculation of $\mathbf{q}_k^*$ and $\dot{\mathbf{q}}_k^*$ from $\mathbf{x}_k^*$ possible and then carry out the direct (or indirect) inversions of the functions F or J; usually the control program uses the inversion results by manipulating the functions $q_i^*(k)$. At high

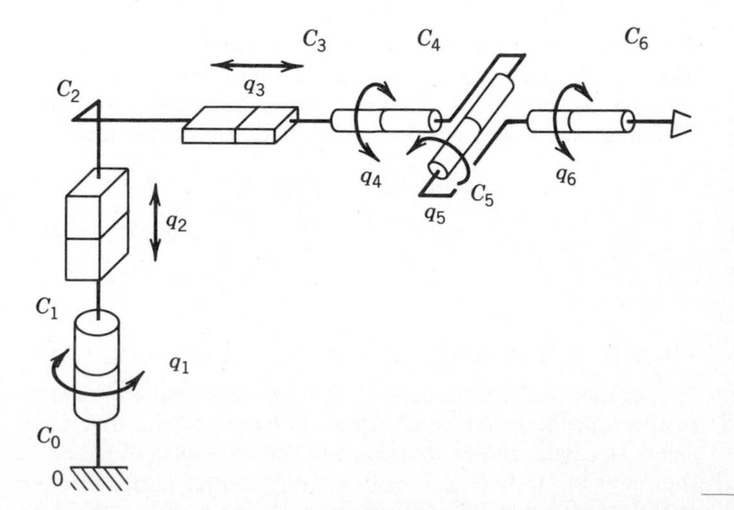

Figure 4. Diagram of the RPPRRR mechanical articulated system.

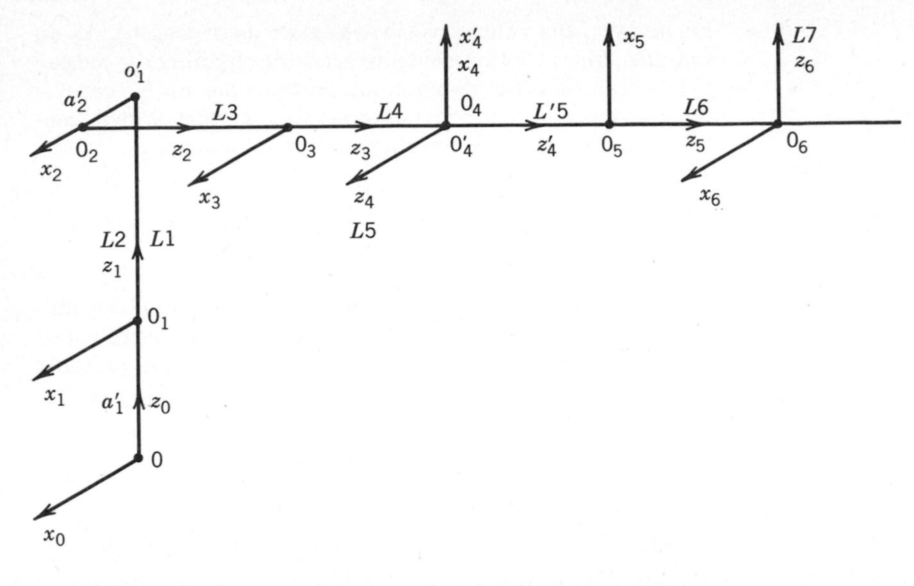

Figure 5. Configuration figure. $q_2 = r_2 = \overline{0_1 0_1'}$ along z_1; $q_3 = r_3 = \overline{0_2 0_3}$ along z_2; L7 is fictitious; $x_0\ x_1\ x_2\ x_3\ x_4\ z_4\ x_6$ are equipollent, and so are $z_0\ z_1\ x_4\ x_5\ z_6$.

speeds the calculations are longer. In such cases the dynamic model $D\psi$ becomes relevant; the matrix D here uses the values $\mathbf{q}\ \dot{\mathbf{q}}\ \ddot{\mathbf{q}}$ directly ($\mathbf{q}$ and $\dot{\mathbf{q}}$) or indirectly ($\ddot{\mathbf{q}}$) provided by the models F and J; the vector ψ components result either from an approximation (14) or from an identification (17) (in this identification Γ and D result from the measurements carried out on the studied MAS). The real time-bound control constraints make the implementation of a real computer control using the complete dynamic model difficult. In practice this model is simplified to reduce the calculation time (2,4,9,13).

THE TAKING INTO ACCOUNT OF ACTUATORS: MODELING AND CALCULATION OF MOTOR CONTROL SIGNALS

The last two paragraphs show how the program to be applied must take into account the specific robot actuators (motors and motion transmission organs), the existing servo-controls as well as the control law which can be set up by the roboticist in a base configuration. Geometry intervenes with the actuators (through motion transmission organs) and in the choice of servo-controls (it also intervenes in parameter optimization).

First consider the case of electrical stepper motors which are in themselves open-loop position servo-controls (2). As an example the motor control signals of the MINIMOVER robot actuators (4,22) can be directly deduced from the registers controlling the stepper motor windings through the following relation:

$$V_i = \mathrm{RCM}_{n,i} - \mathrm{RCM}_{n-1,i} \qquad i = 1, 2, 3, 4, 5, 6 \qquad n = 0, 1, \ldots\ldots, N \tag{7}$$

if it is decided to use N configurations to go from the initial configuration C_0 to the final C_N; V_i is the number of steps $\langle P_i \rangle$ to be made by the ith actuator motor; $i = 6$ correspondings to the gripper opening and closing (it is a pseudo-degree of freedom). The control law and the servo-control are simplified because the relation between the joint variable q_i and the motor control signal V_i is $V_i = K_i\, q_i$ (K_i constant) for $i = 1, 2, 3$; the relation is more complex for the wrist pitch (P') and roll (R') moves resulting from the interacting actuators 4 and 5. Then the TIAA and CL modules are modeled through the following equations:

$$V_i = K_i\, q_i \qquad i = 1, 2, 3$$
$$V_4 = -K_4\,(P' + R') = -K_4\,(q_4 + q_1 - q_5) \tag{8}$$
$$V_5 = K_5\,(P' - R') = K_5\,(q_4 - q_1 + q_5)$$

The procedure STEP $\langle S \rangle$ $\langle P1 \rangle$ . . . . $\langle P5 \rangle$ $\langle P6 \rangle$ is used to go from the configuration C_{n-1} to the configuration C_n; $\langle P_i \rangle$ is the number of steps made by the motor i ($V_i \equiv \langle P_i \rangle$); the sign of P_i determines the rotation direction; the procedure STEP brings the move coordination into effect. The joint variables used are different from the DH parameters; in Fig. 6 these variables are defined with respect to the horizontal plane ($q_1\ q_2\ q_3\ q_4$) and the $y\ 0\ z$ plane (q_5).

Now consider actuators using excited-field direct current motors (induced current/torque motor $= K_m^-$) controlled in current by a convertor voltage/induced current $= K_a^-$), coupled with the MAS through a gearing in the case of a revolute joint (reduction n), through a gearing and a sheave in the case of a prismatic joint (reduction $n = N/R$ where N is the

Table 1. Denavit and Hartenberg Parameter Table (DHPT)

	1	2	3	4	4′	5	6
σ_i	0	1	1	0	0	0	0
α_i	0	$-\frac{\pi}{2}$	0	$-\frac{\pi}{2}$	$+\frac{\pi}{2}$	0	$\frac{\pi}{2}$
a_i	0	a_2'	0	0	0	0	0
θ_i	q_1	0	0	q_4	0	q_5	q_6
r_i	a_1'	q_2	q_3	$\overline{0_3 0_4}$	0	$\overline{0_4 0_5}$	$\overline{0_5 0_6}$
q_i value in Fig. 5	0	$\overline{0_1 0_1'}$	$\overline{0_2 0_3}$	$-\frac{\pi}{2}$	0	0	$\frac{\pi}{2}$

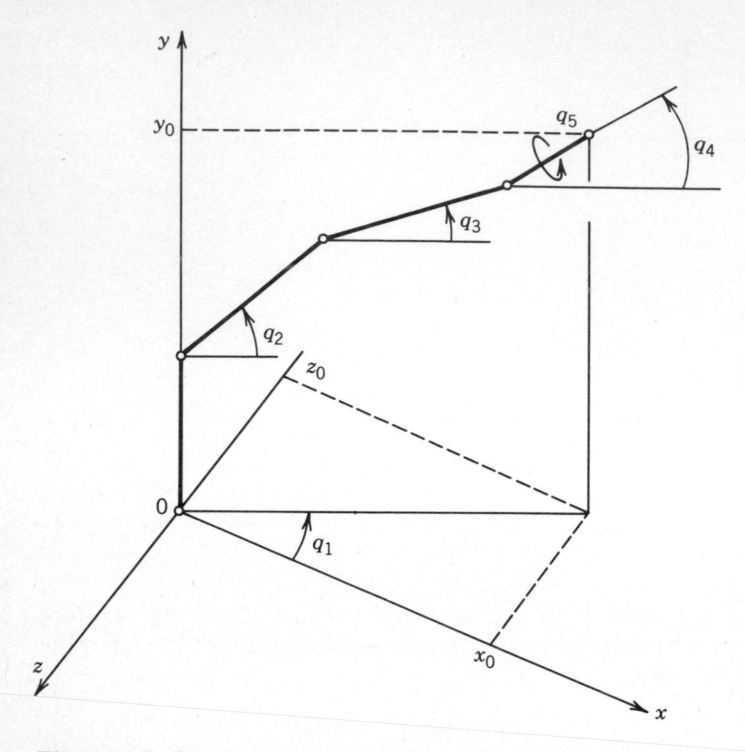

Figure 6. Definition of q_i joint variables used in simplified geometric model.

gearing reduction and R is the sheave radius) (14). For the ith joint the working equation which is explicit with respect to u_i can be written:

$$u_i = \frac{1}{Ka_i\, Km_i}\left[Jm_i\, n_i\, \ddot{q}_i + F_i + \frac{1}{n_i}\Gamma_i\right] \tag{9}$$

where Γ_i is a component of $\mathbf{\Gamma} = D\boldsymbol{\psi} \cdot F_i = Fv_i\, \dot{q}_i\, n_i + Fs_i$ (sig $\dot{q}_i$); Fv_i = viscous friction coefficient; Fs_i = Coulomb friction coefficient; and Jm_i inertia inherent in the motor and the associated reducer.

The $\boldsymbol{\omega}$ dimension can then be increased and $\mathbf{u}$ be written in a linear way with respect to the MAS and actuator dynamic parameters.

$$\mathbf{u}_{n\times 1} = C_{n\times(m+3n)} \cdot \boldsymbol{\phi}_{(m+3n)\times 1} \tag{10}$$

Equation 10 generalizes $\mathbf{\Gamma}_{n\times 1} = D_{n\times m} \cdot \boldsymbol{\psi}_{m\chi 1}$ (2); $m + 3n$ because every actuator implies 3 other dynamic parameters Jm_i, Fv_i and Fs_i, the joints are always taken as uncoupled. As an example, Figure 7 represents an actuator entailing the motion q_i (16); the Coulomb and viscous frictions are neglected; r is the resistance of the motor induced-circuit which is here controlled in tension. The working equation 9 becomes:

$$u_i = k_e\, n_i\, \dot{q}_i + \frac{r\, Jmi\, n_i}{k_c}\ddot{q}_i + \frac{r}{k_c\, n_i}\Gamma_i \tag{11}$$

This modelization matches the ERICC robot actuators depending on axes 1, 2, and 3. These motors are servo-controlled in position and speed; the fourth paragraph gives the contents of the set value q_i^* in terms of the values q_i, $\dot{q}_i$, $\ddot{q}_i$ and Γ_i resulting from MAS programming and modeling (4,16).

TAKING INTO ACCOUNT THE SERVO-CONTROLS: MODELING, CONTROL LAW, SET VALUE CALCULATION

When servo-controls are not entirely programmed the transformation of $\mathbf{u}^*$ into $\mathbf{v}^*$ and then of $\mathbf{v}$ into $\mathbf{u}$ must be distinguished (see Fig. 2). Afterwards the module CL reflects both the model of the used servo-control and the chosen control law.

At low speeds for example, Figure 2**a** becomes as follows for a stepper motor actuator in the MINIMOVER robot (see Fig. 8)

$$\mathbf{u}^* = \mathbf{v}^* = G\,(q^*) \text{ and } \mathbf{q}^* = F^{-1}\,(\mathbf{x}^*)$$

Equation 8 model the modules CL and TIAA; it is relatively straightforward to obtain the reversal geometric model from the direct geometric model and expresssed through the joint variables defined in Figure 6 (for example: $q_5^* = -\text{Arcten}\,\frac{z0^*}{y0^*} - R'^*$; $q_4^* = P'^*$; $q_1^* = \text{Arcten}\,\frac{z0^*}{y0^*}$; etc (4). For an actuator with a direct current motor which is controlled in voltage and servo-controlled in position and speed (such was the case of the already mentioned ERICC robot (16)), the block-diagram of Figure 7 has a new expression (see Figure 9). The set value q_1^* can be expressed as follows:

$$q_i^* = \frac{\alpha 1}{\alpha 0} H\, q_i + \frac{K_1\, k_e\, n_i}{K_0\, \alpha_0}\dot{q}_i + \frac{r\, Jm_i\, n_i}{K_0\, \alpha_0\, k_c}\ddot{q}_i + \left(\frac{r}{k_c\, K_0\, \alpha_0\, n_i}\right)\Gamma_i \tag{12}$$

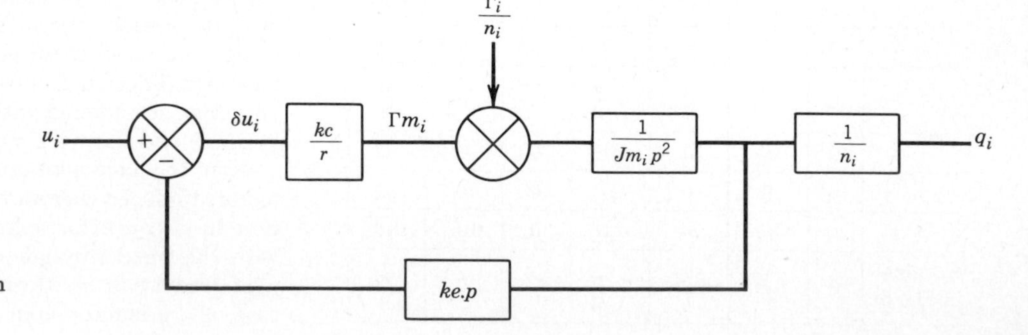

Figure 7. Actuator entailing the motion q_i.

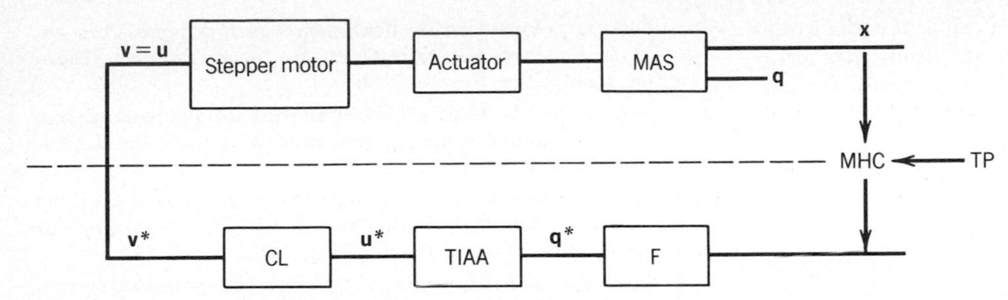

Figure 8. Equation model for stepper motor actuator in the MINIMOVER robot.

The form of this set value is given in terms of q_i, $\dot{q}_i$, $\ddot{q}_i$ and Γ_i whose values are provided by the MAS model and the task programming. Equation 12 takes into account the actuator (motor and motion transmission organ) as well as the adopted control law (PD). The value of the reduction ratio n_i often leads to neglecting the term including Γ_i before the one including Γ_{m_i}; for example, as far as the ERICC robot is concerned and with the values provided by the constructor, the following set value q_1^* connected with the actuator 1 is obtained

$$q_1^* = q_1 + 0.215\,\dot{q}_1 + 1.1\,10^{-4}\,\ddot{q}_1 + 1.1\,10^{-4}\,\Gamma_1 \qquad (13)$$

In general the control algorithm includes several modules; it produces $\mathbf{u}^*$ through task programming, coordinate transformer, actuator model and using data provided by the proprioceptive (perhaps $\mathbf{q}$, $\dot{\mathbf{q}}$, $\ddot{\mathbf{q}}$) and exteroceptive sensors (in the case of an environmental task closed-loop). The context depends on the application and choice carried out by the roboticist: speed range; $\ddot{\mathbf{q}}\,(t)$ evolution between the moments k and $k + 1$; chosen control law $\mathbf{v}^* = \mathrm{CL}\,(u^*)$ (PID in industrial practice, optimal and adaptive in laboratories) (7,12,16). Equation 12 allow the control signals of motors or servo-control set values to be calculated as they are applied at the moment $k - 1$ to go from the situation $\mathbf{x}_{k-1}^*$ to the effector situation $\mathbf{x}_k^*$.

The roboticist can determine the evolution of joint variables between the triplets $(k, \mathbf{x}_k^*, \dot{\mathbf{x}}_k^*)$. Suppose that the control law connecting $\mathbf{q}^*$ with $\mathbf{u}^*$ is known (22) and choose in the generalized space an evolution (such that q_i^k and $\dot{q}_i^k$ at the moment k and q_i^{k+1} with $\dot{q}_i^{k+1}$ at the moment $k + 1$) which will be performed through an acceleration.

$$\ddot{q}_i\,(t) = a_i^k\,(t - t^k) + \ddot{q}_i^k \qquad (14)$$

The change rate a_i^k and the acceleration $\ddot{q}_i^k$ can be expressed in terms of the boundary conditions as follows:

$$\left.\begin{aligned} a_i^k &= \frac{-12}{T^3}\,(\Delta q_i - \dot{q}_i^k\,T) + \frac{6}{T^2}\,\Delta\dot{q}_i \\ \text{and } \ddot{q}_i^k &= \frac{6}{T^2}\,(\Delta q_i - \dot{q}_i^k\,T) - \frac{2}{T}\,\Delta\dot{q}_i \\ (\text{Given: } T &= t^{k+1} - t^k,\ \Delta q_i = q_i^{k+1} - q_i^k \\ &\text{and } \Delta\dot{q}_i = \dot{q}^{k+1}_i - \dot{q}_i^k) \end{aligned}\right\} \qquad (15)$$

The constraints set by the roboticist in every space should be noted: the law $\mathbf{x}_k$ in the task space (task programming itself); the law $\ddot{q}_i^{(t)}$ between two set value triplets $(k, \mathbf{q}_k, \dot{\mathbf{q}}_k)$ and $(k + 1, \mathbf{q}_{k+1}, \dot{\mathbf{q}}_{k+1})$ in the generalized space; the law $\mathbf{u}^*(t) = C^*(t).\,\boldsymbol{\phi}$ taking into account not only the MAS geometry but also the inertias and the actuators in the actuator control space. In the setting up of actuator control signals $\mathbf{u}^*$ neither the part played by the modules concerning exteroceptive data treatment and the taking of these data into account by the decision unit was considered (23) nor the part played by the modules providing the decision taking in programming and task performance with artificial intelligence (24). At the moment these last mentioned modules cannot all be procured for industrial robot configuration (they depend on robot real time control constraints and not all the problems that have arisen have been solved even in simulation). By the computation of $\mathbf{u}^*$ this article does not deal either with the influence

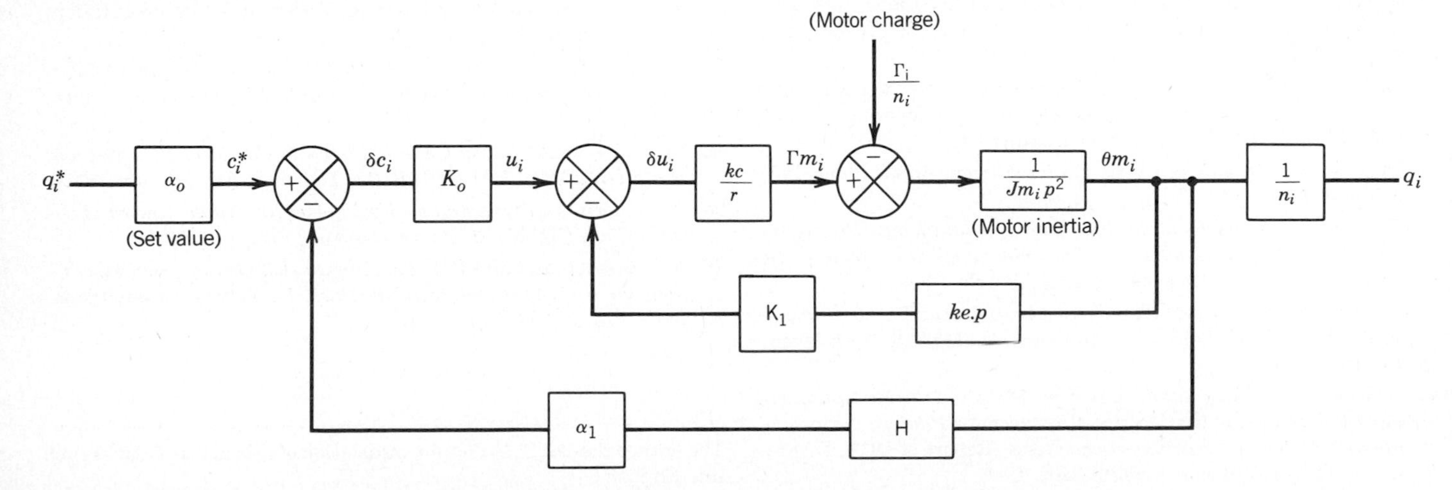

Figure 9. New expression for Figure 7, actuator with direct current motor (16).

of chaotic behaviors which can occur in a computer robot control. These behavior conditions can result from geometric (slacks, elasticities, distortions) or electronic nonlinearities (motor feeding through motor convertors) (24,25).

CONCLUSION

This article has considered the acute problem that arises in robot task performance (the robot is an artificial system acting on the environment in a more or less intelligent way) ie, the actuator control signal computation or in other words the set value computation of the servo-motors driving joint moves. These values are characteristic of the studied system as they depend on the MAS through its geometry (coordinate transformer) and its dynamics (inertial matrix); they also depend on actuators (motors and motion transmission organs), servocontrols and the adopted control law. Every factor can be connected with a module or a subroutine; through this computer-aided architecture the application program is given flexibility allowing it to develop into a valid robot system working in a closed-loop and intelligent as far as the decision-taking is concerned (artifical intelligence module playing a part in programming and task performance).

In every control and task context it proved to be beneficial for the robot purchaser (and consequently the constructor) to choose (or provide) a robot including in its base configuration useful software modules such as: the coordinate transformer, the Jacobian, the actuator model, the control law (PID). The other modules (data processing, AI, elaborate control law) should gradually be provided as options by constructors. All these modules use MAS geometry more or less directly, even if only through the conversational graphic simulation (26) representing a computer aid in system driving as well as staff training.

BIBLIOGRAPHY

1. R. P. Paul, *Robot Manipulators—Mathematics Programming, Control,* MIT Press, Cambridge, Mass., 1981.
2. P. Lopez and J. N. Foulc, *Introduction à la Robotique, Enseignement, Recherche et Développement,* Parts 1 et 2, PSI-Editests, Paris, 1984.
3. D. Viel, *Etude et Réalisation de la partie commande d'un Robot de Conduite Automobile Multieffecteur,* Engineering Report CNAM, Le Havre/Toulouse, France, June 1986.
4. P. Lopez, *Commande des Systèmes Robotiques, Applications,* PSI-Editests, Paris, 1986.
5. C. Laugier, *Un Système de Simulation Graphique de Robots incluant une gestion élémentaire des incidents et des capteurs,* RR n° 421, ENSIMAG, Grenoble, France, march, 1984.
6. S. Morisse, *Commande d'un bras manipulateur par Reconnaissance Vocale,* Mémoire de 3ème Cycle, IMERIR, Perpignan, France, 1986.
7. A. Abi-Ayad, *Coordination des Mouvements d'un Manipulateur dans le formalisme des Matrices Homogènes—Pilotage par calculateur extérieur en position et en vitesse,* Report of DEA, GARI / DGE / INSA, Toulouse, France, Sept. 1985.
8. A. Fournier, *Génération de Mouvements en Robotique—Applications des Inverses Généralisées et des pseudo-inverses,* Thèse d'Etat, Montpellier, France, 1980.
9. M. Renaud and S. Megahed, *Minimisation du Temps de Calcul pour la Commande Dynamique des Robots Manipulateurs,* LAAS-CNRS, Toulouse, France, 1981.
10. M. Al Mouhamed, *Contribution à la Commande automatique en robotique; application à la Commande par microprocesseur des manipulateurs articulés,* Thèse d'Etat, Paris, Jan. 1982.
11. C. Maire, *Les applications des méthodologies adaptatives à la commande des robots manipulateurs,* Report of the third Cycle CNAM, Le Havre/Toulouse, France, March 1986.
12. P. Lopez and C. Maire, *Adaptive Methodologies and Robotics,* FP3 Robot Modelling and Control, The 25th EEE Conference on Decision and Control, Athens, Greece, Dec. 1986.
13. Ph. Coiffet and M. Chirouze, *Elements de Robotique,* Hermes Publishing, Paris, 1982.
14. J. Z. Itturralde, *Commande des Robots Manipulateurs à partir de la Modélisation de leur Dynamique,* Thesis. Toulouse, France, July 1978.
15. F. Tortora, *Commande par microcalculateur d'un portique trois axes à moteurs pas à pas,* Report of DEA, GARI / DGE / INSA, Toulouse, France, Sept. 1986.
16. N. Rekik, *Contribution à la Modélisation et à la Commande Dynamiques de la Configuration Robotique ERICC-GOULD SEL 3275,* Report of DEA, GARI/DGE/INSA, Toulouse, France, July 1986.
17. M. Renaud, *Calcul Intrinsèque Analytique Itératif des Modèles d'un Robot Manipulateur,* Conférence AFCET-IASTED, RAI / IPAR' 86, Toulouse, France, June 1986.
18. B. Gorla and M. Renaud, *Modèles des Robots Manipulateurs—Application à leur commande,* CEPADUES-Editions, Toulouse, France, 1984.
19. C. S. G. Lee, *Robot Arm Kinematics, Dynamics and Control,* IEEE-Computer, 1982, pp. 62–79.
20. P. Lopez, A. Abi-Ayad, H. Donnarel, *Implantation Itérative des Repères et des Paramètres dans la Modélisation Géométrique des Chaînes Articulaires Simples,* IMACS-IFAC Symposium, Lille, France, June 1986.
21. P. Lopez and R. Ferrier, *Generality of Denavit-Hartenberg Formalism in the Homogeneous Modelling of the Joints in a Mechanical Articulated System,* ICIAM 87, SIAM, Paris, 1987.
22. J. N. Foulc, A. Faure, and P. Lopez, *Elaboration des consignes des systèmes actionneurs de robots commandés par calculateurs, Point en Robotique* Vol. 2, Lavoisier Technique et Documentation, Paris, 1985, pp. 37–58.
23. P. Lopez, *Perception Sensorielle Artificielle—Traitement des Informations Extéroceptives,* Conférences DGE / II, Toulouse, France, 1985.
24. Ch. Mira, *Modelling in Chaotic Dynamic Conditions,* Conférence AFCET-IASTED, RAI / IPAR' 86, Toulouse, France, June 1986.
25. A. Crochemore, *Dynamique Chaotique et Robotique,* Report of the third Cycle CNAM, Le Havre/Toulouse, France, 1986.
26. A. Benabdelhafid and P. Lopez, *Simulation Graphique Conversationnelle à l'aide de microcalculateurs,* Le Nouvel Automatisme, Paris, March 1985, pp. 52–57.

The author thanks E. Mallet for translation of this article from French into English.

CONVEYOR TRACKING

WILBERT E. WILHELM
The Ohio State University
Columbus, Ohio

INTRODUCTION

Conveyor tracking requires a robot to perform designated operations while the workpiece is being moved continuously by conveyor (1). The primary advantage of conveyor-tracking applications is that continuous operations may maximize robot utilization and minimize production cycle time, contributing to economic performance and system productivity.

Conveyors may be integrated with robotic stations in several ways. An intermittent conveyor, which may be viewed as a special case of the continuous conveyor, indexes workpieces from one station to another, allowing each robot to work on a stationary workpiece. A composite of continuous and intermittent conveyors may use a barrier to halt the workpiece (2) as it is conveyed to the station. Finally, conveyors may be used to transport workpieces between robotic cells which are otherwise decoupled production stations. Tracking specifically indicates that the robot operates on a moving workpiece and may require different hardware as well as software compared with the applications that involve stationary operations.

Since continuous conveyors are relatively simple mechanical devices, they are less expensive and more reliable than intermittent conveyors (3). In addition, continuous conveyors typically require less repair time should a breakdown occur.

However, line tracking requires intricate control over robot activities to effect complex contouring motions (4). Applications may therefore involve painstaking planning (5) and debugging effort, particularly if a fixed-base robot is used. It is emphasized that contouring motions are controlled routinely in a variety of applications, eg, arc welding, so that conveyor tracking does not require new technologies.

Either fixed-base or mobile robots may be used for conveyor tracking. Gantry (6) and track-mounted (7) are the most commonly used types of mobile robots in tracking applications. Track-mounted robots can move in parallel to the conveyor at synchronized speed (1), reducing the complexity of robot motions and, therefore, programming effort. However, manufacturers of fixed-base robots may provide software to allow them to travel in synchronization with conveyors in line-tracking applications (8). The cost of hardware, not software, will ultimately determine the type of robot used (3).

The purpose of this article is to provide an overview of conveyor tracking and to serve as a source of references for the reader interested in making an in-depth study of the topic. In the Introduction section, line tracking is defined, advantages are itemized, and the types of robots typically used in such applications are described. The body of this article is organized in three main sections. Typical applications are described in the first section, including spot welding, spray painting, assembly, and material handling. The second section presents approaches for analyzing cycle time, and hence the productivity, of single- and multiple-station designs. In addition, several types of simulation models are described along with their applications in designing tracking stations. The last section forecasts trends in the application of line tracking.

APPLICATIONS

Most applications of line tracking described in recent literature have been in the automotive industry and have involved spot welding, spray painting, assembly, and inspection. Many automobile manufacturers have converted to robots to perform these tasks, including General Motors (9), British Leyland's Austin Rover Company, manufacturer of the Montego (10), Vauxhall, manufacturer of the Astra car (11), Nissan (12), and BMW (13).

However, notable exceptions to automotive implementations have been reported in applications such as material handling in the electronics, applicance, and other industries.

Welding

To date, the predominant use of robots in the United States has been in the automotive industry; the majority of these applications have been to spot weld (14) structures to assemble automobiles. Robots offer substantial advantages over human operators in welding applications. Humans are subjected to an unpleasant, if not hostile, environment while welding and must deal with heavy equipment and protective clothing. Robots are not subject to these limiting factors and can make dramatic improvements in "arc-on" time, weld quality, and product uniformity (3).

Robots offer a number of advantages in spot-welding applications, including speed and more accurate positioning of welds, resulting in better appearance and more uniform quality (15). In automotive spot-welding applications, robots may be stationed on opposite sides along the line, allowing one supervisory computer to control all robots. This arrangement permits large numbers of welding programs to be stored and used to process different products and reallocate tasks as necessary. Robots also offer advantages in arc welding, an application that requires the robot to track the weld seam. Seam tracking is a specific type of tracking and is not treated in this paper.

Ref. 16 provides a background in spot-welding fundamentals and describes an approach for planning robotized lines. Data required to complete the analysis include (*1*) parts to be assembled, (*2*) geometrical conformation of parts and the number of stations required, (*3*) distribution of spot welds and the number of robots required to complete them, (*4*) production rate and the number of lines required to achieve the rate, (*5*) desired degree of flexibility, (*6*) basic principles for transferring and positioning the assembly, (*7*) final selection of the robot and its equipment and installation, and (*8*) the environment and the available space. This approach is applied to auto-body assembly, pointing out the importance of maintaining product orientation throughout assembly, assuring high utilization of robots through the use of buffer stocks, minimizing the number of robots, balancing work load among stations, redistributing tasks in case of robot downtime, and providing flexibility to handle new product designs in the future. The use of gantry robots as well as hydraulic polar movement robots installed in overhead positions is discussed.

One hundred and thirty robots are used in the Toyo Kogyo plant to perform 76% of the welding required to assemble the Mazda 626 automobile (17). An innovative development, 3 pairs of "tunnel" robots hang from a common slide, which they traverse horizontally. The arms are grouped to form a tunnel, allowing 18 axes to be managed by one console unit, which can control 62 functions, compared with 51 for normal robot configurations.

Ref. 18 discusses reliability and maintainability issues, emphasizing spot-welding applications. The importance of buffers to reduce the impact of downtime is stressed and the use of backup robots located downstream to take over tasks not completed by upstream robots in repair is suggested.

Spray Painting

Robots make excellent spray painters because they do not require the extensive protective devices that humans do. Once taught, a robot can repeat its skill, tracking product shape with a consistency that humans cannot duplicate.

This consistency allows robots to reduce the amount of coating material used as well as the energy consumed while improving overall quality (19). Current applications include automotive exterior top coat and underbody primer, stains on wood furniture, sound deadeners on appliances, porcelain coating of kitchen and bathroom fixtures, and exterior coating on the space shuttle booster rockets (Figs. 1 and 2). A thorough discussion of topics that must be considered in specifying hardward and software in applying finishes can be found in Ref. 19 (see also PAINTING).

Figure 1. A DeVilbiss TR-3500 spray-finishing robot applies sound deadener to the wheelwell area of a Nissan truck. Courtesy of Nissan Motor Manufacturing Corp., USA.

Figure 2. A total of 40 DeVilbiss TR-3500 spray-finishing robots apply sound deadener, primer, and finish coats to light trucks in Nissan's Smyrna, Tennessee, facility. Courtesy of Nissan Motor Manufacturing Corp., USA.

In spray-painting applications, robots may be made mobile in a variety of ways (eg, mount on a turntable, lift table, or traverse table) and parts may be presented through a variety of motions (eg, rotate or index, convey continuously or intermittently) (19). Close attention must be directed to the interface between the robot and the conveyor to assure synchronized speeds, maintaining the correct gun-to-target position for high quality results.

Some types of errors may be corrected at touch-up stations downstream from the automatic spray line, but robotic lines are especially vulnerable to malfunctions that may force the entire line to shut down. In particular, malfunctions caused by robot–product interaction or by failure to shut off the spray may cripple the line (20).

A complete economic analysis of robotized spray painting, evaluating the effects of robot failures as well as multiple-shift operation in automated systems, can be found in Ref. 21.

The John Deere Company has developed an extensive application of line tracking to spray-paint tractor chassis (22) (Fig. 3). The system achieves an efficiency in excess of 90% on 2-shift-a-day operation and can automatically handle 8 different chassis models that require over 30 paint programs.

At the Toyo Kogyo factory in Japan (17), 2 robots are used to apply weld sealer to the underside of 20,000 Mazda 626 cars monthly; photoelectric sensing is used to allow the robot to track the seam. Two additional robots apply weld sealer inside the car, 2 more apply PVC, 4 others apply underseal, 6 apply primer/surfacer, and 4 apply color coats. In the primer/surfacer booth, robots are used to open the car doors before painting and to close them afterward. In the color booth, the line splits and parallel stations employ 2 robots to select the appropriate color from among 27 colors and spray the car; 5 men are used to do the complicated areas.

Assembly

In general, assembly operations are not done in environments that are uncomfortable, unsafe, or hazardous; however, they

Figure 3. A DeVilbiss/Trafa robot, in a pit beneath a John Deere tractor chassis, coating the underside. Courtesy of DeVilbiss Co.

can be monotonous and lead to quality problems due to low levels of operator motivation. The robot must therefore compete with human operators by doing a faster, more accurate, and more consistent job.

Assembly operations may place severe demands related to precision, repeatability, and the variety of motions required on the capabilities of robots. Grippers may require sophisticated designs to allow them to play roles other than simply holding parts in place (23). In addition, assembly requires complex sensing capabilities, which are just becoming available. Application of robots in the area of assembly has been relatively limited but is expected to be the primary area of growth over the next 10 years (14).

Because of the manipulation required to position parts and the precision necessary to locate parts correctly, assembly applications involving line tracking have apparently been somewhat limited to date. For example, there has been a rapid increase in applications to assemble printed circuit boards and other electronic products, but most applications again involve indexing conveyors which transport subassemblies from one station to another. A variety of applications in Japanese industries, including electronics assembly as well as the use of SCARA-type robots to tighten 80 screws in Yamaha motorcycle engines, are described in Ref. 24. Multistation systems that assemble printed circuit boards, including one continuously moving conveyor application in which a series of 9 robots inserts large (ie, exceptional) components, are described in Ref. 25.

A rotary-pump assembly line, in which 3 robots travel along a circular rail track to perform their tasks, can complete an assembly after one trip around the circular track (26). Electrically driven articulated arms with 5 degrees of freedom are used, and a unique technique that uses a parabolic interpolation curve is applied to effect point-to-point control.

Robots should be selected primarily according to their work envelope, type of drive, rigidity, reliability, and ease of programming (27). Westinghouse Series 5000 gantry robots are used to assemble 140,000 gear boxes annually (28). Power- and free-roller conveyors are used to transfer subassemblies between stations which must deal with 17 product styles, each composed of 14 parts. This process was automated to improve quality and reduce costs.

Material Handling

Robots are used in material-handling applications, including tracking, that involve picking parts from conveyorized parts feeders as well as loading and unloading transmission cases from a moving monorail (29). Furthermore, robots can probe components on a printed circuit board while traveling on a synchronized belt conveyor.

A set of 30 robots perform material-handling functions on washing-machine production lines for both drum and tub manufacturing. The robots also position these components for assembly (30).

Fraser Automation, a division of PickOmatic Systems, has installed a Model FR300–4 robot in a conveyor-tracking installation at the Hydra-matic Division of General Motors. Transmission cases are picked off a moving monorail one at a time, swung around 90°, and placed on a belt that conveys them to an inspection area. According to Ref. 31, the cylindrical coordinate robot moves laterally on a track (the fourth axis) over a distance of 32 in. (813 mm). Over this length, robot acceleration requires 5in. (127 mm); deceleration, another 5 in. (127 mm); and tracking occurs over 22 in. (559 mm). While tracking, the robot extends toward the part, clamps and lifts it from the carrier, and swings it clear of the monorail carriers. The monorail is operated at 22 fpm to allow the robot to be kept busy virtually all of the time, maximizing productivity. Speeds of up to 42 fpm have been used successfully in the plant.

End-of-arm tooling uses toggle-actuated gripper clamps activated by air cylinders. The tooling is flexible and would support other types of tracking, such as picking parts from transfer line pallets and placing them on the moving monorail. Sensors are used to detect the arrival of a part and to initiate synchronized movement of the robot. A conical locator which incorporates an infrared sensor is used by the robot to determine relative position by mating with a matching bore in the part. Two additional locators are used to assure correct part location. The end effector is equipped with an antishock clutch to protect against potentially damaging shock loading. A convergent beam sensor is used to determine the distance to a particular machined surface to guide the conical locator to within $\pm 1/4$ in. (6.4 mm). Monorail carriers are made of lightweight fiber-glass-reinforced plastic and have keels on the bottom to maintain a stable position while the robot picks the parts. The system is foolproofed by several features, including carriers that break at an inexpensive clevis rather than at more costly points and controls which allow a part to recirculate if it is not detected and located correctly by the sensing devices.

Miscellaneous Applications

Robots can be used for a variety of quality-control tasks (32). For example, automobile windows, doors, and trunks can be checked for leaks by injecting helium gas into the compartments and using a robot to manipulate a "sniffer" to detect leaks. The robot travels over some 65.6 yd (60 m) of the car body, acculumlating 400 items of information describing leaks. It tracks a specific path around the auto, maintaining relative motion between itself and the auto body. The conveyor track was specially designed to position the auto accurately for these tests.

A track-mounted ASEA IRb-60 robot checks dimensional accuracy of certain features on welded auto bodies (33).

ANALYSIS FOR DESIGN

A robot can be taught to perform tasks while the workpiece is stationary. Subsequently, control computers can synchronize robot motion with that of the moving conveyor line. The speed at which a robot performs assigned tasks can be adjusted by the controller to adapt to variations in conveyor speed. In addition, tasks may be reassigned to other robots in the line by downloading the appropriate program to control the designated robot. These directives may involve both local and central control computers.

Several other issues are particular to line tracking. All three dimensions of the work zone of the robot must be carefully considered to assure that assigned tasks can, in fact, be accomplished. In addition, tracking inherently involves a nonproductive activity, ie, returning to the "starting" point after assigned tasks have been completed.

Approaches for analyzing potential line-tracking designs are discussed in this section. Certain techniques may be included in comprehensive approaches developed for specific types of applications (eg, for automated assembly systems, see Refs. 1 and 34). Analysis required to design single-station as well as multiple-station applications of conveyor tracking is reviewed. Furthermore, the use of simulation to analyze potential designs is described.

Single-station Cycle Analysis

The productivity of a line-tracking station can be maximized by considering the entire set of activities that must be accomplished at the station by the robot, sensing devices, and computer controller. Activities may be classified into four groups to facilitate analysis. The classification is meant to be representative rather than exhaustive; some of the activities may not be required in certain applications, and other applications may involve additional activities that would have to be included. The classification is based on assigning each task to one of four sets:

T = set of preoperation tasks to sense presence of the workpiece, identify the type of workpiece, and download robot-control programs to the station.

P = set of tasks required to prepare the robot, including tool change and travel necessary to engage the workpiece.

O = set of operations performed by the robot, including, perhaps, additional tool changes.

R = set of tasks required for the robot to disengage a workpiece and travel back to a "ready" point to start the next workpiece.

This detailed classification indicates that certain tasks involve sensors (ie, T), whereas others require the robot (ie, P, O, and R). All tasks require control from some level in the hierarchy of the computer network.

Since the set O represents the only truly productive set of activities, station productivity is maximized by reducing the effects of other sets of tasks. For example, the robot could travel at maximum speed to reduce the time required to return to the "ready" point (see set R). If the control system permits, the set of activities T could be done at the same time that the robot is performing tasks in the R and P sets.

Operations (ie, set O) must be studied in detail to assure that they are performed correctly but expeditiously. The process implemented by the robot is the primary determinant of operation time. For example, in paint-spraying applications, paint must be delivered at a rate that gives a high quality finish without runs or missed areas, and the rate of application depends on the type of paint, its viscosity, and the air pressure used to dispense the paint (35).

After analyzing cycle components, conveyor speed and workpiece spacing must be determined to permit completion of all station activities and to enable the robot to move in synchronization with the line.

Multiple-station Analysis

If a series of stations is to be used, conveyor speed and workpiece spacing must be determined in accordance with the "slowest" station. Variations that allow robots to operate in parallel at a station or to overlap the work envelope of neighboring robots must be planned in detail, especially to incorporate collision-avoidance precautions.

A central controller can monitor an entire line, including the status of parts being fed to robots and line speeds; tasks may be reassigned to maintain high levels of productivity. For example, if a robot goes down, tasks may be reallocated by the central controller to other robots in order to maintain production throughput rates.

In multistation lines, the assignment of tasks to individual robots determines line productivity, since the robot requiring the longest cycle time will determine the throughput rate of the entire line. Task assignment can be viewed as a line-balancing problem, which has traditionally been formulated using one of two objectives: (*1*) given a cycle time, minimize the number of stations required; and (*2*) given a number of stations, minimize cycle time. The first of these problems determines the number of robots to be installed and the second relates to the allocation of tasks in ongoing operations. Algorithms that might be applied to solve each of these problems can be found in Ref. 36.

After balancing the line, robots that can complete assigned tasks in less than the cycle time will incur an idle time each cycle. However, this idle time could be eliminated by reducing the speed of the robot, thus conserving energy and, perhaps, extending the life of the robot. This possibility raises questions concerning the types of robots to be specified. Slow, inexpensive robots could be used in conjunction with faster, more expensive ones, but such a mix complicates operations due to the combination of tooling and control requirements of the different robots. Typically, it is best to specify a single type of robot for as many stations as possible.

In addition, the line may produce a mix of models (ie, of workpieces) requiring more lengthy times for workpiece identification and tool changing. In robotized installations, the time to change from one model to another includes only the time to retool—if general-purpose tooling is not used—and to download necessary computer programs. Of course, provision must be made to supply correct parts to produce the new model.

Issues related to balancing robotized lines are discussed in Refs. 36–39 (mixing robots and human operators on assembly lines).

Analytical models have been developed to assist in the design of flowlines (40). Models are available to design and analyze conveyor systems (41). The references summarize available models and approaches; most require adaptation for robotic applications.

Simulation

Three kinds of simulation models can be used to analyze potential conveyor-tracking designs. Discrete-event (Monte Carlo) models, computer graphics, and prototype installations can all be used to gather pertinent information to assist in design decisions.

Discrete-event simulation is best used to evaluate line productivity. By supplying data that give the task times, precedence relationships among tasks, and resources (eg, robot, tooling, sensors) required to perform each task, the designer can use discrete-event simulation to estimate line performance, including throughput rates, resource utilizations, and interactions among resources. Such an analysis could be used, for example, to select the most appropriate type of robot for the application, evaluate station layout, or determine the best type of tooling to use. Discrete-event simulation is limited, however, in capability to analyze the possibility of the robot colliding with the workpiece and to design detailed aspects of methods to be used.

Computer graphics is better suited to analyzing the potential for collisions and for studying methods details. The McAuto Division of McDonnell Douglas has developed an extensive graphics package to aid designers of robotic workplaces (42). In particular, motion sequences associated with line-tracking applications can be readily simulated with PLACE (positioner layout and cell evaluation), one component of the McAuto program. Other portions of the McAuto program are BUILD, COMMAND, and ADJUST.

It is important to use a prototype cell to simulate production operations, allowing potential problems to be identified and corrected before actual production begins (43). Robots are used in line tracking to assemble engines for the Pontiac Motor Division of General Motors. Advantages of using prototype simulation include the following: production personnel can more easily relate to the physical mock-up than to a computer-graphics display; skilled-trades personnel can easily determine the effect of the design on maintenance activities; potential collision points in the work envelope can be readily identified; program steps can be checked in sequence to assure completeness; "what if" checks can be made, for example, to determine the possibility of pallets becoming stuck. Furthermore, consulting production and skilled-trades personnel can give them a sense of involvement, which can facilitate adoption in the actual workplace. Engineers can check the accuracy and repeatability of the robot and can confirm the adequacy of tooling designs. Actual parts can be used to check for variability in gripping surfaces that might cause jams in production operations. The simulation cell described in Ref. 43 cost about 3% of the project cost and was thought to be well worth the expense, since it avoided lost production time to debug the design and led to high quality and high-rate production.

TRENDS

Trends in robotic applications emphasize that, to date, the automotive industry has been the predominant consumer of robots (14). Line-tracking applications within that industry include welding, spray painting, assembly, and material handling, all of which have been discussed previously. However, it is expected that the automotive industry will soon become saturated, so that future growth in robotic applications is anticipated to be in the appliance and electronics industries. Material handling is expected to be the primary application in these and other industries.

However, several other trends are emerging. Robots are becoming faster and stronger mechanically. Sensor technology is improving rapidly, particularly in the area of vision, so that assembly and inspection applications may increase in the future. Another trend is that integrated manufacturing systems developed by a single supplier are becoming more the norm than is the "island of automation" approach. Combining these trends, it seems likely that line-tracking applications will increase since equipment will be more capable of such activities, a single designer will be more likely to consider all opportunities for productivity improvement, and numerous applications not yet envisioned may evolve in the future.

BIBLIOGRAPHY

1. S. Y. Nof, ed., *Handbook of Industrial Robotics,* John Wiley & Sons, Inc., New York, 1985.
2. T. Owen, *Assembly with Robots,* Prentice-Hall, Inc., Englewood Cliffs, N.J., 1985.
3. C. R. Asfahl, *Robots and Manufacturing Automation,* John Wiley & Sons, Inc., New York, 1985.
4. J. Y. S. Luh, "Design of Control Systems for Industrial Robots," in Ref. 1, pp. 169–202.
5. L. A. Hautau and F. A. DiPietro, "Planning Robotic Production Systems," in Ref. 1, Chapt. 34, pp. 691–721.

6. J. P. Ziskovsky, "Gantry Robots and Their Applications," in Ref. 1, Chapt. 60.
7. H. J. Warnecke and J. Schuler, "Mobile Robot Applications," in Ref. 1, Chapt. 59.
8. "Spotlight: Workmaster Goes Against the Fashion," *The Industrial Robot,* 245–247 (Dec. 1983).
9. F. A. DiPietro, "Automated Body Systems from the Ground Up," *Proceedings of the 13th International Symposium on Industrial robots and Proceedings, Robots 7,* Society of Manufacturing Engineers (SME), Dearborn, Mich., Apr. 17–21, 1983, pp. 11-1–11-13.
10. "Modernization: Robots as an Aid to Recovery," *The Industrial Robot,* 82–85 (June 1984).
11. "Cover Story: Change of Gear for Car Manufacturer," *The Industrial Robot,* 220–225 (Dec. 1984).
12. J. Hartley, "Japanese Scene: Nissan Brings Robots to Assembly Line," *The Industrial Robot,* 253–257 (Dec. 1984).
13. J. Hartley, "Welding: BMW's Alternative Robot Technology," *The Industrial Robot,* 114–118 (June 1983).
14. P. G. Heytler, "Robotics and Factories of the Future," in S. N. Dwivedi, ed., *Proceedings of an International Conference,* Charlotte, N.C., Dec. 4–7, 1984, pp. 69–76.
15. M. J. Sullivan, "The Right Job for Robots," *Manufacturing Engineering,* 51–56 (Nov. 1982).
16. M. Sciaky, "Robots in Spot Welding," in Ref. 1, Chapt. 48.
17. J. Hartley, "Japanese Scene: Emphasis on Cheaper Robots," *The Industrial Robot,* 48–54 (Mar. 1983).
18. G. E. Munson, "Industrial Robots: Reliability, Maintenance and Safety," in Ref. 1, Chapt. 35.
19. T. J. Bublick, "Guidelines for Applying Finishing Robots," *Robotics Today,* 61–64 (Apr. 1934).
20. Ref. 3, p. 261.
21. Ref. 3, pp. 261–266.
22. T. J. Bublick, "Robot Applications in Finishing and Painting," in Ref. 1, Chapt. 76.
23. Ref. 3, p. 266.
24. J. Hartley, "Japanese Scene: Applications Diversity," *The Industrial Robot,* 56–61 (Mar. 1982).
25. J. Hartley, "Japanese Scene: Assembly in Focus," *The Industrial Robot,* 203–210 (Sept. 1984).
26. K. Isoda and M. Takahashi, "Assembly Cases in Production," in Ref. 1, Chapt. 66.
27. T. Csakvary, "Planning Robot Applications in Assembly," in Ref. 1, Chapt. 63, p. 1068.
28. *Ibid.,* pp. 1081–1082.
29. J. A. White and J. M. Apple, "Robots in Material Handling," in Ref. 1, Chapt. 54.
30. A. Ferloni and E. Grassi, "Towards the Automated Factory: An Advanced Washing-Machine Production Line Managed by Robots," *Proceedings of the 12th Symposium on Industrial Robots and 6th International Conference on Industrial Robot Technology,* Paris, June 9–11, 1982), pp. 175–183.
31. R. N. Stauffer, "The Anatomy of a Line-Tracking Robot," *Robotics Today,* **6,** 33–35 (Feb. 1984).
32. M. P. Kelly and M. E. Duncan, "Robots in the Automotive Industry," in Ref. 1, Chapt. 40.
33. J. A. Kaiser, "Robot-Operated Body Inspection System," in Ref. 1, Chapt. 71.
34. R. E. Gustavson, J. L. Nevins, D. E. Whitney, and J. M. Rourke, "Assembly System Design Methodology and Case Study," *Proceedings, Robots 8,* Society of Manufacturing Engineers (SME), Detroit, Mich., June 4–7, 1984, pp. 6-44–6-65.
35. C. Morgan, *Robots: Planning and Implementation,* Springer-Verlag, New York, 1984.
36. R. P. Fisher and C. H. Falkner, "An Assessment of the Applicability of Current Balancing Techniques to Robotic Assembly Lines," Dept. of Industrial Engineering, University of Wisconsin, Madison, Wisc., 1983.
37. Ref. 2, pp. 110–116.
38. T. Owen, *Flexible Assembly Systems,* Plenum Press, New York, 1984, pp. 58–62.
39. Ref. 2, pp. 116 and 144–146.
40. G. M. Buxey, N. D. Slack, and R. Wild, "Production Flow Line System Design—A Review," *IIE Trans.,* **5** 37–48 (Mar. 1973).
41. E. J. Muth and J. A. White, "Conveyor Theory: A Survey," *IIE Trans.,* **11,** 270–277 (Dec. 1979).
42. R. N. Stauffer, "Robot System Simulation," *Robotics Today,* **6**(3), 81–90 (June 1984).
43. R. J. Schwabel, "Simulating Production Critical Assembly Line Applications," in Ref. 34, pp. 4-12–4-18.

General References

E. Kaffrissen and M. Stephans, *Industrial Robots and Robotics,* Reston Publishing Co., Inc., Reston, Va., 1984.

B. R. Sarker, "Some Comparative and Design Aspects of Series Production Systems," *IIE Trans.,* **16,** 229–239 (Sept. 1984).

COST/BENEFIT

Lawrence T. Michaels
Ernst & Whinney
Cleveland, Ohio

Lute A. Quintrell
Price Waterhouse
Cleveland, Ohio

INTRODUCTION

Factory automation can provide large benefits to manufacturing companies. Yet many companies find it difficult to justify and track the benefits. Many executives are willing to invest in automation, but they are unsure of which areas represent the greatest opportunity for improvement, how to use advanced technologies to achieve productivity improvement, and which is the best way to monitor the anticipated benefits.

Excellence in manufacturing has traditionally been associated with quality and customer service. Today's competitive marketplace also requires increased emphasis on cost reduction and throughput improvements. In order to win the competitive edge race, companies must continue to improve quality and customer service, while simultaneously reducing operating costs. Success will require *continuing* improvements in productivity and efficiency. In order to meet these demands, many companies are looking to computer-integrated manufacturing (CIM) technologies to achieve their improvement objectives. Manufacturing excellence is therefore dependent on the effective selection and use of technology. Technology management is a critical strategic issue. In addressing these demands, technology managers must answer three questions. These are

- What improvement opportunities will provide the greatest benefit? What does one invest in first?
- How does one quantify the anticipated benefits and track the actual benefits?
- What supporting systems and new performance measures are needed to achieve the anticipated results?

The failure to address these questions can postpone or prevent the successful implementation of new technologies. Those companies which fail to focus on these questions are risking continued profitability and long-term survival. Future success will depend on mastering the technology management process.

As companies invest in computer-based equipment, managers are finding that the "traditional" measures of cost and manufacturing performance are not satisfactory. Management systems, particularly cost accounting, are hindered by a number of constraints in quantifying and tracking the benefits of new investments in factory automation. Some of the more significant constraints include:

- Product life cycles are becoming shorter; engineering changes are becoming more frequent.
- Production processes are becoming less reliant on direct labor as the primary factor in controlling production.
- Introduction of new technologies rarely replaces the existing method in a one-for-one relationship.

These constraints make it difficult to evaluate new manufacturing technologies and develop a baseline from which to measure cost and productivity improvements (see also ECONOMICS, ROBOTIC MANUFACTURING AND PRODUCTS; ECONOMICS, ROBOT MARKET AND INDUSTRY). The next section of this article will describe how the difficulties of one company can be traced to ineffective cost management practices and inadequate project management procedures.

CASE STUDY—THE C. I. MORGAN (CIM) COMPANY

CIM (a fictitious company) manufactures a range of parts for use in the automotive industry. CIM's products are manufactured in two plants located in the Midwest. The production process includes the use of turning, boring, milling operations, heat treating, chrome plating, and metal stamping. CIM sells to OEMs, as well as to the automotive aftermarket. The company was founded in the early 1900s by Charlie I. Morgan to supply parts to the automotive industry, and since then, the company has shown an excellent record of growth. CIM is viewed by its customers and competitors as a well-established company with a reputation for quality products.

In the mid-1970s, CIM was confronted with the increasingly stringent demands of improved customer service and awareness of the competition's use of improved equipment and processes. CIM's president, John Late, decided that a formal planning process was needed to help meet these challenges. The objective of this process was to improve productivity while maintaining or improving quality. Mr. Late's instructions were clear: "Reduce costs, waste, and nonproductive time. Do what it takes!"

The CIM management team started working on these goals. Individual departments developed their own plans, and many improvement projects were implemented. State-of-the-art equipment was in stalled, including new computer-controlled machines for the manufacturing areas. Management was soon able to report that direct labor costs were reduced, in some departments even more than anticipated. Yet Marty Profit, CIM controller, was perplexed; as labor costs decreased, the cost reports were not showing a similar reduction in product costs.

CIM's management was concerned, and John Late, in particular, was furious. CIM management, unsure why its original effort had failed, doubled its efforts. Late authorized the hiring of a strategic planning manager to help develop a business strategy. The resulting plan was embraced by company management, and the work force was given a briefing. The plan represented a sound understanding of CIM's customers and competitors and of the direction the market was heading. Senior management was committed to improving productivity and willing to provide the necessary resources. CIM had competent people, particularly in design engineering and manufacturing, who were excited by the management commitment. In addition, CIM had extensive information systems, although all were relatively old. It appeared that all was in place to be successful this time.

For middle management, the improvement program goal was the reduction of labor costs through the modernization of equipment. An analysis of CIM's product costs showed that half the costs of goods sold were internal value-added costs. The value-added costs comprised direct labor and overhead; overhead was assessed to each product based on direct labor. Some managers reasoned that if direct labor were reduced, a reduction in overhead would follow. Unfortunately, CIM was to discover that there may be a correlation between output and overhead, but output does not cause overhead to occur.

Once again, individual managers developed their own plans. Budgets and cost savings projections were developed by each manager. The thrust of each of the plans was to address the labor-intensive areas of the plant. Manufacturing management decided to address the turning, boring, and milling areas of the plant first. The manufacturing engineering department designed and implemented a flexible manufacturing system. Conveyors and robots were purchased and installed in the heat treating area. The new equipment worked well, and direct labor was reduced. As these improvements were developed, the size of the engineering and manufacturing support areas increased dramatically.

Much to management's disappointment, product costs increased and overhead went through the roof. John Late and his managers could not explain what went wrong with the strategic business plan. In retrospect, CIM's plan was not at fault; the problem was centered around CIM's approach to selecting modernization priorities.

There were three deficiencies in CIM's approach to modernization; if the steps had been included, CIM would have realized improved benefits of factory automation. First, CIM spent a great deal of effort in understanding the competitive marketplace, but did not spend much time understanding its own manufacturing needs. Solutions were implemented on a first-come first-serve basis; there was no prioritization of improvement programs or needs. Second, each department developed individual solutions to address CIM's mandate. These efforts can best be described as single-dimension solutions looking for a problem. This resulted in many "islands of technology" and incomplete results. Finally, no efforts were directed at developing a benefit-tracking system for the improvements or conducting a sound cost–benefit analysis of the proposed solutions.

The existing cost accounting system was used to measure and report product costs. This system reported product and work order costs, but did not provide any management information regarding manufacturing process cost and performance. In developing the solutions, CIM's managers did not analyze the company cost profile and attempt to understand the impact of new technology on the cost structure of the company.

CIM's management would have been spared the disillusionment of the apparent failure of its efforts if it had improved the technology selection and justification effort. CIM's experience is not unlike many other companies that have attempted to implement factory automation projects. How projects are selected can affect a company's ability to improve productivity, quality, flexibility, or other operational factors.

Fortunately, this case study has a happy ending. The remainder of this article will describe a success formula used by this company and many others to improve the technology management process.

THE RELATIONSHIP BETWEEN TECHNOLOGY AND COST MANAGEMENT

Many companies are finding that in order to justify automation projects, it is necessary to expand the traditional role of cost accounting. Cost accounting has traditionally been associated with the gathering of historical data to support inventory valuation and determining costs-of-goods-sold information. The role of cost accounting needs to be expanded to measure the effectiveness and efficiency of the operating functions of a company. Specifically, the role of cost management needs to be expanded to provide:

- Information regarding controllable/value-added process costs.
- Integration of actual operational and financial data to support the performance measurement process.
- Visibility regarding the cost improvement potential associated with productivity and efficiency improvements.
- The impact associated with manufacturing cost displacement; that is, assurance that a solution is complete, will focus on real opportunities, will provide anticipated results, and will not increase overall costs.

Cost information is needed to establish standards of performance, set priorities, provide the means to establish specific design objectives, and manage performance for the new automated factory environment.

Ingredients for Success

The threats to company survival posed by increased competition and declining productivity are forcing companies to adopt new methods. The benefits associated with robotics, CAD/CAM, new manufacturing information systems, and other advanced technologies are perceived by many as a panacea. However, in order to be successful it is imperative that companies adopt state-of-the-art thinking in order to properly select and use available state-of-the-art technologies.

This begins with the development of a factory automation master plan. The overall strategy and structure of the plan should be consistent with and support business goals and objectives. However, it is also important that this plan identify major opportunities for improvement, be implementation oriented, and foster user participation and sponsorship (genuine ownership) of proposed solutions. In order to assure that the efforts address the goals, the following elements should be part of the overall planning and technology improvement effort:

- Preparation of a master plan that encompasses considerably more from a strategic perspective than just "making the machine run faster."
- Recognition within the planning process of the importance of completing a formal needs analysis and conceptual design prior to proceeding with detailed design and implementation.
- Establishment of a multidisciplined project team and a project scope that encompasses all aspects of company operations.
- Use of cost and performance behavior patterns, which focus on cost and productivity "drivers" and integrated performance measures, as objectives to screen and select technology improvements.

The second key ingredient for success is people. Factory automation requires an investment in people as well as machines. The mix of the right people and skills working on a project will help to ensure that the benefits of factory modernization are realized. Unfortunately, many companies do not use a multidisciplined project team. Instead, they let individual departments develop plans that often overlap, are not realistic for the company as a whole, and may not foster an ownership feeling.

Using a multidisciplined team can ensure proper and complete assessments of improvement opportunities and a thorough understanding of the critical issues and potential barriers and will result in user commitment to *company* (rather than department) solutions. This approach will also help to develop an understanding of the strategic importance of this effort.

The final ingredient in a successful automation project is the use of improved planning methodologies to:

- Develop a functional/process baseline of company operations (cost and performance improvements occur at the process level and result in lower product cost).
- Identify costs incurred in the manufacturing process that are controllable or value-added.
- Use of current cost and performance levels to complete a needs analysis (the objective is to identify areas of the company with high cost and low performance, not to justify a particular technology.

- Focus on cost and productivity "drivers" ("drivers" are those independent variables that control or cause costs to occur and impact throughput).
- Identify and justify programs rather than discrete projects.

By using these methodologies, a project team will be able to evaluate the current resources and methods ("as is" baselines). This will ensure that automation solutions are consistent with the needs and objectives of the company. By doing this, the project team can identify those functional areas that, for example, are cost intensive, are characterized by low performance, or do not support future business objectives. The benefits of this approach are

- Completion of a needs analysis that identifies those areas that represent the greatest potential for improvement.
- Identification and understanding of the key success factors or drivers of cost and performance.
- Selection and integration of system, mechanization, and management technologies to provide comprehensive solutions and avoid islands of technologies.
- Identification of the cost tradeoffs associated with factory automation projects to assure that overall costs are reduced.

Guidelines

Based upon the authors' experience with automation projects, it has been found that there are seven general guidelines that should be used to simplify the modernization process and help ensure success.

Guideline 1

Structure the factory model on a top-down functional basis.

Suggested Approach

- The needs analysis should be conducted on a functional basis. Developing a model of factory functions (ie, a logical grouping of interrelated processes) will improve the needs analysis. Cost reduction occurs at the process level, not the cost center on product level. Reduce the cost of the process/function to reduce the cost of the product.
- Second, a functional approach will help to prevent the occurrence of "islands of technology" that can cause integration problems.
- Last, a functional approach tends to simplify the task of verifying that *overall* costs and performance will be improved.

Guideline 2

A formal needs analysis and conceptual design should precede detailed design and implementation.

Suggested Approach

- Performance should be defined before attempting to improve it (in some companies, new performance measures may be necessary to assess improvement potential).
- High-impact areas in terms of high cost and low performance should be identified—these areas will produce the greatest economic and performance benefits.
- "As is" cost behavior patterns and performance should be used as conceptual design objectives to screen technology alternatives.
- Technologies tend to work together to produce benefits; "single-dimension" projects do not produce comprehensive solutions and will not maximize benefits.

Guideline 3

Multidisciplined project teams should be used. User involvement is critical for a complete assessment of improvement potential.

Suggested Approach

Project team members should include individuals from the following areas to avoid parochial solutions and ensure ultimate acceptance of the technology improvement program:

Product and process engineering.
Quality.
Shop floor supervision.
Production planning.
Data processing.
Cost accounting.

Guideline 4

The success of an improvement program should be evaluated by its results, not by the sophisitication of the technology selected.

Suggested Approach

- It is always cheaper to eliminate something rather than attempt to automate.
- The proven techniques of value analysis/engineering, cellular manufacturing, and just-in-time/synchronized manufacturing in conjunction with increased computerization and mechanization should be used.
- Problems should be solved first, and then automation can be considered for additional benefits.

Guideline 5

The time-phased economics of the viable CIM improvement programs should be determined.

Suggested Approach

- Individual projects may not be justified by themselves, but may be needed to make the entire program work and be fully justifiable.
- Cost–benefit analysis should provide management with several levels of economic analysis, including:

Annualized savings (over the planning horizon).

Return on investment (ROI).

Break-even analysis based on discounted cash flow.

Risk-adjusted savings.

Guideline 6

The information provided by the cost-benefit tracking process must be verifiable and auditable.

Suggested Approach

- Actual cost and performance data should be captured.
- Key cost savings and performance improvements should be tracked individually (all lesser costs and performance factors can be analyzed in aggregate).
- Conflicts with other operational and financial reporting requirements should be resolved.
- Costs and benefits should be tracked over the life of the improvement program process to ensure continued management commitment to the modernization process. Actual cost information should be maintained to "close the loop" and support ongoing modernization efforts.

Guideline 7

The project team, key management personnel, and intended users should be educated.

Suggested Approach

- Traditional approaches to cost management and project selection need to be improved; up-front education and training are needed to produce meaningful results.
- The use of software tools to minimize data handling and provide more time for needs analysis and conceptual design activities should be investigated.

MAKING IT HAPPEN

Installing factory automation will not, on its own, ensure that the anticipated benefits will be realized. Developing new technology to improve the manufacturing process is, in reality, a recognition that the automated factory of tomorrow will require a change in management systems and practices. The present methods of measuring business performance may not be appropriate for a new factory. Managers are recognizing that automation impacts not only direct manufacturing operations, but also the support functions. These include:

Material handling.

Quality control.

Engineering and manufacturing services.

Inventory management and production scheduling.

Cost accounting.

Experience has shown that three basic changes are needed to improve technology management. The first involves expanding traditional performance and cost–benefit tracking procedures. New factory automation technologies will have different development and operating costs than traditional manufacturing processes. The amount of engineering development and support is higher, and maintenance, computer system, and production control costs will also change with the introduction of new factory automation. At the same time, there will be corresponding benefits, for example:

Increased manufacturing flexibility.

Improved product quality.

Reduced manufacturing throughput time.

Improved control over manufacturing process to reduce overall costs.

Second, the traditional role of cost accounting needs to be expanded. In particular, the traditional method of classifying certain costs as overhead needs to be reviewed. The introduction of new technology will increase data availability and improve accuracy. This will allow companies to reclassify and manage many cost elements in a direct manner. Specifically, accounting systems will move from a labor-based system to a machine or process cycle accounting system; this will allow for greater visibility and control of the factors of production.

The third change required is in the area of performance monitoring and cost–benefit tracking. Low productivity may be attributed to a manufacturing operation, even though the *cause* may be found someplace else. A new robot may not achieve the desired results if it is continually inoperative due to poor maintenance or part shortages. The performance monitoring system should be able to identify these problems and provide management with action-oriented recommendations to address the problems in an effective and timely manner.

It is important that performance monitoring and cost–benefit tracking be included in the conceptual design of the automation project. Planning must include *what* is to be measured and *how* the measurement will be accomplished. As programs are developed, a "to be" cost baseline is prepared. This baseline will support benefits analysis, risk management and assumptions, and the development of the required tracking mechanisms. The tracking system should integrate financial and statistical operating data. By accomplishing this task, the key cost and productivity drivers will be monitored in addition to overall cost and performance.

Data collection should be accomplished in a timely fashion, and data should be collected at the lowest practical level of detail (where corrective action can be implemented). The collection of unnecessary data should be avoided. A system should be developed to support the following requirements:

Ledger accounts.

Organizational cost centers.

Job/work orders.

Product.

Functional processes.

Projects or contracts.

By developing a cost–benefit tracking system, many companies have found that they have realized additional benefits that equal or exceed the potential benefits for factory automation. By recognizing the changes that are required to be successful, management can ensure that problems are detected on a timely basis, adjustments made, and the benefits of the investment in factory automation realized.

CONCLUSIONS

The introduction of new technology in manufacturing represents a significant opportunity to improve productivity. As companies address increased competition, investments in factory automation will be imperative if companies want to maintain their competitive edge. The management of factory automation is a strategic issue that must be included in strategic planning. To be successful, management must:

- Allocate the time, resources, and skills to evaluate improvement opportunities.
- Recognize and understand the relationship between factory automation and cost management.
- Anticipate changes in management practices as well as technology.
- Recognize the importance of cost–benefit analysis and tracking.

The goal of every manager involved in factory automation should be to develop a program that is affordable, realistic, and consistent with the future strategic direction of the company. The methods and opportunities will change, but the goal will not. Technology cannot manage itself.

BIBLIOGRAPHY

R. G. Eiler, L. T. Michaels, and W. T. Muir, "The Relationship Between Technology and Cost Management," *Material Handling Engineering* (Jan., Feb., Mar., 1984).

E. L. Krommer and W. T. Muir, "Measuring Operating Performance Through Accounting," *Price Waterhouse Review* (1) 26–31 (1983).

L. T. Michaels, "New Guidelines for Selecting and Justifying Factory Automation Projects," *Proceedings of the Robots 10 Conference,* Chicago, Ill., 1986.

J. G. Miller and T. E. Vollman, "The Hidden Factory," *Harvard Business Review,* 142–151 (Sept. 1985).

W. T. Muir, "Integrated Factory Automation: A Cost/Benefit Perspective," *CIM Review,* 16–20 (Winter 1985).

CYBERNETICS

ANDREW P. SAGE
George Mason University
Fairfax, Virginia

INTRODUCTION AND EARLY HISTORY

Cybernetics is a term that denotes the study of control and communication in, and particularly between, humans and machines. The word cybernetics comes from the Greek word *Kybernetes,* which means "controller" or "governor." The first use of the term is attributed to Professor Norbert Wiener, an MIT professor of mathematics who made many early contributions to mathematical system theory. The first book formally on this subject was titled *Cybernetics* and published in 1948 (1). In this book, Wiener emphasized feedback control as a concept presumably of value in the study of neural and physiological relations in the biological and physical sciences. In the historical evolution of cybernetics, major concern was initially devoted to the study of servomechanisms which later become known as control science and engineering. Cybernetic concerns also involve analogue and digital computer development and the combination of computer and control systems for purposes of automation (qv) and remote control (2–5).

It was the initial presumed resemblance, at a neural or physiological level, between physical control systems and the central nervous system and human brain that concerned Wiener. He and close associates, Warren McCulloch, Arturo Rosenblueth, and Walter Pitts, were the seminal thinkers in this new field of cybernetics. Soon it became clear that it was fruitless to study control independent of information flow, and cybernetics thus took on an *identification with the study of communications and control in humans and machines.* An influence in the early notions of cybernetics was the thought that physical systems could somehow be made to perform better by enabling them to emulate human systems at the physiological or neural level.

Another early concept explored by Wiener was that of homeostasis, which has come to be known as the process by which systems maintain their level of "organization" in the face of disturbances, often occurring over time, of a very large or global nature. Cybernetics soon became concerned with purposive systems, as contrasted with systems that are static over time and purpose.

DEFINITIONS

Cybernetics is a way of looking at things, or a philosophical perspective concerning inquiry, as contrasted with a very specific method (1,6). Fundamental to any cybernetic study is the idea of modeling, and in particular the interpretation of the results of a modeling effort as theories that have normative or predictive value. It is tempting to call cybernetics the study of systems from a cybernetic perspective, but such a circular definition is of little value. There is little explicit or implicit agreement concerning a precise definition for cybernetics. Some users of the term cybernetics infer that the word implies a study of control systems. Other uses of the word are so general that cybernetics might seem to infer either nothing, or everything. Automation, robotics, artificial intelligence, information theory, bionics, automata theory, pattern recognition and image analysis, control theory, communications, human and behavioral factors, and other topics have all, at one time or another, been assumed to be a portion of cybernetics.

Heuristics are also important in the study of cybernetics. Heuristics (qv) are ad hoc methods, or rules of thumb, that assist is resolving an identified task. Cybernetic simulation models are typically heuristic in character and, in this sense, are like the models that humans use, in practice, to resolve problems or issues. A major area of interest in cybernetics is the modeling of human interaction with systems. In this sense, cybernetics is similar to and incorporates artificial intelligence, which may be defined as machine emulation of human abilities. The main concern of artificial intelligence is the ability to develop computer programs that will perceive, learn, and adapt to changing environments in order to enable the computer to accomplish tasks as if it were "intelligent." The notion, at least certainly the centrality, of the human nervous system as playing a role in modern cybernetics has all but vanished today. A much more cognitive perspective is prevalent today.

In work to follow, cybernetics will be needed as the study of the communication and control processes in human–machine interaction for the accomplishment of purposeful tasks. While this is not a universally accepted definition of cybernetics, it is certainly a common, as well as useful, one for this *Encyclopedia*.

A COGNITIVE VIEW OF CYBERNETICS

A purpose of this article is to discuss cybernetics, and the design of cybernetic-based support systems, for purposes such as knowledge support to humans, especially the human–system interactions that occur in such an effort. Thus, the discussions are particularly relevant to knowledge-based system design concerns relative to human–machine cybernetic problem-solving tasks such as fault detection, diagnosis, and correction. These are very important concerns for large-scale systems control applications, such as robotics, at this time. Advances in technology involving computers, automation, robotics, and many other recent innovations, together with the desire to improve productivity and the human condition, render physiological skills that involve strength and motor abilities relatively less important than they have been. In many instances, these physical tasks are now accomplished by robots. These changes diminish the need for physical abilities and increase requirements for cognitive abilities.

The need for humans to monitor and maintain the conditions necessary for satisfactory operation of systems, and to cope with the poorly structured and imprecise knowledge that must be brought to bear on unforeseen occurrences which inevitably occur in modern integrated manufacturing systems, robotic applications, and many other contemporary areas, is greater than ever. Ultimately, these primarily cognitive efforts, which involve a great variety of human problem-solving activities, are translated into physical control signals and control or manipulate some physical process. As a consequence, there are a number of human interface issues that naturally occur between the human and the machines over which the human must exercise cybernetic control.

A number of advances in information technology provide computer, control, and communication systems that enable a significant increase in the amount of information that is available for judgment and decision-making tasks at the problem-solving level. However, even the highest quality information will generally be associated with considerable uncertainty, imprecision, and other forms of knowledge imperfection. Computer, control, and communications technology can assist in human problem-solving tasks by enhancing the quality of the information and knowledge that is available for decision making. When this form of assistance to human problem-solving tasks is provided such that reports to various requests for information are supplied, a classical management information system (MIS) is obtained. An enhancement of a classical MIS system will enable the computer, through the incorporation of various modeling and forecasting algorithms, to respond to "What-if" type questions with "If-then" type responses. Today, the need for decision support is considerably greater than this. The great complexity of tasks makes it desirable for support systems to be able to respond to "What is the best alternative?" type questions with recommendations that are based on the value system of the problem solver and available knowledge. A system that assists in this function is called an *expert system* or augments human performance by the artificial intelligence community (7–9) and or a *decision support system* or *executive support system* by the management science and decision analysis community (10–11). The term *knowledge support system* is often used to denote the integrated use of expert system and decision support system technologies. Through use of knowledge-based systems, the need for another human–system interface is created: one between the human and the computer, with a proper system design to ensure appropriate interaction between the human and the computer (12–14).

Consider interactions between humans and systems at each of these levels as fundamentally cybernetic concerns. First, examine a simple definition for, and for the purpose of, a human–machine cybernetic system. This leads to a discussion of the implications of new task requirements for the human, and needs for human assistance that can potentially be provided by computer support. This in turn leads to a discussion of future prospects for developments in the design of systems for human interactions that employ cybernetic principles.

CYBERNETIC SYSTEMS

A human–machine cybernetic system may be defined as functional synthesis of a human system and a technological system or machine. Human–machine systems are predominantly characterized by the interaction and functional interdependence between these two elements. The introduction of communication and control concerns results in a "cybernetic" system. All kinds of technological systems, regardless of their degree of complexity, may be part of a human–machine cybernetic system: industrial plants, vehicles, manipulators, prostheses, computers, or management information systems. A human–machine system may, of course, be a subsystem incorporated within another system. For example, a decision support system may be part of a larger process control or computer-aided design system which also involves human interaction. This use of the term "human–machine cybernetic system" corresponds, therefore, to a specific way of looking at technological systems integrating engineering and behavioral science concerns, in particular, those of cognitive science and psychology.

Task categories such as controlling, communicating, and the use of these for problem solving describe typical human activities in human–machine cybernetic systems. Of particular concern here are the design of systems to assist in human interaction with the cognitive tasks involved in problem solving and the physiological tasks involved in controlling complex systems of people and machines, and the interfaces between humans and systems that are an integral part of this.

The overall purpose of any human–machine cybernetic system is to provide a certain function, product, or service of reasonable cost under constraint conditions and disturbances that involves and influences the human, the machine, or both. Figure 1 presents a simple conceptualization of human–machine cybernetic systems. The primary "inputs" to a human–machine cybernetic system are a set of objectives that are typically translated into a set of expected values of performance, costs, reliability, and safety. Also, the design must be such that an acceptable level of work-load and job satisfaction is maintained. It is on the basis of these that the human is able to (15,16):

1. Identify task requirements, such as to enable determination of the issues to be examined further and the issues not to be considered.
2. Identify a set of hypotheses or alternative courses of actions that may resolve the identified issues.
3. Identify the probable impacts of the alternative courses of action.
4. Interpret these impacts in terms of the objectives or "inputs" to the task.
5. Select an alternative for implementation and implement the resulting control.
6. Monitor performance so as to determine how well the human and the system are performing.

Many researchers have described activities of this sort in a number of frameworks that include behavioral psychology, organizational management, human factors, systems engineering, and management science (15–23).

Traditional human–machine cybernetic systems analysis focused on skill-based behavior and physiological concerns, but new developments emphasize the integration of these with formal knowledge-based and heuristic rule-based cognitive activities. This has been motivated by copious evidence that humans do not react to task requirements in a way that is easily stereotyped but rather in a fashion that is very much a function of the task, the requirements perceived for the task, the environment into which the task is embedded, and the experiential familiarity of the human with the task, the task requirements, and the environment into which these are embedded (19,23–27). Thus, the focus of attention in systems design is recognized

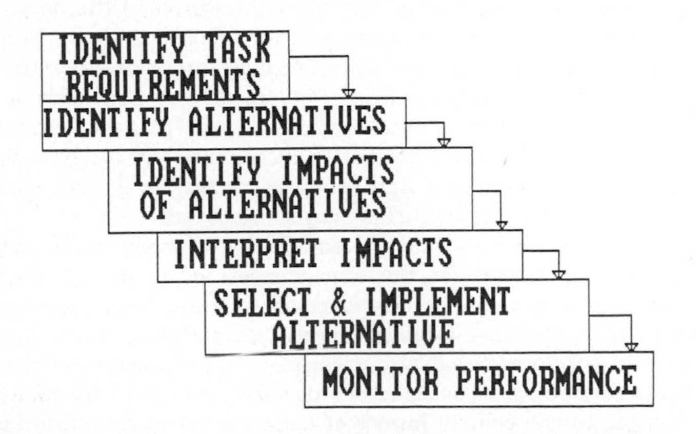

Figure 1. Human–machine cybernetic system (in task performance).

to contain major human–system interaction concerns, and has shifted to systems engineering in a broad sense, particularly to the information science and information technology components of systems engineering.

Many questions can be raised concerning the use of information for judgment and choice activities, as well as activities that lead to the physical control of an automated process. Any and all of these questions can arise in different application areas. These questions relate to the control of technological systems. They concern the degree of automation with respect to flexible task allocation and they also concern the design of computer-generated displays with preprocessing capabilities. Further, they relate to all kinds of human–computer interaction concerns, as well as management tasks at different organizational levels: strategic, tactical, and operational. For example, computer-aiding prospects invade more and more areas of systems design, operation, maintenance, and management. The importance of the shifting from hardware and environmental aspects to software considerations and the integrated consideration of systems engineering that involve human interaction is expressed by the term "software systems engineering."

Human tasks in human–machine cybernetic systems can be condensed into three primary categories that are of principal interest here (28):

1. Controlling (physiological)
2. Communicating (cognitive)
3. Problem solving (cognitive)

In addition, there exists a monitoring or feedback portion of the effort that enables learning. Associated with the rendering of a single judgment and the associated control implementation, the human monitors the result of the effect of these activities. The effect of present and past monitoring is to provide an experiential base for present problem conceptualization. In our categorization of the previous section, activities 1–4 may be categorized as problem (finding and) solving, activity 5 involves implementation or controlling, and activity 6 involves communications or monitoring and feedback in which responses to the question "How good is the process performance?" enables improvement and learning through iteration. Of course, information flow and communication are involved in all these activities.

These three human task categories are fairly general. Figure 2 shows an attempt to integrate them into a schematic block diagram. Controlling shall here be understood in a much broader sense than in control theory. It comprises controlling in the narrower sense, including open-loop vs closed-loop and continuous vs intermittent controlling, as well as discrete tasks such as reaching, switching, and typing. It is only through controlling that outputs of the human–machine cybernetic system can be produced as shown in Figure 2.

Although human functions on a cognitive level can and do play a role in control implementation, their major importance lies in problem-solving activities. Tasks such as fault detection, fault diagnosis, fault compensation or managing, and planning are particularly important in problem solving. Fault detection concerns the identification of a potential difficulty concerning the operation of a system. Fault diagnosis is concerned with

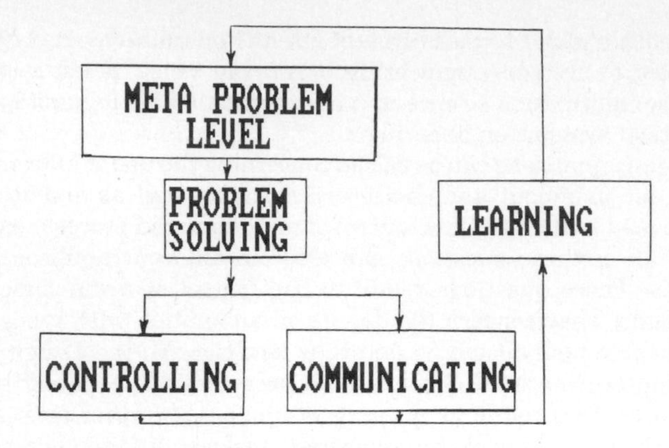

Figure 2. Cybernetic system learning.

identification of a set of hypotheses concerning the likely cause of a system malfunction, and the evaluation and selection of a most likely cause. It is primarily a cognitive activity. Fault compensation or managing is concerned with solving problems in actual failure situations. This may occur through the use of rules that are based on past experience, and the updating of certain rules based on the results of their present application. It is accomplished with the objective of returning the overall system to a good operating state. Fault compensation or managing involves both cognitive and physiological activities. Planning is a cognitive activity concerned with solving possible future problems in the sense of mentally generating a sequence of appropriate alternatives or rules for reaching future states under different foreseeable and unforeseeable conditions (13,26,27).

In all these problem-solving tasks, the skills or rules necessary to accomplish a set of tasks are stored in the knowledge base of a support system after their generation or modification. Often they are generated through a formal reasoning-based effort. A cognitive engine is used to assess the specific skill or rule or formal knowledge-based approach, or combination of these, useful for a specific task. From there, these skills or rules can be utilized in the lower-level processes of controlling, as in Figure 2. This categorization into a *knowledge base,* a *cognitive engine,* and a *interface mechanism* to enable communication between these two and the users of the system, is typical in expert system approaches (7). It illustrates the close relation between expert system and cybernetic system technologies. Both are *information technologies*.

There are many other tasks in human–machine cybernetic systems, such as monitoring and communicating, which can be classified as subtasks or supporting tasks of the two primary activities of problem solving and controlling. Communicating, in a human–machine cybernetic systems context, comprises two very different types of communication. The first one is the verbal communication between the members of the group jointly responsible for aspects of system operation. The second type is "communication" between humans and the technological systems to be controlled and the environment, or between humans and various support systems used to provide cognitive assistance in problem-solving and control tasks. Often this latter form of communication (14,29,30) is called human–computer dialogue. There are many interface concerns between humans and technological systems at the control and communications level. These range from cognitive representations (31–36) to fault management and monitoring (37–40) and the identification of information requirements for problem solving (41).

HUMAN PERFORMANCE LIMITATIONS

A large number of studies (42) show that, in an unaided condition, humans are very inadequate at a number of information-processing tasks. Consequently, associated acts of judgment and choice are often quite flawed. Simply supplying more of the vast amount of information that is now potentially available is not the answer; studies show that the presence of greater amounts of unstructured information may simply make the human problem solver's performance more erratic, and otherwise poorer as well, due to the use of various forms of selective perception to cope with the vastly increased amount of information and associated cognitive overload. In the face of all of this, the human problem solver is likely to resort to time honored and tested methods of judgment that are based on holistic and intuitive behavior. When true expertise is involved, these holistic and intuitive approaches will often be both the appropriate response and the effective and efficient one. There are many types of appropriate skill-based behavior. There is much interest at this time in the design of artificial-intelligence-based systems that capture the skill-based behavior of experts. There is also considerable interest in building decision support systems that assist humans in extrapolating from known and familiar situations using wholistic, skill-based approaches, such as reasoning by analogy.

When the environment (or contingency task structure, as a more general descriptor) has changed and this change is not recognized, then it is very possible that the former "master" becomes no master at all, except perhaps of the art of self deception. In such situations, when the "expert" misjudges the degree of familiarity with the environment, task, or task requirements, then the "expert" is not really an expert at all. The adoption of an intuitive skill-based mode of behavior, when a formal analytical mode is called for, encourages various forms of selective perception, such as failure to seek potentially disconfirming information.

THE DESIGN OF COGNITIVE SYSTEMS

All of this has major implications with respect to the design of systems for the human user, and for robot systems as well. It requires, for the appropriate system design, an understanding of human performance in problem-solving and decision-making tasks. This understanding must be at a *descriptive* level, such that we can predict what humans are likely to do in particular situations, and at a *prescriptive* level, such that we can *aid* humans in various cognitive tasks.

There are many situations in which this need for knowledge-based support for the human operator is strong. With respect to human–machine cybernetic systems, a prototypical robotics application might concern automated systems that require only occasional minor changes in parameter settings for maintenance of satisfactory performance and rare major changes in the control inputs of a structural nature in order to prevent significant malfunction, and perhaps disaster. The need for significant interaction with an operational system will often be very infrequent. The advance notice of the need to interact will often be very short, the importance of the consequence of the human–machine interaction will be very large, the need to interact will occur at times that are very unpredictable, and there will be large amounts of information available when the potential need for operator interaction occurs. A specific description of the situation in which a knowledge-based system is needed will form one of the bases for the determination of information and task requirements for knowledge-based system design. The importance of appropriate knowledge representation to assist the human operator and the need for suitable human–machine interaction and dialogue are apparent. A number of other generic human–machine situations could be cited. These range from strategic planning to manual control of physical equipment, and the need for decision support and associated knowledge representation would be easily recognized in each of them. Unfortunately, the activities that need to take place in order to establish the most appropriate knowledge representation and knowledge support system are not so apparent.

For this reason, it is very important that the designer of human–machine cybernetic systems be aware of various forms of knowledge representation (16,32,34) that are of value to these ends. The form, or frame, of knowledge representation that a person uses is very much a function of the perspective that the person has with respect to the particular issue under consideration. This suggests a contingency task structural approach as being very important. For it is the particular task at hand, the environment into which this task is embedded, and the experiential familiarity of the human problem solver with the task and environment that determines the information acquisition and analysis strategy that is adopted as a precursor to judgment and choice. It is also important to be aware of these concerns as they affect group and organizational behavior. The success of the human in human–machine cybernetic tasks is very much a function of how the human decides how to decide (19–23). Thus, an awareness of a variety of knowledge representations is needed, as is the way in which meta-level knowledge leads to a knowledge representation in terms of the information requirements determined for a particular task, the method of analyzing the acquired information, and the way in which associated facts and values are aggregated to enable judgment.

A special concern here is that the information that is used for judgment and choice is typically not precise clerical and accounting data, but a mixture of this data and information of an imperfect nature. As a consequence, there is a need for the consideration of approaches that allow incorporation of notions of information imperfection (13,27,43). It is an important area for continued research, together with the many other activities related to understanding and improving the large variety of problem-solving tasks associated with human cognitive activities that lead to the effective control of robots and other systems.

Technological advances have changed, and will continue to change, human–machine cybernetic systems for almost all application areas. This is true for industrial plants with integrated automated manufacturing capabilities and for aids to cognitive activities in strategic planning, design, or operational activities. Due to the latter efforts at human support, office automation systems and information systems for observation, planning, executive support, management, and command and control tasks in business, defense, and medicine are similarly influenced by efforts in human–machine cybernetic systems. These involve not only the operation of technological- and management-oriented information systems by highly skilled and knowledgeable personnel, but also cybernetic systems is to provide computer assistance for the maintenance and the design of technological systems.

A problem common to all these applications is the design of appropriate human–computer interaction subsystems. The possibly adaptive task allocation between human and computer, the dialogue design (including the use of natural language), and other software systems engineering aspects are especially important contemporary topics of research and development. With more computerization and higher degrees of automation, much greater attention must be paid to system design for human interaction. Contemporary advanced technologies *potentially* allow a more flexible work organization with higher user acceptance and job satisfaction. However, this potential advantage can only be achieved if the behavioral implications are seriously considered by the designers of computerized human–machine cybernetic systems. Paradoxically, more sophisticated automation systems require greater human knowledge and a better quality of human skills. These are needed so that it becomes possible for humans, at least in a supervisory capacity, to compensate for inevitable physical system limitations that are due to faults and changes in the environment. These will always occur in automated systems; and the potential for catastrophic behavior increases with the degree of automation.

SUMMARY

Contemporary cybernetics interests and needed studies are at the interface of areas of human–machine cybernetic systems, human factors, artificial intelligence, decision support systems, and cognitive science. Human–system interaction concerns that involve models of human problem-solving behavior, models of human behavior in control tasks, displays including visual and auditory presentation, human work load and proficiency, and problem-solving variations with experience and task demands have also been discussed. The implications of all of this for system level design of robotic systems are very significant and are among the most important research frontiers in robotic systems studies.

BIBLIOGRAPHY

1. N. Wiener, *Cybernetics,* The MIT Press, Cambridge, Mass., 1948.
2. W. R. Ashby, *An Introduction to Cybernetics,* Chapman and Hall, Ltd., London, 1956.
3. F. H. George, *Cybernetics,* St. Paul's House, Middlegreen, Slough, UK, 1971.
4. F. H. George, *Cybernetics and the Environment,* Westview Press, Inc., Boulder, Colo., 1977.
5. A. Y. Lerner, *Fundamentals of Cybernetics,* Plenum Publishing Corp., New York, 1976.

6. J. D. Steinbruner, *The Cybernetic Theory of Decision,* Princeton University Press, Princeton, N.J., 1974.
7. A. Barr, P. R. Cohen and E. A. Feigenbaum, eds., *Handbook of Artificial Intelligence,* Vols. I, II, and III, William Kaufman, Inc., Los Altos, Calif., 1981 and 1982.
8. R. Duda and E. Shortliffe, "Expert Systems Research," *Science,* **220,** 261–268 (1983).
9. N. J. Nilsson, *Principles of Artificial Intelligence,* Tioga, Palo Alto, Calif., 1980.
10. J. L. Bennett, *Building Decision Support Systems,* Addison-Wesley Publishing Co., Inc., Reading, Mass., 1983.
11. R. H. Sprague, Jr. and E. D. Carlson, *Building Effective Decision Support Systems,* Prentice-Hall, Inc., Englewood Cliffs, N.J., 1982.
12. S. K. Card, T. P. Moran and A. Newell, *The Psychology of Human-Computer Interaction,* Lawrence Erlbaum Associates, Inc., Hillsdale, N.J., 1983.
13. A. P. Sage, ed., *System Design for Human Interaction,* IEEE Press, New York, 1987.
14. B. Schneiderman, *Software Psychology: Human Factors in Computer and Information Systems,* Winthrop Publishers, Inc., Cambridge, Mass., 1980.
15. A. P. Sage, *Methodology for Large Scale Systems,* McGraw-Hill, Inc., New York, 1977.
16. A. P. Sage, "Methodological Considerations in the Design of Large Scale Systems Engineering Processes," in Y. Haimes, ed., *Large Scale Systems,* North Holland Publishing Co., Amsterdam, 1982, pp. 99–141.
17. R. M. Axelrod, *Framework for a General Theory of Cognition and Choice,* Institute of International Studies, University of California at Berkeley, Berkeley, Calif., 1972.
18. B. H. Kantowitz and R. D. Sorkin, *Human Factors: Understanding People System Relationships,* John Wiley & Sons, Inc., New York, 1983.
19. A. Newell and H. A. Simon, *Human Problem Solving,* Prentice-Hall, Inc., Englewood Cliffs, N.J., 1972.
20. D. A. Norman, *Perspectives on Cognitive Sciences,* Lawrence Erlbaum Associates, Inc., Hillsdale, N.J., 1981.
21. R. W. Pew, *Research Needs for Human Factors,* National Academy Press, Washington, D.C., 1983.
22. H. A. Simon, "On How to Decide What to Do," *Bell Journal of Economics,* **10,** 474–507 (1978).
23. H. A. Simon, *Reason in Human Affairs,* Stanford University Press, Stanford, Calif., 1983.
24. S. E. Dreyfus, "Formal Models VS Human Situational Understanding: Inherent Limitations in the Modeling of Business Expertise," *Office: Technology and People,* **1,** 133–165 (1982).
25. S. E. Dreyfus and H. L. Dreyfus, *Mind Over Machine: The Power of Human Intuition and Expertise in the Use of the Computer,* The Free Press, New York, 1986.
26. J. Rasmussen, "Skills, Rules, and Knowledge: Signals, Signs, and Symbols; and Other Distinctions in Human Performance Models," *IEEE Transactions on Systems, Man and Cybernetics,* **SMC 13,** 257–266 (Mar./Apr. 1983).
27. J. Rasmussen, *Information Processing and Human Machine Interaction: An approach to Cognitive Engineering,* Elsevier North-Holland, Inc., New York, 1986.
28. G. Johannsen, J. E. Rijnsdorp and A. P. Sage, "Human Interface Concerns in Support System Design," *Automatica,* **19**(6), 1–9 (Nov. 1983).
29. A. P. Sage and A. Lagomasino, "Knowledge Representation and Man Machine Dialogue," in W. B. Rouse, *Advances in Man Machine Systems Research,* JAI Press, Inc., Greenwich, Conn., 1983.
30. B. Schneiderman, *Designing the User Interface: Strategies for Effective Human Computer Interaction,* Addison-Wesley Publishing Co., Inc., Reading, Mass., 1986.
31. R. Abelson and M. J. Rosenberg, "Symbolic Psycho-Logic: A Model of Attitudional Cognition," *Behavioral Science,* **3,** 1–13 (1958).
32. R. Abelson, "Differences Between Belief and Knowledge Systems," *Cognitive Science,* **3,** 355–366 (1979).
33. R. Abelson, "The Structure of Belief Systems," in R. C. Schank and K. M. Colby, W. H. Freeman and Co, San Francisco, Calif., 1973, pp. 287–339.
34. J. R. Anderson, *The Architecture of Cognition,* Harvard University Press, Cambridge, Mass., 1983.
35. D. G. Bobrow and A. Collins, eds., *Representation and Understanding: Studies in Cognitive Science,* Academic Press, Inc., New York, 1975.
36. J. F. Sowa, *Conceptual Structures: Information Processing in Mind and Machine,* Addison-Wesley Publishing Co., Inc., Reading, Mass., 1984.
37. W. B. Rouse, "Applications of Control Theory in Human Factors," *Human Factors,* **19** (Aug. and Oct. 1977).
38. W. B. Rouse, *Systems Engineering Models of Human-Machine Interaction,* Elsevier North-Holland, Inc., New York, 1980.
39. T. B. Sheridan and W. R. Ferrell, *Man-Machine Systems: Information, Control, and Decision Models of Human Performance,* The MIT Press, Cambridge, Mass., 1974.
40. T. B. Sheridan and G. Johannsen, eds., *Monitoring Behavior and Supervisory Control,* Plenum Press, New York, 1976.
41. A. P. Sage, B. Galing and A. Lagomasino, "Methodologies for the Determination of Information Requirements for Decision Support Systems," *Large Scale Systems,* **5**(2) (1983).
42. A. P. Sage, "Behavioral and Organizational Considerations in the Design of Information Systems and Processes for Planning and Decision Support," *IEEE Transactions on Systems, Man and Cybernetics,* **SMC-11,** 640–678 (Sept. 1981).
43. L. A. Zadeh, "Outline of a New Approach to the Analysis of Complex Systems and Decision Processes," *IEEE Transactions on Systems, Man and Cybernetics,* **SMC-3**(1), 28–44 (1973).

D

DERIVETERS OF AIRCRAFT WINGS, ROBOTIC

ERNEST FRANKE
JAMES LUCKEMEYER
Southwest Research Institute
San Antonio, Texas

JOHN VRANISH
Naval Surface Weapons Center
Dahlgren, Virginia

INTRODUCTION

In July 1983, the U.S. Navy (North Island Naval Rework Facility) awarded Southwest Research Institute (SwRI) of San Antonio, Texas, a two-year contract to design, develop, and build a robotic deriveter system (RDS) consisting of a robot mounted on a self-propelled vehicle and capable of removing rivets and inspecting rivet holes on aircraft wings. The RDS was conceived in 1979 at North Island and a technical proposal submitted that fall. The proposal was given top priority in fiscal year 1981 and awarded $1.4 million over a two-year development period after it was chosen over 45 other proposals from all three armed forces. The prototype was completed in 1985 and the system is now in production.

The Navy expects rivets to be a significant fastening device in aircraft for many years. It is expecting a $10 million return on investment, primarily from savings in the six naval air rework facilities in the continental United States, at the current work load. The technical paper for the RDS was presented at the Robots V Conference in Detroit in 1980.

High performance aircraft wing structures are held together largely by rivets. From time to time, panels held in place by rivets must be removed for purposes such as inspection for cracks and damage, repair of hydraulic hoses or electrical cables, and modification to new design standards. The skin panels may have an overall area in excess of 1000 sq ft. These panels vary in size and shape and are attached to the frame by up to 4000 rivets for a single upper wing surface in an E-2 or C-2 aircraft. Furthermore, these rivets are not all the same size. The current method of rivet removal is manual drilling and punching of each rivet. This is tedious, slow work and exposes the worker to potential danger from flying metal chips and possible injury due to slipping or drill-bit shattering.

Drilling out the rivets requires concentration, skill, coordination, and strength (to ensure a straight bore down the center of the rivet shaft). The repetitiveness and physical and mental effort involved in this process rapidly lead to fatigue, boredom, and shortened periods of continuous effort with the resultant adverse effect on morale and quality control. As a result, it may take two to three months to get a single wing completely deriveted. Such delays can be costly and critical for aircraft turnaround and availability.

The general requirements for the system as established in the Navy purchase description include:

- The system will be self-contained, with all necessary subsystems mounted on a mobile, battery-powered vehicle.
- The system must use vision or other methods to locate the precise center of rivet heads and to discriminate between common aluminum rivets, steel rivets such as jobolts, and threaded fasteners such as Phillips-head screws.
- Rivet heads must be drilled within 0.005 in. of center with the axis of the drill within 5 degrees of perpendicular to the surface.
- Rivet holes are to be inspected by an eddy current probe to determine if any are cracked or corroded.
- Rivets are to be removed and holes inspected at the rate of at least 120 rivets per hour.
- The system should be easy to operate, and easy-to-teach rivet patterns for new wings and system software should provide support for maintenance and diagnostic operations.

INITIAL CONCEPTS

The initial design concept was based on the use of a central control computer using a data base specifying the approximate locations of rivets on various types of wings. Since the system will be mounted on a mobile vehicle, the position and orientation of the wing work surface will not be fixed in robot base coordinates. The system will use a coordinate transformation to relate rivet coordinates in the data base to the corresponding coordinates in the robot base frame of reference. The appropriate transformation matrix is to be generated by manually moving the robot so that the end effector touches the wing surface at three predefined points and transferring the coordinates of the points to the central control computer that performs the necessary calculations. In addition to performing the necessary mathematical operations, the control computer would be interfaced to direct the operations of the other systems, including:

- The end effector to perform the drilling and punching operations.
- A robot to move the end effector into position.
- A vision system to find and identify the rivets.

A Hewlett-Packard A-700 was selected for the control computer, from previous experience and familiarity with its operation.

To meet the accuracy and cycle-time requirements, the initial design included a multifunction end effector to allow rapid selection of modules required for drilling and punching rivets and for inspecting the rivet holes. A rotating turret configuration was chosen for the end effector with the various modules mounted around the circumference of a 10-in. tool center diameter. To provide the accuracy required for drilling, the end-effector design included provisions to translate the turret vertically and to rotate the turret in steps corresponding to module positions of about 0.001 in.

The requirement of locating rivet centers was met by placing a camera for an IRI P-256 vision system at one of the module positions on the end effector. Vision system software could then analyze an image to determine the rivet centers. Additionally, the fastener type and diameter data could be obtained. To determine surface perpendicularity, the end effector was equipped with a ringlike position sensor which presses against the wing surface. The orientation of the end effector with respect to the wing surface can then be determined with three position sensors attached to the ring.

Key factors in the selection of a robot included load capacity sufficient for the estimated weight (150 lb) of the end effector, a work envelope suitable for the wing surface, rigidity to withstand drilling forces with minimal delection, and capability of interfacing to a host computer. After consideration of several large electric robots (hydraulic robots were excluded by the Navy contract), a Cincinnati-Milacron T3-776 was selected.

Initially, the mobile vehicle was to be obtained from a manufacturer of industrial battery-powered vehicles such as aircraft tow-out tractors or mine vehicles, and modified as necessary. A drawing of the initial concept of the deriveter system is shown in Figure 1.

As the project design progressed, it became apparent that the robot should not be mounted on top of the vehicle, since that restricted the practical work area, did not allow sufficient space for mounting the required subsystems, and necessitated raising the wings several feet from the floor to place them in the work envelope of the robot. Also, this configuration presented several safety problems and did not provide an operator's station with a good view of the operations. The concept was modified to provide a cantilevered platform so that the robot could be mounted as low as possible and the operator's station could be placed behind and above the robot base. It was not practical to modify an existing vehicle to this configuration, so a special-purpose vehicle was designed and built for the deriveter system.

System Operation

The central control computer program includes several modes of operation that are selected from menus by the system operator; the modes include a:

- Teach mode which is used to build a data base for a particular wing or wing section. The data base generated will

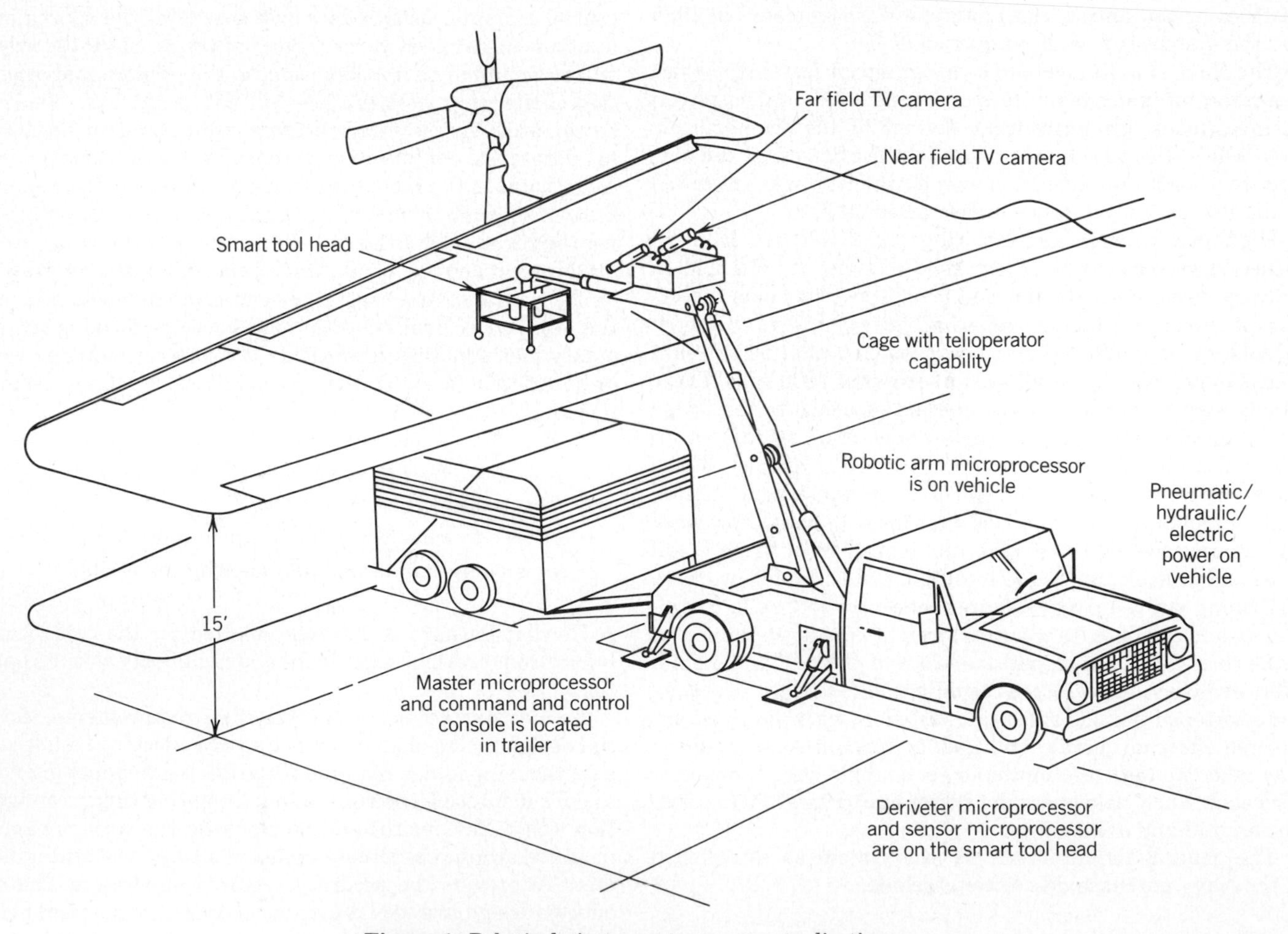

Figure 1. Robotic deriveter systems conceptualization.

contain the coordinates of all rivets to be removed as well as other information such as type, diameter, and length of fasteners originally installed at each location.

- Derivet mode in which the computer uses the information in a wing data base to position the robot to each rivet location and then control the end effector to drill and punch the rivet and inspect the hole.
- Calibration mode that uses standard fixtures to determine calibration parameters automatically for the vision system, position sensors, eddy current system, etc.
- Maintenance mode that allows a technician to operate individual components of the system for troubleshooting.

Initially, the system must be taught the expected pattern of rivets for the types of wings it will be used to derivet. This is done by a two-pass technique in which the robot teach pendant is first used to move the robot to the rivets on a wing manually. The coordinates of each point are transferred from the robot to the computer and stored in the data base for that particular wing type. In addition to location, the data base also contains information such as the expected type and size of each rivet and whether inspection for cracks is to be performed.

The second pass of the teach operation is used to refine the information in the data base. In this case, the system operates under control of the computer using the same operations used for deriveting. The computer directs the robot to move to each rivet location and then commands the vision system to analyze the image in its field of view. The vision system processes the image, locates the center of the fastener, and determines the type of fastener. This information is transmitted to the control computer which updates the data base with this more accurate information.

When a wing is to be deriveted, the operator manually directs the robot to the three reference points on the wing, and the coordinates of the points are transferred to the computer which calculates the transformation matrix to convert the stored rivet location coordinates to the corresponding points on the wing surface. The computer then directs the robot to move to the first location given in the data base and orients the end effector perpendicular to the wing at the correct distance from the wing surface. The end-effector turret is rotated to bring the vision camera over the rivet to obtain an image. The image will not be exactly centered due to robot-positioning error and rivet-location tolerances during manufacture. The system will process the image and determine the rivet type, diameter, and the offset distance from tool center point to the rivet center. This information is transmitted to the computer, which then determines the corrections necessary to position the turret modules directly over the rivet center.

The process of actually removing the rivet then begins as the turret is rotated to center the drill over the rivet head. The rivet is drilled to a depth dependent on the countersink diameter. Next, a punch is used to fracture the weakened rivet at the countersink–shank interface and drive the shank out of the hole. The eddy current probe can then be used to inspect the skin for cracks. If a crack is found, the wing is marked and the crack location entered on a defect log for later printout. The turret is then rotated back to the camera module and the process repeated on the next rivet.

In the automatic derivet mode, the operator first tells the system the name of the data base to be used and then moves the robot to the three wing reference points. This provides the computer with coordinate data for computation of the transformation matrix for the current wing position. The host computer will then retrieve the data associated with a rivet from the data base, move the robot to that rivet, correct for perpendicularity and offset, refine the position using vision, remove the rivet, inspect the hole, and then move on to the next rivet. The system will continue to remove rivets automatically until it reaches the end of the data base or the operator interrupts the process. The operator has the capability of signaling the system to stop after the current rivet is completed (orderly shutdown) or to stop immediately (emergency shutdown).

Vehicle

The original contractual purchase description issued by the Navy specified a trailer that would "provide support and mobility for all the components of the RDS" and would "have the capability of being self-propelled for short travel distances." Thus, the system would be completely self-contained, self-propelled within an aircraft overhaul shop facility, and towable between shops. A jack stabilization system was specified to steady the vehicle during robot operations. Also, it was desired that the entire vehicle package be as small and maneuverable as possible to allow access in various aircraft shop situations. The original plan called for the systems to be built onto a commercial chassis. This would have held down the costs of development, acquisition, and maintenance but did not prove applicable to the special needs of the deriveter concept. A specialized vehicle was designed.

The vehicle was developed along these contractual constraints with only minor changes. The most important exception was the development of an air-bag suspension in lieu of stabilization jacks so that the entire vehicle frame could be lowered to the ground to rest on three contact pads. This, it was felt, would give the robot a solid working platform and lower the robot work envelope, thereby also effectively lowering the aircraft part fixturing and stiffening its structure. The vehicle itself was configured as a forklift with the robot cantilevered ahead and below the front-drive axle and with rear-wheel steering for positioning maneuverability.

External connection for the system was limited to a single 480 VAC 3-Phase power cord "to be contained on a reel." Transformation to 120 VAC power had to be accomplished on board. All hydraulic, compressed air, and vacuum sources had to be available on the vehicle.

Packaging for accessibility and maintenance of the RDS components was of major concern. The following is a list of this equipment divided into the categories of full accessibility (all electronics and controls needed to operate the system) and reduced accessibility (items needing access only for maintenance and repairs):

Full Accessibility	Reduced Accessibility
Robot	Axles/suspension/shocks/steering/tow bar
Robot controller	D-c motor/controller/drive train
RDS electronics	D-c hydraulic pump
Control computer	Battery
Hard disk	Cable reel
Floppy disk	Electrical power distribution
Printer	Circuit-breaker panels
CRT	Junction boxes
Keyboard	Transformers
Vision system	Contactors
Eddy current system	A-c hydraulic power unit
Stepper motor controllers	Air compressor and tank
Interface electronics	Vacuum system and collection tanks
Collision avoidance	Hoses and cabling
Robot programming tools	
RDS operator controls	
Closed circuit TV	
Driving controls	
Tool changer	

The major subsystems required for the RDS operation and packaged on the vehicle are as follows:

- A vacuum collection system consisting of a power unit, dropout tank, and filter tank to remove drilling chips from the end-effector area.
- A compressed air system consisting of a 5-hp (24-cfm) compressor, 30-gal tank, and valving to power the end-effector punches and vortex cooler and to inflate the vehicle air bags.
- A hydraulic power unit with the 15-gal reservoir and servovalving for the end-effector module ram and drill motor.
- An electrical distribution system that includes an isolation transformer, active filter, and wiring precautions for power conditioning to the electronic equipment.

The vehicle frame was designed for stiffness and weight distribution to provide a solid base for the robot reference plane. A pair of 4 × 10 structural steel channels formed the backbone of the frame for the length of the vehicle deck and was subsequently used as troughs for cables and hoses. A webbed structure of ¾-in. plate was welded to one end for the robot cantilever and ⅜-in. plate bulkheads were used to form the axle and wheel wells. The floor in the middle and rear sections was built of 2-in. plate to counteract the weight and center of gravity of the robot when the vehicle was being driven or towed. The frame assembly weighed just over 12,000 lb.

Purchased items were positioned on the vehicle frame to help equalize the weight distribution. The Cincinnati Milacron T3-776 robot selected for the RDS weighs over 5,000 lb (or approximately ⅙ of the estimated 30,000 lb total vehicle and equipment weight). This cantilevered weight was balanced by the floor plate mentioned before and by the robot controller (1,500 lb) and vehicle battery (2,500 lb) placed near the rear of the vehicle. The remaining equipment was distributed in the middle and rear sections for a final estimated weight distribution of 60/40 (F/R).

In addition to that required for the RDS equipment, a substantial amount of vehicle room was required for axles, tires, and suspension. The tires and axles were allowed to have 10 in. of vertical travel so that the frame could rest at floor level for robot operations or be positioned with 8 in. of ground clearance for driving. Parallelogram-type suspension arms made of round stock and rod ends were developed to locate each axle on a pair of 16-in. diameter air bags. Heavy-duty shock absorbers were later added to control the air-bag softness.

The remaining vehicle systems were built using conventional parts. The truck axles used for the vehicle were produced by Rockwell and purchased through a local Chevrolet dealer. The front (drive) axle was mounted with dual tires and connected to a d-c traction motor with a drive shaft, gearbox, and gear belt to get the proper reduction (45.31:1) for a theoretical 6 mph top speed (level ground). The rear (steering) axle included extra linkages for a mechanical connection to a detachable tow bar and a hydraulic cylinder. This cylinder was plumbed to a hydraulic steering valve and d-c-powered pump for power-assisted steering. Disk brakes were purchased with the axles and connected without power assist. A brake disk and mechanical caliper were attached to the drive motor shaft as a parking brake.

The drive motor was selected for a somewhat unconventional 72 VDC electrical system. This minimized the footprint of the battery by keeping the cell count to 36. (Seventy-two volt systems are used on various large electric forklifts, airport baggage tractors, and railroad equipment.) The continuous duty rating of the series wound drive motor is 16 hp, adequate for the level concrete and asphalt surfaces intended for the vehicle (hilly terrain may be traversed under tow). A 72 VDC compound wound motor was connected to a hydraulic pump to power the steering system. This system shares the hydraulic tank required for the deriveter end-effector tools.

END EFFECTOR

The end-effector structure encompasses the motions, tools, and sensors required for computer-controlled rivet removal. The basic design is a tool turret configuration with a translational degree of freedom for initial alignment. Figure 2 shows the overall size and shape of the hardware.

The deriveting operation requires careful centering and perpendicular alignment of the end-effector tool center over the rivet. The robot first places the end effector at a previously taught rivet coordinate (±0.25 in. of the actual rivet center is sufficiently accurate). The contact ring plate position is sensed with three linear variable differential transformers (LVDT) to determine surface distance and perpendicularity. Correction coordinates are passed to the robot until an acceptable iterative position is obtained. At this point, the robot's only function is to hold the end effector in position. With the end-effector camera at tool center position, the vision system determines the offsets required for critical alignment over the true rivet center (within approximately 0.002 in.). These corrections are made using the end-effector 7th axis (vertical adjustment by means of a stepper motor and leadscrew drive) and 8th axis (turret rotation by a stepper motor and gearbox reduction). The 7th axis is then held while the 8th axis rotates the proper tools over tool (rivet) center.

The end-effector tools are of modular design, with each module mounted on guide rods and spring loaded for return. A single hydraulic ram is then used to extend the modules as necessary. The seven modules in their operation sequence are

- The camera module (Fig. 3) is the only one that is not extended by the ram. It consists of a solid-state (CCD) camera with a light source and fiber-optic light ring for uniform illumination of the skin surface.
- The drill module (Fig. 4) includes a hydraulic motor with a gear drive to a precision spindle. A Moorse #1 taper holds a specially developed countersink drill. This drill type eliminates the need for multiple drills to handle various fastener sizes (1/8–1/2 in. in steps of 1/32) by monitoring the drilling depth until the countersink–shank interface is reached.
- There are three punch modules (Fig. 5) and punch sizes to drive out the spectrum of rivet shank diameters. Each module has a pneumatic hammer to provide an impact force on the countersink punch as it is driven against the drilled rivet by the feed ram.
- The eddy current module (Fig. 6) performs an inspection of the drilled rivet hole to detect cracks in the skin panel. A pair of stepper motors are used with a differential drive to rotate the probe about a dynamically adjustable radius. These motions, in conjunction with the module ram feed, scan the eddy current probe over the rivet hole wall. This module also was originally designed to aid in the initial rivet centering operation. A second element built into the tip of the probe may be used for a surface scan to detect the rivet–skin interface and thus the rivet diameter and center.
- The marker module (Fig. 7) is a conventional felt-tip marker pen with a light spring load and a capping arrangement. It is used to indicate crack locations on the aircraft panel itself to aid in correlating defect log information with physical locations.

Figure 2. End-effector robot deriveter.

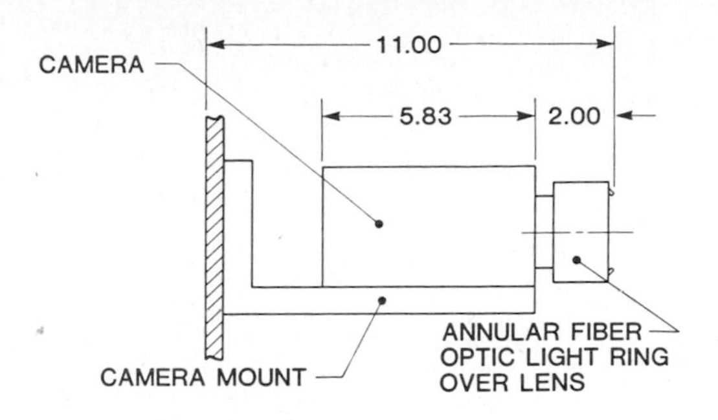

Figure 3. Camera module of an end-effector robot deriveter.

Initial drilling tests uncovered deflection problems with the robot arm at extended working angles. The (untested) solution to this problem is a preload ram that applies and holds a force (somewhat greater than the drilling forces) against the wing surface to take the spring out of the system before the critical rivet alignment is made.

Ancillary drilling equipment includes a vortex gun that uses compressed air to cool the drill bit (all drilling is done without the use of liquid coolants) and a vacuum hose to pull away some of the drilling chips.

The mechanical arrangement of the end effector is fully instrumented for control by the computer system:

- The 7th and 8th axis drives are powered by microstepping motors (25,000 steps/rev). The 8th axis (turret rotation) includes an encoder on the motor shaft. Limit switches are provided for homing the motors and for failsafe protection.

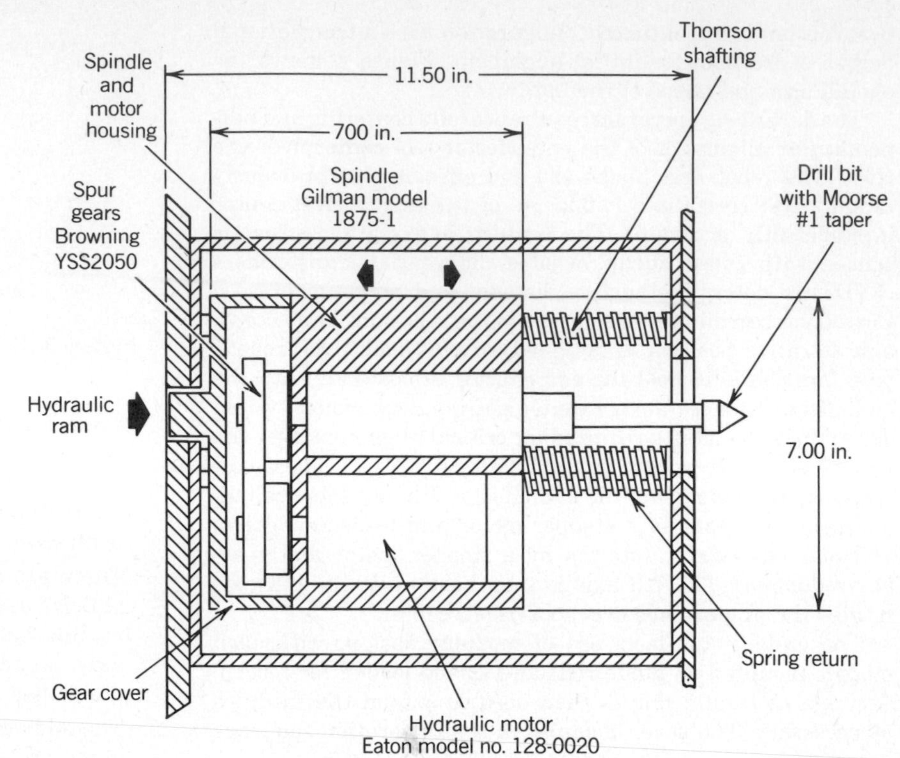

Figure 4. Drill module of an end-effector robot deriveter.

- The hydraulic feed ram is mechanically connected to a linear position potentiometer for feedback to the servovalve control electronics. This position signal is also fed back to the computer for verification. Ram pressure (force) is monitored by the computer.
- The hydraulic drill motor is speed-controlled by a servovalve. Feedback is accomplished with a gear tooth magnetic pickup and a frequency-to-voltage converter. The analogue signal is sent to the servovalve controller and the computer. Speed is software-controlled from 400 to 1500 rpm. Drill-motor pressure (torque) is monitored by the computer.
- The eddy current probe motions are implemented with a pair of microstepping motors. Each has an encoder to close the control loop. Probe leads are connected to an eddy current instrument that outputs four analogue channels for analysis by the host computer.

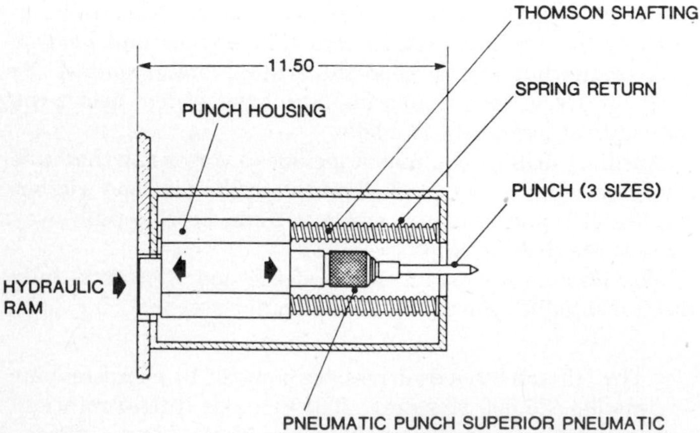

Figure 5. Punch module of an end-effector robot deriveter.

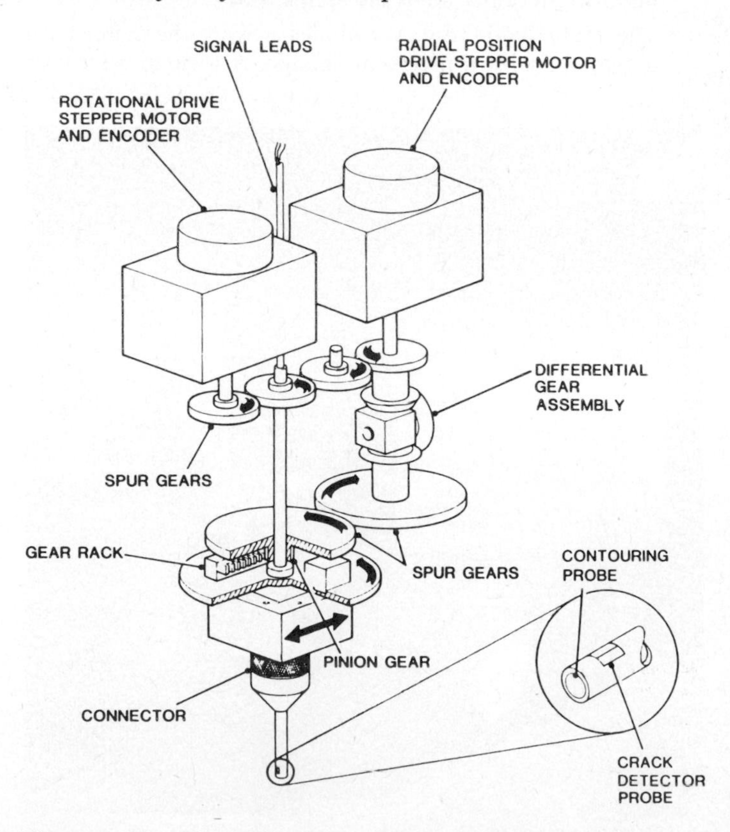

Figure 6. The eddy current module of an end-effector robot deriveter.

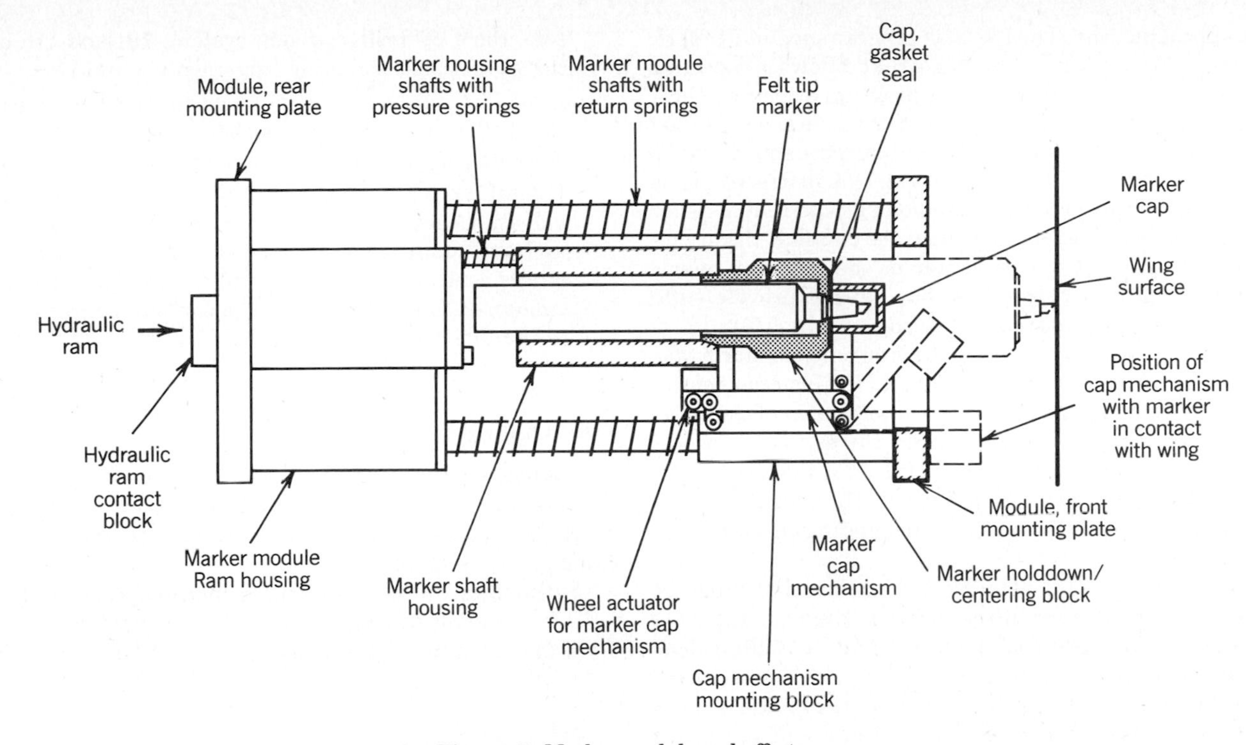

Figure 7. Marker module end effector.

- Solenoid air valves for the pneumatic punches, vortex cooler, and position sensor extension are controlled with digital outputs and a solid-state relay interface.
- Each tool module includes a limit switch to signal its home position. Thus, the module home signal is verified before any turret rotation.

The RDS system includes an automatic tool changer so that deriveting can continue without operator intervention. Software checks are included for drill dullage or breakage and punch breakage. The tool changer includes stations for drill and punch removal and replacement. Switch signals are provided at each station for softward checking of tool presence/absence and appropriate operator prompts at the system console. The vision system is used for tool-change alignment.

ROBOT

A Cincinnati-Milacron T3-776 robot was selected for use in the robotic deriveter system. The 776 is a 6-axis articulated electric robot with a load capacity of 150 lb (10 in. out from the tool mounting plate). The repeatability to any previously taught point is ±0.010 in. The 776 uses Milacron's Version 4 controller which provides controlled path motion, teaching in either rectangular or cylindrical coordinates, and the capability of interfacing to a host computer using serial RS-232 communications.

From the initial conceptual designs, the weight of the end effector was recognized as a key constraint. The load capacity of the 776 was considered barely sufficient, and weight reduction was considered at all end-effector design stages. Even so, the final weight of the end effector will be slightly over 150 lb, but no major robot problems are expected.

The specified repeatability of the robot was considered more than adequate for the project since the vision system will be used to compensate for differences in rivet placement of up to 0.250 in. During testing, the robot operated well within the specified repeatability with the full load of the end-effector weight. The additional drilling load imposed in certain robot orientations did cause a measurable deflection.

The serial RS-232 communications ability of the robot controller was essential to operation with the central control computer. Cincinnati-Milacron uses DDCMP protocol to ensure that transmissing errors are detected and messages interpreted correctly. Both "remote sequence" and "remote point" capabilities are available so that the robot can be completely controlled from the host computer. The DDCMP interface subroutines were written in HP A-700 assembly code for rapid execution.

A Guard-i-Mark capacitive bridge system was included as a collision sensor. The control section is mounted on the vehicle near the robot and coaxial cables connect the bridge to conduit pipes formed into capacitive elements on the robot arm and end effector. Any change in capacitance above a preset amount triggers a relay in the Guard-i-Mark control unit and energizes the "hold-set" input on the robot controller which stops the servomotor drives and locks the brakes on each axis. The Guard-i-Mark relay also generates an interrupt to the HP computer which signals the operator to take corrective actions.

VISION SYSTEM

The vision system used for locating and identifying fasteners is an International Robomation/Intelligence P-256. This is a 68000-based system that has a hardware array processor for very fast convolution operations such as image filtering and gradient approximation. The P-256 is programmed in FORTH, an incrementally compiled language that allows interactive program development and fairly fast program execution time.

Several algorithms are used to find the center of the fastener. The rivet image is first edge-enhanced using Robert's cross operator. A method based on the Hough transform is then used to generate circles centered on all potential fastener edge points. Since all fastener outlines are circular, these generated circles intersect at the center of the fastener image. This technique proved very robust but did not provide sufficient accuracy for this purpose (drilling within 0.005 in. of center), so an algorithm based on the Fourier transform was developed to improve this first estimate of the fastener center. This consists of generating the Fourier coefficients of the curve produced by plotting the distance from the estimated center point to each fastener image edge point as a function of the angle about the center point. This method is applied iteratively to yield an accuracy of 0.001–0.003 in. depending on the quality of the image.

The vision system also includes a classifier that analyzes the patterns on the surface of the fastener head to determine if the fastener is a float-head rivet, a jo-bolt, a Phillips-head screw, etc. This information is compared with the expected type of fastener as stored in the data base to provide a check on the data and the system operation. Details of the vision algorithms used may be found in Ref. 1.

COMPUTER AND INTERFACE

The Hewlett-Packard A-700 computer system provides the primary input for the RDS operator during both teach and automatic deriveting operations. An HP 2623A graphics terminal is used to provide the operator with a menu-based control program allowing selection of the desired operating modes, data bases, etc. A 16.5-megabyte Winchester disk storage system provides on-line storage of all operating programs, the wing data-base information, and an operating log to allow recovery after an error or abnormal operation occurs. A 3.5-in. flexible disk drive (microfloppy) provides 270 kilobytes on each disk for backups of each wing data base. A thermal printer allows listing data-base information generated during the teaching process and generating summary reports listing the rivets removed successfully, the number of cracks found, etc.

The A-700 computer system includes optional floating point hardware to speed up the matrix computations needed to perform the transformations relating the wing coordinates to robot base coordinates. System memory includes 512 kilobytes of parity check random access memory. The A-700 computers use a distributed intelligence I/O structure in which each I/O card contains a microprocessor that controls direct access transfers to/from memory without interrupting the CPU. In addition to the peripherals mentioned, the system includes interfaces to all of the other subsystems of the RDS:

- IEEE 488 interface for the Winchester disk, microfloppy disk, and printer.
- Asynchronous serial RS-232 interface for the system console, robot controller, vision system, 7th and 8th axes of turret motions, and eddy current probe motions.
- Analogue input/output channels to monitor and control the drill speed and ram position, read the LVDT position sensor outputs, and accept the eddy current probe signals.
- Digital input/output channels to control air and hydraulic valves, vacuum and hydraulic motor contactors, vision system illumination, etc, and to monitor the condition of status conditions such as module home switches. Optical isolators are used to couple the computer interface board to the digital devices.

RDS SOFTWARE

The robotic deriveter system software for the HP A-700 is divided into three levels: lower-level device communication routines, mid-level control task routines, and upper-level operating modes of the system.

At the lowest level, the device communication routines provide for sending control commands and receiving responses from the subsystems and modules of the RDS, including:

Robot
Position sensors
Vision system
End-effector 7th axis
End-effector 8th axis
Eddy current rotational drive
Eddy current radial drive
Eddy current module data
Drill module
Punch module
Marker module
Anticollision sensor
Tool changer
Tool calibration

Within the overall operation of the RDS, there are a number of tasks that must be performed to teach new data bases, remove rivets, calibrate tools, etc. These common elements are organized into 16 control task routines, each including the necessary control logic and calls to device communication routines to perform a specific operation. Six of the control tasks deal with robot operations such as robot measurement computation, global robot positioning, perpendicularity and offset correction, and robot position correction using vision. A second group of control tasks is concerned with end-effector and sensor operations, including 7th and 8th axis correction using vision or eddy current, rivet material or type determination, and eddy current skin inspection. Other control task routines perform specific steps in the deriveting operation, including drill, punch, and marker control, etc.

The top level of software includes the user interface and the major operating modes of the system. The user interface is designed to allow operators with minimal computer experience to interact with the system to teach new wing data bases and remove rivets. All system operation options are presented as menu items, and selection is made with function keys. Help screens are provided for additional explanation of the effects of menu selections, and the user is provided with the option of backing up to previous menu screens.

Other high level operating modes include:

Edit Data Base. This allows the operator to modify associated rivet data (but not location coordinates), delete rivets, and merge data bases.

User Utilities. These routines allow copying files between hard disk and floppy, deleting files, and obtaining database directories without having to use the HP operating system.

Calibration Mode. This provides for calibration of the deriveter subsystems including the A/D and D/A converters, robot, vision, eddy current, and position sensors.

Maintenance Mode. This troubleshooting mode allows the user to isolate and operate each of the subsystems independently of the deriveter as a whole.

Diagnostic Mode. Diagnostic tests can be run on each of the subsystems.

SYSTEM PERFORMANCE

Teach

Initial development and testing of the robot, end effector, vision system, and control computer were conducted in a laboratory with the robot and wing fixture bolted to the floor. In September 1984, when the mobile vehicle was completed, the laboratory setup was disassembled and transferred to the vehicle. Initial tests of the complete system were conducted in November 1984.

The two-pass method of teaching approximate rivet locations by moving the robot to points defining rows of rivets works well. Due to the resolution of the robot and operator positioning error, the coordinates obtained during Pass I are accurate to about 0.2 in.

In practice, the operator can use a video monitor to achieve higher positioning accuracy, but the time required increases greatly. It is more efficient to let the operator bring the robot to a position where the rivet is visible in the field of view of the vision system camera and then allow automatic refinement of position and orientation. The time required for the operator to move the robot to a rivet location using the teach pendant, identify the coordinates as individual fastener locations or as points defining a row of rivets, and store the coordinates in robot controller memory is about 30 s per rivet.

The second pass of teach mode (the active refinement of rivet locations) involves moving the robot and nearly always requires several iterations of the Hough transform method to find the rivet. Thus, it is slower than positioning during normal deriveting and requires about 30 s per rivet.

Derivet

In the automatic derivet mode of operation, the information of the data base has been refined to an accuracy limited by the repeatability of the robot (0.02–0.30 in.). Thus, when the system moves to a new position, it is almost always located within about 0.030 in. of rivet center. In this case, the faster Fourier transform method is used to locate the circle center. The time required to locate and classify the image varies from 3 to 6 s, depending on the image quality and on how near center the camera was initially. In some cases, when the image is severely degraded, the Hough Circle-finder algorithm must be used; this requires an additional 15 s.

The time required for drilling varies from 3–5 s for small aluminum rivets to up to 20–25 s for very large steel jo-bolts. Rotating the turret and punching the rivet shank out of the hole requires an additional 4 s. No data on the eddy current inspection is available at this time.

The vision system software has been able to find rivet centers accurately. On the basis of several hundred drilled rivets, it has been found that mechanical considerations introduce the bulk of drill location error. Center location by the vision system has an error of 0.001–0.003 in. Under well-controlled circumstances (drilling similar types of rivets at the same drill feed rate and in nearly the same location with respect to the robot), drilling is performed with a corresponding error. With a required drill depth on the order of 3/16 in., angular errors in drill alignment are not particularly critical. In the course of using the ultrasonic sensor to examine the region around the rivet, the normal to the surface is found to about 1° accuracy, leading to a displacement on the order of 0.00327 in.

By far the largest cause of drill-positioning error is deflection of the fixtured wing and of the robot arm itself when drill force is applied. Errors of 0.03–0.05 in. can occur for a drilling force of 150 lb applied along the wrist axis. This deflection is due primarily to deformation in the elbow joint bearing and is greatest when the loading force is applied at right angles to the horizontal projection of the arm. After consultation with Cincinnati-Milacron, the elbow bearing was tightened to the maximum torque possible without causing excessive wear, but only a slight improvement in deflection was obtained. To eliminate this effect, a system has been designed to apply a preload to the wing before adjusting the end-effector motions to center the image seen by the vision system. In this way, forces on the robot will be nearly constant and very little deflection will occur during drilling.

BIBLIOGRAPHY

1. E. Franke, D. Michalsky and D. McFalls, "Location and Identification of Rivets by Machine Vision," *SME Vision '85 Conference Proceedings,* Society of Manufacturing Engineers, Dearborn, Mich., 1985.

General References

J. A. Torma, "An Innovative C.A.M. Application: The U.S. Navy's Robotic Deriveter system," *ASME PVP Conference,* American Society of Mechanical Engineers, New York, 1984.

DESIGN AND MODELING CONCEPTS

JERZY W. ROZENBLIT
BERNARD P. ZEIGLER
The University of Arizona
Tucson, Arizona

MULTI-OBJECTIVE SYSTEM MODELING

The ability to increase the decision-making capabilities in different environments is related to the scope of our possible intervention into the operation of the systems being modeled. Zeigler (1) classifies the levels of intervention into three broad categories: management, control, and design. Management type of intervention connotes determining policies whose interpretation and execution is then delegated to subordinate levels. Control intervention is an action deterministically related to policy. In contrast, design represents the greatest scope of intervention in that the designer either creates the "real system" or augments, modifies, or replaces a part of existing reality (see also ASSEMBLY, ROBOTIC, DESIGN FOR).

The effects of interventions are uncertain due to the existence of uncontrollable parts in the system and it becomes necessary to encode knowledge about such parts in models of the system. Models represent abstractions of the reality whose primary function is to capture the structural and behavioral relationships in the system. These relationships would be difficult to observe were the models not available.

Thus, the modeling methodology should be an inherent component in computer-aided decision systems in management, control, and design (1–4). The tools and activities prescribed by this methodology enable the decision makers to evaluate (based on the analysis of the models' simulation) the effects of interventions before they are actually carried out. The "best," in terms of performance measures related to the system under evaluation, intervention alternatives are chosen and finally deployed in the real system. The choice of performance measures reflects the objective the decision maker (be it an economist, a designer of a power plant, or a technician supervising a chemical process) attempts to achieve. The nature of the objectives orients and drives the modeling and simulation processes.

It is easy to conceive that any real system could be subject to a multiplicity of objectives in management, control, or design context. Consider an example to illustrate such a situation. Assume that a banking conglomerate is in the process of computerizing operations in several of its branches. Apart from having a computer network designed, the bank is establishing a Computing Services Department to handle and support the new system. Outlined below are some of the aspects relevant to the operation of the new computer system and department.

Objective

1. *Cost requirements* (*required model of*): financing schemes and policies, investment capital, long-term maintenance costs, new personnel hiring and training, etc.
2. *Financial operations* (*required model of*): network operation, likely speed-ups in transaction processing time and decreases in transaction processing costs, growth of service volume, interactions with other computerized institutions.
3. *Customer satisfaction* (*required model of*): customer tastes (eg, popularity of dial-up access, automated teller machines), customer interface (waiting, accessibility, ease of completing a transaction, etc).
4. *Data security:* communication protocols, communication media (phone lines, hardware), security schemes.
5. *Personnel requirements:* system operation, human—machine interface.
6. *Quality control:* network operation, communication schemes, inter- and intrabranch interfaces.

It is rather unlikely that a comprehensive model of a computerized banking network, reflecting all of the above objectives, could be constructed. Even if it were possible to build such a model, its validation, and subsequently, verification and simulation would present a paramount degree of complexity. Instead, envision a collection of partial models, each reflecting a specific objective. This implies that the objectives orient the model building process by helping to demarcate the system boundaries and determine the model components of relevance (1,5). The fundamental formal concept supporting these activities is the *system entity structure*. The entity structure enables the modeler to encompass the boundaries and decompositions conceived for the system (1).

The role of the objectives is equally important in the process of specifying the experimentation aspects for the models that have been perceived for the real system. The key concept in this process is that of *experimental frame* ie, the specification of circumstances under which a model (or the real system) is to be observed and experimented with (1,6,7). The experimental frame definition reflects the objectives of modeling by subjecting the model to input stimuli (which in fact represent potential interventions), observing reactions of the model by collecting output data, and controlling the experimentation by placing relevant constraints on values of the designated model state variables. The data collected from such experiments serve as a means of evaluating the effects of intended interventions.

Generation of meaningful experimental conditions is not a trivial task and requires that the modeler understand the nature of the objectives and their interactions. A frame, similarly to a model, may reflect a single or a complex set of goals. Recall our example of the banking network and the following objectives: customer satisfaction, financial operation, and data security. When a comprehensive model (or partial models reflecting each aspect) is devloped, a relevant set of frames must be available in order to perform the experiment. Notice, however, that the above three objectives may conflict with one another. For example, improving customer satisfaction by providing dial-up access by public telephone lines might decrease the level of data security. On the other hand, purchase of an obsolete computer system may decrease customer satisfaction through the lack of convenient access and not provide the adequate level of security, even though the system might be suited to handle all of the bank's operations. Thus, meaningful trade-off experimental frames have to be specified, and the models must be evaluated within the context of trade-off orderings over the set of objectives.

This article discusses and recognizes the multiplicity of objectives, models, and experimental frames as a *sine qua non* condition of the modern, advanced modeling methodologies. The focus is specifically on the issues concerning the system design, understood here as the use of modeling techniques to procure and evaluate a model of the system being designed. In the ensuing section, the synergism between simulation modeling and system design is underscored.

System Design and Modeling, Synergies

The design aspect in decision-making offers the widest scope of intervention in that the designer develops a model from which a new system will be created. As opposed to system analysis, where the model is derived from an existing, real system, in system design the model comes first as a set of "blueprints" from which the system will be built, implemented, or deployed (8,9). The blueprints might take several forms; they could be simple verbal informal descriptions, a set of equations, or a complex computer program. The goal of such defined system design is to study models of designs before they are actually implemented and physically realized.

At this point, the fundamental question arises: Why are modeling and simulation methodologies needed to support system design? Before this question is answered, the basic elements in the dynamics of the design process must be summarized (9–13):

1. Designs are created by individuals who use the basic problem-solving techniques, namely, problem definition, proposal of a solution, and test of the solution against the problem definition.
2. The problems being addressed are often of a large scale. Thus, there should be methods for decomposing the problems into subproblems easily comprehensible by the designer. Partial solutions could then be generated and integrated using proper aggregation mechanisms.
3. Solutions (designs) are built based on the designer's perception of reality; they result from the transformation of the designer's ideas and knowledge into a blueprint of the system to be created.
4. The attributes of design should be described in comparative measures and applied by using trade-off techniques.
5. The tools, techniques, and methods are currently mostly manual methodologies; automated tools for system design are only now evolving (14,15).

Recall that the primary goal is to locate the system design within the modeling framework. An attempt is made to provide a systematic methodology for a design process supported by adequate formal structures and leading toward future computerization. As depicted in Figure 1, system design is brought into the multifaceted framework with design process being supported by the modeling and simulation techniques in the manner described below.

Modeling is a creative act of individuals using the basic problem-solving techniques, building conceptual models based on the knowledge and perception of reality, requirements, and objectives of the modeling project. The models are design blueprints. This constitutes direct relation to points 1 and 3.

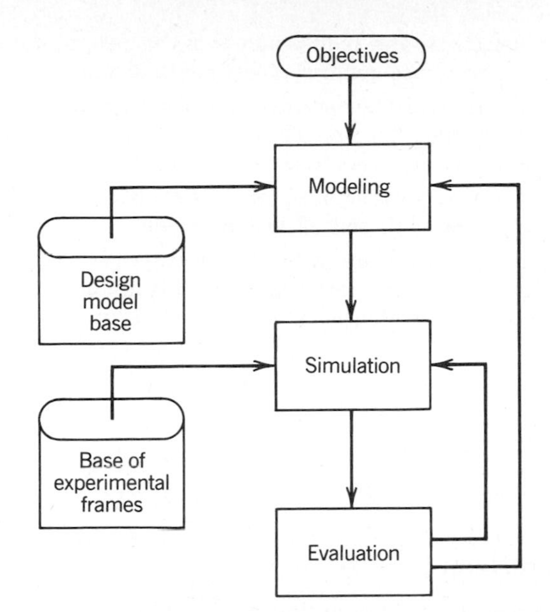

Figure 1. Design in the multifaceted modeling context.

By providing mechanisms for model decomposition, hierarchical specification, and aggregation of partial models (1), the multifaceted modeling approach fully responds to the needs of system design signaled in point 2. Adding the previously mentioned system entity structure, one is now equipped with a facility to generate families of models (of design) in various decomposition aspects.

The experimental frame concept responds to the needs of point 4. This unique structure provides a spectrum of performance evaluation methods, including evaluation of multilevel, multicomponent, hierarchically specified models (13).

Finally, the underlying purpose of the multifacted modeling is that it provides structures implementable in computerized support environments. This is where the possibility for a response to the drastically growing needs for compter-aided design tools is envisioned. The current tools lack an underlying theoretical framework that permits a uniform treatment of system design by providing concepts such as structure and behavior, decomposition, and hierarchy of specification (1,15). The existing architectures are usually conglomerates of various, often incompatible, tools whose coordination poses serious problems and often defies their purpose (4,14,16,17). The representation schemes offered by the multifacted framework are well structured and have formalized operations that can exploit such structures. This significantly reduces the effort of designing expert computer-based environments.

In the following sections, a conceptual framework for model-based design is set up. The proposed methodology utilizes the formal modeling concepts. Several major steps underlie the methodology:

- The system entity structure is a basic means of organizing a family of possible configurations of the system being designed.

- The objectives and requirements of the design project induce appropriate generic experimental frames.
- The design entity structure is pruned with respect to the generic frames. This results in a family of design configurations that conforms to the design objectives.
- The pruned substructures serve as skeletons for generating rules for synthesis of design models.
- Resulting models are evaluated in respective experimental frames and the best design models are chosen on the basis of such evaluations.

FORMAL FRAMEWORK FOR MODEL-BASED SYSTEM DESIGN

This section provides the necessary formal background for the multifaceted system design introduced earlier.

The System Entity Structure

To represent a family of design configurations appropriately, a structure is needed that embodies knowledge about the following relationships: decomposition, taxonomy, and coupling. The *decomposition* scheme allows a representation in which an object (component of a system being designed) is decomposed into components. The structure should be able to operate on and communicate about the decomposition scheme.

Taxonomy is a representation for the kinds of variants that are possible for an object, ie, how they can be categorized and subclassified.

The third kind of knowledge to represent is that of *coupling constraints* on the possible ways in which components identified in decompositions can be coupled together.

A formal object that embodies these three basic relationships is called the *system entity structure*. The system entity structure is based on a treelike graph encompassing the system boundaries and decompositions that have been conceived for the system. An *entity* signifies a conceptual part of the system that has been identified as a component in one or more decompositions. Each such decomposition is called an *aspect*. Thus, entities and aspects should be thought of as components and decompositions, respectively. The system entity structure organizes possibilities for a variety of system decompositions and model constructions.

Both entities and aspects can have attributes represented by the so-called *attached variables* types. When a variable type V is attached to an item occurrence I, this signifies that a variable I.V may be used to describe the item occurrence I. Therefore, whereas an unqualified variable type such as LENGTH may have multiple occurrences in the entity structure, a qualified variable, eg, QUEUE1, LENGTH belongs to one and only one item occurrence, QUEUE1.

The system entity structure satisfies the following axioms (18,19):

1. *Uniformity:* any two nodes that have the same labels, have identical variable types, and isomorphic substructures.
2. *Strict hierarchy:* no label appears more than once down any path of the tree.
3. *Alternating mode:* each node has a mode that is either "entity" or "aspect" (decomposition or specialization); the mode of a node and the modes of its successors are always opposites. The mode of the root is entity.
4. *Valid brothers:* no two brothers have the same label.
5. *Attached variables:* no two variables attached to the same item have the same type.

The entity/aspect distinction can be interpreted as follows: an entity represents an object of the system being designed, which either can be independently identified or is postulated as a component in some decomposition of the system. An aspect represents one decomposition, out of many possible, of an entity. The entities of an aspect represent disjoint components of a decomposition induced by the aspect. The aspects of an entity do not necessarily represent disjoint decompositions.

For a more detailed and formal treatment of the entity structure concept, consult Ref. 1. How the discussed knowledge representation scheme is realized by the concept is discussed below.

An entity may have several specializations. Each specialization may in turn have several entities. The original entity is called a general type relative to the entities belonging to a specialization, which are called special types. Since each such entity may have several specilizations, a *hierarchical structure* results which is called a *taxonomy* (1,20,21).

The alteration property is a salient feature which requires that entities and specializations alternate along any path from the root to leaves. The same property holds for entities and aspects. Specializations have independent existence, just as entities do. A specialization may occur in more than one location. Whenever it occurs, it carries with it all its attributes and substructures. Of course, it may not be meaningful to attach a particular specialization to a particular entity.

Hierarchical decomposition is in many ways analogous to the specialization hierarchy just discussed. The alternation property now requires alternation of aspects and entities. An aspect is a mode of decomposition for an entity just as a specialization is mode of classification for it. There may be several ways of decomposing an object, just as there may be several ways of classifying it. Formally, aspects and specializations are quite alike in their behavior. They each alternate with entities but cannot be hung from each other. A special type of decomposition called a *multiple* decomposition facilitates flexible representation of multiple entities whose number is in a system may vary. (Throughout the illustrations, the multiple decomposition aspect is denoted by triple vertical bars and specializations by double vertical bars.)

Specialization is a concept distinct from that of decomposition. However, there is a way of mapping a specialization hierarchy into an equivalent decomposition aspect which intimately involves the multiple decomposition concept. The transformation, illustrated in Figure 2, is simple: if entities A1 and A2 are special types of entity A, then the multiple component As is decomposable into the multiple components A1s and A2s.

To express the coupling constraints, the following procedure is employed: apply the mapping to remove the specializations to obtain an entity structure containing only entities and as-

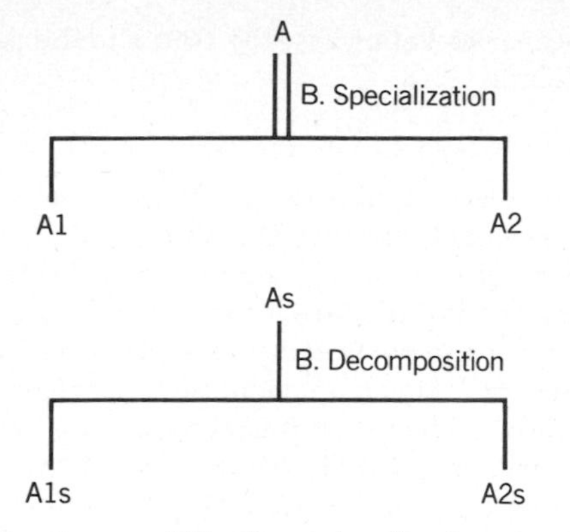

Figure 2. Removing specialization using the decomposition aspect.

pects. Now imagine that models are synthesized by working down the entity structure, selecting a single aspect for each entity and zero or more entities for each aspect. Such a process is called *pruning* of the entity structure (see Entity Structure and Experimental Frame-based Design Model Development). A pruning procedure is also defined as one that operates directly on entity structures with specializations. The coupling constraints must then be associated with aspects, since they represent the decompositions chosen when pruning. Moreover, a constraint must be associated with an aspect that contains all the entities involved in that constraint. What is more, this aspect should be minimal in the sense that there be no other aspect that lies below it in the entity structure which also encompasses all the entities involved in the constraint.

The following example illustrates how the system entity structure can be employed as a representation scheme for a family of design possibilities.

Assume that an aerospace agency is planning to launch a fully automated space station. In the first stages of development, the entity structures representing various configurations for the stations are proposed. One of the possible design entity structures is depicted in Figure 3.

For the sake of brevity, only three aspects, the automation, control, and communications aspects, are presented in the illustration. First, the automation aspect: assume that one of the design objectives is that the station be capable of performing a number of different tasks using a coordinated group of robots (subsequently called robot organization). The tasks, eg, station keeping, maintenance, launching satellites, refueling other space ships, may be performed at different sites called workstations.

The organization, as shown in Figure 3, may employ various types of robots. The two specialized types present in the structure are termed executive and functional robots. It is assumed that an executive robot has managerial skills, (ie, it can coordinate, hire, and fire robots in the task accomplishment process. A functional robot can be coordinated by an executive one and perform tasks for which it has been designed and programmed.

The robots, viewed from the standpoint of their organizational structure, are specialized into adaptive and nonadaptive organizations. The adaptive architecture is further specialized into a hierarchical tree-based architecture and a nonhierarchical distributed adaptive organization. In the tree-based architecture, the robots are recruited from the availability pool and returned to it according to the adaptive reconfiguration strategy. The above scheme, termed "hire/fire," has been proposed by Zeigler and Reynolds (22,23), who are investigating adaptive computer architectures.

The nonhierarchical organization can be visualized as a parallel-serial type of organization whose adaptation scheme may be based on the schemes similar to that of a CPM method. Assuming that a task can be represented by an activity net-

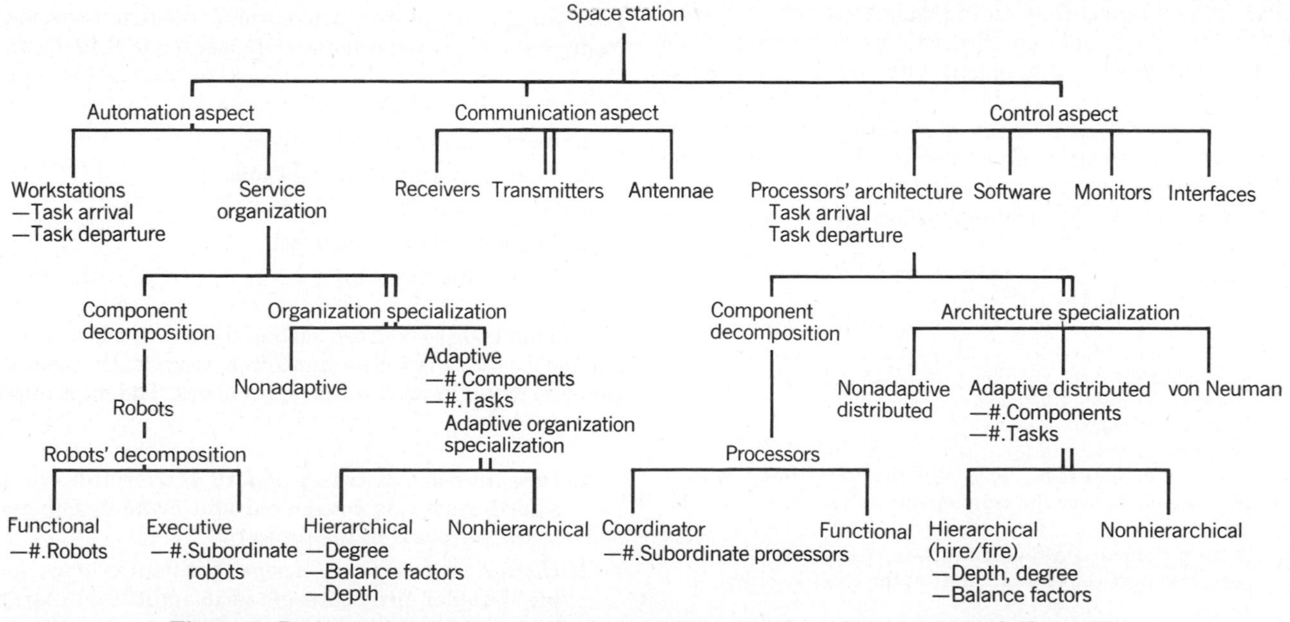

Figure 3. Design entity structure for the space station example. –, denotes an attached variable type.

work, the adaptation consists in shifting the robots from the least to the most critical activities.

In the control aspect, the processors, software, and real-time monitors are the basic components of the station. The processors, like the robots, may have the following architecture specialization types: classic von Neuman, distributed nonadaptive, and an adaptive hire/fire organization. The adaptive architecture requires that functional processors (equivalent in their role to functional robots) and coordinators (equivalent to executive robots) be present.

The Experimental Frame Definition

The system entity concept facilitates the representation of design structures. A means for expressing the dynamics of the design models is also needed. Since the design framework is objectives-driven, the experimental frame concept is used as the other underlying object in system design. The role of experimental frames will be twofold. First, the frame will represent the behavioral aspects of the design objectives and facilitate retrieval of entity structures that conform to those objectives. Secondly, the experimental frame will serve as a means of evaluating the design models with respect to given performance measures.

The conceptual basis for a methodology of model construction in which the objectives of modeling play the key and formally recognized role (therefore called *objectives-driven methodology*) was laid down by Zeigler (1).

The basic process in this methodology is that of defining an experimental frame, ie, a set of circumstances under which a model or real system is to be observed and experimented with. This process comprises the following steps. The purposes (objectives) for which the simulation study is undertaken lead to asking specific questions about the system to be simulated. This in turn requires that appropriate variables be defined so that a modeler can answer these questions. Ultimately, such a choice of variables is reflected in experimental frames that also express constraints on the trajectories of the chosen variables. The constraints on observations and control of an experiment should be in agreement with the modeling objectives. A choice of relevant variables constitutes the first important stage of experimental frame specification. The next step is to categorize the variables into input, output, and run control and place constraints on the time segments of these variables. Formally, the experimental frame specifies the following seven tuple:

$$EF = \langle T, I, O, C, W_I, W_C, SU, W_{SU} \rangle$$

where

- T is a time base
- I is the set of *input variables*
- O is the set of *output variables*
- C is the set of *run control variables*
- W_I is the set of *admissible input segments,* ie, a subset of all time segments over the cross product of the input variable ranges
- W_C is the set of *run control segments,* ie, a subset of all time segments over the cross product of the control variable range.
- SU is a set of summary variables
- W_{SU} $= \{s{:}s{:}I \times 0 \rightarrow SU.range\}$ is the set of *summary mappings*

The I/O data space defined by the frame is the set of all pairs of I/O segments:

$$D = \{(w,r)\colon w \in (T,X),\ r \in (T,Y) \text{ and } dom(w) = (r)\}$$

where X and Y are input and output value sets, respectively.

Since experimental frames should have a meaningful interpretation for both the model and the real system, a concept of restricting the initial state for the model must be provided. The run control variables serve this purpose. They initialize the experiments and set up conditions for their continuation and termination. The set of initial values for the run control variables is called INITIAL. The subset of the control space defined by the termination conditions is called TERMINAL. The set of run control segments is then defined as (for a detailed formal treatment of the experimental frame concept, see Ref. 1):

$$W_C = \{m{:}m{:}\langle t_i,t_f\rangle \rightarrow Z$$
$$\text{and } m(t_i) \in \text{INITIAL},\ m(t) \in \text{CONTINUATION for } t \in \{(t_i,t_f)\}$$

where Z = cross product of the ranges of individual control variables, and t_i and t_f are the beginning and end of the observation interval, respectively.

CHARACTERIZATION OF THE MULTILEVEL, MULTIPHASE SYSTEM DESIGN

System design concepts are found in many disciplines. The paradigms of each discipline underlie the methods for design representation and methodology. In systems theory, the dominant framework is the mathematical representation. In systems methodology, the methods are adopted from operations modeling; in philosophy, the models of thinking play an important role. A wide spectrum of studies on the subject has been documented in the literature (8,9,11,12,24–27).

Common Traits in System Design Methodologies

Reviewing most of the conventional design methodologies leads to the following scheme of reasoning (8,9,12,24,27):

1. State the problem.
2. Identify goals and objectives.
3. Generate alternative solutions.
4. Develop a model.
5. Evaluate the alternatives.
6. Implement the results.

The methods to address each of the above aspects of design depend on the discipline and often vary in the degree that they are found in most of the approaches. The most important traits are

1. *Objectives-driven nature of design.* Everything in the designed system is considered and evaluated in relation to the purposes of the project.
2. *Hierarchical nature of design.* Structures of systems being designed are represented in multilevel hierarchies that express decompositions of the system into subsystems (28).

3. *Design as decision making.* Design is concerned with actions to be taken in the future. Thus, incorrect decisions in the early stages of design may impede all subsequent actions.
4. *Iterative nature of design.* It is widely recognized that the design process should be iterative in that the designer should be able to return to earlier phases of design from any stage of the process. Analogies are often made here with cybernetics and control theory where iteration is continued until a desired equilibrium point is reached (12,28).
5. *Optimization in design.* Design seeks optimization of the whole system with respect to its objectives. Careful consideration must be given if attempts are made to optimize subsystems separately. Appropriate coordination methods must then be used to attain the overall objective (28).

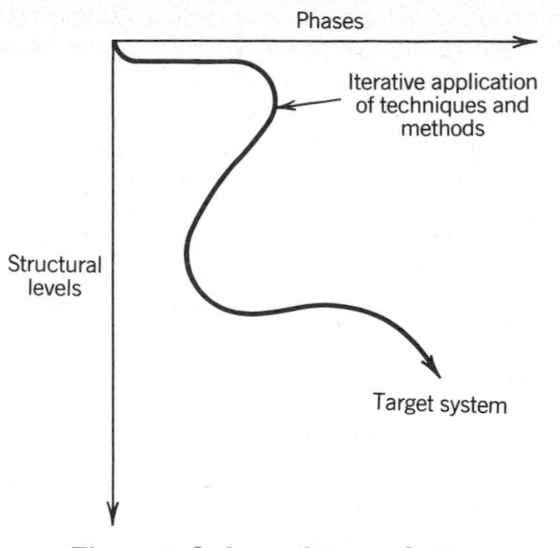

Figure 4. Orthogonal system design.

Orthogonal System Design

In the context of this article the term system design will denote the use of modeling and simulation techniques to build and evaluate models of the system being designed.

The design process is considered a series of successive refinements comprising two types of activities. The first type concerns the specification of design levels in a hierarchical manner. The design levels are successive refinements of the decomposition of the system under consideration. The first, and thus the most abstract level, is defined by the behavioral description of the system. Subsequently, the next levels are defined by decomposing the system into subsystems (modules, components) and applying decompositions to such subsystems until the resulting components are not further decomposable. The atomic system components are represented at the lowest level of the design hierarchy. At each level, the specialization of components into different categories is allowed for. This facilitates the representation of design alternatives.

The second type of design activity is concerned with "horizontal" actions associated with design levels. Such actions include requirements specification, system functional description, development of design models, experimentation by simulation, evaluation of results, and choice of the best design solution.

The design should proceed along both axes of the above characterization. The designer must be able to structure the designs, explore alternative structures, and derive complete specifications and models at any level. Transitions between design levels must be possible and easy to perform.

Such an orthogonal specification of the design process is often called a design matrix. The design process is represented in Figure 4. The space defined by the cross product of phases and levels is in various methodologies filled with different methods, techniques, and tools. However, such methods and tools are often incompatible and there are no underlying structures integrating the design steps at various levels and phases. This, of course, motivates and justifies efforts to solve that problem by applying structures such as the system entity formalism and experimental frame.

To illustrate how the levels of design can be defined, consider the class of computer system designs (16,17,29).

It is common view in the literature (14,16) that the design levels for computer system design are based on the following scheme:

1. *Behavioral level:* general description of system behavior, purpose, and attributes.
2. *Functional system architecture level:* functional system characteristics, partitioning of the system into hardware/software functional specifications.
3. *Hardware/software architecture level:* representation of the system in terms of the main building blocks, (eg, processors, memory, peripherals, etc.)
4. *Module level:* detailed description of the architecture-level building blocks.
5. *Register-transfer level:* the data paths, control sequences, and timing diagrams that implement the modules.
6. *Logic level:* specification of the system in terms of the detailed interconnection of logic components.
7. *Circuit level:* representation by physical circuits.

The phases for this type of design are the horizontal activities that have been specified above.

Although an attempt to refine the definition of design further is not made, nor a description of all its phases, how the multifaceted modeling with its formal objects can unify the design activities and support the construction of environments for expert system design is shown.

ENTITY STRUCTURE AND EXPERIMENTAL FRAME-BASED DESIGN MODEL DEVELOPMENT

This section presents a framework that operationalizes the system entity structure and the experimental frame concept into a systematic design framework. First, the choice of the system entity structure as the underlying object in our design methodology is justified.

Recall that the entity structure represents the following relations:

Decomposition hierarchy of the system being designed. This enables the direct representation of the design levels as discussed earlier.

Taxonomy. This relationship (captured by the specialization aspects) facilitates the classification of design components and constitutes a means of expressing various design alternatives.

Coupling constraints on the possible ways in which components identified in decompositions/specializations can be coupled. This enables the use of the system entity structure as a basis for the hierarchical design model construction.

Again, it is emphasized that the entity structure is the basic means for organizing the family of possible design configurations (structures, architectures). It also serves as a skeleton for the hierarchical design model construction.

It is equally important to understand the role of the experimental frame concept in the design process. The role of frames is twofold. First, an experimental frame is a means of representing the performance measures associated with the behavioral aspects of design objectives. Subsequently, the frame is used in a simulation experiment performed to evaluate the merits of a design model. There is, however, a second important role of the frame concept. A generic form of an experimental frame is employed in the design framework to delimit the design model space given by the system entity structure.

Generic Experimental Frames

A generic experimental frame type represents a general class from which experimental frame specifications can be derived. A generic frame is defined by unqualified generic variable types that correspond to the objectives for which the design study is undertaken.

Design objectives are associated with performance indexes that allow for a final judgment of the design models. A generic frame GEF is defined as the following structure induced by the performance index pi:

$$GEF_{pi} = \{IG, OG, W_{GI}, SU, W_{SU}\}$$

where GEF_{pi} denotes a generic experimental frame for performance index pi and

IG is the set of generic input variable types for pi
OG is the set of generic output variable types for pi
W_{IG} is the set of generic input segment types for pi
SU is the set of summary variables
W_{SU} is the set of standard summary mappings

To illustrate how a generic experimental frame can be defined, let us return to the space station example discussed earlier. Suppose a frame is to be defined that represents the questions concerning the measures associated with the adaptive robot (or computer) architectures perceived for the station (see Fig. 3).

There are two basic measures to be considered. The first one is concerned with the performance measures that are task-oriented, eg, throughput, task transit time, volume of tasks in the system, etc. The other concerns the organization itself, ie, to collect data on the structural properties of the organization.

The following generic frame for the evaluation of tree-based adaptive organizations is proposed:

Generic Frame: Adaptability

{This template is intended to generate experimental frame for collecting observations in adaptive, tree-based organizations}

Generic Input Variables

Task.arrival with range {task identifiers, task type}

Generic Input Segment

Task.workload—a discrete event segment with varying interarrival rate (eg, high, low) and/or varying task types
{One of the objectives of the frame is to observe adaptation patterns to changes in the input segment}

Generic Output Variables

Task.departure with range task identifiers, task types
Number of tasks being processed
Total number of coordinator-type components in the adaptive organization at any given time
Total number of functional-type components in the adaptive organization at any given time
Depth of the tree at any given time
Degree of the tree at any given time
Balance factors (for any or all the nodes)
Number of subordinates for coordinator-type components

Summary Variables

Departure/arrival rate ratio
Average task.transit time
Total number of tasks processed over the observation interval
Average number of components of the organization
Average number of coordinators
Average number of functional components
Number of coordinators/functional components ratio
Average depth of the tree
Maximum depth of the organization tree
Minimum depth of the organization tree

Other examples of generic experimental frame types for various performance criteria are presented in Ref. 30.

The Entity Structure-based Generation of Design Model Structures

Due to the multiplicity of aspects and specializations, the design entity structure offers a spectrum of design alternatives. The procedures that limit the set of design configurations by extracting only those substructures that conform to the design objectives are presented now. Called pruning, this extraction process is based on the following scheme. Assume that an entity structure has been transformed into a structure with no specializations. Then, imagine that the structure is traversed by selecting a single aspect for each entity and zero or more entities for each aspect. All selected entities carry their attributes with them. Also, the coupling constraint of the selected aspect is attached to the entity to which this aspect belongs.

The above process results in *decomposition trees* (1) that represent hierarchical decompositions of design models into components (*design model structures*). The process that extracts the model structures from the design entity structure is called *pruning*.

By pruning the system entity structure with respect to generic frames, the following benefits are obtained:

1. In terms of the contribution to the design process:
   a. a generic frame extracts only those substructures that conform to the design objectives. Thus, a number of design alternatives may be disregarded as not applicable or not realizable for a given problem.
   b. partial models of the design can be formulated and evaluated. This may significantly reduce the complexity which would arise dealing with the overall design model. The generic frame may thus be viewed as an object that partitions the system entity structure into design-objective related classes.
   c. The evaluation of design models constructed from the pruned substructures is performed in corresponding experimental frames. Such frames are generated by instantiating the generic frames used to prune the system entity structure. Hence, automatic evaluation procedures could be employed in the design process.
2. In terms of facilitating the pruning process itself, generic frames automatically determine:
   a. The aspects that are selected for each entity.
   b. The depth of the pruning process.
   c. The descriptive variables of components.

Pruning Algorithms This section presents a suite of pruning algorithms for generating design model structures and begins with the definition of the procedure Prune for pruning pure entity structures, ie, structures in which no specialization relations occur. For entity structures in which specializations are present, procedures for mapping such structures into a set of pure entity structures are provided. The procedure Prune is then applied to the set of pure structures, and relevant design model structures are generated as a result of pruning.

In presenting the algorithms, the concept of a nondeterministic algorithm is employed, ie, the algorithms are allowed to contain operations whose outcome is not uniquely defined but limited to a specified set of possibilities. In the case of pruning pure entity structures, the purpose of the nondeterministic version of the algorithm is to provide a definition for a set of all structures that the deterministic version should produce to be correct.

The case of pruning the specialized entity structures is more complex. The choice of a nondeterministic algorithm is justified in that there is no deterministic procedure that generates the solution in polynomial time.

For the pruning process, it is enough to restrict the generic experimental frame to the generic observation frame (5) ie,

GOF = {IG, OG}

where IG denotes the set of generic input variable types and OG the set of generic output variable types. By defining the observation frame as above, its role is restricted to representing *behavioral* aspects of design objectives. There are also objectives that constrain the structural aspects of the project under consideration. Therefore, in order to realize the strutural constraints, it will be necessary to augment the design model development with a process termed *synthesis rule generation*.

Return to the pruning procedure and define how a generic observation frame generates the design model structures that accommodate behavioral design objectives. Given the generic observation frame, all the substructures that have all the input and output variable types present in that frame are extracted.

First, a nondeterministic version of such a procedure is defined. Assume that the function *choose* selects an aspect for an entity. The algorithm presented in Figure 5 returns a nondeterministically selected decomposition tree that accommodates the generic frame in which the pruning proceeds.

The deterministic version of procedure Prune is based on the depth first tree traversal. In this procedure, every entity in each aspect is searched for occurrences of variable types that are present in the generic observation frame. The entities are attached to the model decomposition tree as the search progresses. At the same time, the algorithm calls itself recursively for each entity being searched. The complete deterministic pruning procedure is given in Figure 6.

To initialize the pruning process, the steps given below are followed:

1. In the system entity structure, choose the entity E_i that represents the model to be evaluated (this entity will label the root of the model structure TE_i).
2. Create a dummy entity DE (with no variables) with a dummy aspect DA in which E_i is a subentity of DE.
3. Call Prune (DE, CV_{GOF}, V_{GOF});

After the procedure has been executed, DE must be eliminated from all the model structures.

The procedure Prune generates a set of design model structures in the form of decomposition trees. Each such structure accommodates the generic observation frame GOF and constitutes a skeleton for a hierarchical model construction. Figure 7 illustrates the results of pruning of the system entity structure with respect to frame GOF.

The limitation of the procedure is that it operates only on pure entity structures, ie, those that do not have specializations. One possible way of dealing with the problem is to map the entity structure that contains specializations into a pure structure using the multiple decomposition concept and the transformation discussed in System Design and Modeling, Synergies. Then the procedure Prune can be applied to the pure structure. However, this approach greatly increases the size of the entity structure and forces the use of multiple entities. This may not be well justified in some design problems. Therefore, the Prune procedure is extended to operate on entity structures that contain specializations.

For the new algorithm, first it is assumed that if there is

```
Nondeterministic Prune(E_j, CV_GOF, V_GOF);
E_j – root of the pure entity structure
CV_GOF – set of variables of the generic frame GOF
          this set is used to check if all the frame variables
          are present in the pruned substructure initially
          CV_GOF = V_GOF
V_GOF set of input and output variable types of GOF
 failure – signals an unsuccessful completion
 success – signals a successful completion

begin
 A_i := choose(E_j);
 for each E_k ∈ A_i do

  begin
   attach E_k with all its variables as a child of TE_j;
   attach coupling constraint of aspect A_i to TE_j;
   { TE_j denotes the node corresponding to E_j in the model
   structure }
   Prune(E_k, CV_GOF, V_GOF);
 end;


   for each node in decomposition tree TE_j do
     begin { verify correctness of choice }
       visit node;
       if v_k ∈ V_GOF then CV_GOF := CV_GOF – v_k;
       { the attached variable types v_k that belong to the
       generic frame GOF are marked as have been found in
       the model structure }
     end; { of for each node }
   if CV_GOF is not empty then failure
     { the substructure does not have all the variables present
     in the frame }

   output Decomposition Tree TE_j;
   { the tree TE_j is a model structure that accommodates the
     frame GOF }
   success;

end. { of Nondeterministic Prune }
```

Figure 5. Nondeterministic prune algorithm.

```
Procedure Prune(E_j, CV_GOF, V_GOF); {Deterministic}
{ This procedure prunes the pure system entity structure and
  returns the model structures that accommodate the generic
  observation frame GOF. Multiple occurrences of a frame
  variable type are permitted in the model structures }

begin
  for each aspect A_i ∈ E_j do
    begin

      for each entity E_k ∈ A_i do
        begin

          attach E_k with all its variables as a child of TE_j;
          { TE_j denotes the root of the model structure being
          built }

          CV_GOF :=CV_GOF – v_k;

          { update the current set CV_GOF by subtracting the
            variable types v_k such that v_k ∈ V_GOF and v_k is
            attached to E_k }

          if E_k has at least one variable type present in V_GOF
          then mark this level in the model structure as the last
                level at which variable types present in the
                frame have been found;

        end; { of for each entity . . . }

      attach the coupling constraint of the aspect A_i to TE_j;

      for each E_k ∈ A_i such that E_k has aspects do
        Prune (E_k, CV_GOF, V_GOF);
        if C_GOF is empty { ie, the frame is accommodated }
        then
          begin

            create a copy of the current model structure rooted
            by TE_j;

            { this copy will serve as a basis for model structure
            construction in the next aspect A_i+1 }

            output the current model structure rooted by TE_j
            without the entities that appear below the level
            marked as the last level with frame variable type
            occurrence;

          end; { of if }

        update the current structure TE_j by cutting off the last
        level entities;

        { thus prepare the structure for pruning in the next
          aspect }

      end; { of for each aspect . . . }

  end. {of prune }
```

Figure 6. Deterministic prune algorithm.

more than one specialization at any given level in the entity structure, such specializations are hung from one another to reduce their number to one at this level. This operation is illustrated in Figure 8.

Secondly, if a general entity that has a specialization is a component in a multiple decomposition of a multiple entity, then the specialization is mapped into the decomposition aspect according to the rules presented earlier. Given these assumptions the extended pruning process is defined.

As depicted in Figure 9, the extended procedure employs three basic modules. The Move module removes all aspects from each level of the design entity structures where specializations occur. The aspects with their full substructures are hung from the entities of the specialization. This is in agreement with the inheritance property that states that the specialized entities inherit the attributes and substructures of the general entity. The algorithm that performs this function is illustrated in Figure 10.

The Move module returns an entity structure that at any given level has either aspects or a specialization. Given such an entity structure, a set of pure entity structures is generated by substituting the specialized entities in place of the general ones. This process is called Split. Notice that a deterministic algorithm to perform the split would have to generate all the combinations of pure entity structures resulting from substi-

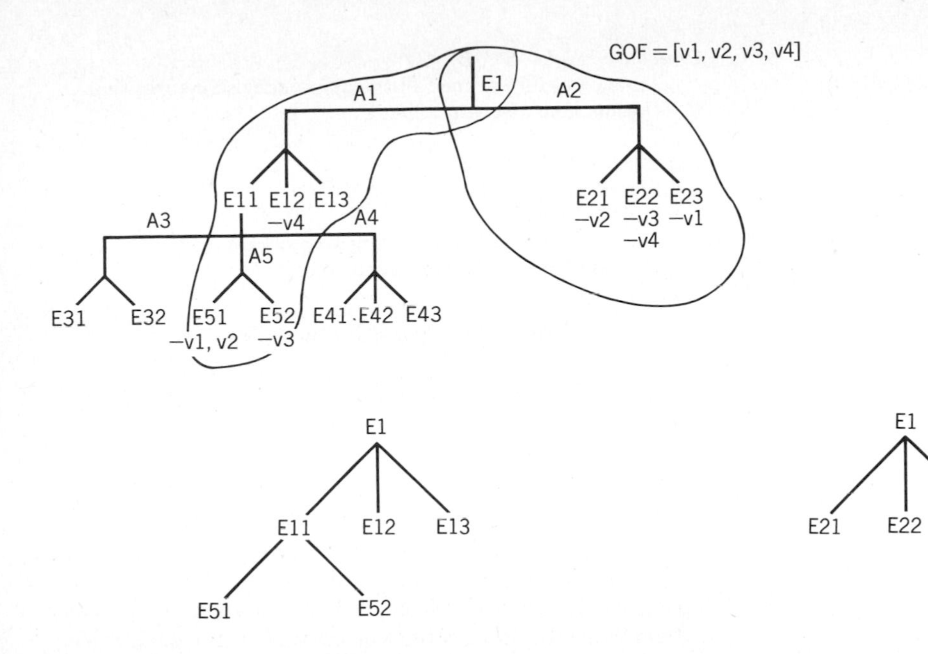

Figure 7. Pruning of the system entity structure in the generic frame type GOF and resulting structures for model construction.

tuting the specialized entities into the general ones. Thus, the problem is computationally hard. Instead, a nondeterministic algorithm that employs the *choose* function for selecting one specialized entity from a set of entities in a given specialization relation is proposed (Fig. 11).

The last module, Prune-Set, employs the deterministic procedure Prune (see Fig. 12).

To illustrate how the extended pruning process works, consider a substructure of the space station design entity structure, depicted in Figure 13. This substructure has both aspects and specializations. The components aspect of the entity Service Organization is first moved down and attached as an aspect to the specialized entities Adaptive and Nonadaptive Organization. This results in a structure shown in Figure 14. The same process is applied again to the specialized entities of Adaptive Organization; the result is depicted in Figure 15. Consequently, an entity structure in which at any level there are either specializations or aspects (never both) is obtained. Finally, the structure is split into pure entity structures with the specialized entities in place of the general ones.

The extended pruning with respect to the generic frame "Adaptability," applied to the station design entity structure, results in the model structures depicted in Figures 16**a** and 16**b**, respectively.

DESIGN MODEL SYNTHESIS

The pruning process described in the foregoing section restricts the space of possibilities for selection of components and cou-

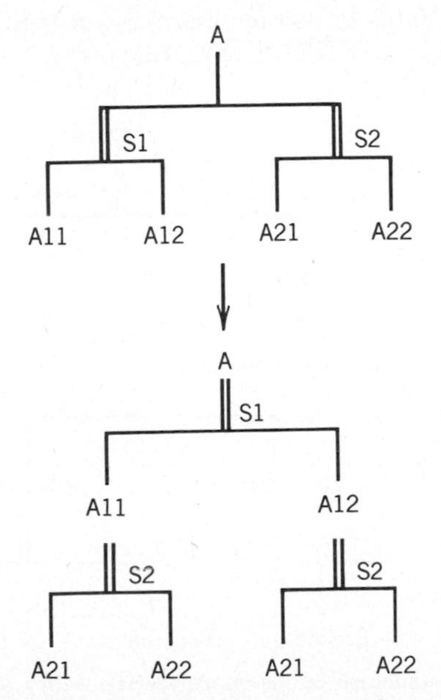

Figure 8. Preparing the entity structure with specializations for pruning.

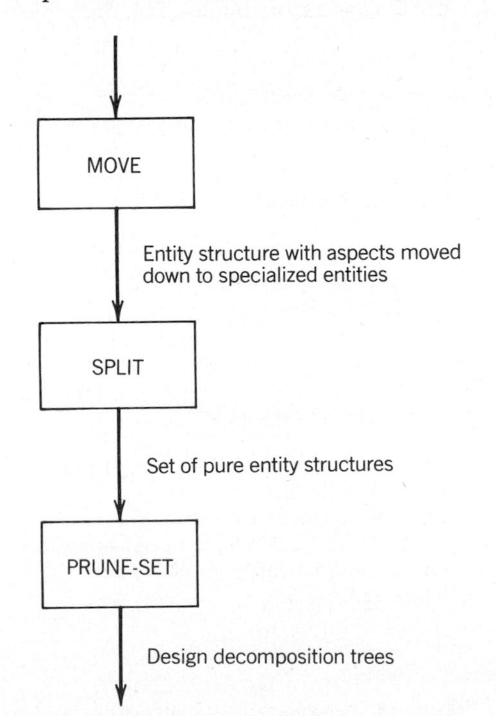

Figure 9. Diagram of extended pruning procedure.

```
Procedure Move(E_j);
    { this module moves all the aspects and variables down to
    the specialized entities }

begin
  if E_j has a specialization S

    begin
      for each aspect A_1 ∊ E_j do

        begin
          for each entity E_k ∊ S do

            begin
              copy all the variables of E_j to E_k;
              attach A_1 with its full substructure and variables as
                an aspect of E_k;
            end;
          delete A_1 from aspects of E_j;
        end;

      for each entity E_k ∊ S do
        Move(E_k);
    end
  else
    begin
      for each aspect A_i ∊ E_j
        for each entity E_k ∊ A_i
          call Move(E_k)
    end;
end. { of Move }
```

Figure 10. The Move algorithm.

```
Procedure Prune—Set({E_j}, CV_GOF, V_GOF);
    { this procedure prunes all the pure entity structures that
      result from algorithm Split }

begin
    for all the pure structures E_j do
      call Prune(E_j, CV_GOF, V_GOF);
      { invoke the procedure Prune to generate a
        set of design decomposition trees }
end. { of Prune—Set }
```

Figure 12. Algorithm Prune_Set.

plings that can be used to realize the system being designed. Thus, it is assumed that the design can now be equivalent to the synthesis of a design model based on the pruned structures and the structural constraints imposed by the project requirements. The following synthesis rule development methodology is proposed:

- Restrict the design domain by pruning the design entity structure in respective generic observation frames.
- Examine the resulting substructures and their constraints. Try to convert as many constraint relations as possible into the active from, ie, into rules that can satisfy them. For those that cannot be converted into such rules, write rules that will test them for satisfaction.
- Write additional rules and modify existing ones to coordinate the actions of the rules (done in conjunction with the selected conflict resolution strategy).

In the following section, a canonical rule scheme for the synthesis problem is presented. See Ref. 30 for a detailed exposition of this methodology.

The constraints imposed by the design requirements can be classified into two basic categories: *convertable to active form,* ie, they can be converted into actions intended to satisfy them, and *passive*. The passive constraints do not guide or motivate any action. They do require satisfaction.

The synthesis problem is conceived as a search through the set of all pruned structures. These are candidates for the solution to the problem.

Assume that for each active constraint, a means of generating such candidates to test against the constraint is present. Call such an operator NEXT_IN_Ci.

```
Nondeterministic Split(E_j, ES);
  ES – entity structure with specializations

begin
 if E_j has aspects then
  begin
    for each aspect A_i ∊ E_j do
      for each entity E_k ∊ A_i do
          Split(E_k, ES);

  end
 else
 if E_j has a specialization then
  begin
    choose(E_i) { an entity in this specialization };
    replace E_j with E_i;
    cut off all the other entities in this
    specialization (including their substructures)
    from the system entity structure ES;
    Split(E_i, ES);

  end;
end. { of Split }
```

Figure 11. Nondeterministic algorithm split.

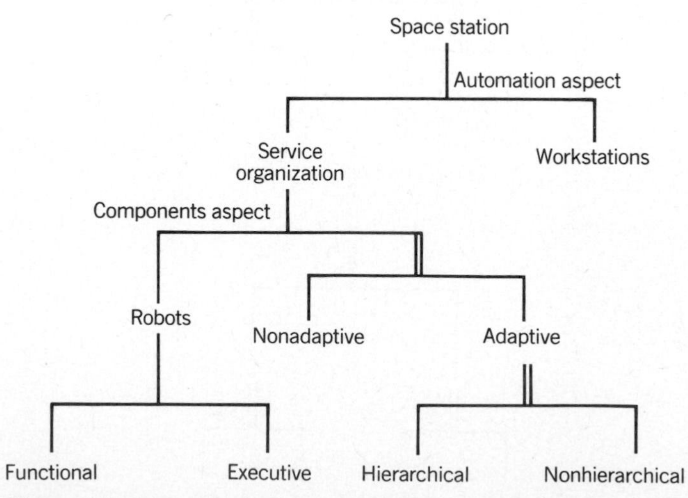
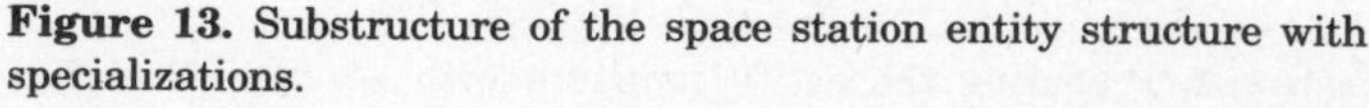

Figure 13. Substructure of the space station entity structure with specializations.

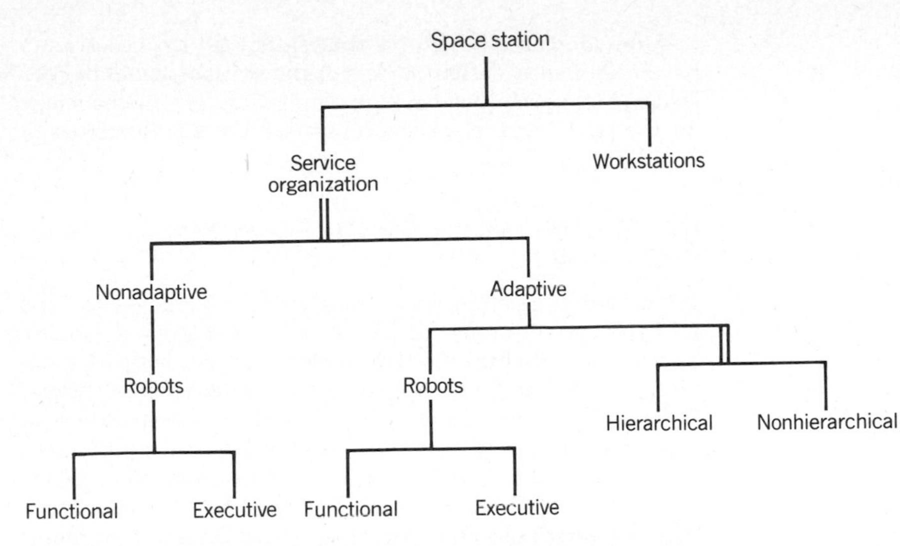

Figure 14. Moving down the aspect of service organization to its specialized entities.

The passive constraints have no corresponding operators and thus can only be tested for their satisfaction. Failure causes backtracking if a state has been reached for which none of the operators can be applied. Instead of applying an operator and then testing if it has consumed more than what remains of an available resource, the application of operators that would bring about the resource depletion is inhibited.

Con is a constraint that should be pretested. An operator, NEXT_IN_Ci, will map a state s into the region satisfying Con if, and only if, Con (NEXT_IN_Ci (s)). To allow the operator to be applied safely, it is necessary to define *applicability* predicate, Ai such that

Ai(s) if, and only if, Con(NEXT_IN_Ci (s))

Thus, a canonical rule scheme for a synthesis problem takes the following form:

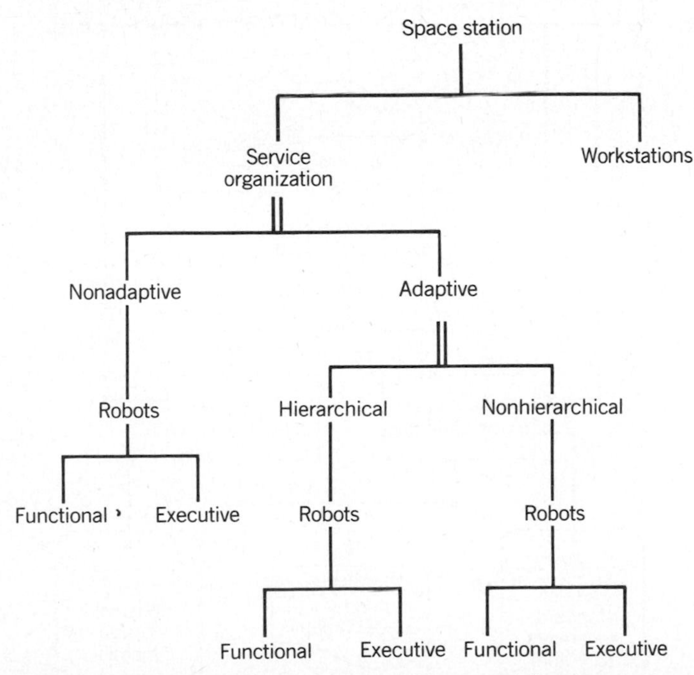

Figure 15. Design entity structure without specializations and aspects present at the same level down any path.

RC If C satisfied on (state)
then Output (state) as the solution
Rl If Cl is not satisfied
Al is satisfied
then state :=NEXT_IN_C1 (state)
....
Ri If Ci is not satisfied
Ai is satisfied
then state :=NEXT_IN_Ci (state)
....
Rn If Cn is not satisfied
An is satisfied
then state :=NEXT_IN_Cn (state)

To illustrate how the above scheme can be applied in model synthesis, use the space station example introduced in Formal Framework for Model-Based System Design. Recall that pruning the space station design entity structure of Figure 3 with respect to generic frame "Adaptability" results in the structures depicted in Figures 16**a** and 16**b**, respectively. Consider the synthesis of the station's model with regard to the automation aspect. For the design model structure of Figure 16, the following constraints are defined:

For the automation aspect of the entity station:

1. TOTAL.ROBOTS.OUTPUT >= TOTAL.STATION.WORK LOAD
2. COMPLEXITY.OF.ROBOT.ORGANIZATION <= MAXIMUM.COMPLEXITY

For the entity robots:

3. NUMBER.OF.ROBOTS IN [1,MAXIMUM].

The first constraint specifies that the robot organization should attain a level of productivity high enough to satisfy the station's work loads. The second constraint is imposed on the level of complexity of the organization. (Different measures of complexity are given in the frame Adaptability.) The third constraint says simply that the number of robots may not exceed a total number of robots available. This constraint can be further refined to apply to functional and executive robots. Similarly, more detailed constraints on the complexity

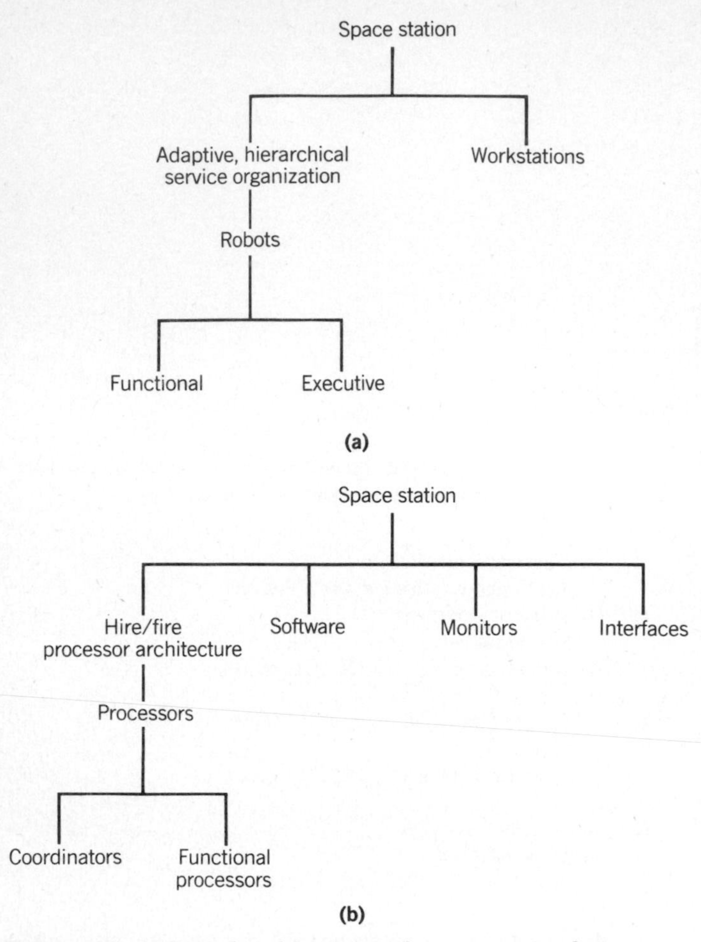

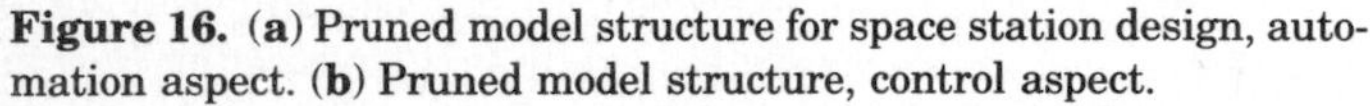

Figure 16. (**a**) Pruned model structure for space station design, automation aspect. (**b**) Pruned model structure, control aspect.

of the organization may be imposed once the reorganization schemes are known.

These constraints are converted into a production rule scheme according to the canonical form. Rule RC is the global constraint checker. Rules RC1 and RC2 are implemented as local constraint satisfiers.

RC if TOTAL.ROBOTS.OUTPUT >= TOTAL.STATION.WORK LOAD
  COMPLEXITY.OF.ROBOT.ORGANIZATION <= MAXIMUM.COMPLEXITY

then "STATION COMPLETED"

RC1 if ROBOT.AVAILABLE
  TOTAL.ROBOTS.OUTPUT < TOTAL.STATION.WORK LOAD

  then HIRE THIS ROBOT
    update COMPLEXITY.OF.ROBOT.ORGANIZATION

RC2 if REORGANIZATION IS FEASIBLE (with respect to restructuring policy)
  COMPLEXITY.OF.ROBOT.ORGANIZATION > MAXIMUM.COMPLEXITY

then
  REORGANIZE THE ROBOTS
  UPDATE TOTAL.ROBOTS.OUTPUT

After candidate structures that satisfy all the constraints have been found, design models of the station should be constructed and evaluated through simulation. The methodology for the model and experimental construction is discussed in detail in Refs. 7 and 30.

ENVIRONMENT FOR INTEGRATED, MODEL-BASED SYSTEM DESIGN

It has been our contention thoroughout the foregoing sections that the system entity structure and the concept of generic frame type constitute the knowledge that can support automatic construction of design models and experimental frames. To explain the argument, the following architecture for an expert system design environment is proposed. As illustrated in Figure 17, the data base of design objectives is one of the major components in the system. It must be well understood that the design objectives drive three fundamental processes in the methodology: first, the retrieval and/or construction of the design entity structure. (Naturally, the designer desires to obtain a family of design representations for a given set of objectives.) Secondly, the objectives serve as a basis for the specification of generic observation frames. Finally, the structural aspects of design generate a set of rules for the design model synthesis.

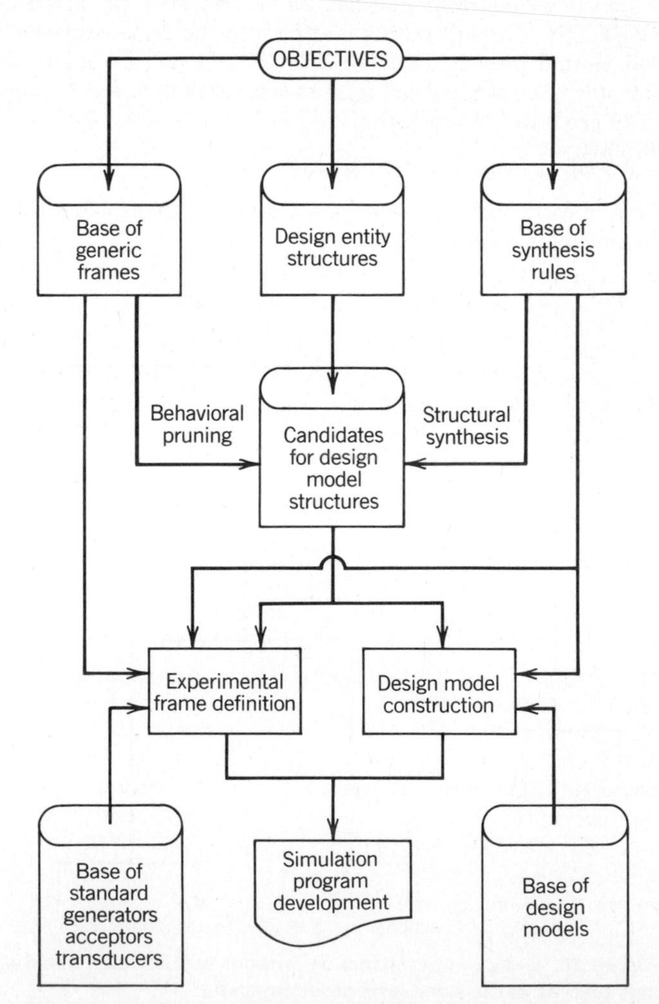

Figure 17. Integrated environment for system design support.

The ultimate purpose of the system depicted in Figure 17 is to analyze and integrate the relationships concerning the objectives specification base, the generic observation frame base, and the design entity structure. Such an integration should result in the formulation of design models and simulation experiments for a problem at hand.

The behavioral aspects of the design objectives are expressed in terms of generic observation frames. Pruning the design entity structure in corresponding observation frames results in substructures conforming to the behavioral objectives.

The substructures are then tested for satisfaction of synthesis rules that are derived from the design structural constraints as presented in Design Model Synthesis. Both behavioral and structural pruning applied to the design entity structure should result in design structures that are candidates for hierarchical model construction. The term candidates implies that some checks for consistency and admissibility (in the sense of conformance to the objectives) should be performed at this stage. If the candidate is inadmissible or no candidates can be obtained by pruning, the process should be reiterated with possible user intervention. The kinds of interventions suggested are modifications or retrieval of the new system entity structure, enhancement of the generic experimental frame, or modification of synthesis rules. The system should construct design models for the structures generated as a result of behavioral and structural synthesis employing the multifacetted model construction methodology presented earlier and in Ref. 1.

The design models should be evaluated through extensive simulation studies in experimental frames induced by the generic frame types. A detailed exposition concerning this aspect of design is presented in Ref. 30.

SUMMARY

In presenting the model-based system design methodology, the main focus is on two major aspects of the design process, the *design model development* and *specification of experimental circumstances* for design simulation.

Based on the formal concepts of the system entity structure and experimental frame, a framework for objectives-driven design model generation has been developed. In this framework, the behavioral aspects of design objectives are reflected in the *generic frame types,* which are prestructures for the experimental frames. The generic experimental frame types serve as a basic means of extracting model structures conforming to the behavioral objectives from the design entity structure. Effective *pruning procedures* have been developed to perform this task. The procedures have been further refined to extract model composition trees from the design entity structures in which specialization relations occur.

The structural aspects of the design objectives represented by a set of constraints have been shown to drive the process called *model synthesis* effectively. A canonical *production rule* scheme has been given for generating model synthesis rules.

These concepts are of a propositional nature. It is stressed that an attempt has been made to lay a foundation on which a design process can be based. A computer-aided expert design environment that internally represents the entity structures and generic frames and has means for dynamically manipulating these structures has been envisoned. Implementation of such a package, in all its generality, may be a long way off. However, efforts are under way to advance the design methodologies further.

BIBLIOGRAPHY

1. B. P. Zeigler, *Multifacetted Modeling and Discrete Event Simulation,* Academic Press, Inc., New York, 1984.
2. M. S. Elzas, "The Use of Structured Design Methodology to Improve Realism in National Economic Planning," in H. Wedder, ed., *Model Adequacy,* Pergamon Press, Ltd., Oxford, UK, 1982.
3. T. I. Oren, "Computer Aided Modeling Systems," in F. E. Cellier, ed., *Progress in Modeling Simulation,* Academic Press, Inc., New York, 1982.
4. D. P. Siewiorek, D. Giuse, W. P. Birmingham, *Proposal for Research on DEMETER: A Design Methodology and Environment,* Carnegie-Mellon University, Pittsburgh, Pa., Jan. 1983.
5. J. W. Rozenblit and B. P. Zeigler, "Concepts for Knowledge-Based System Design Environments," *Proceedings of the 1985 Winter Simulation Conference,* San Francisco, Dec. 1985.
6. T. I. Ören, B. P. Zeigler, "Concepts for Advanced Simulation Systems," *Simulation,* **32**(3), 69–82 (1979).
7. J. W. Rozenblit and B. P. Zeigler, "Entity-Based Structures for Model and Experimental Frame Construction," in M. S. Elzas and co-workers, eds., *Knowledge-Based Modeling and Simulation Methodologies,* North Holland Publishing Co., Amsterdam, 1986.
8. W. A. Wymore, *A Mathematical Theory of Systems Engineering—The Elements,* John Wiley & Sons, Inc., New York, 1967.
9. W. A. Wymore, *Systems Engineering Methodology for Interdisciplinary Teams,* John Wiley & Sons, Inc., New York, 1976.
10. M. Asimov, *Introduction to Design,* Prentice-Hall, Inc., Englewood Cliffs, N.J., 1980.
11. J. C. Enos and R. L. van Tilburg, "Software Design," in R. W. Jensen and C. C. Tonies, eds., *Software Engineering,* Prentice-Hall, Inc., Englewood Cliffs, N.J., 1979.
12. G. Nadler, "An Assessment of Systems Methodology and Design," *Proceedings of The International Conference of The Society for General Systems Research,* Los Angeles, May 1985.
13. J. W. Rozenblit, "Experimental Frames for Distributed Simulation Architectures," *Proceedings of the 1985 SCS Multiconference,* San Diego, Jan. 1985.
14. M. Gonauser and A. Sauer, "Needs for High-Level Design Tools," *Proceedings of the 1983 IEEE Conference on Computer Design,* IEEE, New York.
15. J. W. Rozenblit, *Structures for a Model-Based System Design Environment, Technical Report,* Siemens AG, Munich, 1984 (internal distribution).
16. M. Gonauser, R. Kober, and W. Wenderoth, *A Methodology for Design of Digital Systems and Requirements for a Computer Aided System Design Environment, IFIP WG 10.0,* Sept. 1983.
17. R. Kober and W. Wenderoth, "Problems and Practical Experience in High-Level Design," in Ref. 14.
18. D. Belogus, "Multifacetted Modeling and Simulation: A Software Engineering Implementation," Doctoral Dissertation, Weizmann Institute of Science, Rehovot, Israel, 1985.
19. R. E. Shannon, R. Mayer, and H. H. Adelsberger, "Expert Systems and Simulation," *Simulation,* **44**(6) (June 1985).
20. B. P. Zeigler, Y. V. Reddy, and T. I. Ören, *Knowledge Representation in Simulation Environments,* Academic Press, Inc., New York, in preparation.

21. B. P. Zeigler, "Knowledge Representation from Newton to Minsky and Beyond," *Applied Artificial Intelligence* **1,** 87–107, (1987).
22. B. P. Zeigler, D. Belogus, and A. Bolshoi, "ESP—An Interactive Tool for System Structuring," *Proceedings of the 1980 European Meeting on Cybernetics and Systems Research,* Hemisphere Press, Washington, D.C., 1980.
23. B. P. Zeigler and R. G. Reynolds, "Towards a Theory of Adaptive Computer Architectures," *Proceedings of the Distributed and Parallel Computation Conference,* Denver, May 1985.
24. S. A. Gregory, *The Design Method,* Butterworth Publishers, Ltd., London, 1966.
25. H. D. Hall, *A Methodology for Systems Engineering,* Van Nostrand Publishing Co., New York, 1972.
26. A. P. Sage, *Methodology for Large Scale Systems,* McGraw-Hill, Inc., New York, 1977.
27. W. A. Wymore, *A Mathematical Theory of Systems Design, Technical Report,* University of Arizona, Tucson, Ariz., 1980.
28. M. D. Mesarovic, D. Macko, and Y. Takahara, *Theory of Hierarchical, Multilevel System,* Academic Press, Inc., New York, 1970.
29. P. P. Fasang and M. Whelan, "A Perspective on the Levels of Methodologies in Digital System Design," in Ref. 14.
30. J. W. Rozenblit, "A Conceptual Basis for Model-Based System Design," Doctoral Dissertation, Wayne State University, Detroit, Mich., 1985.

General References

D. K. Baik, "Performance Evaluation of Hierarchical Simulators: Distributed Model Transformation and Mapping," Doctoral Dissertation, Dept. of Computer Science, Wayne State University, Detroit, Mich., 1986.

A. Concepcion, "Distributed Simulation on Multiprocessors: Specification, Design, and Architecture," Doctoral Dissertation, Dept. of Computer Science, Wayne State University, Detroit, Mich., 1985.

L. Dekker, "Concepts for An Advanced Parallel Simulation Architecture," in T. I. Ören, B. P. Zeigler, and M. S. Elzas, eds., *Simulation and Model-Based Methodologies: An Integrative View,* Springer-Verlag, New York, 1984.

J. R. Dixon, *Design Engineering: Inventness, Analysis and Decision Making,* McGraw-Hill, Inc., New York, 1966.

A. Javor, Proposals on the Structure of Simulation Systems," in A. Javor, ed., *Discrete Simulation and Related Fields,* North-Holland Publishing Co., Amsterdam 1982.

R. E. Kalman, P. L. Falb, and M. A. Arbib, *Topics in Mathematical Systems Theory,* McGraw-Hill, Inc., New York, 1969.

T. I. Ören, "GEST—A Modeling and Simulation Language Based on System Theoretic Concepts," in T. I. Ören, B. P. Zeigler, and M. S. Elzas, eds., *Simulation and Model-Based Methodologies: An Integrative View,* Springer-Verlag, New York, 1984.

R. Prather, *Discrete Mathematical Structures for Computer Science,* Houghton Mifflin Publishing Co., Boston, 1976.

Y. V. Reddy, M. S. Fox, and N. Husain, "Automating the Analysis of Simulations in KBS," in Ref. 13.

J. W. Rozenblit, "EXP—A Software Tool for Experimental Frame Specification in Discrete Event Modeling and Simulation," in Proceedings of the 1984 Summer Computer Simulation Conference, Boston, 1984, pp. 967–971.

R. H. Sprague and E. D. Carlson, *Buidling Effecting Decision Support Systems,* Prentice-Hall, Inc., Englewood Cliffs, N.J., 1982.

P. H. Winston, *Artificial Intelligence,* Addison-Wesley Publishing Co., Inc., Reding, Mass., 1984.

B. P. Zeigler, "Structures for Model Based Simulation Systems," in T. I. Ören, B. P. Zeigler, and M. S. Elzas, eds., *Simulation and Model-Based Methodology: An Integrative View,* Springer-Verlag, New York, 1984.

DESIRABILITY OF ROBOTS

Anil Mital
University of Cincinnati
Cincinnati, Ohio

INTRODUCTION

The need to strive for higher productivity is perpetual. Reduction in production costs and greater competition in the international market are two main concerns behind the move of many manufacturers to automate their existing facilities. New technologies, such as FMS, CAD/CAM, and Robotics, are rapidly being implemented in medium and high volume manufacturing. Robots, which are the key supportive elements in automated factories and stand-alone manufacturing cells, are dominating such functions as welding, painting, and loading/unloading.

When first developed, robots were expected to replace workers only in hazardous environments. However, robots have also begun replacing workers in monotonous, highly repetitive, and unstimulating tasks. Existing statistics on robot growth indicate a tremendous increase in the U.S. robot population. With improved capabilities and lower costs, the market could expand to as many as 200,000 units per year (1). By 1990, the world robot population is expected to reach 1 million. The potential market for robots in the United States alone is anticipated to be about 400,000 units per year by that time (2).

Even though new technology has historically created more jobs and led to higher productivity, initial introduction of automation has caused, or is expected to cause, significant displacement (3). In the long run, however, the commercial use of robots is expected to follow this historical trend, although some evidence does raise doubt about a rosy future.

Several different sources forecast that wide-scale usage of robots will lead to worker displacement (see also Human impacts; Employment, impact). According to one study (4), approximately 100,000 jobs may be lost in the U.S. auto industry alone in the 1980s. The Robot Institute of America of the Society of Manufacturing Engineers estimates that 440,000 workers will be displaced by the end of this century and that only 20,000 of these may expect to find another job through attrition or retraining (5). For example, only one fifth of employees laid off by the U.S. auto industries in 1979 has returned to work. A study supported by the Congressional Budget Office (6) predicted a displacement figure of 1 million by the early 1980s. The study conducted by the Ad Hoc Committee on Triple Revolution (7) predicts that the employment displacemet from automation would be so great that it would be necessary for the federal government to provide generous and costly income supports to a large fraction of the work force. Even though the last two studies may be politically motivated, they do not contradict the findings of the others mentioned. On the other hand, the prediction that robots will always lead to worker displacement may not be well-founded. Investigations by the Japanese Labor Ministry (8) and an opinion survey conducted by the Institute of Industrial Engineers (9) contradict the worker-displacement theory.

Displacement predictions aside, there is a psychological fear that robotization will lead to increased unemployment. For instance, Japanese worker resistance has stymied robot introduction in cases where workers felt that robots were taking away the pleasant and easy jobs. The Japanese Labor Ministry, under pressure from the unions (97% of in-house unions and 79% of management believe robots will increase unemployment), has been forced to study the impacts of robots on labor and working conditions (8).

The most serious doubt about robotization stems from the claims of high productivity. Whereas production volume, capacity, and quality are expected to be higher and better and costs of product changeover and human facilities (washrooms, cafeterias, etc) are expected to be lower, there is no assurance that the net associated costs will always be reduced. Increasing costs of U.S. automobiles, despite an increased level of automation, and wage and benefit concessions by the unions is a case in point. As many as 30% of all respondents that are responsible for robot installations felt that they have not achieved any productivity gains (9).

Unless the claims of higher productivity can be verified, it is not possible to state confidently that robots will improve production efficiency. Ineffective modernization, or the lack of it, on the other hand, can lead to feelings of inadequacy or inferiority on the part of the manufacturers and makes it difficult for them to either retain their share of the international market or enjoy new gains.

At the present time, enough doubts exist to question a robot installation. It appears that the decision to replace a human with a robot may not be economical. Prior to any such replacement, a complete economic analysis should be conducted and all costs to the company and the society should be considered. Currently, many firms base these decisions on intuition and fear of being left behind instead of on sound economic replacement analysis.

This article examines the economic desirability of a robot installation from both the company's and society's viewpoint. The effects of robot installations on employment structure, costs involved, changes in cash flows, impact on GNP, etc, are considered. Data from a case study are used to show that robot installations may not always be in the society's best interests (see also ECONOMIC JUSTIFICATION; GOVERNMENT POLICIES).

CONSEQUENCES OF ROBOT INSTALLATIONS—CHANGES IN THE EMPLOYMENT STRUCTURE

Robot installations are, in the simplest form, replacement of manual work by a machine. In the absence of any hard evidence, the consequences of robotization on the employment structure can only be hypothesized. When a worker is replaced by a robot, two scenarios are likely:

1. *Permanent Unemployment.* The displaced worker is relatively older and cannot be retrained. Therefore, the worker does not qualify for newly created jobs, becoming either underemployed or a permanent burden on the society.
2. *Job Relocation.* The displaced worker has some basic skills and can be assigned to a different job, with or without retraining, depending on the skill requirements of the new job.

In both scenarios, the displaced employee experiences changes in income and spending, standard of living, and job satisfaction, along with other social changes. The society also experiences costs (unemployment benefits, welfare, etc), a fact generally ignored by companies automating work places. Conventional equipment replacement analyses rarely consider such costs as taxes paid by the displaced employee, society's expenditure (through the government) on employment benefits, welfare, etc. These costs can be quite substantial, especially if several robot installations are considered. This situation, though beneficial to the company, may not economically benefit the society.

COSTS INVOLVED IN ROBOT INSTALLATIONS

Robotization of a work place requires modifications of the workstation, installation of a robot and accessories, tooling, etc. Many of these costs are obvious and are included in any equipment replacement analysis. Many other costs, however, are not apparent and generally ignored. This usually occurs when only those costs that are intrinsic to the company are considered. The major costs, internal to the company, that are generally considered in economic equipment replacement analyses are:

1. Robot and accessories cost, ie, cost of the robot, test equipment, materials handling system, etc.
2. Installation cost, ie, labor and materials cost for the site, floor and foundation preparations, utilities, interface devices between the robot and fixtures, software development, and factory communication.
3. Rearrangement cost, ie, labor and materials cost for the safety fence, conveyors, etc.
4. Special tooling cost, ie, cost of special end-of-arm devices and changes in the fixture design, clamps, limit switches, sensors, etc.
5. Indirect labor cost, ie, repair and maintenance costs.
6. Operating supplies cost, ie, cost of utilities and services directly attributed to the robot and its support equipment.
7. Maintenance supplies cost.
8. Launching cost, ie, work stoppage costs due to installation.
9. Insurance costs and taxes, eg, sales tax paid on the purchase price of the robot.

In addition to these costs, savings in the direct labor costs and credits due to increased productivity are also considered.

Not counted in a traditional economic equipment replacement analysis are revenues lost and costs endured by the society (taxes) when an employee is replaced by a robot. The following are the major losses and costs to the society resulting from the robotization of a work place:

Social security tax paid by both the employee and the employer
Payroll tax paid by the employer
Federal, state, and city taxes paid by the employee
Unemployment compensation paid by the society
Welfare paid by the society to displaced workers
Retraining costs, if any

In addition to these costs, gains of the society, resulting from corporate federal, state, city, and sales tax, due to increased sales, are also not considered in a traditional economic analysis.

CASE STUDY

To demonstrate the effect of robotization on social costs and to determine the effect of such costs on the decision to install a robot, a case study is presented. The case study concerns the installation of a robot in a medium-sized metal industry located in the midwest. Two different economic analyses are conducted: an initial analysis based on the company's viewpoint and an extended analysis that takes into account all externalities to emphasize the society's viewpoint.

Application Description

The selected application is a gear-hobbing workstation consisting of 4 machines that were individually operated by semiskilled operators. Two sets of machines operated in the shop. Depending on production requirements, either a 4-spindle hobbing machine and shaving machine combination (for low volume) or an 8-spindle hobbing machine and shaving machine combination (for high volume) was used. Major worker func tions involved loading gear blanks, cutting gear profiles, stacking semifinished parts, and trimming the gears on the shaving machine. The clearing and inspection operations, subsequent to shaving, were also manually performed.

The existing workstation was modified so that a single robot could perform all loading/unloading operations. Stacking parts on incoming and outgoing trays, inspection, and overall supervision of the workstation were now performed by only one operator, instead of the four required in the previous setup. The operator who remained was now responsible for (*1*) programming and starting the robot; (*2*) arranging parts on the incoming tray; (*3*) periodic inspection of the finished parts; (*4*) unloading finished parts and stacking them on the outgoing tray; and (*5*) general maintenance of the workstation. Special tools to support the operation were designed and a robot selected. The layout of the modified workstation and specifications of available robots were taken into consideration in making the final robot selection. Using the technical specifications of the robot, a time study was conducted; the operation cycle time was estimated at 88.1 s. From the cycle time, the production capacity (PC) of the robotized plant was determined to be 1315 units per day. The production capacity of the human-operated workstation (PCHO was 800 units per day.

Table 1. Initial Expenditure Due to Robot Installation

Cost Description	Amount, $
Cost of robot (base)	80,000
Special holders and tools	3,000
Installation cost	
Rearrangement cost	1,280
Installation cost	1,500
Feedback and interface devices	5,000
Feasibility study	300
Rearrangement cost	
Conveyor cost	4,362
Fence	670
Feeders and trays	2,000
Special tooling cost	
Grippers	4,000
Special arbor and fixture	2,000
Control locks and safety	5,000
Total expenditure	*109,112*

Costs Involved

The major costs involved in the robot installation were cost of the robot, its accessories, installation, and maintenance, operator training, and other related costs. Table 1 shows the initial expenditure. Annual costs from the company's perspective are given in Table 2.

A savings of $20,000 per year per displacement resulted from the displacement of the three operators. This is shown at the bottom of Table 2. The specific costs to the society and additional revenues generated are discussed later.

Assumptions

In order to conduct an economic equipment replacement analysis, the following assumptions are made:

1. The company is unable to meet the present market demand. The additional capacity, due to robot installation (515 units per day), can be sold without any difficulty.

Table 2. Changes in the Annual Cash Flow of the Company

Description of the Cash Flow	Amount, $/yr
Maintenance and service labor cost for the robot work center	6,000
Operating supplies	3,000
Training cost of the technician[a]	530[b]
Tax and insurance (2.5%)	2,271
Increase in pay due to job upgrading of technician	5,000
Other miscellaneous costs	1,000
Total variable cost/yr	*17,801*
Savings due to labor displacement[c]	60,000

[a] $3,000 is distributed over 10 yr.
[b] Interest rate = 12%/yr.
[c] 20,000/yr/displacement.

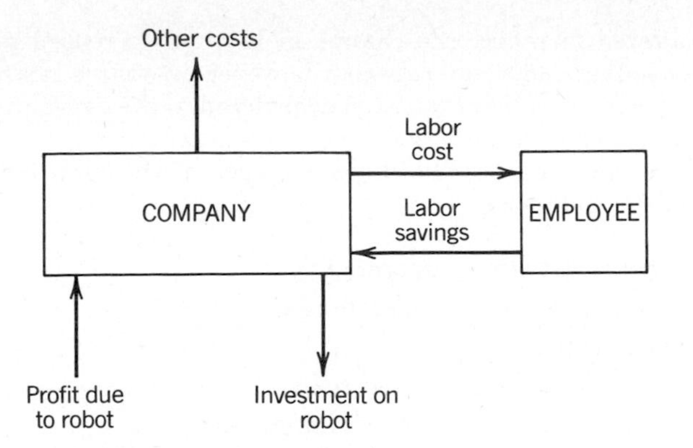

Figure 1. Conventional economic system.

2. The life period of the robot installation is 10 years, with a salvage value of $10,000.
3. The interest rate is 12% per year.
4. All retraining costs are paid by the company.
5. The displaced employees, permanently unemployed, receive their income only from government and welfare agencies.
6. The effect of inflation can be ignored.

Economic Analysis from the Company's Viewpoint

Figure 1 shows the flow chart of a typical economic analysis. Changes in the company's cash flow are estimated by identifying various costs and savings involved. From Tables 1 and 2:

Total fixed cost = $109,112
Total variable cost/yr = $17,801
Total labor savings/yr = $60,000

Using these costs and savings, the return on investment (ROI) is estimated. Table 3 shows the net and aftertax cash flows for the 10 yr study period. The investment is depreciated over 5 years, using the accelerated cost recovery system (ACRS). A corporate tax rate of 46% is assumed to calculate taxes and aftertax cash flow. Using the cash flow in Table 3, the ROI is calculated as follows:

$$[109{,}112 - 10{,}000] - 42{,}199\,[P/Ai,10] = 0 \;\ldots\; \text{pre-tax ROI}$$

or ROI = 40.9%; payback period = 1/0.4096 = 2.45 yr

$$\text{Aftertax ROI} \ldots [109{,}112 - 10{,}000] - 34{,}645\,[P/Fi,1] - 32{,}818\,(P/Fi,2] - 32{,}361\,[P/Fi,3] - 32{,}361\,[P/Fi,4] - 32{,}361\,[P/Fi,5] - 22{,}787\,[P/Ai,5]\,[P/Fi,5]$$

or ROI = 28.76%; payback period = 3.48 yr

Because the company's minimum attractive rate of return (MARR) is 25%, the installation is desirable.

Break-even Analysis The break-even analysis determines the minimum number of extra units that must be produced to justify the investment. Cost and saving (revenue) functions are estimated from the data given in Tables 1 and 2. Annual costs and savings are distributed over the annual production volume (208,000 units = 800 units/d × 5 d/wk × 52 wk).

$$\begin{aligned}\text{Cost function} &= (\text{investment} - \text{salvage value}) + (\text{annual cost/production volume}) \times \text{number of units}\\ &= (109{,}112 - 10{,}000) + (17{,}801/208{,}000)X\\ &= 99{,}112 + 0.086X\end{aligned}$$

$$\begin{aligned}\text{Saving (revenue) function} &= (\text{annual savings/production volume}) \times \text{number of units}\\ &= (60{,}000/208{,}000)X\\ &= 0.228X\end{aligned}$$

Table 3. Cashflow Analysis

End of Year (EOY) (1)	Savings, $/yr (2)	Cost, $/yr (3)	Net Cash Flow, $/yr (4) = (2 − 3)	Depreciation[a], $ (5)	Aftertax Cash Flow[b],[c], (6)
1	60,000	17,801	42,199	14,867	34,645[d]
2	60,000	17,801	42,199	21,805	32,818
3	60,000	17,801	42,199	20,813	32,361
4	60,000	17,801	42,199	20,813	32,361
5	60,000	17,801	42,199	20,813	32,361
6	60,000	17,801	42,199	0	22,787
7	60,000	17,801	42,199	0	22,787
8	60,000	17,801	42,199	0	22,787
9	60,000	17,801	42,199	0	22,787
10	60,000	17,801	42,199	0	22,787

[a] Depreciation, based on the accelerated cost recovery system:
Depreciable amount ($109,112 − 10,000) = $99,112
First-year depreciation = $99,112 × 0.15 = $14,867
Second-year depreciation = $99,112 × 0.22 = $21,805
Depreciation in years 3–5 = $20,813
[b] Aftertax cash flow = net cash flow − tax, where tax = (net cash flow − depreciation) × 0.46.
[c] Includes a corporate tax rate of 46%.
[d] An investment credit of 10% is taken into consideration.

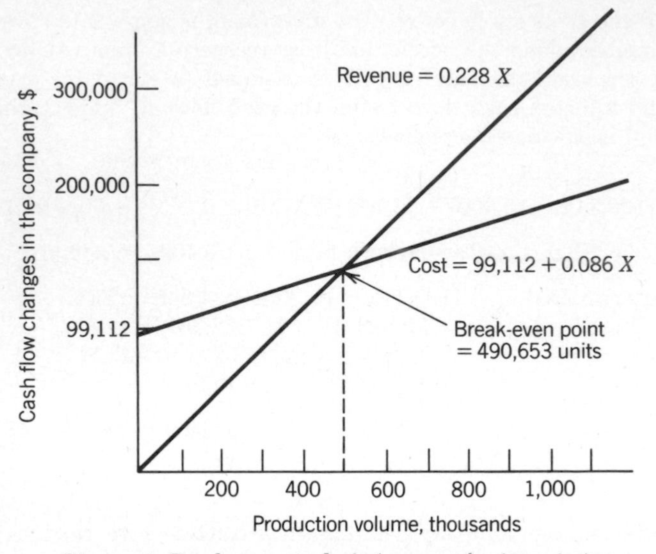

Figure 2. Break-even analysis (company's viewpoint).

At break-even point, the two functions are equal:

$$99{,}112 + 0.086X = 0.228X$$

$$X = 490{,}653 \text{ units}$$

Therefore, at least 490,653 units must be produced to recover the investment. Figure 2 shows the two functions and the break-even point.

Economic Analysis from the Society's Viewpoint

The economic analysis is expanded to include cost and revenue changes in the society's cash flow and to determine if the decision to install the robot is still economically desirable. Figure 3 shows the modified flow chart. The robot installation led to the displacement (permanent unemployment) of three workers. This causes changes in the cash flow to and from the society. Most changes are due to tax losses; additional costs stem from unemployment compensation, welfare, etc. If the employees were relocated to different jobs, tax revenues would be altered to reflect the change in income. Increased profit from selling additional capacity, however, increases tax revenue. These cash flows were not considered in the conventional economic analysis.

The following are the major changes of the cash flow to and from the society:

Federal and state corporate tax
City and corporate franchise tax
Tangible property tax
Sales tax due to increased sales
Federal, state, and city individual taxes
Social security paid by the employees
Payroll tax
Unemployment compensation paid by the society
Welfare

Taxes are computed from tax tables and forms provided by the various tax agencies (because local and state taxes vary widely, the final outcome can vary from location to location). In this case, taxes are calculated on the basis of Ohio's tax laws.

When determining changes in the cash flow of the society, two different cases are examined:

Case 1: Displaced workers are permanently unemployed.
Case 2: Displaced workers are relocated.

Case 1 The displaced workers depend on unemployment compensation and welfare from the society. This change in their status alters the cash flow to the society: various taxes paid by the employees and the employer stop and at the same time, society's expenses increase.

The increased productivity and labor savings, on the other hand, increase the company's profit. This profit then generates tax revenues to the society. The investment (robot) also increases the GNP.

Table 4 lists the additional revenues received by the society as the result of robotization. These are generated from the taxes due to increased savings and sales. The various costs to the society as the result of worker displacement are shown in Table 5. These costs occur mainly due to losses in tax revenue, expenses on unemployment compensation, and welfare. Using the data from Tables 4 and 5, the net loss per year due to displacement amounts to

Net loss = (cost to the society) − (revenue gained by the society)
= \$53,235 − \$29,288
= \$23,947

The break-even analysis is performed from the same information; results are shown in Figure 4:

$$\text{Cost function} = 53{,}235 + (53{,}235/208{,}000)X = 53{,}235 + 0.256X$$

where X is the increased production volume due to robotization (= PC).

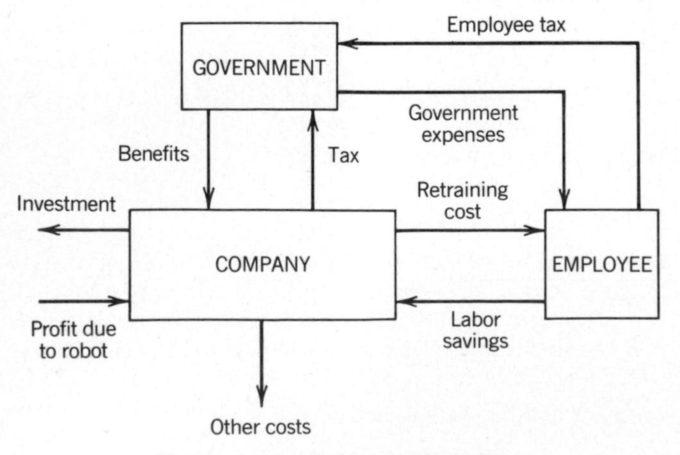

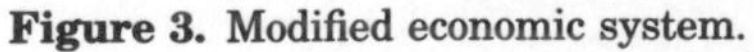

Figure 3. Modified economic system.

Table 4. Changes in the Revenues of the Society

Item	Amount, $/yr
Federal corporation tax (46% of company's profit)[a]	18,930.84
State corporation tax (3.6%)	1,509.75
City tax (2%)	823.08
Corporate franchise tax (CFT)[b]	5,864.88
Tangible property tax (TPT)[c]	507.96
Sales tax (assumed to be 17% of the state tax)	256.53
Indirect gain due to productivity and other sources (assumed to be 5% of the above total)	1,394.65[d]
Total revenue/yr	*29,287.69*

[a] Company's profit = 1.5 (direct labor savings − 2.7 annual costs). Assuming the investment base to be 50% of the expenditure; the extra maintenance was compensated by assuming a factor of 1.2. Or, company's profit = 1.5 × 60,000 − 2.7 × 17,801 = $41,937.30. Profit after tax = 41,937.30 (1 − 0.46) = $22,646.14.
[b] CFT is 5.1% on the first $25,000 of net income plus 9.2% on income in excess of $25,000 plus surtax of 5.6% net income.
[c] TPT is dependent on the investment. Taxable amount for TPT is 34% of the robot cost; tax rate applied is 2%.
[d] Represents gains due to the displacement of three workers.

Table 5. Cash Flow Lost by the Society Due to the Robot Installation

Item	Amount, $/yr
Individual federal income tax (20.7% of company's profit)	4,143.00
State income tax[a]	549.33
City income tax	400.00
Social security paid by the employee (7%)	1,400.00
Payroll tax paid by the employer	1,240.00
Social security paid by the employer (7%)	1,400.00
Unemployment compensation paid by the government to the unemployed	7,000.00
Other government cost (10% of the above)	1,613.20
Total cost/employee	*17,745.00*[b]

[a] State tax = $332.5 plus 4.4% of income above $15,000
= $332.5 + (4.4/100) (20,000 − 15,000)
= $549.33.
[b] Total cost/yr = $17,745 × 3 workers = $53,235.

$$\text{Revenue function} = 29{,}288 + (29{,}288/208{,}000)X = 29{,}288 + 0.141X$$

As shown in Figure 4, the two functions do not intersect. This indicates that the robotized workstation, even though economically beneficial to the company, is not economically desirable from the society's viewpoint if the displaced workers are permanently unemployed.

Case 2 In this case, it is assumed that the displaced workers can be relocated within the company but without retraining, since the new jobs require the same levels of skills. It is also assumed that the three workers, once relocated, will earn the same salary. The net effect of this move would be an increase in revenue. The total revenue in this case would be

Total revenue = $53,235 + 29,288 + tax revenue from the employer due to profits at the new location (RNEW)
= 82,523 + RNEW

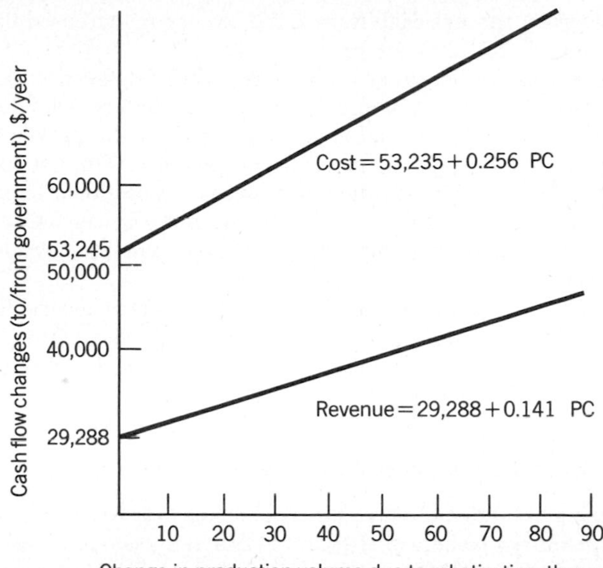

Figure 4. Break-even analysis from the society's viewpoint. Case 1: permanent unemployment.

where RNEW is assumed to be $21,520, the sum of federal, state, local, and sales taxes paid by the employer after the robot installation. This assumption is necessary because the activity at the relocation site for the three displaced employees is unknown. Thus,

$$\text{Total revenue} = \$82{,}523 + 21{,}520 = \$104{,}043$$

The revenue function, therefore, would be

$$\text{Revenue function} = 104{,}043 + (104{,}043/208{,}000)X = 104{,}043 + 0.500X$$

The cost function would remain the same:

$$53{,}235 + 0.256X$$

It is clear that in this case there would be a net annual increase of $51,808 in the society's income. This does not account for the increase due to the larger production volume ($0.249 per unit). This implies that if workers can be relocated, the installation becomes economically beneficial to all.

IMPACT OF ROBOT INSTALLATION

On Worker Displacement

Case 1 depicts a situation where relatively older employees are replaced. Due to their age and basic skill limitations, the three displaced employees became permanently unemployed. For this case, the estimated cash flow to the society is decreased by approximately $24,000 per year.

In the second case, it is assumed that the displaced workers can be relocated to a different location within the company but without any major retraining. Assuming no changes in the pay scale, the net cash flow of the society is increased by approximately $51,808 per year.

In the event the company retrains the three displaced workers for the new jobs, it would spend approximately $24,000. If the new job required higher level skills than the previous job, each worker would be given a pay increase. The net increase in the society's cash flow in this case would be at least $27,808 in the first year and $51,808 annually in the subsequent years (assuming the retraining costs are paid solely by the company).

From the preceding discussion, it is obvious that desirability of the robot installation, from the society's viewpoint, is much greater if the displaced workers are retrained and relocated within the company.

On Producer and Consumer Price Indexes

Producer price index is a measure of changes in prices received in the primary markets of the United States by the producers of commodities in all stages of processing. The consumer price index, often called the "cost of living index," is the statistical measure of the average change in prices in a fixed market basket of goods and services.

Robotization will certainly decrease the "cost per unit" of the product. It is, however, uncertain that this decrease will be passed on to the consumer. If the company maintains the same selling price, the reduction in producer cost will increase the profit margin but will not lower the consumer cost. Examples from the steel and auto industries clearly indicate this.

On GNP

The GNP (gross national product) is the yardstick of an economy's performance. The measure of overall annual flow of goods and services in the economy, it consists of three major components:

1. Personal consumption expenditure (65% of GNP).
2. Goverment expenditure on goods and services (18% of GNP).
3. Gross private domestic investment (17% of GNP).

In the example, the personal consumption expenditure of each displaced worker is expected to change, depending on the new income. If the worker remains unemployed, the reduction in personal consumption expenditure would be

$$[1 - (7{,}000/20{,}000)] \times 100 = 65\%$$

Where $7,000 is the unemployment compensation and $20,000 the original income. This is approximately 3.95 × 10–11% of the GNP (as of April 1984, the GNP was $3,541.6 billion (10). In the example, three workers are displaced, and the GNP is lowered by $11.85 \times 10^{-11}\%$. However, the increase in GNP due to investment ($100, 112 for the robot) is $31.06 \times 10^{-11}\%$. It appears that in this case, there is a net increase in the GNP. However, if the total displacement nationwide ($3 million) and the amount of investment dollars ($5 million) by the year 1990 (11) are considered, the net change in GNP will be a decrease of about 2%. If the workers are relocated instead of being unemployed, the change in GNP will be positive (due to investment).

On Social Factors

The severity of the impact of robot installation on social factors is mainly determined by the level of income. The areas that will be most affected are (*1*) quality of life, (*2*) health and nutrition, (*3*) housing and environment, (*4*) personal and family education, (*5*) welfare and social security, and (*6*) social participation in the society. These factors, although subjective, determine the health of the nation's economy.

CONCLUSIONS

Even though only one case was examined, certain general conclusions can be drawn:

1. Decisions made on the basis of conventional economic analysis may not be sufficient to justify a robot installation. The decision made from the society's viewpoint may be different from that made from the company's viewpoint. The company should compromise some of its profit by incorporating extenalities in the economic analysis.
2. It appears that if the workers are permanently displaced, the decision to install a robot is not economically desirable from the society's viewpoint. However, if the displaced workers are relocated, with or without training, the installation becomes economically desirable from all viewpoints.
3. Without relocation, the net effect of robot installations may be a net decline in the GNP.

POLICY RECOMMENDATIONS

The results of this case study dictate certain changes for the company and the society. Specifically, changes in the tax structure and in companies' profit policy should be incorporated to reduce the adverse consequences of robot installations. The following policy recommendations are suggested (some of these are justified by the results of case study; others, in the author's opinion, should facilitate modernization and enhance the country's ability to compete in manufacturing):

1. There should be an increased sense of responsibility on the part of the employer for continuity of employment of the individual, even if not in the same company. Management needs to oversee retraining and relocation of industrial workers. The training programs should be closely coupled with the future industrial needs. Management should broaden its outlook toward capital budgeting and concern over net earnings and pass some of the profits from robotization to the workers.
2. The society should change the tax structure in favor of advanced manufacturing investments to minimize the impact of unemployment due to robotization. Government policies should influence the direction of growth through special tax incentives, loan guarantees, or other special funding.
3. Risk sharing between government and industry is essential. Expenditures on commercializing technological op-

portunities should be given special tax privileges, and government support of research and development should be sufficient to reinforce domestic industries.

4. National laws requiring portability of pension, seniority, and group insurance are needed. Closer cooperation among industrial firms, community service organizations, and labor organizations in the areas of job counseling, relocation assistance, and employment agencies is needed. The federal government should establish a central facility for sharing information on robotics and employment opportunities.
5. A national plan should be established to encourage the use of robotics and related advanced technologies. Government should rely on the incentives of the free enterprise system to encourage the use of robots in industries where there are sound reasons to do so.
6. The government should not rely on the "trickle-down" effects from military expenditures to maintain technological parities in the economy.

BIBLIOGRAPHY

1. R. U. Ayers and S. M. Miller, *Robotics Applications and Social Implications,* Ballinger Publishing Co., Cambridge, Mass. 1983, pp. 212–222.
2. W. P. Gevarter, *An Overview of Artificial Intelligence and Robotics,* Vol. II, U.S. Government Printing Office, Washington, D.C., 1983, pp. 322–326.
3. E. Appelbaum, "High Tech and the Structural Employment Problems of the Eighties," in E. Collins, ed., *American Jobs and the Changiang Industrial Base,* Ballinger Publishing Co., Cambridge, Mass., 1984.
4. H. Shaiken, "A Robot is After Your Job," *The New York Times,* Sept. 3, 1980, Sect. A19.
5. *Best of Business,* 49 (Apr. 1982).
6. R. U. Ayers, *The Impacts of Robotics on Workforce and Workplace,* Carnegie-Mellon University, Pittsburgh, Pa., 1981.
7. *Overview of Recent Analysis of Technology: Productivity and Employment Trends,* Bureau of Labor Statistics, Washington, D.C., 1982.
8. R. C. Dorf, *Robotics and Automated Manufacturing,* Reston Publishing Co., Inc., Reston, Va., 1983, pp. 6 and 160.
9. "IIE Survey Zaps Robots," *Robot Insider,* 2 (Oct. 7, 1983).
10. *National Economic Indicators-Monthly* (May 1984).
11. P. Deane, *The First Industrial Revolution,* Cambridge University Press, New York, 1979, pp. 90–91.

DYNAMICS

Mo Shahinpoor
University of New Mexico
Albuquerque, New Mexico

INTRODUCTION

Robot manipulators considered in this treatise on dynamic analysis fall in the class of open kinematic chains with servo-controlled joints. Each discrete member of the chain is called a link, attached to neighboring links by means of lower-pair joints (1). Normally, one end of the chain is fixed to a base reference frame and the other end of the chain can move freely in the associated robotic workspace. The free end can be equipped with an end effector, a tool, a hand, or a gripper to manipulate objects and can experience (apply) forces and couples from (to) the objects they manipulate for assembly task processes. In order to understand the dynamics of such robot manipulators, one must first understand the associated kinematics. The kinematics of the robot manipulator, as witnessed in other sections of this encyclopedia, can be completely understood by means of 4×4 homogeneous Denavit–Hartenberg (D–H) (1) transformations.

Some shortcomings of the D–H transformation in dealing with robotic dynamics have been discussed in references 2 and 3, and a new approach based on zero-position description has been successfully utilized. The D–H transformations essentially determine the position of the origin and the rotation of one link coordinate frame with respect to another link coordinate frame. The D–H transformations are also used in deriving the dynamic equations of robot manipulators.

The essential importance of the dynamics of robot manipulators is clearly pronounced in the control of robot motion, forces, and couples. Obviously, the control of robot manipulators is one of the most important subjects in robotics and bears heavily on the dynamics of robot manipulators. Thus, the understanding of the dynamics of robot manipulators is also one of the most important subjects in robotics. Historically, Bejczy (4) presented the first comprehensive treatment on robot arm dynamics, and this presentation parallels his approach and technique in dealing with the dynamics of robot manipulators. The later developments on robot dynamics, as presented in references 5–18, are also utilized.

There are essentially seven different approaches available for the derivation of the governing equations pertaining to robot arm dynamics, as follow:

1. The Lagrange–Euler modeling, L–E (4,6,7,12,14).
2. The Newton–Euler modeling, N–E (5,9).
3. The generalized D'Alembert modeling, G-D (10,11).
4. The bond graph modeling (13,14).
5. The Lagrangian recursive formulation (7).
6. The Newton–Euler recursive formulation (9).
7. The zero reference position formulation (19).

Concentration on the first two techniques is sufficient for this discussion. It is noted that the dynamics equations for robot manipulators are highly nonlinear due to the presence of Coriolis and centrifugal acceleration terms, as will be evident later. In the L–E formation, the real-time control is nearly impossible to achieve due to large computational time, on the order of a few seconds, for point-to-point control. If the nonlinear terms are neglected, then the problem is manageable for real-time control. However, the resulting control is suboptimal and only valid for very slow motion. The N–E formulation is more advantageous than the L–E formulation in both speed and accuracy. The N–E equations can be applied to a robot manipulator link by link and joint by joint either from the base to the gripper (forward recursions) or vice versa (backward recursions). The forward recursive N–E equations transfer kinematic information, such as linear and angular velocities, linear and angular accelerations, and forces and couples applied to the center of mass of each link, from the base reference frame to the gripper frame.

The backward recursive equations transfer the essential kinematic information from the gripper frame to the base frame. Here the computation time is three orders of magnitude smaller than before, on the order of a few milliseconds, thus making it almost possible to achieve real-time dynamic control of robot manipulators in the joint variable space. However, the N–E model is not readily useful for the dynamic control of robot manipulators. A new approach based on the G-D equations that partially circumvents the real-time dynamic control problem in robot manipulations has recently been proposed (11). Brief derivations of the dynamic equations by the L–E and the N–E models are presented below.

THEORETICAL FOUNDATIONS

As discussed before, a computer-based robot manipulator can be modeled as an open kinematic chain with several rigid bodies (links) connected in series by either revolute or prismatic joints. As in the inverse kinematics problem, a set of joint displacements (angles θ_i) are to be found that would place the gripper at a location and orientation (kinematic attitude), specified by T_o^n with respect to the base coordinate. In robotics statics and dynamics, a set of joint forces $\mathbf{f}_i$ and joint torques $\mathbf{T}_i$ are discussed so as to achieve a required set of force $\mathbf{f}$ and torque $\mathbf{T}$ at the gripper. The inverse problem for manipulator dynamics, that is, the problem of computing the joint torques or generalized forces, is discussed.

The dynamic equations for robotic manipulators are formulated by approaches 1 and 2 of the previous section, and finally the recursive formulations are presented briefly. The governing equations are generally nonlinear coupled second-order differential equations.

Deriving the dynamic model of a robot arm by means of the N–E formulation is advantageous because of its speed and accuracy. The derivation is straightforward, but rather messy, and involves a great number of vector cross products. The resulting dynamic equations constitute a set of forward and backward recursive equations that can be applied to the robot links from one end of the arm to the other. In the forward recursive formulation, kinematic information such as the linear velocities, the angular velocities, the linear accelerations, the angular accelerations, the forces, and the couples exerted at the centers of mass of each link is propagated from the base reference frame to the end-effector frame. On the other hand, in the backward recursive formulation such information is propagated from the end-effector frame to the base frame.

The L–E dynamic formulation of robot arm is again straightforward and messy, but systematic. The resulting equations of motion form a set of second-order highly nonlinear coupled differential equations involving a horrendous array of 4×4 homogeneous matrix manipulations. The L–E formulation is computationally inefficient, and real-time control is almost impossible to achieve. Thus, various attempts at approximating the equaitons by neglecting the second-order coupling terms, such as the Coriolis and centrifugal accelerations, have been initiated in the past for slow-moving robotic manipulations.

Presented now are some fundamental definitions for dynamic formulations, and then the governing dynamic equations for robot arms by N–E and L–E approaches are derived.

Preliminary Definitions

The robot arm is considered to be composed of rigid links and subjected to rigid body dynamics formulation. Realistically, all robotic manipulators are indeed made up of flexible structures. However, such treatments are beyond the scope of the present article merely because of their added complexity. Thus, in order to refresh the memory, descriptions of a number of fundamental definitions, such as the generalized robotic coordinates, the general dynamic constraints, the robotic velocity and acceleration in generalized robotic coordinates, the dynamic transformation laws between generalized robotic coordinates, the robotic velocity and acceleration in moving frames, the inertia tensor, the Newton's equation of motion, the Euler's equations of motion, the Lagrangian and the Lagrange's equations, and the Hamilton's principle of least action, which leads to the L–E governing equations for robot dynamics, are presented.

The Generalized Robotic Coordinates

Consider a robot arm as a series of rigid links connected by a series of N joints, either prismatic or revolute. Thus, the independent freedom of motion at each joint could be considered as a degree of freedom associated with the independent actuation of a displacement field q_m, $m = 1, 2, \cdots, N$, in a robot manipulator. On the other hand, the robotic attitude change at the end effector can be described by changes in a six-dimensional coordinate system $x_i = (p_x, p_y, p_z, \theta, \phi, \gamma)$ of the end-effector frame T_o^n such that $\mathbf{p} = (p_x, p_y, p_z)^T$ is the position vector of the origin of the T_o^n frame and θ, ϕ, γ are either the Eulerian angles or the RPY angles of the T_o^n frame with respect to the base frame (x_o, y_o, z_o).

Now from the material covered in the previous sections it is clear that

$$x_i = x_i(q_1, q_2, \cdots, q_N, t) \tag{1}$$

or conversely

$$q_m = q_m(x_1, x_2, x_3, x_4, x_5, x_6, t) \tag{2}$$

where t is the time. A necessary condition for solving for q_m, given equation 1, is that the Jacobian of transformation be nonsingular, that is, that its determinant be different from zero:

$$\left|\boldsymbol{J}(x_i, q_m)\right| = \left|\frac{\partial x_i}{\partial q_m}\right| \neq 0 \tag{3}$$

which implies that for unique solutions $m = b$.

Dynamic Constraints

Dynamic constraints may be presented for a robot manipulator in motion. These constraints are generally expressed in the form

$$\phi_c(x_1, x_2, x_3, x_4, x_5, x_6, t) = 0 \tag{4}$$

or

$$\phi_c^*(q_1, q_2, \cdots, q_N, t) = 0 \tag{5}$$

The Robotic Velocity and Acceleration

For an N-jointed robot manipulator with no constraints, one may assign a Cartesian rectangular coordinate frame at each joint and define a generalized Cartesian joint coordinate $x_i(j)$ pertaining to the jth frame, that is, the $(j + 1)$th joint. Thus, one has the following relationships:

$$x_{i(j)} = \hat{x}_{i(j)}(q_1, q_2, \cdots, q_N, t) \tag{6}$$

The velocity is then defined as

$$\dot{x}_{i(j)} = \sum_{m=1}^{N} \frac{\partial x_{i(j)}}{\partial q_m} \dot{q}_m + \frac{\partial x_{i(j)}}{\partial t} \tag{7}$$

and the accleration is defined as

$$\ddot{x}_{i(j)} = \sum_{m=1}^{N} \left[\frac{\partial x_{i(j)}}{\partial q_m}\right] \ddot{q}_m + \sum_{m=1}^{N} \sum_{n=1}^{N} \left[\frac{\partial^2 x_{i(j)}}{\partial q_m \partial q_n}\right] \dot{q}_m \dot{q}_n + 2 \sum_{m=1}^{N} \left[\frac{\partial^2 x_{i(j)}}{\partial q_m \partial t}\right] \dot{q}_m + \frac{\partial^2 x_{i(j)}}{\partial t^2} \tag{8}$$

As an example, the components of the velocity and the acceleration vectors at the origin of the gripper frame of a cylindrical polar robot manipulator (Figure 1) may be determined as outlined below. It is clear from Figure 1 that at the origin of the gripper frame

$$x = r \cos \theta, \qquad y = r \sin \theta, \qquad z = z \tag{9}$$

$$r = (x^2 + y^2)^{1/2}, \qquad \theta = \tan^{-1}(y/x) \tag{10}$$

The components of the velocity vector are

$$\dot{x} = \dot{r} \cos \theta - r\dot{\theta} \sin \theta \tag{11}$$

$$\dot{y} = \dot{r} \sin \theta + r\dot{\theta} \cos \theta \tag{12}$$

$$\dot{r} = (\dot{x}x + \dot{y}y)(x^2 + y^2)^{-1/2} \tag{13}$$

$$\dot{\theta} = (x\dot{y} - y\dot{x})(x^2 + y^2)^{-1/2} \tag{14}$$

The components of the acceleration vector are

$$\ddot{x} = (\ddot{r} - r\dot{\theta}^2) \cos \theta - (r\ddot{\theta} + 2\dot{r}\dot{\theta}) \sin \theta \tag{15}$$

$$\ddot{y} = (\ddot{r} - r\dot{\theta}^2) \sin \theta - (r\ddot{\theta} + 2\dot{r}\dot{\theta}) \cos \theta \tag{16}$$

The radial and tangential components of the velocity $\mathbf{v}$ and the acceleration $\mathbf{a}$ are given, respectively, by

$$\mathbf{v} = \dot{r}\mathbf{e} + r\dot{\theta}\mathbf{e}_\theta + \dot{z}\mathbf{e}_z \tag{17}$$

$$\mathbf{a} = (\ddot{r} - r\dot{\theta}^2)\mathbf{e} + (r\ddot{\theta} + 2\dot{r}\dot{\theta})\mathbf{e}_\theta + \ddot{z}\,\mathbf{e}_z \tag{18}$$

where $\mathbf{e}_r$, $\mathbf{e}_\theta$, and $\mathbf{e}_z$ are the unit vectors in the radial, tangential, and axial directions, respectively.

In another example, the components of the velocity and the acceleration vectors are determined at the origin of the gripper frame of a spherical robot manipulator (Figure 2) as outlined below.

The origin of the gripper frame is at

$$r = (x^2 + y^2 + z^2)^{1/2} \tag{19}$$

$$\theta = \cot^{-1}\left[\frac{z}{(x^2 + y^2)^{1/2}}\right], \qquad \phi = \tan^{-1}(y/x) \tag{20}$$

$$x = r \sin \theta \cos \phi, \quad y = r \sin \theta \sin \phi, \quad z = r \cos \theta \tag{21}$$

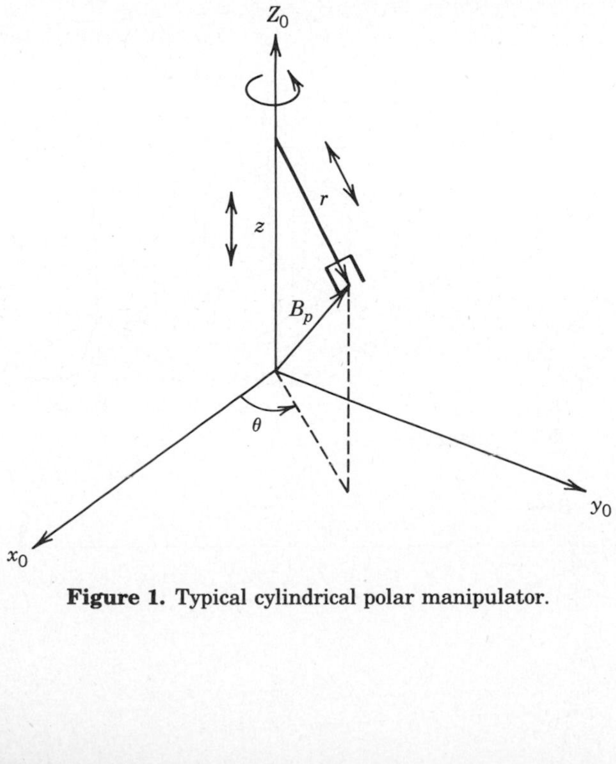

Figure 1. Typical cylindrical polar manipulator.

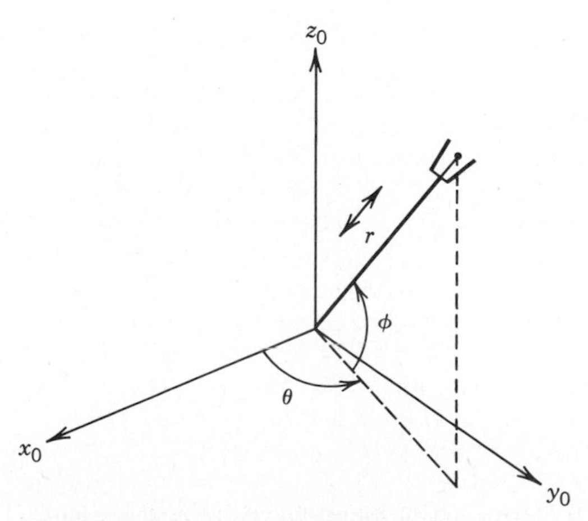

Figure 2. A typical spherical polar manipulator.

The Cartesian and curvilinear components of the velocity vector are, respectively,

$$\dot{x} = \dot{r} \sin\theta \cos\phi + r\dot{\theta} \cos\theta \cos\phi - r\dot{\phi} \sin\theta \sin\phi \tag{22}$$

$$\dot{y} = \dot{r} \sin\theta \sin\phi + r\dot{\theta} \cos\theta \sin\phi - r\dot{\phi} \sin\theta \cos\phi \tag{23}$$

$$\dot{z} = \dot{r} \cos\theta - r\dot{\theta} \sin\theta \tag{24}$$

$$\dot{r} = (x\dot{x} + y\dot{y} + z\dot{z})/(x^2 + y^2 + z^2)^{1/2} \tag{25}$$

$$\dot{\theta} = [(x\dot{x} + y\dot{y})z - \dot{z}(x^2 + y^2)]/(x^2 + y^2)^{1/2}(x^2 + y^2 + z^2) \tag{26}$$

The Cartesian and curvilinear components of the acceleration vector are, respectively,

$$\begin{aligned}\ddot{x} = {} & (\ddot{r} - r\dot{\theta}^2 - r\dot{\phi}^2) \sin\theta \cos\phi + (2\dot{r}\dot{\theta} - r\ddot{\theta}) \cos\theta \cos\phi \\ & - (r\ddot{\phi} + 2\dot{r}\dot{\phi}) \sin\theta \sin\phi - 2r\dot{\theta}\dot{\phi} \cos\theta \sin\phi\end{aligned} \tag{27}$$

$$\begin{aligned}\ddot{y} = {} & (\ddot{r} - r\dot{\theta}^2 - r\dot{\phi}^2) \sin\theta \sin\phi + (2\dot{r}\dot{\theta} + r\ddot{\theta}) \cos\theta \sin\phi \\ & + (r\ddot{\phi} + 2\dot{r}\dot{\phi}) \sin\theta \cos\phi + 2r\dot{\theta}\dot{\phi} \cos\theta \cos\phi\end{aligned} \tag{28}$$

$$\ddot{z} = (\ddot{r} - r\dot{\theta}^2) \cos\theta - (r\ddot{\theta} + 2\ddot{r}\theta) \sin\theta \tag{29}$$

$$\begin{aligned}\ddot{r} = {} & [(x\ddot{x} + \dot{x}^2 + y\ddot{y} + \dot{y}^2 + z\ddot{z} + \dot{z}^2)]\,(x^2 + y^2 + z^2)^{-1/2} \\ & - [(x\dot{x} + y\dot{y} + z\dot{z})^2\,(x^2 + y^2 + z^2)^{-3/2})\end{aligned} \tag{30}$$

$$\begin{aligned}\ddot{\theta} = {} & [(x\ddot{x} + \dot{x}^2 + y\ddot{y} + \dot{y}^2)z - \dot{z}(x\dot{x} + y\dot{y}) - \ddot{z}(x^2 + y^2)] \\ & \times (x^2 + y^2)^{1/2}(x^2 + y^2 + z^2)^{-1} + [(x\dot{x} + y\dot{y})z \\ & - \dot{z}(x^2 + y^2)] \times [-(x\dot{x} + y\dot{y})(x^2 + y^2)^{-3/2}\,(x^2 \\ & + y^2 + z^2)] - 2(x^2 + y^2)^{-1/2}\,(x\dot{x} = y\dot{y} + z\dot{z})\end{aligned} \tag{31}$$

$$\begin{aligned}\dot{\phi} = {} & [(x\ddot{y} - \ddot{x}y)(x^2 + y^2)^{-1} - 2(x\dot{x} + y\dot{y})(x^2 \\ & + y^2)^{-3/2}\,(x\dot{y} - y\dot{x})\end{aligned} \tag{32}$$

The radial and the two tangential components of the velocity and the acceleration vectors are, respectively,

$$\mathbf{v} = \dot{r}\,\mathbf{e}_r + r\dot{\theta}\,\mathbf{e}_\theta + r \sin\theta\,\dot{\phi}\,\mathbf{e}_\phi \tag{33}$$

$$\begin{aligned}\mathbf{a} = {} & (\ddot{r} - r\dot{\theta}^2 - r\dot{\phi}^2 \sin^2\theta)\,\mathbf{e}_r + (r\ddot{\theta} + 2\dot{r}\dot{\theta}^2 \\ & - r\dot{\phi}^2 \sin\theta \cos\theta)\mathbf{e}_\theta + (r \sin\phi\ddot{\phi} + 2\dot{r} \sin\theta\dot{\phi} \\ & + 2r \cos\theta\,\dot{\theta}\,\dot{\phi})\mathbf{e}_\phi\end{aligned} \tag{34}$$

Velocity and Accelerations in Moving Frames

Since a typical robot manipulator is a natural setting for moving and rotating frames where origins are attached to other moving and rotating frames, it would be useful to express the velocity and the acceleration vectors in such frames with respect to each other. Such relationships are further referred to either a fixed or inertial (Newtonian) frame of reference or a noninertial moving frame of reference. Referring to Figure 3, let *OXYZ* be a Newtonian frame of reference and *oxyz* a moving and rotating coordinate frame attached to a point in a robot manipulator. Let $\mathbf{r}(xyz)$, $\mathbf{R}(XYZ)$ be the position vectors of a material point p relative to *oxyz, OXYZ,* respectively, $\mathbf{V}$ be the velocity of the moving origin o, $\mathbf{A}$ be the acceleration of the moving origin o, and $\boldsymbol{\omega}$ be the angular velocity of the *oxyz* frame, all with respect to the base frame *OXYZ*, respectively. Then the acceleration $\mathbf{a}_{(XYZ)}$ is given by

$$\begin{aligned}(d^2\mathbf{R}/dt^2)_{(XYZ)} = \mathbf{a}_{(XYZ)} = {} & \mathbf{A} + (d^2\mathbf{r}/dt^2)_{(xyz)} + (d\boldsymbol{\omega}/dt)_{(xyz)} \\ & \times \mathbf{r}_{(xyz)} + 2\boldsymbol{\omega} \times \mathbf{v}_{(xyz)} + \boldsymbol{\omega} \times (\boldsymbol{\omega} \times \mathbf{r}_{(xyz)})\end{aligned} \tag{35}$$

Note that

$$\boldsymbol{\omega} \times (\boldsymbol{\omega} \times \mathbf{r}_{(xyz)}) = \boldsymbol{\omega}(\boldsymbol{\omega} \cdot \mathbf{r}_{(xyz)}) - \mathbf{r}_{(xyz)}\boldsymbol{\omega}^2 \tag{36}$$

and $(d\boldsymbol{\omega}/dt)_{(XYZ)} = (d\boldsymbol{\omega}/dt)_{(xyz)})$ because $\boldsymbol{\omega} \times \boldsymbol{\omega} = 0$. Now let $\mathbf{h}_{(xyz)}$, $\mathbf{h}_{(XYZ)}$ be the angular momentum vectors of a link in *oxyz, OXYZ* frames, respectively, $\mathbf{R}$ be the position vector of the origin of *oxyz* frame with respect to the *OXYZ* frame, $\mathbf{r}_c$ be the position vector of the center of mass of the robotic link with respect to the *oxyz* frame, and m be the mass of the link. It is then easily established that

$$\mathbf{h}_{(XYZ)} = \mathbf{h}_{(xyz)} + \mathbf{R} \times M\dot{\mathbf{R}} + \mathbf{R} \times M\dot{\mathbf{r}}_c \times M\dot{\mathbf{R}} \tag{37}$$

$$\dot{\mathbf{h}}_{(XYZ)} = \dot{\mathbf{h}}_{(xyz)} + \mathbf{R} \times M\ddot{\mathbf{R}} + \mathbf{R} \times M\ddot{\mathbf{r}}_c + \mathbf{r}_c \times M\ddot{\mathbf{R}} \tag{38}$$

Similarly, the kinetic energy $T_{(xyz)}$ and $T_{(XYZ)}$ are related by the following relationship:

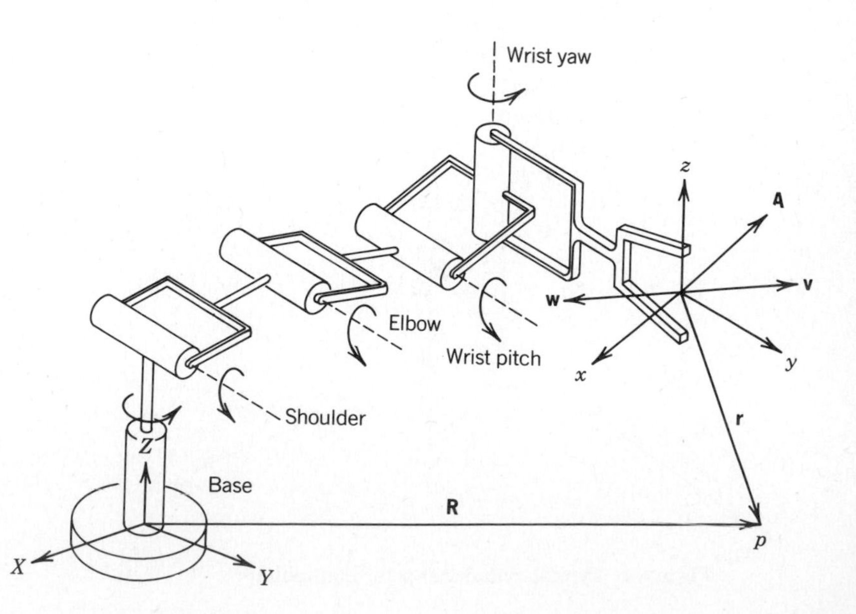

Figure 3. Moving robotic frames for velocity and acceleration.

$$T_{(XYZ)} = T_{(xyz)} + \frac{1}{2} m\,[\dot{\mathbf{R}} \cdot \dot{\mathbf{R}} + \dot{\mathbf{R}} \cdot \dot{\mathbf{r}}_c] \tag{39}$$

ROBOTIC MASS DISTRIBUTION AND THE INERTIA TENSOR

For robot manipulators that are free to move in three-dimensional space, there are a finite number or possible rotation axes. To express the governing dynamic equations about a rotation axis, the mass distribution of the robot about that axis is required. To describe such rotational mass distributions, the notion of moment of inertia tensor is employed.

The inertia tensor ${}^{A}\mathbf{I}$ with respect to a robotic frame A is a 3×3 matrix such that

$${}^{A}\mathbf{I} = \begin{bmatrix} I_{xx} & -I_{xy} & -I_{xz} \\ -I_{xy} & I_{yy} & -I_{yz} \\ -I_{xz} & -I_{yz} & I_{zz} \end{bmatrix} \tag{40}$$

where for a discrete robotic system

$$I_{xx} = \sum_{i=1}^{N} m_i(z_i^2 + y_i^2), \quad I_{xy} = \sum_{i=1}^{N} m_i x_i y_i \tag{41}$$

$$I_{yy} = \sum_{i=1}^{N} m_i(x_i^2 + z_i^2), \quad I_{xz} = \sum_{i=1}^{N} m_i x_i z_i \tag{42}$$

$$I_{zz} = \sum_{i=1}^{N} m_i(x_i^2 + y_i^2), \quad I_{yz} = \sum_{i=1}^{N} m_i y_i z_i \tag{43}$$

and for a continuous robotic system

$$I_{xx} = \textstyle\int\!\!\int\!\!\int \rho(z^2 + y^2)\, dxdydz \tag{44}$$

$$I_{yy} = \textstyle\int\!\!\int\!\!\int \rho(x^2 + z^2)\, dxdydz \tag{45}$$

$$I_{zz} = \textstyle\int\!\!\int\!\!\int \rho(x^2 + y^2)\, dxdydz \tag{46}$$

$$I_{xy} = \textstyle\int\!\!\int\!\!\int \rho\, xy\, dxdydz \tag{47}$$

$$I_{xz} = \textstyle\int\!\!\int\!\!\int \rho\, xz\, dxdydz \tag{48}$$

$$I_{yz} = \textstyle\int\!\!\int\!\!\int \rho\, yzy\, dxdydz \tag{49}$$

where ρ is the mass density of the robot and the integration is taken over the volume and with respect to a frame whose origin either coincides with the center of mass of the robotic link or is attached to the link.

As an example, let us find the moment of inertia of a robotic link with a square cross section with respect to an end coordinate frame $oxyz$ (Figure 4) or the center of mass coordinate frame $o_c x_c u_c z_c$. Note that the expressions 44–49 for this case simplify to the following integrals, first with respect to the coordinate frame xyz, and then with respect to the coordinate frame $x_c y_c z_c$:

$$I_{xx} = \int_0^a \int_0^L \int_0^a (y^2 + z^2)\, \rho\, dxdydz \tag{50}$$

$$= \int_0^a \int_0^L \rho\, (y^2 + z^2)\, a\, dydz \tag{51}$$

$$= \int_0^a \left(\frac{1}{3} L^3 + z^2 L\right) a\, \rho\, dz \tag{52}$$

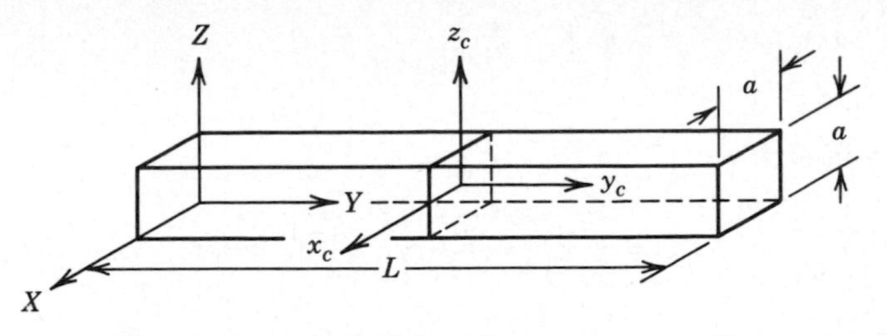

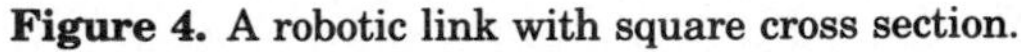

Figure 4. A robotic link with square cross section.

or finally

$$I_{xx} = \rho \left[\frac{a^2 L^3}{3} + \frac{a^4 L}{3}\right] = \frac{m}{3}(L^2 + a^2) \tag{53}$$

Similarly,

$$I_{zz} = \frac{m}{3}(L^2 + a^2) \tag{54}$$

$$I_{yy} = \frac{m}{3}(a^2 + a^2) = \frac{2ma^2}{3} \tag{55}$$

Now

$$I_{xy} = \int_0^a \int_0^l \int_0^a x\, y\, \rho\, dxdydz \tag{56}$$

$$I_{xy} = \int_0^a \int_0^l \frac{a^2}{3} y\, \rho\, dydz = \int_0^a \frac{a^2 L^2}{4} \rho\, dz \tag{57}$$

or finallay

$$I_{xy} = \frac{m}{4} a\, L \tag{58}$$

Similarly,

$$I_{xz} = \frac{m}{4} a^2, \quad I_{yz} = \frac{m}{4} a\, l \tag{59}$$

and thus the inertia tensor with respect to the xyz frame becomes

$$A_I = \begin{bmatrix} \frac{m}{2}(L^2 + a^2) & -\frac{m}{4} aL & -\frac{m}{4} a^2 \\ -\frac{m}{4} aL & \frac{2m}{3} a^2 & -\frac{m}{4} aL \\ -\frac{m}{4} a^2 & -\frac{m}{4} aL & \frac{m}{3}(L^2 + a^2) \end{bmatrix} \tag{60}$$

Now, in order to find the inertia tensor with respect to the center of mass coordinate frame $o_c x_c y_c z_c$, all one has to do is repeat the previous integrals 50–59 with different limits, that is,

$${}^{C}I_{xx} = \int_{-a/2}^{a/2} \int_{-L/2}^{L/2} \int_{-a/2}^{a/2} (y^2 + z^2)\, \rho\, dxdydz \tag{61}$$

$$= \int_{-a/2}^{a/2} \int_{-L/2}^{L/2} \rho(y^2 + z^2)\, a\, dydz \tag{62}$$

$$= \int_{-a/2}^{a/2} \left(\frac{L^3}{12} + z^2 L\right) a\, \rho\, dz \quad (63)$$

or finally

$$^{C}I_{xx} = \phi \left[\frac{a^2 L^3}{12} + \frac{a^4 L}{12}\right] = \frac{m}{12}(L^2 + a^2) \quad (64)$$

Similarly,

$$^{C}I_{zz} = \frac{m}{12}(L^2 + a^2), \quad ^{C}I_{yy} = \frac{ma^2}{6} \quad (65)$$

and thus

$$^{C}I_{xy} = 0, \quad ^{C}I_{xz} = 0, \quad ^{C}I_{yz} = 0 \quad (66)$$

$$^{C}\boldsymbol{I} = \begin{bmatrix} \frac{m}{12}(a^2 + L^2) & 0 & 0 \\ 0 & \frac{ma^2}{6} & 0 \\ 0 & 0 & \frac{m}{12}(L^2 + a^2) \end{bmatrix} \quad (67)$$

Parallel Axis Theorem

Comparing matrix equations 60 and 67 clearly indicates that

$$^{A}\boldsymbol{I}_{zz} = {}^{C}\boldsymbol{I}_{zz} + m(x_c^2 + y_c^2) \quad (68)$$

$$^{A}I_{xy} = {}^{C}I_{xy} + m\, x_c\, y_c \quad (69)$$

$$^{A}I_{yy} = {}^{C}I_{yy} + m(x_c^2 + z_c^2) \quad (70)$$

$$^{A}I_{xx} = {}^{C}I_{xx} + m(y_c^2 + z_c^2) \quad (71)$$

$$^{A}I_{xz} = {}^{C}I_{xz} + m\, x_c\, z_c \quad (72)$$

$$^{A}I_{yz} = {}^{C}I_{yz} + m\, y_c\, z_c \quad (73)$$

which are the familiar expressions of the parallel axis theorem for moment of inertia tensor $^{A}\boldsymbol{I}$ or $^{C}\boldsymbol{I}$.

THE NEWTON'S EQUATION

If the robot arm is considered as a series of rigid links connected by joints, then obviously for each link member a center of mass can be easily defined. Furthermore, an inertia tensor I_{ij} can also be defined, which will be further explained later. In order to cause motion for the links, one must apply forces and couples. Now the Newton's equation of motion states that the force $\mathbf{F}$ and the acceleration $\dot{\mathbf{v}}$ of a body are related to each other by the total mass of the body such that

$$\mathbf{F} = M\, \dot{\mathbf{v}} \quad (74)$$

where M is the total mass of the body. In complex discrete structures, the mass M is normally replaced by a matrix $\boldsymbol{M}$, which is called the mass matrix of the body.

THE EULER'S EQUATIONS

In the same manner, one may want to describe the rotational motion and acceleration of rigid bodies. Under such circumstances, it can be easily shown that $\boldsymbol{\tau}$ is the torque and $\boldsymbol{\omega}$ is the angular velocity vector, and then

$$\boldsymbol{\tau} = \boldsymbol{I}\, \dot{\boldsymbol{\omega}} + \boldsymbol{\omega} \times \boldsymbol{I}\, \boldsymbol{\omega} \quad (75)$$

where $\dot{\boldsymbol{\omega}}$ is the angular acceleration vector and $\boldsymbol{I}$ is the inertia tensor of the body in a body frame whose origin is coincident with the center of mass of the body under consideration.

As an example, let us expand the Eulerian equation of motion of a robotic body if its rotation is the same as the rotation of a coordinate frame *oxyz* embedded in it.

Consider equation 75 and note that

$$\boldsymbol{I}\, \boldsymbol{\omega} = \begin{bmatrix} I_{xx} & I_{xy} & I_{xz} \\ I_{xy} & I_{yy} & I_{yz} \\ I_{xz} & I_{yz} & I_{zz} \end{bmatrix} \begin{bmatrix} \omega_x \\ \omega_y \\ \omega_z \end{bmatrix}$$

$$= \begin{bmatrix} I_{xx}\omega_x + I_{xy}\omega_y + I_{xz}\omega_z \\ I_{xy}\omega_x + I_{yy}\omega_y + I_{yz}\omega_z \\ I_{xz}\omega_x + I_{yz}\omega_y + I_{zz}\omega_z \end{bmatrix} = \begin{bmatrix} a_x \\ a_y \\ a_z \end{bmatrix} \quad (76)$$

Now

$$\boldsymbol{\omega} \times (\boldsymbol{I}\, \boldsymbol{\omega}) = \begin{vmatrix} \mathbf{1}_x & \mathbf{1}_y & \mathbf{1}_z \\ \omega_x & \omega_y & \omega_z \\ a_x & a_y & a_z \end{vmatrix} \quad (77)$$

$$= \mathbf{1}_x(\omega_y a_z - \omega_z a_y) - \mathbf{1}_y(\omega_x a_z - \omega_z a_x) + \mathbf{1}_z(\omega_x a_y - \omega_y a_x) \quad (78)$$

Thus, the governing equations of motion (eq. 75) expand to the following:

$$\tau_x = I_{xx}\dot{\omega}_x + \omega_y\omega_z(I_{zz} - I_{yy}) + I_{xy}(\omega_z\omega_x + \dot{\omega}_y) - I_{xz}(\dot{\omega}_z + \omega_y\omega_x) - I_{yz}(\omega_y^2 - \omega_z^2) \quad (79)$$

$$\tau_y = I_{yy}\dot{\omega}_y + \omega_z\omega_x(I_{xx} - I_{zz}) + I_{yz}(\omega_x\omega_y + \dot{\omega}_z) - I_{xy}(\omega_x + \omega_z\omega_y) - I_{xz}(\omega_z^2 - \omega_x^2) \quad (80)$$

$$\tau_z = I_{zz}\dot{\omega}_y + \omega_x\omega_y(I_{yy} - I_{xx}) + I_{xz}(\omega_y\omega_z + \dot{\omega}_x) - I_{zy}(\dot{\omega}_y + \omega_x\omega_z) - I_{xy}(\omega_x^2 - \omega_y^2) \quad (81)$$

where $\boldsymbol{\omega} = (\omega_x, \omega_y, \omega_z)^T$ is the vector of the rotation of the *oxyz* frame and is related to the Euler's angles ϕ, θ, and ψ by the following equations:

$$\omega_x = \dot{\theta} \cos\psi + \dot{\phi} \sin\theta \sin\phi \quad (82)$$

$$\omega_y = -\dot{\theta} \sin\psi + \dot{\phi} \sin\theta \cos\psi \quad (83)$$

$$\omega_z = \dot{\psi} + \dot{\phi} \cos\theta \quad (84)$$

THE LAGRANGIAN AND THE LAGRANGE'S EQUATIONS

The Lagrangian L is defined as the difference between the kinetic and potential energies of a dynamic system such that

$$L = T - V \quad (85)$$

where T is the kinetic energy and V is the potential energy. If the generalized coordinate is denoted by q_i, it is then clear that

$$L = L(q_i, \dot{q}_i, t) \tag{86}$$

Hamilton's Principle of Least Action

For a dynamic system, the variation of the action integral defined by

$$I = \int_{t_1}^{t_2} L(q_i, \dot{q}_i, t)\, dt \tag{87}$$

is equal to the work done by the external forces. Furthermore, for a conservative system the motion for the system from time t_1 to time t_2 is such that I is an extremum for the path of the motion.

The Lagrange's Equations

Based on the Hamilton's principle, the governing equations of motion for a dynamic system can be shown to be

$$\frac{d}{dt}\left(\frac{\partial L}{\partial \dot{q}_i}\right) - \frac{\partial L}{\partial q_i} = F_i \tag{88}$$

where F_i is the vector of nonconservative generalized forces, that is, not derivable from a potential field.

The generalized momenta of a dynamic system are defined as

$$p_i = \frac{\partial L}{\partial \dot{q}_i} \tag{89}$$

which then from equation 88 implies that

$$p_i = \frac{\partial L}{\partial q_i} + F_i \tag{90}$$

The Hamiltonian H is further defined by

$$H(q_i, p_i, t) = \sum_{i=1}^{N} p_i \dot{q}_i - L \tag{91}$$

The Hamilton canonical equations are then defined as

$$q_i = \frac{\partial H}{\partial p_i} \tag{92}$$

$$\dot{p}_i = -\frac{\partial H}{\partial q_i}, \quad \frac{\partial H}{\partial t} = -\frac{\partial L}{\partial t} \tag{93}$$

N–E ROBOTIC DYNAMICS EQUATIONS

An attempt is made to use the N–E robotic dynamics equation to compute the generalized forces (torques) necessary for a given motion trajectory of a robot manipulator. It is initially assumed that the position, the velocity, and the accelerations of the joints, that is, $(\boldsymbol{\theta}, \dot{\boldsymbol{\theta}}, \boldsymbol{\theta})$ are known, and it is necessary to calculate the joint torques required to cause such time-dependent motions. As mentioned before, recursive-type dynamic modeling is the answer to the above task (see next section), however, first the equations pertaining to the forces and torques acting on a robotic link are presented below.

$$f_i = m_i \dot{\mathbf{v}}_{c(i)} \tag{94}$$

$$\tau_i = {}^{C}\boldsymbol{I}_i \dot{\boldsymbol{\omega}}_i + \boldsymbol{\omega}_i \times ({}^{C}\boldsymbol{I}_i \boldsymbol{\omega}_i) \tag{95}$$

where $\mathbf{v}_{c(i)}$ is the velocity of the center of mass of the ith link and ${}^{C}\boldsymbol{I}_i$ is the moment of inertia of the ith link about its center of mass with respect to frame $\boldsymbol{C}$, which has its origin at the center of mass of link i. Here f_i and τ_i are the net D'Alembert force and couple on the ith link and can be obtained by means of the balance equations for forces and couples applied to the ith link (Figure 5) such that

$$\mathbf{f}_i = \mathbf{f}_{i-1,i} + \mathbf{f}_{i+1,i} + m_i \mathbf{g} \tag{96}$$

$$\tau_i = \tau_{i-1,i} - \tau_{i,i+1} + (\mathbf{p}_{i-1} - \mathbf{r}_i) \times \mathbf{f}_{i-1,i} - (\mathbf{p}_i - \mathbf{r}_i) \times \mathbf{f}_{i,i+1} \tag{97}$$

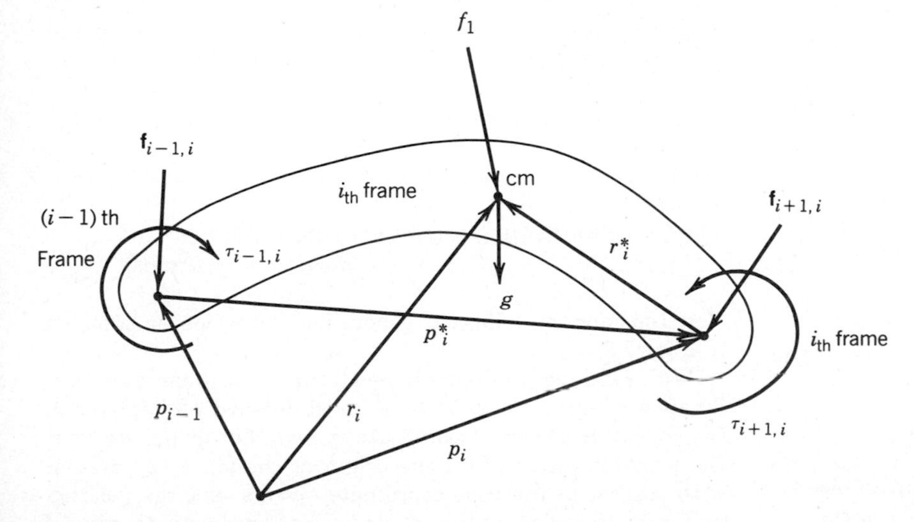

Figure 5. Forces and couples applied to the ith link.

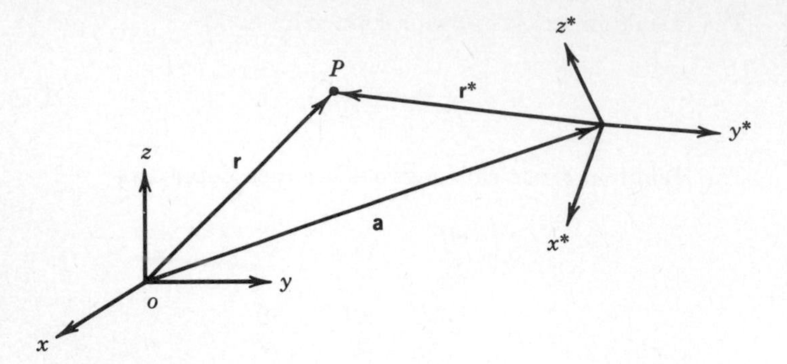

Figure 6. Moving coordinate systems for a robot manipulator.

where $f_{i,j}$ and $\tau_{i,j}$ are, respectively, the force and the couple applied on the ith link by means of the adjoining jth link.

Now consider again two robotic coordinate systems $oxyz$ and $o^*x^*y^*z^*$ as shown in Figure 6.

Let the p be described by a point vector $\mathbf{r}$ with respect to the xyz system and $\mathbf{r}^*$ with respect to the $x^*y^*z^*$ system. The two origins 0 and 0* are separated by a vector $\mathbf{a}$ with respect to the xyz system. It is then clear that

$$\mathbf{r} = \mathbf{r}^* + \mathbf{a}, \quad \mathbf{r}^* = \mathbf{r} - \mathbf{a} \tag{98}$$

where $\mathbf{r} = (x,y,z)^T$, $\mathbf{r}^* = (x^*,y^*,z^*)^T$, $\mathbf{a} \equiv (a_x, a_y, a_z)^T$. If the origin 0* is moving with respect to the origin 0, which is assumed to be fixed, then

$$\mathbf{v} = d\mathbf{r}/dt = d\mathbf{r}^*/dt + d\mathbf{a}/dt = \mathbf{v}^* + \mathbf{v}_a \tag{99}$$

where $\mathbf{v}$ and $\mathbf{v}^*$ are velocities of point p with respect to the xyz and $x^*y^*z^*$ frames, respectively. However, if the coordinate frame $x^*y^*z^*$ is moving and rotating with respect to the frame xyz, then

$$d\mathbf{r}/dt = d^*\mathbf{r}/dt + \boldsymbol{\omega} \times \mathbf{r} \tag{100}$$

where d^*r/dt is the time derivative with respect to the $x^*y^*z^*$ coordinate system and $\boldsymbol{\omega}$ is the angular velocity vector of the $x^*y^*z^*$ coordinate system (Figure 7).

Note that $d^*\mathbf{r}/dt$ is the time rate of change of $\mathbf{r}$ as seen by an observer sitting at the origin of the $x^*y^*z^*$ system, namely, 0^*. From equation 100, it is then noted that

$$d^2\mathbf{r}/dt^2 = \frac{d}{dt}(d^*\mathbf{r}/dt) + \boldsymbol{\omega} \times d\mathbf{r}/dt + d\boldsymbol{\omega}/dt \times \mathbf{r} \tag{101}$$

$$= d^{*2}\mathbf{r}/dt^2 + \boldsymbol{\omega} \times d^*\mathbf{r}/dt + \boldsymbol{\omega} \times [d^*\mathbf{r}/dt + \boldsymbol{\omega} \times \mathbf{r}] + d\boldsymbol{\omega}/dt \times \mathbf{r} \tag{102}$$

Finally,

$$d^2\mathbf{r}/dt^2 = d^{*2}\mathbf{r}/dt^2 + 2\boldsymbol{\omega} \times d^*\mathbf{r}/dt + \boldsymbol{\omega} \times (\boldsymbol{\omega} \times \mathbf{r}) + d\boldsymbol{\omega}/dt \times \mathbf{r} \tag{103}$$

Note that as discussed before

$d^{*2}\mathbf{r}/dt^2 \equiv$ acceleration term as measured by an observer riding with the $x^*y^*z^*$ coordinate system

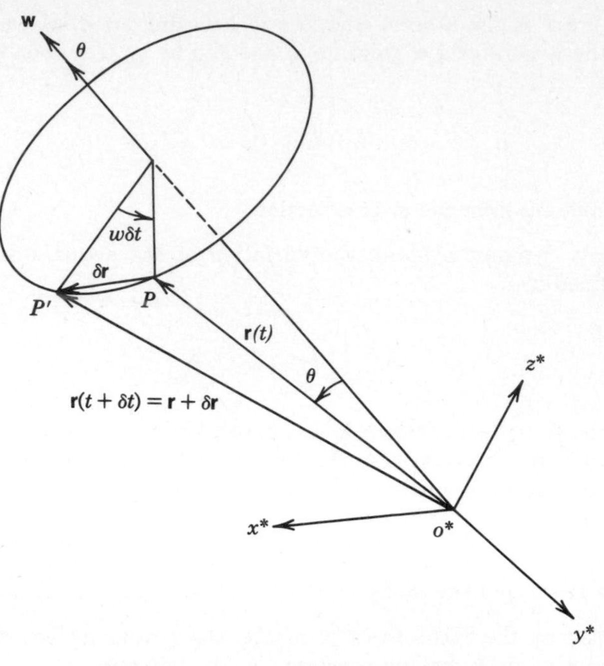

Figure 7. A rotating coordinate system.

$2\boldsymbol{\omega} \times d^*\mathbf{r}/dt \equiv$ Coriolis acceleration

$\boldsymbol{\omega} \times (\boldsymbol{\omega} \times \mathbf{r}) \equiv$ centripetal (toward the center) acceleration.

$d\boldsymbol{\omega}/dt \times \mathbf{r} \equiv$ angular acceleration

Now if the $x^*y^*z^*$ system is an inertial system, that is, its origin 0^* has a uniform velocity with respect to the xyz system, which is assumed to be fixed, then one has

$$m\, d^{*2}\mathbf{r}/dt^2 = \mathbf{f} - m\, \boldsymbol{\omega} \times (\boldsymbol{\omega} \times \mathbf{r}) - 2m\, \boldsymbol{\omega} \times d^*\mathbf{r}/dt - m\, d\boldsymbol{\omega}/dt \times \mathbf{r} \tag{104}$$

where m is the mass associated with the point p and $\mathbf{f}$ is the resultant vector of all external forces applied to the mass m whose center is at p.

If the origin of $x^*y^*z^*$ system is moving with a velocity $d\mathbf{a}/dt$ with respect to xyz where $\mathbf{a}$ is the vector joining 0 and 0^* (Figure 6), then

$$d\mathbf{r}/dt = d^*\mathbf{r}^*/dt + \boldsymbol{\omega} \times \mathbf{r}^* + d\mathbf{a}/dt \tag{105}$$

and further

$$d^2\mathbf{r}/dt^2 = d^{*2}\mathbf{r}^2/dt^2 + \boldsymbol{\omega} \times (\boldsymbol{\omega} \times \mathbf{r}^*) + 2\boldsymbol{\omega} \times d^*\mathbf{r}^*/dt + d\boldsymbol{\omega}/dt \times \mathbf{r}^* + d^2\mathbf{a}/dt^2 \tag{106}$$

Now consider the following generalized situation as depicted in Figure 8.

With reference to Figure 8, recall that an orthonormal coordinate system $(x_{i-1}, y_{i-1}, z_{i-1})$ is established at joint i and (x_O, y_O, z_O) is the base coordinate system. Let us denote by $\mathbf{r}_i$ the position vector of 0^1, the origin of the (x_i, y_i, z_i) system with respect to the base coordinate system. Let the position vector 0*, the origin of the $(x_{i-1}, y_{i-1}, z_{i-1})$ system with respect

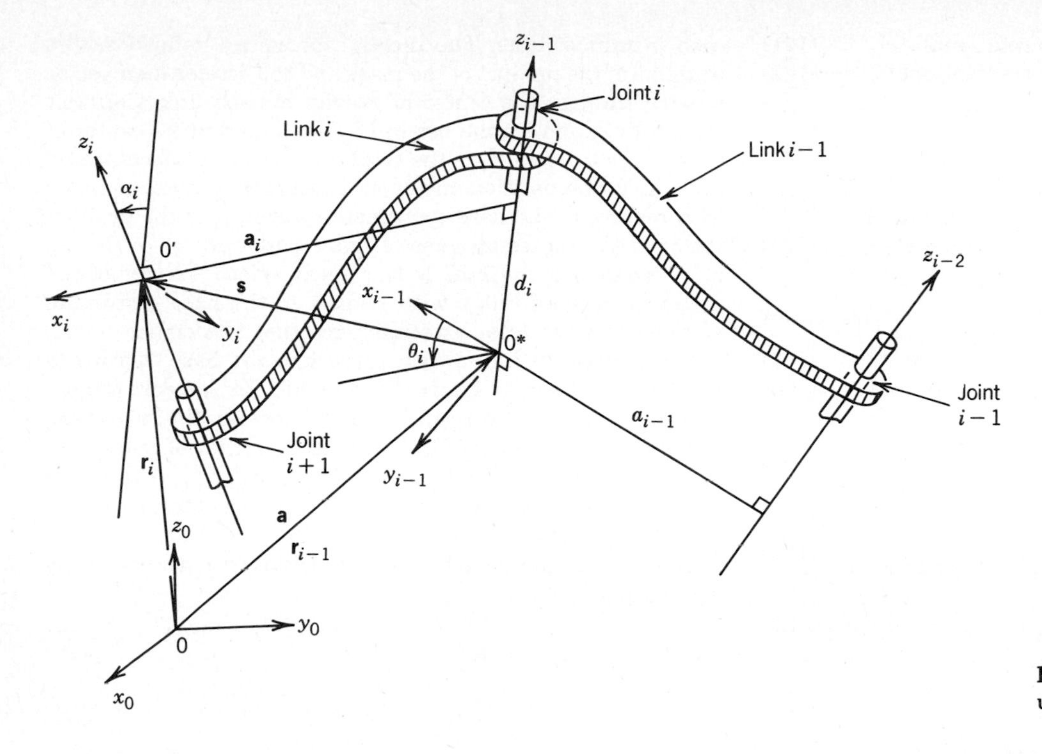

Figure 8. Generalized adjacent links situation for manipulators.

to the base coordinate system, be denoted by $\mathbf{r}_{i-1}$. Call vector $\mathbf{s}$ the one extending from 0^* to 0^1. Note that

$$\mathbf{r}_{i-1} + \mathbf{s} = \mathbf{r}_i \tag{107}$$

Let $\mathbf{v}_{i-1}$ and $\boldsymbol{\omega}_{i-1}$ be the linear and the angular velocity of the coordinate system $(x_{i-1}, y_{i-1}, z_{i-1})$ with respect to the base coordinate system (x_0, y_0, z_0), respectively. The linear velocity $\mathbf{v}_i$ of the coordinate system (x_i, y_i, z_i) with respect to the base coordinate system is

$$\mathbf{v}_i = d^*\mathbf{s}/dt + \boldsymbol{\omega}_{i-1} \times \mathbf{s} + \mathbf{v}_{i-1} \tag{108}$$

where $d^*(\quad)/dt$ denotes the time derivative with respect to the moving/rotating coordinate system $(x_{i-1}, y_{i-1}, z_{i-1})$ and

$$\dot{\mathbf{v}}_i = d^{*2}\mathbf{s}/dt^2 + \dot{\boldsymbol{\omega}}_{i-1} \times \mathbf{s} + 2\boldsymbol{\omega}_{i-1} \times d^*\mathbf{s}/dt + \boldsymbol{\omega}_{i-1} \times (\boldsymbol{\omega}_{i-1} \times \mathbf{s}) + \dot{\mathbf{v}}_{i-1} \tag{109}$$

Let $\boldsymbol{\omega}_i$ and $\boldsymbol{\omega}_s$ be the angular velocity of (x_i, y_i, z_i) with respect to (x_0, y_0, z_0) and $(x_{i-1}, y_{i-1}, z_{i-1})$, respectively. Thus,

$$\boldsymbol{\omega}_i = \boldsymbol{\omega}_{i-1} + \boldsymbol{\omega}_s \tag{110}$$

$$\dot{\boldsymbol{\omega}}_i = \dot{\boldsymbol{\omega}}_{i-1} + \dot{\boldsymbol{\omega}}_s \tag{111}$$

Note that

$$\dot{\boldsymbol{\omega}}_s = d^*\boldsymbol{\omega}_s/dt + \boldsymbol{\omega}_{i-1} \times \boldsymbol{\omega}_s \tag{112}$$

and therefore

$$\dot{\boldsymbol{\omega}}_i = \dot{\boldsymbol{\omega}}_{i-1} + d^*\boldsymbol{\omega}_s/dt + \boldsymbol{\omega}_{i-1} \times \boldsymbol{\omega}_s \tag{113}$$

Note that the angular velocity of the (x_i, y_i, z_i) system with respect to the $(x_{i-1}, y_{i-1}, z_{i-1})$ system, that is, $\boldsymbol{\omega}_s$, is the same as the angular velocity of link i about the z_{i-1} axis. Then

$$\boldsymbol{\omega}_s = \begin{cases} \mathbf{z}_{i-1}\,\dot{q}_i; & \text{if } i\text{th link is rotational} \\ 0; & \text{if } i\text{th link is translational} \end{cases} \tag{114}$$

where $\dot{q}_i$ is the generalized velocity of joint i or the magnitude of angular velocity of link i with respect to the coordinate system $(x_{i-1}, y_{i-1}, z_{i-1})$, which is, incidentally, an orthonormal system. Thus,

$$d^*\boldsymbol{\omega}_s/dt = \begin{cases} \mathbf{z}_{i-1}\,\ddot{q}_i; & \text{if link } i \text{ is rotational} \quad (115) \\ 0; & \text{if link } i \text{ is translational} \quad (116) \end{cases}$$

From equations 110, 112, and 114–116,

$$\boldsymbol{\omega}_i = \begin{cases} \boldsymbol{\omega}_{i-1} + \mathbf{z}_{i-1}\,\dot{q}_i; & \text{if } i \text{ is rotational} \quad (117) \\ \boldsymbol{\omega}_{i-1}; & \text{if } i \text{ is translational} \quad (118) \end{cases}$$

and

$$\dot{\boldsymbol{\omega}}_i = \begin{cases} \dot{\boldsymbol{\omega}}_{i-1} + \mathbf{z}_{i-1}\,\ddot{q}_i + \boldsymbol{\omega}_{i-1} \times (\mathbf{z}_{i-1}\,\dot{q}_i); & i\text{; rotational} \quad (119) \\ \dot{\boldsymbol{\omega}}_{i-1}; & i\text{; translational} \quad (120) \end{cases}$$

Let $\mathbf{s}$ be replaced by $\mathbf{p}_i^*$ where $\mathbf{p}_i^*$ is then the coordinate of the origin 0^1 with respect to the $(x_{i-1}, y_{i-1}, z_{i-1})$ system. If link i is translational with respect to the $(x_{i-1}, y_{i-1}, z_{i-1})$ frame, it then travels in the direction of z_{i-1} with a joint velocity $\dot{q}_i$ relative to link $i-1$. If it is rotational in the coordinate system $(x_{i-1}, y_{i-1}, z_{i-1})$, then it has an angular velocity $\boldsymbol{\omega}_s$ with respect to $(x_{i-1}, y_{i-1}, z_{i-1})$. Thus,

$$d^*\mathbf{s}/dt = \begin{cases} \boldsymbol{\omega}_s \times \mathbf{p}_i^*, & \text{if link is rotational} \quad (121) \\ \mathbf{z}_{i-1}\,\dot{q}_i, & \text{if link is translational} \quad (122) \end{cases}$$

Further,

$$d^*\mathbf{s}^2/dt^2 = \begin{cases} (d^*\boldsymbol{\omega}_s/dt) \times \mathbf{p}_i^* + \boldsymbol{\omega}_s \times (\boldsymbol{\omega}_s \times \mathbf{p}_i^*); & i;\ \text{rotational} \quad (123) \\ \mathbf{z}_{i-1}\,\ddot{q}_i; & i;\ \text{translational} \quad (124) \end{cases}$$

Therefore, equation 108 expands to

$$\mathbf{v}_i = \begin{cases} \boldsymbol{\omega}_i \times \mathbf{p}_i^* + \mathbf{v}_{i-1}; & i;\ \text{rotational} \quad (125) \\ \mathbf{z}_{i-1}\,\dot{q}_i + \boldsymbol{\omega}_i \times \mathbf{p}_i^* + \mathbf{v}_{i-1}; & i;\ \text{translational} \quad (126) \end{cases}$$

and equation 109 expands to

$$\dot{\mathbf{v}}_i = \begin{cases} \dot{\boldsymbol{\omega}}_i \times \mathbf{p}_i^* + \boldsymbol{\omega}_i \times (\boldsymbol{\omega}_i \times \mathbf{p}_i^*) + \dot{\mathbf{v}}_{i+1}; & i;\ \text{rotational} \quad (127) \\ \mathbf{z}_{i-1}\,\ddot{q}_i + \dot{\boldsymbol{\omega}}_i \times \mathbf{p}_i^* + 2\boldsymbol{\omega}_i \times (\mathbf{z}_{i-1}\,\dot{q}_i) \\ \quad + \boldsymbol{\omega}_i \times (\boldsymbol{\omega}_i \times \mathbf{p}_i^*) + \dot{\mathbf{v}}_{i-1}; & i;\ \text{translational} \quad (128) \end{cases}$$

Note that as explained before (equation 118), in equation 128 $\boldsymbol{\omega}_i = \boldsymbol{\omega}_{i-1}$ if link i is translational.

RECURSIVE DYNAMIC EQUATIONS FOR ROBOT MANIPULATORS

In this section D'Alembert's principle is used to derive a set of recursive equations of motion for robot manipulators.

D'Alembert's Principle

The dynamic equilibrium of multibody system is equivalent to a static equilibrium situation of the same system if only the inertial forces are treated as a set of independent forces in equilibrium with all other forces acting on the system at each instant of time. The inertial forces are defined as the negative of the product of the mass and the acceleration vector passing through the center of gravity of each link. Consider the ith link, and let the origin 0^1 be situated at its center of mass (Fig. 9). Then m_i is the total mass of link i concentrated at 0^1, $\bar{\mathbf{r}}_i$ is the position vector of the center of mass of link i with respect to the base coordinate system, $\bar{s}_i$ is the position vector of the center of mass of link i with respect to the (x_i, y_i, z_i) system, $\bar{\mathbf{v}}_i = d\bar{\mathbf{r}}_i/dt$ is the linear velocity vector of the center of mass of link i with respect to the base coordinate system, $\mathbf{f}_i$ is the total external force (net D'Alembert force) vector exerted on link i with respect to the base coordinate system, $\boldsymbol{\tau}_i$ is the total external torque (net D'Alembert torque) exerted on link i with respect to the base coordinate system, $\boldsymbol{I}_i$ is the inertia matrix of link i about its center of mass with reference to the coordinate system (x_0, y_0, z_0), $\mathbf{f}_{i-1,i}$ is the force exerted on link i by link $i-1$, and $\boldsymbol{\tau}_{i-1,i}$ is the torque exerted on link i by link $i-1$.

Now, neglecting the effects of elasticities and viscous damping of the joints,

$$\mathbf{f}_i = m_i\,\bar{\mathbf{a}}_i \quad (129)$$

$$\boldsymbol{\tau}_i = \frac{d}{dt}(\boldsymbol{I}_i\,\boldsymbol{\omega}_i) = \boldsymbol{I}_i\,\dot{\boldsymbol{\omega}}_i + \dot{\boldsymbol{\omega}}_i \times (\boldsymbol{I}_i\,\boldsymbol{\omega}_i) \quad (130)$$

$$\bar{\mathbf{v}}_i = \boldsymbol{\omega}_i \times \mathbf{s}_i + \mathbf{v}_i \quad (131)$$

$$\bar{\mathbf{a}}_i = \dot{\boldsymbol{\omega}}_i \times \bar{\mathbf{s}}_i + \boldsymbol{\omega}_i \times (\boldsymbol{\omega}_i \times \bar{\mathbf{s}}_i) + \dot{\mathbf{v}}_i \quad (132)$$

Clearly, the total external force $\mathbf{f}_i$ and moment $\boldsymbol{\tau}_i$ must be balanced, respectively, by the gravity and link forces from link $i-1$ and $i+1$. Thus,

$$\mathbf{f}_i = \mathbf{f}_{i-1,i} - \mathbf{f}_{i,i+1} + m_i\,\mathbf{g} \quad (133)$$

$$\boldsymbol{\tau}_i = \boldsymbol{\tau}_{i-1,i} - \boldsymbol{\tau}_{i,i+1} + (\mathbf{p}_{i-1} - \bar{\mathbf{r}}_i) \times \mathbf{f}_{i-1,i} - (\mathbf{p}_i - \bar{\mathbf{r}}_i) \times \mathbf{f}_{i,i+1} \quad (134)$$

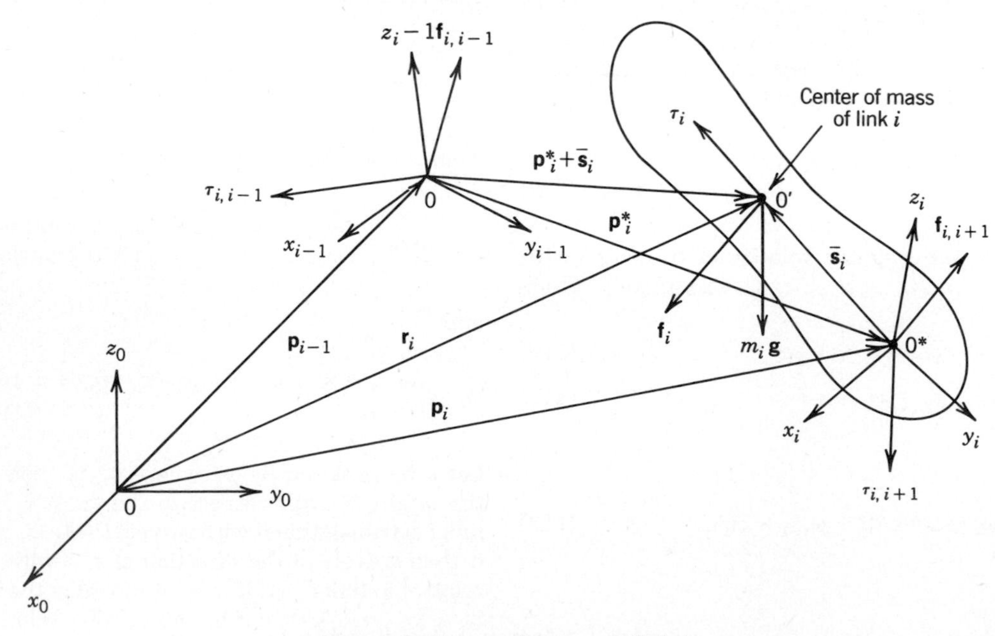

Figure 9. Force and torque on link i.

Note that since $\bar{\mathbf{r}}_i - \mathbf{p}_{i-1} = (\mathbf{p}_i^* + \bar{\mathbf{s}}_i)$, then clearly

$$\mathbf{f}_{i-1,i} = \mathbf{f}_i + \mathbf{f}_{i,i+1} - m_i\,\mathbf{g} \tag{135}$$

$$\boldsymbol{\tau}_{i-1,i} = \boldsymbol{\tau}_{i,i+1} + \mathbf{p}_i^* \times \mathbf{f}_{i,i+1} + (\mathbf{p}_i^* + \bar{\mathbf{s}}_i) \times (\mathbf{f}_i - m_i\,\mathbf{g}) + \boldsymbol{\tau}_i \tag{136}$$

Equations 135 and 136 are called the recursive equations for the dynamic of robotic manipulators. They can be used to compute $(\mathbf{f}_{i-1,i}, \boldsymbol{\tau}_{i-1,i})$, the force and the couple applied to link i by link $i-1$. Note that $\mathbf{f}_{n,n+1}$ and $\boldsymbol{\tau}_{n,n+1}$ are, respectively, the force and the torque exerted by the hand on the object being gripped. Note that if the joint is rotational, then it actually rotates q_i or θ_i rad about the $\mathbf{z}_{i-1}$ axis. Thus, the input torque at joint i is simply the sum of the projection $\mathbf{n}_i$ onto the $\mathbf{z}_{i-1}$ axis and the viscous damping and the elasticity forces in that coordinate system. Thus,

$$\boldsymbol{\tau}_i^* = \begin{cases} (\boldsymbol{\tau}_{i-1,i} \cdot \mathbf{z}_{i-1})\mathbf{z}_{i-1} + [b_i\dot{q}_i + K_i(q_i)]\mathbf{z}_{i-1} & (137) \\ (\mathbf{f}_{i-1,i} \cdot \mathbf{z}_{i-1,i})\mathbf{z}_{i-1,i} + [b_i q_i + k_i(q_i)]\mathbf{z}_{i-1} & (138) \end{cases}$$

where b_i is the viscosity damping coefficient, K_i is the rotational stiffness and k_i the translational stiffness, and $\boldsymbol{\tau}_i^*$ is the joint torque. Clearly, if the base coordinate is bolted on a platform and link *0* is stationary, then $\boldsymbol{\omega}_0 = \dot{\boldsymbol{\omega}}_0 = \mathbf{v}_0 = 0$, $\dot{\mathbf{v}}_0 = \mathbf{a}_0 = (0, 0, g)^T$ where $g = 9.8062$ m/s^2 is the ground-zero gravitational acceleration.

Thus we end up with a set of forward and backward recursion equations. These recursive equations are tabulated in Table 1. For the forward recursions, start with linear velocity, linear acceleration, angular velocity, angular acceleration, and the total forces and moments of each link and propagate from the base to the gripper or the hand. For the backward recursions, start with the torques and the forces exerted on each link propagated from the gripper to the base. Thus, the forward equations propagate the kinematics of each link from the base to the hand, whereas the backward equations compute the necessary torques and forces for each link from the gripper to the base.

RECURSIVE N–E EQUATIONS WITH RESPECT TO LOCAL LINK COORDINATE SYSTEM

Equations 139–148 are in fact with respect to the base coordinate system. It is more efficient to express the dynamic equations of each link with respect to its own coordinate system. Let $\boldsymbol{R}_{i-1}^i$ be the 3×3 rotational submatrix of $\boldsymbol{A}_{i-1}^i$. Recall that this matrix was an orthonormal one, that is, $(\boldsymbol{R}_{i-1}^i)^{-1} = \boldsymbol{R}_{i-1}^i = (\boldsymbol{R}_i^{i-1})^T$ where

$$\boldsymbol{R}_{i-1}^i = \begin{bmatrix} \cos\theta_i & -\cos\alpha_i \sin\theta_i & \sin\alpha_i \sin\theta_i \\ \sin\theta_i & \cos\alpha_i \cos\theta_i & -\sin\alpha_i \cos\theta_i \\ 0 & \sin\alpha_i & \cos\alpha_i \end{bmatrix} \tag{149}$$

$$(\boldsymbol{R}_{i-1}^i)^{-1} = \boldsymbol{R}_i^{i-1} = (\boldsymbol{R}_{i-1}^i)^T$$

$$= \begin{bmatrix} \cos\theta_i & \sin\theta_i & 0 \\ -\cos\alpha_i \sin\theta_i & \cos\alpha_i \cos\theta_i & \sin\alpha_i \\ \sin\alpha_i \sin\theta_i & -\sin\alpha_i \cos\theta_i & \cos\alpha_i \end{bmatrix} \tag{150}$$

Then equations 139–148 are reduced to the ones depicted in Table 2.

Algorithm 1 given for an n-jointed robotic manipulator generates the desired torque or force values for all joint actuators.

Algorithm 1

For the initial conditions,

$$\boldsymbol{\omega}_0 = \dot{\boldsymbol{\omega}}_0 = \mathbf{v}_0 = 0, \quad \dot{\mathbf{v}}_0 = (0,0,g)^T \tag{166}$$

Joint variables: q_i, $\dot{q}_i$, $\ddot{q}_i$, for $i = 1,2,\cdots,n$
Link variables: i, $\mathbf{f}_i$, $\mathbf{f}_{i-1,i}$, $\tau_{i-1,i}$

Table 1. Recursive Equations of Robotics Dynamics

Forward Equations	Equation Number
$\boldsymbol{\omega}_i = \begin{cases} \boldsymbol{\omega}_{i-1} + \mathbf{z}_{i-1}\dot{q}_i, & i \text{ rotational} \\ \boldsymbol{\omega}_{i-1}, & i \text{ translational} \end{cases}$	
$\dot{\boldsymbol{\omega}}_i = \begin{cases} \dot{\boldsymbol{\omega}}_{i-1} + \mathbf{z}_{i-1}\dot{q}_i + \boldsymbol{\omega}_{i-1} \times (\mathbf{z}_{i-1}\dot{q}_i), & i \text{ rotational} \\ \dot{\boldsymbol{\omega}}_{i-1}, & i \text{ translational} \end{cases}$	(139) (140)
$\mathbf{v}_i = \begin{cases} \boldsymbol{\omega}_i \times \mathbf{p}_i^* + \mathbf{v}_{i-1}, & i \text{ rotational} \\ \mathbf{z}_{i-1}\dot{q}_i + \boldsymbol{\omega}_i \times \mathbf{p}_i^* + \mathbf{v}_{i-1}, & i \text{ translational} \end{cases}$	(141) (142)
$\dot{\mathbf{v}}_i = \begin{cases} \dot{\boldsymbol{\omega}}_i \times \mathbf{p}_i^* + \boldsymbol{\omega}_i \times (\boldsymbol{\omega}_i \times \mathbf{p}_i^*) + \dot{\mathbf{v}}_{i-1}, & i \text{ rotational} \\ \mathbf{z}_{i-1}\ddot{q}_i + \dot{\boldsymbol{\omega}}_i \times \mathbf{p}_i^* + 2\boldsymbol{\omega}_i \times (\mathbf{z}_{i-1}\dot{q}_i) + \boldsymbol{\omega}_i \times (\boldsymbol{\omega}_i \times \mathbf{p}_i^*) + \dot{\mathbf{v}}_{i-1} \end{cases}$	(143) (144)
$\bar{\mathbf{a}}_i = \dot{\boldsymbol{\omega}}_i \times \bar{\mathbf{s}}_i + \boldsymbol{\omega}_i \times (\boldsymbol{\omega}_i \times \bar{\mathbf{s}}_i) + \dot{\mathbf{v}}_i$	
$\mathbf{f}_i = m_i\,\bar{\mathbf{a}}_i,\ \boldsymbol{\tau}_i = \boldsymbol{I}_i\,\dot{\boldsymbol{\omega}}_i + \boldsymbol{\omega}_i \times (\boldsymbol{I}\,\boldsymbol{\omega}_i)$	(145)
Backward Equations	
$\mathbf{f}_{i-1,i} = \mathbf{f}_i + \mathbf{f}_{i,i+1} - m_i\mathbf{g}$	
$\boldsymbol{\tau}_{i-1,i} = \boldsymbol{\tau}_{i,i+1} + \mathbf{p}_i^* \times \mathbf{f}_{i,i+1} + (\mathbf{p}_i^* + \bar{\mathbf{s}}_i) \times (\mathbf{f}_i - \mathrm{m}_i\mathbf{g}) + \boldsymbol{\tau}_i$	(146)
$\boldsymbol{\tau}_i^* = \begin{cases} (\boldsymbol{\tau}_{i-1,i} \cdot \mathbf{z}_{i-1})\mathbf{z}_{i-1} + [b_i\dot{q}_i + \mathbf{K}_i(q_i)]\mathbf{z}_{i-1} \\ (\mathbf{f}_{i-1,i} \cdot \mathbf{z}_{i-1})\mathbf{z}_{i-1} + [b_i\dot{q}_i + k(q_i)]\mathbf{z}_{i-1} \end{cases}$	(147) (148)

Table 2. Efficient Recursive Equations for Robotics Forward Dynamics in Local Coordinates[a]

$$\boldsymbol{R}_i^O\boldsymbol{\omega}_i = \begin{cases} \boldsymbol{R}_i^{i-1}(\boldsymbol{R}_{i-1}^O\boldsymbol{\omega}_{i-1} + \mathbf{z}_O\dot{q}_i), & i \text{ rotational} \quad (151) \\ \boldsymbol{R}_i^{i-1}(\boldsymbol{R}_{i-1}^O\boldsymbol{\omega}_{i-1}), & i \text{ translational} \quad (152) \end{cases}$$

$$\boldsymbol{R}_i^O\dot{\boldsymbol{\omega}}_i = \begin{cases} \boldsymbol{R}_i^{i-1}[\boldsymbol{R}_{i-1}^O\dot{\boldsymbol{\omega}}_{i-1} + \mathbf{z}_O\dot{q}_i + (\boldsymbol{R}_{i-1}\boldsymbol{\omega}_{i-1}) \times \mathbf{z}_O\dot{q}_i], & i \text{ rotational} \quad (153) \\ \boldsymbol{R}_i^{i-1}(\boldsymbol{R}_{i-1}^O\dot{\boldsymbol{\omega}}_{i-1}), & i \text{ translational} \quad (154) \end{cases}$$

$$\boldsymbol{R}_i^O\dot{\mathbf{v}}_i = \begin{cases} (\boldsymbol{R}_i^O\dot{\mathbf{v}}_i) \times (\boldsymbol{R}_i^O\mathbf{p}_i^*) + \boldsymbol{R}_i^{i-1}(\boldsymbol{R}_{i-1}^O\mathbf{v}_{i-1}), & i \text{ rotational} \quad (155) \\ \boldsymbol{R}_i^{i-1}(\mathbf{z}_O\dot{q}_i + \boldsymbol{R}_{i-1}^O\mathbf{v}_{i-1}) + (\boldsymbol{R}_i^O\boldsymbol{\omega}_i) \times (\boldsymbol{R}_i^O\mathbf{p}_i^*), & i \text{ translational} \quad (156) \end{cases}$$

$$\boldsymbol{R}_i^O\dot{\mathbf{v}}_i = \begin{cases} (\boldsymbol{R}_i^O\dot{\boldsymbol{\omega}}_i) \times (\boldsymbol{R}_i^O\mathbf{p}_i^*) + (\boldsymbol{R}_i\boldsymbol{\omega}_i) \times [(\boldsymbol{R}_i^O\boldsymbol{\omega}_i) \times (\boldsymbol{R}_i^O\mathbf{p}_i^*)] + \boldsymbol{R}_i^{i-1}(\boldsymbol{R}_{i-1}^O\dot{\mathbf{v}}_{i-1}), & i \text{ rotational} \quad (157) \\ \boldsymbol{R}_i^{i-1}(\mathbf{z}_O\ddot{q}_i + \boldsymbol{R}_{i-1}^O\dot{\mathbf{v}}_{i-1}) + (\boldsymbol{R}_i^O\dot{\boldsymbol{\omega}}_i) \times (\boldsymbol{R}_i^O\mathbf{p}_i^*) + 2(\boldsymbol{R}_i^O\boldsymbol{\omega}_i) \times (\boldsymbol{R}_i^{i-1}\mathbf{z}_O\dot{q}_i) + \boldsymbol{R}_i^O\boldsymbol{\omega}_i \times [(\boldsymbol{R}_i^O\boldsymbol{\omega}_i) \\ \quad \times (\boldsymbol{R}_i^O\mathbf{p}_i^*)], & i \text{ translational} \quad (158) \end{cases}$$

$$\boldsymbol{R}_i^O\bar{\mathbf{a}}_i = (\boldsymbol{R}_i^O\dot{\boldsymbol{\omega}}_i) \times (\boldsymbol{R}_i^O\bar{\mathbf{s}}_i) + (\boldsymbol{R}_i\boldsymbol{\omega}_i) \times [(\boldsymbol{R}_i\boldsymbol{\omega}_i) \times (\boldsymbol{R}_i^O\bar{\mathbf{s}}_i)] + \boldsymbol{R}_i^O\dot{\mathbf{v}}_i \quad (159)$$

$$\boldsymbol{R}_i^O\mathbf{f}_i = m_i\boldsymbol{R}_i^O\bar{\mathbf{a}}_i \quad (160)$$

$$\boldsymbol{R}_i^O\boldsymbol{\tau}_i = (\boldsymbol{R}_i^O\boldsymbol{I}_i\boldsymbol{R}_O^i)(\boldsymbol{R}_i^O\dot{\boldsymbol{\omega}}_i) + (\boldsymbol{R}_i^O\boldsymbol{\omega}_i) \times [(\boldsymbol{R}_i^O\boldsymbol{I}_i\boldsymbol{R}_O^i)(\boldsymbol{R}_i^O\boldsymbol{\omega}_i)] \quad (161)$$

$$\boldsymbol{R}_i^O\mathbf{f}_{i-1,i} = \boldsymbol{R}_i^{i+1}(\boldsymbol{R}_{i+1}^O\mathbf{f}_{i,i+1}) + \boldsymbol{R}_i^O\mathbf{f}_i - \mathrm{m}_i\boldsymbol{R}_i^O\mathbf{g} \quad (162)$$

$$\boldsymbol{R}_i^O\boldsymbol{\tau}_{i-1,i} = \boldsymbol{R}_i^{i+1}[\boldsymbol{R}_{i+1}^O\boldsymbol{\tau}_{i,i+1} + (\boldsymbol{R}_{i+1}^O\mathbf{p}_i^*) \times (\boldsymbol{R}_{i+1}^O\mathbf{f}_{i+1})] + (\boldsymbol{R}_i^O\mathbf{p}_i^* + \boldsymbol{R}_i^O\bar{\mathbf{s}}_i) \times (\boldsymbol{R}_i^O(\mathbf{f}_i - \mathrm{m}_i\mathbf{g})) + \boldsymbol{R}_i^O\boldsymbol{\tau}_i \quad (163)$$

$$\boldsymbol{\tau}_i^* = \begin{cases} \boldsymbol{R}_i^O\boldsymbol{\tau}_{i-1,i}) \cdot (\boldsymbol{R}_i^{i-1}z_0)\boldsymbol{R}_i^{i-1}\mathbf{z}_0 + [\dot{b}_i\dot{q}_i + K_i(q_i)]\boldsymbol{R}_i^{i-1}z_0 & (164) \\ (\boldsymbol{R}_i^O\tau_{i-1,i}) \cdot (\boldsymbol{R}_i^{i-1}\mathbf{z}_0)\boldsymbol{R}_i^{i-1}\mathbf{z}_0 + [b_i\dot{q}_i + k_iq_i]\boldsymbol{R}_i^{i-1}\mathbf{z}_i & (165) \end{cases}$$

[a] Where $\mathbf{z}_O = (0,0,1)^T$ and $\dot{\mathbf{v}}_O = (0,0,g)^T$, and $\boldsymbol{\tau}_i^*$ is the ith joint torque. Note that breaking the terms in equations 151–165 there are multiplications and additions involved with each individual term. These terms are $\boldsymbol{\omega}_i$, $\dot{\boldsymbol{\omega}}_i$, $\dot{\mathbf{v}}_i$, $\bar{\mathbf{a}}_i$, $\mathbf{f}_i$, $\mathbf{f}_{i-1,i}$, τ_i, and $\tau_{i-1,i}$, as depicted in Table 3.

Forward Iterations

1. Set $i = 1$.
2. Compute $\boldsymbol{R}_i^O\dot{\boldsymbol{\omega}}_i$, $\boldsymbol{R}_i^O$, $\boldsymbol{\omega}_i^O$, $\boldsymbol{R}_i^O\mathbf{v}_i$, $\boldsymbol{R}_i^O\mathbf{v}^O$ (equation 167).
3. Compute $\boldsymbol{R}_i^O\bar{\mathbf{a}}_i$, $\boldsymbol{R}_i^O\mathbf{f}_i$, $\boldsymbol{R}_i^O\boldsymbol{\tau}_i$.
4. If $i = n$, go to next step; otherwise, set $i = i + 1$ and return to step 2.

Backward Iterations

5. Set $\mathbf{f}_{n,n+1}$, $\boldsymbol{\tau}_{n,n+1}$ to the required force or torque to carry the load by the gripper.
6. Compute $\boldsymbol{R}_i^O\mathbf{f}_{i-1,i}$, $\boldsymbol{R}_i^O\boldsymbol{\tau}_{i-1,i}$, τ_i given $\mathbf{f}_{n,n+1}$ and $\boldsymbol{\tau}_{n,n+1}$.
7. If $i = 1$, then stop; otherwise, set $i \rightarrow i - 1$ and go back to step 6.

Table 3. Breakdown of Computational Operations in N–E Technique[a]

N–E Equations, Robot	Multiplications	Additions
$\boldsymbol{\omega}_i$	$9(n-1)$	$9n-9$
$\mathbf{v}_i$	$9(n-1)$	$12n-9$
$\dot{\boldsymbol{\omega}}_i$	$12(n-9)$	$12n-9$
$\dot{\mathbf{v}}_i$	$27n$	$24n$
$\bar{\mathbf{a}}_i$	$18n$	$15n$
$\mathbf{f}_i$	$3n$	0
$\mathbf{f}_{i-1,i}$	$9(n-1)+3$	$9n-6$
$\boldsymbol{\tau}_i$	$27n$	$18n$
$\boldsymbol{\tau}_{i-1,i}$	$21n-15$	$24n-15$
Total	$135n-48$	$123n-48$

[a] Where $n \equiv$ number of degrees of freedom for the robot arm.

As an example, for the two-link nonplanar robot manipulator shown below (Figure 10), the dynamic equations of motion may be derived if all masses m_1 and m_2 are concentrated at the outermost end of each link as follows.

Referring to Table 2, the forward dynamics iterations are written out in the following manner:

$$\boldsymbol{\omega}_0 = \dot{\boldsymbol{\omega}}_0 = \mathbf{v}_0 = 0, \quad \dot{\mathbf{v}}_0 = (0,0,-g)^T \quad (167)$$

$$\boldsymbol{R}_1^O\omega_1 = \boldsymbol{R}_1^O(\boldsymbol{R}_0\boldsymbol{\omega}_0 + \mathbf{z}_0\dot{\theta}_1) = \boldsymbol{R}_1^O\boldsymbol{\omega}_0 + \mathbf{z}_1\dot{\theta}_{11} \quad (168)$$

or

$$\boldsymbol{R}_1^O\boldsymbol{\omega}_1 = (0,0,\dot{\theta}_1)^T \quad (169)$$

$$R_i^O\dot{\boldsymbol{\omega}}_1 = \boldsymbol{R}_1^O[\boldsymbol{R}_0^O\dot{\boldsymbol{\omega}}_0 + \mathbf{z}_0\ddot{\theta}_1 + (\boldsymbol{R}_0^O\boldsymbol{\omega}_0) \times \mathbf{z}_0\dot{\theta}_1] \quad (170)$$

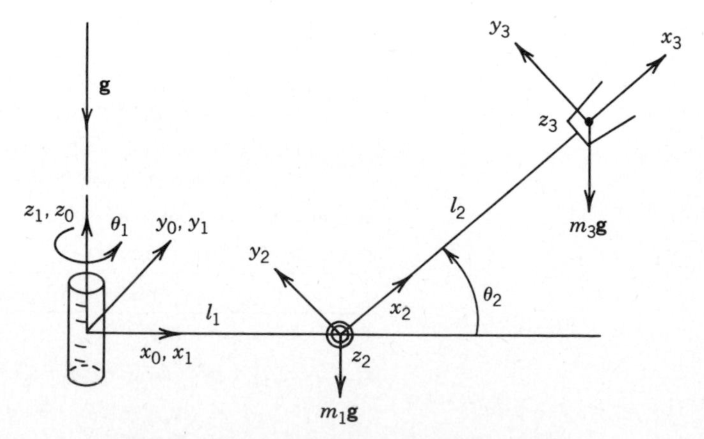

Figure 10. A two-link nonplanar manipulator.

or

$$R_1^O \omega_1^O = (0,0,\dot{\theta}_1)^T \tag{171}$$

$$R_1^O \mathbf{v} = (R_1^O \omega_1) \times (R_1^O p_1^*) R_1^O (R_0^O \mathbf{v}_0) \tag{172}$$

$$= R_1^O(0,0,0,) = (0,0,0)^T \tag{173}$$

Furthermore, as can be seen from the Figure 10,

$$R_0^1 = \begin{bmatrix} C_1 & -S_1 & 0 \\ S_1 & C_1 & 0 \\ 0 & 0 & 1 \end{bmatrix}; \quad R_1^2 = \begin{bmatrix} C_2 & -S_2 & 0 \\ 0 & 0 & -1 \\ S_2 & C_2 & 0 \end{bmatrix}; \tag{174}$$

$$R_2^3 = \begin{bmatrix} 1 & 0 & 0 \\ 0 & 1 & 0 \\ 0 & 0 & 1 \end{bmatrix} \tag{175}$$

Furthermore,

$$p_1^* = (0,0,0)^T, \quad \mathbf{p}_2^* = (l_1,0,0), \quad p_3^* = (l_2,0,0) \tag{176}$$

$$R_1^O \dot{\mathbf{v}}_1 = R_1^O \dot{\omega}_1 \times \mathbf{R}_1^O p_1^* + R_1^O \omega_1 \times [R_1^O \omega_i \times R_i^O \mathbf{p}_i^*] + R_1^O(R_0^O \mathbf{v}_0^O) = (0,0,-q)^T \tag{177}$$

$$R_1^O \bar{\mathbf{a}}_1 = R_1^O \omega_1^O \times R_1^O \bar{S}_1 + R_1^O \omega_1 \times [R_1^O \omega_1 \times R_1^O \bar{S}_1] + R_1^O \mathbf{v}_1^O = (-l_1\dot{\theta}_1^2, l_1\ddot{\theta}_1, -g)^T \tag{178}$$

$$R_2^O \omega_2 = R_2^1(R_1^O \omega_1 + \mathbf{z}_0 \dot{\theta}_2) = R_2^1 R_1^O \omega_1 + R_2^1 \mathbf{z}_0 \dot{\theta}_2 \tag{179}$$

$$R_2^O \omega_2 + \begin{bmatrix} C_2 & 0 & S_2 \\ -S_2 & 0 & C_2 \\ 0 & -1 & 0 \end{bmatrix} \begin{bmatrix} 0 \\ 0 \\ \dot{\theta}_1 \end{bmatrix} + \begin{bmatrix} 0 \\ 0 \\ \dot{\theta}_2 \end{bmatrix} = \begin{bmatrix} S_2 \ \dot{\theta}_1 \\ C_2 \ \dot{\theta}_2 \\ \dot{\theta}_2 \end{bmatrix} \tag{180}$$

$$R_2^O \dot{\omega}_2 = R_2^1 [R_1^O \dot{\omega}_1 + \mathbf{z}_0 \ddot{\theta}_2 + (R_2^O \omega_1) \times \mathbf{z}_0 \dot{\theta}_2] \tag{181}$$

$$R_2^O \dot{\omega}_2 = R_2^1 R_1^O \dot{\omega}_1 + \mathbf{z}_2 \ddot{\theta}_2 + R_2^1 R_2^O \omega_1 \times \mathbf{z}_0 \dot{\theta}_2 \tag{182}$$

$$R_2^O \dot{\omega}_2 = \begin{bmatrix} S_2 \ \ddot{\theta}_2 \\ C_2 \ \ddot{\theta}_1 \\ 0 \end{bmatrix} + \begin{bmatrix} 0 \\ 0 \\ \ddot{\theta}_2 \end{bmatrix} + \begin{bmatrix} S_2 \ \dot{\theta}_1 \\ C_2 \ \dot{\theta}_1 \\ 0 \end{bmatrix} \times \begin{bmatrix} 0 \\ 0 \\ \dot{\theta}_2 \end{bmatrix}$$

$$= \begin{bmatrix} S_2\ddot{\theta}_1 + C_2\dot{\theta}_1\dot{\theta}_2 \\ C_2\ddot{\theta}_1 - S_2\dot{\theta}_1\theta \\ \ddot{\theta}_2 \end{bmatrix} \tag{183}$$

$$R_2^O \dot{\mathbf{v}}_2 = R_2^O \omega_2 \times R_2^O \mathbf{p}_2^* + R_2^O \omega_2 \times [(R_2^O \omega_2) \times (R_2^O \mathbf{p}_2)] + R_2^1(R_2^O \dot{\mathbf{v}}_2) \tag{184}$$

$$R_2^O \dot{\mathbf{v}}_2 = \begin{bmatrix} C_2 & 0 & S_2 \\ -S_2 & 0 & C_2 \\ 0 & -1 & 0 \end{bmatrix} \begin{bmatrix} -l_1\dot{\theta}_1^2 \\ l_1\ddot{\theta}_1 \\ -g \end{bmatrix} = \begin{bmatrix} -l_1C_2\dot{\theta}_1^2 - S_2 g \\ l_1S_2\dot{\theta}_1^2 - C_2 g \\ -l_1\ddot{\theta}_1 \end{bmatrix} \tag{185}$$

$$R_2^O \bar{\mathbf{a}}_2 = R_2^O \dot{\omega}_2 \times R_2^O \bar{\mathbf{S}}_2 + (R_2^O \omega_2) \times [(R_2^O \omega_2 \times R_2^O \bar{\mathbf{S}}_2)] + R_2^O \dot{\mathbf{v}}_2 \tag{186}$$

$$R_2^O \bar{\mathbf{a}}_2 = \begin{bmatrix} S_2\ddot{\theta}_1 + C_1\dot{\theta}_1\dot{\theta}_2 \\ C_2\ddot{\theta}_1 - S_2\dot{\theta}_1\dot{\theta}_2 \\ \ddot{\theta}_2 \end{bmatrix} \begin{bmatrix} l_2 \\ 0 \\ 0 \end{bmatrix} + \begin{bmatrix} S_2\dot{\theta}_1 \\ C_2\dot{\theta}_1 \\ \dot{\theta}_2 \end{bmatrix} \times \begin{bmatrix} 0 \\ l_2\dot{\theta}_2 \\ -l_2S_2\dot{\theta}_1 \end{bmatrix} + \mathbf{R}_2^O \dot{\mathbf{v}} \tag{187}$$

$$R_2^O \bar{\mathbf{a}}_2 = \begin{bmatrix} 0 \\ l_2\ddot{\theta}_2 \\ -l_2C_2\ddot{\theta}_1 + l_2S_2\dot{\theta}_1\dot{\theta}_2 \end{bmatrix} + \begin{bmatrix} -l_2C_2^2\dot{\theta}_1^2 - l_2\dot{\theta}_2^2 \\ l_2S_2\dot{\theta}_1^2C^2 \\ l_2S_2\dot{\theta}_1\dot{\theta}_2 \end{bmatrix} + \begin{bmatrix} -lC_2\dot{\theta}_1^2 - S_2 g \\ l_1S_2\dot{\theta}_1^2 + C_2 g \\ -l_1\ddot{\theta}_1 \end{bmatrix} \tag{188}$$

or

$$R_2^O \bar{a}_2 = \begin{bmatrix} -(l_1C_2 + l_2C_2^2)\dot{\theta}_1^2 - l_2\dot{\theta}_2^2 - S_2 g \\ (l_1S_2 + l_2S_2C_2)\dot{\theta}_1^2 - C_2 g + l_2\ddot{\theta}_2 \\ 2l_2S_2\dot{\theta}_1\dot{\theta}_2 - l_1\ddot{\theta}_1 - l_2C_2\ddot{\theta}_1 \end{bmatrix} \tag{189}$$

The governing equations then become

$$R_1^O \mathbf{f}_1 = m_1 R_1^O \bar{\mathbf{a}}_1 \tag{190}$$

or

$$\mathbf{f}_1 = \begin{bmatrix} -m_1l_1\dot{\theta}_1^2 \\ m_2l_1\ddot{\theta}_1 \\ m_1 g \end{bmatrix} \tag{191}$$

and

$$R_2^O \mathbf{f}_2 = m_2 R_2^O \bar{a}_2 \tag{192}$$

or

$$R_2^O \mathbf{f}_2 = \begin{bmatrix} -m_2(l_1 + l_2C_2)C_2\dot{\theta}_1^2 - m_2l_2\dot{\theta}_2^2 - m_2S_2 g \\ m_2(l_1 + l_2C_2)S_2\dot{\theta}_1^2 - m_2gC_2 + m_2l_2\ddot{\theta}_2 \\ 2m_2l_2S_2\dot{\theta}_1\dot{\theta}_2 - m_2l_1\ddot{\theta}_1 - m_2l_2C_2\ddot{\theta}_1 \end{bmatrix} \tag{193}$$

Now the torque equations become

$$R_1^O \tau_1 = (R_1^O I_1 R_0^1)(R_1^O \dot{\omega}_1) + R_1^O \omega_1 \times [(R_1^O I_1 R_0^O)(R_1^O \omega_1)] \tag{194}$$

$$R_2^O \tau_2 = (R_2^O I_2 R_0^2)(R_2^O \dot{\omega}_2) + (R_2^O \omega_2) \times [(R_2^O I_2 R_0^2)(R_2^O \omega_2)] \tag{195}$$

and since both $I_1 = I_2$ are zero, it then follows that

$$\tau_1 = \tau_2 = 0 \tag{196}$$

Of course, $I_1 = I_2 = \mathbf{0}$ is an idealization because we are assuming that the masses are concentrated at a point at the center of gravities of the robotic structure.

Backward Force Iterations From Link 2

$$R_2^O \mathbf{f}_{1,2} = R_2^3(R_2^O \mathbf{f}_{2,3}) + R_2^O \mathbf{f}_2 = R_2^O \mathbf{f}_2 \quad \text{(see eq. 193)} \tag{197}$$

$$R_2^O \tau_{1,2} = R_2^3[R_2^O \tau_{2,3} + (R_3^O \mathbf{p}_2^*) \times (R_3^O \mathbf{f}_{2,3})] + (R_2^O \mathbf{p}_2^* + R_2^O \bar{\mathbf{S}}_i) \times (R_2^O \mathbf{f}_2) + R_2^O \tau_2 \tag{198}$$

$$R_2^O \tau_{1,2} = R_2^O \mathbf{p}_2^* \times R_2^O \mathbf{f} = \begin{bmatrix} m_2l_1l_2\ddot{\theta}_1 - 2m_2l_2S_2^2\dot{\theta}_1\dot{\theta}_2 + m_2l_2^2C_2\ddot{\theta}_1 \\ m_2l_2(l_1 + l_2C_2)S_2\dot{\theta}_1^2 - m_2gl_2C_2 + m_2l_2^2\ddot{\theta}_2 \end{bmatrix} \tag{199}$$

$$R_2^O \mathbf{f}_1 = R_1^2(R_2^O \mathbf{f}_{1,2}) + R_1^O \mathbf{f}_1 \tag{200}$$

$$R_1^O \tau_{0,1} = R_1^2[R_2^O \tau_{1,2} + (R_2^O \mathbf{p}_1^*) \times (R_2^O \mathbf{f}_{1,2})] + (R_1^O \mathbf{p}_1^* + R_1^O \bar{\mathbf{S}}_1) \times (R_1^O \mathbf{f}_1) + R_1^O \tau_1 \tag{201}$$

After simplifications using equations 197, 199, and 201, one obtains

$$R_1^O \tau_{0,1} = \begin{bmatrix} (m_1 l_1^2 + m_2 l_1^2 + 2m_2 l_1 l_2 C_2)\dot{\theta}_1 \\ * \\ * \\ -2(l_1 + l_2 C_2) m_2 l_2 S_2 \dot{\theta}_1 \dot{\theta}_2 + m_2 l_2^2 C_2^2 \end{bmatrix} \quad (202)$$

So that finally, in the absence of joint friction and elasticity,

$$\tau_1^* = [m_1 l_1^2 + m_2(l_1 + l_2 C_2)^2]\ddot{\theta}_1 - 2(l_1 + l_2 C_2) m_2 l_2 S_2 \dot{\theta}_1 \dot{\theta}_2 \quad (203)$$

$$\tau_2^* = m_2 l_2^2 \ddot{\theta}_2^2 + (l_1 + l_2 C_2) m_2 l_2 S_2 \dot{\theta}_1^2 - m_2 g l_2 C_2 \quad (204)$$

or generally,

$$\boldsymbol{\tau} = \boldsymbol{M}(\boldsymbol{\theta})\, \ddot{\boldsymbol{\theta}} + \mathbf{V}(\boldsymbol{\theta}, \dot{\boldsymbol{\theta}}) + \mathbf{G}(\boldsymbol{\theta}) \quad (205)$$

where

$$\boldsymbol{M}(\boldsymbol{\theta}) = \begin{bmatrix} m_1 l_1^2 + m_2(l_1 + l_2 C_2)^2 & 0 \\ 0 & m_2 l_2^2 \end{bmatrix} \quad (206)$$

$$\mathbf{V}(\boldsymbol{\theta}, \dot{\boldsymbol{\theta}}) = \begin{bmatrix} -2(l_1 + l_2 C_2) m_2 l_2 S_2 \dot{\theta}_1 \dot{\theta}_2 \\ (l_1 + l_2 C_2) m_2 l_2 S_2 \dot{\theta}_1^2 \end{bmatrix} \quad (207)$$

$$\mathbf{G}(\boldsymbol{\theta}) = \begin{bmatrix} 0 \\ -m_2 g l_2 C_2 \end{bmatrix} \quad (208)$$

L–E DYNAMIC MODELING OF ROBOT MANIPULATOR

In this section, the L–E dynamic equations are applied to robot manipulator dynamics, considering them to be nonconservative systems. In deriving the pertinent equations, the D–H transformations of robotic coordinate frames are used (see KINEMATICS for a discussion of these transformations).

Let us recall the L–E equations:

$$\frac{d}{dt}\left(\frac{\partial L}{\partial \dot{q}_i}\right) - \left(\frac{\partial L}{\partial q_i}\right) = \tau_i \quad (209)$$

where

$$L = L(q_i, \dot{q}_i, t) \equiv K - P \equiv \text{Lagrangian} \quad (210)$$

$$K \equiv K(\bar{q}_i, \dot{q}_i, t) \equiv \text{kinetic energy of the whole system} \quad (211)$$

$$P \equiv P(q_i, t) \equiv \text{potential energy of the whole system} \quad (212)$$

$$q_i = \text{the generalized coordinate, ie, } \theta_i\text{'s, etc} \quad (213)$$

$$\tau_i \equiv \text{the generalized force or torque vector} \quad (214)$$

Velocity of the Joints

Consider $\mathbf{r}_0^i$ to be the position vector of a point p fixed with respect to the coordinate frame i (moving with it) and expressed with respect to the base coordinate (x_0, y_0, z_0). Referring to Figure 11, the point $\mathbf{r}_0^i$ has a fixed coordinate with respect to the ith frame, such that the point P as well as other points fixed on link i and expressed in coordinate frame i will have zero velocity with respect to the coordinate frame i. Note that the coordinate frame i is not an inertial frame. Clearly, the velocity of P with respect to the fixed or inertial frame or the base coordinate can be given by

$$\mathbf{v}_i = \frac{d}{dt}(\mathbf{r}_0^i) \quad (215)$$

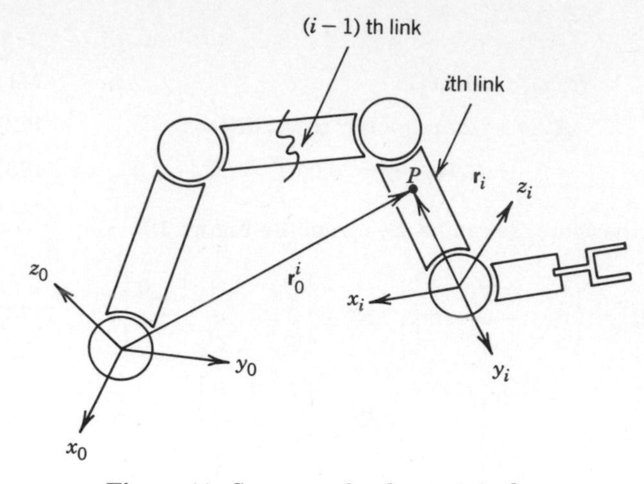

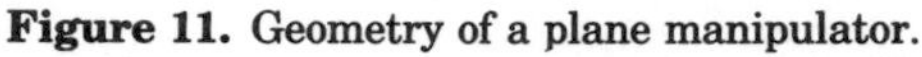

Figure 11. Geometry of a plane manipulator.

Recall that if $\mathbf{A}_{i-1}^i$ is the D–H transformation, then

$$\mathbf{A}_0^i = \mathbf{A}_0^1 \mathbf{A}_1^2 \cdots \mathbf{A}_{i-1}^i \quad (216)$$

and

$$\mathbf{r}_0^i = \mathbf{A}_0^i \mathbf{r}_i, \quad \mathbf{r}_{i-1}^i = \mathbf{A}_{i-1}^i \mathbf{r}_i \quad (217)$$

See Figure 12. Thus

$$\boldsymbol{v}_0^i = \mathbf{v}_i = \frac{d}{dt}(\boldsymbol{A}_0^i \mathbf{r}_i) = \sum_{j=1}^{i} \left(\frac{\partial A_0^i}{\partial q_i} \frac{dq_j}{dt}\right) \mathbf{r}_i \quad (218)$$

where q_j's are the independent variables; that is, θ_i's, $i = 1, 2, \cdots, n$. Thus,

$$\mathbf{v}_0^i = \sum_{j=1}^{i} \left(\frac{\partial \mathbf{A}_0^i}{\partial q_i} \dot{q}_i\right) \mathbf{r}_i, \qquad \text{for } i = 1, 2, \cdots, n \quad (219)$$

It can be shown that

$$\frac{\partial \mathbf{A}_{i-1}^i}{\partial q_j} = Q_i A_{i-1}^i \quad (220)$$

where $\boldsymbol{Q}_i$ is a constant matrix. For example, for $\boldsymbol{A}_{i-1}^i$,

$$\frac{\partial \mathbf{A}_{i-1}^i}{\partial \theta_i} = \begin{bmatrix} -\sin\theta_i & -\cos\alpha_i \cos\theta_i & \sin\alpha_i \cos\theta_i & -a_i \sin\theta_i \\ \cos\theta_i & -\cos\alpha_i \sin\theta_i & \sin\alpha_i \sin\theta_i & a_i \cos\theta_i \\ 0 & 0 & 0 & 0 \\ 0 & 0 & 0 & 0 \end{bmatrix}$$

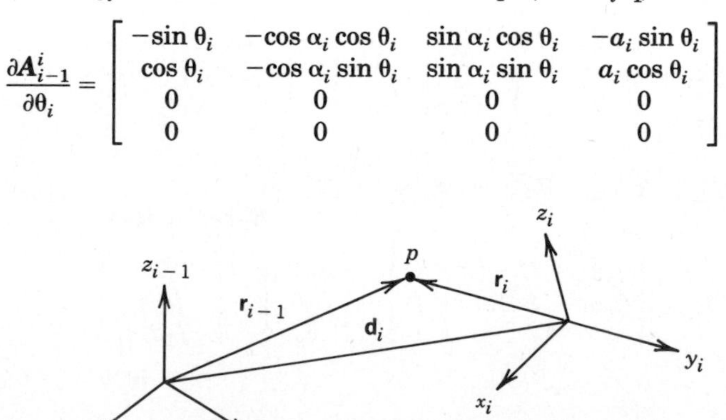

Figure 12. Vectorial representation of joint coordinates.

$$= \begin{bmatrix} 0 & -1 & 0 & 0 \\ 1 & 0 & 0 & 0 \\ 0 & 0 & 0 & 0 \\ 0 & 0 & 0 & 0 \end{bmatrix}$$

$$\begin{bmatrix} \cos\theta_i & -\cos\theta_i \sin\theta_i & \sin\alpha_i \sin\theta_i & a_i \cos\theta_i \\ \sin\theta_i & \cos\alpha_i \cos\theta_i & -\sin\alpha_i \cos\theta_i & a_i \sin\theta_i \\ 0 & \sin\alpha_i & \cos\alpha_i & d_i \\ 0 & 0 & 0 & 0 \end{bmatrix} \tag{221}$$

and thus

$$\frac{\partial \boldsymbol{A}_{i-1}^{i}}{\partial \theta_i} = \boldsymbol{Q}_i \boldsymbol{A}_{i-1}^{i} \tag{222}$$

where for convenience D_{ij} is a quantity defined such that

$$\boldsymbol{D}_{ij} = \frac{\partial \boldsymbol{A}_0^i}{\partial q_j} = \begin{cases} \boldsymbol{A}_0^{j-1} \boldsymbol{Q}_j \boldsymbol{A}_{j-1}; & j \leq i \quad (223) \\ 0; & j > i \quad (224) \end{cases}$$

In a similar manner, the higher-order derivaties of A_0^i are defined; that is,

$$\frac{\partial \boldsymbol{D}_{ij}}{\partial q_k} = \boldsymbol{D}_{ijk} = \begin{cases} \boldsymbol{A}_0^{j-1} \boldsymbol{Q}_j \boldsymbol{A}_{j-1}^{k-1} \boldsymbol{Q}_k \boldsymbol{A}_{k-1}^{i}, & i \geq k \geq j \quad (225) \\ \boldsymbol{A}_0^{k-1} \boldsymbol{Q}_k \boldsymbol{A}_{k-1}^{j-1} \boldsymbol{Q}_j \boldsymbol{A}_{j-1}^{i}, & i \geq j \geq k \quad (226) \\ 0 & i < j \text{ or } i < k \quad (227) \end{cases}$$

To derive the governing dynamic equations for robotic manipulators, the L–E formulation is employed; namely, that for a generally nonconservative systems equation 209 holds. Thus, the Lagrangian $L = K - P$ must be calculated where K is the kinetic energy and P is the potential energy of the manipulator. For a proof of equation 209, refer to any text on "Lagrangian dynamics."

Kinetic Energy E of the Robot Arm

The incremental kinetic energy K associated with a differential mass dm of the manipulator is given by

$$dK = \sum_{i=1}^{n} dK_i \tag{228}$$

where dK_i is the incremental kinetic energy associated with a differential mass dm_i of the ith link of the manipulator and is given by

$$dK_i = \frac{1}{2} \text{Trace} \, (\mathbf{v}_i \, \mathbf{v}_i^T) \, dm_i \tag{229}$$

$$\mathbf{v}_i \, \mathbf{v}_i^T = \begin{bmatrix} v_{1i} \\ v_{2i} \\ v_{3i} \\ 0 \end{bmatrix} [v_{1i} \; v_{2i} \; v_{3i} \; 0] = \begin{bmatrix} v_{1i}^2 & v_{1i}v_{2i} & v_{1i}v_{3i} & 0 \\ v_{2i}v_{1i} & v_{2i}^2 & v_{2i}v_{3i} & 0 \\ v_{3i}v_{1i} & v_{3i}v_{2i} & v_{3i}^2 & 0 \\ 0 & 0 & 0 & 0 \end{bmatrix} \tag{230}$$

and

$$\text{Trace} \, \boldsymbol{A} = \sum_{i=1}^{n} A_{ii} \tag{231}$$

Thus, based on equations 220 and 223,

$$dK_i = \frac{1}{2} \text{Tr} \left[\sum_{p=1}^{i} \boldsymbol{D}_{ip} \dot{q}_p \mathbf{r}_i \left(\sum_{r=1}^{i} \boldsymbol{D}_{ir} \dot{q}_r \, \mathbf{r}_i \right)^T \right] dm_i \tag{232}$$

or finally,

$$dK_i = \frac{1}{2} \text{Tr} \left[\sum_{p=1}^{i} \sum_{r=1}^{i} \boldsymbol{D}_{ip} (\mathbf{r}_i \, dm_i r_i^T) \boldsymbol{D}_{ir}^T \dot{q}_p \dot{q}_r \right] \tag{233}$$

Integrating,

$$K_i = \int dK_i = \frac{1}{2} \text{Tr} \left[\sum_{p=1}^{i} \sum_{r=1}^{j} \boldsymbol{D}_{ip} \left(\int \mathbf{r}_i \mathbf{r}_i^T dm_i \right) \boldsymbol{D}_{ir}^T \, \dot{q}_p \dot{q}_r \right] \tag{234}$$

The integral term is nothing but the inertia moment matrix of the link about the origin of the joint coordinate i, that is, the moment of inertia tensor $\boldsymbol{J}_i$ of the ith link.

$$\boldsymbol{J}_i = \int \mathbf{r}_i \mathbf{r}_i^T dm_i = \begin{bmatrix} \int x_i^2 dm_i & \int x_i y_i dm_i & \int x_i z_i dm_i & \int x_i dm_i \\ \int x_i y_i dm_i & \int y_i^2 dm_i & \int y_i z_i dm & \int y_i dm_i \\ \int z_i x_i dm_i & \int y_i z_i dm_i & \int z_i^2 dm_i & \int z_i dm_i \\ \int x_i dm_i & \int y_i dm_i & \int z_i dm_i & \int dm_i \end{bmatrix} \tag{235}$$

Expressing $\boldsymbol{J}_i$ in terms of the inertia tensor $\boldsymbol{I}$, that is,

$$I_{ij} = \boldsymbol{I} = \begin{cases} \int \left[\delta_{ij} \left(\sum_k x_k^2 \right) - x_i x_j \right] dm, & i = j \\ \int x_i x_j dm, & i \neq j \end{cases}$$

one has

$$\boldsymbol{J}_i = \begin{bmatrix} (-I_{xx} + I_{yy} + I_{zz})/2 & I_{xy} & I_{xz} & m_i \bar{x}_i \\ I_{xy} & (I_{xx} - I_{yy} + I_{zz})/2 & I_{yz} & m_i \bar{y}_i \\ I_{xz} & I_{yz} & (I_{xx} + I_{yy} - I_{zz})/2 & m_i \bar{z}_i \\ m_i \bar{x}_i & m_i \bar{y}_i & m_i \bar{z}_i & m_i \end{bmatrix} \tag{236}$$

where $\bar{x}_i$, $\bar{y}_i$, and $\bar{z}_i$ are the coordinates of the center of mass of link i with respect to the origin of the ith frame. Thus, the total kinetic energy of the robot arm will be

$$K = \sum_{i=1}^{n} K_i = \frac{1}{2} \sum_{i=1}^{n} \sum_{p=1}^{i} \sum_{r=1}^{i} [\text{Tr}(\boldsymbol{D}_{ip} \boldsymbol{J}_i \boldsymbol{D}_{ir}^T) \, \dot{q}_p \dot{q}_r) \tag{237}$$

The Potential Energy of the Robot Arm

The total potential energy due to the weight of the robot manipulator again is given as a sum of all potential energies of the individual links. Thus,

$$P = \sum_{i=1}^{n} P_i = \sum_{i=1}^{n} [-m_i \mathbf{g} \cdot (\boldsymbol{A}_0^i \bar{\mathbf{r}}_i)] \tag{238}$$

where $\bar{\mathbf{r}}_i$ is the location of the center of mass of link i expressed in the ith frame and $\mathbf{g}$ is the gravitational vector.

The Lagrangian ≡ $K - P$

From equations 237 and 238,

$$L = \frac{1}{2}\sum_{i=1}^{n}\sum_{j=1}^{i}\sum_{p=1}^{i}[\mathrm{Tr}(\boldsymbol{D}_{ij}\boldsymbol{J}_i\boldsymbol{D}_{ip}^T)\dot{q}_p\dot{q}_r] + \sum_{i=1}^{n} m_i\mathbf{g}\cdot(\boldsymbol{A}_0^i\,\bar{\mathbf{r}}_i) \quad (239)$$

Recalling that the dynamics equations are to be derived by means of the L–E equations, that is,

$$\frac{d}{dt}\left(\frac{\partial L}{\partial \dot{q}k}\right) - \frac{\partial L}{\partial q_i} = \tau_i, \qquad i = 1,2,\cdots,n \quad (240)$$

it is noted that

$$\begin{aligned}\frac{\partial L}{\partial \dot{q}_k} &= \frac{1}{2}\sum_{i=1}^{n}\sum_{j=1}^{i}\sum_{p=1}^{i}[\mathrm{Tr}(\boldsymbol{D}_{ij}\boldsymbol{J}_i\boldsymbol{D}_{ip}^T)(\dot{q}_p\delta_{jk} + \dot{q}_j\delta_{kp})\\ &= \sum_{i=k}^{n}\sum_{p=1}^{i}\mathrm{Tr}[\boldsymbol{D}_{ip}\boldsymbol{J}_i\boldsymbol{D}_{ik}^T]\dot{q}_p\end{aligned} \quad (241)$$

$$\begin{aligned}\frac{d}{dt}\left(\frac{\partial L}{\partial \dot{q}_k}\right) &= \sum_{i=k}^{n}\sum_{p=1}^{i}\mathrm{Tr}[\boldsymbol{D}_{ip}\boldsymbol{J}_i\boldsymbol{D}_{ik}^T]\,\dot{q}_p\\ &\quad + \sum_{i=k}^{n}\sum_{p=1}^{i}\sum_{m=1}^{i}\mathrm{Tr}[\boldsymbol{D}_{ipm}\boldsymbol{J}_i\boldsymbol{D}_{ik}^T]\dot{q}_p\dot{q}_m\\ &\quad + \sum_{i=1}^{n}\sum_{p=1}^{i}\sum_{m=1}^{i}\mathrm{Tr}[\boldsymbol{D}_{ikm}\boldsymbol{J}_i\boldsymbol{D}_{ip}^T]\dot{q}_p\dot{q}_m\end{aligned} \quad (242)$$

Similarly,

$$\begin{aligned}\frac{\partial L}{\partial q_k} &= \frac{1}{2}\sum_{i=k}^{n}\sum_{j=1}^{i}\sum_{p=1}^{i}\mathrm{Tr}[\boldsymbol{D}_{ijk}\boldsymbol{J}_i\boldsymbol{D}_{ip}^T]\dot{q}_j\dot{q}_p\\ &\quad + \frac{1}{2}\sum_{i=k}^{n}\sum_{j=1}^{i}\sum_{p=1}^{i}\mathrm{Tr}[\boldsymbol{D}_{ipk}\boldsymbol{J}_i\boldsymbol{D}_{ij}^T]\\ &\quad \times \dot{q}_j\dot{q}_p + \sum_{i=p}^{n} m_i\mathbf{g}\cdot\boldsymbol{D}_{ik}\bar{\mathbf{r}}_i\end{aligned} \quad (243)$$

Thus, equation (240) expands to

$$\begin{aligned}\tau_k &= \sum_{i=k}^{n}\sum_{p=1}^{i}\mathrm{Tr}[\boldsymbol{D}_{ip}\boldsymbol{J}_i\boldsymbol{D}_{ik}^T]\dot{q}_k\\ &\quad + \sum_{i=k}^{n}\sum_{p=1}^{i}\sum_{m=1}^{i}\mathrm{Tr}[\boldsymbol{D}_{ipm}\boldsymbol{J}_i\boldsymbol{D}_{ik}^T]\dot{q}_p\dot{q}_m\\ &\quad + \sum_{i=p}^{n}\sum_{j=1}^{i}\sum_{p=1}^{i}\mathrm{Tr}[\boldsymbol{D}_{ikm}\boldsymbol{J}_i\boldsymbol{D}_{ip}^T]\dot{q}_p\dot{q}_m\\ &\quad - \frac{1}{2}\sum_{i=k}^{n}\sum_{j=1}^{i}\sum_{p=1}^{i}\mathrm{Tr}[\boldsymbol{D}_{ijk}\boldsymbol{J}_i\boldsymbol{D}_{ip}^T]\dot{q}_j\dot{q}_p\\ &\quad + \frac{1}{2}\sum_{i=k}^{n}\sum_{j=1}^{i}\sum_{p=1}^{i}\mathrm{Tr}[\boldsymbol{D}_{ipk}\boldsymbol{J}_i\boldsymbol{D}_{ij}^T]q_jq_p - \sum_{i=k}^{n} m_i\mathbf{g}\cdot\boldsymbol{D}_{ik}\bar{\mathbf{r}}_i\end{aligned} \quad (244)$$

Note that the above equation may be simplified further by the following definitions. Let

$$U_{ijk} = \mathrm{Tr}[\boldsymbol{D}_{jk}\boldsymbol{J}_j\boldsymbol{D}_{ji}^T] = \mathrm{Tr}[\boldsymbol{D}_{kj}\boldsymbol{J}_k\boldsymbol{D}_{ki}^T] = U_{ikj} \quad (245)$$

so equation 244 simplifies to

$$\tau_i = \sum_{k=1}^{n}\boldsymbol{M}_{ik}\ddot{q}_k + \sum_{k=1}^{n}\sum_{m=1}^{n}\boldsymbol{M}_{ikm}\dot{q}_k\dot{q}_m + \boldsymbol{M}_i \quad (246)$$

where $\boldsymbol{M}_i$, $\boldsymbol{M}_{ik}$, and $\boldsymbol{M}_{ikm}$ are generalized mass matrices such that

$$\boldsymbol{M}_i = \sum_{j=1}^{n}(-m_j\mathbf{g}\cdot\boldsymbol{D}_{ji}\bar{\mathbf{r}}_j) \quad (247)$$

$$\boldsymbol{M}_{ik} = \sum_{j=\max(i,k,m)}^{n}\mathrm{Tr}(D_{jk}J_jD_{ji}^T) \quad (248)$$

$$\boldsymbol{M}_{ikm} = \sum_{j=\max(i,k,m)}^{n}\mathrm{Tr}(\boldsymbol{D}_{jkm}\boldsymbol{J}_j\boldsymbol{D}_{ji}^T) \quad (249)$$

$$\mathbf{g} \equiv (g_x, g_y, g_z, 0) \quad (250)$$

Equation 245 has also been written in the robotics literature in a more compact vectorial form such that

$$\boldsymbol{\tau} = \boldsymbol{D}(\boldsymbol{\theta})\ddot{\boldsymbol{\theta}} + \mathbf{H}(\boldsymbol{\theta}, \dot{\boldsymbol{\theta}}) + \mathbf{G}(\boldsymbol{\theta}) \quad (251)$$

The numerical computations involved with these equations are extremely large, and for example, they involve approximately 66,394 multiplications and 51,456 additions for a six-axis robot for which $n = 6$.

It must be noted that with regard to numerical calculations, the N–E equations are much more efficient than the L–E equations. For example, for an n-axis robot, the N–E equations require

$(126n - 99)$ multiplications

$(106n - 92)$ additions

while the L–E equations requires approximately:

$(32n^4 + 86n^3 + 171n^2 + 53n - 128)$ multiplications

$(25n^4 + 66n^3 + 129n^2 + 42n - 96)$ additions

For a typical robot of $n = 6$, the computation time for the latter (L–E) is two orders of magnitude more than for the former (N–E).

DYNAMIC EQUATIONS OF MOTION FOR A GENERAL SIX-AXIS ROBOT MANIPULATOR

Equation 246 can be expressed in matrix form by equation 251 such that

$$\boldsymbol{\tau} = \mathrm{M}(\boldsymbol{\theta})\ddot{\boldsymbol{\theta}} + \mathbf{H}(\boldsymbol{\theta}, \dot{\boldsymbol{\theta}}) + \mathbf{G}(\boldsymbol{\theta}) \quad (252)$$

Here $\boldsymbol{\tau}$ is a $\mathbf{6} \times 1$ vector, $\mathrm{M}(\boldsymbol{\theta})$ is a 6×6 matrix, $\mathbf{H}$ is a 6×1 vector involving all coriolis and centripetal acceleration terms, and $\mathbf{G}$ is a 6×1 gravitational vector. For a six-axis robot, and above equations expand to

$$\begin{aligned}\tau_1 &= M_{11}\ddot{\theta}_1 + M_{12}\ddot{\theta}_2 + M_{13}\ddot{\theta}_3 + M_{14}\ddot{\theta}_4 + M_{15}\ddot{\theta}_5 + M_{16}\ddot{\theta}_6\\ &\quad + M_{111}\dot{\theta}_1^2 + M_{122}\dot{\theta}_2^2 + M_{133}\dot{\theta}_3^2 + M_{144}\dot{\theta}_4^2 + M_{155}\dot{\theta}_5^2\\ &\quad + M_{166}\dot{\theta}_6^2 + M_{112}\dot{\theta}_1\dot{\theta}_2 + M_{113}\dot{\theta}_1\dot{\theta}_3 + M_{114}\dot{\theta}_1\dot{\theta}_4\\ &\quad + M_{115}\dot{\theta}_1\dot{\theta}_5 + M_{116}\dot{\theta}_1\dot{\theta}_6 + M_{123}\dot{\theta}_2\dot{\theta}_3 + M_{124}\dot{\theta}_2\dot{\theta}_4\\ &\quad + M_{125}\dot{\theta}_2\dot{\theta}_5 + M_{126}\dot{\theta}_2\dot{\theta}_6 + M_{134}\dot{\theta}_3\dot{\theta}_4 + M_{135}\dot{\theta}_3\dot{\theta}_5\\ &\quad + M_{136}\dot{\theta}_3\dot{\theta}_6 + M_{145}\dot{\theta}_4\dot{\theta}_5 + M_{146}\dot{\theta}_4\dot{\theta}_6 + M_{156}\dot{\theta}_5\dot{\theta}_6\\ &\quad + G_1\end{aligned} \quad (253)$$

$$
\begin{aligned}
\tau_2 = {} & M_{12}\ddot{\theta}_1 + M_{22}\ddot{\theta}_2 + M_{23}\ddot{\theta}_3 + M_{24}\ddot{\theta}_4 + M_{25}\ddot{\theta}_5 + M_{26}\ddot{\theta}_6 \\
& + M_{211}\dot{\theta}_1^2 + M_{222}\dot{\theta}_2^2 + M_{233}\dot{\theta}_3^2 + M_{244}\dot{\theta}_4^2 + M_{255}\dot{\theta}_5^2 \\
& + M_{266}\dot{\theta}_6^2 + M_{212}\dot{\theta}_1\dot{\theta}_2 + M_{213}\dot{\theta}_1\dot{\theta}_3 + M_{214}\dot{\theta}_1\dot{\theta}_4 \\
& + M_{215}\dot{\theta}_1\dot{\theta}_5 + M_{216}\dot{\theta}_1\dot{\theta}_6 + M_{223}\dot{\theta}_2\dot{\theta}_3 + M_{224}\dot{\theta}_2\dot{\theta}_4 \\
& + M_{225}\dot{\theta}_2\dot{\theta}_5 + M_{226}\dot{\theta}_2\dot{\theta}_6 + M_{234}\dot{\theta}_3\dot{\theta}_4 + M_{235}\dot{\theta}_3\dot{\theta}_5 \\
& + M_{236}\dot{\theta}_3\dot{\theta}_6 + M_{245}\dot{\theta}_4\dot{\theta}_5 + M_{246}\dot{\theta}_4\dot{\theta}_6 + M_{256}\dot{\theta}_5\dot{\theta}_6 \\
& + G_2
\end{aligned} \tag{254}
$$

$$
\begin{aligned}
\tau_3 = {} & M_{13}\ddot{\theta}_1 + M_{23}\ddot{\theta}_2 + M_{33}\ddot{\theta}_3 + M_{34}\ddot{\theta}_4 + M_{35}\ddot{\theta}_5 + M_{36}\ddot{\theta}_6 \\
& + M_{322}\dot{\theta}_2^2 + M_{333}\dot{\theta}_3^2 + M_{344}\dot{\theta}_4^2 + M_{355}\dot{\theta}_5^2 + M_{366}\dot{\theta}_6^2 \\
& + M_{312}\dot{\theta}_1\dot{\theta}_2 + M_{313}\dot{\theta}_1\dot{\theta}_3 + M_{314}\dot{\theta}_1\dot{\theta}_4 + M_{315}\dot{\theta}_1\dot{\theta}_5 \\
& + M_{316}\dot{\theta}_1\dot{\theta}_6 + M_{323}\dot{\theta}_2\dot{\theta}_3 + M_{324}\dot{\theta}_2\dot{\theta}_4 + M_{325}\dot{\theta}_2\dot{\theta}_5 \\
& + M_{326}\dot{\theta}_2\dot{\theta}_6 + M_{345}\dot{\theta}_4\dot{\theta}_5 + M_{346}\dot{\theta}_4\dot{\theta}_6 + M_{356}\dot{\theta}_5\dot{\theta}_6 \\
& + G_3
\end{aligned} \tag{255}
$$

$$
\begin{aligned}
\tau_4 = {} & M_{14}\ddot{\theta}_1 + M_{24}\ddot{\theta}_2 + M_{34}\ddot{\theta}_3 + M_{44}\ddot{\theta}_4 + M_{54}\ddot{\theta}_5 + M_{64}\ddot{\theta}_6 \\
& + M_{411}\dot{\theta}_1^2 + M_{422}\dot{\theta}_2^2 + M_{433}\dot{\theta}_3^2 + M_{444}\dot{\theta}_4^2 + M_{455}\dot{\theta}_5^2 \\
& + M_{466}\dot{\theta}_6^2 + M_{412}\dot{\theta}_1\dot{\theta}_2 + M_{413}\dot{\theta}_1\dot{\theta}_3 + M_{414}\dot{\theta}_1\dot{\theta}_4 \\
& + M_{415}\dot{\theta}_1\dot{\theta}_5 + M_{416}\dot{\theta}_1\dot{\theta}_6 + M_{423}\dot{\theta}_2\dot{\theta}_3 + M_{424}\dot{\theta}_2\dot{\theta}_4 \\
& + M_{425}\dot{\theta}_2\dot{\theta}_5 + M_{426}\dot{\theta}_2\dot{\theta}_6 + M_{434}\dot{\theta}_3\dot{\theta}_4 + M_{435}\dot{\theta}_3\dot{\theta}_5 \\
& + M_{436}\dot{\theta}_3\dot{\theta}_6 + M_{445}\dot{\theta}_4\dot{\theta}_5 + M_{446}\dot{\theta}_4\dot{\theta}_6 + M_{456}\dot{\theta}_5\dot{\theta}_6 \\
& + G_4
\end{aligned} \tag{256}
$$

$$
\begin{aligned}
\tau_5 = {} & M_{15}\ddot{\theta}_1 + M_{25}\ddot{\theta}_2 + M_{35}\ddot{\theta}_3 + M_{45}\ddot{\theta}_4 + M_{55}\ddot{\theta}_5 + M_{65}\ddot{\theta}_6 \\
& + M_{511}\dot{\theta}_1^2 + M_{522}\dot{\theta}_2^2 + M_{533}\dot{\theta}_3^2 + M_{544}\dot{\theta}_4^2 + M_{555}\dot{\theta}_5^2 \\
& + M_{566}\dot{\theta}_6^2 + M_{512}\dot{\theta}_1\dot{\theta}_2 + M_{513}\dot{\theta}_1\dot{\theta}_3 + M_{514}\dot{\theta}_1\dot{\theta}_4 \\
& + M_{515}\dot{\theta}_1\dot{\theta}_5 + M_{516}\dot{\theta}_1\dot{\theta}_6 + M_{523}\dot{\theta}_2\dot{\theta}_3 + M_{524}\dot{\theta}_2\dot{\theta}_4 \\
& + M_{525}\dot{\theta}_2\dot{\theta}_5 + M_{526}\dot{\theta}_2\dot{\theta}_6 + M_{534}\dot{\theta}_3\dot{\theta}_4 + M_{535}\dot{\theta}_3\dot{\theta}_5 \\
& + M_{536}\dot{\theta}_3\dot{\theta}_6 + M_{545}\dot{\theta}_4\dot{\theta}_5 + M_{546}\dot{\theta}_4\dot{\theta}_6 + M_{556}\dot{\theta}_5\dot{\theta}_6 \\
& + G_5
\end{aligned} \tag{257}
$$

$$
\begin{aligned}
\tau_6 = {} & M_{16}\ddot{\theta}_1 + M_{26}\ddot{\theta}_2 + M_{36}\ddot{\theta}_3 + M_{46}\ddot{\theta}_4 + M_{56}\ddot{\theta}_5 + M_{66}\ddot{\theta}_6 \\
& + M_{611}\dot{\theta}_1^2 + M_{622}\dot{\theta}_2^2 + M_{633}\dot{\theta}_3^2 + M_{644}\dot{\theta}_4^2 + M_{655}\dot{\theta}_5^2 \\
& + M_{666}\dot{\theta}_6^2 + M_{612}\dot{\theta}_1\dot{\theta}_2 + M_{613}\dot{\theta}_1\dot{\theta}_3 + M_{614}\dot{\theta}_1\dot{\theta}_4 \\
& + M_{615}\dot{\theta}_1\dot{\theta}_5 + M_{616}\dot{\theta}_1\dot{\theta}_6 + M_{623}\dot{\theta}_2\dot{\theta}_3 + M_{624}\dot{\theta}_2\dot{\theta}_4 \\
& + M_{625}\dot{\theta}_2\dot{\theta}_5 + M_{626}\dot{\theta}_2\dot{\theta}_6 + M_{634}\dot{\theta}_3\dot{\theta}_4 + M_{635}\dot{\theta}_3\dot{\theta}_5 \\
& + M_{636}\dot{\theta}_3\dot{\theta}_6 + M_{645}\dot{\theta}_4\dot{\theta}_5 + M_{646}\dot{\theta}_4\dot{\theta}_6 + M_{656}\dot{\theta}_5\dot{\theta}_6 \\
& + G_6
\end{aligned} \tag{258}
$$

Note that in the above equations

$$\sum_{k=\max(i,j)}^{n} \mathrm{Tr}(\boldsymbol{D}_{kj}\boldsymbol{J}_k\boldsymbol{D}_{ki}^T) = M_{ij} \tag{259}$$

$$\sum_{m=\max(i,j,k)}^{n} \mathrm{Tr}(\boldsymbol{D}_{mjk}\boldsymbol{J}_m\boldsymbol{D}_{mi}^T) = M_{ikm} \tag{260}$$

As an example, the explicit forms of the governing dynamics equations for a two-link robot arm are derived as shown in Figure 13.

Note that the table for joint parameters is given by

i	α_i	a_i	d_i	θ_i
1	0	l	0	θ_i
2	0	l	0	θ_2

Now the moments of inertias are $\boldsymbol{J}_1$ and $\boldsymbol{J}_2$, such that

$$J_1 = \begin{bmatrix} \dfrac{m_1l^2}{12} + m_1l^2(1-\alpha_1)^2 & 0 & 0 & -m_1l(1-\alpha_1) \\ 0 & 0 & 0 & 0 \\ 0 & 0 & 0 & 0 \\ -m_1l(1-\alpha_1) & 0 & 0 & m_1 \end{bmatrix} \tag{261}$$

$$J_2 = \begin{bmatrix} \dfrac{m_2l^2}{12} + m_2l^2(1-\alpha_2)^2 & 0 & 0 & -m_2l(1-\alpha_2) \\ 0 & 0 & 0 & 0 \\ 0 & 0 & 0 & 0 \\ -m_2l(1-\alpha_2) & 0 & 0 & m_2 \end{bmatrix} \tag{262}$$

values of D_{111}, D_{211}, D_{212}, D_{232}, A_0^1, A_1^2, A_0^2, D_{11}, D_{12}, D_{21}, and D_{22} are all the same as before, such that

$$\mathbf{D}_{11}\mathbf{J}_1\mathbf{D}_{11}^T = \begin{bmatrix} m_1l^2S_1^2\left(\alpha - \dfrac{1}{6}\right) & m_1l^2S_1C_1\left(\dfrac{1}{6} - \alpha\right) & 0 & 0 \\ m_1l^2S_1C_1\left(\dfrac{1}{6} - \alpha\right) & m_1l^2C_1^2\left(\alpha - \dfrac{1}{6}\right) & 0 & 0 \\ 0 & 0 & 0 & 0 \\ 0 & 0 & 0 & 0 \end{bmatrix} \tag{263}$$

$$\mathrm{Tr}[D_{11}J_1D_{11}^T] = m_1l^2\left(\alpha - \frac{1}{6}\right) \tag{264}$$

Similarly,

$$\mathrm{Tr}[D_{21}J_2D_{21}^T] = m_2l^2\alpha_2 \tag{265}$$

$$
\begin{aligned}
M_{11} = \mathrm{Tr}[D_{11}J_1D_{11}^T + D_{21}J_2D_{21}^T] = {} & m_1l^2\left(\alpha_1 - \frac{1}{6}\right) \\
& + m_2l^2\left(a_2 + \frac{5}{6} + 2\alpha_2C_2\right)
\end{aligned} \tag{266}
$$

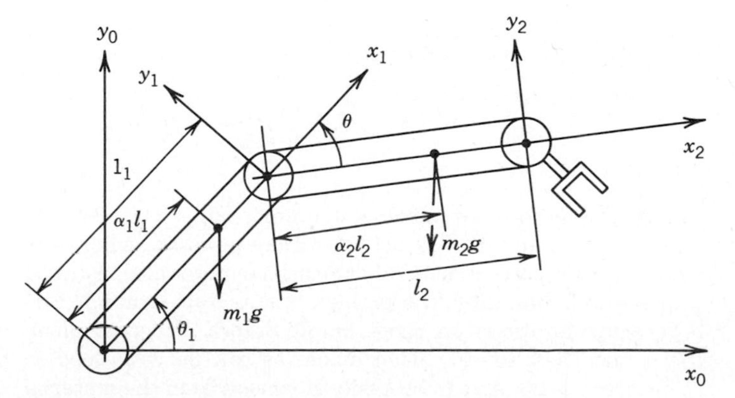

Figure 13. Geometry of a two-link robot with distributed centers of gravity.

$$M_{12} = M_{21} = \mathrm{Tr}[D_{22}J_2D_{21}^T] = m_2l^2\left(\alpha_2 + \alpha_2C_2 - \frac{1}{6}\right) \tag{267}$$

$$M_{22} = \mathrm{Tr}[D_{22}J_2D_{22}^T] = m_2l^2\left(\alpha_2 - \frac{1}{6}\right) \tag{268}$$

$$\mathrm{Tr}[D_{111}J_2D_{11}^T] = 0, \quad \mathrm{Tr}[D_{211}J_2D_{21}^T] = 0 \tag{269}$$

$$M_{111} = \mathrm{Tr}[D_{111}J_1D_{11}^T] + \mathrm{Tr}[D_{211}J_2D_{21}^T] = 0 \tag{270}$$

$$M_{112} = \mathrm{Tr}[D_{212}J_2D_{21}^T] = -S_2m_2l^2\alpha_2 \tag{271}$$

$$M_{121} = \mathrm{Tr}[D_{221}J_2D_{21}^T] = -S_2m_2l^2\alpha_2 \tag{272}$$

$$M_{122} = \mathrm{Tr}[D_{222}J_2D_{21}^T] = -S_2m_2l^2\alpha_2 \tag{273}$$

$$M_{211} = \mathrm{Tr}[D_{211}J_2D_{22}^T] = \alpha_2S_2m_2l^2 \tag{274}$$

$$M_{212} = \mathrm{Tr}[D_{212}J_2D_{22}^T] = 0 \tag{275}$$

$$M_{222} = \mathrm{Tr}[D_{222}J_2D_{22}^T] = 0 \tag{276}$$

Note that $M_1 = m_1g \cdot D_{11}\bar{r}_1 + m_2g \cdot D_{21}\bar{r}_2$. Thus,

$$M_1 = -m_1(0,-g,0,0)\begin{bmatrix} -S_1 & -C_1 & 0 & -lS_1 & (\alpha_1-1)l \\ C_1 & -S_1 & 0 & lC_1 & 0 \\ 0 & 0 & 0 & 0 & 1 \\ 0 & 0 & 0 & 0 & 0 \end{bmatrix}$$

$$-m_2(0,-g,0,0) \times \begin{bmatrix} -S_{12} & -C_{12} & 0 & -l(S_{12}+S_1) & (\alpha_2-1)l \\ C_{12} & -S_{12} & 0 & l(C_{12}+C_1) & 0 \\ 0 & 0 & 0 & 0 & 0 \\ 0 & 0 & 0 & 0 & 1 \end{bmatrix} \tag{277}$$

or

$$M_2 = m_2gC_{12}\alpha_2l \tag{278}$$

Thus, the governing equations are

$$\begin{aligned} \tau_1 = {} & \left[m_1l^2\left(\alpha_1 - \frac{1}{6}\right) + m_1l^2\left(\alpha_2 - \frac{5}{6} + 2\alpha_2C_2\right)\right]\ddot{\theta}_1 \\ & + \left[m_2l^2\left(\alpha_2 - \frac{1}{6} + \alpha_2C_2\right)\right]\ddot{\theta}_2 + \dot{\theta}_1\dot{\theta}_2[-2m_2l^2\alpha_2S_2] \\ & + \dot{\theta}_2^2[-m_2l^2\alpha_2S_2] + lg[m_1\alpha_1C_1 + m_2\alpha_2C_{12} + m_2C_1] \end{aligned} \tag{279}$$

$$\begin{aligned} \tau_2 = {} & \left[m_2l^2\left(\alpha_2 - \frac{1}{6} + \alpha_2C_2\right)\right]\ddot{\theta}_1 + \left[m_2l^2\left(\alpha_2 - \frac{1}{6}\right)\right]\ddot{\theta}_2 \\ & + [m_2l^2\alpha_2S_2]\,\dot{\theta}_1^2 + m_2glC_{12}\alpha_2 \end{aligned} \tag{280}$$

FUTURE OUTLOOK

The future outlook for robotics dynamics will most probably be in connection with flexible robot manipulators, where the dynamics of robotic structural deformation must also be taken into account. Basically, the problem stems from dynamic stability considerations on robot manipulators. Progress along these lines has already been made, as can be witnessed in references 19–26. As can be easily surmised from the material covered in these works, the inclusion of dynamic deformation terms due to flexibility of the arm, will greatly complicate the dynamics equations for robot manipulators.

For further reading on the dynamics of robot manipulators in general, and in particular including the effects of robot flexibility, see references 29–35. In connection with the dynamics of the flexible space shuttle robot arm see references 27 and 28.

BIBLIOGRAPHY

1. J. Denavit and R. S. Hartenberg, "A Kinematics Notation for Lower Pair Mechanisms Based on Matrices," *ASME Journal of Applied Mechanics* **22,** 215–221 (1955).
2. K. C. Gupta, "Kinematics Analysis of Manipulators Using the Zero Reference Position Description," *International Journal of Robotics Research* **5**(2), 13–18 (1986).
3. S. M. K. Kazerounian, "Manipulation and Simulation of General Robots Using the Zero-Position Description, Doctoral Dissertation, University of Illinois, Chicago, Ill., 1984.
4. A. K. Bejczy, "Robot Arm Dynamics and Control," NASA Jet Propulsion Laboratory, Pasadena, Calif., Tech. Memo. 33-669, Feb. 1974.
5. J. Duffy, *Analysis of Mechanisms and Robot Manipulators,* John Wiley & Sons, New York, 1980.
6. R. P. Paul, *Robot Manipulators: Mathematics, Programming, and Control,* M.I.T. Press, Cambridge, Mass., 1981.
7. J. M. Hollerbach, "A Recursive Lagrangian Formulation of Manipulator Dynamics and a Comparative Study of Dynamics Formulation Complexity," *IEEE Transactions Systems, Man and Cybernetics* **SMC-10**(11), 730–736 (1980).
8. D. E. Whitney, "Resolved Motion Rate Control of Manipulators and Human Prostheses," *IEEE Transactions on Man–Machine Systems* **MMS-10**(2), 47–53 (1969).
9. J. Y. S. Luh, M. W. Walker, and R. P. Paul, "On-Line Computational Scheme for Mechanical Manipulators," *Transactions of the ASME, Journal of Dynamic Systems Measurement and Control* **120,** 69–76 (1980).
10. C. S. G. Lee, "Robot Arm Kinematics, Dynamics and Control," *IEEE Computer* **18,** 62–77 (1982).
11. C. S. G. Lee, B. H. Lee, and R. Nigam, "An Efficient Formulation of Robot Arm Dynamics for Control and Computer Simulation," University of Michigan, Ann Arbor, Mich., CRIM Tech. Rep. RSD-TR-8-82, Aug. 1982.
12. J. M. Hollerbach and S. Gideon, "Wrist-Partitioned, Inverse Kinematic Accelerations and Manipulator Dynamics," *International Journal of Robotics Research* **4,** 61–76 (1983).
13. M. Shahinpoor, "Bond Graph Dynamics Modeling of Robotic Manipulators," F. F. Ling and I. G. Tadjbakhsh, eds., *Recent Developments in Applied Mathematics,* RPI, Rensselaer Press, Troy, N.Y., 1983, pp. 176–184.
14. M. Shahinpoor, *A Robot Engineering Textbook,* Harper and Row, New York, 1986.
15. J. W. Burdick, "An Algorithm for Generation of Efficient Manipulator Dynamic Equations," *IEEE Robotics and Automation Conf. Proc.* vol. 1, San Francisco, Calif., April 1986, pp. 212–219.
16. J. J. Uicker, Jr., "Dynamic Behavior of Spatial Linkages," *Transactions of the ASME, Journal of Applied Mechanics* **5**(68), 1–15 (1966).
17. M. E. Kahn and B. Roth, "The Near-Minimum Time Control of Open Look Kinematic Chains," *Transactions of the ASME, Journal of Dynamic Systems Measurement and Control* **93,** series G, 164–172 (1971).

18. K. Kazerounian and K. C. Gupta, "Manipulator Dynamics Using the Extended Zero Reference Position Description," *Proceedings of the 9th Applied Mechanics Conference,* Kansas City, Mo., 1985, pp. 1–11; see also *IEEE Journal of Robotics and Automation* **2**(4), (1986).
19. W. J. Book, "Modeling, Design and Control of Flexible Manipulator Arms," Ph.D. Thesis, Massachusetts Institute of Technology, Cambridge, Mass., Apr. 1974.
20. W. J. Book, "Recursive Lagrangian Dynamics of Flexible Manipulator Arms," *International Journal Robotics Research* **3**(3), 87–101 (1984).
21. O. Maizza-Neto, "Model Analysis and Control of Flexible Manipulator Arms," Ph.D. Thesis, Massachusetts Institute of Technology, Cambridge, Mass., 1974.
22. W. G. Beazley, "The Small Motion Dynamics of Bilaterally Coupled Kinematics Chains With Flexible Links," Ph.D. Thesis, The University of Texas at Austin, Austin, Tex., Aug. 1978.
23. W. Sunada and S. Dubowsky, "The Application of Finite Element Methods to the Dynamics Analysis of Flexible Spatial and Co-Planar Linkage Systems," *ASME Journal of Mechanical Design* **103,** 643–651 (1981).
24. W. Sunada and S. Dubowsky, "On the Dynamic Analysis and Behavior of Industrial Robotic Manipulators With Elastic Members," *ASME Journal of Mechanical Design* **105,** 42–51 (1983).
25. S. J. Derby, "Kinematic Elasto-Dynamic Analysis of Computer Graphics Simulation of General Purpose Robot Manipulators," Ph.D. Thesis, Rensselaer Polytechnic Institute, Troy, N.Y., Aug. 1981.
26. T. Fukuda, "Flexibility Control of Elastic Robotic Arms," *Journal of Robotic Systems* **2**(1), 73–88 (1985).
27. P. C. Hughes, "Dynamics of a Flexible Manipulator Arm for the Space Shuttle," *Proceedings of the AAS/AIAA Astrodynamics Conference,* Jackson Lake Lodge, Grand Teton National Park, Wy., Sept. 1977.
28. P. C. Hughes, "Dynamics of a Chain of Flexible Bodies," *Journal of Astronautical Science* **27**(4), 359–380 (1979).
29. F. A. Kelly and R. L. Huston, "Modeling of Flexibility Effects in Robot Arms," *Proceedings of the 1981 Joint Automatic Control Conference,* American Automatic Control Council, Green Valley, Ariz., June 1981, Paper No. WP-2C.
30. M. H. Raibert and B. K. P. Horn, "Manipulator Control Using the Configuration Space Method," *Industrial Robot* **5**(2), 69–73 (1978).
31. W. M. Silver, "On the Equivalence of Lagrangian and Newton-Euler Dynamics for Manipulators," *Proceedings of the 1981 Joint Automatic Control Conference,* American Automatic Control Council, Green Valley, Ariz., June 1981, Paper No. TA-2A.
32. Y. Stepanenko and M. Vukobratovic, "Dynamics of Articulated Open-Chain Active Mechanisms," *Mathematical Bioscience* **28,** 137–170 (1976).
33. M. W. Walker and D. E. Orin, "Efficient Dynamic Computer Simulation of Robotic Mechanisms," *Transactions of the ASME, Journal of Dynamic Systems Measurement and Control* **104**(3), 205–211 (1982).
34. M. Thomas and D. Tesar, "Dynamic Modeling of Serial Manipulator Arms," *Transactions of the ASME, Journal of Dynamic Systems Measurement and Control* **104**(3), 218–228 (1982).
35. D. E. Whitney, "Deflection and Vibration of Jointed Beams," *Design and Control of Remote Manipulators,* NASA Quart. Rep. NASA-CR-123795, July 1972.

The author thanks Gayla Angel for help in the preparation of the manuscript.

E

ECONOMIC JUSTIFICATION

RICHARD GUSTAVSON
The Charles Stark Draper Laboratory, Inc.
Cambridge, Massachusetts

INTRODUCTION

Industry is always searching for the most cost-effective way of creating its products. Product design is essential for providing good materials and the likelihood of proper fabrication and assembly. Although technology is vital to manufacturing, economics is usually also very significant. How can a rational choice between competing technologically viable alternatives be made? Many factors need to be investigated. The general philosophical issues are discussed elsewhere in this encyclopedia (see COST/BENEFIT; ECONOMICS, ROBOTIC MANUFACTURING AND PRODUCTS) as well as in reference 1 among others. This article deals only with recently developed economic analysis and evaluation techniques that have been found to be particularly useful in a wide variety of industrial applications.

The basic ideas, from engineering economy, involve the time value of money, cost accounting techniques, and cash flow analyses. The present discussion is limited to manufacturing organizations and, more specifically, to individual conditions (a station, cell, system, or factory) and will not be concerned with comparisons to other projects competing for available funds. The techniques described here are general and may be applied to all manufacturing applications. Reference 2 is a useful reference created for industry. It provides a wide variety of forms allowing very detailed cost comparisons; it also contains an excellent bibliography.

Engineers require knowledge in applied microeconomics that they can use as part of proper justification presentations to management. A discussion of minimum attractive rate of return (MARR) is followed by a discussion of fixed and variable cost accounting with applications. Next is a section on quality and the effect of rework on system design and behavior. Then the method of zero present value discounted cash flow (DCF) is explored with particular emphasis on the various ways that a required investment can be justified. Finally, an application exhibits the use of the techniques. A list of symbols and definitions is given at the end of the article under Nomenclature.

MINIMUM ATTRACTIVE RATE OF RETURN (MARR)

Every business has criteria for establishing the desirability of investments. Normally established by accountants, these rules are confusing to engineers. In order to maximize corporate economic status, the MARR for any investment is established. An extensive discussion of the many different ways that the MARR is interpreted is presented in reference 3. Although there is no totally agreed upon definition, most businesses establish at least three categories for proposed investments:

1. High risk (2.5 MARR).
2. Moderate risk (1.7 MARR).
3. Low risk (1.0 MARR).

Every robotic application is subject to one of these economic justification conditions.

How can the MARR be calculated? Few are willing to quantify it; one usable equation (4) can be rearranged as

$$\text{MARR} = r_{\text{LTD}}\left[1 - \frac{E}{\text{LTD} + E}\right] + \left(\frac{r^*}{1-t}\right)\left[1 + \beta\left(\frac{R_m}{r^*} - 1\right)\right]\left[\frac{E}{\text{LTD} + E}\right]$$

This equation can be simplified by defining return on equity as

$$r_E = r^*\left[1 + \beta\left(\frac{R_m}{r^*} - 1\right)\right]$$

then

$$\text{MARR} = r_{\text{LTD}}\left[1 - \frac{E}{\text{LTD} + E}\right] + \frac{r_E}{(1-t)}\left[\frac{E}{\text{LTD} + E}\right]$$

Figure 1 displays results of the calculations for a range of parameter values. It should be noted that well-run companies usually have an $E/(\text{LTD} + E)$ value of 0.9 or higher. Figure 1 reveals that the higher the equity [or the lower the long-term debt (LTD)], the higher the MARR. The MARR is also directly proportional to the effective tax rate t.

For example, in the 1984–1985 era, IBM's financial statements show $t = 0.452$ and $E/(\text{LTD} + E) = 0.924$ on average. Thus,

$$\text{MARR} = 0.076\, r_{\text{LTD}} + 1.686\, r_E$$

If $r_{\text{LTD}} = 10\%$ and $r_E = 15\%$, then the MARR $= 26.1\%$.

In contrast, take another corporation whose values are $t = 0.377$ and $E/(\text{LTD} + E) = 0.813$ on average. Then

$$\text{MARR} = 0.197\, r_{\text{LTD}} + 1.305\, r_E$$

With $r_{\text{LTD}} = 10\%$ and $r_E = 15\%$, the MARR is 21.5%. This company has an easier time justifying an investment in robotics since the MARR is lower.

Although a method for calculating the MARR has been shown, it must be recalled that individual corporate practice, as well as the prevailing economic conditions, will establish the "correct" value to be used in economic justification studies.

UNIT COST ACCOUNTING

An economic justification starts by dealing with the methods through which the cost of a product is established (5). There

$r_{LTD} = 12\%$

Return on equity	Effective tax rate 0%	10%	20%	30%	40%	50%
			$E/(LTD + E) = 0.2$			
8.00%	11.20%	11.38%	11.60%	11.89%	12.27%	12.80%
10.00	11.60	11.82	12.10	12.46	12.93	13.60
12.00	12.00	12.27	12.60	13.03	13.60	14.40
14.00	12.40	12.71	13.10	13.60	14.27	15.20
16.00	12.80	13.16	13.60	14.17	14.93	16.00
18.00	13.20	13.60	14.10	14.74	15.60	16.80
20.00	13.60	14.04	14.60	15.31	16.27	17.60
			$E/(LTD + E) = 0.4$			
8.00	10.40	10.76	11.20	11.77	12.53	13.60
10.00	11.20	11.64	12.20	12.91	13.87	15.20
12.00	12.00	12.53	13.20	14.06	15.20	16.80
14.00	12.80	13.42	14.20	15.20	16.53	18.40
16.00	13.60	14.31	15.20	16.34	17.87	20.00
18.00	14.40	15.20	16.20	17.49	19.20	21.60
20.00	15.20	16.09	17.20	18.63	20.53	23.20
			$E/(LTD + E) = 0.6$			
8.00	9.60	10.13	10.80	11.66	12.80	14.40
10.00	10.80	11.47	12.30	13.37	14.80	16.80
12.00	12.00	12.80	13.80	15.09	16.80	19.20
14.00	13.20	14.13	15.30	16.80	18.80	21.60
16.00	14.40	15.47	16.80	18.51	20.80	24.00
18.00	15.60	16.80	18.30	20.23	22.80	26.40
20.00	16.80	18.13	19.80	21.94	24.80	28.80
			$E/(LTD + E) = 0.8$			
8.00	8.80	9.51	10.40	11.54	13.07	15.20
10.00	10.40	11.29	12.40	13.83	15.73	18.40
12.00	12.00	13.07	14.40	16.11	18.40	21.60
14.00	13.60	14.84	16.40	18.40	21.07	24.80
16.00	15.20	16.62	18.40	20.69	23.73	28.00
18.00	16.80	18.40	20.40	22.97	26.40	31.20
20.00	18.40	20.18	22.40	25.26	29.07	34.40
			$E/(LTD + E) = 1$			
8.00	8.00	8.89	10.00	11.43	13.33	16.00
10.00	10.00	11.11	12.50	14.29	16.67	20.00
12.00	12.00	13.33	15.00	17.14	20.00	24.00
14.00	14.00	15.56	17.50	20.00	23.33	28.00
16.00	16.00	17.78	20.00	22.86	26.67	32.00
18.00	18.00	20.00	22.50	25.71	30.00	36.00
20.00	20.00	22.22	25.00	28.57	33.33	40.00

Figure 1. MARR.

will be fixed costs (factory, robots, conveyors, etc), which are independent of production volume. For each unit produced, there will also be variable costs, which are directly related to the cycle time.

It is often more convenient to use unit cost instead of total cost as a means for comparing alternative methods. Plots of fixed unit cost versus production volume (straight line on a log–log graph) and variable unit cost versus production volume (constant) are shown in Figure 2. The total unit cost will be a curve that is the summation of these two data plots. Many businesses say that manual systems have only variable costs; either the fixed cost is very small in comparison or the investment has been totally written off. Unit cost is defined as

$$C_u = \frac{f_{AC}P}{Q_y} + \frac{T/\epsilon}{3600}[w\bar{L}_H + O_H] \text{ (\$/unit)}$$

where

$$f_{AC} = \text{annualized cost factor} = \left[1 - \frac{v}{(1+r)^H}\right]\left[\frac{r(1+r)^H}{(1+r)^H - 1}\right]$$

The fixed cost portion can be considered the annual repayment of a loan by the manufacturing department to the central account of the business, although no transfer of funds actually takes place. (This is sometimes called capital recovery.) If H is equal to the system economic life (ie, depreciable years) and there is no salvage value, f_{AC} becomes the familiar periodic loan repayment factor.

A system comprises (potentially) many resources, tools, and material-handling devices. Extensive cost analyses have been performed to account for such items as model mix and design changes (6). For present purposes, the various costs are aggregated into one hardware cost. The true cost of a system is a multiple (rho factor) of the hardware cost and must be appropriately cost-accounted. Some labor-intensive industries bury most of this (primarily engineering) cost in overhead, making their burdened labor rate quite high. Robotic systems will be capital-intensive and should be dealt with differently.

When more than one product can be produced by the system, the annualized fixed cost is merely apportioned among the various products; for example,

Product	Common Cost	Particular Cost	Proportion of total
A	$0.62P$	$0.22P$	$0.84P$
B	$0.62P$	$0.16P$	$0.78P$

There is often so little difference between the products (usually addition and/or deletion of parts or processes) that the entire production batch is considered one product for cost purposes.

Variable unit cost is a function of the actual cycle time and the variable cost rate. Cycle time is modified by the actual uptime ratio expected and also possibly by rework. Rate is limited here to two parameters: total labor rate and operating/maintenance rate. The latter term is usually intended to in-

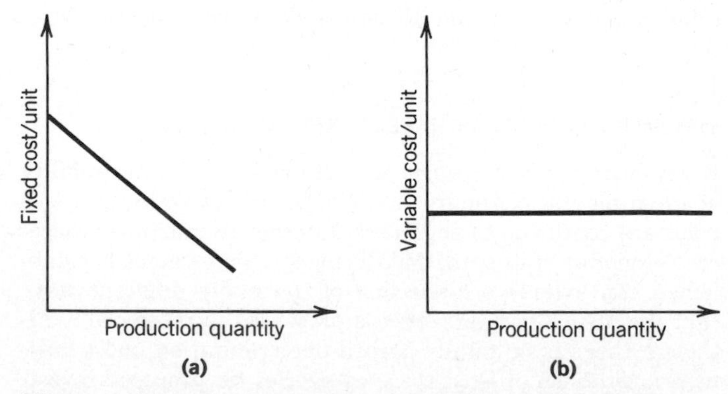

Figure 2. General characteristics of **(a)** fixed and **(b)** variable costs.

Table 1. Assembly System Options General Technology Data

Type	Symbolic Name	Average Resource Price $\bar{P}_R$ ($)	Average Tool and Material Handling Price $\bar{P}_T$ ($/part)	System Uptime Expected ϵ (%)	Minimum Expected Task Cycle Time[a] $\bar{t}$ (s)	Average Loaded Labor Rate $\bar{L}_H$ ($/hour)	Operation/Maintenance Rate for Each Resource Used O_H ($/hour)	Maximum Number of Stations Per Worker m_s	(Total Cost)/(Hardware Cost) ρ
Manual	M05				5	5			
	M10	200	2000	85		10	0.5	0.833	1.2
	M15					15			
	M20					20			
Programmable	P07	7500	3500		2		1	8	2.0
	P15	15,000	5000		2		1.5	7	2.5
	P29	29,000	6500		3		2.0	6	3.0
	P45	45,000	8500	85	4	$\bar{L}_H$	2.5	5	3.5
	P70	70,000	12,000		7		3	4	4.0
	P85	85,000	14,000		7		3	4	4.0
	P120	120,000	18,000		10		4	3	5.0
Fixed Automation	F30		30,000		1.5		1	6	
	F60	0	60,000	90	5.0	$\bar{L}_H$	2	4	1.5
	F90		90,000		10.0		3	2	

[a] Includes tool change, where applicable.

clude only those items directly attributable to the system (power, fluids, floor space, etc), but may incorporate the "institutional cost" (ie, the costs that occur when the business doors are opened).

An example of technology options for an assembly system is shown in Table 1. The values are illustrative and do not necessarily reflect actual cost and/or behavior of the various resources defined. Data from the table applied to a particular assembly are shown graphically in Figure 3 for manual systems, Figure 4 for fixed automation systems, and Figure 5 for programmable (robotic) systems. Similar plots for any set of systems can be created; the crossover production volume point between any two can (usually) be established from

$$Q_{co} = \frac{3600 f_{AC}(P_A - P_B)}{[(T/\epsilon)_B (w_B \bar{L}_{H_B} + O_{H_B}) - (T/\epsilon)_A (w_A \bar{L}_{H_A} + O_{H_A})}$$

An example is shown in Figure 6. Knowing the production batch size required, either A or B can be chosen as the most cost-effective. Some further requirements are discussed in the comparison section.

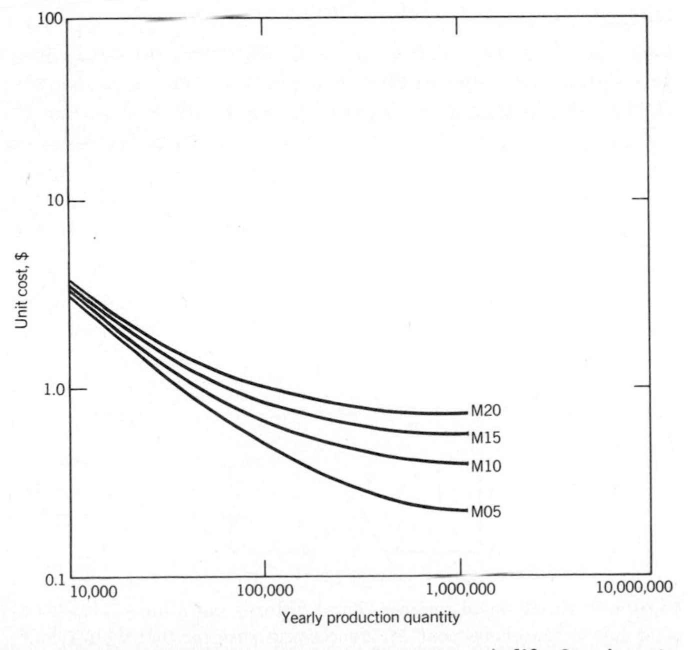

Figure 3. Manual systems. 20 parts 5-yr economic life; 2-yr investment horizon; ACRS depreciation; 35% rate of return; 240 working d/yr; 2 shifts available.

REWORK/QUALITY CONSIDERATIONS

Product quality is of utmost importance. Usually, an automated system (once running properly) produces high-quality

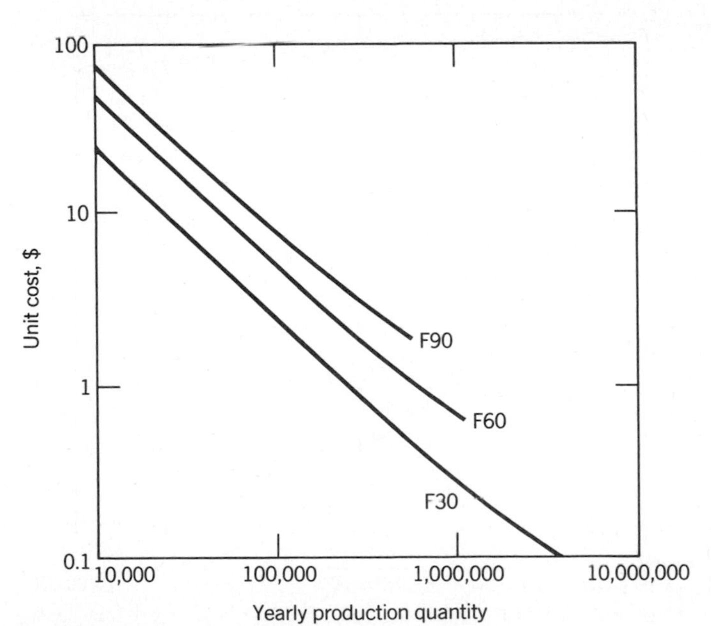

Figure 4. Fixed automation systems. 20 parts; 10 stations. For other conditions see legend for Figure 3.

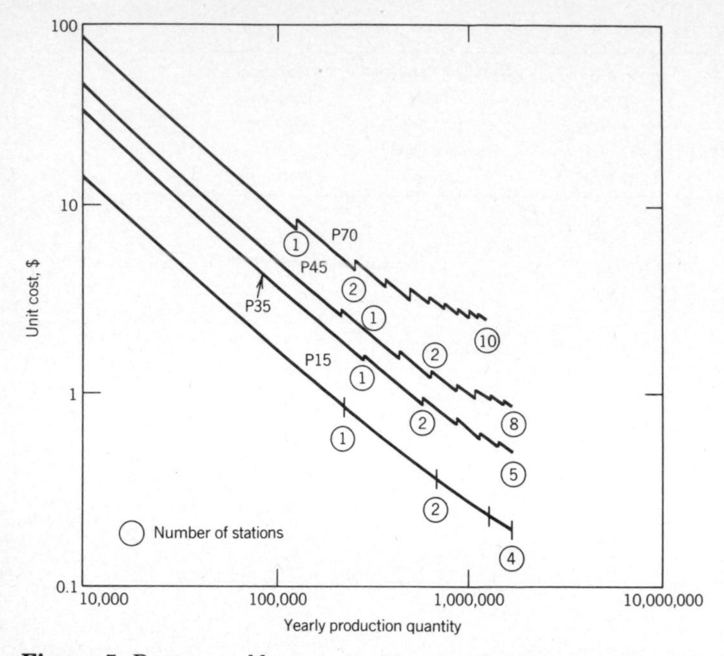

Figure 5. Programmable systems. 20 parts; 3 shifts available. For other conditions see legend for Figure 3.

articles since there is little, if any, recognizable difference between the items. Failures in products created manually occur almost randomly unless there is a design or production flaw. One way of quantifying the effect of quality on production is to increase the actual cycle time inversely as the quality decreases [eg, when yield is 90%, cycle time is 1.11 (T/ϵ)]. The realization that rejected items can be discarded (ie, a new one must be created) or reworked results in additional actual time needed to produce the required batch.

When very low production volume products are considered,

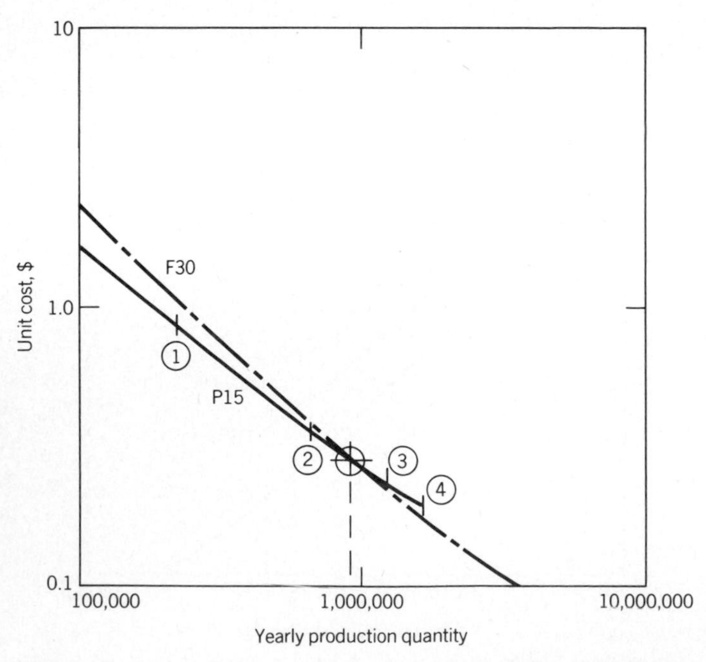

Figure 6. Crossover point example. 20 parts; 3 shifts available. For other conditions see legend for Figure 3.

further significant costs can be incurred due to the extensive rework required to make the products conform to tight specifications. These products are usually extremely expensive and are therefore never scrapped. Figure 7 displays a single-level process that has an assumed constant failure rate for each recycle loop that occurs. Figure 8**a** and **b** exhibits results for a particular cost condition; any particular yield rate shows increasing actual cost as production volume increases until an asymptote is approached. Obviously, the higher the yield (or quality), the lower the total cost.

When the particular process being evaluated is part of a larger system, significant complexity can arise. Figure 9 exhibits a three-level system whose behavior is to be investigated. It is assumed that failed units can only recycle externally to level 1. A schematic of what actually occurs in a poor system (but one that is representative of certain products) for 64 units is shown in Figure 10. If this system can be automated, the behavior could improve to the almost ideal conditions shown in Figure 11. It is the improvement in system yield that lowers the cost, not manpower reduction (as is usually the case).

A comparison of the work required at level 2 (take a vertical cut to the left-hand side of P_2–R_2) shows that the manual case (Figure 10) requires 2.172 + 0.422 + 0.750 = 3.344 time units, whereas the hybrid case (Figure 11) requires 1.031 + 0 + 0.031 + 1.062 time units. This improvement can be further accentuated if we put a weighting factor on each of the process and recycle conditions; Figures 10 and 11 use unity for all stations.

Numerous other failure modes are possible; only one has been illustrated. It shows how the effective cycle time can be reduced by 68% through the addition of automated stations. The variable cost rate is also reduced by almost 65% since there is no reduction in manpower, but a small increase in operating/maintenance rate.

DISCOUNTED CASH FLOW (DCF)

Cash flow is based on the variable costs required for any system. To compare an *a*lternative system (A) with a *b*ase system (B) (7), the first task is to determine the savings (in variable costs) that are expected. In general,

$$S_j = \frac{Q_{y_j}}{3600}\left[(T/\epsilon)_B\,(w_B\bar{L}_{H_B} + O_{H_B})_j - (T/\epsilon)_A\,(w_A\bar{L}_{H_A} + O_{H_A})_j\right]$$

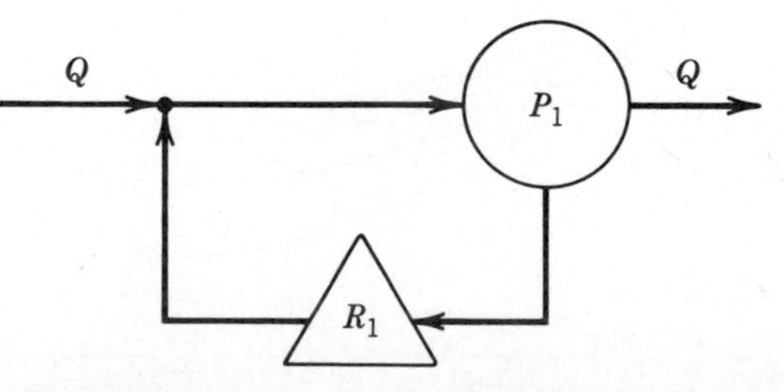

Figure 7. Single-level process. Total failures not allowed. Each unit of Q has a materials cost M. Processing cost for every unit is P_1. Recycling cost for failed units is R_1. Failure rate for each loop is constant. Failed units will recycle until only one unit remains. That unit passes the final loop.

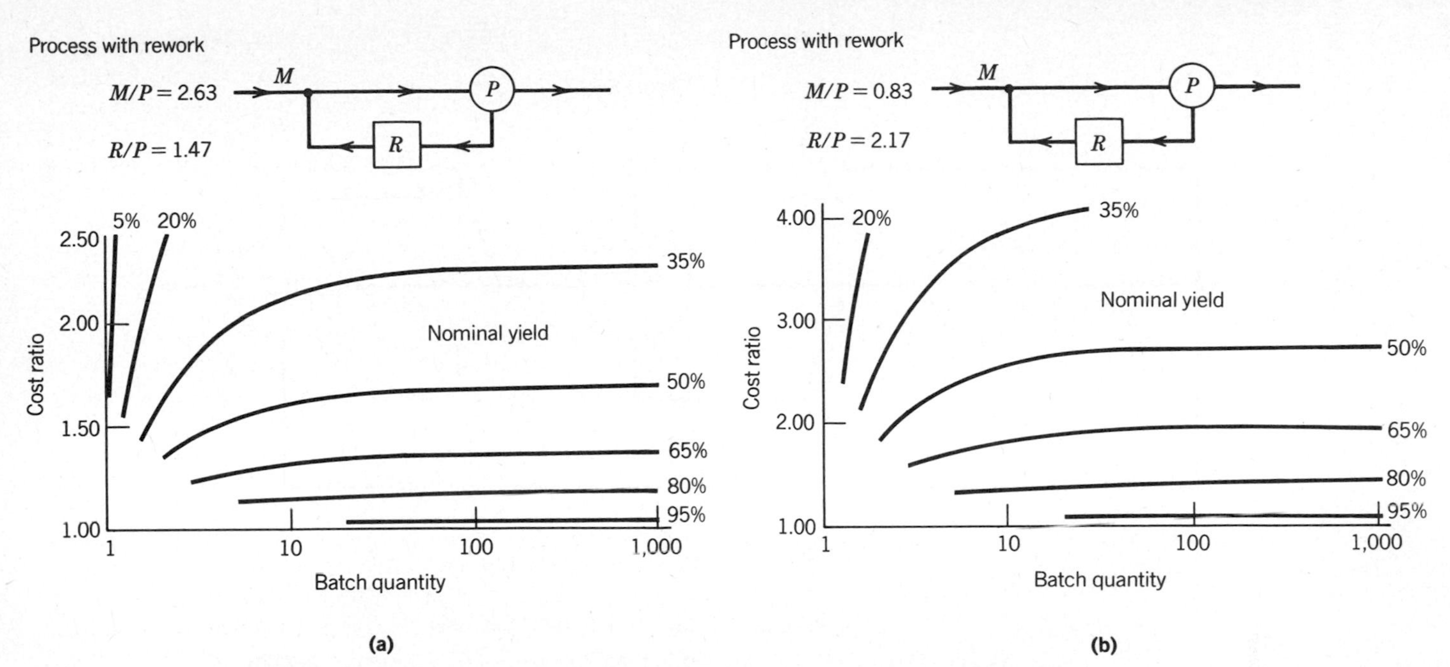

Figure 8. Results for particular cost condition. Two examples **(a)** and **(b)** are shown. M = Material cost; P = processing cost; R = reworking cost.

where the subscript j denotes year and allows for increasing costs and different production volumes. A series of such yearly savings (producing income to the business) over a planning horizon will be termed a savings stream. It can be composed of totally arbitrarily specified values, but could exhibit a constant rate of increase.

The most useful DCF analysis has been found to be the zero present worth method. An investment is made in year 0; some portion of it is depreciable (the inverse of the "rho factor" defined earlier). Each subsequent year's savings must be reduced by the allowable depreciation multiplied by the tax rate, producing a value that is subtracted from the savings to produce a net income. This can be written

$$N_j = S_j - \tau_j \left(S_j - \delta_j \,\mathrm{Ml1}\frac{P}{\rho}\right) = S_j(1 - \tau_j) + \tau_j\delta_j\left(\frac{P}{\rho}\right)$$

Since funds in year j do not have the same value as funds now (year 0), the net incomes must be discounted by

$$\eta_j = \frac{N_j}{(1 + r)^j}$$

The final year of the planning horizon may contain a salvage value for the equipment; it is usually only for accounting purposes since dismantling is unlikely to occur. Zero present worth means that the net outlay in year 0 is equal to the sum of all future discounted net incomes. In general,

$$\sum_{j=0}^{L} \eta_j = 0$$

There are two unknowns: r [internal rate of return (IRR)] and P (the allowable total investment). Most people choose P

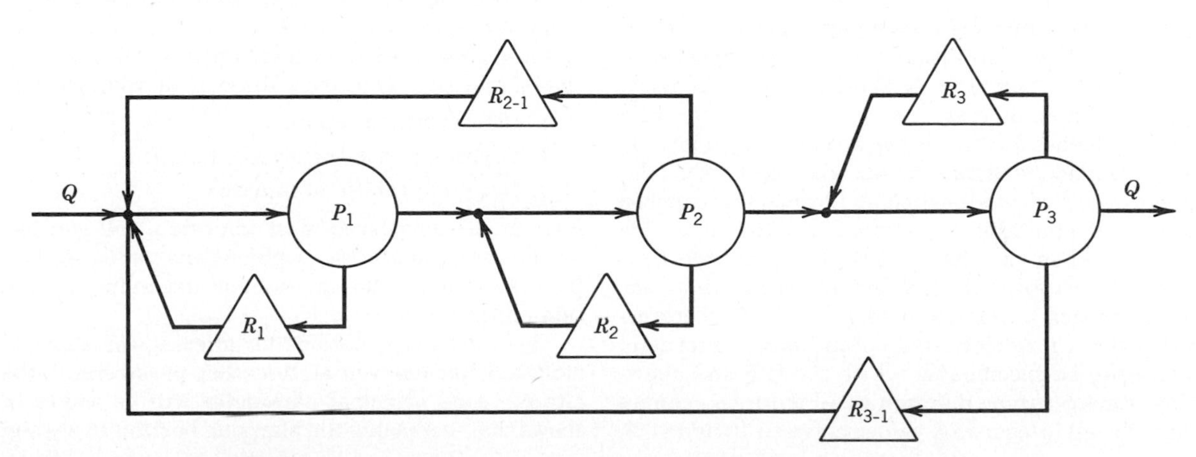

Figure 9. A three-level process. Special case 1. Total failures not allowed.

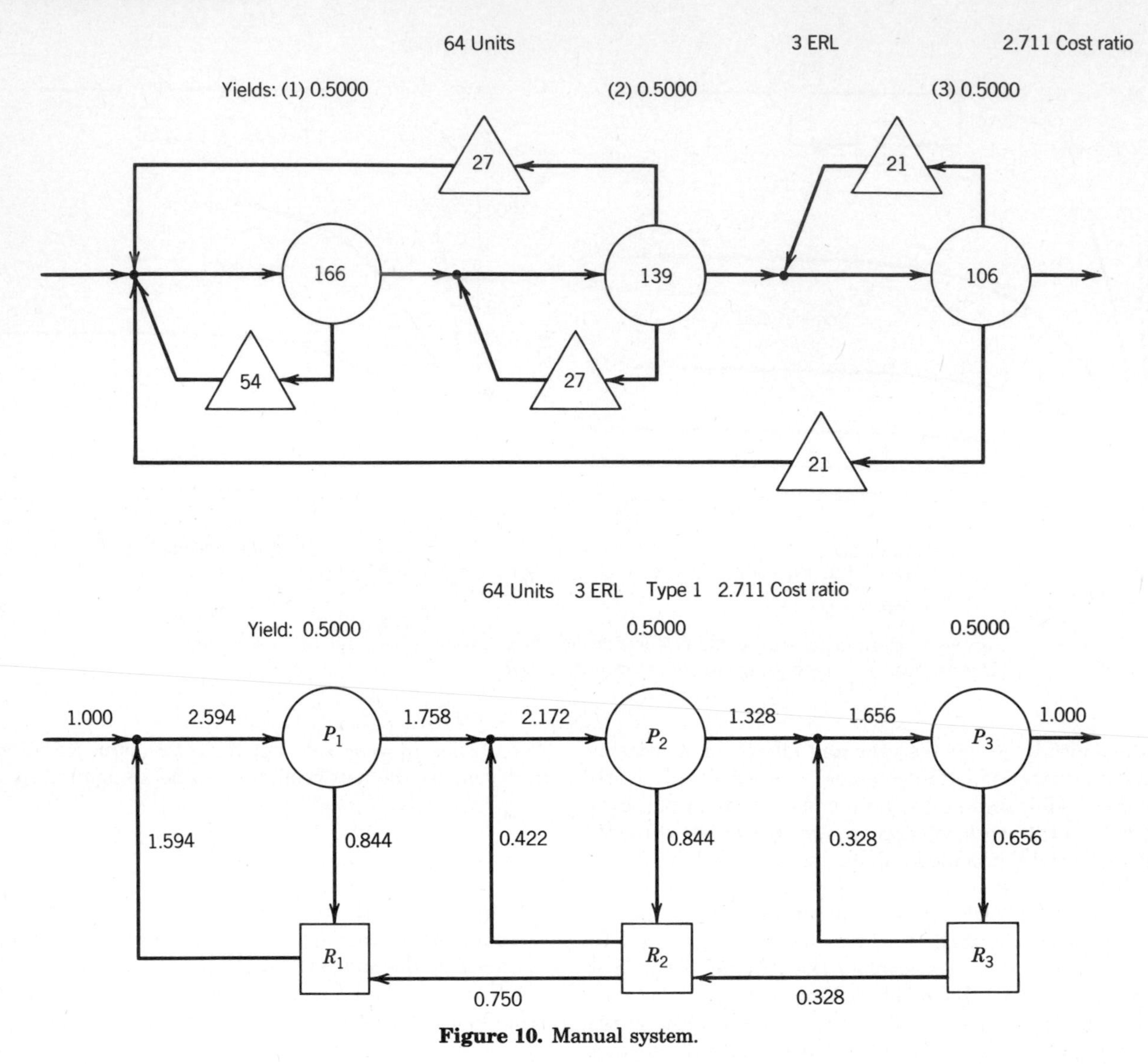

Figure 10. Manual system.

and determine *r;* broader use can be made of specifying *r* and determining *P*.

Refer to Figure 12, which exhibits a particular DCF analysis. The income forecast shows that savings are expected to increase by 15% each year; 5-yr ACRS depreciation is used; a tax credit is available in year 1, and the effective tax rate is constant. The expense forecast is the simplest possible: only one investment is made, without any tax writeoffs for the nondepreciable portion. For this particular income and expense situation, the total allowable investment is 1.198 times the first-year savings when the IRR is 22%. The predicted cash flow table reveals a net profit of 0.444 times the first-year savings and a payback in approximately 1.8 yr. (Payback occurs when the sun of nondiscounted net incomes equals zero; a precise value can be specified for which the IRR and allowable total investment can be determined.) Figure 13 exhibits summary data for all integer IRR values between 10 and 40%. In general, as the IRR increases, the allowable investment decreases, the break-even (or payback) point occurs earlier, and the net profit increases.

For a prescribed investment amount, the company can base the justification on at least three evaluation parameters:

1. IRR (above minimum).
2. Payback period (below maximum).
3. Net profit (above minimum).

Any can be considered most important, all can be weighted equally, or various other combinations can be used to establish the desirability of the investment, depending upon the corporate philosophy.

There are many options for altering the allowable investment amount (as well as the other parameters); the effect of altering each principal parameter will be shown. Figure 14 shows that increasing the planning horizon raises significantly

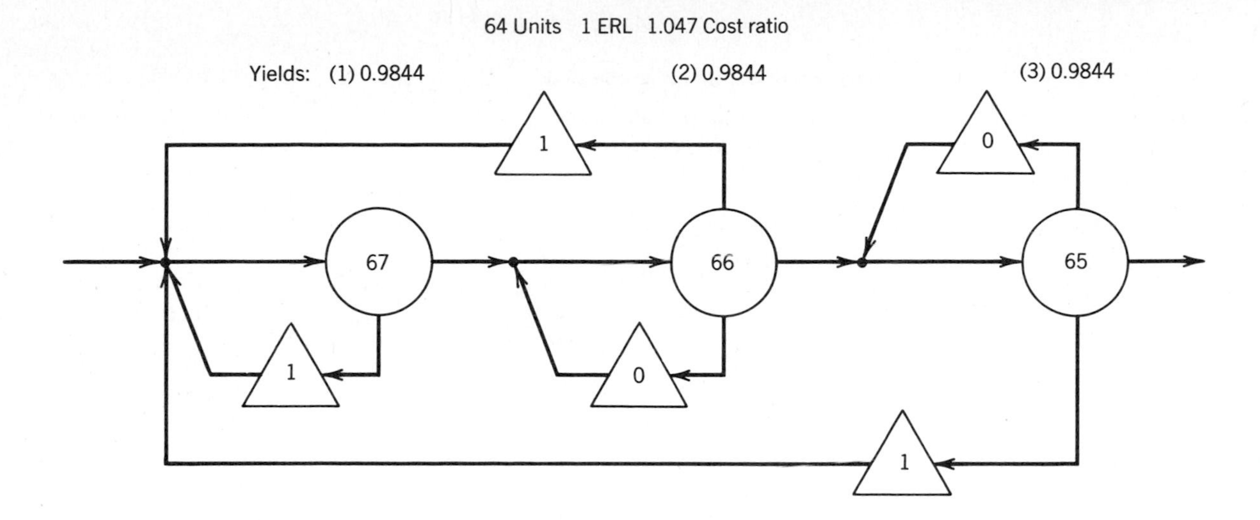

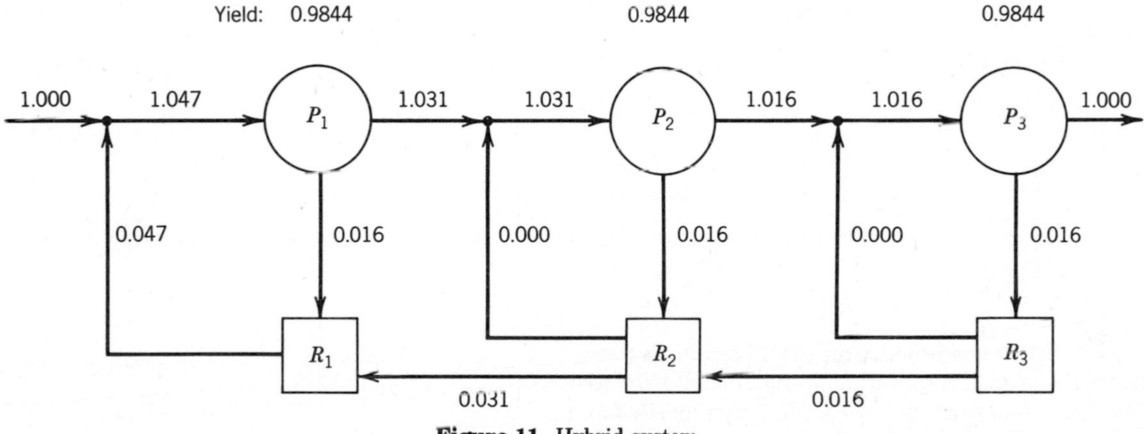

Figure 11. Hybrid system.

the allowable investment. Figure 15 shows that a growth in the rate increase for savings improves the allowable investment. Figure 16 exhibits the fact that the lower the effective tax rate, the higher the allowable investment.

Often there is an option to purchase a "turn-key" system ($\rho = 1$) or to buy only the hardware and perform all necessary engineering and so on in-house ($\rho > 1$); Figure 17 shows that from a purely economic standpoint it is wisest to purchase a system. Other factors, such as company policy or a desire to be current on technology, can certainly limit the depreciable portion (ie, the inverse of ρ) to be used. The nondepreciable portion of the investment (engineering, software, setup, debug, etc) is an expense to the corporation. Depending on circumstances within the business, it may be possible to obtain what is effectively a tax credit for that expense (eg, if other corporate activities are making money). Figure 18 shows that the higher the effective tax rate in year 0 (when the investment is made), the larger the allowable investment. Comparing the 100% depreciable portion behavior to that of the 40% depreciable portion (with a 43% year-0 tax rate), Figure 19 shows that the former is still the significantly better choice economically.

Thus far, only a single investment in year 0 has been examined. When dramatic increases in production volume are forecast, it is deemed necessary to buy only a portion of the ultimate system in year 0 and part in year 1 and/or year 2. Figure 20 displays one set of circumstances where the total savings over each planning horizon period is constant (a fixed amount of income is expected from the product, regardless of how it is distributed). The rates of increase used in Figure 20 are quite arbitrary. Figure 21 exhibits the DCF for case C. Comparison of these data with those in Figure 12 shows that there is a slightly lower allowable investment, the net profit is somewhat higher, and the payback (looking from year 0) is almost twice as long. A vast number of other comparisons could be made, many of which would probably show that spreading out the investment provides the most significant economic benefits.

While the general guidelines and examples above can be

Single Investment—2-yr Horizon

Zero present worth cash flow analysis 07–10–1985

5 yr economic life; salvage value 0% of cost;
0% constant rate of increase in costs

Year	Ratio	Expense forecast Tax rate	Depreciable	Savings	Income forecast Depreciation	Tax rate	Credit
0	100.00%	0.0%	40.0%				
1				100.000	15.0%	43.0%	8.0%
2				115.000	22.0	43.0	
2*				Salvage value	63.0		

Allowable total investment = 119.756
Depreciable investment = 47.902
IRR = 22.00%

Pro forma cash flow

Year	Income	Depreciation	Taxes	Credits	Net	Discounted net
0	−119.756	0.000	0.000	0.000	−119.756	−119.756
1	100.000	7.185	39.910	3.832	63.922	52.395
2	115.000	10.539	44.918	0.000	70.082	47.085
2*	30.179	0.000	0.000	0.000	30.179	20.276
Income totals						
	245.179	17.724	84.829	3.832	164.182	119.756
Net totals						
	125.423	17.724	84.829	3.832	44.426	0.000

Nominal capital recovery = 46.306
Payback in approximately 1.8 yr

Figure 12. A particular DCF analysis. *Accounting method for salvage value, not an actual cash flow.

very useful, specific cases can produce quite different behavior. What is acceptable to one corporation is not necessarily to another. Individual cases must stand on their own merit; the method of evaluation described here will allow reasonable conditions to be found and rated.

COMPARISON OF ALTERNATIVE SYSTEM TO BASE SYSTEM

There are numerous methods that can be used to establish a usable manufacturing system. Most commonly used is the industrial or manufacturing engineer's "bottom-up" approach, which provides technologically viable systems that may be far from cost-optimized. Another technique is to use simulation (8,9) to optimize by repeated trials the systems generated by hand. Synthesis of least-cost systems is possible by linear programming (10) (which produces cost-optimized solutions that may not be physically realizable) or by heuristic methods using elements of dynamic programming (11).

Figure 22 displays two assembly systems to be compared (11). The manual system is the base (B) and the general system (in this case, all robots) is the alternative (A). Recall that the unit cost to assemble is composed of the annualized fixed cost and the variable cost. For this case the variable costs are

$$C_{V_B} = \frac{35.2}{3600}[3.61(19) + 1.50]\,450{,}000 = \$308{,}396$$

$$C_{V_A} = \frac{33.4}{3600}[0.60(19) + 2.40]\,450{,}000 = \$57{,}615$$

Thus, the savings ($C_{V_B} - C_{V_A}$) due to using the alternative is \$250,781/yr. Can this amount of savings justify the necessary investment of \$425,000? For simplification, assume no increase in production or cost. Figure 23 exhibits the fact that a 4-yr planning horizon is required if there is no tax benefit for the nondepreciable investment in year 0. However, when the year-0 tax rate is the same as that for all subsequent years, only a 2-yr horizon is required, as shown in Figure 24.

What if the manual system actually has 15% rework and the automated system has 3% rework? The manual system cycle time becomes 40.5 s/cycle, whereas the automated system actually requires 34.4 s/cycle. This results in a yearly savings of \$295,312 using the alternative, which then allows a 17.8% increase in allowable investment. Other rework percentages produce corresponding savings. There will be cases when an alternative (usually fully or mostly automated) system can only be economically justified through elimination of rework.

Suppose that the production volume increases to 750,000 units in the second year and beyond. Figure 25 displays the least-cost manual system (base) and the least-cost general sys-

Single Investment—2-yr Horizon

Zero present worth cash flow analysis 07–09–1985

5 yr economc life; salvage value 0% of cost;
0% constant rate of increase in costs

Year	Ratio	Expense forecast Tax rate	Depreciable	Savings	Income forecast Depreciation	Tax rate	Credit
0	100.00%	0.0%	40.0%				
1				100.000	15.0%	43.0%	8.0%
2				115.000	22.0	43.0	
2*				Salvage value	63.0		

Rate of return	Allowable investment	Depreciable investment	Approximate break-even	Capital recovery	Net profit
10.0	149.723	59.889	2.18	41.352	24.877
11.0	146.703	58.681	2.14	41.862	26.847
12.0	143.795	57.518	2.11	42.352	28.744
13.0	140.993	56.397	2.07	42.821	30.572
14.0	138.293	55.317	2.04	43.271	32.333
15.0	135.688	54.275	2.00	43.704	34.033
16.0	133.173	53.269	1.97	44.120	35.673
17.0	130.745	52.298	1.94	44.520	37.257
18.0	128.399	51.359	1.91	44.904	38.788
19.0	126.130	50.452	1.88	45.274	40.268
20.0	123.936	49.574	1.85	45.631	41.699
21.0	121.812	48.725	1.82	45.974	43.085
22.0	119.756	47.902	1.80	46.306	44.426
23.0	117.764	47.106	1.77	46.625	45.726
24.0	115.833	46.333	1.75	46.933	46.985
25.0	113.961	45.584	1.72	47.231	48.206
26.0	112.145	44.858	1.70	47.518	49.391
27.0	110.383	44.153	1.67	47.795	50.541
28.0	108.671	43.469	1.65	48.063	51.657
29.0	107.009	42.804	1.63	48.322	52.741
30.0	105.395	42.158	1.61	48.573	53.795
31.0	103.825	41.530	1.59	48.816	54.819
32.0	102.299	40.919	1.57	49.050	55.814
33.0	100.814	40.326	1.55	49.278	56.783
34.0	99.369	39.748	1.53	49.498	57.725
35.0	97.963	39.185	1.51	49.711	58.643
36.0	96.594	38.637	1.49	49.918	59.536
37.0	95.260	38.104	1.47	50.118	60.406
38.0	93.960	37.584	1.46	50.312	61.254
39.0	92.694	37.077	1.44	50.501	62.080
40.0	91.459	36.584	1.42	50.683	62.886

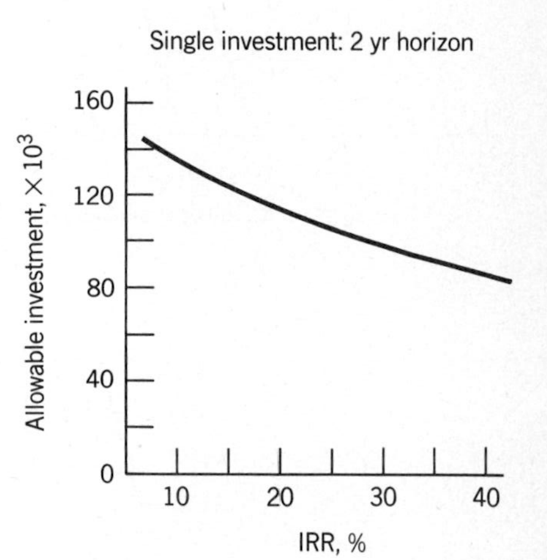

Figure 13. Summary data for all integer values between 10 and 40%. *Accounting method for salvage value, not an actual cash flow.

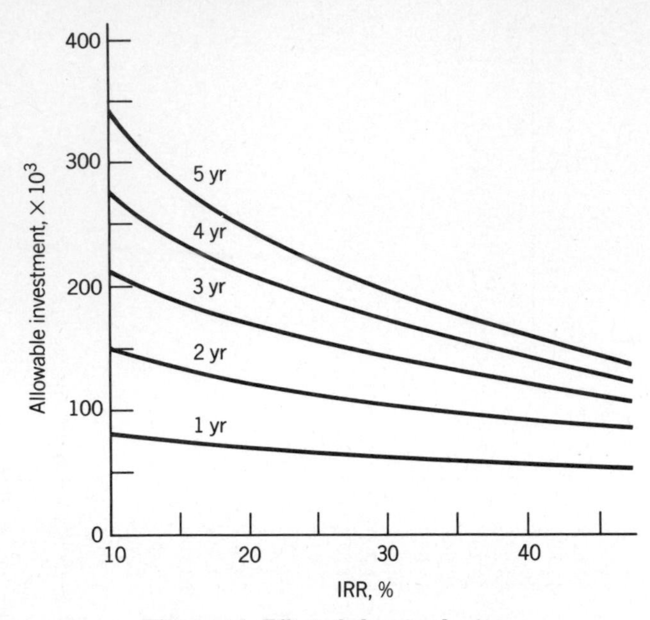

Figure 14. Effect of planning horizon.

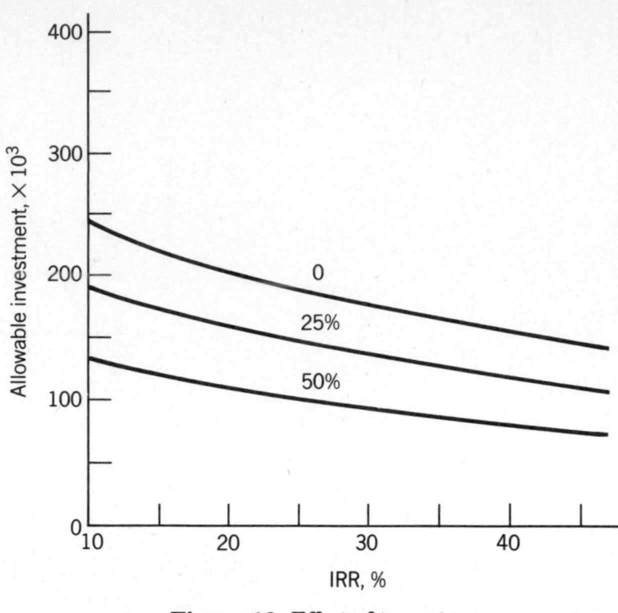

Figure 16. Effect of tax rate.

tem (alternative). Comparison with Figure 22 reveals that one direct person is added to the manual system and two A0 robots are added to the general system. This is a fine example of near-term least-cost systems being part of the ultimate systems. Use of the alternative system yields savings of \$410,744 for 750,000 units/yr. In order to provide the lesser system (Fig. 22), 425,000/575,000 = 73.9% of the total investment in year 0, with the balance in year 1, has to be spent. As before, allowance of year-0 taxes requires only a 2 yr planning horizon (Figure 26). Figure 27 shows that a planning period of over three years is required when there is no tax benefit in year 0.

SUMMARY

Although many components can contribute to the overall justification of automation in general and robots in particular, industry requires that those factors somehow be quantified; no universally recognized scheme has thus far been produced. The methods and techniques shown in this article can generally account for such factors and thus allow engineers and financial people to analyze and interpret the economics of manufacturing mutually. Cost accounting, rework analysis, and DCF are powerful techniques when utilized correctly; they

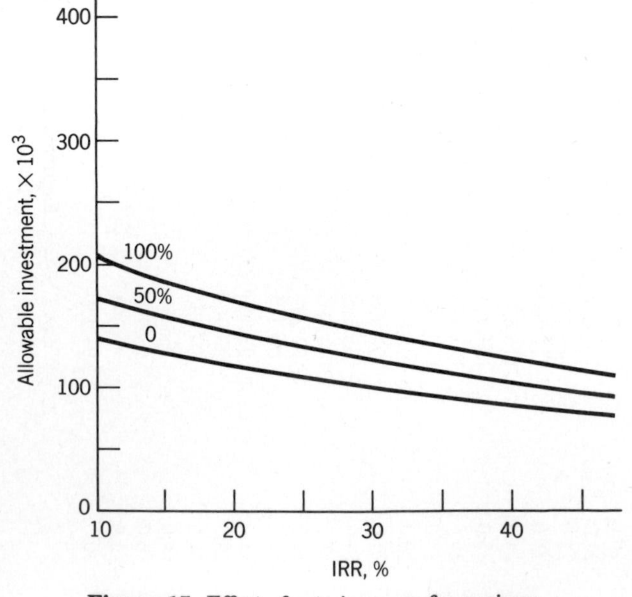

Figure 15. Effect of rate increase for savings.

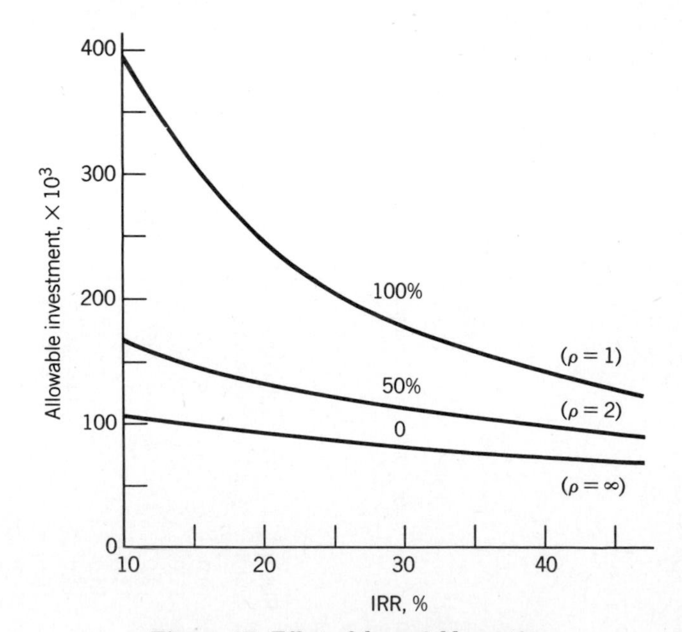

Figure 17. Effect of depreciable portion.

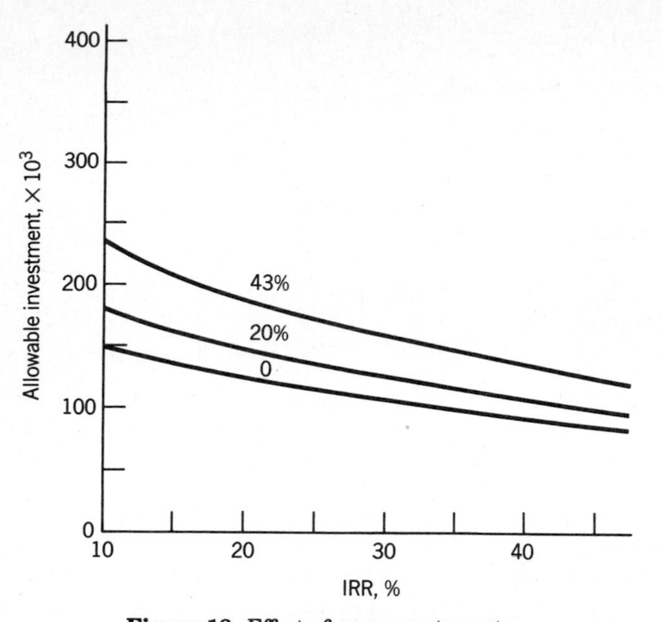

Figure 18. Effect of year zero tax rate.

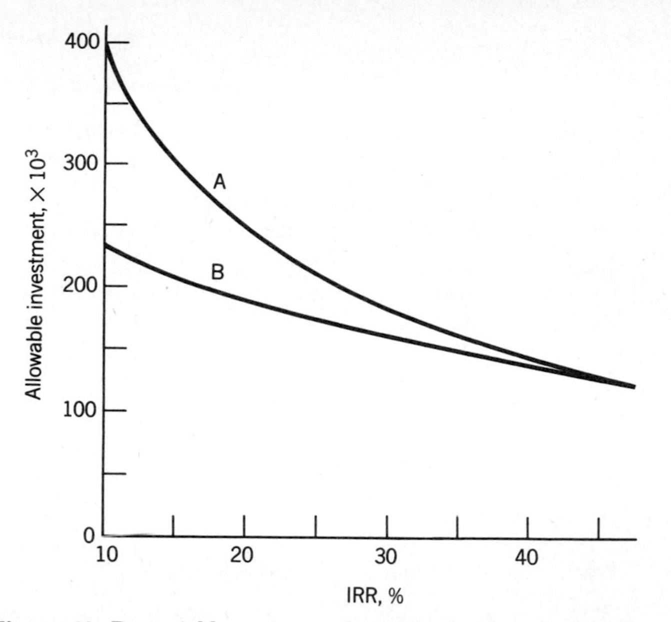

Figure 19. Depreciable portion and year-0 tax rate. A—depreciable portion, 100%. B—depreciable portion, 40%; year-0 tax rate, 43%.

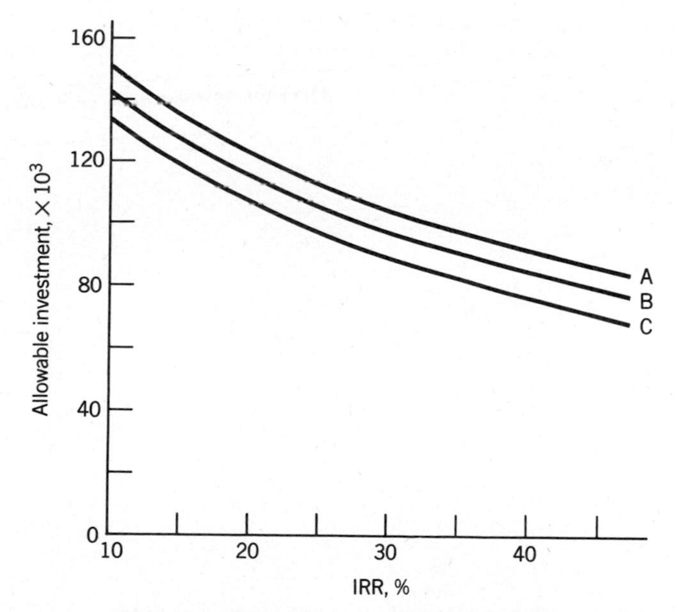

Figure 20. Effect of investment over n years.

Case	Portion of investment (%) Year 0	Year 1	Year 2	Planning horizon (yr)	Annual increase in savings (%)	Year 1 savings
A	100	—	—	2	15	100.000
B	50	50	—	3	30	53.885
C	33.3	33.3	33.3	4	45	28.285

Three-Part Investment—4-yr Horizon

Zero present worth cash flow analysis 07–09–1985

5 yr economic life; salvage value 0% of cost;
0% constant rate of increase in costs

Year	Ratio	Expense forecast Tax rate	Depreciable	Savings	Income forecast Depreciation	Tax rate	Credit
0	33.33%	0.0%	40.0%				
1	33.33	0.0	40.0	28.285	15.0%	43.0%	8.0%
2	33.33	0.0	40.0	41.013	22.0	43.0	8.0
3				59.469	21.0	43.0	8.0
4				86.230	21.0	43.0	
4*				Salvage value	21.0		

Allowable total investment = 103.110
Depreciable investment = 41.244
IRR = 22.00%

Pro forma cash flow

Year	Income	Depreciation	Taxes	Credits	Net	Discounted net
0	−34.370	0.000	0.000	0.000	−34.370	−34.370
1	−34.370	0.000	0.000	0.000	−34.370	−28.172
	28.285	2.062	11.276	1.100	18.109	14.843
2	−34.370	0.000	0.000	0.000	−34.370	−23.092
	41.013	5.087	15.448	1.100	26.665	17.915
3	59.469	7.974	22.143	1.100	38.426	21.161
4	86.230	8.799	33.296	0.000	52.935	23.895
4*	17.322	0.000	0.000	0.000	17.322	7.819
Income totals						
	232.320	23.922	82.163	3.300	153.457	85.634
Net totals						
	129.210	23.922	82.163	3.300	50.347	0.000

Nominal capital recovery = 38.703
Payback in approximately 3.4 yr

Figure 21. The DCF for case C. *Accounting method for salvage value, not an actual cash flow.

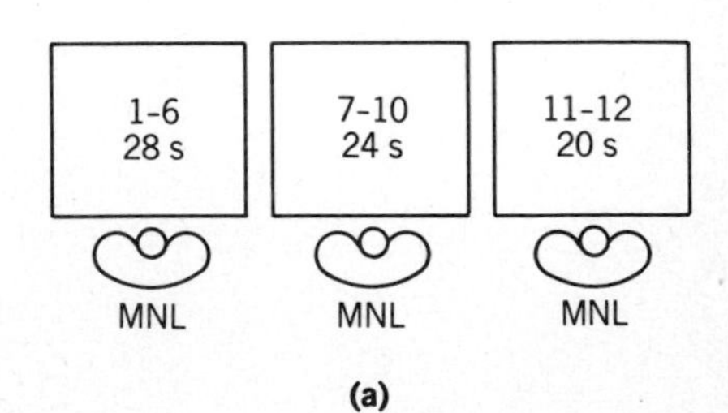

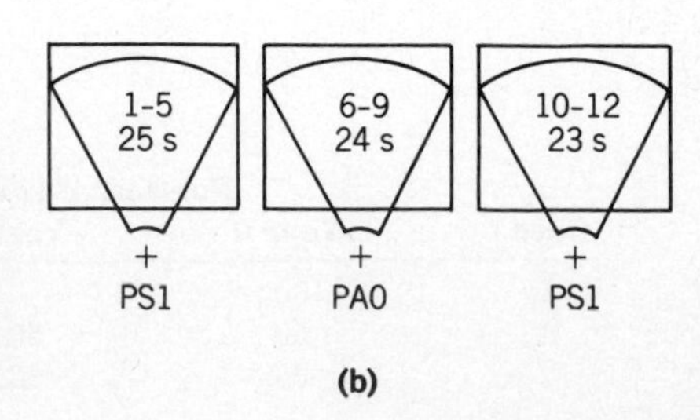

Figure 22. **(a)** The least-cost manual system. 35.2 s usable cycle time; 87.5% system uptime; 1.70 theoretical units/min; 589,091 units actual capacity of system; $6930 total capital investment; rho factor = 1.50; 3.61 workers at $19.00 required; $1.50/h system operating/maintenance rate; 0.764 yr required for 3 shift operation; 240 required for 2.29-shift operation. Units, 450,000; $0.692 each; 1 year; cycle time, 30.8 s; 3 shifts, AF, 0.912. **(b)** The least-cost general system. 33.4 s usable cycle time; 85.0% system uptime; 1.80 theoretical units/min; 620,620 units actual capacity of system; $425,000 total capital investment; rho factor = 2.50; 0.60 workers at $19.00/h required; $2.40/h system operating/maintenance rate; 0.725 yr required for 3-shift operation; 240 d required for 2.18-shift operation. Units, 450,000; $0.493 each; 1 year; cycle time, 28.4 s; 3 shifts; AF, 0.918.

Alternative Versus Base

Zero present worth cash flow analysis 07–11–1985

5 yr economic life; salvage value 0 × of cost;

0 × constant rate of increase in costs

Year	Ratio	Expense Forecast Tax Rate	Depreciable	Savings	Income forecast Depreciation	Tax rate	Credit
0	100.00%	0.0%	40.0%				
1				250.781	15.0%	43.0%	8.0%
2				250.781	22.0	43.0	
3				250.781	21.0	43.0	
4				250.781	21.0	43.0	
4*				Salvage value	21.0		

Allowable total investment = 417.840
Depreciable investment = 167.136
IRR = 22.00%

Pro forma cash flow

Year	Income	Depreciation	Taxes	Credits	Net	Discounted net
0	−417.840	0.000	0.000	0.000	−417.840	−417.840
1	250.781	25.070	97.056	13.371	167.096	136.964
2	250.781	36.770	92.025	0.000	158.756	106.662
3	250.781	35.099	92.743	0.000	158.038	87.032
4	250.781	35.099	92.743	0.000	158.038	71.338
4*	35.099	0.000	0.000	0.000	35.099	15.843
Income totals						
	1038.223	132.038	374.567	13.371	677.026	417.840
Net totals						
	620.382	132.038	0.000	13.371	259.186	0.000

Nominal capital recovery = 151.679
Payback in approximately 2.6 yr

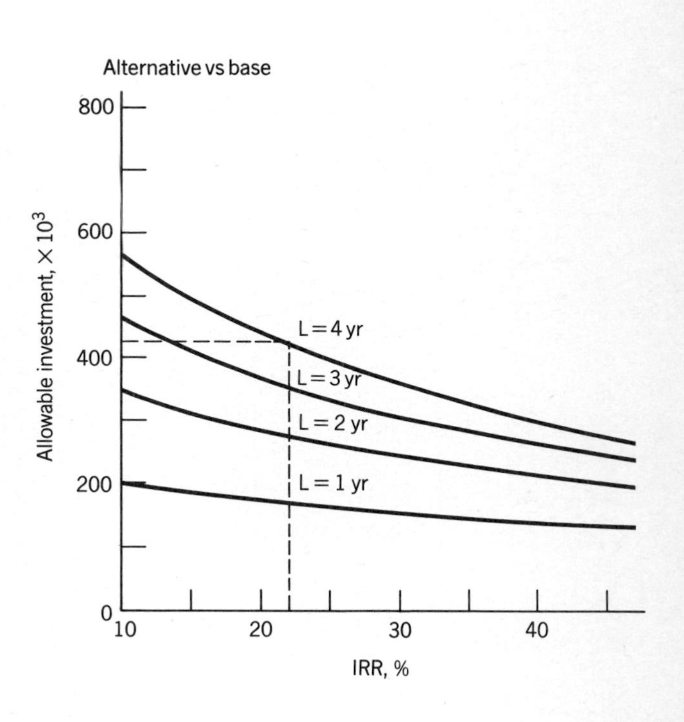

Figure 23. A 4-yr planning horizon is required if there is no tax benefit for the nondepreciable investment in year 0.

Alternative Versus Base

Zero present worth cash flow analysis 07–11–1985

5 yr economic life; salvage value 0% of cost;
0% constant rate of increase in costs

Year	Ratio	Expense forecast Tax rate	Depreciable	Savings	Income forecast Depreciation	Tax rate	Credit
0	100.00%	43.0%	40.0%				
1				250.781	15.0%	43.0%	8.0%
2				250.781	22.0	43.0	
2*				Salvage value	63.0		

Allowable total investment = 426.509
Depreciable investment = 170.604
IRR = 22.00%

Pro forma cash flow

Year	Income	Depreciation	Taxes	Credits	Net	Discounted net
0	−426.509	0.000	−110.039	0.000	−316.470	−316.470
1	250.781	25.591	96.832	13.648	167.597	137.375
2	250.781	37.533	91.697	0.000	159.084	106.883
2*	107.480	0.000	0.000	0.000	107.480	72.212
Income totals						
	609.042	63.123	188.529	13.648	434.162	316.470
Net totals						
	182.533	63.123	78.490	13.648	117.692	0.000

Nominal capital recovery = 164.917
Payback in approximately 1.9 yr
Inversion point IRR = 47.70%

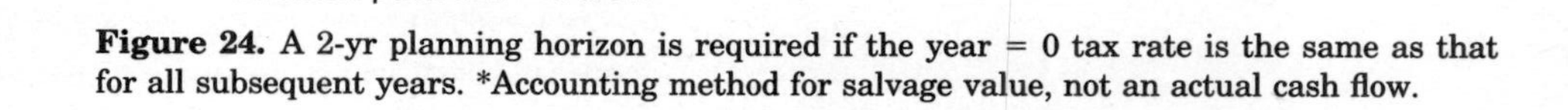

Figure 24. A 2-yr planning horizon is required if the year = 0 tax rate is the same as that for all subsequent years. *Accounting method for salvage value, not an actual cash flow.

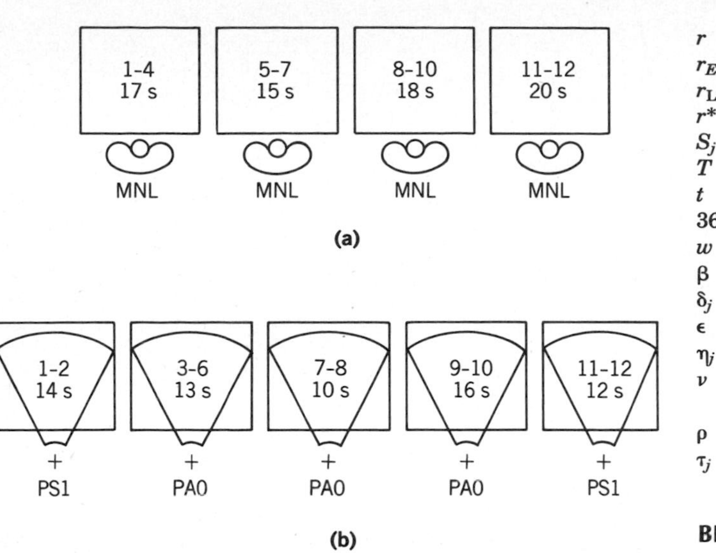

Figure 25. (a) The least cost manual system (base). 26.5 s usable cycle time; 87.5% system uptime; 2.26 theoretical units/min; 782,069 units actual capacity of system; $6930 total capital investment; rho factor = 1.50; 4.82 workers at $19.00/h required; $2.00/h system operating/maintenance rate; 0.959 yr required for 3-shift operation; 240 d required for 2.88-shift operation. Units, 750,000; $0.693 each; 1 year; cycle time, 23.2 s; 3 shifts; AF, 0.966. **(b)** The least-cost general system (alternative). 22.1 s usable cycle time; 85.0% system uptime; 2.71 theoretical units/min; 937,532 units actual capacity of this sytem; $575,000 total capital investment; rho factor = 2.50; 1.00 workers at $19.00/h required; $4.00/h system operating/maintenance rate; 0.800 yr required for 3-shift operation; 240 d required for 2.40-shift operation. Units, 750,000; $0.438 each; 1 year; cycle time, 18.8 s; 3 shifts; AF, 0.875.

can be used to establish that investments in robotics are justified.

Nomenclature

A	Subscript denoting alternative system parameters
B	Subscript denoting base system parameters
C_u	Unit cost to produce
C_{v_A}	Total variable costs for alternative system
C_{v_B}	Total variable costs for base system
DCF	Discounted cash flow
E	Equity
f_{AC}	Annualized cost factor
H	Years for cost accounting amortization of an investment
j	Subscript denoting year in the model(s)
L	Planning horizon years; time frame for a cash flow model
$\bar{L}_H$	Average loaded (or burdened) labor rate ($/h)
LTD	Long-term debt
MARR	Minimum attractive rate of return
N_j	Undiscounted net income in year j
O_H	Operating/maintenance rate ($/h) for the system being investigated
P	Total system cost (hardware, software, engineering installation, debug, etc)
Q_{co}	Crossover production volume; the point at which one system becomes more economical than another
Q_y	Yearly production quantity
Q_{y_j}	Production volume in year j
R_m	Risk attributable to general equity market
r	Minimum (or actual or internal) rate of return
r_E	Return on equity
r_{LTD}	Interest rate for long-term debt
r^*	Risk-free rate of return (probably U.S. Treasury Bill rate)
S_j	Expected savings in year j
T	Theoretical system cycle time (s/unit)
t	(Taxes paid)/(profit before taxes)
3600	Seconds/hour
w	Number of workers in the system being investigated
β	Systematic risk of a stock (eg, value line investment survey)
δ_j	Depreciation rate in year j
ϵ	Uptime (or efficiency) factor
η_j	Discounted net income in year j
ν	Remaining portion (book value) of an investment at the end of year H
ρ	(Total cost)/(hardware cost)
τ_j	Effective tax rate in year j

BIBLIOGRAPHY

1. Y. Hasegawa, "Evaluation and Economic Justification," in S. Y. Nof, ed., *Handbook of Industrial Robotics,* John Wiley & Sons, Inc., New York, 1985.
2. J. R. Canada and R. L. Edwards, *Should We Automate Now?* North Carolina State University, Raleigh, N.C., 1985.
3. J. R. Canada and J. A. White, *Capital Investment Decision Analysis for Management and Engineering,* Prentice-Hall, Inc., Englewood Cliffs, N.J., 1980.
4. R. P. Lutz, "Discounted Cash Flow Techniques," in G. Salvendy, ed., *Handbook of Industrial Engineering,* John Wiley & Sons, Inc., New York, 1982.
5. R. E. Gustavson, "Choosing a Manufacturing System Based on Unit Cost," *Proceedings of the 13th ISIR/Robots 7 Conference,* Chicago, Ill., Apr. 1983.
6. G. Boothroyd, C. Poli, and L. Muech, *Automatic Assembly,* Marcel Dekker, New York, 1982.
7. R. E. Gustavson, "Engineering Economics Applied to Investments in Automation," *Proceedings of the 2nd Assembly Automation Conference,* Brighton, U.K., May 1981.
8. A. A. B. Pritsker, *The GASPIV Simulation Language,* John Wiley & Sons, Inc., New York, 1974.
9. T. J. Schriber, *Simulation Using GPSS,* John Wiley & Sons, Inc., New York, 1974.
10. S. C. Graves and D. E. Whitney, "A Mathematical Programming Procedure for Equipment Selection and System Evaluation in Programmable Assembly," *Proceedings of the 1979 IEEE Decision & Control Conference,* Ft. Lauderdale, Fla., Dec. 1979.
11. R. E. Gustavson, "Computer Aided Synthesis of Least Cost Assembly Systems," *Proceedings of the 14th ISIR,* Gothenburg, Sweden, Oct. 1984.

General References

H. G. Thuesen, W. J. Fabrycky, and G. J. Thuesen, *Engineering Economy,* 5th ed., Prentice-Hall, Inc., Englewood Cliffs, N.J., 1977.

E. P. Degarmo, W. G. Sullivan, and J. R. Canada, *Engineering Economy,* 7th ed., Macmillan and Co., New York, 1984.

E. Grant, W. Ireson, and R. S. Leavenworth, *Principles of Engineering Economy,* 7th ed., John Wiley & Sons, Inc., New York, 1984.

J. A. White, M. H. Agee, and K. E. Case, *Principles of Engineering Economic Analysis,* 2nd ed., John Wiley & Sons, Inc., New York, 1984.

Increase in Production Example

Zero present worth cash flow analysis 07–11–1985

5 yr economic life; salvage value 0% of cost;
0% constant rate of increase in costs

Year	Ratio	Expense forecast Tax rate	Depreciable	Savings	Income forecast Depreciation	Tax rate	Credit
0	73.90%	43.0%	40.0%				
1	26.1	43.0	40.0	250.781	15.0%	43.0%	8.0%
2				410.744	22.0	43.0	8.0
2*				Salvage value	63.0		

Allowable total investment = 598.749
Depreciable investment = 239.499
IRR = 22.00%

Pro forma cash flow

Year	Income	Depreciation	Taxes	Credits	Net	Discounted net
0	−442.475	0.000	−114.159	0.000	−328.317	−328.317
1	−156.273	0.000	−40.319	0.000	−115.955	−95.045
	250.781	250.781	96.420	14.159	168.520	138.131
2	410.744	48.314	155.845	5.001	259.900	174.617
2*	164.637	0.000	0.000	0.000	164.637	110.613
Income totals						
	826.162	74.863	252.265	19.160	593.057	423.362
Net totals						
	227.413	74.863	97.787	19.160	148.785	−0.000

Nominal capital recovery = 228.912
Payback in approximately 2.1 yr
Inversion point IRR = 61.00%

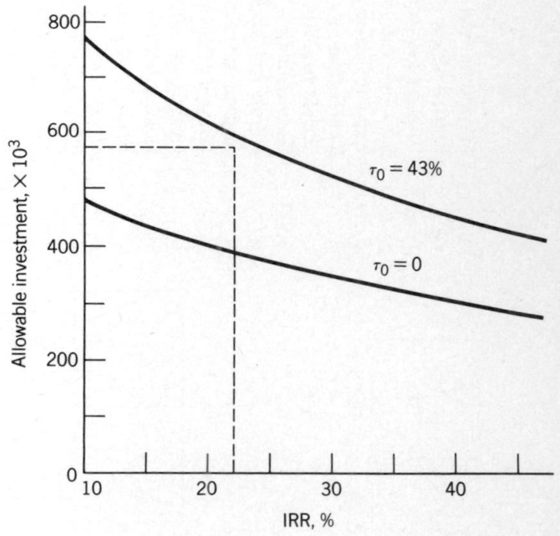

Allowable investment, × 10^3

$\tau_0 = 43\%$

$\tau_0 = 0$

IRR, %

Figure 26. Increase in production example. Allowance of year 0 taxes requires only a 2-yr planning horizon. *Accounting method for salvage value, not an actual cash flow.

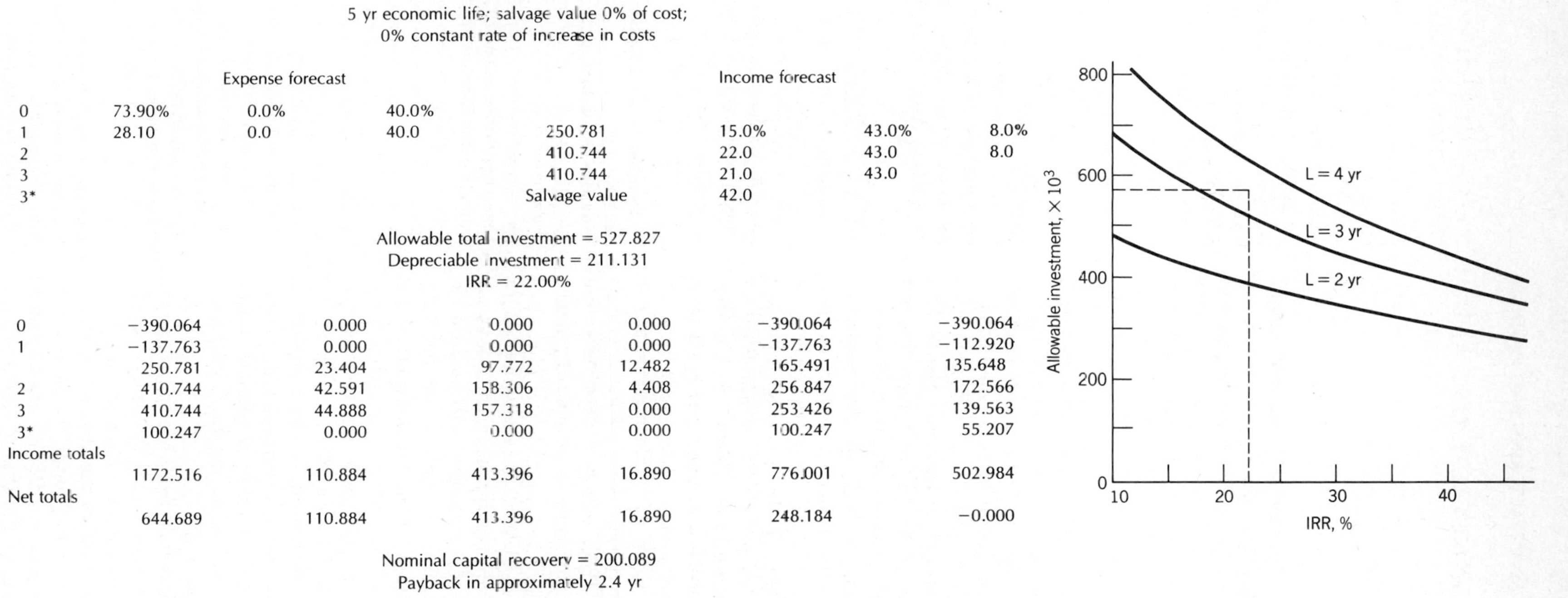

Increase in Product on Example

Zero present worth cash flow analysis 07–11–1985

5 yr economic life; salvage value 0% of cost;
0% constant rate of increase in costs

		Expense forecast			Income forecast		
0	73.90%	0.0%	40.0%				
1	28.10	0.0	40.0	250.781	15.0%	43.0%	8.0%
2				410.744	22.0	43.0	8.0
3				410.744	21.0	43.0	
3*				Salvage value	42.0		

Allowable total investment = 527.827
Depreciable investment = 211.131
IRR = 22.00%

0	−390.064	0.000	0.000	0.000	−390.064	−390.064
1	−137.763	0.000	0.000	0.000	−137.763	−112.920
	250.781	23.404	97.772	12.482	165.491	135.648
2	410.744	42.591	158.306	4.408	256.847	172.566
3	410.744	44.888	157.318	0.000	253.426	139.563
3*	100.247	0.000	0.000	0.000	100.247	55.207
Income totals						
	1172.516	110.884	413.396	16.890	776.001	502.984
Net totals						
	644.689	110.884	413.396	16.890	248.184	−0.000

Nominal capital recovery = 200.089
Payback in approximately 2.4 yr

Figure 27. Increase in production example. A 3-yr planning period is required when there is no tax benefit in year 0.

ECONOMICS, ROBOTIC MANUFACTURING AND PRODUCT

PHILLIP F. OSTWALD
University of Colorado at Boulder
Boulder, Colorado

INTRODUCTION

Do robots influence the cost of products? If so, how much? Are robots and their associated software and equipment replacing labor or merely allowing the skill level of operator to be substituted by that of attendant, with lower wages? What is the economic effect of robots upon material, equipment, and tools? These are the questions addressed in this article (*Market and industry;* COST/BENEFIT). There is no exclusive method that will justify robots for adoption by industry. Economic justification is partially handled by investment strategies such as return on investment, payback, or time value of money methods. Those methods are copiously documented elsewhere (1). But justification is often hampered by these methods (2). Also, many firms do not consider robotic-assisted production because of the difficulty in estimating the savings (3). This article covers the ramifications of the product upon the unit cost, selection of the most likely process, and identification of important cost drivers. Unit cost methods are also considered elsewhere (1,4), and a different emphasis is expected. In some cases, mathematical models of the unit cost are constructed (1,4), but these models are for some fictitious or generic product that is not identified. Robots are difficult to comprehend for economic justification in multimachine systems (5). Unfortunately, short-term payback is sometimes used as the justifying factor, rather than a comprehensive product study for long-term impact. Most models are for stand-alone situations in which the costs, design, and production factors are more clearly identified. These first-order effects are the ones that payback, discounted cash flow, and time value of money principles usually consider.

A specific product may be associated with robotic-assisted production. Once that is determined, quantity variations are allowed. It remains to specify a system that will produce the product. It is incumbent upon the economic model and database to be realistic and accepted by the business firms who would fabricate these products. Also, the cost-effectiveness of robotic-assisted production is a wider problem than the cost of the new system versus another older system, as it considers the choice between a manual, dedicated, or other production system. A production system will have a corresponding performance, which can be estimated. If units of product output are necessary, units of labor, material, capital, and tooling are the grist for the simulation mill.

Manufacturing converts materials from one set of properties to another. The transformation makes the product useful and attractive to the buyer. Market stimulus, specifications, technology, and a host of factors affect the product. Manufacturing in the United States is operated for profit and is dependent upon the success of the product. Robots are important in material conversion, but no more important than machines, tools, factory space, and milieu of manufacturing. Before equipment is economically favored, reductions in labor cost, savings in materials, energy, and so on, and improvements in quality must be apparent.

CLASSIFICATION OF MANUFACTURING AND PRODUCTION SYSTEMS

Manufacturing of durable goods can be broadly classified as mass, moderate, or job lot production (6). In mass production, sales volume is established, and production rates are independent of individual orders. In moderate production, parts are produced in variable quantities and perhaps irregularly over the year. Output is more dependent on single sales orders. Batches may be produced once or periodically. Job lot industries are the most flexible, and their output is directly connected to the customer order. Most often, the job lot industry will not build the product unless the order from a customer is assured.

Although exceptions to this classification can be cited, the manufacturer frequently builds for inventory and may be unaware of the customer in mass production. Mass production examples include automated transfer machines for discrete products, partial and fully automated assembly lines, industrial robots for spot welding, machine loading, and spray painting, automated materials-handling systems, and computer production monitoring. Examples in moderate production in-

Table 1. Characteristics of Production Systems

Workstation Individual bench, machine, or discrete process. Handcraft work, single stations, or numerical control machine tools working independently. This is the most primitive production system. The workstation considered here almost always has labor involvement. Movement between workstations by separate material-moving devices or indirect labor.

Cell Simplest organized production effort. May or may not be computer-controlled or robot-assisted. It is composed of two or more workstations with a higher level of coordinated effort between workstations.

Flexible workstation Volume of production has less effect upon cost, and the use of the computer is characteristic. Reduction of changeover cost between lots or products is a premise of this system. One objective is the reduction or elimination of setup costs.

Programmable automation Emphasis is given to hard- or software programming and convertibility of the equipment is stressed between product models and parts or products. Examples include transfer lines that are computer-controlled. As an example, these lines are able to manufacture a range of slightly different engine blocks in large to small quantities.

Mechanization Dedicated production of large quantities of one product with little model variation. Computers and robots may not have an essential role. Transfer lines having tools and pneumatic, electric, and electronic controls are important. Single or multiple machines that are special, that is, designed with one product in mind, are common. Usually, the integration of production is limited to one special-purpose machine tool or assembly line. Mechanization was developed before computers and robots. Examples include high-volume consumer products such as the popular 12-ounce beverage container, plastic shavers, and many high-volume automobile products.

Continuous flow process Examples include production of bulk product, such as in chemical plants and oil refineries. Features are flow process from beginning to end, sensor technology to measure important process variables, use of sophisticated control and optimization strategies, and full computer control. This technology was developed before the advent of the idealized robot. Consideration of continuous flow technology is avoided. Products that are result of this technology are not considered "durable."

clude books, clothing, and industrial machinery. Job lot industries may provide tools and dies or special equipment, or they may serve as suppliers to other industries. Most often, the job lot industries do not have their own proprietary product.

Production can be classified according to another conceptual model. The meaning of a production system is more limiting than that of a manufacturing system. A production system implies converting equipment such as machines, processes, benches, and tools. Production systems are classified into six categories:

1. Workstation.
2. Cell.
3. Flexible workstation.
4. Programmable automation.
5. Mechanization.
6. Continuous flow process.

These systems are discussed in Table 1.

Figure 1 is a sketch of product diversity, volume, type of industry, and production system. Notice that zones are shared between production systems. Demarcation in terms of quantity and class of production system is probably undefinable, except in specific product cases.

These definitions interlock with the economic evaluation of robots or any other component of the manufacturing system. Admittedly, a manufacturer would not consider these definitions explicitly in any economic analysis. Yet manufacturers are not all things to all products. Some are known for a large volume of products and others are capable at producing small quantities for a specific product. Generalization in manufacturing is not the norm. Specialization is.

"Product" and "producibility" are connected. Economic relationships are difficult to separate. An economic model for robots, or any production system component, cannot be formulated in isolation from the product, market factors, and production. Equally valid are assertions about pieces of the production system, such as robots, that cannot be generalized in any production system. The product, volume, methods of production, and market are related, and the cause and effect of the factors are often obscure. Product, design, quantity, production system, material, tolerance, and so on are primary influences upon cost. In turn, it is the financial health of products that drives the business to seek productive methods.

PRODUCT CONSIDERATIONS

The product, a pressure cover for a ratio transducer, is partially described by Figure 2. The raw material for this product is a magnesium alloy, specification QQM-38. Military specifications such as MIL STD M3174, 7742, and 6021 apply. These standards pertain to surface treatment, condition of drilled, tapped, and other holes, and inspection for cast products. The part weighs about 0.44 lb (02. kg) and has a surface area of 102 in.^{2} (0.066 m^2). The design is similar to a box with five closed and one open surface. Two feet allow for fastening to its higher-level assembly. A cored hole 2.85 in. (72.4 mm) in diameter is a prominent feature.

The one-time-only quantity for this product is ranged from job lot to mass production. This ranging is not an unreasonable assumption. However, the product is not a consumer item and volumes do not approach those levels.

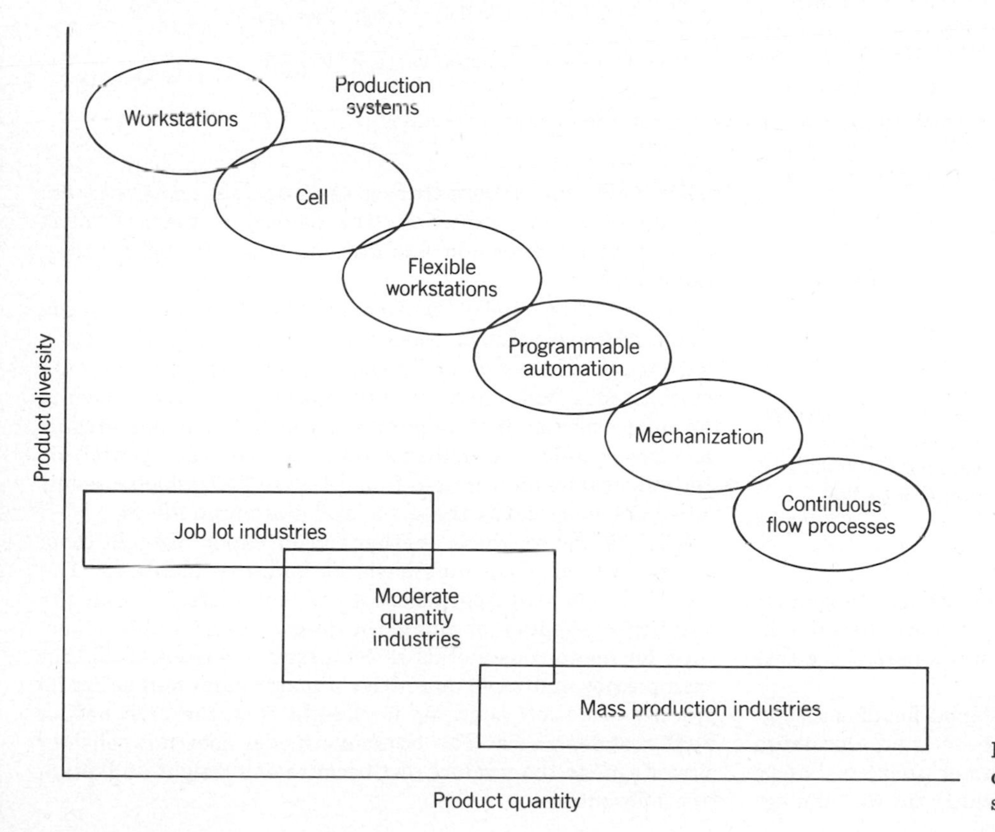

Figure 1. Interaction of product quantity, diversity, type of industry, and production system.

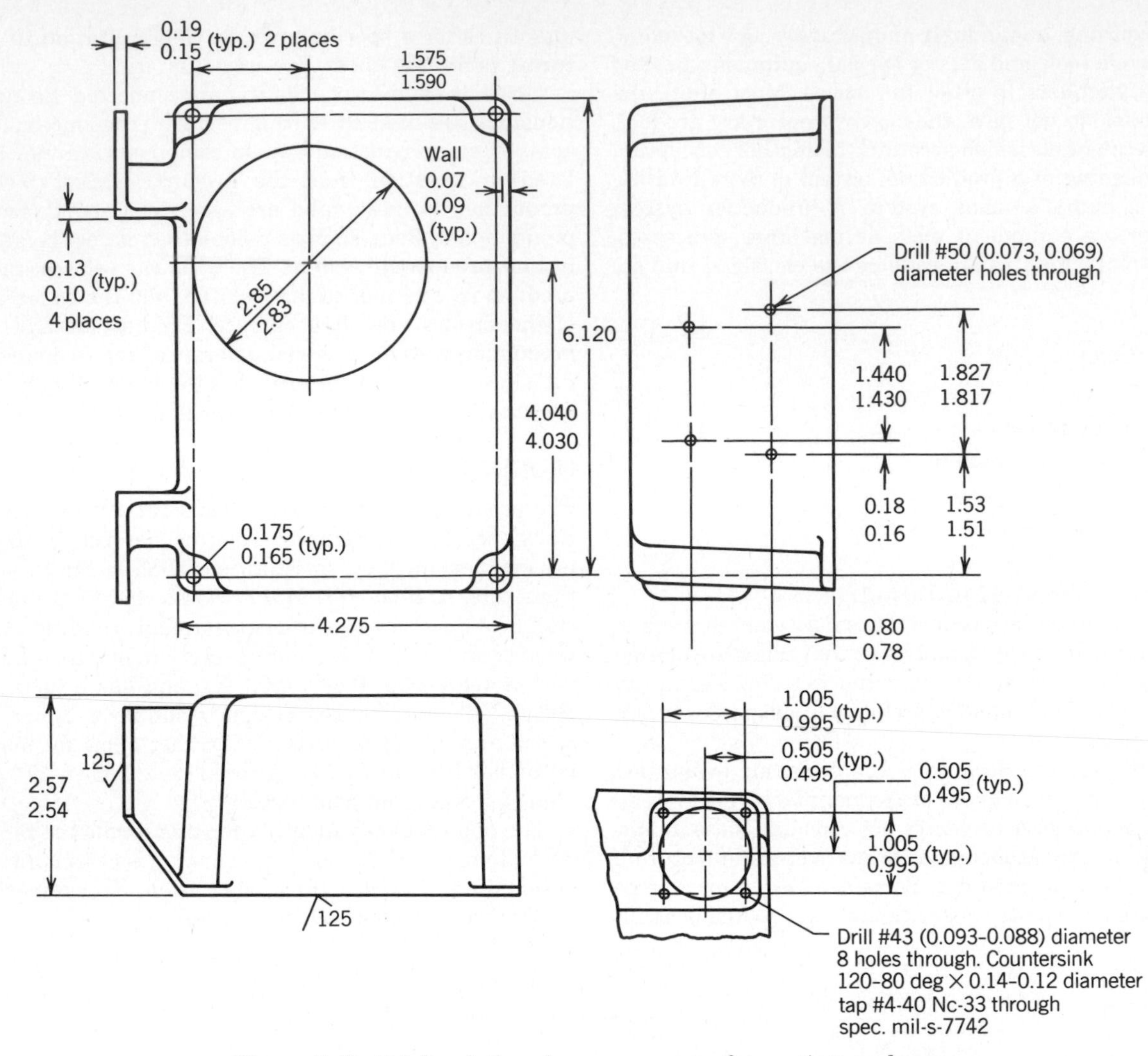

Figure 2. Partial description of a pressure cover for a ratio transducer.

Process Operation Number	Table Number	Process Description
10	2.9	Die cast
20	20.3	Treat with magnesium coat (nearest method)
30	18.2	Deburr flashing
40	9.4, 11.2	Drill holes, tap, ream, etc
50	18.1	Deburr holes
60	20.5	Chrome pickle, Milm3174 (nearest method)
70	22.2	Inspect per Milstd6021—class II B casting

MATERIAL CONSIDERATIONS

An analysis of manufacturing equipment within the context of a product must include considerations of raw material. Unlike in a service economy, material becomes a part of the cost of that product.

Magnesium is not the customary material for die casting. Most suppliers of die cast products use zinc and aluminum alloys, which are more common for consumer products. Safety (magnesium chips are a potential fire hazard), the cost of magnesium pig, and manufacturing and physical properties are considerations in production. The density of magnesium is attractive for lightweight applications (it is 33% lighter than aluminum).

A simulation model invokes a range of quantity for the product. Higher volumes suggest lower unit costs for the magnesium pig. The economic model considers this price break, which ranges from minimum purchase to base rate, which is the point where additional purchasing quantity does not result in a lower unit raw material cost. The cost per pound for this magnesium alloy ranges from $1.29 to 1.37, which is about 260–225% more costly than zinc and aluminum alloys.

The cost for freight is another consideration. Freight costs can be less on a per-unit basis as quantity increases. The cost of freight (transportation of the raw material from the warehouse, supplier, or mill to the die caster) is handled differently for nonferrous metals as compared to ferrous stock. For example, for midrange quantities a magnesium mill shifts its practice of no-cost charging for freight from the FOB mill to FOB customer dock. The economic model does not consider advantages to the product cost because of a freight and quantity differential.

COST RECOVERY POLICIES

The decision to put out a product sets several cost policies in motion. Production at minimum cost is a major objective, albeit minimum cost is an illusory goal. With the product design details in hand and a marketing sales forecast, several manufacturing plans are prepared. If the quantity is low and the firm is job shop, no changes to existing floor space layouts or purchases of new equipment are required. This minimizes preproduction expense. Dies are necessary because die casting equipment requires tooling. As the quantity increases, it is to the die caster's advantage to consider new equipment, tooling, training, and advanced technology for the product. A job shop is interested in advanced technology, but it is the profile of the general business demand and not a specific product that guides these decisions.

Five quantity ranges are evaluated for this product and each production method. Five distinctive production plans are developed. These plans use similar equipment and methods. Of course, the production plans contain common elements, and as quantities increase, the influences of Table 1 are introduced. The processing plans are identified as workstation, cell, flexible workstations, programmable automation, and mechanization.

The term "productive unit cost" recognizes four major cost components: direct labor, direct material, capital, and nonrecurring initial fixed costs. The practice of finding the productive unit cost avoids the accounting application of overhead. A practical problem of overhead accounting is in the allocation of indirect expenses for firms that have multiple products or that share management and factory services. One essential purpose of overhead is cost recovery for pricing products, but its value for equipment selection can lead to dubious choices. Instead, productive unit cost is used to identify the most effective production system. Overhead often obscures equipment selection.

Direct Labor Cost Recovery

The variable costs of production are defined unambiguously as direct labor and direct material. If the product is not made, these costs do not exist. If the product is made, these costs are measured in terms of product quantity. Direct labor has come to mean that labor that operates, watches, controls, ensures, reacts with, and otherwise works with producing equipment. The output of the labor effort can be measured in terms of product quantity. A deburring operator performs direct labor since the effort is attached immediately to the product. A material mover, such as a trucker, crane bay operator, dispatcher, or other, performs indirect labor. The indirect labor cost effort is distributed to a variety of products, and the cost is allocated, so to speak, through overhead practices. The economic model does not consider indirect costs.

Direct Material Cost Recovery

The output weight and the purchase cost of the raw material are primary. Cost additions to this primary cost are scrap, waste, and shrinkage. The runner, sprue, and reject parts are mostly reusable, and the material yield is typically 90–97% for die cast products. This material loss can increase the value-added cost to the product by about 1%.

Capital Cost Recovery

Transformation of raw material into product requires machine tools, ancillary equipment, direct costs, and factory support. The cost model includes a capital portion for equipment. The term capital implies depreciation. This assumes that the equipment cost will eventually be recovered by the price of the product, which includes a portion of the capital. Recovery necessitates time and a variety of products to redeem the full capital that purchased the equipment. Recovery uses the concept that depreciation is over a ten-year life. The cost model presupposes a one-shift operation of 2000 h, and once the lot hours are estimated for the die cast product, they are multiplied by the per-hour depreciation cost for that production system.

Nonrecurring Initial Fixed Cost Recovery

A pick-and-place robot is capitalized equipment able to handle a variety of products. Each product will require a special gripper. This gripper is referred to as a nonrecurring initial fixed cost (6). Nonrecurring initial fixed cost entities are required, especially if general-purpose machine tools are employed. Sometimes the term "tooling" is used to mean nonrecurring initial fixed cost.

The die cast part will require grippers, dies, special tools, fixtures, testing gauges, numerical control tape, and perishable tools such as drills, taps, and deburring tools. Each operation is estimated for these nonrecurring initial fixed costs. Their impact, on the price of the product can be significant.

Lower quantities and manual methods usually dictate a single-cavity mold. As quantities increase, a two-cavity die is used. With a single-cavity mold, each opening and closing cycle will cast one unit. Concurrently, as quantities and die cavities per mold increase, a larger die casting machine is required. Die casting machines are rated in tons, meaning tons of force for the knuckle linkage bringing the die halves together. Die life is assumed to be a maximum of 100,000 units.

In a general overhead allocation scheme, the nonrecurring initial fixed costs are spread to all products despite their quantity, design, or production system. This practice confuses the cost of the product, which may be over- or understated in real dollars. The full initial fixed cost should be assigned to the product. This economic model amortizes the tooling cost by the quantity (6).

COST ESTIMATING PREMISES

The die cast product is estimated using the *AM* (American Machinist) *Cost Estimator* (6), which provides information specifically for direct labor costs and software (7) that organizes the database. A process plan is necessary before estimation. Seven operations are assumed:

10 Die cast.
20 Trim flashing and remove sprue and runner.
30 Treat with magnesium coating.

40 Drill holes, tap, countersink.
50 Deburr holes.
60 Chrome pickle.
70 Inspect.

Concurrent to the estimation step, it is necessary to assume quantities, production equipment, material handling equipment, labor grades and skills, and the elements of work. References 6 and 7 provide the database and encourage consistency in estimating, which is necessary to reach conclusions regarding the economic status of the equipment, process, and product. The *AM Cost Estimator* provides for the selection of the work elements. Once the operation estimate is concluded, the results are setup hours and cycle minutes. Eventually, lot hours are found for the assumed quantity (3,6,7).

In planning for manufacturing, it is necessary to outline the requirements that must be met. Part geometry for easier automatic handling, holding, loading, and unloading and special or commercially available feeders, selectors, loaders, unloaders, transfer devices, indexing equipment, vision systems, and dies are assumptions in the economic analysis. The manufacturing specifications for these devices were detailed only to meet the needs of the estimating program.

In the economic evaluation step, a number of strategies are possible. Consider the situation where there is minimum product variety with large quantity. Automation may be a choice, but full automation may be uneconomical. Instead, partial automation, with the addition of automatic devices until full automation or the point of diminishing economic returns has been reached, is preferred. It is possible to describe levels of automation that range from an automatic fixture at a single machine to full automation. Part variety with low volume may be the dominant constraint, and flexible workstations may be the choice. These varied considerations call for cost estimation to determine a preferred plan.

An example of a nonrecurring cost is the die. The cast product cannot be made unless a die is first produced. Capitalization is analogous to depreciation, and a ten-year term is assumed for the capitalized equipment. This is customary practice within the die casting business.

When calculating the productive cost per unit, there are cost entities that are overlooked. Tax effects, corporate transfer expenses, and overhead terms not specific to the equipment or neutral to the analysis are ignored. Supervision costs, floor space, heating, lighting, power, lubricants, indirect labor such as for material moving and maintenance, and other indirect costs are overlooked. These costs vary with area and company experience. Whether these overlooked cost items are significant to the outcome of the analysis is conjectural. If the analysis recommends a choice of a production system given a product, and if these costs are similar for all systems, their effect on the outcome is neutral. If the purpose of the economic–product model is to estimate absolute cost for pricing purposes, their omission is a serious oversight. Some authors contend (5) that other factors in multimachine justification, such as work-in-process, ergonomics, quality improvements, product, scrap reduction, product recidivism, and material-handling distances that may be favorable to the introduction of robots, are significant but specifics are not provided. The focus here is the evaluation of production systems and their components. Omission of these cost components is not harmful to the analysis.

Other approaches to finding unit cost dwell on fixed and variable cost distinctions (1,4). In references 1 and 4, costs are assumed to be divisible and traceable, and the choice is unambiguous. This position may not be valid, as production systems are often faced with mixed costs, that is, those that have components of both and are not clearly separable. Cost behavior is nonlinear and contains elements of variable, fixed, and semifixed costs. Case-by-case specifics cause the proportions to vary even for the same product. Separation can cause by-product problems in unit cost determination. This approach does not depend upon the separation of fixed and variable costs.

The integration of the robot into individual manufacturing processes requires systems engineering. Although robots are available off the shelf, a robotic system is not. It must be carefully planned and customized to each application. In die casting, the robot represents about 70% of the total robotic component cost. Engineering and peripheral equipment might comprise 50% of the total system cost. In fairness to the die casting industry, robotic types of operation have been in place since the early 1950s. A die casting machine is hazardous, labor intensive, and tedious (8).

Five different production systems were estimated. Space permits only Table 2, which is an estimate of the workstation concept. The quantity for this system is 500. Higher levels of operator skills are usually involved with lower quantities. Some operations for the other production systems are also given. Those operations are selected whenever a drastic change in the method, equipment, tooling, or operator status is made. The workstation estimate may be viewed as the control or standard as compared against the four other methods.

The seven operations, lot hours, unit labor and material cost, and the total operational cost are listed (Table 3). Labor rates are provided. Chicago is the base site and 1986 the year (6).

Table 4 is an example of a die-cast machine with a robot assist. This is the first operation, or operation 10. Despite the improvements that may be introduced into the process, the critical piece of equipment is the die-casting machine. The output of the cold chamber die-cast machine is a casting with runners and sprue, or waste, still attached to the part. In this production method, called programmable automation, the part is removed by a pick-and-place robot, and the die is sprayed for lubrication as the robot arm retracts. Vision equipment verifies that all material is removed from the die. The robot arm rotates 90 deg and presents the casting to the shuttle mechanism. Quenching or air cooling normally follows. It then returns to its ready position to await the nest opening of the die-casting machine. Operation 10 reduces the cycle time of 0.83 min, as found in Table 2 for the workstation concept, to 0.38 min. All of the reduction in cycle time cannot be identified with the robot, however. Midway in the production system switch, a two-cavity die was introduced since the volume now allows for improved tools. The reduction in handling time offsets the increased cost of two cavities over the cost of one. Concurrently, the total setup time increases from 6 to 11 h. The assumptions for the estimates are given internally to the

Table 2. Assumption of Workstation Production System

Lot Quantity, 500 — Material, Magnesium Pig, $1.37/lb; Unit Material Cost, 0.60

Process Operation Number	Table Number	Process Description	Table Time	Adjustment Factor	Cycle, Min	Setup, H
10 Die-Cast machine	2.9	Magnesium die cast				
	2.9.S	Set up for about 102 in.^{2}, 1 part per die, setup does not include robot	6.00			6.00
	2.9.1	Open and close 200-ton press	0.04		0.04	
	2.9.2	Trip ladle metal	0.16		0.16	
	2.9.3	Ladle metal, metal weighs 0.5 lb with runner, sprue	0.06		0.06	
	2.9.4	Shot and dwell	0.17		0.17	
	2.9.5	Guard door, manual	0.10		0.10	
	2.9.6	Eject for part drop	0.04		0.04	
	2.9.6	Remove part	0.09		0.09	
	2.9.7	Clean, lubricate	0.17		0.17	
Total lot hours, 12.92					*0.83*	*6.00*
20 Manual debur bench	18.2	Deburr flash with abrasive belt				
	18.2.S	Set up	0.10			0.10
	18.2.1	Start and stop for the entire lot	0.04	/500	0.00	
	18.2.3C	Pickup, position	0.05		0.05	
	2.9.9	Degate runner, sprue	0.13		0.13	
	18.2.4	Reposition for 3 edges	0.02	3	0.06	
	18.2.5C	Vertical belt for about 13 in.	0.29		0.29	
	18.2.5C	Vertical belt for additional 8 in. of flashing	0.02	8	0.14	
	18.2.7	Blow off	0.25		0.25	
Total lot hours 7.80					*0.92*	*0.10*
30 Magnesium coat line	20.3	Treat with magnesium coat				
	20.3.S	Set up	0.00			0.00
	20.3.3	Bright dip as box of $L + W + H = 12.97$ in. Bright dip as approximate	0.39		0.39	
Total lot hours 3.25					*0.39*	*0.00*
40 Numerical control (NC) drill press	9.4	NC drill press operation				
	11.2.7	Power reaming 2.85-in. hole cleanup, 0.09 in. metal	0.10	0.09	0.01	
	11.2.4	Power drilling 0.19 in.-wall, 1 hole, but for 8-hole total	0.07	0.19	0.01	
			0.01	7	0.09	
	11.2.3	Countersinking	0.03	8	0.24	
	11.2.8	Tapping 40 NC thread for 8 holes	0.12	0.19	0.02	
			0.02	7	0.16	
	11.2.4	Drill 0.175 hole 2.57 in. deep, 4 body holes	0.07	2.57	0.18	
			0.18	3	0.54	
	11.2.4	Drill #50 holes, 4 times, through wall 0.09 in.	0.07	0.09	0.01	
			0.01	3	0.02	
	9.4.1A	Pick up, move	0.08		0.08	
	9.4.2	Clamp, unclamp	0.05		0.05	
	9.4.3A	Manipulation for 3 surfaces by flipping tool on table top	0.01	3	0.03	
	9.4.4A	Machine operation. Start, stop for flip.	0.02	3	0.06	

Table 2. (*continued*)

Lot Quantity, 500			Material, Magnesium Pig, $1.37/lb Unit Material Cost, 0.60			
Process Operation Number	Table Number	Process Description	Table Time	Adjustment Factor	Cycle, Min	Setup, H
	9.4.4B	Position table for 32 holes	1.05		1.05	
	9.4.4B		0.06	13	0.78	
	9.4.4C	Turret index for 6 tools	0.54		0.54	
	9.4.5A	Clean area, medium	0.11		0.11	
	9.4.5A	Clean, lubricate	0.05		0.05	
	9.4.5B	Blow from 8 tapholes	0.21		0.21	
	9.1.S2	Turret machines	0.62			0.62
	9.1.S5	Miscellaneous	0.03			0.03
	9.1.S5	Miscellaneous	0.00	3		0.01
	9.1.S5	Miscellaneous	0.20			0.20
Total lot hours 36.20					*4.24*	*0.86*
50 Drill press deburr holes hand bench	18.1	Drill press deburr holes that are accesible, hand deburr inside difficult ones				
	18.1.S	Set up	0.05			0.05
	18.1.1B	Box-like parts handling	0.07		0.07	
	18.1.1C	Pick up and drop	0.05		0.05	
	18.1.3A	Deburr the big hole	0.03		0.03	
	18.1.3A	Deburr 4 body holes	0.13		0.13	
	18.5.1	Handling, repositioning for hand-deburring of difficult-to-deburr holes	0.07		0.07	
	18.5.2	Tool handling	0.03		0.03	
	18.5.3A	Deburr holes by hand, 8 holes under tab and 4 inside holes	0.59		0.59	
Total lot hours 8.13					*0.97*	*0.05*
60 Chrome pickle	20.5	Chrome pickle, mil Std M 3174				
	20.5.S	Set up	0.05			0.05
	20.5.S	Set up	1.00			1.00
	20.5.1A	String up	0.17		0.17	
	20.5.4A	Electroplating	1.06		1.06	
Total lot hours 11.30					*1.23*	*1.05*
70 Inspect station	22.2	Inspection per mil std 6021 class C casting with sample plans, not 100% checked				
	22.2.S	Set up	0.05			0.05
	22.2.1	Handle	0.10		0.10	
	22.2.3	2.85-μm hole every 2 parts	0.50	0.5	0.25	
	22.2.2	Miscellaneous	0.10	/500	0.00	
	22.2.10	Thread gauging for sample size of 25, but prorated over lot of 500 parts	22.00	/500	0.04	
Total lot hours 3.34					*0.39*	*0.05*

operation. The cost summary for programmable automation assumes 100,000 units. Table 5 shows the economic consequences of a different approach to production. Labor and material costs per unit are provided.

The workstation concept used an NC turret drill press with a table-mounted tumble jig for drilling, countersinking, reaming, and tapping. This operation was revised for the flexible workstation concept. A special drill press with a pick-and-place robot was simulated. Cycle time declines from 4.24 to 0.44 min. Other improvements included a special machine tool hav-

Table 3. Cost Summary, Work Station

Operation Number	Cost Estimator Table Number	Machine, Process, or Bench Description	Lot Hours	Labor Hour Cost, $	Total Operation Cost, $
10	2.9	Hot and cold die cast	12.92	10.82	139.76
20	18.2	Abrasive belt deburr	7.80	10.58	82.53
30	20.3	Chemical clean	3.25	10.21	33.18
40	9.4	Turret drilling	36.20	12.38	448.20
50	18.1	Drill press machine	8.13	7.26	59.05
60	20.5	Electroplating	11.30	10.62	120.01
70	22.2	Table and machine inspection	3.34	12.04	40.15
		Total lot hours	82.94		
		Total operational labor hour cost, $			922.87
		Unit operational labor hour cost, $			1.85
		Unit material cost, $			0.60
		Total direct cost per unit, $			2.45
		Total job cost, $			1224.37

ing a rotary-top table. Side drilling heads with cluster drill units were specified. Tables 6 and 7 provides the details of the operation.

LABOR AND MACHINE CYCLE INTERPRETATION

Figure 3 shows the relationship of direct labor to machine cycle for the quantities and the technology. Direct labor is defined, for the purposes of Figure 3, as those elements of operational work such as "touch," involving actuation, part loading and unloading, and motor and muscle effort. Machine cycle is defined as "at-rest" labor where the operator is watching or observing, or otherwise "attending." The machine process is the controlling factor for the selection. The labor is dejointed in this analysis; that is, if there are multiple machines, two more cavities per die, and simultaneous machine elements, the worker-to-machine cycle is adjusted. The output factor for the cycle is shown by the AM Cost Estimator, Table 3, which gives the dejointed cycle effort. Notice the Adjustment Factor column where division by two suggests multiple part operation. For example, the operator may be unloading two cavities as if the part were a single unit. Division by two re-

Table 4. Assumption of Programmable Automation Production System

Lot Quantity, 100,000

Material, Magnesium Pig, $1.29/lb
Unit Material Cost, 0.57

Process Operation Number	Table Number	Process Description	Table Time	Adjustment Factor	Cycle, Min	Setup, H
10	2.9	Magnesium die cast				
Die-cast machine with robot assist	2.9.S	Set up for about 102 in.^{2}, 2 parts per die	8.00			8.00
	21.3.S	Set up with robot. Die caster has a mechanism that reaches in and removes part.	3.00			3.00
	2.9.1	Open and close 600-ton press	0.12	/2	0.06	
	2.9.2	Trip ladle metal	0.16	/2	0.08	
	2.9.3	Ladle metal, metal weighs 1.2 lb	0.10	/2	0.05	
	2.9.4	Shot and dwell	0.17	/2	0.09	
	2.9.5	Guard door, automatic	0.00		0.00	
	21.3.1	Approach–retract 500 in. lb moment arm to insert to-die-cast part in die cavity for close control of part gripper	0.05	/2	0.03	
	2.9.6	Eject machine sense	0.02	/2	0.01	
	2.9.7	Clean, lubricate with robot spray	0.05	/2	0.03	
	2.9.6	Eject for part drop	0.09	/2	0.05	
Total lot hours, 646.00					*0.38*	*11.00*

Table 5. Cost Summary, Programmable Automation

Operation Number	Cost Estimator Table Number	Machine, Process, or Bench Description	Lot Hours	Productive Hour Cost, $	Total Operation Cost, $
10	2.9	Hot and cold die casting	646.00	10.82	6989.72
20	3.3	Punch Press *M*2nd	147.32	10.24	1508.52
30	20.3	Chemical clean	650.00	10.21	6636.50
40	9.5	Cluster drilling	1254.39	12.38	15,529.35
50	18.1	Drill press machine	451.27	7.26	3276.20
60	20.5	Electroplating	2051.05	10.62	21,782.15
70	22.2	Table and machine	657.05	12.04	7910.88
		Total lot hours	5857.07		
		Total operational productive hour cost, $			63,633.32
		Unit operational productive hour cost, $			0.64
		Unit material cost, $			0.57
		Total direct cost per unit, $			1.20
		Total job cost, $			*120,433.30*

Table 6. Assumption of Flexible Workstations Production System

Lot Quantity, 10,000

Material, Magnesium Pig, $1.31/lb
Unit Material Cost, 0.58

Process Operation Number	Table Number	Process Description	Table Time	Adjustment Factor	Cycle, Min	Setup, H
40	9.5	Cluster drill holes				
Special	9.1.S3	Cluster spindle setup	0.83			0.83
drill	9.1.S3	Cluster spindle, additional 7 spindles	0.08	7		0.56
press with pick-and-place robot	11.2.5	Cluster drilling where 4 holes are simultaneously drilled per approach, there are 3 clusters plus the main 2.85-in. hole.	0.08		0.08	
	21.3.S	Set up for robot moving between 4 drilling heads and input and output conveyor	3.00			3.00
	21.3.1	Approach–retract to pick up and unload parts from bin or conveyor with robot. Pick-and-place robot.	0.05	2	0.10	
	21.3.2	Positioning. Robot manipulator places part in first fixture, and table rotates part to surfaces to the gang drills	0.02		0.02	
	9.5.4	Machine operation to clear drills	0.08		0.08	
	9.5.7	Rapid traverse up and down of drill heads	0.16		0.16	
Total lot hours 77.89					*0.44*	*4.39*

Table 7. Cost Summary, Flexible Work Station

Operation Number	Cost Estimator Table Number	Machine, Process, or Bench Description	Lot Hours	Labor Hour Cost, $	Total Operation Cost, $
10	2.9	Hot and cold die	114.33	10.82	1237.09
30	3.3	Punch press *M*2nd	29.98	10.24	307.03
20	20.3	Chemical clean	65.00	10.21	663.65
40	9.5	Cluster drilling	77.89	12.38	964.28
50	18.1	Drill press machine	153.43	7.26	1113.93
60	20.5	Electroplating	206.05	10.62	2188.25
70	22.2	Table and machine	65.75	12.04	791.63
		Total lot hours	712.44		
		Total operational labor hour cost, $			7265.85
		Unit operational labor hour cost, $			0.73
		Unit material cost, $			0.58
		Total direct cost per unit, $			1.30
		Total job cost, $			*13,025.85*

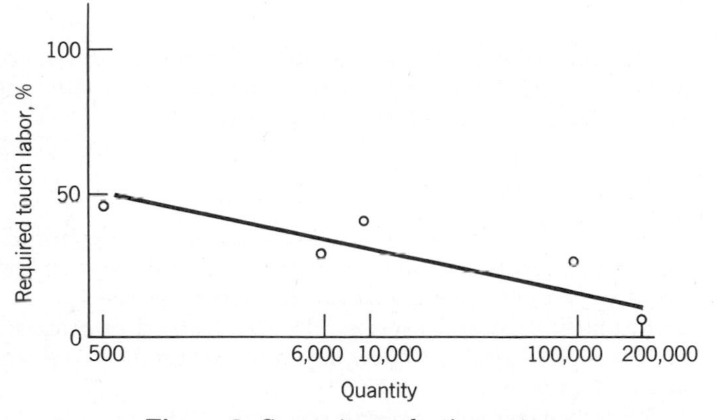

Figure 3. Composite production systems.

duces the work element to a unit output. Also given by Table 8 are the labor and machine hour totals and their percentages. These details are removed from the estimates, similarly to in Tables 4 and 5.

Interpretation from the table and the curve is reasonably certain. Labor control requirements are less for mechanization as compared to other production plans because it is intended for a mechanization policy. These data do not suggest that the machine is unattended for these various plans. On the contrary, machine attendance is assumed for safety intervention and visual quality inspection. Operator attendance is assumed for this product and manufacturing design.

The effect of tools, robots, and ancillary machine tool equipment becomes more important at the middle ranges of volume for this product. Midrange is assumed to be 6000–10,000 total units. Improvement of operations is not really a continuous function. For example, some operations are hard to mechanize, and for this product the random sampling inspection, operation 7, basically did not improve once gauging was introduced at 6000 units. Irregularity in the estimated values, as expressed by cycle minutes and setup hours, is expected. Touch labor improvement is not smooth and continuously declining, but is a step function.

Table 8 deals with the labor time component. However, analysis of labor cost is another matter. If labor is "attending," cost reduction comes from lower skill pay per hour as the job is progressively degraded in terms of skill. Successive improvements in the operation, such as the introduction of robots or specially designed machine tools that might incorporate robots, are examples. In the long term, this effect is likely. Labor cost reduction due to the lower skill level of attending work can be estimated. The attending labor is one labor grade less than the existing or higher labor skill. This labor cost degradation is a conservative assumption. For each operational methods change, a lower labor skill grade is adopted. The workstation quantity of 500 is chosen as the highest level of skill.

INTERPRETATION OF RESULTS OF ECONOMIC MODEL

Space does not allow presenting all economic assumptions and the spreadsheets used for calculating the results. Standard

Table 8. Labor Cost and Work Element Percentage Summary for Selected Technology and Quantity

Lot Quantity	Technology	Labor, H	Machine, H	Labor, %	Machine, %	Unit Labor Cost
500	Workstation	37.9	45.1	46	54	1.85
6000	Cell	240.7	579.3	29	71	1.53
10,000	Flexible workstations	291.3	420.7	41	59	0.71
100,000	Programmable automation	1534.4	4322.6	26	74	0.64
200,000	Mechanization	263.6	11,778.4	2	98	0.60

practice and industry values were used, and estimating data (6) are open to examination. Results are given by several figures that follow. The term "composite" is the adoption of a single and central quantity number representing the range for that technology. The quantities for the production systems are workstation, 500; cell, 6000; flexible workstations, 10,000; programmable automation, 100,000; and mechanization, 200,000. These technologies were also simulated for other quantities to allow cross-comparisons. Quantities are one-time only and are not construed on a per-run or per-year basis.

Composite Direct Costs

Composite direct cost is represented by Figure 4. Direct material cost declines slightly on a per-unit basis. Direct labor cost, on the other hand, declines rapidly once the flexible system is encountered. There is no significant difference in direct labor cost between programmable automation and mechanization systems.

Direct Labor Cost Per Unit

The influence of the setup is more apparent with lower quantities, especially for the workstation and the cell concept. See Figure 5 for this interpretation. Furthermore, direct labor costs for each system level out to a steady-state value. There are labor cost distinctions between the workstation, cell, and flexible workstation, but programmable automation and mechanization systems are similar. Because of greater quantity, programmable automation and mechanization systems are virtually insensitive to a greater absolute setup value as compared to the workstation concept. The workstation method is more handcraft and is susceptible to setup.

Technology results in a greater labor cost reduction than quantity. Note in Figure 5 the reduction in unit cost between the cell and flexible workstation concepts. In orders of magnitude, the direct labor unit cost is reduced by 2 to 1 for technology adaptation. In terms of the prorating of the setup value with respect to quantity, the reduction is about 1.2 to 1 for either the workstation or the cell concept.

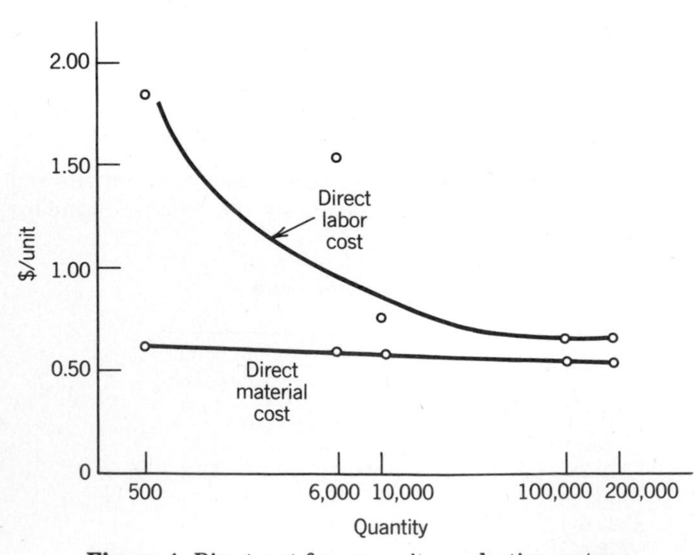

Figure 4. Direct cost for composite production systems.

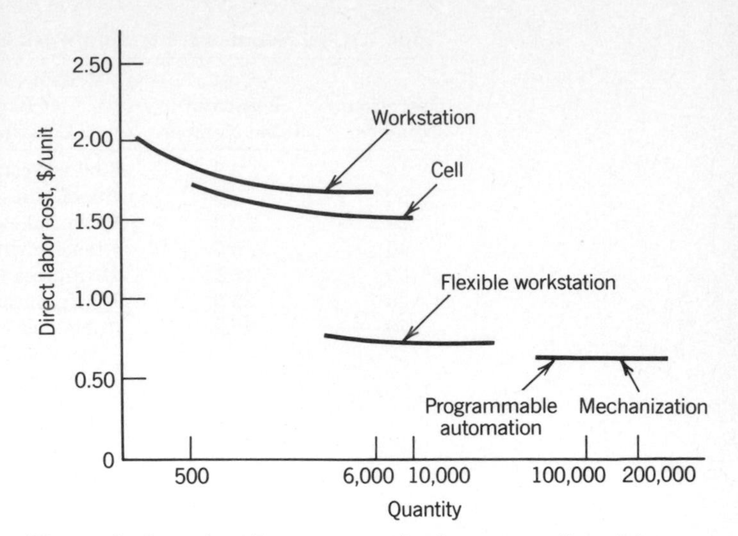

Figure 5. Quantity effects upon production system direct labor cost per unit.

As suggested earlier with a composite graph, the difference between programmable automation and mechanization was nonsignificant for direct labor cost per unit.

Productive Cost Per Unit

The composite curve, Figure 6, has a 4-to-1 order of magnitude reduction in productive unit cost between the workstation and the cell concepts. Reductions are less between the cell and the flexible workstation concepts. The cell method, as designed for this product, did not show any range of comparative economic advantage.

Quantity variations in productive unit cost lead to interpretations other than that suggested by Figure 7. The workstation concept has greater adaptability in its quantity range and is less costly when the quantities overlap with the cell method.

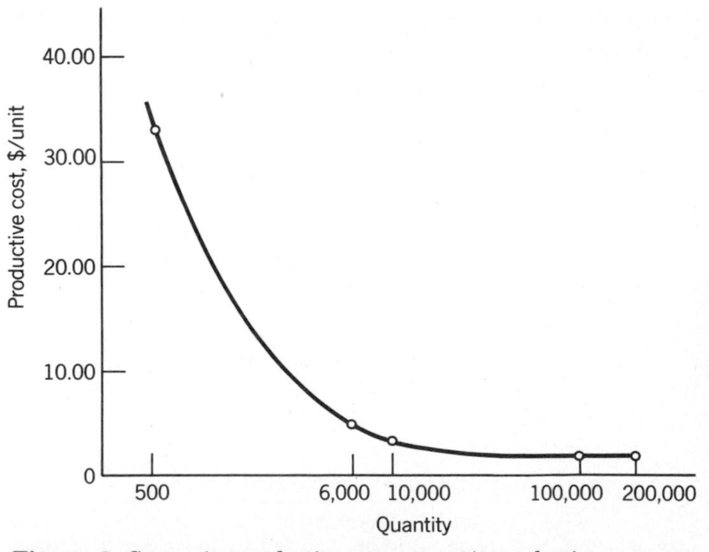

Figure 6. Composite productive cost per unit production systems. Workstation, 500 units; cell, 6000 units; flexible workstations, 10,000 units, programmable automation, 100,000 units; mechanization, 200,000 units.

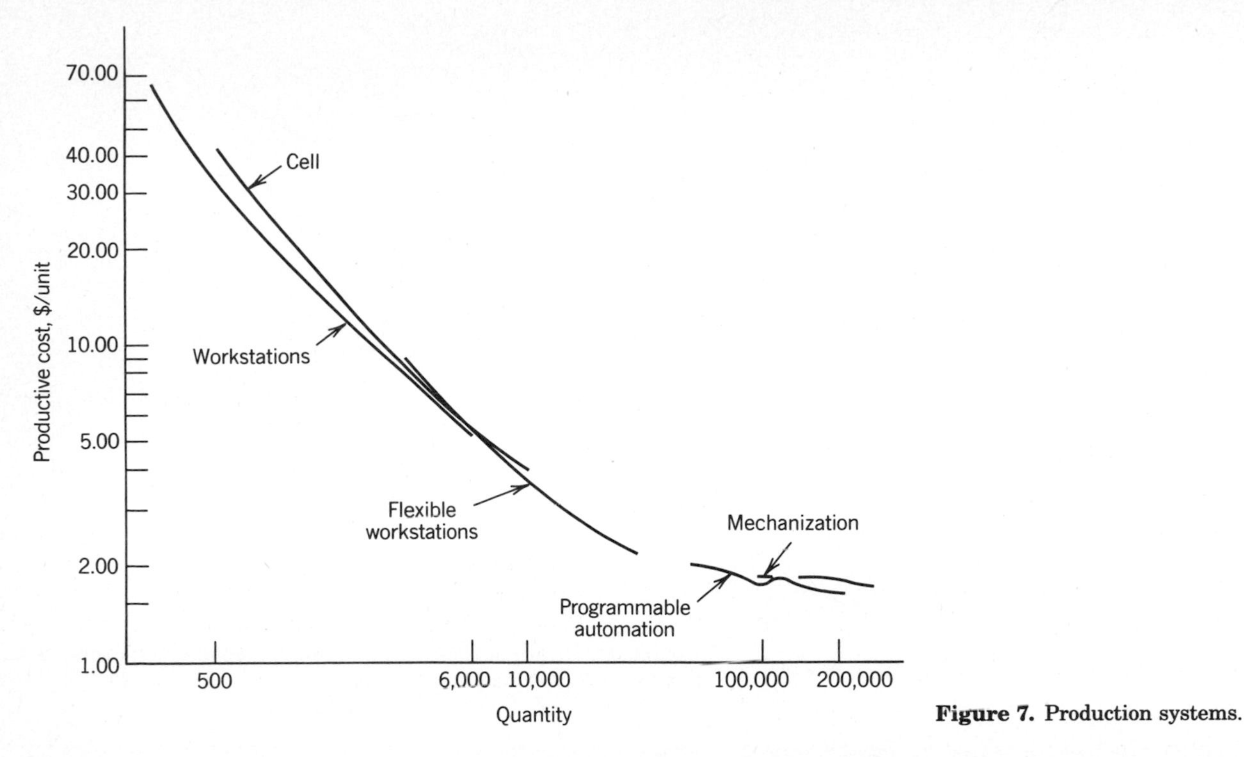

Figure 7. Production systems.

When matched against workstation or flexible workstations, the cell concept costs more over its relative quantity range.

The comparison between mechanization and programmable automation is tied to quantity. Programmable automation is more sensitive to quantity than mechanization, and when varied to 200,000 units, it is slightly less costly, by about 3%. Both methods have a middle hump of rising cost in their quantity range. This phenomenon is due to die life. Both methods require die component replacement for these quantity ranges.

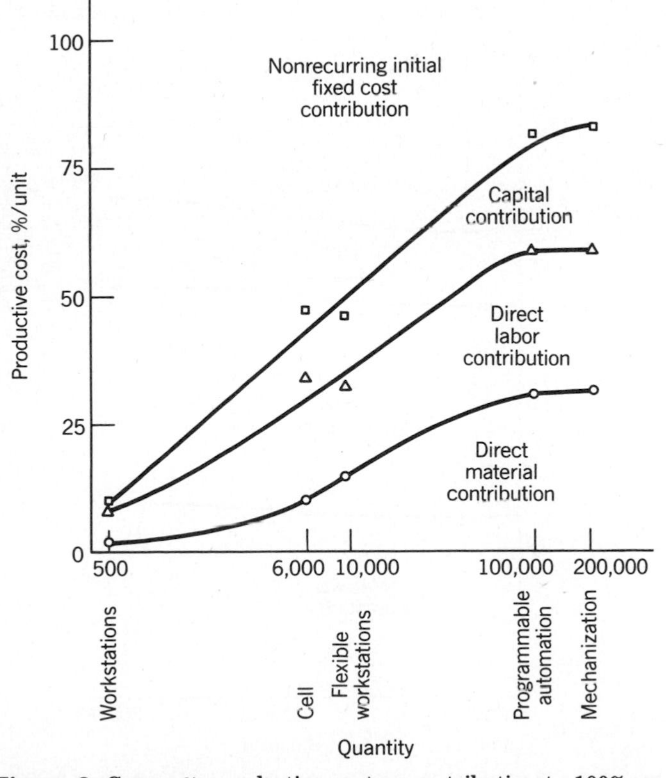

Figure 8. Composite production system contribution to 100% productive cost per unit.

PERCENTAGE CONTRIBUTION TO PRODUCTIVE UNIT COST

Figures 8–13 are concerned with the contribution of the cost elements to the total cost percentage of 100 for a cumulative

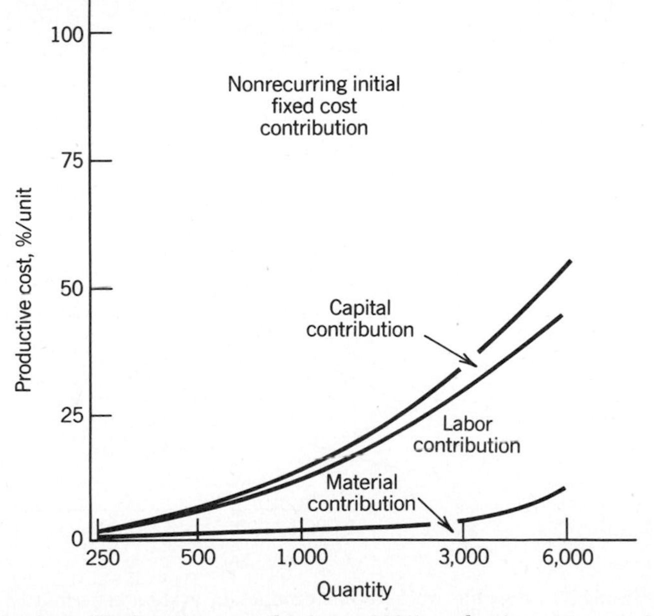

Figure 9. Workstation contribution to 100% productive cost per unit.

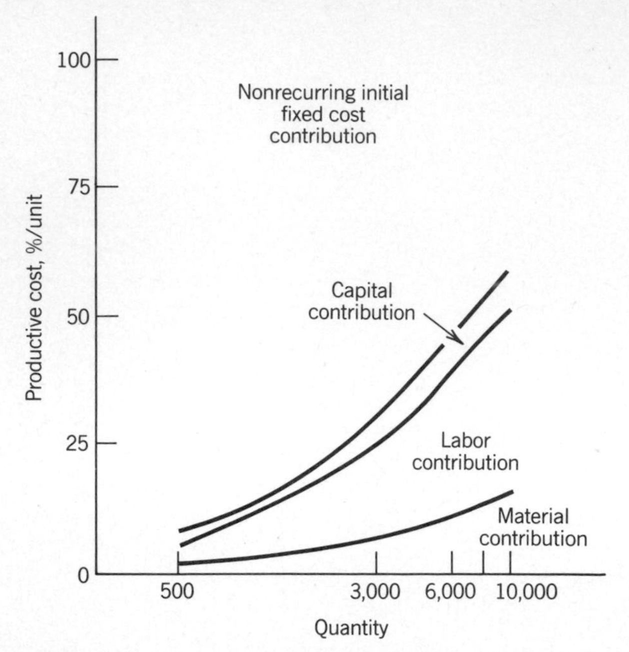

Figure 10. Cell production system contribution to 100% productive cost per unit.

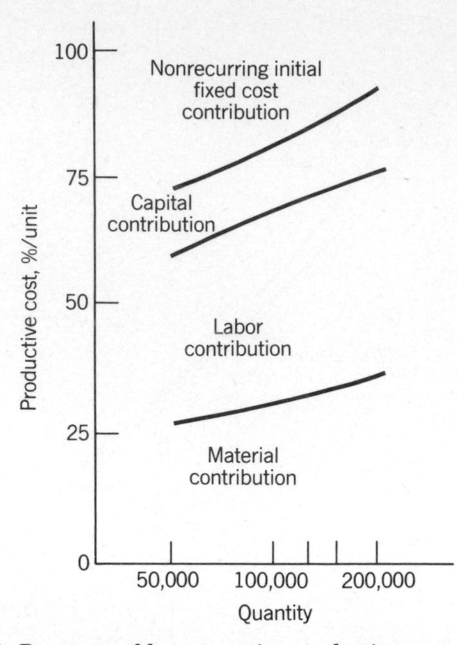

Figure 12. Programmable automation production system contribution to 100% productive cost per unit.

and particular quantity. Each of these figures may be viewed as a layer chart. The percentage difference between neighboring vertical lines is important. Observing Figure 8, at 100,000 units direct material contributes 32%, but at 10,000 units the contribution is 15%. At the intersection of 100,000 units and the direct labor line, the cumulative contribution of direct material and labor is 58%. For direct labor alone, the contribution is 26% (= 58–32).

These charts are interpreted singly and jointly. Direct material is the bottom layer in all instances. Earlier remarks regarding it as a near constant independent of quantity effects should not remove it as a factor in product contribution for lower-quantity production. Direct material contributions are nonfactors for the programmable automation and mechanization concepts since they are identical.

Overall, the major factor in low quantities is the impact of nonrecurring initial fixed costs. Tooling costs, start up, minor tooling, and so on are the most significant here. At 500 units, the tooling costs contribute about 85%. Direct labor is

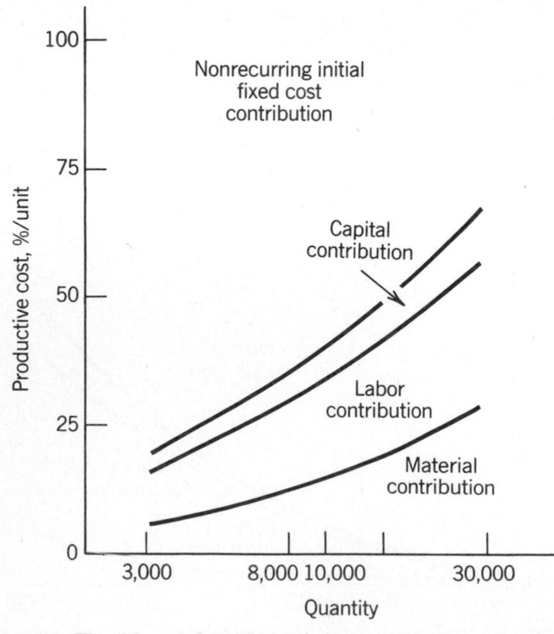

Figure 11. Flexible workstation production system contribution to 100% productive cost per unit.

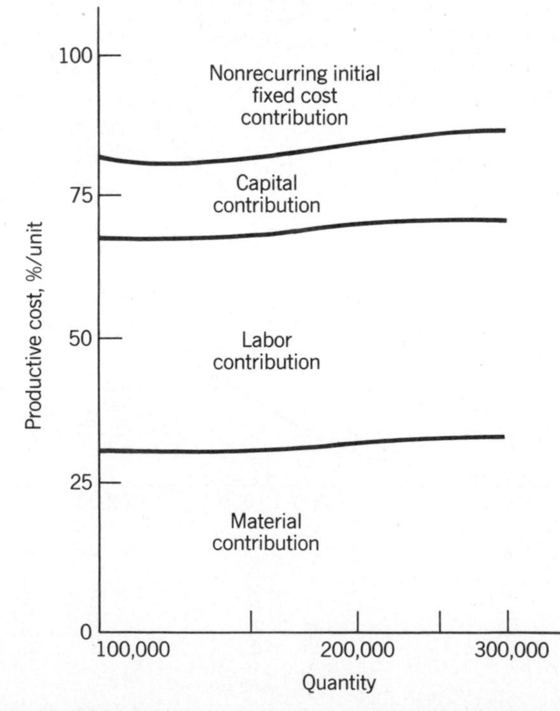

Figure 13. Mechanization production system contribution to 100% productive cost per unit.

the second major contributor. The workstation concept is further analyzed in Figure 9. This concept attempts to minimize capital spending. Even as quantity increases to 6000 units, capital contribution is maximized at about 10%. Labor and tooling costs are the dominant factors and contribute significantly over the quantity range.

The cell system has similar behavior, although the proportions begin to differ from the workstation model. Capital contributes about 5%. This quantity has little change over the span of interest. Tooling and labor are the major factors.

The labor wedge in Figure 11 changes from about 10 to 30%. Increased quantity only magnifies its importance to cost. Nonrecurring initial fixed costs dominate.

The methods and quantity cause a different impression for Figures 12 and 13. Programmable automation results in a more variable unit cost, meaning that direct costs have a greater slope over its range than for the mechanization model. The mechanization concept behaves more rigidly in terms of variable unit cost. Dominant contributors are direct labor and material for both systems.

Analysis on this topic must consider additional questions regarding the assumptions. What would be the effect on the contribution if two shifts were assumed rather than one? Depreciation cost per hour is reduced by 50%, thus compressing the effect of the capital cost component. The major capital model, mechanization, reduces its capital contribution from approximately 15 to 8%. The other components then contribute more to the full unit cost. The observations given earlier about the higher capital cost production systems remain valid, however. The workstation model has a smaller capital component, and two shifts cause a reduction of less than 1% due to the capital cost reduction.

CONCLUSIONS

It is unlikely that a practical production system in die casting technology that is indifferent to robotic influences can be devised. The systems considered here are contemporary and have modern controls, handling, and sensing. Thus, a system with and without robotic attributes is unlikely. Historically, die casting in large volumes used handling devices for the hot casts long before robotics became popular. On the other hand, one can conceive of systems that have more advanced production features. That is the philosophy of this economic simulation.

The economic model gives several answers to the first questions asked. Figure 14 shows the productive cost per unit for the systems. The economic differences are not necessarily unimportant because of the general bunching of the lines in Figure 14. This apparent compression is a result of the use of vertical logarithm scales. Cost differences are important to the success of the product. Sales opportunities are won and lost on pennies. The quantity variations for each system range beyond its practical limits. For example, the workstation concept with quantities appropriate to mechanization and vice versa are considered. Although this is unrealistic, it does confirm earlier statements regarding the potential ability of handcraft systems to respond to quantity demands outside of the original design specification.

For the conditions created for this product, the workstation method is preferred because of its lower cost for quantities of up to 6000 units or so. At that point, flexible workstations are the economical choice. The cell system does not achieve any point of minimum-cost operation. The cell system, on the basis of these simulations, is dominated by other systems.

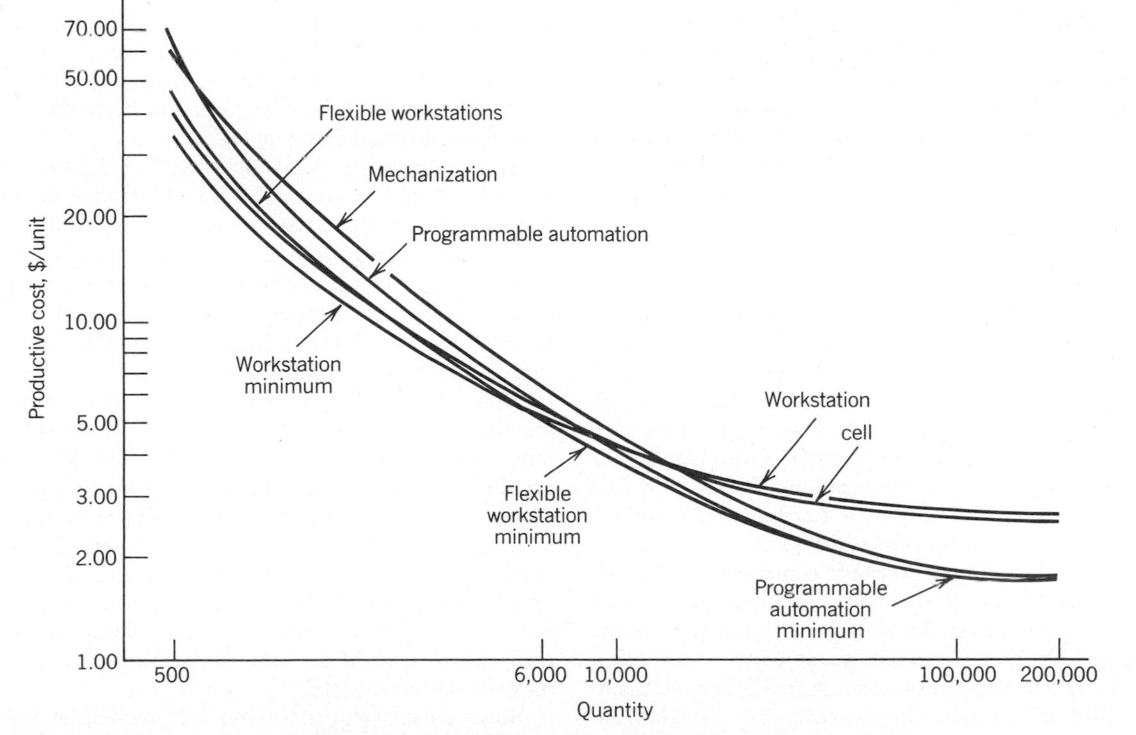

Figure 14. Production systems for die-cast product.

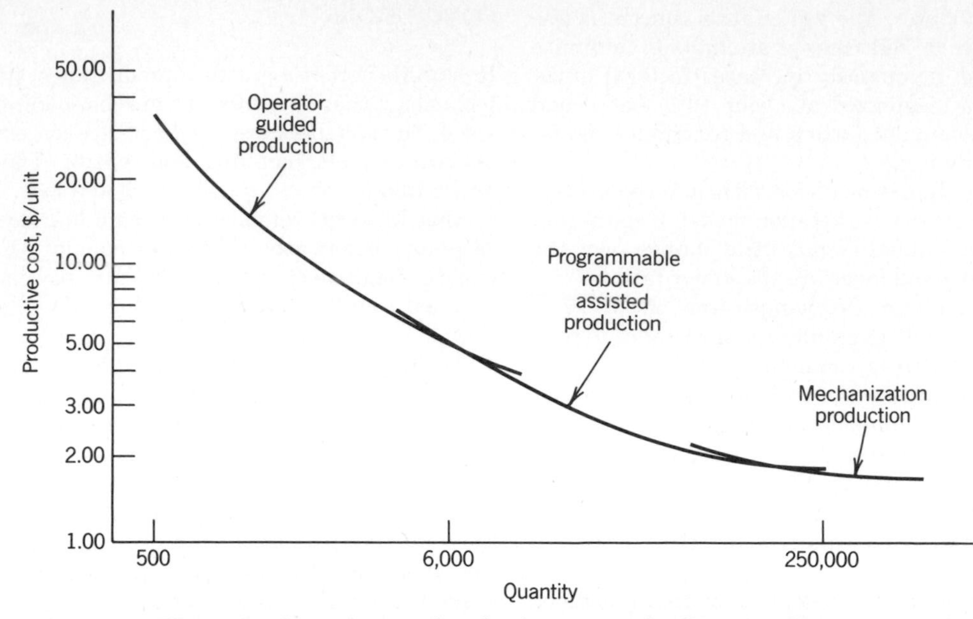

Figure 15. General types of production systems for die cast product.

Flexible workstations, where adaptability is stressed, have a range of minimum cost of up to 50,000 units or so. At that point, the programmable automation method is preferred for up to approximately 250,000 units. For this product, there is little economic difference between mechanization (hard automation) and programmable automation (soft automation) at the higher quantities. The economic costs are very similar for these two technologies, as noted in the practical range of Figure 14. Programmable automation is more sensitive to quantity and would be preferred if the quantity range projected were greater. If the expected market quantity were lower than the design quantity of 100,000 units for programmable automation, other systems, including flexible workstations, would become competitive. Mechanization is in some respects a preferred choice because it seems to be less costly over a quantity range that was not conclusively studied: 300,000 units. The lower-cost output is not unexpected, as hard automation devices are mechanically driven and many of the ancillary machine tools and the automatic handling equipment operate at higher cycle rates.

Figure 15 is a graph of cost versus production philosophy, where titles such as cell, flexible workstation, and so on are ignored. This is more descriptive of the general philosophy of the manufacturing design and its relative economics. This figure concludes that traditional methods, where operators play a major role, are preferred for up to 6000 units. At higher quantities, robotic-assisted systems are least costly, and this trend is extrapolated to 250,000 units. At that point, mechanization is the preferred minimum-cost choice.

As quantity increases, the percentage contribution of direct labor should decline. These effects should be apparent if capital is introduced to replace labor. At best, direct labor percentage remains constant for programmable automation and mechanization. The labor for the workstation indicated greater concentration, as did cell and flexible workstation concepts, although not as much. There may be several reasons for this. It may be that the die casting industry is a mature business field, and for the product studied there was little opportunity for real labor reduction. Die casters state that an overly conservative desire to emphasize personal craftsmanship has played a role in slowing the introduction of automation (8). Lower-cost labor grade substitution as capital level increases is only a partial answer. Another reason is that certain operations may not lend themselves to labor replacement inasmuch as low-cost technology with die cast products may not exist, as with, for example, the deburring technology of magnesium products because of the fire hazard of dust and chips. Furthermore, in the case of coating technology for magnesium die cast products, flow line processes (as contrasted to discrete workstations) already exist, and the labor cost improvement does not offer significant potential.

At any quantity, the opportunity for robots to influence the cost of raw material is not promising for this product. Improvements in yield are more likely to come from metallurgical progress and production cleanliness or recovery of stack losses. The opportunity offered by robotics to reduce the nonrecurring initial fixed costs is unlikely to offer much potential for overall cost reduction. Indeed, an emphasis on setup reduction and just-in-time inventory policies may offer a lower capital risk opportunity. In the higher quantity ranges, where mechanization is common, the tendency to combine operations offers little potential for robot introduction. Midrange quantity provides for capital substitution of labor by robotics. The use of direct labor in a standby status where human observation or intervention may be required limits the cost-effectiveness of robots. It is evident that mechanization offers a cycle reduction in time, and mechanization is developed to assure labor elimination. But the role of robots is dominant in the middle ranges, especially if the manufacturing system is able to eliminate direct labor costs.

Robotic-assisted production will develop along many lines. In Figure 16 a robot is employed to load and unload one die

Figure 16. Robot unloading hot cast from die-casting equipment. Arm has removed cast from quench tank. Courtesy of Unimation, a Westinghouse Company.

casting machine. The quench tank is in the foreground, and the single-cavity die is obscured by the steam. This application underlines the specificity of economic models. The status of the products, materials, labor, capital, tooling indirect changes, software and system engineering, and the various tax codes has a bearing on the economic success of any production system or component. Before a firm is able to justify robotics, its performance must first be estimated. Equipment justification proceeds after product estimates are available.

BIBLIOGRAPHY

1. G. Boothroyd, C. Poli, and L. Muerch, *Automatic Assembly,* Marcel Dekker, New York, 1982.
2. T. Taylor, *The Handbook of Electronics Industry Cost Estimating Data,* John Wiley & Sons, Inc., New York, 1985.
3. P. F. Ostwald, *Cost Estimating,* 2nd ed. Prentice-Hall, Englewood Cliffs, N.J., 1984.
4. R. E. Gustavson, "Choosing Manufacturing Systems Based on Unit Cost," Society of Manufacturing Engineers, Dearborn, Mich., 1983, MS83–322.
5. N. C. Suresh, "Financial Justification of Robotics in Multi-Machine Systems," Society of Manufacturing Engineers, Dearborn, Mich., 1983, MS85–603.
6. P. F. Ostwald, *AM Cost Estimator,* McGraw-Hill, New York, New York, 1985.
7. AM Cost Estimator Software, 2 disks, 640,000 bytes, McGraw-Hill, New York, 1985.
8. J. Canner, "Automatic Die Casting—A Concept Come Full Circle," *Die Casting Engineer* **30** (2), 18–20 (Mar.–Apr. 1986).

General References

J. Canada, "Annotated Bibliography on Justification of Computer Integrated Manufacturing Systems," *The Engineering Economist* **31** (2), 137–151 (Winter 1968).

R. J. Grieve, M. P. Kelly, and P. H. Lowe, "Robots: The Economic Justification," Society of Manufacturing Engineers, Dearborn, Mich., MS83–511.

A. Lewis, P. Watts, and B. K. Nagpal, "Investment Analysis for Robotic Applications," Society of Manufacturing Engineers, Dearborn, Mich., 1983, MS83–324.

P. B. Scott and T. M. Husband, "Robotic Assembly: Design, Analysis and Economic Evaluation," Society of Manufacturing Engineers, Dearborn, Mich., MS83–326.

J. P. Van Blois, "Economic Models: The Future of Robotic Justification," Society of Manufacturing Engineers, Dearborn, Mich., MS83–318.

V. Estes, "Robot Justification—A Lot More Than Dollars and Cents," Society of Manufacturing Engineers, Dearborn, Mich., 1984, MS84–335, 10 p.

W. Levine, *"Automation—It Pays for Itself (Or Buy One, Get One Free),"* G-T85–011, in *Transactions,* The Society of Die Casting Engineers, Inc., River Grove, Ill., 1985, pp. 1–5.

G. Hudak, "Robot Automates Die Casting Cell," in *American Machinist & Automated Manufacturing,* McGraw-Hill, New York, Apr. 1986, pp. 86–87.

B. H. Amstead, P. F. Ostwald, and M. L. Begemean, *Manufacturing Processes,* 8th ed., John Wiley & Sons, Inc., New York, 1987.

ECONOMICS, ROBOT MARKET AND INDUSTRY

Theodore J. Williams
Purdue University
West Lafayette, Indiana

INTRODUCTION

Where is the application of robotic devices or robots economically justified to replace other types of automation or human workers? Under some conditions, their use is automatically justified because no other device can do the task. In other cases they become economic tradeoffs with their competitors, and some special attributes of one competitive system or the other must be exploited to choose the final application system. In still other cases their cost, lack of speed, or some other drawback renders them unable to compete for a particular application. This article reviews the above questions in the context of formulating and answering the vital economic questions of when and where to use robots and robotic devices. Whereas the questions discussed here relate to the use of robots, the discussion presented is completely generic and applies to the application of any device, technique, or system for a particular capability (see also Cost/Benefit; Economic justification; Economics, robotic manufacturing and product).

JUSTIFICATION—THE HEART OF THE ECONOMIC PROBLEM

A major problem that always faces any new innovation or application of even an established technology is the economic justification of using one particular application versus any and all competitive technologies, including manual labor, that might be used for the same application (1). Such justifications must apply to every industrial situation, except perhaps basic research and development, if the company involved is to survive and prosper in competition with its peers. The basic question is as follows: Does the economic gain of the new technology or application include enough additional economic return to pay out the *extra costs* associated with it, within the financial requirements set by the company planning such an installation? Note that it is usually the case that the new technology is, in its totality (procurement, installation, and operating costs), more expensive than the competitive technology chosen for replacement. Obviously, if the new technology is less expensive than the old (including replacement costs of the old), the question of economic justification is moot. Likewise, if required product quality or other factors are obtainable only through use of the new technology, the same conclusions apply.

Such justifications have been insisted on strongly by company management teams because, as discussed later, the new technology may often promise only a relatively small provable economic return at a high technical risk. In addition, this small provable economic return is often associated with a new, different, and unfamiliar technology and faced with a decided shortage of qualified personnel capable of accomplishing such installations.

The economic returns possible from the new technology must also fit within the present, overall economic criterion used by company management for the operation of the plant.

The choices that management usually has for economic gain are listed as follows (2):

1. Maximum production for the sold-out plant.
2. Minimum costs for the plant when operating at less than full production, including minimum fuel and/or minimum raw material usage and reduced manpower.
3. Maximum product quality wherever appropriate.

PRODUCTIVITY—A DEFINITION

Productivity can be defined in many different ways, depending upon the context of its use, such as "units of product made per man hour," "profit per unit of capital investment," "production units per time period for a specific plant," and so on. Therefore, a specific context needs to be stated for clarity. Because a basic need for industry is capital development, productivity is defined as "dollars profit per unit of capital investment." This avoids the calculation of manpower reductions involved in the usual "production per man hour" definition. In addition, it allows for a treatment of the important topics of energy savings, raw materials savings, and quality enhancements for the case of the "non-sold-out" plant where increased throughput is not the desired answer, even though it might be the easiest to attain.

Gains made by new technological applications are best computed as percentage increases or decreases from those operating levels of production, raw material use, energy use, manpower requirements, and so on for the base case plant without the new technology. Unless the rules presented here are expressed in the same percentages, no generality will be possible because of the enormous differences in plant sizes and product values among companies and industries. The final desired payout of the new technology is then obtained by converting the percentages to dollars as a function of plant size and product mix and then computing the ratio of this to the new cost for the desired fractional payout.

The gains from the new technology fall into two categories: those for which an economic factor can readily be applied and those for which an economic factor can be determined only with difficulty or not at all. The former are commonly called tangible benefits and the latter intangible benefits. Unfortunately, many of the latter are often the most attractive to the industrial engineer. As noted above, these benefits are generic and apply to almost any type of new technology used, including robotics. A representative set of both types of benefits is listed as follows:

Tangible Benefits

1. Increased plant throughput or production.
2. Increased product quality.
3. Decreased energy requirements.
4. Decreased raw material requirements.
5. Decreased effluents and pollutants.
6. Manpower savings.
7. Reduced variability of product, less product giveaway.
8. Increased yields or improved product mix.
9. Reduced rework.

Intangible Benefits

1. Increased plant and personnel safety.
2. Better and more timely production, engineering, and management information.
3. Reduced maintenance requirements.
4. Increased flexibility of the system with respect to changes or additions.
5. Increased customer service through faster response and better order monitoring.
6. Better response to emergencies.
7. Warning of impending catastrophic conditions in processing equipment.
8. Better maintenance scheduling.

BENEFITS OF THE USE OF ROBOTS

The major expansion of the use of robots in recent years has been due to a need to increase factory productivity drastically. Because of present union work rules, this has been difficult, if not impossible, to accomplish with the plant labor force. Until recently, the rapid rise in labor costs was beginning to make the capital justification of robot operations as replacements for human workers economically practical in its own right (3–7). However, since the recession of the early 1980s, plant staffs have been drastically reduced. In many cases, wages have also not increased, whereas in others, they have actually been lowered. In addition, the current power of the labor unions is reduced. What effect this will have on future wage progression is not yet clear.

The list below presents commonly claimed advantages and disadvantages for robots as listed in many magazine articles and books (8–13). Again, note the similarity of this list to the one above.

Advantages of Robots

1. Safety of workers.
   a. Hazardous environments.
      i. Toxic fumes.
      ii. High temperatures.
      iii. Radiation.
   b. Hazardous operations.
      i. Loading and unloading of dangerous machinery.
2. Higher productivity.
   a. No need for rest breaks.
   b. 24-h operation.
   c. Higher speed of operation (probably limited to point-to-point robots).
   d. Fewer mistakes resulting in scrapped parts or damaged machines.
3. Production plant flexibility.
   a. Shortened reaction time to required plant changes.
   b. Easier debugging of changes.
   c. Resistance to obsolescence by encouraging updating changes.
4. Ability to work from unusual orientations, such as ceiling or wall mount.

Disadvantages of Robots

1. Difficulty of programming.
2. Relatively slow speed of operation, particularly of continuous-path servo-controlled systems.
3. The accuracy of positioning of robot components is limited by the "play" in robot joints and the flexibility or "bending" of robot limbs. The heavier the load in relation to robot size, the more important is this consideration.

Figure 1 compares the former rapid rate of increase of operating costs of industrial manufacturing operations due to labor versus the corresponding cost for an equivalent capability in robots. It was this drastic increase in personnel costs that appeared to make robotics economically attractive and that promised to lead to a still greater economic incentive for their use in the future (13). However, this prognosis is now unclear because of recent labor changes in industry.

Figure 2 presents an interesting commentary on today's costs. Note that the cost of the mechanical elements of the robot has increased by a factor of about six for essentially the same capability (13). Meanwhile, the controller cost has remained approximately the same, even though the technology involved has changed from that of a relay and transistor design to that of one or more microcomputers with large associated memories and greatly increased capabilities.

Figure 3 continues this economic comparison by relating manufacturing costs for manual labor, robotics, and hard automation as a function of manufacturing volume, thus reemphasizing the place of robotics between manual labor and fixed automation in its use in the manufacturing industries (13).

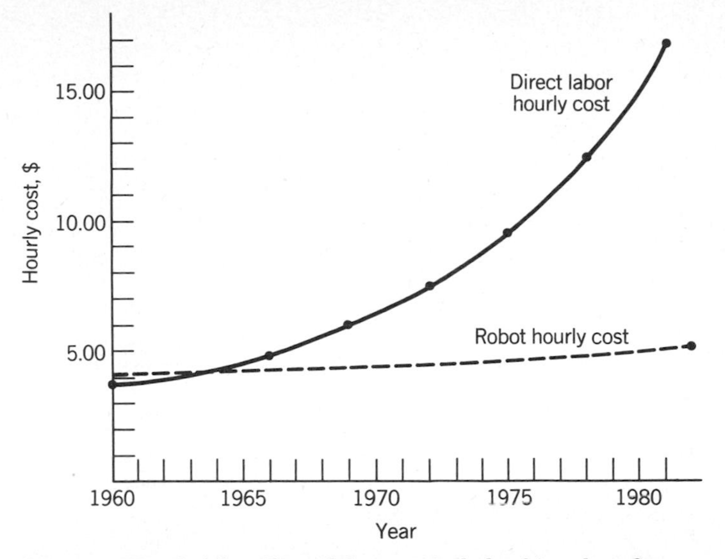

Figure 1. Hourly labor (direct labor cost includes fringe benefits, robot cost includes support) cost of automotive worker versus an industrial robot in the United States, 1960–1981 (13).

APPLICATIONS OF ROBOTS

The major area of application of robots to date has been to handle tasks that have been undesirable, unsafe, or impracti-

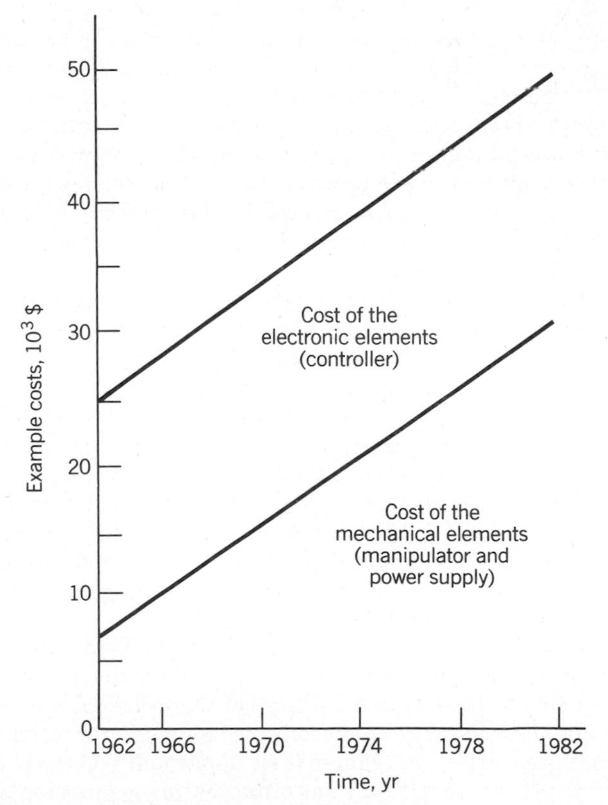

Figure 2. Changes in robotic costs in the last two decades (13).

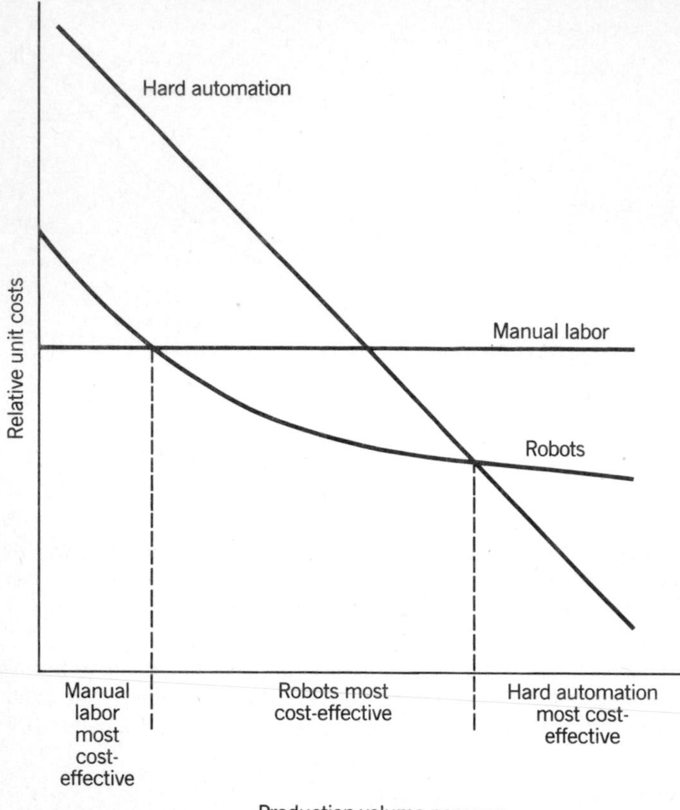

Figure 3. Comparison of manufacturing method unit costs, by level of production (13).

cal for humans to carry out. These include work in hazardous environments, loading and unloading of machines, parts transfer on manufacturing and assembly lines, reactor cleanup, spot welding, spray painting, and light assembly operations (14–16). A second important application has been to increase productivity by improving product quality and reducing rejects in terms of an *assured* repeatability of a correctly completed manufacturing task. Note that the latter is the *fastest growing* application area today.

Thus, the use of robots is most commonly associated with the handling of, or the performing of operations upon, *discrete objects*. There is no foreseen advantage of robots in a plant handling only bulk materials, particularly flowing materials such as liquid, gaseous, or granular materials. Such plants may indeed be completely automated in terms of computer supervision, direct digital control of continuous processes, programmable controller direction of sequencing operation, automated material handling, and so on. But none of these controlling or materials-handling functions are carried out by a *reprogrammable multifunctional manipulator,* that is, a robot, as commonly defined and discussed below (2).

As noted in the common definitions of the field, robots are specifically designed to handle objects, either the discrete objects being worked on in the factory or the tools that are carrying out the necessary operations on them. Such a restriction in definition seriously reduces their applicability to the chemical and petroleum industry and other continuous-process types of industries.

Applications in these latter industries are quickly reduced to packaging operations in the fine chemical, pharmaceutical, and food industries (where packages are relatively small), to plactics molding operations for multiple small lots of products, to the layup of products made from reinforced plastics, or to operations in very severe environments that are dangerous or uncomfortable for human workers.

A major advantage of robots is their flexibility. They are able to compete with so-called *fixed, dedicated,* or *hard* automation, which is difficult and costly, if not impossible, to modify for a new application.

In hard automation, a task is performed by a tool or machine set up so that no human control is required during operations. Hard automation is typically dedicated to one application throughout the life of the tool or machine. The primary disadvantage of hard automation is thus the difficulty of justifying the investment for a batch or small lot manufacturing operation in which changeovers may be required. An additional drawback is the usual need for human assistance in loading and unloading the tool.

Therefore, the limitation of most forms of fixed, or dedicated, automation has been that they can only be used in certain high-volume manufacturing operations requiring only the simplest types of motions. Robots, on the other hand, can be used in any operation where a certain degree of manipulation is required. In other words, a much larger number of tasks formerly requiring manual labor can now be automated. This unique capability to perform tasks requiring the manipulation of objects, combined with the need to reduce direct labor costs and to increase productivity, has resulted in the current enormous interest in robots on the part of manufacturing managers in discrete products plants.

Until recently, the alternative to hard automation was to increase the direct labor content of a manufacturing task. However, a new technology, identified as flexible automation, has been developed to increase the range of tasks that can be performed and to improve the changeover capability of manufacturing tools. In flexible automation, a tool is preprogrammed by a human as in hard automation, although in this case, the workpiece can be manipulated so that a greater number of tasks, such as machine loading and unloading or part transfer, can be performed in each cycle. In addition, a changeover to another job can typically be accomplished by reprogramming rather than reworking or replacing the equipment. Therefore, machinery can be more productively used throughout its useful life. Industrial robots can be classified as a type of flexible automation and are very often components of such systems.

Note that even flexible automation devices are often programmed by the adjustment of mechanical limits, and so on, by humans. The rapidly increasing use of microprocessors or programmed logic controllers to add flexibility to these machines has not yet received enough attention. However, they can have as profound an effect in the discrete manufacturing industries as they can and are having in the continuous-process industries. As in the continuous-process industries, they will probably be responsible for a whole new way of developing and considering automation in these factories. Should this occur, it will limit many of the areas claimed for robotics by their proponents.

THE PARTS OF A ROBOT—A NOTE TO CONSIDER IN THE ECONOMIC JUSTIFICATION

Despite the wide variety of configurations in which they appear, all robots consist of three basic parts: a manipulator, a power supply, and a controller. These are normally sold as a complete package by the robot manufacturer.

The manipulator and its support stand are the basic mechanical elements involved and perform the work assigned to the robot. The power supply provides energy for the manipulator's motion. Last, but far from least, the controller is the robot's "brain" and directs the manipulator's movements (17–19).

Considering the mechanical manipulators and the controller as two separate entities rather than as one whole can be very helpful to the industrial engineer in considering the potential application of robots and robotics to manufacturing plants, particularly in relation to other competitive types of processing units and thus the economic justification for their use on a particular application.

Because of its anthropomorphic qualities, the robot is often considered to be an operator or controller of a process, that is, external to the process. Actually, its manipulator should more properly be considered part of the process or another unit of process equipment separate from the intelligence imparted by its controller. The following questions can then be asked. Is the manipulator (arm and hand) the best method of accomplishing the task at hand? Would the intelligence imparted by the robot's controller perhaps be better applied through another type of processing equipment? Is it the manipulator's capabilities or the controller's intelligence that is needed to accomplish the task at hand? For the situation at hand, another type of mechanism might make a better and less expensive actuator than the robot's manipulator (20).

Almost all industrial process systems would greatly benefit from the amount of intelligence commonly attributed to robotic functions (much more than is actually used in most applications). However, it is often the intelligence involved, not the manipulative ability of the robot's arm, that is most important for many industrial applications, such as most chemical and related processes. Thus, it is very important to keep these two different aspects of a robot separate when evaluating its use in all types of industrial operations. When judged as a mechanical process unit (manipulator arm) with a separate intelligent controller to direct it (as the true economic analysis would view it), the robot assumes an aspect in industrial operations planning somewhat different from that of its more common description in the popular press.

MAKING THE ECONOMIC JUSTIFICATION

The necessary steps to take in making an overall systems engineering investigation and justification of a new manufacturing technology, or even merely of a new piece of manufacturing equipment, are outlined in Figure 4. Note the major emphasis on the economic aspects of the procurement in this analysis (items 2–4, 7–10). It is the economic justification of item 9 that is of greatest concern, using the expected tangible benefits of the type listed earlier. The so-called intangible benefits must be used only as weighing factors since an economic enumeration of their value is, by definition, impossible. Note that most of the advantages of robots listed earlier are of the latter type, except for the higher productivity cited in item 2.

METHODS OF EXPRESSION OF ECONOMIC JUSTIFICATION

There are three basic approaches commonly used in manufacturing firms today to compare alternative project proposals: return on investment (ROI), net present value (NPV), and payback period.

Return on Investment (ROI)

This is probably the most commonly used tool for comparing alternative investments. A series of annual cash flows is developed for each alternative, taking into account both expected annual cost savings and expenses. These cash flows are then compared with an initial investment, or cash outlay, to determine an overall annual rate of return on the investment. This return is then compared with a minimum investment criterion to evaluate the attractiveness of each alternative.

Net Present Value (NPV)

Under this approach, a series of discounted annual cash flows is generated for each alternative over the life of the project (eg, ten years). The discount rate is usually equal to the cost of securing capital for the company, which today may be 20% or even higher. These discounted cash flows are totaled and compared with the initial cash investment. If the sum of the discounted cash flows is larger than the initial investment number, then the difference between the two represents the present value of the alternative to the company. This must be a positive number in order for the alternative to meet the company's investment return criterion.

Payback Period

This is a measure of the time required to recover the initial investment costs for each alternative. For example, if a payback period is three years, this means that the sum of the cash flows during the first three years is equal to the initial investment cost. After three years, the project will then generate positive net dollars.

COSTS FACTORS TO BE CONSIDERED

In the beginning of this article, the benefits of the use of robotics, particularly those that can be expressed in economic terms, were discussed. The alternative in making the analysis is to enumerate the costs involved in the procurement, installation, and operation of the new technology. The following are the major factors to be considered.

Purchase Price of the Robot

The purchase price of a robot is highly variable. The range might extend from \$5000 to \$100,000, depending upon the number of articulations, sphere of influence, weight-handling

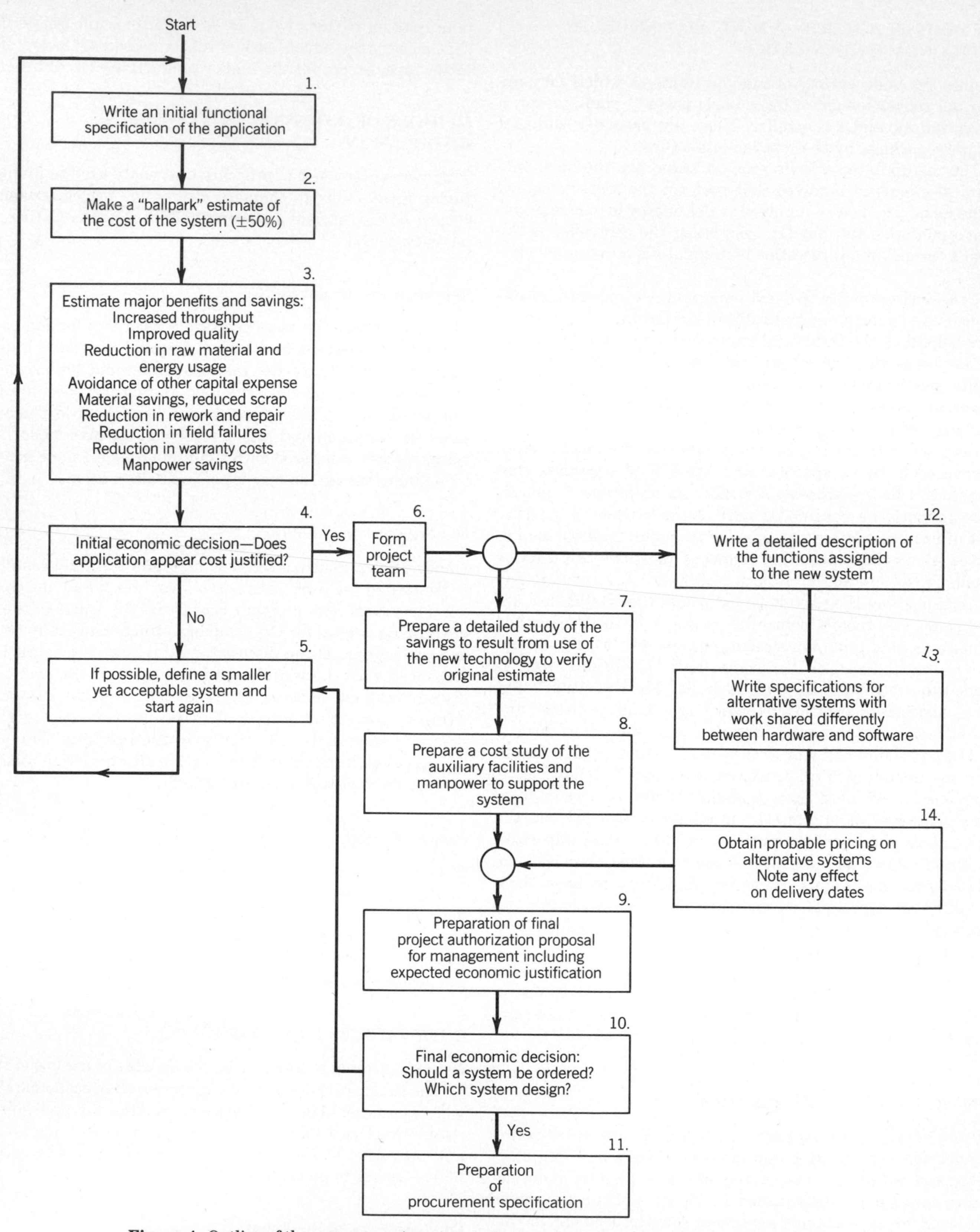

Figure 4. Outline of the systems engineering study phase of the procurement of a new manufacturing technology (21).

capacity, and control sophistication. Generally speaking, the higher-priced robots are capable of more demanding jobs, and their control sophistication assures that they can be adapted to new jobs when original assignments are completed. The more expensive and sophisticated robot also requires less special tooling and lower installation costs ordinarily. Some of the pick-and-place robots are no more than adjustable components of automation systems.

Special Tooling

Special tooling might be nothing more than a $300 gripper for a robot installed at a die cast machine. Or the tooling might include an indexing conveyor, weld guns, transformers, clamps, and a supervisory computer for a complex of robots involved in spot welding automobile bodies. For assembly automation, the special parts presentation equipment may cost well in excess of the robot equipment cost.

Robots are not stand-alone equipment. The interface with the workplace can be critical to success, so that customers often ask that robot manufacturers bid on a turn-key basis. For economic evaluation, the two prime cost factors, *purchase price of the robot* and *special tooling,* are thus combined.

Installation

Installation cost is sometimes charged fully to a robot project, but is often carried as overhead because plant layout changes were already planned. At a model change, there are usually installation costs to be absorbed even if equipment is to be manually operated. There is no reason to penalize the robot installation for more than a differential cost inherent in the automation process.

Maintenance and Periodic Overhaul

To keep a robot functioning well, there is a need for regular maintenance, a periodic need for more sweeping overhaul, and a random need to correct unscheduled downtime incidents. A rule of thumb for well-designed production equipment operated continuously for two shifts is a total annual cost of 10% of the acquisition cost.

There is variability, of course, depending upon the demands of the job and the environment. Maintenance costs in a foundry are greater than those incurred in plastic molding.

Operating Power

Operating power is easily computed as a product of average power drain times the hours worked. Even with increased energy costs, this is not a major robot cost.

Finance Charges

In some cost justification formulas, one inputs the current cost of money. In others, one uses an expected return on investment to establish economic viability.

Depreciation

Robots, like other equipment, will exhibit a useful life. It is ordinary practice to depreciate the investment over this useful life. Since a robot tends to be general-purpose equipment, there is ample evidence that an 8–10-yr life running multishifts is a conservative treatment.

For cost justification formulas straight-line depreciation is commonly used, but the tax schemes of different countries may involve depreciation weighted to early years (eg, in the United States, double-declining balance). Special tax credits to encourage capital investment may also influence the cost justification formula and the buying decision. See the engineering economics texts listed in the General References for a thorough discussion of this topic.

AN EXAMPLE OF THE SIMPLE PAYBACK OR PAYBACK PERIOD

A simple payback example involves a robot at a workstation where 250 days are worked in a full calendar year and where the robot would replace one unskilled human operator whose wages and fringe benefits amount to $11.67/h. Against this saving, the robot would cost $2.75/h to run and maintain a one-shift operational basis, $1.38/h for two shifts, and $0.92/h on a three-shift basis (this assumes that daily costs are the same regardless of work periods used). Capital investment for the robot and its accessories is $66,000. The company normally operates only one 8-h shift per day, but has the option of increasing this to two or three shifts when production demands are sufficient.

The following cases contain the calculations and illustrate the method, by using the simple payback formula:

$$P = \frac{I}{L - E}$$

where P = payback period in years, I = total capital investment in robot and accessories, L = annual labor costs replaced by the robot and other economic gains, and E = annual expense of maintaining the robot.

Case 1. With single-shift operation,

$$P = \frac{\$66{,}000}{11.67(250 \times 8) - 2.75(250 \times 8)} = 3.7 \text{ yr}$$

Case 2. With two-shift operation,

$$P = \frac{\$66{,}000}{11.67(250 \times 16) - 1.38(250 \times 16)} = 1.6 \text{ yr}$$

Case 3. With three-shift operation,

$$P = \frac{\$66{,}000}{11.67(250 \times 24) - 0.92(250 \times 24)} = 1.02 \text{ yr}$$

If only one 8-h shift is to be operated on each of the 250 working days, the payback period amounts to an unsatisfactory 3.7 yr. If, however, sales forecasts and production plans indicate that two shifts could be maintained, then the payback period is reduced to approximately 1.6 yr, which would usually justify going ahead with the project. The three-shift value of 1.02 yr would normally allow proceeding without any doubt.

PRODUCTION RATE IMPACT ON PAYBACK

A robot does more than simply replace a worker. It might work at different speeds, and it can sometimes be included in an automatic system that allows more efficient operation of one, two, or even more pieces of expensive machinery. When the production rate project appraisal method is used, these realities must be taken into account by including the value of capital equipment in the application and the production rate coefficient compared with the manual worker standard. In this more complex example, the production rate payback formula

$$P = \frac{I}{L - E + q(L + Z)}$$

is used.

Assume that the company takes an annual write-down percentage of 10% depreciation in the operating cost budget. This gives an annual depreciation of \$6600, represented as Z in the formula. The production rate coefficient q is the rate by which robotized production either exceeds or lags behind that achieved by a human operator. A rate of 20% above the human rate and a rate 20% below the human rate are to be tested in the calculation, giving values for q of ± 0.2, respectively. The remaining variables are unchanged from the preceding cases.

The actual calculations are given in Cases 4–6.

Case 4a. With single-shift operation, where the robot is 20% slower than the human operator,

$$P = \frac{\$66{,}000}{23{,}340 - 5500 - 0.2(23{,}340 + 6600)}$$
$$= 5.57 \text{ yr}$$

Case 4b. With single-shift operation, where the robot is 20% faster than the human operator,

$$P = \frac{\$66{,}000}{23{,}340 - 5500 + 0.2(23{,}340 + 6600)}$$
$$= 2.77 \text{ yr}$$

Case 5a. With double-shift operation, where the robot is 20% slower than the human operator,

$$P = \frac{\$66{,}000}{46{,}680 - 5500 - 0.2(46{,}680 + 6600)}$$
$$= 2.16 \text{ yr}$$

Case 5b. With double-shift operation, where the robot is 20% faster than the human operator,

$$P = \frac{\$66{,}000}{46{,}680 - 5500 + 0.2(46{,}680 + 6600)}$$
$$= 1.27 \text{ yr}$$

Case 6a. With triple-shift operation, where the robot is 20% slower than the human operator,

$$P = \frac{\$66{,}000}{70{,}020 - 5500 - 0.2(70{,}020 + 6600)}$$
$$= 1.34 \text{ yr}$$

Case 6b. With triple-shift operation, where the robot is 20% faster than the human operator,

$$P = \frac{\$66{,}000}{70{,}020 - 5500 + (70{,}020 + 6600)}$$
$$= 0.84 \text{ yr}$$

It is interesting to compare the results with those obtained from the simple payback period calculation and to see how the consideration of production rates and associated capital utilization can reduce or extend the expected payback period. Assuming that the robot is reliable and has no inherent bugs, the shortened payback result should apply to most applications where the robot can operate continuously without rest periods and with dependable repeatability. Thus, for one-shift operation, the 3.7 yr indicated by the simple study varies between 5.6 and 2.8 yr in the complex example, depending on the production rate coefficient. Similarly, the two-shift result changes from 1.6 yr to a payback period of between 2.2 and 1.3 yr.

RETURN ON INVESTMENT EVALUATION

Payback analysis works best when the overall time scales under consideration are short. Changes in money values owing to inflation or notional or real rates of interest applicable to project financing are then generally ignored. Special-purpose automation, custom built to produce one workpiece, is prone to early obsolescence. Provided that the payback period is suitably short compared with the expected equipment life, such factors as inflation or annual interest rates are unlikely to weigh heavily in the economic justification argument. This is obviously true when payback periods of only one or two years emerge from the calculations.

A Simple Example

An example of an ROI analysis is made by slightly modifying the original simple payback period example to show the effect of varying labor cost. The formula

$$S \times \frac{100}{I}$$

gives the ROI percentage where S = the annual savings resulting from the robot and I = the initial investment. Table 1 presents the results. It is common for companies to require an ROI that is usually 2–3 times higher than the current

Table 1. ROI Analysis Results

	Hours Per Day Operation		
	8	16	24
Robot Costs ($):			
Annual depreciation	6600	6600	6600
Annual upkeep	5500	5500	5500
Total annual robot costs	12,100	12,100	12,100
Corresponding Labor Costs ($):			
$8/h	16,000	32,000	48,000
$10/h	20,000	40,000	60,000
$12/h	24,000	48,000	72,000
$15/h	30,000	60,000	90,000
Annual Cost Savings and ROI ($ ROI, with % in parentheses):			
$8/h	3900 (5.9)	19,900 (30.2)	35,900 (54.4)
$10/h	7900 (12.0)	27,900 (42.3)	47,900 (72.6)
$12/h	11,900 (18.0)	35,900 (54.4)	59,900 (90.8)
$15/h	17,900 (27.1)	47,900 (72.6)	77,900 (118.0)

interest rate available from investment sources to counteract the risk present in a new technology. Thus, 20–30% is often needed. In the example of Table 1, a multiple-shift operation would be required to achieve the expected return.

Figure 5 graphs the resulting ROI versus the hours worked for the wage rates chosen. Note again that, for simplicity, only labor savings are used in the example.

The Time Value of Money

Robots are not special-purpose automation. Their flexibility means that they can be redeployed when a product line changes. Their reliability indicates a long working life, at least 8 yr in the case of many robots. It is not, therefore, unreasonable to regard robots as general-purpose equipment. Since the investment will produce useful work over a period of many years, changes in the value of money, interest payable, and the rate of return on the money invested all provide factors

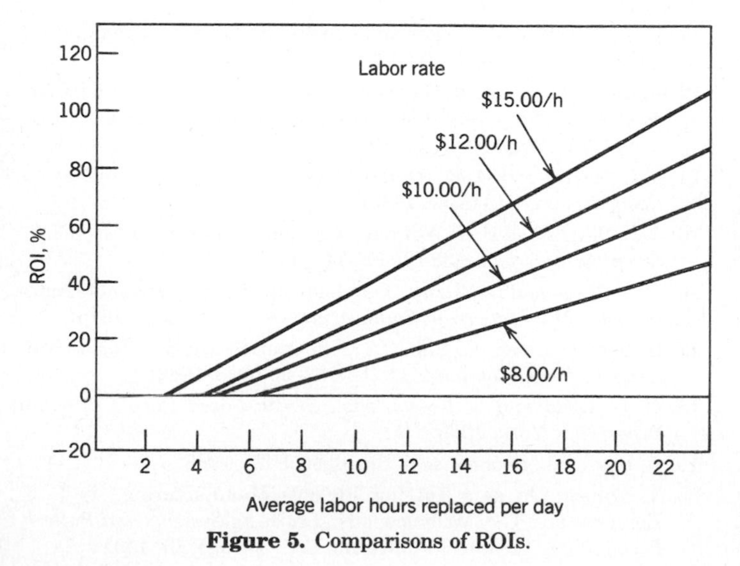

Figure 5. Comparisons of ROIs.

that can be evaluated and considered in deciding if a project is worthwhile. The simplest way to approach this problem is to decide the rate of interest or other return that company policy dictates for its investments and then ensure that the proposed new project at least matches these expectations.

For the ROI calculation illustrated here, the proposed project is the same robot acquisition used in the previous examples, but one additional factor has been introduced. This is the concept of net present value, that is, the current value of the robot installation if the gains in future years are converted to the amount of money that, invested at the required rate of return, would give the same amount of cash at that future date (22, 23). The value of the required investment for each year is found by

$$P = F\left(\frac{1}{(1+i)^n}\right) \tag{1}$$

where P = present value of a future amount, F = receivable in, n = years at interest rate, and i = the interest rate percent.

Thus, in our analysis, F is the amount of gain made each year, and the series

$$P_T = P_1 + P_2 + \cdots + P_n$$

is the total present value of all the gains to be made from the project over the next n years. The number is then subtracted from the cost of the installation to determine whether the project has paid out over the chosen n-year period.

A more common equation derived from equation 1 develops the required annual gain necessary to pay back the original cost P in n years as follows:

$$A = P\left(\frac{i(1+i)^n}{(1+i)^n - 1}\right) \tag{2}$$

Table 2 presents the value in the parentheses of the above equations as a function of the number of years n for a required ROI of 20%.

Table 2. Factors for Compound Interest for Time Value of Money Calculations

n	To Find P, Given F $\left(\frac{1}{(1+i)^n}\right)$	To Find A, Given P $\left(\frac{i(1+i)^n}{(1+i)^n-1}\right)$
1	0.8333	1.2000
2	0.6944	0.6545
3	0.5787	0.4747
4	0.4823	0.3862
5	0.4019	0.3344
6	0.3349	0.3007
7	0.2791	0.2774
8	0.2326	0.2606
9	0.1938	0.2481
10	0.1615	0.2385

As an example of the use of this table, refer to Table 1. For the $15.00/h, one-shift case, the gain of $17,900/yr would require a factor of 0.2712, that is, 66,000/17,900, to pay out by this method, and between 7 and 8 yr to achieve the required 20% ROI.

SALES OF ROBOTICS—THE GROWTH AND FUTURE OF THE ROBOTICS FIELD

The numbers of robots used by the following industrial nations of the world in 1981 were: Japan, 6000; United States, 3500; Germany, 1000; Sweden, 1000; and the United Kingdom, 400; for a world total of about 15,000 (38). Growth in the applications has been rapid, with major investments in the automotive, appliance, and electronics industries in particular.

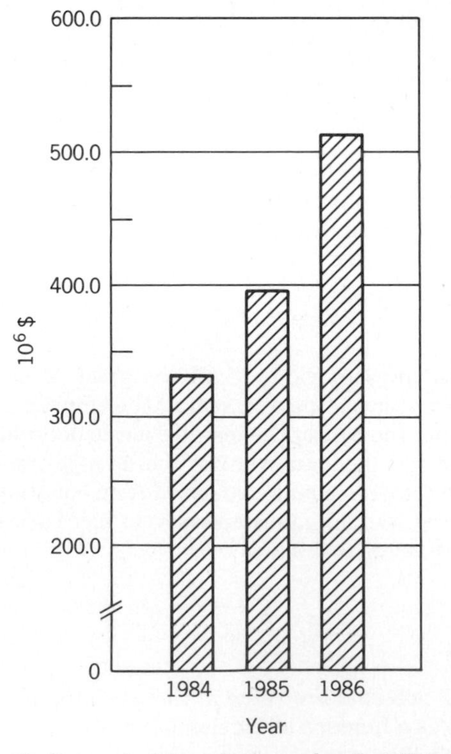

Figure 6. Forecast of the robotic market for 1984–1986.

Ref. 24 presents the recent and expected near future growth of the robotics marketas shown in Figure 6. The actual figures are $333 million for 1984, $395 million for 1985, and $510 million expected for 1986, or an annual increase of 25%, which is expected to continue in the future. However, it was noted that many applications of current robotic technology, such as spray painting and spot welding, are becoming saturated, and future applications will involve more assembly operations, particularly in the automotive and electronic industries.

BIBLIOGRAPHY

1. G. Salvendy, ed. *Handbook of Industrial Engineering,* John Wiley & Sons, Inc., New York, 1982.
2. T. J. Williams, "Robots in the Chemical Industry—An Appraisal and a Note of Caution," *Chemical Engineering* **90** (26), (Dec. 26, 1982).
3. Advanced Technology Advisory Committee, National Aeronautics and Space Administration, *Advancing Automation and Robotics Technology for the Space Station and for the U.S. Economy, Volume I—Executive Summary, Volume II—Technical Report,* Johnson Space Flight Center, Houston, Tex., Mar. 1985.
4. K. S. Fu, guest ed. "Special Issue on Robotics and Automation," *IEEE Computer* **15** (12), (Dec. 1982).
5. A. Garoogian, *Robotics, 1960–1983: An Annotated Bibliography,* CompuBibs, Brooklyn, N.Y. 1984.
6. S. Y. Nof, J. L. Knight, and G. Salvendy, "Effective Utilization of Industrial Robots: A Job and Skills Analysis Approach," *AIIE Transactions* **12** (3), 216–225 (Sept. 1980).
7. R. L. Paul and S. Y. Nof, "Robot Work Measurement—A Comparison Between Robot and Human Task Performance," *International Journal of Production Research* **17** (3), 277–303 (May 1979).
8. N. Andreiev, "Computer Aided Manufacturing—A Stepping Stone to the Automated Factory," *Control Engineering* **29** (7), 51–55 (June 1982).
9. J. F. Engelberger, *Robotics in Practice,* AMACOM, New York, 1980.
10. D. K. Grierson, "The Automated Factory—An Engineering Challenge," *Instruments and Control Systems* **55** (3), 36–41 (Mar. 1982).
11. J. C. Loundes and co-workers, "Aerospace Factory of the Future," *Aviation Week and Space Technology* **117** (5), 40–101 (Aug. 2, 1982).
12. H. H. Rosenbrock, "Robots and People," *Measurement and Control* **15** (3), 105–112 (Mar. 1982).
13. R. J. Sanderson, J. A. Campbell, and J. D. Meyer, *Industrial Robots: A Summary and Forecast for Manufacturing Manager,* Tech Tran Corp., Naperville, Ill. 1982.
14. J. L. Nevins and D. E. Whitney, "Assembly Research," *Industrial Robot* **7** (1), 27–43 (Mar. 1980).
15. J. L. Nevins and D. E. Whitney, "Computer-Controlled Assembly," *Scientific American* **238** (2), 62–74 (Feb. 1978).
16. C. A. Rosen and D. Nitzan, "Developments in Programmable Automation," *Manufacturing Engineering* **75,** 26–30 (Sept. 1975).
17. S. Bonner and K. G. Shin, "A Comparative Study of Robot Languages," *IEEE Computer* **15** (12), 82–96 (Dec. 1982).
18. G. G. Dodd and L. Rossol, eds., *Sensor-Based Robots,* Plenum Press, New York, 1979.
19. A. Pugh, ed., *Robot Vision,* Springer, U.K., 1983.
20. T. Vamos, "AI as a Tool for Flexible Manufacturing," in E. J. Kompass and T. J. Williams, eds., *Learning Systems and Pattern Recognition,* Technical Publishing Co., Chicago, Ill. 1983.

21. T. J. Williams, *The Application of Digital Computers to Industrial Process Control,* Instrument Society of America, Research Triangle Park, N.C. 1984.
22. W. J. Fabrycky, "Manufacturing Integration: A Life-Cycle Engineering Approach," *Proceedings of the NSF Workshop on Systems Integration for Manufacturing,* University of Michigan, St. Clair, Mich., Nov. 6–7, 1985.
23. W. R. Park and D. E. Jackson, *Cost Engineering Analysis: A Guide to Economic Evaluation of Engineering Projects,* 2nd ed., John Wiley & Sons, Inc., New York, 1984.
24. "U.S. Market Report, Industrial," *Electronics* **59** (1), 51–52 (Jan. 6, 1986).

General References

Reference 23 is a good general reference.

L. D. Miles, *Techniques of Value Analysis and Engineering,* 2nd ed., McGraw-Hill, New York, 1972.

J. F. Young, *Robotics,* John Wiley & Sons, Inc., New York, 1973.

G. A. Taylor, *Managerial and Engineering Economy: Economic Decision Making,* 2nd ed., Van Nostrand, New York, 1975.

M. E. Merchant, "The Future of Batch Manufacture," *Philosophical Transactions of the Royal Society of London* **275A,** 357–372 (1976).

H. J. Warnecke and B. Brodbeck, "Analysis of Industrial Robots on a Test Stand," *Industrial Robot* **4** (4), 194–198 (Dec. 1977).

N. A. Barish and S. Kaplan, *Economic Analysis for Engineering and Managerial Decision Making,* 2nd ed., McGraw-Hill, New York, 1978.

G. W. Smith, *Engineering Economy: Analysis of Capital Expenditures,* 3rd ed., Iowa State University Press, Ames, Ia., 1979.

K. K. Humphreys and S. Katell, *Basic Cost Engineering,* Marcel Dekker, New York, 1981.

C. A. Collier and W. B. Ledbetter, *Engineering Cost Analysis,* Harper and Row, New York, 1982.

E. L. Grant, W. G. Irison, and R. S. Leavenworth, *Principles of Engineering Economy,* 7th ed., John Wiley & Sons, Inc., New York, 1982.

J. L. Riggs, *Engineering Economics,* 2nd ed., McGraw-Hill, New York, 1982.

T. Au and T. P. Au, *Engineering Economics for Capital Investment Analysis,* Allyn and Bacon, Boston, Mass., 1983.

L. T. Blank and A. J. Targuin, *Engineering Economy,* McGraw-Hill, New York, 1983.

R. C. Dorf, *Robotics and Automated Manufacturing,* Reston Publishing Co., Reston, Va., 1983.

D. G. Newman, *Engineering Economic Analysis,* 2nd ed., Engineering Press, San Jose, Calif., 1983.

M. W. Thring, *Robots and Telechirs,* John Wiley & Sons, Inc., New York, 1983.

E. P. DeGarmo, W. G. Sullivan, and J. R. Canada, *Engineering Economy,* 7th ed., Macmillan and Co., New York, 1984.

Y. Koren, *Robotics for Engineers,* McGraw-Hill, New York, 1984.

M. Kurtz and R. A. Kurtz, *Handbook of Engineering Economics: Guide for Engineers, Technicians, Scientists and Managers,* McGraw-Hill, New York, 1984.

G. J. Thuesen and W. J. Fabrycky, *Engineering Economy,* 6th ed., Prentice-Hall, Englewood Cliffs, N.J. 1984.

J. A. White, M. H. Agee, and K. E. Case, *Principles of Engineering Economic Analysis,* John Wiley & Sons, Inc., New York, 1984.

R. J. Brown and R. R. Yanuck, *Introduction to Life Cycle Costing,* Prentice-Hall, Englewood Cliffs, N.J. 1985.

EDUCATION, ROBOTICS

JACK LANE
Robotic Integrated Systems Engineering, Inc.
Flint Michigan

INTRODUCTION

Robotics education and training in the United States have expanded greatly since 1980. The expanded usage of robots over this time period and the increased market forecasts for the future have resulted in increased interest in the education and training of personnel for all robotic-related job functions and responsibilities. As a result, efforts to this end have increased in institutions and companies involved in the use of education associated with robots.

This article discusses the current activities toward robotic education and training efforts in the United States. An attempt is made to summarize the activities being pursued at all levels, without elaboration on the detailed course content and subject material being presented.

EDUCATIONAL INSTITUTIONS

Institutions at all levels of formal education are becoming more actively involved in robotics. High schools, colleges with two- and four-year programs, and universities with advanced degree (master's and doctoral) programs are either initiating or expanding existing programs. The increased attention has resulted from industry's need for personnel at all job levels, which are as follows:

1. Users. Management, engineering, maintenance, and research and development.
2. Manufacturers (Robot). Design, engineering, application, maintenance and research and development.
3. Education. Teaching, and research and development.

As a result of the need for the above personnel all over the United States, robotics education should start at an early age and continue to the point where one decides to end one's formal education. This is occurring in the United States, as in other countries, and is discussed below.

High School

Since many colleges and universities in the United States have implemented the study of robotics in their programs, many high schools have looked at how they might introduce their students to the field of robotics. The term "robotics" is a "buzz word," and it appears more and more on television and in the newspaper. Since potential exists for jobs and careers in the robotics field, high schools are being placed in the position of having to meet the demand for knowledge in this field.

In addition, colleges, universities, and user companies are suggesting that instructional programs at the high-school level are needed for preparing students for further education and possible job placement. As a result, several high schools have implemented courses that either act as prerequisites for a good background and are associated with robotics, and some have implemented robotics programs. In the former case, courses are offered in the following areas:

Introduction to robotics.
Computers.
Electronics.
Pneumatics/hydraulics.

In the latter case, some high schools have developed laboratories which normally contain computers and educational robots and provide the students with "hands-on" experience. The laboratory sessions are normally supplemented by lecture courses on introduction to robotics, computers, programming, and introduction to power systems.

Community Colleges

Community colleges all over the United States are offering robotics technician training programs. Many of these have been implemented specifically at the request of manufacturing companies in the immediate vicinity of the colleges. Joint efforts have resulted in the development of programs for the following reasons:

Training technicians for the robot user and supplier.
Upgrading the skills and competencies of skilled trade personnel.
Providing training to hourly personnel to upgrade their job skills for better paying and more secure jobs.

Many educators and other people feel that the community colleges have overreacted to the needs of industry by educating and training more personnel for the robotics industry than are required. Many of these colleges have recently modified their programs to be more generic in terms of where their graduates might utilize their acquired skills and abilities. They have modified their programs to become broader in study, incorporating automation and automatic manufacturing systems and their required affiliated subject areas. As a result, graduates of these programs can now be employed by users and industries that do not use robots or that employ them on a limited basis.

Colleges and Universities

Many colleges and universities with undergraduate and graduate programs have introduced robotics into their educational program in the form of courses, options within a degree program, and degree programs. Although survey information is not available for the year 1986. Past surveys will be quoted to give better insight into the number of institutions and the growth from 1979 to 1983. The first survey, in 1979, indicated that there was a total of 24 colleges in the United States with robotic activity. Three were at the associate level, and 21 at the bachelor's and graduate levels. The RI/SME survey taken in 1982 (ie, two years later) indicated that the total had increased by 400%, with a 900% increase at the associate level and a 340% increase at the bachelor's level. The March 1983 State University of New York College of Technology survey indicates that the dramatic increase in robotics education activity is continuing. The total activity has increased by 180%, with a 207% increase at the associate level and a 170% increase at the bachelor's level. At this point in time, this survey indicates that there are 56 colleges with associate degree programs and 119 with Bachelor's and graduate programs, or a total of 175 with robotics courses. Presently, it is thought that approximately 2000 colleges and universities are offering courses or programs in robotics.

In the United States, there are very few colleges or universities that have degree programs in robotics. Possibly four or five do exist, but this has not been verified.

Some of the universities having graduate programs at either the master's or doctoral level have implemented robotics programs with the intent to produce graduates with more theoretical knowledge so that they might be in better position to enter the teaching profession or to continue research and development interests within various robot-user or robot-manufacturing organizations.

Research opportunities are offered to graduate institutions by federally funded and corporate-sponsored programs. These programs are advancing the field of robotics and are having an impact on the future use of more intelligent and sophisticated robotic systems.

ROBOTICS-USER PROGRAMS

The increased use of industrial robots in the United States by the automotive, appliance, and other industries has resulted in a greater demand for education and training of personnel in all types of job capacities and functions. Management, engineering, maintenance, supervisory, and hourly operating personnel all require training in varying subjects and in different depths. It has been recognized by users of large numbers of robots that to be successfully implemented and utilized the skills of personnel need to be improved and updated. To give insight into what has been done, a description of some of the training efforts being pursued by the U.S. automotive industry is discussed.

Types of Programs

Many different programs are offered by the U.S. automotive companies to their employees. These are offered at either the plant or some other location, depending upon the type of program and program requirements. When equipment and facilities are not available at the plant, various education and training resources that might require travel on the part of the participants are utilized. Many of the automotive companies have developed their own robotics training facilities and have properly equipped them to accommodate training for different programs and vendor-specific equipment.

Since each automotive company has developed its own training programs, it is impossible to discuss the details of each. The following list gives an indication of the types of programs offered without relating them to a particular company. Many diversified programs are being presented.

Types of Robotic Training Programs

Robot basics.
Robot safety.
Robot application programs.
Management overview programs.
Vendor-specific—programming.
Vendor-specific—maintenance.
Pneumatics and hydraulics.
Electronics.
Computers.
New technology—vision, tactile sensing, artificial intelligence, and so on.
Welding controls.

Training Facilities

An important part of the robotics training efforts is the facilities, equipment, and materials that are used for conducting the various robotic programs. The dedication to training at various levels involves a financial commitment by automotive management to the establishment of the required facilities. Since it has been recognized that training is essential, facilities of various types have been established with different priorities and objectives. This article will not elaborate on any specific company's facilities, but will discuss those things that are common and generally used in any robotic training facility.

Course Training Materials

The materials utilized by the automotive industry are mostly materials developed in-house and by the robot manufacturers. They are

Proprietary and vendor-specific manuals and written material.
Audio-visual materials such as slides, films, and video cassette tapes.
Case studies with hardware.

Training Robots

Most of the robotics training performed by the automotive industry is done with the actual robots used in production. These robots are located either at the plant site or in a central robotic training facility. Generic robotic training permits the use of robots other than those to be used in a particular plant's production operations. However, this type of training is usually supplemented by vendor-specific training on a particular manufacturer's robot.

Educational robots are used by some automotive plants for acquainting skilled tradespersons with robot operation, components, and controls. However, these robots are very limited in use at the present time.

Training Classrooms

The classrooms used for conducting training programs are usually well equipped with audio-visual equipment such as videocassette units, overhead projectors, and slide projectors. The classrooms are normally fairly small in size with a seating capacity limited to less than 30–35. The classrooms are usually dedicated to the teaching of various robotic subjects and have robotic components and parts available for observation by the participants.

ROBOT MANUFACTURERS

Robot manufacturers in the United States have all developed training programs to meet the needs of their customers in properly understanding, operating, and maintaining robots and other peripheral equipment used in conjunction with the robots. These programs are offered at the robot manufacturer's location or can be contracted to be performed at the user's facility.

The majority of the robot suppliers have developed a course that covers the robot components and systems, and the interaction of these systems during operation, programming, maintenance, adjustment, and repair. Following this course, specific maintenance training and programming training courses are presented, with the following general format:

Maintenance Training

Course material	Equipment operation manuals with trouble shooting procedures
Audience	Plant maintenance or service personnel
Program duration	One to five days
Equipment	Production robots or laboratory training robots if available
Method of presentation	A combination of classroom and laboratory presentations and exercises.

Programming Training

Course material	Robot programming manuals (vendor-specific)
Audience	Engineering, maintenance, and other technical personnel
Program duration	One to five days
Equipment	Production or laboratory robots and other teaching audio-visual aids
Method of presentation	A combination of classroom and laboratory sessions and exercises

PROFESSIONAL SOCIETIES

The involvement of professional societies in robotics education and training efforts in the United States has been mainly to organize conferences, seminars, and special programs for engineers and management. They have not been involved with the skilled trades or operating personnel, as would be expected.

Professional societies have played an important role in promoting robotics and have largely been responsible for informing industry and the general public about the merits and virtues of robots. They have provided assistance to industry in seeking knowledgeable personnel and programs that have been beneficial to the robot users in continuing educational efforts for engineering and management personnel. Through cooperative efforts on a professional level, they have been responsible in promoting an awareness of the use of robots (see also AUTOMATION, HUMAN ASPECTS).

CONCLUSION

In conclusion, it can be said that robotics education and training programs in the United States have greatly increased in number and quality in the past five years. Everyone concerned with robotics has met the needs that are required for properly understanding, operating, designing, and maintaining robotics systems. The robotics industry has awakened the manufacturing industry and educational institutions to the need for education and training in order to introduce new technology successfully. With more and more emphasis being placed on automation and new technology, the types of educational programs offered will greatly expand in the future.

ELECTRIC POWER INDUSTRY, ROBOTS IN

HARRY T. ROMAN
Public Service Electric and Gas Company
Newark, New Jersey

INTRODUCTION

An exciting and diverse new application area for robots is emerging—the electric power industry. Faced with rising operation and maintenance costs, increasingly stringent personnel safety regulations, and the economic necessity of maintaining and improving already high levels of system reliability, electric utilities are beginning to evaluate robotic technology applications. These application efforts are currently concentrated in two areas: nuclear power plants and the transmission and distribution of electrical energy.

Over the last five years, robot manufacturers and electric utility engineers have begun to focus on the adaptation of conventional industrial robotic technology to the utility industry. This has resulted in automatic machines and preprogrammed robotic devices being developed and used in today's commercial nuclear power plants. This utility–manufacturer dialogue has now matured to the point where utility industry operation and maintenance requirements are providing the economic incentive for manufacturers to research and develop a new generation of robotic devices that is both mobile and interactive with their human operators. In the United States, the research arm of the electric utility industry, the Electric Power Research Institute (EPRI), has played a key role in the identification, assessment, and evaluation of robotic applications. EPRI has also acted as a clearinghouse for the integration of international utility industry robotic experience.

It should be noted that robots being developed for utility use today are not robots in the classical sense of the U.S. definition, ie, as defined by the Robot Institute of America ("a reprogrammable, multifunction manipulator designed to move materials, parts, tools, or specialized devices through variable programmed motions for the performance of a variety of tasks").

With time and future developments, however, these devices will ultimately evolve into machines that fit this definition, but now, in their early stages of evolution, they are direct descendants of the remote manipulator arms developed in the late 1940s to handle radioactive materials (1):

> Robots in the utility industry today might be classified as in the Neanderthal stage of evolution; a stage characterized by only rudimentary intelligence, the use of only simple tools, and the ability to survive in a harsh and hostile environment.

These "robots," as they shall loosely be referred to in this article (and in honor of their future inevitable status), operate in real time and under direct human control. They interact with their environment basically through a camera(s) system which provides the human controller with a visual sense of "being there." Equipped with remote vision capability, a manipulator, and some instrumentation, a trained utility operator can perform a hazardous task from the comfort and safety of a remote-control console.

NUCLEAR POWER PLANT APPLICATIONS

Background

A typical commercial nuclear power plant is a complex facility capable of continuously producing 1,000 MW of electric power. With numerous sophisticated subsystems and safety-related hardware, these modern-day generating stations require hundreds of trained personnel working around the clock to service and maintain over 30,000 valves and many other associated plumbing and mechanical devices. Annually, tens of thousands of hours of operation and maintenance time are required to perform this necessary work as well as comply with stringent governmental regulations that mandate the routine inspection and monitoring of critical plant components and safety systems. Much of this inspection and maintenance requires that the plant itself be out of service or drastically reduced in power output in order to lower radiation down to levels under which humans can safely work.

With the rising cost of commercial nuclear power plant maintenance and continued regulatory pressure to reduce plant-personnel radiation exposure, electric utilities worldwide are actively evaluating the use of robotics technology. Basically, the people managing electric utilities would like robotic devices to:

- Reduce the routine occupational exposure of personnel during the normal course of plant inspection and surveillance
- Reduce occupational exposure of personnel during necessary plant maintenance activities
- Improve existing preventive maintenance practices to reduce personnel exposure

- Provide for in-service inspection, surveillance, and on-line maintenance, which will reduce plant downtime and save costly replacement power
- Have the capability to deal with radioactive leaks or spills effectively, thereby reducing personnel accident exposure levels
- Reduce the amount of radioactive waste materials and expendibles (an area rapidly escalating in importance and cost for utilities using commercially available disposal sites)
- Assist in radioactive waste handling activities, thereby reducing personnel exposure levels

In addition, some nuclear plant robotics advocates would ultimately like to see the construction and decommissioning of plants chiefly accomplished with robotic devices. In the future, robots could be included in the design of standardized nuclear plants for optimal utilization of their capabilities. The major cost-saving and expense factors associated with the use of robotic devices in nuclear power plants are outlined below (2):

Robotic Application Cost-saving Factors

Factors should include all costs associated with not requiring workers to enter controlled radiation areas:
Reduction in exposure (radiation as well as environmental)
Reduction in personnel hours
Reduction in crew size
Improved worker safety
Repeatability of job performance/reduction of inspections
Savings in personnel hours for health physics procedures
Reduction in safety equipment, hardware, and instrumentation
Reduction of radioactive waste
Performs multiple tasks in work environment
Reduction in worker illness and injury costs
Potential for reduction of plant downtime
Robots used for "spill/leak" cleanup

Robotic Application Expense Factors

Purchase cost of robot
Site-specific installation cost
Robot operation and maintenance (O&M)
Robot utilization (single/multiple use)
Operator/technician training
Decontamination cost to move robot around plant
Establishment of robot support staff

Generally speaking, "as low as reasonably achievable" operational radiation exposure for personnel (as specified by the U.S. Nuclear Regulatory Commission (NRC); present rules of the NRC limit workers to a maximum cumulative radiation exposure of 3 REM in 3 mo and no more than 5 REM in a year) is usually accomplished through worker dilution, ie, larger work crews (often supplemented with temporary or transient workers) with less overall dose per worker. The result is high personnel costs. With personnel exposure costs often exceeding $5,000 per man-rem in U.S. power plants, those tasks with high total radiation exposure and/or tasks that occur often with moderate radiation exposure are likely candidates for robotic applications. The cost of replacement power is also a central economic concern, especially with many state regulatory and consumer groups (3):

> Purchased replacement power to substitute for the output of a 1000-MW_e reactor costs an average of $500,000 a day. Robots potentially could contribute to improve plant availability by avoiding delays in scheduled outages and handling some tasks while the reactor is operating.

Recently, the EPRI and the NRC sponsored in-depth studies (4,5) to determine how robotic devices could be used in nuclear power plants and the cost-effectiveness of applications. Both studies basically concluded that robotic devices can be used to reduce personnel radiation exposure significantly and reduce work crew size. Cost savings were found to be substantial. The EPRI study found that for a projected robot cost of $200,000, cost savings for the maintenance and health physics tasks identified in the study could range from $100,000 to $1 million. This same study also cautioned about the limited availability of robotic equipment geared to the nuclear industry and pointed out the need for the industry to fund its own developmental programs directly and help establish the marketplace for robotic designers and manufacturers.

Commercially available robot technology must be modified before use in power plants. The use of robots may also require the modification of plant equipment and maintenance procedures to facilitate the integration of these devices into the plant. Sometimes the simple rethinking of maintenance tasks in a plant can achieve significant savings without the installation of a robot. Moving stationary robotic technology from the factory floor to the electric utility environment will require a great deal of modification. Today's industrial robots will need to be sophisticated, ie, given mobility and senses, and hardened to withstand the often harsh environments that the humans they are designed to replace are exposed to.

The NRC study, which based its conclusions on the examination of three existing nuclear power plants, concluded that commercially available robotics technology will also reduce radiation exposure and save plant operating costs. However, it cautioned that the economic retrofit of robotics technology into existing power plants is a site-specific situation. Benefits can vary significantly among plants due to design differences, operating experience, and site-specific conditions. Some of the conditions that need to be addressed when the retrofit of robotic devices into existing nuclear power plants is being considered appear below (6):

- Each plant contains a large number of radiation areas (rooms) that require an approved permit for entry
- The radiation areas are located at different elevations within the reactor and auxiliary buildings
- Reasonable access is provided for equipment transport to most radiation areas, but some are difficult for even a worker to enter
- There is no defined aisle around much of the equipment and piping within most radiation areas, especially in the lower containment areas of PWRs

- Suit-up rooms, monitor stations, and/or step-off pads are located at the entry to most radiation areas to minimize tracking of contamination into hallways
- Tasks within some of the areas require workers to climb stairs, vertical ladders, spiral staircases, and to maneuver under/over pipes and other obstructions
- Some concern exists regarding the use of wireless signal transmission using radio frequencies within the containment building
- Some areas require entry for surveillance every shift, whereas others are entered monthly or less frequently
- Many of the areas, especially in PWRs, have floor obstructions such as water dams, pipes, and ducts
- The surveillance of some areas must be completed quickly when on critical path for plant outage, and a team of workers is deployed to do the work
- Routine surveillance during power operation is performed simultaneously by a number of worker teams because of the many areas involved
- Many of the areas have very poor lighting, few electrical outlets for power-supply connections, and restrictions on the addition of wall penetrations

The NRC study used a cost-effective approach for robotics application (CARA) technique which recognized the need to examine the feasibility of robot applications on an individual plant area-by-area basis. Boiling water reactors (BWRs) are more likely to employ robotic devices than pressurized water reactors (PWRs) because a BWR generally has more access space than a PWR and is more likely to have higher levels of radiation that can be reduced (see Fig. 1) (7).

A sampling of potential nuclear plant robotic applications that were identified in the EPRI and NRC studies and at a recent EPRI-sponsored conference (8) follows (9) (see also MOBILE ROBOTS; NUCLEAR INDUSTRY, ROBOTS IN):

Radiation monitoring/general health physics task/smear test
Sampling radioactive systems
Small-spill cleanup
Changing cartridge filters
PWR steam-generator repair
BWR recirculation piping repair
Roving robot for:
- Routine radiation monitoring
- Steam-, gas-, fluid-leak detection
- Noise, vibration
- Security/intruder alerts
- Valve/critical instrumentation/equipment status verification and operation
- Infrared scan

Surface cleanup
Condenser tube-sheet inspection and repair
Post-accident monitoring/cleanup/repair
Restricted access/hazardous environment maintenance
Reactor-containment inspection

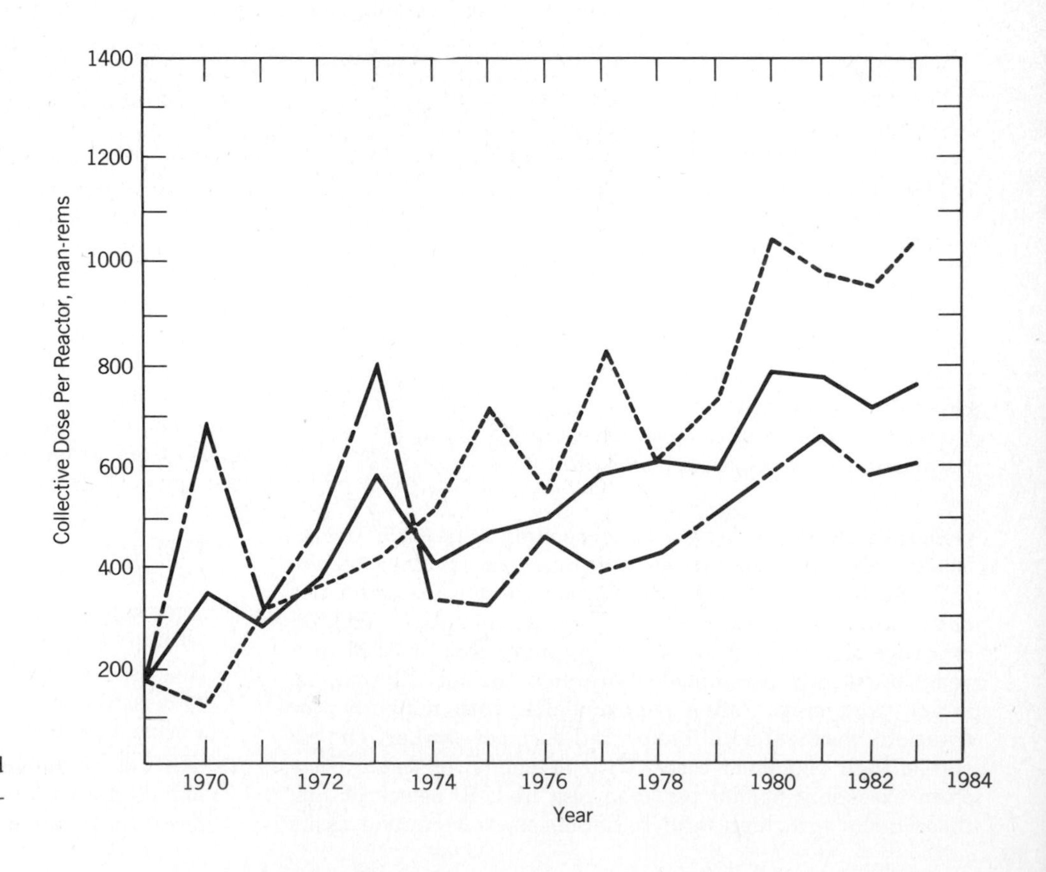

Figure 1. Commercial light water-cooled reactors, 1969–1983: ------- , BWR ____ __ __ ____, PWR; ______, LWR.

- Valve repair/manipulation
  - Steam-gland tightening
  - Steam-gland repacking
  - Bonnet removal
  - Disk lapping
  - Reassembly and bonnet retorquing
- Internal/external pipe inspection/repair
- Equipment and tool transporters
- Pipe/tube x rays
- Painting/coating/cleaning of surfaces
- Sandblasting or chemical cleaning of metal parts and components prior to machining, milling, or repair
- Turbine inspection during operation
- BWR turbine inspection/maintenance
- Snubber inspection
- Worker observation in hazardous areas
- Reactor-cavity cleaning
- Flange unbolting/rebolting for valve maintenance
- Lubrication
- Pipe welding
- Weld grinding/inspection
- Radwaste/drum handling
- Pressurizer surveillance
- Fire fighting/firewatch
- General station security
- Surveillance inside bioshield
- Personnel transport
- Miniature probes for visual inspection in areas inaccessible to fiber optics, eg, pipe inspections
- Drywell observation
- Control rod drive surveillance

An interesting conclusion of the NRC study and the basis of EPRI's funded research is that to perform nuclear inspection and surveillance effectively, a family of robots or a "robotic staff" would be needed. This implies that robots will need to be designed in a modular fashion, capable of quick assembly/disassembly, easy decontamination, and performing more than one task. It is unlikely that any utility will purchase an expensive robot that can only perform a single task. Mature nuclear power plant robot designs can be ultimately envisioned as discrete subsystems carefully designed to work compatibly in various combinations (10,11).

In the future, nuclear plants may have on staff a retinue of robotic subsystems designed to be quickly assembled to do one task, decontaminated, reassembled into a new configuration, and then used again the same day for another plant task.

TODAY'S NUCLEAR PLANT ROBOTIC DEVICES

Over the last 35 years, robotics technology in the form of remote manipulator applications in the nuclear industry has been traditionally used in spent-fuel processing, post-irradiation analysis, waste management, and research and development (12). Variations of this technology are now being applied in commercial nuclear reactors for a variety of applications. For the most part, this technology is represented by dedicated, single-purpose, preprogrammed devices. These first-generation machines often provide little or no feedback to their operator(s) except for perhaps visual input from a remote camera system.

The following EPRI quote (13) (with author-provided references added) excellently summarizes the current state of the art of robotic applications for nuclear power plants:

> A large part of the inspection and repair of steam generators (14) is now performed with automated, preprogrammed robotic devices. Such devices walk along the bottom surface of steam generator tubesheets, for example, cleaning the inner surfaces of the tube ends and inserting plugs to maintain steam generator tube integrity. Other automated probes, armed with computerized diagnostic aids, search the tubes for signs of deterioration. Similarly, ultrasonic detectors (15) mounted on remotely operated scanners, traverse primary coolant piping (16), sending signals to portable computers that can characterize defects. Automatic devices have been developed that cut pipe, weld pipe (17), grind weld contours, and perform operations needed to repair nuclear plant primary piping. Automatic, remotely controlled devices have also been developed for such diverse tasks as processing ion-exchange resins and changing pins in control rod guide tubes. These devices have been developed by reactor manufacturers, nuclear service companies, and EPRI.

A classic example of an existing robotic device (programmable manipulator) for nuclear plant maintenance is Westinghouse Electric Corporation's remotely operated service arm (ROSA) (18–20). First introduced to the electric utility industry in 1983, ROSA (Fig. 2) is actually a family of robotic arms with a common design. These arms are composed of modular, electrically driven actuators (joints) that share the same control system and are operated from a remote-control console. Each actuator is a self-contained unit with its own motor, gear train, brake, and position-feedback elements. Taking advantage of modular actuator and link construction, arms can be built to suit a specific task, with failed arm units easily replaced. ROSA has been used for underwater-reactor internal repair, steam-generator tube plugging and sleeving, eddy-current examination, and tube-end repair. ROSA's highly articulate design allows it to enter, by remote control, the steam channel head of a PWR steam generator through an existing entrance and attach itself to the tube sheet. Here, it can position its operating tool and begin work without requiring operating personnel to enter this high radiation area. When new end effectors are required, it reaches back through the entrance to a shielded, protected worker outside the channel head who attaches a new tool onto it.

In May 1984, ROSA was utilized at a power plant to perform a complete steam-generator tube-end repair; 2200 tube ends were hard-rolled and 1200 tube ends expanded. At this task, ROSA demonstrated a 92% availability, saved 50 working hours in hard-rolling time, and reduced total personnel exposure by approximately 540 man-rem (20). In a European plant sleeving operation, ROSA operated nonstop for 3 d, honing, swabbing, inspecting, inserting sleeves, hard-rolling, brazing, and ultrasonically testing 17 tubes in a steam-generator chan-

Figure 2. ROSA, Westinghouse Electric's remotely operated service arm.

nel head. Total dose to personnel during this entire operation was only 5 man-rem (21).

Babcox & Wilcox Company, another familiar name in electric utility fossil and nuclear steam supply systems, has also developed a remotely operated generator examination and repair (ROGER) device that can be used to examine and repair PWR steam-generator tubes. ROGER, like ROSA, is a computer-controlled remote inspection and repair manipulator. It is a 3-piece, 3 degree-of-freedom arm comprised of 3 mechanical joints connected by structural members, which rotates around a fixed mast in order to reach all of the steam-generator tubes (22). It has the following characteristics (23):

> Installing the equipment does not require entry into the steam generator channel head—an important feature for exposure considerations since fields of over 14rem/h can be experienced in these areas. The arm can carry out a full range of inspection, leak detection and repair operations, including eddy-current testing, helium leak detection, tube plugging and remote welding. Since its commercial introduction in January 1984, Roger has carried out tube inspections and repairs on 18 steam generators (13 recirculating and five once-through), with an estimated personnel exposure reduction of 30–50 per cent.

The Quadrex Corporation also has developed a version of a steam-generator inspection arm called the tube inspection manipulator (TIM). This "zero dose" entry device can perform eddy-current probing, tube profilometry, or light machining of tube ends. TIM can be operated in a manual mode or a master–slave computer indexing mode(s) (24).

Recently, the Belgian Laboratory of the Electricity Supply Industries (Laborelec) has implemented a fully automated polar manipulator (25) for Belgian PWR steam-generator tube inspections. In development since 1977, this device can be preprogrammed to inspect steam-generator tube sheets once it is assembled by 2 trained personnel. Unlike ROSA or ROGER, however, this device must be installed by trained personnel inside the steam generator. ROSA and ROGER are designed for "zero" entry installation and setup. The Belgian device is reported to have a tube-inspection speed as high as 250 tubes/h.

General Electric Company envisions that for each of its BWR-supplied plants there exists a $40 million/yr market in sophisticated services (26):

> At present GE offers a new low level waste disposal process, remote control BWR pipe inspections and elaborate maintenance training facilities. Future markets for the company will be services aimed at extending the life of a unit beyond the currently licensed 40 years, decommissioning (GE is the main contractor on the decommissioning of the Shippingport reactor), and waste disposal.

Robotic devices have wide potential during the decommissioning process, and their specific application will largely depend on the type of decommissioning option selected (safe storage, decontamination, or entombment). The cost of a robotic device may be more easily justified if its later use as a decommissioning tool is also anticipated (27).

Remote handling and computer-controlled operation is built into Canadian "Candu" heavy-water nuclear reactor power plants. To date, Canada has accumulated 22 years' experience using remote-controlled devices (28), generally to facilitate on-line refueling. The Canadian systems use graphic-display monitoring of process status with operator intervention initiated as necessary to correct malfunctions (29):

> The trend of remote handling equipment development is towards increasing use of computer control of sophisticated task-specific systems so the operator can concentrate on managing the operation.

and (23):

> In Canada, the development of remote repair and maintenance has been principally oriented towards the inspection, adjustment and replacement of pressure tubes. Large scale pressure tube replacement has from the inception of the Candu system, been envisioned as a planned refurbishing process to extend reactor life.

An example of this technology is the work Spar Aerospace, Canada, is doing with Ontario Hydro on a large remote manipulator system designed to retube Candu nuclear power reactors completely by remote handling methods. This system utilizes the controls-systems technology developed for the space shuttle's "Canadarm" (30).

MOBILE ROBOTS—THE NEXT GENERATION

As success has been achieved with the preprogrammed robotic and remote manipulator devices described above, utilities and robotic manufacturers are pushing back the frontiers of technology to reveal a new generation of robots, ie, multipurpose, mobile robots, designed to be interactive with their operator and working environment. Mobile robotics technology borrowed from existing applications in such diverse areas as military/ordinance retrieval, nuclear fuel/waste handling, fire fighting, mining, underwater research/retrieval, and ship-hull cleaning is now being researched for adaptation and modification for use in nuclear power plants.

The concept of a mobile robot has particular appeal for use in nuclear plants. Actually, there is an emerging "class structure" to mobile robots (31):

> Just like the utility workers they may soon displace in part, different robots have different strengths and weaknesses. No single robot is best at performing the spectrum of utility maintenance and surveillance tasks. Mobile robots can be divided broadly into two groups: those that do physical work, such as scraping or turning screwdrivers, and those that do strictly surveillance, such as inspecting or collecting data.
>
> The work robots can then be subdivided into two types: so-called "end effector" robots, which have grippers at the end of manipulator arms and perform delicate, nimble jobs, like grasping objects or screwing in bolts; and "work drones," which have heavy attachments for robust jobs, like scrubbing and chipping.
>
> The surveillance robots have none of these characteristics but instead come equipped with more advanced communication and vision systems than the work models.

All mobile robots do have several commonalities: a propulsion or locomotion system, a communication system, and "work packages" (31). Propulsion systems can be wheeled, tracked, or recently, walking- or leg-type devices. Power to these systems comes from on-board batteries or a power cable tether. Communication can be by radio/microwave or coaxial/fiber-optic cable in conjunction with a power tether. Work packages are the "business end" of the robot or what its mission is. For instance, a remote surveillance vehicle could be developed to perform a wide range of routine inspection, surveillance, and maintenance tasks, thereby freeing human workers from radiation exposure. This vehicle, equipped with various sensors/instrumentation, could conceivably monitor radiation levels, read gauges, view restricted areas through camera eyes, monitor temperature and humidity, scan critical equipment for noise or infrared emissions, or measure component vibration. The ideal mobile robot would be designed so that work packages could be conveniently changed to perform a range of similar work tasks utilizing common propulsion and communication systems.

Mobile robotics development is a humbling experience for the designer. Nowhere does it become more evident to the engineer how marvelous the human mind and anatomy are than when attempts are made to replicate their profound sophistication in the form of a machine. The versatility of the human hand, with over 20 degrees of freedom, dwarfs today's best and most dexterous end effector. A person can reach out in 3-dimensional space and pluck an object off a table effortlessly, demonstrating that the response of human graphomotor coordination is truly awesome compared with today's most advanced robot. However, the challenge is not to replicate the human but to provide the means to take the human presence, through the robot, into hazardous environments. The robot is a tool, a means by which a job is effected from a distance (telepresence). A well-designed robot combines the best of human and machine.

The EPRI is providing the focus for U.S. utility development of robotic devices and the refinement and modification of existing robotic concepts. It has sponsored the development of a mobile, untethered industrial remote inspection system (IRIS) (10,32). This device (Figs. 3 and 4) is strictly designed as a remote surveillance and inspection vehicle for hazardous environments. For all practical purposes, this 300-lb, tanklike, dual-tracked vehicle is a rolling instrumentation package, capable of carrying optical, audio, radiological, and environmental sensors into controlled radiation areas. Its unique, high frequency wireless communication system was designed for operation in a crowded, cluttered, physical environment, such as that which may be encountered in an existing nuclear plant. Stereoscopic vision provides the control operator with a 3-dimensional view of vehicle surroundings. IRIS is battery-powered, capable of maneuvering through 6 in. of water, and can climb stairs of not more than 45° in slope. It has a 32-in. vertical clearance. IRIS can also be adapted with a telescoping arm to take multiple smear samples from floors, walls, and equipment surfaces. The device has undergone extensive testing at EPRI's Nondestructive Evaluation Center (NDE) in Charlotte, N.C. and at Duke Power's Catawba Station. It is now offered for sale under the product name SURVEYOR to the utility industry by its producer, Automation Technology Corporation (ATC) of Columbia, Md.

In the long term, EPRI may sponsor research to fabricate and test a maintenance robot based on Odetics, Inc.'s ODEX-

Figure 3. IRIS surveyor.

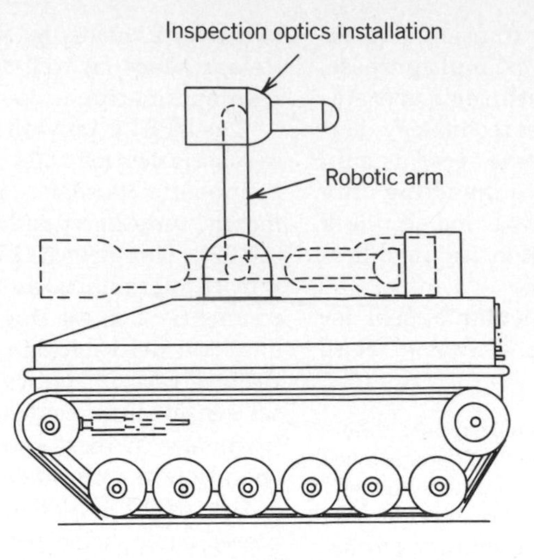

(a)

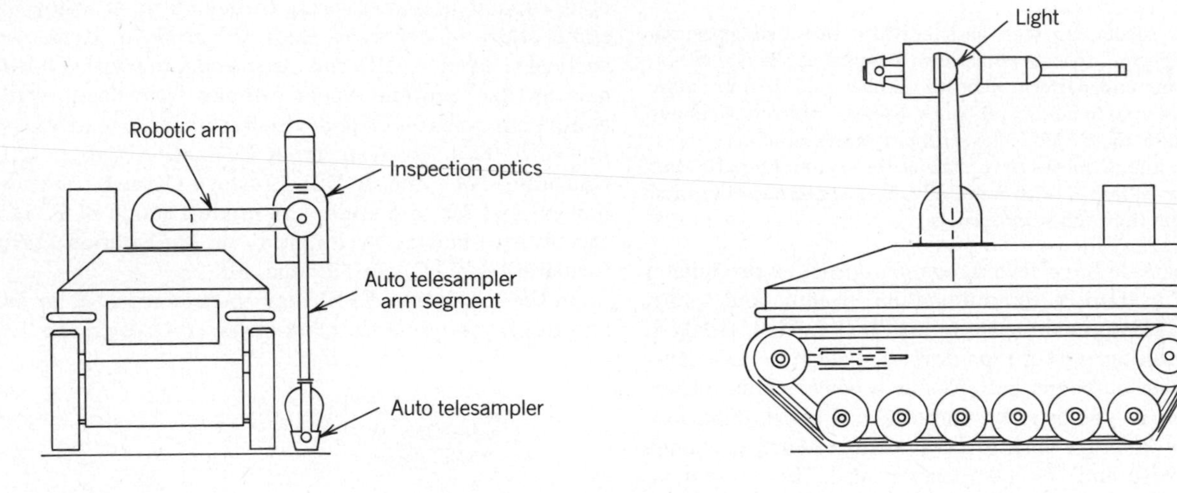

(b)

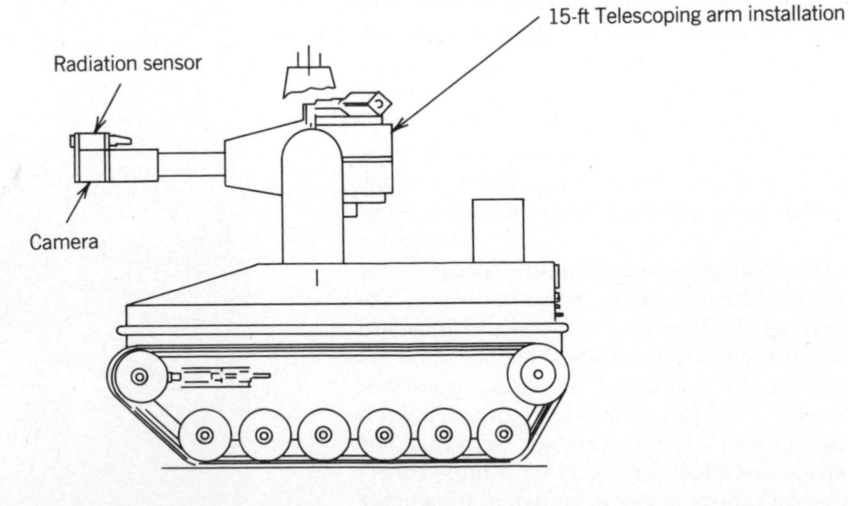

(c)

Figure 4. IRIS surveyor configurations: **(a)** with inspection optics; **(b)** with inspection optics and smear sampler; **(c)** with 15-ft telescoping arm installation.

1 (33,34) functionoid. This 6-legged walking device is suited to a wide variety of tasks which can be accomplished in a cluttered nuclear plant, including climbing stairs. It has an unusually high strength to weight ratio (5–1) which would enable it to transport tools and other heavy maintenance equipment. ODEX derives its unusual mobility from 7 on-board microprocessors; 6 control the legs (one for each leg), and the 7th acts as the integrating control computer. ODEX can modify its omnidirectional profile to fit through doorways or tuck into a squatting position. It is easily operated from a remote console, has no umbilical cord, and is battery-powered.

Recent Canadian experience with an off-the-shelf, 6-wheeled, articulated remote mobile investigation vehicle, equipped with a manipulator arm and camera, suggests that such simple devices can also play an important role in nuclear plants. The vehicle, made by Pedsco Canada and designed for bomb-disposal work and hostage incidents, operated flawlessly (after a few minor modifications) in cleaning up a highly radioactive fuel-element spill where total dose to the vehicle was an estimated 10,000 rem. The device is now being considered for routine nuclear plant radiological surveys, visual surveillance, and materials/components recovery (29).

The Japanese are also moving forward in the development of robotic devices for use in nuclear power plants. Tracked (35) and wheeled–walking (36) mobile vehicles have undergone experimental design and application testing. Plant decommissioning robotic applications are also being considered. A remotely operated robot system has been developed by a consortium of Japanese utility companies and the Toshiba Corporation. The system is designed for inspection inside the primary containment vessel of nuclear power plants (37). It moves around the containment vessel on a multirouted monorail system that runs along upper and lower elevation routes. The robot can enter the primary containment vessel through a personnel airlock and perform various inspection tasks during normal reactor operation. At a recent American Nuclear Society Conference (38), France's national electric utility, Electricité de France, also announced the implementation of a rail-based robotic system for in-service plant surveillance and inspection.

The Japanese Advanced Robot Technology Research Association (ARTRA), established in 1984, is the operational organization for the Japanese Ministry of International Trade and Industry's national project to develop advanced robot technology. A major portion of its work will include the development of a multitask robot for light water reactor maintenance (39):

> It will be directed by Hitachi Ltd., the Toshiba Corp. and Mitsubishi Heavy Industry Ltd. The robot is to be multijoint, multifinger robot that can "walk" on wheels and perform diverse chores in LWR maintenance and repair, such as cutting, welding, part replacement, and in-service inspection, regardless of reactor type.

Table 1 is a compilation of known mobile teleoperated/robotic devices worldwide that are either operational or under

Table 1. Mobile Robots for Nuclear Power Plants

Manufacturer/Supplier	Country	Robot	Type of Locomotion[a]
Automation Technology Corporation	United States	Surveyor	T
Battelle Columbus Laboratory	United States	Rocomp	T
Blocher Motor (CMS Technologies, Inc.)	FRG	MF3	T
Blocher Motor (CMS Technologies, Inc.)	FRG	MF4	T
Carnegie-Mellon University	United States	RRV	W
Cybermation	United States	Kluge	W
GCA/PAR (Martin Marietta Corporation)	United States	Herman	T
Hitachi, Ltd.	Japan	RIS	W
Inspectronics (CEA[b])	France	Ariane	T
Inspectronics (CEA[b])	France	Oscar	T
Inspectronics (CEA[b])	France	Oreste	T
Lawrence Livermore National Laboratory (DOE[c])	United States	Atom	T
Meidensha Electric	Japan	Meirobo	T
Meidensha Electric	Japan	DCR	W
Miti	Japan	MMLW	L
Mitsh-Kobe Ship	Japan	CRDM	W
Mitsh-Kobe Ship	Japan	CRDM	L
NTG Nukleartechnic (Scientific International)	FRG	Unnamed	T
Odetics, Inc.	United States	Odex	L
Pentek, Inc.	United States	Moose	W
Remote Technology Corporation	United States	Surbot	W
Robot Systems International	Canada	Hazcat	W
Rockwell International (DOE[c])	United States	Worm	T
Sumitomo	Japan	ACR	W
ACEF Nuclear Division (TeleOperator Systems Corporation)	Belgium	Telemac	T
Viking Energy	United States	Rod	W
Westinghouse Hanford	United States	Sisi	T

[a] Key; T, tracks; W, wheels; L, legs.
[b] Commissariat à l'Énergie Atomique.
[c] U.S. Department of Energy.

development and that could have application in nuclear power plants (40).

Another robotic application of potential widespread use in nuclear power plants is the pipe crawler. Nuclear-grade pipe-inspection devices by the Quadrex Corporation (24) and Combustion-Engineering (41) use wheeled or track/wheel devices which can negotiate various pipe diameters and perform various inspection routines utilizing cameras and special instrumentation.

THE INFLUENCE OF THREE MILE ISLAND

Three Mile Island (TMI) was an unfortunate accident and a cruel learning experience for the U.S. electric utility industry. From the standpoint of robotics, however, the TMI crisis offered a unique and important opportunity for rapid technological development. TMI provided robotic manufacturers and utility industry personnel with:

1. A wide variety of radiological hazards which served as a realistic basis for various robotic designs (42):

   According to revised estimates, cleanup workers at the Three Mile Island Unit 2 reactor are likely to receive a total collective dose of between 13,000 and 46,000 person rems for the entire cleanup project.

2. An actual test bed for existing and new robotic devices that would feed back robot operating experience to designers and manufacturers.
3. An understanding of the immediate need to work together to adapt/modify existing robotic technology and/or develop new technology specifically for nuclear plant use (43):

   Water reactors are intentionally designed to be repaired by contact means which required personnel entry into the shielded containment building. The TMI accident proved that it is possible to have an accident which prevents entry into the containment building. When this happens, the result is equivalent to having a very large hot cell with no remote viewing or manipulation capability. Hence, no work can be performed within the cell (containment building). TMI engineers have surveyed the nuclear industry for remote equipment that can be used to perform decontamination and cleanup work. They found that commercially available equipment to do this work does not exist.

To prevent such a catastrophic occurrence at other reactor sites, the industry would have to consider upgrading other nuclear plants with robotic or remote handling devices to prevent personnel from being barred access to accident situations. This might not be possible in many cases due to site-specific construction concerns. Furthermore, new nuclear plants would feel the pressure to incorporate robotic applications during design and construction. Robotic design for nuclear plants was now given a significant incentive for development.

An important lesson learned from TMI-2 is that there is a hierarchy of automated "things to do" at a nuclear plant. Some of these things are not necessarily robotic in form, but, when applied in a sequential fashion, lead to an integrated robotic application. Equipment such as hard-wired sensors, radiation sensors, and camera systems can and should be installed in closed radiation areas first, to provide remote surveillance for routine and emergency situations. Had such remote equipment been available at TMI, post-accident data gathering and impact assessment would have been greatly simplified. Robotic equipment should be used as an extension of this basic sensory hardware and integrated with it.

Another insight gained from TMI-2 concerns the need to develop robotic standards (44):

> Cognizance of tasks to be performed in the nuclear industry by remotely-operable devices suggests a need for industry-wide cooperation and effort in developing standards for the use of remote equipment. Development of remote or robotic equipment to perform specific tasks in the maintenance of nuclear facilities and standards for equipment to be worked by remotely-operable or robotic devices would encourage their mass production and progressive improvements in their performance and reliability. This could improve performance of tasks within nuclear facilities and reduce exposures to personnel.

Table 2 details the robotic devices (3,45–47) used at TMI and briefly describes the tasks performed (or to be performed) by them (48). In some of the most radioactive environments, robotic vehicles would be exposed to radiation fields as high as 3000 rad/h (49).

The most widely publicized TMI robotic device is the remote reconaissance vehicle (RRV) known as ROVER (3,50,51). This device (Fig. 5) derives its mobility from a 6 wheel-drive, skid-steer crawler base that has been designed to facilitate the quick change of various payload modules. An on-board reeling mechanism dispenses its power/communication tether. In November 1984, this device twice entered the highly contaminated basement at TMI, the most contaminated area of the reactor building (52):

> Rover was lowered 24 ft. down through a hatch into the dark basement, which had only been glimpsed through tiny optical probes since the March 1979 accident. In eight hours of inspection on the two trips, the tethered robot ventured up to 100 ft in all directions from the hatch. It carried three television cameras (two color and one black and white) and various radiation-detecting instruments.
>
> Operated by two people—one to control its movement via television and one to tend the power tether—Rover waded through 4 inches of radioactive water and as much as 4 inches of sludge. Its six headlights illuminated walls marked by a "bathtub ring" left by the contaminated water. RRV measured average radiation levels from 5 to 35 REM an hour with the highest local reading being 1100 REM an hour near the concrete block wall of an elevator shaft and stairwell, according to Gordon Tomb, public information officer at TMI. In comparison, radiation levels where people regularly work on other floors in the building are around 0.035 to 0.1 REM an hour.

Upon returning from the basement, ROVER was hot-spray decontaminated down to levels below 100 mrem/h and stored on site in a low radiation area. The fact that ROVER could be decontaminated so well is important; robots sent into radioactive environments must be reused or they will never pay for themselves. Tomb verified that ROVER would be extremely expensive to use if it could not be decontaminated enough so

Table 2. Robotic Devices Used at TMI

Robot and Robot Statistics	Robot Use	Description of Task
SISI (Supplied by DOE); first used in 1982; weighs 25 lb (11 kg)	Remote surveillance	This remotely controlled, tracked vehicle was used to inspect and photograph highly radioactive areas in the plant
FRED Used in 1983; weighs 400 lb (181 kg), can lift 150 lb (68 kg), and has an arm extension of 6 ft (1.8 m)	Decontamination	This 6-wheeled, remotely controlled device was fitted with a high pressure water spray that was used to decontaminate walls and floors of selected areas in the auxiliary building
LOUIE (On loan from Westinghouse Hanford); can lift 1000 lb (454 kg) and has radiation-hardened cameras	Radiation monitoring	This device has been used for radiation surveillance and may be modified to operate steam-spray and vacuum decontamination systems
RRV or ROVER (EPRI/Carnegie-Mellon University/General Public Utilities Nuclear/DOE/Ben Franklin Partnership were the developmental sponsors); first used in 1984; weighs 1000 lb (454 kg) and has a unique umbilical cord reeling system; can be quick-spray decontaminated	Remote surveillance	This 2-person-operated remote-surveillance vehicle has entered the dark and once-flooded basement of the reactor-containment building to inspect the highly radioactive sludge left from over 600,000 gal (2,270 m^3) of primary cooling water and other spilled water now removed from the building; another similar vehicle is being prepared to take core-bore-drilling samples in contaminated areas to help determine the depth of radioactivity absorption into concrete surfaces
RWV or remote work vehicle (EPRI/GPU/Ben Franklin sponsoring CMU) to be used in 1985–1986	Sampling	Devices are being developed that can collect liquid/sludge samples and concrete floor/wall core samples and perhaps dismantle minor structural assemblies

workers could modify it to do different tasks. And if the radiation levels ROVER picked up at TMI could be reduced, he noted, well-designed robots used elsewhere should be cleanable (52).

In future missions, the basic RRV design will be modified to deal with obstructions and debris and to collect material samples. Future vehicle missions envisioned include core boring (drilling to determine penetration of contamination into concrete), hydrolazing (water blasting to scour contaminated surfaces), scarifying (scraping to remove contaminated surfaces), material transport of removed contaminants, and remote demolition tasks.

Robot use in radioactive emergency or accident conditions is not new. "Herman," a 1966 vintage device still used today, is a fully mobile manipulator that was designed by Martin Marietta Energy Systems to handle radioactive incidents at nuclear-fuel-related facilities and has seen some duty at commerical nuclear power plants. Although it has not been designated for duty at TMI (53),

> The system is designed to operate as far as 700 feet from its control console, to which it is attached by a trail cable. Its mechanical hand can lift up to 160 pounds and drag up to 500 pounds. It has a reach of up to 11 feet upwards and 4 feet, 2 in. sideways. Two black-and-white television cameras are mounted behind the arm.

Recent work with nuclear emergency mobile teleoperated devices at Oak Ridge National Laboratories (54) suggests that there are 5 critical device design parameters that effectively allow a robot to deal with an accident:

1. *Mobility*. The device should be capable of climbing/descending stairs and maneuvering over and around a wide variety of obstacles, ideally including airlocks. Self-contained power supplies offer greater mobility than tethers.
2. *Reconnaissance*. Visual surveillance utilizing camera systems is necessary to determine the immediate "state" of the situation. High resolution and zoom-camera systems are preferred. Radiation-detection equipment is

Figure 5. ROVER, a remote reconaissance vehicle.

also of immediate need to help plan personnel exposure times.

3. *Availability*. The device should be perferably on-site or able to be put into action in a matter of hours.
4. *Bilateral Manipulators*. A 2-armed manipulator is very useful in an accident situation (rather than a traditional 1-armed device) and provides the operator with a human-like response in what may be a critical time-response situation.
5. *Force Feedback*. Without force feedback, a robotic manipulator could damage valves or other components and thus further worsen the accident situation. It will also speed up operations and ease operator control.

FOSSIL PLANT APPLICATIONS

Robots can, of course, be adapted to fossil plant use, but the compelling cost savings associated with reducing personnel radioactivity exposure in nuclear plants is missing, making a tougher economic argument for the planner. Nevertheless, the strong trend toward fossil plant life extension will dictate more frequent and vigorous maintenance routines as well as on-line automated data-collection processes. With this will come the possibility for robot-based systems and distributed intelligence and muscle around the plant. Also, fossil power plants have their share of hazardous, monotonous, and repetitive tasks that can perhaps justify robot applications (55).

Programmed, automatic devices that are used in nuclear power plants, such as pipe welders and ultrasonic scanners, can certainly be used in fossil plant applications, although to date very little work in mobile robotic and remote handling devices has been done. A list of potential fossil power plant applications follows:

- *In situ* boiler tube welding
- Pipe welding
- Coal pile grooming/loading/unloading
- Condenser tube-sheet inspection and repair
- Roving robot for:
  - Steam, gas, fluid-leak detection
  - Noise, vibration
  - Valve/critical instrumentation/equipment status verification and operation
  - Security/intruder alerts
  - General equipment surveillance
  - Infrared scan
- Fire fighting
- Chemical-spill cleanup
- Restricted access/hazardous environment maintenance
- Internal/external pipe inspection
- Equipment and tool transporters
- Pipe/tube x rays
- Sandblasting or chemical cleaning of metal parts and components prior to machining, milling, or repair
- Slag removal/declogging
- Gas turbine engine-room inspection: surveying, monitoring, and repairing minor problems while in-service
- Removal of hazardous materials such as asbestos

POWER TRANSMISSION APPLICATIONS

Power lines bring electrical energy from the large central generating stations to the utility customers. There are 2 classifications of power lines: the high voltage transmission lines, generally 115 kV and above, that bring bulk energy to large metropolitan load centers by the familiar steel lattice, or polelike, towers; and the distribution lines, generally 69 kV and on down to 4 kV, that bring energy directly to the residential, commercial, and industrial consumers. Both types of power lines have operation and maintenance costs associated with them, and these costs have traditionally been large and escalating (56).

> The maintenance of overhead transmission line systems is presently performed live by means of insulated "hot stick" techniques or by 'bare handing.' When neither 'hot stick' or 'bare hand' operations can be performed, a line outage is often taken, and maintenance is performed with the line de-energized. During inclement weather, maintenance and tasks such as insulator replacement is postponed until the weather improves. Many environmental effects can lead to maintenance delays. These effects include high

wind, temperature extremes, snow and ice, fog, rain, and blowing dust. Delays of several hours due to inclement weather are common, and system outages of many days have occurred to many utilities.

Most utility companies have occasionally had to perform line maintenance during adverse or inclement weather without de-energizing the line. Responding to this high voltage maintenance need, EPRI recently developed a conceptual design for a teleoperator for operations, maintenance, and construction using an advanced technology device called "TOMCAT." Its initial design had 3 objectives based on assisting lineworkers in performing hot-line maintenance tasks:

1. Change out individual insulators in a vertical insulator string.
2. Change out entire insulator strings in an angled or dead-end configuration.
3. Install line spacers or line sleeves in the mid-span of a line.

TOMCAT is a master–slave manipulator that could be installed in place of a line-worker bucket on an aerial truck. It would be capable of servicing energized high voltage lines ranging in potential from 138 kV to 765 kV, with possible upgrade to as high as 1200 kV. TOMCAT would function safely in adverse weather conditions and would not impair the integrity of the power line or associated equipment. The device would be operated from an enclosed remote station on the ground.

If TOMCAT is not to appear a threat to the labor force, it will have to be perceived not as an automated line person, which it is not, but rather as a multifunction tool that will enable the line person to work more effectively, longer, and during a broader range of weather conditions than if present methods were used. In this regard, TOMCAT could actually extend the productive work years of a person working on the line by utilizing older, more experienced but less physically capable personnel as system operators. Also, regardless of technological sophistication, there will always be the need for a human to go aloft for some jobs (57).

TOMCAT benefit–cost analysis centered on the savings it would accrue in regular maintenance of a utility's transmission and distribution system as well as its ability to shorten the time a high voltage line is out of service. TOMCAT can also realize savings in replacement energy if its use can return a major line to service more quickly than is possible without it. EPRI estimates that in some parts of the United States the outage of a 500-kV line can cost a utility $35,000 or more per hour. Analysis of the economic feasibility of TOMCAT indicates that it can be entirely assembled from available off-the-shelf components with only minor enhancements. TOMCAT itself, without the aerial lift truck, costs $300,000–500,000. At $300,000, it is projected that 90% or more of U.S. utilities would experience lower maintenance costs using the device. At $500,000, 50% of utility users would save money.

Other potential applications for TOMCAT that might be developed after the prototypical development of the basic device appear below (58):

- Insulator test and inspection
- Insulator washing and protective coating
- Insulator replacement (individual)
- Insulator replacement (strings) (345 kV, 500 kV, 765 kV, or 1200 kV)
  - Vertical string
  - Dead-end string
  - Compact line strings
- Conductor repairs
  - Minor repairs: attach armor rods
  - Major repairs: attach split repair sleeves or full tension splice
- Conductor hardware repair
  - Spacer replacement
  - Ball marker replacement
  - Wire damper replacement
- Right-of-way clearance
- Substation repair and component replacement
- Underground vault repair or emergency
- Tower and pole maintenance

TOMCAT was recently used by the Philadelphia Electric Company to replace a string of transmission-line insulators. People on the line using the device found it easy to operate and replaced the insulators without trouble. EPRI is now moving toward a full prototype device for actual field testing by the utility industry (59).

Robotic Systems International, Ltd. (RSI) of Sidney, British Columbia, a spin-off of International Submarine Engineering, Ltd., the world's leading manufacturer of remotely operated subsea work vehicles, has also designed a teleoperated device to perform utility-line inspection and maintenance. RSI has developed this human-assisted robot (60) in conjunction with British Columbia Hydro with a contract award from the Canadian Electrical Workers Association. Similar in concept to the EPRI work discussed above, which was performed by the Southwest Research Institute, the RSI work concentrates on the lower-voltage utility-distribution networks consisting of the traditional wood pole lines seen along the streets in cities, town, and rural areas.

Also mounted on the end of the boom of an aerial lift truck, the RSI device consists of several manipulator arms, special tools, and remote video capability. Typical work envisioned to be performed by this device includes insulator replacement, crossarm replacement, conductor repairs, and minor maintenance. Since the RSI device is manipulated from within a truck-mounted control center, it permits improvements in worker safety and increased worker productivity in inclement weather. Preliminary tests of the robotic arm(s) for this device involved British Columbia Hydro workers, who received it favorably since it contributed to safety on the job (61).

Public Service Electric and Gas Company, (PSE&G) has recently identified a number of transmission and distribution robotic applications:

1. Live transmission-line work under adverse conditions:
   - Transmission-line conductor repair
   - Transmission-line insulator replacement
2. Meters: electric and gas
   - Repair or reconditioning
   - Inspection
   - Testing and adjustment
   - Disassembly
   - Pressure testing
   - Painting
3. Materials handling, storage, and reconditioning:
   - Cleaning and restoring electrical connectors
   - Repair of street lights
   - Electrical testing of rubber gloves, bucket lines, and dielectric footwear
   - Storage and retrieve of general-stores items, meters
4. Splicing of underground cable
5. Trenching, plowing, and installation of cable or pipe
6. Remote operations:
   - Unmanned substations and switching stations
   - Automated system operations, load management
   - Meter reading
   - Billing and bill preparation
7. Internal inspection of pipelines
8. Tower-line and tower inspection

These applications were selected from survey questionnaires sent to transmission and distribution personnel in the company. Most of the applications suggested deal with complex but repetitive tasks that currently absorb large amounts of expensive craft labor. Applications such as steel-tower-line inspection might permit anticipation and correction of problems before they actually occur (62):

> There is clearly a need to perform emergency services in a distribution or transmission network under adverse conditions that can be very hazardous to line personnel. Perhaps there is a role for robots here. And it is also possible to envision that, someday in the future, tireless robot workers could man remote substations or switching locations that would be more costly to operate with human operators . . . such robots could detect and correct incipient power-system problems, or quickly alert remote human operators of the need to effect changes, thus maintaining or improving overall service reliability.

The installation of underground electric service cable in residential communities is largely limited by the cost of trenching and the restoration of customer lawns and patios. A "mole-type" robotic device that could bore a tunnel suitable for the insertion of electric service cable(s) would be an ideal solution to this high cost situation. EPRI first approached this problem a few years ago, and now the FlowJet Corporation of Kent, Wash., is developing an electronically guided mole/tunneling device. Water jets are used to "soft tunnel" a path 3–6 in. in diameter up to a 400-ft distance. Field trials of this device are under way (63).

CONCLUSION

The electric utility industry can and will use robotic systems. While not truly robots in the classical sense, these interactive human–machine systems will offset the rising cost of operating and maintaining increasingly complex electric generation and transmission facilities. The driving force for the development of this technology is today's nuclear power plant, where prototypical and commercially available human–machine systems are beginning to demonstrate the ability to reduce human radiation exposure, decrease plant downtime, and perform on-line inspection, service, and diagnostic procedures. Tomorrow's design of standardized commercial power plants will initially incorporate robotic systems, thereby reducing many of the difficulties and site-specific retrofit concerns presently faced by robotic designers.

Fossil-fired power plants also offer the opportunity for robotic applications, although this potential is as yet unexplored. Early nuclear plant robotic activity will naturally spill over to fossil plants as experience with nuclear applications gives plant managers confidence with robotics. While the cost justification for fossil plant applications is admittedly not as great as for nuclear plants, the use of robots in these stations will broadly expand the market potential for all utility robotic devices and hence ultimately lead to "cheap" generic robotic designs, with options available for both fossil and nuclear plant applications.

The use of robotic devices for tasks involving the transmission and distribution of electric power offers yet another challenging aspect of utility applications. Prototypical development of several designs has already been undertaken and will be tested further under actual use conditions.

Two factors will govern how rapidly robotic devices are incorporated into the electric utility industry. First, robotic devices must inevitably become standardized if they are to find wide application. Standardization will expand the market potential for these devices and spur greater productive efforts by robot developers. For standardization to be effective, robot manufacturers and utility engineers must continue their technical exchange, work cooperatively to identify robotic applications, and establish realistic methodologies for determining cost-effectiveness. Of parallel importance will be the development of utility equipment that is itself standardized, so that robotic devices may act upon it in a reliable fashion. This may be more difficult to accomplish than the standardization of robotic devices, since utility equipment is normally designed for long service life.

Second, robotic applications do not stand by themselves unrelated to the rest of the data-gathering and work activities normally involved in a modern utility system. Utility management must see robotics in its true perspective: another link in the chain of technology which automates and integrates diverse utility functions. Today, technology stands firmly at the human–machine interface level. Tomorrow, mobile robotic devices will be desired to provide "action at a distance" in an unstructured environment. Ultimately, the goal is toward intelligent muscle in the form of robots which incorporate artificial intelligence.

It is conceivable that by the year 2000, fully autonomous, artificially intelligent robotic devices will survey and inspect various areas of a power plant and, based on data collected, determine the maintenance to be performed on a piece of equipment *and* carry out the maintenance procedure. Perhaps even a central computer system in charge of a robotic work force will dispatch an intelligent mobile robot to a specific area of the plant in response to an operational alarm, or the robot might be automatically dispatched routinely to inspect and maintain a piece of equipment.

Robotics in its most fundamental concept is a computer-based technology similar to many of the computer-based devices already used by utilities today to monitor and operate large, geographically disperse electric grids. What is so special about robots is their ultimate potential capability to perform the "information–action" sequence autonomously, thus imitating humanity's basic lower-order reasoning processes. This is not a threat to people, but an opportunity. In the next century, robots will help keep the lights on and people will think and create the future.

BIBLIOGRAPHY

1. W. C. Hayes and E. F. Gorzelnik, "Special Report-Robotics: The Age of the Robot is Dawning," *Electrical World, 53 (Mar. 1985).*
2. H. T. Roman, *Robotic Applications in the Electric and Gas Utility Industries, ASME Paper 85-Mgt.-18,* American Society of Mechanical Engineers, New York, Mar. 1985, p. 7.
3. Electric Power Research Institute, "Robots Join the Nuclear Workforce," *EPRI Journal,* 10 (Nov. 1984). The article from which this excerpt is taken is an excellent summary of U. S. electric utility robot technology development.
4. Electric Power Research Institute. *Automated Nuclear Power Plant Maintenance, EPRI NP-3779,* performed under contract by Battelle Columbus Laboratories and available from EPRI, Palo Alto, Calif. 1983.
5. U.S. Nuclear Regulatory Commission, *Evaluation of Robotic Inspection Systems at Nuclear Power Plants, NRC Report NUREG/CR-3717,* performed under contract by Remote Technology Corp. and available from NRC Division of Facility Operations, Office of Nuclear Regulatory Research, Washington, D.C., 1984.
6. J. R. White, *Demonstration Project for Robotic Inspection Systems at Nuclear Power Plants,* Remote Technology Corp. Oak Ridge, Tenn., 1984.
7. U.S. Nuclear Regulatory Commission, *Occupational Radiation Exposure at Commercial Nuclear Power Reactors 1983, NRC Annual Report NUREG-0713, Vol. 5, NRC, Washington, D.C., 1985, p. 8.*
8. *EPRI Mobile Robots Conference for Electricity Generating Plants,* Charlotte, N.C. (Oct. 1984). No formal transactions/proceedings of this conference were published; however, videotape information is available from Electric Power Research Institute NDE Center, Charlotte, N.C.
9. Ref. 2, p. 8.
10. R. L. Horst, T. M. Law, and E. B. Silverman, "A Procedure for Developing A Standardized Robotic Maintenance System for Nuclear Power Plants," *Proceedings of the 1984 National Topical Meeting on Robotics and Remote Handling in Hostile Environments,* Gatlinburg, Tenn., Apr. 1984, American Nuclear Society, Hinsdale, Ill., pp. 173–179.
11. Ref. 2, p. 2.
12. J. R. White and H. W. Harvey, "The Evolution of Maintenance in Nuclear Processing Facilities," *1982 Winter Meeting of the American Nuclear Society,* Nov. 1982, American Nuclear Society, Hinsdale, Ill.
13. J. Taylor, "The Robotics Evolution," *EPRI Journal,* 2–3 (Nov. 1984).
14. E. N. Hayden and T. R. Wagner, "A Review of Nuclear Steam Generator Maintenance Remote Tooling and Robotics Developed By A Major NSSS Service Vender (Westinghouse)," *Proceedings of the International Conference on Robotics and Remote Handling in the Nuclear Industry,* Sept. 1984, Canadian Nuclear Society, Toronto, Ontario, pp. 125–128.
15. E. deBuda, M. D. C. Moles, and W. K. Chan, "A Remote Scanning System for Ultrasonic Inspection," in Ref. 14, pp. 64–67.
16. M. F. Fleming, "Automated Pipe Scanner for Nuclear Power Plants," in Ref. 10, pp. 181–187.
17. W. Cameron and C. Mark, "Remote Machining and Robotic Welding in a Proton Cyclotron," in Ref. 14, pp. 24–29.
18. J. J. Zimmer, J. R. Herberg, and M. D. Hecht, "On Line Robotics in Today's Nuclear Power Plants," in Ref. 14, pp. 21–23.
19. J. J. Zimmer and T. J. Donnelly, *Advancements in Remote Nuclear Plant Maintenance,* Nuclear Services and Integration Div., Westinghouse Electric Corp, Pittsburgh, Pa.
20. Ref. 14, pp. 127–128.
21. Ref. 18, p. 22.
22. L. H. Bohn, "Remotely Operated Generator Examination and Repair (ROGER)," in Ref. 14, pp. 123–124.
23. D. Mosey, "Getting to Grips with Remote Handling and Robotics," *Nuclear Engineering International,* 24 (Dec. 1984).
24. D. P. Worthy, "Overview of Current Nuclear Maintenance Hardware," *Proceedings of American Nuclear Society Executive Conference on Remote Operations and Robotics in the Nuclear Industry,* Atlanta, Ga., Apr. 1985, American Nuclear Society, Hinsdale, Ill.
25. D. Dobbeni and C. van Melsen, "Inspecting Steam Generators with a Fully Automated Polar Manipulator," *Nuclear Engineering International,* 43–45 (June 1984).
26. "General Electric BWR Service Market," *Marketing Notes for American Nuclear Society Organization Members,* **2**(6), 1 (Apr. 1984).
27. R. J. Giordano, "Robotic Applications in Decommissioning Activities," in Ref. 24.
28. *Proceedings of the International Conference on Robotics and Remote Handling in the Nuclear Industry,* Sept. 1984, Canadian Nuclear Society, Toronto, Ontario. This reference contains an excellent series of papers dealing with remote maintenance in Canada's nuclear power industry.
29. Ref. 23, p. 25.
30. "CFFTP Assists Remote Handling Effort at JET," *Fusion Fuels Technology,* **2**(4) (Nov. 1984).
31. M. A. Fischetti, "Robots Do The Dirty Work," *IEEE Spectrum,* 66, (Apr. 1985).
32. E. B. Siverman, "Industrial Remote Inspection System (IRIS™)," in Ref. 10, pp. 255–259.
33. T. G. Bartholet, "ODEX 1-A New Class of Mobile Robotics," in Ref. 10, pp. 261–266.
34. P. Britton, "Engineering the New Breed of Walking Machines," *Popular Science,* 66–69 (Sept. 1984).
35. Y. Shinohara, H. Usui, S. Saito, A. Kumazai, and Y. Fujii, "Experimental Remote Handling Systems," in Ref. 10, pp. 117–121.
36. N. Ozaki, M. Suzuki, and Y. Ichikawa, "Remotely Operated Manipulator Vehicle for Nuclear Facilities," in Ref. 10, pp. 123–130.
37. M. Mizuno, M. Ohnuma, K. Hamada, T. Mizutani, A. Shimada, M. Segawa, and K. Kubo, "Transfer System Development for a

Remote Inspection Robot in Nuclear Power Plants," in Ref. 10, pp. 189–195.
38. J. P. Bémer, "Remote Control Tooling for Maintenance and In-Service Inspection at EDF Plants in France," in Ref. 24.
39. Ref. 26, p. 4.
40. This list, which first appeared in Ref. 31, p. 68, was compiled by Harvey B. Meieran, president of HB Meieran Associates. It does not necessarily represent all of the robots available worldwide.
41. C. W. Ruoss, "Recent Equipment Development for Remote Maintenance in Nuclear Power Plants," in Ref. 24.
42. Ref. 6, p. 3.
43. J. R. White, "Remote Systems in the Nuclear Industry," *ANS Mini-Plenary Session on Human Development and Radiation Protection,* 1981, p. 2.
44. M. D. Pavelek, "Use of Teleoperators at Three Mile Island-2," in Ref. 24.
45. G. M. Taylor, "Remote Handling Systems Help TMI-II Cleanup Efforts," *Nuclear News,* 52–55 (Dec. 1984).
46. "Studies Suggest Robots Worth Serious Thought for Nuclear Plant Use," *Nuclear News,* 5 (Nov. 1984).
47. *A Prototype Remote Reconnaissance Vehicle: Features and Capabilities,* Civil Engineering and Construction Robotics Laboratory, Carnegie-Mellon University, Pittsburgh, Pa., Dec. 1983.
48. H. C. Alexander, "Remote Systems Application at Three Mile Island Unit 2," in Ref. 10, pp. 169–172.
49. Ref. 3, p. 12.
50. Ref. 4, pp. 52–55.
51. Ref. 5, p. 5.
52. Ref. 31, p. 68.
53. "EPRI Demo Highlights Various Systems," 51 (Dec. 1984).
54. C. V. Chester, *Improved Robotic Equipment for Radiological Emergencies," Oak Ridge National Laboratories Report ORNL-6081,* Oak Ridge National Laboratory, Oak Ridge, Tenn., Sept. 1984.
55. Ref. 2, p. 4.
56. Electric Power Research Institute, *Remote Controlled Maintenance Device Feasibility Study," EPRI Report EL-3296,* performed under contract by Southwest Research Institute and available from EPRI, Palo Alto, Calif., 1983, p. 1-1.
57. *Ibid.,* p. 2-11.
58. Ref 59, p. 2-4.
59. Ref. 1, p. 59.
60. Press release, Robotic Systems International, Ltd., Sidney, British Columbia, 1984.
61. "Robotic Arm Tested for Line Work," *British Columbia Hydro News,* **7**(22) (July 18, 1984).
62. Ref. 1, p. 58.
63. Ref. 1, p. 60.

General References

Ref. 28 is a good general reference.

Proceedings of the 1984 National Topical Meeting on Robotics and Remote Handling in Hostile Environments, Gatlinburg, Tenn., Apr. 1984, American Nuclear Society, Hinsdale, Ill.

Proceedings of the American Nuclear Society Executive Conference on Remote Operations and Robotics in the Nuclear Industry, Atlanta, Ga., Apr. 1985, American Nuclear Society, Hinsdale, Ill.

G. D. Friedlander, "Nuclear Maintenance Criteria for the 80's," *Electrical World,* 65–72 (May 1984).

L. J. Sennema, L. Buja-Bijunas, and J. Stephenson, "The Economics of Remote Tooling and Dose Reduction," in Ref. 14, pp. 3–12.

T. A. Mueller and J. F. Tyndall, "Robots in Pipe and Vessel Inspection," *Mechanical Engineering,* 44–49 (Aug. 1984).

Electric Power Research Institute, "Artificial Intelligence; Human Expertise from Machines," *EPRI Journal,* 6–15 (June 1985).

The author gratefully acknowledges EPRI's input to this article, especially the efforts of Dr. Floyd E. Gelhaus, Program Manager, Nuclear Engineering and Operations Department.

ELECTRONICS INDUSTRY, ROBOTS IN

John Dudley
Seiko Instruments U.S.A., Inc.
Torrance, California

INTRODUCTION

Many people's first impression of robots borders along the lines of the popular myths promulgated by such publicly recognized characters as R2D2 and CP30 from the movie *Star Wars*. The truth is that it will be a long time before such anthropomorphic robots are widely available. However, industrial robots operate in many different circumstances within industry and their use in the electronics industry will be highlighted in this article. Robots are used in silicon-chip manufacturing, through-hole printed circuit-board assembly, chip-assembly operations, memory-disk manufacturing, electronic connector assembly, surface-mounted device assembly and discrete parts welding, to name just a few. Four of these applications are discussed in detail, but the key consideration to remember about an industrial robot is that it combines intelligence in the form of a robot controller with a piece of precision equipment and offers the user flexibility. In the fast-paced world of electronic product development, technology advances, and rapid innovation, today's state-of-the-art product can be either modified beyond recognition or simply outdated within a few years. So, a robot user receives flexibility two ways: first, within the context of the robot itself, the system can be readily adapted by reprogramming or retooling to meet production changes relative to the individual workcell and; second, the robot offers the user flexibility in production in terms of overall plant manufacturing strategy.

The robot systems discussed here illustrate applications that are justified by producing or assembling parts with a precision beyond manual assembly capability. Productivity results from faster cycle times and/or improved quality (fewer rejects). These systems have been designed to maximize the functions and operations a robot performs best, yet not necessarily duplicate the way a human might perform these tasks. Conventional productivity measurement, however, typically does not cover one crucial aspect on the production line, ie, work-in-process inventory. Two engineering experts writing in the UK discussed this often overlooked factor in automation project evaluation (1):

Unit assembly cost thus comprises the expenditure (including such costs as design) associated respectively with (one) robotic equipment, (two) reusable nonrobotic equipment and (three) equipment "fixed" to a given product and (not used except when assembling batches of that product). In addition to these three costs, however, there is frequently a fourth "hidden" overhead cost which may nevertheless be substantial.

The major contributing factor to this overhead is often cost of work-in-progress (WIP). Conventional techniques of assembly, whether involving dedicated automation or manual labor, tend to result in a significant value of components and partly finished products simply waiting for assembly. Use of a number of integrated robots for assembly may increase efficient processing of components, resulting in fewer parts being actually "tied up in the system" at any one time. As a result, the WIP inventory will cost less, so less of the company's capital will be tied up. The potential impact of robotics on such assembly overheads requires that such costs be considered separately from the three other costs previously mentioned (1).

Any evaluation of the productivity of a proposed application sooner or later considers cycle time. Again, the two UK engineers write (2):

It is worth noting here that the "maximum end-effector velocity" supplied by manufacturers' specifications may be a poor indicator of actual cycle time.

This is because a robot joint does not, of course, instantaneously reach its slew rate but must instead accelerate to it, maintain it for a time, and then start to decelerate in time to stop at a desired location. If a robot movement is a small one (as is common during assembly), then a given joint may not have reached its slew rate before it has to decelerate again. A robot with a "snappier" response (though perhaps a lower actual slew rate) might be able to perform the same movement more quickly because it in fact reached a higher speed, more rapidly, than the potentially faster alternative. It may even temporarily have reached its slew rate before having to decelerate (2).

The authors go on to point out the indirect factors in selecting a flexible manufacturing system (3):

It is no longer valid to conduct an economic evaluation merely of the direct savings of a robot system—as the level of integration with, and so impact on, the remainder of the factory is increased, so too is the importance of the indirect factors in any economic justification. This integration effect also forces the consideration of strategic as opposed to solely tactical factors in any economic evaluation of robotic systems. This, of course, suggests the need for a detailed understanding of the potential consequences of high technology by those most able to appreciate its likely strategic implications, namely top management.

These four applications illustrate these points:

1. The RCA miniature welding system features extremely precise parts that cannot successfully be produced manually. The system operates as a total assembly line and uses well-designed moves by a "snappy" robot to produce an efficient cycle time.
2. Hewlett-Packard's SMD assembly station demonstrates another application too delicate to assemble manually. The application also demonstrates HP's long-term manufacturing strategy toward the flexibility and efficiency of true mixed-model production.
3. The Deutsch Company application illustrates a unique robot workcell design and again, superior production efficiency over manual methods. It also demonstrates a clear strategic production advantage in harness and connector assembly.
4. Finally, the Cypress Semiconductor workcell demonstrates a strategic onshore business advantage while performing a delicate task with higher yields than manual production.

APPLICATION: RCA MINIATURE WELDING STATION

RCA automated this welding application (4) for a special video tube. The system exhibits multiple subassembly steps to a finished part. The workcell minimizes robot-arm movement in an excellent display of system design. Proper design allows the robot to work at maximum efficiency for parts production with minimum arm movement. The complete application feeds back quality-control information after each stage of operation for robot program decision (see Fig. 1).

Inside an RCA video camera, there is a pickup tube (Vidicon). Deeper in the center of the Vidicon is a small, seemingly innocuous cylinder with a cap protruding from one end. This part, the cathode assembly, the "heart" of the Vidicon, is fabricated from four subminiature metallic parts spot-welded together (see Fig. 2).

When the Vidicon "heart" is assembled manually, the component parts are so tiny that a highly repetitive, tedious assembly task results. RCA decided to automate the assembly even though the parts were small and delicate. However, it was found that these parts were so small and the alignment so critical that manual operations had difficulty producing consistent quality parts. RCA called on Datum Industries, Inc., a Palisades Park, N.J., robotic system integrator to aid in the design and development of their flexible manufacturing workcell.

The Assembly Sequence

The four parts are so small that the robot workstation actually functions as a complete miniature assembly line as the work progresses through three subassembly welding stations. The first subassembly begins when the inner sleeve is fed from a vibratory feeder bowl down a slanted track like a tiny log through a sluice. At the bottom of the track, the sleeve is held at a buffer station, then released when clear to slide over a pin. The pin and sleeve rotate upward to present the sleeve to the robot in a vertical position.

Once the robot has gripped the sleeve, the robot moves to a test station where the robot lowers the sleeve over a pressure sensor to test for part presence. Presence testing is a must in this system to avoid any circumstance where the welding operation might begin without a part being present. Presence testing could have been done internally in the gripper, but because of the parts' sizes, the cost would have been significantly higher than a simple external sensor. Should a sensor

Figure 1. The RCA miniature part subassembly and assembly station for welding subminiature parts.

not detect a part, the robot is programmed to repeat part pickup/presence test. After three pickup attempts, it stops and signals for an operator.

Once the sensor has signaled correct presence, the robot lowers and releases the sleeve onto a custom-tooled rod-shaped grounding weld mandrel. The robot arm next moves to the feeder for the most delicate part, the cap. The vibratory cap feeder orients the caps with the open end facing down. At the escapement, a pin raises the cap into the pickup position. The robot lowers and grips the cap (same gripper used for the sleeve) and moves to an external sensor and tests for presence. The robot then moves back to the previously placed sleeve and lowers and releases the cap over the sleeve with a maximum clearance of 0.001 in. between cap and sleeve. RCA chose the robot because it has the necessary precision to perform this task reliably (see Fig. 3).

The gripper rotates and a "tamp" tool on the end effector taps the cap twice to assure that it is properly seated in position on the sleeve. The spot-welder electrode slides forward horizontally, welds the first spot, and withdraws. The welding-mandrel fixture rotates and the cycle repeats (4 times) at each 90° increment. As each spot is welded, the "tamper" lowers onto the cap and a spring-loaded backup pin projects horizontally from the end effector directly opposite the welding electrode on the sleeve. The robot arm in this instance functions as a mobile fixture to ensure correct relative part position during welding.

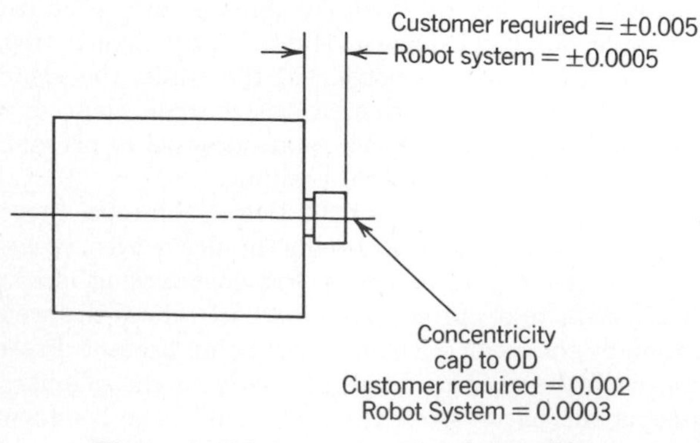

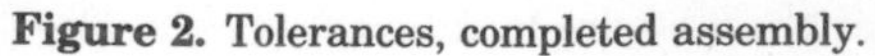

Figure 2. Tolerances, completed assembly.

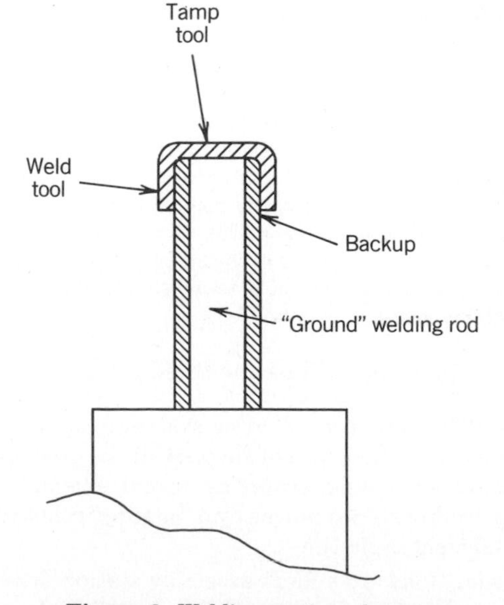

Figure 3. Welding station, sleeve/cap.

After welding, the robot descends and grasps the welded part and returns to the presence testing station. The part is inserted over a hardened pin and tested for a quality weld; if it encounters resistance, the cap weld is judged good. The robot moves the part to a station with an inverter jaw which grasps the part and vertically rotates it 180°. The robot regrips the part and moves and lowers it into the next welding station.

Second Subassembly

The second subassembly operates identically to the first with one major exception. The capped end is oriented downward to accept another component properly, a funnel-shaped sleeve. The robot arm again functions as a mobile fixture, but now the tooled welding electrode rod on the end effector inserts into the open inner sleeve. The robot places a funnel over the first subassembly sleeve at the second welding station. A spring-loaded device shaped like an inverted cone lowers over the funnel and lowers and taps the funnel three times. Another spring-loaded device shaped like a tuning fork straddles the electrode and holds the assembly in place for the weld operation. This "fork" assists the electrode in maintaining correct cone position relative to the inner sleeve (see Fig. 4).

As soon as the funnel is properly placed on the sleeve, the robot signals the welding sequence to begin. This test station again serves the multiple function of testing for part presence from the feeder and testing for presence/QA (quality assurance) once the part is welded.

Third Subassembly

The third subassembly is conceptually similar to the first two. Once the second subassembly is positioned with the electrode rod again serving as a tooling fixture, a cylindrical outer sleeve is welded onto the part. The mobile-fixture aspect of the robot arm during welding changes again slightly. This time, compound spring pressure is applied to both the inner sleeve (of the first subassembly) and the cylindrical outer sleeve during each weld to square the parts properly. Again, the test station checks both incoming part presence and outgoing presence/QA. The robot deposits the finished part in a tray and begins a new cycle.

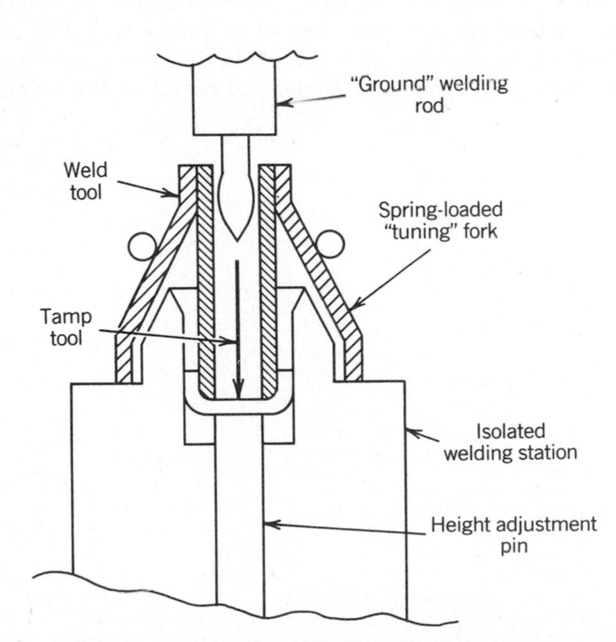

Figure 4. Welding station, sleeve/cap and cone.

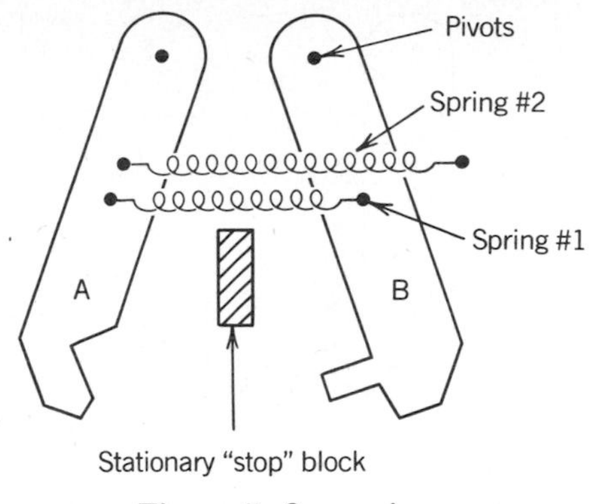

Figure 5. Open gripper.

Summary

The system produces a finished assembly with quality, repeatable welds every 54 s. The overall system, which includes the air-pressure control valves, electrical-relay controls, electrical power for feeder, signaling to welding unit, part sensing, and robot-arm motion, is operated by the Seiko D-TRAN controller.

One of the keys for the success of this application is the 5-part end effector designed by Datum Industries. An advanced gripper design for the cylindrical parts reduces by half any part's tolerance variation which would contribute to overall system tolerance. This gripper design is unlike parallel acting grippers which not only add the full variation due to part's tolerance but also contribute their own "play" to increase tolerance problems further. Datum's unique gripper utilizes a "V"-block design used in machine vises to form a "fixed" finger of the gripper with an adjustable spring-tensioned finger to provide just the minimum pressure necessary to grip the delicate cylindrical parts without permanent deformation (see Figs. 5–7).

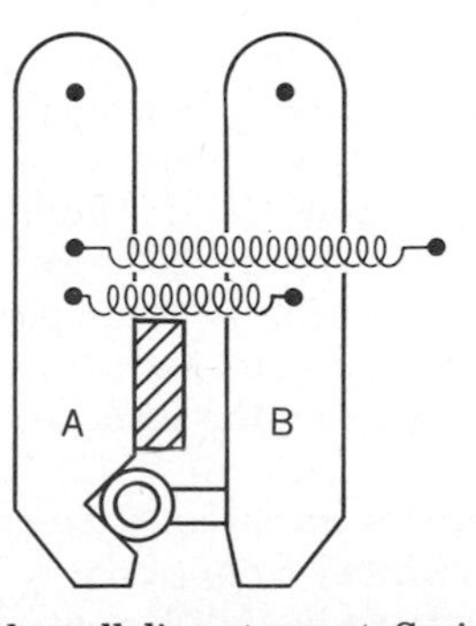

Figure 6. Jaws closed small diameter part. Spring 1 closes jaws (light spring pressure so as not to damage part). Spring 2 causes more pressure on jaw A, forcing jaw A to close on stop block.

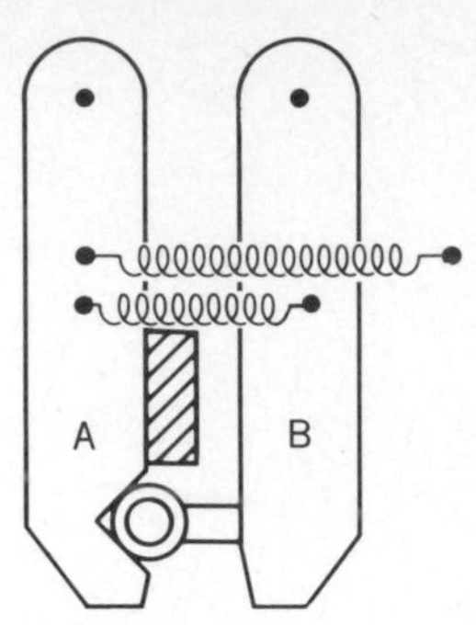

Figure 7. Jaws closed large diameter part. Jaw A is still forced to close on stop block. Lightly spring loaded jaw B compensates for diameter variation and holds part securely. Robot compensates for "off-center" gripping.

APPLICATION: HEWLETT-PACKARD'S PRECISION SURFACE-MOUNT ASSEMBLY

The demand for miniaturized electronic components has surged and new production methods have had to keep pace. Manual assembly of miniature surface-mount devices (SMDs) is a highly repetitive precision task that results in costly parts and uneven quality. To meet SMD printed circuit-board demand for Hewlett-Packard hand-held calculators, HP's Corvallis, Ore., plant selected SMD assembly for their first flexible automation project (5). Flexible automation offered high production volumes at consistent, repeatable quality without sacrificing the flexibility of design changes or system improvements.

The single criterion that distinguishes this automation project in surface-mount technology is precision tolerances. The application places chip resistors, capacitors, inductors, 28-pin SOIC packages, and 3 sizes of an HP package called a quad pack. The quad packs are flat-pack-style devices with leads on all 4 sides that are bent in a "gull-wing" shape.

The techniques and systems described here were developed to accommodate the small distance between quad-pack leads, 0.031 in. centerline to centerline. The allowable misalignment of the centerline of each quad-pack lead to the centerline of the corresponding printed circuit pad is ±0.008 in. Two materials and process-tolerance variations substantially exceeded this tolerance, challenging the HP design team to produce a precision robotic assembly workcell.

To assemble any two parts accurately, the position of the critical features of each of the parts must be controlled. When part drawings of the quad-pack component were first reviewed, the centerline of the body was defined as coincident with the centerline of the center leads. HP built a mechanism that centered the component on its body for robot handling but found that they did not achieve the expected results. HP designers learned from their quad-pack supplier that the centerlines of the body relative to the leads actually has a ±0.007 in. tolerance.

The second tolerance variation concerns the other half of the assembly, the printed circuit board. The pads' positions on the circuit board must be tightly controlled relative to tooling-pin holes for a tooling-pin fixture solution to succeed. This is not common practice by printed circuit-board suppliers because of the expense involved. HP's only specification stated that there was to be no more than a specific amount of hole "breakout" through the annular ring around a plated through hole. The resulting pad-to-hole relationship varies ±0.009 in. When the tolerance typical for tooling-pin fits is added, positional variation increased to at least ±0.012 in., nearly double HP's 0.008-in. maximum misalignment specification.

Board Tolerance Solution

HP adopted the design theme, "Use the capability of the robot to do most of the work." HP engineers used the Seiko D TRAN RT 3000 robot to achieve total system tolerance of ±0.008 in. and solve both tolerance problems. The robot needed to learn the precise location of each printed circuit-board's traces. Trace features on circuit boards printed from the original art work have a repeatable dimensional tolerance of ±0.001 in. Although tooling pins cannot acceptably position boards for part placement, they can position boards within ±0.030 in., sufficient for the robot to then locate art-work features.

Special target traces are now printed on all of HP's circuit boards that use SMDs (see Fig. 8). The target traces (in an annulus design) provide known board positions relative to the component placement positions. The target traces are "seen" by a retroreflective photo detector positioned on the robot end effector. The detector senses the difference in reflectivity between the surface of a trace and the background surface of the printed circuit board. The detector, located in the center of the vacuum pickup head, is concentric with the Alpha (wrist rotation) axis of the robot. Because the detector is concentric with the Alpha axis, the robot can use points found by the detector without concern about the rotational position of the Alpha axis. The detector's binary output is wired to a gripper input of the robot. A specially written software program generates detector-adjustment data using a fixed calibration pedestal. If adjustment is needed, the program displays the number of degrees to turn each adjusting screw to center the detector.

As mentioned, the target design is an annulus. The robot is directed to move so that the photo detector comes to the start position looking at the nominal center of the annulus.

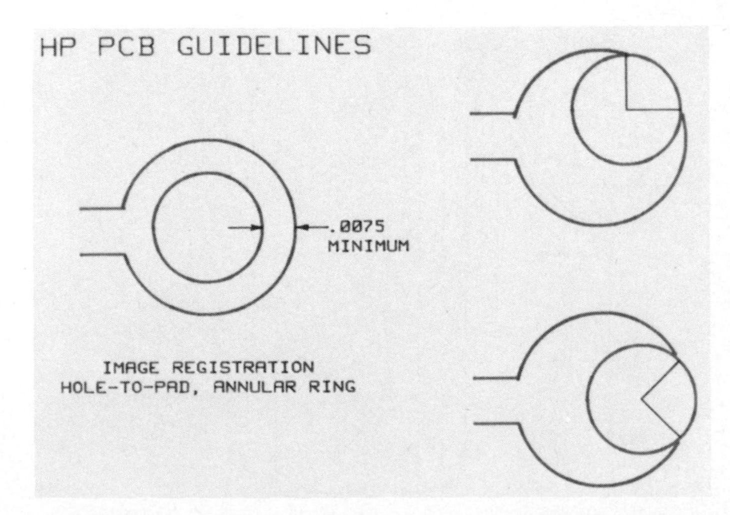

Figure 8. Annular ring design for precision SMD placement.

The robot checks its input from the photo detector to be sure that it does not see a reflective trace. If no trace is seen, the robot assumes it is somewhere in the center, as desired. HP engineers selected motions for the target search that required driving only a single robot axis. This eliminates the "stair-step" motion sometimes caused by multiple-axis moves. Under program control, the robot moves in a clockwise direction while constantly monitoring the input from the photo detector. When the input indicates a trace is present, the robot identifies its current position and remembers it for future calculation. The robot returns to the beginning position and searches in a counterclockwise direction until it finds another trace. The robot computes the average of these two found points and moves to this new third point.

By virtue of the robot's design and its present position, an extension or retraction of the arm, called the R axis, will result in the photo detector traversing a path that bisects the annulus. Therefore, the robot is commanded to extend the R axis until the input from the detector signals the presence of a trace. The robot remembers the location and returns to its secondary starting position, where it is commanded to retract the R axis until a trace is sensed. The average of these last two sensed positions is the center point of the annulus which the robot remembers for later use.

The center point will be difined as the position of the origin of a new coordinate frame which will be used to place components (see Fig. 9). Before the robot can specify a new coordinate frame, it must have knowledge of any point that would be on the x axis of the new coordinate frame. The second annulus on the printed circuit-board art work becomes this point. The robot is commanded to use the same type of search pattern to determine the center of the second target. Utilizing the standard FRAME subroutine in the controller, a new coordinate frame is defined with the center of the first target being the origin and the second target center a point on the x axis. As mentioned, the RT 3000 robot provides the coordinate transformation calculation as a built-in function, but a host computer could also perform the task.

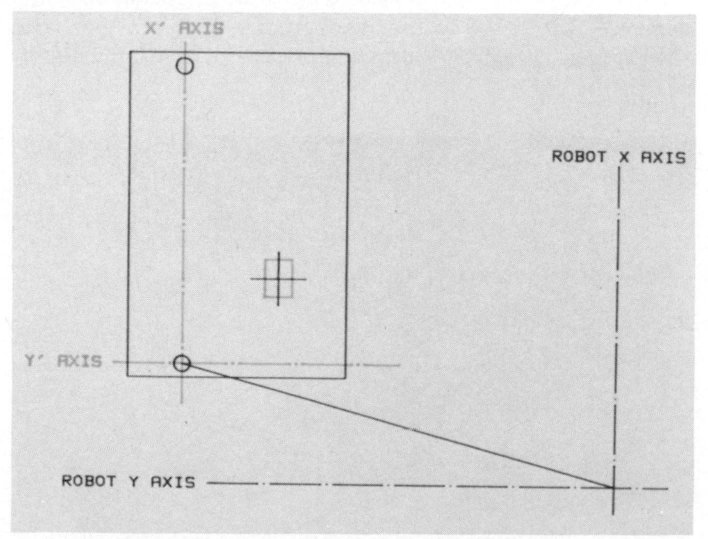

Figure 9. Center point of new coordinate frame.

Figure 10. Positioning equipment prepares for the quad-pack assembly by first centering on the quad pack's body.

Quad-pack Tolerance Solution

Peripheral equipment centers the quad pack on its body for robot pickup (see Fig. 10). This yields ±0.007-in. positioning of the center of the device leads. This tolerance could be improved by a vision system. At the time the system was engineered, both development costs of a vision system and estimated production cycle times were excessive. So a precision nest with slotted "V" grooves to match the nominal position for each lead was developed for the quad-pack device (see Fig. 11). The robot brings the device directly over the nest and lowers the device until the leads are just below the top edges of the nest. The robot releases the device and then lightly pushes it down to ensure that the leads have moved to the bottom of the "V" grooves. The robot regrips the device by vacuum pickup and moves up, lifting the device from the nest. The robot has gained positional control of the device and can proceed with the other placement activities.

Figure 11. A precision tooled nest with "V" grooves to match the nominal positions of the quad-pack leads. The nest allows the robot to be truly centered on the centerline relative to the leads.

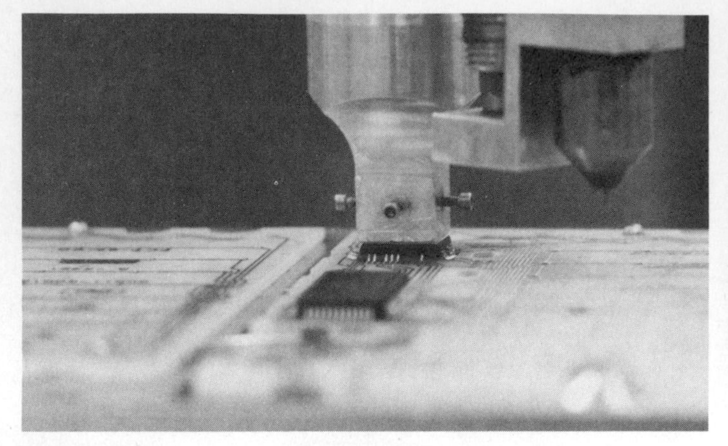

Figure 12. Once the quad pack is positioned, the robot exerts force for 2 s to allow the adhesive to set.

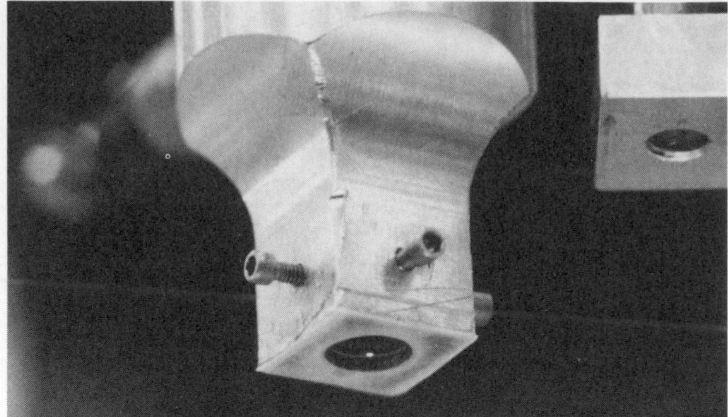

Figure 14. Close-up view of the vacuum pickup and effector equipped with retroreflective detector for SMD assembly.

The nest works if the robot originally controls the device accurately enough to position the leads into the correct "V" grooves. The device must also move enough upon release to translate freely as necessary while it seeks the bottom of the "V" groove. It is difficult to quantify this process, but HP's best engineering estimate is that the tolerance for the centerline of the center leads is reduced from ±0.007 in. to ±0.002 in.

Component Assembly Process

After the robot lifts the quad pack from the positioning nest, the robot takes the quad pack to a cyanoacrylate adhesive dispenser and a drop is placed on the underside of the device. The robot moves to the placement location on the printed circuit board in the coordinate frame that is defined by the centers of the board targets as found during the search. The robot pushes the device down on the board until a force transducer in the end effector senses a predetermined force (see Fig. 12). The robot pauses for 2 s to allow the adhesive to set under force and then releases the device. The adhesive holds the leads in the solder paste and in alignment with the printed circuit-board pads. Gluing quad packs yields 2 unexpected benefits: first, the soldering process works more effectively with difficult-to-solder leads, and second, the PCBs can also tolerate varying volumes of solder paste.

The end effector consists of the vacuum pickup/retroreflective detector previously mentioned plus 2 other applicators (see Figs. 13 and 14). One applicator dispenses epoxy for chip capacitors and resistors, the other cyanoacrylate accelerator. The accelerator is applied to the printed circuit board before quad-pack placement to speed adhesive setting. The epoxy is also dispensed at the placement site and serves the same function as the quad-pack adhesive. Both dispensers retract pneumatically relative to the vacuum pickup when not in use.

Figure 13. View of SMD assembly end effector, which includes the vacuum pickup, epoxy dispenser, and cyanoacrylate accelerator dispenser.

Chip resistors and capacitors are fed to the robot on 8-mm tape dispensers (see Fig. 15). The robot relies on the positional accuracy of the component as fed from the tape because placement accuracy for these devices is less critical. A custom feeder dispenses the SOIC to the robot from an integral precision "V"-grooved nest. These components are placed immediately

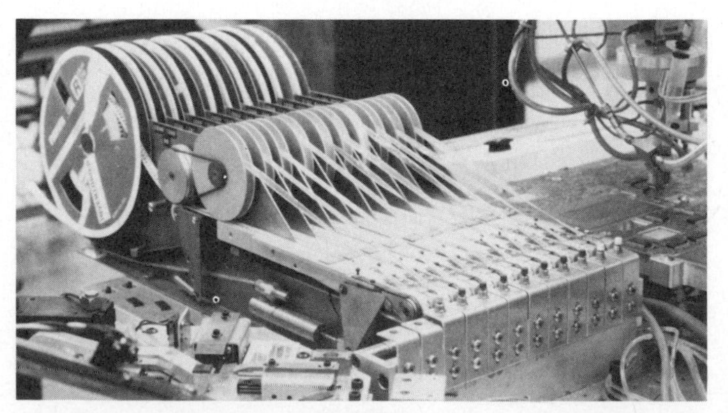

Figure 15. Chip resistors and capacitors are fed to the Hewlett-Packard SMD assembly unit on 8-mm tape.

after pickup at printed circuit-board positions determined by the target searches.

Evaluation

To date, Hewlett-Packard's 4 Seiko RT 3000 robots have assembled over 2.5 million quad packs on 7 different printed circuit-board designs. Unscheduled downtime on the SMD assembly workcells is less than 5% and is primarily associated with peripheral equipment. Unscheduled robot downtime is well under 1%. Soldering defect densities relative to solder joints have fallen to 300 parts per million. Current cycle times are estimated as follows: search for printed circuit-board targets, 6 s; placing a chip resistor or capacitor, 4 s; dispensing a drop of epoxy, 0.75 s; placing and bonding a quad pack, 5.5–8 s, depending on peripheral tooling configuration.

Besides improved quality, the system assembles boards that would otherwise be extremely difficult to build. The robot controller easily accommodates program changes for HP's many short-term part variations that might otherwise have shut down production.

The system was developed in one year and has operated for two years. Based on those three years' operating results plus cash-flow projections for the coming two years, the average annual internal rate of return exceeds 50%. (Internal rate of return is the highest rate at which funds can be borrowed using the cash flow generated by the investment to repay the loan.)

The most important long-range benefits to HP, however, were intangible. SMD assembly was the first flexible automation project at the Corvallis, Ore., facility. HP engineers were forced to understand their manufacturing process, giving them the power to control it. This knowledge, combined with successful operating results, convinced plant management to support development of other robotic applications which will lead to a more automated factory.

APPLICATION: AUTOMATED CONNECTOR ASSEMBLY

High labor costs and assembly uniformity are two major problems in connector assembly. The Deutsch Company, a leading supplier of connectors, decided to prove conclusively that cylindrical connectors could be properly terminated using automation. Deutsch engineers researched their customers' electronic harness assembly methods and found two systems that offered promise. Both approaches begin with a blank spool of wire. A computer-controlled ink-jet marker measures and marks each part number directly on the wire. The wire is rewound onto an empty spool and then mounted adjacent to the harness routing table.

With one method, a worker at the harness table presses a pressure-sensitive strip running along the top edge of the slanted harness routing table. A CRT displays a large-sized number which originates from the same computer that monitored the ink-jet marker. The number correlates to a numbered peg or nail on the harness assembly board. The worker first routes the wire to the appropriately numbered peg or nail on the harness board, wraps the wire around the peg, and presses the pressure-sensitive strip; the CRT then displays the next numbered peg location. The worker repeats these steps until the entire harness is routed. The other approach is identical except that a gantry robot, again directed by the host computer, routes the wire on the harness board. Deutch's research further disclosed that no effective automated method existed to affix connectors to the harness properly. Deutsch decided that the termination part of the assembly could be integrated with present harness routing methods via the host computer which controlled the ink-jet-marking and harness-routing segments of the process.

To help develop the automated termination system, Deutsch contacted BaGaard Automation in Portland, Ore. Deutsch specified that the finished unit must operate at multiple assembly sites within the robot's work envelope. This would allow efficient use of the robot during the assembly operation. For instance, an operator could prepare one assembly site(s) while the robot worked at another. Deutsch also specified that the robot system must select the correct wire, cut it, strip it properly, crimp on the contact and insert it in the correct connector location with no damage to the wire, contact, or connector in a 15-s cycle time, all with only a 3-in. breakout. Concerns about system integration with users' existing computer systems proved groundless due to the flexible communications package associated with the robot's controller (6).

System-design Developments

Several new and unique design developments made the above application possible. Like a carpenter, this robot wears an apron to carry both tools and the parts for assembly. The RT 3000 robot is a cylindrically oriented robot. In order to meet the multiple-station parameter noted previously, an aluminum "apron" was mounted directly on the robot's Theta axis (the rotational axis). Two feeder bowls, which contain the contacts to be applied, are mounted on the "apron." Also mounted on the "apron" are the cutter/stripper unit and two differently sized crimp heads. Since available wire is limited due to the 3-in. breakout, these tools move to the wire in proper sequence (see Fig. 16).

To integrate easily with the harness methods discussed earlier, the design team also had to develop a means for the host computer to monitor specific wire locations at each breakout. This would allow the host computer to command the robot to select the correct wire. A simple device called a wire comb was designed to replace the nail or peg on the harness routing board. Instead of wrapping the wire, the operator or gantry robot simply presses the wire into the next available slot in the wire comb. When the harness routing is completed, the operator removes the harness and the combs. The operator arranges the combs in the proper order at the appropriate assembly site(s) within the robot's work envelope. The host computer then directs the robot to select the correct wire from the comb and proceed through the automatic termination cycle. Several more design innovations made this application possible. These relate to specific segments of the assembly process.

Assembly: The Termination Cycle

The assembly sequence begins once the wire combs from the attached harness are stacked in sequence at the robot work

Figure 16. The Deutsch connector assembly system with "apron"-mounted tools.

site. The robot moves to the tool-change area and changes its own end effector to begin with the properly sized tool. The robot then selects the proper wire from the comb (see Fig. 17). The robot's Alpha axis rotates 180° and presents the wire to the "apron"-mounted tools. The stripper/cutter mechanism positions itself in front of the wire. The robot positions the wire slightly up or down for the correct "cut" hole and inserts the wire end into the "cut" hole (see Fig. 18). The robot retracts the wire from the "cut" hole and repositions slightly up or down, then inserts the wire into the correctly sized "strip" hole.

Innovative design made the cut/strip function possible. No commercial stripper/cutter with vertically aligned holes and a vertically aligned air cylinder existed in combination. Horizontal space constraints below the surface of the apron dictated that the design team develop a new, completely vertically aligned cutter/stripper unit. The stripper strips a wire in less than 80 ms with no damage to the wire. Sensors mounted on the strip unit verify that the wire has been both stripped and stripped to the proper length as the robot retracts the wire from the "strip" hole. Should there be any "aborts" at this or any other stage of the insertion cycle, they are identified and printed on the robot-controller monitor upon completion of the rest of the connector. The robot withdraws the wire from the strip hole. The "apron"-mounted tools move again to present the crimp head to the waiting wire.

The unit has two feeder bowls mounted on the "apron." Each is equipped with dual escapement, pins from one escapement and sockets from the other, both of the same size. Each feeder bowl connects to its appropriately sized crimp head. Once the contact has been air-blasted from the escapement into the crimp head, specially designed sensors verify the presence of the contact. If present, the robot inserts the wire into the crimp head. The contact is automatically crimped and released. The robot removes the wire (with contact crimped) and rotates 180° to the connector. (Note: this application works for any connectors regardless of the number of pins.) The robot positions itself slightly to align precisely in front of the correct connector hole and perfectly perpendicular to the elastomeric grommet (see Fig. 19). The robot inserts the connector. A force test in the robot's R axis automatically tests to assure that the contact does not catch on the connector body and that the connector fingers activate, locking the contact in place.

One major anticipated design hurdle never developed. When manually assembled, cylindrical connectors display a phenomena called "grommet growth." As the grommet is partially filled with contacts, the elastomeric grommet creeps out of position so that the remaining empty grommet holes and the matching connector body holes no longer match. Special moves were programmed into the robot to overcome the problem. However, the robot so accurately aligned and inserted the contacts that no "grommet growth" occurred. The robot could also affix a mated connector to the newly assembled connector and conduct an electrical continuity test.

The system quickly changes its own tooling as necessary to meet specific connector assembly requirements (see Fig. 20). Installation of a single contact requires only 12 s, a vast improvement over the awkward manual insertion methods and accompanying inspection and supervision difficulties.

APPLICATION: PRODUCTIVE SEMICONDUCTOR MANUFACTURING, ONSHORE

Cypress Semiconductor's new state-of-the-art factory advances the semiconductor manufacturing process. One of the first

Figure 17. The robot selects the proper wire for eventual insertion into the connector. The "combs" come directly from the harness assembly board.

pieces of equipment seen on a tour of the new facility is a robot loading lead frames from magazines onto mold plates (see Fig. 21). Robots are not usually found in high volume semiconductor factories. However, in Cypress' case, the robot is a perfect example of its unique business strategy. Cypress' business commitment is to build and market the highest speed C-MOS chips 100% within the United States. Unlike the giant Japanese or U.S. semiconductor manufacturers, Cypress focuses on market niches that are too small for the industry giants and require a technology that discourages custom and semicustom chip manufacturers from competing.

Cypress manufactures a variety of components generally divided between ceramic and encased plastic packages. Because 80% of Cypress' C-MOS volume is in plastic packages, it devoted most of its attention to manufacturing-process advancements in that area. To compete successfully as a stateside manufacturer, it also was committed to producing maximum yields on a "ship-to-order" basis only. This required hands off of all assembly operations and extremely high turnaround times for finished components to compete onshore. Add a mixed-model production line, and the only feasible solution is flexible automation. Only a description of the manufacturing process can explain the mixed-production concept. Cypress is a model facility with design, wafer fabrication, assembly, and test under one roof.

When the 5-in. silicon wafers leave wafer fabrication, they enter the assembly operations area. In assembly, the wafer is sawed into dice, which are inspected for quality and marked accordingly. Dedicated automation equipment attaches the dice to the appropriate lead frame called for by the production schedule. Though Cypress manufactures over 60 different component variations, they all derive from one of eight different lead-frame styles. During this stage of manufacture, called front-end assembly, the variety of lead frames does not demand that equipment be dedicated for each different lead-frame count. The same holds true during the next step, wire bonding, where the lead frame is connected to the die by tiny gold wires. Post-wire-bonding quality assurance ends front-end assembly.

The lead frames then enter packaging and finishing operations in what the industry terms back-end assembly. Cypress' plate molds for packaging devices in plastic following wire bonding were designed with runner systems that would be universal for the eight different lead-frame mold plates. They chose the plate-molding process to package the devices because of its inherent flexibility. The bottleneck in the back-end assembly operation comes in loading the different lead frames onto the mold plates. To retain their high yield/quality orientation of "hands off the components," Cypress had to automate loading lead frames onto the mold plates. Dedicated machines existed for loading the mold plates but would have required substantial capital investment and valuable floor space. Because of the flexible nature of the production line, eight dedi-

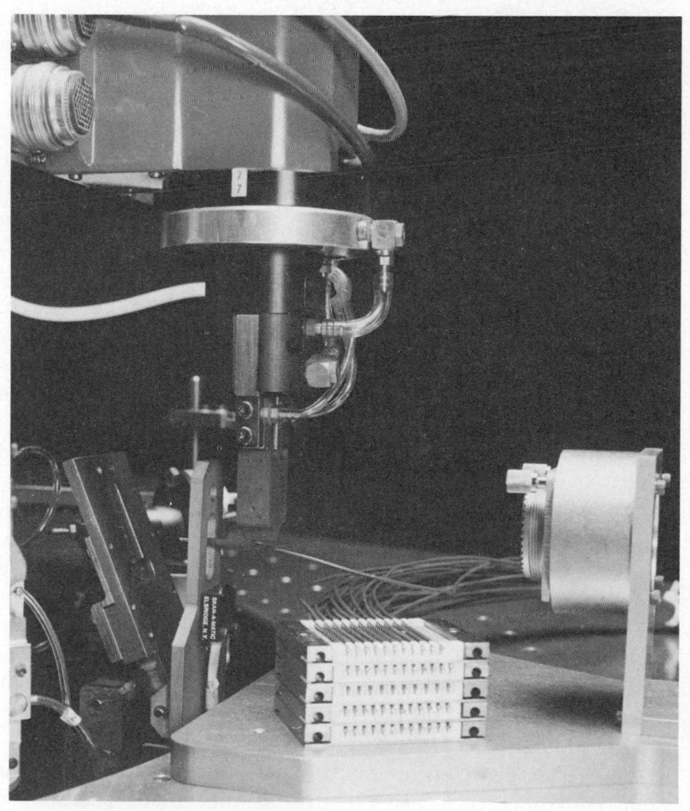

Figure 18. The robot strips a wire in less than 80 ms.

Figure 19. The robot inserts the contact into the connector.

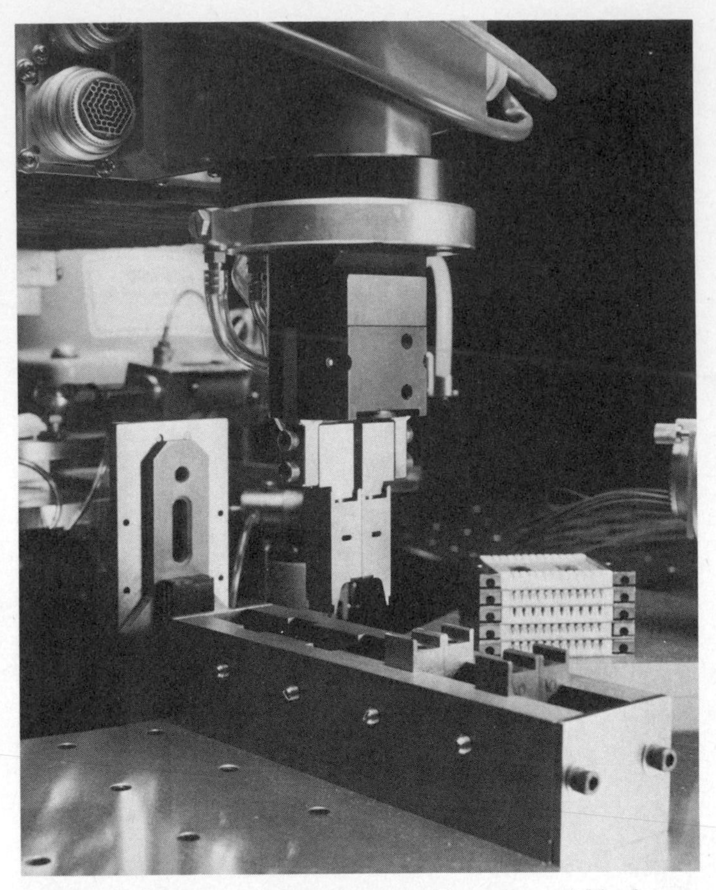

Figure 20. The Deutsch connector assembly system quickly changes its own tooling to perform multiple tasks.

cated machines in the product flow would have created awkward and therefore costly handling problems. The only solution was a flexible plate-loading concept based on a robot (7).

Robot Mold Plate Loading

The lead frames arrive at the robot loading station from wire bonding in a 40-strip magazine. An operator places the lead-frame magazine in the unloader and selects the appropriate model lead frame at the controls and pushes the "ON" switch. The robot first self-adjusts both in its own end-of-arm tooling and then the magazine unloader to the matching size. The robot controller is programmed with multiple feedback loops to detect any difference in lead-frame size from the operator's selection. Any difference stops the robot's arm movement and the robot signals the operator.

The magazine unloader at the rear of the lead-frame magazine slides the topmost lead frame out of the magazine into the robot's work envelope. The custom end-effector tooling of the robot then indexes the lead frame forward on the unloader platform. The robot locates and straightens the lead frame and positions it against a stop on the unloader platform. Once against the stop, the lead frame is accurately located for the robot to pick up and properly place it. The end effector covers the lead frame and picks it up by vacuum.

The robot proceeds using the palletizing function in the robot controller to locate the precise placement position on the mold plate. The end effector descends and deposits the lead frame on tiny pins which correctly position and mount each lead frame on the mold plates. To locate the pins correctly, with their matching holes in the lead frames, requires robot repeatability of 0.001 in. For maximum work-flow throughput,

Figure 21. Cypress Semiconductor robot system loading lead frames onto mold plates.

the lead-frame robot workcell was designed with two complete workstations. While the robot works at one station, the operator unloads finished mold plates and loads a new magazine of lead frames.

Final Back-end Assembly (Mold and Finish)

The mold plates leave the robot workcell and proceed through molding. After curing, the packaged devices are trimmed and formed and each device is laser-marked with its assembly lot code. The leads are finished in a wave solder machine before a final QA in assembly. The devices proceed from assembly into testing, where they are class-sorted by performance capability. The devices are then packaged and shipped. As mentioned before, Cypress manufactures on a "ship-to-order" basis only. The facility carefully matches wafer fabrication capacity, assembly, and test to produce finished product on turnaround times of 2–3 days. This reduces costly work-in-process inventories and eliminates finished good inventories. At full capacity, the plant produces 250,000 devices per week.

Justification

Mold-plate loading was scheduled for automation before ground was ever broken at the facility. As already noted, dedicated equipment exists to handle lead frames but costs about 5 times as much as flexible automation approaches. Furthermore, dedicated equipment would have required extensive factory floor space and disrupted the natural work flow in the assembly process. Automation increased yields because manual handling more easily fractures the delicate gold-wire bonding. Yields decrease ½% of 1% with manual handling, which in a $100-million annual production facility amounts to at least a $500,000 increase in profit due to the flexible automation system.

Summary

Integrated manufacturing is the future in industry. No longer can companies expect to compete globally while retaining enormous investments in work-in-process inventories at offshore facilities. Communications spanning oceans between U.S.-based engineering staffs and offshore manufacturing management must ultimately be counterproductive. So, this plant models the onshore productivity of an integrated factory. As low cost surface-mount devices are used more widely, assembly factories will have to respond with several different package styles within the same factory. Wherever market requirements produce a demand below the justification levels necessary for dedicated automation, flexible automation will play an important role.

BIBLIOGRAPHY

1. P. B. Scott and T. M. Husband, "The Search for Cost Effective Robotic Assembly," reprinted by permission of the Council of the Institution of Mechanical Engineers from *UK Robotics Research 1984,* 11 (1984).
2. *Ibid.,* 12 (1984).
3. *Ibid.,* 14 (1984).
4. A. Month, RCA Corp., Lancaster, Pa., and W. Salvesen, Datum Industries, Inc., Palisades Park, N.J., private communication.
5. J. Altendorf, Hewlett-Packard Co., Corvallis, Ore., private communication.
6. T. Howell, The Deutsch Company, Banning, Calif., and W. Reiersgaard, BaGaard Automation, Portland, Ore., private communication.
7. R. Winslow, Cypress Semiconductor, San Jose, Calif., and M. Akagawa, Intelmatec Corp., Hayward, Calif., private communication.

EMPLOYMENT, IMPACT

ROBERT AYRES
Carnegie-Mellon University
Pittsburgh, Pennsylvania

INTRODUCTION

The subject matter of this article necessarily covers a very large scope, from employment and productivity issues to international trade, structural change, and quality of life. Many of these topics cannot be discussed quantitatively with any realism. Hence, this article is divided into three parts. The first part reviews the capabilities and limitations of robots as a species, compares robot capabilities with human capabilities in ergonomic terms, and identifies the job categories in which robots are now most competitive with humans. The second part deals with the more readily quantifiable questions relating to the penetration of robots in job markets and includes a mathematical forecasting model. It also discusses some of the implications for labor and capital productivity. The third part is more of a literature review touching on some of the other relevant social and economic topics. It is important to emphasize, more than once if need be, that the social importance of various issues may well be in inverse ratio to their quantifiability. In other words, the quantitative results displayed in the second part of the article are much less significant than surface appearances indicate, whereas the reverse may be true of the third part.

ROBOT CAPABILITIES

The Future Economic Role of Robots

To most economists, robots are just another kind of capital equipment, not to be distinguished in any fundamental way from other forms of capital. This attitude may be justified from a very abstract and distant "bird's-eye" view of the matter. However, whereas other forms of capital enhance and complement human labor by making it more productive, robots can be regarded as direct substitutes for some categories of labor. Hence, the first question of some importance is "What is a robot?"

This topic is thoroughly covered from a technical perspective elsewhere in the *Encyclopedia.* Robots have been defined somewhat differently by various authors and groups. The common denominator seems to be that robots, today, are one-armed, two- or three-fingered materials-handling devices or manipulators. They are capable of being instructed to carry out (and repeat) a specified sequence of motions.

Robots are commercially differentiated in terms of speed, payload, range of motion, mobility, number of degrees of freedom, accuracy, memory capacity, teachability, programmability, and control flexibility. The major differences between early robots and later technological generations relate to the latter attributes. Early robots were memory-limited and mechanically or electromechanically taught, programmed, and controlled. Most robots of the current generation are controlled by microcomputers and programmed by higher-level languages. Memory is no longer a significant limitation. The next major technological generation of robots will increasingly incorporate and/or be linked to sensors, eg, to detect obstructions, and self-programming features to avoid the same. Reprogramming will become progressively easier, and correspondingly less specialized engineering will be needed to move a robot from one task to another. Mobility, multiarm coordination, and general-purpose grippers will also be increasingly available.

Nevertheless, in terms of the socioeconomic issues considered in this article, it seems appropriate to define robots not in technical terms as above, but in terms of the kinds of tasks they can do. Even more relevant, perhaps, is a clear understanding of the kinds of tasks that robots of the present (and next) generation *cannot* do. This topic is considered more thoroughly later.

There is a school of thought that puts essentially no limits on the potential capabilities of robots with artificial intelligence (1). People holding this view argue, in effect, that robots and related forms of automation will eventually replace all human workers. There are others, including the author, who believe that humans will continue to hold a comparative advantage over machines for certain classes of tasks for as far into the future as can reasonably be foreseen (2,3). The two views are not even necessarily in conflict. The first emphasizes ultimate technological limits, the second rates of technological progress and market penetration.

For purposes of this article, however, the potential of robots can be considered to be quite narrowly prescribed. To summarize briefly, it is assumed here that for the next 10–20 years (at least), robots will be largely confined to repetitive manufacturing operations in an engineered environment. Other applications within that time frame will be quite specialized, eg, to inimical environments such as outer space, the ocean floor, chemical or nuclear plants, fire fighting, riot control, or warfare. For a more extended discussion, see Ref. 4. Nonmanufacturing applications, especially in space, may someday offer enormous economic benefits. However, meaningful quantitative assessment of those benefits is almost impossible at this time, so further comments are deferred until Other Social and Economic Impacts (see also AUTOMATION, HUMAN ASPECTS).

Task Taxonomy

In the economist's "bird's-eye" perspective, the manufacturing sector (as distinguished from mining, construction, or services) is devoted to the conversion of raw materials into finished and portable products ranging in size from tiny electrical components or fasteners up to ships and in complexity from nails to supercomputers. Basic activities can be subdivided into several basic categories:

- Materials processing (refining, alloying, rolling, etc)
- Parts manufacturing (cutting, forming, joining, finishing)
- Parts assembly and packaging
- Inspection
- Shipping, storage, maintenance, sales, etc

Materials, energy, capital, and labor are said to be "factors of production." As a rough generalization, factors of production are regarded as substitutable for each other, ie, labor or energy inputs can be decreased by increasing capital inputs. (This is not true, of course, for materials actually embodied in the product or for fuel to run machines.) However, upon closer scrutiny, such substitutions are typically possible only at the margin and in a rather special aggregated sense. To clarify this point, consider the role of fixed (physical) capital, disregarding liquid working capital for the moment. Capital plant and equipment are of several distinct kinds:

- Tools, dies, patterns
- Machines tools and fixtures
- Materials-handling equipment, eg, pallets, conveyor belts, transfer machines, pipes, pumps, forklifts, cranes, vehicles
- Containers, eg, shelves, bins, tanks, drums
- Structures and land

Machine tools do substitute for workers insofar as they wield tools such as hammers, drills, punches, saws, milling cutters or grinding wheels, files, or cutting implements similar in function to hand tools as used by human workers. Machine tools are now used almost universally in manufacturing (at least, in developed countries) because they can be faster, stronger, and more accurate and tireless than human workers using hand tools. Motor vehicles are used for transportation for similar reasons. Containers and structures are required to store and protect materials in process as well as shelter tools, machines, and workers from the elements. Clearly, these categories of capital are somewhat complementary; at any rate, capital in one category cannot substitute for capital in another. Traditionally, the substitution of capital for labor has meant the greater employment of machine tools in place of manual tools, and motorized forms of transportation in place of nonmotorized ones. But, until recently, each machine has needed a human operator. In short, machines have been substituted mainly for human muscles but not for human senses and intelligence. In the past, machines and their human operators have been effective complements. The question implicit in the title of this article can now be made explicit: *To what extent can machines be expected to take over other* (*control*) *functions of human workers* in the near future, thus becoming direct substitutes for human labor?

To elucidate this question, a better functional taxonomy of repetitive factory tasks that are directly related to fabrication or assembly of parts is needed. (For present purposes, workers whose jobs are nonrepetitive can be ignored; these include personnel involved with building or machine maintenance, setup, scheduling, inventory, transportation, product design and testing, administration, or sales.) The major generic, ie, repetitive, task categories are

Parts recognition, sorting, and selection
Machine-parts transfer loading/unloading
Tool wielding
Parts inspection
Parts mating, ie, assembly

All of these generic tasks can be accomplished, in principle, either by machines or human workers. The most common patterns of automation in factories today are shown in Table 1. In custom (or small-batch) manufacturing, most control tasks are and will remain largely manual, simply because it is not worthwhile to mechanize any task that is not highly repetitive. The increasing use of programmable machines tools in small shops does not contradict this conclusion. It merely reflects the fact that NC machine tools are becoming easier to program, so that microprocessors are able to control operations that can be entirely committed to memory, in advance. In larger-batch manufacturing, machine-tool loading/unloading is gradually being taken over by robots or programmable feeders, whereas assembly remains largely manual, though machine-assisted. Insensate robots also perform some tool-wielding operations, such as welding, spray painting, or gluing. In mass production, mechanization now extends to virtually all tasks except for magazine or pallet loading, inspection, and assembly, and even these are increasingly machine-assisted.

Although statistical evidence of skill levels over time is scarce, it is likely that the need for intelligence and skill in repetitive factory tasks on the factory floor has generally been decreasing for several decades. In effect, much of the skill formerly needed by a machinist, for instance, has now been embodied in sophisticated machines. This tendency was perhaps anticipated by many writers in the 19th century, but it was empirically confirmed for the first time by a 4-year study at the Harvard Business School by James Bright (5–7). Bright constructed a 17-level "automation ladder" (Table 2) and summarized the results of his observations and conclusions with respect to skill requirements, shown in Figure 1. It should be noted that factory automation today is, on average, several steps more advanced on Bright's ladder than it was in the late 1950s, with many applications exemplifying levels 12–14.

In virtually all cases, the remaining nonmechanized but repetitive factory jobs of today seem to require a significant level of tactile or visual sensory feedback. In fact, it is quite realistic to regard most factory workers in the semiskilled job classifications as "operatives" (Bureau of Labor Statistics terminology) or "machine controllers," to use a term that perhaps better conveys the essence of the human role in the production system.

In the modern manufacturing context, the human factory worker can be modeled as part of an information processing feedback system. (This insight was expressed at least 35 years ago by Norbert Wiener in *Cybernetics* (8), and a number of early workers in "human factors"/ergonomics.) He (or she) receives status information from the machine, the workpiece, and the environment. The worker processes and interprets that information, arrives at certain conclusions, and translates those conclusions either into new control settings for the machine or a new position/orientation for the workpiece. The amount of higher-order (problem-solving) intelligence required by the worker depends on how limited the set of possible responses is and how precisely the criteria for choosing among them can be prespecified. In many cases, the worker need only decide whether the last operation was successful and signal for the next operation to begin. The major difference between jobs requiring semiskilled and skilled workers is that the former jobs involve relatively few and simple choices, each made many times, whereas the latter jobs involve a very wide range of possible choices. Greater intelligence is involved when the range of choice is so wide that each case is likely to be unique in some respects, requiring the worker to extrapolate or interpolate from known and understood situations. (This is the essence of a nonrepetitive job, of course.)

Jobs in the goods-production sectors (agriculture, mining, manufacturing, construction) are classified as *production* or *nonproduction*. The latter category includes office workers, sales personnel, those personnel involved with logistics, etc. Within the "production" category, there is a further division between *direct* and *indirect*. The former refers to workers whose labor is, in some sense, embodied in the product and whose wages are counted as *variable* costs of production. The latter refers to production workers whose labor is needed to keep the plant open and whose wages (or salaries) can be thought of as part of the *fixed* costs of production. (This distinction is somewhat arbitrary, to be sure. In Japan, it is reported that much of the direct labor is regarded as a fixed cost—with important consequences for long-term strategy. On the other hand, academic economists tend to regard all labor as a variable cost.) For the purposes of this article, indirect labor

Table 1. Level of Automation vs Scale

Task Category	Custom	Batch	Mass
Parts recognition and sorting	Manual	Manual	NA[a]
Parts transfer	Manual	Transitional (eg, belt machine)	Mechanized (eg, transfer machine)
Machine loading and unloading	Manual	Mostly manual	Mechanized (eg, feeders)
Tool wielding (including machine operation)	Semimechanized (manual control)	Mostly mechanized (NC) except for supervisors	Mechanized, fixed sequence
Parts inspection	Manual	Manual	Transitional
Parts mating and assembly	Manual	Mostly manual	Transitional

[a] NA = not applicable.

Table 2. Automation Ladder

Levels of Mechanization and Their Relationship to Power and Control Sources

<table>
<tr><th>Initiating control source</th><th colspan="2">Type of machine response</th><th>Power source</th><th>Level number</th><th>Level of Mechanization</th></tr>
<tr><td rowspan="9">From a variable in the environment</td><td rowspan="6">Responds with action</td><td rowspan="3">Modifies own action over a wide range of variation</td><td rowspan="15">Mechanical (nonmanual)</td><td>17</td><td>Anticipates action required and adjusts to provide it</td></tr>
<tr><td>16</td><td>Corrects performance while operating</td></tr>
<tr><td>15</td><td>Corrects performance after operating</td></tr>
<tr><td rowspan="3">Selects from a limited range of possible prefixed actions</td><td>14</td><td>Identifies and selects appropriate set of actions</td></tr>
<tr><td>13</td><td>Segregates or rejects according to measurement</td></tr>
<tr><td>12</td><td>Changes speed, position, direction according to measurement signal</td></tr>
<tr><td colspan="2" rowspan="3">Responds with signal</td><td>11</td><td>Records performance</td></tr>
<tr><td>10</td><td>Signals preselected values of measurement (includes error detection)</td></tr>
<tr><td>9</td><td>Measures characteristic of work</td></tr>
<tr><td rowspan="4">From a control mechanism that directs a predetermined pattern of action</td><td colspan="2" rowspan="4">Fixed within the machine</td><td>8</td><td>Actuated by introduction of work piece or material</td></tr>
<tr><td>7</td><td>Power-tool system, remote controlled</td></tr>
<tr><td>6</td><td>Power tool, program control (sequence of fixed functions)</td></tr>
<tr><td>5</td><td>Power tool, fixed cycle (single function)</td></tr>
<tr><td rowspan="4">From person</td><td colspan="2" rowspan="4">Variable</td><td>4</td><td>Power tool, hand control</td></tr>
<tr><td>3</td><td>Powered hand tool</td></tr>
<tr><td rowspan="2">Manual</td><td>2</td><td>Hand tool</td></tr>
<tr><td>1</td><td>Hand</td></tr>
</table>

on the factory floor is largely concerned with engineering supervision and maintenance. It is essentially nonrepetitive in nature. There are two different kinds of mental activity involved in doing direct manufacturing work, namely:

- *Process Monitoring:* decision-making in response to external sensory data, regarding the state of the workpiece itself, or the state of the machines, tools, processing equipment, and/or the environment. Parts recognition and inspection are examples of "pure" monitoring tasks.
- *Motion Control:* decision-making in response to either internal or external sensory data reflecting the physical state of the worker (but, in a more general ergonomic context, the notion of "workers" will later have to be broadened to include robots) in relation to the requirements of the task in hand. Most manipulative tasks, including assembly, involve motion to some degree.

These may be termed internal control decisions. The more efficient the process, the less control information, of either type, is necessary. The theoretical minimum amount of control information is that which is ultimately embodied in the product by that process. The rest of the information is ultimately lost.

Generic Transfer and Manipulative Tasks

The above distinction between inherently task-related decisions and decisions needed for purposes of worker/operator motion control is helpful in comparing human workers and

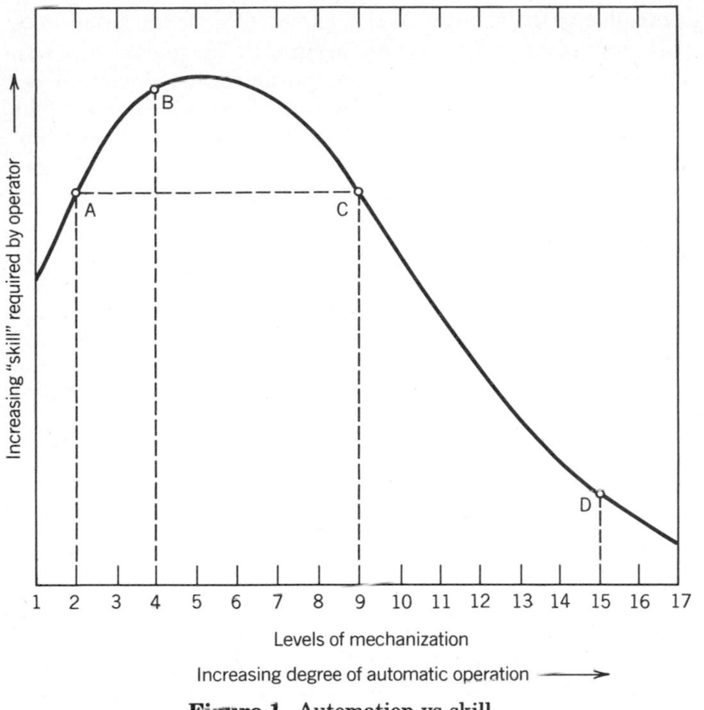

Figure 1. Automation vs skill.

machine tools or robots as manipulators. In the discussion that follows, attention is focused on two interfaces: the *manipulator–tool* interface and the *manipulator–workpiece* interface. In fact, all manipulative tasks can be described in one of three ways:

1. Bringing the workpiece to the tool (typical of manufacturing)
2. Bringing the tool to the workpiece (typical of construction)
3. Bringing workpieces together (assembly and construction)

All three task categories are essentially equivalent from a motion-control perspective.

Until very recently, all of these generic tasks were largely carried out by human workers, although gross (transfer) motions have been increasingly mechanized since prehistoric times and especially since the early 19th century. Fine motions of cutting/forming tools against workpieces have also become increasingly automated. The assembly task is often regarded as being the least susceptible to automation, but automatic assembly of fibers into threads (spinning) and of threads into fabrics (weaving) also goes back to very early times. In fact, programmable automation was first introduced in the silk-weaving industry with the so-called Jacquard loom in the early 1800s.

As noted above, from the decision and control perspective it does not matter whether the object being manipulated is a workpiece or a tool. Thus, there are really only two types of motion worth distinguishing:

1. Motions carrying an object to a specified end-point and orientation, along an unspecified path. (Hereafter, the term "end-point" will always be taken to include an orientation as well as (3-D) position. Hence, 6 variables must be specified to define an end-point, unless some symmetry reduces the orientational information required; for instance, a cylindrical pin can be inserted at any angle around its longitudinal axis.)
2. Motions carrying an object, usually a tool, along a specified path consisting of a dense sequence of specified locations and orientations. Objects can be carried passively, eg, on conveyor belts, on pallets, or in magazines, or they can be actively gripped. It is therefore useful to distinguish passive carriers and active carriers or grippers. In general, passive carriers suffice when the task only demands a change of position, whereas active grippers are needed when a change of orientation is also required. In practice, passive carriers are often used for large-scale intermediate motions, eg, between workstations, but an active gripper is often needed at the beginning and almost always at the end of such an intermediate motion, for purposes of loading/unloading. There are two kinds of active grippers inherently capable of re-orienting an object: (*1*) human arms/hands and (*2*) robot arms/hands.

From the information perspective, it is important to notice that when a workpiece or tool is released by an active gripper (unless it is transferred to another) it often loses position/orientation information. This happens, for instance, whenever a part is dropped into a bin. This loss of information must later be compensated for by a decision-making process involving visual identification of the part (or tool), precise location (and orientation) in 3-dimensional space, along with identification/location of potentially interfering objects, barriers, etc, and the initiation of a motion to bring a gripper to the object, to have it grasp the object, and to carry out the next task. This process is highly information-intensive. Until the present decade, only the human eye–brain–hand combination sufficed. For humans, incidentally, most of the requisite mental activity is unconscious. It has not been clearly understood by human-factors engineers until recent decades that it takes a finite amount of time, requires energy, and induces mental/fatigue to sort and separate parts from a pile. Indeed, the number of errors of identification or location increases with the output rate.

Methods of engineering around the problem of lost information began to be developed as really large-scale manufacturing (mass production) methods evolved early in the 20th century. In the absence of computers, machine vision and robots were used. One possible strategy was to prevent locational information from being unnecessarily lost in the process of transferring discrete workpieces from workstation to workstation. The so-called *synchronous transfer line* was introduced for this purpose. The function of a transfer line is to maintain the relative positions and orientations of parts or workpieces as they progress through a sequence of mechanical operations, so that loading/unloading at each station can be mechanized with a minimum of additional sensory information.

The transfer line was and is a technically successful solution to this problem in some contexts, but the costs of this solution have been quite high, both in terms of capital outlay and in terms of creating effective barriers to product redesign.

It is not a viable solution where workpieces are not both highly standardized and manufactured in very large numbers. (A more extended discussion of flexibility and "the productivity dilemma" can be found in Refs. 9 and 10.)

Another engineering solution to the problem of "lost" location/orientation information is the automatic mechanical parts sorter/feeder (11). Here, the strategy is not to prevent information from being lost in the first place but to use the shape-information embodied in the parts themselves for purposes of selection and orientation. Mechanical details apart, the parts are in effect self-orienting. The (analogue) selection algorithm needed to accomplish this is, in turn, embodied in the design of the feeder device.

The mechanical solutions to the parts-handling problem discussed above play a major role in modern manufacturing. However, because of the inherent complexity of the machines and the fact that a transfer machine must be customized to a particular application, this approach is very capital-intensive and inflexible. Consequently, so-called "hard" automation is limited by economic factors to very large-scale production of very standardized parts/products. In addition, automatic parts sorters/feeders are inherently limited to handling relatively small parts of fairly simple shape that are not easily damaged or broken. Thus, other nonautomatic approaches must also play a role. In this context, "nonautomatic" implies sensory-based control of motion, either by human workers or robots and computers.

Motion Control and Feedback

A brief digression at this point may be helpful for some nonspecialist readers. *Open-loop* (nonfeedback) control is exemplified by the Jacquard loom and by the present generation of NC machine tools. Instructions are prepared in advance, coded, stored (eg, as holes in paper tape or as small magnetized spots on a magnetic tape), and fed automatically to the machine loom or machine tool in real time. The machine operating speed and the rate of instructional input must somehow be synchronized. In an unsophisticated machine such as the Jacquard loom, the cycle time is fixed either by the machine design itself or by the power source; the punched paper tape of instructions is fed mechanically at the same rate as the material being woven. In a more complex modern NC machine tool, there is a buffer (an internal memory) to permit the tape-feed mechanism to operate intermittently and independently of the machine drive as more instructions are needed. In this case, the machine must be able to signal that the last instruction has been executed (whether successfully or not) before receiving the next one.

Such a signal is, in fact, a simple form of feedback, and this is the simplest and crudest form of *closed-loop* (feedback) motion control. The "sensors" involved are mechanical stops or limit switches. The feedback from a stop or limit switch is obviously very limited, however, and most engineers would probably classify a device using limit switches (but no other feedback) as open-loop. True closed-loop control involves more than "on/off" or "go/no-go" decisions. It involves continuous adjustment of the actuating system, eg, tool-drive spindle or the robot arm, by periodically monitoring operating variables such as spindle speed (rpm) and/or arm or tool position. The controller automatically compares actual speed/position data with desired speed/position as specified by the job instructions. Its control decision depends on the measured deviation between the actual speed/position and the specified ideal. The decisions are based on a set of explicit rules (an algorithm) that was previously built in by the programmer. A still more sophisticated future controller will presumably be able to learn from experience and reprogram itself (level 15 or higher on Bright's ladder).

A further distinction is helpful here between *internal* and *external* feedback loops. The kind of feedback described above in the context of a machine tool or robot arm uses only sensory information about its internal state of motion possibly supplemented by temperature, voltage, or other signals, to decide what to do next. It may also use some "force-feedback" signals reflecting interactions at the tool–workpiece interface. But it does not utilize external sensory information about the workpiece on the environment. External sensory information is much more important in practice for controlling animal motion than machine motion, for reasons considered in the next section.

The conventional control system, for a manufacturing process based on information gathered by human eyes and ears and processed by the human brain, can be represented as a simple model, as shown in Figure 2**a.** The (still-primitive) computer, or automated control system, can be represented by a similar model, shown in Figure 2**b.**

Ergonomic (human factors) studies have ascertained that there are two types of human motion control. For gross motions, such as moving the arm toward an object to be grasped, a kind of open-loop (nonfeedback) control suffices. The brain simply commands the muscles to move the arm in a certain direction by a certain amount. The motion-control information required for gross motions is on the order of 5 bits/s. However, fine motions require visual feedback and the control-information requirements rise to around 12 bits/s. This is about the maximum level of mental output for most humans, even for short periods. It requires intense concentration and results in a significant number of errors if concentration lapses. This is a fundamental limitation on humans as workers.

The potential for machines to take over jobs from human workers can now be considered in the light of this fact. Although machines are not yet nearly as good as humans at interpreting visual information, this is not a serious constraint for machines as regards motion control. In short, machines require less information for this purpose. Because machines have many fewer degrees of freedom than humans and are made from stiffer and more rigid materials, open-loop motion control suffices to achieve tolerances that are adequate for the vast majority of industrial applications.

In summary, it can be concluded with considerable confidence that machines can take over from humans essentially any task involving "pure" motion control in an unchanging or predictable environment. They can also take over process-monitoring and control tasks requiring sensory data not directly accessible to human senses. Humans, on the other hand, still have a tremendous advantage in tasks involving significant amounts of visual process monitoring and decision-making. Most real jobs actually involve some combination of the two. Where the decision-making component is dominant and

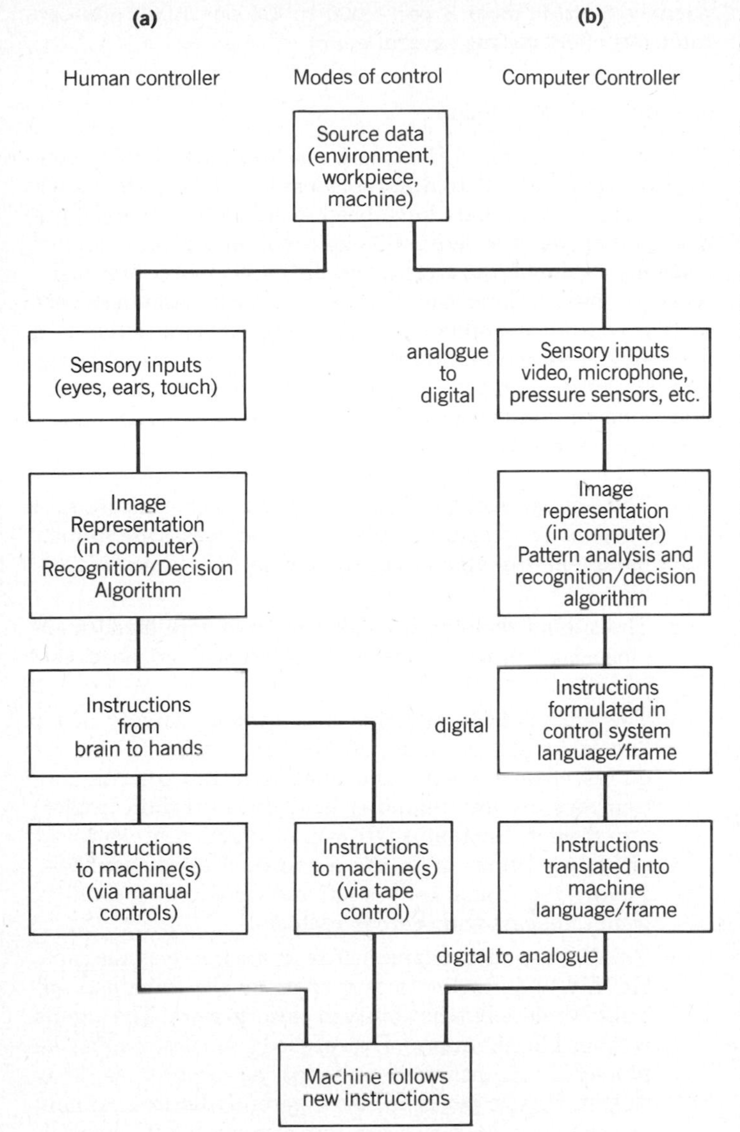

Figure 2. Modes of control.

the motion-control component is essentially trivial (eg, button-pushing), as in the case of a typist, there is no immediate prospect that "smart" machines could substitute for human workers. However, where the decision-making component is relatively minor, as in the case of machine loading and unloading, the reverse is true.

It follows that a quantitative measure of the motion-control (vs monitoring) content of repetitive manufacturing jobs should be a good predictor of their future vulnerability to mechanization. Unfortunately, whereas such a measure could probably be inferred from time–motion data compiled in the past (12), no such study has yet been carried out systematically.

Sensory Processing in Humans

As was pointed out above, some manufacturing tasks, such as parts recognition and inspection, involve sensory perception alone, whereas others require sensory information for motion control. Thus, sensory-information acquisition and processing are critical.

Information inputs to humans in all situations (not only the workplace) is initially conveyed by the five senses: vision (eyes), hearing (ears), touch, taste, and smell. It is known from a variety of kinds of evidence that, for humans, vision accounts for around 90% of total sensory input; touch and hearing account for most of the remainder (see Table 3). This does not imply that vision is the only sense that must be considered, however. On the contrary, there are a number specialized workplace situations in which audio or tactile information is at least as important as vision, such as inserting a bolt in a threaded hole. There are also some industrial tasks in which other senses are critical, including cases where senses not possessed by humans are needed, eg, x rays. The fact that humans cannot "turn off" their eyes and ears means that significant amounts of distracting extraneous information are also processed.

Estimates of the (input) transmission capacity of the human eye vary considerably, depending on the measurement technique used. For instance, combining data on monocular visual acuity and so-called flicker-fusion frequency leads to an estimate of 4.3×10^6 bits/s (13). This implies an estimate of 4.3×10^6 for each optic nerve fiber. However, other data have suggested greater input capacities, eg, 5×10^7 bits/s. Visual input capacity is evidently a function of luminance (light intensity) and may be as high as 10^9 bits/s at high luminance levels (14). The quantity of information that can be extracted from visual or audio inputs (ie, channel capacity) can be estimated in terms of human ability to discriminate among stimuli and respond suitably. Response normally involves a physical action. The discrimination ability varies according to the nature of the sensory input, as shown in Table 4, which assumes no time limitations.

The information that can be processed, ie, discriminated, and responded to per unit time increases in production to the input rate until a maximum is reached, depending on the internal information-transmission capacity of the organism and the nature of the response. Obviously, a single cell can respond to stimuli more quickly than a complex organism, hence, its effective rate is higher. At the other extreme, the response rate of an organization is inherently much slower than that of an individual (see Fig. 3).

Figure 3 indicates clearly that all biological information-processing systems exhibit a saturation phenomenon. When inputs exceed a certain threshold, the processor becomes over-

Table 3. Sensory Input Channels

Sense	Receptors	Nerve Fibers
Vision	2×10^8 (retina)	2×10^6 (optic nerve)
Hearing	3×10^4	2×10^4
Touch	5×10^5	1×10^4
Smell	1×10^7	2×10^4
Taste	1×10^7	2×10^3
Pain	3×10^6	1×10^6 (Pain, Heat, Cold combined)
Heat	1×10^4	
Cold	1×10^5	

Table 4. Sensory Outputs: Discrimination Capability[a]

Sensory Modality and Stimulus	W No. of Levels Which can be Discriminated on Absolute Basis	lnW No. of Bits of Information Transmitted[b]
Vision, single dimensions		
Pointer position on linear scale	9	3.1
Pointer position on linear scale		
Short exposure	10	3.2
Long exposure	15	3.9
Visual size	5–7	2.3–2.8
Hue	9	3.1
Brightness	3–5	1.7–2.3
Vision, combinations of dimensions		
Size, brightness, and hue[c]	17	4.1
Hue and saturation	11–15	3.5–3.9
Audition, single dimensions		
Pure tones	5	2.3
Loudness	4–5	1.7–2.3
Audition, combination of dimensions		
Combination of 6 variables[d]	150	7.2
Odor, single dimension	4	2.0
Odor, combination of dimensions		
Kind, intensity, and number	16	4.0
Taste		
Saltness	4	1.9
Sweetness	3	1.7

[a] Amount of information in absolute evaluations on different stimulus dimensions (15).
[b] Since the number of levels is rounded to the nearest whole number, the number of bits does not necessarily correspond exactly.
[c] Size, brightness, and hue were varied concomitantly, rather than combined in the various possible combinations.
[d] The combination of 6 auditory variables was frequency, intensity, rate of interruption, on-time fraction, total duration, and spatial location.

loaded and performance is degraded. (Essentially the same phenomenon is physically observable as highway congestion.) The output degradation initially takes the form of errors. The error rate increases slowly at first, but as saturation approaches, it rises sharply and the system collapses into "gridlock." The maximum effective output level for humans appears to be in the range of 5–10 bits/s. It is evident, however, that the error rate will tend to be quite high at effective output rates near the maximum. *In order to achieve low error rates, the output rate must also be held to a low level.* This fact has important implications for manufacturing quality control.

Incidentally, all available evidence (16–20) points to the fact that humans are inherently error-prone: even under the best possible working conditions, error rates below 1 per 10,000 opportunities appear to be unattainable. Under highly stressed or life-threatening conditions, errors are far more probable (up to 0.25). An exceptionally intensive inspection system will catch up to 98% of the errors, but an 80% correction rate is more typical. It follows that the probability of uncorrected human error is at least 10^{-6} per operation, and under realistic conditions, significantly greater than that, ie, closer to 10^{-3}. Xerox Corporation reported "success" in sharply reducing its parts reject rate from 8 per 1,000 to 1.4 per 1,000 after an intensive effort lasting several years.

Sensory-controlled Motion

Control strategies for animals and machines both involve sensory information, but to different degrees. It is useful to observe here that animals have been evolving for several hundred million years at least. Efficient motion control is almost certainly a valuable survival attribute for any motile organism, whether prey or predator. Hence, it may be assumed that motion control of animals is now quite efficient, subject to the physical constraints imposed by the characteristics of the animal skeleton and the physiological nature of the nerves and muscles and their chemical-activation mechanisms. Among these constraints, six are worth noting:

1. The animal skeleton has a large number of degrees of freedom as compared with most machines. One human hand and arm alone have more than 20 degrees of freedom.
2. The animal skeleton is weak compared with modern engineering materials, especially in terms of tensile stress/strain.
3. The animal body requires a continuous supply of air and intermittent inputs of food and water. It also requires adequate light and environmental conditioning (temperature and humidity held within certain ranges) for efficient functioning. It cannot function in very cold or hot conditions or in the presence of toxic chemicals. Finally, the animal body is soft and vulnerable to damage from collisions with harder materials.
4. Animals are autonomous self-reproductive systems capable of an enormous variety of metabolic functions not directly related to the ability to perform work. The organs responsible for many of these body functions must be protected, maintained, and carried around as body weight. Muscle power is also inherently limited. Animal power-to-weight ratios are low compared with specialized prime movers made from hard and strong materials.
5. The internal animal skeleton does not permit true rotary motion, although circular cranking motions are possible. Thus, mechanisms depending on circular motion such as roller bearings, gears, pulleys, etc, are impossible for animals with internal skeletons.
6. External feedback information used by most animals for gross-motion-control purposes is primarily provided by reflected light in the solar spectral ranges. (A few species such as bats and whales depend much more on acoustic signals, but this is not particularly relevant for humans.) Humans and anthropoid apes, in particular, also rely extensively on tactile information from the fingers for certain manipulative activities.

A consequence of the first of the physiological constraints of animals, the large number of degrees of freedom, is that control based on internal (proprioceptive) feedback alone is relatively inaccurate, with a tolerance/amplitude ratio of about 7% for arm–hand motions. Greater accuracy is possible as noted pre-

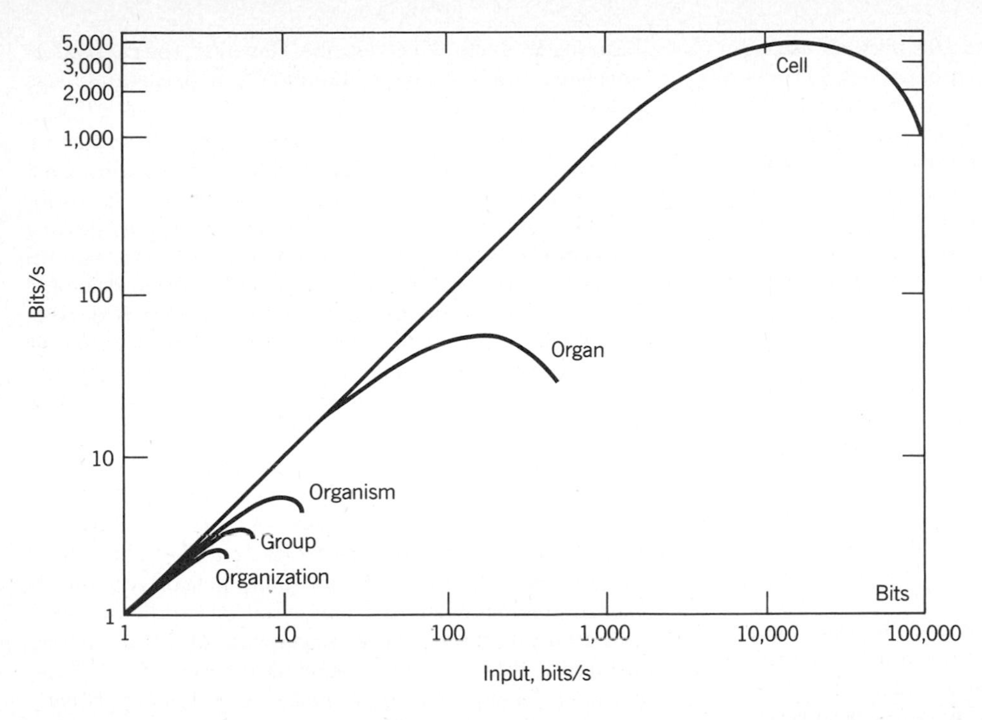

Figure 3. Channel saturation.

viously, but *only* by using external visual feedback (closed-loop) and successive approximations. The minimum time required to complete the motion accurately increases as the logarithm of the tolerance needed. For tasks involving very fine tolerances, humans may be therefore excessively slow compared with machines. By contrast, a nonbiological (robot) arm with only 3–6 degrees of freedom can perform much more accurately than humans with internal feedback controls (or even from memorized instructions). Robot arms can achieve open-loop tolerance/amplitude ratios of 10^{-2}–10^{-3}; higher precision requires closed-loop feedback control. This fact has important implications in choosing between humans and robots to carry out certain tasks requiring precise, repetitive motions.

A consequence of the second and third constraints is that machines are more durable than humans and can work continuously, rather than intermittently. A consequence of the third constraint is that machines (robots) are inherently much better suited than humans at working in remote or hazardous environments. The fourth and fifth constraints imply that machines are inherently much better than humans at performing tasks requiring strength or power or rotary motion. The implication of the sixth constraint is that machines may be more appropriate than humans in situations where the relevant control information is in a nonvisual form. A good example of the latter is submarine navigation, depending on sonar or air traffic control, where the key source of information is radar signals, ie, microwaves. Also one may consider the example of spacecraft reentry, as contrasted with submarine navigation.

Machines, on the other hand, are inherently much less adaptable than humans, especially in situations where the environment or the task is continuously changing in an unpredictable way. As a quick (but not bad) generalization, it can be concluded immediately that machines will almost always be inherently better at routine, repetitive tasks, especially in an unchanging environment. However, there are a number of transitional tasks in manufacturing that humans can still do better than machines, primarily because of current limitations in machine sensing technology. In the past decade, quite a lot of research has gone into the development of the elements of general-purpose computerized sensor-based machine controllers. Significant progress has been made, to be sure. But it is now very clear (though perhaps only dimly understood a decade ago) that the most sophisticated sensor-based computer control system that can be built today is still vastly inferior in input information-processing terms to the human eye–ear–hand–brain combination. It is important to distinguish here between raw input information, such as the optical signals received by the retina of the eye, and the output (control) information sent by the human brain to the hands or feet.

The vision systems of animals comprises an optical focusing device (the lens), a light-sensitive detector (the retina), and a post-processing device (the visual cortex). The retina of a vertebrate contains about 10^6 light-sensitive cells of two distinct types: (*1*) cells that detect both light intensity and peak wavelength for daylight color vision (cones) and (*2*) cells that detect only intensity light for night vision (rods). The retinas of animals, such as birds, that require very high quality daytime vision over wide angles cannot spare much retinal space for night vision and, conversely, animals that hunt at night cannot also enjoy good quality color vision by day.

Within the retina itself, the visual field is processed by about 2×10^7 neurons to determine edges, curvature, and motion. The retina sends a reduced or coded form of this visual input data via the optic nerve to the visual cortex of the brain, where about a billion neurons carry out further processing. The entire system processes about 10 scenes per second, each scene consisting of a matrix of about 1000×1000 picture cells (or pixels) in three colors. By comparison, a state-of-the-art minicomputer requires around 2 min to process one black-and-white scene recorded by a solid-state camera in the form of a matrix of 256×256 binary pixels.

In summary, the color picture recorded 10 times per second by the retina of a human eye initially contains about 50 times as much visual information as that recorded by a Vidicon TV camera (or the equivalent) and it is processed 100–1000 times as fast. Thus, the eye has an overall performance advantage on the order of 5,000–50,000 (1985). Although the above estimates are crude, they serve to make the key point. It seems clear that improved solid-state sensors, higher computational speeds, and computer-memory capacities alone will not quickly bring machine vision to a comparable level; the gap is still much too great. The problem is fundamentally one of inappropriate computer architecture and primitive image-recognition techniques. Image representation and analysis are in principle more suited to parallel-array processors than to the von Neumann-type serial processors utilized by virtually all computers today. Very few parallel-processing computers exist as yet and none are used in commercial vision/taction systems. Indeed, parallel-processing computer architecture is still in its infancy, although several computers with parallel-processing architecture appeared on the market in 1985.

By way of contrast to the computer-controlled "smart" machine, what are the relevant attributes of the human worker? He/she is born with high quality sensory equipment (eyes, ears, and hands), and develops excellent image-representation and pattern-analysis capabilities (brain), utilizing a parallel-processing architecture that is still little understood. These capabilities are innate and not improved significantly by education or training. Inanimate sensors and computers are currently orders of magnitude inferior to the human brain in terms of information processing and interpretation. Even with another decade of rapid progress in electronic technology, the gap will still probably be enormous.

Human vs Machine

Thus, humans are currently able to process and reduce enormous amounts of information, relative to machines (including computers). The advantages of machines, as would be expected, are primarily on the output side, eg, greater operating rate or power (or strength) and tolerance (or precision). It is helpful to consider the above three variables separately:

1. *Rate*. If weight and/or precision of location are not constraining factors, humans can identify and feed or transfer small parts one by one at rates of the order of 1/s. Times for elementary motions have been compiled and published by the Maynard Foundation (12). Transfer machine magazine feeders and rotary bowl feeders can probably achieve consistently higher operating rates than humans for parts of a given size. However, the rate differences are small, perhaps factors of 2 or 3, certainly less than a factor of 10.
2. *Power*. Adult men in excellent physical condition sustain a power output of the order of 250 W or more in short bursts, and 75–10 W for fairly long periods. (A world-class athlete, such as a swimmer or cyclist, may be able to generate 300 or more watts of power output for several hours.) Machines, on the other hand, can be designed to deliver almost any amount of power. In practice, modern machine tools range in continuous effective power from 1 to 100 kW, or even more, depending on the application. Machines can outperform human workers in this regard by at least a factor of 10^2 or 10^3.
3. *Tolerance*. Using hand tools and unaided eyes (or simple lenses), skilled human workers such as seamstresses, jewelers, and watchmakers can work to tolerances up to about 10^{-3} in. (or, perhaps, to 10^{-3} cm). Using mechanical and optical aids such as micrometers and microscopes, tolerances of 10^{-5} cm or better can be achieved by human workers such as engravers. Machine tools or automatic dimensional measuring devices with 1–3 degrees of freedom can be adjusted to move repetitively along paths or to points in space with comparable precision. However, robots with more degrees of freedom tend to be about a factor of 10 less exact in repeating a motion than the most precise machine tools.

Taking into account the above comparisons yields the cost-independent human/machine performance ratios for all three groups of tasks shown in Table 5.

It is clear that the actual choice of a robot vs a human for a particular task such as parts transfer is not determined *a priori* by a single comparison such as that in Table 5. If machines are expensive enough and human labor is cheap, of course human workers will be used. The comparisons in Table 5 are more useful in providing insight as to the order in which various tasks are likely to be taken over by machines. This question is better addressed in terms of a 2-dimensional comparison, as in Table 6, which takes into account all the human–machine differences noted above. From left to right, tasks are ranked roughly in terms of increasing intrinsic difficulty for humans, and from top to bottom in terms of increasing difficulty for machines. Evidently, tasks in the upper-left quadrant are hard for either. The first tasks likely to be automated are therefore in the upper right, the boundary being determined by the cost of labor vs the cost of capital.

Table 5. Human/Machine Performance Ratios for Generic Factory Tasks

	Tasks Category	Measure	Human/Machine Performance Ratio, P
I	Parts transfer machine unloading	Rate	$10^{-1} < P < 1$
II	Parts recognition and selection machine loading; parts mating; inspection	Rate/tolerance	$10^{-1} < P < 10$
III	Tool wielding	Power/tolerance	$10^{-4} < P < 10^{-2}$

Table 6. Intrinsic Task Difficulty: Human vs Machine

Increasingly Difficult for Machines ↓	Increasingly Difficult Tasks for Humans ⟶		
	Pick and place a medium-sized oriented metal part Spot-weld a repetitive pattern Sandblast a wall Spray-paint a simple surface Drive a train on tracks Arc-weld along a seam	Assemble a wooden cabinet, electric motor, or pump, repetitively Pick and place a heavy metal part Solder very tiny wire connectors Spray-paint a complex surface	Pick and place a very heavy metal part in a hot, noisy (toxic) environment Operate a fire extinguisher inside a burning building Laser brain surgery (*ex* diagnosis) Land a spacecraft, good weather
	Pick and place a randomly oriented part (medium size) from a bin Pick unripe fruits/vegetables by size only Wash windows, selectively Wash dishes and glassware individually Inspect eggs in a hatchery	Inspect a printed circuit board for faults Build a brick wall (3-D) Cut coal from a face Operate a farm tractor Assemble a wire harness Finish (eg, lacquer) a cabinet, to order Land a small plane (day, good weather, no traffic) Cut and assemble a suit Identify counterfeit paper money	Assemble a mechanical watch Control air traffic at a busy airport Weld a broken waterpipe from inside Inspect a VLSI chip for faults
	Pick and place very floppy objects, eg, thread, yarn, wire Cut and arrange flowers Harvest ripe soft fruits/vegetables by color or texture Separate crabmeat from shells or clean shrimp Inspect seedlings in a nursery for quality Plant rice seedlings (or similar)	Drive a truck or bus through traffic Deliver a (normal) baby and inspect for faults Dental hygiene Repair lace "Invisible mend" a garment	Land an airliner at night, bad weather Identify a counterfeit "old master" Repair a damaged "old master" High speed auto chase through city traffic Diagnose a medical condition Heart or liver transplant

QUANTIFIABLE IMPACTS ON EMPLOYMENT

Vulnerability of Various Occupational Categories to Automation

The scope of this discussion has already been narrowed to manufacturing jobs in the "direct" category. Having considered the factors that make a particular production job suitable for mechanization, the next question to be answered is "What are the vulnerable job categories and how many workers are currently employed?" It must be recalled that the basic criteria for substitutability are repetitiveness, dominance of motion control over process monitoring, (ie, minimal need for interpretation of sensory data), and simplicity of decision-making. A harsh or unpleasant environment also reduces the efficiency of human wokers and their willingness to work and thus increases the economic attractiveness of machines.

Most writers have ignored the question posed above, recognizing its obvious difficulty. A possible future approach, based on the characterization of repetitive manufacturing tasks in terms of motion-control *vis-à-vis* process monitoring information requirements, was sketchily outlined in Motion Control and Feedback earlier. Lacking systematic data of this sort, the matter can still be approached by asking a group of experts to assess the inherent potential for mechanization of various job classifications in different industries. Two such surveys have been reported. One was done at Carnegie-Mellon University (1981–1982) under the supervision of the author. (The original survey (CMU81) was unpublished. An expanded version was reported in Ref. 2 and appears in Ref. 21. The latter has been further revised in Refs. 22 and 23.) It ultimately covered 24 robot manufacturers and users, primarily in the metalworking sectors (SIC 34–37); SIC refers to the Standard Industrial Classification used by the U.S. Census of Manufacturers. Respondents, all of whom were familiar with robot uses and capabilities, were given a detailed list of manufacturing job classifications and asked to estimate the fraction of each occupational group that could *in principle* be replaced by Level I ("insensate") and Level II ("sensate") robots in their industry. The survey data are given by detailed occupation group in Ref. 24. A summary of the data for (Level-I robots only) is given in Table 7, along with the results from two "Delphi" surveys sponsored by the Society of Manufacturing Engineers (1982, 1985). It can be seen that the results were comparable for most occupations. Although the Ayres–Miller survey, carried out in 1981–1982, did not specify a definite time frame, it is reasonable to suppose that most respondents implicitly assumed that penetration of Level-I ("insensate") robots would be substantially complete by the year 2000, if not sooner. This would not be true for Level-II robots.

A variant on the survey approach was used by Miller in his dissertation (21). He proposed a rule for determining which types of machine tools could be operated (in principle) by robots of Level I and Level II and estimated potential robot penetration on the basis of the *American Machinist Inventory of Machine Tools*. Table 8 shows that the two approaches led to very similar results.

It is clear that other types of automation, such as CAD and FMS, will also reduce the need for workers in various occupational groups. Thus, displacement estimates hereafter

Table 7. Estimates of Displacement by Occupation, %

Occupation	Revised CMU Survey Data for Potential Displacement by Level-I Robots[a]	UM/SME Delphi Survey Results for Expected Displacement by 1995[b]	
		Original Survey, 1982	Updated Survey, 1985
Assembler	8.9	10	12
Inspector	7.5	6	15
Packer	10.8	5	10
Painter	43.5	20	20
Welder and flame-cutter	25.5	20	15
Machinist and machine operator	14.3	13	10
Tool and die maker	1.3	2	

[a] Ref. 21.
[b] Ref. 25.

Table 8. Comparison of Estimates of the Potential for Robot Substitutions in the Operation of Metalcutting Machine Tools, %

Level I	Level II	Refs.
9.4–15.7	46.7	21[a]
14.3	40.8	2,21[b]

[a] Figures are Miller's estimates of the percentage of all metalcutting machine tools that could be operated by a Level-I and a Level-II robot, based on data from Ref. 26.
[b] Figures are derived from survey respondents' estimates of the percentage of metalcutting machine-tool operating jobs that could be performed by a level-I and a Level-II robot.

are likely to be understated. It is also clear that the manufacture and installation of robots will add some additional production jobs in the "robot" sector (unfortunately, many of these new jobs appear likely to be in Japan, not the United States). This would only compensate at best, for a fraction (perhaps 1 in 5) of the job losses.

Manufacturing Jobs

The percentage of the work force employed in the production of goods (agriculture, manufacturing, and mining) has steadily declined from nearly one-third of the work force in 1959 to barely over one-quarter in 1977. This trend is likely to continue into the foreseeable future (see Table 9). There were about 20 million manufacturing workers in 1980.

As the service sectors themselves have increased in importance, the "service content" of the manufacturing sector itself has also increased. Nearly 30% of the work force in manufacturing in 1980 was in "white-collar" managerial, professional, clerical, sales, or supervisory activities, up from 25% in 1960. Therefore, about 14 million workers were classified as "production workers" in manufacturing sectors in 1980.

Table 9. Work Force Employed in Goods and Service Sectors, %[a]

Percentage of Work Force Employed in	1959	1968	1977	1990 (Projected)
Agriculture, mining and manufacturing	34.4	30.5	25.2	23.2
Government, construction, utilities, transportation, trade, and services	65.6	69.5	74.8	76.8

[a] Ref. 27.

Slow growth in manufacturing employment in the 1970s contributed to the substantially higher unemployment rates in that decade than in the previous one. The economy was consequently unable to provide employment for all of the women wanting to enter the labor force or the "baby-boom" generation born in the late 1950s and early 1960s. The significance of robotization in the 1980s and 1990s is that it will add to the combination of factors that have retarded growth decline in employment in manufacturing.

Nearly half of all the 14 million production workers in manufacturing (and for that matter, nearly half of all manufacturing workers) are concentrated in the metalworking industries. This group includes (again, SIC refers to the Standard Industrial Classification used by the U.S. Census of Manufacturers):

- Primary metals (SIC 33)
- Fabricated metal products, except machinery, and transportation equipment (SIC 34)
- Machinery, except electrical (SIC 35)
- Electrical and electronic machinery, equipment, and supplies (SIC 36)
- Transportation equipment (SIC 37)

Table 10 shows the breakdown of U.S. metalworkers by major occupational classification. Through the mid-1980s, at least 80% of all industrial robots were used within the metalworking industries. Statistics from Japan confirm that about 85% of all robots sold in that country went to the metalworking sectors. Table 11 shows the total displacement that would result in that sector if the percentage shown in column 1 of Table 7 were correct, based on 1980 employment. It is also clear from occupational titles and descriptions that most of the routine, repetitive jobs that currently lend themselves to automation and robotization are performed by the semiskilled (nontransport) operatives. These jobs accounted for about 40% of total manufacturing employment in 1980.

Table 10. Production Worker Employment in Industries SIC 34–37, 1980

Occupation	Employment by Occupation, 1980	Production Employment, %
Assemblers	974,410	19.1
Inspectors	216,800	4.2
Maintenance and transport workers	341,740	6.7
Material handlers and laborers	479,960	9.4
Metalcutting machine operators	1,073,380	21.0
Metalforming machine operators	337,841	6.6
Miscellaneous craft workers	507,980	10.0
Other machine operators	785,660	15.4
Tool handlers	383,490	7.5

In the past, operative jobs in manufacturing have provided entry-level positions to immigrants, to blacks migrating from the rural south to the north, and women and other minorities. The more than 8 million manufacturing operatives in 1980 included a greater than proportionate share of blacks and other nonwhites than in the overall civilian labor force, as shown in Table 11. There is some reason for concern that a major expansion of robotization and related advances in computer-controlled manufacturing processes would have a disproportionate impact in creating job losses among groups that

Table 11. Distribution of the Manufacturing Work Force by Sex and Race, 1980, %[a,b]

Total employed persons: 97,270,000

	W	NW	Totals
M	51.7	5.8	57.5
F	37.1	5.4	42.5
Totals	88.8	11.2	100.0

Skilled Workers

Machine jobsetters: 658,000

	W	NW	Totals
M	88.2	7.9	96.1
F	3.2	.7	3.9
Totals	91.4	8.6	100.0

Other metalworking craftworkers: 638,000

	W	NW	Totals
M	89.2	6.9	96.1
F	3.3	.6	3.9
Totals	92.5	7.5	100.0

Semiskilled Workers

Motor-vehicle-equipment operatives: 431,000

	W	NW	Totals
M	65.7	14.6	80.3
F	15.1	4.6	19.7
Totals	80.8	19.2	100.0

Other durable goods manufacturing operatives: 4,166,000

	W	NW	Totals
M	55.7	8.5	64.2
F	30.2	5.6	35.8
Totals	85.9	14.1	100.0

Nondurable goods manufacturing operatives: 3,390,000

	W	NW	Totals
M	35.0	6.6	41.6
F	47.3	11.1	48.4
Totals	82.3	17.7	100.0

Manufacturing laborers: 961,000

	W	NW	Totals
M	68.0	16.3	84.3
F	13.0	2.7	15.7
Totals	81.0	19.0	100.0

[a] Ref. 28.
[b] Key: M = male; F = female; W = white; NW = nonwhite.

are already substantially disadvantaged in the nation's labor markets.

The Rate of Diffusion of Robots

The total number of industrial robots installed in the United States between 1960 and 1985 is on the order of 25,000, of which 80% have been purchased since 1980. By the arguments given in the last two sections, it would seem that the ultimate potential "market," defined in terms of jobs displaced, must be, at least on the order of 40% of the direct production labor employed in manufacturing, or about 5.6 million jobs. There is no reason to suppose that this is an upper limit in the long run, given continuous progress in robot capabilities. However, if it is assumed that each robot can replace no more than 2–3 human workers, it is clear that the smallest reasonable estimate of the potential industrial robot market is around 2 million units. It follows that the present level of market penetration is not much over 1% at most, and probably less. Under the circumstances, the small amount of experience accumulated to date is not likely to be very representative of what can be expected in the next 2 decades.

One possible basis for near-term forecasting, therefore, is to fit recent historical trends to an exponential function of time and extrapolate that function. Statistically significant deviations from the exponential model will likely appear before the 10% saturation level is reached (ie, 200,000 to 250,000 robots in service), but this is unlikely to occur until 1990 or later.

A second approach is to assume that the rate of market growth is determined by a balance between two contrary forces. On the one hand, robot manufacturers gain both experience and customers by reducing prices, provided the market is price-elastic enough such that, sales, and gross profits, increase faster than unit prices decrease. On the other hand, if prices fall too fast, consumers may decide to wait. The balance between waiting/not waiting is a function of the consumers discount rate and his/her expected benefit (or the return on his investment). It has been shown that the profit-maximizing market growth rate k is given by

$$k = -\sigma\,(\tau - \delta)$$

where τ is the target real rate of return on a risky investment, δ the discount rate, and σ the price elasticity, assumed to be a (negative) constant. This quantity is defined hereafter. The discount rate δ is normally equated to the average rate of return on safe long-term financial investments, so the term $\tau - \delta$ can be interpreted as a "risk premium." Typically, the premium might be 0.05–0.15 for a large firm, depending on the perceived risk.

The price elasticity over an interval in time is defined as

$$\sigma = \left(\frac{P}{Q}\right)\left(\frac{\Delta P}{\Delta Q}\right) = \left(\frac{P}{\Delta P}\right)\left(\frac{\Delta Q}{Q}\right) \tag{2}$$

where P is the unit price (of robots), Q the quantity sold, ΔP the price change in period Δt, and ΔQ the change in demand during that period. If prices and production quantities are known at successive times, σ can be estimated. Thus, combining equations 1 and 2,

$$k = -\left(\frac{P}{\Delta P}\right)\left(\frac{\Delta Q}{Q}\right)(\tau - \delta) \tag{3}$$

To summarize, k can be estimated directly from time-series sales data alone or, indirectly, from a combination of sales and price data at two (or more) points in time.

Using the first of these methods, the best fit to historical U.S. data (Fig. 4) indicates a growth rate of 24% per annum for the installed robot population, corresponding, of course, to a somewhat larger growth rate in sales. Projecting this rate of growth gives an expected robot population of roughly 135,000 by 1995 (Fig. 5).

OTHER SOCIAL AND ECONOMIC IMPACTS

There are a number of other indirect economic impacts, especially in the medium-to-longer term. Some of these that can be estimated, given a credible forecast of the rate of robot introduction by sector. These effects include:

1. Growth of employment and output in the robot manufacturing sector and its suppliers.
2. Reduced labor costs to industrial robot users.
3. Cost savings passed on to customers, especially through lower costs of captial goods.
4. Increased productivity of the economy as a whole.
5. Changes in patterns of international trade and competition.
6. Faster technological change in "mature" industries as increased flexibility reduces disincentives.

Robot Manufacturing

The first topic is, superficially, a straightforward application of input–output modeling techniques. Such a study, for robots per se was carried out by Howell (29), based on a more comprehensive study by Leontief and Duchin (30). The results strongly depend on a number of ancillary assumptions, however. (Howell calculates a "net" job loss by 1990 ranging from 167,000 to 718,000.) Apart from this wide range of outcomes, the study is predicated on an unchanging robot production technology and no decrease in robot production costs, which is highly unrealistic. The only firm conclusion that can be drawn from Howell's analysis is that, for reasonable assumptions about robot costs and the marginal rate of substitution between robots and workers, growth of a robot-manufacturing industry cannot be expected to create more than a small fraction of the manufacturing jobs taken over by robots. This is only to be expected, given that robots are primarily intended to replace human labor.

Productivity Effects

The history of technological change suggests strongly, however, that more new jobs are likely to be created in the economy *as a whole* than would be lost to direct substitution effects. The mechanisms at work include effects 2–4 on the above list. In macroeconomic terms, the savings enjoyed by robot users translate into increased productivity and greater eco-

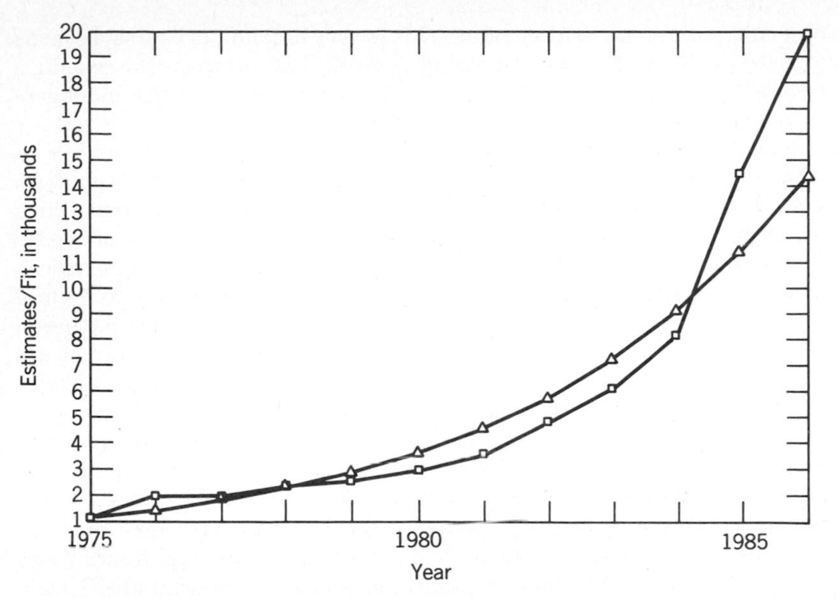

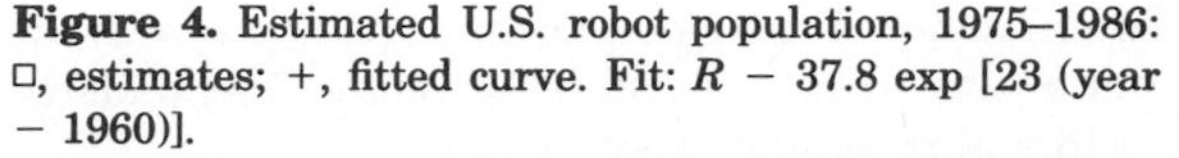
Figure 4. Estimated U.S. robot population, 1975–1986: □, estimates; +, fitted curve. Fit: $R - 37.8 \exp [23 (\text{year} - 1960)]$.

nomic output per unit of combined inputs (labor and captial). This increment of increased output is potentially available for distribution as real wealth among the population as a whole, as wages paid to labor, profits to investors, and interest paid to savers.

Historically, most increases in productivity have gone to raise the wages of labor (contrary to the predictions of Marx) and returns on captial investment have actually been declining for many decades. However, there is a theoretical possibility that the pendulum might swing the other way for a time due to long-term shifts in sociopolitical relationships. Certainly, there is no guarantee that labor will continue to enjoy most of the fruits of increased productivity over the next two or three decades. However, the major point that needs to be emphasized here is that the choices are social and political—society can allocate these benefits as it chooses.

In microeconomic terms, the mechanism that creates new jobs is simply that reduced costs of production are immediately translated into increased profits to producers or reduced prices to customers. If the market is cartelized, equilibrium prices and profits are higher than if it is competitive, but the mechanism of "recycling" works either way. Increased profits may be reinvested directly or passed on as dividends or higher wages. Some of it is captured as taxes. Ultimately, however, the money goes back into circulation as spending. The net effect of reduced production costs in *any* sector is to increase final demand in *every* sector.

Transitional Effects

The benefits of productivity growth are not, in general, distributed uniformly. In the first place, there are winners and

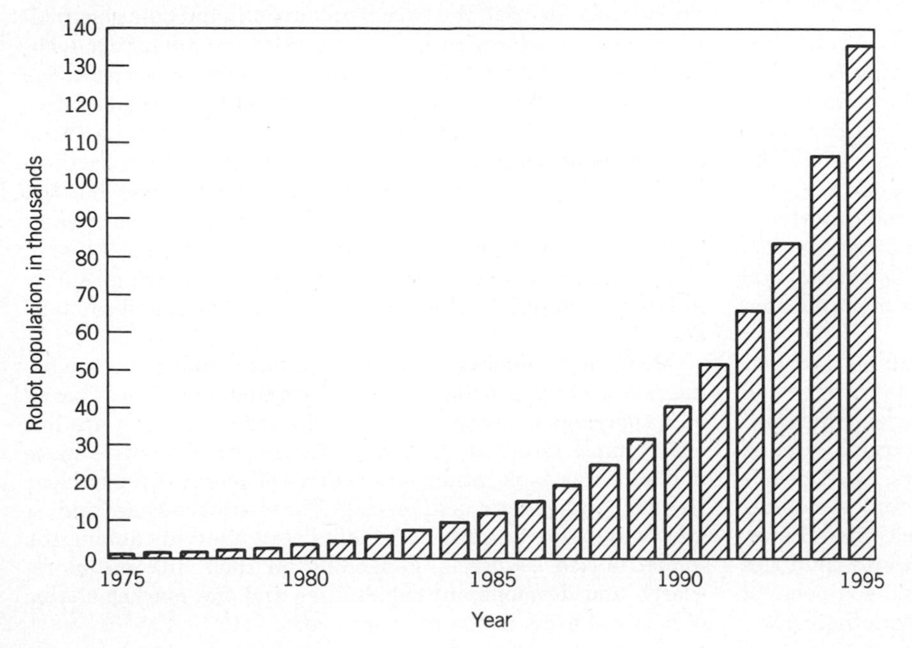

Figure 5. Projected U.S. robot population. Fit: $R - 28.94 \exp [24 (\text{year} - 1960)]$.

losers in any process of change. The biggest winners would be in the robot-manufacturing industry itself, as noted above. The semiskilled workers directly replaced by robots, especially in the metalworking industries, are the most likely to be immediate losers. Whereas the "productivity effect" noted above will presumably result in some incremental growth in demand for the products of the industry, the trend of declining blue-collar employment is likely to continue. In fact, a few decades hence, the manufacturing industry in the United States could employ as few workers as the farms do today, while maintaining current output levels.

The bad news for some workers in mature industries such as automobiles and appliances could well be good news for other workers whose jobs are now threatened by massive migration of components manufacturing to Japan and low wage countries such as Korea, Taiwan, and Brazil. Between 1975 and 1985, many times more U.S. workers lost jobs due to the inability of their employers to compete with imports than due to any massive invasion of robots. The underlying causes of this phenomenon may include

- Worldwide overcapacity in most industries
- Wage differentials favoring foreign producers
- Capital-cost differentials favoring foreign producers, especially Japan
- Unfair trade practices such as export subsidies and "dumping"
- Undervalued Japanese yen
- Overvalued U.S. dollar needed to finance the federal budget deficit because of low domestic savings
- U.S. focus on short-term financial gains rather than long-term investments

The relative weight to be attached to each cause is a matter of continuing dispute, which suggests that the phenomenon is not really well understood. Under the circumstances, it is not clear what government policies, if any, would be effective in reversing the trend.

What is clear is that if the present trend continues, the United States will lose much of its manufacturing base over the next decade. Eventually, this would reduce worldwide overcapacity. In the long run, too, the United States cannot expect capital inflows to balance current account deficits. To be an importer, it (like any country) must also be an exporter. It is not feasible for the United States to be a net exporter of mineral resources (except for coal), so imported petroleum and metals must be paid for; it can export its surplus agricultural production at the right price, but this requires a much cheaper dollar.

With a cheap enough dollar, the United States can also be a competitive exporter of "high tech" products with a small direct labor component. This is because the salaries of managerial and technical personnel are much less variable around the world than the wages of unskilled workers. Such people are relatively mobile and are unwilling to work for low salaries if they can earn much more by emigrating to the United States. Thus, to compete in "high tech" fields, third-world countries must either restrict emigration of scientists and engineers or offer them a standard of living commensurate with their international market value. This is very difficult, of course, because it results in a very large, painful gap between the earnings of the unskilled and semiskilled and those of the managers and engineers.

International trade equilibrium would therefore assign labor-intensive "low tech" manufacturing to low wage countries and "high tech" manufacturing to high wage countries. (Such an equilibrium would also require a continuing flow of capital investment from the developed countries to the least developed countries that do not yet have an export industry to finance their transition to the "low tech" stage.) Given this picture of the long-term trends in international trade, the United States can expect further losses of domestic manufacturing industry over the next two decades, mainly to countries such as Korea, Taiwan, the Philippines, Mexico, and Brazil. The areas affected will be relatively standardized materials, components, and complete consumer products such as household appliances and items of clothing. However, the United States may continue to dominate in some fields, such as commercial aircraft and large computers, and can continue to be internationally competitive in other areas, such as process-control equipment, telecommunications equipment, construction and mining equipment, agricultural equipment, and specialized machine tools.

Thus, in the long run, robots and computer-integrated manufacturing (CIM) will not reverse the present trends in international trade. Countries with low wages can continue to use human workers, whereas a high wage country might use a robot, but if robots are needed for other reasons (eg, for quality control), they can be installed as easily in a plant in Korea or Brazil as in the United States. The greater flexibility of robots and CIM as compared with traditional "hard" automation can pay off, however, in terms of lowering the costs of innovation. Therefore, a somewhat higher rate of technical change can be expected in the auto industry, in particular, as U.S. and European firms seek to compensate for their labor-cost disadvantages by increasing the pace of new model introduction.

A major structural realignment is taking place in international trade. In brief, the "mass-production" and bulk commodity industries—characterized by standardized production technologies—are migrating increasingly rapidly away from the older industrialized countries (Europe and the United States) to countries with cheaper labor or resources. Japan has been a major beneficiary of this movement (especially in textiles, steel, shipbuilding, consumer electronics, and autos), but the first four of these industries have already begun to move away from Japan to still lower cost locations such as Korea, Taiwan, Hong Kong, and Singapore. Japan, in turn, is moving rapidly into the "high tech" fields, such as machine tools and computers.

Meanwhile, important new science-based industries are being created by technology. The "information industries" based on a marriage of computers and telecommunications are but one example. Other areas of major future growth include space and undersea exploration, new sources of energy (from nuclear to solar) and genetic engineering. These science-based industries will, however, provide benefits very unevenly among the industrialized countries, depending on their differential research and development capabilities and the market shares of national firms in the relevant areas.

Moreover, the increased application of computers to manufacturing (CAD/CAM, robotics) will unquestionably displace workers in some industries and occupations and cause frictional adjustment problems (see UNMANNED FACTORIES). The metalworking sectors appear to be the most vulnerable at present. These problems will be amplified because industries tend to be concentrated in certain areas. In the United States, for instance, the metalworking industries are particularly concentrated in the so-called Great Lakes states, ie, Michigan, Ohio, Indiana, Illinois, and Wisconsin. In Europe, the Ruhr Valley in the FRG, the Alsace-Lorraine in France, and the Midlands of the UK are corresponding areas of industrial specialization.

Most of the potentially affected workers are also members of trade unions, which are politically powerful, especially in Europe. One common response in Europe has been for each country to try to preserve domestic jobs by subsidizing exports (especially of steel)—which is tantamount to exporting unemployment. Many noncompetitive firms in Italy, France, and the UK have been partially or totally nationalized and subsidized. Most countries have adopted a variety of nontariff protectionist policies, from explicit "quota" agreements to "safety" regulations designed to discriminate against imports. The Japanese have been particularly adept at excluding foreign competition in their domestic markets.

In a period of rapid worldwide economic growth, such abuses and distortions might be bargained away without undue hardship. However, in a period of prolonged recession, even depression in some sectors, it is likely that protectionist pressures will continue to grow.

BIBLIOGRAPHY

1. N. J. Nilsson, "Artificial Intelligence, Employment & Income," *The A. I. Magazine,* 5–14 (Summer 1984).
2. R. U. Ayres and S. M. Miller, *Robotics: Applications and Social Implications,* Ballinger Publishing Co., Cambridge, Mass., 1983.
3. R. U. Ayres, "A Schumpeterian Model of Technological Change," *Journal of Technological Forecasting and Social Change,* **27**(4) (1985).
4. Ref. 2, Chapt. IV.
5. J. R. Bright, *Automation and Management,* Harvard Business School, Boston, Mass., 1958.
6. J. R. Bright, *Harvard Business Review* (July–Aug. 1958).
7. J. R. Bright, *The Relationship of Increasing Automation and Skill Requirements,* National Commission on Employment and Economic Progress, Washington, D.C., 1966, Appendix, pp. 201–221.
8. N. Wiener, *Cybernetics,* The MIT Press, Cambridge, Mass., 1948.
9. W. Abernathy, *The Productivity Dilemma,* Johns Hopkins University Press, Baltimore, Md., 1978.
10. R. U. Ayres, *The Next Industrial Revolution: Reviving Industry through Innovation,* Ballinger Publishing Co., Cambridge, Mass., 1984.
11. G. Boothroyd, C. Poli, and L. E. Murch, *Handbook of Feeding and Orienting Techniques for Small Parts, (Automation Project),* University of Massachusetts, Amherst, Mass., undated (ca 1983).
12. H. B. Maynard, G. J. Stegmerten, and J. L. Schwab, *Methods-Time Measurement,* McGraw-Hill, Inc., New York, 1948.
13. H. Jacobson, "The Informational Capacity of the Human Eye," *Science,* **113** 292, (1951).
14. C. R. Kelley, *Manual and Automatic Control,* John Wiley & Sons, Inc., New York, 1968.
15. E. J. McCormick, *Human Factors in Engineering,* McGraw-Hill, New York, 1970.
16. L. V. Rigby and A. D. Swain, "Effects of Assembly Error on Product Acceptability and Reliability," *Proceedings, 7th Annual Reliability and Maintainability Conference,* American Society of Mechanical Engineers, New York, July 1968.
17. D. Meister, *Human Factors: Theory and Practice,* John Wiley & Sonc, Inc., New York, 1971.
18. D. Meister, "Reduction of Human Error," in G. Salvendy, ed., *Handbook of Industrial Engineering,* John Wiley & Sons, Inc., New York, 1982.
19. A. D. Swain, *Design Techniques for Improving Human Performance in Production,* available from author, 712 Sundown Place SE, Albuquerque, N.M. 87107, 1977.
20. A. D. Swain and H. E. Guttmann, *Handbook of Human Reliability Analysis With Emphasis on Nuclear Power Plan Applications, Final Report NUREG/CR-1278-F SAND80-0200 Rx,AN,* Sandia Laboratories, prepared for U.S. Nuclear Regulatory Commission, Washington, D.C., 1983.
21. S. M. Miller, Ph.D. thesis, Carnegie-Mellon University, Pittsburgh, Pa., 1983.
22. S. M. Miller, "Impacts of Robotics and Flexible Manufacturing Technologies on Manufacturing Costs and Employment," in P. R. Kleindorfer, ed., *The Management of Productivity and Technology in Manufacturing,* Plenum Press, New York, 1985, Chapt. 3.
23. S. M. Miller, *Impacts of Industrial Robotics: Effects on Labor and Costs in the Metalworking Industries,* University of Wisconsin Press, Madison, Wisc., in press.
24. Ref. 2, Table 5.9.
25. D. N. Smith and P. Heyther, *Industrial Robots: Forecasts and Trends,* 2nd ed., Society of Manufacturing Engineers, Dearborn, Mich., 1985.
26. *The 12th American Machinist Inventory of Machine Tools,* as reported in *American Machinist* (1978).
27. Bureau of Labor Statistics, Dept. of Labor, Employment Projections for the 1980s, Government Printing Office, Washington, D.C., 1979, p. 32.
28. Ref. 2, Chapt. 5.
29. D. R. Howell, "The Future Employment Impact of Industrial Robots," *Technological Forecasting and Social Change,* **28,** 297–310 (Dec. 1985).
30. W. Leontief and F. Duchin, *The Impacts of Automation on Employment,* Oxford University Press, New York, 1985.

END-OF-ARM TOOLING

RONALD D. POTTER
Robot Systems, Inc.
Norcross, Georgia

INTRODUCTION

End-of-arm tools, also called end-effectors, give robots the ability to pick up and transfer parts and to handle a multitude of differing tools to perform work on parts. Robots have been fitted with grippers to load and unload parts from a variety of machines and processes, such as forging presses, injection-molding machines, and die-casting machines. They have also been fitted with tools that perform work on parts, such as spot welding guns, drills, routers, grinding and cutting tools, and other types of tools that help fabricate and form parts, such as arc-welding torches and ladles for pouring molten metal. Tools for assembling parts, such as automatic screwdrivers and nutrunners, have also been attached to robots, as have tools for performing finishing operations, such as paint spray guns and special inspection devices, such as linear variable, differential transformers (LVDTs) and laser gauges, that perform quality control functions. There are very few limitations on the type of hand or tool that can be attached to the end of a robot's arm. Unlike the human hand, which is fairly standard with five fingers and a relatively uniform size, a robot's hand can be any size or shape and is normally a unique, one-of-a-kind device designed for a specific application.

End-of-arm tooling is a critical part of an industrial robot system, as it is the part of the system that actually links the robot to the workpiece. The success or failure of an application is very dependent on how well the end-of-arm tooling is conceived, designed, and implemented. In most industrial robot applications, the end-of-arm tooling must be custom-designed to match the process requirements. Defined by the part and process, end-of-arm tooling cannot be viewed separately from the other system elements. Throughout the design stages, the interrelationship of all system components with the part and process must be considered. Since the possibilities for end-of-arm tooling are even more diverse than the types of manufacturing processes and machines that exist, it is very difficult to generalize or attempt to restrict the imagination in developing end-of-arm tooling, as the possibilities are infinite. However, the development should follow a systematic approach to ensure that the optimum robot system results.

SYSTEMATIC APPROACH TO DEVELOPING END-OF-ARM TOOLING

Development of end-of-arm tooling for a particular application should occur at a specific time as one of the sequential steps in the development of the entire robot system. These sequential steps ensure that optimum productivity and efficiency are achieved in the system. Two general rules should be kept in mind when developing a robot system and end-of-arm tooling.

1. Do not attempt to mimic human operations. A human operator, when performing an industrial task, cannot be realistically compared to a robot. Although a human has much more sophisticated sensory capabilities than a robot (ie, sight, hearing, tactile senses, etc), a robot does not possess the inherent physical limitations of a human in other areas. A robot, not equipped with a relatively small five-fingered hand, has much greater capabilities than a human in handling heavier weights for longer periods of time in harsher environments. Thus do not limit the capabilities of the robot system by simply trying to duplicate human capabilities.
2. Do not select the robot first and then try to fit it with an end-of-arm tool and put it to work; select the most appropriate robot for the application as another of the sequential steps in overall system development.

The following systematic approach details the timing of end-of-arm tooling development and robot selection in the system development, as shown in Figure 1.

Sequential Steps in Developing a Robot System, Including Timing of End-of-Arm Tooling Analysis

1. Understand the process thoroughly. Consider what modifications must be made to the process to automate it with any generic robot. Look for ways of improving the efficiency and productivity of the process by altering the present method of manual operation.
2. Analyze the production equipment used in the process. Consider what modifications must be made to the equipment to automate it with any generic robot. For example, provisions may have to be made for automatic clamping, sensors for malfunctions, removal of protective guards, interfaces to controls for automatic start/stop, changeover, relocation, and clearances.
3. Analyze the sensors and peripheral equipment that are required to produce an automatic system. Sensors in their various forms provide the paths of communication between all elements of the system, including the end-of-arm tooling. Define all the various conditions that must be sensed in the system and make provisions for them. Peripheral equipment, such as parts presentation devices, holding fixtures, conveyors, and inspection stations, can be provided to assist the robot in performing the task. Divide up the tasks and do not make the robot do everything. Use peripheral equipment to simplify the design of the end-of-arm tooling and overlap actions to optimize cycle time considerations. As the tasks are divided between robot and peripheral equipment, and sensor requirements are determined, the performance requirements of the end-of-arm tooling can be defined.
4. Conceptualize the end-of-arm tooling. At this point, the robot make or model has not been selected.

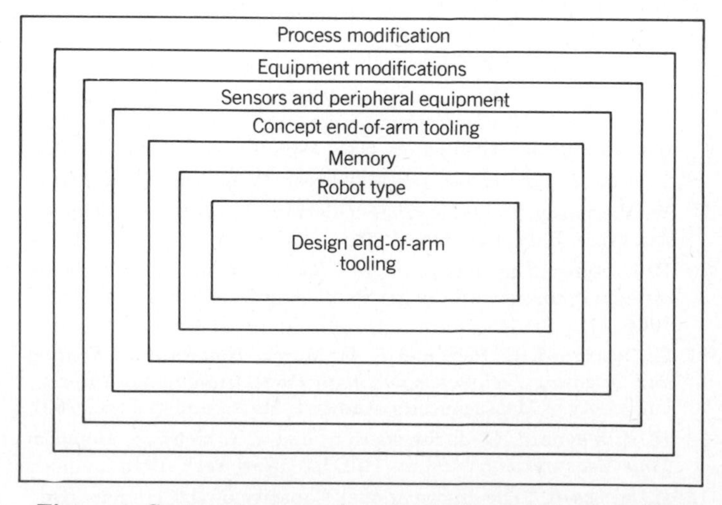

Figure 1. Systematic approach to developing end-of-arm tooling.

5. Analyze the memory type and capacity required for the system. In addition to the robot controller, a programmable controller is normally used to control peripheral equipment and sequence external events. Considerations should be made concerning batch run sizes, number of steps per program, number of programs, changeovers, and so on, and an appropriate system controller should be selected.
6. Analyze the robot type and options best-suited for the application. The selection of a particular robot make or model should be made based on the following technical criteria: (*1*) type of robot (nonservo point-to-point, servo-controlled point-to-point, servo-controlled continuous path); (*2*) work envelope; (*3*) load capacity (including end-of-arm tooling weight); (*4*) cycle time; (*5*) repeatability; (*6*) drive system (pneumatic, hydraulic, or electric); (*7*) unique hardware or software capabilities. In addition, other nontechnical considerations such as cost, reliability, and service should be considered in selecting a robot for a particular application.
7. Final concept and preliminary design of end-of-arm tooling to match the selected robot, peripheral equipment, and other system elements. At this point the preliminary concept of the end-of-arm tooling should be analyzed in relation to the robot tool-mounting plate and work envelope and modified accordingly.

The preceding sequence of events allows the end-of-arm tooling to be conceptualized and designed only after proper consideration has been given to required modifications in the process and manufacturing equipment, analysis of sensor and peripheral equipment needs, and robot selection.

Definition of End-of-Arm Tooling

End-of-arm tooling is defined as the subsystem of an industrial robot system that links the mechanical portion of the robot (manipulator) to the part being handled or worked on. An industrial robot is essentially a mechanical arm with a flat tool-mounting plate at its end that can be moved to any spatial point within its reach. End-of-arm tooling in the form of specialized devices to pick up parts or hold tools to work on parts is physically attached to the robot's tool-mounting plate to link the robot to the workpiece.

ELEMENTS OF END-OF-ARM TOOLING

End-of-arm tooling is commonly made up of four distinct elements, as shown in Figure 2, which provide for (*1*) attachment of the hand or tool to the robot tool-mounting plate, (*2*) power for actuation of tooling motions, (*3*) mechanical linkages, and (*4*) sensors integrated into the tooling.

Mounting Plate

The means of attaching the end-of-arm tooling to an industrial robot is provided by a tool-mounting plate located at the end of the last axis of motion on the robot. This tool-mounting plate contains either threaded or clearance holes arranged in a pattern for attaching tooling. For a fixed mounting of a gripper or tool, an adapter plate with a hole pattern matching the robot tool-mounting plate can be provided. The remainder of the adapter plate provides a mounting surface for the gripper or tool at the proper distance and orientation from the robot tool-mounting plate. If the task of the robot requires it to automatically interchange hands or tools, a coupling device can be provided. An adapter plate is thus attached to each of the grippers or tools to be used, with a common lock-in position for pickup by the coupling device. The coupling device may also contain the power source for the grippers or tools and automatically connect the power when it picks up the tooling. An alternative to this approach is for each tool to have its own power line permanently connected, and the robot simply picks up the various tools mounted to adapter plates with common lock-in points.

Power

Power for actuation of tooling motions can be either pneumatic, hydraulic, or electrical, or the tooling may not require power, as in the case of hooks or scoops. Generally, pneumatic power is used where possible because of installation and maintenance, low cost, and light weight. Higher-pressure hydraulic power is used where greater forces are required in the tooling motions. However, contamination of parts due to leakage of hydraulic fluid often restricts its application as a power source for tooling. Although it is quieter, electrical power is used less frequently for tooling power, especially in part-handling applications, because of its lower applied force. Several light

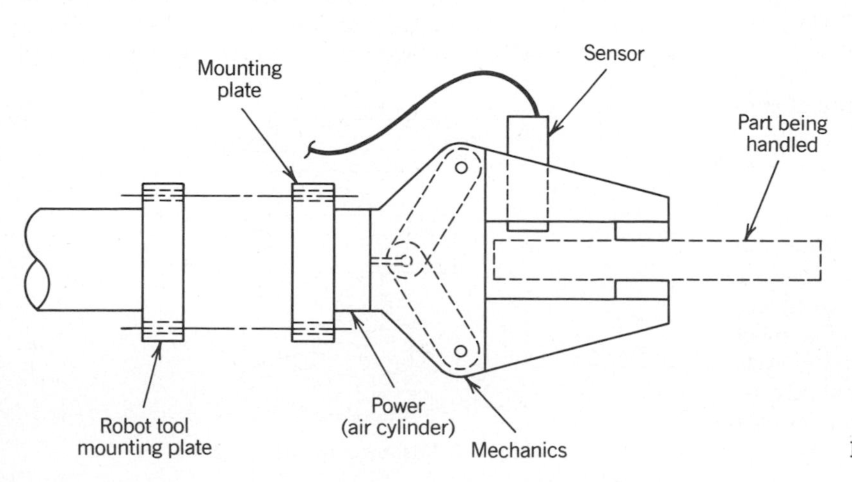

Figure 2. Elements of end-of-arm tooling.

payload assembly robots utilize electrical tooling power because of its control capability. In matching a robot to end-of-arm tooling, consideration should be given to the power source provided with the robot. Some robots have provisions for tooling power, especially part-handling robots, and it is an easy task to tap into this source for actuation of tooling functions. As previously mentioned, many of the robots are provided with a pneumatic power source for tooling actuation and control.

Mechanics

Tooling for robots may be designed with a direct coupling between the actuator and workpiece, as in the case of an air cylinder that moves a drill through a workpiece, or use indirect couplings or linkages to gain mechanical advantage, as in the case of a pivot-type gripping device. A gripper-type hand may also have provisions for mounting interchangeable fingers to conform to various part sizes and configurations. In turn fingers attached to grippers may have provisions for interchangeable inserts to conform to various part configurations.

Sensors

Sensors are incorporated in tooling to detect various conditions. For safety considerations sensors are normally designed into tooling to detect workpiece or tool retention by the robot during the robot operation. Sensors are also built into tooling to monitor the condition of the workpiece or tool during an operation, as in the case of a torque sensor mounted on a drill to detect when a drill bit is dull or broken. Sensors are also used in tooling to verify that a process is completed satisfactorily, such as wire-feed detectors in arc-welding torches and flow meters in dispensing heads. More recently, robots specially designed for assembly tasks contain force sensors (strain gauges) and dimensional gauging sensors in the end-of-arm tooling.

TYPES OF END-OF-ARM TOOLING

There are two functional classifications of end-of-arm tooling for robots: grippers for handling parts and tools for doing work on parts. In general, there is more effort required in developing grippers for handling parts for machine loading, assembly, and parts transfer operations than for handling tools. It is generally more difficult to design special-purpose grippers for part handling than it is to attach a tool to the end of a robot's arm.

The following sections describe the five basic types of end-of-arm tooling, including grippers and tools.

Attachment Devices

These devices are simply mounting plates with brackets for securing tools to the robot tool-mounting plate. In some cases attachment devices may also be designed to secure a workpiece to the robot tool-mounting plate, as in the case of a robot manipulating a part against a stationary tool where the cycle time is relatively long. In this case, the part is manually secured and removed from the robot tool-mounting plate for part retention.

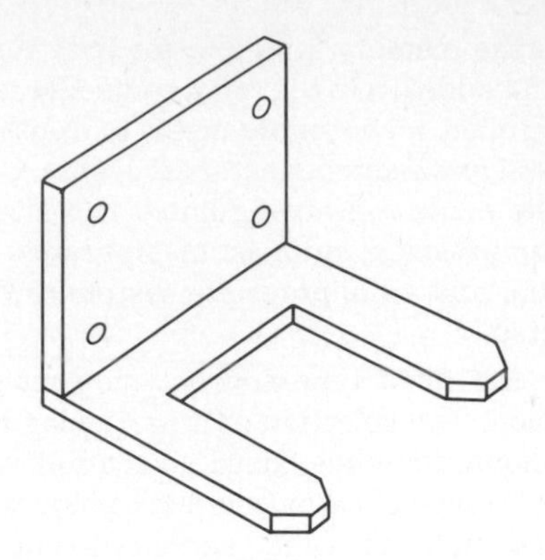

Figure 3. Support and containment device.

Support and Containment Devices

As shown in Figure 3, lifting forks, hooks, scoops, and ladles are typical examples of this type of tooling. Again, no power is required for the tooling, as the robot simply moves to a position beneath the part to be transferred, lifts to support and contain the part or material, and performs its transfer process.

Pneumatic Pickup Devices

The most common example of this type of tooling is a vacuum cup (Fig. 4) that attaches to parts to be transferred by a suction or vacuum pressure created by a venturi transducer or a vacuum pump. Typically used on parts with a smooth surface finish, vacuum cups are available in a wide range of sizes, shapes, and materials. Parts with nonsmooth surface finishes can still be picked up by a vacuum system if a ring of closed-cell foam rubber is bonded to the surface of the vacuum cup, which conforms to the surface of the part and creates the seal required for vacuum transfer. Venturi vacuum transducers are relatively inexpensive and are used for handling small, lightweight parts where a low vacuum flow is required. Vacuum pumps, quieter and more expensive, generate greater vacuum flow rates and can be used to handle heavier parts. With any vacuum system, the quality of the surface finish of the part being handled is important. If parts are oily or wet, they tend to slide on the vacuum cups. Therefore some type of containment structure should be used, in addition to the vacuum cups, to enclose the part and prevent it from sliding on the cups. In some applications a vacuum cup with no power

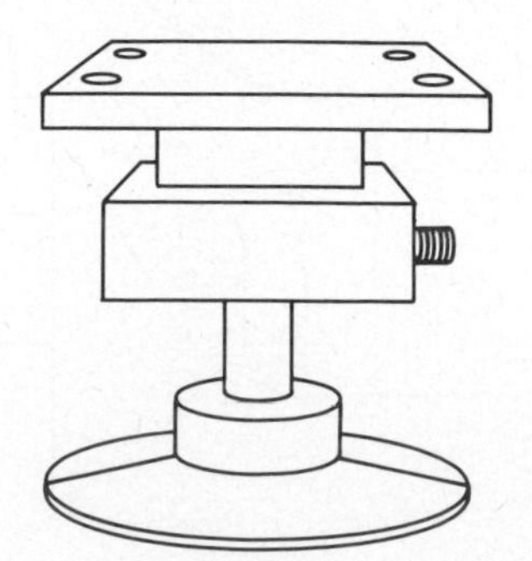

Figure 4. Vacuum cup pickup device.

source can be utilized. By pressing the cup onto the part and evacuating the air between the cup and part, a suction is created capable of lifting the part. However, a stripping device or valve is required to separate the part from the cup during part release. When a venturi or vacuum pump is used, a positive air blow-off may be used to separate the part from the vacuum cup. Vacuum cups have temperature limitations and cannot be used to pick up relatively hot parts.

Another example of a pneumatic pickup device is a pressurized bladder. These devices are generally specially designed to conform to the shape of the part. A vacuum system is used to evacuate air from the inside of the bladder so that it forms a thin profile for clearance in entering the tooling into a cavity or around the outside surface of a part. When the tooling is in place inside or around the part, pressurized air causes the bladder to expand, contact the part, and conform to the surface of the part with equal pressure exerted on all points of the contacted surface. Pneumatic bladders are particularly useful when irregular or inconsistent parts must be handled by the tooling.

Pressurized fingers, shown in Figure 5, are another example of pneumatic pickup devices. Similar to a bladder, pneumatic fingers are more rigidly structured. They contain one straight half, which contacts the part to be handled, one ribbed half, and a cavity for pressurized air between the two halves. Air pressure filling the cavity causes the ribbed half to expand the "wrap" the straight side around a part. With two fingers per gripper, a part can thus be gripped by the two fingers wrapping around the outside of the part. These devices also can conform to various shape parts and do not require a vacuum source to return to their unpressurized position

Magnetic Pickup Devices

These devices comprise the fourth type of end-of-arm tooling and can be considered when the part to be handled is of ferrous content. Either permanent or electromagnets are used, with permanent magnets requiring a stripping device to separate the part from the magnet during part release. Magnets normally contain a flat part-contact surface but can be adapted with a plate to fit a specific part contour. A recent innovation in magnetic pickup devices uses an electromagnet fitted with a flexible bladder containing iron filings, which conforms to an irregular surface on a part to be picked up. As in the case of vacuum pickup devices, oily or wet part surfaces may cause the part to slide on the magnet during transfer. Therefore containment structures should be used in addition to the magnet to enclose the part and prevent it from slipping. Three other concerns arise in handling parts with magnets. If a metal-removal process is involved in the application, metal chips may also be picked up by the magnet. Provisions must be made to wipe the surface of the magnet in this event. Also, residual magnetism may be imparted to the workpiece during pickup and transfer by the magnetic tooling. A demagnetizing operation may be required after transfer if this is detrimental to the finished part. If an electromagnet is used, a power failure will cause the part to be dropped immediately, which may produce an unsafe condition. Although electromagnets provide easier control and faster pickup and release of parts, permanent magnets can be used in hazardous environments requiring explosion-proof electrical equipment. Normal magnets can handle temperatures up to 140°F (60°C), but magnets can also be designed for service in temperatures up to 300°F (150°C).

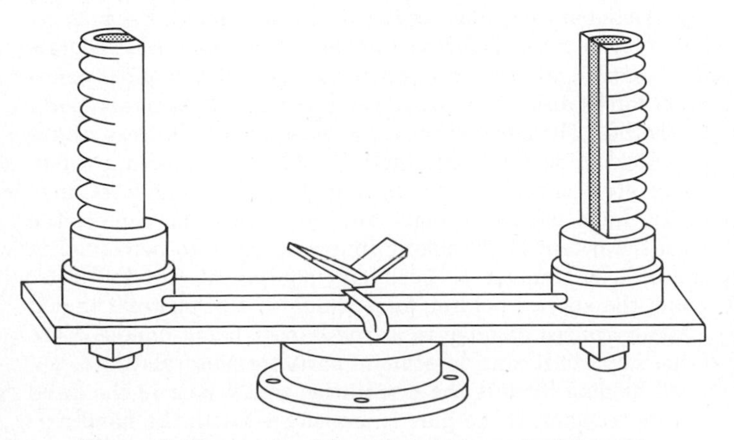

Figure 5. Pneumatic pickup device (pressurized fingers).

MECHANICAL GRIP DEVICES

Mechanical grip devices, the fifth category, are the most widely used type of tooling in parts-handling applications. Either pneumatic, hydraulic, or electrical actuators are used to generate a holding force which is transferred to the part by linkages and fingers. The most commonly used power source for finger closure actuation is a pneumatic cylinder the bore and stroke of which are selected in relation to the available operating to provide an optimum amount of holding force on the part. The grip force of the fingers may be varied by regulating the pressure entering the tooling actuators. The grip force may be further reduced by mounting soft conforming inserts in the fingers at the point of contact with the part to be handled. Some recent innovations in standard commercially available grippers contain sensors and features for controlling the amount of grip force exerted on the part, in addition to dimensional gauging capabilities incorporated in the gripper. In these hands, the fingers can close on parts of various sizes until a predetermined force is attained and stop at that point. Still other standard grippers are equipped with strain gauges in the fingers to detect the position of parts within the gripper and adjust the robot arm accordingly to center the fingers around the part.

The motions of the fingers in a mechanical grip device vary, and an appropriate finger motion can be selected to best suit the part shape and its constraints at the pickup and release stations. The simplest finger motion involves a gripper with one stationary or fixed finger and one moving finger, commonly referred to as a single-action hand (see Fig. 6). This hand functions by having the robot move the open hand to a position around the part and having the moving finger close on the part, clamping it against the stationary finger. This hand requires that the part be free to move during the pickup process to allow for clearance of the stationary finger around the part. If a part is fixtured tightly during pickup or release, this hand is not the most appropriate choice. By placing a V-block insert in either of the fingers, this single-action hand can center the part during the grip process and locate it accurately.

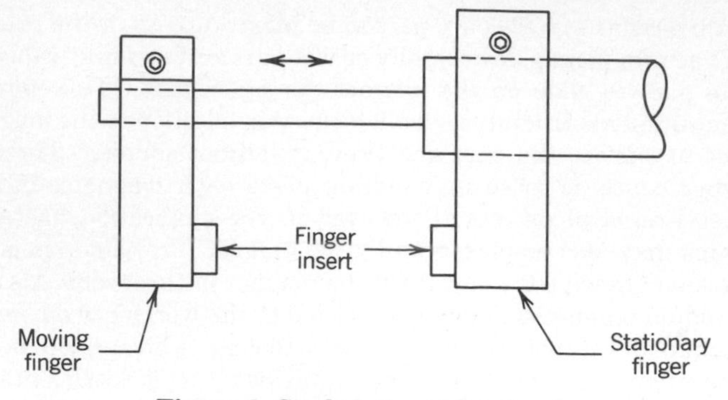

Figure 6. Single-action pickup hand.

Where parts are tightly fixtured in the pickup or release position, a double-action hand can be utilized. This hand contains two moving fingers that close simultaneously to hold a part. The motion of the fingers can be either pivoting or parallel. Pivoting fingers have greater limitations, as they must be designed to match the part shape at the contact points at a specific angular orientation of the fingers. Since parallel motion double-action hands move in a straight-line motion in closing on a part, they do not have the angular orientation consideration to meet. Thus they can also handle a wide range of part sizes automatically. By placing V-block locators in the fingers as inserts, the double-action hand can center the parts not consistently oriented at the pickup station, if they have provisions for movement. On tightly fixtured parts, such as in the unloading of a lathe, both fingers have clearance around the part in the open position and simultaneously close to center on the part.

Some mechanical grip devices contain three moving fingers that simultaneously close to grasp a part or tool. These hands are particularly useful in handling cylindrical-shaped parts, as the three-point contact centers round parts of varying diameters on the centerline of the hand. Machine-loading operations, where round parts are loaded into chucks or over mandrels, are best suited for use of a three-fingered centering gripper. Still other mechanical grippers contain four moving fingers that close simultaneously to center a square or rectangular part on the centerline of the hand.

With most of the hands described, the position of the fingers can be reversed to allow gripping of the internal surfaces of parts, if required.

General Design Criteria for End-of-Arm Tooling

Although robot end-of-arm tooling varies widely in function, complexity, and application area, there are certain design criteria that pertain to almost all robot tooling.

General Guidelines. First, end-of-arm tooling should be as lightweight as possible. This primarily affects the performance of the robot. The rated load-carrying capacity of the robot, or the amount of weight that can be attached to the robot tool-mounting plate, includes the weight of the end-of-arm tooling and that of the part being carried. The load that the robot is carrying also affects the speed of its motions. Robots can move at faster rates carrying lighter loads. Therefore, for cycle time considerations, the lighter the tool, the faster the robot is capable of moving. The use of lightweight materials, such as aluminum or magnesium for hand components, and lightening holes whenever possible are common solutions for weight reduction.

Second, end-of-arm tooling should be as small in physical size as possible. Minimizing the dimensions of the tooling provides for better clearances in workstations in the system. In relationship to the robot, most load-carrying capacities of robots are based on moment of inertia calculations of the last axis of motion, that is, a given load capacity at a given distance from the tool-mounting plate surface. Therefore, by minimizing the size of the tooling and the distance from the tool-mounting plate surface. Therefore, by minimizing the size of the tooling and the distance from the tool-mounting plate to the center of gravity of the tooling, robot performance is enhanced.

At the same time, it is desirable to handle the widest range of parts with the robot tooling. This minimizes changeover requirements and reduces cost for multiple tools. Although minimizing the size of the tooling somewhat limits the range of parts that can be handled, there are techniques for accomplishing both goals. Adjustable motions may be designed into the tooling so that it can be quickly and easily manually changed to accommodate different-sized parts. Interchangeable inserts may be put in tooling to change the hand from one size or shape part to another. The robot may also automatically interchange hands or tools to work on a range of parts, with each set of tooling designed to handle a certain portion of the entire range of parts. This addresses weight and size considerations and reduces the total number of tools required.

Maximizing rigidity is another criterion that should be designed into tooling. Again, this relates to the task performance of the robot. Robots have specified repeatabilities and accuracies in handling a part. If the tooling is not rigid, this positioning accuracy will not be as good and, depending on part clearances and tolerances, may cause problems in the application. Excessive vibrations may also be produced by attaching a nonrigid or flimsy tool on the tools-mounting plate. Since robots can move the tooling at very high rates of speed, this vibration may cause breakage or damage or damage to the tool. Providing rigid tooling eliminates these vibrations.

The maximum applied holding force should be designed into the tooling. This is especially important for safety reasons. Robots are dynamic machines that can move parts at high velocities at the end of the arm, with only the clamp force and frictional force holding the part in the hand. Because robots typically rotate the part about a fixed robot base centerline, centrifugal forces are produced. Acceleration and deceleration forces also result when the robot moves from one point to another. The effect of these forces acting on the part makes it critical to design in an applied holding force with a safety factor great enough to ensure that the part is safely retained in the hand during transfer and not thrown as a projectile with the potential of causing injury or death to personnel in the area or damage to periphery equipment. On the other hand, the applied holding force should not be so great that it actually causes damage to a fragile part being handled. Another important consideration in parts transfer relating to applied holding force is the orientation of the part in the hand during transfer. If the part is transferred with the hand axis parallel to the floor, the part, retained only by the frictional force between the fingers and part, may tend to slip in the hand, especially at programmed stop points. By turning the hand axis perpendicular to the floor during part transfer, the required holding force may be decreased, and the robot may be able to move at higher speed because the hand itself acts as a physical stop for the part.

Maintenance and changeover considerations should be designed into the tooling. Perishable or wear details should be designed to be easily accessible for quick change. Change details such as inserts or fingers should also be easily and quickly interchangeable. The same type of fastener should be used wherever possible in the hand assembly, thereby minimizing the maintenance tools required.

ESTABLISHING DESIGN CRITERIA FOR END-OF-ARM TOOLING

As previously mentioned, the development of end-of-arm tooling concept and design should proceed only after analysis of the part to be handled or worked on and the process itself has been thoroughly completed. This analysis can be divided into two phases: a preengineering, data collection phase and a design phase. The preengineering phase involves analysis of the workpiece and process with emphasis on productivity considerations, whereas the second phase involves the actual design of the tooling to best meet the criteria developed during the first phase.

Preengineering and Data Collection

Workpiece Analysis. The part being transferred or worked on must be analyzed to determine critical parameters to be designed into the end-of-arm tooling. The dimensions and tolerances of the workpiece must be analyzed to determine their effect on tooling design. The dimensions of the workpiece determine the size and weight of the tooling required to handle the part. It also determines whether one tool can automatically handle the range of part dimensions required, whether interchangeable fingers or inserts are required, or whether tool change is required. The tolerances of the workpieces determine the need for compliance in the tooling. Compliance allows for mechanical "float" in the tooling in relation to the robot tool-mounting plate to correct misalignment errors encountered when parts are mated during assembly operations or loaded into tight-fitting fixtures or periphery equipment. If the part tolerances vary so that the fit of the part in fixture is less than the repeatability of the robot, a compliance device may have to be designed into the tooling. Passive compliance devices, such as springs, may be incorporated into the tooling to allow it to float to accommodate very tight tolerances. This reduces the rigidity of the tooling. Other passive compliance devices, such as remote center compliance (RCC) units, are commercially available. These are mounted between the robot tool-mounting plate and the end-of-arm tooling to provide a multiaxis float. RCC devices, primarily designed for assembly tasks, allow robots to assemble parts with mating fits much tighter than the repeatability that the robot can achieve. Active compliance devices with sensory feedback can also be used to accommodate tolerance requirements.

The material and physical properties of the workpiece must be analyzed to determine their effect on tooling design. The best method of handling the part, by vacuum, magnetic, or mechanical-grip pickup, can be determined. The maximum permissible grip forces and contact points on the part can be determined, as well as the number of contact points to ensure part retention during transfer. Based on the physical properties of the material, the need for controlling the applied force through sensors can also be resolved.

The weight and balance (center of gravity) of the workpiece should be analyzed to determine the number and location of grip contact points needed to ensure proper part transfer. This also resolves the need for counterbalance or support points on the part in addition to the grip contact points. The static and dynamic loads and moments of inertia of the part and tooling about the robot tool-mounting plate can be analyzed to verify that they are within the safe operating parameters of the robot.

The surface finish and contour (shape) of the workpiece should be studied to determine the method and location of part pickup (ie, vacuum on smooth, flat surfaces, mechanical grippers on round parts, etc.). If the contour of the part is such that two or more independent pickup means must be applied, this can be accomplished by mounting separate pickup devices at different locations on the tool, each gripping or attaching to a different section of the part. This may be a combination of vacuum cups, magnets, and/or mechanical grippers. Special linkages may also be used to tie together two different pickup devices powered by one common actuator.

Part modifications should be analyzed to determine if minor part changes that do not affect the functions of the part can be made to reduce the cost and complexity of the end-of-arm tooling. Often, simple part changes, such as holes or tabs in parts, can significantly reduce the tooling design and build effort in the design of new component parts for automation and assembly by robots.

Part inconsistencies should be analyzed to determine the need for provision of out-of-tolerance sensors or compensating tooling to accommodate these conditions.

In tool-handling rather than part-handling applications, the workpiece should be analyzed to determine the characteristics of the tool required. This is especially true for the incorporation of protective sensors in the tooling to deal with part inconsistencies.

Process Analysis. In addition to a thorough analysis of the workpiece, the manufacturing process should be analyzed to determine the optimum parameters for the end-of-arm tooling.

The process method itself should be analyzed, especially in terms of manual versus robot operation. In many cases physical limitations dictate that a person perform a task in a certain manner whereas a robot without these constraints may perform the task in a more efficient but different manner. An example of this involves the alternative of picking up a tool and doing work on a part or picking up the part and taking it to the tool. In many cases the size and weight-carrying capability of a person is limited and forces him to handle the smaller and lighter weight of the part or the tool. A robot, with its greater size and payload capabilities, does not have this restriction. Therefore it may be used to take a large part to a stationary tool or to take multiple tools to perform work on a part. This may increase the efficiency of the operation by reducing cycle time, improving quality, and increasing productivity. Therefore, in process analysis, consider the alternative of having the robot take a part to a tool or a tool to a part, and decide which approach is most efficient. When a robot is handling a part, rather than a tool, there is less concern about power-line connections to the tool, which experience less flexure and are less prone to problems when stationary than when moving.

Because of its increased payload capability, a robot may also be equipped with multifunctional end-of-arm tooling. This tooling can simultaneously or sequentially perform work on a part that previously required a person to pick up one tool at a time to perform the operation, resulting in lower productivity. For example, the tooling in a die-casting machine unloading application may not only unload the part, but also spray a die lubricant on the face of the dies.

The range and quantity of parts or tools in the manufacturing process should be analyzed to determine the performance requirements for the tooling. This will dictate the number of grippers or tools that are required. The tooling must be designed to accommodate the range of part sizes either automatically in the tool, through automatic tool change, or through manual changeover. Manual changeover could involve adjusting the tool to handle a different range of parts, or interchanging fingers, inserts, or tools on a common hand. To reduce the manual changeover time, quick disconnect capabilities and positive alignment features such as dowel pins or locating holes should be provided. For automatic tool change applications, mechanical registration provisions, such as tapered pins and bushings, ensure proper alignment of tools. Verification sensors should also be incorporated in automatic tool change applications.

Presentation and disposition of the workpiece within the robot system affect the design of end-of-arm tooling. The position and orientation of the workpiece at either the pickup or release station will determine the possible contact points on the part, the dimensional clearances required in the tooling to avoid interferences, the manipulative requirements of the tooling, the forces and moments of the tooling and part in relation to the robot tool-mounting plate, the need for sensors in the tooling to detect part position or orientation, and the complexity of the tooling.

The sequence of events and cycle time requirements of the process have a direct bearing on tooling design complexity. Establishing the cycle time for the operation will determine how many tools (or hands) are needed to meet the requirements. Multiple parts-handling tools often allow the robot to increase the productivity of the operation by handling more parts per cycle than can be achieved manually. The sequence of events may also dictate the use of multifunctional tooling that must perform several operations during the robot cycle. An example of this is in machine unloading, where the tooling not only grasps the part, but also sprays a lubricant on the molds or dies of the machine. Similarly, robot tooling could also handle a part and perform work on it at the same time, such as automatic gauging and drilling a hole.

The sequence of events in going from one operation to another may cause the design of the tooling to include some extra motions not available in the robot by adding extra axes of motion in the tooling to accommodate the sequence of operations between various system elements.

In-process inspection requirements will affect the design of end-of-arm tooling. The manipulative requirements of the tooling, the design of sensors or gauging into the tooling, and the contact position of the tool on the part are all impacted by the part-inspection requirements. The precision in positioning the workpiece is another consideration for meeting inspection requirements.

The conditional processing of the part will determine the need for sensors integrated into the tooling as well as the need for independent action by multiple-handed grippers.

The environment must be considered in designing end-of-arm tooling. The effects of temperature, moisture, airborne contaminants, corrosive or caustic materials, and vibration and shock must be evaluated, as must the material selection, power selection, sensors, mechanics, and the provision for protective devices in the tooling.

SUMMARY

Design Tips for End-of-Arm Tooling

The following list presents some tips that are useful in designing end-of-arm tooling (EOAT).

1. Design for quick removal or interchange of tooling by requiring a small number of tools (wrenches, screwdrivers, etc) to be used. Use the same fasteners wherever possible.
2. Provide location dowels, key slots, or scribe lines for quick interchange, accuracy registration, and alignment.
3. Break all sharp corners to protect hoses and lines from rubbing and cutting and maintenance personnel from possible injury.
4. Allow for full flexure of lines and hoses to extremes of axes of motion.
5. Use lightweight materials wherever possible, or put lightening holes where appropriate to reduce weight.
6. Hardcoat lightweight materials for wear considerations, and put hardened, threaded inserts in soft materials.
7. Conceptualize and evaluate several alternatives in EOAT.
8. Do not be penny-wise and pound-foolish in EOAT; make sure enough effort and cost is spent to produce production-worthy, reliable EOAT and not a prototype.
9. Design in extra motions in the EOAT to assist the robot in its task.
10. Design in sensors to detect part presence during transfer (limit switch, proximity, air jet, etc).
11. For safety in part-handling applications, consider what effect a loss of power to EOAT will have. Use toggle lock gripper or detented valve to promote safety.
12. Put shear pins or areas in EOAT to protect more expensive components and reduce downtime.

13. When handling tools with robot, build in tool inspection capabilities, either in EOAT or peripheral equipment.
14. Design multiple functions into EOAT.
15. Provide accessibility for maintenance in EOAT design, and quick change of wear parts.
16. Use sealed bearings for EOAT.
17. Provide interchangeable inserts or fingers for part changeover.
18. When handling hot parts, provide heat sink or shield to protect EOAT and robot.
19. Mount actuators and valves for EOAT on robot forearm.
20. Build in compliance in EOAT of fixture where required.
21. Design action sensors in EOAT to detect open/close or other motion conditions.
22. Analyze inertia requirements, center of gravity of payload, centrifugal force, and other dynamic considerations in designing EOAT.
23. Look at motion requirements for gripper in picking up parts (single-action hand must be able to move part during pickup; double-action hand centers part in one direction; three or four fingers center part in more than one direction).
24. When using electromagnetic pickup hand, consider residual magnetism on part and possible chip pickup.
25. When using vacuum cup pickup on oily parts, a positive blow-off must also be used.
26. Look at insertion forces of robot in using EOAT in assembly tasks.
27. Maintain orientation of part in EOAT by force and coefficient of friction or location features.

Future Considerations

To date, most of the applications of industrial robots have involved a specially designed hand or gripper. Current research is ongoing to develop more flexible general-purpose grippers that can adapt to a variety of sizes and shapes of parts. The state-of-the-art is nowhere near duplicating the complexity of the human hand. However, with increased sensory feedback integrated into robot tooling, more sophisticated tasks are being completed by the robot. This trend will continue in years to come.

BIBLIOGRAPHY

General References

G. Lundstrom, B. Glennie, and B. W. Rooks, *Industrial Robots Gripper Review,* IFS Publications, Bedford, England, 1977.

T. Okada and S. Tsuchiya , "On a versatile finger system," *Proceedings of the 7th International Symposium on Industrial Robots,* Tokyo, Japan, Oct. 1977, pp. 345–352.

K. Mori and K. Sugiyama, "Material handling device for irregularly shaped heavy works," *Proceedings of the 8th International Symposium on Industrial Robots,* Stuttgart, FRG, May 1978, pp. 504–513.

H. Van der Loos, "Design of three-fingered robot gripper," *Industrial Robot* **5** (4), 179–182 (Dec. 1978).

I. B. Chelponov and S. N. Kolpashnikov, "Mechanical features of grippers in industrial robots," *Proceedings of the 13th International Symposium on Industrial Robots,* Chicago, Ill., Apr. 1983, pp. 18, 77–90.

Reprinted from S.Y. Nof, ed., *Handbook of Industrial Robotics,* John Wiley & Sons, Inc., New York, 1985.

ERGONOMICS

SHIMON Y. NOF
Purdue University
West Lafayette, Indiana

INTRODUCTION

Ergonomics in Greek means "The natural laws of work." Traditionally, it has meant only the study of the various aspects of humans in working environments aimed at optimizing the work system. This is the focus of the articles ERGONOMICS WORKPLACE DESIGN and ERGONOMICS, HUMAN–ROBOT INTERFACE. Today, however, when robots are being introduced to the working environment, there is also a need for a new kind of ergonomics, that can be termed Robot Ergonomics.

Robot Ergonomics is defined as the study and analysis of relevant aspects of robots in working environments. It is used to provide tools for the purpose of optimizing overall performance of the work system. In a broad sense, this includes analysis of robot work characteristics, work methods planning, workplace design, performance measurement and integrated human and robot ergonomics. Robot work in general should be optimized with respect to objectives such as minimize the time and cost per unit produced, maximize safety and maximize the quality most effectively. The article ERGONOMICS, ROBOT SELECTION covers the important subject of selecting robots for given tasks.

The purpose of this article is to provide summary of some work that has been carried out over recent years in robot ergonomics, to outline the scope, and provide foundation for further work in this new subject area. The article is organized along the five major subareas of robot ergonomics (see Fig. 1). Finally, robot ergonomics is compared to traditional ergonomics.

ANALYSIS OF WORK CHARACTERISTICS

For a given job requirements, one must first decide whether a human or a robot should be employed for the job. If a robot then the best combination of robot models and work-methods has to be selected. In any consideration it is necessary to know the detailed characteristics and skills of current industrial robots, as well as those of humans.

A series of Robot-Man charts was developed (1) to aid in determining quickly whether a robot can perform a job under consideration. Also, the charts can serve as a guideline and reference for robot specification. This issue is further discussed in ERGONOMICS, ROBOT SELECTION. The full charts (updated in

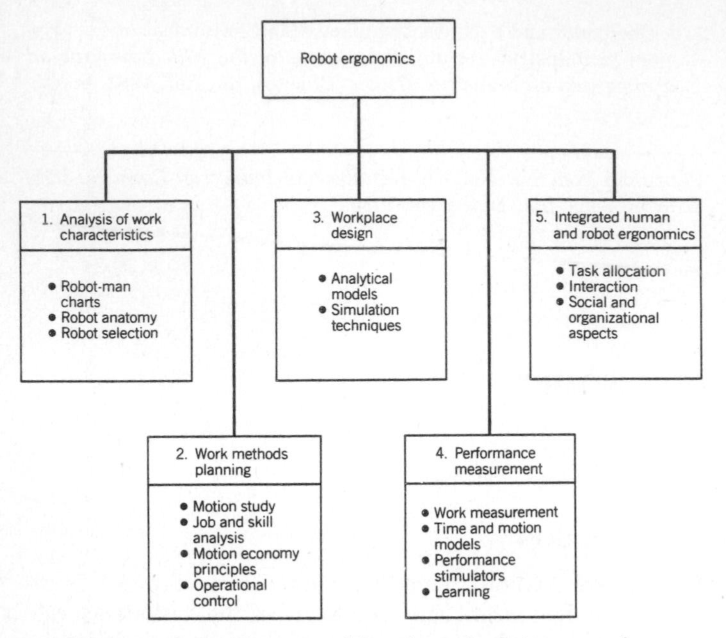

Figure 1. General scope and functions of robot ergonomics.

Ref. 2) provide appropriate numerical range values for each item of three main categories of work characteristics: physical properties, mental and communicative abilities, and energy considerations. The information in the charts can be utilized effectively to select jobs that robots can do well.

Since robot technology advances rapidly, the charts have to be updated periodically. Specialized charts can be prepared for particular areas of application, where attention can be focused on application-specific characteristics. Ref. 3 carried the Robot-Man charts a step further by adding the characteristics of automation as another dimension of consideration.

ROBOT ANATOMY

A thorough anatomy of industrial robots and their controls (4) provides a close examination of the basic robot structure and reveals their current limitations, particuliary in sensor ability and task interactions. In order to determine the current abilities of industrial robots, literature describing 282 models was analyzed in two forms (5): frequency distributions of properties and motion-velocity graphs.

The properties studied are robot size, arm structure, lift capacity, actuator type, degrees of freedom, control mode, repeatability, weight, and velocity. Each property was analyzed for Japanese, U.S. and European, and combined robot populations. Motion-velocity graphs are regions of maximum movement and velocity combinations for common arm and wrist motions. Individual points inside the region are generated for individual robot models. for instance, a Cybotech V-80 arm can rotate up to 270 degrees right-to-left at up to 1 rad/sec. The regions are developed by connecting extreme perimeter points. Both the frequency distributions and the motion-velocity graphs provide important ergonomic information. For example, it was observed that 1982 Japanese robots concentrated more on rectangular models (52% of all their models) where as the U.S. market had more articulated models (48% of all U.S. Models).

A subsequent survey of 1984 models (6) including 676 Japanese and U.S. market robots provides, in addition to updated distributions general trends in robot work characteristics, statistical correlation between various robot properties and regression equations of significant robot parameters. For example, it was found that Japanese 1984 models moved to majority of spherical arms (38%) compared to only 5% of rectangular ones. In the U.S. market articulated robots continue to be in the lead (54%). Conditional distributions provide additional insight as to how current robot models are designed (Fig. 2). This type of ergonomic study may be more useful for knowledge of general trends of how robot populations are designed, and what are their work abilities. Figure 3 shows typical configuration of robotic workplaces.

WORK METHODS PLANNING AND WORKPLACE DESIGN

The work method determines how well limited resources in the work systems are being utilized, and its planning has to be carried out both at the macro and micro levels. Robot work-method planning techniques that have been developed and applied include motion study, job and skill analysis, motion economy principles, and operational control. One of the early motion studies of robot work was carried out in Japan (7). Memo-motion cameras were used to record the details of free-forging group operations performed by operators. Robotization plans were developed in response to poor working conditions, and the motion study supplied the necessary information about work requirements, constraints, and flow. Basically, however, this study did not differ from traditional motion studies.

Robot oriented job and skill analysis was developed by Nof, Knight and Salvendy, (1). Traditionally, such methods have been applied for cost-production programs in manual and human–machine work environments and for effective personnel selection and training. The job analysis focuses on what to do; the skill analysis focuses on the "how." In revising the method and adopting it for robot-oriented analysis, the features that distinguish robot and human operators were considered.

As before, a task is broken down to elements, specified each with their time and requirements. However, specification of left/right hand operations is replaced by a general limb column, since a robot may be designed to have any number and variety of limbs, eg, arms, grippers, special tool-hands, etc. Other features include new considerations of computer memory and programming details that influence the required robot design. With humans most work-related decisions may be self-explanatory and trivial, however, robot decision and control must be completely specified. In a case study described in Ref the method of a mechanical assembly by a robot is analyzed and improved by rearranging the work elements. The job and skill analysis provided for that case lead to the development of alternative workplace designs.

Operational control of robots can also be considered part of the robot work-methods planning. Consider, for instance, the planning of optimal bin-picking sequences, or the optimal sequencing of components insertion by a robot. Such problems were studied under the Assembly Plan Problem formulated and solved by Drezner and Nof (9). The general problem involves several ergonomic tasks: to find the optimal location of each component in a robotic assembly cell; to find the optimal sequence of picking and of inserting the components. A number of variations to the problem exist depending on whether assembled components are repeated or not, excess of bin cells, and whether different assembly configurations are required. Beyond the optimal planning of physical cell organization and assembly tasks sequencing, the assembly problem is associated with other ergonomic functions. The design of the robot selected to perform the planned assembly may be specified differently, depending on the optimal method that results; the times of robot motions that are carried out during the assembly must be provided by performance measurement (discussed later); motion economy principles, discussed next, can guide the robot selection and overall cell planning. Other work on the robot workplace design is done by analytic models, (10) and by simulation (11).

MOTION ECONOMY PRINCIPLES

Principles of motion economy have traditionally guided the development, troubleshooting, and improvement of work methods and workplaces, with special attention to human operators. In developing such principles for robot work, some of the principles can be adopted, but new principles, unique to robots, must also be added.

The robot-motion economy principles can be applied as guidelines for the initial planning of robotic installations, where the planned tasks have to satisfy the recommendations made by the guideline. The guidelines can also serve in analyzing existing robotic facilities, where tasks are checked after changes made by accumulated requirements to determine whether they still follow the guidelines.

Human-motion economy principles are divided into three main groups: principles concerning motions, eg, two hands should not be idle at the same time; principles concerning the design of the work-cell, eg, chair type and height to permit good posture equipment, eg, tool should be prepositioned whenever possible. In robot-motion economy principles (2,12), there are four main groups of principles: (*1*) robot motions, (*2*) design of the robotic work cell, (*3*) tooling and equipment, and (*4*) sensors. The latter is added because sensors for robots can be selected and optimized.

The first principle had to be adapted for the robot case. The revised principle states: in a multi-armed cell, two or more arms should not be idle at the same time, unless this is unavoidable due to machine sequencing.

The second principle, concerning chair design, is irrelevant for robots. On the other hand, a unique robot principle is to minimize the number of arms necessary for a prospective robot application. The third example above, stating that tools should be prepositioned whenever possible, is valid for both humans and robots.

Robinson (12) investigated robot-motion economy principles and found that in principles concerning motion, five manual principles are irrelevant, three can be adapted, and four new principles can be developed for robots. In workplace design he found that three manual principles are irrelevant, five can be adopted, and three new robot principles are needed. In tooling and equipment, three manual principles irrelevant, two can be adapted, and one new robot principle is needed. Finally, he proposed five new sensor related principles. One example of a new robot principle concerns arm-joints utilization.

ARM-JOINTS UTILIZATION PRINCIPLE

This principle is stated as follows: structure tasks to make greater use of arm-joints which, when in motion, require less effort of the drive-system. As a result of this guideline, wear of the robot is reduced, leading to better consistency and accuracy, and to less maintenance.

The principle was investigated in three ways (12): by kinematic analysis and by experimental analysis of the work effort by two robots, PUMA and Cybotech H-80. The PUMA is a small articulated robot and the H-80 is large and cylindrical. The third method was an investigation of performance time by the RTM method. In all studies, two tasks were compared, bin picking and kitting, each performed with a planar bin, ie, a bin placed horizontally in front of the robot, and with a vertical bin, ie, a bin placed vertically in front of the robot, (see Fig. 4). In the bin-picking task, the robot picks specified parts from bin-cells, depositing each part in the drop point. In kitting, on the other hand, the robot travels through a given set of cells, picking kit-components into a magazine and finally depositing the whole kit at the drop point. Both tasks are typical of many industrial robot assignments.

Based on this research it was found that:

1. The horizontal work orientation, namely, with planar bin, is *statistically* preferred for less robot effort, for both robot types and for both tasks.
2. The horizontal work orientation is *always* preferred for the cylindrical robot, H-80, because of its work envelope.
3. Some vertical positions are better for the articulated robot, PUMA, because of its kinematic structure.
4. In terms of performance time, there is no significant difference, overall, between the two bin orientations. However, pick-patterns that require different distances and orientations do have a significant effect on time. For instance, in the horizontal bin orientation when bin cells are closer to the point where parts are dropped, the cycle time required is obviously shorter.

The contribution of such robot ergonomic knowledge and guidelines to task-oriented robot selection, method planning, and performance evaluation can be significant. Other motion economy principles for robots analyzed by Robinson concern the selection of single vs multiple gripper for machine loading/unloading, and selection of the degree of robot sensor sophisti-

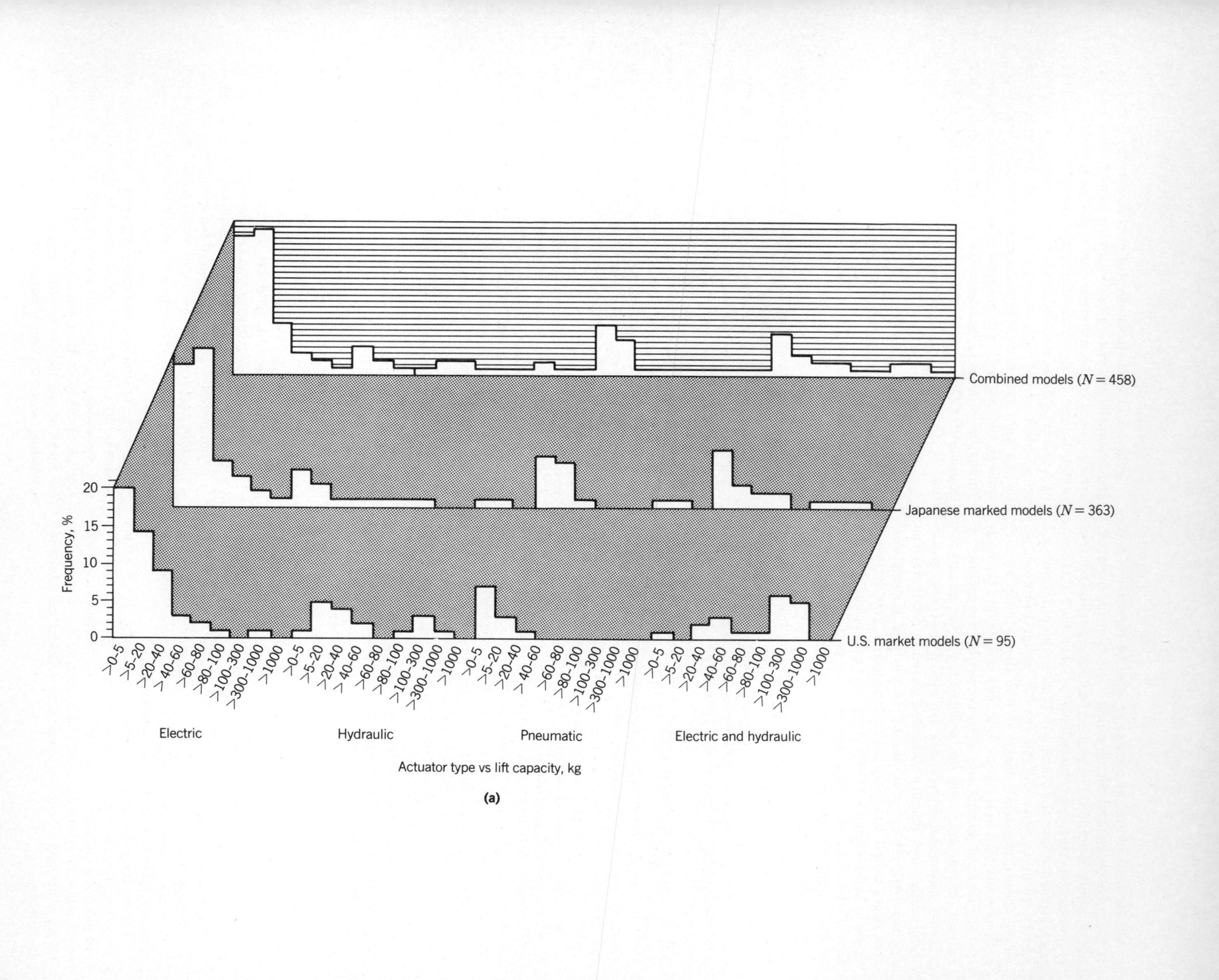

Combined models ($N = 458$)
Japanese marked models ($N = 363$)
U.S. market models ($N = 95$)
Frequency, %
20
15
10
5
0
>0-5
>5-20
>20-40
> 40-60
>60-80
>80-100
>100-300
>300-1000
>1000
>0-5
>5-20
>20-40
> 40-60
>60-80
>80-100
>100-300
>300-1000
>1000
>0-5
>5-20
>20-40
> 40-60
>60-80
>80-100
>100-300
>300-1000
>1000
>0-5
>5-20
>20-40
>40-60
>60-80
>80-100
>100-300
>300-1000
>1000
Electric
Hydraulic
Pneumatic
Electric and hydraulic
Actuator type vs lift capacity, kg

(a)

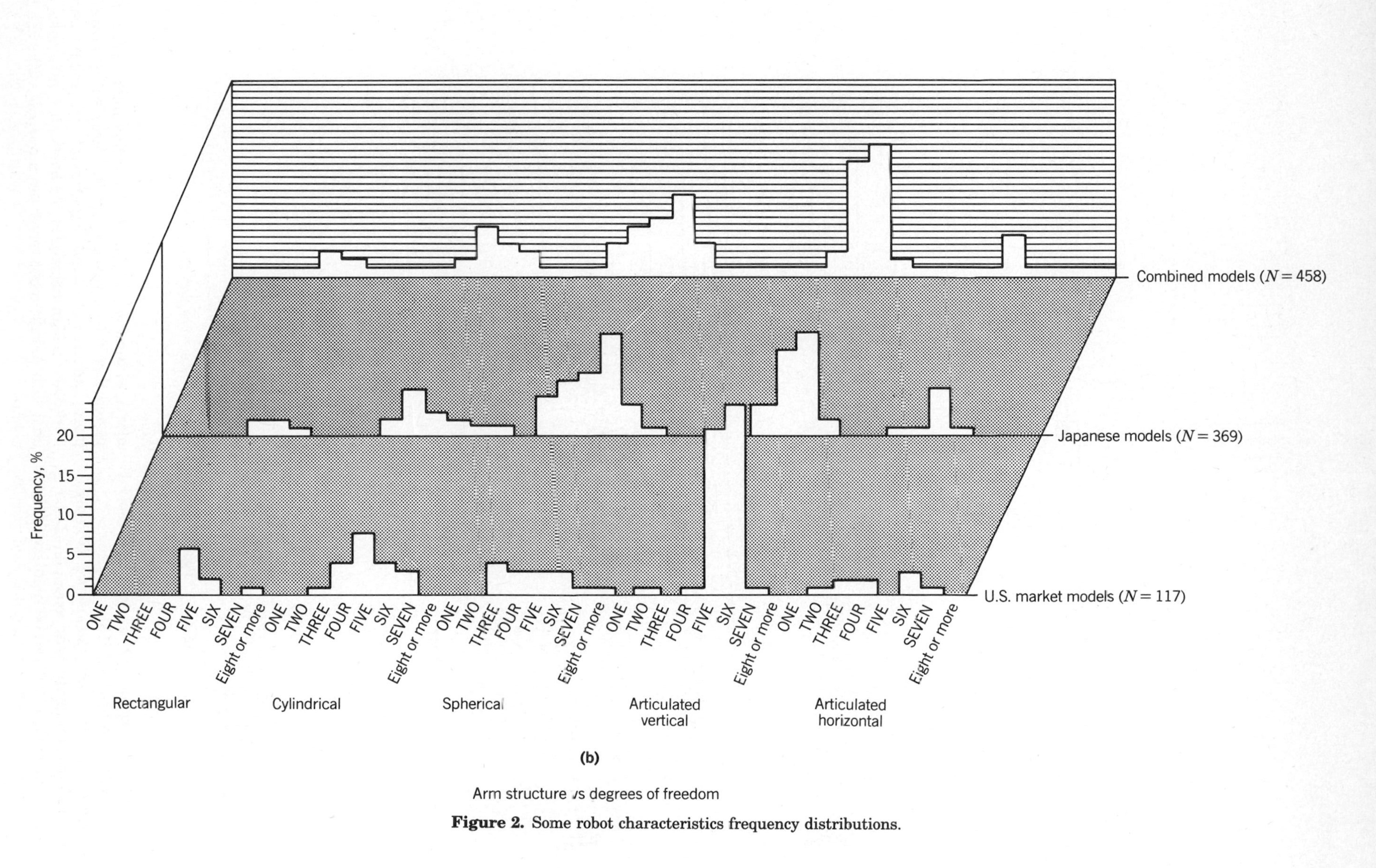

(b)

Arm structure *vs* degrees of freedom

Figure 2. Some robot characteristics frequency distributions.

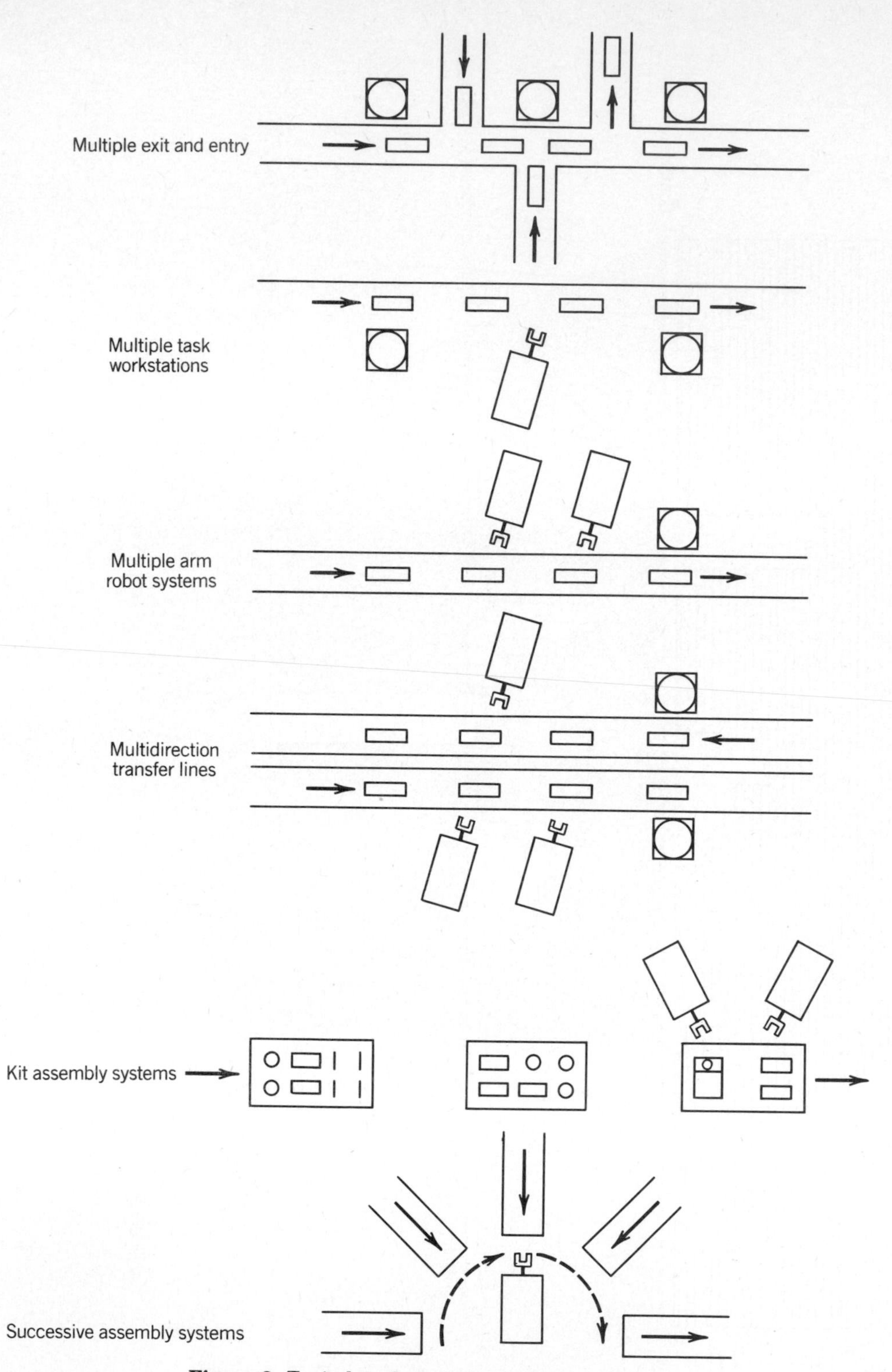

Figure 3. Typical configurations of robotic workplaces.

cation. Both analyses were carried out using a flow-control simulator for robotic cells, SINDECS-R (13).

Although the motion economy principles serve as general guidelines, their modeling and analysis as exemplified above offer robot ergonomic knowledge that can be highly valuable. Task-oriented robot selection, method planning and performance evaluation can benefit from this information. It is expected that future research in this area will continue to explore, analytically and experimentally, robot work and workplace design principles. In particular, principles associated with specific robot types and industrial tasks can be very useful.

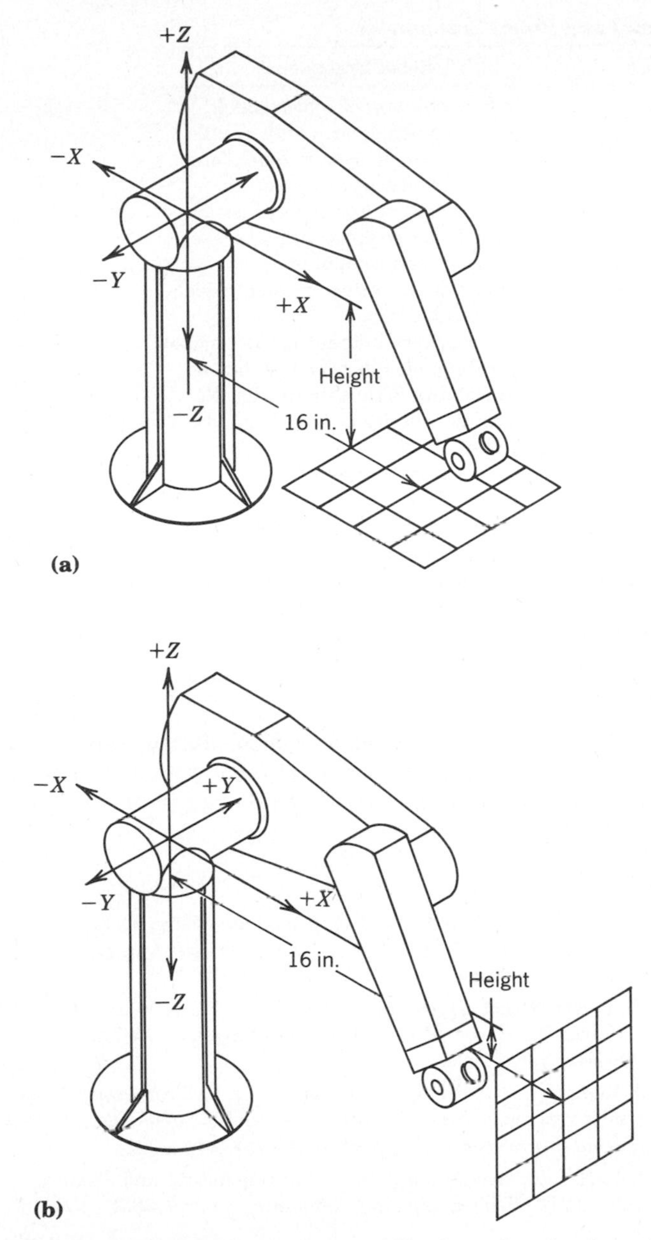

Figure 4. (**a**) Graphic representation of horizontal work orientation. (**b**) Graphic representation of vertical work orientation.

PERFORMANCE MEASUREMENT

Robot performance measurement is also considered as part of robot ergonomics according to its definition. Included here are work measurement, performance prediction, and performance evaluation. Robot designers have long been concerned with planning robot motions in an optimal way (14). Early studies (15) investigated the minimum cycle-time problem of a robot moving among obstacles. Alternative combinations of simultaneous motions of different joints were analyzed; Birk and Franklin (1974) reduced cycle-time by proper placement of workpieces at the workstation, which follows one of the motion economy principles discussed above is discussed in Ref. 16. Others performed time amd motion studies in order to evaluate, select and compare alternatives for robot work systems designs (17–20).

THE RTM METHOD

The Robot Time and Motion method (RTM) (8,21) for predetermination of robot cycle times is based on standard elements of fundamental robot work motions. It is analogous to the Motion Time Measurement technique (M+M). Both enable users to estimate the cycle time for given methods without having to first implement the work method and measure its performance. But RTM differs in that its data and performance models have to be tailored to particular robot models or families. MTM users must consider human individual variability and pacing effects, however, RTM can rely on the relative consistency of robots and apply computational models based on physical parameters of the robot.

The RTM system was developed experimentally for several robot models such as T^3, PUMA, IBM RS-1, and others. It is basically comprised of three major components. (*1*) The 16 standard elements, including movement elements, sensing elements, gripper and tool elements, and delay elements; (*2*) a library of robot computational performance models; and (*3*) the RTM analyzer, which computes performance measure directly from user-specified work methods. The RTM research and developments are summarized well in Ref. 2.

Although a method such as RTM can provide useful performance measures for given combinations of work methods carried out by certain robot models, it is limited in terms of evaluating the interaction with other components of which a work system consists. For this purpose, the performance simulator can be appropriate. Analysis and evaluation of robotic work with emphasis on performance has been done (13,22,23). As indicated above, the latter simulator has been used in investigations of several motion economy principles for robots. RTM was applied to generate time data input for those analyses. In the context of robot ergonomic, performance simulators will continue to play an important role in analyzing and evaluating relationships between ergonomic variables and robotic work system design.

INTEGRATED HUMAN AND ROBOT ERGONOMICS

Work in this area has been performed mainly by human ergonomists, with the general objective of considering human ergonomic needs when designing robot applications. Examples of this approach are works contained in Refs. 24 and 25.

Although industry has tended to separate employees from robots for safety reasons, a vital area of ergonomic study is the integration of humans and robots in work systems. Ref. 25 identifies the typical jobs human workers will perform in a robotic work environment. Out of eight, seven job-types are performed side-by-side or separately, while one prospective job is indicated as "synergy": combination of humans and robots in operations such as assembly or supervisory control of robots by humans. The objective of the ergonomic study in this case is to plan a work system with the required degree of integration to best utilize the respective advantage of humans and robots working in concert.

Table 1. Some Issues of Comparison between Traditional and Robot Ergonomics

Area	Human Ergonomics	Robot Ergonomics
1. Work abilities (anatomy, physiology, psychology, etc)	Study population of people Select and train	Study population of robot models Specify (eg, single arm with double gripper, single eye, no touch/force sensor) and teach or program
2. Job and skill analysis	Based on work abilities Right/left hand analysis	No psychological/sociological issues Columns per required limbs, senses, memory and computer
3. Predetermined time standards	Data gathered by experimentation Must consider human individual variability and pacing	Same; can also utilize motion models to calculate times Must consider different motion models for different robot models, but no variability within the model. No pacing, velocity is programmed.
4. Motion economy principles	Worker related principles Equipment related principles Time-reduction principles	Similar Same or very similar Same Additional principles, eg Robot specification/design principles
5. Learning	Learning by worker's experience Learning by organization's experience	Same, plus "learning" by artificial intelligence

Another relevant direction concerns robot learning. Robot learning can be found in three main areas in industry:

1. Human operators learning to accept and work with robots
2. Organizations learning to introduce and utilize robots effectively
3. Robots learn to improve their own operation by adaptive or learning control.

Some research has started in the above areas (see discussion in Ref. 2), which has significant potential for productivity gains.

CONCLUSION

Techniques and knowledge developed by robot ergonomics research have been identified and described as necessary for planning effective, optimized robot work. Traditional ergonomic tools can be extended to develop human-robot work systems. On the other hand, new techniques have evolved from traditional ergonomics for the purpose of planning the work of robots themselves. Basically, robots offer a new dimension in work design by allowing the "operator" to design (ie, the robot) itself. Table 1 summarizes comparison issues between traditional and robot ergonomics. Five ergonomics areas are analyzed, namely, work ability, job and skill analysis, predetermined time standards, motion economy principles, and learning. Throughout this comparison, it is shown that even though similar principles of ergonomics are applied, robot oriented techniques do require certain adaptations. It is expected that further research and practice will lead to the development and refinement of additional, innovative robot ergonomic techniques and knowledge.

In recent years it has often been argued that the bottleneck to reaping the true productivity promise of robotics is not in the robot machine structural or control design; rather, the difficulty and challenge are in correct robot work system planning. For ergonomics experts, this challenge offers an important opportunity.

BIBLIOGRAPHY

1. S. Y. Nof, J. L. Knight, and G. Salvendy, "Effective Utilization of Industrial Robots-A Job and Skill Analysis Approach, *AIIE Transactions,* **12**(3), 216–225 (Sept. 1980).
2. S. Y. Nof "Robot Ergonomics: Optimizing Robot Work," in S. Y. Nof, ed., *Handbook of Industrial Robotics,* John Wiley & Sons, Inc. New York, 1985.
3. J. Kamali, C. L. Moodie, and G. Salvendy, "A Framework for Assembly Systems: Humans, Automation and Robots," *International Journal of Production Research* **20**(4), 431–448 (1982).
4. J. Y. S. Luh, "An Anatomy of Industrial Robots and Their Controls," *IEEE Transactions on Automatic Control* **28**(2), 133–153 (Feb. 1983).
5. S. Y. Nof and E. L. Fisher, "Analysis of Robot Work Characteristics," *The Industrial Robot,* 166–171 (Sept. 1982).
6. R. A. Pennington, E. L. Fisher, and S. Y. Nof, "Survey of Industrial Robot Characteristics: General Distributions, Trends, and Correlations, in S. Y. Nof, ed., *Robotics and Material Flow,* Elsevier, 1986.
7. Y. Hasegawa, "Analysis of Complicated Operations for Robotization," SME paper no. MS79-287, 1979.
8. S. Y. Nof and H. Lechtman, "Analysis of Industrial Robot Work by the RTM Method," *Industrial Engineering Journal,* (April 1982).
9. Z. Drezner and S. Y. Nof, "On Optimizing Bin Picking and Insertion Plans for Assembly Robots," *IIE Transactions* **16**(3), 262–270 (Sept. 1984).
10. S. C. Sarin and W. E. Wilhelm, "Prototype Models for Two-dimensional Layout Design of Robot Systems, *IIE Transactions,* **16**(3), (1984).
11. R. N. Stauffer, "Robot System Simulation," *Robotics Today,* 81–90 (June 1984).

12. A. P. Robinson, "Principles for Robot Work Design," M. S. Thesis, School of Industrial Engineering, Purdue University, West Lafayette, Ind., 1984.
13. A. P. Robinson and S. Y. Nof, SINDECS-R: A Robotic Work Cell Simulator, *Proc. 1983 Winter Simulation Conf.* pp. 350–355.
14. M. Brady and co-eds., *Robot Motion Planning and Control,* MIT Press, Cambridge, Mass., 1982.
15. D. L. Pieper and B. Roth, "The Kinematics of Manipulators Under Computer Control," *Proc. 2nd Int. Conf. Theory of Machines and Mechanisms,* Warsaw, Poland, Sept. 1969.
16. J. R. Birk and D. E. Franklin, "Picking Workpieces to Minimize The Cycle Time of Industrial Robots," *Industrial Robot* **1**(5), 217–222 (Sept. 1974).
17. W. L. Verplank, *Research on Remote Manipulation of NASA/Amers Research Center,* NBS Special Publication 459, U.S. Department of Commerce, 1976, pp. 91–96.
18. J. Vertut, *Experience and Remarks on Manipulator Evaluation,* NBS Special Publication 459, U.S. Dept of Commerce, 1976, pp. 97–112.
19. P. F. Rogers, "Time and Motion Method For Industrial Robots," *Industrial Robot,* **5** 187–192 (Dec. 1978).
20. A. S. Kondoleon, *Cycle Time Analysis of Robot Assembly Systems, SME Paper, MS79-286,* 1979.
21. H. Lechtman and S. Y. Nof, "Performance Time Models for Robot Point Operations," *International Journal of Production Research,* **21**(5), 659–673, (1983).
22. T. Kuno, F. Matsunari, H. Moribe, and T. Ikeda, Robot Performance Simulator, *Proc. 9th ISIR,* Washington, D.C. March 1979, pp. 320–330.
23. D. J. Madeiros, R. P. Sadowski, D. W. Starks, and B. S. Smith, "A Modular Approach to Simulation of Robotic Systems," *Proc. 1980 Winter Simulation Conf.,* pp. 207–214.
24. L. Noro and Y. Okada, "Robotization and Human Factors," *Ergonomics,* **6**(10), 985–1000 (1983).
25. G. Salvendy, "Review and Reappraisal of Human Aspects in Planning Robotic Systems," *Behaviour and Information Technology,* **2**(3) 263–287 (1983).
26. H. M. Parsons and G. P. Kearsley, Robotics and Human Factors: Current Status and Future Prospects *Human Factors,* **24**(5), 535–552 (1982).

This article is based in part on research grants from the National Science Foundation Program on Production Research and Technology for "Advanced Industrial Robot Control"; Purdue University's Computer Integrated Design, Manufacturing, and Automation Center for "Design Data Base Technology in Computer Integrated Production Systems." The author wishes to thank his graduate students and colleagues with whom work reported and referenced in this article was developed.

ERGONOMICS, HUMAN–ROBOT INTERFACE

JOHN G. KREIFELDT
Tufts University
Medford, Massachusetts

INTRODUCTION

Ergonomics or engineering psychology, human factors or human factors engineering, are terms used interchangeably to describe the unique considerations and approaches that pertain to designing equipment, products, environments, workplaces, jobs, services, and so on for humans. A good definition of human factors based on reference 1 is cited below.

Human Factors

Focus

1. The design and analysis of man-made objects, products, equipment, facilities, and environments that people use.
2. The development of procedures for performing work and other human activities.
3. The provision of services to people.

Objectives

1. Enhance the effectiveness and efficiency with which people perform work and other human activities.
2. Maintain and enhance certain desirable human values such as health, safety, and satisfaction.

Approach

1. The systematic application of relevant information about human abilities, characteristics, behavior, and motivation in the execution of such functions.

In this classical sense, many of the considerations and principles that pertain to good design for the "human–robot" system and environment are of longstanding interest and found in many, if not most, of the "prerobot" human–machine systems (particularly in remote manipulation or "teleoperator" systems) and therefore, experience and knowledge gained in those applications can be readily drawn upon. On the other hand, because new "machinery" or systems such as robotics always bring somewhat new concerns, not all of the pertinent human factors problems have been solved.

The basic concerns of human factors (qv) in the workplace are for the productivity, safety, and well-being of the worker. Robotic systems are, in this respect, another arena for these concerns.

From the point of view of the human factors system designer, the robot demystified is an engineering system compromised of essentially an electro- (or hydraulic- or pneumatic-) mechanical system coupled to a computer and designed to accomplish certain tasks. Because this description would cover many items (eg, automatic pilots), for the purposes of this article, the definition of an industrial robot is a "reprogrammable, multifunctional manipulator designed to move parts, tools, or specialized devices through variable programmed motions for the performance of a variety of tasks" [Robot Industries Association (RIA)]. It could also be added that the manipulator sometimes has at least vaguely anthropomorphic (human-looking) characteristics.

Unlike strict and complete automation, in which the human element is removed totally, robotic systems are often a type of semiautomation in the sense that humans are still involved in the tasks to a degree that their performance has a large bearing on the total system performance in terms of desirable measurement criteria such as overall cost, speed of production, safety, and so on. Thus, the joint activities of humans and robots should be as effective as possible.

Nine ways that humans participate in robot systems have been identified (2):

Surveillance (monitoring).
Intervention.
Maintenance.
Backup.
Input.
Output.
Supervision.
Inspection.
Synergy (task sharing or task interaction).

Surveillance is necessary to the extent that absolute and complete trust that nothing will go wrong with the robotized operation is lacking. *Intervention* pertains to the multiple activities that include setup, startup, shutdown, (re)programming, teaching, on-line editing, and taking corrective actions.

Maintenance involves both hardware and software and can encompass troubleshooting, repair, calibration, and equipment substitution.

Backup substitutes human (manual) operations for robotic operations during emergencies such as breakdown when it is necessary to keep production going.

Input pertains to such activities as a human supplying materials to, or orienting parts for, robots such as might be needed in the absence of automatic part orientation mechanisms.

Output tasks may engage a human in removal of robot-finished goods, retrieval of dropped tools or parts, and so on.

Supervision could comprise managing of both humans and robots in shared tasks, managing multiple robots, planning operations, and dealing with emergencies.

Inspection activities would typically be found in quality control during processing and in quality assurance of the end products.

Synergy describes a harmonious interaction of human and robot to perform some set of tasks. The human may, at a higher level, act as the "brains" while the robot acts as the "brawn" in a command/feedback task-sharing concept termed "supervisory control." At a lower level, the human may do that manual part of a task not economically or state-of-the-art robotizable while the robot does the other part.

PARADIGMS FOR HUMAN–ROBOT SYSTEMS

The human–robot system shares many aspects of previously encountered human engineering systems, and these aspects are expressed in several different paradigms, which illuminate the essential features of these systems. These systems are *man–machine* (now human–machine) systems and *user–product–task* systems.

Human–Machine Systems

The human–machine system is one of the earliest engineering-based paradigms (3) and is based upon the notion of cyclic or closed-loop information flow through a system whose purpose is to produce some desired action or output. The system can take purposeful or corrective action to bring the actual output (performance) into agreement with the desired one (the input) by comparing the actual and desired output and modifying the current output value appropriately. When the corrective action is taken by a human (as, for instance, the driver of an automobile), this philosophy of control is symbolized by the block diagram in Figure 1. This paradigm points out important aspects of such an information and control system.

A *display* means whereby the human can be apprised of the status of the desired behavior (input) and the status of the output and thereby take corrective action through means of a controller. A *controller* whereby the human can input desires and have them changed into a form that can then act upon the *object to be controlled,* which is the action producer. The actual output or behavior of the controlled object is made known to the human through *feedback* of the behavior to the *display*.

Various technical means may be used to display information to the human and to transform the human's actions into a form suitable for directing the controlled object. This particular paradigm emphasizes the "human–in-the-loop" form of control. When the object to be controlled is a robot, the *controller* could be a keyboard, teach pendant, and so on, and the *display* could be the human "eyeball" alone or a CRT screen with pertinent robot action output displayed, and so on.

Both the display and the controller exist for human use, not for robot use, although obviously they must be designed for both the human and, for example, the robot to use since one end of each mechanism is the human end and the other end is the robot end. That is, there are several *interfaces* that are also important from a design point of view. The display must be designed to match the sensory (usually visual) and cognitive characteristics of the human, and the controller must be designed to match the effector (usually motor action) characteristics of the human. For example, letters that are too small to be seen on the display under the working conditions surpass human visual capabilities, whereas poorly encoded information (eg, attempting to signal which of 12 axes of a robot is overheating by using a different warning tone for each axis)

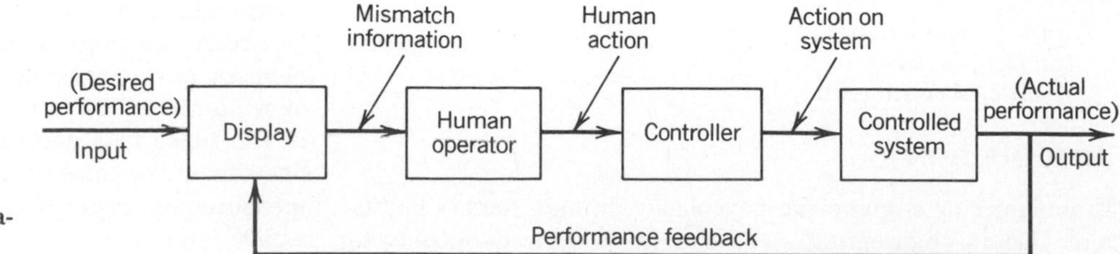

Figure 1. Human–machine system paradigm.

surpasses the human's information processing cognitive abilities. Or again, if the controller contains a computer that takes more than 0.2 s to turn the keyboard (or joystick) human input into form suitable to drive the robot, then a serious degradation in the human's ability to control the robot in a smooth, continuous fashion (should that be necessary as a requirement for precision work) can be expected (4). More on the human control ability can be found in texts such as reference 5.

Studies and knowledge of human "input," "output," and cognitive characteristics are the foundations of human factors engineering and many pertinent principles, facts, and so on are contained in numerous handbooks and texts as referenced. The body of human factors knowledge is very extensive and grows continuously, and it is just as applicable in human–robot systems as it is in other areas.

The User–Tool–Task Paradigm

When the actual use of a "tool" to perform some task is more pertinent than the concept of "information flow," then there is a different paradigm that is more applicable to the problem. The user–tool–task paradigm as shown in Figure 2 has been developed for these purposes (6). This paradigm focuses attention on the fact that the designer can take design action on only three design objects: *user, took,* and *task.*

Design action for the *user* of the tool means training, instruction, and selection.

The *tool* is the obvious focus of design and usually receives the "engineering" attention.

The *task* represents the various elements and subelements describing exactly what has to be accomplished. Modification of the task is often as important as the "tool" design.

The *environment* is also a legitimate design object to the extent that it affects the worker's ability to perform tasks adequately.

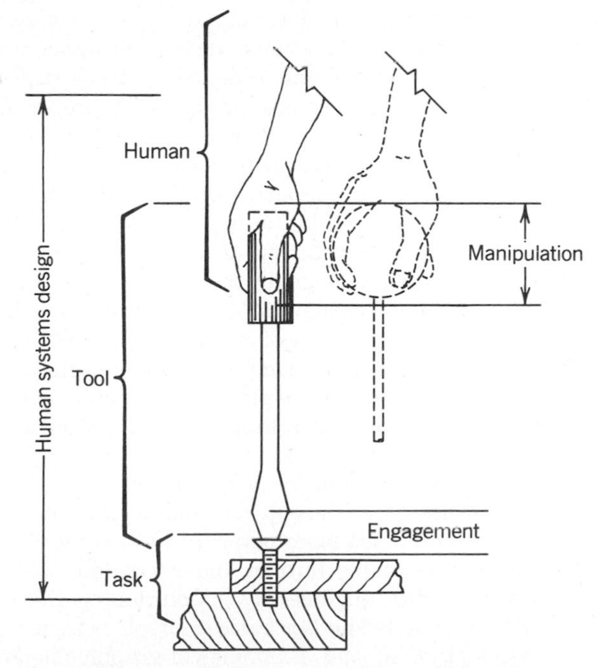

Figure 2. The user–tool–task paradigm (6).

In addition to the three (or four) design objects, there are also two *interfaces* that "connect" the user to the tool and the tool to the task.

The *manipulation* interface encompasses the various human factors concerns pertaining to a good "fit" of the tool to the human's manipulative abilities.

The *engagement* interface encompasses the engineering concerns pertaining to how well the "action end" of the tool is matched to the various task requirements.

Finally, there are several *influences* or feedbacks/feedforwards that have very strong influences on how well the user and tool will "fit together" and how well the tool and task will interact.

The most important of these influences is the *task* influence on *manipulation.* The task demands determine, to a large extent, how the user will manipulate the tool. Inasmuch as the task changes but tools do not automatically keep up with these changes, the user is forced by the task demands to alter the manipulation of the tool in order to accomplish the task. This is one of the reasons, for example, that finger-contoured handgrips are rarely successful. For a particular hand orientation, the contours might fit the hand, but as the task changes and the tool does not, the user might need to alter a grip on the tool. In this case the contours no longer fit the fingers. Many principles of tool design can be found in references 7 and 8.

The word "tool" should not be taken literally inasmuch as it stands for the device that couples the worker to the task or even to another device. For example, the handgrip on a remote manipulator can properly be looked at as a tool by which the worker accomplishes the task of manipulating the manipulator itself. This view has definite (human) engineering design implications for the design of the hand control, just as there are engineering design implications for the design of the manipulator's end effector based upon what it must manipulate. In the user–product–task paradigm, the display as a special-purpose device is often not paramount inasmuch as direct viewing is usually sufficient.

Where the work being performed can be affected by environmental conditions such as lighting levels, noise or sound, temperature and humidity, and so on, *environmental engineering* becomes important. This aspect of human engineering must be addressed.

ENVIRONMENTAL ENGINEERING

There are a number of environmental variables known to have an effect on the human's ability to perform the required tasks efficiently, safely, and with a sense of well-being. Some of these are lighting, noise and sound, temperature and humidity, vibration, atmosphere, and radiation.

There are professional societies with special publications and journals addressed to the complexities of each of the above. The human engineer may draw upon the results of these during the course of an environmental evaluation and/or design. Several of these societies are referenced below.

Lighting

The handbook of the Illumination Engineering Society (9) contains practical information regarding the proper design of the visual environment in terms of the tasks to be performed and the room variables. The study of illumination for human use is one of the oldest of the human factor areas and is intimately connected with the visual characteristics of the eye and the human's cognitive and motor ability. For example, it has been well established that the quickness with which a letter (or small task detail) can be perceived is directly related to the amount of illumination on the detail (or letter), which has definite implications for the speed with which the human can react to, say, an emergency. It is the case that the brighter a light (eg, a warning light), the quicker the response (10) up to a limit of course.

The lighting and view perspectives of a task are also crucial if the human is to perform well. Shadows, glare, (especially on CRT screens) confusing depth clues, and so on, all reduce human performance and well being. For example, when the operator must perform control actions using TV monitors (as in certain highly hazardous environments), the placement of the cameras and light sources is not intuitive and not necessarily what would be imagined if the robot or manipulator had "eyes" through which the remote operator could see (11).

Noise and Sound

Excessive levels of noise and sound can cause hearing damage and in some cases work degradation (12). Even if a noticeable effect cannot be discerned in the short term, uncomfortable acoustic levels or characteristics of the sound can definitely degrade the worker's well-being in terms of a pleasant workplace. The effect of a pleasant workplace should not be overlooked, because one way the worker can compensate for unpleasantness is by taking frequent breaks. This has an effect on productivity (13).

Because of the potential for hearing loss produced by excessive levels of sound, the Occupational Safety and Health Administration (OSHA) sets allowable durations for exposure to different noise levels. For example, an 8 hour daily exposure is allowed to levels of 90 dBA (decibels read on the A scale of a sound level meter) (29 CFR 1910.95) (14) (about the equivalent of a boiler room).

Various methods exist for reducing the level of noise in an environment ranging from passive to active. Passive means include room deadening by sound absorbers (ceiling tiles, carpets on floors and walls), sound absorbing machine mountings, increasing the distance between the worker and noise source, replacement with quieter machines, and so on (15). Active means include ear plugs and ear protectors. Robots may not care about their acoustic environments, but workers assigned to the same area definitely will.

As is the case for the eye and light, the human can respond faster to a louder sound than to a softer one, and surprisingly, the response time to an auditory signal can be faster than to a visual one (10). There are many characteristics peculiar to the human ear and auditory senses (16) that are important when designing the worker's auditory environment. The ability of the worker to hear necessary signals or communicate orally with other human or robot workers (perhaps equipped with voice recognition systems) can be of paramount importance, particularly in terms of worker safety, and human factors studies of verbal communication in noisy environments may be drawn upon for this purpose (17).

OTHER ENVIRONMENTAL VARIABLES

Other environmental variables may also affect the productivity, well-being, or safety of the worker in a robotic environment. A fairly comprehensive introduction to such human factors concerns as temperature/humidity, vibration, atmosphere, and radiation may be found in references 18 and 19.

Although the robotic environment may be imagined to be of a "gentle" nature because of its advanced technological connotations of artificial intelligence, computers, and so on, in fact, it can be just the opposite, given the uses to which the robot may be put. When designing the environment the robot inhabits, the human occupant must not be treated as an afterthought.

THE WORKPLACE

In some instances of human operators working together with robots, the arrangement is one of work or task sharing. In such cases, the robot may be treated as an extension of the human (often into environments unsuitable for or unreachable by humans), and the robot may be directed from a console or workstation designed for this purpose. As in the human–machine paradigm, an important issue in this case is that of display and control: display of information regarding the task and robot performance and communication of control intentions by some means that take into account human capabilities.

Primary display means include CRTs and TV monitors, although simple windows for direct viewing may be pertinent. In addition to the basic vision human factors pertaining to, for example, letter size, contrast, color selection, eye fatigue, and so on, there are also the cognitive issues of how to display information for humans (20). Concepts such as the display integration of different informational pieces, as opposed to their separate display, have found wide use in areas such as aircraft cockpits (21). It has been found that with proper display integration, the human is able to absorb information more quickly and with less error and to perform tasks otherwise difficult or impossible. With computers coupled to TV or CRT displays, many forms of "display aiding" are possible. It is often possible to make a prediction of the future course of some activity that does not change too rapidly over time. Activity such as the course of movement of a remotely controlled arm used for handling large pieces of machinery (hence "slowly changing") can be predicted by computer means quite accurately over the short term and displayed in suitable fashion for the human to use while controlling the arm, primarily to avoid collision between the arm and objects in its environment. This prediction can be updated frequently so that it is never "stale" and inaccurate. This concept, referred to as a predictor display (22, 23), is one of the more powerful ways of providing necessary information to the human and exemplifies one of the principles of human engineering—unburdening the human's mental load, just as another basic principle is to unburden the human's physical load. Examples of graphic displays for controlling (remote) manipulators that provide sensory "feel" information are given in reference 24.

The human operator or controller of a robot must utilize some means of signaling the control intentions to the robot. In some cases, simple buttons and switches may suffice for discrete actions (even though the robot may take complicated actions subsequently). Where "steering" actions are to be imparted, various types of joysticks, trackballs, or other hand controllers, largely developed for aircraft and more recently for remote manipulators such as in the space shuttle (25), are available. These hand controllers often attempt to combine, within the scope of one hand and its fingers, multiple function buttons for discrete controls that may be actuated, although the unit itself may have multiaxis capability. That is, the unit may have the capability to signal up–down, left–right, back–forth, and so on, all at controllable speeds, depending on how far the unit is moved in the appropriate direction.

Where continuous steering-type actions are not required, and where a computer intermediates between the human and the robot, a computer keyboard or other similar multiple-switch device is a natural choice for human input to the robot. The basic engineering type of human factors (eg, forces, sizes, spacings, etc) of switches and keyboards has been well researched (26, 27). The principle of grouping switches by function to minimize error and increase response speed (another early human factors principle) has recently received renewed emphasis for the complex switch boards of large installations, such as at nuclear power facilities (28). The same principles are applicable to robot installations, particularly where the human must control more than one robot, or to the hand controller unit for a single robot.

WORKSTATION DESIGN

Similarly to the previous issues, the physical design of the workstation for the human is a well-researched human factors area. The actual design of the workspace is dictated largely by anthropometric and biomechanical aspects, that is, by the human's own static and dynamic geometry and strength characteristics, as though the human were an articulated engineering machine that must be integrated into the confines of a generally fixed layout of table- (or console) tops, panels, displays, controls, and so on. Again, a basic human factors principle is to fit the design to the human rather than expect the human to adapt to the design. Human adaptation almost invariably comes at a price ranging from dissatisfaction through loss of productivity to possible fatalities.

Anthropometric data form the single largest collection of human factors data and are continually enlarged. Reference 29 supplies examples of such data as well as examples of proper design techniques and several applications. For workstation design, there are certain design aids ranging from flat articulated manikins in various aspects (side, front, top) that may be slid easily around a drafting board layout of the workspace (30) to elaborate three-dimensional computer models with realistic motion constraints on the joints. The entire workspace, together with a user, may be modeled with such computer aids. Several computer-based anthropometric aids have been compared (31).

In laying out the workspace, another human factors principle requires that it be suitable for the "smallest to the largest" forseeable user, which is usually interpreted to encompass the 5th percentile female to the 95th percentile male. (In at least one case, this requirement of good practice is codified as a federal regulation for the design of commercial motor vehicles (32).

It may be surprising to find that although the 5th and 95th percentile manikins (or data) are reasonably similar in proportion, the workspace designed for each need not be proportionally the same. That is, one cannot simply scale up or down the same geometric layout and expect it to fit the intended user. Each feature of the workspace must be considered in terms of the individually sized user and the user's various pertinent limbs.

In addition to the basic engineering layout of a workspace satisfying the user's geometric constraints, it is important to account for the well-being of the user. Figure 3 (33) is an example of a workstation designed to be both "well fitted" to the users as well as aesthetically pleasing. This two-person workstation is designed to allow two persons to perform remote guidance of a manipulator under camera observation but could as easily be used to guide robot manipulators.

TASK SHARING

One of the primary modes of interaction of human workers with robots, and one which will become increasingly important, is that of task sharing or task interaction. In this mode, the human and robot are both in the "manual" mode in order to accomplish some task rather than in the "supervisory" mode in which the human proposes and the robot disposes.

There are two general modes of interaction in task sharing: sequential and parallel. More complex interactions may be analyzed as some combination of these main modes.

Sequential Task Sharing

Figure 4 presents a generalized paradigm of sequential task sharing (34). This mode is marked by an alternation of task

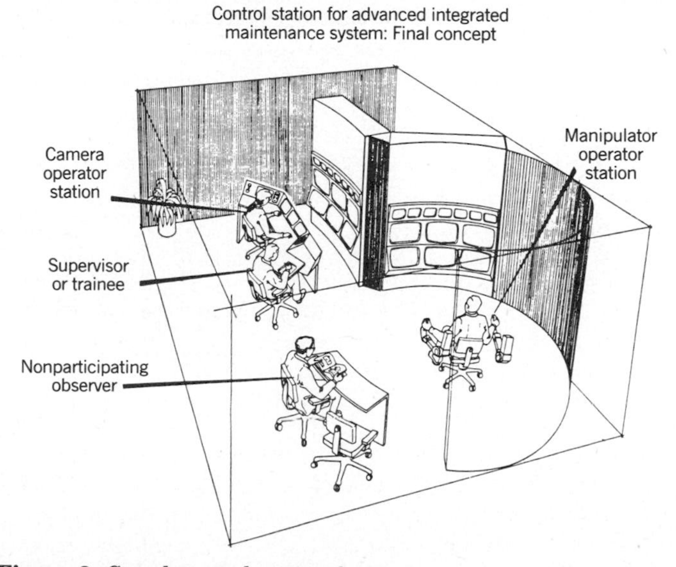

Figure 3. Complete workstation design for a remote control facility (31) (Oak Ridge National Laboratory).

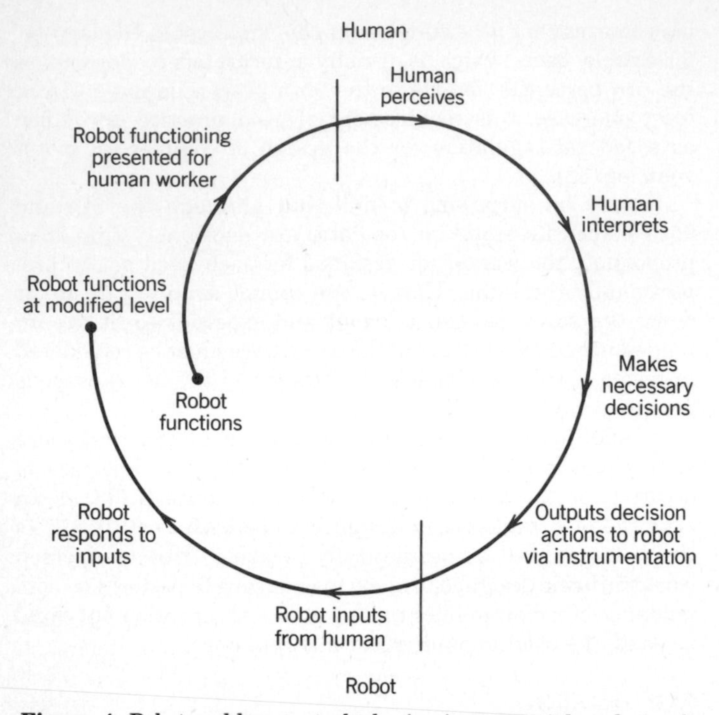

Figure 4. Robot and human task sharing in sequential mode (modified from reference 34).

elements in which the robot operates and then is essentially idle while the human operates. The human, in turn, is essentially idle while the robot operates. In sequential operation, the robot (or other machine) usually performs the tasks at which it excels, such as those demanding speed, very high accuracy, and so on, and the human performs those uniquely human functions such as pattern recognition, decision making, and so on. This mode is very common in many human–machine systems characterized as "semiautomated." Because of the usually great disparity between the tasks and often short duration of each, it is usually impossible to keep one operator profitably busy while the other is in operation. Thus, some tasks that have been proposed for semiautomation actually turn out to take longer than when they are performed entirely by a human worker. Tasks being considered for semiautomation (ie, with a human in the loop) in order to realize time savings should be carefully analyzed first to see if the proposed time savings will materialize. This analysis may be performed using simulation as well as analytical techniques familiar in human factors engineering such as SAINT (35), which is a computer-based time-line simulation program.

Parallel task sharing forms the alternative to sequential task sharing. Figure 5 presents this concept pictorially. In parallel task sharing, both robot and human operate in a continually busy mode while working on the same job. There is little idle time for either. Although there are many examples on assembly lines of parallel operation between two humans, such as jointly lifting, positioning, and then bolting one assembly piece to another, there are fewer such examples so far in human–robot interactions. This is probably because of the great disparity between them in nearly all respects: sensory, physical, dynamic, kinematic, in strength, intelligence, and so on. The major form of parallel task sharing between human and robot presently seems to be of the type where the interaction is not critical. This is the case, for example, where the human aligns a batch of pieces while the robot grasps and further manipulates them, or where the human steers the robot while it performs some task not directly dependent on the steering, as when the human steers a robot arm clear of obstructions while it grasps a retrieved part that has fallen in, say, some highly hazardous environment (36).

Most human–robot task sharing (and much semiautomation) seems to be of the sequential type, although this is more due to an inability to balance the capabilities of the two operators (human and robot) than desirability. Part of this inability to balance the capabilities of each is attributable to the effects of *work pacing* when planning tasks for humans and for robots.

WORK PACING

There are two general modes of work pacing for humans: *self-pacing* and *forced pacing*. In general, humans work better under self-pacing.

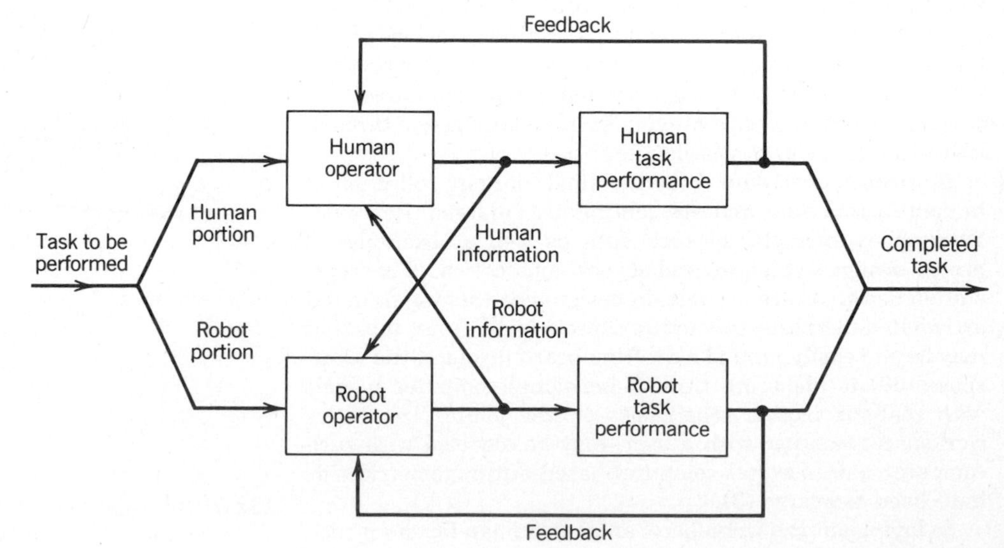

Figure 5. Human and robot in a parallel task-sharing mode.

Self-Pacing

When an operator is allowed to work at a rate and rhythm of his/her own choosing, the work pace is said to be self-paced. This occurs when, for example, a new task element or job is not undertaken until the worker completes the previous one and there is no time pressure except that which is self-imposed to complete it. The worker can adjust the work pace to suit the immediate needs. A worker cutting material in a garment factory may work in a self-paced mode, although piecework wages often introduce a (self-imposed) type of forced pacing. This mode is often more "humane" in that the human worker can adjust the work to personal considerations, which may change with time over the course of the work cycle, and may include pausing to talk, pausing to retrieve some tool, speeding up to work off a backlog, and so on. Self-pacing follows the sound human factors principle of adjusting the work to the worker.

Forced Pacing

The opposite of self-pacing is forced pacing, in which the worker works at a (rhythmic) rate, usually undeviating and set by external (to the worker) conditions, as in assembly line work in which a new piece arrives every X seconds without deviation and without sensitivity to the worker's immediate conditions. (The classic Chaplin movie "Modern Times" illustrates this type of pacing in the extreme.) There is often the connotation of "sweatshop" conditions associated with forced pacing. However, with successful labor–management relations, the pace of a forced-pace job can be set at some humane and productive level. The pace of the assembly line in an automotive plant is often a critical issue in labor–management negotiations.

Naturally, every actual job is some mixture of both types of pacing, although whenever a human worker must work in an automated environment with or without robot workers, there is a decided likelihood of encountering forced-pace work conditions just because this approach is natural for machine-dominated work planning. Rhythmic work at an unvarying pace over long periods is natural for machines, programs, and planning work, but it is much less natural for humans. When these types of demands are imposed on humans, problems of one sort or another can be expected to develop, leading to less than the desired productivity or safety. One of the difficulties in task sharing between robots and humans (in addition to those mentioned previously) is in matching these two distinctly different modes of pacing. The robot can work in an undeviating, rhythmic mode for an arbitrarily long duration without rest or missing a beat. The human cannot. When the two are integrated, the unfortunate result is usually a demand that the human worker adapt to the (robot's) work place rather than a plan in which both can work in their natural modes. A decided challange in work planning is this integration of the two different pacing modes in the same work.

Results from elementary queueing theory suggest some of the difficulties of this integration. Imagine a robot and a human working side by side doing identical tasks that consist simply of picking a cylindrical piece from a storage bin of arriving pieces and placing it in a hole in a rotating carrier in another location. The carrier indexes whenever the insertion is completed. In both cases, the pieces arrive at each storage bin at random intervals rather than in a regular fashion. Further assume that both human and robot workers "pick and place" the same number of parts per minute on the average, but that the robot never deviates from this rate whereas the human fluctuates around this rate in a random fashion. Under these simple conditions, at any given time there will be more waiting parts in the human's holding bin than in the robot's even though they both pick and place at the same average rate. Furthermore, even if the parts arrive at an undeviating rate, this will still be true.

There is another fundamental difference between humans and robots that complicates interactive task sharing and that can be exemplified by the same example above. The time required by a robot to pick a peg from a holder and place it into a hole in a carrier can be decomposed into times for the three sequential task elements: pick–transport–place. The time for each of these three elements is essentially independent of the others in the sense that the time required to transport the peg is not dependent on the time required to pick or place the peg but rather simply on the velocity and distance of the move. Thus, the time to transport and place the peg for a robot would be

$$\text{time} = \text{time (transport)} + \text{time (place)}$$

where

time (transport) = distance/average velocity
time (place) = proportional to tightness of fit between hole and peg

On the other hand, when a human is performing such a task rapidly, the total time is not decomposable into completely independent times. The time to transport and place the peg depends upon the distance of the move and inversely on the tolerance of the fit between peg and hole according to a version of the so-called Fitts law (37):

$$\text{time} = A + B \times \log(\text{distance}/\{\text{hole diameter} - \text{peg diameter}\})$$

where the tightness of fit is measured by the difference between the hole and peg diameters such that the greater the difference, the faster the total time. For humans, it appears that the movement (transport) time is influenced by the final destination tolerances of placement.

Thus, the human and the robot workers have different movement laws, so that integrating their work cycles must be carefully considered, particularly since at present the robot worker is not sensitive to human presence or subtleties in behaviors, which means that the human will need to do the "adapting" to the robots.

Work Planning

It is not a given that robots and humans will necessarily coexist to perform tasks jointly. The need for coexistence will arise when no other way can be found to perform the required tasks safely, economically, or otherwise efficiently using only automation (including robots) or only human workers (38). In fact, it has happened in the past that the limited capabilities of robots (compared to humans) has required certain assembly tasks to be carefully studied and redesigned in order that a robot could perform it, but the paradoxical result was that after the task redesign, humans could perform it even better or more productively. The often unconscious (and sometimes misplaced) dependence on human adaptability, intelligence, and trainability largely causes the task and tool redesign approach to achieving improved work to be slighted.

Notwithstanding the above comments, where humans and robots must work together, certain basic planning must be performed.

Task Analysis

A careful and detailed study must first be made of the task elements (or subtasks) together with the presumed times necessary or desired for accomplishing each task element. The sequential and parallel characteristics of the overall task must be defined and expressed in some time-line flow form so that the organization is clear. The control mode (robot-programmed or human operator) can then be assigned based on the feasibility of each assignment. However, practically, tasks will be assigned to humans by "default" when a programmed method of performance is not feasible because of cost, technical feasibility, and so on. It is also the case that the mode of interaction will usually be sequential (Fig. 5), although if the work can be planned to be executed in parallel fashion (human and robot active simultaneously), it can be accomplished more efficiently.

Figure 6 and Table 1 are an example of "human-in-the-loop" involvement in a proposed remote handling control station for nuclear waste cask handling (34). The overall concept is shown in Figure 7, in which the task times and sequences given in Figure 6 pertain to Area B. In this area, a large jib crane and robotic arm are used to carry out the tasks and subtasks listed in Figure 6. The human operator's function is mainly surveillance of the system's preprogrammed operation. However, operators will remotely enter the control loop for six specific functions that are not accomplished by the robot in preprogrammed modes.

Human intervention into robot operation is part of a planned sequence, to meet unplanned contingencies (eg,

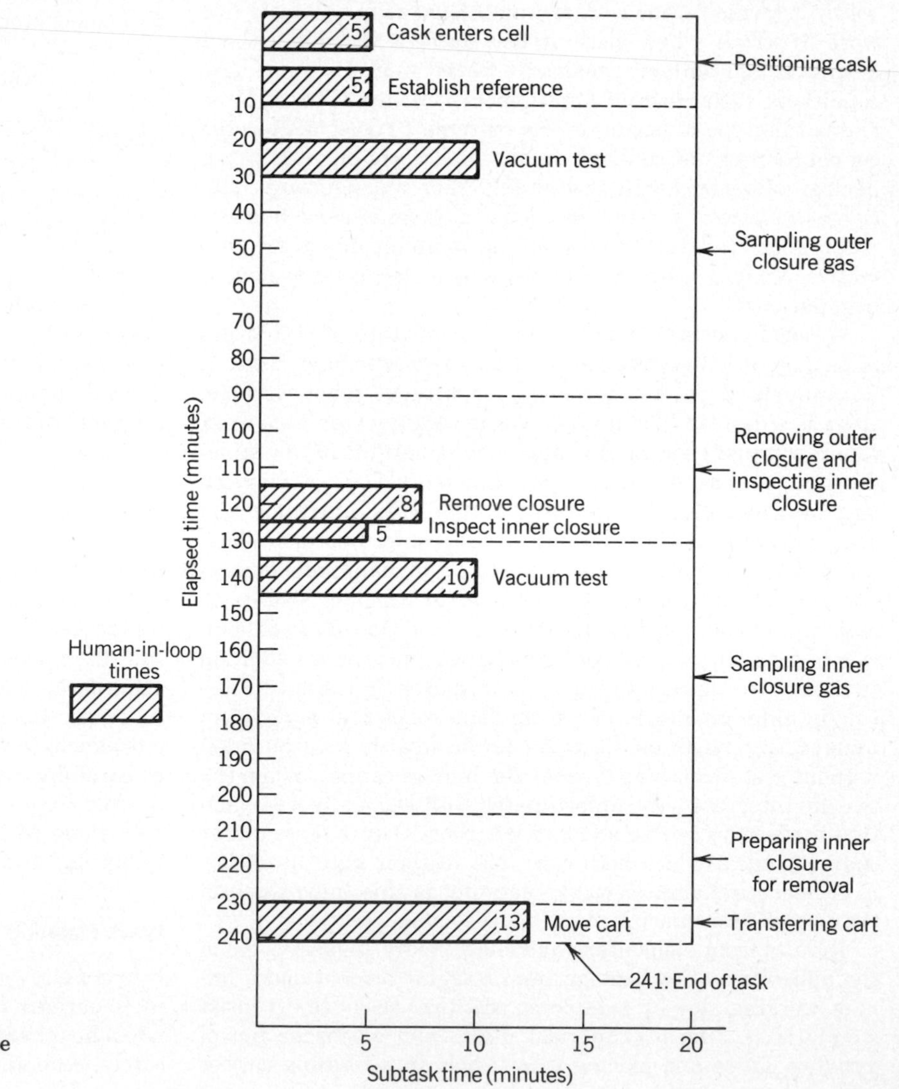

Figure 6. Human-in-the loop robot and remote waste handling (36).

Table 1. Subtasks and Normal Control Modes for Area B

Subtask	Task time, min	Elapsed time, min	Control mode
1. Position cask cart on arrival from			
Area A	5	5	Operator
Attach wrench	2	7	Programmed
Use robot to establish a datum	5	12	Operator
2. Sample gas from outer closure			
Remove gas seal plug	4	16	Programmed
Install gas sample adapter	4	20	Programmed
Vacuum testing seal	10	30	Operator
Loosen gas sample plug	4	34	Programmed
Take sample	44	78	Programmed
Close gas sample plug	4	82	Programmed
Remove gas sample adapter	4	86	Programmed
Replace gas seal plug	4	90	Programmed
3. Remove outer closure			
Prepare outer closure for removal	20	100	Programmed
Position lifting fixture	5	115	Programmed
Position and use crane to remove closure	8	123	Operator
Use camera to inspect inner closure	5	128	Operator
4. Sample gas from inner closure	78	206	
Remove gas seal plus	4	16	Programmed
Install gas sample adapter	4	20	Programmed
Vacuum testing seal	10	30	Operator
Loosen gas sample plug	4	34	Programmed
Take sample	44	78	Programmed
Close gas sample plug	4	82	Programmed
Remove gas sample adapter	4	86	Programmed
Replace gas seal plug	4	90	Programmed
5. Prepare inner closure for removal	22	228	Programmed
6. Transfer cart to unloading cell	13	241	Operator

dropped tool or piece), or unplanned and undesired (accident). In order to make human intervention effective (and safe), particularly if it is to be a rare event, the operator must be kept "in the loop," at least cognitively, throughout every step of the operation. Therefore, a CRT or other interface is needed to apprise the operator of the *past, present,* and *next* major steps of the operation at a glance. This is based on human factors experience that human performance is generally improved if a history, present status, and preview of inputs (and next action) are appropriately displayed. Where possible, information must be integrated in a meaningful whole, rather than depending on the operator to integrate it. Reference 36 may be consulted for guidelines to designing the human–computer interface for such a purpose.

SAFETY IN THE ROBOTIC WORKPLACE

The safety of the worker in any workplace must be of paramount importance. Mechanization in the workplace brought with it machines that posed hazards to the worker. As a result, there is now a considerable body of literature pertaining to safety design, warnings, and instructions that extends over at least 100 years (39). For example, the first patent on a machine guard was granted in 1886. It was recognized that the machine and workplace designer must take into account that human workers have definite limitations and peculiarly human characteristics (no matter how undesirable in the workplace they may be). That is, the human cannot be assumed to be perfect any more than material can be assumed by a designer to be perfect. The realistic characteristics must be part of the initial design considerations. Humans make mistakes, can be distracted, misjudge their own abilities, forget, tire, stumble, and so on. They are quite unrobotlike.

Robots Versus Traditional Automation

Robotization of the workplace brings several new opportunities for injury to the worker (38).

Robot Danger Zones. Robots are designed to have great flexibility in spatial movement in order to be flexibly programmable for different tasks. As a result, workers often fail to make an instant judgment of the danger zone extent and may approach it too closely, particularly when the robot is "waiting" but looks to be inoperative. This is compounded by the fact that the robot reach (hazardous area) is not limited to its front.

Teaching. The majority of robots need their movements to be preprogrammed. This often requires that the worker be

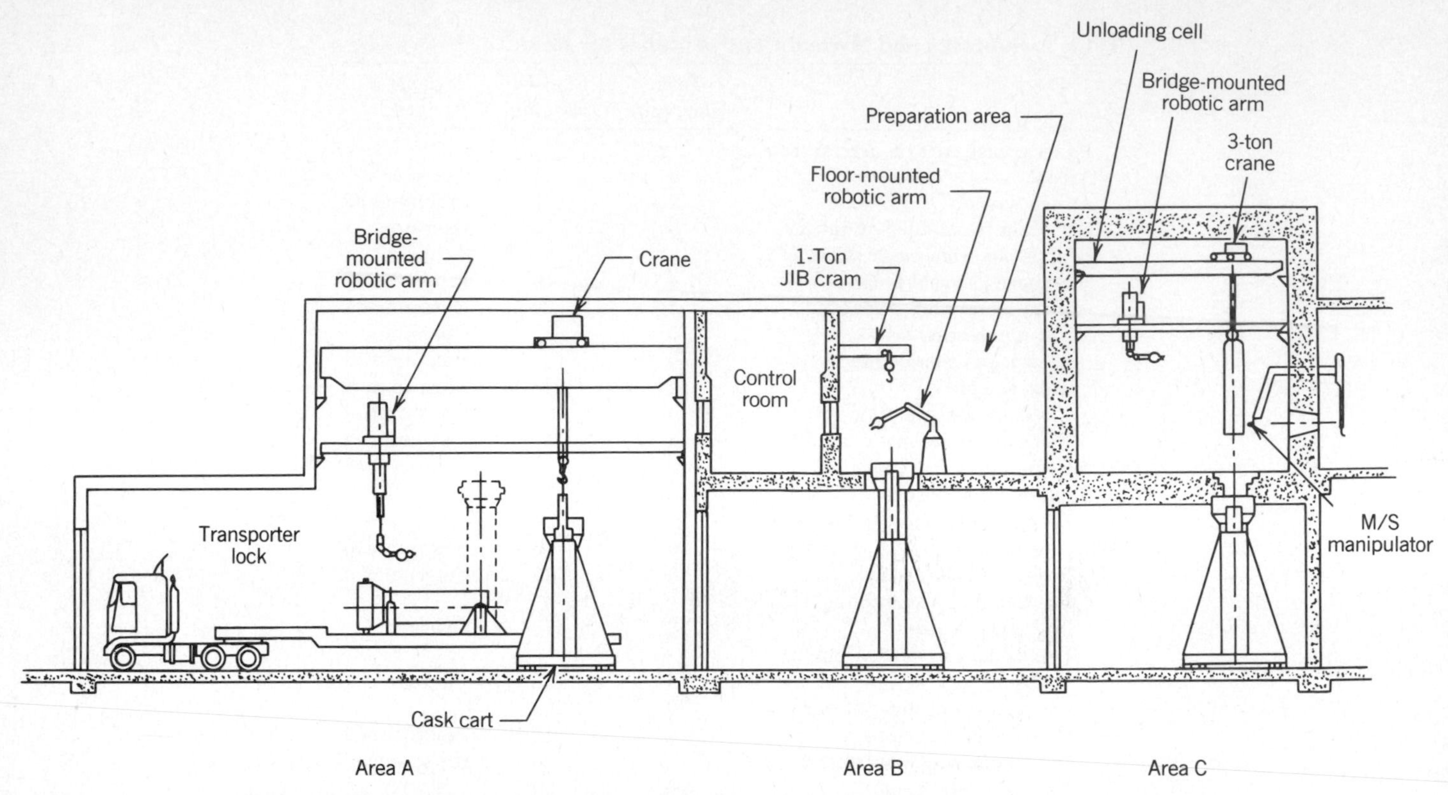

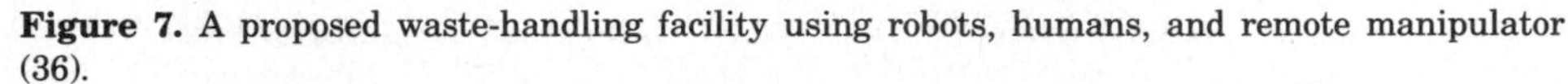

Figure 7. A proposed waste-handling facility using robots, humans, and remote manipulator (36).

physically near the robot and within its hazardous reach area while the power is on.

Malfunctions and Control Errors. "Noise" or undesired inputs into the system can cause the movements of the robot to be different from the preprogrammed ones or can otherwise cause it to deviate from its expected movement or schedule.

Human Characteristics. Workers may approach a robot too closely because:

1. The danger zone is invisible.
2. They trust the design to limit the robot's motion.
3. The anthropomorpic ("R2D2") structure may foster a misplaced trust.

There have been a number of safety devices and practices devised over the years for traditional workplaces and machines that are applicable to robotic environments. Specific devices and practices are recommended in reference 40:

1. Robots should be installed with flexible stopping devices that physically limit their range of motion to the specific task. This helps remove the possibility of undesired motion into an unexpected space.
2. The robot area should be painted to distinguish it from safe areas as a visible reminder to the worker and others (eg, visitors and sightseers).
3. The robot area should be enclosed by an iron fence to prevent intruders physically.
4. The fence gate should be interlocked so that opening it renders the robot inoperative.
5. A presence-sensing device (eg, a safety mat on the floor of the hazardous area) should be installed to render the robot inoperative if the hazard zone is intruded.
6. "Start" buttons that power up the robot should be distant from the workpiece.
7. Any workpiece (eg, "jig") attached to the robot arm should be able to be repaired outside the fence, that is, outside or at the limit of the robot's reach.

Good human factors practice requires that any "safety" device or procedure be designed based on actual human characteristics and that it be "fail safe." Devices that hinder the worker and/or reduce productivity are candidates for being bypassed if possible.

Teaching and Maintenance

Because workers and maintenance personnel must often be physically near the robot when it is powered, special procedures are recommended for ensuring the safety of the personnel (40) (see also SAFETY, OF OPERATORS):

1. The worker engaged in teaching should turn a switch on the robot controller to "teach" and hang a name tag on the switch to give others an indication of the robot mode and the identity of the person near the robot. This is similar to a "man aloft" sign.

2. The worker should remove a safety plug at the fence door when going inside. The robot should move at a slower movement for teaching speed when a safety plug is inserted into it.
3. Another worker should observe the first worker's teaching behavior from outside the robot working area and assist him in the teaching task—the "buddy system."
4. An *emergency stop* button should be installed in the teaching console, and the design of the teaching console should be the same for different makes to reduce the opportunity for errors. Standardization can greatly help promote safety in the workplace, just as standardization of automobile (and aircraft) controls and operation permits easy driver transfer from one vehicle to another.
5. The buttons on the teaching console should be of the nondefeatable "dead man" type so that in the event of an accident the robot power will be removed. This "dead man" principle should be followed elsewhere where appropriate.
6. A clear signal should indicate that the robot is in the "wait" mode. During some portion of its work cycle, the robot may be quiescently waiting for further inputs to reinitiate movement—inputs invisible to the worker. This may mislead the worker into assuming that it is safe to approach the robot.

There are three simple rules for coexisting with robots:

1. *Stay out of reach and throwing range of a robot.* Not only can the robot arm swing about in a flexible manner, it is also possible that an object held in the robot "hand" can slip out of its grasp while it is swinging, thus becoming a lethal flying object.
2. *Never turn your back on a robot.* Two reported deaths happened when a robot, in each case, struck the worker in the back.
3. If you must approach a robot, *kill it before it kills you.* That is, render it inoperative by removing (or restoring) its power from a safe position.

SUMMARY

Thus, in summary, there exists a long history of well-founded human factors design principles and experiences directly applicable to human–robot interaction design on a variety of bases. However, the ultimate procedure, after the design is conceptualized and before it is implemented, is to "test, test, and test again" until the performance specifications are assured.

This article does not purport to be a manual for designing robot–human work or workspaces. Rather, it is an attempt to alert the reader to the contributions that human factors engineering can and has made to this topic as well as to some of the important human factors considerations necessary to ensure a safe, efficient, and satisfying human–robot relationship.

BIBLIOGRAPHY

1. E. McCormick and M. Sanders, *Human Factors in Engineering and Design,* McGraw-Hill, New York, 1982.
2. H. Parsons and G. Kearsley, "Robotics and Human Factors: Current Status and Future Prospects," *Human Factors* **24**(5), 535–552 (1982).
3. H. P. Birmingham and F. V. Taylor, "A Design Philosophy for Man–Machine Control Systems," *Proceedings of the IRE* **XLII**(12), (1954).
4. W. R. Ferrell, "Remote Manipulation with Transmission Delay," *IEEE Transactions on Human Factors in Electronics* **HFE-6**(1), 24–32 (1965)
5. T. Sheridan and W. Ferrell, *Man–Machine Systems,* M.I.T. Press, Cambridge, Mass., 1974.
6. J. Kreifeldt and P. Hill, "Toward a Theory of Man Tool System Design: Application to the Consumer Product Area," *Proceedings of the Human Factors Society 18th Annual Meeting,* 1974.
7. S. Konz, *Work Design,* Grid Publishing Inc., Columbus, Oh., 1979.
8. L. Greenberg and D. Chaffin, *Workers and Their Tools: A Guide to the Ergonomic Design of Hand Tools and Small Presses,* Pendell Publishing Co., Midland, Mich., 1980.
9. J. Kaufman and H. Haynes, *IES Lighting Handbook, Reference Volume 1981,* Illuminating Engineering Society of North America, New York, 1981.
10. B. J. Underwood, *Experimental Psychology,* Meredith Publishing Co., 1966.
11. M. Clarke, J. Garin, and P.-A. Andrea, "Development of a Standard Methodology for Optimization of Remote Visual Displays for Nuclear Maintenance Tasks," *Proceedings of the Human Factors Society 25th Annual Meeting,* 1981, 454–456.
12. E. Eggleton, *Ergonomic Design for People at Work, Vol. 1.,* Lifetime Learning Publications, Belmont, Calif., 1983.
13. E. Tichauer, "Ergonomic Aspects of Biomechanics," in *The Industrial Environment—Its Evaluation and Control,* National Institute for Occupational Safety and Health, 1973.
14. 20CFR 1910.95 Title 29, *Code of Federal Regulations,* Pt 1910, Section 95, "Occupational Noise Exposure."
15. C. R. Asfahl, *Industrial Safety and Health Management,* Prentice-Hall, Inc. Englewood Cliffs, N.J., 1984.
16. F. Geldard, *The Human Senses,* John Wiley & Sons, Inc., New York, 1972.
17. K. Kryter, "Speech Communication," in Vancott and R. Kinkade, eds., *Human Engineering Guide to Equipment Design,* U.S. Government Printing Office, Washington, D.C., 1972.
18. J. Parker, Jr. and V. West, *Bioastronautics Data Book,* 2nd ed., NASA, SP-3006, 1972.
19. P. Fanger, *Thermal Comfort,* Danish Technical Press, Copenhagen, Denmark, 1970.
20. W. Simcox, "Cognitive Considerations in the Design of Graphic Representations," Ph.D. Dissertation, Department of Engineering Design, Tufts University, Medford, Mass., 1983.
21. S. Roscoe, *Aviation Psychology,* The Iowa State University Press, Ames, 1980.
22. J. Kreifeldt and T. Wempe, "Pilot Performance During a Simulated Standard Instrument Turn With and Without a Predictor Display," NASA, TMX-62,201, NASA-Ames Research Ctr., Moffett Field, Calif., 1973.
23. W. Simcox, and J. Kreifeldt, "Prosthetic EMG Control Enhancement Through Application of Man–Machine Principles," *Proceedings of the 13th Annual Conference on Manual Control,* M.I.T., Cambridge, Mass., 1977.

24. A. Bejczy and G. Paine, "Displays for Supervisory Control of Manipulators," presented at the 13th Annual Conference on Manual Control, M.I.T., Cambridge, Mass., 1977.
25. A. Lippay, "Multi-axis Hand Controller for the Shuttle Remote Manipulator System," *Proceedings of the 13th Annual Conference on Manual Control,* M.I.T., Cambridge, Mass., 1977.
26. D. Alden, R. Daniels, and A. Kanarick, "Keyboard Design and operation: A Review of Major Issues," *Human Factors* **14**(4), 275–293 (1972).
27. D. Lang, "Ergonomic Aspects of Keyboards: A Summary of Current Literature on Ergonomic Standards and Recommendations," for GenRad, Inc., Concord, Mass., Internal Rep., 1983.
28. R. Kinkade and J. Anderson, *Human Factors Guide for Nuclear, Power Plant Control Room Development,* EPRI NP-3659, Research Project 1637–1, 1984.
29. Roebuck, Jr., K. Kroemer, and W. Thomson, *Engineering Anthropometry Methods,* John Wiley & Sons, Inc., New York, 1975.
30. S. P. Rodgers, *Anthropometric Data Application Mannikin,* Santa Barbara, Calif., 1976.
31. M. Dooley, "Anthropometric Modelling Programs—A Survey" *IEEE Computer Graphics and Applications* **2**(9), (Nov. 1982).
32. Title 49, Federal Motor Carrier Safety Regulations, Section 399.205, Transportation, Chapt. III, Federal Highway Administration, 1981.
33. M. Clarke and J. Kreifeldt, "A Control Room Concept for Remote Maintenance in High Radiation Areas," *Proceedings of the Human Factors Society 28th Annual Meeting,* 1984, pp. 230–233.
34. A. Zeller, "The Slow Speed Demon, Part I," *Flying Safety Magazine* **15**(2), 16–20 (1959).
35. G. Chubb, "Saint, A Digital Simulation Language for the Study of Manned Systems," in J. Moraal and F.-K. Kraiss, eds., *Manned System Design,* Plenum Press, New York, 1980.
36. M. Clarke, J. Kreifeldt, and J. Draper, "Man–Machine Interface for a Nuclear Cask Remote Handling Control Station," U.S. Department of Energy, Westinghouse Hanford Co., HEDL-7465, TTC-0499, July 9, 1984.
37. P. Fitts and J. Peterson, "Information Capacity of Discrete Motor Responses," *Journal of Experimental Psychology* **67,** 103–112 (1964).
38. W. Seering, "Who Said Robots Should Work Like People?" M.I.T. *Technology Review* 5–13 (Spring 1986).
39. G. Marshall, *Safety Engineering,* Brooks/Cole Engineering Div., 1982.
40. M. Nagamachi, "Human Factors of Industrial Robots and Robot Safety Management in Japan," *Applied Ergonomics* **17**(1), 9–18 (1986).

ERGONOMICS, ROBOT SELECTION

ROBERT M. WYGANT
Western Michigan University
Kalamazoo, Michigan

INTRODUCTION

More companies are beginning to realize that industrial robots are capable of performing many of the functions that have traditionally been done by human operators. However, one of the problems facing the purchaser or designer of robots is how to determine when a robot should be used to perform a specified task. A second problem arises when the analyst is forced to select a particular robot model from the hundreds currently available.

Traditionally, the field of ergonomics, often referred to as human factors, has been used in the design and analysis of work tasks and equipment to optimize the relationship between the human operator and the work environment. An integral part of ergonomics is the study of human capabilities and limitations. If the robot is to replace the human operator in the work system, it is imperative that the work analyst also understand the capabilities and limitations of robots and apply this information in the same way ergonomics is used in designing for human use (see ROBOTS, WORK PLANNING; ROBOTS, WORKPLACE DESIGN).

One of the difficulties in comparing humans and robots is the lack of standardization throughout the robotics industry and the availability of an easy-to-use data base of robot specifications. The need for a standardized method to compare robot systems which can be used to assist engineers in robot applications is identified in Ref. 1; mechanical robot components, electric drive systems, and controller parameters are reviewed. After comparison charts are designed with all of these categories and parameters, a subjective evaluation can be made using a grading scale to compare all of the selected robot systems. The method of grading is left up to the user. No recommendations are made on how robot specifications are to be standardized.

Researchers at New Jersey Institute (2) have developed a computerized robotics data base and retrieval system. Information is available on vertical and horizontal reach, maximum load, memory and address capability, programmable steps, repeatability, degrees of freedom, cost and power considerations, end-effectors, coordinate system, type of control, and typical applications. There are 220 robots from 90 manufacturers included in the data base.

A commercially available robotic directory called *ROBOT-CALC 1* was recently introduced by Robot Calc, Inc. (3). It is a computerized data-base directory of industrial robots, manufacturers, robotic terminology, and criteria for justification. It was originally designed by robotics engineers to aid robot users in their robot search. Robot selection is based on user-defined requirements, such as welding applications, painting/finishing, materials handling, machining processes, reach, drive, axis, payload, and repeatability.

A decision model for selecting alternative robotic systems has been presented (4). The model consists of two phases: (*1*) a technology-needs identification phase in which task requirements are determined, and (*2*) a needs-to-availability matching phase in which technology needs identified in the first phase are matched to available market technologies. Two modeling approaches are discussed as possible mechanisms for the decision model: integer programming and a rule-based expert system.

A robot selection model was developed at Virginia Polytechnic Institute (5) to illustrate some of the tangibles and intangibles that should be considered when selecting a robot, and to assist the user in making an appropriate decision, based on the firm's objectives, preferences, and constraints. An interactive computer program consists of two parts. The first part uses specifications defined by the user to select a feasible set from a data base of commercially available robots. The second part of the program assists the user in determining which particular robot should be selected by using a linear multi-attribute decision model. The user must enter the desired attributes and rate each attribute's importance on a scale of one to ten or may rank each attribute and make pair-wise trade-off decisions.

This article compares the major physical features of humans and robots which are important in the evaluation and selection of robots. A model for robot performance is presented and the methods for predicting the time a robot takes to complete a task are reviewed.

ROBOT–MAN CHARTS

A method to assist engineers and analysts in the selection of humans vs robots in performing an industrial task has been developed (6). This method is in the form of Robot–Man Charts, which contain three main types of work characteristics (7):

1. Physical skills and characteristics, including manipulation, body dimensions, strength and power, consistency, overload/underload performance, and environmental constraints. Table 1**a** provides details of this category. Typical ranges of maximum motion capabilities (TRMM) are given for several categories of body movement and speed, and arm and wrist motions.
2. Mental and communicative characteristics: the Robot–Man Charts contain mental and communicative system attributes for robots and humans (Table 1**b**).
3. Energy considerations: a comparison of representative values of energy-related characteristics, such as power requirements and energy efficiency, for robots and humans, is given in Table 1**c**.

GENERAL

One of the areas in which a human has a decided advantage over a robot is sensing. Humans are capable of sensing a tremendous amount of information. It is estimated that a human

Table 1. Robot–Man Charts[a]

Characteristics	Robot	Human
	a. *Comparison of Robot and Human Physical Skills and Characteristics*	
1. Manipulation		
A. Body	a. One of four types: Uni- or multiprismatic Uni- or multirevolute Combined revolute/prismatic Mobile	a. A mobile carrier (feet) combined with 3-DF[b] wristlike (roll, pitch, yaw) capability at waist.
	b. Typical maximum movement and velocity capabilities: Right–left traverse 5–18 m at 500–1,200 mm/s Out–in traverse 3–15 m at 500–1,200 mm/s	b. Examples[c] of waist movement: Roll: ≃ 180° Pitch: ≃ 150° Yaw: ≃ 90°
B. Arm	a. One of four primary types: Rectangular Cylindrical Spherical Articulated	a. Articulated arm comprised of shoulder and elbow revolute joints
	b. One or more arms, with incremental usefulness per each additional arm	b. Two arms, cannot operate independently (at least not totally)
	c. Typical maximum movement and velocity capabilities: Out–in traverse 300–3,000 mm 100–4,500 mm/s Right–left traverse 100–6,000 mm 100–1,500 mm/s Up–down traverse 50–4,800 mm 50–5,000 mm/s Right–left rotation 50–380°[d] 5–240°/s Up–down rotation 25–330° 10–170°/s	c. Examples of typical movement and velocity parameters: Maximum velocity: 1,500 mm/s in linear movement Average standing lateral reach: 625 mm Right–left traverse range: 432–876 mm Up–down traverse range: 1,016–1,828 mm Right–left rotation (horizontal arm) range: 165–225° Average up–down rotation: 249°

Table 1. (*continued*)

Characteristics	Robot	Human
C. Wrist	a. One of three types: Prismatic Revolute Combined prismatic/revolute Commonly, wrists have 1–3 rotational DF[b]: roll, pitch, yaw; however, an example of right–left and up–down traverse was observed	a. Consists of 3 rotational DF[b]: roll, pitch, yaw
	b. Typical maximum movement and velocity capabilities: Roll 100–575°[e] 35–600°/s Pitch 40–360° 30–320°/s Yaw 100–530° 30–300°/s Right–left traverse (uncommon) 1,000 mm 4,800 mm/s Up–down traverse (uncommon) 150 mm 400 mm/s	b. Examples of movement capabilities: Roll: ≃ 180° Pitch: ≃ 180° Yaw: ≃ 90°
D. End-effector	a. The robot is affixed with either a hand or a tool at the end of the wrist. The end-effector can be complex enough to be considered a small manipulator in itself	a. Consists of essentially 4 DF[b] in an articulated configuration; five fingers per arm each have three pitch revolute and one yaw revolute joints
	b. Can be designed to various dimensions	b. Typical hand dimensions: Length: 163–208 mm Breadth: 68–97 mm (at thumb) Depth: 20–33 mm (at metacarpal)
2. Body dimensions	a. Main body: Height: 0.10–2.0 m Length (arm): 0.2–2.0 m Width: 0.1–1.5 m Weight: 5–8,000 kg	a. Main body (typical adult): Height: 1.5–1.9 m Length (arm): 754–947 mm Width: 478–579 mm Weight: 45–100 kg
	b. Floor area required: from none for ceiling-mounted models to several square meters for large models	b. Typically about 1 m^2 working radius
3. Strength and power	a. 0.1–1,000 kg of useful load during operation at normal speed; reduced at above normal speeds	a. Maximum arm load: <30 kg; varies drastically with type of movement, direction of load, etc
	b. Power relative to useful load	b. Power: 2 hp ≃ 10 s 0.5 hp ≃ 120 s 0.2 hp ≃ continuous 5 kc/min subject to fatigue; may differ between static and dynamic conditions
4. Consistency	Absolute consistency if no malfunctions	a. Low b. May improve with practice and redundant knowledge of results c. Subject to fatigue: physiological and psychological d. May require external monitoring of performance
5. Overload/underload performance	a. Constant performance up to a designed limit and then a drastic failure b. No underload effects on performance	a. Performance declines smoothly under a failure b. Boredom under local effects is significant
6. Environmental constraints	a. Ambient temperature from −10 to 60°C b. Relative humidity up to 90%	a. Ambient temperature range 15–30° C b. Humidity effects are weak

Table 1. (*continued*)

Characteristics	Robot	Human
	b. Comparison of Robot and Human Mental and Communicative Skills	
1. Computational capability	a. Fast, eg, up to 10 Kbits/s for a small minicomputer control	a. Slow, eg, 5 bits/s
	b. Not affected by meaning and connotation of signals	b. Affected by meaning and connotation of signals
	c. No evaluation of quality of information unless provided by program	c. Evaluates reliability of information
	d. Error detection depends on program	d. Good error detection correction at cost of redundancy
	e. Very good computational and algorithmic capability by computer	e. Heuristic rather than algorithmic
	f. Negligible time lag	f. Time lags increased, 1–3 s
	g. Ability to accept information is very high, limited only by the channel rate	g. Limited ability to accept information (10–20 bits/s)
	h. Good ability to select and execute responses	h. Very limited response selection/execution (1/s); responses may be "grouped" with practice
	i. No compatability limitations	i. Subject to various compatibility effects
	j. If programmable, not difficult to reprogram	j. Difficult to program
	k. Random program selection can be provided	k. Various sequence/transfer effects
	l. Command repertoire limited by computer compiler or control scheme	l. Command repertoire limited to experience and training
2. Memory	a. Memory capability from 20 commands to 2,000 commands, and can be extended by secondary memory such as cassettes	a. No indication of capacity limitations
	b. Memory partitioning can be used to improve efficiency	b. Not applicable
	c. Can forget completely but only on command	c. Directed forgetting very limited
	d. "Skills" must be specified in programs	d. Memory contains basic skills accumulated by experience
		e. Slow storage access/retrieval
		f. Very limited working register: ≃ 5 items
3. Intelligence	a. No judgment ability of unanticipated events	a. Can use judgment to deal with unpredicted problems
	b. Decision making limited by computer program	b. Can aniticipate problems
4. Reasoning	a. Good deductive capability, poor inductive capability	a. Inductive
	b. Limited to the programming ability of the human programmer	b. Not applicable
5. Signal processing	a. Up to 24 input/output channels, and can be increased, multiasking can be provided	a. Single channel, can switch between tasks
	b. Limited by refractory period (recovery from signal interrupt)	b. Refractory period up to 0.3 s
6. Brain–muscle combination	a. Combinations of large, medium, and small "muscles" with various size memory, velocity and path control, and computer control can be designed	a. Fixed arrangement
7. Training	a. Requires training through teaching and programming by an experienced human	a. Requires human teacher or materials developed by humans
	b. Training does not have to be individualized	b. Usually individualized is best
	c. No need to retrain once the program taught is correct	c. Retraining often needed owing to forgetting
	d. Immediate transfer of skills ("zeroing") can be provided	d. Zeroing usually not possible
8. Social and psychological needs	a. None	a. Emotional sensitivity to task structure—simplified/enriched; whole/part
		b. Social value effects

Table 1. (*continued*)

Characteristics	Robot	Human
9. Sensing	a. Limited range can be optimized over the relevant needs	a. Very wide range of operation (10^{12} units)
	b. Can be designed to be relatively constant over the designed range	b. Logarithmic: Vision: 1. Visual-angle threshold: 0.7 min 2. Brightness threshold: 4.1 μμℓ 3. Response rate for successive stimuli ≃ 0.1 s Audition: 1. Threshold: 0.002 hyn/m^2 Tactile: 1. Threshold: 3 g/mm^2
	c. The set of sensed characteristics can be selected; main senses are vision and tactile (touch)	c. Limited set of characteristics can be sensed
	d. Signal interference ("noise") may create a problem	d. Good noise immunity (built-in filters)
	e. Very good absolute judgment can be applied	e. Very poor absolute judgment (5–10 items)
	f. Comparative judgment limited by program	f. Very good comparative judgment
10. Interoperator communication	Very efficient and fast intermachine communication can be provided	Sensitive to many problems, eg, misunderstanding
11. Reaction speed	Ranges from long to negligible delay from receipt of signal to start of movement	Reaction speed ¼–⅓ s
12. Self-diagnosis	Self-diagnosis for adjustment and maintenance can be provided	Self-diagnosis may know when efficiency is low
13. Individual differences	Only if designed to be different	100–150% variation may be expected
	c. Comparison of Robot and Human Energy Considerations	
1. Power requirements	Power source 220/440 V, 3 phase, 50/60 Hz, 0.5–30 KVA; limited portability	Power (energy) source is food
2. Utilities	Hydraulic pressure: 30–200 kg/cm^2 Compressed air: 4–6 kg/cm^2	Air: oxygen consumption 2–9 L/min
3. Fatigue, downtime, and life expectancy	a. No fatigue during periods between maintenance	a. Within power ratings, primarily cognitive fatigue (20% in first 2 h; logarithmic decline
	b. Preventive maintenance required periodically	b. Needs daily rest, vacation
	c. Expected usefulness of 40,000 h (about 20 one-shift years)	c. Requires work breaks
	d. No personal requirements	d. Various personal problems (absenteeism, injuries, health)
4. Energy efficiency	a. Relatively high, eg, 120–135 kg/2.5–30 KVA	a. Relatively low, 10–25%
	b. Relatively constant regardless of workload	b. Improves if work is distributed rather than massed

[a] Ref. 7.
[b] DF = degrees of freedom.
[c] Where possible, fifth and ninety-fifth percentile figures from Ref. 8 are used to present minimum and maximum values. Otherwise, a general average value is given.
[d] A continuous right–left rotation is available.
[e] A continuous roll movement is available.

may receive as much as 1 billion bits of information per second via sensory receptors (9). The quantity of information that a human can receive depends on the sensory modality (eg, eyes, ears, touch) and the nature of the stimulus.

The rate at which information can enter permanent storage is relatively small, estimated to be about 0.7 bits/s. In contrast, the amount of information that can be retained in the human brain is estimated to be between 100 million and 1 million billion bits (10). As rough as these estimates may be, it is clear that the human brain has a tremendous capacity to store and process information.

The controller of a robot may process information at a very

fast rate. For a small microcomputer, this may be 10 bits/s. Without secondary storage, most robots are limited to less than 2000 commands, and in some cases can only store one program at a time.

Unlike a human, a robot has very limited sensing abilities. However, this is rapidly changing. The proliferation of research that has been done in the last few years has shown that the technology is now available to expand the capabilities of robots with the addition of sensing devices that will provide information to the controller from the surrounding environment.

One of the fundamental requirements of robot operation is the need to verify the position of the robot. Sensor information to verify position is provided in different ways, depending on the type of robot. Mechanical stops are used on nonservo robots and motion is verified at the mechanical stop limit. For servo-controlled robots encoders, resolvers and potentiometers are used to provide a continuous indication of the robot joint.

Vision systems are being used to determine part size and location at various workstations. Most of these systems are limited to two dimension black-and-white imaging, which puts restrictions on the type of objects that may be identified and the work environment in which sensing takes place. Proximity sensors and strain gauges are being used in attempts to provide end-of-arm tooling that will have the ability to "feel" objects. Up until now, very few robots with sensing capabilities have actually been used successfully in industry. As new technology develops further, the gap between human and robot sensing abilities is expected to diminish.

HUMAN WORK ENVELOPES

The first concept of a normal work area for humans was given by Maynard (11). It was assumed that the operator was seated at, or standing by, the desired worktable or bench. The sketch was of inner and outer semicircles of the right and left hands. Several modifications were made to add dimensions to the work areas since the original recommendations of Maynard were without dimensions. Farley (12) provided dimensions for normal and maximum working areas based on average male and female workers at General Motors Corporation. These new workplace layouts are cited in many texts and are widely used in workstation design for industrial applications.

There have been numerous studies done on the functional body dimensions of humans similar to the one shown in Figure 1. This diagram shows the single-arm grasping reach of a male operator in the sitting position. To accommodate the differences in body dimensions between individuals, most anthropometric data are specified to cover 95% of the population. A frequently used reference for this type of information is Ref. 14. Squires (15) recognized that the elbow does not move in a regular circular path. It was suggested that the elbow does not stay at a fixed point but moves out and away from the body as the forearm pivots. Research was conducted to determine design parameters or dimensions for industrial workplace layouts for various operating positions and operators. To describe the work envelope for the individual operator, the concepts developed by Farley and Squires were used. For determining the limits of reach and clearance requirements, the dimensions of smaller (5th percentile) and larger 95th

			Percentiles, in. (cm)		
Angle, deg	N	Min	5th	50th	95th
L165					
L150					
L135					
L120					
L105					
L90	11			14.00 (35.6)	18.75 (47.6)
L75	16			18.00 (45.7)	21.50 (54.6)
L60	20	17.00	17.50 (44.5)	20.50 (52.1)	24.50 (62.2)
L45	20	18.25	19.50 (49.5)	22.75 (57.8)	26.75 (67.9)
L30	20	20.25	21.50 (54.6)	24.75 (62.9)	28.25 (71.8)
L15	20	22.50	23.50 (60.0)	26.75 (67.9)	29.75 (75.6)
0	20	25.00	25.50 (64.3)	28.75 (73.0)	31.75 (80.6)
R15	20	27.25	28.00 (71.1)	30.50 (77.5)	34.00 (86.3)
R30	20	29.00	30.00 (76.2)	32.00 (81.3)	35.75 (90.8)
R45	20	30.50	31.00 (78.7)	33.50 (85.1)	36.25 (92.0)
R60	20	31.50	32.00 (81.3)	33.75 (85.7)	36.25 (92.0)
R75	20	31.50	32.25 (81.9)	34.00 (86.3)	36.50 (92.7)
R90	20	31.75	32.25 (81.9)	34.00 (86.3)	36.00 (91.4)
R106	20	31.50	31.75 (80.6)	33.50 (81.3)	35.75 (90.8)
R120	19		30.50 (77.5)	33.00 (83.8)	35.50 (96.1)
R135	9				34.50 (87.6)
R150					
R165					
180					

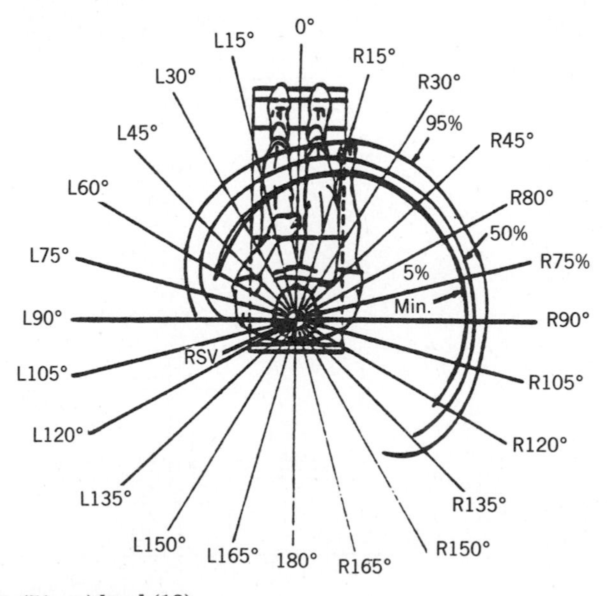

Figure 1. Grasping reach of a male, 20-in. (51-cm) level (13).

percentile) operators were used, respectively. Adjustments were also made in the anthropometric data for clothing, shoes, slump posture, and other allowances. Additional information on work-space layouts may be found in Refs. 15 and 16.

ROBOT WORK ENVELOPES

Research carried out on the work space of robots is quite limited. One of the reasons is that to analyze an open-loop system with multiple degrees of freedom is a complicated task that lends itself to computer-graphics technology rather than mathematical modeling. Ref. 17 presents a discussion and listing of computer graphics and robots.

Roth's study (18), published in 1975, was the first to evaluate the work space of some manipulator mechanisms. Based on homogeneous transforms, a general expression for describing the work space of a robot is included (18). The influences of geometric parameters of robot linkages on shapes of tori are discussed analytically. By using a series of specified planes to cut the surface of a torus, the boundary of work space can be formulated by numerical algorithms and drawn by computer.

When attempting to determine the work envelopes of robots, it becomes obvious that there are no standards that have been accepted and utilized throughout the robotics industry. Manufacturers differ on how they specify the various axes and working envelopes. To eliminate some of the confusion over terminology and dimensions, robot structures used in this article are based on those shown in Figure 2–6 (19). The axes are designated as closely as possible to those that are used in conventional machining and NC program languages. Each of the X, Y, and Z axes have both a minimum and a maximum dimension which clearly define the work envelope in relation to the centerline of the robot base. By standardizing these dimensions, it will be possible for all engineers and analysts to "talk the same lanaguage" when evaluating or specifying a particular robot for a given application.

The axes of movement for a rectangular-type robot, sometimes referred to as a Cartesian coordinate robot, are shown in Figure 2. This type of robot moves in the vertical and horizontal axes in simple straight-line motions, forming a rectangular work space. To extend the range of these robots, they are frequently mounted on a track or overhead gantry.

The work envelope that is shown in Figure 3 is for a cylindrical robot. The vertical and horizontal motions are similar to the rectangular robot except for the base movement. Instead of a straight-line motion, the cylindrical robot rotates about the base, forming an arc.

With a spherical robot, the reach and sweep (rotation about the base) motions are the same as those of the cylindrical robot. The vertical motion results from a pivoting action of the robot arm and forms an arc in the vertical plane. The pattern of a spherical robot is indicated in Figure 4.

A fourth type of robot is the jointed-arm robot. As shown in Figure 5, this type of robot forms a horizontal swing pattern similar to those of the cylindrical and spherical models. A unique motion in the vertical and horizontal reach planes is formed by the combination of angles that are possible between the various joints of the arm.

Information from 334 different robot models (19) indicates that the maximum vertical motion of robots is from 1 in. (2.5 cm) to 192 in. (488 cm). The range of movement in the Y axis, or horizontal reach, is a minimum of 3 in. (7.6 cm) to a maximum of 480 in. (1219 cm). With the exception of rectangular models, the X axis is usually specified in degrees. The maximum swing of all of the robots listed in the data base is

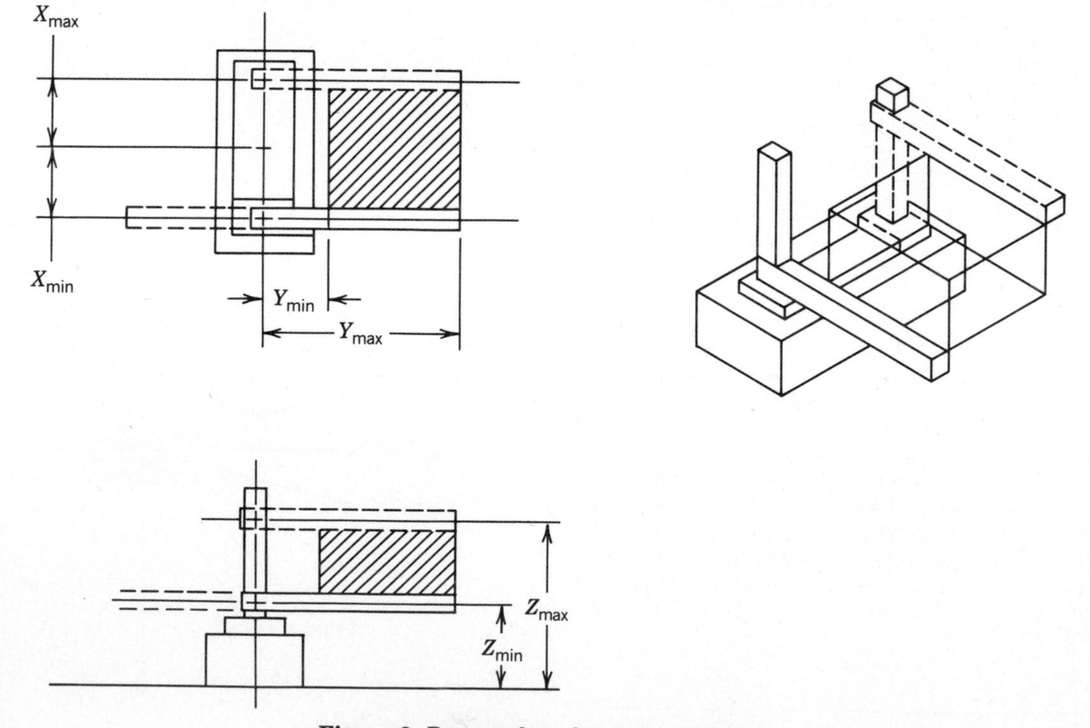

Figure 2. Rectangular robot work envelope.

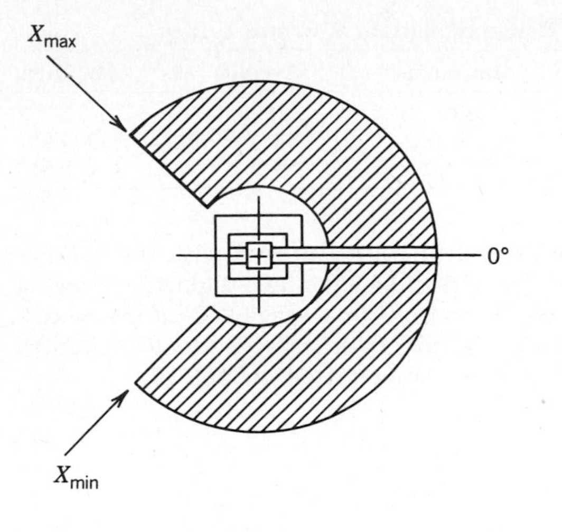

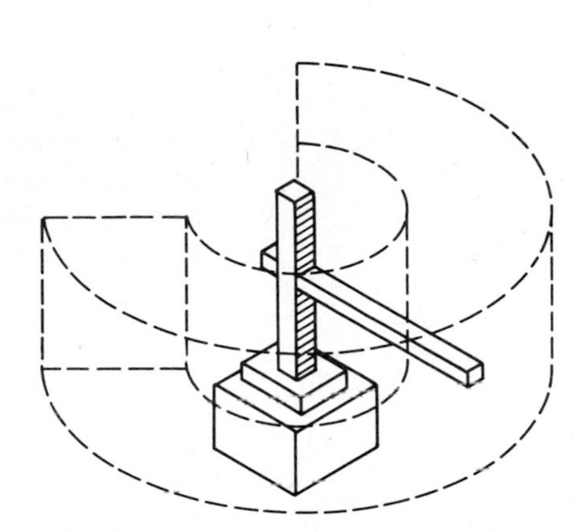

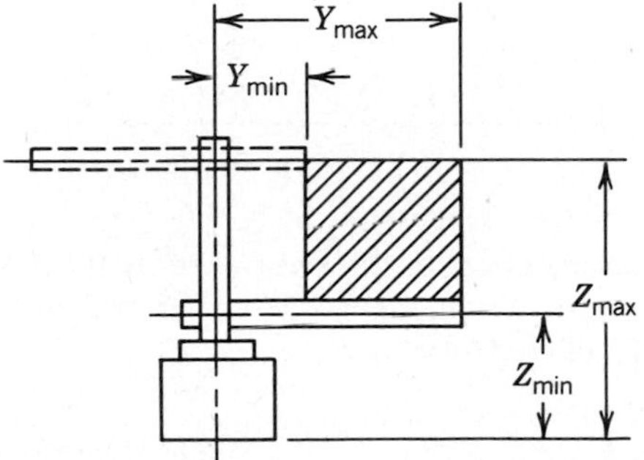

Figure 3. Cylindrical robot work envelope.

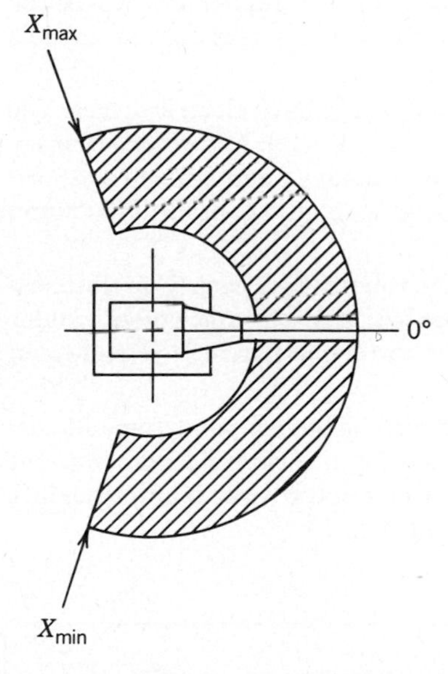

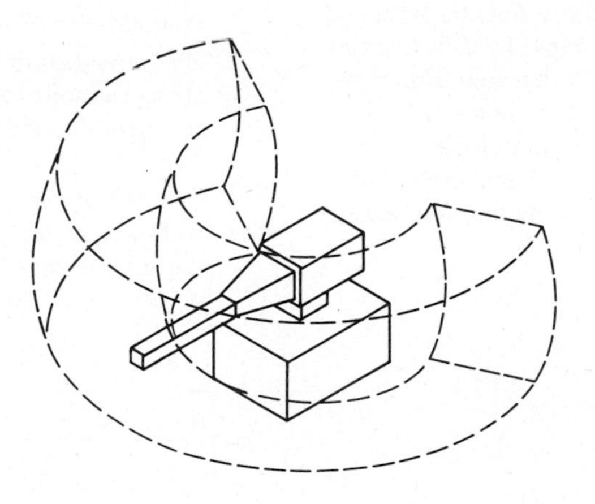

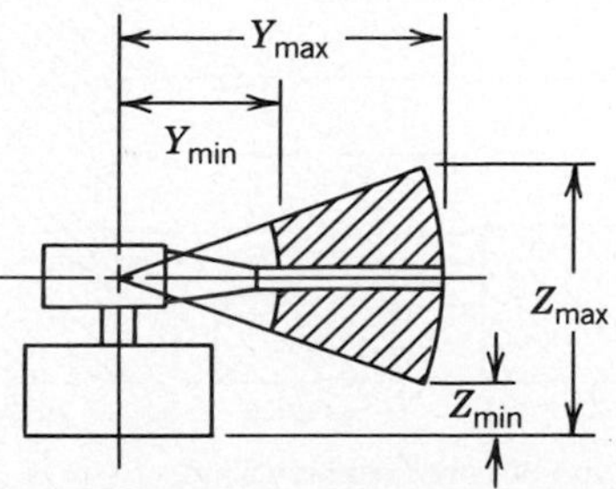

Figure 4. Spherical robot work envelope.

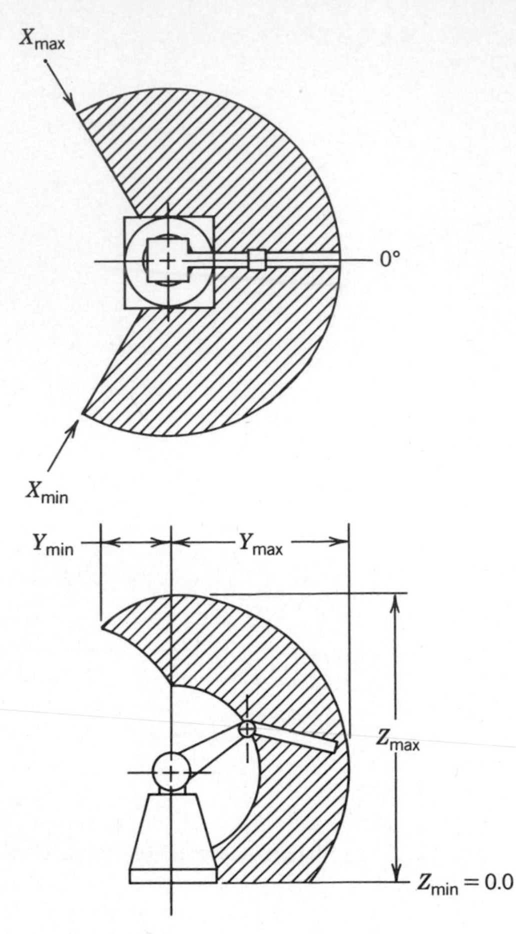

Figure 5. Jointed-arm robot work envelope.

432° in each direction. The average is 120° clockwise and 120° counterclockwise around the base. These data are summarized in Table 2.

Additional manipulation abilities of robots may be extended beyond the limitations of the three main axes of reach (Y axis), swing (X axis), and vertical stroke (Z axis) by including wrist movements. Thirty-eight percent of all robots have provisions for controlling 6 axes of movement. These motions are shown in Figure 6. When comparing the range of movement of a human vs that of a robot, the human-wrist flexion and extension (pitch) are limited to 189° compared with the average 200° for a robot. The wrist abduction and adduction (yaw) of a human average 74° and robots average 269° of movement. The roll motion of robots averages 341°; the human is limited to 190°. These factors are summarized in Table 3.

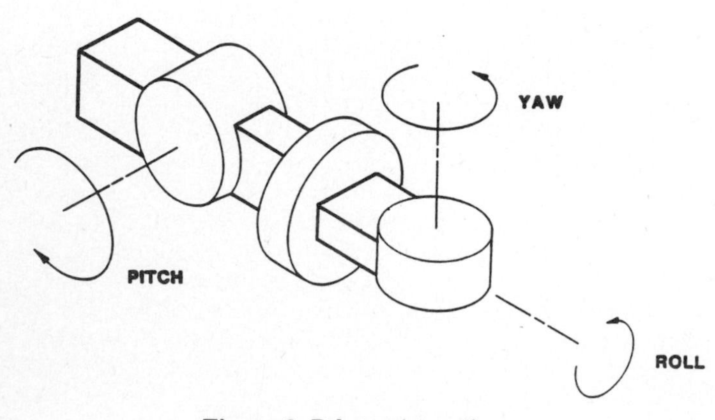

Figure 6. Robot-wrist motions.

Table 2. Robot Range of Motion: X, Y, and Z Axes

Axis	Maximum	Average	Minimum
X, deg	432	120	5
Y, in. (cm)	480 (1219)	66 (168)	3 (7.6)
Z, in. (cm)	192 (488)	58 (147)	1 (2.5)

Table 3. Robot-Wrist Range of Motion

Movement	Maximum	Average	Minimum
Pitch, deg	747	200	55
Yaw, deg	776	269	60
Roll, deg	768	341	90

Additional information in the data base includes information on type of construction, method of power or driver, repeatability, and type of control.

1. More than half (52%) of all models available in the United States have arm type of construction (Fig. 7). One in every five of the arm-type robots are SCARA or assembly models. One-third of all rectangular models are gantry robots.
2. Fifty-eight percent of the robots have electric drives. The electric drive provides a good combination of accuracy and load capacity. Only 14% have pneumatic drives, primarily because of the need for high repeatability in many applications (Fig. 8).
3. The repeatability, or the repetitive accuracy or closeness of agreement of repeated position movement under the same conditions at the same location, is shown in Table 4.
4. Two out of every three models have a point-to-point control system for programming a series of points without regard for coordination of axes, in which intermediate points are not controlled (Fig. 9).

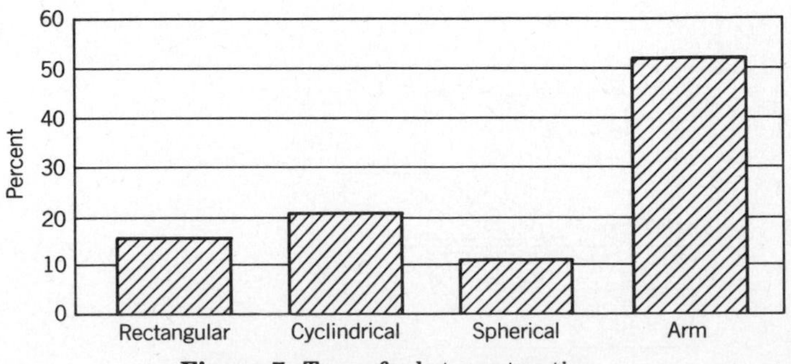

Figure 7. Type of robot construction.

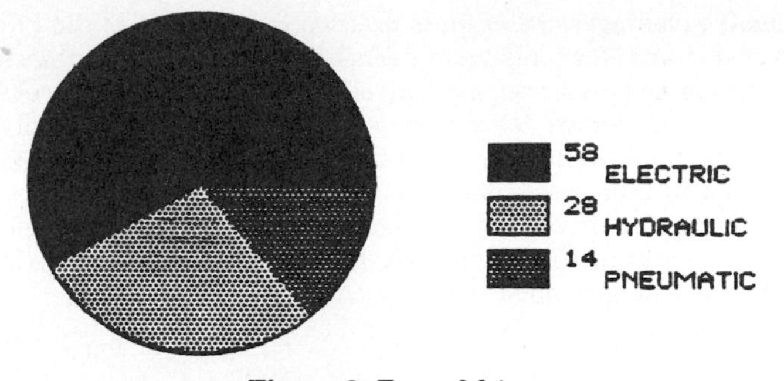

Figure 8. Type of drive.

Table 4. Repeatability

in.	cm	Percent of Total
0.0005	0.0013	5
0.001–0.005	0.0025–0.0140	29
0.006–0.010	0.0141–0.0269	33
0.011–0.015	0.0270–0.0399	3
0.016–0.025	0.0400–0.0650	13
0.026–0.050	0.0651–0.1299	12
0.051–0.500	0.1300–1.2700	5

STRENGTH AND ENDURANCE

When comparing the strength of robots with that of humans, the mechanical advantage of the robot far exceeds the limits of human operators. The load capacity of robots ranges from 1 mL to over 2500 lb (1134 kg). However, most of the current models are designed to handle loads of less than 60 lb. Only 9% of the industrial robots are designed to exceed 225 lb (27 kg) (see Table 5).

Typically, the human operator works 8 hours per day with some time provided so that the workers may recover from fatigue and tend to their personal needs. Many researchers (20–23) have investigated fatigue, but unfortunately there is no agreement on what, if any, allowance should be used. Traditionally, physiological fatigue allowances are recommended when the energy demands of the job, for 8 hours, exceeds one-third (33%) of the aerobic capacity of a person (24). In reality, the majority of industrial companies in the United States add an allowance factor for personal needs, fatigue, and delays which amounts to 15% of the hours worked.

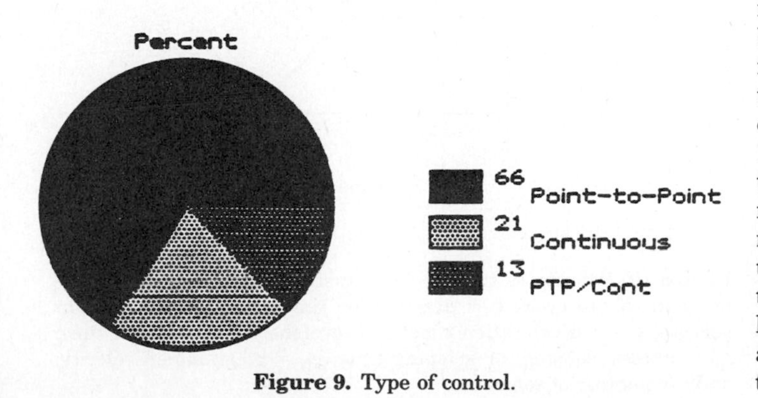

Figure 9. Type of control.

Table 5. Robot Load Capacity

Capacity, lb (kg)	Number of Robots	Percent of Total
1–60 (0.45–27)	179	63
61–110 (28–50)	38	13
111–170 (50–77)	26	9
171–225 (78–102)	18	6
226–450 (103–204)	12	4
451–2500 (205–1134)	13	5

Compared with the 8 hours per day (less approximately 1.2 h for allowances) that the human operator works, many companies are operating robots 24 hours per day with no time out for breaks, etc. With proper maintenance, companies are realizing 95–98% "up time" on their robot applications.

A MODEL FOR ROBOT MOTION PERFORMANCE

In attempting to evaluate robot performance, many analysts have had to rely on the recommendations of sales representatives or use "rule-of-thumb" methods. In a special report on industrial robots (25), a statement is made that "suppliers of robots say you should allow one second for each move or change in part orientation, another one second to grasp the part, and yet another second to release it." Another suggestion is that "if the robot is to pick up a part, swing 90 degrees, set down the part, and return to start, a comfortable cycle time would be 10 to 15 seconds" (26). Contrast this estimate with the claim by Adept Technology that Adept robots pick up a 1-lb (0.45-kg) load, move 1 in. (2.54 cm) up, 36 in. (9 cm) over, 1 in. (2.54 cm) down, and then back in 1.8 s. It is recommended that robots should be programmed at the same speed as the distance moved. This would result in the move time always being 1 s plus the time for acceleration and deceleration (27). It is also claimed that "it only takes ¼ second for most robots to reach terminal velocity no matter what velocity you have programmed and another ¼ second is required to stop." Obviously, these methods of predicting robot performance times are not accurate enough to make a sound selection of a robot for an industrial application.

Based on robot performance studies at Purdue University, four possible modeling approaches to determine robot performance times have been suggested (28). The first method is similar to predetermined time systems that are used for human work methods. The idea of describing robot motion times by using predetermined time systems is not new. Hassiegawa first suggested this concept in 1972 when he stated that predetermined time standards could be assigned to different types of robots. Ref. 29 cites the work of Nevins and co-workers in 1974 and Whitney in 1975, who used times studies, including the MTM system of predetermined times, to analyze various robot design configurations. The major advantage of this method is its simplicity. Once tables have been developed with times for the various motions, the analyst can determine the time it takes to perform a task by describing the motions and looking up the times in the tables. With proper training, this approach provides both a consistent and accurate estimate of the task time.

A second approach to modeling performance times is the use of regression analysis to analyze observation values. This method can be used to evaluate the relationship of one or more independent variables to a single continuous dependent variable. Regression analysis was used in developing time values for a system called Robot Time and Motion (RTM) (28). Analogous to MTM, RTM is based on standard elements of fundamental work methods. The SPSS linear regression technique to develop models to estimate performance time for point operations such as spot welding and riveting can also be used (29).

When engineering data is available for individual robot models, performance times can be modeled as motion control based on velocity, acceleration, and deceleration values. The main difficulty with this method is that a different model must be developed for each specific robot type and make. A second problem is that reliable data are not readily available.

Robot motions can also be modeled using the robot joint and link velocities to calculate path geometry. A discussion of a procedure for obtaining the dynamic equations of motions for single-arm robotic systems based on Kane's dynamic equations can be found in Ref. 30. Euler parameters, partial velocities, angular velocities, relative coordinates, and generalized speeds are used in the development of governing equations. A thorough discussion of kinematic and dynamic analysis is found in Ref. 31. It describes the points of an object in terms of a coordinate system and then defines the object in space in relation to location and orientation of the object's coordinate system. By defining a series of robot-arm end positions, a task can be described as a sequence of manipulator moves and actions. The structural task description is in terms of homogeneous coordinate transforms.

Simulation is another area that has been used to evaluate robot motion performance. A simulation program has been developed (32) that allows the user to vary angular, length, and velocity constraints interactivity and immediately study the resulting variations in the dynamic motion as limited by the defined accessible work space. McDonnell Douglas Automation Company has developed two robotic animation systems, PLACE and ANIMATE. Robot kinematic routines must be derived and encoded into the software (33). A model by IBM simulates the motions of the IBM 7535 SCARA robot. Accurate cycle time estimates have been realized because the robot's control system was designed to account for the dynamics of the model (34). A summary of robot system simulation is given in a special report published in *Robotics Today* (17).

GENERALIZED MODEL OF ROBOT MOTION

There are two basic methods to formulate a model of a process. One method is to derive the relationship between variables analytically, using fundamental assumptions and principles such as Newton's laws. The other method is to formulate a model based on data gathered from actual observation of the process.

Researchers differ in their methods for obtaining the equations in describing the motion of robots. Some investigators have suggested that Lagrange's equations present the most efficient procedure for developing equations of motion. Others maintain that using Newton's laws provide distinct computational advantages. Still others prefer a combination of the Lagrangian and Newtonian approaches (35). However, regardless of the method preferred, most investigators are currently seeking methods which (*1*) are easy to formulate, (*2*) are easily converted into computer algorithms and codes, and (*3*) lead to efficient numerical solutions (30, 31).

The profile of velocity vs time for a typical robot motion pattern is shown in Figure 10. Using this model, the cycle time for a programmed task is

$$t = \sum_{n=1}^{i} t_{ci} + t_{ai} + t_{vi} + t_{-ai} + t_{oi} + t_{pi} \quad (1)$$

where n = the number of segments in the cycle.

During the initial part of the segment, shown as t_c, the robot does not move while the controller calculates position and switches to the next command. The time for t_c varies according to the model of robot being used. Experiments on the Prab model E (36) indicate that this time averages 98 ms. Studies (37) reveal that this time is 115 ms for the Cincinnati model T3.

The total time per segment of motion without sensing or process time is

$$t = t_c + s/v - t_a + t_o \quad (2)$$

Given the time t_a necessary to accelerate the arm from rest to maximum velocity, the time t_m to move from position to position must be greater than 2 t_a for equation 2 to be applicable. Figure 11 shows velocity vs time when t_m is not greater than 2 t_a.

It has been suggested that it only takes ¼ s for most robots to reach terminal velocity no matter what velocity has been programmed, and another ¼ s to stop (38). The laws of physics are such that a given mass, such as the robot arm, can only be accelerated within a certain time given a reasonable power system. The robot reaches the desired velocity in an acceleration or deceleration stage during a fixed period of a ¼ s (39). Experiments with the Prab model E (26) show that acceleration and deceleration time averages 553 ms, with a standard deviation of 16 ms.

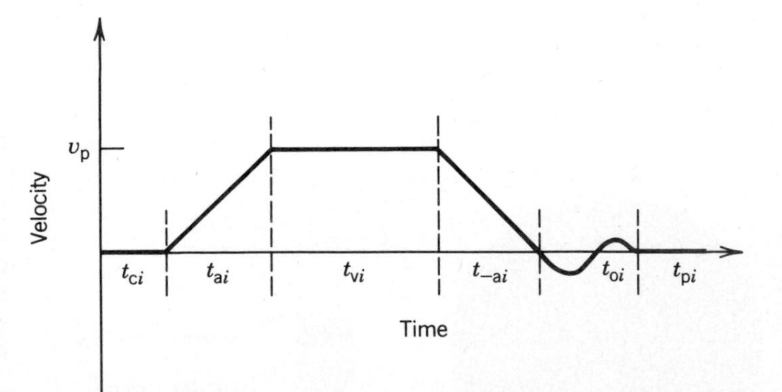

Figure 10. Velocity vs time for an industrial robot, where t_c = time for controller to cycle; t_a = acceleration time; t_v = time at constant velocity; t_{-a} = deceleration time; t_o = overshoot or oscillation time; t_p = process, sensing, or gripping time; v_p = programmed velocity; and i = number of segments.

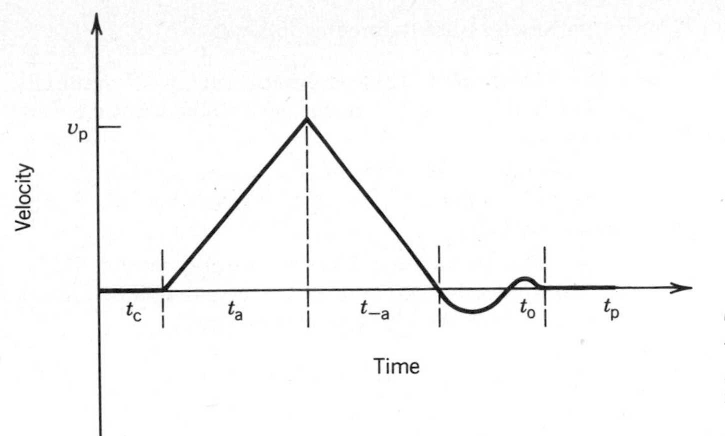

Figure 11. Velocity vs time when total move time is less than 2 t_a.

When a robot accelerates and decelerates at high rates, or when violent changes in direction take place, oscillations or overshoot may occur. Some researchers have chosen to ignore overshoot as a factor to consider in cycle time. Instead, they recommend that the robot should be programmed to avoid this type of motion.

ROBOT TIME SYSTEMS

The first predetermined time system used in the study and evaluation of manual work elements was Motion Time Analysis, was introduced by A. B. Segur and Company in 1924. Many systems followed, including Work Factor in 1938 and Methods Time Measurement (MTM), which was released by the Maynard Engineering Council in 1948. Partly because the MTM system was turned over to the MTM Association for public use, it soon became the most widely used system of predetermined times. Since the first introduction in 1948, a family of MTM systems has evolved, including MTM-1, MTM-2, MTM-3, and several specialized versions. These systems are now available in computerized form for both main-frame and personal computers.

In the 1970s, a new concept of predetermined time systems was developed. Called MOST (Maynard Operation Sequence Technique) (40), the system is 40–50 times faster than MTM-1 and 10 times faster than MTM-2. Computer versions of this system are available for the VAX and IBM PC computers; most recently, it was programmed on the APPLE computer.

A comprehensive survey of work measurement for robots was conducted in 1981 (37). This research found that some of the earliest studies indicated that a methodology similar to MTM seemed appropriate for developing robot standards. A manipulator and Ames arm using the same rules as MTM have been modeled (41).

A method called Robot Time and Motion (RTM) was introduced by Paul and Nof in 1979. Since that time, the system has been undergoing development at Purdue University. Like MTM, RTM is based on standard elements of fundamental work motions (42). The RTM methods consists of eight elements of task performance by robots:

Symbol	Element
Rn	Reach
SE	Stop on error
SF/ST	Stop on force/touch
Mn	Move
GR	Grasp
RE	Release
VI	Vision
TI	Process time

Three of the elements in the RTM system, SE, SF/ST, and VI, have limited application. Most robot controllers control position error, and the time of this calculation is very small. When this happens, the element SE is not required. The element for stop on force/touch is only applicable when the robot model has force compliance capability. Vision systems have been receiving a lot of attention in the past few years, but their use in industry is still limited.

Recently, a computerized method to predict cycle times for robots was introduced. Named Robot Factor (ROFAC), it was developed with Nof of Purdue and has been tested on the Cincinnati T3 and a Unimation Puma robot (34).

The use of metalanguages similar to MTM and RTM that would involve the description by one metalanguage of a manual task, the translation of the metalanguage to a second metalanguage that describes the task performed by a robot, and finally the transform of the second metalanguage into a program language for the operation of the robot has been suggested. A new method, ROBOT MOST, enables the user to input a robot task in the same terms that are used in as input to MOST for manual work elements. The use of a common language for both manual and robot operations eliminates the need to learn two different predetermined time systems and provides a simple and fast technique to compare the performance of a human vs that of a robot.

MOST Work Measurement Systems

The MOST system uses a sequence of basic motions to describe the fundamental activities that are generally found in the movement of objects. The basic MOST sequence models as described in Ref. 44 are

General Move Sequence: for the spatial movement of an object freely through the air

Controlled Move Sequence: for the movement of an object when it remains in contact with a surface or is attached to another object during the movement

Tool Use Sequence: for the use of common hand tools

General Move Sequence

The General Move follows a fixed sequence that includes the follow activities:

1. Reach with one or two hands a distance to the object(s), either directly or in conjunction with body motion.
2. Gain manual control of the object.

3. Move the object a distance to the point of placement, either directly or in conjunction with body motion.
4. Place the object in a temporary or final position.
5. Return to work place.

The sequence model is shown in a series of letters that represent each of the various subactivities. The General Move Sequence is

ABG BPA

The parameters are defined in the manual as follows:

A = *Action Distance*. This parameter covers all spatial movement or actions of the fingers, hands, and/or feet, either loaded or unloaded. Any control of these actions by the surroundings requires the use of other parameters.

B = *Body Motion*. This parameter refers to either vertical, ie, up and down, motions of the body or the actions necessary to overcome an obstruction or impairment to body movement.

G = *Gain Control*. This parameter covers all manual motions (mainly finger, hand, and foot) employed to obtain complete manual control of an object and to relinquish that control afterward. The G parameter can include one or several short-move motions whose objective is to gain full control of the object(s) before it is to be moved to another location.

P = *Place*. This parameter refers to actions at the final stage of an object's displacement to align, orient, and/or engage the object with another object(s) before control of the object is relinquished.

Each of the sequence-model parameters are assigned time-related index numbers based on the motion of the subactivity. The time units, identical to those used in basic MTM, are called time measurement units (TMU). One TMU is equivalent to 0.00001 h. The time value in TMUs for the sequence models is calculated by adding the index numbers and multiplying the sum by 10.

The General Move Sequence can be thought of as three separate phases:

GET	PLACE	RETURN
A B G /	A B P /	A

Controlled Move Sequence

The Controlled Move also follows a fixed sequence similar to the General Move. In the Controlled Move, instead of the PLACE phase, three different subactivities are included:

1. Move the object over a controlled path.
2. Allow time for a process to occur.
3. Align the object following the Controlled Move or at the conclusion of the process time.

The Controlled Move Sequence is shown as

A B G M X I A

The new parameters are defined as follows:

M = *Move Controlled*. This parameter covers all manually guided movements or actions of an object over a controlled path.

X = *Process Time*. This parameter occurs as that portion of work controlled by processes or machines and not by manual actions.

I = *Align*. This parameter refers to manual actions following the Controlled Move or at the conclusion of process time to achieve the alignment of objects.

The three phases of the Controlled Move are

GET	MOVE OR ACTUATE	RETURN
A B G /	M X I /	A

A complete discussion of all sequence models can be found in Ref. 40.

ROBOT MOST

A computer program, ROBOT MOST, has been developed that can be used to compare the performance times of a human operator and a robot for the same task. The robot analysis is part of the ROBOTMGR.2A program that was developed by the author (36).

ROBOT MOST uses the same sequence models as described in the previous section. Since the models are the same for both manual and robot performance, information for the task or element description needs to be entered only once.

ROBOT MOST Index Values

Action Distance (*A*). The most common motion control for robots is point-to-point control, where the path of the robot arm moves in several axes at the same time. Generally, all axes move at the maximum programmed velocity. Whichever axis has the smallest distance to move will reach its position first and then wait for the others. An example of point-to-point travel is shown in Figure 12. The time for the robot action distance (*A*) is the maximum value of the motions in the X, Y, and Z axes.

Body Motion (*B*). Since the robot moves in the vertical direction simultaneously with the horizontal motions, no time is added for the Body Motion unless the motion cannot be performed by a robot (eg, pass through a door or climb on or off a work platform). In a case where the robot cannot perform the motion, the same index value that is used for the manual Body Motion will be used for robot Body Motion.

Gain Control (*G*). To gain control of an object, the robot must move to the object, in some cases orient the end-of-arm, and close the gripper. Since the actual path of

Figure 12. Path generated by point-to-point control.

the robot arm is not known, most motions are programmed to a position just clear of the object; a second motion is then made to gain control to ensure that the gripper does not come in contact with the object before it is ready to grasp. The time to close the gripper includes 100 ms for cycle time and the actual close time, which varies by robot model. The orientation time also varies with the particular robot model that is being analyzed.

Place (*P*). To place an object, the robot must move to the object, orient the end-of-arm as required, and open the gripper. The method of calculating the index value for Place is the same as used in Gain Control.

Move Controlled (*M*). For a robot, this motion is simply a single move or a series of short moves that are calculated in the same way as Action Distance.

Process Time (*X*). Process time is the same for both manual and robot sequence models.

Align (*I*). Since most robot models are capable of aligning or orienting the object with the desired accuracy during the Move Controlled activity, no time is allowed in the robot sequence model for Align.

Robot Data Manager

The ROBOTMGR.2A program utilizes four data bases:

- ROBOTDAT.1: Main data base with specifications on various robot modes
- ROBOTDAT.2: Supplementary information on the models listed in ROBOTDAT.1
- ROBOTDAT.3: Time–performance data for robot models that can be analyzed using ROBOT MOST
- ROBOTDAT.4: Continuation of ROBOTDAT.3

Examples of information that is included in these data bases are shown in Figures 13 and 14. The computer program allows

RECORD NUMBER = 1

COMPANY	ACCUMATIC MACHINERY
MODEL	NUMAN I
TYPE	ARM
DRIVE	HYDR
MOTION	PTP
REPEAT	0.015
VELips	24
VELdps	
LOAD	75
X-min	140
X-max	140
Y-min	24
Y-max	48
Z-min	24
Z-max	136
PITCH	90
YAW	180
ROLL	280
RM	N

Figure 13. Example of information for each record in ROBOTDAT.1.

RECORD NUMBER = 1

COMPANY	ACCUMATIC MACHINERY
MODEL	NUMAN I
TYPE	ARM
DRIVE	HYDR
MOTION	PTP
# AXES	6
SERVO	YES
TEACH	NO
STORE	300
INTER	YES
APPLIC	
WEIGHT	1500
SPACE	16
COST	75000
YEAR	1984

Figure 14. Example of information for each record in ROBOTDAT.2.

the user to change data titles, change or add data, display data on the monitor, and to print data as a hard copy for each of the data bases.

The computer program provides two additional options that assist the user in selecting a robot for a particular task. These options are listed as Robot Selection and ROBOT MOST Analysis.

Robot Selection

This routine will select robot models from the ROBOTDAT.1 and ROBOTDAT.2 data bases that are capable of performing a given task as specified by parameters that are entered by the user. The most common limiting parameters are maximum horizontal reach, maximum load, repeatability, and whether or not continuous-path motion is required. After the information for the limiting parameters has been entered, the computer will search the data base and print out the record number, manufacturer, and model for each robot capable of performing the desired task.

An example of robot selection for a simple task of unloading boxes from a conveyor and placing them on a pallet is shown in Figure 15. The maximum reach that is required is 60 in. (152 cm) with a maximum load of 50 lb (23 kg). The desired repeatability for this task is 0.5 in. (1.27 cm). Continous-path motion control is not required. The data input for this task is shown in Figure 16 and a partial listing of robot models that are capable of performing this task in Figure 17.

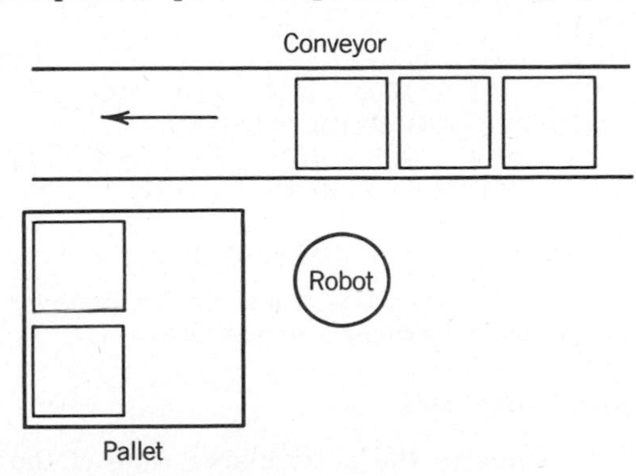

Figure 15. Example of robot selection, unload conveyor.

This routine will select robot models from the the data base that are capable of performing the task as limited by the parameters you will enter.

ENTER THE MAXIMUM HORIZONTAL REACH (inches) ? 60

ENTER THE MAXIMUM LOAD INCLUDING END-OF ARM-TOOLING (pounds) ? 50

ENTER THE REQUIRED REPEATABILITY (decimal inches) ? 0.5

DO YOU REQUIRE CONTINUOUS PATH MOTION (Y or N) ? N

Figure 16. Computer input for robot selection, unload conveyor.

76	CYBROTECH	G80
78	CYBROTECH	H80
79	CYBROTECH	V80
84	DEVILBISS	TR-3500W
87	FMC	A-450
90	GCA	DKB 1440
92	GCA	DKB 3200

Figure 17. Computer output for robot selection, unload conveyor.

```
DRILL 0.5 IN. HOLE IN CASTING
            Manual Sequence Model                    TMUs
LOAD CASTING INTO FIXTURE
  A1    B3    G1    A3    B0    P1    A1             100
DRILL HOLE
                                                     160
  A0    B0    G0    M0    X16   I0    A0
UNLOAD CASTING AND ASIDE TO SKID
  A1    B0    G1    A1    B6    P1    A1             110
TOTAL TIME FOR THIS OPERATION USING A
  HUMAN OPERATOR IS                                  370 TMUs
                          TOTAL MINUTES =            0.22

DRILL 0.5 IN. HOLE IN CASTING
             Robot Sequence Model                    TMUs
LOAD CASTING INTO FIXTURE
  A3    B1    G3    A3    B1    P3    A1             150
DRILL HOLE
                                                     160
  A0    B0    G0    M0    X16   I0    A0
UNLOAD CASTING AND ASIDE TO SKID
  A1    B1    G1    A3    B3    P3    A3             150
TOTAL TIME FOR THIS OPERATION USING
  A ROBOT IS                                         830 TMUs
                          TOTAL MINUTES =            0.50
```

Figure 18. A typical comparison of performance times for a human operator and a robot performing a similar task.

ROBOT MOST ANALYSIS

This option compares the performance time of the average human operator with the time of a selected robot model. The program is designed to prompt the user to give all the information necessary to calculate the performance times. After the task description and element information have been entered, the computer will calculate the human performance time using the MOST system which was described previously. Parameters from the ROBOTDAT.3 and ROBOTDAT.4 data bases are used to calculate the robot time using ROBOT MOST. An example of the output from the ROBOT MOST analysis is shown in Figure 18. Whenever the user specifies a task that cannot be performed by a typical robot, a message is displayed on the screen. The message identifies the motion that should not be assigned to a robot without some modification in the task or special capabilities that are not customarily found on standard robot models.

BIBLIOGRAPHY

1. J. D. Laney, "A Suggested Standard for Robot Comparisons," *Robots 8 Conference Proceeding,* Robot Institute (RI) of the Society of Manufacturing Engineers, Detroit, Mich., 1984.
2. K. J. McDermott, "An Industrial Robot Data Base and Retrieval System," in Ref. 1.
3. *ROBOT CALC 1,* Robot Calc, Inc., South Bend, Ind., 1985.
4. O. Z. Maimon and E. L. Fisher, "Analysis of Robotic Technology Alternatives," *Proceedings of the 1985 Annual IE Conference,* Institute of Industrial Engineers (IIE), Atlanta, Ga., 1985.
5. M. S. Jones, "A Robot Selection Model," in Ref. 4.
6. S. Y. Nof, J. L. Knight, Jr., and G. Salvendy, "Effective Utilization of Industrial Robots—A Job and Skills Analysis Approach," *AIIE Transactions,* **12,** 216–225 (1980).
7. S. Y. Nof, "Robot Ergonomics: Optimizing Robot Work," in S. Y. Nof, ed., *Handbook of Industrial Robots,* John Wiley & Sons, Inc., New York, 1985, pp. 550–556.
8. W. E. Woodson, *Human Factors Design Handbook,* McGraw-Hill, Inc., New York, 1981.
9. E. J. McCormick and M. S. Sanders, eds., *Human Factors in Engineering and Design,* 5th ed., McGraw-Hill, Inc., New York, 1982, p. 45.
10. *Ibid.,* p. 46.
11. S. Konz, *Work Design: Industrial Ergonomics.* 2nd ed., Grid Publishing, Inc., Columbus, Ohio, 1983, pp. 257–302.
12. R. R. Farley, "Some Principles of Methods and Motion Study as Used in Developing Work," *General Motors Engineering Journal,* **2**(6), 20–25 (1955).
13. K. W. Kennedy, *Reach Capability of Men and Women: A Three-Dimensional Analysis, AMRL-TR-77-50,* National Technical Information Service, Springfield, Va., 1978.
14. H. P. VanCott and R. G. Kinkade, eds., *Human Engineering Guide to Equipment Design,* U.S. Government Printing Office, Washington, D.C., 1972.
15. B. Das and R. M. Grady, "Industrial Workplace Layout and Engineering Anthropology," in T. O. Kvalseth, ed., *Ergonomics of Workstation Design,* Butterworth Publishers, Ltd., London, 1983, pp. 103–128.
16. R. M. Grady, "Design of Industrial Workplace Layout Through the Application of Engineering Anthropology," Masters of Engineering Report, Texas A&M University, College Station, Texas, 1979.
17. R. N. Stauffer, "Robot System Simulation," *Robotics Today,* **6,** 81–90 (June 1984).
18. Q. Huang, "Study on the Workspace of R-Robot," in Ref. 1.
19. R. M. Wygant, *Robot Data Manager,* Dept. of Industrial Engineering, Western Michigan University, Kalamazoo, Mich., 1985.

20. L. Brouha, *Physiology in Industry,* Pergamon Press, Inc., New York, 1960.
21. "American Industrial Hygiene Association: Ergonomics Guide to Assessment of Metabolic and Cardiac Costs of Physical Work," *American Industrial Hygiene Association Journal,* **32,** 560–564 (1971).
22. E. Grandjean, *Fitting the Task to the Man,* Taylor & Francis, London, 1981.
23. D. W. Karger and W. M. Hancock, *Advanced Work Measurement,* Industrial Press, Inc., New York, 1982.
24. A. Mital and R. L. Shell, "Determination of Rest Allowances for Repetitive Physical Activities that Continue or Extended Hours," *Proceedings of the 1984 Annual IE Conference,* Institute of Industrial Engineers (IIE), Atlanta, Ga., 1984.
25. "Vision Systems—Bringing Sight to Automation," *Modern Material Handling,* 64–69 (Apr. 9, 1984).
26. C. Reed, *"Multi-Part Handling," Proceedings of the 2nd International Robot Conference,* Inter Robot West, Long Beach, Calif., 1984.
27. L. A. Hautau, "Robotic Process Planning—A Total Systems Approach," *Robots 7 Conference Proceedings,* Robot Institute (RI) of the Society of Manufacturing Engineers, Detroit, Mich., 1983.
28. H. Lechtman and S. Y. Nof, "Performance Time Models for Robot Point Operations," *International Journal of Product Research,* **21,** 659–673 (1983).
29. R. P. Paul and S. Y. Nof, "Work Methods Measurement—A Comparison Between Robot and Human Task Performance," *International Journal of Product Research,* **17,** 277–303 (1979).
30. R. L. Huston and F. A. Kelley, "The Development of Equations of Motions of Single-Arm Robots," *IEEE Transactions on Systems, Man, and Cybernetics,* **12,** 259–265 (May–June 1982).
31. R. P. Paul, *Robot Manipulators: Mathematics, Programming, and Control,* The MIT Press, Cambridge, Mass., 1981.
32. R. Mahajan and J. S. Mogal, "An Interactive Graphic Robotics Instructional Program," in Ref. 27.
33. S. J. Kretch, "Robotic Animation," in *Computers in Engineering 1982,* Vol. 2, American Society of Mechanical Engineers, New York, 1982.
34. R. L. Hershey, A. M. Letzt, and S. Y. Nof, "Predicting Robot Performance with ROFAC—A Computerized Decision Making Aid," *Proceedings of the 1983 Fall IE Conference,* Institute of Industrial Engineers (IIE), Atlanta, Ga., 1983, pp. 585–587.
35. R. C. Dorf, *Robotics and Automated Manufacturing,* Reston Publishing Co., Inc., Reston, Va., 1983.
36. R. M. Wygant, "Robots vs Humans in Performing Industrial Tasks," Ph.D. thesis, Dept. of Industrial Engineering, University of Houston, Houston, Texas, 1986.
37. H. Lechtman, "Robot Performance Models Based on the RTM Method," M.S. thesis, School of Industrial Engineering, Purdue University, Lafayette, Ind., 1981.
38. Ref. 27, p. 111.
39. Ref. 37, p. 14.
40. K. B. Zandin, *MOST Work Measurement Systems,* Marcel Dekker, Inc., New York, 1980.
41. J. W. Hill, *Study of Modeling and Evaluation of Remote Manipulator Tasks with Force Feedback, Final Report, Contract NSA 7-100,* SRI International, Menlo Park, Calif., Mar. 1979.
42. S. Y. Nof and H. Lechtman, "Now It's Time for Rate-Fixing for Robots," *Industrial Robot* (June 1982).
43. P. B. Scott and T. M. Husband, "Robotic Assembly: Design, Analysis, and Economic Evaluation," in Ref. 27.
44. *MOST University Program Student Users Manual,* H. B. Maynard and Co., Pittsburgh, Pa., 1983.

ERGONOMICS, WORKPLACE DESIGN

Martin G. Helander
State University of New York at Buffalo
Buffalo, New York

INTRODUCTION

This article provides an overview of ergonomic issues, which are important in the design of human/robot systems. The focus will be on job and workplace design as well as safety.

Ergonomics problems are systems problems, as shown in Figure 1. There are two major components of the system: the environment and the operator. The environment is composed of two subcomponents: the task and the machines. The type of machine used is of main importance for this article. Of concern also are the ergonomic implications of robots and other automated equipment. Task allocation between humans and robots is also an important ergonomic issue. Often, because of technological limitations in robot performance, the human operator is left to perform leftover tasks. Since the operator's task often depends on what the robot is doing, some trends in the use of robots and how humans are needed for both supervising and assisting robots and the effect of task allocation on job satisfaction are also discussed.

The operator component of the system consists of three major subsystems: perception, information processing, and muscular response.

Accurate perception is particularly important for robotic safety. Many accidents occur because of failure and difficulties in perceiving and anticipating the movement of the robot arm. Measures to increase the perceptibility of robot arm movements are discussed.

Information processing and decision making are important for tasks such as programming and maintenance of robots and will be discussed in greater detail in connection with these tasks.

Finally, following the decision making, the operator performs tasks such as interaction with terminals and manual materials handling. These tasks are not specific for robotic workplaces, rather they are common for all industrial environments, and some general ergonomic design principles are therefore summarized.

If the environmental demands exceed operator capabilities, there may be consequences such as accidents, decreased job satisfaction, or biomechanics problems. These types of problems are discussed throughout the article.

Ergonomics Defined

Ergonomics and human factors engineering are fairly close synonyms. Both are concerned with the design of the interface between humans and machines. The term "ergonomics" is used more in Japan and Europe, whereas "human factors engineering" is used in the United States. In the past, ergonomics had a bias toward work physiology and workplace design and human factors toward psychology and systems design. Today the terms are used more or less synonymously. Most of the research support in the United States has come from the Department of Defense; this is also true for research on robotics. In Europe, however, with its interest in workplace applications, resources have come from departments of labor or simi-

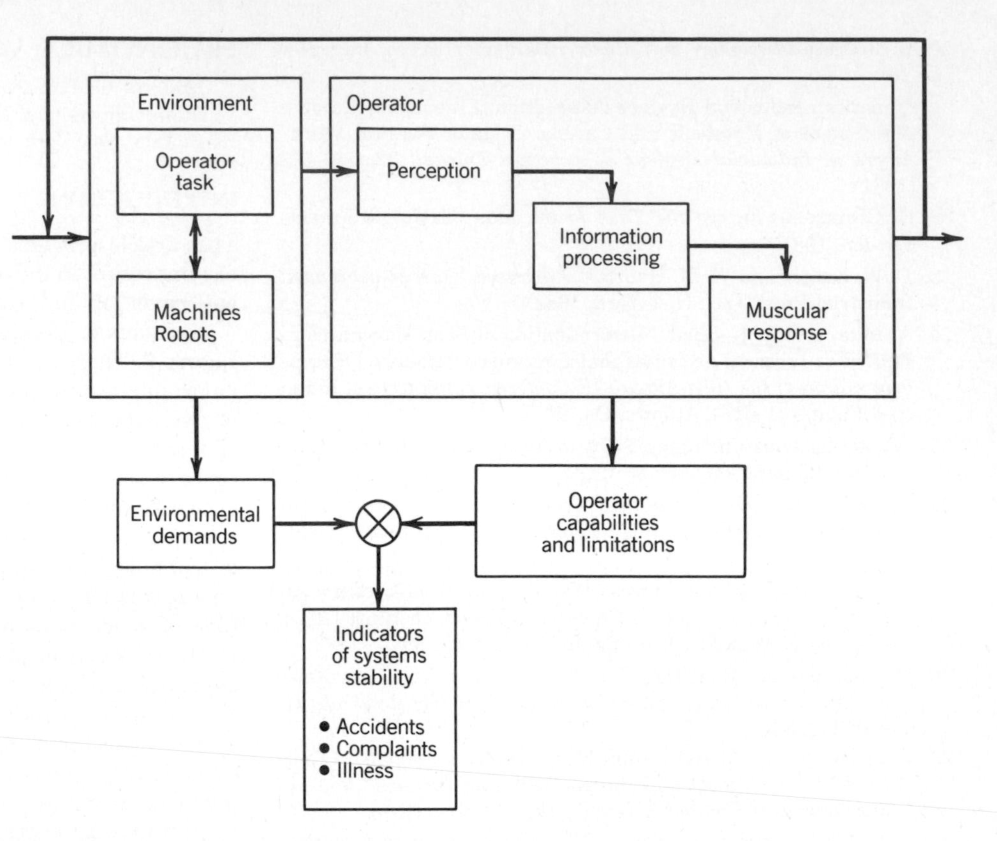

Figure 1. A systems approach to analysis of ergonomic problems in human–robot interaction.

lar institutions. Ergonomics uses knowledge from several sciences, including psychology, sociology, medicine, and engineering. It includes not only research and applications, but also addresses the adaptation in both sides of the human–machine relationship. Humans may adapt to machines by training and selection, and machines may be adapted to humans by improving the design of hardware and software.

For complementary information on human factors in robotics, see references 1–5. There are several textbooks in human factors that are appropriate as a general background for further studies of this topic (6–9).

ROBOT TASKS AND WORKER TASKS

In the past, the most common robot applications were welding, followed by machining and painting. In the future, robots will perform more complicated tasks. At General Motors, for example, welding robots are presently the most common, but there are predictions that by 1990 assembly robots will dominate (10). Many companies have difficulties understanding how robots can be used. One study investigated the introduction of 237 robots in 37 U.S. companies (11). In 19% of the cases, management found an application after the decision to buy a robot.

The major motive for investing in industrial robots is to enhance productivity. In addition to increasing productivity, robots do have the potential to relieve human operators from tasks that are difficult or hazardous. Such tasks are sometimes referred to as "4-D" work since they are dangerous, difficult, dirty, or disappointing. For the early applications of robots in welding and spray painting, workers were indeed relieved from hazardous tasks. It was claimed (12) that the U.S. industry was only "mildly interested in shielding workers from hazardous working conditions" through the use of robots. A questionnaire study was used to investigate the motivating factors for installing 577 robots in 23 U.S. corporations (13). The most important factor was to "reduce labor costs" followed by "relieve workers of tedious and/or dangerous jobs," "increase output rate," and "improve product quality."

A Japanese study found that for most robot installations it was possible, in retrospect, to find working environment arguments for the robots. Although most of the installations were made to improve productivity, 77% of the installations claimed worker benefits due to mechanization of hard work, 82% due to elimination of harmful working conditions, 89% due to elimination of dangerous working conditions, and 92% due to mechanization of monotonous work (14).

The U.S. Department of Defense sees additional merits in using robots and other intelligent systems. In addition to reducing the hazards, there are also reduced requirements for manpower, operator skills, and training. The research sponsored by the Department of Defense is likely to produce civilian spinoff effects, not all of which are desirable. For example, the emphasis on reduced skill levels and training costs would have negative carryover effects for workers if unconditionally applied in industry (15).

In designing an appropriate working environment, it is necessary to understand what types of tasks people do. Operator tasks are largely determined by the requirements of the system. For example, a human operator may be able to supervise 20 welding robots, but only 4–5 assembly robots. This is because in the latter case there are additional maintenance

Table 1. Common Robot and Human Tasks

Robot	Human Operator
Welding	Monitoring/supervising
Painting	Programming
Assembly	Maintenance
Machining	Intervening/helping
Material handling	Input/output
Forging	Leftover tasks
Surface cleaning	Replace robot during failure

tasks, such as clearing feeding devices and picking up things that are dropped (eg, end-effectors and screws).

It is easy to automate and roboticize tasks that are understood and can be modeled. The human supervisor can then take care of less structured tasks (16). However, in many cases the outcome of robotics, despite good intentions, is less desirable work for humans. Many tasks are characterized by shorter cycle times and monotonous work simply because it is difficult to suit both humans and robots (17). Several of the tasks in Table 1 provide examples of this and will be commented upon in greater detail below.

The first three tasks in Table 1 (monitoring, programming, and maintenance) are all reasonable from a work satisfaction point of view. The remaining tasks (intervening, input/output, leftover tasks, and replacing robot) lack many of the attributes that constitute a satisfactory task. These tasks are explained below.

Monitoring

Monitoring involves the detection of events that require operator attention. A model of monitoring, or supervisory control (16), is given in Figure 2. There are two major parts of the system in Figure 1: the computer/controller and the user interface. To the computer/controller are attached sensor(s) and actuator(s) and to the user interface display(s) and control(s). The operator has two options in observing the task. It may be done indirectly by observing displays or directly by observing the robot.

This type of monitoring is different from that, say, in a power plant or a steel plant, where much of the information comes from measured process parameters that are best displayed on a CRT. In a manufacturing environment, the operator is looking for events that are usually so obvious that it is not necessary to use indirect observation; the operator can obtain much more detailed information by simply observing the robot. A CRT may then be used for confirmation of the information.

Monitoring is usually a passive and boring task since the operator rarely needs to do anything except when things go wrong (18). At that stage, many tasks require highly skilled operators who can perform the necessary corrections of the system. Because of the monotonous nature of a monitoring task, it has proven difficult to retain qualified operators. It is therefore advisable to make the task more appealing by expanding the scope of the task to include additional elements—so-called job enrichment.

Programming

There are three common ways of programming a robot: through a computer terminal, by use of a teach pendant, and by manually leading the robot arm. Operator errors in programming always lead to frustration and, in the case of robots, to safety hazards. It is therefore important to understand what the difficulties are and how a task can be made easier. The manual lead through is commonly used for programming spray painting robots. It is thereby possible to imitate the fluid movement of manual spray painting. However, compared to a spray gun, the robot arm is more difficult to move and there is no visual feedback from paint hitting the surface. It may therefore be difficult to achieve a satisfactory product finish, even for programmers who are experienced at manual spray painting (19). Some manufacturers use a special lightweight robot arm for teaching the robot. It is easier to move and makes the task a little easier.

Few manufacturers of robotic systems have yet addressed the ergonomic design of teach pendants (20). This is an important issue since teach pendants have become very complex. Some have more than 50 different controls. A teach pendant should be designed so that it can be held comfortably with either hand and operated with the other. The controls should be positioned in functionally related groups so that different groups of controls are used for different subtasks. Color coding and shape coding of controls should also be considered. Preferably, it should be possible to operate the controls sequentially (horizontally or vertically) within each group. The direction

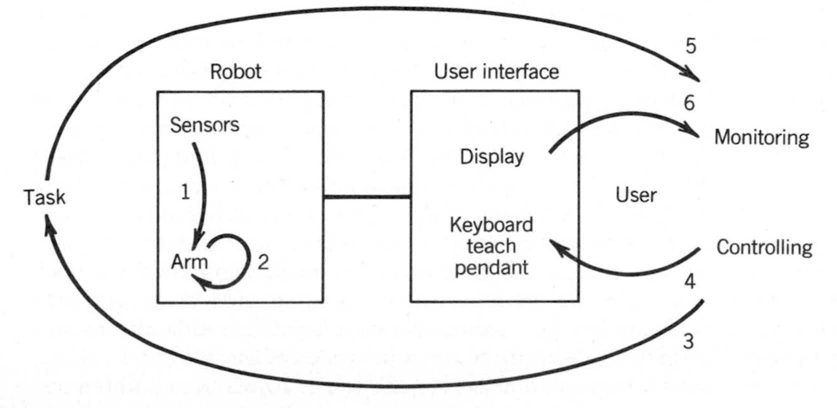

Figure 2. A model for monitoring and controlling robots. 1, Robot controls task autonomously with feedback from sensors. 2, Robot controls task autonomously without feedback. 3, Operator controls robot arm directly by moving it. 4, Operator controls robot arm indirectly through keyboard or teach pendant. 5, Operator monitors task directly. 6, Operator monitors task indirectly through display.

controls on a teach pendant may be confusing to use, especially when the programmer moves around the robot so that the frame of reference changes. To indicate the direction of movement, it may not be meaningful to use "left–right," arrows, or an *xyz* coordinate system. It may be easier to use numbers that have no previous meaning to the operator, for example, 1–4, 2–5, and 3–6.

There is a considerable amount of literature addressing these types of problems (7, 9). Often, however, design recommendations have limited applicability, which would warrant research addressing the specific design problems for teaching pendants.

Programming through a terminal is more difficult because the task is controlled and observed indirectly through displays and controls (see Figure 2). This increases the likelihood of making errors. According to one study, industrial workers average 1 error in 1000–10,000 discrete acts such as moving a control, putting a part in place, or reading a number (21). This could be compared with the much greater rate of errors in using a teach pendant. In one investigation, ten undergraduate engineering students used a teach pendant to move objects with a PUMA robot; the error rate was about one error in ten discrete acts (22).

There are also errors in human–computer interaction. One investigation of error rates in word processing found that novice users of a text editor made mistakes in 19% of the commands and experienced users in 10% (23).

Although these studies are not fully representative of the problems in programming robots, there is at least one generalized conclusion. Direct manipulation of the robot arm is much less prone to errors than indirect manipulation through a teach pendant or software. More research is needed to improve the interface of both teach pendants and software so that they are easier to use.

Maintenance

Maintenance of robots is a complex task requiring knowledge of mechanics, electronics, hydraulics, and computers, and it may be difficult to recruit personnel with such broad experience. There is, however, an increasing use of automatic test equipment and computerized diagnostics, which simplifies maintenance. One study reported that it once took an average of 35 min to locate and repair a fault in a printed circuit board used in an NC machine. With automatic test equipment, the same board can be diagnosed in 80 s, representing a 96% time saving (24).

Maintenance personnel and programmers work in close proximity to robots and are highly exposed to safety hazards. In fact, these operators incur a majority of the accidents with robots. It is therefore important that they receive safety training.

Intervening/Helping

Robots often need help, particularly to perform complex tasks. In robotic assembly, for example, gravity feed devices must be cleared of obstructions, and objects that the robot drops must be picked up from the floor or replaced by the human operator. The operator is actually serving the robot, and the interventions are paced by the robot. From a job satisfaction point of view, it would be better if the roles were reversed. This would require a different approach to the organization of manufacturing. Most of us have made the reflection that it is tedious to serve or help other people at work if repeatedly required to do so. Robots may also need plenty of assistance, and just waiting to help is very tedious. It may be better to do the task oneself—from both the productivity and job satisfaction perspectives.

Input/Output

The input and output to a process may represent tasks that are difficult to automate. An automated card-stacking machine, for example, places electrical components on cards. The input represents putting empty cards in frames, which are fed into the process, and the output consists of removing the finished cards from the process. These are tedious and repetitive tasks created by the technical limitations of the robot (or the automated machine process). Job enrichment techniques (see Table 4) could help in increasing job satisfaction.

Replacing

Finally, operators must sometimes replace robots during failure. This has implications for both product design and workplace design. Products must be designed so that they can be assembled easily by both robots and human beings, and the workplace must also be appropriate for both robots and human operators. This entails the use of adequate illumination, noise protection, easy access to tools, proper seating arrangements, and appropriate working height.

Leftover Tasks

Leftover tasks are those that remain after robots (or automated devices) have performed what they can do. One representative example is visual inspection. Despite the developments in machine vision, human operators are much better at pattern recognition than robots. It may therefore be the case that most of an assembly is performed by a robot, but the quality is controlled by a human operator. From a job satisfaction point of view, this is not a good arrangement. Visual inspection is a highly monotonous, repetitive, and often boring task, and it may be better to expand the operator's task by integrating some of the mechanical assembly. These issues are discussed further below.

Ever since the time of the industrial revolution and the introduction of machinery, workers' jobs have been redesigned because of technical innovations. In the beginning, there was a relatively high concern for the human operator. Early inventions such as the Spinning Jenny and the Spinning Mule were designed so that, rather than eliminating human skills, they required an extension of the operator's skill beyond what was required to perform the original task (25). But later developments in industrial mechanization usually substituted for human skills. For many of the tasks illustrated above, the only elements left for human operators are those difficult to automate or those required for supervision of the system. In effect, the human operator serves the robot. It has been pointed out that maybe production systems could be arranged in the opposite way (robot serving human), thereby enhancing job satisfaction (25). One example, discussed later, is given in Figure 5.

ALLOCATION OF TASKS BETWEEN HUMANS AND ROBOTS IN MANUFACTURING

The allocation of tasks between humans and robots is one of the most important, yet neglected, elements in the planning of automated manufacturing. In the past, products were designed predominantly for manual assembly. Recently, a new methodology of design has emerged, the intention of which is to facilitate automatic or robotic assembly; these products are said to be designed for automation (DFA). Principles of DFA have emerged in industry in the form of compilation of empirically developed rules and guidelines (see Table 2).

In most cases, it is desirable to have the flexibility to switch between robotic and human assembly. It is then important to ensure that principles for DFA not interfere with human principles for assembly. The increased flexibility is important since product life may be as short as 2–3 years. Under such circumstances, it may be difficult to justify the additional costs of setting up automatic assembly. In an example from IBM, it was pointed out that, using CAD/CAM equipment, product development took only 6 months (27). This was much less than the 1–1.5 years necessary for implementing the automated assembly, which is due to long delivery times for automated equipment. Since the product lifetime is limited to 2–3 years, it may therefore be practical to use manual labor, at least during a transition period before the automated process is installed. Manual and automated workcells will then coexist on an assembly line, and as the manufacturing processes change, the manual workcells will gradually be replaced by robot workcells.

Table 2. Principles of Design for Automation[a]

Design for unidirectional assembly, preferably top-down.
Reduce the number of sizes of screws.
Design for insert and snap assembly.
Design chamfers for self-alignment.
Eliminate parts that are difficult to feed automatically, eg, springs, washers, fragile parts, etc.
Eliminate parts requiring extremely tight tolerances.
Eliminate parts that are difficult to handle, either too bulky or too small.
Combine parts to reduce the number of assembly steps.
Eliminate cables, wires, and other flexible parts.

[a] Ref. 26.

Figure 3 suggests a systematic approach to task allocation between humans and robots. There are four major steps: resource analysis, design of products, task analysis and allocation, and evaluation and implications of product design.

Resource Analysis

To some degree, the services and skills supplied by humans and robots are interchangeable. Assuming flexibility in choice,

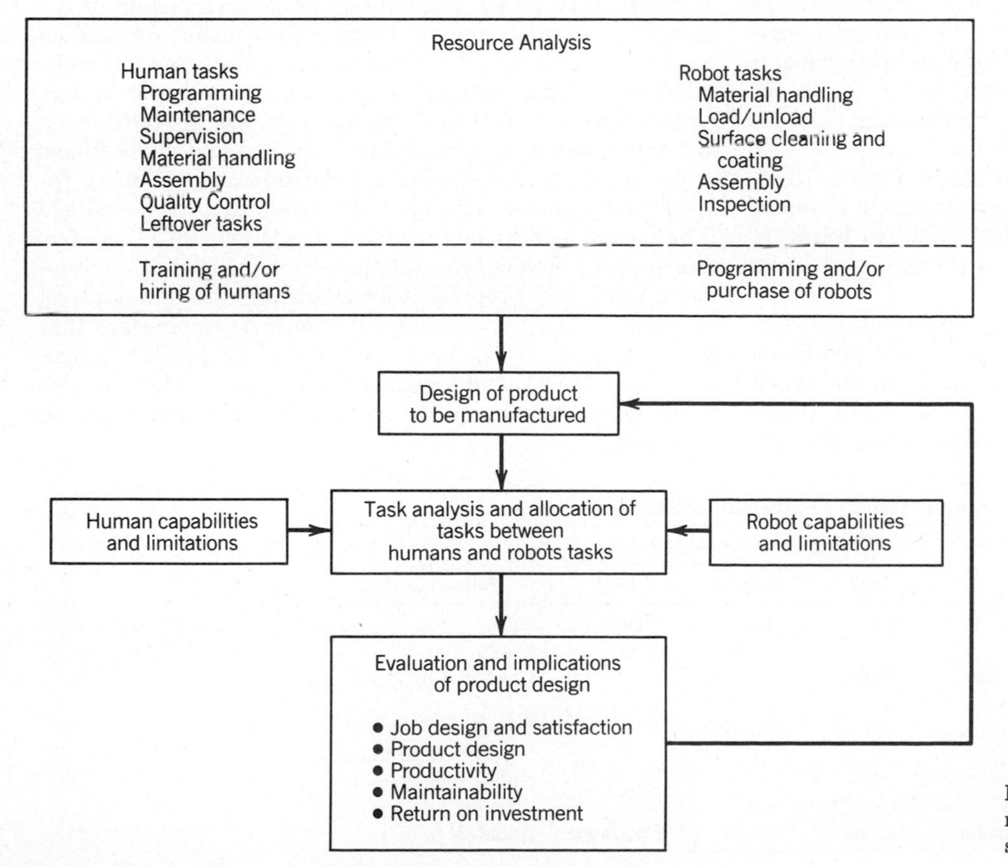

Figure 3. Task allocation between humans and robots is affected by product design.

it is feasible to train existing employees, hire new employees, program existing robots, or purchase new robots. By analyzing existing restraints, it should be possible to propose an optimum mix of humans and robots. The choice depends on both organizational and economic resources, such as the qualifications and skills of the existing work force.

Design of Products

The design of products may well be the most complex step in planning automated manufacturing since it has important ramifications for manufacturability, utilization of resources, standardization of parts, and formation of product families. This involves important strategic decisions, and it is not surprising that most manufacturers keep postponing such decisions.

Task Analysis

Task analysis, the next step in Figure 3, breaks down the components of the manufacturing process into subtasks and further into task elements. By assessing the overall requirements of the task, it is possible to specify the kind of robot that is necessary. Table 3 provides a framework for task analysis that may be used to specify robotic features and performance levels necessary for industrial assembly (27). When it comes to human capabilities, it is less appropriate to specify different performance levels since the variability among individuals is fairly minor, at least for the types of skills involved in repetitive assembly. Table 3 compares primarily materials-handling capabilities for robots and humans, characteristics that traditionally have been investigated by industrial engineers. Information processing requirements are more difficult to assess. There has been little research, and it is not yet meaningful to specify human information processing capabilities for comparison with those of robots.

Lifting capability depends upon the frequency, direction, and distance of lifts. For design of industrial tasks, it is appropriate to assume a maximum limit for repetitive lifting of about 25 lb (28). For heavy tasks, it is therefore generally advantageous to use robots.

Robotic repeatability accuracy is fairly comparable with human performance, although some robot task completion times may be a bit shorter. Human repeatability accuracy and speed of hand movements are governed by Fitts' law (29), which postulates that the time T to move the hand to a target depends on the relative precision required; that is the ratio between the distance D to a target with width W, according to the following formula:

$$T = K\,(D/W + 0.5)$$

For a human operator, the reach is limited to 15–25 in., depending upon work posture and size of operator (30). This is considerably less than robotic reach, but not necessarily a limitation since human operators are free to move around and thereby compensate for reach limitations.

For repetitive tasks, human memory is usually not a restricting factor, and comparison with robotic memory is therefore not particularly relevant. For nonrepetitive tasks, such as implied by the ever-shifting tasks in a flexible manufacturing systems, human learning is more difficult and the performance more vulnerable and error prone.

The human operator has great versatility of hand and arm movements, about 25 deg of freedom of motion. Although this exceeds by far the capacity of robots, there are many biomechanical limitations of the joints, ligaments, and tendons that make it uncomfortable to assume certain work postures. To maximize comfort, all joints should utilize only the midrange of motion (31). This restriction effectively reduces the degree of freedom of motion.

Evaluation and Implications of Product Design

After tasks have been allocated between humans and robots, there must be evaluations of the product design in terms of economic implications, feasibility, and job satisfaction. If the results are unsatisfactory, it is necessary to modify the product design. If so, the resource requirements must again be scrutinized, a new task analysis and allocation must be performed, and so forth. We will first comment on the job satisfaction and then address the feasibility and economic implications.

Job satisfaction depends on several different conditions. Table 4 lists a number of important factors that have emerged from several independent studies (eg, reference 25). These factors are not unique to the automated manufacturing environment; rather, they represent job characteristics that apply in all types of industrial settings. It should be emphasized that all of the issues in Table 4 are affected by product design, allocation of tasks, and design of the workplace. They are also fairly simple to manipulate so that job satisfaction can be increased.

Table 3. Robot and Human Performance Characteristics

Type of Performance	Robot Performance Levels			Human Performance
	Low	Medium	High	
Load capacity, lb	<15	15–100	>100	<25
Reach, in.	<20	20–50	>50	15–25
Repeatability, in.	>0.05	0.05–0.02	<0.02	Governed by Fitts' law
Memory, points	<300	300–1000	>1000	Usually not a restriction for tasks that have been trained
Degrees of freedom	2 or less	3–5	6 or more	About 25

Table 4. Criteria for Job Satisfaction

Does the job provide:
- Opportunities to cooperate with others
- Opportunities to talk to others
- Opportunities to learn new things
- A range of experiences
- Performance feedback
- Control over work pace
- Use of own judgment and decision making

The choice between humans and robots is not a simple one. It has come as a surprise to many design engineers that once the product has been designed for automatic assembly, it is also easier to assemble manually. In other words, most of the design principles that make it easier to assemble a product by a robot also facilitate manual assembly. The choice of robotic or manual assembly is, therefore, not necessarily straightforward. In many instances, when a product has been redesigned for automation, the manual assembly work has also been simplified and minimized to the extent that capital spending on automation can no longer be justified.

There are many examples of this within IBM (27). One is a paper pick mechanism for a copying machine. In this case, the original product had 27 parts. These were too many to be assembled by a robot, and the product was redesigned. The redesign incorporated several principles to facilitate robotic assembly; for example, symmetrical parts were used so that the robot would not have to turn parts around, notches were used to guide the insertion of parts, and there was a reduction in the number of screws used. After redesign, the number of parts was reduced to 14, 13 of which could be assembled by a robot. The remaining part had to be inserted by a human operator. The surprising outcome was that due to the redesign of the product, the manual assembly had become so simple that it did not pay off to use automation. This product is presently assembled manually.

Another example is a computer terminal product manufactured by IBM. All internal cables were eliminated so that the parts simply snapped together. In this case, the manual assembly time became so short that automation could not be justified. Economics and expediency considerations substantiated the manual operation, although the highly repetitive task could be predicted to result in poor job satisfaction.

One may speculate about what the design principles are that simplify manufacturing. In most cases, it seems that human assembly is simplified by the same principles as simplify robotic assembly. Humans and robots are both aided by principles such as screws with attached fasteners, springs with closed ends that cannot tangle, symmetrical parts that may be inserted in any of several directions, reduction of the number of screws, and so forth. There are, however, important differences. For example, robots prefer to insert screws vertically; horizontal insertion is much more difficult since there must be special end effectors to prevent screws from falling. In contrast, humans prefer horizontal insertion, which offers biomechanical advantages.

Design for automation implies the use of certain design principles that simplify automatic assembly. Ideally, products should be designed so that they can be assembled by humans and robots alike. Thereby, it is possible to maintain full flexibility in the choice of manufacturing method. There are already models that can predict automated assembly times (32). Research is now needed to develop models that can predict human assembly times. Without such information, it is impossible to assess the economic implications of either method of assembly.

It seems ironic that the introduction of robots has clarified the important issue of modeling, predicting, and simplifying human assembly. Although methods engineering and principles for economizing human motion have a long tradition within industrial engineering, the implications of product design on ease of assembly are virtually unknown. There is a substantial amount of information in the human factors literature, such as laboratory studies of reaction times and perceptual motor skills. Surprisingly, this research tradition was never applied in the more practical context of industrial assembly.

SAFETY AND INTERACTIVE ASPECTS

Few statistics have been published on robot accidents. It is therefore difficult to evaluate the safety of robot workplaces in detail. A study conducted in Sweden analyzed 15 accidents occurring with 270 robots (33). A common cause was "pushing the wrong controls." The accident rate was estimated to be 1 accident per 45 years. This may be compared to industrial presses, previously the most hazardous industrial machines, with an accident rate of 1 accident per 50 machine years.

A study in Japan analyzed 18 near-accidents caused by industrial robots (26). Summarizing those results and the Swedish study, the authors stated:

- Most manufacturing companies are aware of the dangers of robots.
- The accident rate of robots is significantly higher than that for other automated machines.
- A majority of accidents occur during teaching, testing, and maintenance.

As a result of this investigation, the authors commented that only when robots themselves are able to detect the approach of humans and actively avoid accidents will safety in the workplace be ensured.

The first fatal robot accident in Japan occurred in a manufacturing cell (34). There were four machines and a robot that moved parts between the machines. At the time of the accident, the robot was transporting a workpiece from a conveyor belt to one of the machines. In order to adjust a door movement of the machine, a worker stepped into the robot work area without taking off the safety chain, which would have cut the energy to the robot. Instead, he switched off the robot movement. After the repair work, the robot operation movement was reactivated. As the worker was watching the door movement of the machine, the robot arm moved forward and killed the worker. Obviously, this accident would have been avoided if proper safety procedures had been observed and the chain had been taken off.

Not all robots are unsafe. For example, a Seiko robot used for assembly of watches handles only small objects and is not

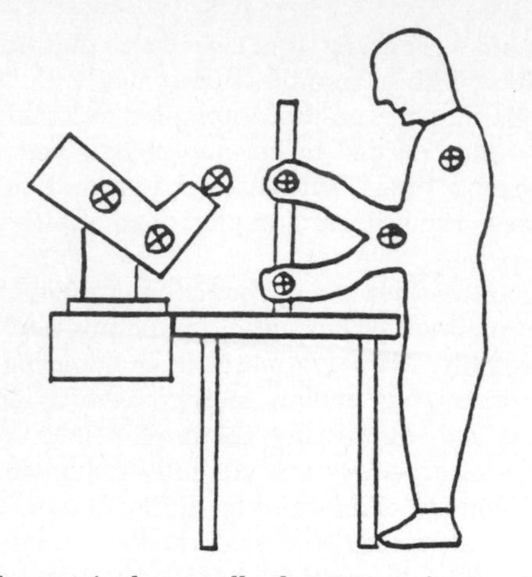

Figure 4. Human-sized or small robots are easier to work with and cause fewer safety problems than larger robots (12).

much of a threat. In contrast, a Cincinnati Milacron T3 robot, which is larger than humans, poses much more of a threat. It is therefore unlikely that large, medium-size, and small robots would require similar types of safety precautions.

It was proposed that robots be human-sized (12) (see Fig. 4). Human-sized robots are more comfortable to work with than large robots since it is easier to get a visual overview of the robot and the working environment. Working with smaller-sized robots therefore gives a better sense of control.

Situations where the worker would do a job in tandem with a robot were foreseen in reference 12. For example, a robot could be used as a flexible fixture or tool by holding a workpiece and moving it through difficult positions to make it easier for the operator to assemble (see Fig. 5).

For such situations, it may be important that the robot move in a humanlike fashion. Rather than moving in a jagged, nervous fashion according to *xyz* coordinates, the motion envelope could be smooth and humanlike (as for a spray painting robot).

Hence, there are situations, other than maintenance and programming, in which the operator may have to work close to a robot. For any task, it is important that the operator can anticipate the movements of the robot arm and any possible safety hazards.

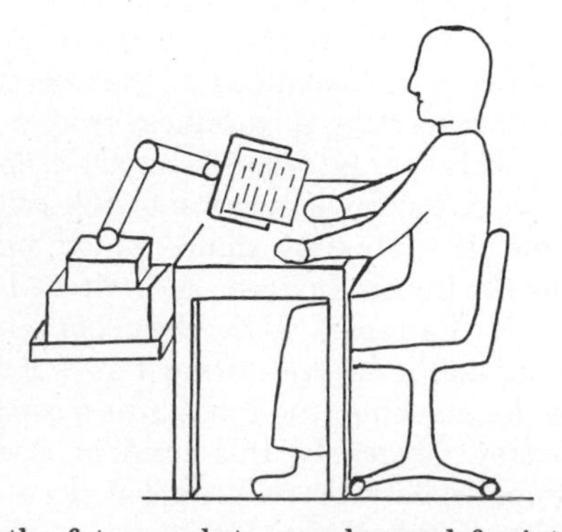

Figure 5. In the future, robots may be used for interactive tasks. The figure shows a robot used as a flexible fixture or tool to aid the human operator in assembling the product.

Perceptibility of Robot Arm Movement

Robots are machines capable of complex movement patterns, and it is difficult for operators to understand those patterns. As long as the robot arm is moving at normal speed, its power is evident, and workers and onlookers keep their distance. But when it stops or moves at a slower speed, they may move in for a closer look. This may pose a great danger since the robot may simply be stopping as a natural sequence of the program and could take off again with great force a moment later.

It has been pointed out that there are so many different stop conditions that it is difficult for operators to distinguish among them (26) (see Table 5). Emergency and temporary halts may be induced by the robot operator during the work. Malfunction and runaway halts are due to machine failure, and a condition halt is induced by software. Even for the experienced operator, it may be difficult to understand why the robot has stopped and if it is safe to approach the robot. To help the operator, a visual and auditory warning signal could be mounted on top of the robot to indicate the status of the robot and its potential danger.

To perceive safety hazards, the operator should always be able to anticipate the robot arm movements. It is important that the velocity of the robot arm be low enough that operators can perceive and avoid it. The results of several experiments that investigated the perceptibility of robot arm movements as a function of speed have been reported (30, 34). In the first experiment, a robot arm moved at different speeds from 100–500 mm/s. Subjects found that it was possible to make "work corrections" as long as the robot did not move more than 200 mm/s.

In a second experiment, the robot dropped a component part under the mechanical arm. Subjects were then asked to estimate the hazard in picking it up while the robot made a brief pause. Five robot pauses were investigated, 0, 1, 2, 3, and 4 s, and the robot moved at one of five speeds varying from 140 to 460 mm/s. The 3- and 4-s pauses were judged to be completely safe regardless of arm speed. It was concluded that in real life even the longer pauses were unsafe. The test subjects therefore misjudged the situation, and the robot should have been turned off. Hence, there is a high possibility of unsafe acts occurring with long waiting times.

In a third experiment, subjects were told to place themselves at the closest safe distance from the robot arm while the robot arm approached at speeds varying between 5.5 and 18.1 in. /s (140 and 460 mm/s). For the lower speed, the median distance was 0.5 in. (13 mm) and for the highest speed 7.6 in. (195 mm). Again, with the slower speed, operators had

Table 5. Causes for Robot Stoppage[a]

Emergency halt induced by emergency control.
Temporary halt induced by pause control.
Malfunction halt due to abnormality.
Runaway halt due to machine failure.
Condition halt for machine recycling.
Apparent halt due to fixed-point position control.
Halt due to work termination.

[a] Ref. 26.

difficulties in assessing the real hazards and positioned themselves too close to the robot.

The Robot Industries Association (RIA) recently published an ANSI safety standard for robots. This standard stipulates that in addition to the faster operating speed, robots shall have a "slow speed" for programming that shall not exceed 10 in./s (250 mm/s).

People do not behave in single acts or even series of consecutive acts, but in coordinated activity to achieve goals. Hence, when unfamiliar with a situation, a person may well set in motion a train of actions to achieve this goal without reflecting that some of these acts are unsafe. It is therefore important to train personnel in the operation of robots and procedures necessary to safely perform various tasks. Such training is also mandated in the ANSI standard.

Safety Devices

Sensors may be used to sense the presence of operators intruding in the robot workplace so that the robot will be automatically turned off. The most common types of sensor include automatic gates, floor mats, and photoelectric beams.

A photoelectric beam or an automatic gate is usually mounted on poles surrounding the workplace. Opening the gate or walking through the beam initiates a safety response. Typically, the robot is shut down, but there are other alternatives such as alerting the intruder by visual or auditory warnings, including the use of spoken messages derived through computer speech synthesis. It was suggested that programmers and maintenance personnel who must function within a robotic enclosure might wear a device that would transmit a signal to the robot's safety sensor, indicating human presence (35). Similar systems are used in air-traffic control by using transponders or radar reflectors on aircraft.

It was suggested that operators who enter the workplace detach a safety plug off the gate and pocket it (36). Thereby, other workers could not close the door by mistake and thus restart the robot. The motion of operators may be detected through so-called presence-sensing devices, which use ultrasonic, microwave, or infrared sensors. Touch-sensitive sensors, such as touch-sensitive skin or pressure-sensing gauges in the robot's joints, can also be used.

Signs in the form of catchy slogans or exhortations to be cautious are rarely effective in preventing accidents (35). For a sign to be effective, it should simulate what can happen directly. It could, for example, depict in graphics and words either the behavior that would cause an accident or the behavior that would avoid an accident.

The ANSI standard recommends that the type of safeguarding shall correspond to the type and level of hazard. This is in agreement with our observation that large and small robots require different measures. The standard mandates that intrusion by personnel be restricted by the use of one or several safeguarding devices, including presence-sensing devices, barriers, perimeter guarding, awareness barriers, and awareness signals.

ERGONOMICS IN THE WORKPLACE

Most principles of good ergonomic design apply equally to robotic workplaces and other types of environments. There are basic human requirements for work posture, biomechanics, and materials handling. This section is only a brief review of these issues and references other information sources.

Choice of Work Posture

The designer of a workplace has to make a choice among sitting, standing, and sit/stand work postures (30). Sitting workplaces are best if fine assembly or writing tasks dominate and all items can be easily handled within easy reach. A standing work posture is more appropriate for tasks with items weighing more than 10 lb. This is also appropriate for tasks with an extended reach envelope or where downward forces must be applied, as in packaging, for example. Sit/stand workplaces use a standing work posture with a high chair that can be used for occasional sitting. This is good for jobs that could be done sitting were it not for extensive reaching. It is also a good compromise for jobs with several tasks that may require either sitting or standing. Both standing and sit/stand work postures make it easier for operators to move around, and therefore add more flexibility (30).

Depending upon the task, the work surfaces should be put at different heights. For precision work with supported elbows, the work height should be approximately 2 in. above elbow height. For light assembly work, it should be about 4 in. below elbow height, and for heavy work about 8 in. below elbow height (see Fig. 6).

Guidelines for Arrangements of Controls and Other Workplace Items

Several items compete for space in the work area. There are parts to be assembled, controls, and hand tools. By doing a task analysis, it is possible to divide a task into subtasks. Controls, tools, and parts can then be arranged in the workplace according to subtask so that they can be handled in a spatially sequential order.

It is also helpful to analyze how frequently the various items are used. The most frequently used items are called primary items, and the not so frequently used items are called secondary items. The primary items should be placed closest to the operator in the primary reach envelope, and the second-

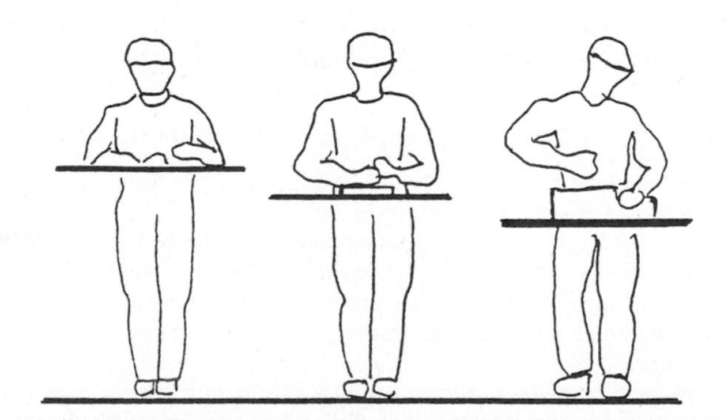

Figure 6. For precision work, the work height should be approximately 2 in. above elbow height. For light assembly work, the hands should be about 4 in. below elbow height, and for heavy work, about 8 in. below elbow height. The lower height makes it easier to apply greater force due to increased leverage.

Table 6. Guidelines for Allocation of Component Parts, Controls, Hand Tools, and Other Items

Keep the number of items to a minimum.
Distinguish between primary, secondary, and emergency items.
Locate primary items closest to the operator and secondary items farther away.
To prevent accidential activation, place emergency items away from other frequently used items.
Establish subtasks and arrange items sequentially within each subtask.
Arrange the items so that the operator can adjust posture frequently.
If one hand must reach several items in sequence, arrange the items to allow for continuous movement through an arc.
When there is only one major control that at times must be operated by either hand or both hands, place it in front of the operator.
Handedness is important only if task requires skill and dexterity.

ary items can be positioned farther away, where convenient reach is not so important. It is sometimes helpful to form a third category, consisting of emergency items such as emergency stop controls. These should be within easy reach, but placed away from other frequently used items, thereby preventing accidental activation. These and some other guidelines are summarized in Table 6.

Manual Materials Handling

Several guidelines for ergonomic principles in materials handling are summarized in Table 7. A majority of back injuries in the manufacturing industry are due to lifting. Especially hazardous is when the operator is bending to lift items from the floor while twisting/turning the body to the side. These types of combined movements can be avoided by a combination of measures.

If possible, lifting and lowering should be restricted to take place only between knuckle height and elbow height for a standing operator. (The knuckle height is the height from the floor to the knuckles.) For items located below knuckle height, the operator must bend his/her back. Lifting above shoulder height is more difficult than lower lists because the corresponding muscles are much weaker. In addition, it is more difficult to balance the load, and the risk of dropping items increases.

The "straight back–bent knees" method for lifting has been preached by industry for many years. Recent research has shown that the method is applicable only to heavy compact objects that can be held close to the body. Bulky objects are more difficult to lift this way because the arms must be held further out than if the operator were to lift with straight legs and bent back. The physiological cost (in calories per foot pound of lifted items) is also greater. It is difficult to enforce the straight back–bent knees method, even for the types of lifting where it is appropriate.

Table 7. Rules for Improving Manual Materials Handling

The predominant movement should be horizontal. Pushing and pulling is preferred over lifting and lowering.
Material delivered to the workplace should be located at convenient height, ideally between knuckle height and elbow height.
Lifting and lowering should take place between knuckle height and shoulder height. Otherwise, injuries due to overexertion are likely to occur.
The straight back–bent knees method for lifting is only applicable for heavy, compact objects. It is difficult to use for bulky objects.
Make sure that the material is light, compact, and has handles so that it is easy to grip.
Bins and containers for material must be easily removed so that operators do not have to dive into a container to reach material.

A better approach is to redesign the job and work environment so that heavy lifting is eliminated. Although this may at first glance appear to be a very costly proposition, there are several simple measures that would be helpful. An example is the use of storage racks located at a convenient height. The motto is "Don't put it on the floor so you won't have to pick it up again." Other inexpensive measures are to use boxes with handles and storage bins that are easily accessible.

CONCLUSION

This article has provided a systems overview of several types of ergonomic problems in the use of robotics.

One important issue is task allocation. Much research must be done in this area to develop models that can allocate tasks to optimize both productivity and job satisfaction. Such models must eventually be incorporated in CAD/CAM so that engineers can readily use the information. It may seem ironic that presently it is easier to predict robotic assembly time than human performance. Until such information is available, it will be difficult to make decisions of whether or not to automate.

One of the major problems in robotic safety is the lack of accident data. Without such information, it is difficult to understand how robotics safety can best be improved. Over time, as more robots are used and more accidents reported, this problem will be solved. It is important to collect the right type of information. Such efforts should be coordinated by federal agencies. Meanwhile, ergonomics research could investigate issues such as perceptibility of robot arm movements and operator risk-taking behavior. Such information can be collected both in controlled laboratory settings and in the field and would complement the accident statistics (see also ERGONOMICS, WORK PLANNING; ERGONOMICS, ROBOT SELECTION).

BIBLIOGRAPHY

1. O. Ostberg and J. Enqvist, "Robotics in the Workplace: Robot Factors, Human Factors, and Humane Factors," in H. W. Hendrick and O. Brown, Jr., eds, *Human Factors in Organizational Design and Management,* North-Holland, Amsterdam, the Netherlands, 1984.
2. K. Noro and Y. Okada, "Robotization and Human Factors," *Ergonomics* **6,** 985–1000 (1983).
3. H. M. Parsons and O. P. Kearsley, "Robotics and Human Factors: Current Status and Future Prospects," *Human Factors* **24,** 535–552 (1982).
4. G. Salvendy, "Review and Reappraisal of Human Aspects in Planning Robotic Systems," *Behavior and Information Technology* **2,** 263–287 (1983).
5. A. J. Macek, "Human Factors Facilitating the Implementation of Automation," *Journal of Manufacturing Systems* **1,** 195–206 (1981).

6. G. Salvendy and M. J. Smith, *Machine Pacing and Occupational Stress,* Taylor and Francis, London, 1981.
7. E. J. McCormick and M. S. Sanders, *Human Factors in Engineering and Design,* 5th ed., McGraw-Hill, New York, 1985.
8. E. Grandjean, *Fitting the Task to the Man,* Taylor and Francis, London, 1980.
9. W. E. Woodson, *Human Factors Design Handbook,* McGraw-Hill, New York, 1981.
10. R. Mittelstadt, "Robotics—Thoughts about the Future. *Proceedings of the Robots 8 Conf.,* Society of Manufacturing Engineers, Dearborn, Mich., 1984, pp. 1:12–2:21.
11. J. Fleck, "The Adoption of Robots," *Proceedings of the Robots 7 Conf.,* Society of Manufacturing Engineers, Dearborn, Mich., 1983, pp. 1:41–1:51.
12. J. F. Engelberger, *Robots in Practice,* American Management Association, New York, 1980.
13. R. U. Ayres and S. M. Miller, "Robotics Realities: Near-Term Prospects and Problems, *Annals of the American Academy of Political and Social Science* **470,** 28–55 (1983).
14. Japanese Industrial Safety and Health Association. *Prevention of Industrial Accidents due to Robots,* Tokyo, Japan, 1983.
15. O. Ostberg, *Review of Workplace Aspects of Robot-Based Production,* draft report. National Institute for Occupational Safety and Health, Cincinnati, Oh, 1985.
16. T. B. Sheridan, "Modeling Supervisory Control of Robots," in A. Morecki and K. Kendzior, eds., *Proceedings of Symposium on Theory and Practice of Robots and Manipulators,* Elsevier, Amsterdam, the Netherlands, 1977, pp. 894–1106.
17. R. Langmoen, "Automatic Assembly of Electric Heaters," *Proceedings of the 12th International Symposium on Industrial Robots,* Paris, France, 1982.
18. J. Sharit, "Supervisory Control of a Flexible Manufacturing System," *Human Factors* **27,** 47–60 (1985).
19. D. E. Jarvis, Review of Applications and Operation Experience in the Automobile Industry with Devillbiss Trallfa Spray Painting Robots," in *Robots in the Automotive Industry,* IFS Publications, Bedford, England, 1982.
20. H. G. Shulman and M. B. Olex, "Designing the User-Friendly Robots: A Case History," *Human Factors* **27,** 91–98 (1985).
21. L. V. Rigby and A. D. Swain, "Effects of Assembly Error on Product Acceptability and Reliability," *Proceedings of the 7th Annual Reliability and Maintainability Conference,* American Society of Mechanical Engineers, New York, 1968, pp. 3.12–3.19.
22. K. Ghosh and C. Lemay, "Man/Machine Interaction in Robotics and Their Effect on the Safety at the Workplace," *Proceedings of the Robots 9 Conference,* Society of Manufacturing Engineers, Dearborn, Mich., 1985.
23. B. Schneiderman, "Correct, Complete Operations and Other Principles of Interaction," in G. Salvendy, ed., *Human–Computer Interaction,* Elsevier, Amsterdam, the Netherlands, 1985.
24. A. Macek, L. Heeriga, B. Somberg, D. Sauer, D. Robbins, and J. Howard, *Human Factors Affecting ICAM Implementation.* Materials Laboratory, Wright-Patterson AFB, Oh., Rep. AFWAL-TR-81-4095, 1981.
25. H. H. Rosenbrock, "Engineers, Robots and People," *Chemistry and Industry* **19,** 756–759 (1982).
26. N. Sugimoto and K. Kawaguchi, "Fault Tree Analysis of Hazard Created by Robots," *Proceedings of the Robots 7 Conference,* Society of Manufacturing Engineers, Dearborn, Mich., 1983, pp. 9:13–9:28.
27. M. G. Helander and K. Domas, "Task Allocation Between Humans and Robots in Manufacturing," *Material Flow* **3,** 175–185 (1986).
28. National Institute of Occupational Safety and Health, *Work Practices Guide for Manual Lifting,* U.S. Dept. of Health and Human Services, Cincinnati, Oh., 1981.
29. A. T. Welford, *Fundamentals of Skills,* Methuen, London, 1968.
30. Eastman Kodak Co., *Ergonomic Design for People at Work,* Lifetime Learning Pub., Belmont, Ga., 1983.
31. N. Corlett, "The Human Body at Work: New Principles for Designing Workspaces and Methods," *Management Services,* May 1–8, 1978.
32. G. Boothroyd, *Design for Assembly Handbook,* Department of Mechanical Engineering, University of Massachusetts, Amherst, Mass., 1982.
33. J. Carlson, L. Harms-Ringdahl, and U. Kjellen, *Industrial Robots and Accidents at Work,* Royal Institute of Technology, Stockholm, Sweden, 1979.
34. M. Nagamachi, "Industrial Robot Safety Management in Japanese Enterprises," *unpublished paper,* Department of Industrial Engineering, Hiroshima University, Hiroshima, Japan, 1984.
35. H. M. Parsons, "Human Factors in Robot Safety," Paper presented at Robots East Exposition, Essex Corporation, Alexandria, Va., 1985.
36. M. Nagamachi and Y. Anayama, "An Ergonomic Study of Industrial Robots (1)—The Experiments of Unsafe Behavior in Robot Manipulations," *Japanese Journal of Ergonomics (in Japanese)* **19,** 259–264 (1983).

General References

B. K. Ghosh and M. G. Helander. "A Systems Approach to Task Allocation of Human–Robotic Interaction in Manufacturing," *Journal of Manufacturing Systems,* **5,** 41–49 (1986).

M. Helander, "Documentation from Workshop on Human Factors in Robotics," presented at the International Conference on Industrial Ergonomics, Human Factors Society, Toronto, Canada, 1984.

EXPERT SYSTEMS

William B. Gevarter
National Aeronautics and Space Administration
Ames Research Center
Moffett Field, California

INTRODUCTION

Expert systems is probably the "hottest" topic in artificial intelligence (AI) today. Prior to the last decade, AI researchers tended to rely on nonknowledge-guided search techniques or computational logic when trying to find solutions to problems. These techniques were successfully used to solve elementary problems or very structured problems such as games. However, the search space of complex problems tends to expand exponentially with the number of parameters involved. For such problems, older techniques had generally proved inadequate and a new approach was needed. This emphasized knowledge rather than search and has led to the field of knowledge engineering and expert systems. The resultant expert systems technology, limited to academic laboratories in the 1970s, is now becoming cost-effective and is beginning to enter commercial applications.

WHAT IS AN EXPERT SYSTEM?

An expert system is an intelligent computer program that uses knowledge and inference procedures to solve problems that are difficult enough to require significant human expertise for their solution. The knowledge necessary to perform at such a level, plus the inference procedures used, can be thought of as a model of the expertise of the best practitioners of the field.

The knowledge of an expert system consists of facts and heuristics. The "facts" constitute a body of information that is widely shared, publicly available, and generally agreed upon by experts in a field. The "heuristics" are mostly private, little-discussed rules of good judgement (rules of plausible reasoning, rules of good guessing) that characterize expert-level decision making in the field. The performance level of an expert system is primarily a function of the size and quality of the knowledge base that it possesses(1).

It has become fashionable today to characterize any large, complex AI system that uses large bodies of domain knowledge as an expert system. Thus, nearly all AI applications to real-world problems can be considered in this category, though the designation "knowledge-based systems" is more appropriate.

THE BASIC STRUCTURE OF AN EXPERT SYSTEM

An expert system consists of:

1. A knowledge base (or knowledge source) of domain facts and heuristics associated with the problem.
2. An inference procedure (or control structure) for using the knowledge base to solve a problem.
3. A working memory, "global data base," for keeping track of the problem status, the input data for the particular problem and the relevant history of what has thus far been done.

A human "domain expert" usually collaborates to help develop the knowledge base. Once the system has been developed, it can also be used to help instruct others in developing their own expertise in addition to solving problems.

It is better, though not yet common, to have a user-friendly natural language interface to facilitate the use of the system in all three modes: development, problem solving, and instruction. In most sophisticated systems, an explanation module is also included, allowing the user to challenge and examine the reasoning process underlying the system's answers. Figure 1 is a diagram of an idealized expert system. When the domain knowledge is stored as production rules, the knowledge base is often referred to as the "rule base," and the inference engine as the "rule interpreter."

An expert system differs from more conventional computer programs in several important respects. In an expert system ". . . there is a clear separation of general knowledge about the problem (the rules forming a knowledge base) from information about the current problem (the input data) and the methods for applying the general knowledge to the problem (the rule interpreter)."

In a conventional computer program, knowledge pertinent to the problem and methods for using this knowledge are combined, so it is difficult to change the program. In an expert system, " . . . the program itself is only an interpreter (or general reasoning mechanism) and (ideally) the system can be changed by simply adding or subtracting rules in the knowledge base (2)."

THE KNOWLEDGE BASE

The most popular approach to representing the domain knowledge (both facts and heuristics) needed for an expert system is by production rules (also referred to as "SITUATION-ACTION rules" or "IF-THEN rules"). Not all expert systems are rule-based. The classic network-based expert systems MACSYMA, INTERNIST/CADUCEUS, Digitalis Therapy Advisor, HARPY and PROSPECTOR are examples which are

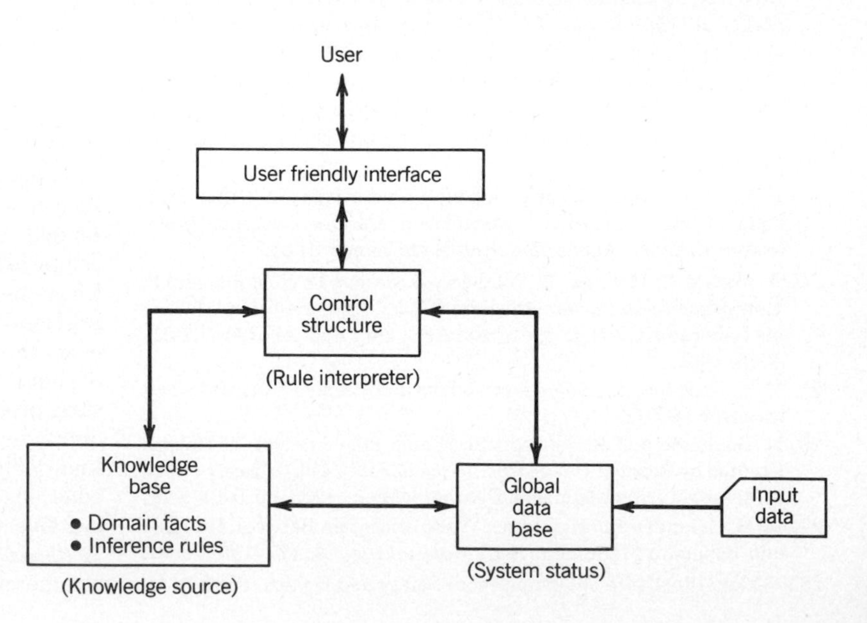

Figure 1. Basic structure of an expert system.

not. The basic requirements in the choice of an expert system knowledge representation scheme are extendibility, simplicity and explicitness. Thus, rule-based systems are particularly attractive (3). Often a knowledge base is made up mostly of rules which are invoked by pattern matching with features of the task environment as they currently appear in the global data base. However, representing a knowledge base in part as a hierarchical collection of objects is now coming into vogue.

THE CONTROL STRUCTURE

In an expert system a problem-solving paradigm must be chosen to organize and control the steps taken to solve the problem. A common, but powerful approach involves the chaining of IF-THEN rules to form a line of reasoning. The rules are actuated by patterns in the global data base. The application of the rule changes the system status and therefore the data base, enabling some rules and disabling others. The rule interpreter uses a control strategy for finding the enabled rules and for deciding which of them rules to apply. The basic control strategies used may be top-down (backward-chaining or goal driven), bottom-up (forward-chaining or data driven), or a combination that uses a relaxationlike convergence process to join these opposite lines of reasoning at some intermediate point to yield a solution. Virtually all the heuristic search and problem solving techniques that the AI community has devised have appeared in the various expert systems.

ARCHITECTURE OF EXPERT SYSTEMS

One way to classify expert systems is by function (eg, diagnosis, planning, etc). However, examination of existing expert systems indicates that there is little commonality in detailed system architecture that can be detected from this classification. A more fruitful approach appears to be to look at problem complexity and problem structure and deduce what data and control structures might be appropriate to handle these factors. The use of the techniques in four classic expert systems is illustrated in Tables 1–4. Table 1–4 outlines the basic approach taken by each of these expert systems and shows how the approach translates into key elements of the Knowledge Base, Glodal Data Base and Control Structure. An indication of the basic control structures of the systems in Tables 1–4, and some of the other well known expert systems, is given in Table 5.

Table 1. Characteristics of Example Expert Systems (Dendral)[a]

		Key Elements of		
Purpose	Approach	Knowledge Base	Global Data Base	Control Structure
Generate plausible structural representations of organic molecules from mass spectrogram data.	1. Derive constraints from the data. 2. Generate candidate structures. 3. Predict mass spectrographs for candidates. 4. Compare with data.	Rules for deriving constraints on molecular structure from experimental data Procedure for generating candidate structures to satisfy constraints Rules for predicting spectrographs from structures	Mass spectrogram data Constraints Candidate structures	Forward chaining Plan, generate and test.

[a] Institution: Stanford University.
Authors: Feigenbaum and Lederberg.
Function: Data Interpretation.

Table 2. Characteristics of Example Expert Systems (AM)[a]

		Key Elements of		
Purpose	Approach	Knowledge Base	Global Data Base	Control Structure
Discovery of mathematical concepts.	Start with elementary ideas in set theory. Search a space of possible conjectures that can be generated from these elementary ideas.	Elementary ideas in finite set theory. Heuristics for generating new mathematical concepts by modifying and combining elementary ideas.	Plausible candidate concepts.	Plan, generate, and test.
	Choose the most interesting conjectures and pursue that line of reasoning.	Heuristics of "interestingness" for discarding bad ideas.		

[a] Institution: Stanford University.
Authors: Lenat.
Function: Concept Formation.

Table 3. Characteristics of Example Expert Systems (R1)

Purpose	Approach	Key Elements of: Knowledge Base	Global Data Base	Control Structure
Configure VAX computer systems (from a customer's order of components).	Break problem up into the following ordered subtasks: 1. Correct mistakes in order. 2. Put components into CPU cabinets. 3. Put boxes into unibus cabinets and put components in boxes. 4. Put panels in unibus cabinets. 5. Lay out system on floor. 6. Do the cabling. Solve each subtask and move on to the next one in the fixed order.	Properties of (roughly 500) VAX components. Rules for determining when to move to next subtask based on system state. Rules for carrying out subtasks (to extend partial configuration). (Approximately 1200 rules total)	Customer order. Current task. Partial configuration (System state).	"MATCH" (data driven) (no backtracking)

[a] Institution: CMU.
Authors: McDermott.
Function: Design.

Table 4. Characteristics of Example Expert Systems (MYCIN)[a]

Purpose	Approach	Key Elements of: Knowledge Base	Global Data Base	Control Structure
Diagnosis of bacterial infections and recommendations for antibiotic therapy.	Represent expert judgmental reasoning as condition-conclusion rules together with the expert's "certainty" estimate for each rule. Chain backwards from hypothesized diagnoses to see if the evidence supports it. Exhaustively evaluate all hypotheses. Match treatments to all diagnoses which have high certainty values.	Rules linking patient data to infection hypotheses. Rules for combining certainty factors. Rules for treatment.	Patient history and diagnostic tests. Current hypothesis. Status. Conclusions reached thus, far, and rule numbers justifying them.	Backward chaining thru the rules. Exhaustive search.

[a] Institution: Stanford University
Authors: Shortliffe
Function: Diagnosis

Table 5 represents expert-system control structures in terms of the search direction, the control techniques used, and the search space transformations employed. The approaches used, and the search space transformations employed. The approaches used in the various expert systems are different implementations of two basic ideas for overcoming the combinatorial explosion associated with search in real, complex problems. These two ideas are (*1*) Find ways to efficiently search a space, (*2*) Find ways to transform a large search space into smaller manageable chunks that can be searched efficiently.

There is little architectural commonality based either on function or domain of expertise. Instead, the design of complex expert systems may best be considered as an art form, like custom home architecture, in which the chosen design can be implemented from the collection of available AI techniques in heuristic search and problem solving.

In addition to the techniques indicated in Table 5, also emerging are distributed knowledge and problem solving approaches exemplified by the MDX expert system (4) and the object-oriented programming language, LOOPS (5).

EXISTING EXPERT SYSTEMS

Table 6 is a list, classified by function and domain of use, of many of the pioneering expert systems. There is a predominance of systems in the medical and chemistry domains following from the pioneering efforts at Stanford University. Stanford University dominates in the number of these systems, followed by MIT, CMU, BBN and SRI, with several dozen efforts scattered elsewhere.

The list indicates that the major areas of expert systems development have been in diagnosis, data analysis and interpretation, planning, computer-aided instruction, analysis, and automatic programming. However, the list also indicates that a number of pioneering expert systems exist in quite a number of other functional areas. In addition, a substantial effort is

Table 5. Control Structures of Some Well-Known Expert Systems

			Control Structure																
			Search Direction				Control								Search Space Transformations				
System	Function	Domain	Forward	Backward	Forward and Backward	Event Driven	Exhaustive Search	Generate and Test	Guessing	Relevant Backtracking	Least Commitment	Multilines of Reasoning	Network Editor	Beam Search	Multiple Models	Break into Subproblems	Hierarchical Refinement	Hierarchical Resolution	Meta Rules
MYCIN	Diagnosis	Medicine		x			x												
DENDRAL	Data interpretation	Chemistry	x					x											
EL	Analysis	Electric circuits	x						x	x									
GUIDON	Computer-aided instruction	Medicine				x													
KAS	Knowledge acquisition	Geology	x										x						
META DENDRAL	Learning	Chemistry	x					x											
AM	Concept formation	Math	x					x											
VM	Monitoring	Medicine				x	x												
GAI	Data interpretation	Chemistry	x					x											
R1	Design	Computers	x				x									x			
ABSTRIPS	Planning	Robots		x													x		
NOAH	Planning	Robots		x							x					x			
MOLGEN	Design	Genetics			x				x	x	x					x	x		x
SYN	Design	Electric circuits	x												x				
HEARSAY II	Signal interpretation	Speech understanding			x						x	x						x	
HARPY	Signal interpretation	Speech understanding	x											x					
CRYSALIS	Data interpretation	Crystallography				x		x											x

under way to build expert systems as tools for constructing expert systems.

CONSTRUCTING AN EXPERT SYSTEM

To construct a successful expert system:

- There must be at least one human expert acknowledged to perform the task well.
- The primary source of the expert's exceptional performance must be special knowledge, judgment, and experience.
- The expert must be able to explain the special knowledge and experience and the methods used to apply them to particular problems.
- The task must have a well defined domain of application.

Using present techniques and programming tools, developing a major expert system takes about two years. Two to five people usually work on major projects.

SUMMARY OF THE STATE-OF-THE-ART

The state of the art in expert systems is characterized by (6):

- Narrow domain of expertise

Because of the difficulty in building and maintaining a large knowledge base, the typical domain of expertise is narrow. The principal exception is INTERNIST, for which the knowledge base covers 500 disease diagnoses. However, this broad coverage is achieved by using a relatively shallow set of relationships between diseases and associated symptoms. (INTERNIST is now being replaced by CADUCEUS, which uses causal relationships to help diagnose simultaneous unrelated diseases.)

- Limited knowledge representation languages for facts and relations
- Relatively inflexible and stylized input-output languages
- Stylized and limited explanations by the systems
- The use of commercial expert system building tools for construction

At present, it usually requires a knowledge engineer to work with a human expert to extract and structure the information to build the knowledge base.

- Single expert as a "knowledge czar."

Table 6. Some Pioneering Expert Systems by Function

Function	Domain	System[a]	Institution
Diagnosis	Medicine	PIP	MIT
	Medicine	CASNET	Rutgers University
	Medicine	INTERNIST/CADUCEUS	University of Pittsburgh
	Medicine	MYCIN	Stanford University
	Medicine	PUFF	Stanford University
	Medicine	MDX	Ohio State University
	Computer faults	DART	Stanford University/IBM
	Computer faults	IDT	DEC
	Nuclear reactor accidents	REACTOR	E G & G Idaho Inc.
Data analysis and interpretation	Geology	DIPMETER ADVISOR	MIT/Schlumberger
	Chemistry	DENDRAL	Stanford University
	Chemistry	GAI	Stanford University
	Geology	PROSPECTOR	SRI
	Protein crystallography	CRYSALIS	Stanford University
	Determination of causal relationships in medicine	RX	Stanford University
	Determination of causal relationships in medicine	ABEL	MIT
	Oil-well logs	ELAS	AMOCO
Analysis	Electrical circuits	EL	MIT
	Symbolic mathematics	MACSYMA	MIT
	Mechanics problems	MECHO	Edinburgh
	Naval Task Force threat analysis	TECH	Rand/NOSC
	Earthquake damage assessment for structures	SPERIL	Purdue University
	Digital circuits	CRITTER	Rutgers University
Design	Computer system configurations	R1/XCON	CMU/DEC
	Circuit synthesis	SYN	MIT
	Chemical synthesis	SYNCHEM	SUNY Stonybrook
Planning	Chemical synthesis	SECHS	University of California, Santa Cruz
	Robotics	NOAH	SRI
	Robotics	ABSTRIPS	SRI
	Planetary flybys	DEVISER	JPL
	Errand planning	OP-PLANNER	Rand
	Molecular genetics	MOLGEN	Stanford University
	Mission planning	KNOBS	MITRE
	Job shop scheduling	ISIS-II	CMU
	Design of molecular genetics experiments	SPEX	Stanford University
	Medical diagnosis	HODGKINS	MIT
	Naval aircraft Ops	AIRPLAN	CMU
	Tactical targeting	TATR	RAND
Learning from experience	Chemistry	METADENDRAL	Stanford University
	Heuristics	EURISKO	Stanford University
Concept formation	Mathematics	AM	CMU
Signal interpretation	Speech understanding	HEARSAY II	CMU
	Speech understanding	HARPY	CMU
	Machine acoustics	SU/X	Stanford University
	Ocean surveillance	HASP	System Controls Inc.
	Sensors on board naval vessels	STAMMER-2	NOSC, San Diego/SDC
	Medicine—left ventrical performance	ALVEN	University of Toronto
	Military situation determination	ANALYST	MITRE
Monitoring	Patient respiration	VM	Stanford University
Use advisor	Structural analysis computer program	SACON	Stanford University
Computer-aided instruction	Electronic troubleshooting	SOPHIE	BBN
	Medical diagnosis	GUIDON	Stanford University
	Mathematics	EXCHECK	Stanford University
	Steam propulsion plant operation	STEAMER	BBN
	Diagnostic skills	BUGGY	BBN
	Causes of rainfall	WHY	BBN
	Coaching of a game	WEST	BBN
	Coaching of a game	WUMPUS	MIT
		SCHOLAR	BBN
Knowledge acquisition	Medical diagnosis	TEIRESIAS	Stanford University
	Medical consultation	EXPERT	Rutgers
	Geology	KAS	SRI

Table 6. (*continued*)

Function	Domain	System[a]	Institution
Expert-system construction		ROSIE	Rand
		AGE	Stanford University
		HEARSAY III	USC/ISI
		EMYCIN	Stanford University
		OPS 5	CMU
		RAINBOW	IBM
	Medical diagnosis	KMS	University of Maryland
	Medical consultation	EXPERT	Rutgers
	Electronic systems diagnosis	ARBY	Smart Sys. Tech.
	Medical consultation using time-oriented data	MECS-AI	Tokyo University
Consultation/intelligent assistant	Battlefield weapons assignments	BATTLE	NRL AI Lab
	Medicine	Digitalis Therapy Advisor	MIT
	Radiology	RAYDEX	Rutgers University
	Computer sales	XCEL	CMU/DEC
	Medical treatment	ONCOCIN	Stanford University
	Nuclear power plants	CSA Model-Based Nuclear Power Plant Consultant	Georgia Tech
	Diagnostic prompting in medicine	RECONSIDER	University of California, San Francisco
Management	Automated factory	IMS	CMU
	Project management	CALLISTO	DEC
Automatic programming	Modeling of oil well logs	ΦNIX	Schlumberger-Doll Res.
		CHI	Kestrel Inst.
		PECOS	Stanford University
		LIBRA	Stanford University
		SAFE	USC/ISI
		DEDALUS	SRI
		Programmer's Apprentice	MIT
Image understanding		VISIONS	University of Mass.
		ACRONYM	Stanford University

[a] Refs. 2, 3, 5–9.

The ability to maintain consistency among overlapping items in the knowledge has is limited. Therefore, though it is desirable for several experts to contribute, one expert must maintain control to ensure the quality of the data base.

- Fragile behavior

Most systems exhibit fragile behavior at the boundaries of their capabilities. Thus, even some of the best systems come up with wrong answers for problems just outside their domain of coverage. Even within their domain, systems can be misled by complex or unusual cases, or for cases for which they do not yet have the needed knowledge or for which even the human experts have difficulty.

- Delivery of developed expert systems on personal computers is becoming commonplace.

Despite a rough beginning, there have been notable successes. A methodology has been developed for explicating informal knowledge. Representing and using empirical associations, five of the pioneering systems (DENDRAL, MACSYMA, MOLGEN, R1 and PUFF) have been routinely solving difficult problems and are in regular use. The first three all have serious users who are only loosely coupled to the system designers. DENDRAL, which analyzes chemical instrument data to determine the underlying molecular structure, has been the more widely used program. R1, which is used to configure VAX computer systems, has been reported to be saving DEC twenty million dollars per year, and has been followed up with XCON. As indicated in Table 6, dozens of pioneering systems were constructed. These early systems have been followed by hundreds of others destined for daily use.

FUTURE TRENDS

- Medical diagnosis and prescription
- Medical knowledge automation
- Chemical data interpretation
- Chemical and biological synthesis
- Mineral and oil exploration
- Planning/scheduling
- Signal interpretation
- Space defense
- Air traffic control
- Circuit diagnosis
- VLSI design
- Equipment fault diagnosis
- Computer configuration selection
- Speech understanding
- Intelligent computer-aided instruction
- Automatic programming

- Signal fusion—situation interpretation from multiple sensors
- Military threat assessment
- Tactical targeting
- Intelligent knowledge base access and management
- Tools for building expert systems

There appear to be few domain or functional limitations in the ultimate use of expert systems. However, the nature of expert systems is changing. The limitations of pure rule-based systems became apparent. Not all knowledge can be readily structured in the form of empirical associations. Empirical associations tend to hide causal relations (present only implicitly in such associations). Empirical associations are also inappropriate for highlighting structure and function.

Thus, the newer expert systems are adding deep knowledge having to do with causality and structure. These systems will be less fragile, thereby holding the promise of yielding correct answers often enough to be considered for use in autonomous systems, not just as intelligent assistants.

The other change is a trend towards an increasing number of non-rule based systems. These systems, using semantic networks, frames and other knowledge representations, are often better suited for causal modeling and representing structure. They also tend to simplify the reasoning required by providing knowledge representation more appropriate for the specific problem domain.

Some of the future opportunities for expert systems are

- *Building and Construction*

  Design, planning, scheduling, control

- *Equipment*

  Design, monitoring, control, diagnosis, maintenance, repair, instruction

- *Command and Control*

  Data fusion, intelligence analysis, planning, targeting, communication

- *Weapon Systems*

  Target identification, adaptive control, electronic warfare

- *Professions*

  (Medicine, law, accounting, management, real estate, financial, engineering) consulting, instruction, analysis

- *Education*

  Instruction, testing, diagnosis, concept formation and new knowledge development from experience

- *Imagery*

  Photo interpretation, mapping, geographic problem-solving

- *Software*

  Instruction, specification, design, production, verification, maintenance

- *Home Entertainment and Advice-giving*

  Intelligent games, investment and finances, purchasing, shopping, intelligent information retrieval

- *Intelligent Agents*

  To assist in the use of computer-based systems

- *Office Automation*

  Intelligent systems

- *Process Control*

  Factory and plant automation

- *Exploration*

  Space, prospecting, etc

It thus appears that expert systems will eventually find use whenever symbolic reasoning with detailed professional knowledge is required. In the process, there will be exposure and refinement of the previously private knowledge in the various fields of applications.

On a more near-term scale, the next few years will increasingly see expert systems with thousands of rules. In addition to the increasing number of rule-based systems one can expect to see an increasing number of nonrule-based systems (10). Much improved explanation systems that can explain why an expert system did what it did and what things are of importance, can also be expected.

By the 1990s, intelligent, friendly and robust human interfaces and much better system building tools can be expected.

Somewhere around the year 2000, systems which semiautonomously develop knowledge bases from text will appear. The result of these developments may very well herald a maturing information society where expert systems put experts at everyone's disposal. In the process, production and information costs should greatly diminish, opening up major new opportunities for societal betterment.

BIBLIOGRAPHY

1. E. A. Feigenbaum, *Knowledge Engineering for the 1980's,* Computer Science Dept., Stanford University, 1982.
2. R. O. Duda, "Knowledge-Based Expert Systems Come of Age," *Byte,* (9) 238–281 (Sept. 1981).
3. B. G. Buchanan and R. O. Duda, "Principles of Rule-Based Expert Systems," Heuristic Programming Project Report No. HPP 82–14, Dept. of Computer Science, Stanford, Calif., Aug. 1982. (To appear in M. Yorit., ed., *Advances in Computers,* Vol. 22, Academic Press, New York.
4. B. Chandrasekaran, "Towards a Taxonomy of Problem Solving Types," *The AI Magazine,* **4** (1), 9–17 (1983).

5. M. Stefik, J. Alkins, R. Balzer, J. Benoit, L. Birnbaum, R. Hayes-Roth, and E. Sacerdoti, "The Organization of Expert Systems, A Tutorial," *Artificial Intelligence,* **18,** 135–173 (1982).
6. B. G. Buchanan, *"Research on Expert Systems,"* Report No. STAN-CS-81-837, Stanford University Computer Science Department, Stanford, Calif., 1981.
7. A. Barr and E. A. Feigenbaum, *The Handbook of Artificial Intelligence,* Vol. 2, W. Kaufman, Los Altos, Calif., 1982.
8. *IJCAI-81—Proc. of The International Joint Conference on AI,* Vancouver, Aug. 1981.
9. *AAAI-82—Proc. of the National Conference on AI,* CMU and the University of Pittsburgh, Pittsburgh, Penn. Aug. 18–22, 1982.
10. W. B. Gerarter, "The Nature and Evaluation of Commercial Expert System Building Tools," *Computer* **20** (5), 24–41 (1987).

F

FABRICATION AND MACHINE LOADING APPLICATIONS

William Uhde
UAS Automation Systems
Bristol, Connecticut

When robots were first developed, it was assumed that machine loading and unloading would be their primary applications (see also Machine loading/unloading). Reality was somewhat different, due in large part to the complexity of the interface between machine and robot. Automatic fixturing and sequence control are required before robot servicing is feasible. Machines work in concert with other machines to produce a product, and advances in the technology of integration of these machines, such as group technology, FMS, and CIM, all increase the efficiency of robots in a work cell environment.

GENERAL TYPES OF SYSTEMS

In a typical job shop, most of the process time of a part from start to finish is spent in storage. Some process float is necessary to assure high machine utilization factors, but having a part rest in storage for weeks when total machining time is measured in minutes is not good conservation of capital. It is important to first identify the type of system that is required. Process conditions can range as follows:

1. A single machine performing all the machining operations on a part with short lot sizes and wide variations in parts.
2. High volume, limited part changes with several machines performing duplicate functions.
3. Intermediate volume, several part numbers that can be assigned to a limited number of families with machines performing consecutive operations on the parts.

Other factors important in evaluating system concepts are machine cycles, electrical interfaceability, capital value of machinery, mechanical reliability, and mechanical means to accept automatic loading equipment.

WHEN NOT TO CONSIDER ROBOTS

If a machine does much work on a part over a long period of time and there are no overriding health or safety problems to consider, robot automation is probably not cost-effective. Transfer pallets may serve well where parts are very large and consecutive machine operations are required. A robot alternative is a large, far-reaching intelligent gantry robot working as a transfer crane, if part to part variations are not too great. When volume is low, part to part variation is great and there is no family resemblance, or total machine utilization is low, automation is not cost-effective.

MACHINE SPECIFICATIONS

Even if automation is not immediately implemented for a machine being purchased, careful planning is required so that it can be automated when the right volume is reached or the part becomes suitable for automatic processing.

Cutting machine tools require large coolant pumps so that parts can be flooded during cutting, allowing chips to wash downward. Automatic chip removal equipment is required to get scrap into waste bins so that the machine does not have to be stopped for scrap removal.

Automatic tool changing is desired, so that sharp cutting points are available without stopping for setup. An alternative is fully equipped turrets holding duplicate tools. Hollow spindles in horizontal cutting machine tools are useful so that coolant has access "behind" the part or can wash chips out of a bored hole past the boring bar.

Automatic chucking or fixturing is required so that parts can be loaded and unloaded through I/O commands with the robot loader. Sufficient I/O controls are needed in the machine tools so that other equipment can be interfaced to make an operation completely automatic.

In-process or interfaceable post-process gauging is necessary to keep parts in tolerance.

It is discouraging to find how often a part family finally reaches a volume suited to robot automation only to find expensive machine tools not equal to the process change. The only alternative is to rebuild the machine tool to make it perform or to sell it and replace it with a machine that can be automated. The last alternative is losing the business to someone who has the right technology.

PARTS WASHING

Another element in the system is to clean the part and prepare it for the next operation or to protect it during storage. Flow-through washers are employed for high volume work, but batch washers are effective where robot work cycles allow the time to load and unload these units. It is also required in some cases for the robot to have the intelligence to easily palletize product in oriented batch-washing operations.

DEBURRING EQUIPMENT

Occasionally, the machine tool's cutting tool leaves a burr which must be removed before more machining takes place. Deburring equipment can take the form of gun drills with reaming heads to clean up interior holes, or chemical deburring machines can be supplied to remove rather substantial burrs in a short period of time.

INSPECTION

Post-process and in-process gauging is required not only to eliminate bad parts from the process so that more processing time and space are not spent on them, but to prevent bad parts from being made in the first place.

More CNC machine tools are being equipped with in-process gauging systems, which either correct the cutting or grinding action while parts are made or put adjustments into the machine for the next part.

Basic inspection is also necessary when processing parts in order to avoid fixture or part damage. In some cases optical or mechanical switches can be used in the robot tooling to make sure the part is where it should be. If the robot is very active in utilizing its range of motions it might be more reliable to put part sensing on the fixture, rather than in the tooling.

CONTROLS

There is little doubt that companies investing in automatic handling equipment are also looking to a future when the reporting of floor conditions will be nearly instantaneous.

The robotic industry is developing along two avenues. One is to pack as much sophistication, memory, and input-output (I/O) capability into a unit as possible, to the point of adding additional multi-tasking CPUs for management functions. The other is to provide plenty of I/O capability in a robot but to rely on a properly sized programmable logic controller to provide the management capabilities.

Both techniques are used to great success, now that computers are becoming hardened to their industrial environment. A subset of this approach is to employ a workcell multitasking computer instead of a PLC, and this approach is also economical where concurrent operational and management functions are required.

PARTS STORAGE

Robots are becoming so sophisticated that a wide range of part storage options are open to the user. If the robot selected is restrictive of memory or is difficult to program, single-part presentation positions are provided by blue steel conveyor, table top chain conveyor, or accumulation roller conveyor. Belt conveyors can be used if parts are light or do not have to be accumulated.

Magazines for medium-sized parts or bulk storage for large parts can be used where parts can mix or roll against each other. These are typically used before machining is done, and one oriented pickup location can be provided with this equipment. Bowl feeders are used if parts are very small. If the robot selected has advanced programming capabilities and the accuracy to make palletizing feasible, overhead power and free chain conveyor with multiple part racks or pallets with oriented pockets for parts can be used as effective economical alternatives (see ECONOMIC JUSTIFICATION). Palletized product is particularly valuable where part families are very broad in dimension and weight and where lot sizes are relatively small.

LINE BALANCING

When combining several consecutive machine operations to make a continuous process, simulation of the total process is helpful. Many of today's simulation programs are based on General Purpose System Simulation (GPSS) technology, which generally assumes an orderly linear flow from work center to work center. However, robot-based systems are almost never truly linear in their operation. Floating storage can be used for a range of parts or for several different operations.

The robot can follow several different parallel paths when servicing parts, machines, or processes. Therefore, more sophisticated and flexible simulation programs are required. Fortunately, these programs are becoming available. Their primary thrust in machining operations is to assist in determining through-put by evaluating equipment capability, reliability, and availability and desired buffer or in-process part storage.

ROBOT TOOLING

Robot tooling takes many forms when working with a wide range of machine tools. The most basic set of tooling consists of mechanical grippers that grasp interior or exterior surfaces by parallel motion or by toggle action (see Fig. 1). A double set of grippers is used in order to cut machine service time so that a green part is ready for loading in the "hand" while the other gripper is grasping the worked piece.

Multiple spindle machines require more sets of tooling. Sometimes the hands are independently activated. The robots are then able to reject unsuitable parts on an individual basis, or to individually process parts to an output conveyor or inspection or rework station.

In several instances total exchange of part gripper devices is required. This is true when the part goes through a large change during machining, or when the grip point changes between inside and outside part surfaces and the clearances inside machines are so tight that all required tooling cannot be carried permanently on the robot arm to all locations (see Fig. 2).

Sometimes multiple pocket fingers are used to work between inside and outside surfaces or multiple surfaces of a part at different times during the process. A typical condition is where a part diameter is cut down in a lathe and the gripping surface is markedly different during pickup. The smaller pocket allows positive location after machining. The same gripper actuator can be used both in delivery and removal of the part.

SAFETY IN A MACHINING CELL

The typical robotic work cells uses several techniques to keep the work area safe. Since most robots in machinery centers are large and cover a large service volume, the area is best served by a 6–8 ft wire mesh fence with entry gates that are interfaced to the robot control in order to shut it off when a gate is opened. The high fence keeps the parts from leaving the area if dropped by the robot.

The robot is allowed to operate in the teach mode, and with electric robots acceleration is electrically limited, and in hydraulic robots the machines use low flow control valving as well as electrical velocity limitation devices. The machine tools in the work center are to be put into a manual operation mode.

Safety posts to limit robot motion, usually in the rotary mode, are used to block out certain areas from robot motion, but these posts can also act as pinch points while the robot

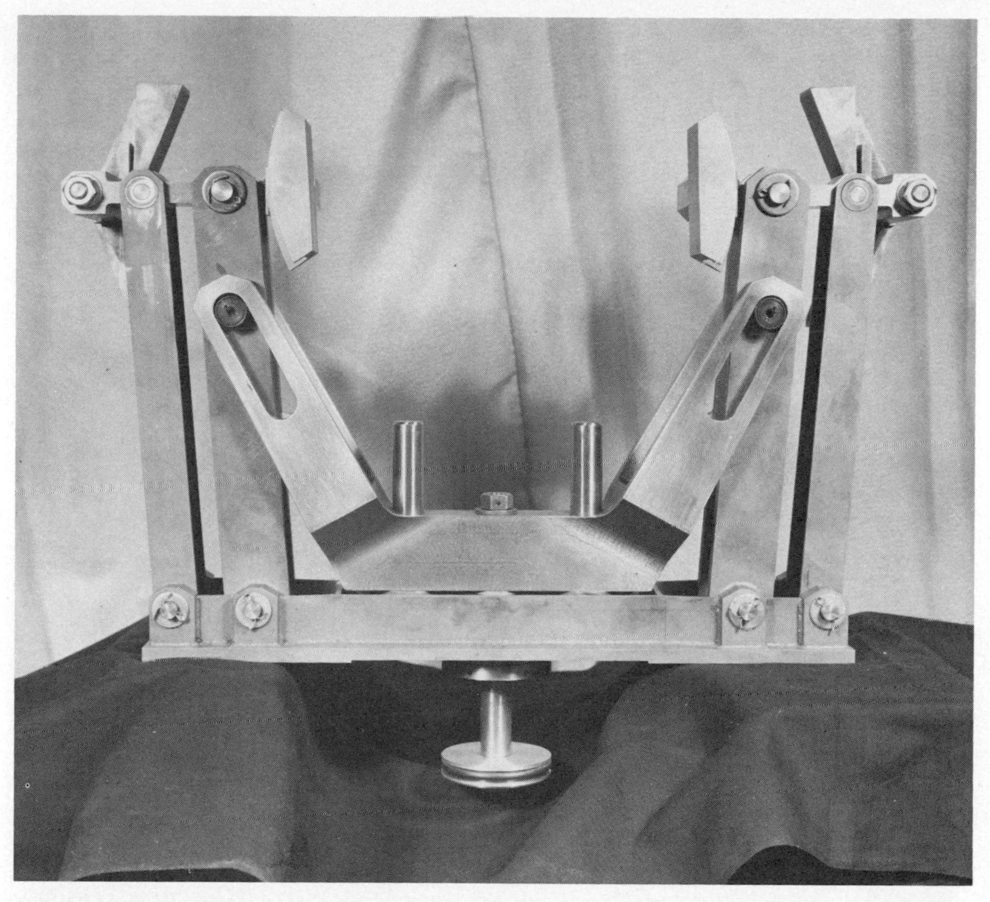

Figure 1. Dual activating hand to grip both ID and OD surfaces. Courtesy of Unimation and UAS Automation Systems.

is moving toward them. End-of-arm tooling is usually painted OSHA orange, and robot arms are usually painted in bright colors to highlight the working end of the robot. To keep abreast of the latest technology when planning the work cell, it is best to consult the most up-to-date safety articles in the robot trade magazines and OSHA regulations.

EXAMPLES

Brake Assembly and Machining Line

A truck brake-shoe assembly is constructed of two major parts that must be assembled by riveting (see Fig. 3). The assembly is then machined so that the brake will run true.

The large three-axis robot in the work cell processes the backing plate for the brake from the input conveyor to a press die. A bracket is added from a second input conveyor and placed in the press die over the backer plate. The small four-axis robot takes rivets from a feeder bowl and places them through the holes in the bracket. Five rivets are added one at a time. The tooling in the small robot grasps the rivets by the heads. The same gripper aligns itself with two holes in the bracket, which allows the placement of the rivets to be accurately done.

After the press assembles the parts into one unit, the large robot moves the brake assembly to a multispindle drilling machine and then to a chamfering machine. The large robot is single-tooled and therefore works its way backwards through the operation, clearing stations before putting new parts in place for work. The 2 robots working together process a part every 30 seconds.

Interface signalling is limited to limit switches and machine "cycle one" control outputs and is completely directed by the robots' controllers, with the large robot controller serving as the master.

Suture Needle Forming and Machining Center

This work center is organized around the material handling of 12 or more small parts at a time (see Fig. 4). One operator controls the activities of 2 robots.

Needles are spread into slots from a bulk storage device that passes over a plate. A three-axis robot picks up a row of needles with a multiple pocket hand and passes the product through a grinding operation. After grinding, polishing compound is applied and the needles are put in a polishing machine. The parts are passed to a six-axis robot that feeds the parts to a curling machine, which somewhat resembles a rolling mill. After being formed, the parts are transferred to the unload station where an operator takes a sample and uses an optical comparator to evaluate the finish. If any one part is below standard, the appropriate machine is checked for operability and adjustments made as required.

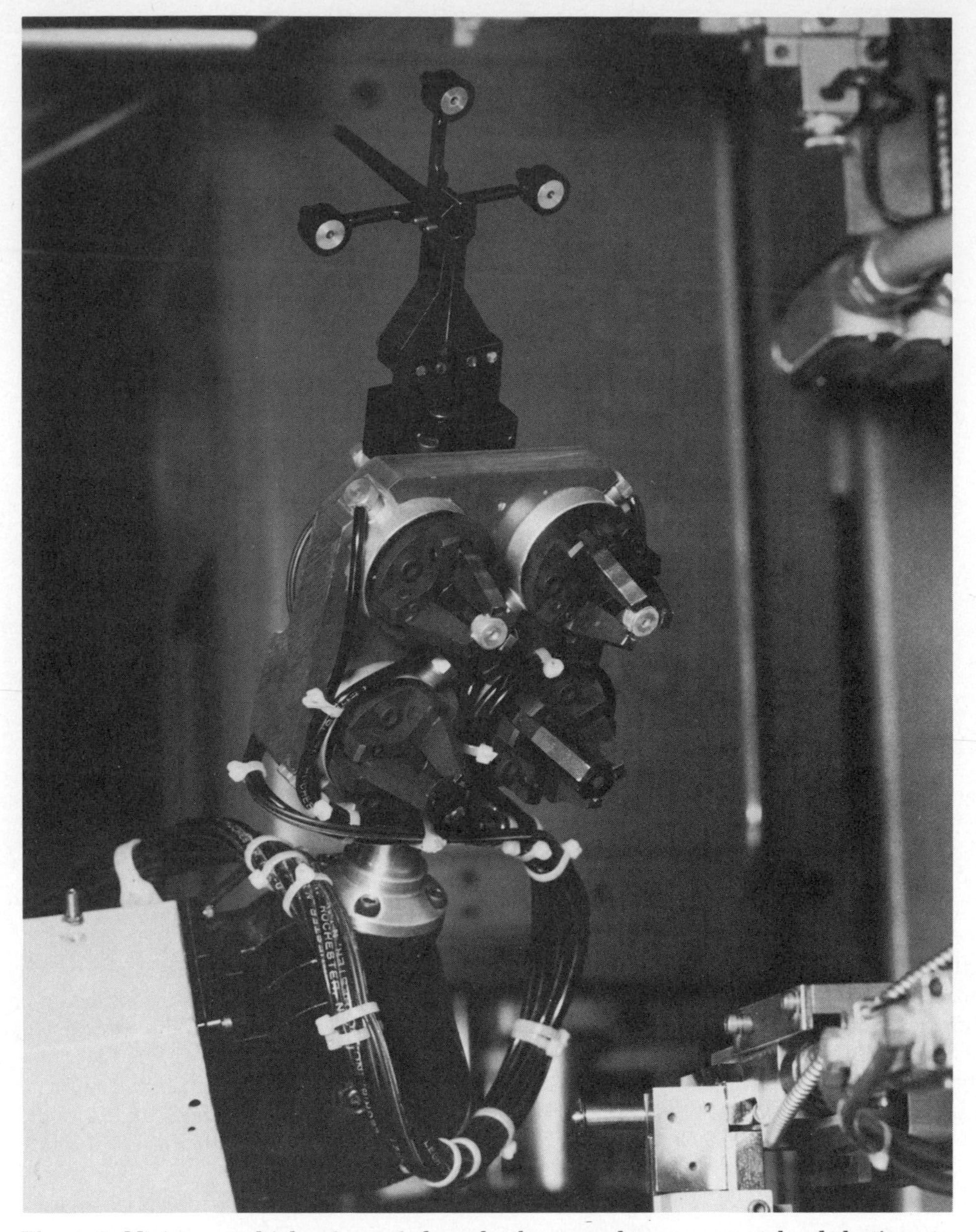

Figure 2. Miniature multiple grippers independently actuated to process metal and plastic parts, mounted to a Puma 550. Courtesy of Unimation and UAS Automation Systems.

Automobile Wheel Machining Line

When large machining lines consisting of fifty or more machine tools are assembled into one line, high speed feed-through loaders are used at each machine to maximize machining time. Up to ten of these machines may be dedicated to a single operation on one product. Two or three models of parts may be produced on one line at different times. Blue steel conveyor is employed to transfer the parts from operation to operation, and elevators with storage silos are used so that enough storage can be provided to serve between operations. Robots are employed in a unique way in these operations.

In order for blue steel conveyor to distribute a balanced load to each machine tool in a group, diverters are required to divide the product into individual lanes. After the product is machined, blenders on these lanes bring back the product to one lane. Hitch lifts are used to achieve enough height so that gravity conveyor can reach the machines furthest down the line.

Robots eliminate the need for these lifts, diverters, and blenders by acting as intelligent distribution devices, placing the product where it is needed and gathering it from finished product lanes and placing it in common inspection lanes. Another advantage is that more than one product can be processed at the same time.

The robot also reports source data to a programmable logic controller (PC) so tracing of the machining source can be done at the inspection station. The robot serves in additional ways.

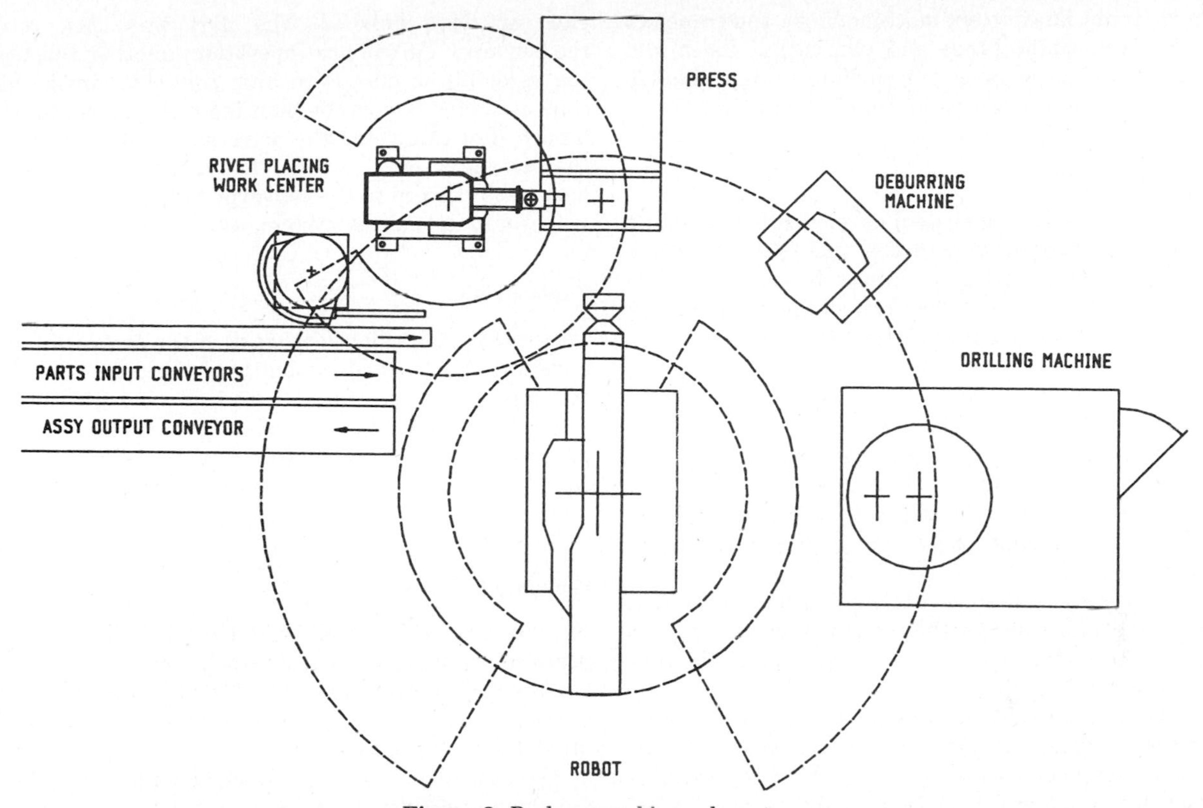

Figure 3. Brake assembly work center.

Buffer storage silos hold additional product when downstream inspection operations are not functioning. When downstream operations start up again, the robot feeds product to the downstream lanes directly from the machine tools so that up-to-date inspection data can be fed back to a PC that has the ability to shut down machine tools not producing parts to tolerance. On a time-available basis, the parts in the storage silo holding previously machined uninspected product are gradually brought back into the system. The parts are inspected for in or out of tolerance dimensioning, but the PC does not shut down machine tools when parts are out of tolerance because the data are too old.

Because the robot is a programmable unit, future product changes and production volume changes for a mixed product line can be accommodated without the need to rebuild the conveyorization feeding the work area.

The robot has complete control over which lane gets which product. At the present installation, the robot processes parts

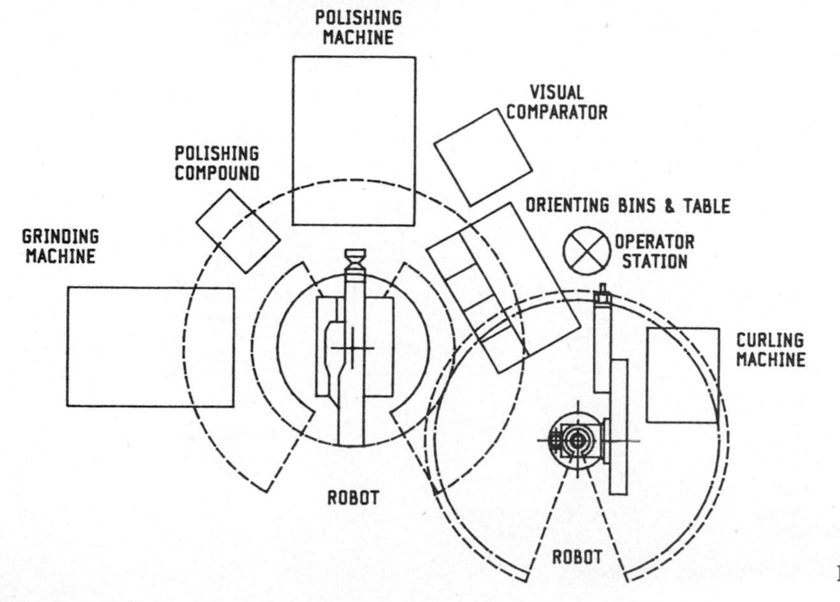

Figure 4. Suture needle forming and machining center.

from up to two input lanes to six machine lanes and transfers product from six machine lanes and two output lanes and also controls feeding a two lane intermediate storage silo. Up to four parts at a time are processed from lane to lane.

Machining a Copier Drum

The second robotic machining system developed for a product was built by Xerox Corporation in the mid-1970s. The line included fabrication by brazing of three subassemblies. Two special center-driven CNC lathes performed the basic turning operations on the copier drum and end caps (see Figs. 5, 6, and 7).

The tube selection of the drum was placed by an operator on a recirculating carousel-style pallet conveyor. The first CNC lathe turned both ends of the tube and bored out the interior so that it could accept the end caps. The tube was transferred to a brazing machine where the end caps, fed from storage, met the tube. After brazing, the assembly was returned to the conveyor.

One robot performed all tube handling operations to and from the conveyor and through the machines. A two-gripper mechanical hand minimized load and unload time at the conveyor and at the machines.

At the next operation another robot processed the assembly to the second center drive lathe where the end caps were finish-machined with reference to the tube outside diameter. The part was next fed to a grinder where bearing surfaces in the end cap were finished. The part was then returned to the conveyor. At the next operation, another robot processed the assembly to an OD turning center to finish the outside tube diameter. The robot then fed one end cap to a nest in a broach that completed the drive slot. This fixture supported the bearing surfaces so that the drum would run true. The part was returned to the conveyor for delivery to the operator who would unload the system, after inspecting the assembly.

Control Unit

The process controller kept track of the status of the machine tools and robots. It also monitored the parts and their stage of completion. The controller also directed the robot program selections for six different styles of parts plus alternate path programs to avoid machines withdrawn from service. Parts would be recirculated until completely machined. All machine interface functions were tied through the controller.

Planetary Pinion Gear Line

The machinery required for the manufacture of a family of gears consisted of a broach, which sized the billet interiors, two vertical chuckers, for ID and OD turning operations, three gear shapers to rough cut the teeth, and three shavers to finish the teeth (see Fig. 8).

The best line design was determined to be the grouping of the same operations in work cells with each machine tool of

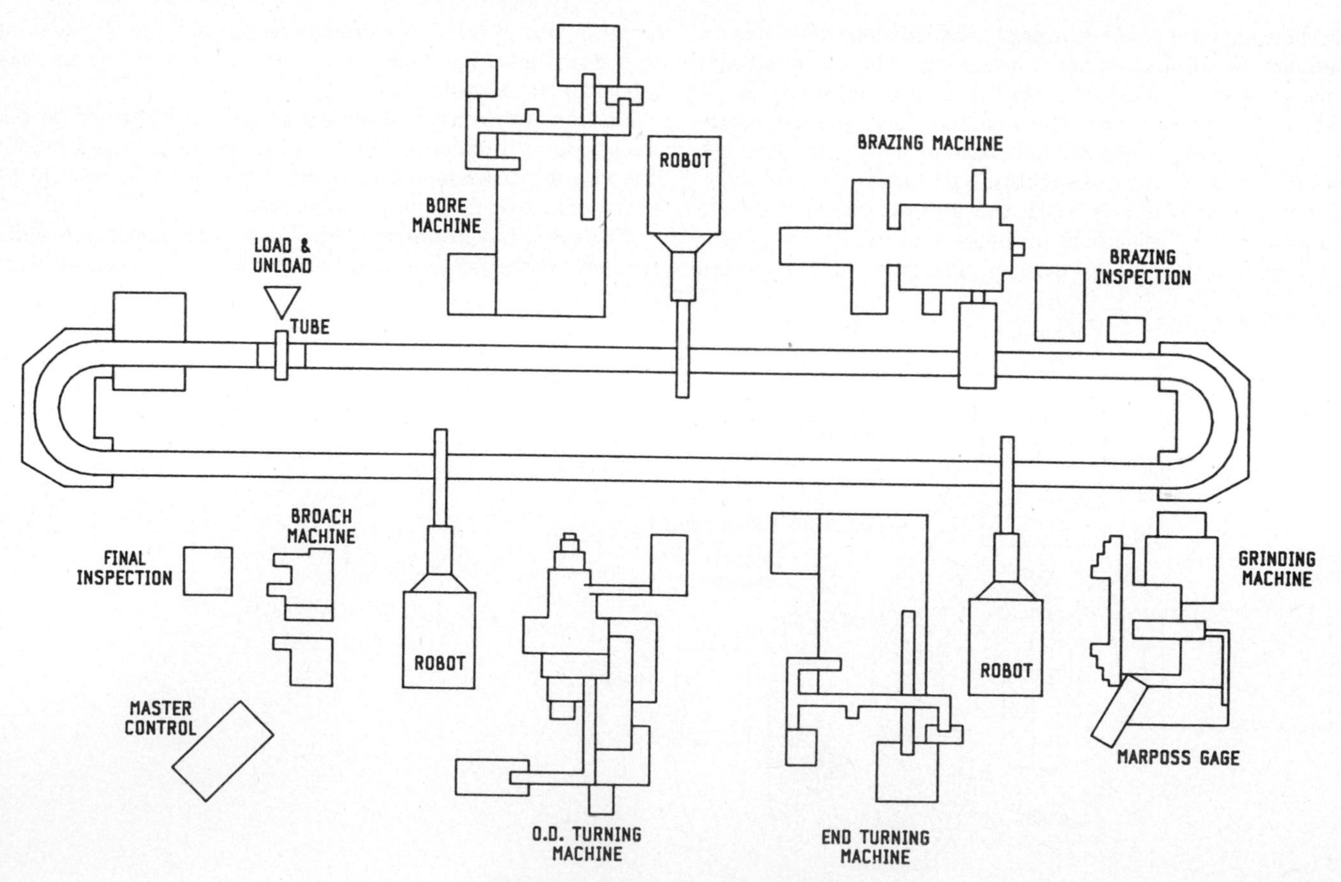

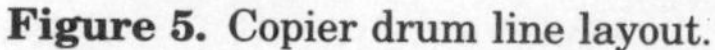
Figure 5. Copier drum line layout.

Figure 6. Turning cell with double-tooled robot. Courtesy of Unimation and VAS Automation Systems.

similar function working in parallel with the others. A part arriving at a common input point could be fed to any one of the machines in the cell and there was only one exit position to the next operation.

The one exception was the turning operation where one machine did rough-turning and one machine did finish-turning. An intermediate storage silo recirculated parts that were only rough-turned and the robot loaded it when the finishing machine was out of service. When the rough-turning machine was out of service the robot processed parts by drawing from this storage silo and loading the finish-turning machine.

THE PROCESS PROGRESSION

Broaching and Turning

The front end of the line consisted of a bulk feeder which oriented the billets and fed them to a pick and transfer device

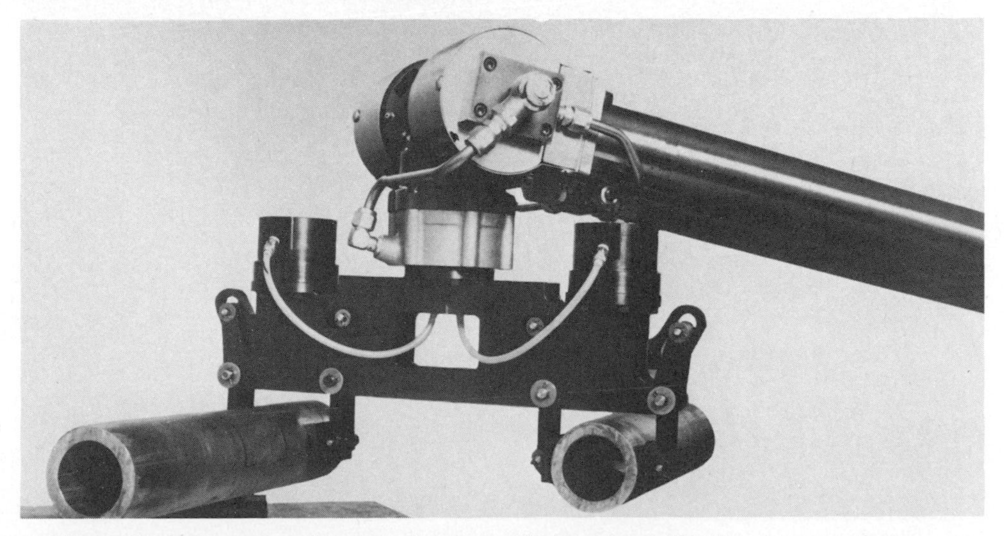

Figure 7. Double-actuated tooling. Courtesy of Unimation and UAS Automation Systems.

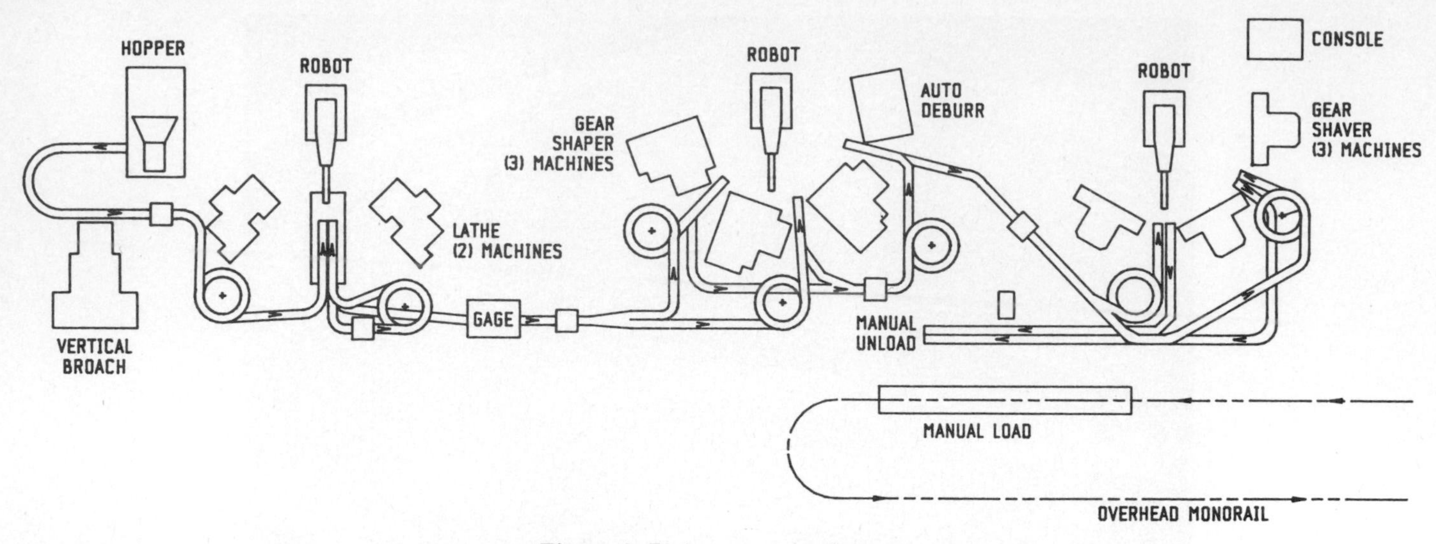

Figure 8. Pinion gear manufacture.

that moved the parts through the broach (see Figs. 9 and 10).

The turning work cell required a robot that would service two turning centers. The first machine rough-turned the inside and outside diameter of the billet, and the second machine finish-turned the diameters. However, it was suspected that downtime for tooling changes might be substantial, particularly the waiting for manpower to perform service.

The robot program options allowed it to service machines independently, and unfinished parts would be recirculated through the work cell by using a recirculating storage silo. If a machine tool suffered a major breakdown, the remaining machine tool would be retooled to make a completely turned part. The robot would perform the new task simply by changing its program sequences. If a machine tool could not complete its cycle, a programmable controller circuit would time out

Figure 9. Vertical turning work center. Courtesy of Unimation and UAS Automation Systems.

Figure 10. Double cam-actuated OD tooling for turning operations. Courtesy of Unimation and UAS Automation Systems.

and the robot would be signalled to change sequence automatically to bypass the inoperable condition, using intermediate storage silos as required.

Because of sequence variations, the double-pocketed hands had to be individually activated through the robot control logic, and because forging surfaces are inaccurate and irregular, the hands were set in a floating frame. This allowed for seating two parts at a time in collet chucks that had restrictive clearances when open.

Rest stand pockets in front of each machine allows parts to be accurately held between chuckings, since one chucking is not enough to turn both faces and both diameters. Rest stands also allow for rapid interchange of parts in order to maximize machining time, since the robot does not have to return to a conveyor for replenishment until it gets some "free" time.

It can now be seen that although the robots used for this application do not have to be smart with regard to management capability, program flexibility is required to minimize downtime due to malfunctions.

Gear Tooth Cutting and Shaving

The gear shaping and shaving work centers both work in an identical manner (see Figs. 11 and 12). Each robot loader must make a decision as to whether a machine is serviceable or not and make the appropriate loading decision.

The robot tooling must be designed around the limitations of the machine tool. In this case the shaper spindle and supports stop in an undefined area in the vertical plane and the hand tooling must find the fixture. The shaving machine required hand tooling that would roll the gear into mesh with the shaving cutter with all the attendant sensing equipment to assure that mesh was completed properly.

A deburring machine was found to be required later after shaping operations were complete, and this machine was added to the shaper work center. It required robot programming modifications at the shaper work center. If hard automation had been planned, another work center would have been required to accommodate the change.

Conveyors

The conveyorization selected for in-process storage and recirculation for a family of five different gears was roller-bearing "blue steel" flexible-banded conveyor. Storage elevators and helical silos were used to store several hours' worth of parts between work cells without using excessive floor space.

PC controller activity in the work cells initiated the robots' program selections, monitored status at different points in the line, and also kept part counts and scrap counts in the post-process gauging stations.

Overall increased productivity showed a gain of 25% over manual. Although these robots are not the most skillful available on the market today, they are a long way from pick and place devices, which could not hope to provide the flexibility in operation that this system demanded.

High process in line automation would also have been prohibitively expensive for a system producing 200 pieces per hour.

Pump Body Machining

Figure 13 is a flexible machine tool cell. Hydraulic pump cover castings, up to 5 in. square and 3 ½ in. thick and weighing

Figure 11. Gear shaping and deburring work center. Courtesy of Unimation and UAS Automation Systems.

Figure 12. Gear shaping work center. Courtesy of Unimation and UAS Automation Systems.

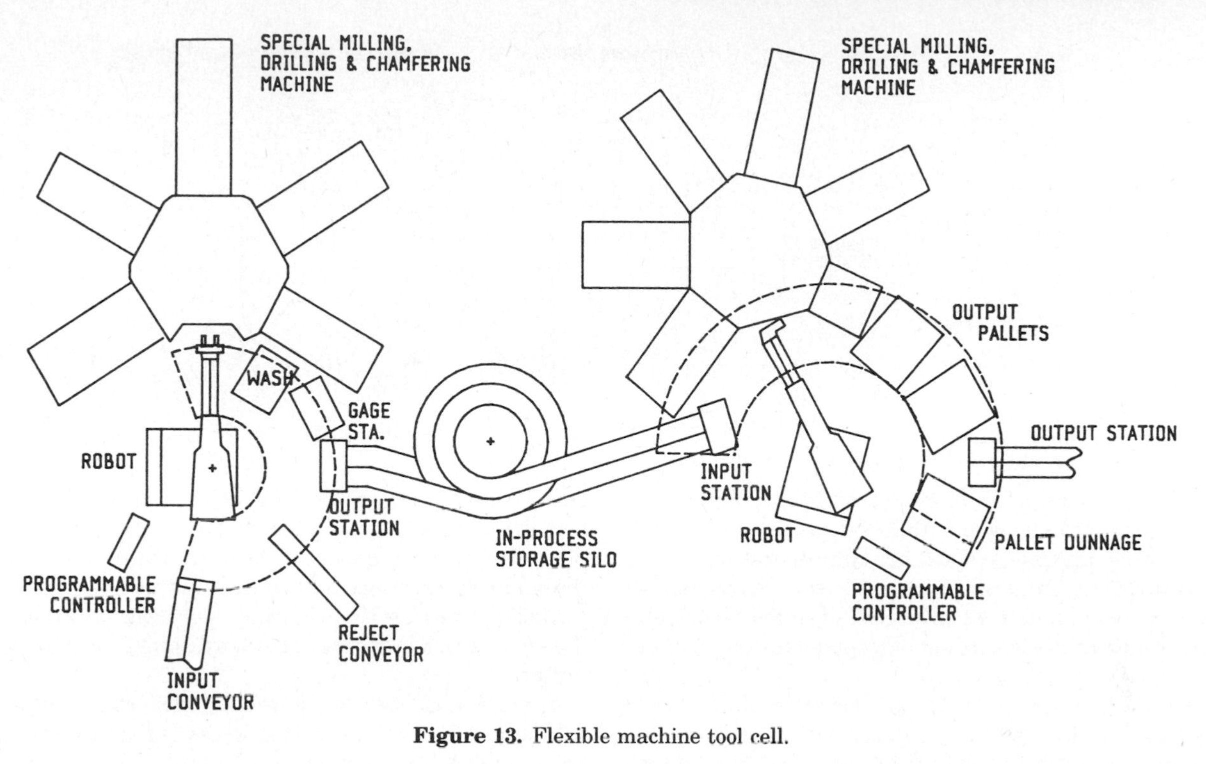

Figure 13. Flexible machine tool cell.

from 3–5 lb are machined in 2 consecutive rotary milling, drilling, and reaming centers.

Twenty-five different part numbers are run through the system in batches of several hundred pieces. A washer and gauge are also included in the first cell. Rather than palletize product in cargotainers, which would destroy orientation and eliminate the ability to automatically load and unload the machines, it was decided to provide in-process oriented storage utilizing a blue steel storage tower with a bearing rollway.

Pallets roll on the bearings and a series of cavities and pins provides accurate rest points for the 25 different products between work centers. To avoid using storage elevators on the pallet return loops, the robot tooling lifts empty pallets to sufficient height for the return. A two-level inner and outer track coil arrangement in the same silo saves floor space. Two hundred pallets provide in-process storage. A second storage silo at the input to the system operates in the same manner, so that only part-time manual servicing of the system is needed.

The robot palletizes acceptable output in one of two pallets, which are located on lift tables. The self-leveling lift tables limit the teaching of palletizing programs to one level, an advantage with 25 different parts.

The first robot has two hands, independently actuated, so that rejected parts can be independently placed on the reject conveyor. Wide stroke fingers are used so that parts and pallets can be handled by the same hand. Since all machinery in this cell processes two parts at once, the programming sequence starts from the last operation forward. To minimize loading time at the rotary machining centers, staging tables are used to prealign parts for loading. (see Fig. 9, Vertical Chucker Work Center, Planetary Pinion Gear Line).

PROCESS FLOW

In work center 1, the robot starts with the unloading of the gauging station, rejecting parts that are unsuitable for further machining. Then the washer output is loaded to the gauge, and then the rotary machine output station is unloaded to the washer. Parts are loaded into the rotary machine from the staging tables, and it in turn is recharged from the input conveyor.

In work cell 2, the robot has a mechanical hand which uses through-pins to align the part accurately for machine tool loading. The vacuum gripper is held up out of the way until needed for dunnage placement, which supports the parts being palletized.

Self-leveling lift tables make sure that dunnage is at a common presentation height and the parts are always loaded to the same level. With 25 different part numbers to palletize, this procedure saves much memory and teaching time while maintaining accuracy of placement.

The controls consist of a programmable logic controller that directs the programs of the robots and the order in which they occur. If pallets must be removed from the system, this is accomplished by the robot under the direction of the controller. System status is also displayed on the controller panel board and a part count is maintained.

FLEXIBLE ROBOT-BASED SHAFT MACHINING SYSTEM

A flexible system is shown in Figure 14. An intermediate electric motor facility required a shaft machining line that encompassed a wide range of lengths and diameters with lot sizes as low as 30 pieces.

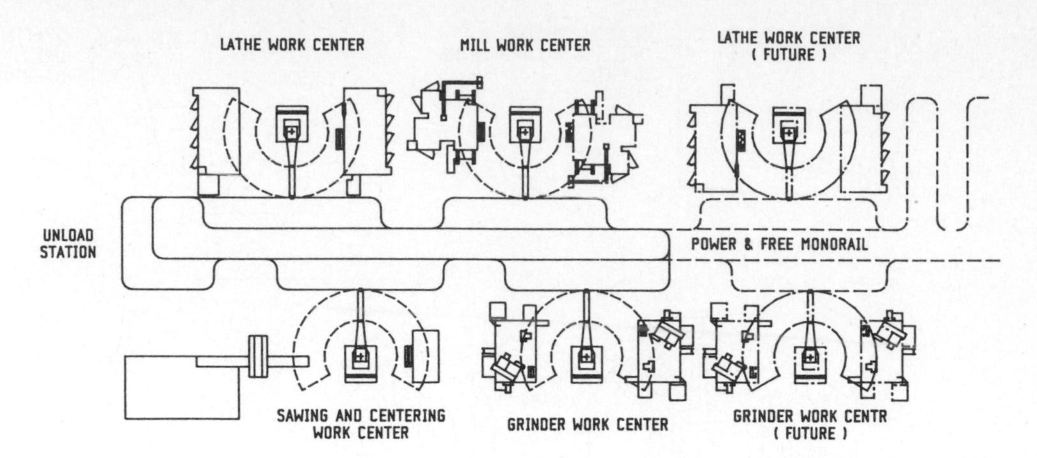

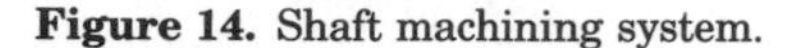
Figure 14. Shaft machining system.

Raw shafting stock was to be delivered from bulk storage directly to a saw. Billets would then be processed through centering, turning, milling, and soft grinding operations. Standard CNC control machine tools were to be used. Several different production schedules were projected over five years and the automation equipment was to be designed for expandability.

The initial production requirement over two shifts would be provided by one saw, one face and centering machine, two lathes, one mill, and two grinders. Two years into the project, an additional mill was to be added plus two more lathes and two more grinders.

The parts vary from 1 in. diameter to 4 in. diameters, the length from 9 in. to 28 in., and they weigh as much as 60 lb. The light parts consist of 50% of the total production quantity, and the heaviest parts have 10% of the total volume. However, machining time at the lathes for the largest part is 10 times as long as for the lightest part. Their machining time is most critical for production throughput.

The first system concept developed was for a pallet-based system with oriented dunnage. Up to 40 pieces could be held in a pallet. A holding area for parts stored between machining operations consisted of several recirculating pallet roller conveyors. Pallet orientation conveyors were concepted for each work cell and an automatically guided vehicle (AGV) processed parts between work cells. There was a strong resemblance to a pallet-based FMS machining system. The cost was too high.

The least expensive part delivery system was a rack-based monorail system, but it was not flexible. However, a flexible version of this conveyor is a power and free monorail conveyor. In this system three rack configurations suited all parts. Pick and place robots with palletizing capabilities were required to serve this conveyor. Work load and time cycles were such that the machines in the system are divided into cells by function. Lathes, grinders, and mills have their own work centers. Saw, face, and centering operations are assigned to another center.

In-process gauging is accomplished at the lathe center, and the grinding center has automatic wheel compensation and dressing.

Conveyor Arrangement

Simulation techniques applied to the system resulted in a 53-rack storage purchase. Total conveyor length required was in excess of 160 ft. The conveyor chain strength is 1000 lb and 2 drives are used. A total of 5 processing stations allow part presentation for all operations. An unload station is employed to prepare parts for heat treating, and empty racks return to the saw face and centering work center.

Racks awaiting a machining operation will recirculate on the main track, waiting for an available location at a work center.

Several racks can be stored at each work cell so that no part starvation occurs unless excessive downtime occurs on an upstream work center. Each work center has a location device that accurately locates a part at its left end so that machines can be properly loaded. Therefore parts can be allowed to shift horizontally while riding the hooks on the racks, without affecting the accuracy of part-processing at each work cell.

Conveyor Operation-Controls

The PC is utilized for storing rack number and destination for the 53 racks in the system. Two racks are stored upstream of the unload location and 2 racks downstream of the unload position. When a robot has completed a rack of parts, it relays a signal to the controller that a rack is being released to the recirculating conveyor. The next rack is brought into location and orienting units lock the parts in place. The robot then supplies parts to the machines.

The loaded rack is released to the recirculating conveyor and a new destination is assigned. It recirculates until a spot becomes available at the destination. It then goes into storage at the destination and awaits processing.

Progression of Product

The logic of the control is set so that empty conveyors are recycled into location at the saw/centering work center for loading with product.

The robot assigned to the saw and centering machines charges the system with product. Empty racks are delivered from the unload station to the workstation as they become available. When a rack has been filled, the product enters the recirculating portion of the system and remains there until requested by the appropriate lathe work center.

The output of this center is delivered to a drop location at one of the lathe centers. The output rack moves from the lathe work center to the vertical machining center. The output rack from the vertical machining center is delivered to the grinders.

The output rack from the grinders is delivered to the output station.

Product Support

Each of the 53 racks store from 8 to 24 pieces, depending upon weight and size of the product. The rack capacity is 500 lb. The racks store the parts horizontally on hooks.

Since the saw and centering station has a low utilization factor because machining cycles are short, racks can be rapidly made up to supply the work centers with product. A maximum of 2½ hr storage is kept in advance of any machine center.

The 2 lathes in 1 work center are processing the same product at the same time. Five racks are assigned to the lathes when small product is run. Since 70% of all production involves the lighter part series, it is normal to expect two lathes to run this product at least 70% of the time. Storage density decreases with heavier part weights, but cycles are longer and 5 to 6 racks are required to store 2½ hours' worth of large parts for 2 lathes. The mill areas require from 5 to 12 racks for 2½ hours' worth of storage. The grinding work center is set up to run 1 part at a time. Five racks to 11 racks of storage are required for 2½ hours' worth of storage. Additional racks allow some float for downtime periods at the work cells.

Manning Requirements

Parts are accumulated after grinding so that the output station requires an operator service time of only 10%. After grinding, the parts are sent to the output station where they are gathered until a sufficient number accumulate in order to keep manpower busy off-loading product for about an hour.

This leaves approximately seven racks as a float to start accumulating a second product for the lathe operation and to provide rack float for downtime periods between turn, mill, and grind areas. The system will manage different parts running at the same time. Manpower is required at the input-output station for the following purposes:

To direct machine setup personnel to various machines for part changeover.

To initiate a new part lot; person is to log parts into and out of the system.

To call up inventory and status report data for record keeping and action.

To keep the system operating in accordance with management part supply requirements.

To provide direction for maintenance personnel regarding malfunctions in the system.

To palletize system output for downstream operations, to control input for downstream operations, and to control input bar stock to the accumulator storage in front of the saw.

Production Line Expansion

When planning for higher production, line expansion must be implemented. The present production line can be expanded by the addition of a lathe cell and a grinder work center with associated doubling of material handling storage, which should allow a processing capacity of 150,000 parts per year during a 14 hr day. An alternative method is to leave the first line intact and build a second line to parallel the first. The output of the saw would also be conveyorized to the second face and centering work cell, and a conveyor package would be utilized to distribute product to a copy of the first line. Several advantages are gained by building a second line: flexibility is realized in the layout of the plant because the machines can be placed where there is room, the conveyorized line can be shifted along its length to accommodate available space; the first line can stay intact, producing at full capacity while the second line is being constructed; if necessary, the second line can be placed in another facility, if so desired, with the purchase of an additional saw; and several production decisions can be made in future years, depending upon market needs.

ECONOMICS

There is no doubt that automation applied in the right environment greatly increases productivity. An old Kearney and Trecker study showed that automation replacing manual activities at a CNC machine increases its utilization from a low of 30% to 80% and higher. If demand for the machine is on a multishift basis, the increase in productivity alone usually justifies the automation.

Direct labor savings on a machining system occur in several areas. First is the direct savings realized from a reduction in machine operators. Also realized is a savings in material handlers moving the parts from machine to machine. If the system connects several machining operations together and localized productivity and status reporting are tabulated in a PC or small computer, there is a savings in expediting and management support services.

Indirect savings occur from the more efficient use of capital in the reduction of in-process inventory and the reduction in physical numbers of machines with attendant floor space to produce the same amount of product.

BIBLIOGRAPHY

General References

"Machine Cells," *Robotics Today,* 69–71, (Feb. 1982).

"Safety," *Robotics Today,* 61–65, (Oct. 1983).

"Safety," *Robotics World,* 32–35, (Dec. 1983).

J. Engelberger, *Robotics in Practice,* Amacon, New York, 1981.

R. Maiette, "How to Plan a Robotic Machining Cell," *Tooling and Production,* 1 Paper MS83–885, (July 1983).

W. Uhde, "Flexible Robot Based Shaft Machining System S.M.E.," Mach-Tec Conference Proceedings, Dec. 1983.

FACTORY OF THE FUTURE—A CASE STUDY

L. J. George
Anil Mital
University of Cincinnati
Cincinnati, Ohio

INTRODUCTION

The factory of the future (FOF) is the integration, by a central facility, of equipment used in manufacturing and materials handling, and controlling the process so that a given product or component is produced with the least human intervention.

Robots are an integral part of manufacturing and materials handling equipment and can be used in a wide variety of applications. Robots performing assembly, painting, welding, and materials handling are now routinely encountered.

The work discussed here evaluates the role of robots in the factory of the future. Factors are discussed such as how and where robots should be used, types of robots that should be used (classified by geometry, control configuration, and programming methods), standardization of software and programming for robotic application, implementation of robotic technology from the point of view of organizational structure, maintenance, operation, and product and process design.

The factory of the future to manufacture computer printers in the United States (actual case study) is discussed to alert potential decision makers and implementors of the factory of the future concept to the stumbling blocks that may be encountered in the design, development, and implementation of robotic systems. In this article, emphasis is placed on the discussion of the case study. Conclusions are generalized whenever possible.

ROLE OF ROBOTS IN THE FACTORY OF THE FUTURE—IMPLEMENTATION OF FLEXIBLE AUTOMATION

The implementation of robots in an automated factory environment has to go through a logical sequence. A typical sequence is:

1. Conception and evaluation of process alternatives, defining alternative manufacturing processes to create the net end result.
2. Financial justification of the automated process. Can the capital expenditures be justified on manpower savings and productivity improvements?
3. If robots are the choice, performance of feasibility studies and a preliminary analysis to determine if the robot in a particular application can perform the task within the capabilities and cycle time requirements to meet throughput.
4. Examination of product designs, materials handling system designs, process and tooling designs, manufacturability concerns.
5. Analysis of software, programming, and hierarchial control (communication) issues.
6. Evaluation of robot reliability and maintainability issues.
7. Analysis of robot operator and maintenance training issues.
8. Examination of robot safety concerns and safety device implementation.

CASE STUDY DESCRIPTION

This case study describes a factory for assembling computer printers (Fig. 1). The printers are a high-volume low-cost product comprising approximately 50 component parts (Fig. 2).

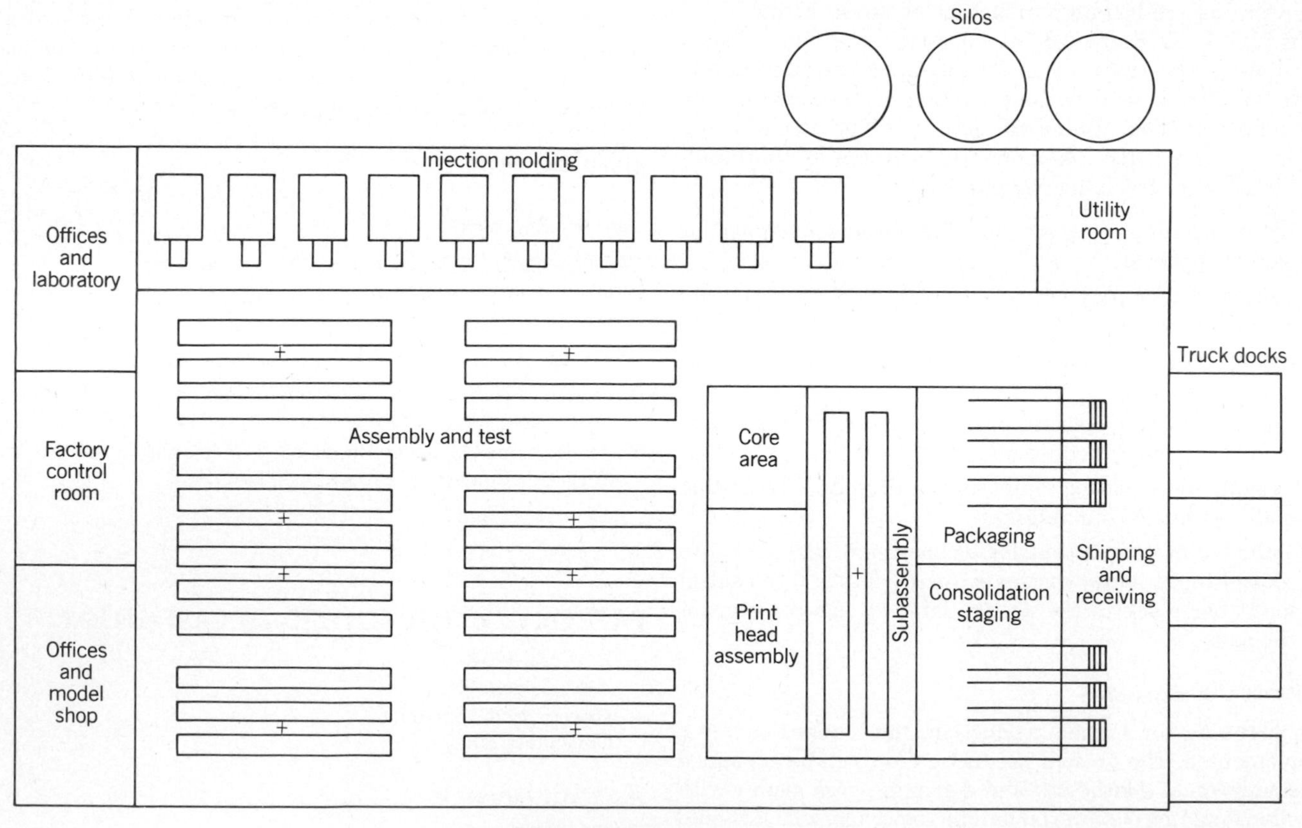

Figure 1. Factory layout.

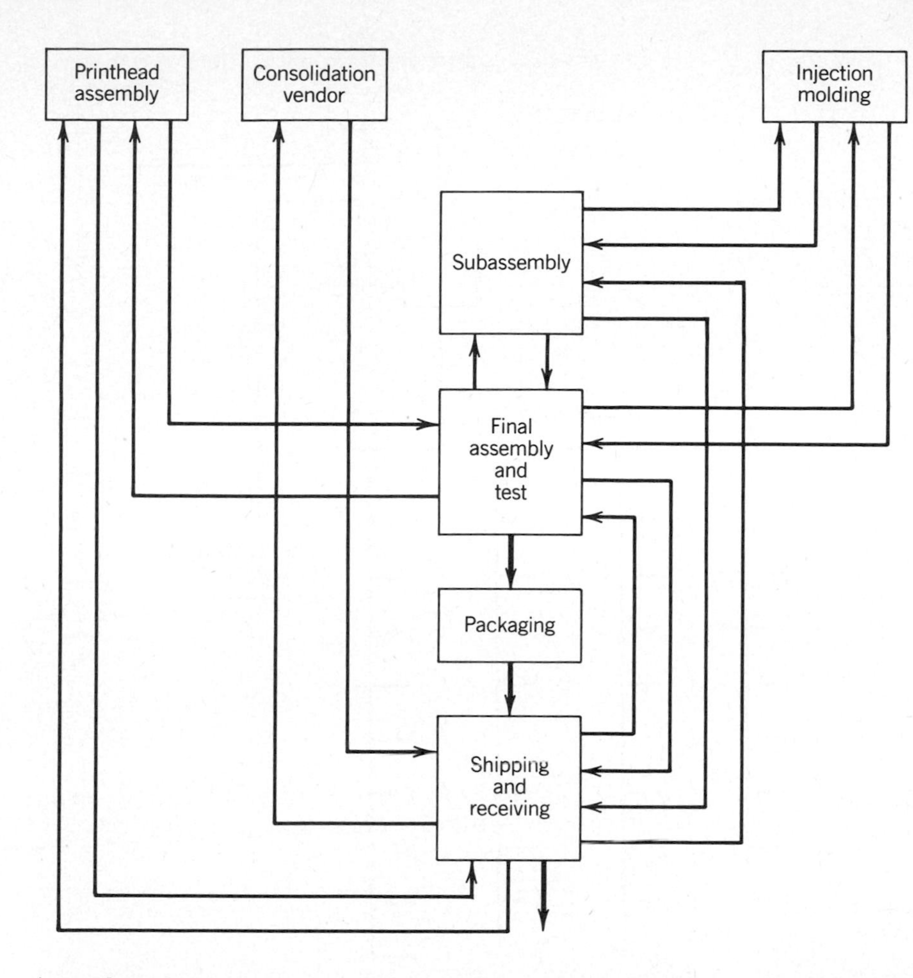

Figure 2. Material flow chart. —, Product flow; — tote flow.

Annual production volume for this product is approximately 1 million units and the price of the printer is approximately \$400. The process starts with plastic raw material being fed from silos into 20 injection-molding machines for molding into individual plastic parts. The molded plastic parts are robotically loaded into matrix totes for conveyance to 8 sequential robotic assembly lines where they are stored in automated storage and retrieval systems for use in the assembly process. The assembly and testing of the product are done on a sequential robotic assembly line (Fig. 3), where component parts are

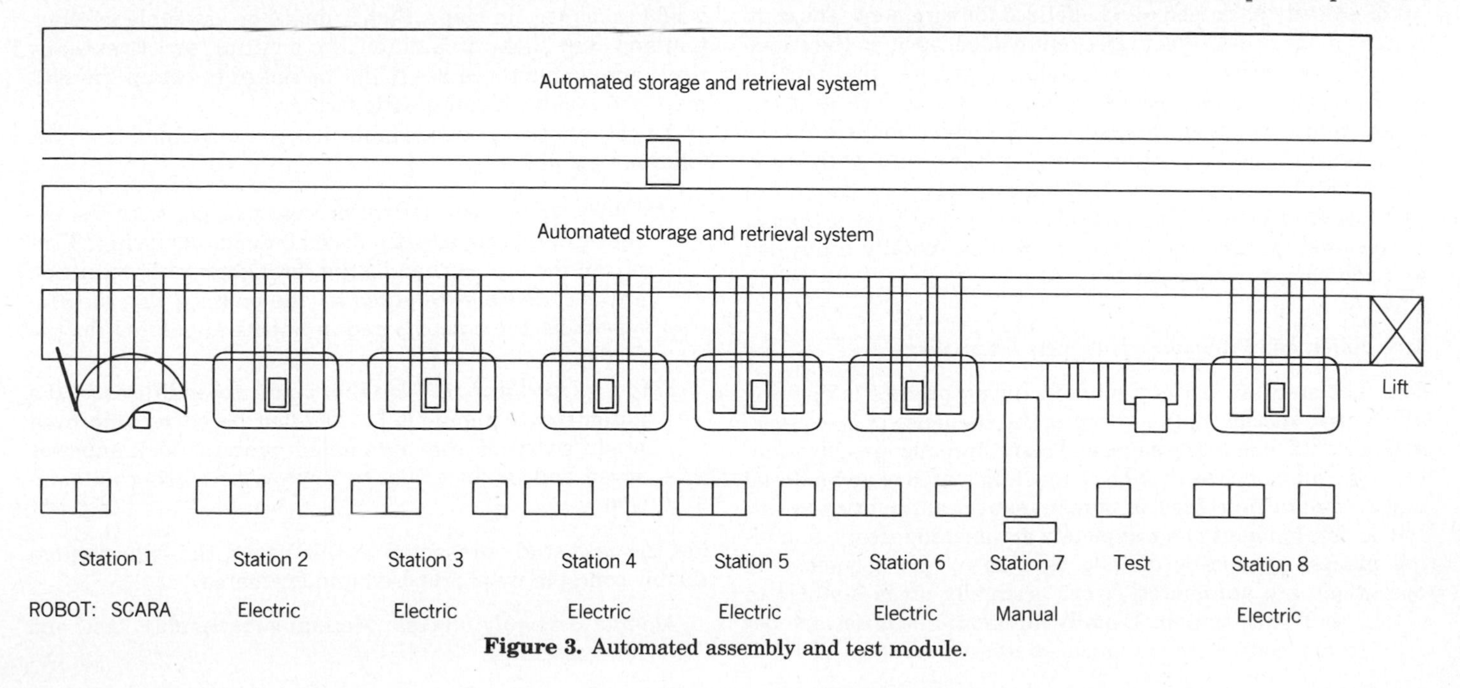

Figure 3. Automated assembly and test module.

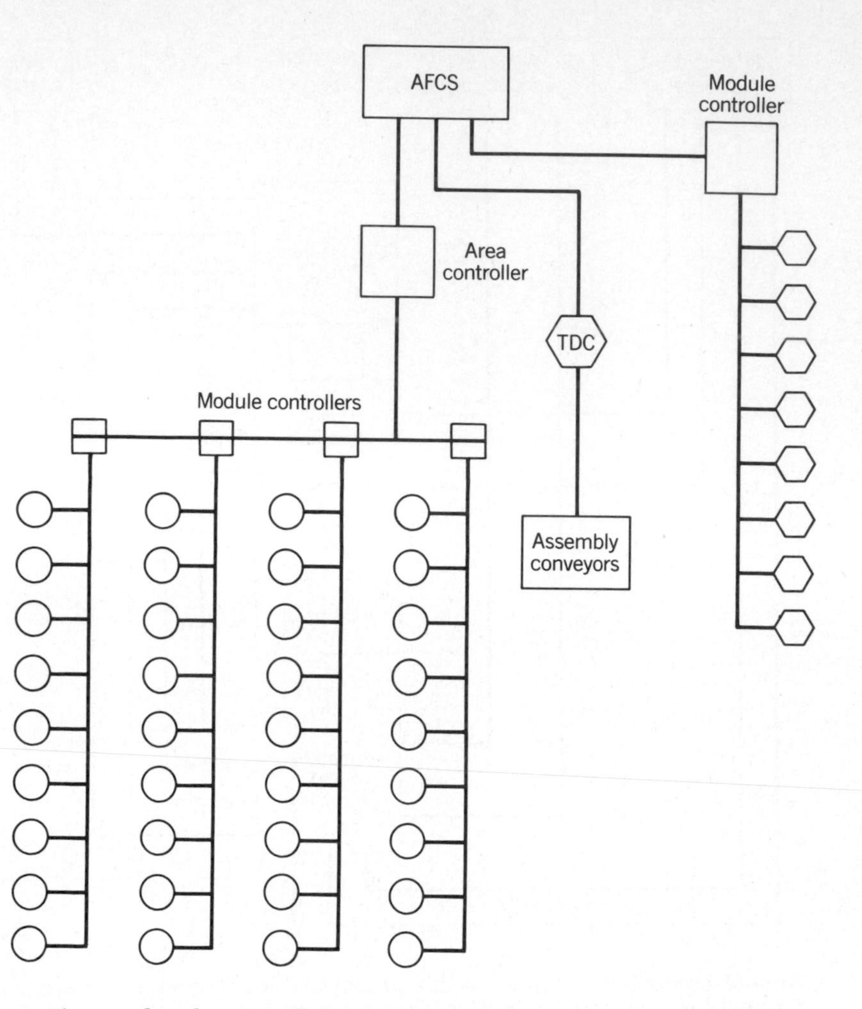

Figure 4. Factory communication network. ⬭ Automated storage and retrieval system controller; ○ robot workstation controller.

assembled on the product as the product is moved from station to station. Print quality testing is performed prior to the assembly of covers. The finished printer is transferred by overhead conveyor to the packaging area where the printers are automatically packaged and palletized for shipment. The component parts that are not injection-molded, such as the transformer, are received at a consolidation vendor, the function of which is to load the respective matrix totes with these parts. These totes are then shipped to the factory for use in the final assembly of the printer. Subassemblies, such as the tractor, feedroll, and pressure roll subassembly, are assembled in-house and loaded into totes for use in the final assembly of the product. The entire factory is hierarchially controlled by a network of computers (Fig. 4).

Conception and Evaluation of Process Alternatives

The first step toward any manufacturing process is defining alternative methods of performing the sequence of operations which would lead to the same end result (product or subassembly). The alternatives could be a mix and match of any manual and automatic processes, or pure manual or automatic process. In the development of a completely automated factory, a manual process is usually defined as a base, and sequences of operations are automated either gradually or as a whole to obtain total automation. Usually automatic factories are derived from reconfiguring products, processes, and facilities that already exist. "Islands of automation" are eventually linked to obtain a totally automated facility. For example, in an automobile assembly plant with a sequence of assembly, welding, and painting operations, it is conceivable that automation would take place in steps. Step 1 might be assembly automation and step 2 might be automatic painting, etc. Eventually these automated processes might be linked to obtain the ultimate end result, the automatic factory.

The alternatives, once established, are evaluated from the following standpoints:

Doability. Is the alternative process feasible from the design criteria that are stipulated for manufacturing? That is, can the process handle the throughputs of production at the cycle times required for the process? Can the process meet the capacity requirements stipulated for the factory?

Financial justification. Can the capital expenditures for the automated equipment be justified based on improved productivity of the automated process. Is manpower saved and product quality improved by using automation?

For the case study presented in this paper, the following assembly concepts were proposed and evaluated:

Manual assembly process. Manual subassembly, test, and final assembly.

Hybrid assembly process. Completely automated subassembly and manual testing and final assembly.

Sequential robotic assembly process. Completely automated subassembly, final assembly, and testing.

Single station robotic assembly process. Same as the sequential robotic assembly process, but instead of a sequential process to assemble this product, a tray of component parts and subassemblies was built up for assembly at a single robot station.

Manual Assembly Process. The component parts are picked from totes manually and placed on a parts tray (Fig. 5). The

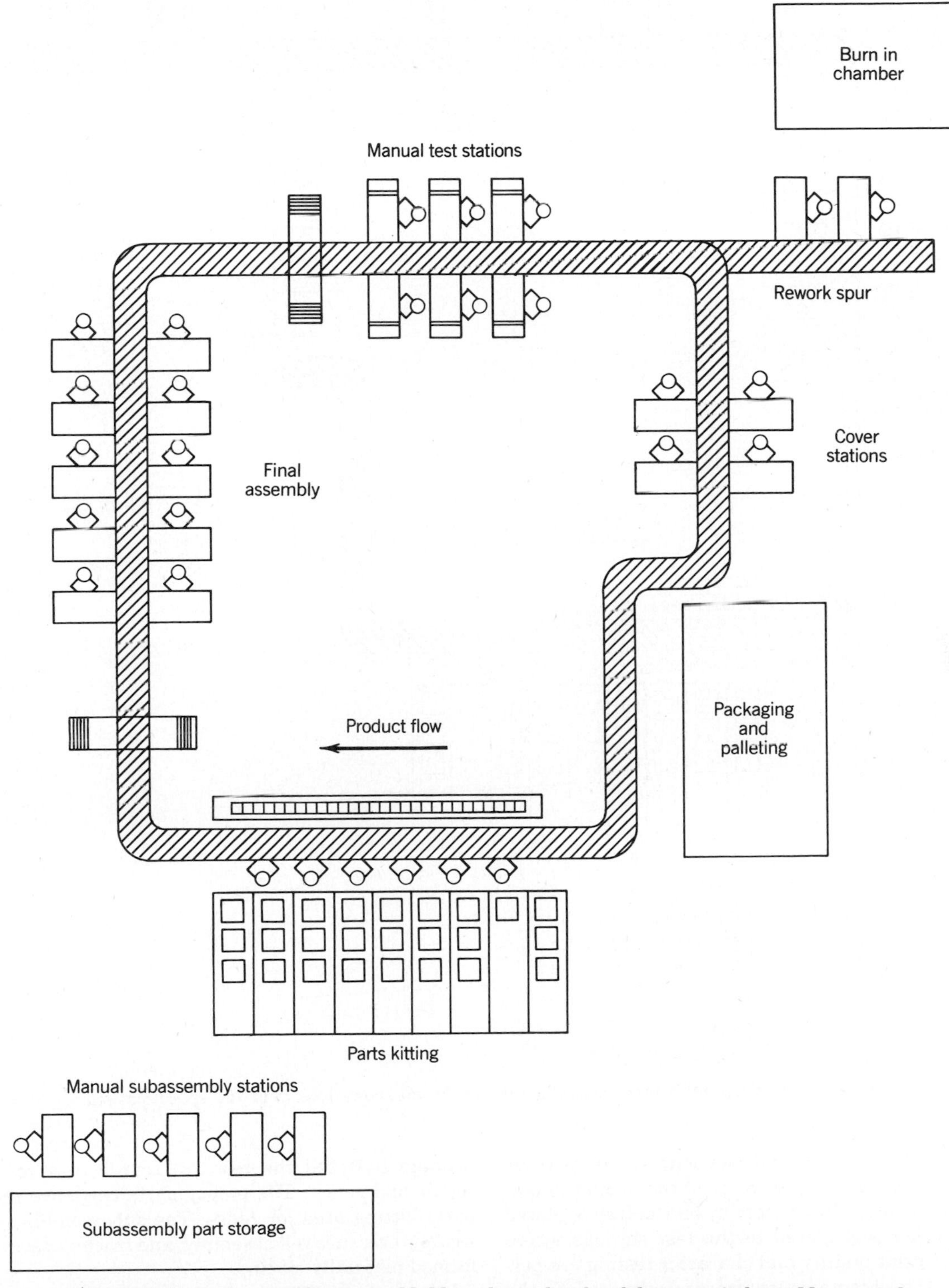

Figure 5. Manual assembly process-M. Manual case lost breakdown: capital cost-M1; manual cost/year-M2; space cost/year-M3.

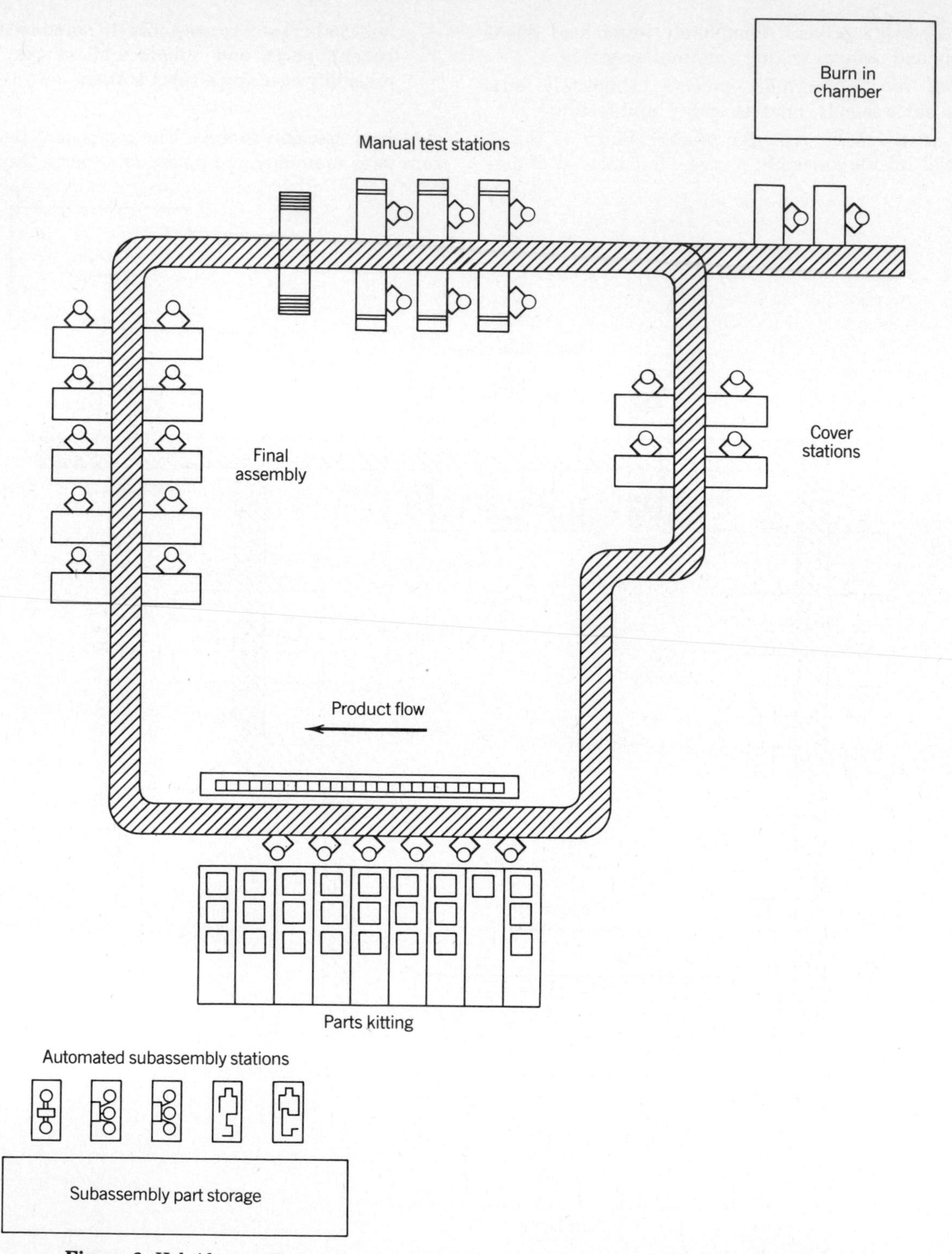

Figure 6. Hybrid assembly process-H. Capital cost-H1; manpower cost/year-H2; space cost/year-H3.

tray is conveyed to the final assembly stations where the trays are taken off the conveyor and assembly of the printer is performed. The completed product assembly on the tray is placed back on the conveyor and moved to the test stations where manual testing of print quality and character testing are performed. Finally, the accepted printers are conveyed to the cover stations where the covers are snapped on, and the complete printer is packaged and palletized for shipment. The printers that fail the test are routed to a rework spur for repair and retest. The empty parts trays are recycled to the parts kitting area for reuse. The subassemblies—feedroll assembly, pressure roll assembly, and tractor assembly—are performed manually off-line.

Hybrid Assembly Process. The printer assembly portion of the hybrid process is the same as in the manual process. The only difference is that the subassemblies in this case are totally automated (Fig. 6).

Sequential Robotic Assembly Process. The sequential robotic assembly process consists of six electric (seven axis) robot assembly workstations, one electric (scara) robot workstation, one manual workstation, and an automated test station. The component parts are both stored in bulk and matrix totes (Fig. 3) in the automated storage retrieval system. The bulk-stored parts are bowl-fed or magazine-fed at the individual robot workstations. The first robot (SCARA) places the base of the product on the product pallet. The product pallet is then indexed from station to station and the product is sequentially assembled. The operator at the manual station performs a printhead to platen gap check, installs the printhead, and greases two shaft bearings; the product is then conveyed to the automated tester, where it is tested, and on to the cover station for assembly of covers. Products that fail the test bypass the cover station and are conveyed to a rework area. The individual bowl feeders and magazines at each of the workstations are filled by an operator who monitors the entire assembly line.

Single Station Robotic Assembly Process. In the case study, since the majority of robots (six of seven) in the sequential assembly process were identical, the alternative of assembling the entire product (except for covers) at a single robot station was investigated. The component parts were robotically assembled from a kit similar to the parts tray in the manual process and fed to individual robot workstations for assembly. The test and cover assembly processes were identical to the sequential robotic assembly.

On completion of alternative conception and evaluation, comparisons should be made between alternatives on other merits and demerits before the final choice is made. For example, a comparison of the sequential robot assembly process and the single station robot assembly process gave rise to several advantages and disadvantages, shown in Table 1.

Financial Justification of the Automated Process

Any automated process or capital expenditure has to be financially justified (see ECONOMIC JUSTIFICATION). In the financial justification of stand-alone robot systems, the total system is justified based on single operation savings. Usually the following factors are considered: labor savings, improved product quality, hazardous environments, improved throughput rates, material savings, and space savings.

To illustrate the justification of a stand-alone robot system, let us look at a typical painting application.

Assuming that the robot replaces one operator/shift for a two shift operation/day, the capital justification approximating the return on investment and the payback period is approximately as follows:

CAPITAL BREAKDOWN

Item	Cost, $
Robot and controller costs	75K
Spray guns	2K
Spray booth modifications	3K
Fixturing and location of product in booth	15K
Programming costs	2K
PLC, sensors, interfacing, product identification device	4K
Installation costs	5K
Design and feasibility study	4K
TOTAL	110K

Table 1. Comparison of Sequential Robot Assembly and Single Station Robot Assembly

Advantages	Disadvantages
Sequential Assembly	
Reduces assembly cycle times	Requires higher capital investment
Reduces in-process inventory	Increases the possibility that one breakdown will interrupt the whole line
Reduces congestion of tools and parts at each workstation	Harder to replicate, less flexibility in handling product mix
Allows for faster changeovers	
Single Station Assembly	
Minimizes the seriousness of a breakdown since no other stations are affected	Requires long cycle times to complete each assembly, leads to high throughput times which cause high inventory levels, additional floor space required
Requires lower capital investment	Creates congestion of material and tooling at each station
Eliminates line balancing problems	Requires additional delays for changeovers, instructions, training, etc.
Increases flexibility; easier to replicate, hence easier to handle product mix	

Savings. The robot is justified based on manpower and material savings. The robot replaces one operator/shift, two operators/day.

In order to calculate manpower savings (a), let us assume the operator hourly wage is $9. The annual wage of each operator is $9 × 8 hrs × 250 or $18K. Fringes and benefits are assigned as 30% of wage rate. The total cost for each operator per year is $18K × 1.3 = $23.4K, so the total manpower savings per year is $46.8K. As for material savings (b), the overall material (paint) savings is $7.5K/yr.

$$\text{Total savings (a + b)} = \$46.8 + \$7.5 = \$54.3\text{K}$$

The return on investment analysis (before tax)

$$\text{Capital cost} = \$110\text{K (1)}$$

$$\text{Savings} = 54.3\text{K (2)}$$

$$\text{Payback period} = \frac{(1)}{(2)} = \frac{110}{54.3} = 2.02 \text{ yrs}$$

$$\text{Return on investment} = \frac{1}{\text{payback period}} = \frac{1}{2.02} = 49\%$$

Assuming a corporate minimum attractive rate of return of 25%, the robot installation is economically justified.

In an automated factory, the justification process is more complex. In this case a broader perspective must be considered, namely, evaluation of economic alternatives that would lead to the same end result, the manufacture of the end product.

In the case study as stated earlier, the following alternatives were considered:

1. Manual assembly process, consisting of manual, subassembly, final assembly, and testing.
2. Hybrid assembly process, a combination of completely automated subassemblies and manual assembly and testing.
3. Automatic assembly process, a totally automated scenario which consists of automatic subassemblies, robotic final assembly, and automated testing.

The costs associated with these three cases are shown in Table 2.

The financial justification was carried out as follows: A return on investment analysis was carried out between the hybrid and manual options using incremental capital (H1-M1), incremental manpower (H2-M2), and space (H3-M3) savings. It was found that the incremental capital was justified based on the decision rule ROI (H-M) ≥ MARR. This procedure was then repeated for the incremental capital (A1-H1) for the automatic option. Using the same decision rule ROI (A-H) ≥ MARR, the capital expenditure for the automatic case was justified.

The following guidelines were adhered to in the comparisons: (*1*) depreciation method of all capital equipment was the ACRS method; (*2*) equipment life for planning purposes was taken as 5 years; (*3*) investment tax credit was assumed to be 10% on all capital expenditures; and (*4*) corporate tax rate of 46% was assumed.

It should be noted that since the financial justification of the project occurs during the planning stages, the accuracy of the numbers could vary considerably from the observed end result.

Table 2. Capital, Manpower, and Space Requirements

	Manual	Hybrid	Automatic
Capital[a]	M1	H1	A1
Manpower cost[b]	M2	H2	A2
Space cost/year	M3	H3	A3

[a] Manual case is least capital-intensive.
[b] Manpower requirements are determined by operations analysis on all assembly, testing, and material handling operations in the process, using predetermined motion time study (in this case MTM) and a two shift operation.

Feasibility Study, Performance Evaluation, and Preliminary Analysis for Robot Selection

Once the process has been decided on and the decision made to robotize, the next step is identification of potential robots that could be used for individual applications based on cost and performance. The process must be broken down into the individual workcells and specifications for robots at each workstation should be defined. Factors that typically make up robot specifications are

1. Requirements of manipulator arm (number of axes, repeatability, payload, work envelope, etc).
2. Programming requirements (continuous path programming or point-to-point programming).
3. End effector, tooling requirements (spray guns, welding torches, vacuum grippers, etc).
4. Software requirements (error recovery, data communication to a hierarchial controller, multitasking capabilities, etc).
5. Sensory requirements (interfacing between the robot controller and sensors in the workstation such as vision systems, cell controllers, etc).

Compared to robots in stand-alone workcells, the robots used in an automatic factory would be characterized by high levels of intelligence, flexibility, and communication capabilities. The next step is to compare the robots in the marketplace in order to find the robot that best fits the specifications for the particular application. With this in mind, let us investigate a typical airless painting application in which the robot is required to spray the interior of a container with a sound deadener. Robot specifications for this application are

Manipulator. 6 axis minimum (3 on the arm and 3 on the wrist); gun flipper attachment required for 90 degree rotation of spray gun.

Controls. Hydraulic drives not critical as the material to be sprayed is water-based and does not require an explosion proof environment.

Interfacing requirements. A minimum of 8 inputs and 8 outputs for tie-in to sensors and peripheral equipment in the workstation.

Programming. Both point-to-point and continuous path required with application program editing capabilities. Storage for up to X minutes of programming time and Y number of program points.

Other. Speed not critical as cycle time of application is 1 minute (relatively long for application). Payload not critical as all painting robots considered can handle the weight of an airless spray gun. Repeatability within ± (0.25) in.

Documentation. Robot manuals (programming, maintenance), schematics (mechanical and electrical).

Other equipment required. Maintenance table to mount the robot for the purpose of removing it in the case of a breakdown.

Five robots vendors are investigated and compared, assuming that all have demonstrated, through feasibility studies, that their robots can perform the task at hand (Table 3). If the decision is to be based purely on cost, the choice would be between vendor 3 (V3) and vendor 4 (V4). But here again, factors such as quantity discounts, vendor loyalties, vendor standardization, and lead times for delivery would have to be considered before the final decision could be made. It should be noted that standardization of hardware is key from the standpoint of ease of maintainability, spare parts, and operation. In the case study, the final decision was to purchase robots internally as no other robot available in the market had the desired capabilities. The robot (Fig. 7) was selected for its large work envelope (approximately 70 cu ft—7 ft × 5 ft × 2 ft), error recovery capabilities, user friendly programming, sensory capabilities (tactile sensing—strain gauges in gripper, LED part detection), multitasking and communication capabilities of the software that could be run on the controller.

Table 3. Robot Vendor Selection Matrix (Painting Application)[a]

Manufacturer	Model	Geometry Type
Vendor 1	V1	Articulated arm
Vendor 2	V2	Articulated arm
Vendor 3	V3	Articulated arm
Vendor 4	V4	Articulated arm
Vendor 5	V5	Articulated arm

Robot	Repeatability, in.	Speed	Axes	Payload	Cost, $
V1	±0.025	N/A	6	N/A	140K
V2	±0.02	N/A	6	N/A	80K
V3	±0.15	N/A	6	N/A	75K
V4	±0.02	N/A	6	N/A	75K
V5	±0.10	N/A	6	N/A	75K

[a] All numbers for cost and repeatability are approximate. The above matrix was developed for an airless finishing application in which the speed of the robot was not a gating factor and the payload (weight of airless gun) was well within the payload capabilities of these robots.

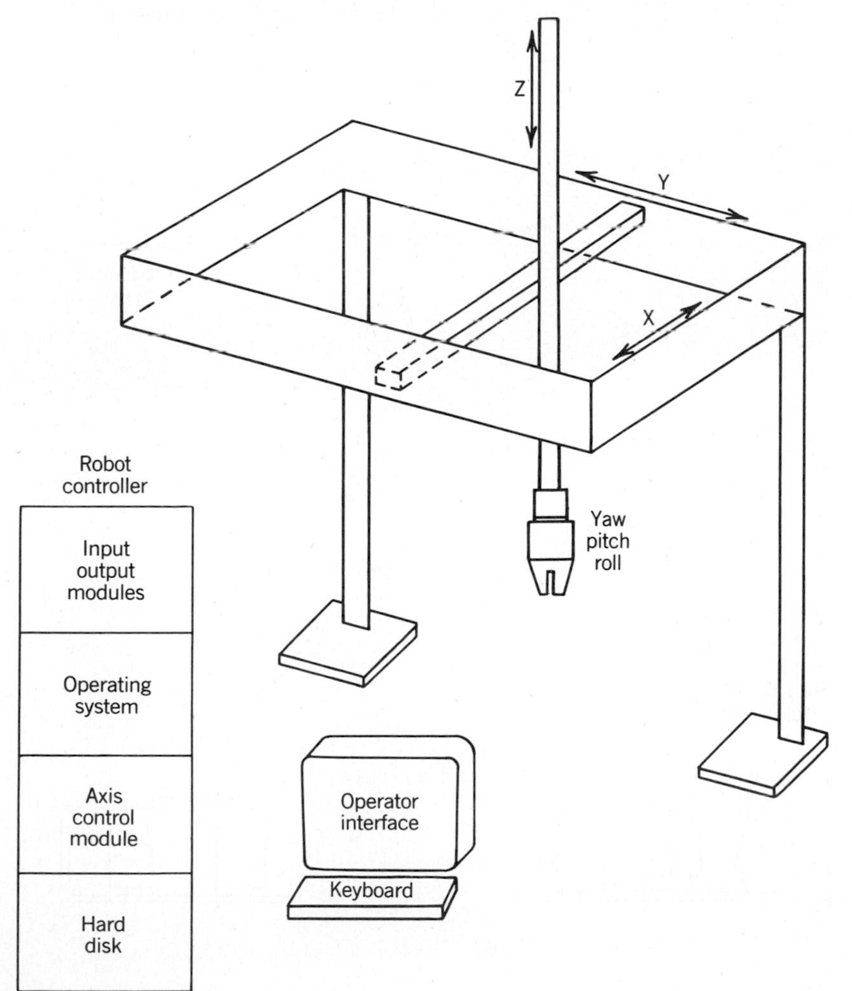

Figure 7. Electric assembly robot.

Product Designs, Material Handling, System Designs, Process Tooling Designs, and Manufacturability Concerns for Robotization

In an automatic factory, selection, installation, and programming of the "shelf robots" usually represent a relatively small degree of complexity when compared to the other factors that are involved in making the project a success, such as product designs for automation, material handling (qv) system designs, process tooling designs, and other manufacturability concerns. It is in these areas that a high level of technical innovation is required. In the case study, the following were accomplished:

1. Product was designed for automation (Figs. 8 and 9). There were no screws or fasteners to hold component parts together as inserting screws represents one of the most arduous robotic assembly operations. All parts either snapped or twisted into place.
2. Design engineering group and automation process designers worked together from initial design concept to finalized product to minimize costly design changes in product and process during implementation and to evaluate alternative component part designs during robotic feasibility analysis.
3. Tolerance of individual component parts was maintained within the accuracy and capabilities of the robots to be used.
4. The number of parts to be assembled was minimized (a total of approximately 60 parts), thus resulting in reduced materials handling and cycle time and increased space savings and throughput. In the case study, a molded side frame took the place of 20 other parts.
5. Chamfers or tapers were provided in component part mating areas to facilitate assembly.
6. Projections, holes, or slots that could cause entanglement in feeders were avoided.
7. Adequate totes (Fig. 10) for component part transportation were provided.
8. Component part inserts (Fig. 11) in totes were designed to accommodate critical tolerances and orientation for parts picked up by robot.
9. Adequate space was provided between the side of the tote and parts for robot end-effector access to part.
10. Locating notches and clamping devices for parts totes and product pallet for accurate location of totes and pallet in robot work envelope were provided.
11. Magazines' and bowl feeders' capacity at individual robot workstations was designed for 2 hours of production.
12. Flexibility was built into robot end-effectors (Fig. 12) for multiple parts handling without tool changes.
13. Automated storage-retrieval system was designed to hold adequate capacity for 2 days of production.

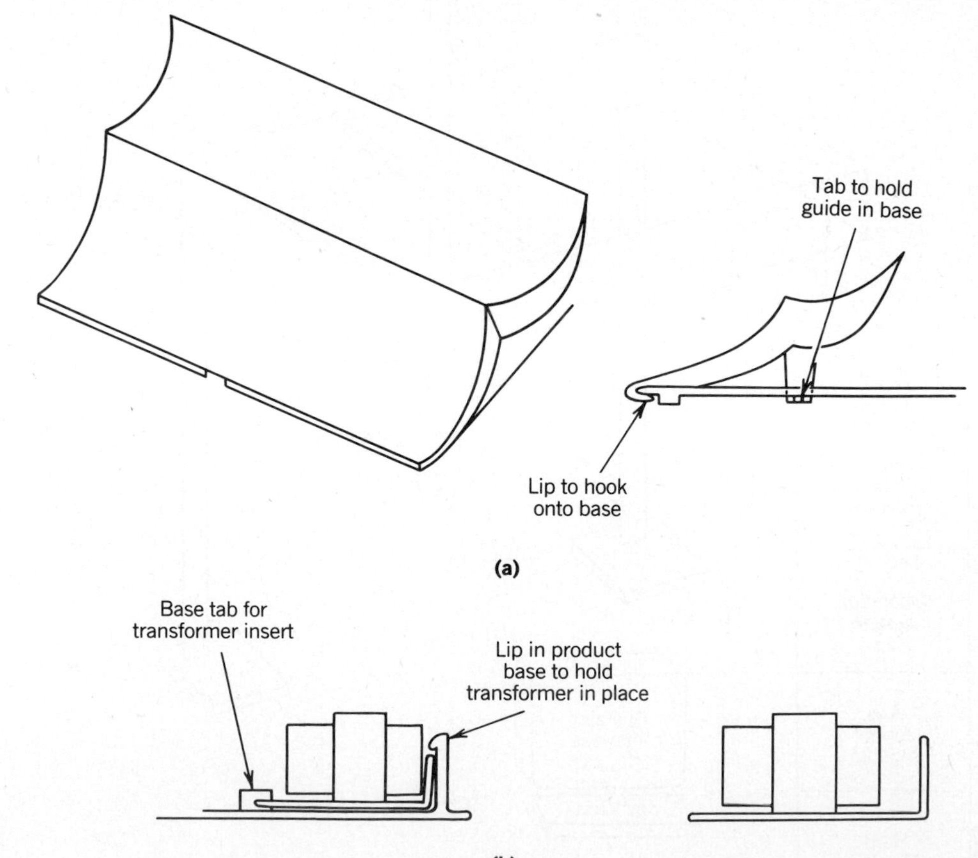

Figure 8. Design for automation examples. (**a**) Front panel guide. (**b**) Transformer snap.

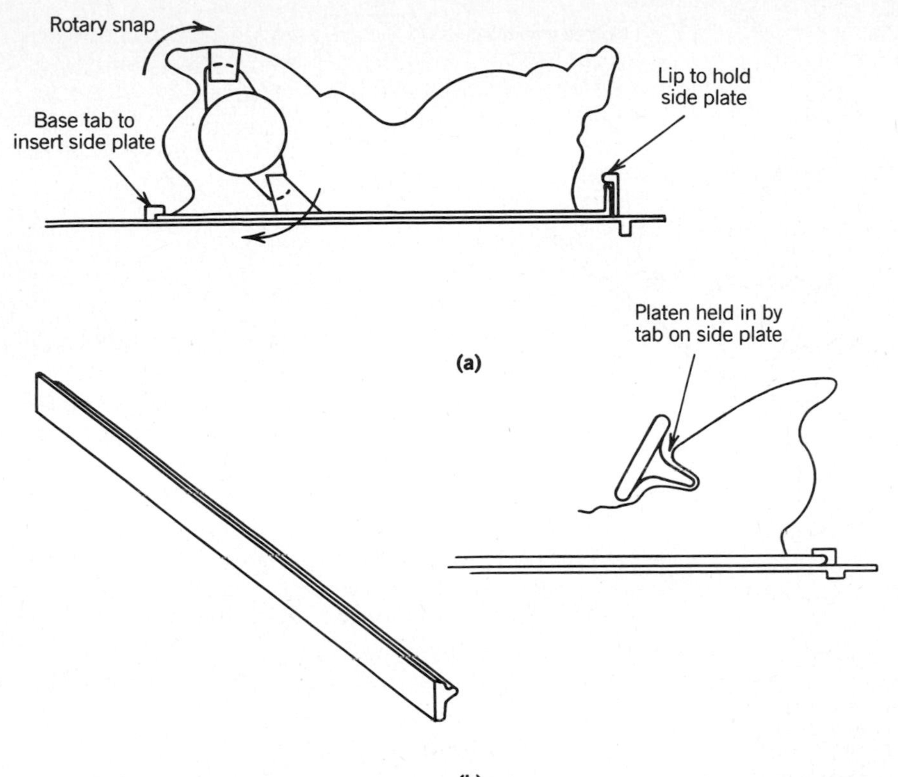

Figure 9. Design for automation examples. (**a**) Paper feed motor assembly. (**b**) Platen snap.

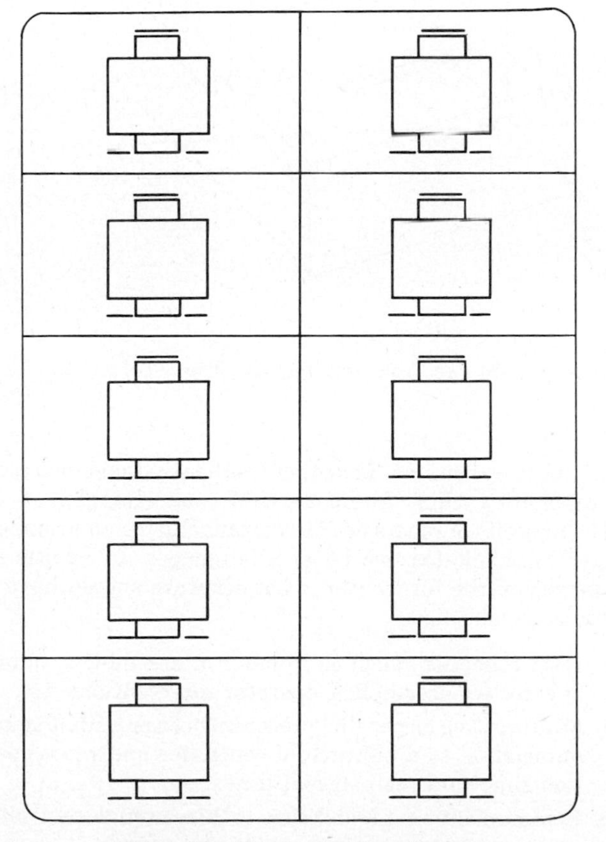

Figure 10. Matrix tote (transformer).

14. The automated line was balanced for an individual station cycle time of 2 minutes. The parts breakdown by station was given.
15. The bowl feeders' and parts magazines' locations in individual stations were configured for optimal robot cycle times.

Software, Programming, and Hierarchial Control Issues

Software and programming represent an integral part of any robotic application. In an automated factory not only is software at cell level critical, but up-down communication in a computer control hierarchy plays an important role for inventory management; data collection (downtime and error recording, throughput summaries); and storing and downloading of production schedules and robot application programs, and downloading of part set up information in NC machines, etc (Table 4).

Software Overview of Present Day Robots. Programming capabilities of present day robots can be classified in the following manner:

Level 1 Joint control languages
Level 2 Path control languages
Level 3 Task-oriented languages
Level 4 World model languages

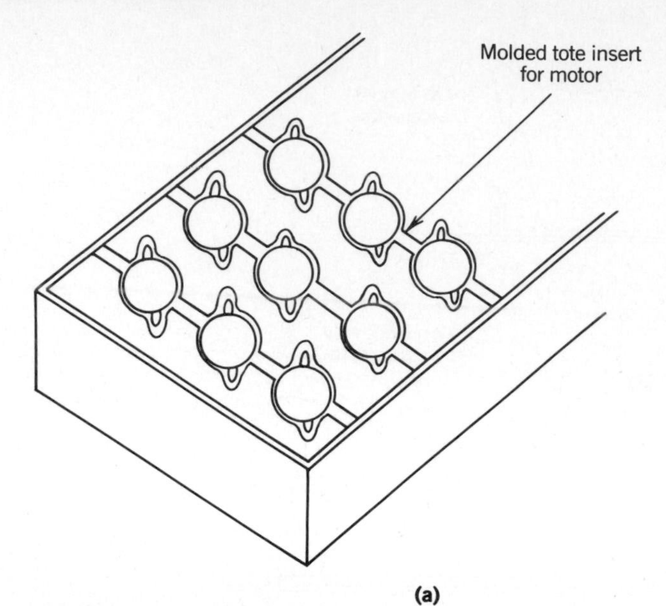

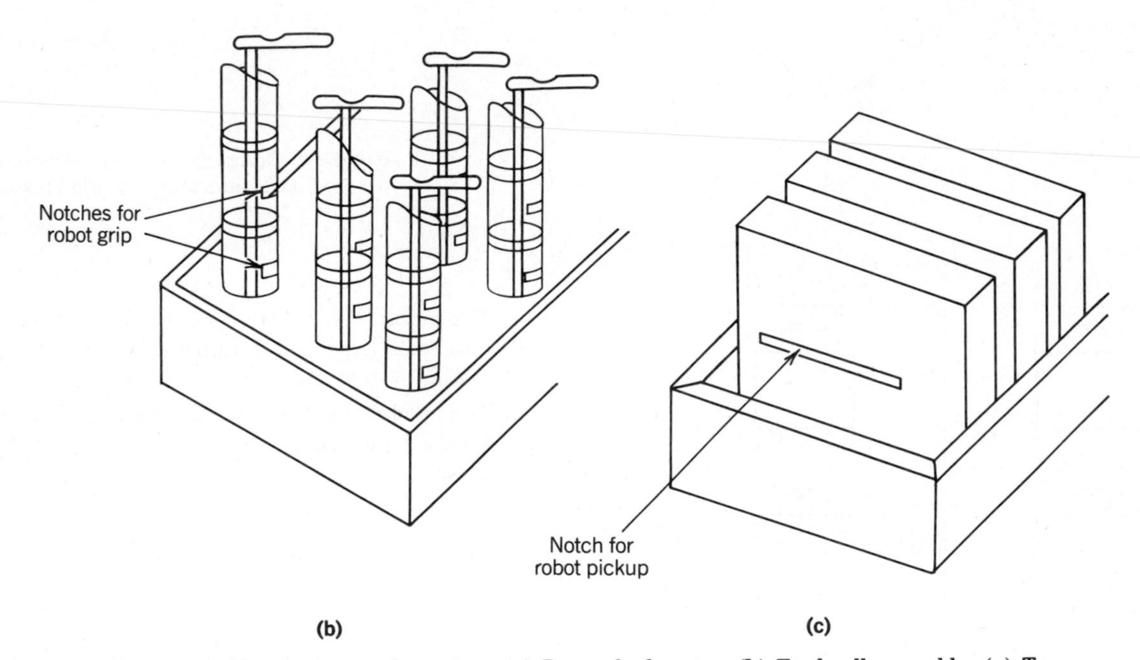

Figure 11. Matrix tote configurations (**a**) Paper feed motor; (**b**) Feed roll assembly; (**c**) Top cover assembly.

Level 1—Joint Control Languages. Characteristics of this level require teach pendant of the robot to position each axis in workspace. Switch settings define robot operation. Examples are Rhino Robot Language and Armbasic for the Microbot.

Level 2—Path Control Languages. This level uses teach pendant of the robot to command direction of motion of robots and effector. CRT/keyboard is used to enter commands that define robot cycle logic. Examples include Val (Unimation) and Cincinnati Milacron T3 language.

Level 3—High Level Languages. Structured programming constructs are used to define robot operations. Examples are AML (IBM) and Rail (Automatix).

Level 4—World Model Languages. This level uses statements that define a task. Each statement must be decomposed to define robot operations. Learning frequently takes place during automatic cycling. Autopass, IBM is an example.

The majority of robots used in manufacturing environments in the United States use Level 2 languages. In addition, in the factory of the future, the robot software should have the following capabilities:

1. Data collection, such as number of assemblies, number of defective assemblies, operator interventions, etc.
2. Multitasking or parallel processing of tasks, that is, communciation to a hierarchial controller and robot operation simultaneously in real time.
3. Communication capabilities to hierarchial controllers, computer-integrated manufacturing.

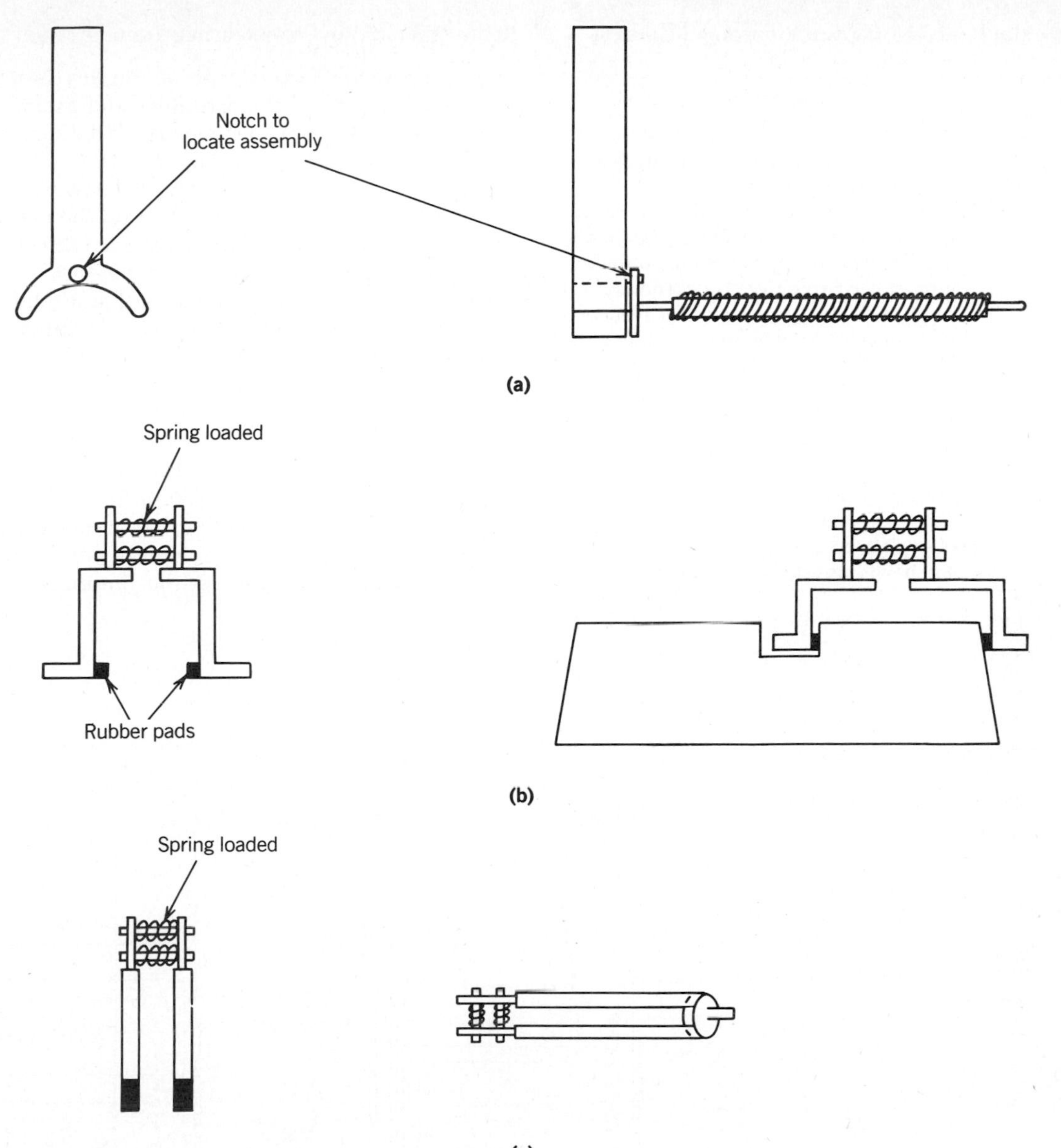

Figure 12. End-effector attachments. (**a**) Helix motor assembly attachment. (**b**) Top cover pickup attachment. (**c**) Knob pickup attachment.

4. High degree of sensory intelligence, error recovery capabilities.

With those criteria in mind, part of the application program for station 2 in the sequential robotic assembly process (Fig. 13) was developed (Table 5). The emphasis was placed on error recovery, communication, and operator intervention requirements.

The electric robot at station 2 assembles the following component parts on the computer printer: transformer, front panel guide, right side plate, end of form flag, eccentric bushing, and ground plane. The application program was set up in the following manner:

Program description. Assemble transformer, right side plate, eccentric bushing, ground plane, end of form flag, and front panel guide on the product.

Primary segment. Contains sequential manipulator moves and input/output functions to assemble component parts on the assembly.

Swap segments. Contains manipulator moves and functions to pick up right side plate from alternate feeder.

Error recovery segment. Contains manipulator moves for all possible error contingencies in the assembly sequence, such as a dropped part and improper part pickup and assembly. This segment also contains error message information and operator actions required to correct these errors on occurrence.

Table 4. Hierarchial Control Diagram Functional Control Levels

Level	Major Functions
1. Site/facility management	Planning, engineering, financial, site services, distribution, interplant interface, management information
2. Plant/shop management	Shop scheduling, parts planning, order management, inventory control, maintenance support, controlling ECs/MCs, network support, assisting resource planning, problem tracking, management reports
3. Module/cell management	Module scheduling, maintenance support, implementing EC/MCs, warehousing control, shipping control, maintaining process data, alerting to problems, management reporting, receiving control
4. Element/area control	Workstation control, work-in-process control, parts distribution control, out-plant material movement
5. Tool/equipment control	Tool/equipment control (eg, sensing, measuring, handling, transporting, etc)

Robot Operator and Maintenance Training Issues

In an automated factory, one of the key issues is the skill level and training of the operators and maintenance technicians. Another key consideration is the delineation of responsibilities between operators and technicians.

In the startup stage, due to lack of knowledge of the process, operators may be totally dependent on the maintenance technician to fix even routine problems such as reorienting a part in a feeder, ejecting a completed product assembly from a workstation, etc. With adequate training and instruction of the operators, this problem can be alleviated. The following should be considered: Operators should be assigned responsibilities based on competence levels. For example, operators with high mechanical and electronic aptitudes can be assigned the task of overseeing robot operation while personnel with lower aptitudes can be assigned routine tasks such as routing process monitoring or inspection tasks. Next, it should be recognized by management that during the startup phase an overdependence of operators on maintenance and engineering personnel will exist. This can be minimized by monitoring

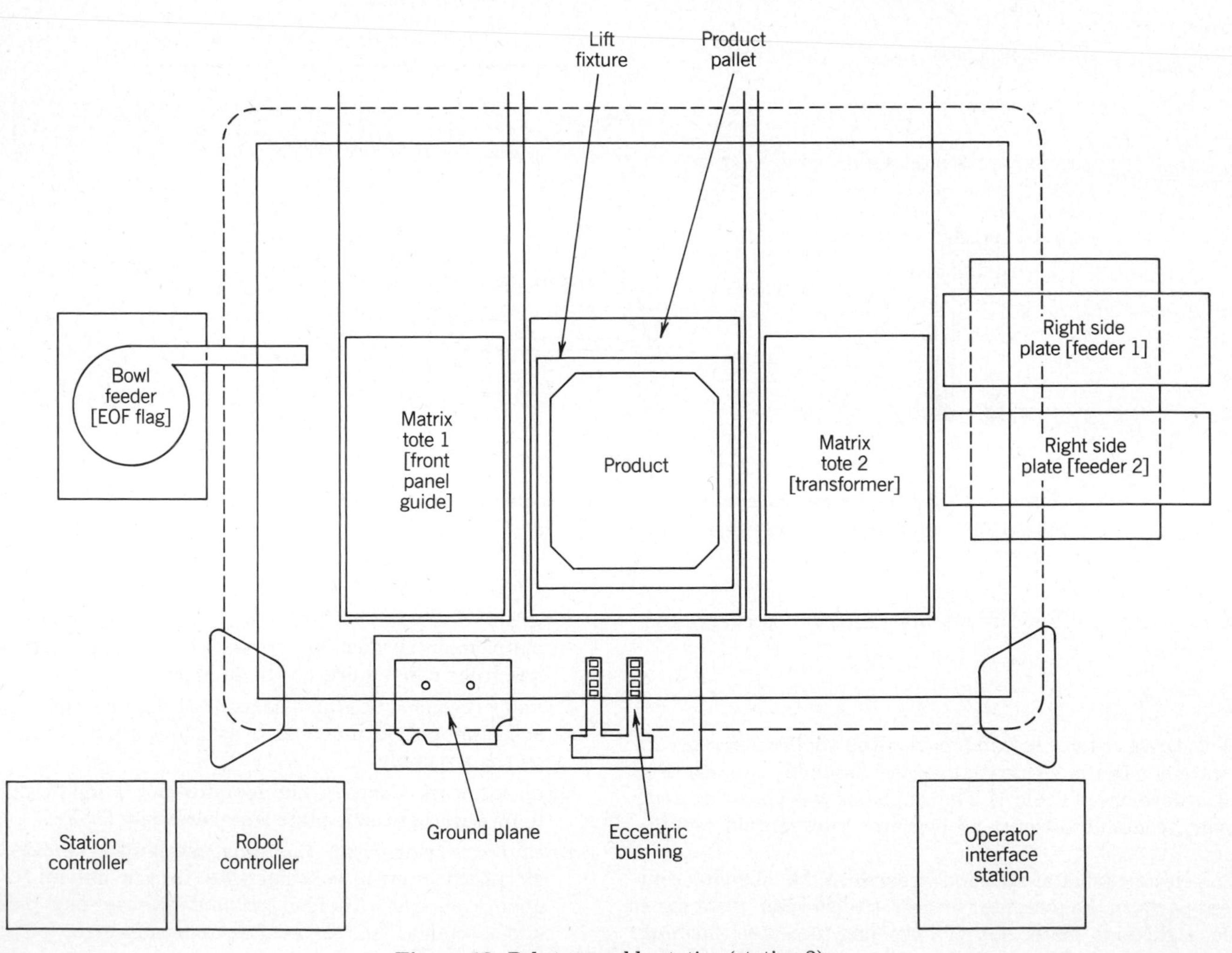

Figure 13. Robot assembly station (station 2).

Table 5. Application Program Listing

Record Number	Record Description
Primary Segment 1—Pick up Side Plate from Feeder 1	
Record 1	Move to safe Z over side plate magazine 1
Record 2	Dummy record-check shuttle out on side plate feeder 1
Record 3	Move down to pick up side plate
Record 4	Orient gripper around side plate
Record 5	Grip side plate
Record 6	Move up with side plate
Record 7	Dummy move—check strain gauge < × grms, jump to record 500
Primary Segment 2—Assemble Side Plate on Product	
Record 26	Move over product
Record 27	Move down with side plate
Record 28	Pitch
Record 29	Insert backend of side plate into tab on base of product
Record 30	Move down to make snap
Record 31	Dummy move—check S.G. for correct insertion of side plate jump to record 508 [if S.G. < × grms]
Record 32	Open grip
Record 33	Move manipulator arm to safe Z position
Swap Segment 1—Pick up Side Plate from Feeder 2	
Record 251	Move to safe Z over side plate feeder 2
Record 252	Dummy move-check shuttle out
Record 253	Move down to pick up side plate
Record 254	Orient gripper around side plate
Record 255	Grip side plate
Record 256	Move up with side plate
Record 257	Dummy move—check strain gauge for side plate presence S.G. < × grms—jump to record 511
Error Recovery Segment—Manipulator Moves and Operator Intervention Messages	
Record 501	Retry side plate pickup, dummy move, check side plate feeder shuttle out
Record 502	Move down to pick up side plate
Record 503	Orient gripper around side plate
Record 504	Grip side plate
Record 505	Move up with side plate
Record 506	Dummy move—check S.G. for side plate presence S.G. < × grms operator intervention—error message—check orientation of side plate in feeder 1
Record 507	Jump to record 26
Record 508	Retry side plate assembly on product, move Z down for proper installation of side plate in product
Record 509	Dummy move, check strain gauge S.G. < × grms operator intervention, check assembly of side plate on product
Record 510	Jump to record 51, assembly of transformer
Record 511	Retry side plate pickup from feeder 2, dummy move, check side plate 2 shuttle out

operators initially "on the job," giving them leeway in fixing routine problems until a "comfort level" is reached and the operator can perform independently. A compromise has to be reached on how much this leeway will be as an operation error could result in extensive damage to a manipulator or peripheral equipment. Finally, a significant problem in operator training is caused by the fact that classroom training is seldom supplemented with hands-on experience. This is critical as problem recognition and the speed with which the operator fixes process problems come through continual hands-on experience.

Typical operator responsibilities are startup and shutdown of the workstation at the beginning and the end of the shift; clearing routine jams of parts, product, tooling in robot workstations (operator interventions); and notifying maintenance technicians of equipment malfunctions—the more skilled and experienced the operator, the higher will be the level of his troubleshooting skills. Operator training documentation should typically consist of a functional overview of the automated process equipment description (with schematics), by workstation, of individual parts, processes carried out; listing of all possible malfunctions and corrective actions (operator interventions) required, by workstation; safety procedures and devices, by workstation; and routine startup and shutdown procedures, by workstation. The following operator intervention procedure was developed for the case study:

Step	Action
1	Read error message on operator interface screen, eg, knob jammed in feeder track. Remove knob from track and ensure knob is oriented correctly in track; robot will retry.
2	Press intervention key on robot teach pandant. Will freeze motion of manipulator.
3	Shut down robot power by pressing power off button. Ensure that power-on light on manipulator is off.
4	Enter robot work envelope, remove defective knob from track, and ensure that track is clear and feeder is operating correctly.
5	Ensure robot arm will not collide with any objects in the workplace when power is turned on.
6	Start up robot.
7	Monitor assembly of a few subsequent operations to ensure that the problem has been corrected.

Robot Reliability Issues

One of the key factors for successful operation of a completely automated factory is the continuous problem-free operation of components (hardware and software) in the factory. The reliability of a piece of equipment can be defined as the probability that the product will give satisfactory performance for a specified period of time when used under specific conditions. There exists a simple relationship between the reliability of an equipment and its mean time between failures.

$$R = e^{-T/MTBF}$$

where R = reliability factor.

Availability is the probability that at any given point in time the equipment will be ready to operate at a specified level of performance.

$$A = \frac{\mathrm{MTBF}}{\mathrm{MTBF}} + \mathrm{MTTR}$$

where A = availability factor.

MTTR = mean time to repair

MTBF = mean time between failures

Typically, an automatic factory requires an extensive integration of a variety of vendors' hardware and software. Incompatibilities between these components can lead to extensive downtime in the production cycle. Robot vendors usually specify, as part of their reliability data, an MTBF for a robot system. But in most cases this does not represent the ultimate true application environment of the robot and hence can be misleading. Typically robot vendors do both parameter and extended life tests on manufactured systems. Parameter testing is testing to verify accuracy, repeatability, and payloads under different conditions of loads and speeds. The extended life test is usually a cycling of the robot over a continuous period of time (eg, 100 h). Within the extended run the manipulator arm is programmed so that the axes are subject to varied acceleration, velocity, and stress conditions. The drawback with such testing is that it does not represent the true application environment in which the robot is ultimately to be used, thus reducing the "comfort level" of the ultimate customer as to the acceptance and reliability of the system when in production. From a vendor's standpoint, it is not economically feasible to customize reliability testing by purchaser. As a result, it is in the best interest of the end user to develop a generic acceptance test for purchased robots. The key criterion to be tested should be the analysis of the unusual motions, or functions, that the robot would have to perform in its application and the simulation of this application over a significant period of time. (Table 6 lists the reliability program for a material handling simulation.)

Robot Safety Issues in the Factory of the Future

Safety of personnel is also a major concern in an automated environment (see also SAFETY OF OPERATORS). As a result of the large number of accidents and injuries caused by robots in the United States and abroad, robot safety has received repeated attention in the past few years.

In an automated factory environment, with a large number of robots in operation simultaneously, safety becomes even more critical. This is especially true in the "bring up" of robot systems when manufacturing personnel might inadvertently step into the work envelope of a robot. During application teaching or debugging, it is critical that personnel involved in maintenance, installation, and operation understand the unpredictabilities and capabilities of a moving robot arm in case of a malfunction, such as an encoder or resolver breakdown. Thus a safety awareness needs to be created and emphasized repeatedly. The key to robot safety is to understand the most effective way for protection and still have easy access to the robot work area for intervention and maintenance purposes.

In the case study, the following equipment and personnel safety measures were implemented. Hard hats were mandated

Table 6. Reliability Program Listing—Material Handling Simulation

Step No.	Command	Function
Program: To simulate pick and place of blocks		
10	Speed 300	Set speed to 300
20	Move T1	Move over block pickup
30	Move T2	Move down to block pickup
40	Do output + OGI-200	Pick up blocks with vacuum end-effector
50	Speed 100	Slow down robot to speed 100
60	Move T1	Move up with blocks
70	Move T3	Move over block setdown
80	Move T4	Move down with blocks
90	Do output + OGO-200	Set blocks down
100	Move T3	Move up
110	Move T4	Move down to block pickup
120	Do output + OGI-200	Pickup blocks
130	Move T3	Move up with blocks
140	Move T1	Move over block setdown
150	Move T2	Move down to set down blocks
160	Do output + OGO-200	Release blocks
170	Speed 300	Increase robot speed
180	Move T1	Move up
190	Move T5	Move to dial indicator
200	Move T6	Hit dial indicator
210	Move T5	Move back from dial indicator
220	Got to 10	Repeat program

for all personnel entering the manufacturing area for protection of material flow on overhead conveyors. Safety mats, safety shields and light curtains were strategically placed to protect personnel from robot arms, mechanical fixtures and feeders, shuttles, and moving material handling devices such as conveyors. A key switch was provided to override these devices for maintenance purposes. Emergency power-off buttons and emergency power-off cords were installed at strategic locations to deenergize robots and moving devices. (It is advisable to have an operator monitor the maintenance technician during the period work is done within the work envelope. The operator should have easy access to the emergency power-off button during this period.) Due to the small number of operators monitoring operation in the factory, the following strobe light configuration was used to indicate robot workstation status:

Color of Strobe	Conditions
Amber	Parts low condition of feeders, magazines
Blue	Parts out condition of feeders, magazines
Red	Malfunctions such as parts jams, problems in product infeed, outfeed, improper clamping, etc—operator intervention required
Green	Safety mats, shields overridden for robot station maintenance—override mode
White	Station operational

Finally, it should be emphasized that safety cannot be compromised to accelerate installation and production schedules as an equipment-related injury could have serious implications and cause setbacks in production schedules and employee and management morale.

SUMMARY AND CONCLUSIONS

Despite drawbacks that have been encountered in the implementation of robots in the United States and abroad, robots are still a viable alternative capable of providing the flexibility necessary in any automatic factory. With proper planning during the initial stages, robots can play a very important role in the factory of the future. The key issues that should be kept in mind are

1. Do not robotize for the sake of robotization. Thoroughly evaluate each application and weigh the pros and cons from all relevant standpoints, such as flexibility for future needs, feasibility, financial justification, etc.
2. Develop the skills that are required for robotization success—programming, mechanical and electrical design, for example.
3. Successful robotization comes from innovative technology and a harmonic integration between the elements of people, equipment, materials, and information which constitute an automatic factory.

This provides some general guidelines for planning and implementation of highly automatic or completely automatic factories. The guidelines proposed are the result of a successful implementation of complete automation in assembly and should prove useful in similar applications.

FINISHING

Atlas J. Hsie
SUNY College of Technology
Utica, New York

Danny McCoy
B&D Sencoy Inc.
Utica, New York

INTRODUCTION

It has long been a desire of industry to be able to fully automate the very labor intensive "off-hand" process of deburring, buffing, polishing, grinding, and similar mechanical surface preparation processes, herein to be called finishing. The utilization of robotics offers yet another alternative to automation for these boring, repetitive, often hazardous tasks.

The use of robotics has not been the "cure all" that many had hoped it would be. There are problems of adaptation, integration, and understanding not only of the robots' capabilities but also of the tooling/media for the process. (Tool refers to metal cutters. Media refers to buffing wheels, brushes, nonwoven nylon products, abrasive belts, etc). Material-handling criteria and the general manufacturing capabilities in which the robotic system is to be used to perform the finishing process are also considerations. These problems are being overcome and the use of robotics is starting to increase in the areas of finishing.

In discussions with the major robot manufacturing companies, the number of installations is approximated at 400–500 world wide. The bulk of these is being claimed by one manufacturer at approximately 300 installations. A breakdown of the number of units in each individual area, such as buffing, deburring, or polishing, was not available, nor the number of installations used for just loading and unloading parts into some other automated finishing unit. The vast majority of all units installed are in Europe, where automation of these finishing operations is undertaken seemingly with more desire to use state of the art technology.

Robots are able to improve the finishing operations by providing more consistency in the finished parts, often processing parts faster than by hand; are able to work multiple shifts untended; are able to utilize more powerful tools for faster finishing; and they can work in the noisy, dusty environments that humans find unsuitable and/or hazardous.

TYPICAL APPLICATIONS

The typical installation cannot be readily defined. There has been a certain degree of proprietary information surrounding many of these installations. This is because the end user or the robot manufacturer wants information about the process kept from the competition. Thus, many of the installations are not reported in trade publications, nor are provided by the robot companies to customers inquiring about such processes, other than to say yes or no to the question have you installed a robotic system to do "this type of finishing."

There are some generalities that are evident from reported installations and from looking at the capabilities of the robots themselves. Robots used for these types of operations are usually of the five and six axes variety in order to obtain the required orientation between the tool/media and area to be processed. These robots, in most cases, can not hold accuracies, within their entire work envelope, of better than ± 0.005 in. (0.125 mm) to ± 0.010 in. (0.25 mm) with older units having less accuracy. Repeatability of better units is in the range of ±0.002 in. (0.05 mm) to 0.005 in. (0.125 mm).

When you combine the robot accuracies with the inaccuracies usually present in the part, the varying load conditions usually present in processing (causing varying deflections in the robot arm), and the cantilevering loads from the tooling/media units or part/fixture combination, it becomes apparent that tolerances better than approximately ± 0.015 in. (0.375 mm) are difficult to obtain without special care. Some precision deburring operations are being performed but this is the exception rather than the norm.

Electrically operated robots are the units of choice over those using hydraulic power sources. The load capacities are normally less than approximately 50 lb (223 N) (1). If robot mounted drive units for the finishing tool/media is used, they are usually under 5 hp (3 kW) (1). These are considerably over the typical loads and power of drive units employed in manual deburring but are average to light for buffing and polishing operations.

The work envelopes have been such that the robot has a reach of 48 in. (1,200 mm) or less from the base of the robot. One of the exceptions has been units used to drill and deburr fiberglass components, with each unit being a gantry type robot having a working envelope of approximately 10 ft (3.0 m) by 3 ft (0.9 m) × 4 ft (1.2 m) and load capacity in excess of 100 lb (446 N).

Robots have been used for various deburring operations in the automotive and aerospace industries and others. They have also been used in the foundry industry for grinding gates, risers, and flash as well as various chamfering of internal and external edges. Many parts have also been processed for a variety of industries using media such as various brushes, buffing wheels, nonwoven nylons, hard tooling (carbide bits, tool steel cutters, etc), bonded grinding wheels, and speciality products to obtain the desired results. The use of robots has not been limited to any industry, type of part, material, etc. The only limits seem to be the accuracies required and the ability to adapt the proper tooling/media to the robotic system. The latter has been seen as the most troublesome for most of the robot manufacturers and system house engineers.

CONTROL METHODS FOR ROBOTS

It is not the intent of this article to review all the various control methods used with robotics but to briefly mention those that have the most applications in the finishing processes.

The manual method of programming a robot is similar to the setting up of other machine tools. It is the procedure for the simpler robots which use mechanical stops, cams, switches, relays, etc, as the memory and control unit. This type of robot is little used for finishing operations. It is very precise, however, and can be used for selected finishing of given areas without the necessity of following an edge or contour.

Walkthrough Method

This method allows the programmer to physically grasp the robot "hand" or item mounted on the end of the robot assembly and guide it through the motions desired. The locations and speed are recorded and can be played back for the robot to repeat the motions. The speed can be varied as needed. In some cases a separate robotlike device that duplicates the robot configuration is used for programming, where the actual robot is too large for a person to manipulate or the production robot is not to be taken off line for programming. This method is not utilized for many finishing operations but is utilized more in the painting areas. The robots usually used in conjuction with this type of programming method are not as rigid nor as accurate as needed for finishing operations.

Leadthrough Method

This method uses a teach pendent or joy-stick type of control to power drive the robot to its desired locations. It generally has the capabilities of driving each joint individually or all joints simultaneously, in a world coordinate system (X,Y,Z) where the Z axis is vertical. With more sophisticated control systems one has the capability of having a tool coordinate drive capability with the X,Y,Z orientated so that the Z axis passes through the centerline of the flange of the last joint of the robot arm. This, in conjuction with a tool offset command (allowing the X,Y,Z to be offset to the tip of a known tool attached to the flange) allows control at the tip of the tool/media in power driving the robot through/to all the desired locations and motions as if the tool were operating through the finishing process. This type of system for programming the robot is the most used system with the method utilizing the tool coordinate and tool offset capabilities as being most desirable.

Off-line Programming

Off-line programming relieves the problems of taking production robots out of service while programming takes place for new parts or for modifications of old programs (see PROGRAMMING, OFF-LINE LANGUAGES). This method is available with those robots usually having the more sophisticated languages and advanced control methods as described in leadthrough methods above.

This programming has had some major drawbacks and rarely is as effective as the user would like. The problems center around the inability of the robot to go to any defined locations in space with the accuracy normally desired. This is a result of the tolerances built into the robot, the part tolerances, and the location of the part and tool/media in reference to the robot. The inaccuracies of these tolerances, in total, cause the interface location between the area of the part to be processed and the tool/media used for the processing not to be at the location programmed into the controls. This almost always requires on line correction to compensate for the tolerance buildup.

In those cases where accuracies are not very exact and some form of location compensation system (sensors, force feedback, etc) is used, it can be near 100% suitable with little or no on line corrections. Again the accuracies of the particular job determines the effectiveness of the off-line programming. The more exacting the task the less effective the off-line programming, the less exacting the task the more effective the programming. With the cost of this feature dropping over the years and the advantages of the more sophisticated controls normally present in robots used for finishing, the added cost of this feature is well worth the price for the overall time savings for most applications.

METHODS FOR INACCURACY CORRECTION

Currently there are three approaches to offset the robot inaccuracies, workpiece variations, and spacial relationship inaccuracies of all components within the work envelope, to obtain more precise processing results. They are identified as compliant approach, force feedback control methods, and "fine-tuned" robot method.

The compliant approach is basically a method whereby some type of "forgiveness" or "float" is built into the system using springs, air, or resilient compound (rubber, urethane, etc). This allows movement usually in preset directions and normally requiring a preset controlled amount of pressure to obtain the desired motion. The amount of "float" is determined by either/and/or the amount of tolerance variation in: the part, the location of the part in the work envelope, the tool/media used, the robot inaccuracies, the tool/media wear during processing, the location of the tool/media in the work envelope, etc. Proper design is necessary to provide the correct travel or "float" under the appropriate pressure.

The requirement of controlled rigidity is necessary under the correct cutting speeds and travel rates to prevent chatter. This is particularly true for the harder types of tooling/media such as carbide bits, bonded wheels, etc. This can be a major problem in that chatter is not only detrimental to the finishing process, but is also highly destructive to the robotic arm assembly.

Most compliant devices operate in one direction which is usually perpendicular to the interface of the tool/media and the parts surface to be processed. Devices are available to operate in up to three directions simultaneously for special multicontouring processes (1).

Consistent edge breaks and chamfering processes can be performed holding tolerances of 0.010 in. (0.25 mm) or better, using compliant devices even when total tolerance variables of the system exceed 0.100 in. (2.5 mm) (1).

Currently many robotic manufacturers have ongoing research and development in the areas of compliant devices for a variety of motions, pressures, etc. There are also joint ventures to develop these compliant devices between robot manufacturers and companies devoted to the design of fixturing, grippers, and special "tooling" for robotic use.

The term "fine-tuned" robot refers to a robot that has had adjustments and/or modifications to the electrical resolvers to increase the accuracy of the unit within a given area of its work envelope. This is a time-consuming process and often slows the operational speed of the robot. It also can make it necessary to reprogram all locations from all programs once an accident occurs or, in some cases, once mechanical maintenance occurs. This procedure is used when an existing electrical robot is not as accurate in a given area of its work envelope as desired. It is a little used procedure and is not recommended except as a last resort to obtain the accuracies needed.

The force-feedback control methods are control systems that allow corrections in the path of the robot to be altered as a direct result of sensing an increase in a preset load condition. These systems have been under development for some time by the makers of robots, educational institutions, and others. Operational units have been employed for a number of years in certain robotic installations with success.

The use of these systems in robotic finishing has not proven as successful as hoped. The major drawback centers around one of the unique problems of finishing processes—speed. Finishing operations, especially deburring, require travel rates that are usually at the top of most suitable robots' speed range. The existing force feedback systems do not have the capability of processing all the pertinent information within a time frame necessary to make them work within the economical travel rates required for the finishing process. Thus in the typical finishing process of deburring, if an increase in force is sensed as the result of a bigger burr, by the time the system responds to make a change to correct for the resultant deflection of the tool/media and/or robot arm, the tool/media has gone past the bigger burr.

The other problems with these systems are as follows:

- Pressure sensing is not sensitive enough for many precision finishing operations.
- Dynamic responses of the robot system is difficult to define for all positions within the work envelope.
- Robot deflections under various operating loads and positions affect sensor data adversely.
- Where parts are three dimensional curvilinear geometries, having surface conditions and tolerances, it is difficult to conceive the general algorithms necessary for robot logic to be effective (1).

Many of these current problems are being overcome by the use of larger, faster computers employed to process the very large quantities of data necessary for finishing processes within a suitable time frame. Computer programs are also being developed to assist in solving the other design problems mentioned.

Although these systems are somewhat costly, adding to the cost of the continuous path robots they are normally used with, they can solve many of the tolerance problems currently prevalent, particularly for those finishing processes that are slower such as coarse brushing, rough grinding, degating, etc. The alternative to this system, that can work as well in many situations, is the use of compliant devices, normally available at lower cost.

In some applications a combination of using a compliant

device and the force feedback system can utilize the traits of both to obtain the necessary control. An application where a part has short, closely spaced undulations as well as large tolerances in part location/orientation within the work envelope could benefit from this combination. The compliant device is used for processing the short undulations and the force feedback system is used to compensate for the larger location/orientation problems. It is not known if this type of combination has been used in a robotic finishing application.

PART MANIPULATION VERSES TOOL/MEDIA MANIPULATION

To move the part via the robot to the tools/medias for processing or to move the tools/medias to the part depends on a number of factors. Again each job has to be studied closely to evaluate which of the two choices is best suited for the particular set of conditions you have. Some general guide lines and assumptions can be of assistance.

1. The greater the mass to be moved, the larger and more costly the robot.
2. The larger the robot, the slower and more costly it will be.
3. The greater the projection of part or tool/media past the end flange, the greater the work envelope required and the larger the robot.
4. The greater the number of surfaces/sides required to be processed on a part, the greater the chances of requiring multiple fixtures, grippers, or auxiliary part moving devices which also increases the number of interlocks, sensors, input/output ports to the controller, etc.
5. Multiple tools/medias require costly, often specially designed, quick change couplings if they are to be moved by the robot.
6. If the tools/medias are to be stationary and the part moved by the robot, avoid having the tool/media and part interface in such a position that requires the operator to perform upside-down programming. This increases programming costs and potential for accidents.
7. The greater the number of quick tool/media changes or regrasping of the part the slower the overall cycle time becomes. This can be a significant part of the overall cycle time.
8. Higher powered stationary supports for tool/media drive units can greatly shorten processing time, but require a larger more costly robot to operate under the increased loads imposed.
9. Stationary supports for tool/media drive units need to be built properly to resist deflections under loads. The greater the accuracy required, the more rigid the supports must be and the greater the cost.
10. Auxiliary motion control devices, such as precision multi axes welding tables, can be less costly for complex part orientation than the cost of multiple grippers and the value of the time lost in regrasping the part. This becomes more evident as the number of parts to be processed increases.

TOOL/MEDIA WEAR COMPENSATION

The change in the dimensional size of tools/medias used for finishing operations has a significant impact on the quality of processing performed. The same is true for tools that dull rather than have an easily measurable size change.

Size change in medias such as buffing wheels, grinding wheels, nonwoven nylons, brushes, etc, can be compensated for in a number of ways. One method is to use compliant devices that have enough travel to compensate for any wear that takes place. Another method is to use a force feedback system measuring the loads imposed during processing and correcting the path of the robot to maintain the desired pressure. Measurement systems that use proximity switches, potentiometer units, photoelectric cells, etc, to directly measure the size of the media and through software makes a tool offset, can correct for the change. Power load changes for simple processing operations, where the power requirements remain constant except for a reduction in power load as the result of size change of media, has also been used for certain finishing processes. Again software allows for a tool offset to correct the path of the robot automatically. A variation of the power load system uses a separate program that moves the media to a known hard surface at which point the robot performs a search routine until an increase in the power load is sensed indicating that the media has contacted the surface. The location is determined and through software a tool offset is introduced to correct the robot's path. Where the media does not wear a significant amount and close tolerances are not a problem, a simple counter can be utilized to count the number of parts processed. When a preset number of parts have been processed the software automatically introduces a predetermined tool offset to correct the robot's path. This method is predicated on an analysis that the media wears at a constant rate and its size change, after processing a given number of parts, can be predetermined.

Tools/medias that become dull without an easily measurable change in size must be dealt with via either those systems that monitor loads applied on the tool/media (either interface forces or power consumption) or by utilizing a counter system and changing the tool/media after a predetermined number of parts are processed. In the case of those finishing processes that require very light loads, it is very difficult to monitor interface forces and power consumption effectively given that there usually are changes in these forces as a result of variations in the parts (burr configuration changes or flash changes, etc) that can be greater than the forces resulting from a dull tool.

In all cases where computer software is relied upon, to make corrections in the robot's path as a result of tool/media wear, the selection of the exact tool/media wear system and the robot system must be closely coordinated, for not only compatability, but also for ease of operation, cost factors, and suitability for the application.

When multiple tools/medias are used, it may be necessary to use a combination of the wear systems above to obtain the best overall installation.

GENERAL CONSIDERATIONS

When converting from an "off-hand" process to an automated one, there are a number of problems that become evident which can jeopardize the entire installation. These problems are often not discovered until you are well into the project and most often occur during the robot test run and quotation phases. These problems can be classified under: part problems, educational problems, and cost justification problems (2).

Part Problems

The "off-hand" finishing processes, particularly buffing, polishing, and grinding, have often been used to correct upstream tolerance problems. The "off hand" processes also allow for greater variations in many parts particularly if it is one of the first material removal operations such as in deflashing castings.

When automation is attempted, these tolerance problems and many other unknown tolerance problems (that "off hand" operators corrected for, unbeknown to others) cause major problems in fixturing, repeatability, and final quality of finishing. It is the overlooked part variations that occur most often to defeat the automated process attempts.

Educational Problems

Few of the personnel responsible for the robotic finishing project have an adequate working knowledge of various finishing operations. This is understandable given that almost no schools provide educational course work in these areas nor is there in-house training other than in the same processes that the particular company has used for years (3). This lack of knowledge is even more prevalent among most of the robotic manufacturers and system houses responsible for providing the customer with a production robot installation. The end result often is an attempt to adapt the existing "off-hand" finishing process complete with the exact same tooling/media to a robot system. What gets overlooked are the much better and faster finishing processes that are available and easily adaptable to the robot. In defense of those involved in these robotic finishing systems, this is a new technology as a result of the marriage of the finishing processes and the robot, thus, little experience and history is available to draw upon. On the other hand, few have taken enough interest to seek out the available information on the subject of finishing that has been available for many years.

One other side to the educational problem is the lack of understanding by the manufacturers' personnel regarding the tolerance problems of the parts. The personnel responsible for the robot finishing project are all too often totally unaware of part configuration and out of tolerance problems and also the impact they can have in causing the robotic system to fail (2).

The educational problem can be alleviated by contacting various engineering groups (such as the Society of Manufacturing Engineers), vendors of various finishing products, and library research for books and articles on finishing and robotics. The personnel should become more familiar with the parts they want processed and also engage in discussions with the "off-hand" operators who can supply an excellent working knowledge of part variations and problem areas.

Cost Justification

Cost justification is one of the major problems of automating the finishing processes (see ECONOMIC JUSTIFICATION). Two of the main reasons for a lack of cost justification of many installations have been the low priority in the overall manufacturing operations of these finishing processes, and the stubborn tenacity of management to adhere to the questionable practices of the traditional cost-justification methods currently used by most companies.

Finishing operations often have unique factors that can have a greater impact on the overall prosperity of a company, long and short term, than many of the other manufacturing operations that may take place on the same part. These factors are the quality of the finishing processes such as buffing, polishing, and deburring, which can greatly affect the amount of sales of the product; poor quality of finishing, which causes costly rework both internally and externally (in the form of service calls and warranty work); the inconsistency and sometimes poor quality of "off-hand" finishing, which can preclude many downstream automation projects; the many injuries that occur in the finishing departments; insurance claims regarding joint injury as a result of operators using hand-held power tools in finishing processes. If companies want to remain competitive in the world marketplace, the finishing processes, which are the most labor intensive areas in many companies, must be automated.

In the nontraditional cost-justification process a dollar value would be placed on all the above factors as well as other subjective factors. This nontraditional approach simply adds more factors, which have subjective values, to the less subjective traditional factors, such as direct labor and materials.

The treatment of overhead costs should be reanalyzed to determine if the unique problems of finishing cause a disproportionately low value to be used. This can occur when such factors as the cost resulting from the often high rate of accidents in finishing is not fully proportioned to the finishing area. This could adversely impact the justification of automation.

From the incomplete list of the nontraditional factors mentioned, just one of them, such as increased market share from better quality, could have enough value to offset the entire cost of the automation. With the current and expected world competition, the nontraditional cost justification practices give a more realistic value to all factors and reflect the impact of the unique value that the finishing processes have in many companies.

SYSTEM CONSIDERATIONS AND FEATURES

Because of the uniqueness of problems faced by robotic finishing systems, there are a number of features that can increase the productivity of the system, reduce operating costs, and increase the ease of using the overall system. The following features can accomplish this depending on your particular application.

1. Electrically operated robots are preferable.
2. Capability to operate the robot in "tool mode" wherein the X,Y,Z orientation can have the Z centerline passing through the flange or center of the last robot arm joint.
3. The ability to utilize a "tool offset" command whereby the X,Y,Z orientation can be moved from the location in tool mode to the tip or edge of a tool/media mounted on the robot.

4. High level of language sophistication, which reduces the complexity and number of commands necessary for the more demanding programming requirements of finishing tasks.
5. Programming language that uses common word definitions instead of symbols or uncommon meanings of words.
6. Off-line programming capabilities.
7. Five or six axes robot unless finishing operations are very simple.
8. Continuous path capabilities.
9. Circular interpolation for following circular segments via programming commands.
10. Duplication of sequences. Basically the ability to duplicate a series of movements and move them into any plane in space. Used to eliminate the requirement of teaching a given series of movements for processing one area of a part and then duplicating the teaching process of the same movements at another location.
11. Particular attention must be paid to the problem of some robots' controllers to introduce unpredictable and undesirable motions while the tool/media is being moved in a straight line or just from point to point. This can result in disaster.
12. Ease of robot controls, sensors, and other auxiliary devices to be interfaced.
13. The more accurate your requirements for finishing the more rigid and accurate the robot, fixtures, tool/media spindles, etc, must be.
14. A larger than usual safety factor for load capacity of the robot must be used to offset the reactions of loads that can be imposed by the torque of drive motors, motion of robot, etc, during an accident.
15. Extra safety devices for many types of finishing processes to protect operators and/or passerbys from flying debris from normal finishing operations, such as grinding, deburring, and polishing, and from flying "projectiles" resulting from broken media, such as grinding wheels, or in rare cases, broken parts that failed under the normal forces of finishing.

TOOLING/MEDIA USED

The types of cutting tools and medias used for robotic finishing are the same as those used for both "off-hand" and automated finishing processes. The exceptions are only in the sizes of the tools/media, which are not as large nor powered by as large drive units as you may find in certain buffing or grinding operations using wheels in excess of 12 in. (300 mm) dia and powered by 15 hp (9 kW) or larger motors.

Abrasive belt units that have been used are usually floor mounted as opposed to being mounted on the robot. This allows you to take advantage of the economics of longer belts and efficiencies of more powerful drive units.

Robots have been utilized with almost every type of finishing tool or media used for finishing either in production applications or in testing for customer acceptance pending sales of complete installations. This includes using the robot to manipulate a part immersed in vibrator bowls filled with abrasive finishing media, mounting abrasive blast finishing nozzles on the robot (both dry and wet methods being used), utilizing robot mounted water jets and abrasive water jets, lasers mounted to robots, robotic shot peening processes, and the more traditional robotic applications such as using carbide burrs, grinding wheels, buffing wheels, wire brushes, nonwoven nylon wheels, reciprocating files, etc. Basically, if someone has a finishing tool/media that can be used by hand and in hard automation, someone has probably considered using it or has tried it with a robot.

FUTURE TRENDS

The robot will be used more extensively for finishing processes in the near future. Those trial cases and now rare or unusual installations will become more commonplace. Some of the newer technologies such as abrasive water jet cutting and lasers will find many more applications in the finishing areas than are currently being looked at as possible applications.

It will only be a matter of time before the major factors of part problems, education, and cost justification are overcome and widespread use of robotics for the boring, hazardous, repetitive, costly finishing operations becomes routine.

BIBLIOGRAPHY

1. *Materials, Finishing, and Coating,* Vol. 3, of *Tool and Manufacturing Engineer's Handbook,* 4th ed., Society of Manufacturing Engineers, Dearborn, Mich., 1985.
2. D. McCoy, *Robotic Usage for Deburring, Polishing, and Buffing, Tech. Paper MR85–464,* Society of Manufacturing Engineers, Dearborn, Mich.
3. D. R. Sayre and R. R. Myers, *Coatings and Finishing Courses Offered in North America,* Kent State University, Kent, Ohio, July 14, 1981, Appendix I.

FLEXIBLE MANUFACTURING CELLS AND SYSTEMS

Yoh-Han Pao
Mariann Jelinek
Case Western Reserve University
Cleveland, Ohio

SOME BASIC DEFINITIONS

Flexible manufacturing is manufacturing capable of producing a range of products of varied design and of different materials, or utilizing basic production process in different combinations or sequences. Although the terms "cells" and "systems" are sometimes confused, a flexible manufacturing cell (FMC) is simply one distinct element within a more inclusive flexible manufacturing system. The cell is a single piece of equipment or several related pieces (eg, a turning center, milling machine, and robot) capable of automatically performing a related series or sequence of activities (eg, drilling a series of holes and finishing them) on a single workpiece, changing tools and repositioning the workpiece as required. The cell is responsible for fairly autonomous performance of a segment or stage of the activities required to produce the product. Because it is flexible, it can perform a range of such specified tasks. The flexible manufacturing system may include several such cells, and will certainly include the associated computer-controlled design, supervision, and monitoring equipment that integrates an entire factory. FMCs offer a logical first step for learning about flexible manufacturing and implementing it.

The adaptability of such manufacturing arrangements to many products stands in sharp contrast to the practices of the past, when, except for custom work, manufacturing generally consisted of rigid production practices aimed at producing a single product. Changeover time and costs were high, encouraging long, standardized production runs in order to lower costs. Manufacturing is a complex, detail-ridden activity, and never more so than today: more is known about even older processes and meticulous control is essential to quality production. In the past, such control was achievable only by ruthlessly controlling variation and resisting change, so that human minds and paper systems could keep track of the details. Today, computers offer the option of both variation and control by allowing one to master the increasing details of manufacturing across varied products. But while change is now feasible, managers must still comprehend what they seek to accomplish, and system designers must understand the capabilities they seek to coordinate and define.

This discussion will outline the objectives of flexible manufacturing systems (FMS), provide an overview describing the levels of control and equipment configurations that comprise FMS, and characterize the operations of FMS in some detail. A number of associated enabling technologies that contribute to making FMS feasible are discussed, as are some key technical and managerial issues facing FMS at present. Because such a system constitutes a fundamental shift from the practices and assumptions of the past, the strategic consequences of flexible manufacturing for the firm are also discussed.

INTRODUCTION

Modern manufacturing is often a highly complex operation involving many steps from the conception of a design through its detailing, the specification of often elaborate, science-based production steps and their implementation to accomplish the transformation of the designer's idea into a finished physical product. Knowledge of science and materials has increased, and thus the complexity of manufacturing processes has also increased markedly in recent times. Advances in computers, robotics, automated processes, sensors, information transfer (telecommunications), new engineered materials, and materials handling have added both to manufacturing complexity and to the potential for manfacturing control. Enormous amounts of information—about materials, processes, and the status of activities at the instant they occur—are now routinely available through computers and can be used to direct, monitor, and control manufacturing. These technologies are driving manufacturing toward increasingly flexible, responsive systems and cells, run by computer, in order to master the detailed complexity involved.

In the past, before computers, manufacturing management typically lacked the information needed to manage these processes. The information might simply be unavailable; it might be available only to a highly trained and experienced technician operating the equipment; the information needed might be available only after extensive analysis, perhaps in the lab, days after it was originally generated; or information from one step in the process might really be needed to adjust operations at an earlier stage. At every step of the way from product concept to the tool point, compromises were required and repetition and error were endemic, because information was hard to get, costly, unreliable, scattered, slow, or all of these.

Many manufacturing practices necessarily evolved to deal with these realities. For instance, steps required to transform the designer's concept were separated, and specialist departments designated as responsbile for each step—for design, product engineering, manufacturing engineering, actual manufacturing operations, and quality control. Such separation allowed specialists to develop in-depth knowledge of the fine points of each separate specialty. Rigid rules and procedures for manufacturing practices were instituted, and workers were expected to follow rules precisely. For excellent reasons and with unimpeachable logic, manufacturing managers typically resisted change. It was so difficult to control a complex manufacturing process that managers sought long production runs of standardized products, to amortize the heavy costs of special-purpose equipment (that performed a limited range of activities precisely), or the cost and trouble of specifying the elaborate rules and procedures needed to insure effective production.

Given these circumstances, it is easy to see that manufacturing managers in the past had every reason to fix the details, insist on long, standardized production runs, and use inventories of raw materials, work in process, or finished goods as buffers against uncertainty or problems. Making manufacturing rigid made it predictable and controllable. All of this made sense when information was scarce and unreliable, when little was known about the mysteries of production or materials, and past experience was the best guide to practice. It did not seem much of a hazard, either, when competition consisted mostly of local producers with similar approaches, similar equipment, and similar products.

The practices that permitted successful manufacturing management before reliable information became readily available were essentially strategies to subdue or contain what might go wrong. Uncertainty and its consequences could be decreased and managed by using inventory to decouple successive segments, thus buffering the various activities from problems elsewhere; reducing each task to its smallest components, so as to simplify, standardize, and control operations; reducing variety by manufacturing standardized designs; lengthening product life cycles to minimize changeovers and the need to design or implement new processes; encoding control data and knowledge of procedures in conveyors, pipes, or hard automation equipment that performed a single task with precision; and utilizing statistical control methods, such as forecasting, intermittent batch scheduling, inventory management, and materials requirements planning to fix the future and compensate for variations in demand levels, process and movement times, and materials availability.

The net result was that traditional manufacturing, and especially traditional automated manufacturing, embodied a clear set of tradeoffs between flexibility and cost, control and improvement, cost and quality. Control could be achieved, although this severely limited flexibility in product design. A given level of quality could be attained, but once achieved, improvements or variations were extremely costly. Incremental changes or innovations were difficult to incorporate. Productivity could be achieved through the use of special tooling, complex materials-handling equipment, and rigidly implemented procedures. Large factories, where procedures could be implemented once and for all in so-called hard automation, seemed desirable. But initial investments were very high, and consequently only long production runs and standardized, unchanging product designs produced over an extended product life cycle could garner the desired low unit costs. This is the essence of economies of scale. Since changes in product design, process equipment, or production schedules were expensive, they were to be avoided. The result was a highly productive, rigid factory that did one thing very well. Such a factory is essentially inflexible. Changing the product means expensive changes in the special-purpose equipment. Only by such rigidity could high reliability, high volume, and thus low unit cost be attained.

The traditional economy-of-scale approach seemed sufficient as long as there were few sources of supply and thus little competition. Scale economies are often significant, in comparison to the costs of small-batch manufacturing in highly flexible but costly manual systems of very low productivity. Productivity seemed to demand rigidity; flexibility seemed to demand low productivity and thus high cost. Today these premises no longer hold. In recent decades, an increasingly global market, vastly increased scientific knowledge about materials, manufacturing processes, and product design, and computers have overturned every one of these assumptions of the past. Many competitors around the world can duplicate manufacturing know-how and seek to compete in worldwide markets, so keeping up is important. Information about production is now readily available, often at little expense. Unlike information held by a highly skilled, expensively trained craftsperson, computer-derived and transferred information can be instantaneous as well as highly accurate, readily transferred and shared. Computers vastly exceed human information management capability as well. The result is that flexibility can be combined with low cost.

Manufacturers face a rapidly changing, technologically challenging, and information-rich manufacturing environment. The best practices from around the world constitute the new competitive standard. These practices are learnable and transferrable around the globe. They are no longer "black art," but based on science. Science-based knowledge about materials and processes is often bolstered by massive amounts of up-to-the-minute, tool-point data. Other information too has an impact: global shortages or gluts of commodities (eg, oil), embargoes, cartels and protectionism, currency fluctuations, and the like will affect all competitors. Intensified competition encourages rapid product design change, and customer preferences can quickly respond to widespread information about new developments.

Flexible manufacturing uses such information flows, responding with change instead of rigidity. Traditional manufacturing and hard-automated systems are unable to respond to large variations in product demand, continuous competitive pressure on cost and quality, and rapid technological change, the conditions characteristic of competition today. While there are exceptions, labor-intensive small-batch manufacturing is generally not cost-competitive with more automated production. Perhaps more seriously, manual production often fails to attain quality standards in highly demanding, highly complex situations. Some manufacturing processes are so complex and demanding that they are impossible by manual means. Examples include many semiconductor manufacturing processes and some ultraprecise metal-cutting, flame-cutting, and stamping operations. It is typically because of requirements for tight tolerances, high reliability, waste control, and speed that such activities are impossible by manual methods.

Until recently, highly flexible yet automated manufacturing was not possible, but advances in computer technology and computer applications technology have produced a wide array of computer numerically controlled (CNC) machine tools, robots, robot vehicles, sensor systems, gauges, feedback and control mechanisms, materials handling devices, and other needed advances. Perhaps most importantly, computer communications networks and powerful methods for handling information have been developed. As a result, manufacturing management now involves coping effectively with potential information overload, not information scarcity as in the past. Indeed, manufacturing has moved from an effective focus on materials movement and transformation to an imperative focus on information as its central organizing concept. FMS can be thought of as an information-processing system, using design information to operate a series of peripheral devices that constitute the factory in order to transform information and raw materials into desired finished goods. Although the use of computer-controlled machines is essential, the central challenge is getting the right data to the right location at the right time. Comparatively speaking, all other aspects of the task are easy.

This article will briefly outline the key information flows that affect flexible manufacturing and necessarily connect it to the rest of the business enterprise. Having located flexible manufacturing within the organization, flexible manufacturing systems and the elements that compose them, flexible manufacturing cells, will be defined. The required hierarchy of control will be described, along with the technological developments that support it. Approaches to flexible manufacturing will be outlined, highlighting not only the technical specifics, but also the essential strategic perspectives.

OVERVIEW OF FLEXIBLE MANUFACTURING

A flexible manufacturing system is first and foremost a system, that is, an integrated and articulated series of tools, equipment, control mechanisms, monitors, inspection equipment, robots, materials handling and storage devices, and so on constituting an interactive manufacturing capability. It is not a single piece of equipment, nor is it the equipment without its control function; it is not even the equipment needed to manufacture the current product or product line. Instead, the FMS is best conceived of as an integrated manufacturing capability, an information-processing system designed to transform a range of design concepts into reality, utilizing a range of intelligent subsystems, modules, or nodes (automated welding, CNC machining, inspection, assembly, and the like). Specified raw materials will be transformed by the system in accordance with designated operations, within specified tolerances, to produce specified outputs. En route, information will be fed back to control each operation to insure output results.

The U.S. National Bureau of Standards Center for Manufacturing Engineering has a research facility intended to demonstrate flexible automation for small-batch parts manufacture. Since the overwhelming majority of manufacture in the United States even today consists of small-batch manufacture (under 50 pieces), the practical necessity of such a focus is obvious. The Automated Manufacturing Research Facility (AMRF) shows a keen appreciation of the information-processing and control aspects of manufacturing, and places an accompanying emphasis on measurement, in-process monitoring, and quality control. In the fall of 1986, the floor plan of the AMRF was as shown in Figure 1. The AMRF's system architecture—the fundamental conceptual model underlying its design—is based on a classical hierarchical command structure (shown in Fig. 2).

There are five levels of control identified in the AMRF model, each with a specified time horizon and range of operation. The levels of control and their time horizons provide a conceptual frame for understanding and managing the operation of the factory over time. In other words, these levels and time horizons offer a dynamic model of manufacturing operations and their changes in response to new product, new process, and new material demands. Longer time-horizon elements will change least readily and thus least often; shorter time-horizon elements will be most responsive to change. Together, they comprise a complete and integrated manufacturing operation with the following organization:

Facility-Level Control (time scale: several mos to several yrs).

Manufacturing engineering. Provides computer-aided design to determine geometrical specifications and bill of materials for assemblies, parts, tools and fixtures, process planning, and specifications for all manufacturing operations.

Information management. Provides user-data interfaces for support of business functions such as cost estimating, customer billing, order handling, inventory, accounting, personnel management, and payroll.

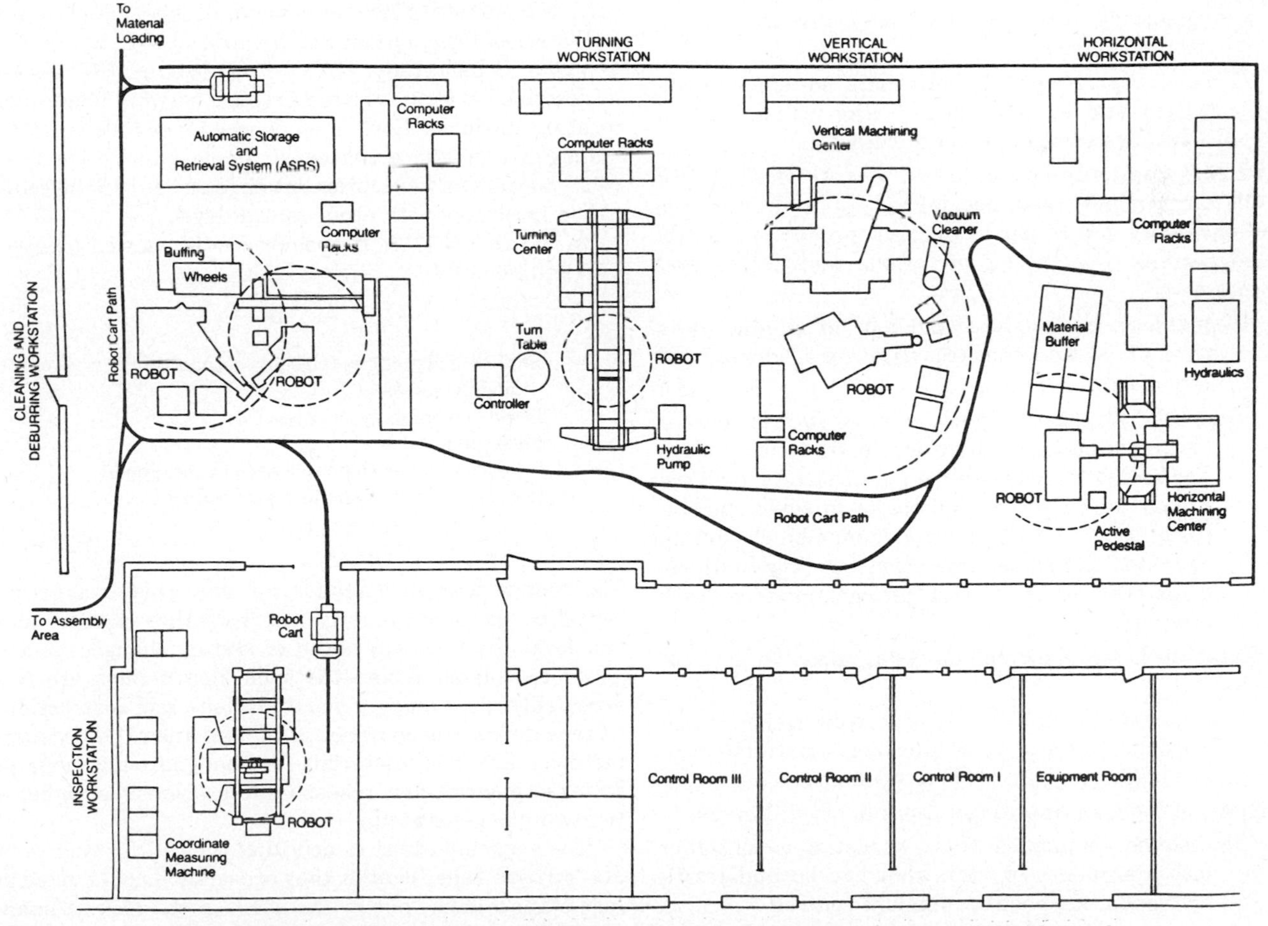

Figure 1. Layout of the Automated Manufacturing Research Facility. Courtesy of the National Bureau of Standards.

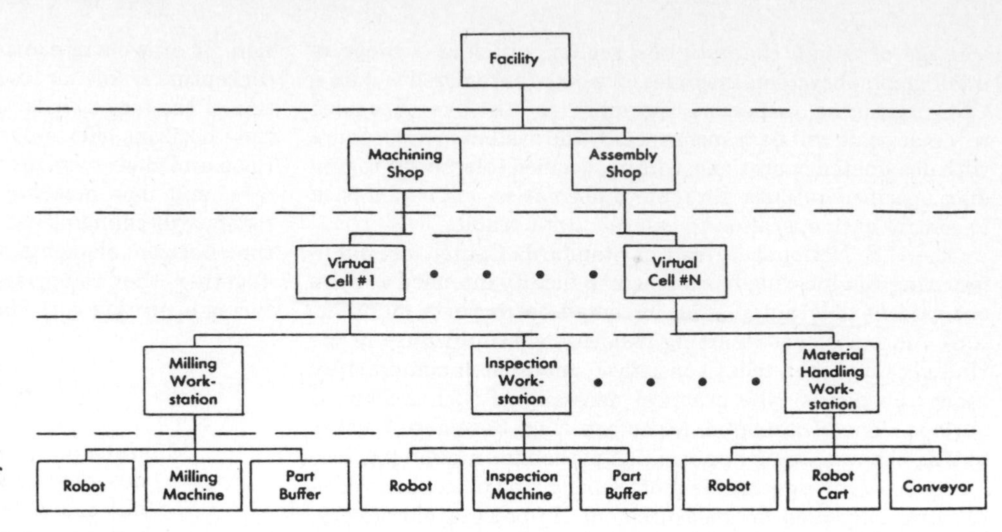

Figure 2. The AMRF's system architecture. Courtesy of the National Bureau of Standards.

Production management. Handles major project management, long range scheduling, production resource requirements, production planning, and release of work orders; identifies need for additional capital investment; determines excess production capacity; summarizes quality performance data.

Shop-Level Control (several wks to several mos).

Resource manager. Allocates workstations, tools, and materials; monitors stock and inventory (uses group technology).

Task manager. Schedules job orders, equipment utilization and housekeeping; tracks utilization and schedules preventive maintenance.

Cell-Level Control (several hrs to several wks). (AMRF cells are "virtual" cells, not defined by fixed groupings of machinery, but rather by dynamic production control that permits time-sharing of various workstation-level processors.)

Scheduling. Sequences batch jobs of similar parts through workstations (constituting a "virtual" cell sequence).

Support. Supervises support services such as materials handling and calibration, task decomposition and analysis, and resource requirements analysis; prepares requisitions; reports job progress and status of shop control; makes dynamic batch-routing decisions; schedules operations at workstations; dispatches tasks to workstations; monitors task progress.

Workstation-Level Control (several mins to several hrs).

Operates and monitors a typical workstation, which consists of a robot, a machine tool, a materials storage buffer, and a control component.

Equipment-Level Control (millisecs to several mins).

Translates commands from workstation controller into a sequence of tasks that can be understood by (vendor-supplied) equipment controls.

Each level will exercise control at its own level and communicate with other levels (receiving instructions and transmitting instructions or data as required) in order to integrate control at all levels. Currently the highest level of control implemented within the AMRF is at the cell level, but the hierarchy of the AMRF model offers a useful framework for discussing flexible manufacturing systems. The AMRF seeks to demonstrate a highly flexible, generic FMS. Higher levels of control have been demonstrated in practical systems aimed at more limited flexibility (with occasional gaps in the integrated activities), such as Ingersoll's FMS in Rockford, Ill., the John Deere works in Waterloo, Iowa, or General Electric's Large Steam Turbine Works in Schenectady, N.Y. An idealized AMRF system can be described in terms of a hierarchical network of intercommunicating modules. Each module would include an activities manager (handling activities at the designated level) and its own planning and execution intelligence, as well as communication interfaces with other control levels.

In the abstract, every module can be viewed as executing a repetitive cycle:

→ IN —

Receive instructions (objectives, constraints, context, module status)
Decide on strategy
Make plan
Generate instructions to attached equipment
Generate instructions to other modules

— OUT →

The control process demands not only that instructions received be executed properly, but that other modules and control levels be kept appraised of status, changes, and newly generated plans. Where the same instructions are received repeatedly, previously generated plans and instructions can be repeated and need not be generated anew. Where improved methods, different materials, or changed equipment permit better sequences, new responses to previously specified directions can be generated.

It is very important to note that the cells of such a system are "virtual" cells. That is, they are defined not by fixed groupings of machinery, but by the needs of the task at hand. The composition of virtual cells is determined at the shop-control level. In practice, the dynamic reconfiguration characteristic of a full-blown flexible manufacturing system will most certainly be severely constrained by transfer path and other logistic considerations, including previous commitments to machine loading and shop scheduling. However, the ability to vary workpiece path and manufacturing operation sequence is a key component in manufacturing flexibility. Efficient management of this ability clearly requires extensive optimizing and scheduling capability. Actual plant layout of machines or workstations eventually exerts considerable influence on the freedom with which virtual cells can be configured. Some form of group technology will clearly facilitate efficient operations, but in doing so will also impose some inefficiencies on operations very different from those used as the basis for grouping.

The potential layouts for FMS equipment cover a range of options. In the past, such layouts were typically specified on the basis of simple convenience or conceptual clarity. Layout can offer significant advantages where group technology concepts are used to design efficient workpiece paths for families of workpieces. Virtual cells attempt to exploit both the simplicity or conceptual clarity of earlier physical layout designs and the efficiency of specified workpiece paths in a flexible, programmable layout. Thus these layouts constitute the physical context within which virtual cells would be defined. Some prototype layouts for FMS equipment are illustrated in Figure 3. In schematic outline, layout options include:

Random layout, in which a number of machines or workstations are laid out in a rectangular shop (or some other specific, convenient arrangement). This scheme offers modularity and flexibility in the configuring of virtual

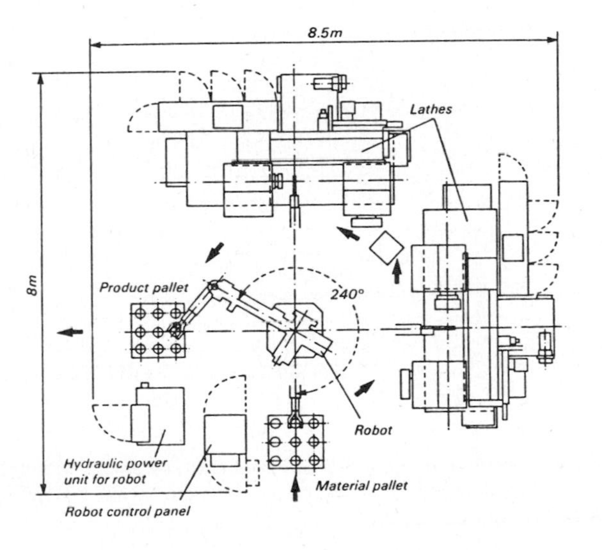

(a)

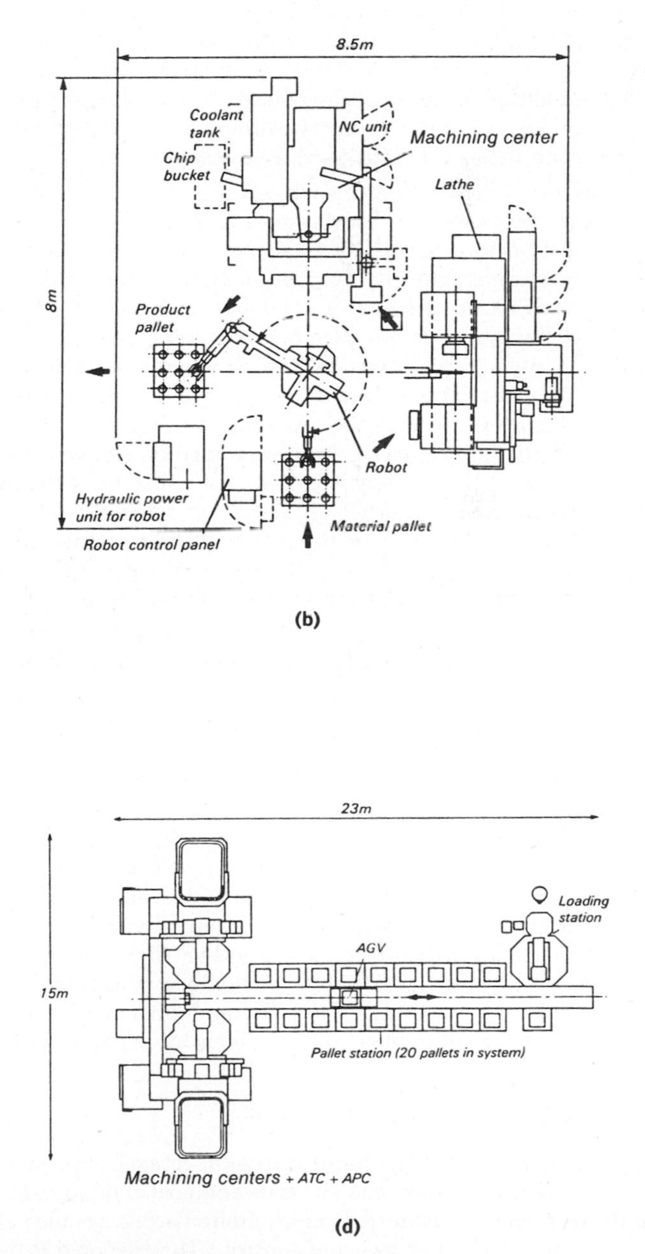

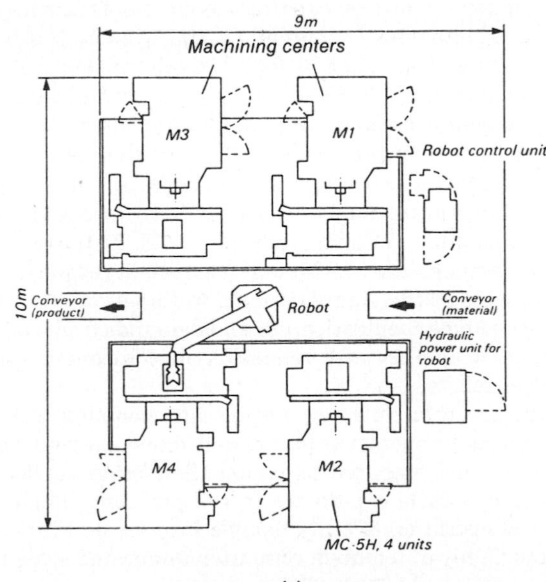

(c)

(d)

Figure 3. Some simple FMS equipment layouts: (**a**) two lathes and robot; (**b**) machining center, lathe, and robot; (**c**) four machining centers, one robot; (**d**) two machining centers, pallet transfer (1). Courtesy of Elsevier-North Holland and IFS (Publications) Ltd.

cells with the disadvantage that transfer patterns can be longer and more complicated than necessary. The longer and more complicated the transfer patterns, the more dependent such a system is on computer control for both efficient operation and simply keeping track of work in process, individual machine operation, and error analysis.

Functional layout, in which machines or workstations are arranged according to function, such as turning, milling or boring, and grinding. The workpieces flow through the shop from one end to the other (assuming no loops or repetitions are required in operations). This is a typical job-shop layout.

Modular layout, in which identical modules perform similar processes in parallel. This design is likely to result in some redundant capacity, but can be an alternative to the functional design. Its redundancy gives it the capacity to cope with rush orders or unexpected problems. Modular layout is an alternative to ever-larger factories and ever-greater-capacity equipment, since each module constitutes a minifactory, in essence.

Cellular layout, in which each cell is dedicated to a specified group of products. This system, an extension of the group technology concept, is likely to give the best match of machining capacity to process time of the workpieces. Each cell is product or product-family specific. Its efficiency and relative ease of supervision offer significant benefits in tracking the flow of workpieces, identifying problems, and reducing queuing and waiting time. It also lends itself to the gradual introduction of FMS for different workpieces, since each module can be self-contained. In emergencies, there is scope for machining workpieces in another cell from that originally intended. Typically, each cell's product will differ rather significantly from another's, in contrast to the general similarity of product output from the modular layout.

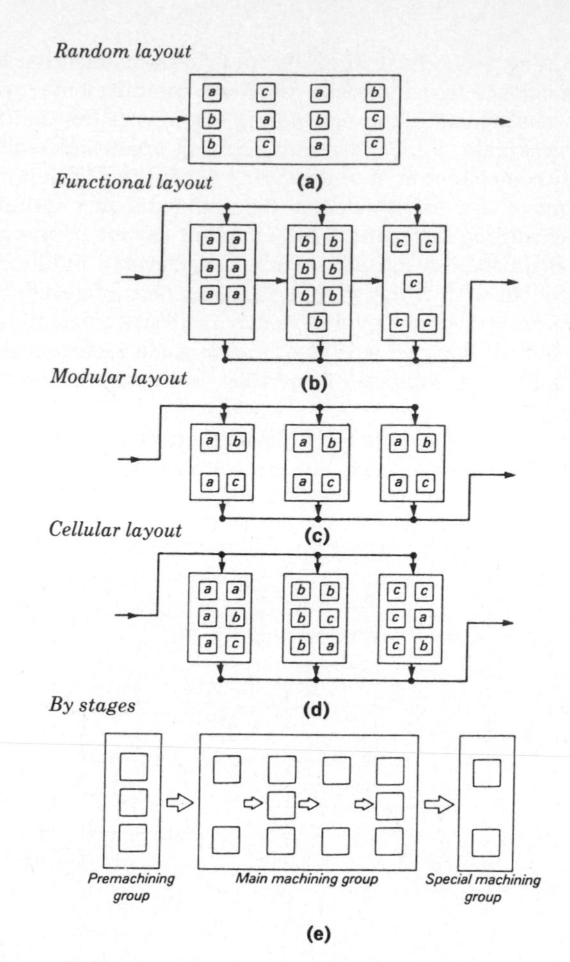

Figure 4. FMS equipment layout options: (**a**) random; (**b**) functional; (**c**) modular; (**d**) cellular (**e**) by stages (2). Courtesy of North Holland and IFS (Publications) Ltd.

Figure 4 contrasts these layout schemes and their output.

Virtual cell operation provides the freedom for creating a temporary grouping or sequencing of operations at need, reconfigurable simply by programmed instruction. It would be misleading to suggest that virtual cells offer wholly unconstrained sequencing, since, as noted, physical machine layout and its associated logistics, sheer information complexity, and prior machine loadings impose efficiency constraints. The virtual cell need not be constrained nearly to the extent that physical equipment layout is, however, and of course virtual cells will far more easily differ from one another than can physical equipment layout. Nevertheless, some form of cellular layout appears to offer maximum flexibility and efficiency, and least cost in terms of unduly complex or lengthy workpiece paths.

ELEMENTS OF TECHNOLOGY FOR FMS

Currently the cell is the highest level of control implemented within the AMRF, representing the vanguard of highly flexible FMS technology. (Somewhat more limited-scope systems elsewhere embody higher levels of control.) The National Bureau of Standards' effort was undertaken as a demonstration project, intended to utilize existing, available technology integrated in a highly flexible manufacturing system. It is broader, perhaps, than what might be found in practice because a variety of vendors' manufacturing equipment, computer controls, and equipment controllers were deliberately included. Though it is easy to understand conceptually that the essence of flexible manufacturing is information management, it is extremely difficult to integrate such a variety of hardware and software subsystems into an effective, efficient FMS. At the same time, many subsystems are continuing to evolve in response to such developments as the Manufacturing Automation Protocol (qv), growing customer sophistication and demands from customers for imporved quality and reliability in components such as industrial robots.

There are relatively few large-scale, complex, fully integrated FMSs in operation today, and many systems that are in place are not operated to exploit their inherent flexibility. Thus there is little experience in the planning, implementations, and operation of truly flexible large-scale FMSs. There are some highly automated, computer-integrated systems that operate with great cost-efficiency within a much narrower scope of the manufacturer's current product mix. Such systems were typically planned and installed by very large companies with significant resources. They can be very expensive—the more so when they unwittingly include obsolete assumptions or expectations about the nature of the manufacturing task, or fail to include crucial new realities. For smaller companies, the AMRF offers a more reasonable approach: it copes with a variety of vendor-supplied hardware and software. The AMRF modular system also permits addressing FMS in stages, implementing FMS first at the machine and cell levels, taking care to insure against bottlenecks. Significant gains by this approach are available to small users.

The ARMF also demonstrates a wide range of associated enabling technology and equipment necessary to support a truly viable FMS. Some of these enabling technologies and equipment are well-developed predecessors; others are new; and a few are very recent indeed. A key advantage of such a demonstration model as the AMRF (or of individual companies' efforts) is to highlight just where additional development is required. Among others, the required enabling technologies include

- CNC, or computer numerical control of machinery
- Special-purpose CNC machinery and devices
- Robotics, including both general and special-purpose programmable manipulators, grippers, and positioning devices
- Sensor technology: computer vision, proximity detection, coordinate measurement, and so on
- ATC, or automatic tool changes, permitting computer control of which tool is utilized
- AGV, or automatic guided vehicles (essentailly robot carts), capable of computer control
- ASRS, or automatic storage and retrieval systems, where raw materials, parts, subassemblies, or finished goods can be reliably stored and retrieved automatically, at need
- Computer communications mechanisms and communications protocols, so that various pieces of equipment can communicate with each other and with their governing control system
- Database management systems, to keep track of voluminous amounts of information and permit ready access to it as needed
- Systems planning, operation and integration methodology and software, to provide the conceptual frameworks for joining the elements of the FMS
- Machine intelligence or artificial intelligence (AI) in support of the elements of FMS
- Developments in computer memory, computation speed, information storage and access speed, and data-handling capabilities to support the detailed monitoring and information management of manufacturing

The steady growth in computer capabilities, with the cost per operation declining and the information capacity increasing during the past 30 years, has clearly had a major impact. So too has the development of computer communication networking capability. In FMS terms, for instance, these translate into practical machine-intelligence availability that permits both local program storage and operation monitoring, instruction downloading from a supervisory computer, and uploading of selected or summarized machine operation information.

The full range of capabilities of an FMS described in the NBS AMRF is at the leading edge of a rapidly expanding technical capability: it is virtually impossible to accurately describe the state of the art at any moment, because it is changing so rapidly. Nevertheless, the modular, hierarchical control-based conceptual design used by the AMRF and outlined here reflects capabilities widely recognized as desirable, and widely sought in many research and applications efforts besides those at the AMRF. The concept is important, and although its full-blown form pushes available capabilites, it is closer than is often realized in the United States. In Japan, existing factories (with capabilities similar to those in the United States) are operated with generally greater flexibility, deliberately seeking to expand the range of control. So-called lights-out operations, or unstaffed overnight manufacture, have also been achieved. In contrast, U.S. firms often utilize potentially flexible equipment in nonflexible modes: long production runs and standardized products with little change are the result of management decisions, not production equipment limitations (3). Any individual FMS will require decisions, configurations, connections, and substantial effort to achieve flexibility, but the problem is generally not a hardware problem. Technically, the requisites for highly flexible manufacture, if not completely flexible manufacture, exist today.

The fully figured FMS suggested by the AMRF is still not available "of the shelf"; however, it can be described on the basis of available technology and concepts, with very few gaps. In the fully integrated FMS, intelligence will be widely distributed throughout the entire system. It is useful to think of the system as an organization of interacting experts, sharing information and control in response to programs of instructions, objectives, and current status information. At some levels, generally higher levels of control, the machine experts will act primarily in support capacity, as advisors, downloading instructions and programs, directions and directives. At lower levels of control, machine intelligence will generally act autonomously, executing instructions, reporting status or summaries upward, and unilaterally intervening for safety or problem resolution. Thus, for instance, while executing a specified operation, a lower-level machine would automatically stop in response to a tool condition indicating imminent breakage, or a mispositioned workpiece.

Especially when acting autonomously, the intelligent modules act as strategists as well as tacticians, planning how to act as well as executing specified action sequences. Given a task, the typical module generates a plan for carrying out the task and generates instructions to other modules without addressing the details of how those instructions will be carried out. In addition, it generates in detail the action steps needed for its own use (as other modules will, for their use on their tasks). The module will also specify the feedback it needs and expects, evaluate results, and report them to all appropriate modules. Thus a cell executing an instruction to "Make Part A" will generate a plan for making the part, drawing on several subordinate machine capabilites, scheduling their operation, and utilizing expected machine-operation feedback. When one lower-level machine experiences feedback identifying an imminent tool break, for instance, the lower-level machine will stop, change tools, and then proceed, all the while informing the cell of its status and permitting effective rescheduling. The cell's machine intelligence might reprioritize activities on the delayed workpiece.

Such descriptions help both to set a target for the broader system flexibility sought and to outline the sorts of machine intelligence and detailed capability needed to achieve it. A broad range of activities contributes. For example, extensive effort on such apparently disparate details as monitoring chip formation and control (4) and machine vision capabilities, scheduling software, and automated materials-handling capability, interface standards and sensors must all be elaborated to produce a range of products or components: each element must be flexible if the system as a whole is to be flexible.

CHARACTERIZATION OF FMS

A great deal has been written about what should be achievable with computer-supported flexible manufacturing systems. Although there are many instances of impressive accomplishment, the evidence suggests that present-day FMSs are flexible only if the seeming flexibility is incorporated both into the initial design and into management thinking about operational philosophy. First initial design contraints are dealt with.

Repertoire and Envelope Flexibility

A CNC machining center can make ten consecutive copies of the same design, or it can make one each of ten different designs in any desired order, provided that the ten different designs are all in the software of that workstation or available to it. As long as higher levels of control remain essentially rigidly programmed and difficult to modify, it will be difficult to attain true flexibility to shift from preplanned activities to cope with unforeseen developments. Within its range or repertoire of programs, the machining center can be said to be flexible; this is referred to as repertoire flexibility.

Anecdotal reports of successful FMS implementations indicate that outstanding successes have been attained under two circumstances. Impressive gains can be attained with preprogrammed flexibility at the workstation or cell level—the repertoire flexibility noted above. Impressive gains can also be obtained in cases where an entire plant is carefully designed and implemented with a high degree of computer integration and preprogrammed flexibility. This expanded flexibility aims at controls and methods of exploiting the range of capabilities inherent in the equipment itself. This is called envelope flexibility, or flexibility within the physical limits of the equipment, not within a preprogrammed set of part recipes. Envelope flexibility goes beyond the readily available software-based repertoire flexibility. Indeed, the lesson is that present-day FMSs have repertoire flexibility, so long as the scope of flexibility is carefully designed and prescribed ahead of time. Beyond this, it seems that the hardware and equipment inflexibility of older, hard-automated equipment has been replaced by software inflexibility as a current manufacturing limit. Current research efforts aimed at modular approaches, specified interface protocols like MAP and TOP, and especially concerted efforts to develop broader machine intelligence, including true automated process planning, all address this problem by expanding software flexibility.

The concepts of repertoire flexibility and envelope flexibility touch on a paradox of FMS: what is flexible at one level appears fixed at another, and the trade-off between stability and flexibility is central to comprehending what sorts of manufacturing flexibility can be attained at reasonable cost. Such trade-offs are also very much a matter of relativity and control: simply hand tools are highly flexible but wholly inadequate to submicron tolerances; programmable manufacturing equipment is vastly more flexible than hard automation, but is in fact limited in its flexibility. Repertoire flexibility indicates the degree of limitation as constrained by control software, whereas envelope flexibility suggests more absolute limits entailed by physical equipment capabilites. The distinction also highlights the source of limitations and thus the appropriate focus of attention to transcend the limits. Repertoire flexibility offers a substantial improvement in programmable control and flexibility over the manual and hard-automated systems of the past; envelope flexibility multiplies this flexibility, bringing managerial and strategic concepts into play as ultimate sources of limits to flexibility.

Strategic Capabilities

Within this framework, the capabilities of a fully featured FMS can also be characterized from the perspective of underlying strategy and managerial philosophy. It has been suggested that traditional manufacturing practices and their rigidity emphasize economies of scale, while computer-integrated, flexible manufacturing emphasizes economies of scope (5). As computer-integrated manufacturing and its capacities for flexibility are developed, factories will emphasize the end target, tending toward AMRF goals of efficient, economical production in lot sizes down to one unit, and interconnection between equipment produced by a wide variety of vendors. In short, factories will look beyond repertoire flexibility toward envelope flexibility.

There is however a serious difficulty facing FMS and other advanced technologies. Managers have been overimpressed and sometimes oversold on such computer-based technologies as CNC, robotics, CAD/CAM, sensors, automated storage and retrieval, and computer-aided engineering (CAE) as instant solutions to manufacturing problems. For every component, there exist effective, economically operating examples. In addition, there are operating systems combining these components and achieving outstanding results. However, for most components, the successful systems are those keyed very closely to application-specific designs aimed at achieving flexibility within a carefully specified and often quite limited repertoire.

Managers are told that "the technology exists," and it does. However, in many instances, its flexibility will be limited by the available software and supporting technology, which may be experimental or costly, incomplete or inappropriate for the specific application at hand, or may require extensive reworking to be useful. For smaller manufacturers especially, for whom purchase of an entire new system at once is not feasible, and who must utilize existing equipment from a variety of vendors, retrofit and interface problems may be staggering. The difficulties of such interfaces, and smaller manufacturers' need for adaptive solutions, are the focus of much effort at the ARMF. This work highlights the importance of a host of enabling technologies and software interfaces.

ENABLING TECHNOLOGIES

CNC

Computer-controlled machines are the building blocks of FMS. The first computer-controlled machines were computer numerically controlled (CNC) replacements for machine tools operated formerly by numeric tapes of paper or mylar. NC equipment is still widely used and has advantages of insensitivity to electromagnetic disturbance in factory environments. However, tape production is relatively slow and error-prone, in contrast to electronic data transfer. Moreover, direct computer links are clearly crucial for the highly integrated factories of the future.

Considering machine tools alone, the localized kernel of an FMS, a flexible manufacturing cell, can be implemented with a number of CNC machine tools and associated automatic, computer-controlled equipment for changing tools, loading and unloading workpieces, monitoring processes and inspection (both in process and on an intermittent basis). Figure 5 illustrates such cells. Rapid advances in CNC machine tool technology in recent years have been largely due to improvements in accurately controlled variable-speed motor drives and advances in microcomputer technology. Microcomputer advances include the now widespread availability of 16-bit and 32-bit microcomputers and buses, with corresponding increases in addressable memory, availability of large-scale memories and permanent memories, accompanied by dramatic decreases in cost per function, cost per memory storage unit, and cost for capabilities of a given scale. Capabilities formerly available only in very large and very expensive equipment

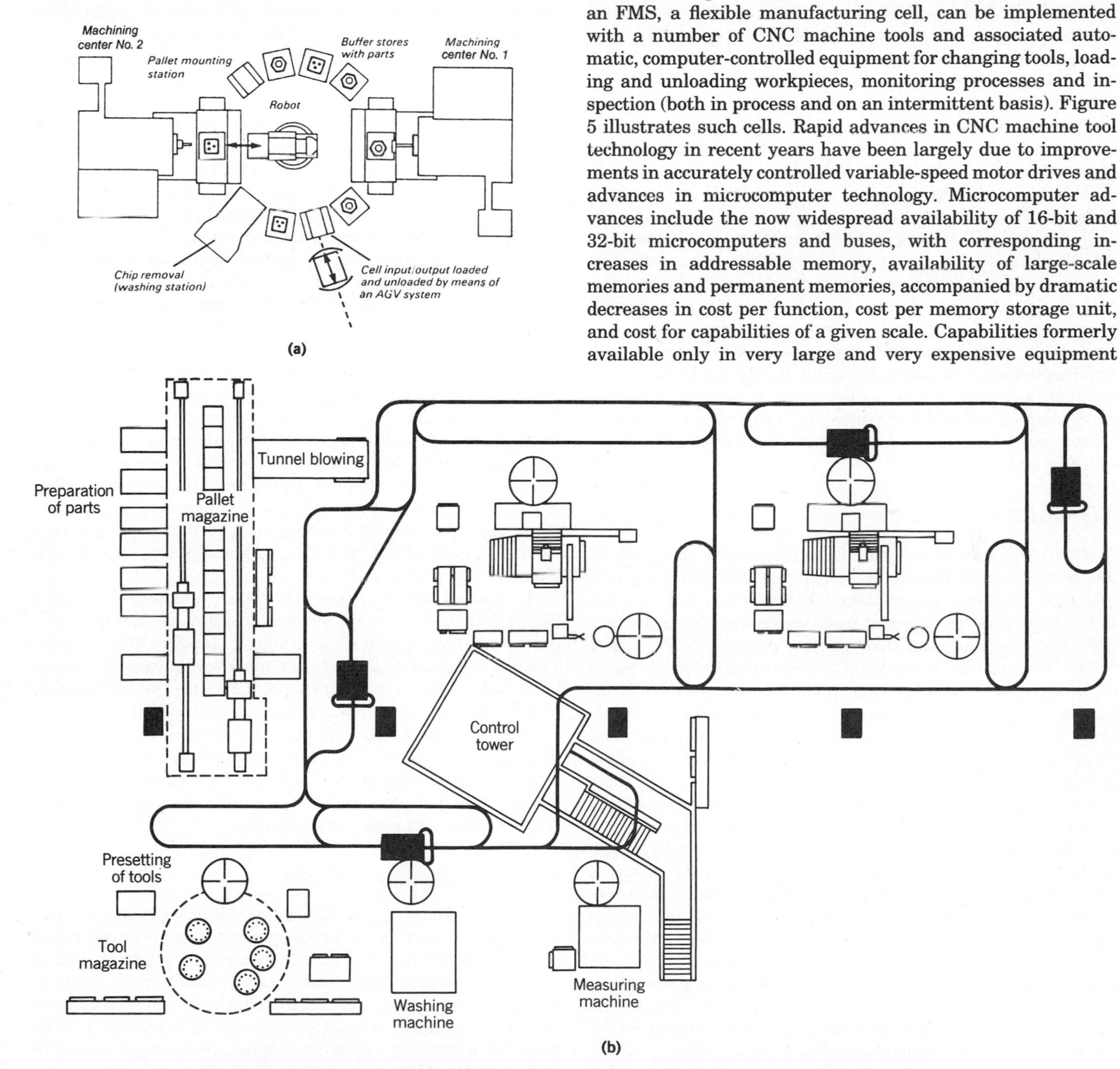

Figure 5. FMS equipment layouts: (**a**) cell layout: two machining centers, pallet load/unload robot, buffer stores, washing station, and cell input/output (6); (**b**) Citroen's FMS: automatic delivery of workpieces and tools to two machining centers, robot washing of workpieces, and CMM (7). Courtesy of North Holland and IFS (Publications) Ltd.

are now routinely available in desktop (or smaller) computers at prices low enough to make them accessible even to very small manufacturers.

The power and quality of software have also improved substantially. As but one example, computer-aided design software is now available for prices ranging from a few hundred to a few thousand dollars, depending on how elaborate the program, and a computer to run it may be bought for well under $10,000. Formerly, such capabilities required special workstations, huge mainframe computers, and very expensive programming. (Of course, special workstations and expensive programming are still available; however, they perform far more elaborate tasks, eg, incorporating automated error-checking and computer-aided engineering). The point is that the explosive development of computer capabilities, memories, and software has put substantial computer controls within financial reach of small users. What remains to be done is to tie together these capabilities.

Machine tools controllable in CNC format include a wide range of lathes, drilling, tapping, milling, broaching, and grinding machines, flame cutters, and the like. Other manufacturing, inspection, and assembly operations are also computer-controllable. Current efforts in CNC focus on providing intelligence at the local cell control level, and on interfacing with other parts of the manufacturing system. The range of computer-controllable activities is rapidly increasing as computer, memory, and software advances make it possible to specify more complex activities. For instance, even rather complicated grooves or splines or elaborate shapes can now be cut under CNC control.

CNC Workstation

A closer look indicates some of the complexity involved. As has been noted, a flexible manufacturing workstation might generally consist of one or more CNC machine tools, a robot to carry out pick-and-place or positioning tasks, an automatic tool changer (ATC), and perhaps an automatic pallet carrier (APC). Necessarily, the peripheral equipment would also be computer-controlled, although input into the cell might be manual. Some companies have produced embryo FMS operations, constituting neither a workstation nor a cell but a small FMS facility. In light of the potential for even a single cell to operate reliably and its capacity for continuous utilization, this is a nontrivial achievement. One such system is illustrated in Figure 3**a.** It consists of two CNC lathes mounted in a rectangular configuration, at 90° to one another, with one pallet on the third side of the rectangle for workpieces and a second pallet on the fourth side for finished, machined components. A robot, located between the pallets and the machine tools, moves workpieces from the material pallet to the first lathe for operation. A turnover fixture or table between the two lathes allows the single-handed robot manipulator to adjust its grip on workpieces if needed. The robot loads and unloads the lathes in order to keep them functioning smoothly, removing finished components to the product pallet.

APC

For unmanned machining over an eight-hr shift, the material pallet would have to provide enough workpieces to occupy the equipment during that period, and the product pallet would have to accommodate the finished components. Two methods are available for achieving this. Either a number of pallets can be loaded onto an automatic pallet carrier that is then available to the robot parts server; or robot carts may be used to ensure that empty materials pallets are replaced by full ones at appropriate times. These robot carts or automatic guided vehicles (AGVs) could run on rails or might be guided by magnetic tape, fluorescent tape or wire, or laser beams (see also AUTOMATED GUIDED VEHICLES). Figure 5**b** shows AGVs servicing two cells, utilizing a pallet magazine.

ATC

In addition to the equipment shown in Figure 3**a,** there needs to be a magazine of tools available to the equipment and a means for taking the appropriate tool from the magazine, inserting it in tool holders, and finally mounting it on the CNC machines. Of course, the parallel demounting and return to the magazine are also needed. Such a magazine or index of tools will be required for unmanned operation, where tools may wear or break, as well as for performing multiple operations requiring different tools. There are many variants to the implementation of automatic tool changers, but the basic requirement for ATCs is that the system must not run out of tools during the unmanned machining period, whether that constitutes a sequence of operations during a staffed shift or an entirely unmanned shift. The ATC could be a rather massive automated machine with a large number of tools serving a number of CNC equipments, or it might be effected by use of a magazine of detachable tools mounted near or even within the machine using them, with tool change accomplished by the machine itself or by means of some form of separate robot to change tools as required. Figure 5**b** also shows a tool magazine.

As manual tool changing is a major weakness in existing FMSs, attention is being directed to limit or eliminate this time-consuming activity, eg, through standard preset tools. This imposition of standards, an apparent limitation on flexibility substitutes arrangement or structure for information and allows for vastly more efficient programming and operation of the FMS. As such, it facilitates flexibility, despite the limitation it imposes. For instance, a convention specifying which tool shall reside in which slot and where rarely used or special-purpose tools shall reside will enormously expedite programming and program revision.

CMM—Monitoring and Inspection

Other components of an FMS can include coordinate measurement and tool-wear monitoring systems. Computer-controlled inspection machines are similar in concept to CNC machine tools. They require rigid construction with alternate positioning and sensors, accurately controlled drives, and good controllers. In particular, coordinate measuring machines (CMM) must be located in vibration-isolated and temperature-controlled installations. (Other manufacturing equipment, especially for highly precise operations such as integrated circuit or microprocessor manufacture, where device architecture requires line-widths of a micron or less, must also be vibration- and temperature-controlled.) Laser inferometric measurement techniques may be required for calibration, although actual inspections might be carried out with electromechanical touch-trigger probes. The touch-trigger probe illustrated in Figure 6 is typical of such devices, which are available from many vendors.

Some advantages of using CMMs in FMSs include the following:

1. No special fixtures or gauges are required, since the finished parts can be measured on the pallet on which they were manufactured, which can be automatically loaded to present parts for inspection in the same way that parts are loaded for machining or assembly operations.
2. CMMs do not need part alignment since all orientations of parts can be detected automatically and all measurements are relative to one another.
3. Human error in execution of measurement and data interpretation is eliminated. However, programming errors are possible. (Some highly integrated systems derive machine instructions and measurement parameters directly from CAD/CAE part geometry, eliminating this source of error.)
4. Automatic collection of data, generation of results, and preparation of reports are all possible. Such reports can be in electronic form, alerting supervisory systems to problems and shutting down production automatically, or they can be in hardcopy, printed form.
5. CMMs can be used for scribing and marking parts and for production layout (eg, in sheet metal operations).

The advantages of CMMs and other forms of computer-controlled inspection and measurement are quite analogous to the advantages of other computer-controlled operations: accuracy, replicability, reliability, speed, and reduction of wasted time and materials. Computerized laser-guided inspection of automobile body panels, for instance, is highly accurate and extremely fast: highly skilled humans require 45 minutes, while the automated inspection detects defects as small as $1/100$ inch in just over a minute. With such speed and accuracy, 100% inspection is clearly feasible and economical—an important point as quality becomes ever more important.

Procedures used by CMMs for locating center points and for determining lines, planes, and angles are sketched in Figure 7. The probe paths and calculation routines for interpreting readings, as well as the instructions for positioning the probes, the number of readings required, and the like are details that must be programmed. Since the routines for measuring a circle or for measuring perpendicularity, for instance, are identical each time they are performed (although the size of the circle or the orientation of the intendedly perpendicular lines will vary), the routines are obvious candidates for local machine intelligence, with the advantages in speed and simplicity this offers.

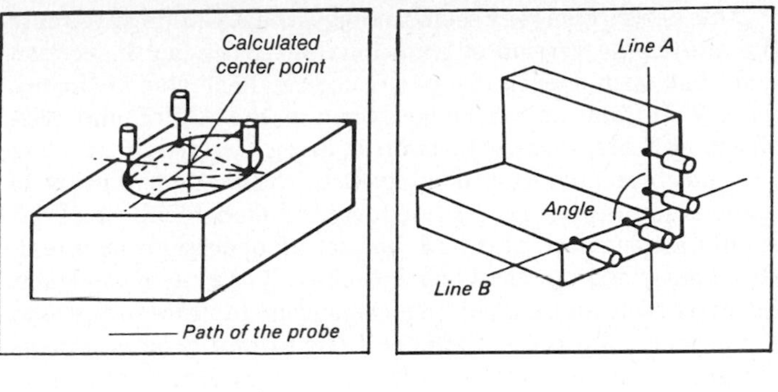

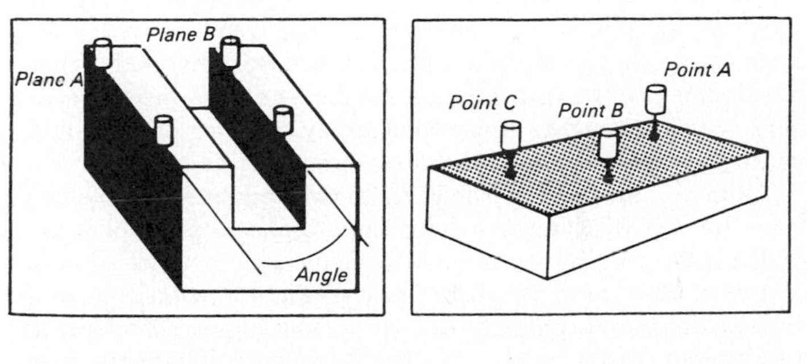

Figure 7. Coordinate measuring machine (CMM) measurement routines (9). (**a**) Multipoint circle measurement routing. Using three or more measured points, it automatically calculates the center point and the diameter of the best fit circle. (**b**) Perpendicularity measurement of two lines, with a canned cycle. Perpendicularity is defined as the tangent of the calculated angle. To calculate the angle, lines A and B have to be measured by touching them at two points. (**c**) To be able to calculate the parallelism of two planes, the machine has to determine the angle between the two planes. (**d**) The best fit plane can be determined by a minimum of three measured points. Courtesy of North Holland and IFS (Publications) Ltd.

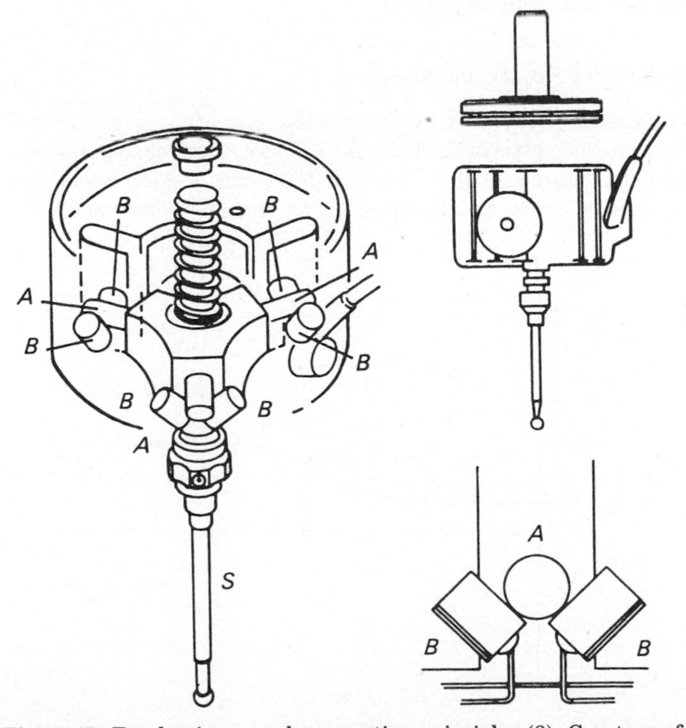

Figure 6. Touch-trigger probe operating principles (8). Courtesy of North Holland and IFS (Publications) Ltd.

This description may seem to suggest that CMMs are simply the automated version of traditional after-the-fact inspection and that such inspection is the logical next step to insure quality. In fact, as extensive efforts at the AMRF and elsewhere indicate, more inspection is being directed at insuring that machine operation is controlled, rather than waiting to discover faults after production has concluded. Ideally, an FMS should inspect each part and monitor all operations on a real-time basis, feeding results back to the cell or system controller for correction and embedding measurement traceability into the production process. FMSs will rely primarily on in-process measurement and control, rather than on intermittent inspection of discrete parts and process results. Inspection and measurement can, of course, be incorporated at the tool point to control machine operation, rather than simply recording data. Local machine intelligence can control operations to correct machine adjustment in response to tool wear or spot worn tools before they produce out-of-tolerance workpieces. Further development of both inspection machines and in-process sensors is being sought, and a vast array of other sensors and controllers are being incorporated into machine tools.

Optical inspection methods for measuring part dimensions have been available for a long time, and the principles are well known. Examples include comparing silhouette dimensions (comparators) or using laser triangular measurements to measure minute changes in distances or angles. Interference techniques are also routinely used to measure distances either in ultra-accurate calibration procedures or, less accurately, in layout procedures. When combined with machine intelligence, these older technologies can be utilized to improve the performance of FMS.

Acoustical Emission

Inspectors and controllers are needed both to check on product characteristics and to monitor machine operation. Tool breakage is an example: a broken tool must be stopped and removed. The workpiece on which it broke may have to be discarded. Clearly, tool breakage should be avoided as much as possible. One way of reducing tool breakage is to operate the FMS with machines running at 65–75% of rated cutting speed. This may not be feasible or economical, however. Consequently, monitors have been developed to sense breakage using spindle current (in lathes) or, more generally, power, thrust, or noise levels generated by equipment. Acoustic emission (AE) systems are used most in drilling applications. It is claimed they can detect damage in a tool as small as 3-mm diameter and stop the operation before breakage occurs.

AE can also be used in other ways to avoid tool breakage and marring of workpiece surfaces. In unmanned machining, where metal-cutting machines must operate for long periods of time without operator intervention, chip control is a major problem. Chip formation is not usually a problem on machining centers except in drilling or boring, since in milling the cutting is inherently interrupted. However, in single-point turning, long, continuous, unbroken chips can be generated, causing serious difficulties. Long chips can become entangled with the tool or the workpiece, damaging surface finish of the part or interfering with workpiece or tooling changes. Chip removal is also an important problem in untended operations. Long chips make automated chip disposal impossible.

Researchers at the AMRF have reported on a method for monitoring chip form by detecting the burst of high frequency AE signal that accompanies the material deformation or breakage of chip formation. The root-mean-square (rms) of the high frequency signal is compared against a "floating" threshold that accounts for such continuous (but changing) factors as tool speed, feed, depth of cut, tool geometry (including chip breakers or grooves), tool wear, and the like, all components of regular tool noise, which can change over time (Fig. 8). The high frequency burst of AE signal accompanying chip form is compared with the threshold. Examples of these signals are displayed in Figure 9. The factors contributing to noise are among those that can shift chip form from acceptable broken pieces, which are easy to remove, to potentially disastrous long, unbroken chip pieces. By monitoring the frequency of bursts of AE noise and the rms-fast signal accompanying it, long chips can be detected before they cause damage. AE has also been used to detect insert fracture, tool wear, and chip tangles, but not as reliably as in monitoring chip form.

While these examples, like the AMRF, emphasize metal cutting and discrete parts manufacture, the enabling technologies and the concepts of flexible manufacturing are highly adaptable to a wide range of manufacturing activities. Among other activities presently operated with flexible controls are semiconductor and integrated circuit design and manufacture; a wide variety of assembly tasks, welding operations, flame cutting, and water-jet cutting; as well as foundry and molding. Process production—continuous production of products like bread, textiles, basic metals, and chemicals or refined petroleum products—also readily lends itself to computer controls, as does batch manufacture of chemicals. The exact sensing, information acquisition and analysis, feedback-based control and flexibility that constitute the heart of flexible manufacturing are differently applied in different situations, but the concepts are broadly applicable.

Automated Storage and Retrieval

Another development affecting FMS operation is automated storage and retrieval (ASRS). An ASRS can increase the potential time frame during which an FMS can operate, since raw materials, tools, and finished goods can all be automatically

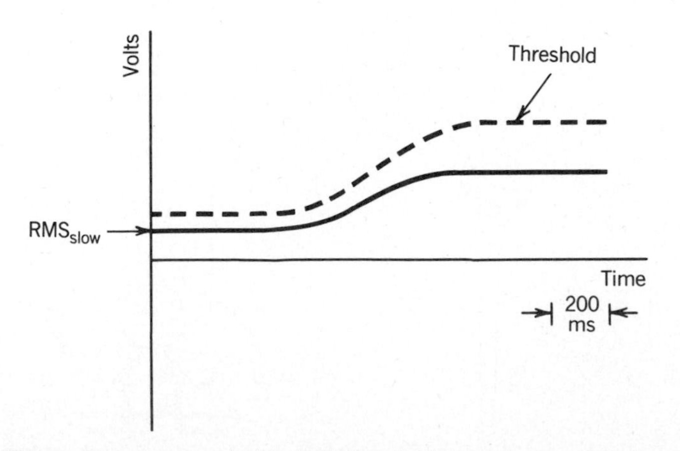

Figure 8. Acoustic emission (AE) threshold level change (10). Courtesy of National Bureau of Standards.

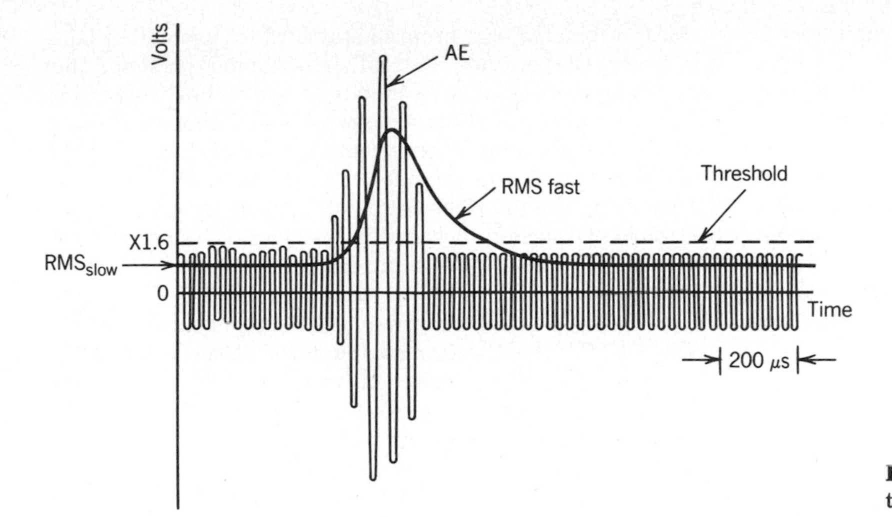

Figure 9. AE burst and threshold (11). Courtesy of National Bureau of Standards.

stored and retrieved at need by computer-controlled equipment. Thus an empty materials pallet can signal the computer of the need to retrieve full pallets from the ASRS; a full product pallet can cue the system to remove the completed work to the ASRS and return with another empty product pallet for further output. The benefits of such operation are high rates of machine utilization and productivity; the hazards are continued production of unneeded product or of product that should have been adapted or adjusted to new customer needs. The ASRS constitutes a large buffer storage area, but using it is likely also to mean extended travel paths and time for parts, tools, and finished products. Figure 10 shows an ASRS.

Apple Computer's MacIntosh production line at Cupertino, Calif. was an early example of ASRS in conjunction with automated and manual production activities. Many more such ASRS installations exist around the world. Computer controls can ensure that items do not languish unrecognized in inventory, that they are used before they age, and that items in storage are accurately tracked. One potential hazard of ASRS approaches is that inventory may be unwittingly increased "just in case," because storage is an easy, convenient solution to production difficulties or forecasting problems. In contrast, the "just in time" approach demands detailed attention to coordination and timing of manufacturing operations and parts and components delivery, and minimizes inventory. North American manufacture, responding to the limitations of older human and paper information systems, has quite rationally relied on "just in case" inventory for generations. Some inventory will continue to be needed as an appropriate buffer to smooth production operations. The danger of ASRS is that it may simply encourage continued dependence on this expensive, inefficient method. Ideally, ASRS would be combined with stringent inventory control and a highly coordinated manufacturing operation designed to minimize materials holdings and throughput time.

Robots

It is sometimes said that FMS would not be possible without robots. Because this entire work addresses robotics this article will not dwell on them at any length, but it is nevertheless worth observing that robots come in many varieties. Defini-

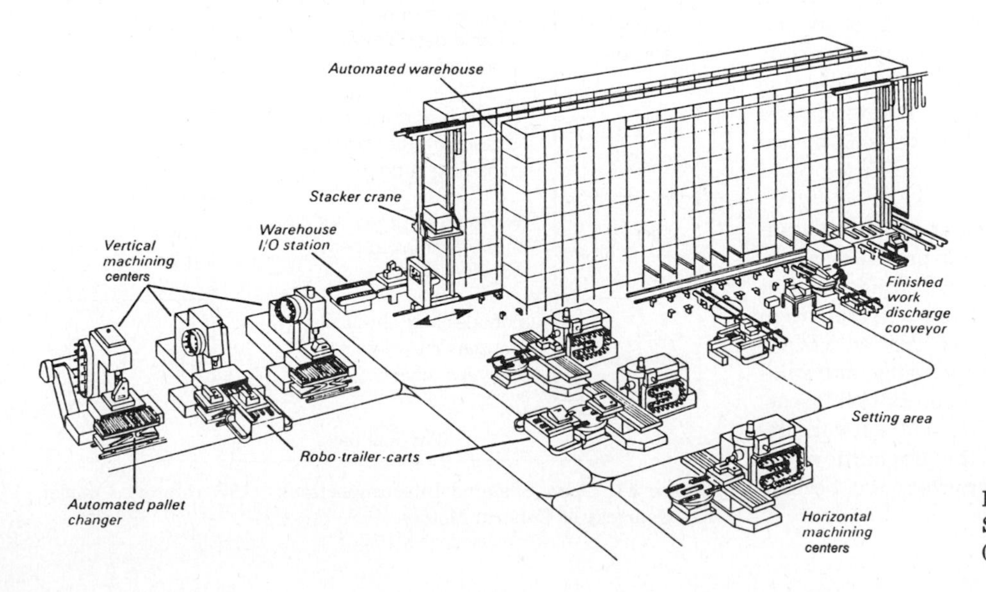

Figure 10. Automated Storage and Retrieval System (ASRS) integrated with manufacturing (12).

tions differ, too: in Japan, for example, many computer-controlled devices are called robots that would be excluded from that category in the United States on account of insufficient autonomy or machine intelligence. Within an FMS, robots provide important connectors to integrate sequences of operations by moving workpieces from one machine to another, or from one series of operations to another. Among important robot types widely utilized in FMS installations are the following:

1. Robot vehicles (AGVs, or automatic guided vehicles) are widely used to move pallets, tools, or workpieces from place to place.
2. Pick-and-place robots have typically anthropomorphic "arms" with end-actuators of various sorts, including grippers, suction wands and clamps. These robots handle relatively lightweight loads. Figures 3 and 5 illustrate typical pick-and-place installations, in which the robot serves parts (and perhaps tools) to one or several machines, loading and unloading them under computer control to coordinate cell operation.
3. Gantry robots, capable of handling much heavier objects than the anthropomorphic robots noted above, may incorporate highly versatile multiple axes of rotation to position heavy workpieces for other machine operations.

Advances in robotic grippers have important implications for FMS operations. Grippers with three or more independent "fingers," good force and control, some sense of touch and movement-sensing capabilities would facilitate better automatic assembly procedures. Advances in robot controls and machine intelligence offer great promise in extending FMS capabilities from repertoire flexibility to envelope flexibility by helping to fully exploit existing machine capabilities.

COMMUNICATIONS PROTOCOLS

In computer-controlled automated manufacturing, communication is a central issue. Equipment provided by diverse vendors must interface smoothly, with no loss or degradation of information. Such machines include the entire gamut of FMS equipment, from highest-level computers to cell supervisors, machine controllers, AGVs, ASRS, CAD and CAE, robots and CMMs and sensors. Accurate, timely, and comprehensibly shared information is the mode of integration that allows the innumerable details of manufacturing to be monitored and coordinated and optimized for efficient, effective, safe operation. In addition, business and management activities must also communicate with manufacturing.

In addition to the NBS activities aimed at outlining standard interfaces between a variety of vendor-provided equipment, mentioned earlier, there are two protocol standards that should be mentioned here: the Manufacturing Automation Protocol (MAP) and the Technical and Office Protocol (TOP). MAP focuses on engineering, design, manufacturing, and similar technical data, while TOP addresses business task areas. Both standards depend on further hardware and software development, and both will evolve further as the pragmatic needs of the automated, computer-integrated environment become better understood.

MAP originated as a protocol standard for use within General Motors Corporation. General Motors managers found that networking costs to connect shop-floor equipment from different vendors were prohibitively expensive and time-consuming, as each vendor's proprietary protocol and interface specification differed. MAP was intended to overcome these difficulties by specifying the interface characteristics at various levels of control. MAP is generally based on the Open Systems Interconnection (OSI) protocol of the International Standards Organization, but it also includes recommendations based on protocols defined by a number of other organizations (among others, ANSI, CCITT, the Electronics Industries Association [EIA], IEEE, the Instrument Society of America [ISA], and the National Bureau of Standards). Because General Motors so profoundly affects industrial manufacturing in the United States, MAP had to be widely accepted by GM suppliers. As a result, MAP is now a public protocol, being developed and supported under the guidance of committees from various national and international standards organizations, and supported by a large number of equipment vendors and software houses.

The seven-layer structure of the OSI of the reference model is shown in Figure 11. The OSI model presumes modular networking support software by functionality, with each module taking the form of a layer in the model. Each layer is responsi-

LAYERS	FUNCTION	LAYERS
User program	Application programs (not part of the OSI model)	Server machine
Layer 7 application	Provides all services directly comprehensible to application programs	Layer 7 application
Layer 6 presentation	Restructures data to/from standardized format used within the network	Layer 6 presentation
Layer 5 session	Name/address translation, access security, and synchronize and manage data	Layer 5 session
Layer 4 transport	Provides transparent, reliable data transfer from end node to end node	Layer 4 transport
Layer 3 network	Performs message routing for data transfer between nonadjacent nodes	Layer 3 network
Layer 2 data link	Improves error rate for messages moved between adjacent nodes	Layer 2 data link
Layer 1 physical	Encodes and physically transfers messages between adjacent nodes	Layer 1 physical
	Physical link	

Figure 11. Open Systems Interconnections (OSI) reference model (13). Courtesy of General Motors.

ble for providing specified networking services to the layer above. Network services are provided by programs resident in the layer and by accessing the services available from the layer below. In theory, any layer could be replaced, as a replacement layer which provides the same services in a different way would not affect the user's perception of network operation. Thus networks can be tailored to specific needs while enjoying the economies and benefits of common components.

Applications programs reside at the highest level or layer. At the lowest layer are the physical media over which data is transmitted. The seven layers of the module can be classified into three basic categories: end-node specific, routing, and adjacent-node specific, as follows:

Layer 7: (Application)

Layer 6: (Presentation)

[end-node specific]

Layer 5: (Session)

Layer 4: (Transport)

[routing]

Layer 3: (Network)

Layer 2: (Data link)

[adjacent-node specific]

Layer 1: (Physical)

The application, presentation, session, and transport layers deal with application-level services, data format standardization, address translation, access security, and the end-to-end processing of messages. Protocols at these levels must be consistent for pairs of communicating end-user programs. The network layer is concerned with how to route messages between the two (or more) communicating applications in the presence of intermediate nodes. Protocols at this layer must be consistent throughout a given network. The definition and implementation of the network layer is crucial to the success of MAP as a nonproprietary, multivendor network standard.

The data link and physical layers are concerned with the physical communication of messages between adjacent nodes. Protocols at these levels must be consistent within the connection of each pair of adjacent, communicating network nodes. The use of MAP standard protocols at these levels will allow for the interconnection of nodes from different vendors. Thus one vendor's robot, another vendor's lathe, and a third vendor's AGV should be able to communicate effectively through the MAP standard. The relationship between matching pairs of protocol layers is shown in Figure 12.

The distribution of functions to the layers is not absolute. As an example, addressing is a multilayer problem which includes locally identified names processed by the session layer, global addresses processed by the transport and network layers, and physical layer addresses processed by the data link and physical layers.

Each layer provides services to layers above it, and two types of information are passed between layers in providing

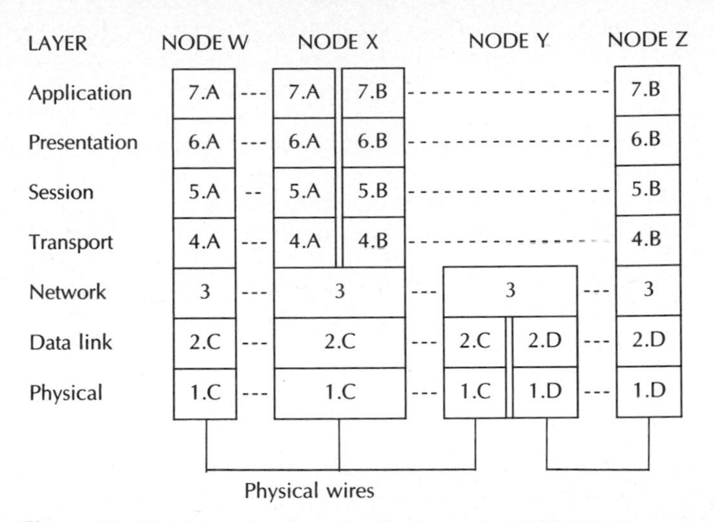

Figure 12. Matching of communication layers in MAP (12). Courtesy of General Motors (13).

these services: control information and data. Services required to process messages utilize control information as their basis. Each layer uses control information required to provide its services and transmits remaining control information on to lower layers, until all control information is used.

Data passed between layers is generally transported transparently. (The presentation layer constitutes an exception: its job is to reformat data passed to it.) Each layer processes data by means of a control block before requesting services from the next lower layer. Control block information is interpreted by the receiving layer. As data is passed to successively lower layers, its size increases (see Figure 13). As data is passed to successively higher layers, the control blocks are removed.

The objective of the MAP is to define a standard protocol that can be implemented on all types of computers, terminals, and programmable peripheral devices. This will allow users to plug into the MAP network and be assured of compatability; it will allow vendors to design devices to set interfaces and anticipate compatibility. TOP is similar to MAP but differs in specifications because of the different data needs and device characteristics involved. The TOP and MAP relationship is shown in Figure 14.

Applications that occur in both MAP and TOP include product definition data, exchange of such data, exchange of graphic files and data base access. To handle some exchange tasks, specifications such as P-DES (Product Data Exchange Specification) are being worked out. Data generated in the design of a product have clear implications for management decisions such as inventory management, scheduling, and cost calculations. There is no question that network protocols are desirable. However, the situation facing TOP designers is somewhat different from that of MAP. The use of distributed data bases is far more fluid, and there is no incentive at present to induce vendors to agree on specifications to indicate how distributed data bases should be structured, accessed, or modified.

IBM is looking into data base standards, particularly through its Standard Query Language and various standardized data structures. Other vendors have their own query languages. Still others believe that some form of artificial intelli-

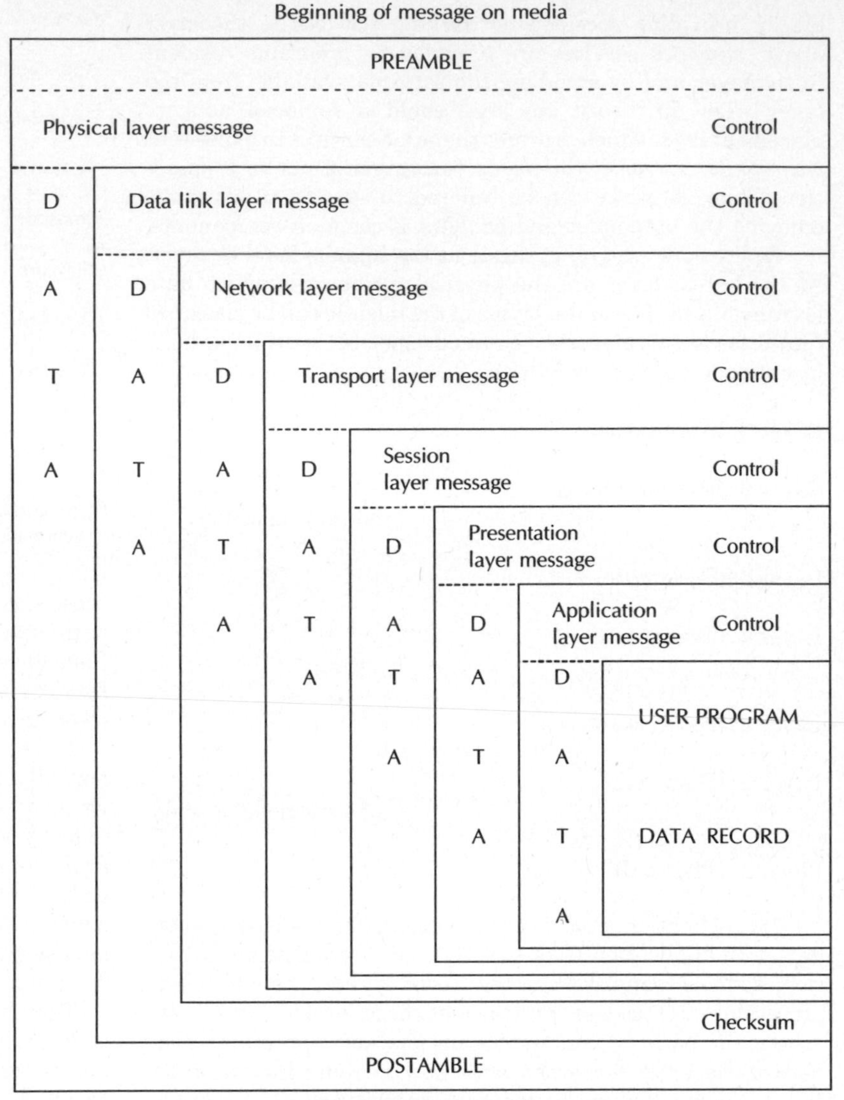

Figure 13. Nesting of layer protocols within MAP (13). Courtesy of General Motors (13).

gence needs to be introduced to provide an expert system to supervise the network, managing information in all the different data bases.

Several other applications areas are also in need of standardization. Standardized protocol specifications should be clearly and unambiguously stated, yet specifications should also be flexible and modular, as the OSI reference model suggests: any layer should be replaceable by a new layer that performs the same function, as long as the input and output formats are compatible with the system specification and the new layer performs the same function as the old.

Current MAP and TOP efforts include extensive development efforts, specifications of both interim and end-result characteristics, and consideration of migration paths and implementation procedures. MAP and TOP are both intended to provide "rules of the road" to permit effective communication among a wide range of vendor hardware and software products. MAP operation has been demonstrated in prototype displays, although development continues. The importance of these standards and interfaces for permitting widespread interconnection is of central importance in practically implementing FMS, particularly for small users and those with some equipment already installed.

INTELLIGENT SYSTEMS TECHNOLOGY

At this point it is clear that both detailed specification according to exhaustive standards and flexibility are desirable for an FMS. Ordinarily it would seem that these are contradictory demands and that it would be wellnigh impossible for autonomous computer-controlled machines to be able to perform in detail according to meticulously prescribed procedures and yet also be flexible enough to modify these procedures appropriately and accurately in accordance with changes in instructions and circumstances. For example, in a highly automated flexible manufacturing system, going from machining one product to another entails much more than changing from one CNC program to another. In any one case, decisions might have to be made regarding a host of related matters including rescheduling of machines, tool changes, and so on.

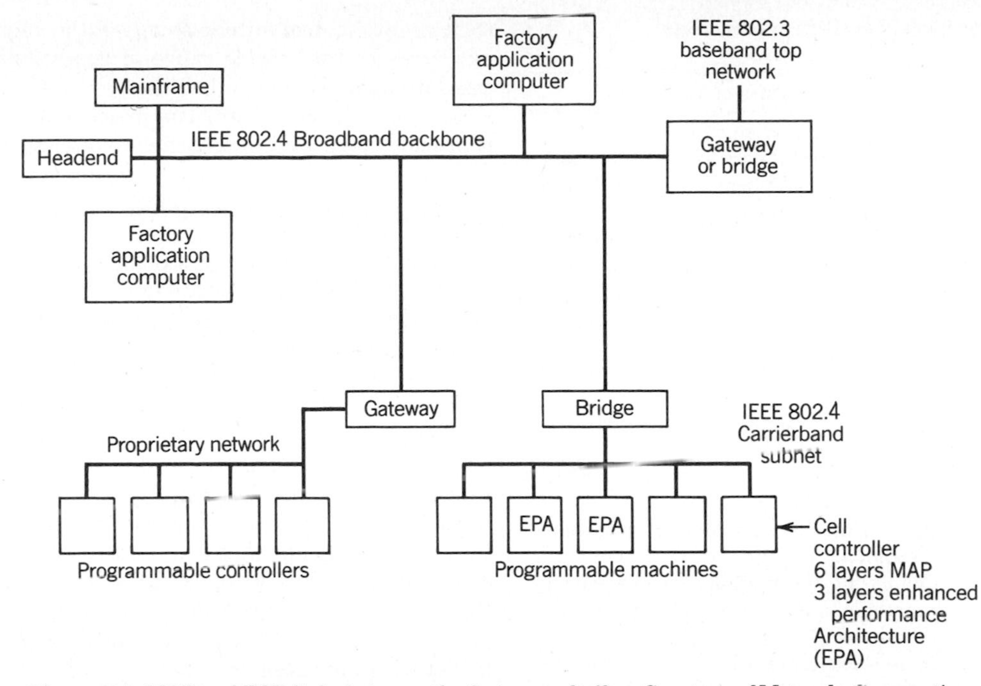

Figure 14. MAP and TOP links between the factory and office. Courtesy of Motorola Corporation.

Two types of situations arise that are particularly difficult for machines to cope with. In some circumstances, the situation is so complex that no optimum solution can be obtained through analytical means, eg, dynamic programming. However, humans can rely on their knowledge and judgment to provide acceptable solutions that satisfy constraints reasonably well. In other cases, even though the situation remains quite simple in all functions, there are so many related decisions to be made that it is not possible for any fixed algorithm or any set of such algorithms to cope with all the circumstances. Here again, human experts can rely on their knowledge and judgment to adapt their procedures in accordance with changes in task objectives or circumstances even when they cannot produce calculated or analytically derived rationales for the (generally successful) changes they generate.

Until recently, it was not possible to implement computer-based systems that could bring knowledge and judgment to the performance of tasks in the manner that human experts do, ie, by judgment-based and noncalculative, nonalgorithmic means. However, it now appears that AI technology, especially expert systems technology, can indeed provide the software required to support such judgmental systems, and thus aid in the implementation of highly automated, flexible manufacturing systems capable of responding to a wide range of problems too complex for simple repertoire flexibility.

AI technology is the integration of the tools and techniques of AI research with many other developments in computer technology to yield a new, more flexibly judgmental dimension to computer programming and control. With this new technology, it is being demonstrated that expert systems can be written that begin to duplicate the problem-solving proficiency of human experts. At present, such systems implemented within carefully defined problem domains have demonstrated performance equal to or better than that of human experts. Examples include some carefully specified medical diagnostic situations, scheduling, geological analysis and trouble-shooting, and maintenance procedures for locomotives. One such system in a manufacturing setting involves trouble-shooting in the operation of complex equipment for manufacturing fluorescent lamps; here, the diagnosis process may involve 40 or 50 steps to bring the equipment back into the desired performance range. Although AI technology is still at a very early stage of development, these practical expert systems do clearly indicate the effectiveness of these approaches for a variety of real and complex manufacturing (and other) settings. They suggest the possibility for many broader applications in the future.

Although the present tendency is to call all such systems expert systems, after the first such systems that drew on human expertise to derive implicit judgment rules, in some circumstances there are no human experts capable of solving the specific task problems at hand. In other circumstances, the performance of the expert system may well be far superior to that of the existing human expert. This is especially so where many factors combine to affect performance: the expert system keeps track and exercises control in such fashion as to permit explicit testing of hypotheses, enabling the expert system to build upon its initial knowledge base (whether human beings set up the experiments or the system itself does).

A list of the tasks that AI technology systems can presently address, with varying degrees of success, is given in Table 1. As mentioned previously, highly competent expert systems have been developed and fielded for certain types of tasks, such as diagnosis, troubleshooting and scheduling and resource allocation. A number of such systems are in operational use for commercial purposes, actually performing diagnoses, assisting in factory operation, or interpreting data. Other tasks, such as understanding and learning, are presently handled in rather primitive fashion by AI systems.

Table 1. Tasks that Artificial Intelligence Systems Can Address

Category of Task	Nature of Task
Allocation of resources	Allocating limited amount resources so that constraints are satisfied to a certain extent and benefits attained are at an acceptable level.
"Routine" processing	Processing routine tasks where nearly all procedures are seemingly fixed, but actually, in practice, judgment is often required.
Planning	Stipulating sequences of events that are to happen if desired results are to be obtained.
Diagnosis	Inferring underlying reasons for observed system malfunctions.
Interpretation of signals and images	Inferring object or situation descriptions from signal or image data.
Design	Configuring objects under specified constraints.
Monitoring	Comparing observations with known situations to infer status of system performance.
Debugging	Diagnosis accompanied by plans for specific remedial action. Generally used in the context of computer software, but applicable to all procedural matters.
Repair	Similar to debugging, but generally for hardware problems, emphasizing details of executing remedial action.
Instruction	Planning presentation of materials and concepts accompanied by diagnosis debugging and repair of student misunderstandings.
Learning	Inferring correct generalizations or extensions of observed phenomena. Inferring specific consequences of observed phenomena (including some not formerly seen).
Text understanding	Recognizing aspects of text in terms of known situations.
Predictive monitoring	Inferring likely consequences of observed situations.
Control	Monitoring, interpreting, predicting, and repairing system behavior.

AI technology is in its infancy, but it does bring a new dimension and highly interesting additional capabilities to the sorts of problems involved in implementing flexible manufacturing systems. As greater familiarity with AI and expert systems evolves, and as still more powerful computers and computer approaches are developed, one can expect AI and expert systems to move quite definitively from status as advanced development, laboratory efforts to more general use as practical management tools. From *tour de force* examples of experimental systems developed in special LISP machine environments, for instance, AI can be expected to move into delivery environments suitable for factory floor, business, and management use.

Flexible manufacturing systems will ultimately be judged by their ability to perform as highly automated systems, capable of autonomous operation for long stretches of time and their ability to deal autonomously with unforeseen contingencies across a fairly wide range of operational circumstances. In addition, all this must be achieved in a manner comprehensible to human overseers. The merit of AI technology is that it provides tools and approaches to address these needs.

FMS PHILOSOPHY AND BENEFITS

A key consideration facing any manager or engineer contemplating a flexible manufacturing cell or FMS should be the objectives sought. In short, why bother with the myriad of detailed specifications, the expensive equipment, the effort to integrate the different components? To be economically viable, FMS must deliver measurable, discernible contributions to the firm, enabling it to deliver better products to the marketplace, more cheaply or more quickly than the competition. These products may incorporate new features or incorporate those features long before others can duplicate them. Alternatively, FMS may enable a firm to deliver a far higher standard of quality at a low price, or at a price only slightly higher than the competition's. FMS promises the advantages of computer controls and automation along with the flexibility to incorporate improvements, new materials, and product or process changes without having to scrap the plant. Manufacturing in any case requires control of the details; the only questions concern how they are to be controlled and what constraints the control effort entails. The options for achieving control appear limited, and most entail unacceptable costs. Is control to be achieved by means of expensive, single-purpose hard automation? This option exposes the manufacturer to high risks when customer tastes change, technology advances, or new constraints compel changes in product design, materials, or processes. Today, it is estimated, the average product life cycle is only three years (and in many industries, it is far less). This is often insufficient time to warrant heavy expenditures for dedicated plant. Is the control to be attained by means of human-resource intensive, expensive operator skills? Even highly skilled workers are less reliable than automatic equipment and, with manual equipment, must typically spend much time, effort, and waste material getting equipment set for production. Maintaining an adequate staff of skilled operators is also difficult. Or can control be obtained by means of flexible, programmable production equipment? Such equipment, capable of substituting machine attention, quality, speed, and reliability to transcend human limitations, seems within reach. Moreover, as flexible control improves, unmanned operation expands the production capacity of such equipment. As experience proves, much careful attention is required before firms actually enjoy the potential benefits this equipment promises, but competitors' adoption of flexible manufacturing may require its adoption nevertheless.

Evaluating such options requires looking behind the obvious, to the true limitations inherent in the choices. Human beings do some things very well and others far less well. Repetition and meticulous detail control, massive real-time data acquisition, and calculated analysis are not key human skills. By contrast, they are skills readily attainable through computer controls, and highly amenable to automation. As labor costs decrease—and today they are almost always below 20% of cost and typically well under 10% of cost—considerations beyond potential labor savings have become essential for effective evaluation of proposed automation investments. Virtually no automated system can be justified economically simply on the basis of elimination of direct labor: today, direct labor almost always constitutes a minimal portion of total costs. Further, emphasis on labor often translates to a lack of attention to the key skills and benefits machines offer, thus contaminating choice and evaluation processes. Quality and flexibility, which can be translated into the ability to adjust manufacturing process in response to real-time data, timely design, and specification changes, or shifting needs, are far more important than labor savings, in the long run.

Markets in many industries are increasingly worldwide markets, so the standard of competition has moved from local minimums to worldclass standards. This tougher competition demands vastly more competent, better-controlled manufacturing. Quality is the central, imperative issue, with price not far behind. All too often, because they cannot effectively control manufacturing to worldclass standards, U.S. manufacturers have been forced to compete on price alone. Unfortunately, lacking effective manufacturing control, manufacturers cannot achieve either effective quality or effective price. Flexible manufacturing facilities enable a manufacturer to compete on product features, new concepts, and continuous improvements simultaneously, rather than on price alone.

Flexibility means that jigs, tools, machines, and layout need not be scrapped with each change. This lowers cost, but it also enables manufacturing personnel to build their experience with existing facilities as well as enabling them to use this knowledge in new ways, for new products. Flexibility also creates potential for firms to respond to changes in demand by addressing other markets, utilizing equipment for multiple purposes and thus spreading the risk that any single market will turn down. The benefits of implementing FMS often include fewer shop-floor employees—in some cases dramatically lower numbers—but its key advantages are not to be found in labor cost reductions. Instead, FMS offers above all far more strategic and long-term advantages, most notably quality and flexibility. FMS permits a firm to enjoy lower risk by enabling it to access other product markets, specialty niches, and varied customer bases, product types, and sometimes varied raw materials as well. Most importantly, by creating possible options, FMS ensures a wider array of future strategies.

SOME EXAMPLES OF FLEXIBLE MANUFACTURING

Some examples will illustrate how manufacturing control, cost, and quality relate to flexible manufacturing systems. These examples are drawn from especially demanding environments, but parallel examples could be offered in many different industries.

IBM manufactures mainframe computer displays in Research Triangle Park, North Carolina. The IBM facility incorporates a host of improvements over its predecessor, including product redesign, process redesign, and much closer integration of designers with manufacturing personnel. New manufacturing technology, emphasizing flexible controls, includes automated design and automated assembly. The new layout reduced manpower from 130 to 10 and reduced required manufacturing floorspace from 51,000 sq ft to 9000 sq ft. Data processing support costs per unit were reduced from $50 to $1.26 (achieved in part by reducing the number of subassemblies with over 50 component parts to fewer than 12). Daily output of display units was increased from 800 to 1200, and assembly time was reduced from several hrs to just one min.

Equipment includes three Cincinnati Milacron robots that remove component parts from pallets and place them on an automated conveyor, which feeds parts to the assembly stations; eight IBM 7540 robots and a GMF robot that assemble the displays; and two product assembly lines for the manufacture of the IBM mainframe displays, including the 3178, 3179, 3191 and 3196. Design and assembly are automated to make the manufacturing operation a continuous flow process implementing a just-in-time approach, and minimizing the amount of time materials and work spend in production. This effort cost IBM some $10 million, and the company declines to state exactly how much dollar benefit it has achieved. In strategic terms, IBM's improvements constitute a significant increase in capacity of an existing facility, providing "extra" space "for free" in addition to the other benefits of the redesign efforts. The new operation also allows IBM to incorporate improvements or to bring new products to market much more quickly and with greater confidence in their quality than before. Capacity, quality, and cost improvements aside, the facility enables IBM to force competitors to match its expertise and its quality and price capabilities at a faster, forced pace than before—formidable competitive advantages in a fast-moving market.

Texas Instruments (TI)'s Miho, Japan semiconductor manufacturing operation won the Demming Prize for manufacturing quality in 1985. TI has long been known for manufacturing expertise and attention to quality, but achieving the Demming Prize is an outstanding testimony to excellence in an extremely demanding environment. TI has sister plants in Miho and in Richardson, Texas, each equipped with "the cleanest rooms in the world" and highly automated in order to ensure product quality. Tighter control in manufacturing was essential to competition and survival in the key DRAM (Dynamic Random Access Memory) market. TI believes position in DRAMs is essential for survival in other important semiconductor markets, such as EPROMs (Electronically Programmable Read Only Memories) and application-specific integrated circuits (ASICs), very large scale ICs, microcontrollers, microprocessors, and other advanced products.

TI's advanced plants cost $100 million each, with 70% of the cost allocated to equipment. These new plants will produce product architecture ranging from the 256K DRAM's 1.25 microns down, with the coming 4Mb DRAM expected at 0.8 microns.

The new plants achieve the following specifications:

Class 1 clean room (no more than one particle per cubic foot larger than 0.2 microns): TI's facilities are believed to be the cleanest in the world.

On-line statistical control, via TI Professional computers, is used to minimize process variations. This translates to improved yields.

Equipment is bulkhead mounted, so that servicing is accomplished from outside the clean room, thus reducing potential contamination. Together with a modular approach, this arrangement will also allow rapid change from 256K to 1 Mb and 4Mb DRAM production.

The production floor is vibration-isolated, constructed to be independent of the building's walls. Vibration measurements on the production floor are lower than 0.01 microns. (This is essential to ensure accurate production of submicron-integrated circuitry.)

Ion implant utilizes a pick-and-place robot to move silicon wafers into the ion implanter, and AGVs pick up and deliver wafers. This eliminates human handling and thus reduces contamination potential, improving output yields.

TI continues to use and further develop automated design and circuit-test capabilities, reducing design time by reducing errors. Early discovery and elimination of such errors is essential both for ultimate product quality and for acceptable cost: it is very much more expensive to correct errors later on. In addition, many IC designs for advanced products are now too complex for unaided human design. Reduced design error and tighter manufacturing control translate into improved yields even on highly complex (and thus more lucrative) products. Advanced manufacturing concepts (including high-level control, clean room conditions, modular design, and design for a suite of related, automatically produced products) give both cost and quality advantages by increasing product yield in highly complex processes otherwise subject to risk of contamination and waste. It is debatable whether demanding, new generation submicron ICs and highly complex designs are achievable at all without automated design, test, and manufacturing of this sort.

Traditional Machining

One more traditional example concerns General Electric's (GE) Large Steam Turbine works in Schenectady, New York. GE managers found that it was not only cheaper but also quicker to manufacture replacement parts on demand in their flexible machining operation than to place them in inventory and retrieve them. GE also eliminated blueprints—and thus the cost, space, and effort required to generate, store, file, and maintain blueprint files—by designing turbines on the computer. Like TI, IBM, John Deere, Ingersoll, and a host of others around the world, GE also experienced lower error rates, faster turnaround, and "better" designs, as well as greater design output.

Another more traditional example is to be found in Volkswagen's Wolfsburg factory, where auto frames are automatically welded. Some 150 simultaneous, precisely positioned welds ensure the dimensional integrity of the automobile and also facilitate quality production down the line. Because the frame's shape is accurate, hinges and doors are more reliably mounted and work as designed; engines and drive-train members fit properly, enabling faster and easier assembly and better performance for the customer. VW's line produces random sequences of different models (including, for instance, left and right hand drives for different markets) on the same setup. Unique barcodes identify each vehicle, permitting coordination of parts delivery to the line and allowing VW to track the vehicles automatically through production and ultimate disposition.

CONCLUSION

Flexible manufacturing systems are factory systems capable of producing a range of products of differing design, different materials, or utilizing basic production processes in differing combinations and sequences. Examples are to be found in a wide range of industries, from metal cutting through electronics, plastics, batch chemicals, and a host of consumer products parts manufacture and assembly. To fully exploit this new manufacturing capability, however, old ways of thinking about manufacturing and about strategy must be updated. New bases for equipment evaluation that emphasize what the new equipment and practices offer that the old equipment and practices do not must be be developed.

The key advantages of FMS consist primarily in the far more precise control and far greater capability for responding to changes in demand, design, priorities, or circumstances. These joint advantages, control and flexibility, overcome long-perceived tradeoffs enshrined in traditional manufacturing. A great many subsidiary technologies (eg, CNC, CMM, CAD and CAE, AI and enhanced communications standards and protocols, and computers, among others) contribute to achieving the benefits of flexible manufacturing on the factory floor. Subsidiary management practices (eg, IBM's product and process redesign, linking designers and manufacturing staff more closely, or TI's emphasis on quality and flexibility) also help.

FMS development, especially toward more integrated control and expanded flexibility, can be expected to continue. Today, substantial capability—sometimes in cells rather than systems—is available and realizes important returns. Though substantial effort may be required to ensure effective coordination and communication across different vendors' equipment and among different elements of the manufacturing process—from design to tool point, from technical to managerial activities—the potential benefits are significant. Keys to success include judiciously evaluating FMS, its associated technologies, and the effort required to implement it successfully in terms of its ability to enable the company to support customers and respond to change more effectively.

BIBLIOGRAPHY

1. J. Hartley, *FMS at Work,* IFS Publications, Kempston, Bedford, UK and North-Holland, Amsterdam, 1984, pp. 39–41.
2. *Ibid.,* p. 48.
3. R. Jaikumar, "Postindustrial Manufacturing," *Harvard Business Review* **64,** 69–76 (Nov.–Dec. 1986).
4. K. Yee, D. S. Blomquist, D. A. Dornfeld, and C. S. Pan, "An Acoustic Emission Chip-Form Monitor for Single-Point Turning," Working Paper, National Bureau of Standards, Automated Manufacturing Research Facility, Gaithersburg, Md.
5. J. D. Goldhar and M. Jelinek, Plan for Economies of Scope," *Harvard Business Review* **61,** 141–148 (Nov.–Dec. 1983).
6. P. Ranky, *Design and Operation of Flexible Manufacturing Systems,* IFS Publications, Kempston, Bedford, UK and North-Holland, Amsterdam, 1983, p. 47.
7. Ref. 1, p. 57.
8. Ref. 6, p. 225.
9. Ref. 6, p. 229.
10. Ref. 4, p. 4.

11. Ref. 4, p. 3.
12. Ref. 6, p. 250.
13. General Motors, MAP, Part 1.

General References

N. L. Hyer and U. Wemmerlov, "Group Technology and Productivity," *Harvard Business Review* **62,** 140–149 (July–Aug. 1984).

J. M. Martin, "Research that Packs a Punch," *Managing Automation* **1**(6), 49–55 (Oct. 1986).

W. Rauch-Hindlin, "Revamped MAP and TOP Mean Business," *Mini-Micro Systems* **19,** 95–110 (Nov. 1986).

FLEXIBLE ROBOTS, CONTROL OF

S. H. Wang
University of California
Davis, California

INTRODUCTION

Industrial robots today are built with heavy rigid links. The motion of its end-effector can be accurately determined from its joint angles and hence controlled by actuators at each joints. However, because of its heavy weight, today's industrial robots can lift objects of no more than five percent of its own weight (1). Heavy links also limit the speed of robot motion. For the next generation high-performance robot, lighter and more flexible materials will be used for its construction. Unfortunately, the complexity of modeling and control of flexible structures is increased considerably. This article surveys several techniques for synthesizing an open-loop control signal to move a flexible robot in high speed and cause minimal residual vibrations at the end of the movement. One has to either avoid exciting any of the resonant modes of the structures, or to cancel any structural vibration before stopping.

Several methods of generating the input function, which is represented as a finite sum of sine or other functions with coefficients chosen to minimize the residual vibrations are discussed later. These techniques are restricted to systems modeled by linear differential equations and can be used for minimizing the residual vibrations only. A more general technique is also presented. This technique is based on the closed-loop simulation to generate the input function. It can be used to control a robot to move precisely along a continuous path. Furthermore, it can also be applied to nonlinear systems.

SERIES EXPANSION OF INPUT FUNCTION

Ref. 2 proposes a procedure for synthesizing input torque waveform for reorientation of spacecraft. The shaped torque waveform is assumed to be

$$u(t) = \sum_{n=1}^{M+1} e^{-\sigma_n t}\left[g_{1,n}\cos(\omega_n t) + g_{2,n}\sin(\omega_n t)\right] \quad \text{for } 0 \le t \le T$$

where $(M + 1)$ is the number of modes of the structure. The coefficients $g_{1,n}$ and $g_{2,n}$ are chosen to satisfy the initial and terminal conditions for the motion of the system. One can use the free parameters ω_n and σ_n to minimize the performance index

$$J_1 = \int_0^T u(t)^2\, dt$$

so that $u(t)$ will tend to be small over the interval $[0,T]$. Another candidate performance index J_2 reflects the accuracy of the terminal conditions with respect to changes in modal parameters,

$$J_2 = \sum_{m=1}^{M+1} \left(\frac{\partial y_m(t = T)}{\partial v_m}\right)^2$$

so that the mode amplitude $y_m(t = T)$ at the end of the maneuver should be unaffected by small changes in each of the mode eigen-frequencies.

A multiswitch bang-bang function to achieve time-optimal control has been derived (3). Then a series of harmonics of ramped sinusoids is used to approximate the time-optimal control without causing excessive structural vibrations. A finite Fourier series expression for the forcing function to attenuate residual dynamic response in elastic systems has been used (4). The approach is based on selecting the Fourier coefficients to depress the envelope of the residual response spectrum in desired regions. High-speed low-vibration motions for linear systems have also been obtained (5,6) using finite trigonometric series.

In the follow sections, a method which is more general and is applicable to nonlinear systems is presented.

CLOSED-LOOP SIMULATION

The approach to be presented here is quite different from the previously known methods. The authors' method (10) is based on the closed-loop simulation to generate the open-loop control input. Based on this method, it is relatively easy to synthesize an open-loop control input to move a robot arm along a desired trajectory. Furthermore, the method can be readily applied to systems modeled by nonlinear differential equations.

An example from ref. 3 illustrates the method. A three-axis Cartesian manipulator is modeled as lumped masses and springs as shown in Figure 1. This three mass model is a reasonable representation of a flexible 3-axis Cartesian robot while a force u is applied to the vertical link. The flexures of the two horizontal links are modeled by two springs.

The dynamical equations are

$$\begin{bmatrix} m_1 & 0 & 0 \\ 0 & m_2 & 0 \\ 0 & 0 & m_3 \end{bmatrix}\begin{bmatrix} \ddot{x}_1 \\ \ddot{x}_2 \\ \ddot{x}_3 \end{bmatrix} + \begin{bmatrix} k_1 & -k_1 & 0 \\ -k_1 & (k_1 + k_2) & -k_2 \\ 0 & -k_2 & k_2 \end{bmatrix}\begin{bmatrix} x_1 \\ x_2 \\ x_3 \end{bmatrix} = \begin{bmatrix} 0 \\ 0 \\ 1 \end{bmatrix} u$$

or

$$M\ddot{x} + Kx = bu \tag{1}$$

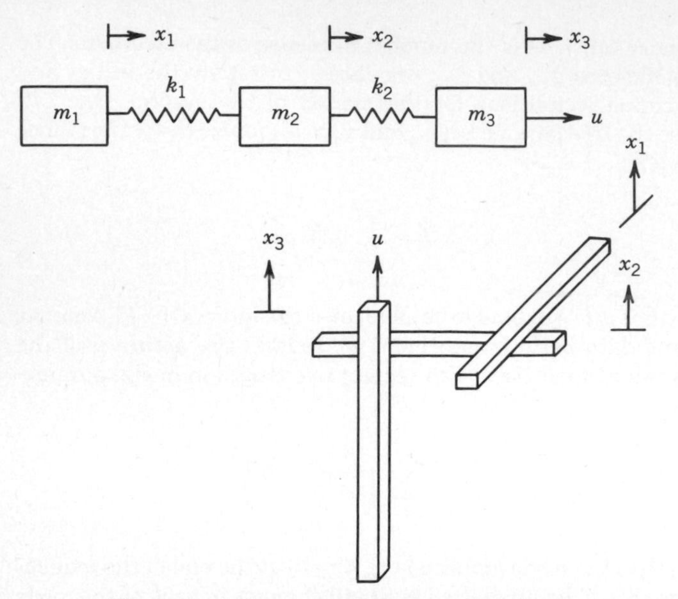

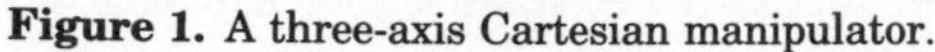

Figure 1. A three-axis Cartesian manipulator.

$m_1 = 7.9$ kg (17 lb) $m_2 = 9.3$ kg (20 lb) $m_3 = 10$ kg (22 lb)

$k_1 = 25{,}280$ N/m (25.3×10^6 dyne/cm)

$k_2 = 67{,}000$ N/m (67×10^6 dyne/cm)

$\omega_2 = 10$ hz $= 62.86$ rad/s $\omega_3 = 20$ hz $= 125.98$ rad/s

Equation 1 can be rewritten as a set of first-order differential equations:

$$\frac{d}{dt}\begin{bmatrix} x \\ \dot{x} \end{bmatrix} = \begin{bmatrix} 0 & I \\ -M^{-1}K & 0 \end{bmatrix} \begin{bmatrix} x \\ \dot{x} \end{bmatrix} + \begin{bmatrix} 0 \\ M^{-1}b \end{bmatrix} u \tag{2}$$

To design a proportional-integral-derivative (PID) controller, add an integrator to integrate the error signal $(r - x_1)$, ie,

$$\dot{z} = r - x_1 \tag{3}$$

Combining equations 2 and 3,

$$\frac{d}{dt}\begin{bmatrix} x \\ \dot{x} \\ z \end{bmatrix} = \begin{bmatrix} 0 & I & 0 \\ -M^{-1}K & 0 & 0 \\ -1\,0\,0 & 0\,0\,0 & \end{bmatrix} \begin{bmatrix} x \\ \dot{x} \\ z \end{bmatrix} + \begin{bmatrix} 0 \\ M^{-1}b \\ 0 \end{bmatrix} u + \begin{bmatrix} 0 \\ \cdot \\ \cdot \\ 0 \\ 1 \end{bmatrix} r$$

or

$$\frac{d}{dt}w = \hat{A}w + \hat{e}u + \hat{b}r \tag{4}$$

It can be shown (7) that the pair of matrices $(\hat{A},\hat{e})$ is completely controllable. Hence, based on the state-feedback pole assignment theory, one can find a row vector f to assign any set of desired poles to the matrix $(\hat{A} + \hat{e}f)$. In other words, by choosing appropriate f_is in

$$u = f_1x_1 + f_2x_2 + f_3x_3 + f_4\dot{x}_1 + f_5\dot{x}_2 + f_6\dot{x}_3 + f_7z = fw \tag{5}$$

the closed-loop system poles can be assigned.

Combining equations 4 and 5,

$$\frac{d}{dt}w = (\hat{A} + \hat{e}f\ )w + \hat{b}r$$

In our case, the closed-loop system poles are chosen to be $(-12 + 2i, -12 - 2i, -14 + 2i, -14 - 2i, -15 + 2i, -15 - 2i, -20)$. The complete feedback configuration is shown in Figure 2. For further details on the design of PID (or robust servomechanism) controllers, refer to Ref. 7.

POINT TO POINT MOVEMENT

First consider the problem of moving the mass m_1 a distance of one meter. The velocities at the initial point and the final point are both equal to zero. Apply a unit-step function at the reference input r. Simulation results are shown in Figure 3. The output x_1 does approach 1 after one second. Also, record the open-loop input function $u(t)$. One can then use the open-loop input function $u(t)$ to drive the system and get the desired output x_1 as in Figure 3. Of course, it must be assumed that the system is accurately modeled by equation 1.

MULTIPLE POINTS TRACKING

In this section, a more general problem of synthesizing open-loop input function for the system output to track a sequence of points or an entire continuous function is considered. In the example, the following desired points are chosen

t, s	0	2	4	6	8	10	12	14	16
x_1, m	0	0.75	2.7	5.3	7.5	8	8	8	8
(ft)		2.5	8.9	17.4	25	26	26	26	26

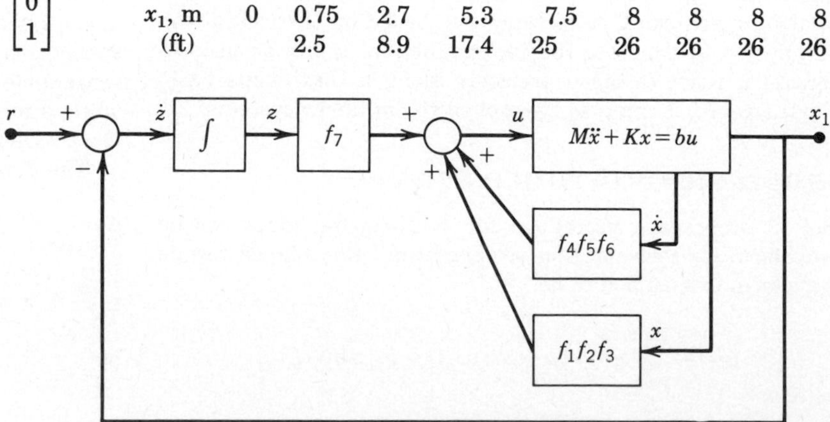

Figure 2. Closed-loop configuration.

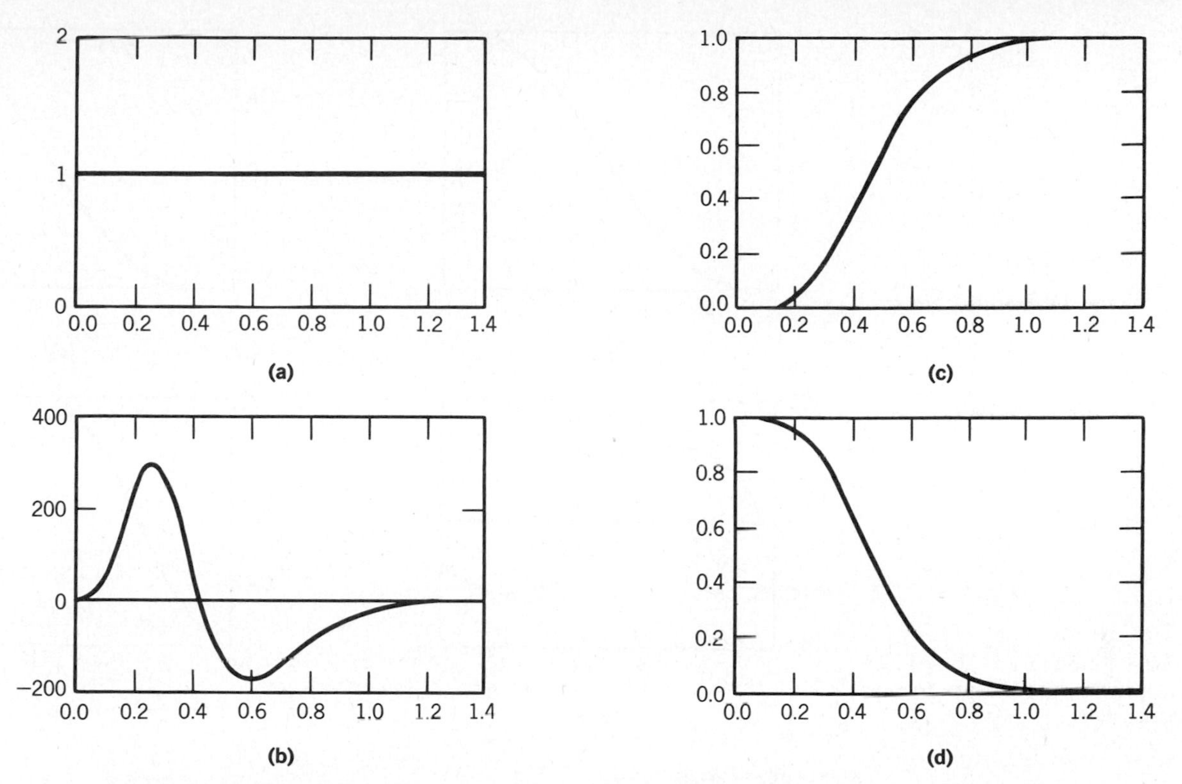

Figure 3. Unit-step response. **(a)** Closed-loop input r. **(b)** Open-loop input u, **(c)** Output X1. **(d)** Output error.

Then define the desired output function $x_1^d(t)$, $0 \le t \le 10$, by using cubic spline interpolation through the set of given points, as shown in Figure 4.

Initially, a reference input $r(t) = x_1^d(t)$ is applied to the system, the corresponding output x_1 and output error, $x_1^d(t) - x_1(t)$ are shown in Figure 5. Also record the open-loop input $u(t)$ in the same figure.

As can be seen, the output error may be too large. In the next section, a new algorithm, called Computed Reference Error Adjustment TEchnique (CREATE), (8,9), to reduce the output error is applied.

COMPUTED REFERENCE ERROR ADJUSTMENT TECHNIQUE (CREATE)

As discussed earlier, when the reference input $r(t)$ is set to be equal to the desired output $x_1^d(t)$, the output $x_1(t)$ usually differs from the desired output $x_1^d(t)$. Based on the error, $x_1^d(t) - x_1(t)$, one can modify the reference input to reduce the output error.

Step 1. Let $i = 1$.

Step 2. Let the reference input $r^i(t)$ be the same as the desired output $x_1^d(t)$, $t \in [t_0, t_f]$.

Step 3. Apply the reference input $r^i(t)$ to the system. Compute the error in the output, $e^i(t) \equiv x_1^d - x_1^i(t)$, where $x_1^i(t)$ is the output of the system due to input $r^i(t)$.

Step 4. If the error $e^i(t)$ is small enough, stop. Otherwise, go to the next step.

Step 5. Compute the reference input $r^{i+1}(t) \equiv r^i(t) + e^i(t)$, $t \in [t_0, t_f]$ for the next iteration.

Step 6. Let $i = i + 1$ and go to step 3.

Based on the above procedure, the new reference input $r^2(t) = r^1(t) + e^1(t)$, where $e^1(t) = x_1^d(t) - x_1^1(t)$ is obtained. The resulting input, output and error are shown in Figure 6. The results from the third iteration are shown in Figure 7. Substantial reduction in the output error has been achieved in three iterations.

PIPELINE IMPLEMENTATION

For real-time application, the CREATE algorithm as stated may require too much computation time to be useful. Hence,

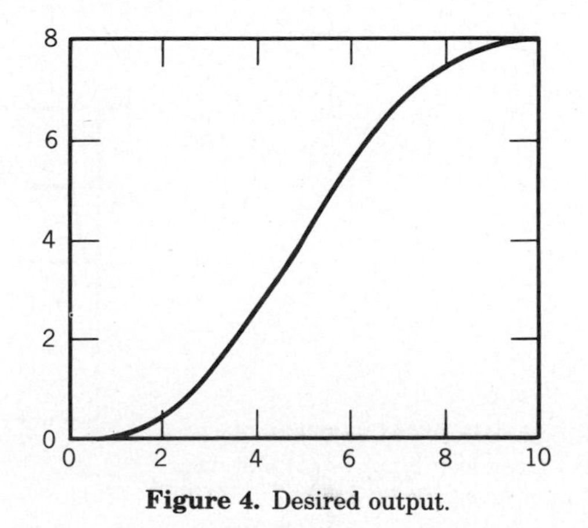

Figure 4. Desired output.

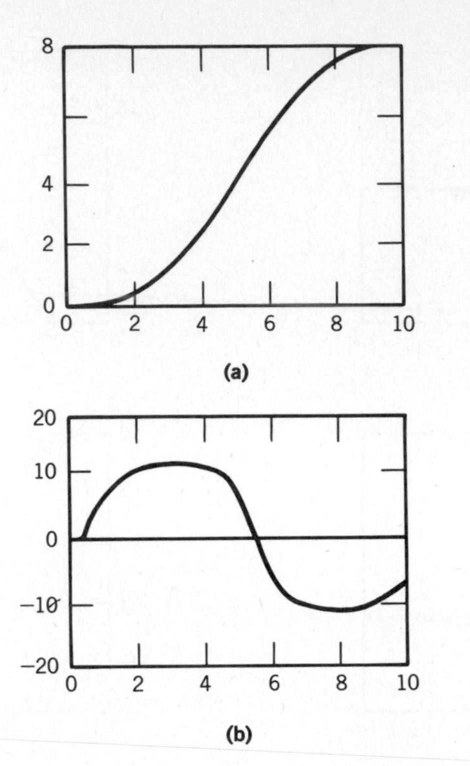

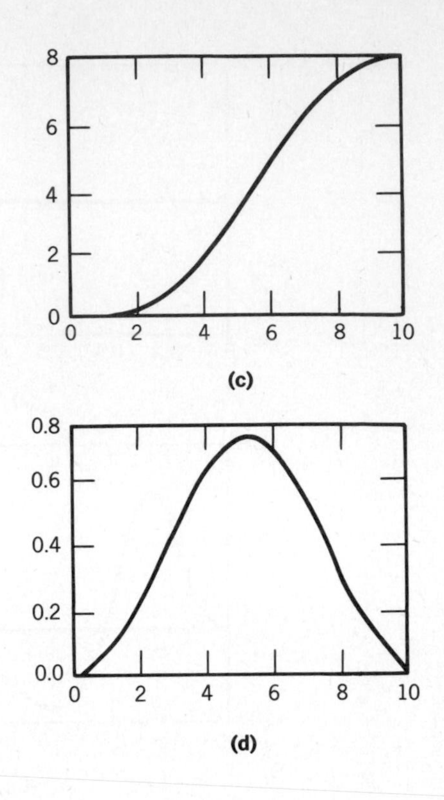

Figure 5. First iteration. **(a)** Closed-loop input r. **(b)** Open-loop input u. **(c)** Output X1. **(d)** Output error.

the following pipeline implementation to overcome this difficulty is proposed. Simulate the closed-loop system S in Figure 2 via hybrid computer. Build n-copies of S and connect them as in Figure 8.

From the n-th copy of S, the desired open-loop control signal u, which can be applied to the physical system is obtained.

CONCLUSIONS

Several different open-loop control techniques for flexible robot arms are discussed. In particular, the last method seems to be rather promising for the precision control of a flexible arm to move along an entire continuous path. The degree of control

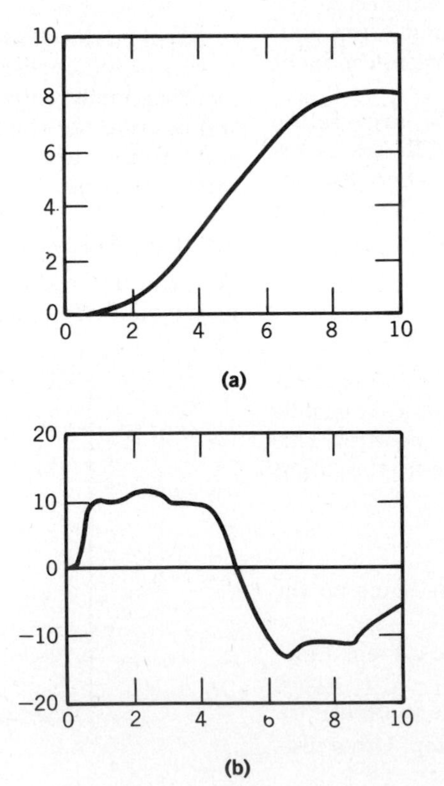

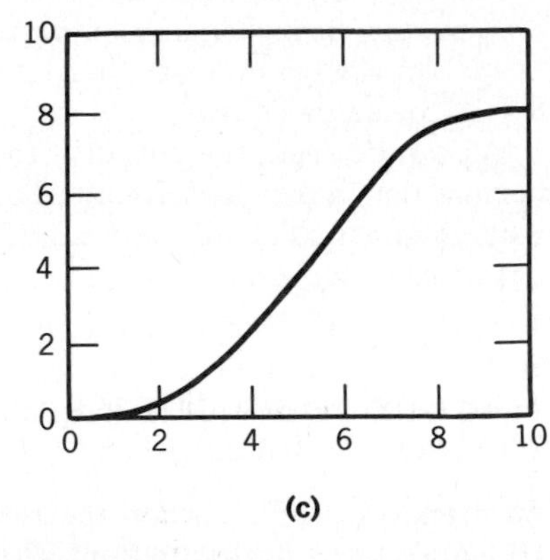

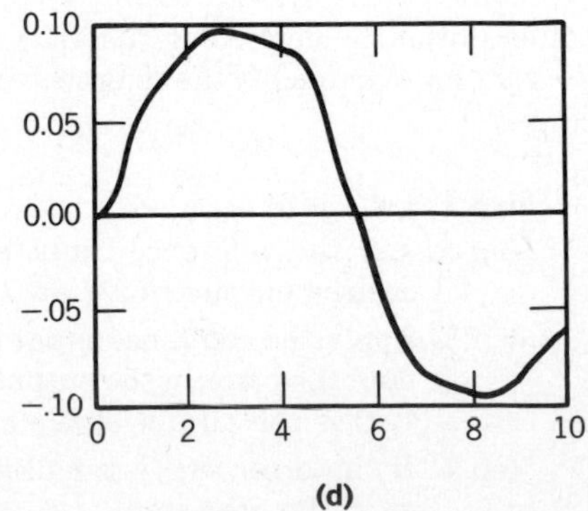

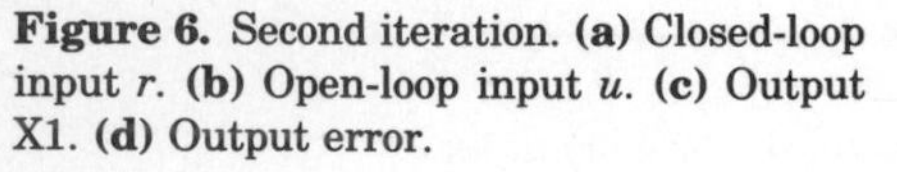
Figure 6. Second iteration. **(a)** Closed-loop input r. **(b)** Open-loop input u. **(c)** Output X1. **(d)** Output error.

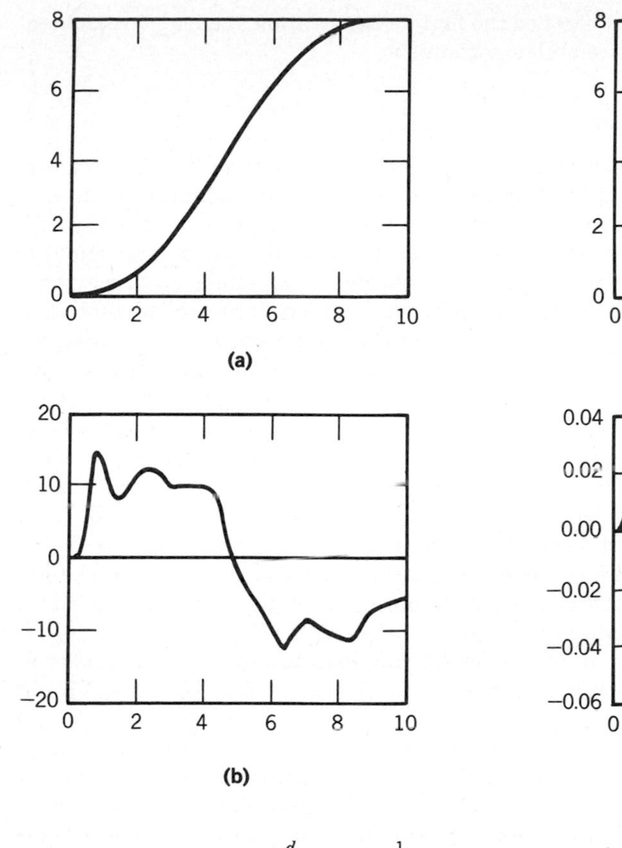

Figure 7. Third iteration. **(a)** Closed-loop input r. **(b)** Open-loop input u. **(c)** Output X1. **(d)** Output error.

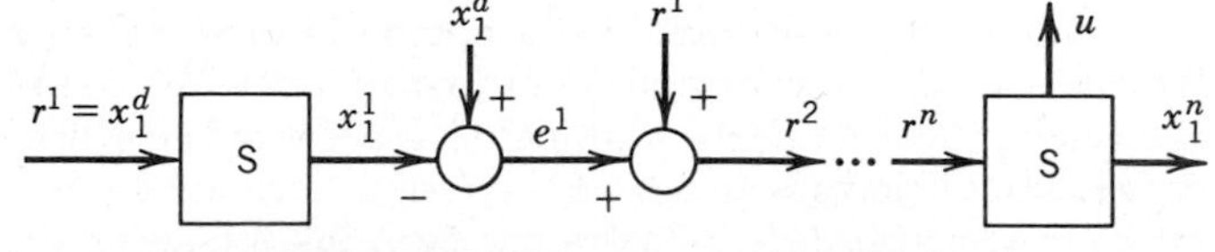

Figure 8. Connection on n-copies to S.

on the residual vibration effect can be adjusted by appropriately selecting the precision points at the end of the desired motion. Since the robust servomechanism theory can be applied to nonlinear systems, the last method can be readily extended to systems modeled by nonlinear differential equations. One of the fruitful research topics is to design the open-loop control input such that the system output is insensitive to system parameter variations.

BIBLIOGRAPHY

1. W. J. Book, T. E. Alberts, and G. G. Hastings, "Design Strategies for High-Speed Lightweight Robots," *Computers in Mechanical Engineering,* 26–32, (Sept. 1986).
2. C. J. Swigert, "Shaped Torque Techniques," *Journal of Guidance and Control* **3**(5), 460–467, (Sept.–Oct. 1980).
3. P. Meckel and W. Seering, "Active Damping in a Three-Axis Robotic Manipulator," *Journal of Vibration, Acoustics, Stress, and Reliability in Design, Trans. of the ASME* **107,** 38–46, (Jan. 1985).
4. D. M. Aspinwall, "Acceleration Profiles for Minimizing Residual Response," *Journal of Dynamic Systems, Measurement, and Control* **102**(1), 3–6, (March 1980).
5. J. L. Wiederrich and B. Roth, "Dynamic Synthesis of Cams Using Finite Trigonometric Series," *Journal of Engineering for Industry, Trans. ASME, Series B* **97**(1), 287–293, (Feb. 1975).
6. J. L. Wiederrich, "Residual Vibration Criteria Applied to Multiple Degree of Freedom Cam Followers," *Journal of Mechanical Design, Trans. of the ASME* **103,** 702–705, (Oct. 1981).
7. C. A. Desoer and Y. T. Wang, "Linear Time-Invariant Robust Servomechanism Problem: A Self-Contained Exposition," Memo No. UCB/ERL M77/50, Electronics Research Laboratory, University of California, Berkeley, Calif., Aug. 1977.
8. S. H. Wang, "Computed Reference Error Adjustment Technique (CREATE) for the Control of Robot Manipulators," *Proceedings of 22nd Annual Allerton Conf. on Communication, Control, and Computing,* Monticello, Ill., Oct. 3–5, 1984, pp. 874–875.
9. S. H. Wang and I. Horowitz, "CREATE—A New Adaptive Technique," *Proceedings of Conf. on Information Sciences and Systems,* Johns Hopkins University, pp. 620–622, March 1985.
10. S. H. Wang, T. C. Hsia, and J. L. Wiederrich, "Open-Loop Control of A Flexible Robot Manipulator," *International Journal of Robotics and Automation* **1**(2), 54–57 (1986).

This research was supported in part by the MICRO research grant jointly funded by the FMC Corporation and the State of California, and in part by the Engineering Research Program in the Mechanical Engineering of the Lawrence Livermore National Laboratory through grant No. 9339805.

FOOD PROCESSING, ROBOTS IN

STEWART J. KEY
JAMES P. TREVELYAN
University of Western Australia
Nedlands, Western Australia

INTRODUCTION

Agriculture for food production is undoubtedly one of the world's oldest industries. It is also likely to be one of the last industries to be affected extensively by robots. The reasons are obvious. Robots have been designed to work in the well-ordered environment of factories where every part is of uniform size and shape and made from materials able to withstand rough treatment and handling without damage. Agriculture on the other hand involves the processing and handling of plants and animals subject to natural variations in form, variations which humans are accustomed to. Contemporary commercial robots find it hard to deal with these variations and consequently all of the robot applications in agriculture are in the research phase. New approaches to robot control are required before commercial applications in agriculture become commonplace.

For the purposes of this article the food processing function is deemed to start once the harvesting of the produce is complete.

The food processing industry has experienced rapid growth in the last decade, but applications of robotics in the industry are relatively few. This is due to the essential character of the food processing chore whether the food be fruit, biscuits, spaghetti, or chocolates. Vast numbers of essentially similar products are processed. These products are usually processed in a standardized fashion. Therefore hard automation in the form of specially designed machines can offer great productivity gains over the soft automation of robotics. The cost of design of special purpose machinery can be absorbed because of the large product runs, and the speed of processing required far exceeds the handling capacity of the most advanced commercial robots. However several of the capabilities that have evolved to solve problems in the industrial robotic arena have found application in the food processing industry; for example, repetitive inspection jobs that previously were not possible to automate can now be tackled with machine vision.

A survey of the published literature concerning robots and robotic research over the last decade shows practically no mention of agriculture and food processing. Recently, due to efforts by organizations like the American Society of Agricultural Engineers (ASAE), reports of work being undertaken internationally have been published through conferences held in the United States.

ROBOTS IN AGRICULTURE

Applications of robotics in agriculture fall into four basic categories: nursery, cultivation, harvesting, and husbandry. While certain other areas of application, such as aids for animal and plant reproduction, have been advanced as possible and worthwhile, applications in these four areas will indicate the flavor and direction of work to date. Commercial application of robotics in these areas is some years away. Several have only just reached the first demonstration stage of development in the research environment.

Nurserying

The tasks of nursery personnel are similar whether the aim is to supply plant stock for vegetable production or to market cut flowers. The structure of the working environment is well-suited to some forms of automation in that a large number of repetitive tasks are performed in a standardized way on a highly concentrated product in ordered, controlled surroundings. However the scale of the operation, the profit margins involved, and the technical skill level of the industry have resulted in the use of unskilled labor remunerated on a piece-work basis. Lower-priced simple robots now offer a means of automation that is well-suited to the needs of nursery owners. Such robots can be used to move a seeding head over seedboxes; transplant from seedboxes into retail pots; handle large numbers of plants to give the correct exposure and to better utilize hothouse space; fertilize based on the measured height or growth of the plant; and spray with insecticide based on visual clues on the plant leaf profiles (1).

Ways in which robots may help to relieve the drudgery of transplanting have been investigated (2). This research addressed the problem of taking a plant out of a seedbox and feeding it to a plough planter (see Fig. 1). A very simple manipulator arm was used. The project demonstrated that with a simple software strategy and sensing, delicate plants could be conveyed. Note that the slow conveying rate required for this application is well-suited to the speed of a robotic system. Work has also been conducted on the tasks of inventory control and yield estimation of tree seedlings (3,4). Here opto-electrical or computer vision systems have been used. Seed-piece preparation in the sugarcane industry has been investigated using RF absorption detection to accurately locate seed nodes and direct cutters for immediate seed-piece preparation and replanting in the field (5).

Cultivation

Considerable work is proceeding on the automated control of tractors to pull cultivating implements. The degree of automation ranges from driver aids to improve fuel consumption to autonomous driverless tractor systems similar to mobile robots developed for industrial applications. The use of robotic arms to cultivate plant beds or groves has not been actively researched at this stage.

Harvesting

Harvesting of fruits, crops, vegetable produce, fish and animal products has attracted widespread interest and attention from farm mechanization specialists. It is in this area of agriculture that high costs are incurred by the use of expensive capital equipment or expensive labor or both. Automated harvesting of oranges, apples, white asparagus, melons, grapefuit, grapes, strawberries, mushrooms, milk, and wool are under investigation (1,6–14). Research into wool harvesting is discussed in this article (see ROBOTS IN AGRICULTURE).

In 1984, the Australian sheep industry earned approxi-

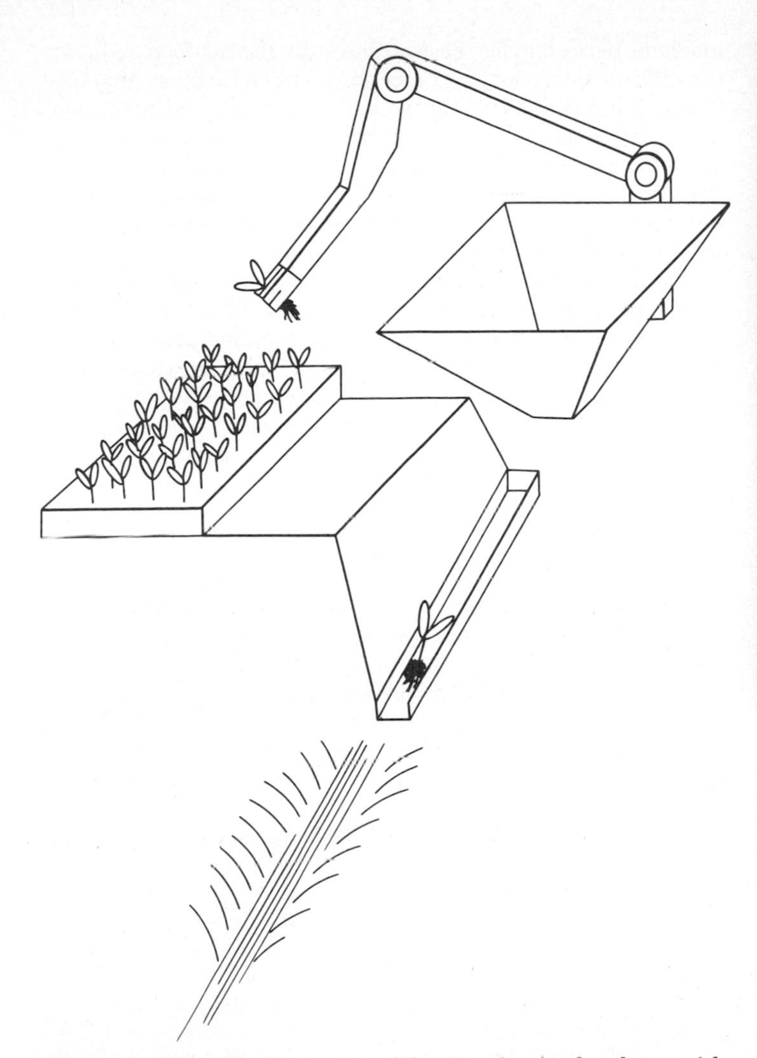

Figure 1. Schematic illustration of the use of a simple robot to pick seedlings from a seed box and feed them into the down shoot of a rolling wheel planter.

mately 13% of the Australian export income. Over the last 20 years production costs have risen to a point where the profit margin from wool has been reduced by two thirds. Further there is increasing concern that because manual shearing is physically arduous and conducted in remote locations it will be unable to compete with other less demanding trades for young entrants. These influences have generated the need in the industry for alternative harvesting techniques aimed at reducing labor costs.

Research at the University of Western Australia has demonstrated over the last seven years that the automated shearing of a sheep by robot is technically feasible. This has required the design, manufacture, and commissioning of three special-purpose research robots that have the unique kinematic and dynamic characteristics to perform the task of shearing. Shearing is achieved by first placing the sheep upside down in a sheep manipulator that catches and restrains the sheep (see Fig. 2). A highly dexterous robot is then used to guide the cutter over the sheep following a shearing blow strategy that is decided in advance. Before shearing, the sheep's weight and several girth measurements are used to statistically predict its surface shape. Surface maps are created that are used to navigate over those areas to be shorn. During shearing, sensors locate the skin of the sheep and enable the control computer to direct the motion to maintain the cutters just lightly touching the sheep's surface. The robot has sufficient dynamic response to react to movements of the sheep such as breathing or twitching due to being tickled by the cutters. As the robot proceeds over the area to be shorn it learns more about the surface of the sheep and adjusts the shearing strategy to optimize the way the surface is shorn. On several occasions the sheep is moved automatically on the manipulator to present previously hidden surfaces of the sheep to the robot for shearing. For example, the sheep is initially loaded on its back and later automatically rolled onto its side so that the back fleece wool may be removed.

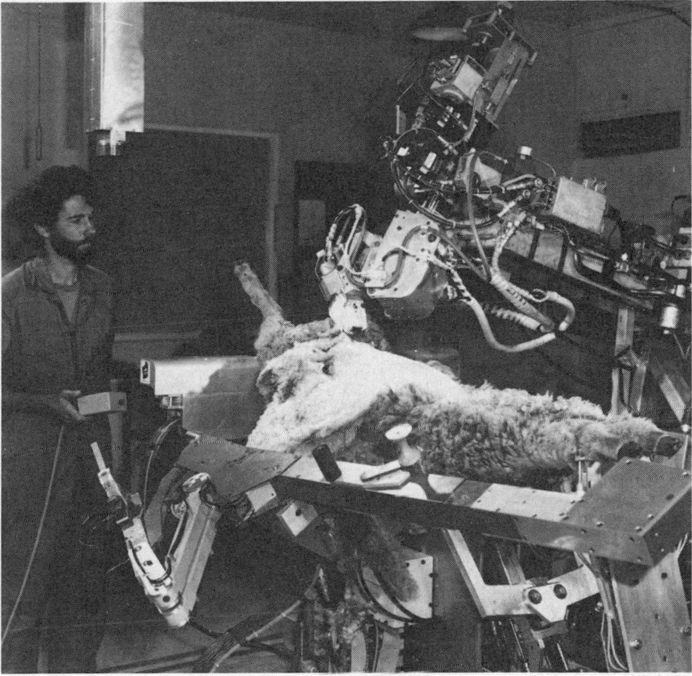

Figure 2. The experimental laboratory testing of the ORACLE sheep shearing robot at the Department of Mechanical Engineering, University of Western Australia. The sheep is held in this case on its back in the automated sheep manipulator while shearing is in progress on the belly. The operator holds a button box that will stop the robot at any time to allow for further investigation of results by the operator. The test may then be aborted or continued.

Estimates of the eventual commercial productivity and capital cost of the machinery required to support the introduction of this technology into the marketplace have been made. The system is likely to be a mobile plant capable of moving from property to property. Woolgrowers will hire the services of the plant in a similar manner to the present conventional practice of shearing contracting teams. It is likely that the eventual cost to the woolgrower will be equivalent to present manual shearing costs.

Husbandry

Care of the crop, pasture, or animal other than for direct production may be considered husbandry. Animal husbandry considerations are certainly built into the work on the automated sheep shearing robot. However work on the care of vines for correct growth in France (15) and prevention of insect damage through chemical spraying (16) have either been proposed or researched to the prototype stage. Techniques of vision processing have been utilized to aid in the research of chemical spray effectiveness, and ultrasonic transducers initially refined for guidance of industrial vehicles are being used to help control the height of boom sprays for broad acre chemical applications. Autonomous lawnmowers designed initially for green keeping have been proposed and tested (17). These may also have applications in the grove environment. With the general acceptance and success of fruit harvesting by robot, it is envisioned that automated pruning of groves will receive research input, thereby allowing the capital-intensive machines needed for harvesting to be part utilized during a more prolonged period of the year.

ROBOTS IN FOOD PROCESSING

As mentioned previously, the food processing industry is in most cases better-suited to the use of hard automation than to the use of robots. However several interesting research programs are being conducted in the areas of packaging, fault detection, sorting, and product preparation.

In the confectionery industry (18–20) a vision system has been coupled to a conventional simple robot to pick the correct chocolate from a conveyor, reorient it, and place it into the correct chocolate from a conveyor, reorient it, and place it into the correct location in the retail box. This approach incorporates detection of faulty products and has been used to direct the arm to decorate selected product lines with melted chocolate before packaging.

Fault detection is a major task, especially in fruit packaging. Correct detection of bruises on the outside of apples, for example, has the potential greatly to increase the price received by packers for the finished product (21,22). Two research groups have independently developed a method to detect bruises. The fruit is aligned and spun in front of a line camera operating through a selective filter. Good reliability has been achieved. A similar technique utilizing visual processing of a line scan camera image is under development for the on-line rejection of potato chips that have holes in them at the point of cut of a chipper machine (14).

Sorting of agricultural produce into size lots for different market requirements has traditionally been performed by mechanical means. With the advent of computer vision, several new approaches to the sorting problem are possible. Cucumbers, for example, are sorted in Japan according to their straightness, length, and girth using a computer vision system (23).

In general meat products require considerable processing before sale. Recent experimentation in deboning of bacon backs by robot has commenced (24). The position of the end of the bone is sensed and a wire hoop cutter is slipped over the end of the bone. The hoop is then drawn up the bone, cutting it away from the flesh. Semi-automated cutting of pig carcasses into eight primal cuts has been demonstrated (25). The machine receives the carcass upside down on a conveyor. The operator digitizes several key points on the carcass and the machine performs the desired cuts. In the seafood industry the difficult task of oyster processing, which involves breaking the shell hinge and cutting the adductor muscle without damaging the meat, is under investigation using computer-controlled machinery (26). Careful analysis of the geometrical variation in the position of the muscle and size of the shells has been required, coupled to sensing of the position of the meat to achieve a reliable demonstration.

These projects illustrate the worth of the application of segments of mainstream robotics research effort to the generally better structured environment of the food processing industry. The application is warranted only for those tasks where a unique ability is required to adapt the standard process to each product as it is being processed.

BIBLIOGRAPHY

1. G. W. Krutz, "Future Use of Robots in Agriculture," *Proc. First International Conference on Robotics and Intelligent Machines in Agriculture,* Tampa, Fla., Oct. 1983, p. 15.
2. H. Hwang, "The Implementation of a Robotic Manipulator on a Mechanical Transplanting Machine," *Proc. AgriMation 1 Conference ASAE,* Chicago, Ill., Feb. 1985, p. 173.
3. D. R. Devoe, "Image Processing to Inventory Tree Seedlings in Nursery Beds," *Proc. AgriMation 1 Conference ASAE,* Chicago, Ill., Feb. 1985, p. 215.
4. G. A. Kranzler, "Opto-Electronic Inventorying of Tree Seedlings in Nursery Beds," *Proc. AgriMation 1 Conference ASAE,* Chicago, Ill., Feb. 1985, p. 223.
5. L. A. Jakeway, "Machine Vision and RF Absorption Detection Techniques for Robotic Production of Sugarcane Seedpieces," *Proc. AgriMation 1 Conference ASAE,* Chicago, Ill., Feb. 1985, p. 312.
6. E. G. Tutle, "Image Controlled Robotics in Agricultural Environments," *Proc. First International Conference on Robotics and Intelligent Machines in Agriculture,* Tampa, Fla., Oct. 1983, p. 84 [apple harvesting].
7. A. Grand d'Esnon, "Robotic Harvesting of Apples," *Proc. AgriMation 1 Conference ASAE,* Chicago, Ill., Feb. 1985, p. 210.
8. J. H. Pejsa and J. E. Orrock, "Intelligent Robot Systems: Potential Agricultural Applications," *Proc. First International Conference on Robotics and Intelligent Machines in Agriculture,* Tampa, Fla., Oct. 1983, p. 104.
9. A. Manimalethu, "Agricultural Robot Application," *Proc. 13th ISIR,* Chicago, Ill., Apr. 1983, p. 1076.
10. S. J. Key and J. P. Trevelyan, "Automated Sheep-Shearing Project," *Proc. National Conference and Exhibition on Robotics,* Melbourne, Australia, Aug. 1984, p. 37.
11. J. P. Trevelyan, "Skills for a Shearing Robot: Dexterity and Sensing," *Proc. 2nd International Symposium of Robotics Research,* Kyoto, Japan, Aug. 1984, p. 317.
12. P. Baylou, "Computer Control of an Agricultural Locomotive Robot," *Proc. 2nd International Conference on Automated Guided Vehicle Systems,* Stuttgart, June 1983, p. 243.
13. D. Ordolff, "A System for Automatic Teat-Cup Attachment," *Proc. AG ENG 84,* Cambridge, Mass., Apr. 1984, p. 222.
14. R. M. Devaeminck, "Vision Systems and Robotics in Food Processing," *Proc. AgriMation 1 Conference ASAE,* Chicago, Ill., Feb. 1985, p. 27.
15. F. Sevila, "A Robot to Prune the Grapevine," *Proc. AgriMation 1 Conference ASAE,* Chicago, Ill., Feb. 1985, p. 190.

16. D. E. Guyer, "Potential for Computer Vision as a Spray Controller," *Proc. AgriMation 1 Conference ASAE,* Chicago, Ill., Feb. 1985, p. 156.
17. K. Loebbaka, "The Research and Development of PEGASUS: An Autonomous Lawnmower," *Proc. AgriMation 1 Conference ASAE,* Chicago, Ill., Feb. 1985, p. 277.
18. A. T. Cronshaw, "Feasibility Study for the Application of Vision to Chocolates," *Proc. 1st International Conference on Robot Vision and Sensory Controls,* Stratford-upon-Avon, UK Apr. 1981, p. 233.
19. A. J. Cronshaw, "Software Developments for the Visual Recognition of Chocolates," *Proc. 2nd International Conference on Assembly Automation,* Brighton, UK, May 1981, p. 191.
20. A. J. Cronshaw, "Automatic Chocolate Decoration by Robot Vision," *Proc. 12th ISIR,* Paris, June 1982, p. 249.
21. R. W. Taylor, "Development of a System for Automated Detection of Apple Bruises," *Proc. AgriMation 1 Conference ASAE,* Chicago, Ill., Feb. 1985, p. 53.
22. Y. Ikeda and R. Yamashita, "On the System Evaluating the Shape of Farm Products via Image Processing Techniques," Research Report on Agricultural Machinery No. 12, Kyoto University, Kyoto, Japan, 1981, p. 83.
23. W. F. McClure, "Agricultural Robotics in Japan: A Challenge for U.S. Agricultural Engineers," *Proc. First International Conference on Robotics and Intelligent Machines in Agriculture,* Tampa, Fla., Oct. 1983, p. 76.
24. J. C. Vickery, "Sensing and Cutting Methods for the Automatic Deboning of Bacon Backs," *Proc. 14th ISIR,* Gothenburg, Sweden, Oct. 1984, p. 653.
25. P. T. Clarke, "Automatic Breakup of Pork Carcasses," *Proc. AgriMation 1 Conference ASAE,* Chicago, Ill., Feb. 1985, p. 183.
26. F. W. Wheaton, "Use of Biological Properties in Food Processing Automation," *Proc. AgriMation 1 Conference ASAE,* Chicago, Ill., Feb. 1985, p. 38.

General References

S. J. Key and D. Elford, "Animal Positioning, Manipulation and Restraint for a Sheep Shearing Robot," *Proc. International Conference on Robotics and Intelligent Machine in Agriculture,* Tampa, Fla. Oct. 1983, p. 42.

S. J. Key, "Productivity Modelling and Forecasting for Sheep Shearing Machinery," *Proc. AgriMation 1 Conference ASAE,* Chicago, Ill., Feb. 1985, p. 200.

K. A. McDonald, "Computer Recognition of Lumber Quality, *Proc. 2nd Conference on Automated Inspection and Product Control,* Chicago, Ill., Oct. 1976, p. 51.

M. Ong, "An Approach to the Mechanical Design of a Sheep Shearing Robot," *Proc. 2nd Wool Harvesting Research and Development Conference,* Sydney, Australia, Aug. 1981, p. 161.

D. Shetly, "Automatic Sizing of Fish using Fluidic Sensors," *Proc. 5th Conference on Automated Inspection and Product Control,* Stuttgart, June 1980, p. 235.

W. D. Shoup, "A Model for Establishing Opportunity Values of Robotics/Automation in Food Processing." *Proc. Agrimation Conference,* Chicago, Ill., 1985, p. 69.

J. P. Trevelyan, "Software for Automated Sheep Shearing," *Proc. 51st ANZAAS Conference,* Brisbane, Australia, 1981.

J. P. Trevelyan, S. J. Key, and R. A. Owens, "Techniques for Surface Representation and Adaptation in Automated Sheep Shearing," *Proc. 12th ISIR,* Paris, June 1982, p. 163.

J. P. Trevelyan, M. Ong, and P. D. Kovesi, *"Motion control for a Sheep Shearing Robot," 1st International Symposium on Robotics Research,* MIT, Cambridge, Mass., 1983, p. 175.

FOUNDRY APPLICATIONS

Sten Larsson
ASEA AB
Vasteras, Sweden

Robots have been used in die casting much longer than in gray iron foundries. Die casters have been using them for more than 20 years to unload die casting machines because robots can easily handle hot castings while humans wait for the castings to cool. In these operations the robots are basically simple manipulators, and pneumatic rather than electronic robots are most often used. They remove the hot castings from the die casting machines and place them on a conveyor or into a container.

In 1979 in the United States alone, 298 robots of all kinds were in use for foundry and die-casting work, according to the Society of Manufacturing Engineers (SME). By 1983 more than 600 robots were in use in metal casting. The SME projects that around 900 robots will be used in the United States in Metal-casting applications by 1985, a growth rate of 250% in five years.

Recent developments provide new applications of robots in foundries. In this article fettling, cutting grey-iron castings, and refractory brick handling are described.

FETTLING

Manual cleaning, fettling, of castings is one of the most arduous and hazardous jobs in industry. It is becoming increasingly difficult to recruit personnel for this work, and the turnover of people in cleaning departments of foundries is substantially higher than in foundries as a whole.

Increasing knowledge of the harmful effects of vibrating hand tools on blood vessels, nerve fibers, and on bones in hands and arms has led to an increasing demand for automatic aids for cleaning castings, all over the world. Instead of lowering the harmful effects of vibration by reducing the size of hand tools, and thereby their efficiency, it is now possible to use industrial robots in place of people in foundry cleaning departments. Additional robotic applications (often for the same robots) are now being found in the same foundries which are turning to robotics for the first time.

Advantages of Robotic Fettling

Robotic cleaning of casting provides economic as well as environmental benefits to foundry managers. The robot operates nonstop and is screened off from the workers, who are thus protected from a job that in the long term carries severe risks of injury. Instead, the manual workers concentrate on supervising the robot, checking the quality of cleaned castings, and, when required, perform robot maintenance.

ASEA's Industrial Robot Division, Vasteras, Sweden, has developed a completely automated installation for cleaning gray and ductile iron castings. The system is now being marketed worldwide. The foundry fettling installation went into operation in late 1982 at Volvo Komponenter, Arvika, Sweden. The robotic cleaning installation works together with a handling and hopper system and uses four separate types of tools, all under adaptive computer control.

No specialized computer knowledge is needed for programming the ASEA IRB 60 robot used in the system. An operator familiar with cleaning castings can manually set those points to be searched by the robot, and then the operator enters the point in the computer program. The robot automatically compensates for tool wear. It searches for the condition of tool edges before it starts an operation. The program also features adaptive control to sense the size and position of risers and external flash, and locates cavities for internal grinding.

When a tool has become so worn that it must be changed, the robot does the job. The tool attachments are designed so that the new tool automatically takes up the same position as the old one. The cleaning installation at Volvo has four tools, as explained later. The complete system is controlled by ASEA's SII electronic robot controller which is based on two Motorola 68000 16-bit microprocessors and floppy disk program storage.

Reducing the Cleaning Cycle

One of the problems initially faced in the Volvo foundry in Arvika was the weight of the gearbox housing castings that had to be handled by the robot. Together with the risers, the gearbox housing weighs more than 132 lb (60 kg), which is a maximum handling capacity of the six-axis ASEA IRB 60 robot.

The problem was solved by making the robot first use a hydraulically operated cutting tool to remove the risers. This operation cuts the weights of the casting to a more manageable 121 lb (55 kg).

Castings are fed to the robot by a roller conveyor parts-handling and hopper system. The hopper holds up to 96 parts at a time. Castings are automatically transferred to the robot and returned to the hopper after cleaning.

"The robot and hopper system lets us operate the installation unmanned at night," reports the Project Manager at Volvo Komponenter. "We load the hopper in the evening after the final shift, and the robot works all night. By morning we have a supply of cleaned castings equal to the production of an eight-hour shift."

The robot at Volvo works with four tools: a cutting wheel, a grinding wheel, a chisel, and a rotary file. In the first step of the cleaning operation, risers are cut from the gearbox as the robot presses the hydraulic cutting wheel against the cast part.

During the next three steps, the robot lifts the 121 lb (55 kg) gearbox and holds it up to the stationary tools. Outer edges of the gearbox are ground, and burrs inside are removed, partly with the chisel, and partly with the rotary file.

The computer program for cleaning each component lasts 7 min—3 min less than required for manual cleaning. The reduction in cleaning time is 30%, and this represents a sizable cost saving.

When a new type of casting is to be cleaned, the robot program is changed. A new program is loaded from a tape cassette into the robot controller, which takes 15 s. At the same time the robot gripper module is changed to permit handling the new casting.

CONSISTENT CLEANING QUALITY

"We are going to work hard on automating our foundry," declares the Volvo Project Manager, "Four or five robot lines for cleaning will be needed before the end of this decade to maintain production—and to keep us competitive. Manual cleaning requires six employees per line, and they are not able to provide the same consistent, high quality as the robot. Equally important, we have totally eliminated personnel hazards such as back injuries from lifting heavy castings, as well as getting our people out of the hot, dusty atmosphere of the cleaning room; and we have solved a high personnel turnover and absentee problem in our foundry."

PLASMA ARC CUTTING OF GREY IRON CASTINGS

High-speed plasma arc cutting has been applied by researchers at the University of Rhode Island to cleaning gates, visors, and sprue from large gray iron engine block castings. A 20,000°F (11,000°C) plasma arc is guided by a six-axis ASEA IRB 60 electric robot to cut sections of ⅛ to ½ in. (0.32 to 1.3 cm) thickness, at speeds between 30 and 70 in./min (1.3 and 3 cm/s). Arc cutting of gray iron was not used until recently because it created a thick, hard layer of white iron carbide. A much thinner layer that can be rapidly removed results, however, under controlled arc parameters.

ROBOTIC REFRACTORY BRICK HANDLING

Another new area being explored by both foundries and steel mills is the robotic handling of refractory bricks, since the same robots that clean casting are not busy when the primary melters are down for relining. European brick makers already have extensive experience with brick handling robots because the manual handling of brick, like cleaning castings, is not a popular job. A worker can wear out three pairs of gloves in a single week handling brick in a brickyard.

In Sweden, the brickmaker Hoganas AB placed a robot online in 1980 to stack bricks on pallets. The robot handles 8000 bricks per 8-h shift. When the job was done by human hands, maximum capacity per shift reached only 5000 bricks.

The brick is handled by an ASEA IRB 60 robot with a capacity of 132 lb (60 kg). Programming for brick handling takes just a few minutes, as the programs are usually quite simple. Once the right moves are in the robot's memory, it performs them with near-perfect repeatability through thousands of working cycles. The unit stacks the bricks to accuracies of a few thousandths of an inch—well beyond the capability of unassisted human hand-eye coordination.

Sensors enable the ASEA robot to detect problems. When it "sees" or "feels" a problem, the robot stops what it is doing immediately and calls for help by an alarm system. When the condition is corrected, the robot resumes work where it left off.

Brick handled by the robot at Hoganas measures 10 × 5 × 5 in (250 × 130 × 130 mm). Each brick weighs about 20 lb (9 kg). Conveyors carry the brick directly to the palletizing station from a tunnel kiln. On the conveyor, they are stacked 12 in. (300 mm) high in five rows. Approximately 10 in. (250 mm) of air space is left between each row of bricks.

The robot's gripper handles five bricks at a time, using five movable suction cups. When the bricks are picked up, a vacuum sensor checks to see that they actually number five and are securely held. When the bricks are stacked on the pallet and released, an optical laser sensor verifies that all five bricks have, in fact, found their way to the proper location on the pallet.

The brick-handling robot is mounted on a 3-ft (1-m) high pedestal, giving it the working height needed to handle the top bricks in the stacks on both conveyor and pallet. If there had been too little space for floor mounting, the robot could have been inverted and mounted overhead. Robots work just as hard upside down as they do right-side up.

BIBLIOGRAPHY

General References

J. Shimogo and co-workers. "A Total System Using Industrial Robots for Electric Arc Furnace Operation," *Proceedings of the 3rd International Symposium of Industrial Robots,* Zurich, Switzerland, May 1973, pp. 359–374.

S. Synnelius, "Industrial Robots in Foundries," *Industrial Robot* **1**(5) 210–212 (Sept. 1974).

M. Mori and co-workers, "Applications of Robot Technology for Tapping Work of Carbide Electric Furnaces," *Proceedings of the 7th International Symposium on Industrial Robots,* Tokyo, Japan, Oct. 1977, pp. 293–300.

M. B. Tomasch, "Materials Handling: Key to Foundry Mechanization," *Foundry Management and Technology,* **106,** 26–27 (July 1978).

J. Kerr, "Britain's First Robot Fettling Shows What Can Be Done, *Engineer* **248,** 15 (March 22, 1979).

A. I. Alves, "Thoughts and Observations on the Application of Industrial Robots to the Production of Hot P/M Forgings," *Robotics Today,* 30–31 (Spring 1980).

A. Ferloni, "ORDINATORE: A Dedicated Robot That Orients Objects in a Predetermined Direction," *Proceedings of the 10th International Symposium on Industrial Robots,* Milan, Italy, March 1980, pp. 655–658.

This article is reprinted from S. Y. N. of, ed., *Handbook of Industrial Robotics,* John Wiley & Sons, Inc., New York, 1985.

FUTURE APPLICATIONS

Lewis J. Pinson
University of Colorado
Colorado Springs, Colorado

INTRODUCTION

It is both challenging and dangerous to discuss future applications in any technical area given the rapid growth and development of new capabilities that characterize such areas. Yet it is fascinating to be in a position of dreaming about future applications and trying to tie those dreams to today's reality and to tomorrow's perceived reality. The role is much like that of the science-fiction writer with only a few more limitations on the exercise of imagination.

Robotics has finally emerged from the realm of science fiction into the world of industry and research laboratories. It is a reality yet it is still embryonic and capable of exploding in many directions. The current state of capability in robotics has been characterized as similar to the state of computers only a few years ago. It was difficult then to predict the impact that computers would have on lives today and it is difficult to predict now the impact that robotics will have on lives in the next ten to fifteen years. Yet, that is the purpose of this article.

In predicting future applications it is reasonable to look momentarily at current applications and to examine trends based on current experience. To a large extent that is the methodology that will be used in this article. To a lesser extent it is impossible to not venture out on a limb occasionally and examine possible applications that are entirely new (see also Applications of robots; Factory of the future; Futurism and robotics; Research programs).

FACTORS AFFECTING FUTURE APPLICATIONS

The actual directions taken by new technologies are determined by complex interactions of a number of factors. Among these are the following:

Hardware Development

The single most important factor contributing to the existence of working robotics systems today has been the development of fast, powerful, and inexpensive computer systems. Other hardware components important to robotics include the mechanical arms and their drive motors, a variety of sensors, and interface components for connecting each part of the robotics system to the computer.

Parallel architectures for computers, faster and more powerful microprocessors, smaller size and lower power requirements will characterize future computing hardware for applications to robotics. This more powerful hardware, coupled with better software, will make possible the implementation of more intelligence in a small package.

Vision sensors currently have more capability than can be used by a robotics system; however, higher resolution imaging sensors are currently being developed by a number of manufacturers. High resolution tactile array sensors that are currently under development will significantly impact the capability of future robotics systems. Auditory sensors coupled with improved speech understanding will provide robots with the capability to understand human speech. This accomplishment alone will revolutionize the way that humans interact with machines and open up new areas of activity aimed at exploiting this capability.

Software Development

The heart of any computer system is its software. As new and more sophisticated hardware becomes available it places new demands on the developers of software tools, languages, and methods to fully exploit the hardware capability. There is usually a period during which a wide variety of incompatible ideas are evaluated and pared down to those that seem to work best. This is certainly true for robotics software. As evidenced by the existence of over twenty robot languages, a good solution to robotics software development methodology is still needed.

The future applications of robotics depend very strongly on software that provides control, sensor signal processing, pattern matching, and intelligent decision making. Further an effective environment for developing and integrating components of a robotics software system is needed. This environment must provide software development tools, simulation and emulation capability, and easy porting of emulation software to a real robotics system. Several new paradigms for software development, eg, object-oriented methodology and parallelism offer promise in this area.

In the area of specific applications software, pattern recognition and speech understanding are two areas that offer quantum improvements in robot system capabilities. Improved algorithms coupled with faster hardware will make possible the realization of these quantum improvements.

Supporting Disciplines

The list of supporting disciplines for robotics includes a large percentage of all the areas of human endeavor. Of course the technical disciplines are heavy contributors to robotics technology. Of particular importance are the areas of engineering, computer science, and materials research with underlying support of all the sciences.

In nontechnical areas, robotics draws from the disciplines of education, psychology, and sociology for ideas on learning, intelligence, and the needs of society for the development of useful robots.

Economic Support

Development of any complex technological system requires a commitment for economic support that goes beyond immediate returns. Fortunately, robotics draws on technical areas that have a support base in other applications already in place. However, a need still exists to support integration of these technical areas into working robotics systems.

Much of the support for robotics is still in the form of research money to develop and verify new principles. This support comes from government agencies and industrial research and development programs. As some of the hardware and software advances described above become available, intelligent robotics systems will move from the research laboratory into development and production. Then the laws of supply, demand, innovation, and competition come into effect and the robotics revolution will have arrived.

Social Acceptance

One of the problems faced by expanded robotics development and application at this time is the fear of job loss to a robot. The same fear was expressed about computers. As with most technological advances, it should be expected that robots will create a whole new set of job categories while replacing some of the jobs that are not well suited to humans anyway. Certainly there is more reason to believe this positive point of view.

Another fear attached to intelligent robots is caused by the delicate insecurity that humans have about machines being smarter than they are. Again a positive outlook is more likely to represent reality. Maybe in the attempt to give "intelligence" to machines, a better understanding of human intelligence will develop. How can this be bad?

And finally, humans tend to accept more readily those things that they know. Robot toys and domestic robots that enter the home and provide first hand interaction will be the ones that achieve acceptance for the broader class of robots.

Safety

One of the most important considerations with respect to social acceptance and new applications for robotics is their safety. If all robots can achieve compliance with the three laws of robotics as stated by Isaac Asimov, then there will be no problem. However, it is nontrivial to ensure that such laws can always be met. This will be particularly true for robots with general mobility.

There are currently efforts under way to establish robot safety standards. The success of these efforts and their general applicability to future robots as well as existing robots will be a key factor in the future of all robotics systems.

CURRENT APPLICATIONS

The first major application of robots in this country has been in the automotive industry. Robots have been used in the automotive industry for a variety of tasks from assembly of parts to welding and spray-painting to laying of complex adhesive patterns. For the most part these have been "dumb" robots. They are computer-controlled manipulator arms that are preprogrammed to perform a specific task without sensory input or decision-making ability.

Additional applications of robotics have been in the assembly of electronic circuit boards and electronic component systems. In most cases these applications have been a combination of ordinary factory automation methods and of robotics. The flexibility of the robotics systems will cause them to slowly replace the ordinary factory automation methods. Again there has been relatively little intelligence in these robots.

Most other current robotics applications have been more in the area of research and prototype development than in widespread use. Robots for palletizing, mowing lawns, repairing satellites, reading and playing music, sentry duty, and performing surgery are under development and have been demonstrated with varying degrees of success and completeness.

ROBOT CAPABILITIES

The applications for future robotics systems will be driven to a great extent by their capabilities. Therefore, before looking at specific applications it makes sense to predict the capabilities of future robots.

Much of the future belongs to intelligent robots; however, there will still be a need for dumb robots as well. A dumb robot is one that has few if any sensors, is preprogrammed to perform a specific task, has no intelligent decision-making ability, and generally has limited if any mobility.

An intelligent robot may be described as one that has sophisticated sensors, sophisticated signal processing capability, intelligent decision-making ability, and general mobility. Alternative descriptions of intelligent robots have appeared in the literature. Further it is not essential that a robot have all the above features to be classified as intelligent.

Sensors

One of the most important capabilities that humans use in performing intelligent tasks is receiving sensory information from their surroundings. Human senses provide a rich variety of sensory inputs that are critical to many of the activities performed.

Much effort has been expended to provide a similar sensory environment for robots. Vision, which is the most important sense (it is estimated that up to 80% of sensory information is received by vision) can be provided to robotics systems using available state of the art imaging sensors. In many cases these sensors provide spatial resolution equivalent to that of the eyes and response in spectral regions beyond that of the human eye. Imaging sensors with greater than 1.5 million pixels are available now and higher resolution imaging sensors are under development. Inclusion of on chip processing is the goal of some sensor designs. This offers the possibility of significantly enhanced utilization of the acquired imagery.

Acoustic sensors have consisted primarily of active devices for determining approximate distance to objects. Passive acoustic devices can provide the sense of hearing for robots if methods for understanding the acoustic signals can be developed. Perhaps a breakthrough in speech recognition methods will spur a new effort to develop smart acoustic sensors with enhanced capability.

There is a new interest in the development of tactile sensors in arrays that approximate the sense of touch in humans. Many of the precision tasks performed by humans are critically dependent on this sense of touch. The state of the art in tactile arrays is not close to that in vision sensors in terms of spatial resolution or dynamic range of individual sensors. Mechanical microcircuits may someday provide sophisticated tactile sensing for robots; however, this capability is not projected for the next few years. The intelligent robot of the next decade will undoubtedly have sophisticated tactile sensors.

Chemical sensors that approximate the senses of smell and taste may be developed for future robotics applications. There are even existing chemical sensors that will find their way into applications involving robots.

Signal/Image Processing

One of the most significant bottlenecks in the development of intelligent robots is that of image processing. Imaging sensors are available that provide high resolution, multi-spectral imagery yet people are unable to process the imagery fast enough to be useful in most real-time applications. Further, it is not always clear how to process an image to extract useful information.

One of the most useful tasks for an intelligent robot to perform is pattern recognition, yet there is still a struggle with the development of algorithms that recognize patterns with high accuracy and in a reasonable time. The human pattern recognition process cannot be imitated because it is not understood. This most valuable skill of the intelligent robot is the key to new applications. Faster processors, better algorithms, and an increased understanding of pattern recognition processes in humans will provide a quantum leap in improved capability of intelligent robots.

Intelligent Software

Intelligent software makes the brain of the intelligent robot work. It takes processed sensory information, databases, and knowledge and makes decisions. Unfortunately, all available intelligent computer languages still are dependent on programmed rules. It is dependent upon interpretation whether or not they exhibit true intelligence.

As more is learned about knowledge, how to represent it, and how to use it, this capability can be provided to a robot through appropriate languages. Since most ideas about knowledge and intelligence are expressed in language, the development of natural language understanding for computers should make it easier to provide similar capabilities for a robotics system.

Mobility

Limited flexible motion is one of the underlying principles of a robotics system. In current applications this implies a fixed but flexible workspace within which a robot arm is free to move.

The addition of a capability for general mobility is the next logical step in the development of more useful robotics system. This opens up a new spectrum of tasks that can be performed by the robot. Currently, mobile robots do not have the intelligence to properly navigate in a general sense. They are constrained to follow specified paths either as defined by tape or wires in the floor, or they are ultimately controlled by humans through teleoperation.

The provision of general mobility and navigation for a robot depends on the development of new image processing, sensor processing, pattern recognition, and decision-making software with faster hardware for its implementation. A better understanding of very simple pattern matching and decision-making methods employed by biological specimens may provide another key to enhanced robot capability. For example the simple cockroach has no trouble navigating a complex path at high speed when surprised with the lights in a dark room. No one would claim that the cockroach is of high intelligence. Yet humans cannot duplicate this task with even the most sophisticated sensors/signal processors/intelligence software.

APPLICATIONS

Semiconductor Industry

There are available now, robots for clean-room applications. In the near future it is reasonable to expect that robots will be used extensively in microcircuit fabrication operations. They have high accuracy, high repeatability, high cleanliness rating (one company claims a Class 1 rating for its top of the line clean-room robot), and high resistance to the hazards in most microcircuit fabrication processes.

The major inhibitor to their application is likely to be initial cost since most microcircuit fabrication facilities are very expensive. Their modification to incorporate robotics is not likely to be easy nor cheap. In the long run, however, increased productivity achievable with robotics should be an overriding economic factor.

Manufacturing, Assembly, and Inspection

Manufacturing and assembly processes have been among the first for application of robotics. This area is expected to continue as a major application area for robotics. It will take advantage of increased capability of robots for manufacture, assembly, and inspection of finished products. Robots coupled with the Manufacturers Automated Protocol (MAP) system can achieve on-demand manufacture of components and products with a near one hundred percent acceptance rate.

The addition of intelligence and improved pattern matching capability will make possible one hundred percent inspection of manufactured products. Higher productivity and reduced costs will keep the manufacturing industry competitive.

Materials Handling

In addition to the usual materials handling that is a part of any manufacturing process, a very significant application area for robotics is the handling of mail and packages that are transferred from point to point each day in this country. The overnight delivery services are handling up to a million parcels each night. Much of the actual sorting and re-routing of these parcels is still done largely by humans. Errors are typically made on about five percent of the parcels. That is they get sent to the wrong place on the first try. Some even get lost.

Experience has dictated a number of checks that are made in the process of parcel transfer. Robots could be used to provide many of the actual parcel-handling operations and to provide the validity checks. Widespread use of robots for these operations will likely depend on further development of improved image processing/pattern recognition methods as well as improved manipulator designs. Containerization which seeks to maximize the fill of an aircraft for package transfer is an intelligent process that humans do very well; it is still not known how to program a robot to do as well.

Automated reading of address zip codes is possible now at high rates if the numerals conform to a well specified standard. However, if handwritten characters are to be recognized, the success rate drops off rapidly with variations in style of the characters. Optical character recognition schemes for recognizing handwritten characters cannot compete with humans. Much of human ability is based on the context within which characters are presented. If context recognition can be provided for machines, then robots will sort more of the general mail.

Domestic

Acceptance by humans of new technology is always enhanced by the friendly, inexpensive version of the technology performing a simple but unattractive task that is a part of everyday life. In that vein, the robot vacuum cleaner will do more to enhance acceptance of robotics than all the other applications put together. Interestingly enough the robot vacuum cleaner is not a simple robot. It requires sophisticated sensing, pattern recognition and control. However, it will be a reality within the next fifteen years.

Space Applications

Robots are a natural for space applications. In particular an intelligent robot could perform many tasks that are extremely hazardous and expensive for humans to perform. In the past and the present these tasks are often not performed at all because of the cost and the risk to human life.

Satellite repair robots and space construction robots provide an extension to human capability in space applications that will be of great interest in future space exploration. Unmanned space probes with intelligent robots aboard may provide a method for expanding knowledge about distant and hazardous regions of space.

Medicine

The accuracy and precision of robot manipulators coupled with intelligent image processing software will make surgery a more precise art than is now possible. It is not advocated that robots will replace surgeons or other health care personnel. Rather, the robot will become a valuable assistant in the delivery of health care. In fact, social acceptance of robots will probably be slowest in this area of all those discussed.

TRACKING PROGRESS IN ROBOTICS

It is far easier to track the progress of a technological area than it is to predict future directions. There are an increasing number of resources that deal with what is happening in robotics. This section presents a representative list of resources for keeping up with new developments in robotics. It is not intended to be complete and apologies are made to those resources that are not included.

Resources for tracking progress in robotics can be grouped into the following categories *1.* newspapers, *2.* trade journals, *3.* conferences, *4.* exhibitions, *5.* technical journals, and *6.* educational seminars. Often conferences, exhibitions, and educational seminars are held concurrently and sponsored by technical societies or groups with an interest in robotics.

Listed below are selected specific resources in each of the six categories.

Newspapers. There are an increasing number of weekly newspapers that provide information relating to developments in robotics. These are potentially the best source of up-to-date information about robotics. In addition to announcements of technical achievements, these papers include articles about the industry, its economic status, and business news about the major players in robotics. One such weekly newspaper is *Automation News*.

Trade Journals. These journals are typically published monthly and contain more in-depth articles about robotics than the newspapers. The articles are written for a general technical audience and tend to be informative without getting lost in technical details. Most of these journals also publish a list of recent activities, product developments, and business activities. Some of the robotics trade journals are *Robotics Engineering* (formerly *Robotics Age*), *Robotics Today; Robotics World;* and *Applied Artificial Intelligence Reporter*.

Conferences. There is an increasing number of conferences on robotics held each year and sponsored by several technical and industrial organizations. The purpose of these conferences is to present technical papers on subjects related to robotics. The following is a partial list of these conferences:

1. Robots xx, where the 'xx' refers to the number of the annual conference. This conference is held in conjunction with one of the largest robot expositions in this country. Robots 11 was held in Chicago in April 1987.
2. IEEE Conference on Robotics and Automation. This conference is held annually.
3. SPIE Conference on Intelligent Robotics and Computer Vision. This conference is held in Cambridge, Mass. each autumn. It recently added a parallel conference on Mobile Robots.
4. SME Conference on Robotics: The Next Five Years and Beyond was held in Scottsdale, Ariz. in 1986. The SME sponsors a number of robotics related conferences throughout the year.
5. Vision xx where 'xx' is the year is closely related to robotics because of the close tie of vision sensors to intelligent robotics.
6. There are many other conferences held each year in specialty areas that significantly impact the future of robotics. These include conferences on image processing, pattern recognition, software development environments, and numerous other specialties.

Exhibitions.

1. See Conferences, no. **1.** It is typically held in conjunction with a technical conference as well.
2. SPIE International Symposium is held each August in San Diego. Although its main focus is not robotics, many of the equipment exhibits are of imaging sensors/image processing equipment that do have direct impact on robotics systems.
3. Vision xx where 'xx' is the year is a conference and equipment exhibition that features imaging sensors and image processing equipment.

Technical Journals. A number of technical societies have added journals that deal specifically with the area of robotics. Among these are *Journal of Robotics Systems* (John Wiley & Sons); *IEEE Journal of Robotics and Automation;* and *Robotica* (Cambridge University Press). Related IEEE Publications are *Computer Magazine; Transactions on Pattern Analysis and Machine Intelligence;* and *IEEE Spectrum*.

Educational Seminars and Publications. Numerous short courses are now available through universities, technical societies, and consulting firms that present various aspects of robotics and robotics related topics. Additionally a number of tutorial publications are available through various technical societies.

CONCLUSION

The objective in writing this article has been to stimulate ideas rather than make bold prognostications about what future robots will actually be doing. A secondary objective has been to provide some comments about the current applications of robots and the factors that will strongly influence future applications of robotics.

Robotics is a fascinating technology area that should have a bright future in at least some of the specific applications mentioned above.

FUTURISM AND ROBOTICS

Alan L. Porter
Frederick A. Rossini
Georgia Institute of Technology
Atlanta, Georgia

ROBOTICS IN THE YEAR 2000, A DELPHI FORECAST

Rapid robotic development with far-reaching implications is suggested throughout the *Encyclopedia*. What future developments can be expected, and when, are of great interest. Moreover, the extent to which various robotic capabilities will be widely applied in the commercial market is vital to those committing resources to robotics. This article attempts to forecast the directions and magnitude of commercial application of particular robotic capabilities in the year 2000 (see also Future applications).

METHOD

Technological Forecasting

Technological forecasting continues to grow in importance (see also Technological forecasts). Changing technology affects business, government, education, and leisure. Moreover, the pace of change has never been greater than during the last quarter of the Twentieth Century: the world is moving beyond the Industrial Age into the Information Age. Changing information technologies, in particular, bring opportunities, and threats, to many enterprises. It follows that being able to anticipate the character and magnitude of technological changes is now a necessary activity.

Technological forecasting is, however, a precarious business. No crystal balls provide exact prognostications of what is to come. (As has been said, forecasting is difficult, especially about the future.) Yet, a respectable body of forecasting methods has emerged (1–4). Judiciously applied, these can substantially narrow the uncertainties in making decisions about research thrusts, capital investments, and choices of systems.

The hundreds of specific forecasting techniques fall into, perhaps, five categories: extrapolative, expert opinion, modeling, scenario, and monitoring. Extrapolative methods rely upon time series data and the assumption of continuity into the future. Expert opinion draws on the tacit understanding by a group of experts of complex causal influences and development patterns. Models, especially computer models, depict and explore the interactions of various factors of interest. Scenarios provide a "storyline" to portray how a complex set of factors might emerge at a future time. Monitoring scans the environment for relevant information. Choice of forecasting methods depends on the situation in which the methods are to be applied. More important than methodological sophistication or precision, the core assumptions must be right (5).

This article forecasts developments in robotics to the year 2000. Many others share a sense of the importance of forecasting developments in robotics. Predicasts (6) recently compiled some 136 forecasts concerning robotics. These range in time horizon from 1990 to 2000, covering topics such as sales, number in use, employment, and worker displacement. With the possible exception of aggregate sales and number in use, however, adequate time series data do not exist for rapidly changing robotic technology. *Predicasts Basebook* for 1986, in fact, provides no historical time series under any of the Standard Industrial Classification (SIC) headings pretaining to robots.

Expert opinion appears the method of choice for this forecast. Time series data are not sufficient for substantial extrapolation; no particular target warrants detailed modeling; no dramatic overall choice would be served well by scenarios (eg, alternative national energy strategies).

Expert opinion methods require experts. A natural set of such experts is the group of authors of this volume. The Delphi method provides a way to bring expert opinion to bear on issues without excessive cost (eg, as in convening committees), yet with a measure of feedback and the opportunity to revise judgment (7–9).

In essence, Delphi consists of repeated survey rounds accompanied by feedback of information on group response in the previous round.

The driving questions behind this forecast of the development and application of robotics in the United States for the year 2000 are

- What are the major technical changes to be expected?
- To what extent are these likely to be adopted by industry by the year 2000?

The intent is to understand the approximate magnitude of these changes and the relative degree of uncertainty associated with these changes. The aim is not to pinpoint scientific breakthroughs nor to achieve precision in the timing and extent of adoption.

APPROACH

The starting points for this forecast are a general causal model that attempts to identify the factors most influential on robotic innovation and a "morphological analysis" that characterizes the important robot attributes (10). The assumption is that the "push" of research and development generates new technical capabilities in robotics and supporting technologies, such as microprocessors. The market (both private and public sector) "pulls" for the introduction of new and improved robotic products. However, the linkage of new robotic technology to markets requires engineering and management skills, organization, capital, etc. This enterprise is affected by societal preferences, economic influences, and governmental incentives and constraints. In one sense, this implies that technological forecasting of robotic developments requires social, institutional, and economic forecasting as well. Our operating presumption is that the experts implicitly integrate such considerations into their judgments. These broader socio-economic influences are noted later with respect to the forecasting of certain robotic features.

Morphological analysis focuses on a single issue (robotic development); it defines all the important parameters critical to that issue and the states that each each parameter can take (11). One then examines the various state combinations to identify potential configurations. For present purposes, specific configurations are less important than characterizing the essential parameters. Table 1 [based on 10] provides the initial set of parameters.

Guided by this framework, the authors reviewed recent robotic forecasts (12–17) to identify those changes perceived important by experts. A preliminary survey form was devised, pilot-tested, and then sent to the authors of the *International Encyclopedia of Robotics*. The results of this exploratory prioritization round are presented in Table 2. Usable returns were received in time from 66 of 138 authors (48%). The purpose of this round was to focus the inquiry for succeeding Delphi rounds, and this it did. However, the results turned out quite interesting in their own right. In particular, it is interesting to note that many items are judged not to be changing significantly or to be changing in ways considered highly important to robotic development. Comments are selected to indicate specific development possibilities or to show contrasting perspectives (see Table 2).

Taking into account the priorities indicated by the expert respondents, the main Delphi forecast instrument was constructed. This was mailed to the same body of experts. Responses were tabulated and statistical response profiles composed for each item.

A final found Delphi consisted of the same items as the main round, with the addition of information on corresponding items from the 1985 Society of Manufacturing Engineers forecast of robotic developments (12) and an analysis of the experts' own responses from the main round on certain items. The final round profiles (shown in Fig. 1) were generally similar to the previous round estimations. However, for 19 of the 24 items, the variance of the responses decreased. This suggests a convergence of expert opinion.

We compared respondents associated with academia (N=21) with those from industry (N=21)—(seven could not

Table 1. Robot Attributes

Class	Parameters	States
Physical characteristics	Size	Transistor size, hand-held calculator size, human size, truck size
	Mobility	Immobile, wheels (manually or self-propelled), track, buried wire, laser code readers, free roving
	Energy sources	Hydraulic, electrical, pneumatic
	Medium of operation	Normal atmosphere, under water (high pressure), vacuum (space), extreme temperatures, zero gravity, radiation, contaminated air
Sensory capabilities	Vision	None, presence vs nonpresence, binary, grayscale, 2-D silhouette (edge or area based), 3-D depth perception
	Tactile	None, touch sensors, stress sensors
	Other	None, temperature, chemical sensors (smell/taste), radiation, radar
Manipulative	Dimensions of movement	One, two, three, six
	Grippers	Hook, parallel jaw, two-fingered, multifingered, coordinated multigrippers
	Force	Friction, physical constraint, attraction, support, pressure, suction, magnetic
	Reach/control	Point-to-point, continuous path, pick and place
Information processing capabilities	Locus of control	Central computer, self-contained
	Programming modes	Keyboard, lead through, walk through, voice-special commands, voice-English, speech-two way, self-adaptive learning
	Processing	Single function, multiple sequence, multiple simultaneous
	Self fault detection	None, yes, yes and repair

be classified into these categories). Of the twelve items that related to technological advance, the academics were more optimistic on six and the industrialists were more optimistic on six. Furthermore, on only one of these (Item 7) was the difference statistically significant at a 5% probability level. A pattern of close agreement held for most items, leading to the conclusion that the academic and industrial experts surveyed perceive the future prospects of robotics quite similarly.

The following section explores the specific final round item forecasts by the combined 49 responding authors of this *Encyclopedia* as of July 1986.

FORECASTS

Sales

Three complementary robot sales measures are addressed for the U.S. in the year 2000: industrial sales, total sales, and sales by type of application.

Table 2. Exploratory Round Assessments of Changing Robotic Technologies[a]

Class/parameter	Rate of Change (Mean)	Importance of Change (Mean)	Specific Advances (Selected Comments)
A. *Technical advances*			
PHYSICAL CHARACTERISTICS			
Servocontrol technologies	Lo Med Hi 1 2 3	Lo Med Hi 1 2 3	Fully digitized control loops. Hydraulic machines will make a comeback due to cost reductions in servo hydraulic components and need for speed/power. Smaller, stronger, more powerful, and accurate direct drive motors.
Mechanical miniaturization	1 2 3	1 2 3	Will follow the technological trend.
Alternative materials (increase payload weight; soft touch)	1 2 3	1 2 3	Medical uses (blood vessels, . . .) will provide incentive and money. Motors constructed as integral part of robot frame. Composite materials (ie, graphite, Kevlar; ceramics)
Systems resistant to hostile environment	1 2 3	1 2 3	Military and space applications, materials, plastics, coverings, sealed units, clean room, and corrosive environments.
SENSORY CAPABILITIES			
Vision, 3-D (depth perception)	1 2 3	1 2 3	Will reduce peripheral equipment costs once vision systems are reduced. "Smart" vision sensors (ie, "smart" computers connected TV cameras.)
Outside visible spectrum (infrared, uv)	1 2 3	1 2 3	Infrared for vision in dirty and dusty environments. Application of semiconductor lasers.
Sound (acoustic)	1 2 3	1 2 3	Hopefully will eventually respond to spoken directions.
Proximity (sonar/radar)	1 2 3	1 2 3	Optical radar will be used for vision not proximity. This has been judged as expensive way of measuring proximity.
Force (tactile)	1 2 3	1 2 3	Widespread use of mode-switched force control. High resolution, reliable, no hysteresis.
Radiation sensitivity	1 2 3	1 2 3	Robot should be capable of more efficient environment analysis in a myriad of unexpected situations.
Temperature	1 2 3	1 2 3	Needs to be done for manufacturing purposes, but does not yet have high priority.
Chemicals	1 2 3	1 2 3	
END-EFFECTOR CAPABILITIES			
Manipulation (gripper design)	1 2 3	1 2 3	Advances needed: doorknobs; key locks; switches; dials; phones. More general purpose grippers needed. Standardization.

Table 2. *(continued)*

Class/parameter	Rate of Change (Mean)	Importance of Change (Mean)	Specific Advances (Selected Comments)
Modification (cut, grind, weld)	Lo Med Hi 1 2 3	Lo Med Hi 1 2 3	Factory uses closer to NC machines tools.
Force, load capacity	1 2 3	1 2 3	Trend will be increased payload.
Force, precision	1 2 3	1 2 3	Force control (sensing and servocontrolled grippers).
Reach/control, accuracy	1 2 3	1 2 3	Redundant manipulators and the redundancy inherent in coordinating multiple robots will have an impact on these categories.
Reach/control, repeatability	1 2 3	1 2 3	Ideal: robots with machine tool capabilities. Automatic assembling robot. Direct drive motor important in reach/control, speed coupled characteristics.
Reach/control, speed	1 2 3	1 2 3	
Coordinated multiple effectors	1 2 3	1 2 3	Should be able to coordinate multiple functions at once. Already being done.
INFORMATION PROCESSING CAPABILITIES			
Computer hardware capabilities	1 2 3	1 2 3	Multiprocessor systems with distributed control and adaptive algorithms.
Computer software capabilities	1 2 3	1 2 3	Standardization; integrates with other systems.
Computer software incorporating AI	1 2 3	1 2 3	Task-level programming.
Real time processing (adapt to environment	1 2 3	1 2 3	Necessary for robots in unstructured environments. Higher performance hardware will enable grater sophistication in real time control algorithm.
Image processing (scene interpretation)	1 2 3	1 2 3	Connected with new methods of recognition.
Self-fault detection	1 2 3	1 2 3	Fault detection simulators, on line fault detection. Formulation of control algorithms not AI.
Self-fault repair	1 2 3	1 2 3	Emphasis will and should be on not needing repair. The technology to achieve automatic fault handling is available now, but will not be rapidly applied due to the increasing reliability of hardware.
Process planning (integration with other units)	1 2 3	1 2 3	Manufacturing automation protocol (MAP) is essential. Process planning integration is bound to happen. Standard interfaces and languages.
TECHNICAL SYSTEMS ISSUES			
Force mechanisms (friction, attraction, suction)	1 2 3		Sufficient force to subdue and take into custody a violent, human, armed subject without injury (eg, a police tactical robot.)

Table 2. *(continued)*

Class/parameter	Rate of Change (Mean)	Importance of change (Mean)	Specific Advances (Selected Comments)
Mobility (stationary, tracted, wheels, free-roving)		Lo Med Hi 1 — 2 ▲ — 3	Free roving robots are necessary if robots are ever to be widely used outside factories. High mobility needed: stair climbing, room searches, etc.
Sensing (discrete sensors to imaging systems)		1 — 2 — ▲ 3	Multiple-design integrated-sensing systems needed.
Controls (point to point or continuous path)		1 — 2▲ — 3	Accuracy rather than path is most important.
Locus of control (self-contained to central control network with other machines)		1 — 2 ▲ — 3	Inclusion in overall factory systems is vital.
Operational programming (walk through, off-line, real-time)		1 — 2 ▲ — 3	Off-line programming is going to be the biggest breakthrough in robotics history. Need methods for arm calibration.
ADOPTION ISSUES			
Degree of integration into systems (vs. stand alone uses)		1 — 2 — ▲ 3	Technology integrators are the key to making robots work. Both will develop.
Sophistication level (more advanced robots operate with simpler models or replace them)		1 — 2▲ — 3	The simpler the better. Emphasis on doing more tasks and integrating with other robots.
Application domain (importance of uses outside industry, such as military, domestic)		1 — 2 ▲ — 3	The household domain is the most difficult, but most important to spur advances. Food preparation, health settings, office. Robots must be successful in unstructured environments before their full benefits can be realized. Military use will supply large research funds.
Safety		Lo Med Hi 1 — 2 ▲ — 3	Of importance, but not as important as profit and quality.
Labor (unemployment, supervision)		1 — ▲ 2 — 3	Robot will not have a significantly adverse effect on employment levels.
Reorganization of production to make full use of robots		1 — 2 — ▲ 3	Also, the redesign of products to facilitate automation of manufacturing and assembly. Robot use must match products, not vice versa. This is not the problem. High need but the lunkheads will not move.
Demand stimulus (market demand relatively given or expandable by new technology)		1 — 2 ▲ — 3	Not enough stimulus.
Marketshare (U.S. exports/imports share in U.S.)		1 — ▲2 — 3	Exports/imports very alarming for future. Not the real issue.
Industry structure (type and number of companies, integration, producers also users)		1 — 2▲ — 3	Too much fragmentation and confusion; not enough customer support; not enough product development (performance and reliability); not enough R&D; robot companies are too small to afford it.

[a] See Methods section for discussion of this survey. Responses for rate of change and importance of change are the means (66 respondents). Specific advances presents selected comments, sometimes chosen for counter perspectives, sometimes for highlighting a particular change.

ITEM — **Final Round Response Profile**

1. Estimate the number of *industrial* robots to be sold in the U.S. in the year 2000. For comparison, use 4200 as an estimate of the number sold in 1984.
[SME estimate for the year 2000 = 20,000.]

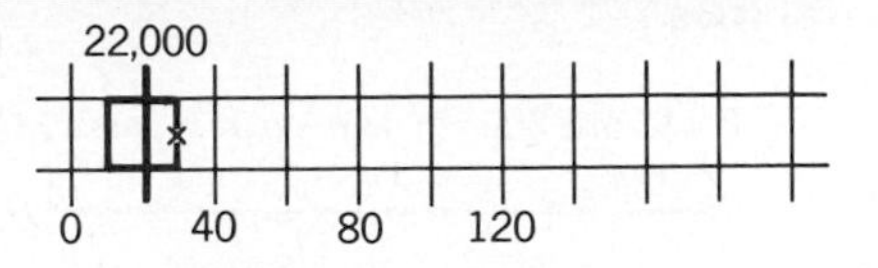

2. Estimate the number of robots sold in the U.S. in the year 2000 for *all* uses (industrial, military, household, etc)

3. Estimate the distribution of industrial robot sales in the U.S. in the year 2000 by *application*. For comparison, use the following 1985 estimates.

A. Welding (spot and arc) and painting (coating)—46% in 1985
[SME estimate for 1995 = 26%]

B. Material handling and machine tending—32% in 1985
[SME estimate for 1995 = 30%]

C. Assembly and inspection—16% in 1985
[SME estimate for 1995 = 36%]

D. Processing (drilling, grinding, etc) and other—6% in 1985
[SME estimate for 1995 = 8%]

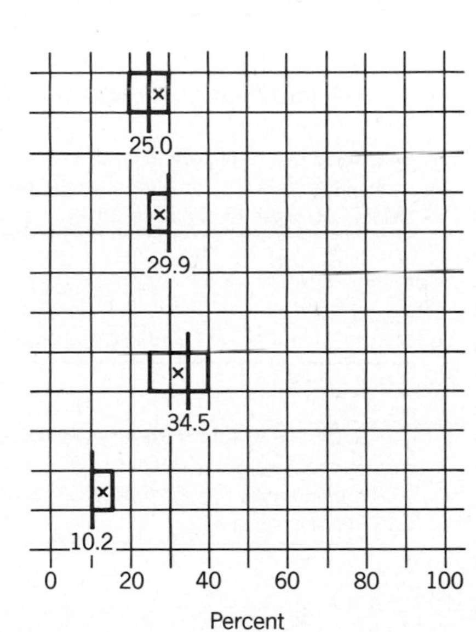

[NOTE: The %'s estimated for Items 3A–3D above should sum to 100.]

4. Estimate the percent change in cost from 1985 to 2000 for comparable capabilities (in constant dollars) for

A. Programmable, servo-controlled, single-arm robot

B. Gray scale vision system

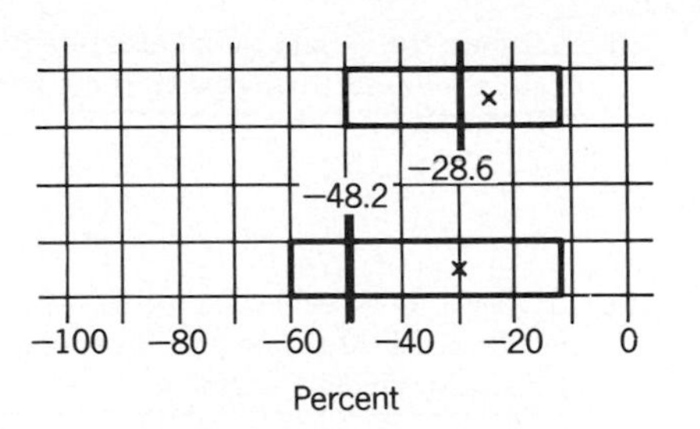

[NOTE: Please indicate plus(+) or minus(−) for Items 4A and B above.]

[The following series of questions (#5–16) all pertain to the percent of *industrial* robots sold in the U.S. in the year 2000:]

5. Estimate the % sold for use in a facility designed expressly to make full use of robotics.

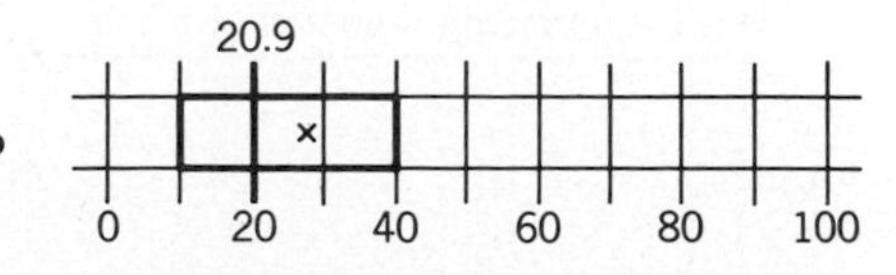

Figure 1.

[Questions 6, 7, and 8 concern mutually exclusive *vision* capabilities.]

6. Estimate the % sold with a basic, binary scale machine vision system.

7. Estimate the % sold with advanced machine vision systems (eg, gray scale, 3-D, and/or color)

8. Estimate the % sold without vision systems

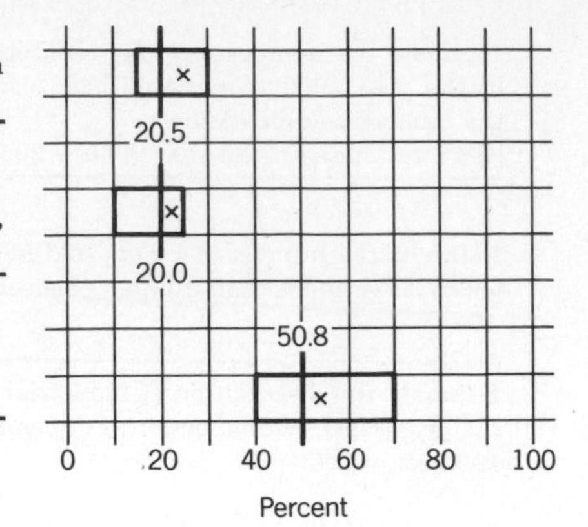

[Items 6, 7, and 8 above should sum to 100%.]

9. Estimate the % of robots sold with imaging systems for real time scene analysis (eg, to pick a part from a bin of mixed parts).
[SME estimate for 1995 = 18%]

10. Estimate the % sold with tactile (touch or stress) sensors.

11. Estimate the % sold with any sort of *non-contact, non-vision sensing devices* (eg, proximity, range).
[SME estimates for 1995 = 40% for Japan and 30% for Canada and Western Europe]

12. Estimate the % sold with two (or more) coordinated arms.
[SME estimate for 1995 = 15%]

13. Estimate the % sold to be integrated into systems (eg, work cells, flexible manufacturing) with some central computer controls.

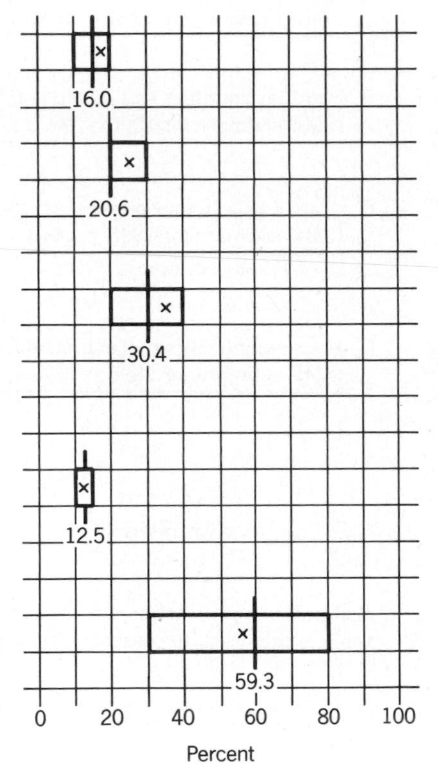

14. Estimate the % sold compatible with the Manufacturing Automation Protocol (MAP) or a direct successor standard for integration of manufacturing systems.

15. Estimate the % sold that will be programmed substantially off-line.
[SME estimate for 1995 = 50%]

16. Estimate the % sold using microprocessors as distributed controllers for particular sensors or joints.

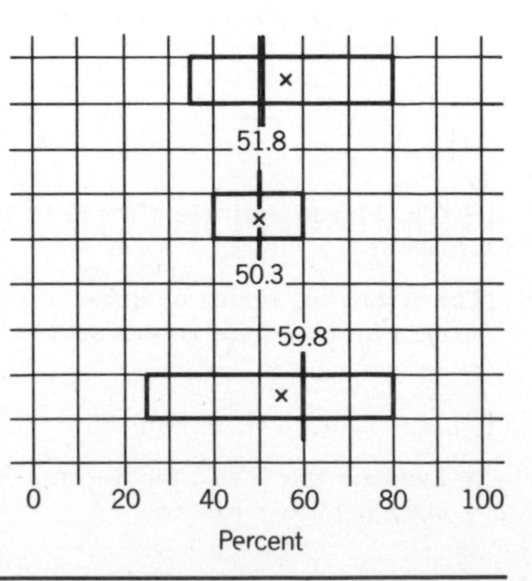

Figure 1. *(continued)*

17. Estimate the % of robots sold in the U.S. in the year 2000 for *all* uses (industrial, military, household, etc) with some degree of self-propulsion (mobility).
[SME estimate for 1995 for *industrial* robots = 29%]

18. Estimate the % of mobile robots sold in the U.S. in the year 2000 for *all* uses capable of real time navigation in a dynamic environment (ie, if all the mobile robots estimated in #17 have such capability, answer 100% to #18.

19. What will be typical reach accuracies (in inches) of large capacity (100 lbs or more) industrial robots sold in the U.S. in the year 2000? Take 0.04 inches as a 1985 estimate.
[SME estimate for 1995 = 0.02 in.]

20. What is the probability of having a standardized *off-line* programming language for robots in extensive use in the year 2000? (express as %)

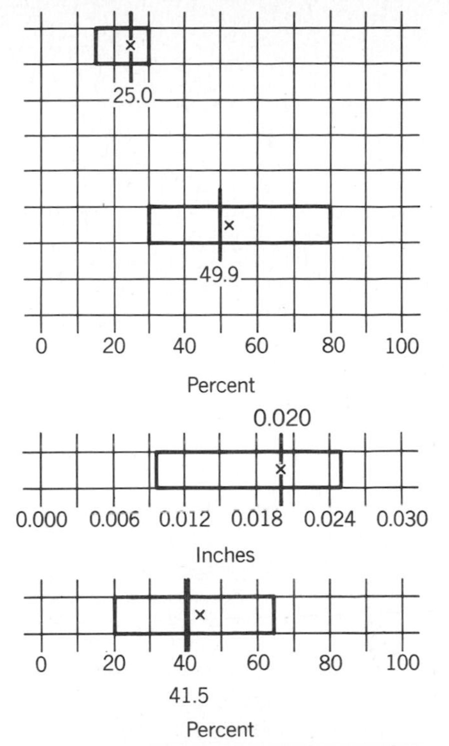

Figure 1 (*continued*). Final round response profile. The twenty items presented to the experts (authors of the *International Encyclopedia of Robotics*) for their judgment are reproduced here. To the right, the 25th and 75th percentile responses are shown as the edges of a "box." That is, half of the 49 responses received fall within this range. For instance, on Item 1, 25 respondents estimate between 15,300 and 29,900 industrial robots will be sold in the U.S. in the year 2000. The line through the middle of the box designates the median (50th percentile) response, also indicated numerically. For instance, on Item 1 this middle response is 22,000. The mean value of the responses received is shown by an "X." Additional information was provided to the respondents on the final round only for certain items. The source was the 1985 Society of Manufacturing Engineers/University of Michigan study entitled *Industrial Robots: Forecasts and Trends* (designated below as "SME estimates"). Except for Item 1, these estimates pertain to the year 1995.

Industrial sales are projected to grow from 4,200 units in 1984 to 22,000 (median estimate) for 2000 (Fig. 2). As shown for Item 1 in Figure 1, the mean estimate (29,200) is considerably higher than the median estimate. This reflects a skewed response distribution, with some very high estimates (up to 150,000). However, only 26% of the respondents favored a value as high as 29,000.

The *interquartile range* (*IQR*) describes the middle half of the responses (the edges of the boxes drawn in Figure 1 and 2). For industrial sales, the 25th percentile is 15,300; the 75th percentile, 29,900. The IQR gives a good indication of the variability of responses. It is less sensitive to outlier estimates than the standard deviation. For Item 1, the standard deviation is 26,900, considerably more than the IQR. However, for 18 of the 24 items, the standard deviation is less than the IQR, so generally the IQR indicates a wider range than does the standard deviation.

These results can be translated into a compound annual growth rate by assuming constant growth through 2000. The median industrial robot sales estimate implies an 11% compound growth rate from 1984 to 2000. The IQR is quite narrow; 8.4–13% annual growth. The SME forecast of 20,000 unit sales for the year 2000 implies a comparable 10% annual growth. *Predicasts* reports five forecasts of unit sales that appear independent of each other. The only year 2000 projection reflects 9.7% annual growth. Projections from 1985 to 1995 that appear in the *Predicasts* compilation suggest annual growth rates of 10.3, 23, 24, and 27.4%. Present results suggest that those high-growth estimates may be excessively optimistic. On reflection, the best estimate for the range of industrial robot sales growth appears to be 9–13% compounded annually.

Presumably most robots sold over this time period will remain in operation. The industrial robot population will become quite substantial by the Year 2000, suggesting strong prospects for robot service and used robot sales. A demand for less expensive and less sophisticated robots should complement the development and marketing of robots with highly sophisticated capabilities. Given the diverse roles robots can play, there is good reason to expect many robotic capabilities to be represented in the workplace.

Total sales are forecast to be 50,000 units in 2000 (Fig. 1, Item 2 and Fig. 2). However, the uncertainty in this estimate far exceeds that for industrial robot sales. The IQR extends from 33,000 to 80,000. The explanation for this range comes

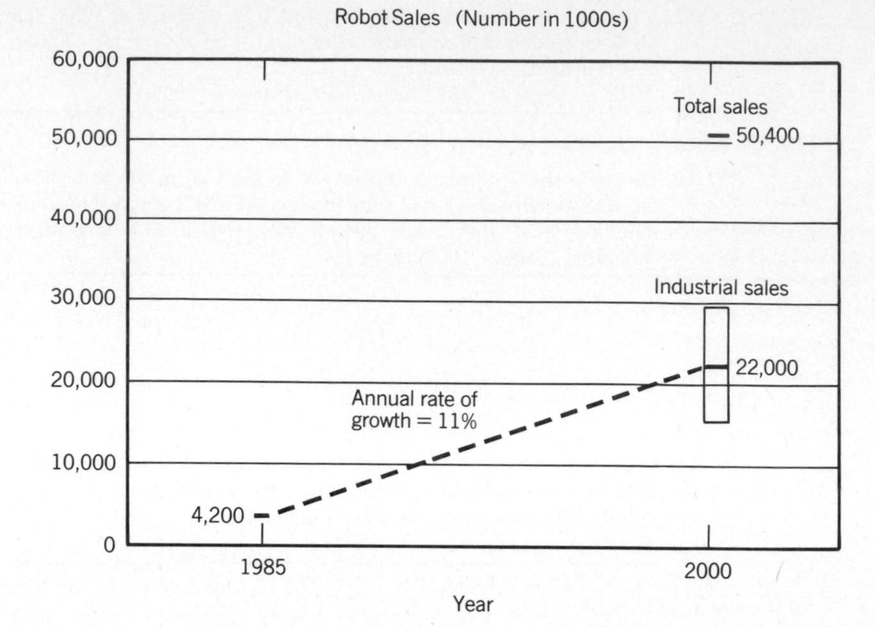

Figure 2. Robot sales.

from consideration of the three components that make up "total robot sales": industrial, military, and household. Our median industrial estimate (22,000) is smaller than the combined military and household shares (28,000) in the total of 50,000. Given that sales today are virtually all industrial, that implies greater growth rates in the nonindustrial arenas. Predicasts' compilation for military robots locates only one forecast for approximately 20% annual sales growth from 1985 to 1994. Their lone household forecast only runs through 1990, yet it postulates sales of 750,000 units! That apparently ridiculous forecast makes one wonder just what a "household robot" is, since the sensory sophistication and ability required for domestic tasks appears to exceed that for industrial tasks. Comments of two of our respondents convey some views of those markets:

> "Industry will be the primary user, military next; household is a long way from being developed as a market."
>
> "Many cars, trains, etc, will be sufficiently automated to be considered robots. If household robots become popular so that tens of thousands are sold, then millions will be sold."

These estimates suggest that development of household and military robotics may increase in importance after 1990 and should be monitored.

Our emphasis in the present forecast, however, lies with industrial application. (Indeed, with the exception of Items 17 and 18 addressing robot mobility, all the Delphi estimates pertain directly to industrial application.) Within industry, changes in the *application* of robotics are of considerable interest. The profile projected by our experts for the year 2000 contrasts with that of 1985 (see Fig. 3):

A. Welding and painting; substantial drop from 46 to 25% of all applications.
B. Material handling and machine tending; little change from 32% in 1985 to 30% in 2000.
C. Assembly and inspection, dramatic increase from 16 to 35%.
D. Processing and other, increase from 6 to 10%.

One might note also (Fig. 1, Items 3A–D) that the IQRs for these estimates are relatively tight; the relative importance of application areas appears quite likely to be reordered.

Cost

Cost estimates require careful interpretation. For instance, Predicasts reports several forecasts of changes in wholesale

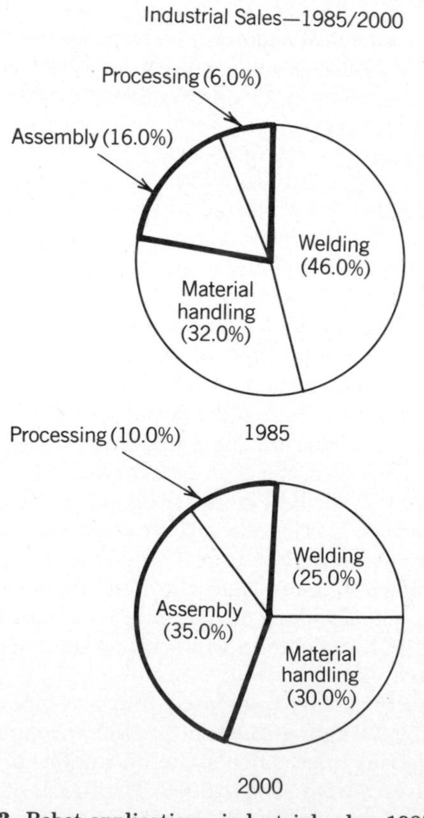

Figure 3. Robot applications, industrial sales, 1985/2000.

prices. One forecast for industrial robot prices through 1995 projects a 4% annual increase; another, a 3% annual decrease. The key here appears to be the associated changes in capabilities. The wholesale price in the former forecast may well be projected to increase in conjunction with increasing robot sophistication.

Items 4A and B posed the cost issue to our experts under the assumptions of comparable capability and constant dollars. Their median projection through 2000 is a for a 29% drop in cost for a basic, programmable, servocontrolled, single-arm robot, and a sharper 48% drop for a gray scale vision system per se (shown in Fig. 4). These translate to annual cost decreases of about 2 and 4% respectively. For comparison, forecasts compiled by Predicasts anticipate an annual 1% annual decline in cost for materials handling robots through 1995 and an 8% annual decline for a robot with vision through 1990. These projections imply that mechanical aspects will show only modest cost improvements, whereas aspects dependent on computer processing, in particular, and sensing will evidence substantial cost improvements. In the long run, therefore, the tremendous price gap between basic and sophisticated capability robots should narrow. For example, in the Predicasts forecast for robots with vision just noted, the price is projected to drop from $150,000 to $100,000 between 1985 and 1990. In the material-handling robot forecast, the price is projected to drop only from $35,000 to $31,000 between 1985 and 1995. Also, as one of our respondents pointed out, "This is hardware and software that will reduce drastically in price as it is mass produced."

Systems and Software

The popular image of robots as stand-alone hardware items is falling ever farther from reality. Industrial robots are increasingly being integrated into production systems. Accomplishing such integration places increasing demands on software for control.

Several items in the present Delphi focus on systems and software issues. Perhaps the most significant issue is the extent to which robots are intended to be integrated into work systems. Item 13 (Fig. 1) predicts 59% will be sold for some form of work cell with central computer controls in the year 2000. Item 5 goes a step beyond to address the issue of sale for use in a facility designed expressly to make full use of robotics. The median estimate is that 21% will be targeted for such an automated facility. The industrial robot will become a significant component in computer-integrated manufacturing.

Integration of manufacturing systems implies a great need for software development. Item 14 inquires as to the percent of robots sold in 2000 that will be compatible with the manufacturing automation protocol (MAP) or a direct successor standard for integration of manufacturing systems. Our respondents estimate 52%. Some noted that this is not an "either/or" issue. MAP may not prove to be a generally valid option, or several major standards may emerge.

Programming for robots must take place at levels ranging from individual motions to integration within systems. Two items concern the extent that programming will be done off-line. In Item 15, our experts estimated that 50% of the robots sold in 2000 would be programmed substantially off-line. In Item 20, they estimated a 42% probability that a standardized off-line programming language would be in extensive use for robots in 2000.

With the exception of Item 15 (substantial off-line programming), all of these systems and software items show a broad response range. So, on the one hand, development of effective robotic industrial systems is considered both important and probable. On the other, there is considerable uncertainty as to the form and extent to which this will be put into practice.

Configuration and Capabilities

Robots, by definition, employ manipulative capabilities. Increasingly, however, these capabilities are complemented by sensory features. We posed a number of queries to our experts on the development of robot sensation.

Locus of control is an aspect subject to change with the

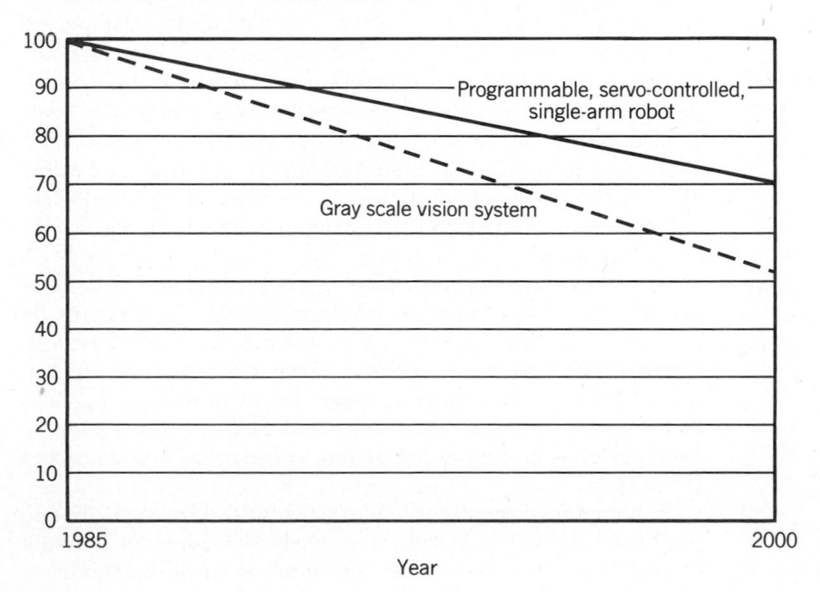

Figure 4. Changes in cost, constant capabilities and dollars.

continuing development of microprocessors. Our respondents estimate that 60% of the robots sold in 2000 will utilize microprocessors as distributed controllers for particular sensors or joints (Item 16).

Reach accuracies are expected to improve. Large capacity industrial robots sold in 2000 should be twice as accurate as their 1985 counterparts (Item 20; accuracy of 0.02 in. (0.06 cm) compared to 0.04 in. (0.1 cm) as a 1985 estimate).

Given expected improvements in control, it seems reasonable to anticipate more complex robot manipulative capabilities. Item 12 posed one such feature; the presence of more than one arm on a robot. Respondents were in strong agreement that such robots would be sold, but would constitute only 12.5% of robot sales in 2000.

Turning to robot sensation, Items 6–8 addressed vision capabilities. Responses divided as follows:

- Robot sold without vision systems (median estimate, 51%: mean, 54%)
- Robot sold with a basic, binary scale vision system (median, 20%: mean, 24%)
- Robots sold with an advanced machine vision system (eg, gray scale, 3-D, and/or color) (median, 20%; mean, 22%).

Almost half of the industrial robots sold in 2000 are expected to have vision systems. As indicated in Figure 1, the IQRs are relatively tight.

We also asked about other sensory capabilities. Our respondents anticipate some 21% of industrial robots sold in the year 2000 to have tactile (touch or stress) sensors. They estimate about 30% will have some sort of noncontact-, nonvision-sensing devices (eg, proximity, range). The IQR for both items (10 and 11) is relatively tight. This suggests that our experts take nonvision sensing very seriously. Sensory packages incorporating multiple capabilities are likely to be significant.

One vital use of vision is for imaging systems that can perform real time scene analysis. Our experts predict some 15% of the industrial robots sold in 2000 will have this feature.

The advent of advanced sensory capabilities will, in turn, support robot mobility. Items 17 and 18 address mobility for all robot uses (not just industrial). Our experts anticipate 25% of all robots sold in the U.S. in 2000 will have some degree of self-propulsion (IQR, 15–30%). They estimate that around 50% of these mobile robots will have real time navigation capabilities for a dynamic environment (IQR, 30–80%).

DISCUSSION

Stepping back from the specifics of this forecast, one senses an expectation of steady progress in robotics. This pertains to both advances in technical capabilities and to the adoption of robots by users. The authors of the *Encyclopedia* who participated in this forecast expect through the year 2000:

- about 11% annual growth in industrial robot sales,
- increases in military/household robot sales,
- steadily decreasing costs of robotic systems (decreasing faster in computer-based areas than in mechanical ones),
- increasing incorporation of robotics into integrated industrial systems,
- standardization of programming,
- steadily improving control and manipulation,
- widespread development of robotic vision systems,
- significant use of other (and multiple) sensory systems, and
- real time imaging and mobility applications.

This is not a prediction of spectacular growth. The annualized rates of change are actually rather modest. In contrast, forecasts made about 1979–1981 called for 35% annual growth rates (18–20). When robot sales dipped below expectations (during the subsequent recession), forecasts were adjusted downward. Forecasts were scaled down to a 25% annual growth level (21). In 1984, robot sales rebounded for 58% growth over 1983 (22). *The Wall Street Journal* (23) reports marked decrease in robot sales expectations and sizeable layoffs by robot manufacturers. Using the U.S. industrial robot sales estimate of 4200 for 1984, 25% annual growth would lead to sales of 150,000 in the year 2000; 35% growth would lead to 500,000. In our forecast, 11% leads to 22,000.

Moderate, steady growth in robotics will cumulate to produce major impacts on industrial (and ultimately military and household) operations. Perhaps the most intriguing question raised by the present forecasts concerns the extent of household application. If this begins to accelerate, the implications will be immense.

This forecast has relied on expert judgement because of the complex array of factors that affect robotic development (as discussed in the Introduction). At this point, it seems useful to examine certain of these factors explicitly, to reflect on their implications for the future of robots in the U.S.

Focusing on industrial innovation (ie, adoption of robots), four key influences deserve attention: technological capabilities; foreign competition; economic activity; and worker displacement and political pressures.

This forecast paints a positive picture for several important technological capabilities (vision, controls, etc). One can identify contributing developments, in particular, computing power, that should enhance robotic capabilities. More specific technical areas (such as grippers, force mechanisms, intelligent programs) are too fine-grained for this article, but their prospects are illustrated throughout the *Encyclopedia*. Are there any competing technologies apt to sidetrack robotics? None stand out as obvious threats. In sum, the technological climate seems extremely supportive of continuing steady robotic expansion.

An interesting feedback loop exists between robot technological capabilities and sales. Robot sales "pull" for and provide capital for further investment in technological development. Further technological development provides improved capabilities, which induce expanded sales. In certain areas, particularly household applications, one could imagine such a positive feedback cycle either not happening or becoming a potent force for change.

Robotic technology is likely to move ahead in steps. At one level (say, the variable sequence robot), development is likely to accelerate to a point, then taper off as target capabilities are achieved or physical limits are approached. This can be depicted as a classic "S-shaped" or logistic growth curve. However, another level (say, an intelligent robot with certain sensory capabilities) may become feasible, improve rapidly, then, too, taper off in the rate of improvement. Such a sequence of technological steps can be visualized as a series of S-shaped curves that build on one another. Such a growth profile seems likely for robotics. One should expect an "envelope" of spurts and slowdowns progressing over time as the fine-grained pattern of a long term upward trend.

Such a technological progression will not be singular in nature. At any given point in time, one is likely to see different levels of robotic technology working together in a complementary fashion. Furthermore, different emphases may dominate in different market niches. A few years ago, the electrical machinery industry emphasized point-to-point, servo-controlled robots, but the fabricated metals industry relied more on nonservo's and continuous-path, tape-controlled robots (24).

Foreign industrial competition provides an unmistakable driving force for U.S. automation. While Japan leads in the adoption of robotics, Western European nations, such as France, have demonstrated coordinated industry-government commitment to advancing robotics and automation (10). Foreign competition can be met either by building steep trade barriers to protect American industry or, more effectively, by improving U.S. productivity. Another aspect of foreign competition pertains to production of robots *per se*. One intriguing prospect is the fusing of efforts, as in the case of IBM marketing Japanese-made robots in conjunction with IBM computers. The main point for the future of robotics is that foreign competition assures continuing U.S. commitment to develop and use robot technology.

Although long-term momentum should be sustained by continuing technological improvements and foreign competition, economics will dictate short term fluctuations in robotics. As noted, the 1980s have already witnessed significant peaks and valleys in U.S. robot sales growth. The state of the economy, as such, certainly affects capital investment. Changes in tax treatment, major military programs, and the fortunes of particular sectors of the economy will all make their mark on robotic development over the coming decade and a half. In addition, the structure of the robotics industry itself deserves attention. Consolidation of firms could augment R&D capabilities. Integration of machining capabilities with information processing skills should boost robot development prospects.

The interplay of automation and economic activity may hold the key to the extent of robotic applications in the long term. Labor displacement is the key concern in the use of robots (and other facets of automation). Of course, robots are introduced for many reasons besides labor saving. Increased accuracy, reliability, and capabilities beyond those of humans (eg, strength, resistance to hazardous environments to name a few). Nevertheless, prospects for major dislocations cannot be ignored. These are not likely to be evened out by gains in employment of technicians, programmers, or maintenance workers to support robotics. Nor is industrial demand likely to rise so dramatically that it will counterbalance productivity gains. The implications of robot-displacement depend on the employment context. Should jobs be plentiful, there will be manageable relocation problems. Should unemployment continue a long term increase, possibly toward extreme levels (25), strong political pressures could be brought to bear to curb automation. Government restrictions, costly retraining requirements, and popular resentment could curtail robot growth.

In sum, we have presented a forecast of selected robotic parameters through the year 2000. This forecast does not pretend to provide fine detail on what will occur. It does pose change vectors worthy of serious consideration by those charting robotic development and application plans.

BIBLIOGRAPHY

1. J. P. Martino, *Technological Forecasting for Decision Making,* 2nd ed., North Holland, New York, 1983.
2. A. L. Porter, F. A. Rossini, S. R. Carpenter, and A. T. Roper, *A Guidebook for Technology Assessment and Impact Analysis,* North Holland, New York, 1980.
3. J. R. Bright, *Practical Technology Forecasting,* The Industrial Management Center, Austin, Texas, 1978.
4. H. Jones and B. C. Twiss, *Forecasting Technology for Planning Decisions,* Petrocelli Books, New York,1978.
5. W. Ascher, *Forecasting: An Appraisal for Policy Makers and Planners,* The Johns Hopkins University Press, Baltimore, 1978.
6. Predicasts, Inc., *Predicasts Forecasts,* Cleveland, Ohio, July 31, 1986.
7. H. A. Linstone and M. Turoff, *The Delphi Method: Techniques and Applications,* Addison-Wesley, Reading, Mass., 1975.
8. H. Sackman, *Delphi Assessment, Expert Opinion, Forecasting and Group Process,* Rand Corporation, Santa Monica, Calif., Report R-1283-pl, April, 1974.
9. K. R. Nelms and A. L. Porter, "EFTE: An Interactive Delphi Method," *Technological Forecasting and Social Change* **28,** 42–61 (1985).
10. A. L. Porter, F. A. Rossini, J. Eshelman-Bell, D. D. Jenkins, and D. J. Cancelleri, "Industrial Robots—A Strategic Forecast Using the Technological Delivery System Approach," *IEEE Transactions on Systems, Man and Cybernetics,* **SMC-15,** 521–527, (1985).
11. F. Zwicky, *Morphology of Propulsive Power,* Society for Morphological Research, Pasadena, Calif., 1962.
12. D. N. Smith and P. Heytler, Jr., *Industrial Robots Forecast and Trends,* Society of Manufacturing Engineers, Dearborn, Mich. and University of Michigan, Ann Arbor, Mich., 1985.
13. M. P. Groover, M. Weiss, R. N. Nagel, and N. G. Odrey, *Industrial Robotics,* McGraw-Hill, New York, 1986.
14. Georgia Institute of Technology, *Robotics for the 21st Century,* Conference, Atlanta, Ga, Feb.1986.
15. V. D. Hunt, "The Future of Robotics" in *Industrial Robotics Handbook,* Industrial Press, Inc., New York, 1983.
16. S. M. Miller, *Impacts of Robot Use on Manufacturing Labor and Costs* to be published by The University of Wisconsin Press.
17. B. Everett and R. L. Jenkins, *Robotic Technology in Ship-building Applications,* Naval Sea Systems Command, Washington D.C. July 1985.
18. *Technology Assessment: The Impact of Robotics,* report to the National Science Foundation, Grant ERS-76–00637, Eikonix Corporation, Cambridge, Mass., 1979.

19. Bache, Halsey, Stuart, and Shields, Inc., private communication, New York, 1980.
20. Department of Engineering and Public Policy and School of Urban and Public Affairs, *The Impacts of Robotics on the Workforce and Workplace,* Carnegie-Mellon University, Pittsburgh, Pa., June 1981, p. 4.
21. Predicasts, Inc., *Predicasts Forecasts,* No. 95, B-161, Cleveland, Ohio, April 30, 1984.
22. G. Kaplan, "Industrial Electronics," *IEEE Spectrum* **22,** 68–71 (Jan, 1985).
23. *The Wall Street Journal,* 1, (Sept. 18, 1986).
24. J. J. Obzrus, "Robots Swing into the Industrial Arms Race," *Guilton's Iron Age,* 51 (July 1980).
25. A. L. Porter, "Work in the New Information Age," *The Futurist* **20,** 9–14 (1986).

The authors wish to thank Professor Stephen L. Dickerson for his assistance in clarifying critical robotic parameters. They also thank Michalina Bickford and her associates at John Wiley for administering the Delphi rounds.

G

GANTRY ROBOTS

Elan Long
CIMCORP Inc.
Aurora, Illinois

Gantry robots are built in a wide variety of configurations although they are all distinguished by a box-like structure which gives them particular flexibility in application and performance. The gantry robot's ability to straddle a workstation or assembly line and to travel long distances makes it particularly cost effective. One gantry robot can replace four or more pedestal robots in a work area. Improvements in machine design have allowed gantry robots to move beyond simple material-handling operations to precision machining and other processing applications.

EVOLUTION

The advantages of overhead operation were recognized by many people who installed floor-mounted robots overhead on tracks or beams to perform operations that could not be accomplished from floor level. The most significant source of gantry-robot technology was in the nuclear industry where remote manipulators translated human motion into mechanical motion in a hostile environment. As gantry-robot technology has advanced, applications have broadened from strictly material-handling operation to precise machining and process operations.

Overhead Cranes

Overhead cranes and hoists have been widely used in manufacturing plants for many years to assist in moving heavy or awkward parts, subassemblies, and machinery. A gantry crane moves on wheels that roll along the floor. Gantry robots are driven along elevated rails similar to bridge cranes where wheels roll along an overhead track. The nomenclature of robotics has not been consistent with that of cranes as this style of robot is referred to as "gantry robot" or even "overhead gantry robot" rather than "bridge robot." Standard industrial cranes and hoists actually played only a small role in the development of gantry robot technology.

Track-Mounted Pedestal Robots

The conceptual benefits of working from above were recognized by users of standard pedestal-based or floor-mounted robots in the early 1980s. With the development of electric-drive robots came the ability to mount standard robots above the work area on either stationary frames or on slides. The slides added an extra axis to the robot's motions, allowing it to traverse the work area as required. Applications for slide-mounted pedestal robots included automotive sealant application, ink-jet marking of lay-up patterns for composite-material aircraft parts, turbine-blade handling and inspection, and welding.

Nuclear Industry Remote Manipulators

The most significant influence on the development of gantry-robot technology was the nuclear-power industry. Teleoperators or remote manipulators have been in use in the nuclear industry since the mid-1950s. Early devices were fully manual master-slave remote arms but these developed into servo-driven manipulators. The purpose of these remote devices was to allow the human operator to perform tasks from outside of the hostile environment of radioactive materials.

Teleoperators were not designed to be programmed for repeated performance of an operation. Instead, designers focused on translating human motions accurately into mechanical motions. The operator provided all of the required decision-making capability in the unstructured environment where results were not always predictable, operations were not regularly repeated, and intelligent responses to unexpected events were critical. The teleoperator systems extended people's reach and manipulation abilities into the remote environment (see Fig. 1). Early pioneers in the application of remote manipulators to nuclear facilities were Argonne National Laboratory, General Mills, and PaR Systems (now a division of CIMCORP Inc.).

In 1981, PaR Systems began applying its teleoperator-system technology to industrial robotics and introduced a gantry-robot system. In the early 1980s, large gantry robots were developed and marketed by companies such as PaR Systems, Mobot, Niko, and Durr Industries. As gantry-robot technology developed and applications were proven other companies such as GMF, Cincinnati Milacron, Cybotech, and American Robots also offered gantry robots as part of their product lines.

Gantry Robots for Electronic Assembly

Small gantry robots or tabletop gantry robots were developed for electronic assembly operations. IBM introduced the 7565 Manufacturing System in 1983 (see Fig. 2). A similar model was offered by Nova Robotics. These small gantry robots operated at speeds up to 40 inches per second (102.5 cm/s) and handled payloads up to 5.0 lb. (2.26 kg) in a six-axis configuration. Although their structure made them inherently stiffer than other assembly robots, the small gantry robots met with limited commercial success. An assembly robot is required to have a straight-line vertical motion to work on parts located in a horizontal plane. The gantry, Cartesian, and SCARA configuration robots all meet this technical criteria. The small gantry robot did not offer the precision of the SCARA robot which was in a later stage of development, nor was it as easily applied to existing work layouts. SCARA robots could be added onto an assembly line without necessarily changing the flow

Figure 1. Gantry-type remote manipulator installed in a nuclear facility.

of operations but the tabletop gantry required the work to be brought into its work envelope.

DESIGN AND OPERATION

Coordinate Systems

Gantry robots are basically Cartesian coordinate systems. Standard industrial models range from two- to six-axis configurations and additional axes of motion may be added, particularly in the end-effector design. The first three axes are Cartesian x-y-z planes; rotational motions are alpha, beta, and gamma corresponding to yaw, pitch, and roll in cylindrical coordinate systems. Controllers for gantry robots ideally have the ability to work in Cartesian, cylindrical, and world (or tool) coordinates to facilitate programming of positions. Although gantry robots of more than three degrees of freedom use all three coordinate systems, they are generally designated as Cartesian or Cartesian-cylindrical robots.

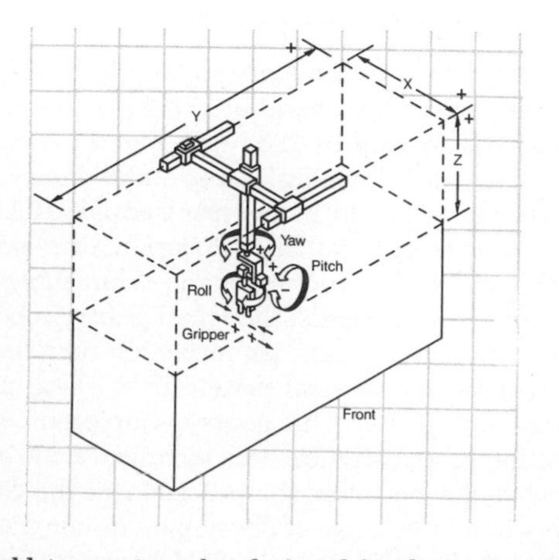

Figure 2. Tabletop gantry robot designed for electronic assembly operations.

Structural Components

Gantry robots are designed with a structure that is box-shaped. The structure consists of four beams joined to form a rectangle and supported in the air by four or more legs. Additional support legs are added in pairs when the length of the structure might cause the beams to sag. The structure is an integral but passive component of the system. Mounted along the longitudinal or x-axis of the structure are the side rails or runway. Motion along the x-axis is the result of travel by the bridge assembly along the side rails. The bridge also supports the carriage assembly (see Fig. 3). In some gantry robots the bridge assembly is replaced by a single beam or monorail. These robots are called half-bridge or monorail robots.

The carriage assembly provides transverse or y-axis motion as it travels along the bridge rails. The carriage also acts as the support structure for the z-axis or robot arm. The vertical motion of the robot is accomplished by raising or lowering the arm, which may be a single mast or a set of telescoping tubes. The single mast or monomast structure is a simpler construction and is more generally applied. The telescoping tube design is used where headroom is limited since a mast construction requires clearance above the structure equal to the z-axis travel. In nuclear facilities telescoping tubes were required because travel distance down to fuel rods was very long and head room expansion was not possible. The monomast design is more rigid and provides tighter tolerances in both repeatability and accuracy specifications.

Multiple Bridge Robot Systems

Multiple bridge gantry robots have two or more bridge and carriage assemblies which share one structure. The bridges travel along the same side rails for x-axis motion and each is theoretically able to cover the entire work envelope. In practice, work area for each robot arm may be limited by hardware

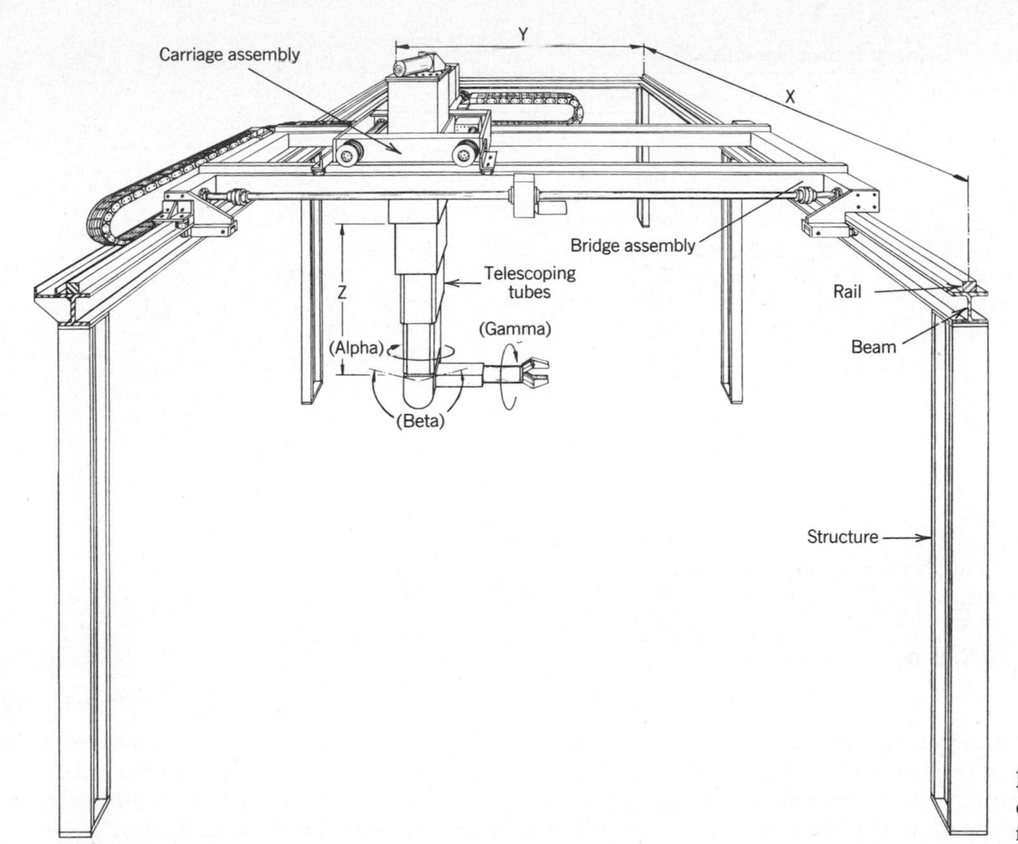

Figure 3. Gantry robot with bridge and carriage construction and telescoping tubes for vertical travel.

stops along the side rails, by software limits to axis travel, or by sophisticated collision-avoidance software which tracks the position and direction of each arm and prevents two bridges or arms from interfering with each other. Collision-avoidance is a complex control problem because of the possibilities for work pieces in grippers or tool-tips to collide when the bridges are separate or for bridges to hit when the robot arms are extended in opposite directions. The multiple bridge gantry robot is most useful in situations where separate operations are performed in areas within the work envelope or where there is a need to work on two sides of a workpiece simultaneously such as an automotive assembly line.

General Robotic Developments Applied to Gantry Robots

The cylindrical movements are varying constructions depending on number of axes and manufacturer. A three-roll wrist similar to that used in some pedestal-mounted robots is often applied in a heavy-duty construction for gantry robots.

A slip-clutch wrist was developed in 1986 by a gantry-robot manufacturer in response to the particular collision problems of gantry robots due to their overhead construction. Because gantry robots operate with many potential obstructions within their work envelopes, operator errors often result in collisions between the arm and some other machine or device. Collisions result in major damage to the arm mechanism because of the power of the x- and y-axes driving it into an obstacle. The slip clutch allows the arm to give way while immediately shutting the robot arm down electrically, and to rearm retaining its original positioning relationships.

Another robotic development that is particularly significant for gantry robots is the quick-change end-effector. Gantry robots are often used for multiple tasks within a work cell. Their flexibility can be limited, however, by the ability of the end effector to perform varied operations. For handling heavy payloads or performing multiple functions, a very strong or complex and heavy end-effector is required. These end effectors are difficult to design and build and they reduce the effective payload of the robot. A special-purpose end-effector can weigh over 600 lb (272.2 kg), reducing the useful capacity of the robot by the same amount. With quick-change end-effectors, one robot is able to pick up one of many single-use tools from a rack or stand within the work area and perform operations such as drilling, routing, tapping, and palletizing. Many requirements for multipurpose end-effectors are eliminated and the useful performance profile of the robot is improved.

Performance Characteristics

The basic gantry robot has the simplest geometry and control system of any robot. The first three axes follow machine tool geometry and are easily comprehended by programmers familiar with numerically controlled machinery. The rectangular work envelope is defined by the robot structure but is not identical to the inside dimensions of the structure. The work envelope volume is smaller than the inside dimensions of the structure because the distance from the centerline of the mast to the outside edge of the carriage assembly is larger than the length of the arm mechanism from mast to faceplate. If the application requires reaching the structure, an extension must be added to the faceplate. The structure adds rigidity to the robot which results in closer tolerances and higher re-

Table 1. Typical Gantry Robot Specifications[a]

		Degrees of Freedom			
		3	4	5	6
Configuration					
x-axis	Travel, ft	10 ~ 36	10 ~ 36	10 ~ 36	10 ~ 36
	Speed, ips	36	36	36	36
y-axis	Travel, ft, in.	6,8 ~ 19,8	6,8 ~ 19,8	6,8 ~ 19,8	6,8 ~ 19,8
	Speed, ips	36	36	36	36
z-axis	Travel, in.	0 ~ 59	0 ~ 59	0 ~ 59	0 ~ 59
	Speed, ips	20	20	20	20
01-axis	,°		345	400	345
	,°/s		60	60	60
02-axis	,°			180	180
	,°/s			60	60
03-axis	,°			360	360
	°/s			60	60
Repeat-ability	Linear, in.	±.004	±.004	±.004	±.004
	Rotational, rad		±.0006	±.0006	±.0006
Mast dead vertical lift capacity, lb					
Max speed 20 ips		300	160	130	100
Max speed 40 ips		150	110	80	50

[a] Courtesy of CIMCORP Inc.

peatabilities. The straight-line motion of the gantry robot eliminates the need to perform special path or trajectory computations for operations with close tolerances between obstacles or reaching into an opening such as a machine tool door or chuck. The performance characteristics of gantry robots also allow them to be successfully programmed directly from a CAD system and to carry out precise freehand operations such as drilling and routing without external fixturing or templates. Typical gantry-robot performance specifications are shown in Table 1; performance specifications for a highly accurate gantry robot are shown in Table 2.

The most significant performance characteristic of the gantry robot is that it achieves the same level of performance across the work envelope. The bulk of the gantry robotic mechanism moves with the arm so that there is less cantilevered posture as the arm extends to reach the work. Only the lighter assemblies close to the tooling are extended out from the fully supported position. The result is less vibration, less performance degredation, and more even performance than that of pedestal-mounted robots in full extension. The exception would be where tools or grippers are mounted cantilevered or off the centerline of rotation but the effect would still be minimized compared to a pedestal-mounted robot where the gripper extension would compound the arm extension.

Table 2. High Precision Gantry Robot Specifications[a]

		Degrees of Freedom			
		3	4	5	6
Configuration					
x-axis	Travel, ft, in.	15,7–32,11	15,7–32,11	15,7–32,11	15,7–32,11
	Speed, ips	40	40	40	40
y-axis	Travel, ft, in.	5,6–14,2	5,6–14,2	5,6–14,2	5,6–14,2
	Speed, ips	40	40	40	40
z-axis	Travel, in.	39–63	39–63	39–63	39–63
	Speed, ips	30	30	30	30
01-axis	,°		330	330	330
	,°/s		114	114	114
02-axis	,°			210	210
	,°/s			114	114
03-axis	,°				900
	,°/s				114
Repeat-ability	Linear, in.	±.006	±.006	±.006	±.006
	Rotational, rad		±.00035	±.00052	±.00052
Accu-racy	Linear, in.	±.014	±.014	±.014	±.014
	Rotational, rad		±.0005	±.0014	±.0014
Mast dead vertical lift capacity, lb					
Max, speed 30 ips		860	650	470	420

[a] Courtesy of CIMCORP Inc.

The interaction of speed, travel, reach and payload capacity affect operating performance. One gantry system installed with a 500 ft (170 m) x-axis travel, 40 ft (13 m) y-axis travel, and 25 ft (8 m) z-axis travel had an effective payload of 15 tons (13.6 Mg). The size and capacity of the system resulted in a corresponding reduction in repeatability to ± 0.25 in. (6.4 mm).

The effective payload of a gantry robot with more than three axes is dependent on the torque of the axes being used for rotational movement. The highest payloads are possible in the dead-lift position where no rotational movement is used and all axes are normal to the floor. The payload for a five-axis robot with 56 in. (143.6 cm) vertical travel and maximum z-axis travel speed of 24 inches per second (61.5 cm/s) might have a payload capacity in any orientation of 230 lb (104.3 kg). The same robot would have a vertical dead-lift payload capacity of 410 lb (186 kg). Reducing the z-axis travel speed of that robot from 24 in./s (61.5 cm/s) to 12 in./s (30.8 cm/s) would have the effect of increasing the payload capacity to 820 lb (372 kg) in the dead-lift orientation or approximately four times its payload in a rotational orientation. The reduction in speed would not have a noticeable improvement in lift in a rotational orientation because the limiting factor is the axis torque rather than the speed of motion.

Thus the performance specifications for any one gantry robot may be very broad depending on the actual motions required. For example, a gantry robot may be used for loading 50 lb (22.7 kg) castings into the chuck of a maching center, unloading the machined castings to a pallet, and then lifting a full pallet weighing 900 lb (408 kg) and moving it to the next transport station. For this reason gantry-robot manufacturers often specify positioning repeatability for each axis individually. The repeatability of the entire system depends on both static and dynamic errors which may be eliminated or augmented according to which axes are actually required to move to the desired position. The actual positioning repeatability can be calculated based on the movements required to perform the operations. In determining cycle times for automatic operations the dynamics of distance and speed and accuracy must be considered. Key tradeoffs are speed vs distance traveled, positioning accuracy vs speed of travel, and distance traveled vs positioning accuracy.

Performance of Multiple Bridge Systems

Multiple-bridge gantry robots may not have identical performance specifications with otherwise identical single bridge systems. When multiple bridges move concurrently they may create dynamic errors which result in performance degradation compared to a single system. It is difficult to calculate the exact theoretical performance characteristics as the arms actually operate in a factory environment. The impact of motion by one arm on the performance of another are affected by the speed, motion, resonance, and weight carried by the other arm or arms.

Safety Considerations

Gantry robots create special safety concerns. Because the moving parts are overhead they are not always visible to plant personnel. They may be "lost" visually in the ceiling, pipes, and beams of the plant. The structure does not create a natural barrier to the robot's work area and must be closed off with safety fencing, electronic beams, or other devices which can shut the robot down when a worker enters the work area accidentally or without following proper safety procedures. The most dangerous safety threat comes as a result of carelessness by people familiar with the robot, usually those working in the envelope for installation, integration, or maintenance. Gantry robots have very large work envelopes. People may move within the work area assuming that the robot arm will stay in another portion of the area only to find that the robot suddenly enters a new segment of its program which directs it to a new workstation in a different area. Caution must always be exercised when working within the structure of a gantry robot.

Many manufacturers provide a series of safety mechanisms with their equipment. Colored flashing lights are mounted on the controller and on the robot structure to indicate various states the robot may be in such as system power on/off, arm power on/off, brakes engaged/disengaged, program executing, etc. These lights are visible to maintenance personnel who may be on or above the gantry structure making repairs or adjustments; they are also indicators to plant workers that the equipment is in operation or could suddenly begin to move. Hardware stops, software limits, and collision-avoidance software discussed earlier are used to allow maintenance people to service equipment in one area of the workcell while the robot and the rest of the equipment continues in operation.

APPLICATIONS OF GANTRY ROBOTS

Gantry robot applications can be divided into two general categories: material handling and process applications. Material handling includes parts transfer, machine loading, palletizing, materials transport, and assembly applications. Process work performed by gantry robots includes welding, painting, machining operations such as drilling, routing, cutting and milling, inspection and nondestructive testing, application of sealants and adhesives, and parts marking (see Figs. 4–6 for some examples). The versatility of gantry robots is the result of the flexibility of configurations, heavy payload capacity, inherent structural accuracy, and ability to reach many points without performance degradation. For many of the applications mentioned, gantry robots perform at least as efficiently and effectively as pedestal robots. Applications described in detail are some of those where gantrys have particular advantages.

Machine Loading

Gantry robots for machine loading range from two-axis monorail to six-axis heavy-duty constructions. The straight-line motion is particularly useful in loading machine tools with chucks. The rotational movement is used for orientation of parts and placement in the machine tool; Cartesian motions are used for palletizing of completed parts and movement of full pallets. The ability of some systems to handle both parts and tool head changing makes them economically beneficial. A human operator will have a clear view of the tools being loaded from overhead allowing him to tend many machines

Figure 4. Four-axis gantry robot with telescoping tubes performs tool-changing operation on flexible machining center.

Figure 5. Six-axis monomast robot paints side stripes on an automobile.

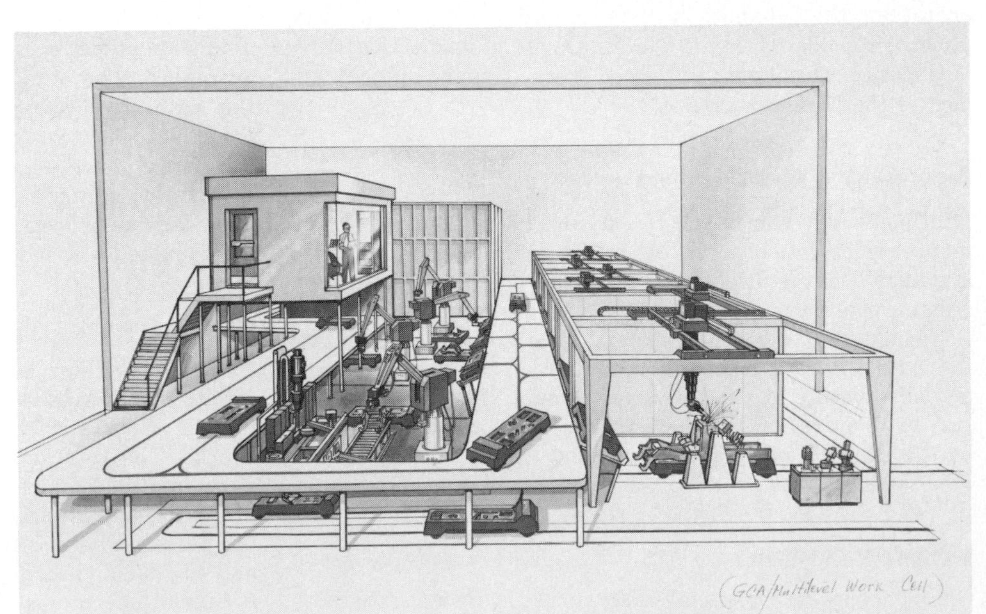

Figure 6. Conceptual drawing of robotic assembly center with gantry robots performing welding of subassemblies in off-road equipment manufacture.

Figure 7. Operator teaches clean-room gantry robot program for transporting of cassettes of silicon wafers between process steps.

without obstruction. Travel speeds over longer distances enable the gantry robot to load multiple machine tools without slowing production. Gantry robots in machine loading operations frequently perform additional tasks such as vacuuming chips from machined parts, gaging or inspecting parts, or changing tools and jaws in the machines (see also MACHINE LOADING/UNLOADING).

Parts Handling

Gantry robots can be incorporated into in-line operations for machining or processing. A simplified straight-line work flow is possible because of the flexible travel of the robot. At Cummins Engine Company, gantry robots handle engines from 290 to 1200 horsepower as they move through in-line machining cells.

Demilitarization of Munitions

GA Technologies has applied gantry robots in remote demilitarization of obsolete chemical agent munitions. The robots unpack munitions from palletized boxes by cutting metal strips on the pallet, removing a box from the pallet, prying open the lid of the box and removing the cardboard packing material. The robots spread the boxes open and remove the two munitions cannisters in each box, placing them on holding racks. The racks are filled with six cannisters of up to 200 lb (90.7 kg) and the loaded racks are moved into and out of cyrogenic baths. The robots performance capabilities meet the requirements of 1500 lb (680.4 kg) payload for the filled racks, the large work envelope of 50 ft (153.8 m) by 16 ft (49.2 m) by 8 ft (24.6 m) for both unpacking pallets and servicing the bath chambers, and the dexterity to perform the unpacking operations described. The GA Technologies installation is more efficient than humans in protective suits, does not require the extensive manpower and equipment required to do the operations manually and provides complete remote operation and maintenance while meeting system reliability goals.

Clean-room Transport

Robotic applications in fabrication of integrated circuits are of interest to semiconductor manufacturers because of the economic and quality effects of humans in the clean room environment (see Fig. 7). A gantry robot was tested transporting cassettes of silicon wafers between process devices. The gantry robot was designed with bellows encasing the y- and z-axis drive gears, a traveling vacuum system for removing drive-gear particles on the bridge, a payload vertical offset to prevent particles from falling onto the wafers from the z-axis mechanism, externally exhausted pneumatic actuators, and non-shedding paint. The gantry robot was compatible with Class 10 cleanroom standards. The tests showed that the human operator produced more particles and larger particles than the robot, resulting in approximately three times more area of the wafers damaged by particles than in wafers handled by the robot. Alternative approaches to cleanroom transport have included track-mounted robots and small arm mechanisms mounted on automatic guided vehicles. The economics of these approaches are not as beneficial as the gantry robot because they require floor space which is very expensive in a clean room (see also CLEAN ROOM APPLICATIONS).

Aircraft Inspection and Painting

The U.S. Air Force has purchased two gantry robot systems to inspect military aircraft and locate damage and defects in the aircraft control surfaces and in the internal honeycomb structure (see Fig. 8). Previous inspection techniques required dismantling sections of the aircraft and testing for damage caused by moisture or stress. The two new systems will use x-ray and neutron-ray systems to inspect the aircraft without

Figure 8. Large-scale gantry robot system for automated x-ray inspection of military aircraft. The system is designed to detect damage to air-control surfaces and internal honeycomb structures without dismantling the aircraft.

removing any components. The side rails of the robot will be incorporated into the building structure which will house the robot systems. Similar large-scale gantry robots have been designed for painting aircraft. The planes can be parked under the gantry structure and painted automatically without operators in the paint room.

Welding Bulkhead of Ships

Large structural sections of ships are now being welded automatically using gantry-robot technology (see Fig. 9). CAD descriptions of the bulkheads of ocean liners can be downloaded directly to the robot controller for path descriptions of the sections to be welded. The multiple-arm welding gantry robot uses two bridges to allow the arms to move independently over the work envelope. The area covered by the gantry creates a work envelope which is 50 ft (153.8 m) by 40 ft (123.1 m) by 10 ft (30.8 m).

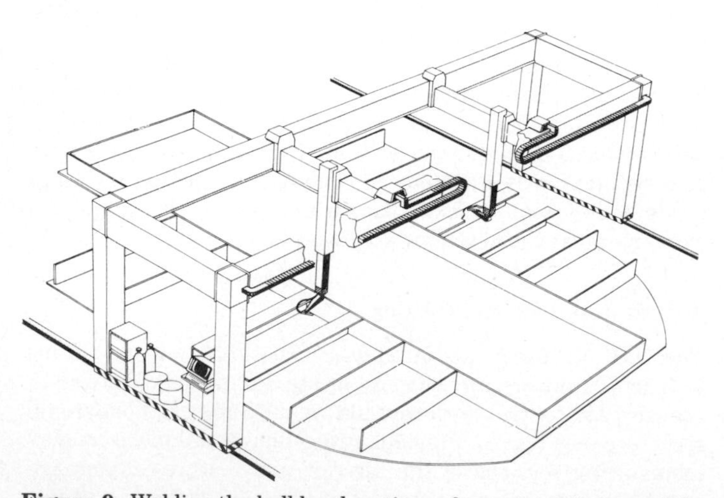

Figure 9. Welding the bulkhead sections of a cruise ship is possible using direct downloads of CAD data with gantry robots. The multiple bridge gantry robot has two independent arms on a single structure.

FUTURE CONSIDERATIONS

Overhead gantry robots have become the fastest growing segment of the robotics industry. Market share was estimated at less than 5% of units shipped in 1985 but is projected to grow to a total of about 30% of the installed base by the late 1990s. Gantry robots are used today as material-handling devices, transport mechanisms, inspection tools, assembly devices, welding systems, and machining centers. The refinement and commercialization of gantry robots has been called one of the most significant developments in robotics. The continued efforts to improve performance in the areas of precise positioning, smooth acceleration/deceleration, and overall reliability and maintainability will increase the industrial acceptance and the application potential of gantry robots. Currently gantry-robot cells are being set up which allow manufacturers to place a sheet of material in the gantry's work envelope and begin automatic cutting, trimming, drilling, milling, assembly and finishing operations which completely manufacture a part or subassembly using quick-change tools and programmed subroutines. The flexibility of gantry robots opens new methods of manufacturing in many industries. Industries which were not previously able to consider significant automation such as aircraft repair, shipyards, and semiconductor fabricators will continue to discover new possibilities for robotics.

BIBLIOGRAPHY

General References

A. J. Sturm and M.-Y. Lee, Stiffness Discussion on Gantry Robots's New Application as a Process Tool," *Proceedings of the 22nd AIM Communicating Manufacturing Technology Conference,* St. Louis, Mo., 1985.

T. Owen, *Assembly with Robots,* Prentice-Hall, Inc., Englewood Cliffs, N.J., 1985.

R. C. Dorf, *Robotics and Automated Manufacturing,* Reston Publishing Co., Inc., Reston, Va., 1983.

H. L. Martin and W. R. Hamel, "Joining Teleoperation with Robotics for Remote Manipulation in Hostile Environments," *RI/SME ROBOTS 8 Conference Proceedings,* Detroit, Mich., 1984.

H. J. Warnecke, "Mechanical Design of Robots," in S. Y. Nof, ed., *Handbook of Industrial Robotics,* John Wiley & Sons, New York, 1985.

J. P. Ziskovsky, "Gantry Robots and Their Applications," in S. Y. Nof, ed., *Handbook of Industrial Robotics,* John Wiley & Sons, New York, 1985.

M. J. Sullivan, "The Use of Industrial Robots in the Lay Up of Composite Material for Aircraft and Other Products," *RI/SME ROBOTS 8 Conference Proceedings,* Detroit, Mich., 1984.

P. Lammineur and O. Cornillie, *Industrial Robots,* Pergamon Press, Oxford, UK, 1985.

J. Hartley, *Robots at Work: A Practical Guide for Engineers and Managers,* North-Holland Publishing Co., Oxford, UK and IFS Ltd, Kempston, Bedford UK, 1983.

J. McElroy, "Increasing Productivity Without Automation," *Automotive Industries* (March 1984).

R. N. Stauffer, "Advancements in Robot Design," *Robotics Today* (Aug. 1985).

T. J. Slattery, "Computer Integrated Manufacturing Special Report," *Machine and Tool Blue Book,* Dec. 1985.

T. J. Slattery, "Gantry Robots Bridge the Gap in Overhead Operations," *Robotics World* (Sept. 1985).

R. C. Asfahl, *Robots and Manufacturing Automation,* John Wiley & Sons, Inc., New York, 1985.

T. E. Klotz, "Robotic Loading of Machine Tools," in S. Y. Nof, ed., *Handbook of Industrial Robotics,* John Wiley & Sons, New York, 1985.

R. S. Muka and S. D. Mayer, "Particulate Contamination During Robotic Wafer Transport," *Microcontamination* (Nov. 1985).

S. D'Agostino, "Robot Applications in Electronic Assembly," *RI/SME ROBOTS 8 Conference Proceedings,* Detroit, Mich., 1984.

L. J. Kamm, "Gantry Robots for Machine Loading," *RI/SME ROBOTS 9 Conference Proceedings,* Detroit, Mich., 1985.

F. Mason and N. B. Freeman, "Special Report: Turning Centers Come of Age," *American Machinist* (Feb. 1985).

R. N. Stauffer, "Gantry Robots Facilitate Handling in In-line Workcells," *Robotics Today* (June 1986).

R. H. Leary, J. M. McNair, and J. F. Follin, "Remote Demilitarization of Chemical Munitions," *RI/SME ROBOTS 8 Conference Proceedings,* Detroit, Mich., 1985.

M.-O. Demaurex and L. Horvath, "High Precision Industrial Robots Suited to High Force Work," *RI/SME ROBOTS 8 Conference Proceedings,* Detroit, Mich., 1985.

GARMENT AND SHOE INDUSTRY, ROBOTS IN

G. E. Taylor
P. M. Taylor
University of Hull
Hull, United Kingdom

INTRODUCTION

As with any other manufacturing industry the prime objective of clothing production is the satisfaction of market demand. In almost all cases the market's requirements are dictated by fashion and hence the industry is characterized by small-batch production runs involving various sizes, colors and different types of raw material. This seems unlikely to change in the foreseeable future—indeed there is continually increasing demand for better fit, style, and quality and for faster response to fashion change. Such an industry has relatively little need for hard automation, but is clearly a prime candidate for the use of all kinds of flexible automation, including robots and thus the particular problems involved are attracting worldwide attention (1–4). Despite this there is currently little penetration by completely automatic assembly systems although many semi-automatic machines, especially those which de-skill the more complex operations, may be found in the developed countries. There are a number of reasons for this lack of full automation, these include factors involving capital costs and payback periods in a market which frequently works with very narrow profit margins, but also a number of materials handling and assembly problems specific to this industry for which solutions are only just being proposed.

Shoe and garment production share a number of common features. The materials used are generally nonrigid and frequently nonhomogeneous, examples range from leather to knitted or woven fabrics of different weights and formed from different types of fibers. Each product comprises a number of parts which must first be cut and then joined together by fusing or sewing and, particularly in the later stages of the assembly, complex manipulation is likely to be necessary to achieve the final three-dimensional shape from its two-dimensional components. These are the features which cause particular difficulty in the application of robots and will be dealt with in detail in this article. There is also an increasing number of first generation robots, especially in the shoe trade involved in traditional robotic tasks such as the loading and unloading of injection—molding machinery (eg, in the production of shoe-sole units) and spray painting (eg, fashion heels). These are not radically different from similar applications in other industries and hence are not considered within the scope of this article (see Painting).

PLY SEPARATION

In the garment industry the most fundamental problem is that of fabric handling. Further, as fabric panels are usually cut in stacks and stacks of partially completed garments are often used as buffers between production cells, fabric-ply separation is a key part of this requirement. Separation techniques must not damage or mark the fabric in any way (this may, for example, preclude the use of intrusive devices such as pins, with fine, lightweight fabrics) and cycle times frequently need to be of the order of two or three seconds or even less. Vacuum is not usually an option because of fabric porosity. Reliability must be high, if the required cycle time is three seconds then even 95% reliability is not acceptable since it implies operator intervention once per minute on average. A particular difficulty is that when stacked and subsequently cut (frequently while under vacuum) the fabric plies tend to cling together. The cohesive forces are caused by the interlinking of fibers from one ply to the next, electrostatic forces and there may also be additional pronounced cohesion along ply edges due to imperfect cutting. When plies cling together in this way there is a tendency for the gripper to separate more than one at once. This leads to a requirement to detect multiple picks and the necessity of an ability in the system to recover from

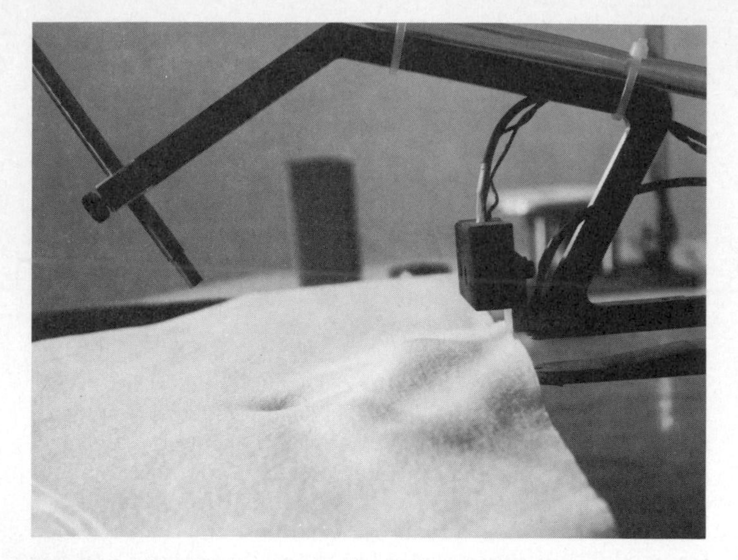

Figure 1. Completely nonintrusive pin-based device developed at the University of Hull.

such errors, usually by aborting and restarting the pick and place sequence.

Despite the constraints listed in the previous paragraph many ply separation devices have been invented (5) although few are, as yet, in common use within the industry. The most well known commercial devices are the Clupicker (6) which uses a pinching mechanism and the Polytex gripper (type NG, Polytex AG, Switzerland) which inserts angled, hollow pins into the first layer of fabric, then blows air down these pins creating a bubble of air between the topmost ply and the remainder. Other proposed pin-based techniques use (*1*) opposing pins (7) and (*2*) a cylindrical comb (8). A completely nonintrusive device has been developed at the University of Hull, UK (9). This comprises an angled air jet fixed at the end of long, parallel jaws as shown in Figure 1. The vibrations caused by the horizontal air flow are considerably increased in amplitude and reduced in frequency by the flow of air through the porous top ply. This flow also causes an air bubble under the top ply which propagates to the stack edge, breaking it up so that the top ply flaps freely and the opened parallel jaws can be inserted. Various sensors are located in the jaw to detect faults such as multiple pick. The devices described all work reliably over a given range of fabrics, but none works satisfactorily for all fabrics. A detailed discussion of the fabric handling problem may be found in Ref. 10 (see also APPAREL INDUSTRY, ROBOTS IN).

As well as initial pick-up of the cut ply, single plies and completed or partially completed sub-assemblies must be transported from one place to another during the assembly process. The most commonly used mechanical methods employ conveyors, sliding plates or wheels. By introducing cameras or other sensors the position and orientation of the transported piece can be determined when high accuracy is required. Another effective means of transportation, assuming automated ply separation is established, is to use the separation mechanism as a transport device, again sensory feedback is necessary to ensure fast and accurate transfer.

Leather and the majority of other materials used in the shoe trade are much less porous to air than fabrics used for garments and thus vacuum can be used successfully for most pick and place operations.

JOINING BY FUSING

A good example of joining by fusing in the context of garment manufacture is the initial part of the shirt-collar assembly process. In one common design the basic collar comprises three parts, a woven polyester/cotton outer and two ready-coated lining panels as shown in Figure 2. The parts are presented to the assembly station in three stacks. The individual panels must be de-stacked, perhaps inspected and then placed accurately (within 0.04 in. (1 mm)) on top of each other as shown in the figure. The completed assembly is then slid onto a conveyor and into a fusing press. An appropriate pressure is applied at a given temperature for a specified time to achieve good bonding. The bonded panels are then passed through a series of sewing operations, some of which may already be implemented in a semi-automatic way by loading the pieces into jigs and making a predetermined stitch pattern in an open loop manner. As far as complete automation is concerned the remaining problems are ply separation, accurate placement (not always straightforward with a floppy piece of material), perhaps automatic inspection and monitoring and control of all the machinery elements to be incorporated into the work-cell.

Shoe soles are usually attached to the lasted uppers by means of an adhesive which is applied to the sole and upper and is then reactivated prior to assembly. In this case the adhesive is applied in the shoe factory and the bottom surface of the lasted upper must be correctly prepared by a roughing operation. Once the adhesive has been reactivated the sole must be placed into the correct position on the upper. The first stage of this is called spotting and is performed by bringing the toe area of the sole precisely (better than 0.04 in. (1 mm) accuracy) onto the toe area of the upper and squeezing the two together. This is then repeated at the heel end. Some stretching or compressing of the compliant sole may be necessary. The spotted items are then placed in a press for full adhesion. The essential features of this process are the application of the adhesive, the picking of the sole and lasted upper, their alignment (in three dimensions!), the application of pressure and the monitoring and control of all the elements of the work cell. Although at first sight there are many differences between this process and the shirt-collar assembly, the crucial and most difficult part of each process as regards achieving complete automation is the same, namely correct alignment of the constituent parts and it seems likely that similar techniques can be used in both of these and, indeed, a multitude of other, similar joining operations. Some possible solutions are discussed in Ref. 11.

JOINING BY STITCHING

The joining of two fabric panels by stitching may comprise the following tasks: separate the fabric panels from a stack, unless they are already separated, place one panel down in a work area, place the second panel on top of the first and slide the complete sub-assembly under a sewing head. Variations on this are (*1*) to have two feeders to the sewing head and (*2*) to load the panels into jigs which are then fed into an

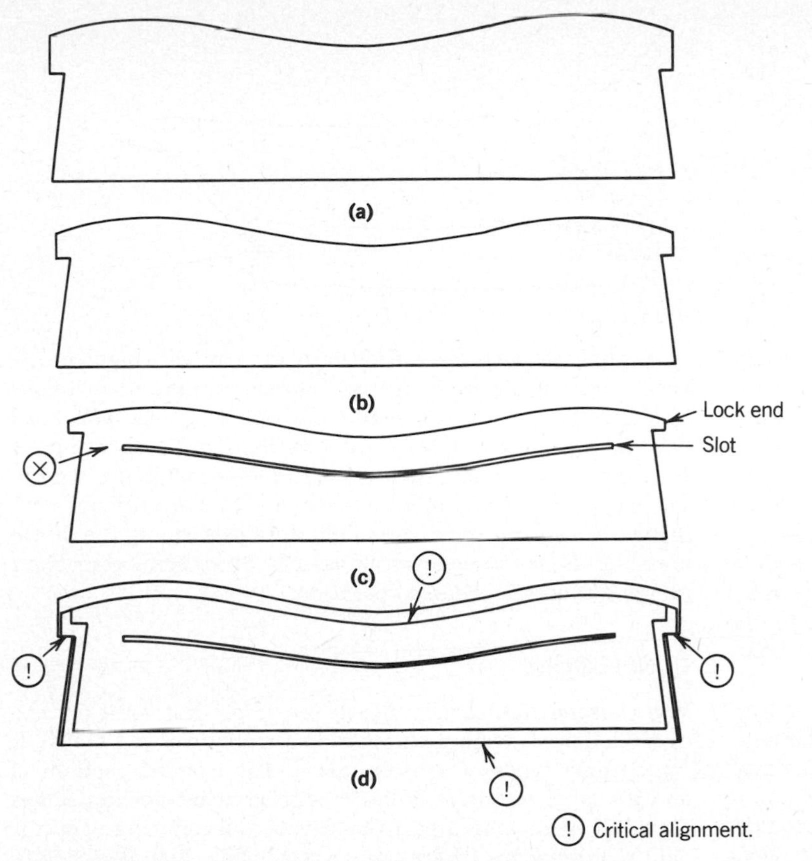

Figure 2. Collar components. (**a**) outer ply, (**b**) half ply, (**c**) inner ply, (**d**) assembly, ! critical alignment.

automatic sewing machine. The key elements here are ply separation, accurate and reliable pick and place and perhaps coordination of movement control with the sewing head. The sewing machines themselves must be monitored for correct operation (thread tensions, misstitching, etc) and to ascertain when bobbins need replacing. All elements of the work cell again need to be integrated together for fully automatic operation.

The operations involved in sewing together pieces of leather are almost identical although leather is stiffer and less porous than most fabrics and hence there may be additional difficulties such as needle breakage unless the material is kept stationary when the needle passes through it.

COMPLEX MANIPULATION OF PARTS

In many of the joining operations necessary to the production of a garment there is an exact match between the edges to be joined and open access to the seam (this is true, for example, of the side seams of a jacket or a dress). However there are certain operations where much more dextrous and complex handling of the parts is required and this clearly presents a much greater challenge if it is to be successfully automated. Two examples are considered here.

Concealed Gusset

This construction is commonly found in ladies and children's briefs and in men's underpants. To give a smoother finish and prevent chafing, part of the garment comprises a double thickness of fabric with the seams joining the two pieces turned to the inside (Fig. 3). A robot based technique for this particular assembly problem is currently being investigated at the University of Hull, UK (12).

Arm Insertion

This operation is typical of many where opposing curves are sewn together to produce a three-dimensional garment (Fig. 4). Skilled manipulation by the operator is normally required to fit the parts together and feed them under the sewing head, but some advances have been made with machines that allow different rates of feed for the two parts.

INTEGRATION OF WORKSTATION TASKS

As has been indicated in previous sections, once the basic handling techniques have been evolved considerable further effort is needed to integrate the various parts of what seems likely to be, in many instances, a complex workcell. A number of demonstrations of such workstations have been described in the literature.

During the first Japan International Apparel Machinery Exhibition, October 1984 (13), there were five examples of robots in action. Most equipment appeared to be at the development stage, but all demonstrated the principles well. A particularly complex work cycle performed by the Brother robot involved pick up, transportation, and orientation of a single ply which was stitched by a lockstitch topsew machine and had buttons attached before being transported to a stacking area.

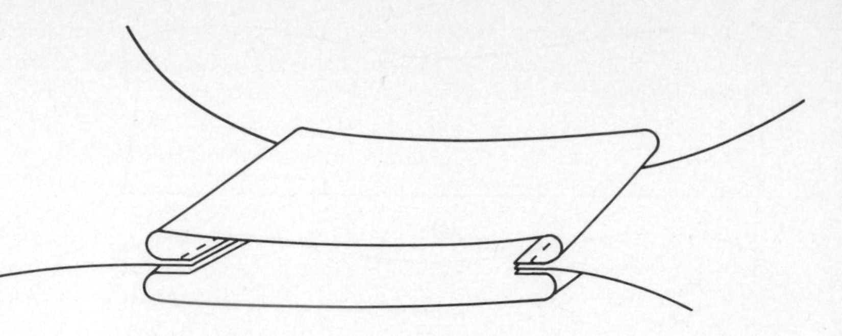

Figure 3. Concealed gusset construction.

Work cycle time was more than 50% slower than for the equivalent manual sequence, but anticipated improvements in robot dynamics and processor speed should remove this discrepancy.

Work from the Charles Stark Draper Laboratories, Cambridge, Mass., was shown on display in the Cobo Hall exhibition (14). Here the robot carried out a sequence of folding, sewing and reopening operations on the sleeve of a man's two-piece tailored suit using a gripper able to contour itself to match the curve of a pattern piece. The vision system, robot and gripper control and sewing machine all reported to and received instructions from an IBM-PC (see APPAREL INDUSTRY, ROBOTS IN).

Over the past few years work has been carried out at Durham University, UK, aimed at establishing principles for automation in garment assembly (8). A workstation has been developed to produce a subassembly which forms part of one of the popular styles of men's briefs. The work cycle includes fabric ply separation, pick and place from a stack of parts, positioning and alignment and synchronized sewing with microprocessor controlled monitoring of sewing-machine yarn for quality control purposes. A number of sensors included in the workstation feed information back to the master control system to enable successful implementation of each part of the subassembly.

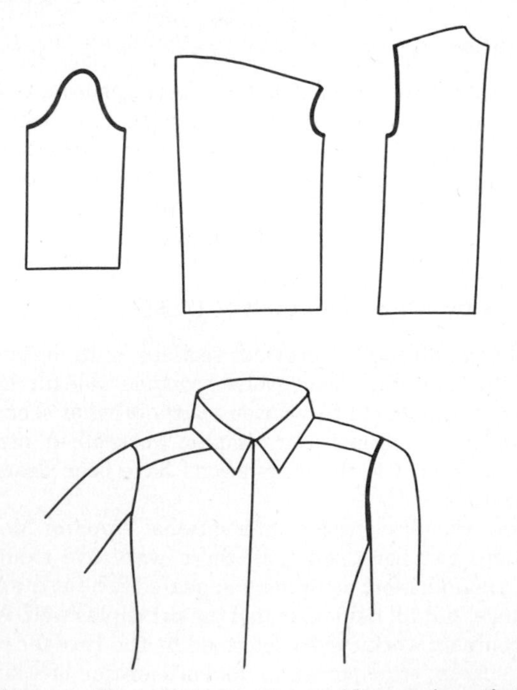

Figure 4. Three-dimensional assembly, arm insertion.

A collaborative project, funded by the United Kingdom Science and Engineering Research Council, is being undertaken by the University of Hull and Corah plc (12). The aim here is to demonstrate a robotic workstation for the assembly of the fronts, backs and gussets of both men's and ladies' briefs. The garment subassemblies are stitched by conventional sewing machines with robotic manipulators performing the fabric handling. Sensory feedback is used to allow error correction in order to obtain reliable operation.

CONCLUSIONS

The clothing industry is clearly an important area for the introduction of flexible automatic assembly workcells including various types of sensory robot. The current position is one of a labor intensive industry with growing mechanization aimed at increasing production levels and reducing operator skill requirements. There are currently few examples of complete automation except in areas of high volume production such as socks and tights where changing fashion affects color and pattern rather than basic style. For such products hard automation is appropriate, available and, indeed, widespread in use, many examples may be found in the trade journals (15). However, a number of countries are now putting considerable research effort into solving the problems necessary before completely flexible automatic clothing design and assembly becomes a reality. These include an extensive Japanese project based on the government research laboratories, MITI, European Economic Community Funding via the BRITE iniative (16) and funding in the UK via the Science and Engineering Research Council's ACME iniative (17). In addition to academic institutions many of the major textile and clothing machinery manufacturers are involved in such projects. Research funded ranges from noncontact body measurement through fabric handling to various demonstrator projects displaying automatic techniques for producing complex subassemblies. Looking to the future a vision of apparel merchandising starting with a video fashion show for each customer and finishing with the delivery to that customer of an automatically "made to measure" product has been proposed (18).

BIBLIOGRAPHY

1. J. Hasegawa, "From Automation to Robotics in Textile Industry" *The Textile Institute, World Textiles: Investment-Innovation-Invention,* Annual World Conference, London, May 1985.
2. J. A. Krol, "The Global Outlook on Marketing and Research," in Ref. 1.

3. P. J. Tredwin, "Computerised Garment Manufacture," in Ref. 1.
4. C. Walter, "The Application of Robotics in Context of Garment Making Up Automation," *Knitting International,* April 1985.
5. J. M. Murray, "Single Ply Pick Up Devices," *Bobbin* (Dec. 1975).
6. I. Dlaboha, "Cluetts Clupicker Enhances Robots Ability to Handle Limp Fabric," *Apparel World, Knitting Times* (July 1981).
7. J. K. Parker, R. Dubey, F. W. Paul, and R. J. Becker, "Robotic Fabric Handling for Automating Garment Manufacturing," *Journal of Engineering for Industry,* **105/21** (Feb. 1983).
8. J. E. L. Simmons, "Research Developments in Automated Garment Assembly," Press Release, University of Durham, UK, April 1986.
9. D. R. Kemp, G. E. Taylor, and P. M. Taylor, "An Adaptive Sensory Gripper for Fabric Handling," *4th IASTED Symposium on Robotics and Automation,* Amsterdam, June 1984.
10. P. M. Taylor and S. G. Koudis, "Automated Handling of Fabrics," *Science Progress,* (71) (1987).
11. P. M. Taylor, G. E. Taylor, P. Bowden, I. Gibson, and S.G. Koudis, "Sensory Robotic Systems for the Garment and Shoe Industries," *16th Industrial Symposium on Industrial Robotics,* Paris, June 1986.
12. P. M. Taylor, "A Garment Assembly Demonstrator," ACME Directorate Project Summaries, Science and Engineering Research Council, Swindon, UK, 1986.
13. J. D. Tyler, "Systems Approach is New Theme at First Japanese Garment Assembly Exhibition," *Knitting International* (Jan. 1985).
14. L. Macarthour, "Water Jet Cutting seen at Robots 9," *Automation News,* (June 1985).
15. "Colosio introduce the versatile Stella," *Knitting International,* (Sept. 1985).
16. "New Garment Project Leads for BRITE Supported Research," *Apparel International,* (Sept. 1986).
17. *ACME Directorate Project Summaries,* Science and Engineering Research Council, Swindon, UK, 1986.
18. W. Rainsford Evans, "Made to Measure by Computer Integrated Manufacturing," *Knitting International,* (March, 1986).

GOVERNMENT POLICIES

PHILIP L. BEREANO
University of Washington
Seattle, Washington

THE CONCEPT OF GOVERNMENT POLICY

"Government policy" is a loose label given to a set of government activities (including inaction as well) related to a particular subject. Usually these actions can be thought of as "means," although the "end" may not be clearly articulated. Often, the first portion of a statute passed by a legislature will recite some objectives, but these may be part of a ritualistic obeisance rather than the explicit goals. (For example, in the years after World War II the Congress passed legislation proclaiming that each citizen was entitled to full employment and to satisfactory housing, although neither goal has been pursued literally during the last four decades.)

"Policy analysis" is a commonly used term, yet it is rarely defined with precision. It may be said to encompass "different sectors of human thought including legal theory, welfare economics, moral philosophy and decision analysis," and employ "techniques and principles of problem-solving variously referred to as policy science, cost-benefit analysis, operations research, systems analysis, and decision theory (1)." Central to good policy analysis are value issues and epistemological concerns about the origins, methods, and limits of human knowledge. One typology for such analysis includes the steps of: goal clarification (based on societal values and the interests of involved parties), goal realization status (the events in legislation, budgeting, etc. related to such goals), analysis of relevant societal conditions, projecting developments, and identifying policy options (2). If one uses different paradigms concerning the relationship between technological change and social change, involving different possible social change mechanisms, institutional arrangements, and value constructs, one can conduct a full exploration of technological options and alternative states of society (3).

General Considerations for Policy Analysis

Behind the idea of governmental policy is the belief that social change can be understood and directed.

> . . . A theoretical model of the way in which society operates is important as a basis for purposeful social policies. Often such policies assume, implicitly or explicitly, cause-and-effect relationships. . . . A social change (whether it be the introduction of a new technology, an alteration of political power, a new economic program, or the introduction of a "social technology" such as no-fault insurance) will have consequences that may lead to great benefits and/or harms to different sectors of society. An accurate social theory is thus essential for planning and policy formation and implementation, especially if distributional and second-order effects are to be recognized.
>
> However, there is no commonly accepted social theory, or ideology, in the United States. The well-known American attitude of pragmatism is, in practice, essentially a non-theory despite the serious attention given to it by John Dewey and others. Instead of the elaborate social constructs associated with mercantilism, laissez-faire capitalism, theoretical Marxism, or philosophical anarchism, the presently dominant American ideology of corporate liberalism stresses the functioning of institutional systems and the achievement of change through small increments of reform. . . . (4)

The available models of social change tend to fall into two major categories: equilibrium models based on mechanistic or functionalistic concepts in which society is seen as a system of interdependent parts which naturally seek to achieve an equilibrium based on common values which are imputed to the population; and conflict models which are based on views of society as a dynamic system constantly changing, often based on the work of Hegel in which social change is the natural tendency of social systems—every situation, the "thesis," develops an opposing tendency, the "antithesis," and these are in a continuing struggle resulting in a "synthesis" which itself is a new thesis in the dynamic (5).

The introduction of a major new technological phenomenon, such as robotics, brings with it concomitant social, economic, and cultural impacts which may engender a felt need for government policies to guide the development of the technology and to mitigate or enhance these impacts. Since the consequences of robotics in domestic households are likely to be insignificant (6), policy concerns have been directed mainly towards industrial applications for robotics. In order to explore this phenomenon, a policy model is necessary. Although the dominant view in the U.S. is that policies can be formulated and implemented without any explicit ideological content, much of the academic literature in this country, and the overall view in European societies, recognizes that policies embody specific values and deal with the distributional or equity aspects of guiding social change (ie, who wins and who loses) and thus are political (7).

The ideological elements that are relevant to a discussion of governmental policies regarding robotics concern the "technological imperative" (determinism) and the notion of progress, the role of competition (especially in regard to industrial developments in Western Europe, the USSR and Japan), and the basic tenets of capitalist ideology—strongly expressed in recent years under the conservative Reagan Administration—which sees technological initiatives as falling primarily in the domain of the private sector and urges limited government involvement in national economic life. In each of these elements, contradictory values and desires exist. Technological determinism is associated with a view that whatever can be done technologically will be done because it contributes to progress (8). Despite whatever adverse consequences a new technology may present, a widely shared belief in this society is that new technologies ought to be applied and that our technical genius can deal with any of the "negative externalities" (adverse effects). Often this view is accepted as a matter of faith, although an underlying reason seems to be related to competition—the fear of falling behind.

> On one point standard economic analysis is correct: there is no stopping the introduction of robots. Even if it should turn out that employment is adversely affected and the rate of profit on capital is approximately the same after the introduction of robots as it was before, firms cannot avoid trouble by never adopting the new technology. So long as some firms (in this case, Japanese or European firms) are going to introduce the new technique, U.S. firms will have to do likewise or find they are unable to compete in world markets. In the absence of competition, firms have sometimes delayed the introduction of cost saving techniques to protect the value of existing capital. In the presence of competition, delay risks loss of sales and, hence, of jobs. U.S. firms will ultimately have to adopt robotics and related technologies. If they do so later rather than sooner, they will lose whatever advantages accrue to having been among the leaders (9).

At the same time that fostering competition is linked to free trade and consumer benefits, pressures arise for tariff restrictions and other policy devices for protecting U.S. manufacturing jobs. The contradictions within conservative capitalist values regarding governmental policies for new technologies were perhaps most clearly illustrated when, in 1984, President Reagan vetoed legislation sponsored by Republicans as well as Democrats which would have provided federal money to support research, development, education and training activities for industrial robotic technologies (see below); he called it improper government activity.

At the root of the conflicts in these policy elements is the peculiar U.S. desire to believe that everybody can benefit from a public policy, an unwillingness to face the fact that virtually all policies are political because they involve the bestowing of advantages or the transfer of opportunities and power from one group to another. However, there has been an increase in recent scholarship denying that a "win-win" condition has ever existed in regard to technologies for industrial production (10).

> . . . The introduction of machines became as much part of the day-to-day tactics of the class-struggle between labor and capital as it was part of the overall strategy. These tactics inevitably included the need for increased social control on the part of capital, and the authoritarian relationships that this implied became crystallized in the machines that were introduced.
>
> To see this process at work, we need go little farther than a remarkable volume, *The Philosophy of Manufactures,* written by a Scottish academic, Andrew Ure, an arch-apologist for the whole factory system, and published in 1835. Ure describes vividly how manufacturers, oppressed by militant unions and "unable to control workers by reducing wages," were led to use technological innovations for this purpose. . . .
>
> To Ure, the message was clear. "This invention confirms the great doctrine already propounded, that when capital enlists science in her service, the refractory hand of labor will always be taught docility". . . . (11)

This more radical, and, indeed, more realistic view recognizes that policies result from the contention of political groups in a society. Similarly, it is argued that the technological formats which are actually researched, developed, and deployed are expressions of the goals of powerful interests rather than weak ones (12).

Ideally, any governmental policy dealing with the effect of robotics in manufacturing would be but a small portion of an overall industrial policy which would also encompass such areas as workers' economic security (full employment policies, unemployment insurance, job-seeking services, etc), plant closure policy, tariff schedules, and the like. Socialist planned economies claim to have such a policy structure, and, indeed, a few capitalist countries, most notably Sweden, may have something approaching this level of comprehensiveness. Nevertheless, the United States has no overall industrial policy and even within specific subtopics policy formulation may be scattershot.

Policies Regarding New Industrial Technologies

Because of the U.S. cultural value that sees technological change as essentially "inevitable," most discussions of the relationship between technological change and social change start with the new technological development as if it were a given and investigate its impacts (see eg, Ref. 13). Although much of this essay is organized along similar lines, it should be stressed that any realistic view of the process must lead to the conclusion that technologies themselves are produced by economic, social, and political (as well as technical) forces. And thus, in order to better analyze policy development regarding a new technological innovation, it is necessary to try to understand the forces and desires which have led to the innovation itself. Looking at the literature on the "first" Industrial Revolution may help to better identify the forces which are at work during the current era resulting in similarly massive technological change. These forces might be: a desire to rationalize production—increase productivity, facilitate planning of investments and contract obligations, etc; a desire on the part of management to exercise control over the workforce; a desire to "deskill" the steps of the work process by the use of machinery, so that the labor rate that must be paid for each such step is reduced (this is a motivation explored by Adam Smith in his famous example of pin manufacturing in *The Wealth of Nations*); and, in a related fashion, to depress wages through greater competition for jobs by the "reserve labor force" (the unemployed) who now, with but little training, can perform the lower skilled function (see Refs. 10 and 11). These four aspects of a motivation for industrialization have an underlying ideological theme, an inherent antagonism between the interests of management and labor. If, however, one uses a model more along corporatist lines, asserting a community of interests between the two groups, then only the first motivation—a desire to increase productivity—makes sense, and then the interesting policy question would be the division of the benefits of increased productivity as between the two sectors.

In analyzing the effects of a category somewhat broader than robotics, programmable automation, the Office of Technology Assessment of the U.S. Congress offered the following motivations for developing policy in this area: the immaturity of the technology and the limited experience of its application in order to fully realize its potential; international competition in which foreign governments are encouraging use of such technologies; the risk of growth in unemployment; the risk of adverse effects on the psychological aspects of the work involvement, which may both diminish productivity and constitute new health problems; and the implications for education, training and retraining at all levels at a time when the capacities and resources of the instructional system in the United States are particularly strained (14, pp. 367–370). The federal government has broad interests in robotics and associated automation because of the implications of the development and use of these technologies for military purposes and in reducing the costs of defense products, a generalized responsibility for the economic well-being of the citizenry (standard of living), a concern about employment levels and workplace equity issues such as occupational safety and health, and a deep ideological strain in the U.S. culture to be "Number One" in technological leadership.

> Existing Federal programs reveal an ample precedent for Federal involvement in the development and use of [these technologies]. In particular . . . the U.S. Government already has a major role in funding [automation] research and development, and it offers tax incentives for capital investment that may motivate adoption of [automated technologies]. Moreover, it is involved in study and regulation of occupational safety and health impacts generally; it measures employment trends and relates them in limited degree to technological and economic development; and it funds and shapes education, training, and retraining activities (15).

Whereas the various U.S. states may foster "high-tech" industrial policies, these usually involve competition for a limited number of industrial facilities; only federal policy can affect the development and use of robotics nationwide. Thus, federal policies must be aimed at coordination of activities among the various regions, coordination between the interests of manufacturers and employees, and coordination in regard to the allocation of educational and training resources. This must be done despite substantial uncertainties: uncertainties regarding the rate of technological development, the rate of deployment of new technologies, foreign utilization of new technological innovations, forecasts of economic growth and health, and the like.

In general, the federal government can pursue one of four basic strategies in regard to coordinating activities regarding the development and the use of new production technologies: do nothing, or continue current activities; emphasize the development and application of the new technologies; emphasize human resources by stressing attention to education and training, potential problems in the work environment, and the creation of new and meaningful jobs; or adopt policies which emphasize both technology and human resource concerns (14).

Two Possible Policy Models

The main contending models for developing policy on robotics are distinguished by believing either that automation is being developed solely for productivity gains or that it is also deployed in order to increase management's control over workers and the workplace. The dominant view in the United States is clearly the former; it can be found in popular scholarly works (16), industry materials, and underlies most governmental reports (eg, see Ref. 17).

> The analytical methodology employed by economists to analyze the effects of the introduction of robots on wages and employment or workers in the affected industries and upon employment generally tends to trivialize what is a complex question, and offers little policy guidance. This does not *appear* to create any difficulties, mainly because the analysis leads the economists to the conclusion that, apart from a short-term need for retraining, workers face no problems. Thus, standard analysis suggests that robotics: 1) will have a positive impact on wage levels, 2) will probably tend to reduce rather than increase unemployment in the long run, and 3) will stimulate total employment even in the industry introducing the robots [citation to Congressional document omitted]. The analysis on which these conclusions are based can only be characterized as glib and superficial (18).

The more radical thesis (10) begins with the notion of the *division of labor* which, as can be seen in Adam Smith's example of pin making, relates to taking a complex task and, by a process of *reduction,* breaking it down to small component tasks which can later be strung together. (Division of labor as a concept must be distinguished from *specialization,* which entails the recognition of a worker's skill or excellence in performing a task which may be either complex or simple—unfortunately, the literature frequently does not acknowledge this distinction.) The division of labor facilitates *deskilling* of the workers because the reduced sub-tasks do not require as much skill to perform as the overall complex task. But these two factors, division of labor and deskilling, enable the additional separation of the mental and the manual portions of an industrial process, so that increasingly planning and the exercise of judgment are performed in the management offices rather than on the shop floor where now work has been more *routinized*. Routinization in turn, enables mechanization or *automation* to be facilitated because mechanical means can be created which can do small and discrete tasks which are of a repetitive nature not requiring much judgment or skill in their performance. The end result of this process of *"rationalization"* of work is a claim of increased productivity, but an inherent effect is also the transfer to management of ever more control over the work process and the workers. Degradation of work is also inherent in the process of mechanization and automation (19). Other scholars have suggested that at early stages of mechanization increasingly higher worker skill levels will be demanded but as automation increases this relationship peaks and required skill levels decline (20, pp. 201–221).

Whereas the dominant thesis sees the social organization of production as being based merely on the requirements of technological innovations, the more dialectical model understands that social arrangements can occur prior to the development of technology that will then make these organizations more effective. At the beginning of the Industrial Revolution, when there were hardly any capitalists and very little machinery, factories were already being organized.

> It seems possible to identify four main reasons for the setting up of factories. The merchants wanted to control and to market the total production of the weavers so as to minimize embezzlement, to maximize the input of work by forcing the weavers to work longer hours at greater speeds, to take control of all technical innovation so that it could be applied solely for capital accumulation, and generally to organize production so that the role of the capitalist became indispensable. Factories provided the organizational framework within which each of these could be achieved. Thus although machines were present in the early factories, they were seldom the *reason* for setting up a particular factory. The factory was a managerial rather than a technical necessity. It imposed a new discipline on the whole production process, and was described by Charles Fourier as a "mitigated form of convict prison" (21).

Objectives for Government Policy

The concept of productivity is essentially an efficiency measure, comparing the output from one particular factor in the production process to the total input. If productivity is understood as being composed solely of this *quantitative* notion certain policy directions are indicated. Others will be suggested if efficiency is also considered to have a *qualitative* aspect which relates to the maintenance of management prerogatives over the work process (in this view, trying to increase productivity is similar to the process faced by an engineer or mathematician using partial differentials when a parameter is dependent on more than one variable). Clearly modern capitalism has little toleration for policy proposals that could greatly increase quantitative productivity and efficiency by dramatically restructuring the relations of production to give workers much more power. The most vivid modern illustration of this point is the refusal of the management of the British firm Lucas Aerospace to accept workers' proposals for a conversion of production to a score of socially useful and needed products (such as kidney dialysis units) when the government of the U.K. greatly cut back its purchases of armaments in the 1970s; protecting management's prerogatives was held more important (22). Managers believe that they can gain competitive advantage over other firms, not only if their costs are minimized but also if they are better able to discipline their workers, avoid strikes, and maintain order. This overall thesis has been called "the ideology of industrialization" (11).

The formulation of policy will be dependent upon the way that the impacts of a technological development are categorized, perceived, and treated (23). Whether an impact receives serious consideration may depend upon whether it is intended or inadvertent, a primary consequence or an indirect one, associated with a major value in the culture and, of course, the political power of the groups affected by it. For robotics a major direct goal is increased productivity. An indirect (but intended) one is to improve competitive position. Impacts which are usually indirect (and *perhaps* are unintended) include effects on the quality of work, job creation/elimination, unemployment, training and retraining, and worker control by management. Thus, one would be interested in asking not only what are the actual impacts on the labor force but also what impacts on the labor force are intended, ie, what are the goals and objectives which the policy should be pursuing. Regarding the workplace, one would need to understand both the *content changes* and *compositional changes* that occur. Content changes in work indicate whether the job involves more or less skill, diversity, and worker autonomy. Compositional shifts mean that certain jobs may be created or eliminated in a particular sector in the economy. This distinction serves as a useful tool in sorting out the theoretical and empirical debates that have gone on concerning the relationship between technological innovation and work, and which set the stage for development of policies. The two predominant views in these debates can be characterized as the upgradating position and the deskilling thesis.

The upgrading view (16) characterizes the contemporary world as one where information and efficiency exist as the guiding principles. Noting the decrease in the number of traditional blue-collar workers and the increase in information and service industries, the proponents of this view claim that the quality and character of worklife have improved. Office work, manipulating numbers and symbols, is considered to involve a greater number of human facilities than blue-collar work, is more autonomous, creative, and fulfilling. Automation of factory work is seen to be similar so that where continuous processing technologies are used workers are said to require greater skills and expertise to operate the equipment; instead of being simply laborers workers are viewed as technicians, and the work is generally less strenuous and dirty. This view, in short, argues that the world of work has become more interesting, "professional," and socially interactive—content and compositional shifts have unfolded in a way that enlarge the content of jobs and create more jobs offering a wider range of opportunity. The upgrading view is consistent with the policy model that sees workers and management united in the interest of pursuing quantitative efficiency.

The deskilling view has been set forth above within the discussion of concepts of productivity. This perspective corresponds to the policy model where employees and managers have antagonistic interests and qualitative efficiency is required so that shifts in the composition of work put more people in jobs where they are subject to greater domination. Management, with the aid of new mechanization, continuously reduces the necessary expertise and autonomy on the shop floor. In addition, explicit control mechanisms may be built in to the new equipment by management; for example, word processors and data entry terminals have the capacity to count the number of keystrokes of the workers in order to more closely monitor them. Machines may change the content of jobs so that they become more routine and repetitive. Insofar as fewer skills are necessary to accomplish the same task with the aid of a machine, the change of the new technology may be associated with wage reductions, the replacement of talented employees by those of fewer skills who are willing to work for less money, or the deunionization of the workforce.

In the first decades of the 19th century, British craftsmen smashed the new machines which were being introduced into the textile manufacture. Although these "Luddites" are often considered to have been antitechnology know-nothings, in actuality they were involved in a struggle between emerging industrial classes in which their wages were being lowered, and their jobs were being lost to the new machinery and cheap child labor which could perform the simplified skills which the new equipment permitted. That technology causes displacement and job loss is thus a phenomenon of long recognition. Similarly well-known is the fact that technology is a critical component in the process of creating jobs. This viewpoint was central to the 1966 report of the National Commission on Technology, Automation, and Economic Progress (20). Since, presumably, productive work is intrinsically good (and inherently necessary in a capitalist society), if one believes that automation, including robotics, has the vastly predominant effect of creating jobs, then the appropriate government policy is to do nothing; if one believes that such technological developments lead to considerable job loss (even if temporary), the conclusion about policy is apt to be quite different, and one will advocate any number of possible government activities to mitigate this negative consequence. The most reasoned view of the overall situation, in the opinion of this author, is that technological change both creates and destroys job opportunities; the problem for policy analysis, formulation, and implementation via government programs has to do with tremendous uncertainty about the numbers of jobs being created and lost (how do these compare in magnitude?), the rate of technological change and sequencing (is this a rapid process?, are jobs being created before other jobs are being destroyed or vice-versa?), and the comparability of skills between those jobs which are being lost and those created (how readily can workers who are being displaced find employment in the new slots? what sorts of retraining, if any, might be necessary?) (24). Arguments can be made that increased productivity itself does not always lead to job displacement. It can lower costs which would enhance competitiveness of the producer, raise consumption and create more jobs. Whatever their effect on productivity, robots themselves often perform only a portion of a job task rather than the complete function; as a result, they are more likely to lead to situations where the work process is restructured rather than where whole jobs are lost.

In sum, one may wish to distinguish between the following automation situations:

1. Replacing many unskilled workers by a few skilled workers plus automated machinery (a reduction in labor cost and some upgrading).
2. Replacement of skilled workers by unskilled workers plus automated machinery (deskilling automation).
3. Increasing output by using the same number of workers in skill levels with new automated machinery, such as by speeding up the assembly line.
4. In cases of expanding production and growth, the use of machinery to reduce future new hires.

Such a typology suggests that there may not be any simple overall expression of the labor force impacts of automation, and that government policies ought to vary depending upon the particular effect(s) of specific automation. Unfortunately, such sensitivity has not been characteristic of governmental programs.

Choice of Policy Tools

After government has decided on what objectives it wishes to pursue, and has some reasonable theoretical or empirical belief in a mechanism of social causation (a belief that the programmatic activities it undertakes will produce desired results), there still remains a choice to be made from among an array of policy tools or instruments which can be employed to effect change. Normally these choices are made by a legislature. Policy tools consist of:

- Direct regulation (with a variety of possible sanctions for transgressions).
- Subsidies (through direct payments or transfer of goods and services, or indirectly through a variety of tax mechanisms such as increased depreciation, tax deductibility or credits, etc).
- Grants (outright transfers of monetary support).
- Charges or fees.
- Loans (perhaps at reduced interest).
- Research programs.
- The use of the government's purchasing power to impose conditions in its contracts.

In general, the policies chosen could seek to inhibit the utilization of robots, promote their utilization, guide and control the effect of the impacts which result from a "natural" deployment rate, or do nothing on the part of the government. For example, regarding the workplace environment and protection of workers' safety and health, the government could seek to raise the existing safety and health standards in situations where robots are deployed, develop new safety and health regulations for categories of workers or workplace situations that are not presently covered, establish standards or specifications for the goods the government purchases which would favor those made by automatic processing, sponsor research and development activities in industry to increase the versatility of robots and to decrease their costs (of both purchase and operation), promote robots by means of industrial tax incentives which could allow rapid write-offs of the robotic machinery itself or provide tax credits to enable industry to more easily acquire this type of equipment, remove whatever policy and programmatic barriers might exist which are hampering the diffusion of robots (perhaps workmen's compensation, product liability laws, labor union contracts, and programs and practices regarding worker training and employment), and finally the government might purchase or build robots itself for use in government facilities and programs (this is currently an important issue with regard to outer space activities). The choice of tools and programs will, to a very important degree, depend upon the policy model which one believes. This discussion now turns to look at what actual policies the U.S. government has in pursuing and regarding robots.

SPECIFIC U.S. POLICIES REGARDING ROBOTICS

Interest Groups, Studies, and Legislation

Parties with a major stake in government policies regarding robotics certainly include the manufacturers of robotic machinery and component parts, industrial corporations likely to be interested in using the new equipment in their operations, workers (particularly machinists and others involved in fabrication processes), and government agencies which have a mission directly relevant to the technological capabilities. The Robotic Industries Association is a North American trade association organized "to promote the use and development of robotics and related technologies." It was begun in 1974 and currently represents more than 330 United States robot manufacturers, distributors, corporate users, accessory equipment and suppliers, consultants, service companies and research organizations (25). The Association reflects some of the general principles discussed above: a belief in the beneficence of "progress;" a desire to increase "productivity," which usually is manifested as a desire to cut labor costs; and non-specific fear of competition, of being left behind if one fails to do what everybody else is doing. The Association's Governmental Relations Committee conducted a survey in 1985 to ascertain what legislative issues the members would like the organization to address. These may be fairly said to represent the industrial policy agenda, in order of importance:

- Work towards an extension of the research and development tax credit which was due to expire.
- Initiate specific tax reform for the robotics community.
- Increase government sponsored research and development funding.
- Work for more general research and development/industrial policy legislation.
- Participate in general tax reform.
- Work for simplification of export administration.
- Increase government sponsored education programs.

The RIA recognizes that there exists divergence between the policy agendas of the manufacturing and user firms. For example, users favor general, rather than robot-industry specific, policy. As of this writing, general tax reform has been enacted greatly reducing industry tax breaks; also, as will be discussed below, legislative attempts to obtain greater government support for research and development activities have been stymied by Presidential veto.

The general position of the International Association of Machinists, the union representing most of the organized workers in industries which would be affected by robotics, may be indicated from its 1981 paper on "Rebuilding America" which urged the inclusion of an "economic and social clause" in corporate charters, relevant legislation, foreign trade agreements or treaties, and private commercial contracts in international trade which would guarantee human rights in the workplace, the right to organize unions and bargain collectively, parity wage standards, safety, health, and environmental protections, and prohibitions on discrimination. Situations of economic dislocation would require impact analyses and plans for economic reconstruction. New technology would be closely monitored and regulated to protect workers' privacy, ease job loss, and facilitate job transition. R/D programs would search for new products and markets, creating new employment opportunities (26, 27)

Several studies have been done by U.S. government agencies investigating the consequences of new automated production technologies and suggesting government policies. The National Science Foundation contracted for one of the first of these, completed in September of 1979 (6). This study was based on the premise that "the capability of replacing most human labor, either in industry or the home" made robotics a subject of national policy importance; however, the study found that "it does not appear that there will be the surge of highly versatile robots one would have expected" and therefore it concluded that no major impacts would be likely through 1990 and that there would be no mass displacement of workers over any time period. The study investigated deployment of robots in three major arenas: domestic, the nuclear reactor industry, and manufacturing industry generally. Despite any number of popular articles in the press to the contrary, the contractor found that the domestic usage of robots is likely to be minimal because of both cost and technical limitations. A likely role for robots in the nuclear industry was foreseen, based on health and safety hazards; new automated machinery was seen as useful and likely in the areas of fuel extraction and fabrication, the operation of power plants, fuel reprocessing, and the areas of decommissioning and repair. The large amount of money already invested in nuclear activities makes the use of robots in this arena cost feasible. The bulk of the report (28) concerns manufacturing. The study paid particular attention to investigating the constraints on the deployment of robots. It noted that robots likely to be available to industry in the 1990s are a long way from the general-purpose systems that are common in science fiction. Robots cannot become truly competitive alternatives to human beings for general-purpose operative functions unless there are significant low-cost advances in robots with sufficient capabilities. The conclusion was that robots are not sufficiently flexible, nor are they likely to be inexpensive when great flexibility is developed, to be competitive for the vast majority of operative jobs. Since robotic deployment is likely to be slow (although steady)

> considering historical trends and our present perspective on robots, it seems probably [*sic*] that the impact of robots will be similar to the general impact of automation over the last two decades.

> Present robots do not represent a revolutionary change in the nature of automation but rather a predictable, although significantly qualitative, increase in capabilities. These increases are going to allow automation to expand its domain of application, and the impacts in these new domains are expected to be similar to those previously associated with automation (29).

Although the authors conclude that robots will not cause massive numbers of people to lose their jobs, there will be some job loss and also a newly created demand for different types of jobs; thus labor displacement is seen as an area requiring some policy intervention. This study concludes that "federal actions may be warranted to insure that there are adequate, equitable and satisfactory means of shifting patterns of employment (30)."

Congress' Office of Technology Assessment has also done several policy studies on industrial automation, including robotics. Its first major venture investigated the economic and social aspects of the production and use of programmable automation technologies, and included their impacts on products manufactured, the structure and competitive behavior of manufacturing industries, the numbers and skill mix of people employed in manufacturing, the working conditions in manufacturing jobs, and the education and training requirements implied by growth in these new technologies (14). As part of the assessment process, OTA first released a technical memorandum on selected labor, education and training issues (9). This memorandum focused on questions of methodology and on a critical concern for whether there was sufficiently widespread appreciation of the existing delivery system for education and training. The document covered the working environment including issues of managerial control, boredom and the circumstances of work, and offered an overview of industrial relations in the U.S., especially regarding the legal and regulatory framework (31).

The subsequent study investigates federal policy (both existing and options for new initiatives) along four dimensions: technology development, diffusion, and use; employment policy (excluding training); work environment; and education, training and retraining (32). The study concludes that computer-based automation in manufacturing can improve productivity, product quality, and working conditions. It might have an enormous long-term impact on the number and kinds of jobs available, but will not generate massive nation-wide unemployment over the next decade (although it may aggravate regional unemployment in the Northeast and Midwest). Automation will gradually alter the mix of occupations and skills needed by manufacturers and may, as a result, limit the mobility of manufacturing employees. Its principal employment effects will be limited to certain occupations, particularly metalworking machine operators. In the long-term, world demand will rise for engineers, technicians, maintenance personnel, and management staff. The impacts of automation on the work environment, particularly the psychological aspects of work, will depend on how the technologies are implemented. Although the report recognizes that automation has had some negative effects (such as decreasing employee autonomy and creativity), it sees the possibility that automation will improve jobs by increasing the variety of tasks and challenges. (Consistent with its use of the dominant model, the report does not fully investigate how the realization of such positive impacts will come about if management keeps decisions about automation within its sole prerogative.) The assessment also suggests that new approaches to education, training and career guidance will be necessary because of these technologies. The study covers a wide range of policy options for Congressional consideration. It discusses the continuation of current federal roles, and suggests instead that further actions could be taken in the areas of: strengthening research and development and the evolution of standards, facilitating mobility among occupations and jobs, guarding against downgrading of working conditions and the work environment, and making the instructional system more responsive to demands from the spread of the new technologies. In sum, the report concludes that the overall success of automation-related policy will depend upon the health of the economy and the broader context of macro-economic policy. Although it strongly urges that planning for the future deployment of these technologies be begun now, later events show that the Reagan Administration tends to see such activities as ideologically improper.

In a subsequent related study (24), OTA reported on the relationship between new manufacturing technologies and structural unemployment. This report highlights again that the most vulnerable workers are those who are unskilled or semi-skilled production workers, and that nearly half of all workers displaced between 1979 and 1984 worked in manufacturing industries, especially those hard hit by international competition. Its main conclusion is that existing federal policies, as embodied in programs such as Title III of the Job Training Partnership Act of 1982, have not provided adequate mechanisms to deal with this situation, estimating that no more than 5% of those eligible for assistance are now being served. The policy options that are discussed for Congressional action include making employment services more readily available to workers, enhancing vocational and basic education training, improving labor market and occupational information, encouraging employers to offer workers chances at both vocational and basic education while they are still employed, and stimulating research on the effects technological changes have on jobs. OTA is continuing to do analyses of technology and economic transition in the U.S.

In 1984, both Houses of the U.S. Congress passed the Manufacturing Science and Technology Research and Development Act (originally S.B. 1286, subsequently folded in as a Title of H.R. 5172). Regarding the policy needs engendered by new automation and robotics technologies, the Senate Committee on Commerce, Science, and Transportation reported that two major policy arenas needed attention: assistance in research to develop new technologies that will meet the goals of higher productivity, flexibility and product quality; and exploring ways to increase utilization of the new manufacturing technologies that are already available (33). The bill reported by the Senate had five major provisions: to establish centers for manufacturing research and technology utilization at consortia (composed of non-profit research institutions with entities such as states, industry, companies or associations, etc); to authorize the Department of Commerce to award grants and contracts for research purposes; to permit the Department to identify approaches for enhancing industry utilization of advance manufacturing technologies and to identify the effects of such technologies upon workers, including the potential need for retraining of displaced workers; to direct the Department to analyze the long-term ability of certain industrial sectors which are sensitive to technological developments to remain competitive; and to set up an advisory committee to assist to achieving the above goals. The report of the House of Representatives Committee on Science and Technology indicated significant concern with the decline of U.S. manufacturing industries and especially the way in which new manufacturing technologies have contributed to heightened competition from foreign firms (34). The new technologies are specifically praised because of their ability to promise substantial improvement in quality of product and labor savings which can return these industries to world pre-eminence. Indeed the only concern of workers which is addressed is that their fears have a negative effect on the adoption of these technologies (perhaps somewhat ironic from a Democratically-controlled committee). While recognizing that the primary responsibility for carrying forth these policies lies with private industry, the House Committee saw an important role for the federal government in supporting programs of research and development involving firms, universities, industry associations, and selected national laboratories, and by upgrading education and training efforts for engineers and technicians. Accommodating the fear of the Reagan Administration that the bill passed by the Senate might place the federal government in the position of "selecting technology winners and losers," the House tried to adopt a narrower approach to eliminate any semblance of government intervention (35). Nonetheless, on October 30th President Reagan vetoed the bill which embodied these program proposals for the Department of Commerce and also contained provisions regarding support by the National Science Foundation for engineering research and education as well as authorization of programs for the National Bureau of Standards. The President's message of disapproval was predicated on two grounds. The first was opposition to "significant federal expenditures with little or no assurance that there are any benefits to be gained," especially in light of the Administration's attempts to revitalize the general economy which, it is believed, will sufficiently stimulate American industry. The second reason is that the bill

> . . . would establish a new program providing Federal financial support for a variety of research, development, education and training activities, whose purported purpose would be to improve manufacturing technologies, including robotics and automation, these activities would total $50 million during fiscal years 1985–1988 and represent an unwarranted role for the Federal government. The decisions on how to allocate investments for research on manufacturing technologies are best left to American industry. It is highly doubtful that this Act and resulting Federal expenditures would improve the competitiveness of U.S. manufacturing.
>
> The new role for the Federal government contemplated by [the Bill] could also serve as the basis for a Federal industrial policy to influence our Nation's technological development. This Administration has steadfastly opposed such a role for the Federal government (36).

The Chair of the House Committee stated at a public hearing a year later that the Committee members

> were disappointed that this law was vetoed by the President, but we also have to recognize that this is an area in which the country's future economy is going to go, and, so although it may have been vetoed last year, we expect similar legislation to be passed eventually on the national level. . . . (37).

The House Committee is also investigating legislation regarding vocational retraining at community colleges to be subsidized by federal funding.

In the Fall of 1986, major new tax reform legislation was enacted but the implementing regulations by the IRS have not yet been promulgated and the effects of all of the provisions have not yet been sorted out. However, in general terms, the legislation reduces or eliminates much of the tax law subsidies which were going to industry, as part of a larger policy determination to shift the tax paying burden from individuals to corporations (which had been paying increasingly smaller percentages of taxes under previous tax law changes). Thus, it is likely that, to the extent previous tax credits and credits for research costs in the area of robotics have helped to stimulate the development and deployment of these technologies, tax incentives for robotics will probably not exist in the future.

Agency Programs as Embodiments of Existing Government Policy

In addition to the general technology assessment on robotics mentioned above, the National Science Foundation, through its Division of Policy, Research and Analysis, Technology Assessment and Risk Analysis Group, has funded a few studies which are relevant to robotics (38). And, under recent Congressional authorization, it has established a small number of centers for engineering excellence at engineering colleges around the nation, including one at the University of California, Santa Barbara, which focuses on manufacturing automation. In the late 1970s, based on the research the Foundation had been sponsoring on advanced manufacturing and assembling technologies at several academic institutions, NSF mounted a demonstration project at the Westinghouse Corporation which floundered because of a lack of clear objectives and theoretical underpinning as regards the proper role for the government and how to weigh the importance of the various impacts.

> Federal policy toward manufacturing technology for new products or production processes is piecemeal at best. Relevant programs principally address research and development, although macro-economic policies and more specific programs, such as tax credits, may indirectly stimulate technology change in manufacturing by encouraging capital investment. Only in the area of defense procurement does the Federal government actively coordinate product and process technology development and application (39).

As already shown, federal policy on diffusion or increasing utilization of these technologies has been minimal, mainly for ideological reasons. The federal activities that have existed are small and mainly directed toward research and development activities with some concern for standards which would increase the ease of having these technologies used and applied. The actual research and development activities of the federal government have been mainly carried out by those agencies which themselves are users, or potential users, of robotics—the National Aeronautics and Space Administration and the Department of Defense (there has also been a small amount of activity in the government's nuclear energy programs). In regard to NASA, there has been very little policy analysis investigating NASA's activities. The space-station program will, however, be a major potential arena for using robotic technologies. There have been specific Congressional mandates in some NASA appropriations bills which have directed the agency to examine how robotics can be used on the space station program and how to stimulate the necessary R&D for such technologies (40). These activities are tied to a still-unresolved controversy over the extent to which the space station should be "manned" as opposed to operated by automated machinery.

> . . . The Senate Bill requires . . . a NASA automation report. This report and the ongoing NASA-sponsored automation study are expected to be completed no later than April 1, 1985.
>
> The Committee also expects the [space station] contractors to devote a significant portion of their effort pursuing the study of automation and robotics technologies identified in the [Advanced Technology Advisory Committee] report with the objective of advancing the state-of-the-art in these technologies and increasing their terrestial application.
>
> The Committee acknowledges the need to pursue a manned space station; however, the Committee believes that NASA needs to pursue the areas of automation and robotics more vigorously. Consequently, the bill language is intended to assure that such advanced technologies are indeed made an integral part of the planning development for a manned space station (41).

Subsequently, at the request of the Senate Committee on Commerce, Science, and Transportation, the Office of Technology Assessment convened several workshops and developed a staff paper (42). The motivation for this work was the "intense national and Congressional interest in the U.S. industrial competitiveness, and specifically an interest in the extent to which the space station can become the showcase of advance technology for industrial automation. . . ." (43). OTA concluded that "the space station could be an appropriate vehicle to advance the state-of-the-art in [automation and robotics], provided that objectives are carefully chosen and early expectations are not unreasonably high (44)." However, at projected NASA funding levels and space station plans, "no significant advances in the state-of-the-art in automation and robotics are likely to result from NASA activities, and the eventual evolution of the space station . . . will be hampered (45)." One has the sense that there is a conflict between the space bureaucracy and important members of Congress in this area, since NASA maintains the position that advancing terrestrial robotics is not a goal of its space station program.

The Department of Defense has a number of relevant activities, principally those of the Department of the Army, which it carries out directly with contractors, and those of the Department of the Navy, which are in large measure carried out jointly with the National Bureau of Standards in the Department of Commerce. Either items are used in such enormous quantities by the Armed Services that even a small reduction in the cost per unit would amount to a large savings, or items are so specialized and used in such small numbers that their costs, and the costs of spare parts for repairs, are very high. Often the operational life of the combat unit (such as a ship) is greater than the life of the manufacturing companies which make the components, thus exacerbating the problem of repair or replacement. Thus, there is a generalized belief that what is good for industry is good for the Defense Department; in particular, if all the information about making a specialized part in a factory operation could be represented by a memory file or automated program, the fabrication of another unit, even many years later, would be facilitated. However, the DOD's production needs are so specialized that some scholars argue that it will never develop engineering for cheap production of many multiples of a product (the typical industrial situation).

The Army's Manufacturing Technology Program has been hard hit by recent budget reductions (of about two-thirds), apparently on the belief of high-level departmental personnel that the program was not doing enough to improve the general productivity of their contractors. The program has been limited to "captive facilities" such as Army depots, arsenals and ammunition plants, or other government-owned facilities even if they are operated by a contractor, but can no longer operate by investing in facilities which are contractor owned. The motivation of the Army in these investments is to reduce the cost of its weapons systems. There are not too many situations in which safety and protection of workers from hazard plays a role, except in regard to demobilizing chemical weapons. The economic potential comes from the elimination of labor and the production of less scrap and less need for reworking fabricated items; also, these items would be able to be made with higher quality, and the batches or runs would have greater repeatability and accuracy. A list of the projects current in mid-1986 includes such stated problems as "is labor intensive and creates harsh working environment," "higher overhaul/rebuild costs," "is time-consuming and labor intensive," "inefficient and costly," "requires much hand labor," and "limited productivity." Contractors can still be hired by the Department of the Army to do development work, but the actual utilization ("primary and first use") must be by the Army; afterwards, the process is available for industrial utilization.

The Navy is largely a silent partner in the activities of the National Bureau of Standards, funding about one-half of NBS activities. It has entered into an arrangement with the state of South Carolina and several private firms to establish a laboratory facility in the Charleston shipyard to work on creating an automated "manufacturing cell."

The National Bureau of Standards, in its Center for Manufacturing Engineering, conducts two sorts of programs relevant to robotics. The first has to do with the more traditional concern of NBS for questions of measurement and standardization. The basic question the Center is trying to address concerns interface standards for automated equipment, on the belief that the future of automation for U.S. industry will be heavily dependent upon the standards enabling simple interconnection of the units together. The driving force behind this work is competition with Japan and Germany. The mainstay of U.S. industry is smaller companies of less than 50 employees which cannot afford capital investment, specialized technical employees, and the like, and thus are unable to rapidly adopt automation technologies. The goal of the Bureau is to bring automated manufacturing to the state that exists for home entertainment technologies, ie, all the various units can hook together simply. The bureaucracy is trying to foster a "hierarchical architecture" in which the pieces of manufacturing technology break down commands received from machines "above" into simpler portions and pass them on. That concept is seen as the basic unit for automated factories of the future, but the Bureau (and its parent Department of Commerce) do not want the standards to stifle technological development. A related set of activities concerns the traditional measurement function which is at the heart of the Bureau's historic work, and relates to measurements for the output of robotic systems (quality control for example) as well as the design of robotics. The theory here is that if you can control the manufacturing system sufficiently (in terms of movement, compensation for variable operating temperatures, and the like) then you do not have to measure the output of the system because its degree of replicability will be so high that it will function as if the system were monitoring itself.

In addition to these activities, a sub-project of the Bureau's Center is the Automated Manufacturing Research Facility (the activity which the Navy is particularly interested in). The Facility is a kind of consortium of government researchers, industrial firms, and university people (usually on NBS grants) which provides a basic array of modern manufacturing equipment and systems for purposes of experimentation with new standards, and studying new methods of measurement and quality control. The ultimate goal appears to be to create a Year 2000 fully automated machine shop, but to do it by focusing on aspects of the mission of the National Bureau of Standards, ie, its traditional concerns for problems of measurement and standardization. Although the Bureau has no authority to impose these standards and measurements, they have been adopted in the past because private industry considers them unbiased and authoritative, and the prospects for the future also look good. As of September 1986, three industrial standards have already been developed based on this research: a standard method for exchanging graphics data between otherwise incompatible computer-aided design systems, a standard for the characterization for computerized coordinate measuring machines, and a standard for the method of measuring surface texture. Seven other potential standards are being developed. Although the highest echelons of the Department of Commerce have visited the Facility and are very supportive, the program exists at the very edge of toleration of the Reagan Administration ideology against government subsidization of private industry. The distinction that is made is that NBS will focus on research, leaving development of technology to the private sector. Although there are some development spinoffs, the official literature resolutely maintains the distinction—the AMRF is not a demonstration or a prototype of the factory of the future.

Since it has been shown that one of the main forces pushing for development of government policies on robotics is international competition, it should not be surprising that the International Trade Administration of the Department of Commerce has been looking at new manufacturing technologies and the position of U.S. industry in relationship to the world market. In 1985 a study was published on "flexible manufacturing," and this year studies are in progress on both CAD/CAM and robotics (46). The ITA researchers find the structure of the robotics industry bothersome; of the ten major producers (and the definition of "producer" may affect one's analysis) only three are independent of foreign materials and parts and could stand alone during a national emergency. Most of "U.S. production" largely consists of taking basic equipment manufactured in Germany or Japan and adding on elements which enable the robot to be applied to specific functions domestically. This is especially true in regard to uses within the auto industry, probably the largest arena of current application for robots in the United States. Although many of the companies show a profit, often this comes from systems service operations, eg, where robots are combined with automated machinery to provide an entire manufacturing cell on a turnkey basis. Given this view of the domestic industry as mainly consisting of software and applications, and a projection that hardware is not, and is not likely to be, a main aspect of it even in the future, the conclusion is that the industry itself is not too competitive. Thus the report will probably recommend, as policy suggestions: (*1*) that existing research and development efforts be maintained (in NASA, NBS, NSF, and DOD); (*2*) since robotics is already being applied, the government need not get involved in these activities (which would be consistent with the present Administration's ideological biases), but should expand efforts to promulgate accurate information about the industry and the equipment's capabilities; (*3*) the government should investigate conducting its research and development activities through limited partnerships as a way to strengthen the domestic industry; (*4*) programs to expand upon current NBS activities developing robotic industry standards should be created as a way of reducing costs; (*5*) continue studies on the employment effects of robots, provide accurate factual documentation and publicize valid statistics, offer counseling assistance to laid-off workers and to potential robotics workers (although it is estimated that 45,000 new jobs will be created in manufacturing robots, the jobs lost are estimated to exceed this and be at far lower skill levels than would be required in the robot assembly industry); and (*6*) provide funding to educational institutions for the purposes of training skilled workers to go into the industry.

Methodological Concerns

Government policies on robotics are much more likely to be based on ideologies (what those in power would like to have happen) than on factual analyses because of serious methodological problems: inadequate theoretical bases, conflicting and imprecise definitions, lack of a clear analytical methodology for ascertaining impacts and mechanisms to realize the positive effects and mitigate the negative ones, and a data pool consisting of spotty and not always reliable statistics (47). Earlier in this article, it was discussed how a lack of a demonstrable underlying theory renders policy work implausible and suspect. Not only is the dominant U.S. view opposed to explicitly discussing theory (perhaps because this is associated with foreign intellectuals and "-isms"), there is substantial lack of agreement on even general theoretical conceptions.

> Despite the important role of industrial relations in the U.S. economy, the analysis of industrial relations tends to be relatively imprecise and experiential. As one participant in the OTA labor markets and industrial relations workshop put it, there seem to be

more "ad hoc-eries" than true theories for explaining industrial relations phenomena. Further complicating an industrial relations issue are the differences in approach taken by different analysts. For example, most labor economists and so-called industrial relationists tend to regard workers and managers as having opposing interests, with workers striving to minimize work effort and maximize compensation, and managers striving to minimize cost and maximize production. Most organizational behaviorists and organizational development specialists tend, by contrast, to regard workers and managers as sharing basically similar interests that stem from their association with the same organizations. The former group tends to focus on the setting of wages and other internal "economic" issues, while the latter group tends to focus on job satisfaction and performance supervisory relationships, and job designs (48).

For example, according to one scholar "simple textbook models of supply and demand are not sufficient for analyzing industries that are highly unionized, capital intensive, and in which firms are able to exercise market power. An alternative analysis which poses very different questions and possibilities, is a prerequisite to research" (49).

In addition, there is little agreement on the definition of the phenomenon which is the subject. One study offers "a pragmatic definition" of robots: simply, a robot is viewed as a means of mechanizing jobs which are performed by the so-called operative worker, specifically, in the batch manufacturing industries. Batch processes are where the distinctions between a robot and general automation devices are evident, "automation can be viewed as the process of mechanizing some task or process currently being carried out by a human worker. Automation consists of rationalizing the task (expressing it as an explicit procedure), controlling relevant factors in the work environment, and fabricating a physical system capable of performing the task in that environment. In this study "the robot is viewed as an aspect of automation having characteristics which can be described as programability and adaptability. These characteristics enable a specific robot to automate jobs which require certain types of flexibility" (50). The Office of Technology Assessment examined a broader category, "programmable automation technologies," which were described as including "computer-aided design (CAD); computer-aided manufacturing tools, eg, robotics, numerically controlled (NC) machine tooled, flexible manufacturing systems (FMS), and automated materials handling (AMA); and computer-aided techniques for management, eg management information systems (MIS), and computer-aided planning (CAP). When systems for design, manufacturing, and management are used together in a coordinated system, the result is computer-integrated manufacturing (CIM)" (51). In other words, "programmable automation, which weds computer and data-communications capabilities to conventional machine abilities, increases the amount of process control possible by machines and makes possible the use of single means of equipment and systems from multiple applications" (52). Key aspects of robots seem to be that they are capable of physical manipulations and that they are computer programmable; although the robotics trade association usually includes in the definition of the term that the machine be autonomous, other authors and workers in the field allow the term to be used for situations where there is some human control.

Definitional issues abound as well when discussing the impacts of robotics/automation on the workforce. As already noted above, the dominant view is that the mechanization of industrial and office work will "upgrade" jobs by requiring a better trained and more highly educated working population. One problem with this view is that it uses average values, a statistical process which thus can mask any increasing polarization of skill levels which might be occurring within the category being observed (ie, both upgrading and deskilling could be occurring simultaneously). The definitions of the occupational categories themselves, however, are far greater barriers to meaningful policy analysis. The categories used by researchers are those of the U.S. Bureau of the Census and the U.S. Bureau of Labor Statistics, originally done in the 1930s; the original labels "Craftsmen," "Operatives," and "Laborers," soon became—both in official terminology and in common parlance—the categories of "skilled," "semi-skilled," and "unskilled" labor. The claimed distinction in skill levels between categories "was based not upon a study of the occupational tasks involved, as is generally assumed by the users of the categories, but upon a simple *mechanical* criterion, in the fullest sense of the word. . . . By making a connection with machinery . . . a criterion of skill, it guaranteed that with the increasing mechanization of industry the category of the "unskilled" would register a precipitous decline, while that of the "semi-skilled" would show an equally striking rise. This statistical process has been automatic ever since, without reference to the actual distribution of "skills." (53) In terms of examples, in the fantasy world of the statistics a truck driver is assumed to have greater skills than a gardener and a parking lot attendent more than a lumberman, because their jobs are connected with more elaborate machinery, no matter how trivial that connection may be nor how few actual skills are required to control or watch the machine.

A similar arbitrariness exists with categories of industrial activities as well, such as the distinction between "manufacturing" and "service" occupations. ". . . [F]or example, restaurant labor which cooks, prepares, assembles, serves, cleans dishes and utensils, etc, carries on tangible production just as much as labor employed in many another manufacturing process; the fact that the consumer is sitting nearby at a counter or table is the chief distinction, in principle, between this industry and those food-processing industries which are classified under manufacturing." (53) Perhaps problems such as these are behind the OTA conclusion that "historically, attempts to forecast detailed changes in occupational employment have met with limited success" (54).

Forecasting of labor force impacts from automation appear to be based on either of two methodologies—called "engineering" and "economic" by OTA—which are actually very limited models and reflect a lack of understanding of the relationships between the mechanization processes and the consequences to the labor force (55). Finally, policy analysis is severely hampered by the spotty amount and quality of systematically assembled data, the statistics. And attempts to make statistical comparisons over time are particularly prone to error because of both problems of poor data and shifting category boundaries.

"Moreover, what data may exist (eg, in case studies) may have little general value because early programmable automation applications have been limited in number compared to applications of other types of systems and equipment, and they have been tailored to individual company needs. Early applications also are likely to be different from later applications involving more sophisticated equipment and systems, especially since future applications are expected to feature greater computer integration of production and other company activities" (56).

Despite these problems, however, government policies have been promulgated, and will be promulgated in the future. They will have enormous implications for the robotics industry, the consumer, the worker, manufacturing firms, and so on. Close examination of policies and programs in other industrialized countries, mounting pilot projects and case studies here, and careful use of assessment and evaluation mechanisms are probably the best ways one can proceed sensibly to understand government policies for robotics.

BIBLIOGRAPHY

1. L. H. Tribe, "Policy Science: Analysis or Ideology?," *Philo. and Pub. Affairs,* **2** (1), 66–67 (Fall 1972).
2. M. R. Berg, K. Chen, and G. J. Zissis, "Methodologies in Perspective," in S. Arnstein and A. Christakis, eds., *Perspectives on Technology Assessment,* Science and Technology Publishers, Jerusalem, 1975, pp. 40–42.
3. R. S. Ahmed and A. Christakis, "A Policy-Sensitive Model of Technological Assessment.", *IEEE Trans. on Systems, Man and Cybern.,* **SMC-9,** 450–55, Sept. 1979.
4. P. L. Bereano, ed., *Technology as a Social and Political Phenomenon,* John Wiley & Sons, Inc., New York, 1976, p. 53.
5. P. L. Bereano, ref. 4, pp. 54–60.
6. *Technology Assessment: The Impact of Robots,* report to the National Science Foundation under grant ERS7600637, Eikonix Corp., Burlington, Mass., Sept. 30, 1979, pp. 81–86.
7. P. L. Bereano, ref. 4, pp. 148–157.
8. P. L. Bereano, ref. 4, pp. 8–10.
9. U.S. Congress, Office of Technology Assessment, *Automation and the Workplace: Selected Labor, Education, and Training Issues, A Technical Memorandum,* U.S. Government Printing Office, Washington D.C., March 1983; p. 62.
10. D. Noble, *The Forces of Production,* Alfred A. Knopf, New York, 1984; H. Braverman, *Labor and Monopoly Capital,* Monthly Review Press, New York, 1974 is perhaps the earliest influential scholarly work exploring this more radical view.
11. D. Dickson, *The Politics of Alternative Technology,* Universe Books, New York, 1975, pp. 79–80.
12. P. L. Bereano, "Technology and Human Freedom," *Science for the People,* **16**(6), 17–21 Nov./Dec. 1984.
13. Ref. 9, pp. 5–8.
14. U.S. Congress, Office of Technology Assessment, *Computerized Manufacturing Automation: Employment, Education, and the Workplace,* OTA-CIT-235, U.S. Government Printing Office, Washington D.C., April 1984.
15. *Ibid.*, p. 369.
16. D. Bell, *The Coming of Post-Industrial Society,* Basic Books, Inc., New York, 1973.
17. Ref. 9, pp. 75–80.
18. Ref. 9, pp. 60–61.
19. H. Arendt, *The Human Condition,* University of Chicago Press, Chicago, 1958.
20. National Commission on Technology Automation, and Economic Progress, *The Employment Impact of Technological Change,* U.S. Government Printing Office, Washington, D.C., 1966, Appendix vol. II.
21. Ref. 11, pp. 73–74.
22. M. Cooley, *Architect or Bee?, The Human/Technology Relationship,* South End Press, Boston, 1980, pp. 81–129.
23. A. M. Lee and P. L. Bereano, "Developing Technology Assessment Methodology: Some Insights and Experiences, *Tech. For. Social Change,* **19,** 15–31 (1981).
24. U.S. Congress, Office of Technology Assessment, *Technology and Structural Unemployment: Reemploying Displaced Adults,* OTA-ITE 250, U.S. Government Printing Office, Washington D.C., February 1986; see esp. Chapts. 4, 8.
25. Robotics Industries Association, *1985–86 Annual Report,* 2, available from RIA at 900 Victors Way, P.O. Box 3724, Ann Arbor, Mich 48106.
26. International Association of Machinists and Aerospace Workers, 1300 Connecticut Ave. NW, Washington, D.C. 20036, communication.
27. Ref. 9, pp. 88–96 and Appendix B "Industrial Relations," pp. 53–59.
28. Ref. 6, pp. 98–223.
29. Ref. 6, p. 216.
30. Ref. 6, p. 231.
31. Ref. 9, pp. 53–59.
32. Ref. 14, pp. 373–397.
33. U.S. Congress, *Senate Report 98–431* (May 8, 1984).
34. U.S. Congress, House of Representatives, *Report 98–1078* (Sept. 25, 1984).
35. *Ibid.*, pp. 10–11.
36. "Federal Expenditures for Manufacturing Technologies," Memorandum of Disapproval of H. R. 5172, Oct. 30, 1984, *Weekly Compilation of Presidential Documents,* **20** (4), 1696–1697 (Nov. 5, 1984).
37. U.S. Congress, House of Representatives, Hearing, *The Role of Automation and Robotics in Advancing United States Competitiveness,* U.S. Government Printing Office, Washington D.C., No. 62, October 7, 1985, P. 2. In a related fashion the National Institute for Occupational Safety and Health had planned to do a study of the impacts of office automation, with the participation of activist employee groups, but the Administration held it up. OTA, however, not subject to Executive Branch pressures, did a report along these lines.
38. W. Leontief and F. Duchin, *Automation, the Changing Pattern of U.S. Exports and Imports, and Their Implications for Employment,* Institute for Economic Analysis, New York University, March 1985.
39. Ref. 14, p. 373.
40. U.S. Congress, *Senate Report 98–506* (June 7, 1984).
41. *Ibid.*, pp. 65–66.
42. U.S. Congress, Office of Technology Assessment, *Automation and Robotics for the Space Station: Phase B Considerations,* undated (early 1986).
43. *Ibid.*, p. 2.
44. *Ibid.*, p. 3.
45. *Ibid.*, p. 5.
46. J. Mearman and co-workers, "A Competitive Assessment of the U.S. Robotics Industry," U.S.G.P.O. #003-009-00499-3, 1986, and "A Competitive Assessment of the U.S. Computer Aided Design and Manufacturing Systems Industry," U.S.G.P.O. #003-009-00498-5, 1986.

47. Ref. 4, pp. 404–414.
48. Ref. 9, p. 53.
49. Ref. 9, p. 61.
50. Ref. 6, p. 98.
51. Ref. 14, p. 33.
52. Ref. 9, p. 4.
53. H. Braverman, ref. 10, p. 429.
54. Ref. 9, p. 12.
55. Ref. 9, pp. 14–19.
56. Ref. 9, pp. 11–12.

GRIPPERS

KAZUO TANIE
Mechanical Engineering Laboratory, MITI
Ibaraki, Japan

INTRODUCTION

Generally the gripper for industrial robots is for used special purposes—a device to handle limited shapes of objects and limited functions. This kind of gripper makes designing easy and also keeps machinery costs relatively inexpensive, but versatility and dexterity are reduced. In some applications the simplification of gripper function may be more important than versatility and dexterity from the point of view of economics. In others, however, the gripper will be required to handle and manipulate many different objects of varying weights, shapes, and materials. The universal gripper, actually robot hands, will be suitable in such cases.

Currently, the development of a universal gripper and the investigation of manipulation using it are under way. There are no practical universal grippers or hands at present. Only an outline of recent developments on universal grippers is described, with gripper functions and related design factors.

FUNCTIONS OF GRIPPERS AND RELATED FACTORS

Human-hand grasping is divided into six different types of prehension: palmar, lateral, cylindrical, spherical, tip, and hook, which were identified by Schlesinger (Fig. 1) (1). Crossley classified manipulation functions by human hand into nine types: trigger grip, flipping a switch, transfer pipe to grip, use cutters, pen screw, cigarette roll, pen transfer, typewrite, and pen write (2).

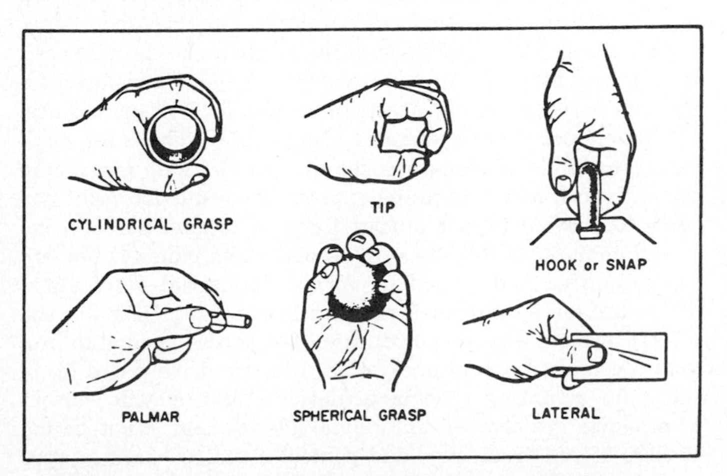

Figure 1. The various types of hand prehension (Schlesinger).

There are several factors relating to these variations of function. Important factors are the number of fingers, the number of joints for each finger, and the number of degrees of freedom of a hand.

A human arm including a hand has five fingers: the thumb, the index finger, the middle finger, the third finger, and the little finger. The whole arm structure has 27 degrees of freedom (DF), 20 of which are for the hand (3). Each finger except the thumb has three joints, and each can produce 4 DF motion. The thumb has two joints with 3 DF. There is 1 DF in the palm. To approximate a subset of the human grasp and hand manipulation, the relation between gripper structure and its function must be considered.

To achieve minimum gripping function, a gripper needs two fingers connected to each other using a joint with 1 DF for its open-close motion. If the gripper has two rigid fingers, it has only the capability of grasping objects of limited shapes and is not able to enclose objects of various shapes. Also, this type of gripper cannot have manipulation function because all degrees of freedom are used to maintain prehension.

There are two ways to improve the capability to accommodate the change of object shapes. One solution is to put joints on each finger. The other is to increase the number of fingers up to a maximum of five. The manipulation function will also emerge from this. To manipulate objects it is usually necessary that the gripper have more fingers and joints driven externally and independently than does the gripper used only for grasping objects. The more fingers, joints, and degrees of freedom a gripper has, the more versatile and dexterous it can become. Table 1 shows approximate relations between the numbers of fingers and the functions of grippers.

GRIPPER CLASSIFICATION

A gripper can be designed to have several fingers, joints, and degrees of freedom, as mentioned before. Any combination of these factors gives different grasping modalities to a gripper. Also, a gripper can be designed to include several kinds of drive methods. A discussion of classifications according to

Table 1. Approximate Relations between the Numbers of Fingers and Joints and the Functions of Gripper[a]

Type of Finger	Functions		
	Grasping	Shape Accommodation	Manipulation
2fG-Rf[b]	O	X	X
2fG-Af[c]	O	O	X
3fG-Rf	O	O	X
3fG-Af	O	O	O
5fG-Rf	O	O	X
5fG-Af	O	O	O

[a] O: Some of the grippers can involve the function. X: None of the grippers can involve the function.
[b] nfG-Rf (n = 2, 3, and 5) means n-finger-type gripper with rigid fingers.
[c] nfG-Af (N = 2, 3, and 5) means n-finger-type gripper with articulate fingers.

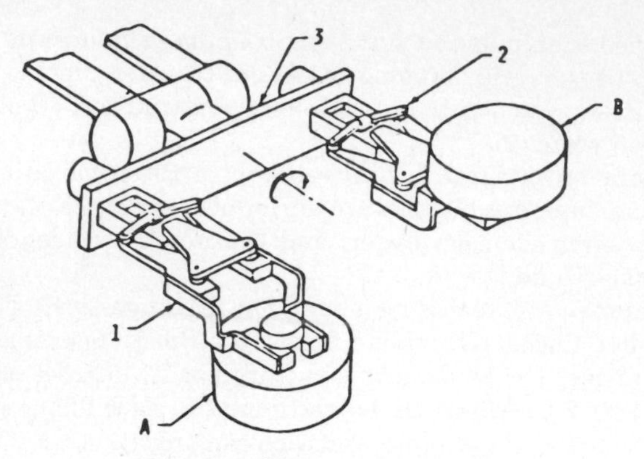

Figure 2. Multigripper system.

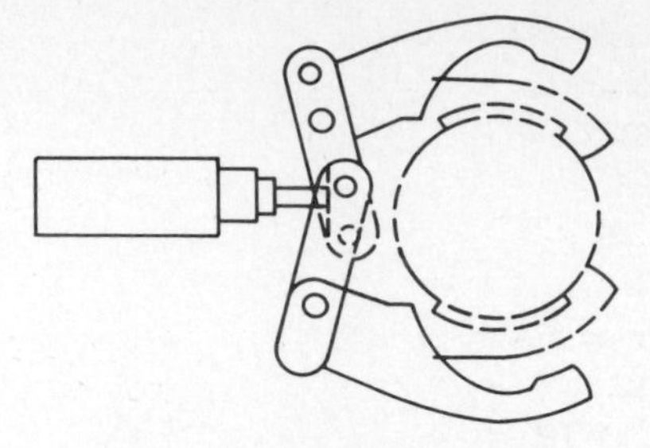

Figure 3. External gripper.

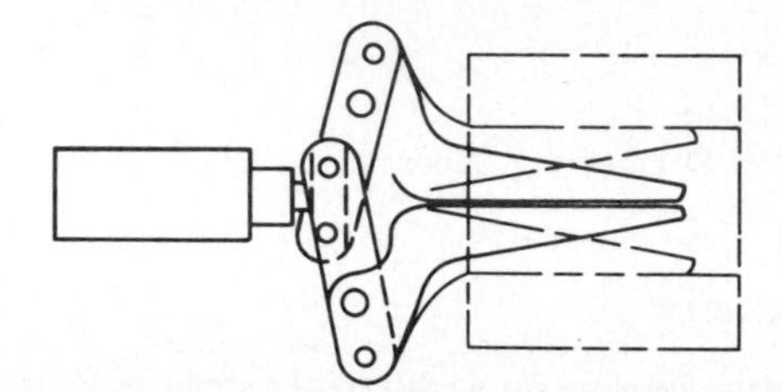

Figure 4. Internal gripper.

grasping modalities follows later. The drive system is described in the next section.

In general, grippers can be classified according to type of grasping modality as follows: 1. mechanical finger; 2. special tool; and 3. universal finger. Modality 1 corresponds to grippers with fingers designed for a special purpose. This category includes less versatile and less dexterous finger grippers with fewer numbers of joints compared with modality 3. However, they economize in the device cost.

This type of finger can be subject to finger classifications, for example, the number of fingers, typically two-, three-, and five-finger types. For industrial applications, the two-finger gripper is the most popular. The three- and five-finger grippers, with some exceptions, are customarily used for prosthetic hands for amputees.

Another classification is the number of grippers, single or multiple, mounted on the wrist of a robot arm. Multigripper systems (Fig. 2) enable effective simultaneous execution of more than two different jobs. Design methods for each individual gripper in a multigripper system are subject to those of single grippers.

Classification of the mode of grabbing results in external and internal systems. The external gripper (Fig. 3) is used to grasp the exterior surface of objects with closed fingers, whereas the internal gripper (Fig. 4) grips the internal surface of objects with open fingers. There are two finger-movement classifications: translational finger grippers and swinging finger grippers. The translational gripper can move its own fingers, keeping them parallel. The swinging gripper involves a swinging motion of fingers.

Another classification may be possible according to the number of degrees of freedom included by gripper structures. Typical mechanical grippers belong to the classification of 1 DF. A few grippers can be found with more than 2 DF.

Modality 2 is a special-purpose device for holding objects. Vacuum cups and electromagnets are typical devices in this class. In some applications the objects to be handled may be too large or too thin for finger grippers to grasp them. Here this gripper has a great advantage over the others.

Modality 3 is comprised of multipurpose grippers of usually more than three fingers and/or more than one joint on each finger which provide the capability to perform a wide variety of grasping and manipulative assignments. Almost all grippers in this category are under development.

DRIVE SYSTEM FOR GRIPPERS

In typical robot systems there are three kinds of drive methods: electric, pneumatic, and hydraulic. Pneumatic drive can be found in gripper systems of almost all industrial robots. The main actuator systems in pneumatic drive are the cylinder and motor. They are usually connected to on-off solenoid valves which control their directions of movement by electric signal. For adjusting the speed of actuator motion, air-flow regulation valves are needed. A compressor is used to supply air (maximum working pressure, 10 kg/cm^2) to actuators through valves.

The pneumatic system has the merit of being less expensive than other methods, which is the main reason that many industrial robots use it. Another advantage of the pneumatic system relates to the low degree of stiffness of the air-drive system. This feature of the pneumatic system can be used effectively to achieve compliant grasping, which is necessary to one of the most important functions of grippers: to grasp objects with delicate surfaces carefully. On the other hand, the relatively limited stiffness of the sytem makes precise position control difficult. Air servo valves are being developed for this purpose but are not practical enough for widespread use.

The electric-drive system is also popular. There are typically two kinds of actuators, d-c motors and step motors. In general, each motor requires appropriate reduction gear systems to provide proper output force or torque. Direct-drive torque motors (DDM) are commercially available (4) but are too expensive to be used in normal industrial application. There are few examples of robot grippers using DDM. In the electric system a servo power amplifier is also needed to provide a complete actuation system. Electric drive has a lot of merit for actuating robot articulation. First, a wide variety of products are commercially available. Second, constructing flexible signal-processing and control systems becomes very easy because they can be controlled by electric signals, and this enables the use of computer systems as control devices. Third, actuators in electric systems, especially those using d-c motors, can be used for both force and position control. However, there are some drawbacks of the electric system as follows:

1. It is a little expensive compared with the pneumatic system.
2. The transient response is lower than those in pneumatic and hydraulic systems.
3. It is less stiff than hydraulic systems.
4. It can not be used in an explosive environment because of its spark and heating.

Hydraulic drives used in robot systems are electrohydraulic drive systems. They have almost the same system configuration as pneumatic systems, though their features are different from each other. A typical hydraulic drive system consists of actuators, control valves, and power units. There are three kinds of actuators in the system: piston cylinder, swing motor, and hydraulic motor. To achieve position control using electric signals electrohydraulic conversion devices are available. For this purpose electromagnetic or electrohydraulic servo valves are used. The former provides on-off motion control, and the latter is used to get continuous position control. Hydraulic drive gives accurate position control and load-invariant control because of the high degree of stiffness of the system. On the other hand, it makes force control difficult because high stiffness causes high pressure gain, which has a tendency to make the force control system unstable. Another claimed advantage of hydraulic systems is that the ratio of the output power per unit weight can be lower than in other systems if high pressure is supplied. Facts show this drive system can provide an effective way to construct a compact high-power system.

Outside of the foregoing three types of drive, there are a few other drive methods. One method uses a springlike elastic element. A spring is commonly used to guarantee automatic release action of grippers driven by pneumatic or hydraulic systems. Figure 5 shows an example of a spring-loaded linkage gripper using a pneumatic cylinder (5). Gripping action is performed by means of one-directional pneumatic action, while the spring force is used for automatic release of the fingers. This method considerably simplifies the design of the pneumatic or hydraulic network and its associated control system.

The spring force can also be used for grasping action. In this case, the grasping force is obviously influenced by the spring force. To produce a strong grasping force, it is necessary to use a spring with a high degree of stiffness. This usually causes the undesirable requirement for high-power actuators for the release action of the fingers. Therefore the use of spring force for grasping action is limited in low-grasping-force grippers for handling small machine parts such as pins, nuts, and bolts.

The reason the spring force can be used for a one-directional motion of the pneumatic and the hydraulic actuator is that the piston can be moved easily by the force applied to the output axis (piston rod). The combination of a spring and electric motor is not viable because normal electric motors include a gear-reduction system which makes it difficult to transmit the force inversely from the output axis.

Another interesting method uses electromagnets. The electromagnet actuator consists of a magnetic head constructed with a ferromagnetic core, conducting coil, and actuator rod made of ferrous materials. When the coil is activated, the magnetic head attracts the actuator rod, and the actuator displacement is locked at a specified position. When the coil is not activated, the actuator rod can be moved freely. This type of actuator is usually employed with a spring and produces two output control positions.

Figure 6 shows a gripper using the electromagnetic drive. The electromagnetic actuator produces the linear motion to the left along the L-L line. The motion is converted to grasping action through the cam. The releasing action is performed by the spring.

The actuator displacement that this kind of actuator can make is commonly limited to a small range because the force produced by the magnetic head decreases according to increase of the actuator displacement. Therefore this drive method can be effectively used only for gripping small workpieces.

In the design of effective gripper systems, the selection of drive system is a very important problem. Selection depends on the kinds of jobs required of the robot. Briefly, if a gripper has some joints that need positional control, an electric or hydraulic system is a better choice. If not, a pneumatic system is better. For robots required to work in a combustible environment, for instance a spray-painting environment, pneumatic or hydraulic systems are suitable. If force-control function is

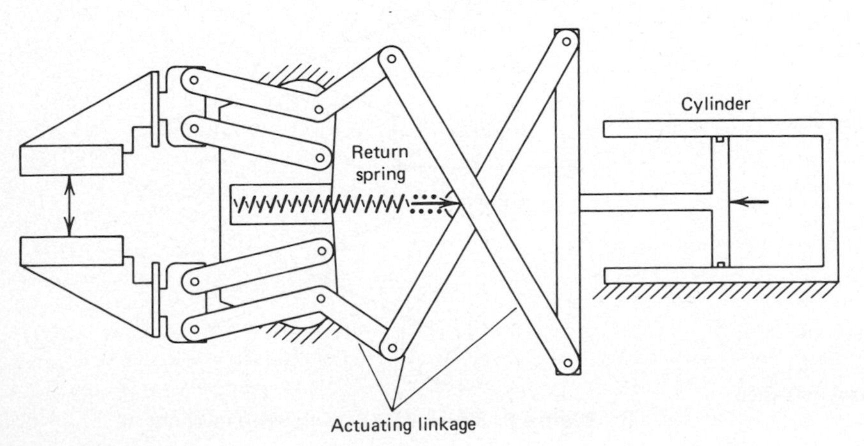

Figure 5. Spring-loaded linkage gripper. Courtesy of Dow Chemical Co., Ltd.

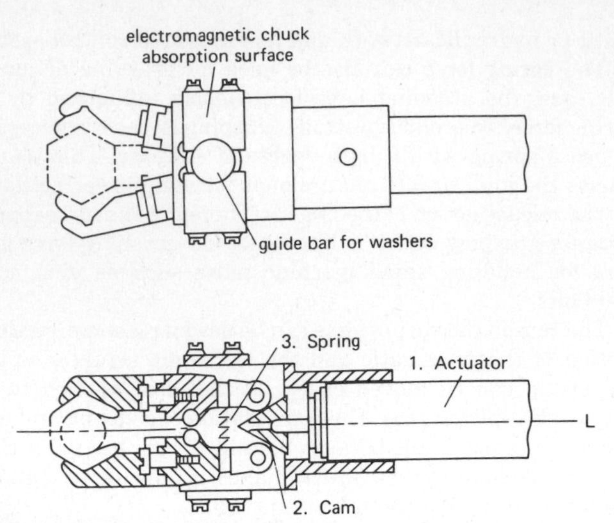

Figure 6. Gripper using an electromagnetic drive. Courtesy of Seiko-seiki Co., Ltd.

needed at some joints, for example, to control grasping force, electric or pneumatic systems are recommended.

MECHANICAL GRIPPERS

Several kinds of gripper functions described earlier can be realized using various mechanisms. From observation of the usable pair elements in gripping devices (6), the following kinds are identified: (*1*) linkage, (*2*) gear-and-rack, (*3*) cam, (*4*) screw, (*5*) cable and pulley, and so on. The selection of these mechanisms is affected by the kind of actuators to be employed and the kind of grasping modality to be used. In the past, many gripper mechanisms have been proposed. However, fewer mechanisms have been put into practical use. The following sections explain the practical gripper mechanisms.

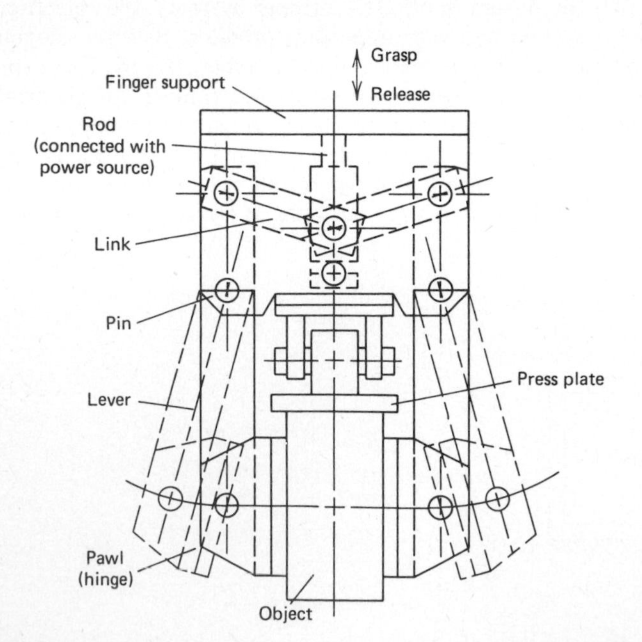

Figure 7. An example of swing gripper. Courtesy of Yasukawa Electric Manufacturing Co., Ltd.

A number of other possible gripper mechanisms can be found in Ref. 6, if needed.

Mechanical Grippers with Two Fingers

Swinging Gripper Mechanisms. This is the most popular mechanical gripper for industrial robots. It can be designed for limited shapes of an object, especially cylindrical workpieces. Figure 7 shows a typical example. Mechanisms to be adopted depend on the type of actuators used. If actuators are used that produce linear movement, like pneumatic or hydraulic piston cylinders, the device contains a pair of slider-crank mechanisms.

Figure 8 shows an example of a pair of slider-crank mechanisms commonly adopted by grippers using pneumatic or hydraulic piston cylinder drive. When the piston 1 is pushed by hydraulic or pneumatic pressure to the right, the elements in the cranks, 2 and 3, rotate counterclockwise with the fulcrum A2, respectively, when y is less than 180°. These rotations make the grasping action at the extended end of the crank elements 2 and 3. The releasing action can be obtained by moving the piston to the left. The motion of this mechanism obviously has a dwell position at = 180°. To obtain an effective grasping action, $\gamma = 180°$ must be avoided. An angle y ranging from 160 to 170° is commonly used.

Figure 9 shows another example of a mechanism for swing ing grippers that uses the piston cylinder, the swing-block mechanism. The sliding rod 1, actuated by a pneumatic or a hydraulic piston, transmits motion by way of the two sym-

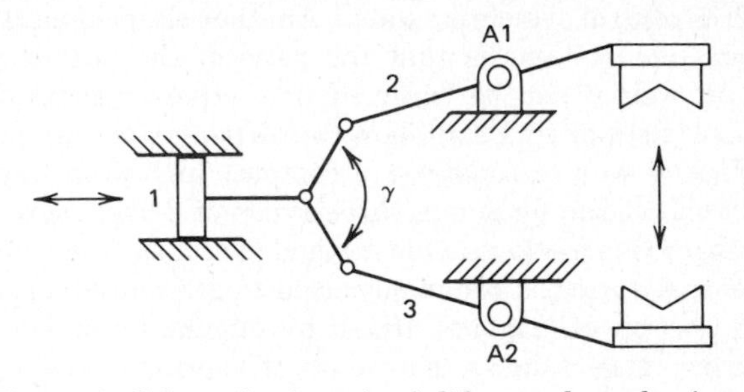

Figure 8. Schematic of a pair of slider-crank mechanisms.

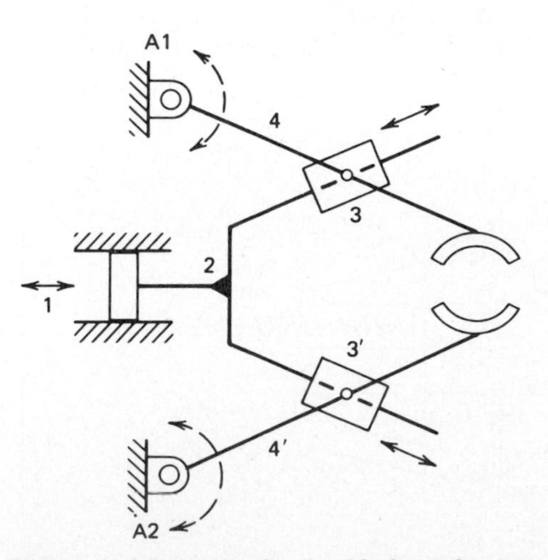

Figure 9. Schematic of swing-block mechanism.

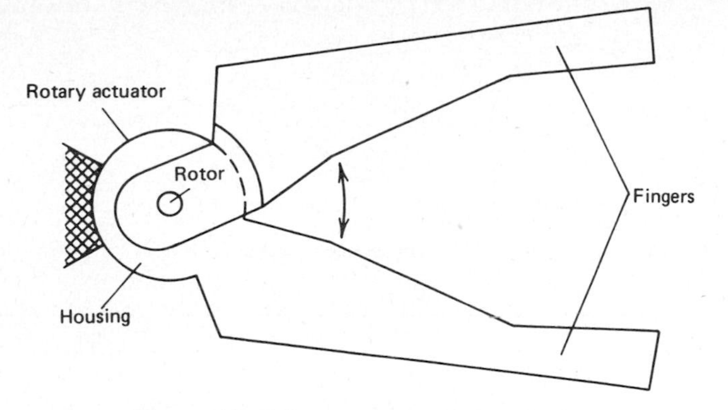

Figure 10. Gripper using a rotary actuator.

metrically arranged swinging-block linkages 1–2–3–4 and 1–2–3′–4′ to grasp or release the object by means of the subsequent swinging motions of links 4 and 4′ at their pivots A1 and A2.

Figure 10 is a typical example of a gripper using a rotary actuator in which the actuator is placed at the cross point of the two fingers. Each finger is connected to the rotor and the housing of the actuator, respectively. The actuator movement directly produces grasping and releasing actions. This gripper mechanism includes a revolute pair.

Translational Gripper Mechanisms. Translational mechanisms are used widely in grippers of industrial robots. The mechanism is a little complex compared to the swinging type.

The simplest translational gripper uses the direct motion of the piston cylinder. Figure 11 shows this type of gripper using a hydraulic piston cylinder. As depicted in the figure, the finger motion corresponds to the piston movement without any connecting mechanisms between them. The drawback to this method is that the actuator size decides the gripper size. This can sometimes make it difficult to design the desired size of gripper. The method is suitable for the design of wide-opening translational grippers.

Figure 12 shows a translational gripper using a pneumatic or hydraulic piston cylinder, which includes a dual-rack gear mechanism and two pairs of the symmetrically arranged parallel-closing linkages. This is a widely used translational gripper

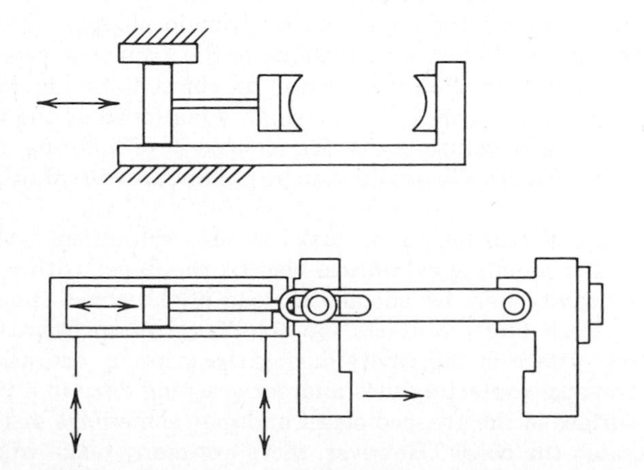

Figure 11. Translational gripper using a cylinder piston.

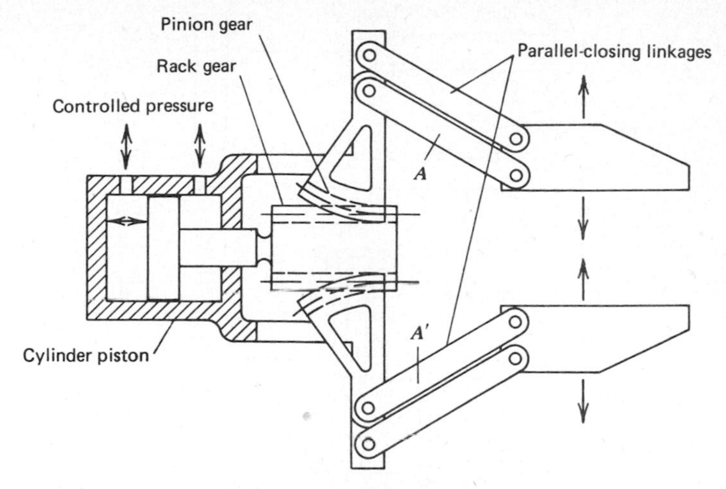

Figure 12. Translational gripper including parallel-closing linkages driven by a cylinder piston and a dual-rack gear.

mechanism. The pinion gears are connected to the elements *A* and *A*′, respectively. When a piston rod moves toward the left, the translation of the rack causes the two pinions to rotate clockwise and counterclockwise, respectively, and produces the release action, keeping each finger direction constant. The grasping action occurs when the piston rod moves to the right in the same way. There is another way to rotate the two pinions. Figure 13 shows the mechanism using a rotary actuator and gears in lieu of the piston cylinder and rack.

Figure 14 shows two examples of translational gripper mechanisms using rotary actuators. Figure 14**a** consists of an actuator and rack-pinion mechanism. The advantage of this kind of gripper is that it can accommodate a wide range of

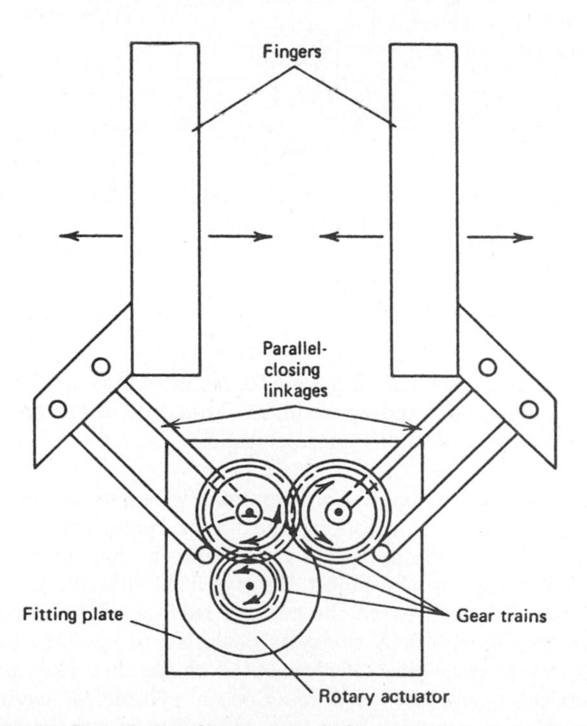

Figure 13. Translational gripper including parallel-closing linkages driven by a rotary actuator and gears.

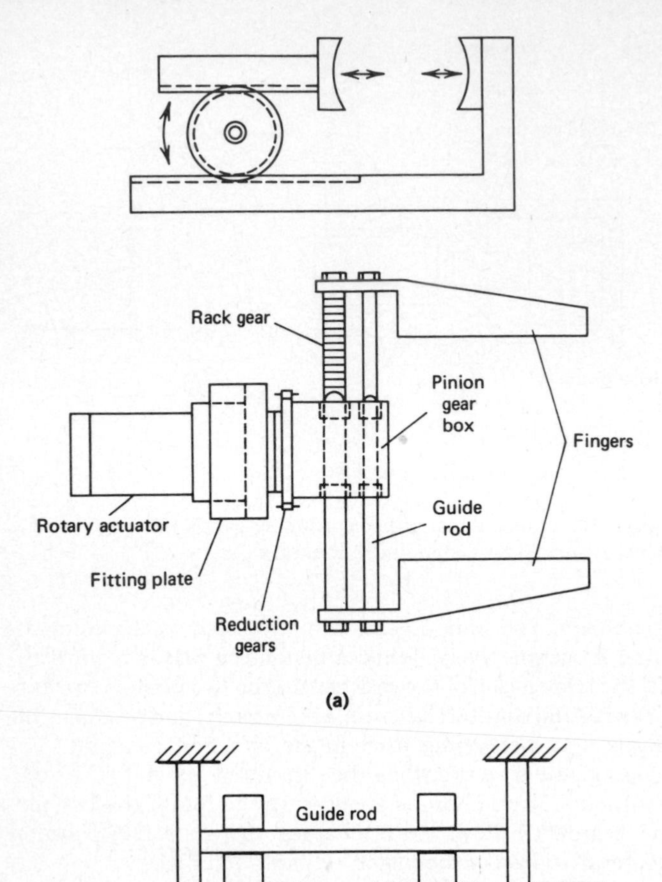

Figure 14. **(a)** A translational gripper operated by a rotary acuator with rack-pinion mechanism and an example of this type of gripper. Courtesy of Tokiko Co., Ltd. **(b)** A translational gripper using a rotary actuator and ball-screw mechanism.

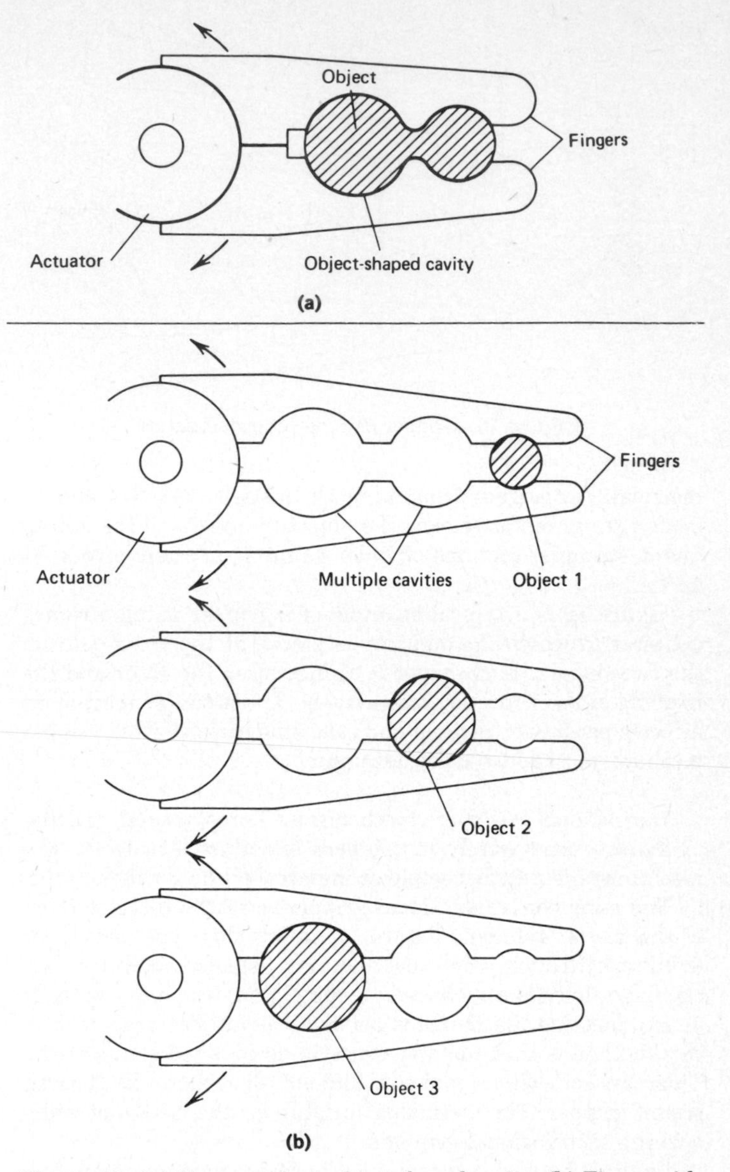

Figure 15. **(a)** Finger with an object-shaped cavity. **(b)** Finger with multiple object-shaped cavities.

dimensional variations. Figure 14**b** includes two sets of ball-screw mechanisms and an actuator. This type of gripper enables accurate control of finger positions.

Consideration of Finger Configuration. When using rigid fingers for mechanical grippers, the finger configuration must be contrived to accommodate the shape of the object to be handled. To grasp the object tightly, it is effective to make object-shaped cavities on the contact surface of the finger as shown in Figure 15. A cavity is designed to conform to the periphery of the object of a specified shape. For example, if cylindrical workpieces are handled, a cylindrical cavity is made. The finger with this type of cavity has the merit of grasping single-size workpieces more tightly with wider contact surface. The establishment of wide contact surface is important to prevent the grasping force from localizing.

Structure will limit the capability of the gripper to accommodate change in the dimension of an object to be handled. Versatility of the gripper can be slightly improved by the use of a finger with multiple cavities for objects of differing size and shape. Figure 15**b** shows examples of fingers with multiple cavities.

In manufacturing, many tasks involve cylindrical workpieces. For handling cylindrical objects, the finger with a V-shaped cavity may be adopted instead of the object-shaped cavity. Each finger contacts the object at two spots on the contact surface of the cavity during the gripping operation. The two-spot contact applies a larger grasping force to a limited surface of the grasped object and may sometimes distort or scratch the object. However, there are many tasks where this problem is not significant, and the device has a great

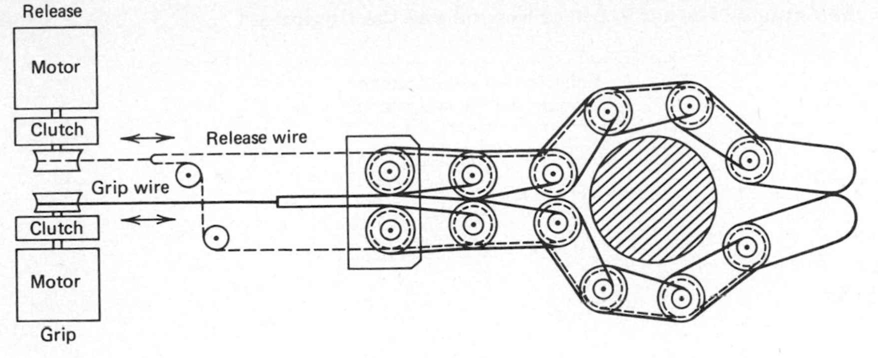

Figure 16. The soft gripper (7). Courtesy of S. Hirose and Y. Umetani.

advantage over the gripper with object-shaped cavities. One advantage is that it can accommodate a wide range of diameter variations in the cylindrical workpiece, allowing the shape of the cavity to be designed independent of the dimensions of the cylindrical object. Another advantage is that it is easy to make, resulting in reduced machinery costs.

To provide the capability to completely conform to the periphery of objects of any shape, a soft-gripper mechanism has been proposed, which is shown in Figure 16 (7). The adjacent links and pulleys are connected with a spindle and are free to rotate around it. This mechanism is manipulated by a pair of wires, each of which is driven by an electric motor with gear reduction and clutch. One wire is called a grip wire, which produces the gripping movement. The other is a release wire, which pulls antagonistically and produces the release movement from the gripping position. When the grip wire is pulled against the release wire, the finger makes a bending moment from the base segment. During this process of wire traction the disposition of each gripper's link is determined by the mechanical contact with an object. When the link i makes contact with an object and further movement is hindered, the next link, $(i + 1)$, begins to rotate toward the object until it makes contact with the object. This results in a finger motion conforming to the peripheral shape of the object. In this system it is reported that the proper selection of pulleys enables the finger to grasp the object with uniform grasping pressure.

Calculation of Grasping Force or Torque. The maximum grasping force or grasping torque is as important a specification for gripper design as the geometrical configuration. The actuator output force or torque must be designed to satisfy the specification of the maximum force. How hard the robot must grasp the object depends on the weight of the object, the friction between the object and the fingers, how fast the robot is to move, and the relation between the direction of movement and the fingers' position on the object.

The worst case is when the lines of action of the gravity and the acceleration force to the grasped object are parallel to the contact surface of the fingers. Then friction alone has to hold the object. Therefore this situation is assumed in evaluating the maximum grasping force or grasping torque.

After the maximum grasping force or torque has been determined, the force or torque that the actuator must generate can be considered. The calculation of those values requires the conversion of the actuator output force or torque to the grasping force or torque, which depends on the kind of actuator used and the kind of mechanism employed. Table 2 shows the relation between the actuator output force or torque and the grasping force in grippers that include the various kinds of mechanisms and actuators described above.

Mechanical Hands with Three or Five Fingers

Three-Finger Hand. The increase of the number of fingers and degrees of freedom will greatly aid the improvement of the versatility of grippers. However, this also complicates the design process. Although design methods for this type of gripper have still not been established there are a few examples that have been put into practical use.

The simplest example is a gripper with three fingers and one joint driven by an appropriate actuator system. The main reason for using the three-finger gripper is its capability of grasping the object in three spots, enabling both a tighter grip and the holding of a spherical object of differing size keeping the center of the object at a specified position. Three-point chuck mechanisms are typically used for this purpose. Figure 17 gives an example of this gripper. Each finger motion is performed using a ball-screw mechanism. Electric motor output is transmitted to screws attached to each finger through bevel gear trains which rotate the screws. When each screw is rotated clockwise or counterclockwise the translational motion of each finger will be produced, which results in the grasping-releasing action.

The configuration of the grasping-mode switching system using three fingers (8) is shown by Figure 18. This includes four electric motors and three fingers and can have four grasping modes, as shown in Figure 19, each of which can be achieved by the finger-turning mechanism. All fingers can be bent by motor-driven cross-four-bar link mechanisms, and each finger has one motor.

The finger-turning mechanism is called a double-dwell mechanism. This mechanism transfers the state of gripper progressively from three-jaw, to wrap, to spread, and to tip prehension. The gears for fingers 2 and 3 are connected to the motor-driven gear directly, whereas the gear for finger 1 is connected to the motor-driven gear through a coupler link. Rotating the motor-driven gear in three-jaw position, finger 1 rotates, passes through a dwell position, and then counterrotates to reach the wrap position. Similarly, finger 1 is rotated out of its spread position but is returned as the mechanism assumes tip prehension. Finger 2 is rotated counterclockwise 60° from its three-jaw position to the wrap position, then coun-

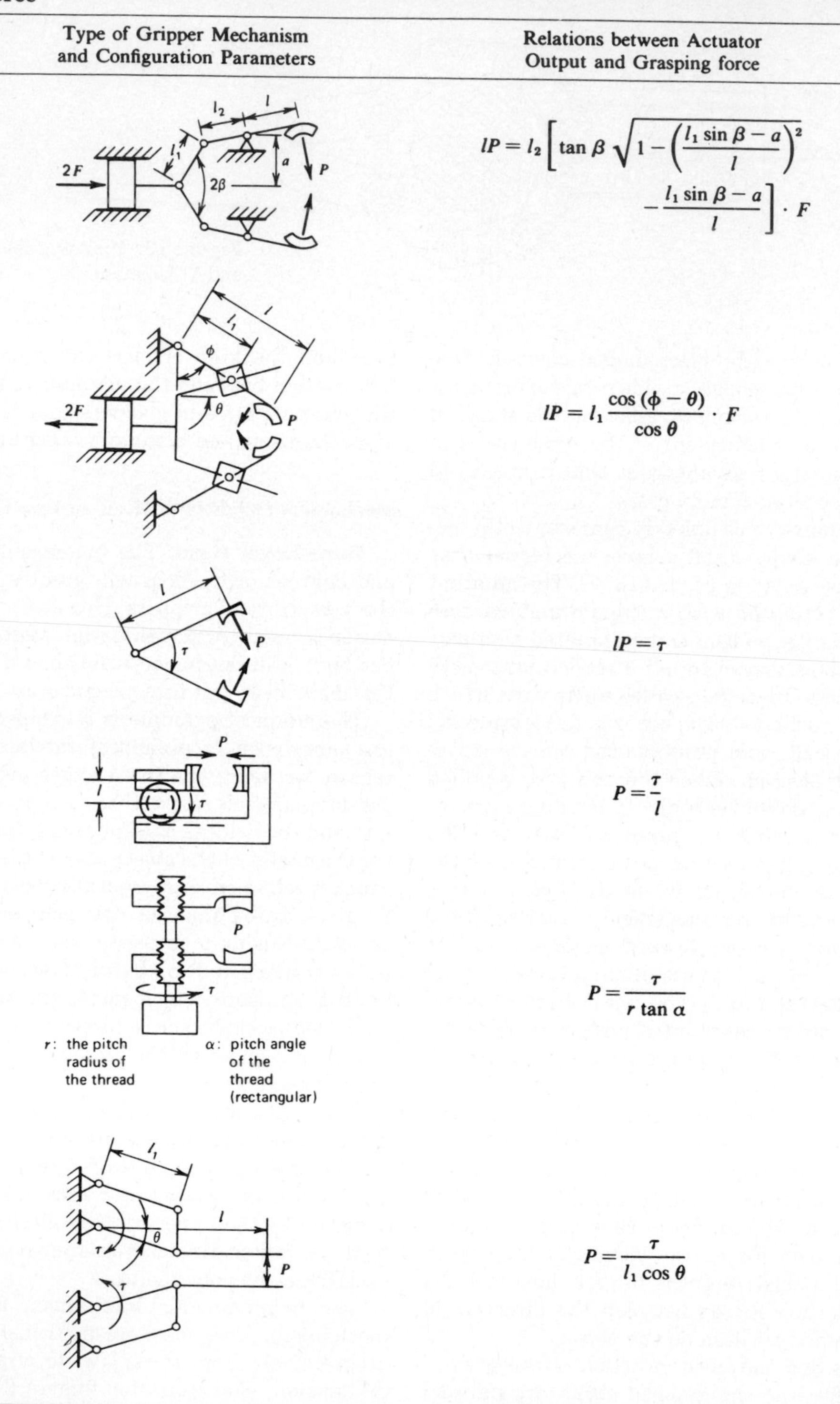

Table 2. Relations between the Actuator Output Force or Torque and the Grasping Force

Type of Gripper Mechanism and Configuration Parameters	Relations between Actuator Output and Grasping force
	$lP = l_2 \left[\tan\beta \sqrt{1 - \left(\frac{l_1 \sin\beta - a}{l}\right)^2} - \frac{l_1 \sin\beta - a}{l} \right] \cdot F$
	$lP = l_1 \frac{\cos(\phi - \theta)}{\cos\theta} \cdot F$
	$lP = \tau$
	$P = \frac{\tau}{l}$
r: the pitch radius of the thread; α: pitch angle of the thread (rectangular)	$P = \frac{\tau}{r \tan\alpha}$
	$P = \frac{\tau}{l_1 \cos\theta}$

terclockwise 120° into the spread position, then counterclockwise 150° into the tip position. Finger 3 rotates through identical angles but in a clockwise direction. A multiprehension system of this type is effective for picking up various-shaped objects.

Five-Finger Hand. A small number of five-finger hands have been developed in the world, with only a few for industrial use. Almost all of them are prosthetic hands for amputees. In the development of prosthetic arms, cosmetic aspects are more important to the mental state of the handicapped than

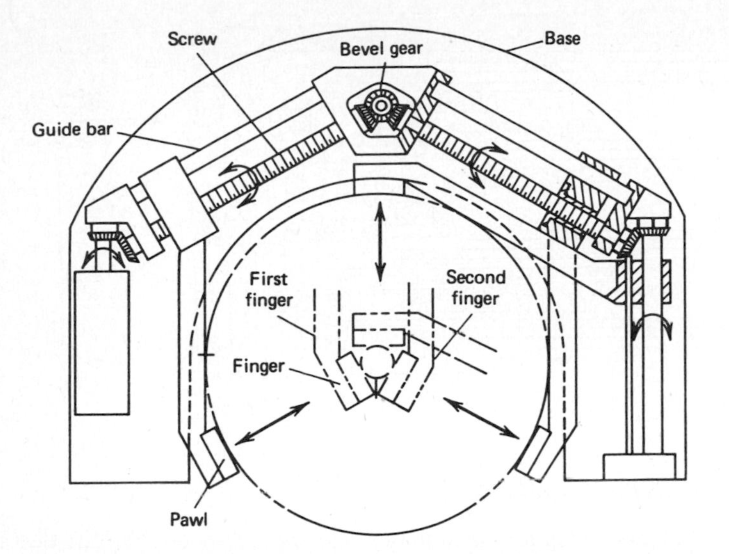

Figure 17. Gripper using three-point chuck mechanism. Courtesy of Yamatake Honeywell Co., Ltd.

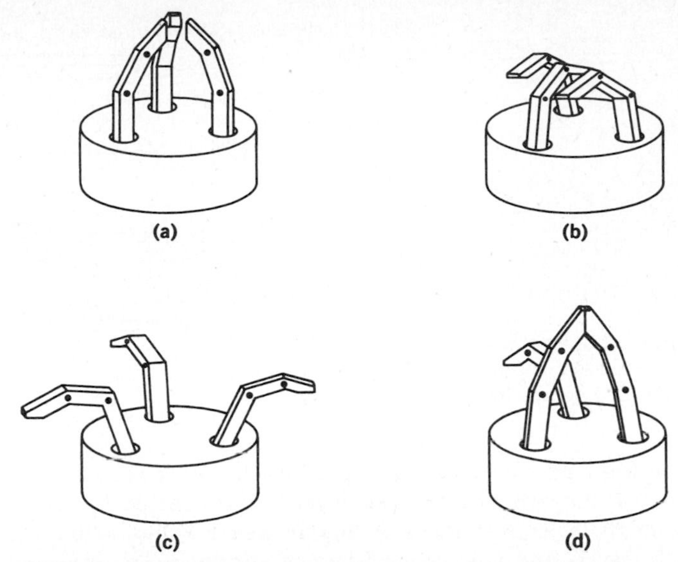

Figure 19. Mechanical equivalent of prehensile modes (8). (a) Three-jaw position; (b) wrap position; (c) spread position; (d) tip position. Courtesy of F. Skinner.

functions. This requires anthropomorphism in the design of prosthetic hands. For industrial use, the function is more important than cosmetic aspects of the gripper. Therefore anthropomorphism is beyond consideration in the design of industrial grippers. This is why there are only a few five-finger industrial grippers. Nevertheless, five-finger grippers that have been developed so far for prosthetic use are described in the following part of this section because they include many mechanisms that will be effective in the design of industrial grippers.

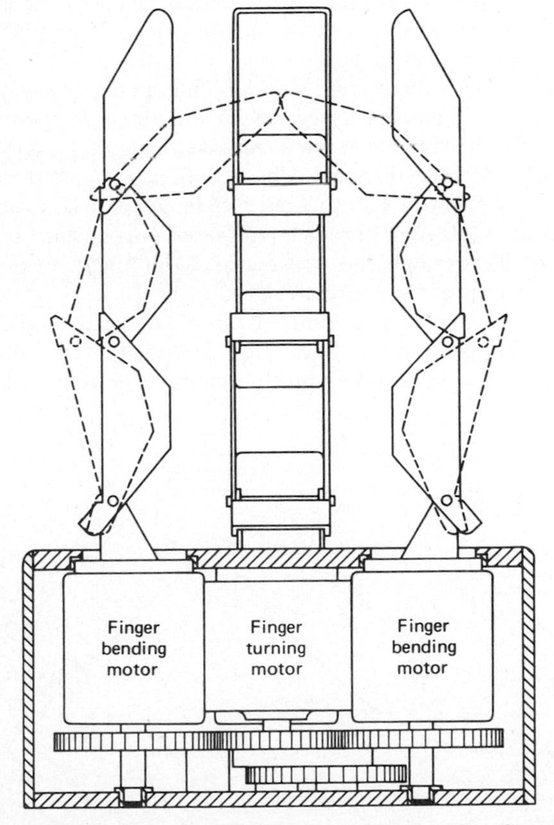

Figure 18. Multiple prehension gripper system (8). Courtesy of F. Skinner.

A handicapped person must produce control signals for operation of a prosthetic arm. The number of independent control signals available determines how many degrees of freedom the prosthetic device can have. Typical models of a five-finger gripper for prostheses have only one degree of freedom. Each finger is connected to a motor by appropriate mechanisms.

To ensure that the gripper holds the object with the equilibrium of the forces between the fingers and the object, the arrangement of fingers must be carefully considered. In typical five-finger hands, the thumb faces the other four fingers and is placed equidistant from the index finger and middle finger so the tips to the fingers can meet at a point when each finger is bent (see Fig. 20).

If each finger connects to the drive system rigidly, finger movements are decided by the motion of the drive system. The finger configuration cannot accommodate the shape change of grasped objects. To remedy this problem, motor output can be transmitted to each finger through flexible elements.

Figure 21 shows examples of this type of gripper (9). There are two pairs of fingers: the index and the middle fingers,

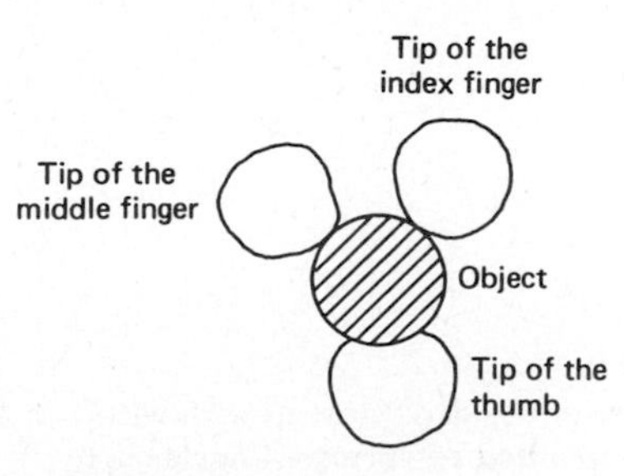

Figure 20. Three-point pinch.

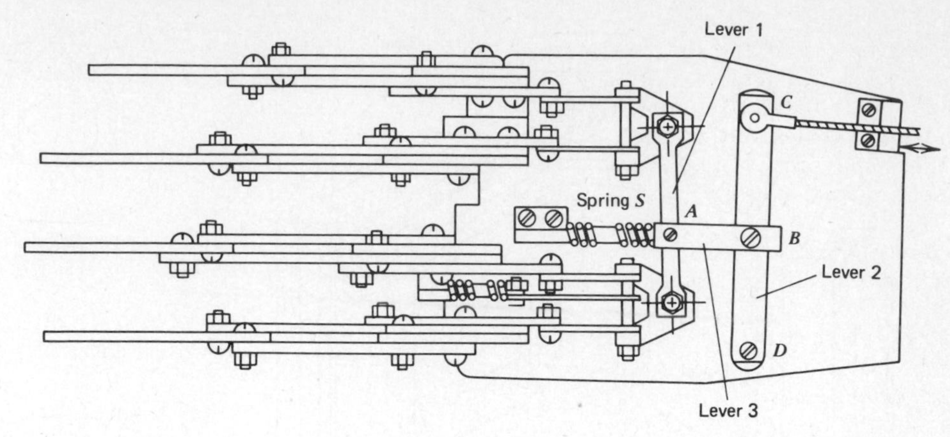

Figure 21. Prosthetic hand equipped with accommodation mechanism. Courtesy of Mechanical Engineering Lab. MITI, Japan.

and the ring and the little fingers. Each pair is connected to lever 1 through a pivot. Also, lever 1 is connected to lever 2 through lever 3 at joints A and B. Lever 2 has a fulcrum point at D and is supported by the spring S. The grasping operation is executed by pulling lever 2 at the end C. Assuming that the index and middle fingers touch the object, movements of these fingers will stop, but the third and little finger can still continue to move till these fingers touch the object because lever 2 can rotate at joint A. This causes the finger to accommodate the shape of the object.

It is possible to move each finger independently, if the cross-four-bar link for each finger is driven by a different motor. Usually this type of gripper requires small motors to be installed in the finger.

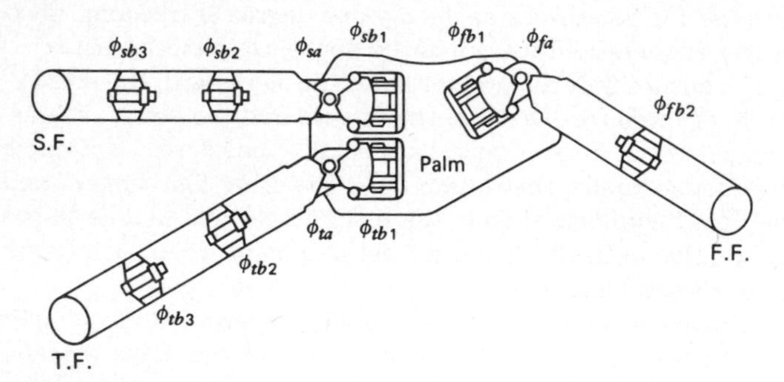

Figure 22. Versatile finger system (10). Courtesy of T. Okada.

UNIVERSAL GRIPPERS

Universal grippers with many degrees of freedom like the human hand have been researched by several investigators. Increasing the degrees of freedom causes several problems. One difficult problem is how to install in the gripper the actuators necessary to activate all degrees of freedom. This requires miniature actuators that can produce enough power to drive the gripper joint. Commercially available actuators are too large to attach at each joint of the finger. The use of SME (shape memory effect) actuators is one solution, but it is not practical at present. The most frequent solution to this problem is to use cable and pulley mechanisms that enable a motor to be placed at an appropriate position away from the joint.

There are two examples of universal grippers. Each used cable pulley mechanisms and d-c motor drives. Figure 22 shows an example of the gripper, which includes three fingers: a thumb, an index finger, and a middle finger (10). Usually three fingers provide enough functions for universal grippers. Each finger contains two or three segments made of 17-mm brass rods bored to be cylindrical. The tip of each segment is truncated at a slope of 30° so that the finger can be bent at a maximum angle of 45° at each joint—not only inward but also outward. This makes the workspace of the finger more extensive than that of the human finger.

The thumb has three joints. Each of the index and middle fingers has four joints. Each joint includes 1 DF, which is driven using a wire-pulley mechanism and electric motors. A

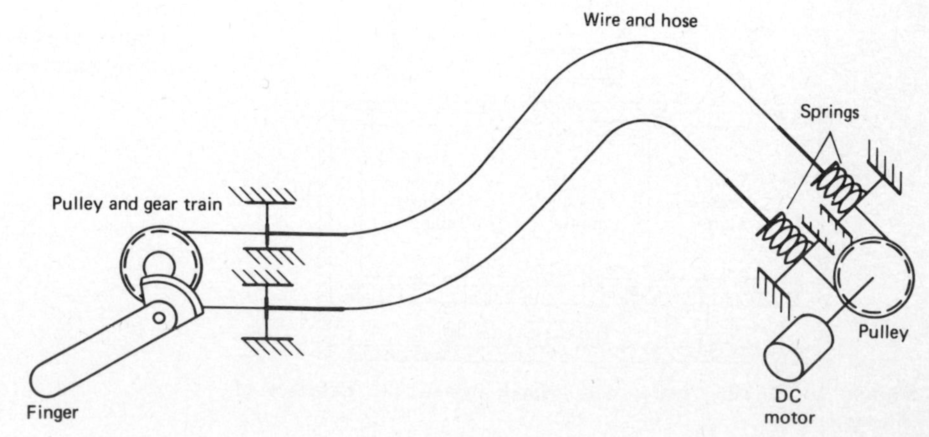

Figure 23. Wire-pulley drive system with wire-guiding hoses supported by springs. Courtesy of S. Sugano and co-workers.

pulley is placed at each joint, around which two wires are wound after an end of each wire is fixed on the pulley. The wire is guided through coil-like hoses so that it cannot interfere with the complicated finger motion. Using coil-like hoses is effective in protecting the wire and also in making it possible to eliminate relaying points for guiding wire. To make the motions of the finger flexible and to make the gripper system more compact, the wires and hoses are installed through the finger tubes. The system in Figure 22 is reported to perform some manipulative tasks, screwing a bolt, bar-and sphere-turning tasks through the use of the playback control method. Using coil-like hoses reduces the complexity of the wire-guidance mechanism, but it raises the problem of friction between hoses and wires. To resolve this problem it is effective to support the hoses with flexible elements like springs (11). Figure 23 shows the construction of a wire-pulley drive system with wire-guiding hoses supported by springs. If a hose whose ends are rigidly fixed is bent, the hose will be extended. This reduces the cross-section area of the hose, which increases the friction. In the system of Figure 23, the spring, instead of the hose, will be extended when the hose is bent. Therefore the cross-section area of the hose will not be affected by the change of hose configuration, which prevents friction from increasing.

To achieve stable prehension using universal grippers, each joint for the finger must be moved cooperatively. For this purpose a stable prehension control algorithm has been proposed based on potential method (12).

BIBLIOGRAPHY

1. G. Schlesinger, *Erstzglieder und Arbeitshilfen,* Part II, Springer, Berlin, Germany, 1919.
2. F. R. E. Crossley and F. G. Umholts, "Design for a Three-fingered Hand," in E. Heer, ed., *Robot and Manipulator Systems,* Pergamon Press, Oxford, UK, 1977, pp. 85–93.
3. A. Morecki, J. Ekiel, and K. Fidelus, "Some Problems of Controlling a Live Upper Extremity and Bioprostheses by Myopotential External Control of Human Extremities," *Proceedings of the Second Symposium on External Control of Human Extremities,* Belgrade, Yugoslavia, 1967.
4. H. Asada and T. Kanade, *Design of Direct-Drive Mechanical Arms,* Robotics Institute, Carnegie-Mellon University, Pittsburgh, Pa., 1981.
5. O. L. Sheldon, "Robots and Remote Handling Methods for Radioactive Materials," *Proceedings of the Second International Symposium on Industrial Robots,* Chicago, Ill., pp. 235–256.
6. F. Y. Chen, *Gripping Mechanisms for Industrial Robots, Mechanism and Machine Theory,* Vol. 17, No. 5, Pergamon Press, Oxford, UK, 1982, pp. 299–311.
7. S. Hirose and Y. Umetani, "The Development of a Soft Gripper for the Versatile Robot Hand," *Proceedings of the Seventh International Symposium on Industrial Robots,* Tokyo, Japan, 1977, pp. 353–360.
8. F. Skinner, *Design of a Multiple Prehension Manipulator System, ASME Paper, 74-det-25,* 1974.
9. Y. Maeda, A. Fujikawa, K. Tanie, M. Abe, T. Ohno, K. Tani, F. Honda, T. Inanaga, T. Yamanaka, and I. Kato, "A Hydraulically Powered Prosthetic Arm with Seven Degrees of Freedom (Prototype-I)," *Bulletin of Mechanical Engineering Laboratory,* No. 27, Tsukuba, Japan, 1977.
10. T. Okada, "On a Versatile Finger System," *Proceedings of the 4th International Symposium on Industrial Robots,* Tokyo, Japan, 1977, pp. 345–352.
11. S. Sugano, J. Nakagawa, Y. Tanaka, and I. Kato, "The Keyboard Playing by an Anthropomorphic Robot," *Preprints of Fifth CISM-IFTomm Symposium on Theory and Practice of Robots and Manipulators,* Udine, Italy, 1984, pp. 113–123.
12. H. Hanafusa and H. Asada, "Stable Prehension By a Robot Hand with Elastic Fingers," in M. Brady and co-eds., *Robot Motion,* MIT Press, Cambridge, Mass., 1983, pp. 337–359.

General References

J. F. Engelberger, *Robotics in Practice,* Kogan Page Ltd., London, 1980.

I. Kato, *Mechanical Hand Illustrated, Survey Japan,* Tokyo, Japan, 1982.

S. Timoshenko and D. H. Young, *Engineering Mechanics,* McGraw-Hill, New York, 1956.

This article is reprinted from S. Y. Nof, ed., *Handbook of Industrial Robotics,* John Wiley & Sons, Inc., New York, 1985.

H

HANDICAPPED, ROBOTS FOR

Tomović Rajko
Popović Dejan
Faculty of Electrical Engineering
Belgrade, Yugoslavia

INTRODUCTION

Medical robotics, in the broadest sense, may be identified with artificial organs. Artificial organs are mechanisms designed to replace body parts (heart, kidney, pancreas, arm, hand, leg, etc) or improve the body functions (pacemaker, cochlear prosthesis, functional electrical stimulation, etc). The field of artificial organs can be subdivided in various ways (eg, function, structure). The most general division of artificial organs breaks them down into two categories: orthoses and prostheses. The term orthosis is used for artificial organs that compensate for functional deficiencies without anatomical substitution of natural body parts. The prosthesis, however, compensates for lost body activity both anatomically and functionally.

Motor deficiencies are a frequent cause of human handicaps. They result in the reduction of manipulation and locomotion capabilities and require external support to improve functions of paralyzed or lost extremities. External support for motor deficiencies may be passive or active. Clearly, robotics for the handicapped deals with active assistive devices for the compensation of reduced or lost functional motions.

To understand better the full scope of robotics for the handicapped, a structural diagram of the biological system responsible for the control and execution of functional motions is shown in Figure 1. The system comprises skeletal parts, muscular actuators, and control mechanisms. It must be kept in mind that all approaches in medical robotics that neglect the strong interaction of the above-mentioned elements in the performance of functional motions led to mechanistic solutions usually rejected by the handicapped.

The skeletal system represents a complex kinematic chain. Segments of the mechanism can be approixmated by rigid bodies. The number of independent parameters determining the positions of the skeleton is over 350. Joints, which link segments of the skeleton, are cylindrical, spherical, and limited in motion range (1).

The muscular system can be considered the actuator part of the biomechanism. The muscular system is made up of about 700 muscle pairs. Muscle contractions generate torques at the joints. Contractions are isomorphic or anisomorphic, resulting in a relative state of rest or the displacement of neighboring segments. Muscles act in pairs (agonist, antagonist) over one or two joints (2).

In robotics, the two aspects of the biomechanism are merged in the hardware of the assistive device. In dynamical considerations, however, factors such as tendons, ligaments, variable mass geometry during contractions, and so on must be taken care of.

The neural system is responsible for the control of functional movements. The hierarchical nature of natural neural control is recognized, but many aspects of the actual system operation are unknown. The highest level, the cortex, is responsible for motion planning. The cortex receives sensory information and interacts with the spinal cord, basal ganglia, and cerebellum directly, and with lower motor neuron and effectors through the spine. Two distinct control mechanisms of the central nervous system (CNS) are identified. The first one deals with the preprogrammed control of motion without sensory feedback. The second CNS mechanism depends on the stretching and loading of muscles, so it is sensory driven. The latter CNS mechanism consists of the so-called short-latency segmental stretch reflex and the longer-latency segmental and suprasegmental functional stretch reflex. The lowest level of control consists of simple single reflexes.

Medical robotics for the handicapped can be divided in the following way (3): prosthetics, orthotics, and environmental control. This division reflects specific features of each field in terms of performance requirements and synthesis methods. Devices such as wheelchairs, special beds, and so on are not discussed here.

In certain aspects, medical robotics has been ahead of research in the industrial field. For instance, the importance of sensory feedback, the need for dexterous multifingered hands, and leg locomotion was first emphasized in rehabilitation engineering. On the other hand, widespread applications of externally powered assistive systems are still an exception rather than current rehabilitation practice. A general explanation for the slow acceptance of active prostheses and orthoses by the patient may be that robotics for the handicapped implies

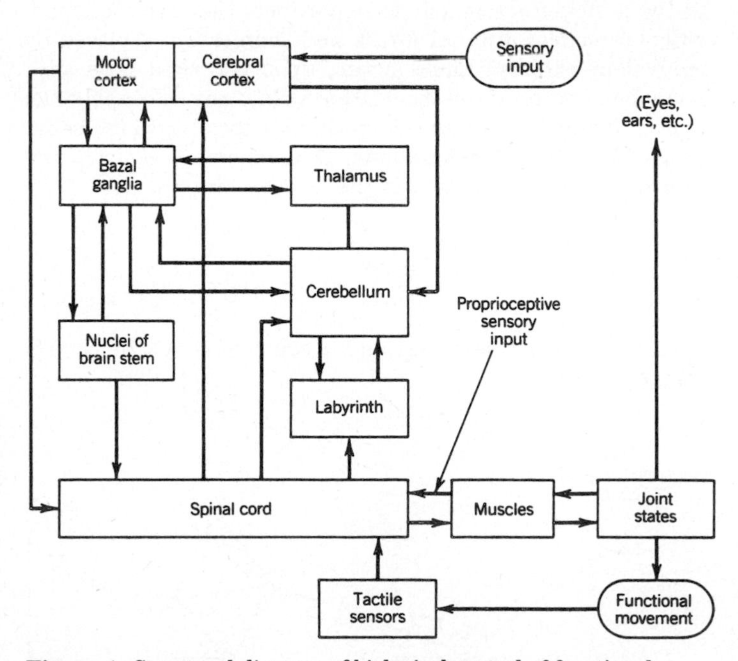

Figure 1. Structural diagram of biological control of functional motions.

a complex and delicate human–machine interaction. As a matter of fact, any active assistive device interacts with the patient in three ways: control, power, and force transmission.

Special attention must be paid to the human–machine interaction at the control level in rehabilitation engineering. If the patient has the impression that the machine dominates his/her behavior, such a solution will be rejected. The lack of sensory feedback is another cause for rejection of active assistive systems. In many instances, mere compensation for motor functions by a machine, without sensory feedback, is not sufficient motivation for a patient to accept an external device as a part of the body.

Medical robotics must meet other demanding requirements as well. Simplified fitting procedures, high equipment reliability, easy repair, compactness, and so on are decisive factors in the evaluation of assistive systems. Thus, robotics for the handicapped represents a highly challenging field of bioengineering.

PROSTHETICS

The limb prosthesis replaces the lost part of the body both functionally and morphologically. Locomotion and manipulation are two basic functions to be performed by artificial extremities. Because of different task requirements, it is convenient to deal with hand, arm, and leg prostheses as separate topics. Another distinction refers to amputation levels, so there are hand prostheses, below-elbow, above-elbow, and total arm prostheses, as well as below-knee, above-knee, and total leg prostheses (see PROSTHESES).

The mechanical model of a prosthesis can be represented as a system of rigid bodies linked by joints (Figure 2). Electrical, hydraulic, or pneumatic actuators apply torques at the joints. The number of joints is determined by the type of prosthesis. The design procedure for the prosthesis is based on the principle of segment decomposition. Using the decomposition principle, internal forces and couples are replaced by equivalent external forces acting upon the rigid body with nonholonomic constraints. At this point, one can apply the known methods of classical mechanics (eg, Lagrange equations, Newton–Euler equations, D'Alembert's principle) and derive equations of motion in the Cauchy form (4):

$$\dot{X} = f(X,U)$$

$$X = \{x_1,x_2, \cdots ,x_n\}, \qquad U = \{u_1,u_2, \cdots ,u_m\}$$

where x_i are state variables and u_i components of the control vector.

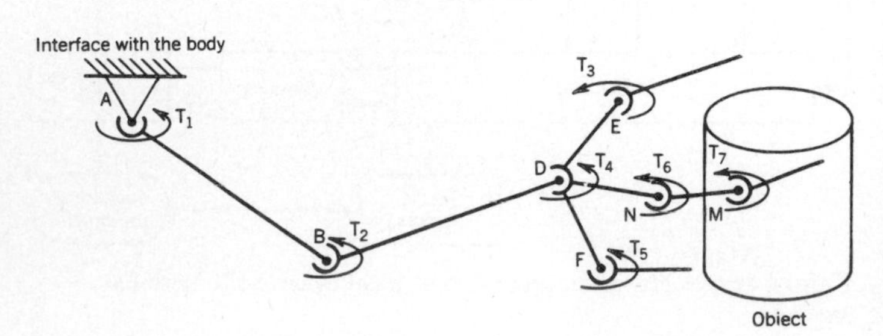

Figure 2. Mechanical model of the extremity prosthesis.

Equations of motion of the kinematic chain are nonlinear and nonstationary. Nonlinearities appear in the form of trigonometric functions and squares of the first derivatives of the state variables. Additional difficulty is caused by dynamic indetermination at the closing of the kinematic chain (grasping action, meeting obstacles, double support phase of gait, use of crutches, etc).

Hand Prostheses

Active hand prostheses were the first electronic devices designed for rehabilitation purposes (Figure 3). The development work in this area of medical robotics started in the early 1960s (5,6). In terms of mechanics, the human hand represents a five-fingered, articulated, open-chain structure. The design of the mechanical equivalent of the human hand is very difficult to implement. For instance, if each joint of an articulated five-fingered hand is to be controlled, the prosthesis must be equipped with 19 motors. Such a solution is evidently out of the question in rehabilitation.

In order to overcome the above difficulty, two approaches to the design of an active hand prosthesis have been used.

The first reduces the five-fingered, articulated structure to a two-fingered, nonarticulated gripper. Such a solution requires no more than one actuator and one control site, which are essential. The functional capabilities of the gripper-type prosthesis are rather limited, but this is the price paid for control simplicity. The control source for such a hand prosthesis is usually a myoelectrical signal derived from the convenient muscle group. Hand prostheses of this type are commercially available (eg, the Swedish myoelectrical hand, Wienaton). The myoelectrical control site is used to control the closing–opening action and the grasping force adjustment.

The effort to preserve the five-fingered, articulated hand structure and still use one actuator led to an interesting solution in terms of control and mechanical design. The basic intention was to preserve the potential of the human hand for automatic shape adaptivity without increasing the number of voluntary control sites or motorized degrees of freedom. To acomplish this goal, some compliance and synergy features for automatic shape adaptation of the human hand were incorporated in the mechanics of the prosthesis (7). This anthropomorphic hand prosthesis was controlled by touch and slip sen-

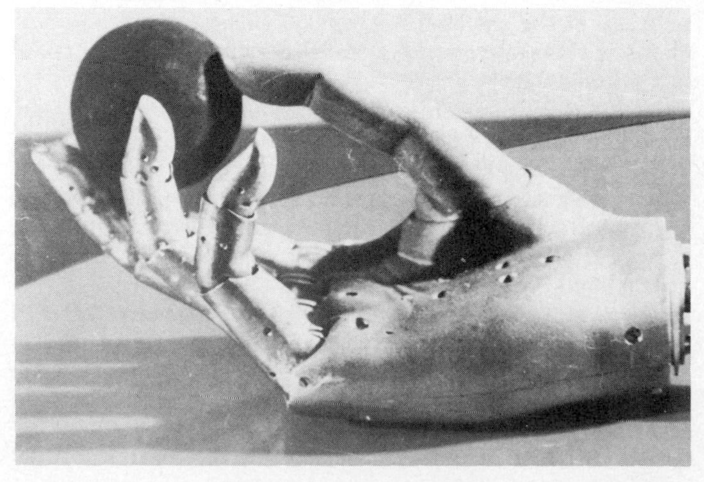

Figure 3. Belgrade hand.

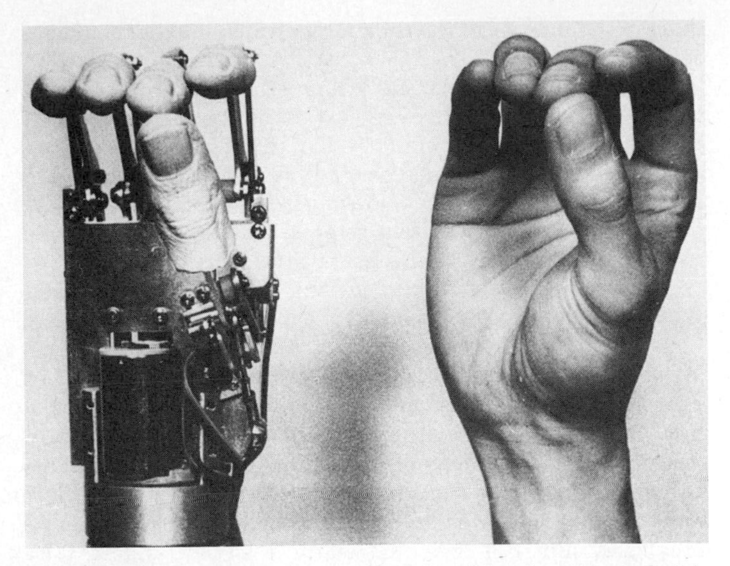

Figure 4. The hand on the multifunctional hand with myoelectric control.

sors (8). Hand prostheses of this type are not commercially available.

This research was done at Waseda University (WIME hand, Figure 4).

When it comes down to the day-to-day use of an active hand prosthesis, a high rejection rate by the patients is observed because electronic technology is much more demanding and less reliable than the solutions used in passive prostheses. In general, the acceptance rate of assistive devices depends very much on nonengineering factors such as social status, profession, education, and so on. The level of amputation (below-elbow, above-elbow, unilateral, bilateral) also affects the acceptance rate of active prostheses to a high degree (see HANDS).

Arm Prostheses

The complexity of external motor control and the lack of sensory feedback indicates the use of an active hand prosthesis. During the rehabilitation of one-sided or two-sided above-elbow amputees, these complex difficulties become much more pronounced. An increased number of degrees of freedom of the arm prosthesis requires a corresponding extension of control sites. But it is hard to meet such requirements since coordination of more than two to three external control sites for the execution of functional motions is beyond the amputees' capabilities. Because of this constraint, the use of active above-elbow prostheses, especially for two-sided amputations, is rather limited.

The simplified mechanical model of the arm has seven degrees of freedom (Figure 5). It can be considered a redundant manipulator with an adaptive terminal device. Redundancy implies several options of target approach and hand orientation for the same grasping task. From the mathematical point of view, redundancy implies the nonuniqueness of solutions of system dynamics. An arm prosthesis, which always goes together with a hand prosthesis, is shown in Figure 6. Because of control site constraints, above-elbow prostheses are built with a reduced number of degrees of freedom. Usually, no

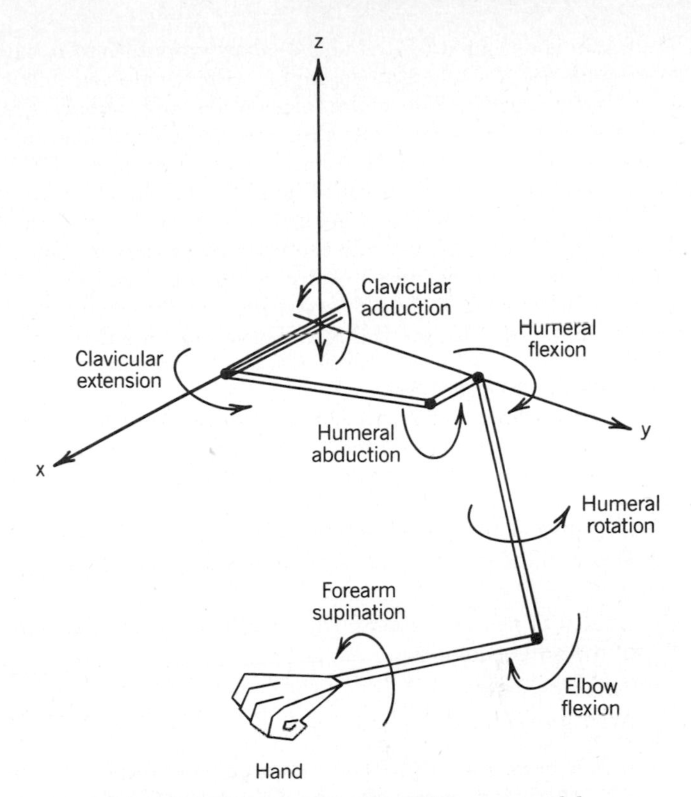

Figure 5. Degrees of freedom of the arm model.

more than four degrees of freedom are available (Boston Arm, Temple Arm, Utah Arm, Swedish Hand).

In order to get better insight into the operation of an arm prosthesis, the Utah Arm is described in more detail. The

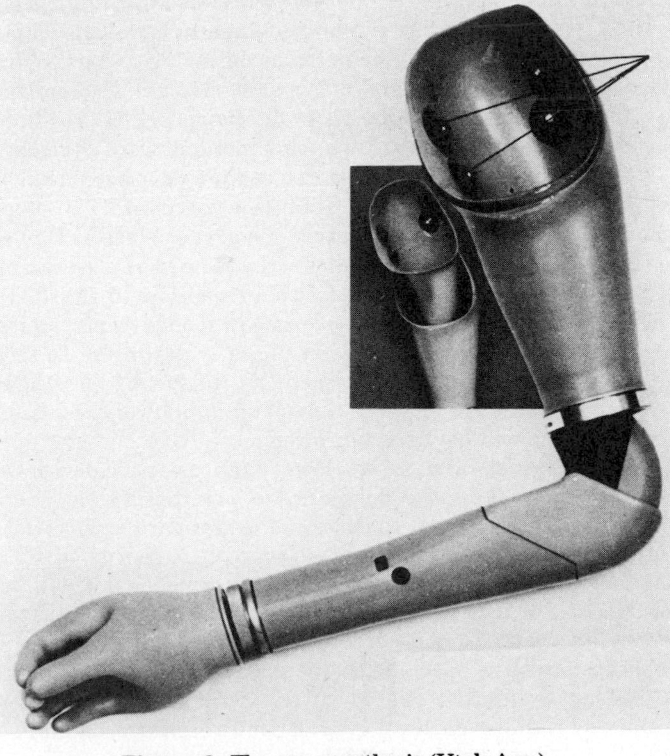

Figure 6. The arm prosthesis (Utah Arm).

Utah Arm is designed for the above-elbow amputation. It has two electrodes for electromyographic (EMG) measurement fixed in the socket. The electronics and power supply are mounted in the interior of the prosthesis so that it operates as a self-contained device. The control system processes EMG inputs and generates outputs for humoral rotation, elbow flection, forearm supination, and gripper operation. The prosthesis cosmetics is satisfactory and the training process is simple, so the device is easily accepted by amputees. Extensive research efforts at M.I.T. in the development of the Boston Arm have preceded the design of this assistive machine (9).

Above-knee (A/K) Prostheses

The requirements for the A/K prosthesis were set up 30 years ago (10). They are still valid. The following requirements are of special interest for active A/K prostheses:

- Minimum power consumption.
- Resemblance of assisted and nonassisted walking patterns.
- Ability to flex and extend the prosthesis at any moment of the gait sequence.
- Jerk minimization.
- Ability to change the gait mode.

An A/K prosthesis cannot be unambiguously classified as passive or active. Such solutions as the polycentric knee mechanism (11), the polycentric knee mechanism with hydraulic value (12), or the A/K prosthesis with friction-type breaking (13) satisfy some of the above performance requirements. The logically controlled A/K prosthesis with an hydraulic valve represents a further bridge between purely passive and fully controllable assistive devices (14). All of the above-mentioned prostheses satisfy the stance phase requirements as well as the minimum power consumption principle. However, the amputee is not able to flex extend the knee in the stance phase once it has been flexed. Extension in the swing phase requires additional metabolic energy. Gait asymmetry is also present.

In order to overcome the above deficiencies, A/K prostheses having fully controllable knee joint motions were developed. In the last decade, several research centers have been involved in this kind of scientific and applied activity (M.I.T., Boston, Faculty of Electrical Engineering, Belgrade, Waseda University, Tokyo, etc). Detailed studies of the knee-joint performance in the stance phase are part of this research (15,16). It has also become evident that for several gait modes (ramp, stairs), active knee control in the swing phase is desirable. To meet these requirements, self-contained, multimode A/K prostheses were introduced (17,18). The control philosophy of these assistive systems will be discussed later.

The operation of an active, lower-limb, assistive device will be explained taking the Belgrade A/K prosthesis as an example. Its basic hardware consists of the standard U.S. skeletal prosthesis. The knee joint is fitted with a special actuator having four controllable states: flection, extension, rigid, and loose (17). In this way, stiffness control at all gait instances including ballistic motion, is available. The prosthesis is equipped with a special passive foot that allows for dorsal and plantar flection. The power supply is sufficient for three hours of walking on level ground, and the microcomputer is fixed in the interior of the prosthesis, which makes the device self-contained.

The control of the Belgrade A/K prosthesis is nonnumerical. It is based on proprioceptive and exteroceptive pattern recognition and pattern matching (17). For this purpose, a sensory system generating the necessary information is an integral part of the prosthesis. The following types of artificial sensors are used: insole with matrix of pressure transducers, angle measurement sensors, and sensors of the vertical line.

The ankle joint plays an important role in the normal walk. It can generate high torques and a fast response when needed. Consequently, a powerful and fast actuator is needed if an active ankle is desired. In many instances, well-designed foot prostheses without electronics can do the job. As a matter of fact, they are the best solution for simple gait patterns. In more complex environmental conditions (ramp, stairs), an active ankle prosthesis can improve the gait considerably. Several solutions have been proposed to integrate the active ankle prosthesis into the lower-extremity assistive system (19). Waseda University's A/K leg prosthesis WLP-7, shown in Figure 7, is a satisfactory solution in this respect.

ORTHOTICS

The orthotic assistive system operates in conjunction with a natural limb whose control is impaired or lost. Thus, the biomechanical requirements for operation of an orthosis are more complex than those for a prosthesis.

The classical orthotic approach relies on the application of an external skeleton supporting the impaired functional mo-

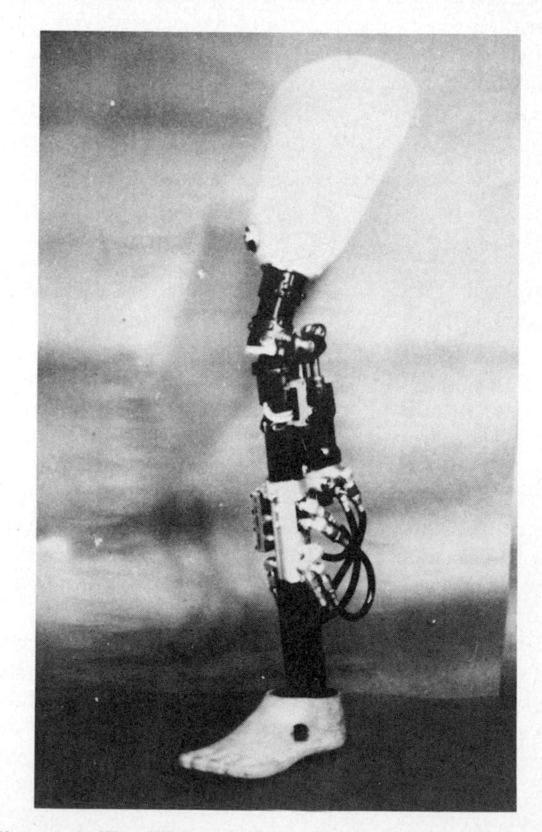

Figure 7. The WLP-7 A/K prosthesis of Waseda University, Tokyo.

tion. Such an arrangement involves force transfer over the human–machine interface as well as external control adapted to the patient's need and body attitude. The force transfer may exercise excessive pressure on some regions of the patient's body surface. Machine control must take into account the specific nature of each disease since motor deficiencies are caused by many different factors. In addition, paralysis produces spasticity, contractions, blood circulation problems, tissue damage, reduced elasticity of the skeleton, and so on.

The modern orthotic approach relies on functional electrical stimulation (FES). In many instances, FES can produce sequences of controlled muscle contractions leading to functional motions such as standing, ambulation, and hand actions. FES rehabilitation technology is not effective in all cases of paralysis. In order to extend the scope of FES, methods combining FES and active exoskeletons, the so-called hybrid assistive systems (HASs), have been proposed. Advantages of HASs will be discussed separately.

Active Exoskeleton

Models of active exoskeletons take the form of kinematic chains with joints and viscoelastic forces (20). A structural diagram of the model is presented in Figure 8. Modeling the exoskeleton with viscoelasticity is necessary since the orthosis acts upon the body of the paralyzed through muscles and interface elements. The general solution for the above model was derived on the basis of Newton–Euler equations (21). Elastic coupling of the active orthosis means that the exoskeleton, like natural joints, must have polycentric links so that the two kinematic chains can operate in parallel. This problem does not exist with conventional orthoses because the three-point support fixes the joint in the desired position.

The first type of exoskeleton was built as a rigid, anthropomorphic system designed to carry the patient using stored, prescribed trajectories (22). In the next phase, soft exoskeletons were introduced, which made possible a distributed force transfer (23,24). Figure 8 presents the French soft-suit orthosis. All proposed exoskeletons are meant to assist only the locomotion cycles. Gait stability and posture control are maintained without machine assistance by involving the patient's upper extremities (bars, crutches, canes). Even in this form, existing systems are cumbersome and unfit for field applica-

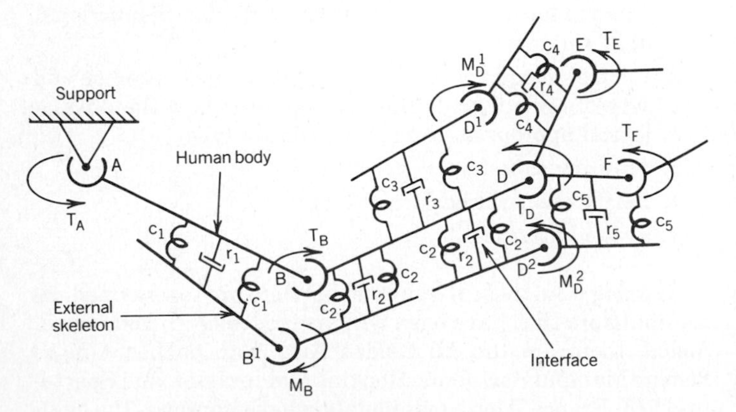

Figure 8. Model of the active skeleton.

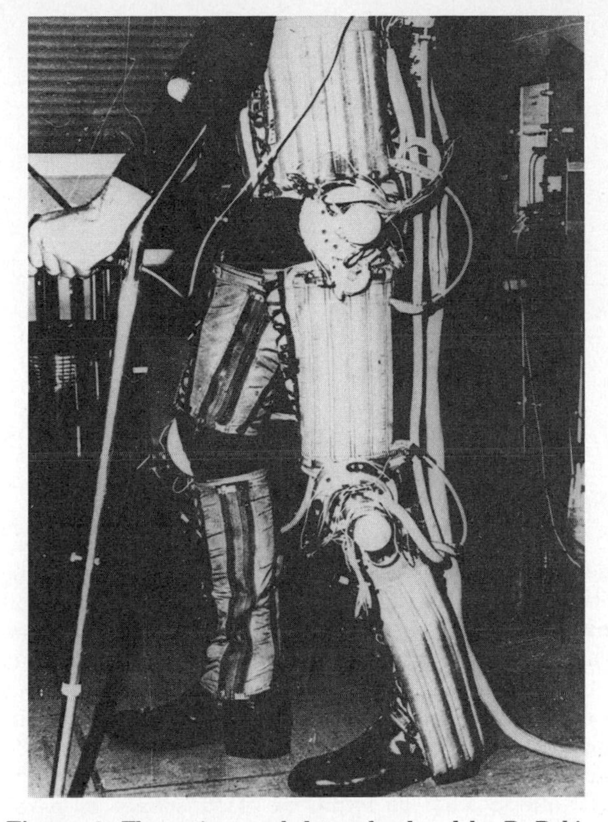

Figure 9. The active exoskeleton developed by P. Rabishong, Montpellieur, France.

tion. In the present form, they are used as clinical rehabilitation devices.

In order to improve the exoskeleton approach and add to it new desirable features such as modularity, self-fitting, and self-adaptive joint center matching, a new type of active orthosis was introduced (Figure 9) (20). On the basis of this orthotic device, assistive systems with a partial external power supply are built. Namely, the active external support is used in only those phases of bipedal locomotion (rigidity control of the joint, extension and flection support) that are less power demanding. Actuators of such an active orthosis are state controlled rather than trajectory controlled. Partial external power supply assumes, obviously, that motor deficiency is incomplete.

Exoskeletons are used less as complete upper extremity orthotic devices. They are mainly applied to improve impaired hand motions. Active arm orthosis for heavy motor deficiency has been developed for research purposes (25,26). It is a three-degree-of-freedom orthosis with joystick control. Namely, the motor deficiency in question prevents arm motions from counteracting gravity in the so-called large joints. Hand finger motions are preserved so that they can control the joystick.

Hybrid Assistive Systems (HASs)

FES has a great advantage over exoskeletons in that it attempts to regenerate functional motions by evoking external biological reflex mechanisms and internal synergies over which voluntary control has been lost because of damages in the neural networks (27). Ultimately, FES paves the road to

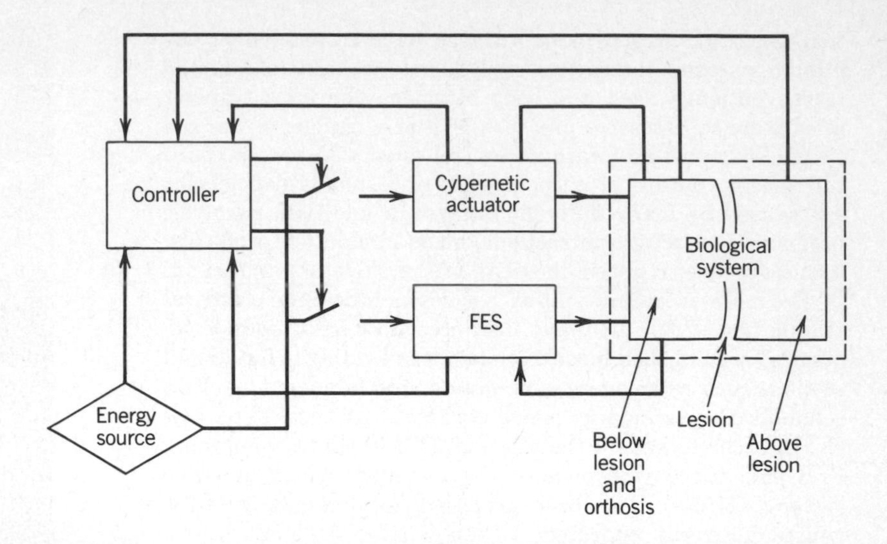

Figure 10. The structural diagram of HAS.

a neural prosthesis. The multichannel stimulation significantly extends the range of externally initiated functional motions (28–32). By multichannel electrical stimulation, sequences of muscle groups are activated, following patterns of normal functional motions. The application of FES enables the patients to reproduce the motor function using their own muscles as actuators and their own metabolic energy (33,34). In addition to the orthotic function, FES produces other therapeutic effects such as improved blood supply to paralyzed parts, spasticity reduction, and so on. However, many technical difficulties, especially related to implantable stimulators, must be overcome before the potential of this rehabilitation approach may be fully exploited.

The idea that FES should be supported by externally controlled orthoses in order to improve and extend the application field of FES was presented quite early (35). It took some time before hardware and software tools for HAS were developed so that clinical feasibility studies of this rehabilitation method could be started (36–38).

It must be kept in mind that FES by itself cannot restore in a satisfactory way functional motions in all instances of paralysis. With total lesions or denervated muscles, application of FES is not yet possible. Therefore, in cases of severe lesions, HAS may be the only solution.

A structural diagram of an HAS is shown in Figure 10. The hardware of the Belgrade HAS is explained in more detail (Figure 11). It consists of the following components:

1. A six-channel stimulator with surface electrodes. Each channel is activated independently. The pulse rise is exponential with the adjustable time constant. The output voltage of each channel is also adjustable. Pulse frequency and pulsewidth are affected by the controller. In the case of the patient in question, three muscle groups are stimulated: gluteus medius, quadriceps, and peroneus.
2. Assistive hardware consists of the self-fitting modular orthosis (SFMO) attached to lower extremities by trousers. The soft interface improves the pressure distribution on the skin due to external power transfer. The orthosis can be equipped with state-controlled actuators (flection, extension, rigid, loose) at hip and knee joints.
3. Bionic sensors reproducing elementary proprioceptive and exteroceptive information serve as input to the microprocessor.
4. The control is sensory driven. The sensory patterns, recognized by the controller, are matched to the discrete changes of joint states. Such a control has been given the name "artificial reflex control."

ROBOTICS FOR ENVIRONMENTAL CONTROL

Severely handicapped patients are unable to interact directly with the environment in terms of manipulation and locomotion. A remotely controlled manipulator, adapted for control purposes to the remaining minimal motor capabilities of the patient, can act as a flexible machine, strong enough for physical impact on the environment. In this way, the basic needs of the bed-confined patient to interact physically with the environment can be met. Special effort was made to define the minimal class of environmental interactions that should be at the disposal of an immobilized patient. It turns out that about 12 daily activities can significantly increase the patient's autonomy (39). Medical manipulators are actually evaluated in these terms (40).

The main features of a medical manipulator are

1. Manipulator performance efficiency and reliability measured in terms of speed, accuracy, and range of functional ability.
2. Control adequacy (number of control sites, number and type of control modalities, considerations of the physiological appropriateness of the control type).
3. Training time.
4. Safety of operations.
5. System logistics.

Starting from these requirements, there are several medical manipulators that have been designed, developed, and tested. Among them are the All Golden Arm, Rancho Los Amigos Remote Manipulator, Johns Hopkins Manipulator, and Spartacus IRIA, France. These manipulators have improved the quality of life of severely handicapped people.

Because of the minimal residual voluntary motor control

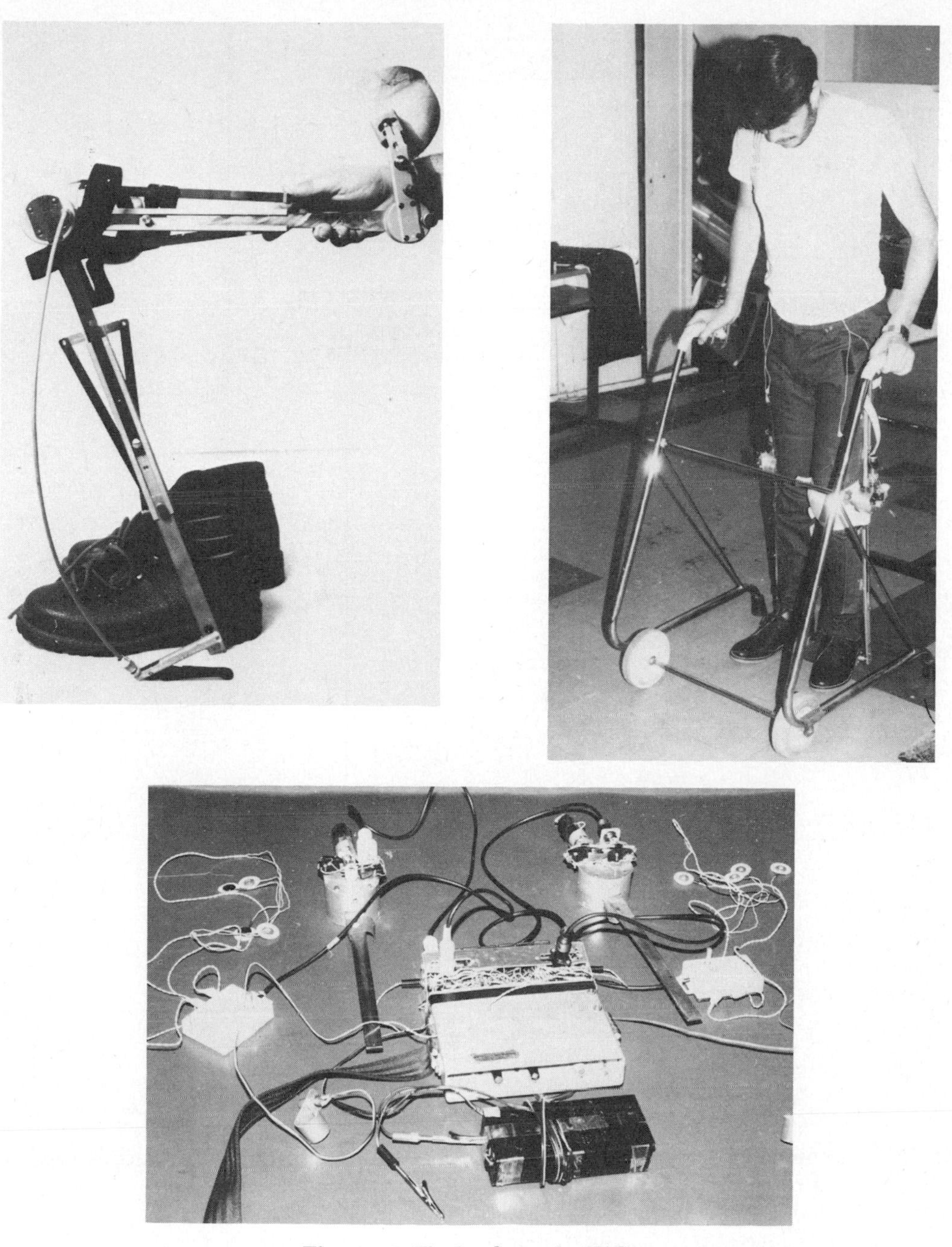

Figure 11. The hardware for HAS.

of the severely handicapped, human–machine control interaction becomes a crucial issue in the design of medical manipulators. In this regard, several options have been explored: voice control, ultrasound control, infrared rays, head displacement, and the joystick. Specific features of medical manipulators are displayed in Figure 12. Robot hardware for both medical and industrial manipulators is the same.

The lack of feedback from the manipulator to the patient requires considerable training effort before satisfactory proficiency in the use of the equipment is reached. Good medical manipulators must take into account the special human–machine interaction problems due to the state of the severely handicapped (41). Experiments have shown that semisequential control (allocating, in succession, control sites to different subsets of control variables) best meets the needs of the severely handicapped (42–44).

CONTROL METHODS FOR MEDICAL ROBOTICS

In the early stages of medical robotics, control methods of production processes were transferred in a straightforward way to rehabilitation engineering. Hand and arm prostheses used to be controlled by elementary methods such as on–off, position and velocity, and trajectory control. Later on, predictive, adaptive, and optimal control methods found application in rehabilitation engineering as well. The first control systems in robotics for the handicapped were of the open-loop type. Open-loop control assumes complete knowledge of the

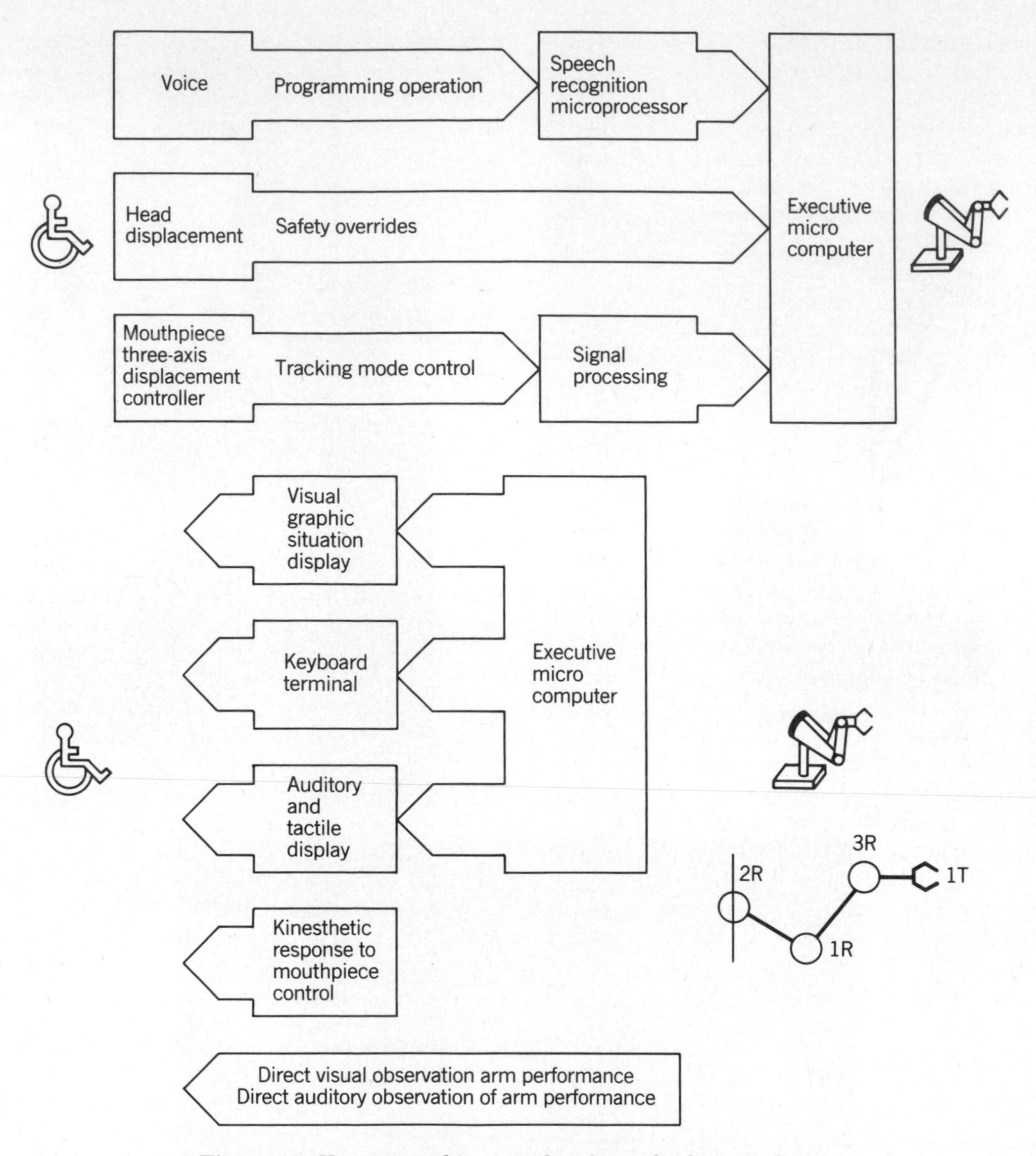

Figure 12. Human–machine interface for medical manipulators.

system behavior in all environmental conditions. The behavioral analysis involves several research fields, such as neurophysiology, anatomy, biomechanics, and so on. Within these research fields, motion analysis represents a highly challenging task. Photogrammetry, TV measurements, goniometric systems, Selspot, and other techniques have been developed (1–3) in order to collect data on body motions. Special EMG techniques, implants, surface electrodes, signal processing, and filtering, have been developed for the same purposes. Nonetheless, the data available deal only with simplified models of human behavior, which makes the application of open-loop controllers impossible in rehabilitation.

By introducing feedback, the exercise moves to the nature and quality of sensors. Two different feedback approaches exist. One is based on myoelectric activity as a source of control signals. The second one is based on artificial sensors. Myoelectric control uses surface or intramuscular EMG measuring and signal processing. By application of pattern recognition methods or correlation techniques, functional motions can be created if an adequate interface is used. The use of multiple electrodes and special filtering techniques results in several degrees of freedom regulation (knee flection, knee extension, dorsal flection). By the development of an anatomically based muscle model, the vector myogram, which represents the concurrent myoelectric activity of numerous muscle groups, can be evaluated. Off-line measurements on healthy and handicapped humans were used to develop an optimization procedure (9,15,45) controlling several degrees of freedom simultaneously (eg, humoral rotation, elbow flection, wrist pronation/supination, thumb rotation, finger prehension).

Medical robotics has stimulated, on the other hand, the development of control methods of its own that are now helping to improve machine performances in other fields of robotics. Touch, position, pressure, and slip sensors are essential elements of prosthetic and orthotic devices of all kinds. What actually matters is not just the output of sensors, but their overall properties. Medical robotics, as well as the new generation of smart robots, is in great need of distributed, matrix-type, sensory systems with high resolution. Such sensors must be mounted on soft, flexible, supporting surfaces that can cover objects of arbitrary form.

Sensory information derived in the above way must be adequately preprocessed before it is used for control purposes. In medical robotics, pattern-driven control is often more appropriate than error-driven control. Patterns involved in medical robot control are of visual, proprioceptive, exteroceptive, and acoustic natures. Consequently, pattern recognition algorithms for the above types of sensory information must be available. Looking at the control aspects of medical robotics in a general way, one can say that the nonnumerical nature of biological control should be reflected in machine control as well.

Pattern-driven and pattern-matching control of active assistive systems is consequently a highly desirable feature of human–machine interaction in the execution of functional motions. In order to apply such an approach to robot control, it was necessary to outline general methods for the synthesis of nonnumerical, pattern-matching controllers. The closest equivalent to pattern-matching control in biological systems is reflex action. Inspired by the role of reflexes in the execution of natural functional motions, a new method of robot control called artificial reflex control (ARC) was proposed. An outline of this nonnumerical control method is given below.

ARC control of assistive systems avoids solving of equations of motion of the kinematic chain. Thus, *a priori* knowledge about plant dynamics is not available in the form of state-space equations. Artificial intelligence has provided an alternative solution. The required *a priori* information about system behavior is now represented in the form of a knowledge base containing human expertise on the execution of functional motions at the reflex level.

The development of such a computer knowledge base relies on the following elements:

1. Formal identification of human reflex actions involved in the execution of the desired functional motion.
2. Knowledge representation in the machine.
3. Appropriate organization of the knowledge base.

Formal identification of human reflex actions is the most difficult part of the control synthesis. Namely, working at the reflex level, one cannot encode human expertise by interrogation and person-to-person communication. Instead, observation and records of natural functional motions serve as sources of expertise. The idea is to identify invariant and singular events occurring permanently in the execution of classes of natural functional motions (walking, stair climbing, grasping, manipulation, etc). In this context, the term identification means

1. Recognition of invariants of a class of functional motions in terms of proprioceptive, exteroceptive, and visual patterns.
2. Identification of pattern-matching invariants, that is, invariant mappings of sensory patterns on joint states.

Expert systems of this kind have been successfully applied to control active A/K prostheses, as explained earlier (17). By combining artificial and natural reflex control, the background for the development of HASs was laid down (35–37). On the basis of ARC, expert systems for industrial manipulators with dexterous multifingered hands are now being developed (46,47).

BIBLIOGRAPHY

1. V. M. Zaciorskii and co-workers, *Biomehanika dvigatelnoga aparata čeloveka* (in Russian), Fizikultura i sport, Moskva, USSR, 1980.
2. A. D. Winter, *Biomechanics of Human Movement,* John Wiley & Sons, Inc., New York, 1979.
3. *Advances in External Control of Human Extremities (ECHE) I–VIII (Proceedings of the International Symposia on ECHE I–VIII,* Dubrovnik, Yugoslavia, 1962–1984), Yugoslav Society for ETAN, Belgrade, Yugoslavia.
4. M. Vukobratović, and V. Potkonjak, *Dynamics of Robots and Manipulators,* monograph, Springer-Verlag, Berlin, FRG, 1984.
5. R. Tomović, "Human Hands as a Feedback System," *Proceedings of the I IFAC Congress Moscow,* Butterworth's Scientific Pub. London, 1961, pp. 624–628.
6. R. Tomović and G. Boni, "An Adaptive Artificial Hand," *Transactions on Automatic Control* **AC-7,** 3–10 (Apr. 1962).
7. M. Rakić, "An Automatic and Prosthesis," *Medical Electronics and Biological Engineering* **2,** 47–55 (1964).
8. R. Tomović and Z. Stojiljković, "Multifunctional Terminal Device with Adaptive Grasping Force," *Automatica* **11,** 567–571 (Nov. 1975).
9. W. R. Mann, "Cybernetic Limb Prosthesis," *Annals of Biomedical Engineering* **9,** 1–43 (1981).
10. E. M. Wagner and J. G. Catranis, "New Developments in Lower-extremities Prosthesis," in *Human Limbs and Their Substitutes,* NAS, 1954, Chapt. 17.
11. A. Capozzo and co-workers, "A Polycentric Knee–Ankle Mechanism for Above-knee Prostheses," *Journal of Biomechanics* **13,** 231–239 (1980).
12. C. W. Radcliffe, "Biomechanical Basis for the Design of Prosthetic Knee Mechanisms," *Proceedings of the Rehabilitation Engineering International Seminar REIS '80,* Tokyo, Japan, 1980, pp. 68–88.
13. F. Aoyama, "Lapoc System Leg," *Proceedings of the Rehabilitation Engineering International Seminar REIS '80,* Tokyo, Japan, 1980, pp. 59–67.
14. R. Tomović and co-workers, "Bioengineering Actuator with Nonnumerical Control," *Proceedings of the IFAC Conference on Control Aspects of Prosthetics and Orthotics,* Ohio, 1982, pp. 55–63.
15. W. C. Flowers and R. W. Mann, "An Electro-hydraulic Knee-torque Controller for a Prosthesis Simulator," *Transactions of the ASME, Journal of Biomechanical Engineering,* 3–8 (1977).
16. W. C. Flowers, D. Rowell, M. Tanquary, and H. Cone, "A Microcomputer-controlled Artificial Joint," *Proceedings of the 3rd Symposium on Theory and Practice of Robots and Manipulators,* 1978, pp. 133–146.
17. R. Tomović, "Control of Assistive System by External Reflex Arc," *Proceedings of the 8th International Symposium on ECHE (ETAN),* Dubrovnik, Yugoslavia, 1984, pp. 7–21.
18. K. Koganezawa and I. Kato, "A Development of A/K Prosthesis Adaptable to Voluntary Walking Period," *Proceedings of the 8th International Symposium on ECHE (ETAN),* Dubrovnik, Yugoslavia, 1984, pp. 343–355.
19. A. P. Kuzhekin and co-workers, "Subsequent Development of Mo-

torized Above Knee Prosthesis," *Advances in ECHE VII,* Yugoslav Society for ETAN, Belgrade, Yugoslavia, 1981, pp. 525–530.

20. D. Popović, "Assistive Systems for Rehabilitation of Impaired Locomotion," Ph.D. dissertation, Faculty of Electrical Engineering, Belgrade, Yugoslavia, 1981.
21. K. Ito and co-workers, "Computer Aided Dynamic Analysis of Multilink Systems," in *Research Report on Automatic Control Laboratories,* Vol. 27, Nagoya University, Nagoya, Japan, 1980, pp. 9–30.
22. M. Vukobratović, *Legged Locomotion Robots and Anthropomorphic Mechanisms,* monograph, Mihajlo Pupin Institute, Belgrade, Yugoslavia, 1975.
23. P. Rabishong and co-workers, "AMOL Project," *Advances in ECHE V,* Yugoslav Society for ETAN, Belgrade, Yugoslavia, 1975, pp. 33–48.
24. D. Hristić and co-workers, "New Model of Autonomous 'Active Suit for Distrophic Patients,'" *Advances in ECHE VI,* Yugoslav Society for ETAN, Belgrade, Yugoslavia, 1978, pp. 33–42.
25. D. Hristić and co-workers, "Active Arm Orthosis and Experience with Its Practical Use," *Advances in ECHE VIII,* Yugoslav Society for ETAN, Belgrade, Yugoslavia, 1984, pp. 33–40.
26. J. R. Allen and co-workers, "Orthotic Manipulators," *Advances in ECHE III,* Yugoslav Society for ETAN, Belgrade, Yugoslavia, 1970, pp. 261–270.
27. L. Vodovnik and co-workers in T. Hambrecht and J. Reswick, eds., *Functional Electrical Stimulation Application in Neural Prothesis,* Marcel Decker, New York, 1977, pp. 39–54.
28. A. Kralj and co-workers, "Gait Restoration in Paraplegic Patients: A Feasibility Demonstration Using Multichannel Surface Electrodes FES," *Jouranl of Rehabilitation Research and Development* **20,** 3–20 (1983).
29. E. B. Marsolais and R. Kobetic, "Functional Walking in Paralyzed Patients by Means of Electrical Stimulation," *Sin. Orthotic Rel. Research* **17s,** 30–36 (1983).
30. J. S. Petrofsky and co-workers, "Feedback Control System for Walking in Man," *Computer Biology and Medicine* **14,** 135–149 (1984).
31. P. H. Pekham and co-workers, "Controlled Prehension and Release in the C-5 Quadriplegic Elicited by FES of the Paralyzed Forearm Musculature," *Annals of Biomedical Engineering* **8,** 369–388.
32. D. R. McNeal and G. A. Bekey, "Closed-loop Control of the Human Leg Using Electrical Stimulation," *Advances in ECHE VIII,* Yugoslav Society for ETAN, Belgrade, Yugoslavia, 1984, pp. 113–126.
33. T. Bajd and A. Kralj, "Standing-up of a Healthy Subject and a Paraplegic Patient," *Journal of Biomechanics* **1s**(1), 1–10 (1982).
34. A. Kralj and co-workers, "Electrical Stimulation Providing Functional Use of Paraplegic Patient Muscles," *Medical Progress Through Technology* **7**(1), 3–9 (1980).
35. R. Tomović and co-workers, "Hybrid Assistive Devices for Rehabilitation," *Advances in ECHE IV,* Yugoslav Society for ETAN, Belgrade, Yugoslavia, 1972, pp. 7–19.
36. D. Popović, *Technical and Clinical Evaluation of the Self-fitting Modular Orthoses (SFMO),* NIHR, Washington, D.C., Reports on Research Grant, 1983–1985, p. 432.
37. L. Schwirtlich and D. Popović, "Hybrid Orthoses for Deficient Locomotion," *Advances in ECHE VIII,* Yugoslav Society for ETAN, Belgrade, Yugoslavia, 1984, pp. 23–32.
38. B. Andrews and T. Bajd, "Hybrid Orthoses for Paraplegics," *Advances in ECHE VIII Supplement,* Yugoslav Society for ETAN, Belgrade, Yugoslavia, 1984, pp. 56–61.
39. J. Bruett, "Twelve Activities of Daily Living Scales," *Physical Therapy* **49,** 857–862 (1969).
40. K. Corker and co-workers, "The Devlopment of Criteria for Evaluation and Prescription of Remote Medical Manipulators," *Advances in ECHE III,* Yugoslav Society for ETAN, Belgrade, Yugoslavia, 1970, pp. 121–135.
41. H. H. Kwee, "Quelque repres de commande pour tetraplegique," IRIA, Le Chesuay, France, Spartacus Report HL005.
42. V. Paeslack and H. Roesler, *Medical Manipulators for the Seriously Disabled,* Heidelberg, FRG, Annual Report Stiftung Rehabilitation, 1975.
43. J. Vertut, J. Charles, P. Coiffet, and P. Pettit, "Advance of the New MA-23 Force-reflecting Manipulator System," *Proceedings of the Second CISM–IFTOMM Symposium,* Ro-mansy, Warsaw, Poland, 1976.
44. D. E. Whitney, "Resolved Motion Rate Control of Manipulators and Human Prostheses," *IEEE Transactions on Man–Machine Systems* **MMS-10,** 47–53 (1969).
45. C. B. Jacobsen, "Control Systems for Artificial Arms," Ph.D. dissertation, Department of Mechanical Engineering M.I.T., Cambridge, Mass., Jan. 1973.
46. R. Tomović and G. Bekey, "Robot Control by Reflex Actions," *Proceedings of the 1986 Conference on Robotics and Automation,* Vol. 1, 1986, pp. (1)240–(1)248.
47. R. Tomović, G. Bekey, and W. Karplus, "A Strategy for the Synthesis of Grasp with Multi-fingered Robot Hands," Robotics Institute, University of Southern California, Los Angeles, Calif. Internal Report, 1986.

HANDS

William Palm
University of Rhode Island
Kingston, Rhode Island

INTRODUCTION

Most robots consist of an arm for moving objects and a device that is generically called an end-effector, which is attached at the end of the arm. The end-effector can be a special-purpose tool like a drill or grinder, or it can be a device that holds tools or other objects (see End-of-arm tools). The most common end-effector is a gripper that has two or more rigid fingers (see Grippers). Although a gripper is capable of grasping many types of objects, it depends on the robot arm to move the object. The term hand is used in robotics to mean an end-effector that can grasp an object and manipulate it within the hand by changing its grasp in an adaptively controlled manner. Sometimes the phrase dexterous hand is used to emphasize those capabilities not achievable with a gripper. As will be seen, robot hands need not look like the human hand, but many robot hand designs have been based on the human model. A hand is articulated if it has fingers with joints. The terms versatile gripper and universal gripper are sometimes used to describe dexterous hands.

The Need for Dexterous Hands

Robot capabilities are currently severely limited by the rather crude end-effectors available. Successful future applications of robots will often require a hand to manipulate the workpiece. In assembly tasks the robot hand may be required to operate near or within rigid obstacles and may have to provide finer positioning accuracy than is available from the robot arm. Also, in many situations it is either impossible or uneconomical to present the workpieces in a predictable, prescribed orientation, and the robot acquires the workpiece in a random orientation (1–4). Several possibilities exist for solving this problem, including a grasping station, a reorienting station, the use of more than one arm, and a robot hand that can manipulate the workpiece.

Many assembly applications require workpiece motions that are difficult if not impossible to achieve without in-hand manipulation. For example, recent studies have demonstrated the usefulness of active-compliance control strategies for the assembly of bayonnet-type electrical connectors and for insertion operations in disk drive assembly (5). With active-compliance control, the control system plays the role of an elastic member with programmable stiffnesses in several axes that can be adaptively changed in the course of the assembly process. The required workpiece motions must be rapid and sometimes precise. It is cumbersome to achieve them by moving the entire arm, because arm motions are slower, less precise, and difficult to perform in an environment crowded with obstacles. An analogy can be made with the human hand-arm system used to control a pencil for writing; the arm is not used for the fine motions, which instead are produced by the fingers and wrist.

A wrist can be used to produce some of the required workpiece motions, but these are somewhat limited, for example, to rotational motions. Even if a workpiece translation capability is added to the wrist, the device is still incapable of altering its grasp on the workpiece. The ability to change the grasp is the distinctive characteristic of in-hand manipulation. This ability is important for several reasons. Many workpieces cannot be picked up with the same grasp that can be used to perform the assembly operation; also, the grasp may need to be altered as the assembly process proceeds. Applications requiring such a solution occur in many mechanical assemblies and in the assembly of circuit boards containing odd-form components such as switches, connectors, transformers, and large capacitors. These components are often imprecise in their external geometry. They are therefore difficult to acquire and insert with hard automation, and sensor-based manipulation is required. In general it will not be possible to design industrial products to eliminate completely the need for dexterous hands. In addition, the use of robots in unstructured environments such as exploration and remote maintenance, repair, and inspection will require such hands (6).

Some robot system designers have attempted to increase workpiece handling capability by equipping the robot with an array of end-effectors that are either located on a rotatable carousel on the end of the arm or stored in a rack and equipped with quick-connect/disconnect mechanisms. This solution may be necessary in applications where a variety of manufacturing processes such as grinding, drilling, deburring, or welding are performed on the workpiece. However, in many applications, such as assembly, the primary need is to impart prescribed motions and forces to the workpiece and to be able to change the grasp on the workpiece during the assembly process without losing control of it. For such applications, designing a single hand for certain classes of assembly processes is feasible and desirable for economic and physical reasons. The extra time required to switch hands will often make this solution uneconomical (7). Also, the cost or weight of an array of special purpose end-effectors can be prohibitive. Finally, in many assembly applications, the workpiece must be maneuvered in tight confines, where it is not feasible to change end-effectors.

History and Status of Dexterous Hands

The earliest uses of mechanical hands were as prosthetic devices to replace lost limbs. This application is quite old, with one such design reported as early as 1509 (8). Most prosthetic hands are capable only of grasping objects and not of manipulating them. Later applications of mechanical hands were in remote manipulation of hazardous materials. It is expected that more applications of these teleoperators will occur in space and underwater exploration.

Most of the manipulating hand designs developed thus far have utilized multiple articulated fingers that mimic the design and function of human fingers (9–15). An exception is the hand design described by Birk (1), which can achieve and detect the orientation of cylindrical-type objects about one axis. Two ways of improving the manipulation capability of an articulated hand are increasing the number of joints on each finger and increasing the number of fingers. Crossley (16) designed a four-fingered hand that included a thumb with a retractable fingernail. The Utah/MIT Hand (11,17,18) has four fingers including a thumb, and each finger has four joints. But most articulated hands developed thus far have three fingers with typically three joints per finger. The best-known of these is perhaps the Stanford/JPL Hand (19–22). To date most of the work on multi-fingered articulated hands has been concerned with three aspects: (*1*) the development of suitable actuators and mechanical transmissions (11,18,23), (*2*) the problem of stable grasp configurations (20, 24–31), and (*3*) the interpretation of finger feedback information (9, 32–36). However, significant control problems need to be solved before such hands can achieve gross motion reorientation of workpieces under closed-loop control. Another approach is to abandon the anthropomorphic model and to design hands for specific functions or for specific classes of workpiece shapes. Researchers at the University of Rhode Island have taken such an approach for cylindrical workpieces (37). The Rhode Island Hand for Cylindricals is described in a later section.

HAND FUNCTIONS AND HAND COMPONENTS

The human hand has several functions. The principal ones are grasping objects, manipulating the objects with both fine and gross motions, controlling the forces applied by the object on the external environment, and sensing various quantities such as force, weight, shape, orientation, movement, texture, and temperature. The hand components needed to perform these functions include the kinematic structure formed by the palm and fingers with their joints; the tendons and muscles to provide actuation; the skin; and, of course, the brain and nervous system. In order to treat robot hands, it is necessary to discuss these same functions and the artificial counterparts to the human hand's components. Artificial skin and associated tactile sensors are discussed in another article (see SENSORS,

TOUCH, FORCE, AND TORQUE MEASUREMENT). The kinematic design requirements for a robot hand, possible actuation systems, and control methods are discussed here.

Grasp

In order to treat the kinematic design requirements, one must first consider the function of grasping in various situations. Schlesinger (38,39) identified the six types of prehension that characterize the grasps used by the human hand. The cylindrical and spherical grasps are employed when grasping objects like bottles and balls. The tip grasp is used when holding an object such as a pill beween two fingertips. The palmar grasp is that used when holding a pen while writing. The lateral grasp holds an object such as a plate between the thumbpad and the side of the index finger. Finally, the hook or snap grasp is used to hold a suitcase handle. All six grasps can be achieved with the thumb and only two other fingers. The remaining fingers provide additional stability (due to increased frictional area) and strength (due primarily to the muscle extending along the outside of the hand between the small finger and the wrist). The palm of the hand serves primarily as a supporting structure for the fingers and touches the workpiece in only the cylindrical and spherical grasps.

Manipulation

Grasps can be considered static functions, but manipulation requires movement of the hand's components. The manipulation functions of the human hand are categorized by Crossley (16) as shown in Table 1. Note that manipulation types A and B require the motion of only one finger; the difference between the two is that type A exerts a force on an object held in the hand while the type B exerts a force on an object external to the hand. Type H is similar to type B, except that type H requires multiple motions. Type C is actually a transfer from a tip grasp to a cylindrical grasp, and type G is the reverse. The remaining manipulation types involve motion while retaining the same type of grasp. For example, writing with a pen uses only a palmar grasp, but requires that the grasp be adjusted dynamically in order to move the pen.

Table 1. Manipulation Functions of the Human Hand[a]

Type	Action	Comment
A	Gripping a trigger	Index finger produces motion while thumb and fingers 3, 4, 5 grasp.
B	Flipping a switch	Similar to A except that thumb and fingers 3, 4, 5, rest.
C	Transferring a pipe to grip	The inverse of A, with motion provided by fingers 3 and 4.
D	Using cutters	A grasp of varying width; it requires no independent finger motions.
E	Screwing a pen top	Tip of index finger rolls object down thumb, which is held stiff.
F	Cigarette rolling	Same as E except that lateral surface of index finger is used.
G	Transferring pen	Thumb and index grasp, then finger 3 hooks under object and pushes away.
H	Typewriting	Selecting and pushing an array of buttons independently.
J	Penwriting	Can be achieved either by bending the fingers or, less accurately, by moving the forearm while keeping the fingers rigid.

[a] Modified from Crossley (16).

HAND DESIGN

Degrees of Freedom

The number of degrees of freedom (DOF) of a mechanism is the number of independent ways (translations or rotations) the device can move. A variable representing every DOF must be specified to describe the position and orientation of the device. The human hand has 20 DOF. There are 4 DOF in each finger (2 DOF in the knuckle and 1 in each of the remaining two joints), 3 DOF in the thumb, and 1 in the palm. In a given task any DOF utilized to maintain prehension cannot be used to achieve workpiece manipulation. There are several ways to increase the DOF of an articulated robot hand. One is to increase the DOF in each joint; another is to increase the number of joints. Finally, the number of fingers can be increased.

Types of Contacts

A joint is a permanent connection between two links to limit their relative motion. Joints are typically either revolute (rotational) or prismatic (sliding). When contact occurs between the surfaces of the links of an articulated hand and the object being manipulated, the object can be treated as one of the links. The nature of the contacts is very important in hand design and depends on the shapes of the mating surfaces and their frictional properties. The relative motion between two objects in contact at one or more locations depends on the shape of the contact area, the relative locations and orientations of the contact areas, and the effect of friction in each contact area (22). In general, each surface can have point, line, or plane contact. Two surfaces in contact have nine possible contact types. Three of these are unstable and therefore transient: point-point, point on a line, and line on a point. Two of the remaining six types are symmetric, so there are actually only four types of stable contacts: point-plane, line-plane, plane-plane, and line on a nonparallel line. The latter is essentially a point contact.

Contact types can also be classified according to the relative freedom of motion between the two bodies that is allowed by the contact. In a 1 DOF contact, only one parameter is needed to specify the relative motion, and so on up to a 5 DOF contact. The DOF of a contact depends partly on whether friction is significant.

Kinematic Synthesis

Mechanism design in general and robot hand design in particular may be attempted in several ways. Number synthesis considers only the DOF in the joints and contacts connecting the links. Type synthesis considers the type of relative motion allowed by each connection and the resultant motion of each link due to the net effect of all such connections. Finally, dimensional synthesis considers the effect of mechanism dimensions (such as link lengths) on the link motions. Number synthesis is usually the easiest to apply, and the first thorough kinematic analysis of articulated hands was done by Salisbury using number synthesis in conjunction with mobility and connectivity analysis (22). The mobility of a mechanism is the number of independent parameters needed to specify the position of every body in the system. The connectivity between two bodies in a mechanism is the number of independent parameters needed to specify the relative positions of the bodies. Since the palm is often used as the reference point, the connectivity measure normally employed is that between the palm and the grasped object.

TWO PROMISING DESIGNS

The Stanford/JPL Hand

Salisbury (19,20,22) limited his analysis to a maximum of three fingers with at most three joints per finger, and he assumed that all the contacts in a given design allow the same DOF. The eight ways that a three-link finger can touch an object are shown in Figure 1. The range of possible mechanisms was also limited by considering only those designs that have a connectivity of 6 with the joints active (for general tasks), and that have a connectivity of zero with the joints locked (in order to completely constrain the grasped object). Thirty-nine potentially acceptable designs were found with this method.

The 39 designs included ones with 3, 4, and 5 DOF contacts at the fingertips. The 33 designs with 5 DOF were eliminated because it would be impossible for them to exert moments on cylinders and spheres without excessive reliance on a sufficient amount of friction. Of the six remaining designs, the one having three fingers with three joints per finger was chosen because it was the only one of the remaining designs in which a more secure grasp was achieved from the extra joints by keeping the friction constraints active. The last two axes were chosen to be parallel, and the second axis was chosen to be perpendicular to the first in order to allow the fingers to curl around objects. The placement of the fingers relative to one another was made to accommodate a 1 in spherical workpiece.

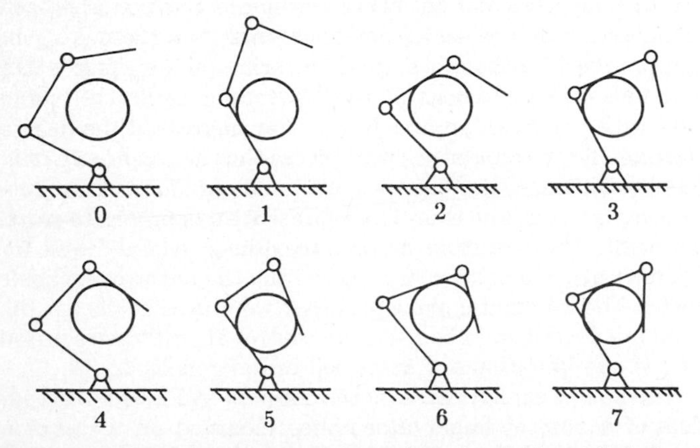

Figure 1. The eight contact configurations in which a 3-link finger can touch an object. Redrawn from Ref. 22.

The resulting design is the Stanford/JPL Hand, shown in Figure 2. Teflon-coated cables in flexible conduits transmit the tension-cable actuator forces to the joints. Four cables are used for each finger; each cable has a tension sensor composed of strain gauges. Position and velocity feedback is obtained from incremental optical encoders on the shafts of the motors that drive the cables. Each joint controller uses a 6502 8-bit microprocessor. A VAX 11/750 performs joint space interpolations, reads sensor data, and sends setpoint commands to the joint controllers. An LMI LISP machine performs top-level control for coordination.

The Utah/MIT Hand

The Utah/MIT Hand (11,17,18) evolved from biomedical research directed toward the design of prosthetic devices and was therefore designed to be anthropomorphic in size, geometry, and performance. The latest version, shown in Figure 3, has four fingers, including an opposing thumb. In order to decrease the system's complexity, the small finger of the human hand was not replicated. In addition, the 2 DOF that are present in the human knuckle were implemented as sepa-

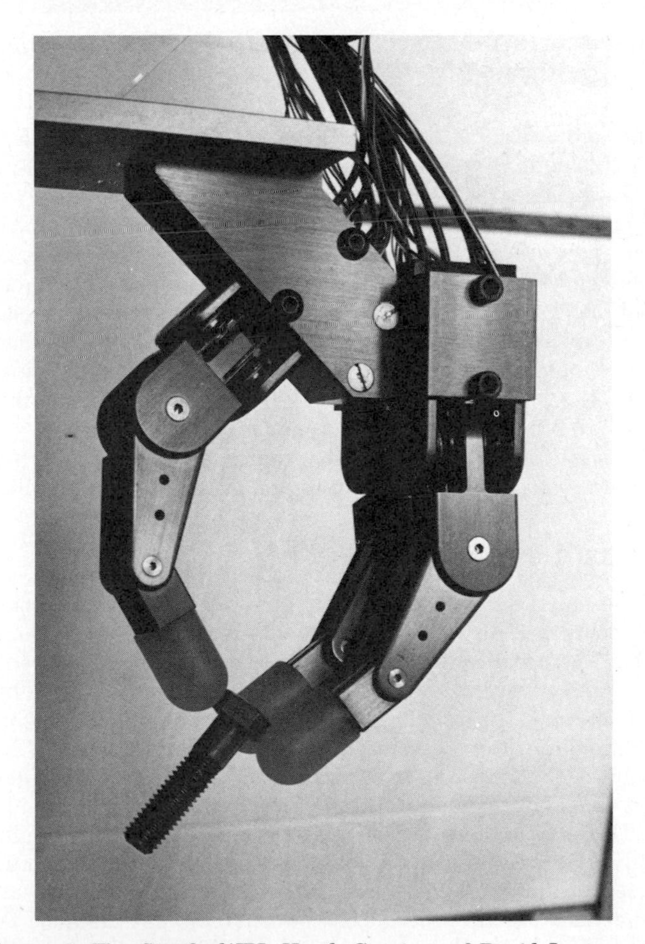

Figure 2. The Stanford/JPL Hand. Courtesy of David Lampe and Ken Salisbury of MIT.

Figure 3. The Utah/MIT Hand. Courtesy of M. Milocek and S. Jacobsen of the University of Utah.

rate joints because of the routing requirements imposed by the rectangular cross-sections of the tendons.

Each finger has 4 DOF. Each DOF is actuated by a pair of antagonist tendons and requires a pair of actuators. The actuators are electropneumatic devices consisting of a glass cylinder with a carbon-graphite composite piston. Each joint contains a Hall effect device for position sensing, and each tendon has a strain gauge device for tension sensing. Feedback control of the joints is obtained with a hybrid analogue/digital controller having five Motorola 68000 single board computers, one for each of the fingers and one for coordination.

Current research with the Stanford/JPL and Utah/MIT Hands is focused primarily on improving their controllability.

CONTROL OF DEXTEROUS HANDS

While much work has been done on the control of robot arms, the control of dexterous robot hands is a largely unexplored area. The hand control problem is extremely difficult because the hand must not only acquire and hold the workpiece but also provide the required precise motions and forces while providing position, force, or tactile feedback information. This requires sophisticated control algorithms for integrating sensory feedback, task models, and force/motion control.

The control of a hand with articulated fingers like the human hand is a difficult problem because the hand's structure is serial and not parallel and because its operation is vitally dependent on the precise control of forces (21). Because the finger forces act through relatively small surface areas, the fingers must be properly coordinated to maintain a secure grasp while manipulating the workpiece, and this cannot be done with only local controllers at the joints. A high level controller is required for the fingers to move cooperatively. This is so difficult a problem that no even partially-successful results have been reported to date, with two exceptions (40, 41). No articulated robot hand is currently capable of reliably performing workpiece reorientation involving gross motions under closed-loop control.

In addition to the force-control/stable-grasp issue, another difficult aspect of the control problem is the need to generate the trajectories (joint angles vs time) that each joint controller needs to perform the coordinated motion. These trajectories are highly dependent on the specific workpiece geometry, the initial grasping configuration at acquisition, and the desired final orientation of the workpiece. It is extremely tedious for the control system designer to specify these trajectories, even if all of the possible geometries that can occur can be identified. What is needed is an automatic method for generating the trajectories for different initial and final workpiece states, and for dealing with variations in workpiece geometry. This is known as implicit or object level programming.

NONANTHROPOMORPHIC APPROACHES TO HAND DESIGN

Apparently, a hand will be more controllable if it can be designed to minimize the need for active control to maintain grasp stability. Also, if the hand's kinematics are selected to match the manipulation requirements of a generic class of workpiece shapes, the trajectory generation problem will be simplified. For example, many industrial workpieces have the same generic geometry, such as a cylindrical shape with a hole in the center. During assembly, the center hole may either be inserted onto a peg or screwed onto a threaded stud. To handle these pieces, a hand must provide small translations in three directions and small rotations about three axes, for alignment purposes. Also, large rotations may be required to flip the piece or to provide multiple revolutions for threading.

The Rhode Island Hand for Cylindricals

If complete generality is not sought, a hand's kinematics can be simplified and the hand can be control-configured for ease of achieving gross motions of the workpiece. Such an approach promises to produce useful devices sooner than the more general approach, although a smaller variety of workpieces can be handled. Researchers at the University of Rhode Island have taken such an approach (3). They addressed the design of a hand for manipulating workpieces that are approximately cylindrical, since these shapes are commonly found in industrial applications and therefore represent an appropriate starting point. The resulting design, the Rhode Island Hand for Cylindricals, is a drastic departure from the anthropomorphic model. The kinematic principles that were used to design the hand are described in Ref. 41; the control algorithms required for the manipulation are described in References 42–44.

The hand's configuration is represented by Figure 4. It consists of a pair of finger assemblies mounted on a common headframe. Each finger is driven by its own power screw and can translate independently. This allows both grasping and

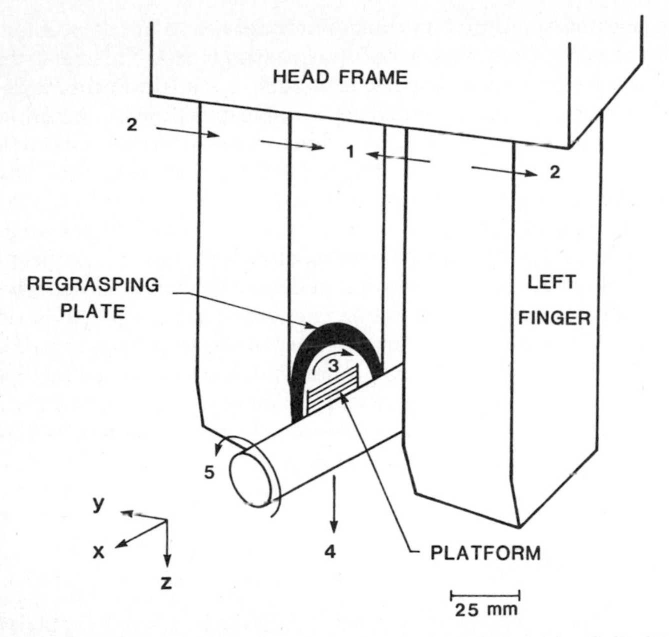

Figure 4. Modes of operation of the Rhode Island Hand for Cylindricals.

translation of the workpiece along the y axis. These operations are shown as motions (1) and (2) (Fig. 4). Each finger contains a rotating platform that has an independently-powered translating belt. The multiple parallel lines shown in the figure represent the belt's ribs. The platform rotation and belt translation are shown in Figure 5. Each finger also has a pneumatically-actuated plate for grasping the workpiece while the belt platform reorients itself. These plates contain two pairs of optical through-beam sensors that are used to sense the presence of the workpiece and its orientation with an algorithm that is presented in Ref. 43. The emitters are on one plate; the detectors are on the other. Thus the hand is capable of six independent motions: translation of each finger along a common axis (two motions), coupled rotation of the belt platforms (one motion), translation of the belts in the same direction and in opposite directions (two motions), and coupled actuation of the grasping plates (one motion).

These motions can be used to manipulate a cylindrical workpiece as follows. Once the workpiece has been grasped by the belt platforms, it can be rotated about the axis of rotation of the platforms. This rotation is about the y axis and is shown as motion (3) in Figure 4. Because the belts can translate on the platforms, they can be used to translate the workpiece in any direction normal to the platform rotation axis. This requires that the platforms be oriented at the same angle, and the mechanical drive employed in the hand ensures that this is so. When the belts translate in the same direction, the workpiece translates in that direction namely, the direction perpendicular to the belt ribs. This is shown as motion (4) in Figure 4. When the belts are translated in opposite directions, the cylindrical workpiece is rotated about its major axis (motion (5)). Thus the cylinder can be rotated about its major axis, rotated about a line joining the finger contact points, and translated in the x, y, and z directions. The hand thus manipulates cylinders in 5 DOF. The only basic motion that the hand cannot provide is a rotation about the z axis. However this type of motion is usually available from the wrist rotation of the robot arm.

The grasping plate is used when the platform angle is such that the belts cannot translate the workpiece in the desired direction. For example, the platform angle as shown in Figure 4 cannot produce a translation of the cylinder along the x axis. To accomplish this, first the grasping plates are actuated

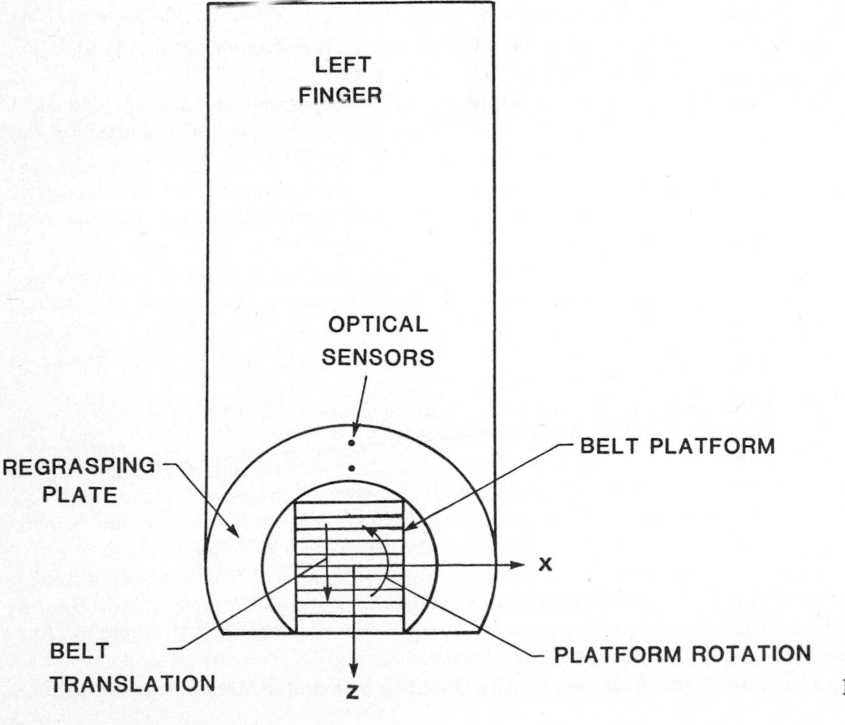

Figure 5. A view of one finger of the Rhode Island Hand.

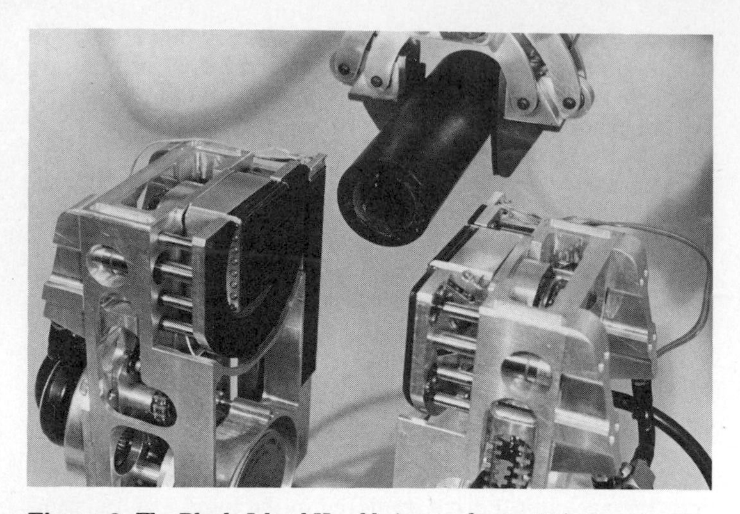

Figure 6. The Rhode Island Hand being used as a stationary reorientation device to assist a robot with a simple gripper. Courtesy of the University of Rhode Island.

to hold the cylinder while the platforms rotate by 90°. The plates then release, and the cylinder is gripped by the platforms. The desired translation can then take place. The mechanical design allows the plates to push the platforms back slightly, enough away from the workpiece to allow platform rotation to occur.

Figure 6 shows the hand being used as stationary reorientation device, that is, as a "robot's helper" or "second hand." In this application a robot with a simple gripper acquires an object that must be reoriented before the robot can perform the next task. The robot's helper takes the object from the robot, reorients it, and hands it back to the robot in the desired orientation. The Hand for Cylindricals has been successfully tested in such an application. Current work is aimed at improving the hand's capability for manipulating noncylindrical workpieces. This will require compliance-based control of the grasping force to allow the fingers to "float" while the workpiece is rotated by the belts. Since the control code is now heavily dependent on workpiece geometry, another research goal with this hand is to develop means for automatic programming, in which the workpiece model is composed of different primitive shapes. A high level planning algorithm would then generate the control code (44).

SUMMARY

Since 1982 significant progress has been made in the theory, design, and control of dexterous hands. Salisbury established a formal basis for understanding the mechanics of grasping and the use of force control, and demonstrated these concepts with the Stanford/JPL Hand. In addition, the state of the art of finger actuation and control techniques has been greatly advanced by the Utah/MIT Hand. These two hands have become the recognized standards, and several other research groups are attempting to apply and to improve them.

Despite the extensive analysis that led to the Stanford/JPL Hand, it probably does not represent the last word in articulated hands. Salisbury's designs were based on instantaneous analysis and thus ignored dynamic effects. Also, the analysis focused on small-motion control and did not treat the problem of obtaining large motions of the grasped object. This requires repositioning of the fingers, and such a transition might require more fingers to maintain a stable grasp while the other fingers are repositioned. The integration of other sensor types, such as tactile sensors at the fingertips, is an area that will receive more attention in the future.

The lack of high-level control algorithms required for gross motions and workpiece reorientation will remain a significant obstacle to the implementation of dexterous hands in the foreseeable future. New actuation methods that allow the hand size to be reduced are vitally needed. Also, in light of the controllability problems and other limitations imposed by imitating the human hand, one will probably see increased efforts in the design of nonanthropomorphic hands, as illustrated by the Rhode Island Hand for Cylindricals.

BIBLIOGRAPHY

1. J. Birk, "A Computer-Controlled Rotating-Belt Hand for Object Orientation," *IEEE Transactions on Systems, Man, and Cybernetics* **SMC-4**(2), 186–191 (Mar. 1974).
2. J. Birk, "A Computation for Robots to Orient and Position Hand-Held Workpieces," *IEEE Transactions on Systems, Man, and Cybernetics* **SMC-6**(10), 665–671 (Oct. 1976).
3. J. Birk, R, Kelley, and H. Martins, "An Orienting Robot for Feeding Workpieces Stored in Bins," *IEEE Transactions on Systems, Man, and Cybernetics* **SMC-12**(2), 151–160 (Mar./Apr. 1982).
4. R. Kelley, J. Birk, H. Martins, and R. Tella, "A Robot System Which Acquires Cylindrical Workpieces from Bins," *IEEE Transactions on Systems, Man, and Cybernetics* **SMC-12**(2), 204–213 (Mar./Apr. 1982).
5. G. Chaoui and W. Palm, "Active Compliance Control Strategies for Robotic Assembly Applications," in M. Donath and M. Leu, eds., *Robotics and Manufacturing Automation,* American Society of Mechanical Engineers, New York, 1985, pp. 43–50.
6. R. R. Schreiber, "Research in Robotics: Some Critical Issues," *Robotics Today,* 66–71. (Aug. 1984).
7. F. Skinner, "Multiple Prehension Hands for Assembly Robots," *Proceedings of the 5th International Symposium on Industrial Robots,* Chicago, Ill., Sept. 1975, pp. 77–87.
8. D. S. Childress, "Artificial Hand Mechanisms," *Proceedings of the Mechanisms Conference and International Symposium on Gearing and Transmissions,* San Francisco, Calif., Oct. 1972.
9. R. Bajcsy, M. McCarthy, and J. Trinkle, "Feeling by Grasping," *Proceedings of the 1st IEEE International Conference on Robotics,* Atlanta, Ga., Mar. 1984, pp. 461–465.
10. H. Hanafusa and H. Asada, "A Robot Hand with Elastic Fingers and Its Application to Assembly Processes," in M. Brady and co-eds., *Robot Motion: Planning and Control,* MIT Press, Cambridge, Mass., 1982, pp. 338–359.
11. S. C. Jacobsen, J. E. Wood, D. F. Knutti, and K. B. Biggers, "The Utah/MIT Dextrous Hand: Work In Progress," in M. Brady and R. Paul, eds., *Robotics Research (The First International Symposium),* MIT Press, Cambridge, Mass., 1984, pp. 601–653.
12. D. Lian, S. Peterson, and M. Donath, "A Three-Fingered Articulated Robotic Hand," *Proceedings of the 13th International Symposium on Industrial Robots/Robots 7 Conference,* Chicago, Ill., Apr. 1983, pp. 18-91–18-101.
13. T. Okada, "Object-Handling System for Manual Industry," *IEEE*

Transactions on Systems, Man, and Cybernetics **SMC-9,** 79–89 (Feb. 1979).

14. T. Okada, "Computer Control of Multi-Jointed Finger for Precise Object-Handling," *IEEE Transactions on Systems, Man, and Cybernetics* **SMC-12**(3), 289–298 (May/June 1982).
15. P. K. Wright, "A Manufacturing Hand," *Robotics and Computer-Integrated Manufacturing* **2**(1), 13–23 (1985).
16. F. R. E. Crossley and F. G. Umholtz, "Design for a Three-Fingered Hand," *Mechanism and Machine Theory* **12,** 85–93 (1977).
17. K. B. Biggers, S. C. Jacobsen, and G. E. Gerpheide, "Low Level Control of the Utah/MIT Dextrous Hand," *Proceedings of the 1986 IEEE International Conference on Robotics and Automation,* San Francisco, Calif., Apr. 1986, pp. 61–66.
18. S. C. Jacobsen, J. E. Wood, D. F. Knutti, K. B. Biggers, and E. K. Iversen, "The Version I Utah/MIT Dextrous Hand," in H. Hanafusa and H. Inoue, eds., *Robotics Research* (*Second International Symposium*), MIT Press, Cambridge, Mass., 1985, pp. 301–308.
19. J. K. Salisbury and J. J. Craig, "Articulated Hands: Force Control and Kinematic Issues," *The International Journal of Robotics Research* **1**(1), 4–16 (Spring 1982).
20. J. K. Salisbury and B. Roth, "Kinematic and Force Analysis of Articulated Mechanical Hands," *ASME Journal of Mechanisms, Transmissions, and Automation in Design* **103,** 35–41 (Mar. 1983).
21. J. K. Salisbury, D. Brock, and S. Chiu, "Integrated Language, Sensing, and Control for a Robot Hand," *Proceedings of the 3rd International Symposium on Robotics Research,* Gouvieux, France, Oct. 1985.
22. M. T. Mason and J. K. Salisbury, *Robot Hands and the Mechanics of Manipulation,* MIT Press, Cambridge, Mass., 1985.
23. J. K. Salisbury, "Design and Control of an Articulated Hand," *International Symposium on Design and Synthesis,* Tokyo, Japan, July 1984.
24. J. M. Abel, W. Holzmann, and J. M. McCarthy, "On Grasping Planar Objects with Two Articulated Fingers," *Proceedings of the 1985 IEEE International Conference on Robotics and Automation,* St. Louis, Mo., Mar. 1985, pp. 576–581.
25. B. S. Baker, S. Fortune, and E. Grosse, "Stable Prehension with a Multifingered Hand," *Proceedings of the 1985 IEEE International Conference on Robotics and Automation,* St. Louis, Mo., Mar. 1985, pp. 570–575.
26. M. Cutkosky, "Mechanical Properties for the Grasp of a Robotic Hand," Technical Report #CMU-RI-TR-84-24, The Robotics Inst., Carnegie-Mellon Univ., Pittsburgh, Penn., Sept. 1984.
27. H. Hanafusa and H. Asada, "Stable Prehension by a Robot Hand with Elastic Fingers," in M. Brady and co-eds., *Robot Motion: Planning and Control,* MIT Press, Cambridge, Mass., 1982, pp. 323–335.
28. W. Holzman and J. M. McCarthy, "Computing the Friction Forces Associated with a Three Fingered Grasp," *Proceedings of the 1985 IEEE International Conference on Robotics and Automation,* St. Louis, Mo., Mar. 1985, pp. 594–600.
29. H. Kobayashi, "On the Articulated Hands," in H. Hanafusa and H. Inoue, eds., *Robotics Research* (*Second Interntional Symposium*), MIT Press, Cambridge, Mass., 1985, pp. 293–297.
30. D. M. Lyons, "A Simple Set of Grasps for a Dexterous Hand," *Proceedings of the 1985 IEEE International Conference on Robotics and Automation,* St. Louis, Mo., Mar. 1985, pp. 588–593.
31. J. K. Salisbury, "Active Stiffness Control of a Manipulator in Cartesian Coordinates," *Proceedings of the 19th IEEE Conference on Decision and Control,* Albuquerque, N.M., Dec. 1980, pp. 83–88.
32. R. Bajcsy, "Shape from Touch," Lecture Notes, Univ. of Penn., Computer and Information Science Dept., Philadelphia, Penn., 1983.
33. T. Okada and S. Tsuchiya, "Object Recognition by Grasping," *Pattern Recognition* **9**(3), 111–119 (1977).
34. J. K. Salisbury, "Interpretation of Contact Geometries from Force Measurements," *Proceedings of the 1st IEEE International Conference on Robotics,* Atlanta, Ga., Mar. 1984.
35. J. K. Salisbury, "Interpretation of Contact Geometries from Force Measurements," in M. Brady and R. Paul, eds., *Robotics Research* (*First International Symposium*), MIT Press, Cambridge, Mass., 1984, pp. 565–577.
36. R. Tomovic and Z. Stojilkovic, "Multifunctional Terminal Device with Adaptive Grasping Force," *Automatica* **11,** 567–570 (1975).
37. D. Edson, "Giving Robot Hands a Human Touch," *High Technology,* 31–35 (Sept. 1985).
38. G. Schlesinger, "Der mechanische Aufbau der kunstlichen Glieder," *Ersatzglieder und Arbeitshilfen,* Part II, Springer, Berlin, 1919.
39. K. Tanie, "Design of Robot Hands," in S. Nof, ed., *Handbook of Industrial Robotics,* John Wiley & Sons, Inc., New York, 1985, Chapt. 8.
40. R. S. Fearing, "Implementing a Force Strategy for Object Re-Orientation," *Proceedings of the 1986 IEEE International Conference on Robotics and Automation,* San Francisco, Calif., Apr. 1986, pp. 96–102.
41. P. Datseris and W. Palm, "Principles on the Development of Mechanical Hands Which Can Manipulate Objects by Means of Active Control," *ASME Journal of Mechanisms, Transmissions, and Automation in Design* **107,** 147–156 (June 1985).
42. W. Palm, D. Martino, and P. Datseris, "Coordinated Control of a Robot Hand Possessing Multiple Degrees of Freedom," in D. Hardt and W. Book, eds., *Control of Manufacturing Processes and Robotic Systems,* American Society of Mechanical Engineers, New York, 1983, pp. 253–266.
43. W. Palm and P. Datseris, "Pose-Seeking Algorithms for the Control of Dexterous Robot Hands," *Proceedings of the 1985 International Conference on Robotics and Automation,* St. Louis, Mo., Mar. 1985, pp. 582–587.
44. W. Palm, "Expert Systems for the Control of Dexterous Robot Hands," *Proceedings of the 1985 ASME International Conference on Computers in Engineering,* Boston, Mass., Aug. 1985, pp. 177–182.

HEURISTICS

R. R. Gawronski
University of West Florida
Pensacola, Florida

WHAT IS HEURISTICS?

Depending on the field of application and the author's view, several definitions of heuristics may be found in the literature. Some of these definitions are very simple, perhaps oversimplified. To illustrate, "A heuristic is a 'rule of thumb' used in problem solving" (1), or similarly, heuristics are "popularly known as rules of thumb, educated guesses, intuitive judgments or simply common sense" (2). There is no separate article on heuristics in the *Encyclopedia Brittanica,* but it is mentioned in the essay devoted to "Thought and Thought Processes" in connection with computer simulation of thinking:

". . . the search is organized heuristically in such a way that the directions most likely to lead to success are explored first" (3). Definitions found in psychological literature are usually more optimistic and general, for example, "heuristic: leading to the discovery of, or the search for, new thought formulations or conclusions" (4).

Originally the term heuristics came from the Greek word "heuriskein" (to discover). It was used by Socrates when he tried to develop in his disciples the ability to discover ideas. Teachers have used heuristics to describe a method of teaching students independent thinking and invention.

In modern usage, heuristics is any set of methods for solving a complex problem or reaching a difficult goal without preliminary evidence that these methods will lead to the solution of the problem.

Unlike algorithmic methods, heuristic methods do not always lead to the desired result. In other words, "The heuristic procedure consists in choosing a method of attack which seems promising, while keeping open the possibility of changing to another if the first seems not to be leading quickly to a solution" (5). Such a definition of heuristics, typical in psychology, needs the introduction of a measure of "certainty" (feasibility or possibility) of the applied method, but not necessarily probability in the mathematical sense. This measure may be based on previous experiences, intuition, analogies, or other sources of certainty that seem to justify the selection of the method. Some authors introduce for this purpose so-called heuristic evaluation functions (1). Application of fuzzy set theory (6) opens some new possibilities for a more exact description of this "certainty" measure.

Problems are usually solved using sets of rules, conventions, and procedures that, if properly applied, lead to an outcome. As problem complexity increases, procedures to effect a solution are less certain and outcomes problematic. An appropriate example to illustrate this point is the game of chess. Even if it were theoretically possible to verify which set of decisions is the best, it is practically impossible to test all combinations of possible solutions (10^{120} for chess). Human beings solve many such problems using "heuristic methods." Indeed, human beings apparently make many such creative acts in the solution of the heretofore insolvable. A crucial element in the exposition of artificial intelligence (AI) will almost certainly involve a process to permit machines to move towards the solution of complex tasks without precise mathematical formulation for every step in the process.

In conclusion heuristics may be defined as a method of solving complex problems that cannot be solved by mathematically founded algorithms by applying an intuitively selected sequence of rules that promises to be most effective in achieving a given goal. This sequence of rules is usually selected on the basis of experience and successful solutions of similar problems. Heuristic methods form a basic part of AI. Several authors define AI as an application of heuristic methods for the solution of complex problems that cannot be solved with the help of exact algorithms (2,6,7). Other authors (1,8,9) stress that in the larger sense of AI there are several methods to be employed, including the heuristic search as a method of intuitive, partially justified, selection of a successive step in a decision tree leading to the final goal.

Heuristics is, in most cases, a necessary but intermediate step in the development of more sophisticated systems. (10,11). When partial solutions are found and verified, the next step is the determination of appropriate assumptions and precisely defined procedures that may be mathematically described. The scientific goal is to develop an algorithmic solution to a given complex problem. Once solved, heuristics may again be employed in the partial solution to a more complex problem. Heuristic methods are and always have been crucial steps in the discovery of new solutions and making the unknown known. Therefore, heuristics is closely related with learning as a process of improving an activity using iterative procedures and intermediate results.

HEURISTIC METHODS

Although heuristics is vaguely defined, it is possible to outline some basic methods, rules, and occasional tricks that may be useful in the practical applications in AI, and especially in robotics.

Let one assume that a system for the solution of a complex problem must be designed. Two typical examples may be given:

1. A robot must find the shortest path (trajectory) from the existing position to a given point behind several obstacles. That is an easy task, even for a dog, but not for a robot.
2. A robot must be designed to assist in feeding a badly disabled person (tetraplegic). During the learning period, this assistance must be improved to minimize the complaints of the disabled person.

To describe the general methodology employed to solve such problems, assume that a "metasystem" exists, that is, a larger system embedding the system that must be designed. The metasystem may contain higher-order computers, and people. Before direct application of heuristic methods, some preliminary work must be performed in the metasystem. First, a physical system enabling *symbolic representations* of all processes connected with the problem must be constructed. The idea of a "physical symbol system" was introduced in AI in the early 1960's (12) and then developed in connection with the so-called general problem solver (GPS).

A symbol is a sign (or a set of signs) represented by any physical signal, pattern, or set of values that has a concrete semantic meaning. It means that a generator may exist that develops a specific signal in response to a concrete event or situation. This signal may be appropriately (for example, binary) coded and introduced to the computer as an entity.

More concrete description of the symbol system may be given for a class of symbols denoting physical objects and their images. The basic idea of this description is given here to illustrate the concept of representation.

Assume that the distribution of the intensity of light "I" over a bounded plane $\langle x,y \rangle$ is described by

$$I(x,y) \epsilon S_x$$

where S_x a space of all possible images. In the metasystem, there exists a human being, or a classifier (a decision layout using a given set of rules and a threshold element), which is able to divide the space S_x into three disjoint subclasses S_{r1},

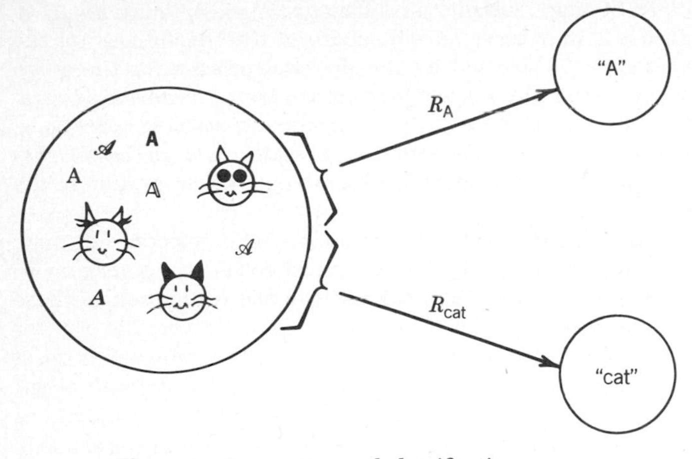

Figure 1. Recognition and classification process.

S_{r2}, and S_0 (see Figure 1). To the first class, a symbol "S_r" (for example, "A" or "cat") is assigned. To the second class, a negated symbol "$-s_r$" (for example, "not A" or "not a cat") is assigned. To the third class, a symbol "?" ("I do not know") is assigned (this class may be empty). If the result of classification is repeatable, this procedure is called recognition of the image representing the object having the name "s_r" (for example, "A" or "cat"). The process of assigning a separate code to every sign "s_r" ($r = 1, \cdots, R$, where R is the number of different symbols) creates a vocabulary of symbols (here, names) denoting appropriate classes of pictures or objects represented by these pictures. Design and physical realization of a classifier using multilayer threshold elements have been described (13).

Similar but more complex procedures may be used to define symbols representing relationships between objects of the external world and the activity (dynamic processes) of objects. "The Physical Symbol System Hypothesis" was introduced as follows: "A physical symbol system has the necessary and sufficient means for general intelligent action" (14).

Signs, symbols, and different methods of coding may also be considered from the point of view of the theory of information as a process of communication between systems (15). Application of the fuzzy set theory (16) introduces more flexibility in the representation of the goal and external world in the system by using a measure of "membership" for every element. A well-defined method for mathematical operations on fuzzy sets was developed, but the selection of the "membership coefficient" usually requires an heuristic approach.

Representation of the external world by the physical symbol system requires a grammar, that is, syntactic rules controlling the manipulation and admissible relations between symbols. These rules should also reflect possible relations between objects represented by symbols. An interesting approach to the methodology of representation using linguistic methods was presented recently in reference 5.

Symbolic representation—usually constructed in the metasystem—enables the introduction of the *goal* for the system under consideration. In more complex situations, several goals may be introduced, but in these cases a hierarchy of goals must be established. Rules that change the hierarchy of goals may also be introduced, depending upon the external situation. For example, if a robot's visual system recognizes a danger for the cooperating people, first of all it should try to avoid the accident. A description of that goal may utilize both symbolic and numeric expressions; that is, the final destination may be described by the name of an object close to the destination (eg, table), direction (eg, left or right) and distance from the object (numeric variable and unit).

If the system is equipped with a sensing and recognition subsystem, the state of the environment is introduced to the system. Combining with the results of measurements of the internal state of the system, a current state of the problem may be found and compared with the goal.

Symbolic representation of the set of possible operations must also be introduced from the metasystem. For example, for a robot the designer defines operations

$$O_j \epsilon \mathbf{O}, \text{ where } \mathbf{O} \text{ is a possible set of operations } j = 1, \cdots, J$$

Which may be performed physically by the robot or symbolically—and/or numerically—by the computer. Selection and design of this set of operations is usually a very difficult problem that may be solved by heuristic methods.

Every operation "O_j," that is, moving a robot, measuring or calculating a distance, or grasping a tool, changes the current state of the system "S_k" to a new state "S_{k+1}." Therefore,

$$(S_{k+1})_j = O_j\,(s_k), j = 1, \cdots, J$$

where J is the number of possible operations from the state "k."

This assumes that the system may have a finite, or countable, number of states. For example, it is true if the system may be described by symbolic expressions only. If continuous variables are introduced, more complex description is necessary.

All possible states, and operations performed by the system, may be presented in the form of a graph. A segment of a graph is presented on Figure 2. Every *state* of the system is represented by a node (circle); every line connecting two nodes represents an *operation* transferring the system from a "source" ("parent") state to the "successor" ("child") state. The number of successors is called branching degree "J." An optional evaluation element and a *decision* element are associated with every node. Evaluation elements perform processing of the data, which may be used for the decision if necessary. For example, a sonar system compares the distances to all surrounding obstacles. The decision element selects an operation—or set of operations—that transfers the system to the next state. Every node may generate up to "J" nodes–successors. After "l" steps, up to "l^J" steps may be generated.

Assuming that the number of levels and branches is finite, two kinds of graphic representation are possible (Figure 3**a** and **b**):

1. A net, when it is possible to reach some nodes (ie, D or G in Figure 2**a**) using different operations (ways).
2. A tree, if there is a starting node and every node has only one parent (Figure 3**b**).

If some nodes are repeated in the graphic representation, it is possible to represent any net as a tree. An example is given in Figure 3**b.**

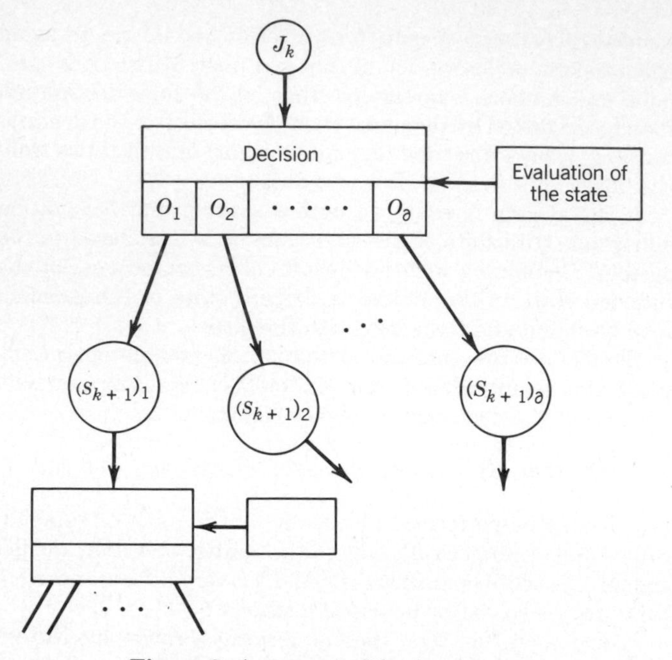

Figure 2. A segment of the decision net.

The main problem is caused by the exponential explosion of possible states. Therefore, if no reliable algorithm for the univocal selection of successive steps exists, a search procedure for the best (or only admissible) solution must be designed. Pure random—sometimes called blind—selection of branches and verification of the result may be used only for a small number of *"I"* and *"J."* Random selection is not considered to be an heuristic method.

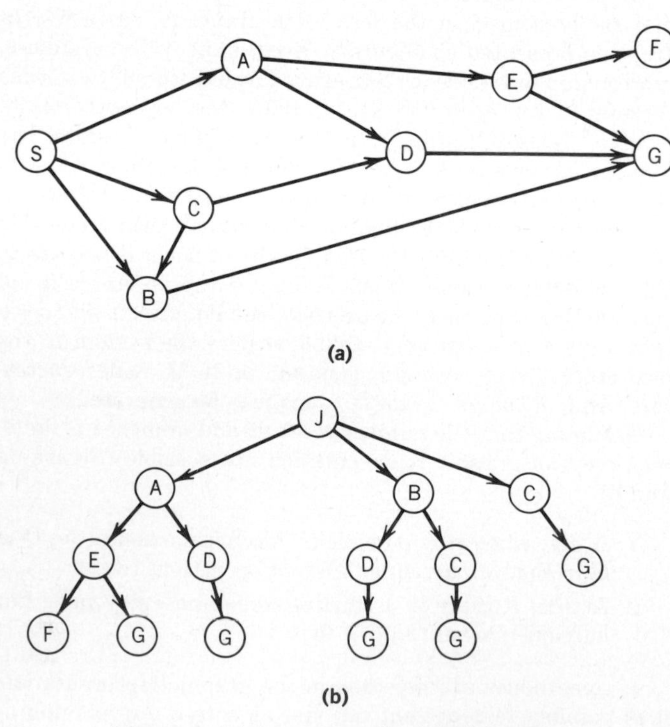

Figure 3. **(a)** A state net. **(b)** A decision tree.

To be more specific, a coefficient *"$c_j(goal)_k$"* may be introduced. It may serve as a measure of the "usefulness" of the operation *"j"* selected by the decision process for the given state *"k"* and with respect to the given "goal." Symbolic description of the goal, state of the environment, possible operations, results of every operation, and *evaluation of coefficients "$c_j(goal)_k$"* is generally called knowledge representation in the system.

The simplest method would be the selection of max $\{c_j(\text{goal})_k\}$ at every stage. But in most cases, it is a very inefficient way to reach the goal because the evaluation of these coefficients—even if it is reliable enough—is possible only for some nodes close to the current one. The problem is most difficult when the values of $c_j(\text{goal})_k$ are closer to each other. In such cases, it is necessary to find some rules for successive decisions to continue the process. Consequently, several methods of *heuristic search* have been developed. The first two methods are called "systematic searches." They are used when poor information about $c_j(\text{goal})_k$ is available.

Depth-first Search. Starting from the initial node *S* (Figure 3) the first branch, for example the left one, is selected. Other branches are neglected as long as we can continue downstairs and a successor state is generated, that is, *"E"* in Figure 3**b.** Then the next successor, that is, *"F"* is verified. The procedure is continued until the goal is achieved or the end of the branch is reached. In that case, the procedure requires a return to the nearest previous node where an unexplored branch exists.

Breadth-first Search. In this procedure, every successor of the starting node is verified for the goal first (*A,B,C* in Figure 3**b**). Then the procedure goes to the next level and verifies all successors at this level (***E,D,C,G*** in Figure 3**b**).

Depth-first search is good when dead branches (branches without success) are not very long. Breadth-first search may be applied when the number of branches at every level is not very large. The heuristic decision of the one method that should be applied requires some evaluation of the possible structure of the tree.

Subsequent procedures require the utilization of some methods to estimate the "distance" to the goal or some other method of evaluation of the improvement of the situation after every operation.

Hill-climbing Search. This is a procedure based on local optimization. For every node *"k,"* a heuristic measure $c_j(\text{goal})_k$ of the remaining distance must be evaluated. The branch with maximum gain is selected. Then the procedure is repeated at the next node. The hill-climbing search may be an efficient procedure if no danger of local maxima, ridges, or plateaus exists.

Best-first Search. All successors from the starting node and next considered nodes are called "open nodes." They are sorted by the estimated remaining distance $c_j(\text{goal})_k$. The operation (branch) with the best estimate is selected first—independent of the level.

A group of search procedures exists that optimize the effort *"e"* necessary to reach the goal.

Branch-and-bound Search. If, in the tree representing possible states, some nodes correspond to the same state *"s"* (*D* in Figure 2**a** and **b**) then the node with minimum effort *"e_k"* from the starting state *"S"* to the current node *"k"* is selected and extended one step, where the sorting procedure is repeated. The procedure is terminated when the goal is reached and the shortest incomplete path is longer than the path to the goal. An optimal path may be selected in this manner.

Underestimate of Remaining Effort. Some improvement may be introduced to the branch-and-bound procedure if an underestimation of the effort (distance) from every open node to the goal may be calculated. For every open node, a sum of "used" effort e_k and underestimated future effort to the goal "c_k" is calculated. All open nodes are sorted with respect to this sum: $e_k + c_k$. A successor with a minimum sum is selected for the continuation of the procedure.

Search Procedure A^*. This is an efficient combination of several procedures: branch-and-bound search, underestimate of the remaining effort, and rejection of all branches longer than the minimum-effort path between any two open nodes (dynamic programming principle).

Several more or less sophisticated search procedures may be found in the literature. A clear and comprehensive presentation of different search procedures is given in reference 2. Selection of the best procedure is another heuristic problem usually solved in the metasystem.

Some general approaches may be used for the improvement of heuristic decisions. Some of them are used unconsciously by the designer, and others are based on both theoretical and practical experience. The most common approaches are as follow.

Analogy is probably the most important source of ideas and constructions. Successful solution of similar problem does not mean that the same solution may be applied to another complex problem. Nevertheless, without other evaluations, an analogue sequence of operations may help in reaching the goal. In robotics, the main source of analogue solutions is biology (17), and to some extent psychology. Some examples will be given in the next section.

Application of simplified models may be a successful clue if the simplification does not go too far and if it is possible to find and analyze exact algorithmic solutions. For example, the construction of a control algorithm for a multilink manipulator meets essential difficulties resulting from strong nonlinearities of the system. Linearization of the manipulator permits the construction of a satisfactory control system, which may be improved in subsequent iterations (11,18). It is possible to construct a flexible control system that starts from simplified models and automatically detects the necessity of the development of the model using additional search procedures.

Introduction of the *hierarchical structure* of the system and procedures using *partial solutions* is a typical method when the direction of the final solution cannot be determined. This method is used, for example, when a large number of data must be considered. In modern recognition systems, a hierarchical structure of data processing is used. At lower levels of the system, some local properties of images (vertices, ends, crossings, and others) are detected. Then, at higher levels some general features of the image are discovered (11,19,20).

Learning processes are normally excluded from heuristic methods. Nevertheless, they are closely connected to heuristics. In the broader sense ". . . learning is any process by which a system improves its performance" (21). Therefore, heuristic methods play a crucial role in the construction of efficient learning algorithms. The definition presented in reference 22 and accepted in psychology is even more general: "Learning is a relatively permanent change in a behavioral potentiality that occurs as a result of reinforced practice." This definition accepts learning without improvement of performance. It is possible that more developed AI systems will acquire such a property as a "by-product" of complexity.

Four basic learning situations can be discerned (23):

1. Rote learning, in which the environment provides information exactly at the level of the performance task, and thus, no hypotheses are needed.
2. Learning by being told, in which the information provided by the environment is too abstract or general, and thus, the learning element must hypothesize the missing details.
3. Learning from examples, in which the information provided by the environment is too specific and detailed, and thus, the learning element must hypothesize more general rules.
4. Learning by analogy, in which the information provided by the environment is relevant only to an analogous performance task, and thus, the learning system must discover the analogy and hypothesize analogous rules for its present performance task.

Application of learning algorithms in AI and especially in robotics is the most promising tool for the solution of many complex problems that cannot be solved analytically.

Psychology and behavioral neurophysiology (22,24,25) are the most fruitful areas for concepts concerning learning algorithms. There is no consensus about the physiological basis for learning processes. The old Pavlovian theory of association is generally accepted as oversimplified, but it spawned a better-developed connectivity theory. This theory assumes that variable connections between nerve cells are responsible for basic learning processes (24,25). Another theory assumes that some changes of proteins and neurotransmitters (biochemical substances responsible for the relations between nerve cells) form the physiological basis for memory and learning (26). It is evident that learning in higher animals and in humans is a very complex hierarchical process involving different structures of the brain (24) (not only the cortex, as Pavlov believed) and several stages with different time constants (27).

Selection of appropriate ideas for the construction of a learning algorithm is a typical heuristic procedure that must be supported by a more exact description of the algorithm and experiments using computer modeling. An important problem that usually must be solved by modeling and successive improvement is the convergence of the algorithm to the expected goal. Some special clues may be applied to increase the speed of convergence of the algorithm to the goal. For example, the introduction of a preliminary training period or adaptation of the gain of rewording loops may give positive results (27,28). In robotics, the investigation and modeling of psychomotor learning processes are a very promising direction.

HEURISTIC METHODS IN ROBOTICS

Two possible applications of heuristics in robotics may be discerned. The first concerns the design phase of an "intelligent" robot. Many solutions observed in the motor systems of animals influenced the mechanical construction of robots and manipulators. A typical example is the construction of the dexterous artificial hand (29). It is believed that the design of control systems for a higher-generation robot will be based on many neurophysiological results on the structure and processes in the motor control systems of higher animals.

The second possible application of heuristic methods in robotics is more controversial. Is it possible to construct a *cognitive robot* supplied with a flexible control system having some degree of "intelligence" to make optimal decisions in a concrete situation? Pessimists say no. Optimists say yes. The last part of this section is written from a "cautiously optimistic" point of view. Undoubtedly, a cognitive robot would meet situations where its internal library of algorithms contained no algorithm suitable to respond to them. The next action of such a cognitive robot would have to be based on some heuristic decisions using methods described in the previous section, or methods that will be discovered in the "real brain." This is probably the highest level of flexibility that can be expected in an intelligent robot with existing knowledge.

Comparison of the structure of the motor control system in higher animals and that of an "intelligent robot" is presented on Figure 4. In both cases, several hierarchical levels may be identified. The lowest level is formed by the mechanical system with actuators. Engineering intuition and experience in robot design are strongly supported by biological analogy. The design of a flexible mechanical construction (ie, elastic manipulators) will initiate several new problems that may involve heuristics. Also, the construction of multilegged robots is based on intuition and analogy.

The construction of the next level is usually based on classical control theory. Nevertheless, some concepts coming from neurophysiology may be very fruitful. The first one concerns the application of multiloop feedback with variable gain and decoupling cross connections (30,31). The second one is even more interesting and needs more investigation. In contrast to the typical robot, the lowest level of motor control in animals is supplied with a very great number of "interoreceptors," that is, receptors placed in muscles, tendons, and joints. They enable "on-line" measurement of positions, speed, and forces for every degree of freedom. The result of measurement is most probably used at every level of the muscle control system. This complex measurement system, including tactile sensors and vision, strongly influences the general structure of the robot control system. If the term heuristics is understood in the broader sense, the design procedure evidently must use

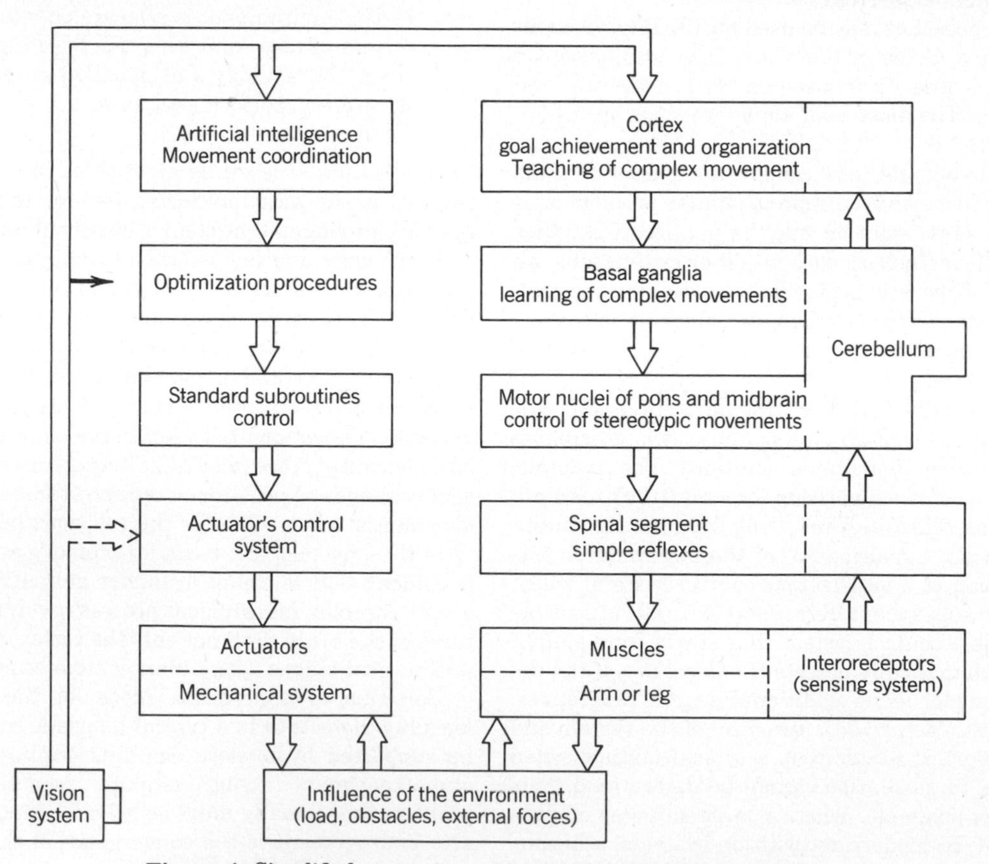

Figure 4. Simplified comparison of technical and biological systems.

heuristic methods. It is likely that complex computer programs using search procedures presented in the previous section can be utilized for the selection of optimal or quasi-optimal structures of the control system.

Much less is known about the structure and function of the next levels of the motor control system of the brain. It is accepted (24–26) that, at all control levels but the lowest, learning procedures play a crucial role in the formation of stereotypical skill movements. Analytical calculations of the control signal for fast, skilled movement encounter essential difficulties. The application of learning algorithms using special training procedures (27) may be in many cases the best solution. Decisions concerning parts of the job that must be done by the hardware structure and parts that must be realized by software are the next heuristic problems.

Practical application of the optimization theory for the on-line optimal control of dynamic movement has not been well developed. Complex mathematical description, variable influences of the environment, and variable objective functions hinder the application of classical optimality theory. The general goal will determine the performance function to be selected, for example, minimum time of movement, minimum energy, or minimum error of trajectory or final position. A higher-level decision center of a cognitive robot should select the final performance index and an appropriate learning procedure for the optimization level of the control system.

The planning of complex movement, usually composed of several stereotypical movements, is the task of the highest level of the complex control system. It is convenient to give a symbolic name to common stereotypical movements. Decomposition of the current goal to a set of simpler movements may sometimes be done with the application of linguistic methods (5) or with the application of invented learning procedures (2). It is difficult to expect direct application of the neurophysiological results at this level, but studying current results of biological investigation may generate several ideas that may be applied in robotics.

BIBLIOGRAPHY

1. E. Charniak and D. McDermott, *Artificial Intelligence,* Addison-Wesley Publishing Co., Reading, Mass., 1984, p. 261.
2. J. Pearl, *Heuristics,* Addison-Wesley Publishing Co., Reading, Mass., 1984, p. vii.
3. *Encyclopedia Brittanica,* 15th ed., Vol. 28, p. 643.
4. J. P. Chaplin, *Dictionary of Psychology,* A. Laurel, ed., New York, 1975, p. 235.
5. A. Bonnet, *Artificial Intelligence,* Prentice-Hall International, Englewood Cliffs, N.J., 1985, p. 18.
6. C. Negoita, *Expert Systems and Fuzzy Systems,* Addison-Wesley Publishing, Co., Reading, Mass., 1985.
7. A. Barr and E. A. Feigenbaum, eds., *The Handbook of Artificial Intelligence,* Vols. 1, 3, Heuris Tech Press, Stanford, Calif., and William Kauffman, Inc., Los Altos, Calif., 1981/1982.
8. N. V. Findler and B. Meltzer, eds., *Artificial Intelligence and Heuristic Programming,* American Elsevier Pub. Co., New York, 1971.
9. N. J. Nilsson, *Principles of Artificial Intelligence,* Tioga Pub. Co., Palo Alto, Calif., 1980.
10. R. B. Barneji, *Artificial Intelligence—A Theoretical Approach,* Elsevier North-Holland, New York, 1980.
11. R. R. Gawronski, "Learning, Hierarchy and Parallel Processing as a Tool for the Improvement of Control Quality" *Proceedings of the IEEE Workshop on Cognitive Control,* Troy, N.Y., Aug. 26–27, 1985.
12. A. Newell, J. Shaw, and H. A. Simon, "Report on a General Problem Solving Program for a Computer" *Proceedings of the International Conference on Information Processing,* UNESCO, Paris, France, 1960, pp. 256–264.
13. N. J. Nilsson, *Learning Machines: Fundamentals of Trainable Pattern-Classifying Systems,* McGraw-Hill Inc., New York, 1965.
14. A. Newel and H. A. Simon, "Computer Science as Empirical Inquiry: Symbols and Search, in J. Haugland, ed., *Mind Design,* M.I.T. Press, Cambridge, Mass., 1981.
15. R. J. McElience, *The Theory of Information and Coding,* Addison-Wesley Pub. Co., Reading, Mass., 1977.
16. A. Kandell, *Fuzzy Mathematical Techniques with Applications,* Addison-Wesley Publishing Co., Reading, Mass., 1986.
17. R. Gawronski, *Bionics—The Nervous System as a Control System,* Elsevier Publishing Co., Amsterdam, the Netherlands, 1971.
18. M. Vukobratovic, V. Pontiak, and D. Stokic, *Scientific Fundamentals of Robotics 1 and 2,* Springer-Verlag, New York, 1982.
19. J. F. Gerrison, "Theory and Model of the Human Global Analysis of Visual Structure, Part I, *IEEE Transactions on Systems, Man and Cybernetics* **SMC-12**(6), 805–817 (1982).
20. J. F. Gerrison, "Theory and Model of the Human Global Analysis of Visual Structure, Part II," *IEEE Transactions on Systems, Man and Cybernetics* **SMC-14**(6), 847–862 (1984).
21. R. Michalski, J. G. Carbonel, and T. M. Mitchell, *Machine Learning: An Artificial Intelligence Approach,* Morgan Kauffman Publishers, Los Altos, Calif., 1986.
22. G. A. Kimble and N. Garmezy, *Principles of General Psychology,* Ronald Press, New York, 1968.
23. P. H. Winston and R. H. Brown, *Artificial Intelligence: An MIT Perspective,* M.I.T. Press, Cambridge, Mass., 1979, pp. 328, 329.
24. J. Konorski, *Integrative Activity of the Brain, an Interdisciplinary Approach,* University of Chicago Press, Chicago, Ill., 1967.
25. C. D. Woody, *Memory, Learning and Higher Function,* Springer-Verlag, New York, 1982.
26. L. R. Squire and N. Butters, *Neurophysiology of Memory,* Guilford Press, New York, 1984.
27. R. R. Gawronski, "Two Stage Learning Algorithm for Optimal Control of a Manipulator," *Proceedings of the SME Conference on Robotics Research: The Next Five Years and Beyond,* Aug. 14–16, 1984.
28. B. Macukow and R. Gawronski, *Modelling the Movement of a Two-Link Manipulator, Computers in Industry* (3), 187–198 (1982).
29. J. K. Salisbury, "Kinematic and Force Analysis of Articulated Mechanical Hands, *Journal of Mechanics, Transactions on Automation in Design* **105,** 35–42 (1983).
30. R. R. Gawronski, "On Structures of the Muscle Control System," *Proceedings of the III International Symposium on External Muscle Control,* Dubrovnik, Aug. 25–30, 1969, pp. 23–34.
31. R. R. Gawronski and B. Macukow, "The Adaptive Neuronlike Layer Net as a Control Learning System," in *Progress in Cybernetics and System Research,* Vol. III, Advance Pub. Ltd., London, 1977.

HOSPITALS AND NURSING HOMES, ROBOTS IN

HIROYASU FUNAKUBO
Medical Precision Engineering Institute, University of Tokyo, Tokyo, Japan

INTRODUCTION

The problem of an aging population is no longer limited to specific countries or regions, but is becoming a worldwide phenomenon. As can be seen from Table 1, the speeds at which populations are aging in different regions are not necessarily the same. It took 115 years for the population in the 65-and-over age bracket to rise to 14% of the total population in France and 85 years in Sweden. It must be noted that social systems also change in response to demographic changes. Unlike in the above-mentioned European countries, the same demographic changes will occur in Japan in a span of only 26 years. The population in the 65-and-over age bracket is expected to reach the above levels in Japan by 1996. This means that unless Japan makes drastic changes in its social systems in response to these demographic changes, serious social friction may develop in many different parts of the society.

Major changes in medical, social welfare, and health care fields as well as systems must be realized quickly in order to keep the above-mentioned friction to a minimum and thus maintain the vigor and energy of not only the individuals, but also this sector of society. Finally, the time necessary to modify the social structures in function of social needs has to be as short as possible.

Measures from the medical as well as physical and mental viewpoints are determined primarily by the progress made by research and development in these fields. It is, however, needless to say that behind the progress in these fields lie technological developments in measuring, treatment, and diagnosis.

An aging population also means a corresponding rise in the number of handicapped persons. It must not be overlooked that the increasing number of aged, handicapped persons means a proportional decrease in the number of young, able-bodied persons to look after the aged. This means that the type of technological support necessary to overcome this problem must be identified.

Increasingly, a society of this type will have to orient its research toward a wide variety of fields, such as diagnosis, treatment, measurement, preventive care, and rehabilitation, among others. The utilization of robots is spreading rapidly and is being viewed with great hopes.

A variety of robots have been commercialized for industrial applications. There is a strong possibility that such robots can be modified and play a vital role in these nonmanufacturing fields also. The robots include those designed for microsurigical operations, which will be able to take over some of the workload from surgeons, those designed to assist bedridden persons, who cannot be looked after adequately because of shortages of young people, and seeing-eye robots for those with vision problems.

This section describes the current prototype robots actively being tested at hospitals and nursing homes for medical applications and for assisting disabled persons.

Table 1. Comparison of Speed of Population Aging in Different Countries[a]

	Number of Years Required for Percentage of Population 65 and Over to Reach Following Levels:		
	7%	14%	Years
Japan	1970	1996	26
France	1865	1980	115
FRG	1930	1975	45
Sweden	1890	1975	85
UK	1930	1975	45
United States	1945	2020	75

[a] Statistics prepared by the Japanese Ministry of Welfare on the basis of data provided by the United Nations. Except for Japan, time is shown in 5-year units.

Robots for Medical Applications

Figure 1 shows a microsurgical robot that is used to automatically trephinate corneas for corneal transplants. In order to accurately trephinate a cornea of soft 0.5-mm tissue without damaging the pupil lying behind the cornea, it is necessary that the trephine be accurately positioned vertically above the specified position and stopped immediately upon completion of the cut. At the same time, it must hold the cornea firmly to prevent it from falling into the eye. Moreover, the robot should be designed so that the eye being operated on is clearly visible to the surgeon. This is the reason why the cutting axis protrudes at an angle from the side. Microcomputer-controlled micromotors are provided to drive the arms horizontally, drive the cutting axis vertically, rotate the cutting axis, oscillate the cutting axis, and rotate the cutting trephine tool. This allows automatic positioning of the trephine at a position perfectly perpendicular to the cornea. Accurate positioning over the eye is controlled by the visual sensory device. This is not shown in the figure. Measurements taken by the device are sent in the form of control signals to control the motors.

Figure 1. Microsurgical robot for corneal transplant.

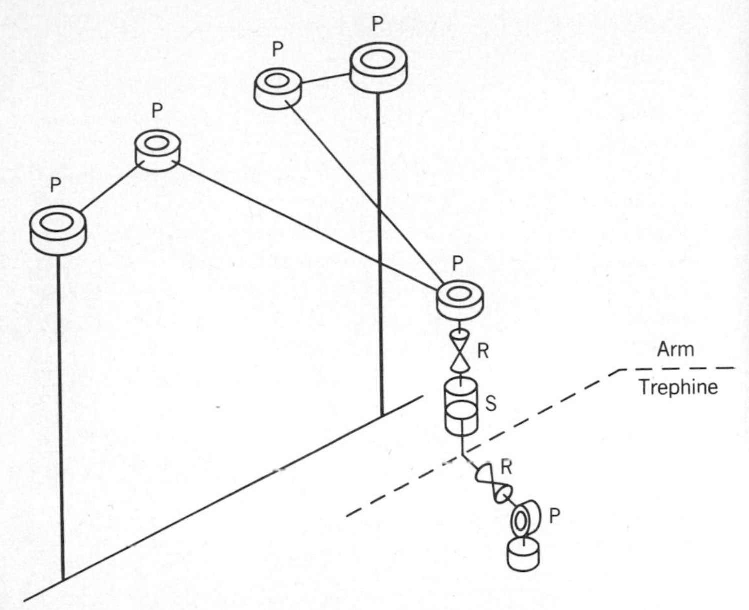

Figure 2. Microsurgical robot configuration.

Figure 3. Experimental removal of pig's cornea using a microsurgical robot.

Operation of this mechanism is illustrated in Figure 2. Figure 3 shows an example of an actual trephination being performed on a pig's cornea (see also SURGERY, ROBOTS IN).

ROBOTS FOR ASSISTING HANDICAPPED PERSONS

Robots for Bedridden Patients

An entire robot system is shown in Figure 4. Helping bedridden patients in their routine activities places a great burden, both physical and mental, on those responsible for it. Apart from this, it is also desirable that physically handicapped persons do their own daily chores as much as possible. Therefore, this system has been designed so that a patient can command it with whatever functions, such as touch, voice, head move-

Figure 4. A system for helping bedridden patients with their routine chores.

ment, or whistle, remain. Upon receipt of a command by any of these means, the system displays a menu on the screen, and the menu is also transmitted to the patient by a voice synthesizer to avoid misunderstanding of the menu's contents. From this the patient can select the desired operation by a simple command signal. A central processing unit (CPU), not shown in the figure, issues commands to the various system devices in accordance with the selected menu. The system is configured from modules to automatically control curtains, doors, windows, air conditioners, TVs, lighting, modules to store daily necessities such as bread, milk, magazines, and so on. It is provided with a window from which the product ordered from the CPU can be retrieved, an automatic cart to deliver the product retrieved from the window to the patient's bedside, and two manipulators to carry the food to the patient's mouth. Display devices, storage units, the automatic cart, and the manipulator have their own CPUs to issue instructions and control operations. All commands are transmitted telemetrically to eliminate the need for complex wiring in the patients room. As shown in Figure 5, the robot has a total of nine micromotors: three for shoulders, two for arms, one for

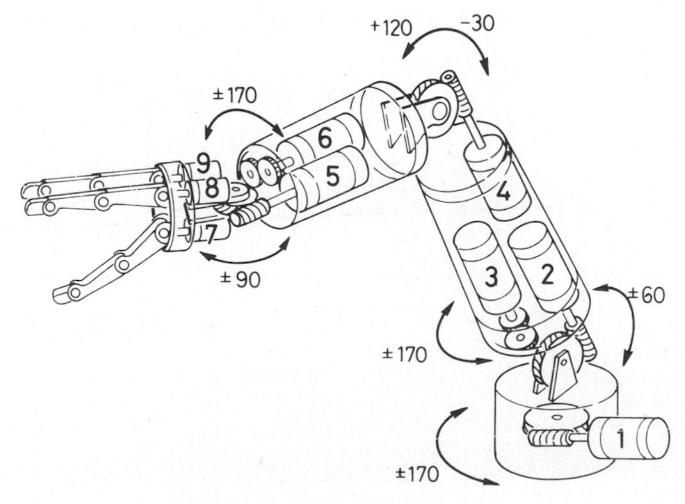

Figure 5. Manipulator structure.

Figure 6. Example of a pair of manipulators in operation.

the wrist, and three for fingers. Two sets of micromotors are controlled by a 16-bit microcomputer to provide 18 different freedoms.

As shown in Figure 6, a tray is held horizontally by the two arms and carried to the specified point, and motions of the two arms are coordinated automatically. The robot has a payload of 2 kg and weighs 6 kg. It has been improved considerably and has been provided with vision, as shown in Figure 7. This project was granted by the Science and Technology Agency, Japanese Government, between 1980–1982.

Spartacus Robot for Tetraplegic Patient

The French Spartacus pilot project had the medical objective to demonstrate the feasibility of applying robotic technology to solve some of the functional problems of tetraplegic patients. This study was successfully terminated at the end of 1980 by the development of a prototype "telethesis," the MAT1, a manipulator still under evaluation in the Occupational Therapy Department of the Raymond Poincare Hospital near Paris (see Figure 8) (see also HANDICAPPED, ROBOTS FOR).

The telethesis consists of the following elements:

1. An articulated mechanical arm with six degrees of freedom, powered by electrical torque motors, and carrying

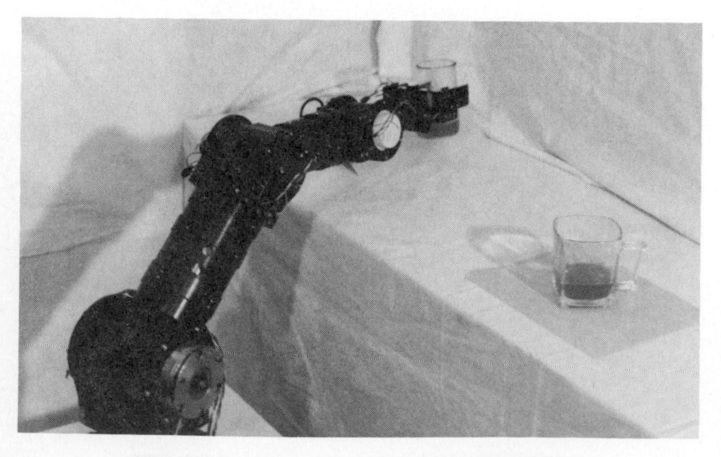

Figure 7. Modified version of a manipulator.

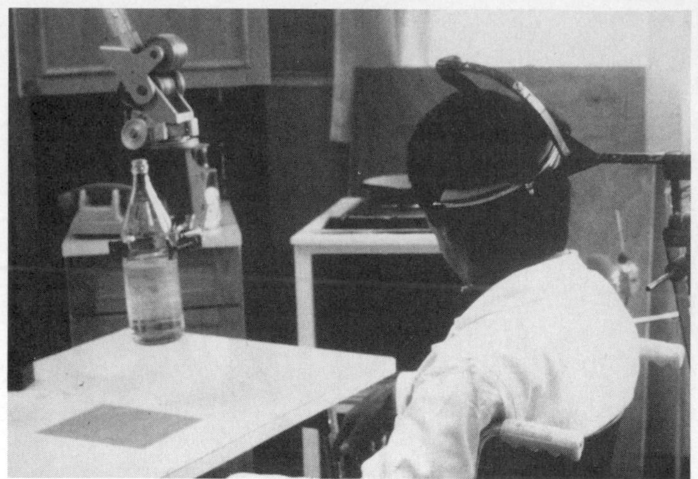

Figure 8. Thelethesis in action. Courtesy of E. Lesigne, CEA, France.

a powered gripper at the end. The arm, studied by CEA, has been specially designed to work on a table, to reach objects stored on shelves, and to retrieve objects fallen to the floor.
2. A variety of control transducers, selectable to suit the individual patient. In particular, a head movement transducer with two or three degrees of freedom has proven to be well suited for severely damaged tetraplegic patients. For those who still have some arm movement, a roller transducer is a very convenient complement.
3. A microprocessor system with special interfaces to establish an ergonomic, individually adaptable programmable link between the user and the mechanical arm.

Transport Cart Installing a Robot

A photograph of a robot designed specifically for picking up objects from a patient's bedside and carrying them to specified destinations is shown in Figure 9.

Robot Designed to Climb Stairs

The robot shown in Figure 10 has been specifically designed to enable patients who have lost the use of their legs to climb up and down staircases. The design assures that the patient's chair remains horizontal and that the center of gravity is maintained in the region of the legs. As is clear from the figure, it is a six-legged robot.

Shifting a Patient

The job of lifting a patient from a bed and transferring him/her to a wheelchair is not only physically demanding and difficult, but is likely to cause backaches and other problems in the persons physically handling such patients.

This robot approaches the bed under the manual control of a nurse. Precise positioning of the robot for docking is achieved by means of an omnidirectional wheel system. The robot stretches out its arms, inserts them into a special board

Figure 9. A manipulator mounted on a wheelchair.

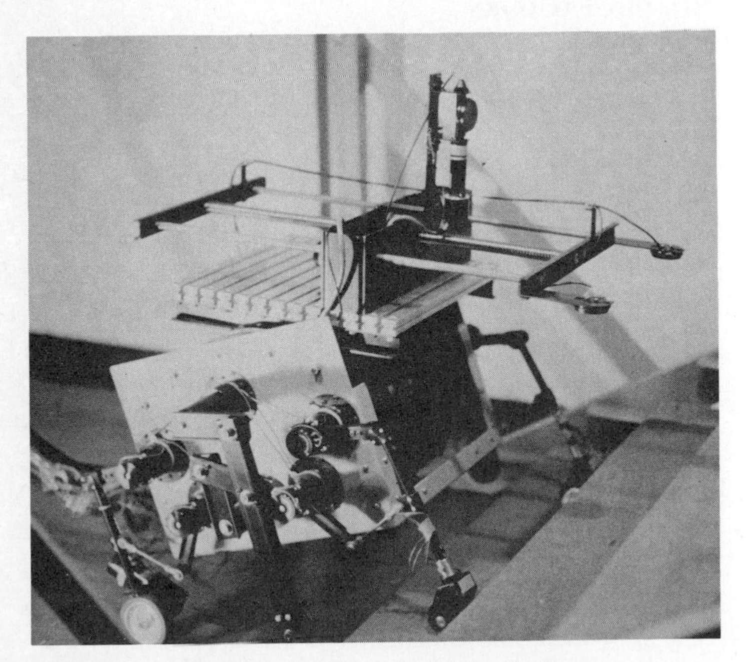

Figure 10. A six-legged robot for climbing stairs.

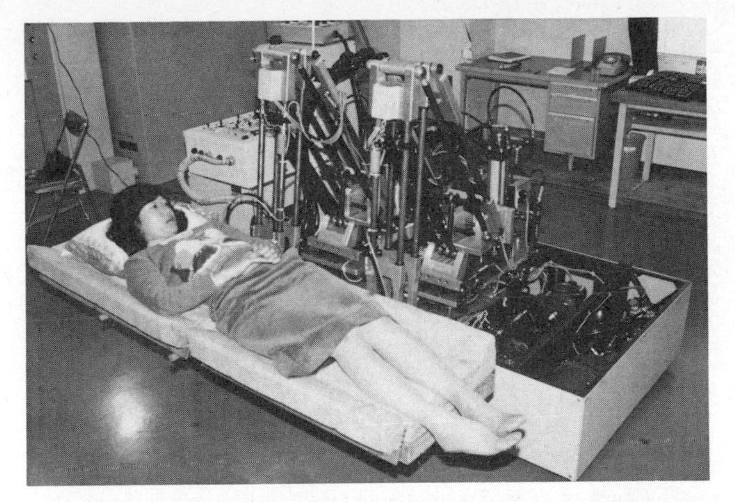

Figure 11. Robot MELCOM transfers a patient from a bed to a stretcher. Courtesy of E. Nakano, Mechanical Engineering Laboratory, MITI, Japan.

under the mattress, lifts, turns around, and puts the patient on a wheel stretcher (see Figure 11).

Seeing-eye Robot

Figure 12 shows a robot that acts as a seeing-eye robot and facilitates safe nagivation both indoors and outdoors. The robot features a scanner that scans the surroundings for obstacles,

Figure 12. Seeing-eye robot MELDOG. Courtesy of S. Tachi, Mechanical Engineering Laboratory, MITI, Japan.

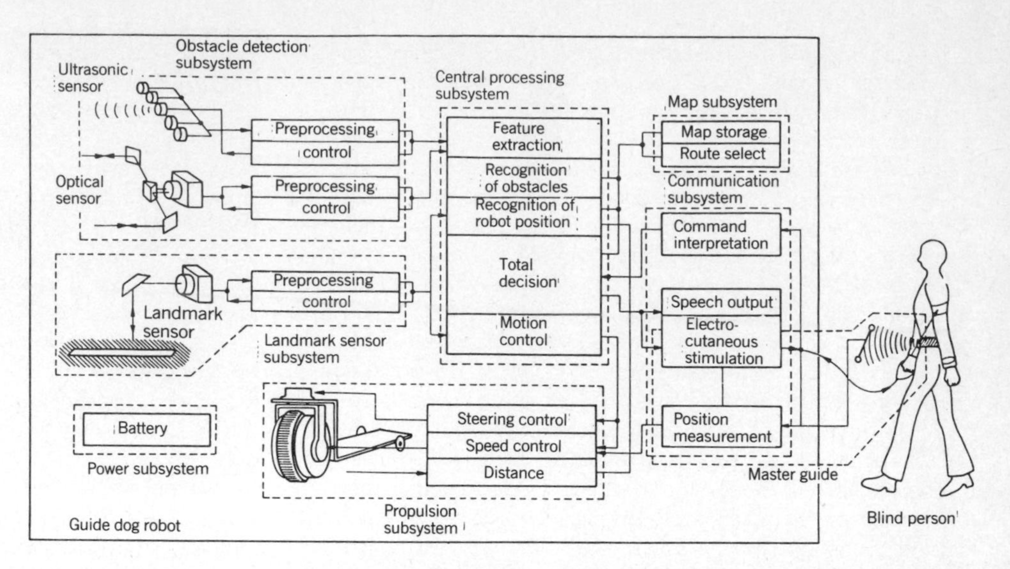

Figure 13. Schematic diagram of MELDOG. Courtesy of S. Tachi, Mechanical Engineering Laboratory, MITI, Japan.

enabling the user to navigate them. Figure 13 is a block diagram of a seeing-eye robot.

FUTURE PROSPECTS

By developing increasingly precise robots for medical applications, some of the workload of overworked surgeons may be alleviated. This type of help is not only desirable, but will become absolutely necessary.

Robots designed to assist the handicapped in their routine chores must perform many different complex jobs, unlike industrial robots, which are designed to carry out a specific operation. These robots must therefore be customized to match the requirements of the operating environment, the type of handicap, the intellectual potential of patients, their economic capacity, and their household and surrounding environment. The robots must also be highly intelligent, compact, light, and energy efficient. They must be designed so that even the handicapped can control them easily. It should be pointed out that semiautomatic robots are preferable to fully automatic ones.

The most important point, which cannot be overlooked, is that robots must be absolutely safe, and the patient must not be hurt accidentally by robots under any circumstances. Robots will therefore have to be provided with duplicate or triplicate hardware and software backup systems.

INDEPENDENCE, MUTUAL HELP, AND TECHNOLOGICAL ASSISTANCE

Unlike industrial robots, the above-mentioned robots are intelligent, compact, have superior precision functions, are much safer, and are easy to use. Machines, the end results of technology developed by humans, must not in any way contribute to the deterioration of a person's natural physical and mental abilities or increase the degree of a disability.

In a society, individuals must be asked to do everything independently as much as possible. Family and neighbors should provide mutual assistance whenever necessary. These are the very foundations of human society. Although robots will eventually invade it, they must not be allowed to affect human relations or the independence of the person. Although the above-mentioned robots are considered to be desirable for persons with advanced disabilities because of their high intelligence levels and advanced functions, they should never be used by persons with minor disabilities just because they are convenient. This will have ill effects on society. Also, they will affect human relations, and there is a possibility that they will destroy the very fabric of society in the end.

However, who uses what kind of robot is a human decision and does not lie with the robot. Therefore, only society can be blamed for the end results.

It is needless to say that everyone must be aware of the fact that a decision on the use of household robots requires utmost care. This is because technological developments will continue unabated, irrespective of their effects on human society.

HUMAN FACTORS

H. M. Parsons
Essex Corporation
Alexandria, Virginia

INTRODUCTION

The term human factors has always been somewhat ambiguous, but one way to define it is to ask what those who apply this label to their endeavors do. Like others who may not use the label, they investigate or arrange relationships between the processes or products of technology and those who use them; they try to fit the technology to its users or vice versa. This article examines robotic processes from that perspective.

Users can be considered as individuals or in the aggregate. Human factors researchers have concentrated on the former, leaving aggregates to economists and sociologists, though human factors practitioners have tried to extend research findings through design guidelines and collective training. This article emphasizes those human factors methods that relate robots to individuals, notably human factors engineering, accident prevention, and training. It is granted, however, that occupational effects of robotics, including job alterations and eliminations and industrial relations, can be construed as human factors considerations, as can broader social and economic impacts. Indeed, these are among the objectives of the Human Factors and Safety Division of Robotics International, a component association of the Society of Manufacturing Engineers.

In the United States, interest in human factors had two origins. One was within the robotics community, which produced the Robotics International division. The other was in the human factors community, primarily in a few universities such as Purdue and M.I.T. If the growth of robotics has been both recent and relatively gradual, that of human factors in robotics has been even more so. It developed somewhat earlier and faster in Japan and Sweden (where, as elsewhere overseas, human factors is known as ergonomics); limited human factors experiments and some accident surveys have been conducted in each of these countries. On the other hand, considerable human factors research has occurred in the United States in the design of teleoperators, under the sponsorship of the Navy, NASA, Department of Energy, and Electric Power Research Institute. The U.S. Army through its Human Engineering Laboratory has funded several modest human factors studies in robotics as well as in battlefield telerobotics. The Air Force sponsored a brief investigation of human factors in a robotic manufacturing project, apparently the sole such investigation within a U.S. factory. The National Bureau of Standards has developed robotic safety techniques, and the National Institute of Occupational Safety and Health (NIOSH) has investigated two robotic accidents. Only one report has been published describing a robot manufacturer's human factors analysis to improve a human-robot interface, and apparently only one robot manufacturer (in Sweden) has conducted a human factors design experiment. In short, not a great deal has occurred, though human factors involvement in robotics was urged some years ago (1–3).

HUMAN FACTORS ENGINEERING

This major component of human factors or ergonomics includes the allocation of tasks or functions between humans and equipment, the design of the hardware in that equipment, and the design of the software.

Function/Task Allocation

An important human factors requirement is the consideration of which tasks or functions to allocate to a robot and which to a human. Three overall functions can be considered in industrial robotics: operation, maintenance, and programming. Maintenance is generally allocated to the human, except for diagnostics in computer programs. Programming of operating systems is also largely a human function, computer-aided in such operations as compiling/interpreting from a higher-order language into machine language, kinematic transformations, and diagnostic aids. Creating an application program is performed by a human with the aid of the robot itself or stored computer files such as those for computer-aided design. Operation consists of operating the robot at a robot controller panel or computer terminal, or performing operations in an industrial process in lieu of a robot. It is the latter that has been primarily examined with respect to function/task allocation. One of the pioneers in industrial robotics (4) raised this issue in applying the term "symbiosis" to relationships between robots and those working with them. Others suggested joint robot-human activities in industrial assembly operations (5) and called attention to "the human supervision of industrial robots" (6). The respective performance capabilities of workers and robots with quantitative data were analyzed (7), and this analysis, which is explicitly directed toward helping designers allocate tasks and functions, was updated (8), though it is not apparent that this information has actually been used as yet, except in a laboratory study of a pump assembly (9). The notions of symbiosis and synergy figured prominently in perhaps the earliest comprehensive review of human factors in robotics (2), which gave numerous examples from industry, as had ref. 10.

Missing has been any more recent or systematic examination of actual task allocation and combination in factories. Indeed, very few studies have been published with human factors data collected about robotics in factory settings. Robotic drilling and routing for an aircraft's large, contoured metal parts have been observed (11). Maintenance activities in two robot installations in French automobile plants have been analyzed as part of a study of shift work (12). In England, researchers visited a large number of plants and interviewed workers and executives about automation, including robots (13), and 16 companies in three European countries were surveyed for "the implications for health, safety and quality of working life of the introduction of robots into manufacturing technology" (14); the report included "allocation of robot-related tasks," selection and training for these, "robot-operator interface," "residual human tasks," and "design of work." A before-and-after questionnaire study of workers' reactions to the installation of a robot in a manufacturing plant has also been conducted (15). More investigations of robot-worker interactions in industry are needed to find out how functions and tasks are actually allocated and shared, as well as for other human factors objectives. Otherwise one is left with anecdotal accounts, which are inadequate for understanding the impact of robotics or for guidance in shaping the course of this new technology. One such account has come to this author from a former robot welder in an automotive plant. He said he used a pocket flashlight to help him align the end-effector's tool center point with the required location on the workpiece in programming the robot and had to view that location from only a few inches away.

Combinations of human and machine operations constitute the essence of telerobots, stationary or mobile. Just what form such combinations should take, ie, how tasks and subtasks should be allocated, has aroused human factors interest, though not as much as has the design of the interfaces in the combinations. An extensive research program in undersea telerobotics, funded by the U.S. Office of Naval Research at M.I.T., investigated task mixes between robot operator and computer, the tasks being measured in terms of unpredictability. The concept of "supervisory control" that emanated from the program is a "hierarchical control scheme whereby a teleoperator or other device having sensors, actuator, and a computer, and capable of autonomous decision-making and control over short periods and restricted conditions, is remotely monitored and intermittently operated directly or reprogrammed by a person" using another computer (16). These authors noted that operator and computer could either share control by being active at the same time or trade control back and forth.

The potential kinds of combinations will merit increasing human factors study as robotics is applied to service activities such as sales, transportation, law enforcement, health, and recreation. As a noted robotics expert has written, "At present fully autonomous robots cannot cope with the difficult environmental conditions" in these services, in the home, and out-of-doors, but "a partially controlled robot (hybrid teleoperator robot) can be developed within the present state of the art that would be economically viable for the majority of these service tasks. . . . Subroutines controlling often-used manipulative procedures would be called up by the operator, to be implemented autonomously by the robot/teleoperator, fitted with available sensors. Seldom-occurring tasks would be performed by the operator. . . ." (17) The design issue will be to analyze and allocate the tasks. This author suggested further that "dangerous, arduous, and repetitive physical manipulation of objects and control of simple manipulative actions would be performed by our new 'slaves'; the target of these 'slaves' will be progressively to minimize human detailed control as we learn to improve our robot systems."

Design of Equipment

Effective operation, maintenance, and programming of robots require appropriate (ie, ergonomic, user-friendly, and human engineered) design of workplaces and equipment, controls and displays, and application programs in computer software. The operation of an industrial robot is relatively simple. The robot's controller panel has a limited number of switches and indicators to be properly configured and coded. However, even these may present problems, especially in an emergency, if only because the panels' control elements and their arrangement and coding differ among robot manufacturers (18–20); perhaps some are better than others—they have not been evaluated. Nor (with rare exceptions) has design for maintenance been analyzed, especially with regard to easy access, anthropometric aspects, and communications among maintenance personnel. Maintenance design becomes especially important when multiple robots work together and when there are numerous interfaces with other equipment, such as conveyors, machine tools, assembly feeders, and sensors. Two or more maintenance personnel may have to function together at different locations. Safety considerations (discussed later) may place special demands on workplace design for maintenance.

But it is programming that challenges human engineering for industrial robots. The principal equipment units are a teach pendant (portable control panel) and a computer terminal; for programming continuous paths, as in spray painting, controls are attached to the robot arm or to a surrogate arm. A major design problem has been the division of programming between the pendant and the terminal; the former is carried inside the robot work envelope (bounded by a barrier or fence, with a gate), whereas the latter is usually next to the controller housing a microprocessor outside the enclosure. As indicated earlier in the anecdotal account, to program with sufficient precision the locations to which the robot's end-effector will move automatically, the programmer must go very close to the alignment of the end effector (eg, tool center point) with the workpiece on which it will operate, to achieve the necessary visual discrimination, operating the pendant controls to move the robot to acquire the desired alignment. Hence the need for the portable pendant. But since numerous additional steps have to be programmed, involving speeds, trajectories, pauses, other equipment, sensors, decisions, and repetitions, the programmer must have available many control elements and displays other than those just for moving the robot to program the positions and orientations at the end of each motion or motion component. The design choice may lie between adding control and display components to the pendant (thereby unduly increasing its size and weight) and performing the additional programming at the terminal. In the latter approach the programmer will be spending a lot of valuable time rushing back and forth between pendant and terminal.

The pendant's controls should be designed so they can be operated quickly and without error. Since exclusive space is not available for all the required functions, more than one of these may be assigned to a control element, with selection by another element. What functions are assigned, and how? In what way are control elements grouped and coded? What kinds of elements are better, for example, buttons or joystick? What use might be made of a menu system with "soft" buttons? Decisions have been made about these design choices by robot manufacturers, along with design decisions about pendant shape and weight and inter-element spacing and the size and contents of pendant displays. Yet only recently has a data base been assembled showing how major robot manufacturers have designed their pendants (18–20). This survey revealed (*1*) the actual complexity of this seemingly simple device; (*2*) the variety in designs—no two are the same; (*3*) the variations in solving the pendant vs terminal dilemma; and (*4*) the absence of explicit application of human engineering principles other than some degree of functional grouping, some use of menus, and some modest color coding. Feedback from users has apparently been relatively infrequent or at best sought in no systematic fashion. Apparently only one experimental test (showing that a joystick is faster than pushbuttons) has been conducted—by a Swedish manufacturer; no other pendant time data have been published except for one report that also presented the only data on errors (with undergraduate engineering students as subjects) (21). Another researcher examined teach pendant designs and said they occasioned errors but gave no data (22). Some Japanese surveys showed a high incidence of operational error with robots, but it is not clear how many were due to mistakes in operating teach pendants. One researcher observed that "Informed criticism of the detailed design of the teach controls and some recommendations for improvements await a serious investigation into this problem as it exists in the field, where all the relevant circumstances of operation may be taken into account" (14). The only published human factors intervention to improve the programmer-robot interface concerned not a teach pendant but controls on a surrogate or teaching arm for spray painting (23). As a result of that intervention, the programmer no longer needed three hands to perform the task.

The survey of teach pendant designs (18–20) also revealed much diversity among robot manufacturers in the designs of application programs, ie, in the names and categorization of commands, the structures of menus, and error prevention and recovery. One programming manual listed 201 different errors; another 133 error messages; a third 96 errors (among 241 error and status messages); a fourth 106 system fault and error messages; and a fifth 40 messages to indicate errors in command entry to the controller, as well as 5 categories of error codes with 477 "nonfatal" warnings and 97 "fatal" errors. The teach pendant display might show a single error, the last 10 errors, or the last 30. The manuals contained virtually no guidance for recovering from errors, other than, in one instance, "Correct the condition that caused the error" and resume operations by pressing the reset button, and in another instance the information that a CLR ERR button clears the error. The only case in which a specific design feature was mentioned for preventing errors was a requirement to press both the record and shift keys to record a location and to press both the delete and shift keys to delete one. In the manuals and in interviews with representatives of some of the robot manufacturers, there was no indication of any familiarity with research and guidelines published in recent years to improve software design with respect to command terminology, error feedback, or menu structure. That research and those guidelines have been directed at helping computer end users rather than programmers. In industrial robotics it would be the application programmers on the factory floor who need such help, in this case from programmer-friendly software. For the most part, they are not professional programmers.

It would appear that detailed examinations of future robot application programs might suggest human factors improvements. In any case such programs might receive more attention to error management. Errors in programming can result in expensive mishaps, even accidents.

Human-machine interfaces have at times been studied more energetically in telerobots, though some, eg, for commercial undersea mobile telerobots and bomb disposal units, seem to have lacked human factors interventions. Research has been reported for interfaces in space, naval undersea, battlefield, and nuclear power systems, due presumably to governmental requirements. The interface components have been primarily the controls to move the manipulator (and the vehicle if the telerobot is mobile), the displays or other feedback methods to show the human operator what the manipulator is doing and the environment in which it is operating, and the workstations where these are situated.

In the early days of stationary teleoperators in nuclear facilities, control assumed a master-slave configuration in which the controlling device mimicked the joints and links of the manipulator. Investigators studied how much longer remote control took than direct hands-on control and how force feedback might increase dexterity. As mobile telerobots entered the scene, smaller control devices, eg, joysticks, were introduced that spatially limited operator stations could accommodate and that could also more easily control vehicles. The design of these devices has occupied a number of investigators. More recently research attention has been given to telerobot control by voice. Because direct visual access to the manipulator, its environment, or its vehicle was no longer available as it was with the early stationary teleoperators, the design of closed circuit television became important, attracting human factors research in stereo displays for depth perception. In addition investigators asked how an operator might acquire force feedback when there was no mechanical linkage and if feedback could be obtained through a sensory modality other than proprioception or pressure.

SAFETY

Although safety has long been a problem in industrial life, the advent of robots has dramatized it, especially for those unfamiliar with what goes on in factories. Robots also created a new kind of hazard. Unlike hard automation, their motions may be varied in space and time and thus less predictable. Unlike humans (most of the time), they may move unexpectedly and strike a person without warning; their anthropomorphic nature, it has been suggested, tends to make workers less cautious. However, there have been relatively few robot-related deaths or injuries; five robotic homicides were reported in Japan by the end of 1985 and two in the United States by the end of 1986, and increasing attempts have been made in these and other countries to prevent accidents (24). A Swedish survey (25) suggested that robot accidents had been occurring at a rate of 1 per 45 robots each year, and a subsequent analysis (26) showed that hand, arm, and finger injuries were the most numerous. Case studies and surveys in Japan indicate that accidents are most likely to occur during programming and maintenance. In the United States, the General Motors Corporation, with 2500 robots in operation, in 1985 had experienced "one serious injury requiring sutures, one case resulting from a cervical strain, and one minor injury case" during the 23 years it had been using robots (27). An investigation of a robot-caused death was carried out by the U.S. National Institute for Occupational Safety and Health (28). Unfortunately, near accidents, the investigation of which can be regarded as a human factors technique, have rarely been reported to guide safety endeavors.

Such endeavors have been described at length (29–31). They have included access prevention (visual or auditory warnings and fences and guardrails), access/presence/proximity detection by perimeter devices and interior sensors, robot control through speed limits, and emergency stop buttons as well as other features. Safety endeavors have also included procedural methods, such as requiring two persons to be present during maintenance, one inside and one outside the enclosure, and special safety training. Safety standards have been adopted in Japan, the United States, and the USSR and guidelines in the UK and Sweden; an international standard has been under discussion. Among topics presenting some difficulty for agreement have been the definition of the space around the robot regarded as hazardous and how the robot's motions should be interrupted in an emergency. There has been human factors participation in standards writing and in many of the accident prevention methods, as well as in some of the accident investigations and surveys. More such participation can be expected in the future as can participation in injury litigation (as expert witnesses). But perhaps the greatest human factors contribution to robotic safety in industry can come from preventing errors that lead to accidents, by improving the design of pendants and software.

Other contributions have occurred in fault tree analysis

(32), in gathering "critical incident" data (33), and in conducting experimental research. In a report of a series of experiments to try to find out how various robot pauses and velocities affect worker caution behavior (34), the aim was to help establish a maximum velocity during programming with a pendant, a major interest among standards writers. Recently NIOSH has supported similar research in the United States (35).

In another Japanese experiment, researchers forced a teach pendant error that unexpectedly made a robot move (36). These investigators obtained reaction times before a stop button was pressed, for various robot speeds, thereby estimating how close a worker might be to the robot before it hit him. More of this kind of research, as well as the error experiments mentioned earlier (21), seems definitely warranted.

TRAINING

Though the distinction is somewhat arbitrary, robotics education may be said to pertain to college and university courses and robotics training to vocational schools, community colleges, courses given by robot manufacturers, and training at a plant installing robots given by the manufacturing user or by the robot vendor. Installation-specific training, whether at a plant or at the vendor's premises, is what holds human factors interest primarily. College and university education in robotics for engineers might benefit from including some human factors considerations, as might coursework in vocational schools and community colleges for technicians (mostly, it appears, for maintenance tasks).

Plant-oriented training has been discussed from the viewpoint of a robot manufacturer (37) and from that of the manufacturer using robots (38). Relatively little has been published about specific factory training and retraining, though it has been extensive; one such program has been described (39). An important issue is maintenance cross-training of mechanical craftspeople in electronics and electrical craftspeople in mechanics. The former seems more difficult than the latter. Another is training of manual employees to perform robot programming with teach pendants and computer terminals. The more these and associated software are designed to be programmer-friendly, the easier it will be, presumably, to train workers to use them, and the less training will be required. This relationship, familiar to human factors practitioners, has not been noted in the robotics literature.

In safety training, which should be ongoing, new methods could be introduced or emphasized, eg, accident simulation or avoidance, at a robot location, by videotape, or with computer-aided design. In general, hands-on training seems essential, though it may be difficult to arrange on the factory floor. Factory-oriented training, for particular robots, depends on having user manuals and documentation that are matched to the learning capabilities of the students: the craftspeople and technicians. The documentation should also be performance-oriented, ie, describe the task performance and then the devices or software commands for accomplishing it, rather than the devices or commands and then what to do with them. It is not clear that either this requirement or the need to match the verbal presentation to the educational levels of the students has been fully understood by those who write robotics manuals for operation, maintenance, and programming. Human factors inputs through task analysis and preliminary tryouts might help improve this documentation as well as training content and presentation.

OTHER ASPECTS

Organizational Impacts

One organizational impact is the design or redesign of jobs (40). Job design in robotics depends, as it does elsewhere, largely on skill requirements, which also indicate what training is needed and, to some extent, who can be trained for various jobs. What skills are required? What skills can be combined in a job, for example in robot maintenance? In military telerobotic systems, can some maintenance be performed along with operations by the same individual? In a factory, who should do the programming of the robot with the teach pendant? (This issue may fade somewhat as improved machine vision makes pendant programming less consequential.) How various might be the tasks for an employee to make the work more challenging, with greater prospects for advancement? Are there situations in industrial robotics suitable for team activity? These kinds of questions often fall under the heading of human resources, a subset of human factors. Public knowledge about what is happening in this regard is in short supply, though there has been speculation based on presumed transfer of concepts to robotics from other manufacturing settings and equipment. How valid is such transfer in the absence of empirical evidence?

Another organizational impact of robotics may involve management (40). Factory automation, including robotics, can alter relationships between design and manufacturing engineers. Computer-integrated manufacturing (CIM) may bring about some management restructuring. What role may robotics play in this respect? Anecdotally, at least, robots have affected management in two ways. A management may direct that robots be acquired without sufficient analysis of the need for them. Or a management may require a quicker return on investment than may be desirable for corporate prosperity in the long run.

A third kind of organizational impact of industrial robotics concerns industrial relations (40). Though industrial unions have not been resisting the introduction of robots, they are understandably worried that their members might lose their jobs and not acquire new ones. Job loss or change are two of the most significant effects of technology on individuals as well as on aggregates of workers. Job loss is a major stressor as well as, too often, a major misfortune. Though this has not been the kind of stressor that human factors scientists have customarily examined, it and other aspects of industrial relations related to robotics constitute important human factors considerations from the broader viewpoint noted at the start of this article.

Social and Economic Impact

One area under this very broad heading is overall occupational changes resulting from robotics, including shifts in employment and unemployment patterns. These have been comprehensively discussed (41–43) by researchers who emphasized the "job twist," the shift in job qualifications that has been developing. It seems clear that public perceptions of such changes, based on unfamiliarity with industrial life and on hyperbole from publicists and entrepreneurs, have been exaggerated. Perhaps the science and practice of human factors will help reduce the unfamiliarity and hype, at least a little.

Another approach to social impact is to view an industrial plant as a "sociotechnical system," ie, as a social as well as a technological entity, to see how each affects the other. An unanswered question is how the actual social impact of robots in a plant can be distinguished from that of other forms of industrial automation or of nonindustrial automation (eg, agricultural or office). This topic calls for rigorous inquiry and analysis rather than the kind of speculation it has so far received.

CONCLUSION

All of these aspects as well as human factors engineering and human factors considerations in safety and training could be taken into account in planning and introducing a robotics installation, perhaps more systematically and effectively than they have been in the absence of the human factors label, profiting from experience in other human-machine contexts. Robot installation, a time-consuming and relatively expensive operation, can and often does involve some "test and evaluation" by specialists who bring into the act technicians, programmers, and engineers who will be associated with the robots in production. Such specialists may identify themselves as industrial engineers, many of whom have been trained in human factors/ergonomics in recent years.

BIBLIOGRAPHY

1. H. M. Parsons and G. P. Kearsley, "Robotics and Human Factors: Current Status and Future Prospects," *Human Factors* **24,** 535–552 (1982).
2. K. Noro and Y. Okada, "Robotization and Human Factors," *Ergonomics* **26,** 985–1000 (1983).
3. G. Salvendy, "Review and Appraisal of Human Aspects in Planning Robotic Systems," *Behavior and Information Technology* **2,** 263–287 (1983).
4. J. F. Engelberger, "Man-Robot Symbiosis," *Proceedings of the 4th International Symposium on Industrial Robots,* Tokyo, 1974, pp. 149–162.
5. D. A. Thompson, "The Man-Robot Interface in Automated Assembly," in T. B. Sheridan and G. Johannsen, eds., *Monitoring Behavior and Supervisory Control,* Plenum Press, New York, 1976, pp. 385–391.
6. T. B. Sheridan, "On the Human Supervision of Industrial Robots," *Proceedings of the 3rd Conference on Industrial Robot Technology and the 6th International Symposium on Industrial Robots,* University of Nottingham, UK, 1976, pp. 45–52.
7. R. L. Paul and S. Y. Nof, "Work Methods Measurement—A Comparison between Robot and Human Task Performance," *International Journal of Production Research* **17,** 277–303 (1979).
8. S. Y. Nof, "Robot Ergonomics: Optimizing Robot Work," in S. Y. Nof, ed., *Handbook of Industrial Robotics,* John Wiley & Sons, New York, 1985, pp. 549–604.
9. S. Y. Nof, J. Knight, and G. Salvendy, "Effective Utilization of Industrial Robots: A Job and Skills Analysis Approach," *AIIE Transactions* **12,** 216–225 (1980).
10. J. F. Engelberger, *Robotics in Practice,* Amacom Division of American Manufacturing Association, New York, 1980.
11. J. M. Howard, "Human Factors Issues in the Factory Integration of Robots," Technical Paper MS82-127, Society of Manufacturing Engineers, Dearborn, Mich., 1982.
12. F. Daniellou, *L'Impact des Technologies Nouvelles sur le Travail en Postes dans l'Industrie Automobile,* Conservatoire National des Arts et Metiers, Laboratoire Physiologie du Travail-Ergonomie, Paris, 1982.
13. N. Corlett and co-workers, personal communication.
14. M. Edwards, "Robots in Industry: An Overview," *Applied Ergonomics* **15**(1), 45–53 (1984).
15. L. Argote and P. S. Goodman, *Human Dimensions of Robotics,* Technical Paper MM84-640, Society of Manufacturing Engineers, Dearborn, Mich., 1984.
16. T. B. Sheridan and W. L. Verplank, *Human and Computer Control of Undersea Teleoperators,* Technical Report for the Office of Naval Research, M.I.T., Cambridge, Mass., 1978, Sect. 1, p. 1.
17. C. A. Rosen, "Robots and Machine Intelligence," in ref. 8, pp. 21–28.
18. H. M. Parsons, "Human-Machine Interfaces in Industrial Robotics," *Proceedings of the Annual Conference of the Human Factors Association of Canada,* Vancouver, Canada, 1986, pp. 189–192.
19. H. M. Parsons, "Data Base of Industrial Human-Robot Interfaces," International Conference on Intelligent Robots and Computer Vision, SPIE Symposium, Cambridge, Mass., 1986.
20. H. M. Parsons and A. S. Mavor, *Human-Machine Interfaces in Industrial Robotics,* Report for the U.S. Army Human Engineering Laboratory, Essex Corp., Alexandria, Va., 1986.
21. K. Ghosh and C. Lemay, "Man/Machine Interactions in Robotics and their Effect on Safety at the Workplace," *Proceedings of the Robots 9 Conference,* Society of Manufacturing Engineers, Dearborn, Mich., 1985, pp. 19.1–19.8.
22. G. J. Levosinski, "Teach Control Pendant for Robots," *Proceedings of the 1984 International Conference on Occupational Ergonomics,* Human Factors Association of Canada, Toronto, Canada, 1984, pp. 599–603.
23. H. G. Shulman and M. B. Olex, "Designing the User-Friendly Robot: A Case History," *Human Factors* **27,** 91–98 (1985).
24. H. M. Parsons, "Human Factors in Industrial Robot Safety," *Journal of Occupational Accidents* **8,** 25–47 (1986).
25. J. Carlsson, L. Harms-Ringdahl, and U. Kjellen, *Industrial Robots and Accidents at Work,* Royal Institute of Technology, Stockholm, 1979.
26. J. Carlsson, *Robot Accidents in Sweden,* National Board of Occupational Safety and Health, Solna, Sweden, 1984.
27. K. E. Lauck, "Development of a Robot Safety Standard," paper presented in the Robot Safety Workshop of the Robotic Industries Association, Robots East Exposition, Boston, Mass., 1985.
28. L. M. Sanderson, J. W. Collins, and J. D. McGlothlin, "Robot-Related Fatality Involving a U.S. Manufacturing Plant Employee: Case Report and Recommendations," *Journal of Occupational Accidents* **8,** 13–23 (1986).
29. M. C. Bonney and Y. F. Yong, eds., *Robot Safety,* IFS Publications, Springer-Verlag, New York, 1985.
30. G. E. Munson, "Industrial Robots: Reliability, Maintenance, and Safety," in ref. 8, pp. 722–758.
31. P. M. Strubhar, ed., *Working Safely with Industrial Robots,* Robotic Industries Association, Ann Arbor, Mich., 1986.

32. N. Sugimoto and K. Kawaguchi, "Fault Tree Analysis of Hazards Created by Robots," *Proceedings of the 13th International Symposium on Industrial Robots and Robots 7,* Society of Manufacturing Engineers, Dearborn, Mich., 1983, pp. 9.13–9.28.
33. M. Nagamachi and co-workers, "Human Factor Study of Industrial Robot (2): Human Reliability on Robot Manipulation," *Japanese Journal of Ergonomics* **20,** 55–64 (1984).
34. M. Nagamachi and Y. Anayama, "An Ergonomic Study of the Industrial Robot. 1. The Experiments of Unsafe Behavior on Robot Manipulation," *Japanese Journal of Ergonomics* **19,** 259–264 (1983).
35. M. Helander, personal communication.
36. N. Sugimoto and co-workers, *Collection of Papers Contributed to Conferences Held by the Machinery Institute of Japan,* No. 844-5, 1984.
37. R. J. Rosato, "Training Strategies for Robotic Implementation," in T. M. Husband, ed., *Education and Training in Robotics,* IFS Publications, Springer-Verlag, New York, 1986, pp. 69–79.
38. J. D. Lane and R. S. Richards, "Robotics Training in the U.S. Automotive Industry," in ref. 36, pp. 81–90.
39. R. W. Basey, "Training for New Technology. Robots and Control Systems, Operation and Maintenance," in ref. 37, pp. 91–101.
40. H. M. Parsons, "An Overview of Human Factors in Robotics," *Proceedings of AUTOFACT 86,* Society of Manufacturing Engineers, Dearborn, Mich., 1986, pp. 3.11–3.17.
41. R. U. Ayres and S.M. Miller, *Robotics: Applications and Social Implications,* Ballinger Publishing Co., Cambridge, Mass., 1983.
42. R. U. Ayres and S. M. Miller, "Socioeconomic Impact of Industrial Robots: An Overview," in ref. 8, pp. 467–496.
43. H. A. Hunt and T. L. Hunt, *Human Resource Implications of Robotics,* W. E. Upjohn Institute, Kalamazoo, Mich., 1983.

HUMAN IMPACTS

NOEL PERRIN
Dartmouth College
Hanover, New Hampshire

The impact of robots on human life and welfare can be considered in three categories: demonstrable effects that have already occurred, probable effects during the remainder of the twentieth century, and likely effects thereafter. Naturally, there is no complete agreement on the future effects. They are, however, the most significant.

Because the robotic age began so recently and the number of existing robots is so small (roughly 1 industrial robot per 10,000 human beings), the current impact is slight. It is also almost entirely beneficial.

From the beginning of the industrial revolution, progressively more advanced tools have served two main purposes. One is simply to increase productivity. In the earlier part of the machine age, that purpose was overwhelmingly dominant. Human impacts were not of much concern, except to the occasional philanthropist or Luddite. In addition, the effect of the new factories on workers was often, perhaps even mostly, negative. Former craftsmen found their jobs deskilled and their working conditions worse than those of their grandparents.

Gradually, the other main purpose of mechanization grew in importance—to spare humans first from dangerous work, then from physically degrading work (loading bananas onto ships, for example, from which the Honduran stevedore wound up with a permanently deformed shoulder), and finally from boring and repetitive work. The advantage of folding-and-stuffing machinery is not merely that it can prepare mailing envelopes faster, but that it saves human workers from the ennui of an essentially mindless task.

The robot is the logical culmination of the industrial revolution. Here at last is the truly autonomous tool. Process—as far as human workers are concerned—is reduced to a minimum; product is expanded to a maximum. Even the relatively simple models now in existence can and do save human beings from dangerous, physically degrading, and boring, repetitive work. As regards danger, a familiar example is the spray painting robot now used by most automobile factories. For human workers, an 8-h shift of spray painting constitutes hazardous duty since it is nearly impossible to avoid breathing a few of the fine droplets of paint. An industrial robot, of course, has no lungs or need to breathe and may spray paint in perfect safety, one shift after another. Other examples abound: police robots that deactivate bombs, undersea robots, space robots, and robots such as the French RM3 designed to wash windows on tall buildings. Concerning degradation, the loading and palletizing robots now present in many factories spare workers the jobs of being mere human beasts of burden. As for rote work, assembly robots, such as the two-robot team in Japan that can put together about 10,000 tubes of lipstick during one 8-h shift, are just beginning to appear in quantity. It is estimated that the two Japanese robots save about 20 human beings from doing almost wholly uninteresting work.

Because the present number of robots is so small, there seems to be no corresponding disadvantage to offset these advantages. The common fear with new technology is that it will produce human redundancy in the form of unemployment. Robots have not done this so far. In Japan, where their numbers are at present largest, robots have unquestionably eliminated tens of thousands of human jobs—but they have not produced unemployment. Japanese unions are structured by company rather than by craft; employees whose previous jobs have been taken over by robots can expect simply to be reassigned within the company. So far, such work has always been available. There has therefore been little or no worker resistance to automation in Japan.

In the United States and other countries where unions are organized by type of work, the potential for robot-caused unemployment is obviously higher. There has nevertheless been very little of it so far. The large corporations, which employ most of the robots, have generally chosen to reduce their human work force by attrition rather than by firing people. According to one study, only 5% of robot-displaced workers in the United States have actually wound up unemployed, and the study predicts that this low figure will continue to prevail in the future (1,2).

It can be argued, of course, that this is too narrow a focus and that there are young unemployed people who *would* have been hired if robots did not exist. The counterargument is that so far the development of robots has created at least as many new jobs as it has eliminated old ones. These new jobs, to be sure, almost invariably require higher levels of training, perhaps even of intelligence, than the jobs currently being eliminated by robots, so it is not clear that a given unemployed person who has lost the opportunity for an old-style job will be able to benefit from the availability of a new-style one. Even granting this, it remains the case that the net effect of robots so far has been almost entirely beneficial.

The short-term outlook is also bright, although there are several dark spots looming. Robots will continue to expand production and to take over tedious and dangerous work. Increasingly, they will be able to leave the special environment of the factory floor. They are scheduled, to replace the migrant workers who pick citrus fruit in the 1990s and will probably replace most other farm workers as well. (A sheep-shearing robot is now in the testing state, and a robot farm tractor was patented in 1985.) They are just beginning to enter the building trades. Before the year 2000, they are expected to saw boards, hammer nails, lay bricks, and so on. More important, robots, which already do much of the assembly of computers, will be able to manufacture other robots and thus drastically reduce their own cost. Historically, computing costs have dropped about 20% a year, and in the most recent year for which statistics are available, the cost of industrial robots dropped by the same percentage. As robots build other robots, prices may drop even faster. That will produce first a rapid increase in the number of robots (the Japan Industrial Robot Association currently predicts that the number of Japanese robots will reach 1 million just before the year 2000 (3), and its director believes the United States will reach the million mark even sooner), and second a gradual and continuing drop in the cost of manufactured goods of all kinds. One of the few U.S. government officials to give serious thought to the future of robots has predicted that manufactured goods will eventually "cost only slightly more than the raw materials and energy from which they are manufactured" (4).

Before the twentieth century ends, however, robots will be doing more than working in factories and on farms and building sites. The household or domestic robot will join the industrial robot as a major presence. Domestic robots are not likely to be the android figures of science fiction or larger versions of the toy robots currently on sale. That is, they are not likely to be general-purpose creatures, as human beings are. Instead, most of them will be "smart" versions of household appliances, such as refrigerators, stoves, vacuum cleaners, and so on, all coordinated by a robot majordomo. This probability is discussed in some detail in the book *Silico Sapiens* (5). The principal human impact of domestic robots will, of course, be a vast reduction in the amount of housework performed by human beings. Cooking, for example, will become a sort of optional hobby; in the middle class, cleaning is not likely to remain a human activity at all.

By the year 2000, there will also be growing numbers of service robots. They will work in fast-food restaurants and as clerks in stores. CheckRobot, a prototype replacement for human cashiers in supermarkets, is currently in limited production. In short, before the century ends, robots will be a major presence on this planet, although there will still be many more human beings than robots. In advanced industrial societies, robots will have taken over nearly all the dangerous, degrading, and boring work. They will also have become sufficiently inexpensive and reliable to have begun to penetrate less industrialized societies.

Two problems seem to be inherent in the changes that are now occurring. One is that robots do not hold much short-term promise for the third world, but rather seem to pose a threat. This is true whether the penetration is extensive or slight. A primary characteristic of third-world countries is the presence of a youthful and rapidly growing population. Typically, half of all the people in such a country are 15 years old or younger.

If robots penetrate such a society extensively, which would mean their widespread domestic use by the upper classes and a more limited use in factories and large-scale farming, they will simply produce increased unemployment among the great mass of unskilled (technologically speaking) and uneducated people, and probably a further drain of capital. However, the penetration before the year 2000 is likely to be slight because of a lack of electricity and maintenance facilities. In that case, unpaid robot labor in the industrialized countries will be competing with low-paid labor in the third world and, of course, winning the competition. The present steady movement of factories to cheap human labor regions will be halted and perhaps even reversed; the lowest labor costs will be in the regions of highest technology. Either way, the probable effect of robotization will be to widen the existing rift between rich and poor nations and to devalue still further the principal asset of most of the poor nations—their large potential work forces.

The other problem is one that will affect the developed nations themselves, the ones with the most robots and highest technology. There is, of course, no way to limit robots to dull and dangerous jobs, or even to the kind of activity one calls "work." Before the year 2000, robots will not only be holding some of the interesting jobs, they will also be doing a good deal of playing. Examples of interesting work that robots (or in some cases, expert programs) will take over in whole or in part are photography, medical diagnosis, and engineering design. "Smart" cameras already exist in profusion, although they are not mobile and require human transporters. However, even the most clever camera operates in the traditional mode in that it must be physically present at the scene it is intended to record. In the next stage, which is just beginning now, physical presence is no longer required, and industrial photographers are beginning to lose work to computer simulation. A shipping company, for example, wishing a photograph of one of its ships in a particular harbor, can sail the ship there and hire a photographer. Or, it can utilize existing separate pictures of the ship and the harbor and have a computer create a true, in-depth single picture. Similarly, an expert program currently exists that can perform some aspects of diagnosis as well as the average internist. Computer-aided design is just beginning to metamorphse into computer design, as in the case of General Electric turbines. "What took the senior designer six weeks to do manually and two weeks to complete on an interactive graphics system is now accomplished by the computer in a mere five minutes" (6).

In no case are human beings likely to be eliminated from any interesting line of work in the near future, but the number able to be photographers, physicians, or designers on a commercial basis *is* likely to be reduced. Insofar as human beings derive meaning and a sense of personal worth from their work (which so far in human history has been to a very large extent indeed), some loss is necessarily involved.

As for play, the current test case is chess. More than 20 years ago, Norbert Wiener, the founding father of cybernetics, made a gloomy prediction about that game. "I find it to be the general opinion of those of my friends who are reasonably proficient chess players that the days of chess as an interesting human occupation are numbered. They expect that within from 10 to 25 years, chess machines will have reached the master class, and then . . . [chess] will cease to interest human players" (7).

It is still too soon to say if Wiener's prediction is current. A small number of chess programs *have* reached the master class, and human interest in the game has not perceptibly diminished. On the contrary, it may have increased. There is widespread interest in the kind of inexpensive home machine that, although no Grand Master, nevertheless plays better chess than 99+% of all human beings, and that can be set to play at various skill levels. An ambition to be able to play the machine at successively higher levels seems to animate and interest many human beings.

If and when a robot becomes the best player on earth (as happened in backgammon when the robot called Gammonoid beat the newly crowned world backgammon champion four games to one in 1979), there may indeed be a falling off of human interest. As robots eventually become masters of most or all sports, there may be a general loss of human interest and pleasure. On the other hand, human beings may simply choose to ignore their mechanical superiors. Humans have a known capacity for selective attention. There remains intense and even growing interst in marathon racing, even though at any time since the day of the battle of Marathon a horse could have covered the distance in a shorter time than the fastest runners. In recent times, numerous machines have come into existence that can traverse the course 10 or 20 times faster than a human being.

The unknown element is that robots, although not at all necessarily humanoid in form, have historically developed as replacements for human beings rather than as human tools or as a clearly distinct separate species, and hence may threaten our diversions in a way that animals and earlier machines have not. By the end of the century, it should be clear whether robots will be regarded as rivals.

Finally, there are the effects that robots are likely to have in the twenty-first century. There is no present agreement on what the robots themselves will be like, let alone their human impact. Two views, however, predominate: that robots will have achieved what is sometimes called human equivalence and that although they will continue to expand their abilities rapidly, they will remain mere machines into the indefinite future. If the first view is correct, there will probably be a new form of human existence quite unlike any that has ever previously existed and possibly the supercession of human species' dominance on this planet. If the second view is correct, the changes will be less dramatic. What would then be likely to occur is the evolution of a leisure society, not unlike some that have existed in the past. It would obviously be on a far grander scale, and there would be a couple of key differences, one of them being universality. All, or nearly all, human beings would play the roles that in previous leisure societies were played only by members of the upper class; robots would serve as the working class and, to some extent, as management. This would eventually become true worldwide, so that the distinctions between first, second, and third worlds would become minimal, at leat in terms of living standards.

Both views deserve examination.

Human equivalence is something robots now conspicuously lack. Artificial intelligence (AI) in control of a mechanical body can command such speed and accuracy of numerical thought and such tireless movement of end effectors as no human being can hope to equal. Thanks to insertable programs, they can instantly "learn" skills that a human being might require years to pick up, or be unable to acquire at all. Offsetting these strengths, current robots lack common sense, imagination, good eyesight, and the ability to travel with speed on other than predetermined paths. They also lack, in all but a few trivial ways, any awareness of their own existence and exhibit no trace of emotion.

One school of thought holds that in about 20 years, robots will have acquired almost all these abilities, be able to do almost anything that human beings can, and still retain their vastly superior speed at abstract numerical thought, their tirelessness, and their virtual immortality. (Because machine parts can be replaced ad lib, machines can continue to be repaired indefinitely. Even now there exist working machines several hundred years old, including a few primitive automata.) There is another branch of the school, whose members believe that robots will indeed acquire all these abilities, but after a much longer period of time.

If robots really do acquire human equivalence, and hence become self-aware beings with the ability to reproduce themselves unaided, the impact on humans is apt to occur in one of two forms. One possibility, imagined more than 40 years ago by the English science-fiction writer Arthur Clarke, and in a different form by his American contemporary Isaac Asimov, is that robots will emerge as independent beings, superior to our race, but well disposed toward it. The residual effect of their programming and the human-centeredness of their original design—that is to say, their evolutionary history—will make them a race of guardian angels, or even demigods. They may even be able to make human beings into immortal cyborgs. The final role of human beings will then be that of any group existing in an achieved paradise: to enjoy living in a protected, essentially unchanging environment and to do very little. There would be very little that needed doing. As Hans Berliner (who wrote the winning backgammon program) has said, "Eventually, almost all important tasks will be better done by machines than by people" (8). Human beings would be free to devote their lives to learning and maximizing awareness. The effect might be somewhat like going to college for life instead of for a few preparatory years.

The other likely possibility, if robots achieve human equivalence, is that they will not remain human-centered. "Advanced robots will eventually be capable of their own view of the world," Joseph Deken has written (9). The world view and the emotions (if any) of a crystalline intelligence are, of course, virtually unimaginable, although generations of science-fiction writers have made the attempt. The one point on which they agree is that independent machines need not feel benevolent toward the organic creatures who developed them and with whom they find themselves sharing the universe. The handful of roboticists and AI experts who foresee self-conscious, highly intelligent robots and who have considered the human implications, tend to take the same view. A well-known joke, commonly attributed to either Marvin Minsky or Robert Jastrow, goes, "If we're lucky, they may decide to keep us as pets." Deken, quite seriously, proposes that the robots are kept as unwilling slaves, precisely in order to avoid dominance by them. "They are our tools and creations, to be kept in place as a subservient species by whatever methods we find necessary," he has written (10). If robots turn out to be equal or superior to men in general ability, however, it may be impossible to keep them subservient; as the emergence to equality occurs, it may also seem unjust. In the 1980s, it is already a strongly held view among some AI students that to disconnect a self-conscious machine would constitute murder.

It is nearly impossible to predict, then, what the consequences for humanity would be if robots were to emerge as the second highly intelligent life form on this planet. However, whether good, bad, or some unguessable combination, robots will have consequences of immense significance.

The other long-range possibility, and the one more widely held by roboticists themselves, is that robots will not achieve self-consciousness, come to feel emotions, or even learn to hink in the full human sense of that term. Should this be e case, there will be no question of using ruthless methods keep them subservient, any more than there is with, say, dozers. There may, however, be problems in adapting to ure society. This leisure would not be complete if robots to fail to achieve human equivalence, as people would retain much of the work of raising their children and also hold many professional, service, and managerial ere is indeed one school of thought that holds that a re society will not evolve in the twenty-first century ew jobs for human beings will open up as fast as nes are filled by robots. "I think there is an almost otential for growth in interpersonal work," O. B. as written, "and I believe the interpersonal field ost likely to absorb workers displaced by machines oduction and service economices." He goes on to although there have long been "counselors, lawyers, ts, rehabilitation experts, politicians, diplomats, rs, and artists," there have never been enough, "and y access to their services has traditionally been limited he wealthy" (11). In the twenty-first century, their services ll be available to everybody, he thinks, and will utilize almost the entire labor pool.

There are several objections to this prediction. One is that clearly not all human beings are temperamentally suited to be lawyers, diplomats, psychiatrists, and so on. Many are better suited to working with their hands in the kinds of jobs that robots will have taken over almost entirely. Another is that the demand for interpersonal workers is not entirely limitless. For example, of the types Hardison lists, there is currently a sufficient number of lawyers and diplomats, and a generous surplus of artists. The number of people seeking to work as actors or musicians or trying to sell their writing or their painting far exceeds market capacity; this discrepancy can only be further increased by advances in robotics. The very primitive robots that impersonate the U.S. presidents at Disney World replace about 40 human actors; the machine called the synclavier is said to have displaced a substantial fraction of all U.S. commercial musicians.

However, the main objection is that it seems to underestimate both the ingenuity of engineers and the speed with which mechanical developments can now occur. Growth in computing power over the past 40 years has consistently outrun predictions, so that statements made as recently as the 1960s already have a quaint air. Z. Rovenskii, writing a "philosophical essay" on robots for a Soviet audience in 1960, concluded, "A cybernetic device neither now nor in the future can be regarded as a brain" (12). Such a conclusion is infinitely less clear now. As recently as 1967, a U.S. robotics expert could doubt that during his lifetime robots would master any game more complicated than checkers. "It's one thing to hope for a machine able to play chess, and quite another to make it actually do so" (13). To go from playing chess to assisting in therapy is no enormous leap. Even without consciousness, robots will be able to do many interpersonal jobs, although the personalities they use will, of course, be those designed for them by human beings.

By the early decades of the twenty-first century, human leisure is apt to be considerable. There will not be enough full-time jobs to go around. One often-discussed possibility is to redefine full-time work to mean a 20-hour week, or eventually even a 10-hour week. The idea is both to preserve the concept of paid jobs for all who seek them ("full employment") and to utilize the workplace, as has been traditional, for the distribution of income. Such ideas miss the main point, however. There is no necessity to distribute income through the workplace. Nor is there any need to distribute it through some form of investment, such as what Albus imagines when he proposes that displaced human workers be given the opportunity "to own the equivalent of one or two robots" (14). A much more direct system could be used. As long ago as 1888, Edward Bellamy imagined, in *Looking Backward,* a system (the book shows it operative in the year 2000) in which every human being would simply receive a credit card at birth, good for a fixed income throughout life. A variant on such a system seems to be coming into existence in The Netherlands. "We believe you must give people a basic wage, and let them choose whether or not to work," one of the aldermen of Amsterdam said recently (15).

The main point, of course, is whether work is a burden from which human beings rightly strive to free themselves or whether it is the central and necessary ingredient for a satisfactory life. "Work" is defined here as an activity in which there is considerable external compulsion to perform. It thus stands in opposition to freedom. A life consisting of nothing but assigned work is commonly referred to as slavery; a life consisting wholly of leisure is a life of perfect freedom. No human being doubts that slavery is undesirable; the question of freedom is more complex.

There is no clear evidence that human beings can or cannot endure a life of perfect freedom. Leisure classes in past aristocratic societies have tended to apply compulsion to themselves, so that their members did in fact work often, generally as warriors and political leaders. They merely avoided work performed primarily for money. In simpler modes of human organization, such as in tribes of hunter–gatherers, there seems to have been no sharp distinction between work and play. Such tribes still exist in small numbers, and they demonstrate a great capacity for enjoying leisure. In one tribe that has been closely studied, the hunters sometimes lounge about for months at a time. "During these periods, visiting, entertaining, and especially dancing are the primary activities of men" (16).

On the other hand, it is the hunt punctuating this leisure that gives the hunters their status, and it is a combination of danger and skill that they perceive to be the essence of the hunt (along with the fact that they are providing the tribe with necessary food). If robots were to free hunters from the necessity of hunting, it is not clear that they would relish the additional time for visiting and dancing. Indeed, the experience of people such as the Eskimos points in the opposite direction. Numerous groups of Eskimos have been freed from their traditional, arduous methods of hunting and are currently living more or less Bellamy style—that is, on a fixed income from the government. One result is a great deal of drunkenness. The Eskimos have not, of course, been prepared or trained for lives of perfect leisure, any more than lottery millionaires (a majority of whom return to work within a year or two of winning), and their lives might be very different if they had access to enough counselors, therapists, rehabilitation experts, and teachers. What seems more likely is that things like danger and arduousness can be eliminated from human life only at a severe cost in meaning.

It is possible, to be sure, that danger and arduousness can be sufficiently provided through play. The sport of mountain climbing, for example, provides generous doses of both, and climbing is more common now than it was in the days before machines such as helicopters provided an effortless way of reaching mountaintops. It has indeed been the practice of leisure classes, past and present, to seek out dangerous and arduous sports, from big game hunting (before the big game big gun rendered it too safe and easy) to hang gliding. However, it is not clear that all human beings can be satisfied by playful or voluntarily chosen danger, any more than all can be satisfied by therapeutic jobs. To some human beings, the playful and voluntary seems insufficiently serious, even insufficiently real.

Some evidence from sources other than Eskimos and hunter–gatherers is available as well. One striking piece of it is a study that tracked several hundred male Bostonians over a period of about 35 years—from the time they were boys just entering adolescence until they became middle-aged men (17). The study began in the 1940s and continued into the 1980s. It included original interviews of the boys, their parents, and their teachers and reinterviews of the grown men at ages 25, 31, and 47.

Those men who had done chores and held part-time jobs while they were growing up turned out to live longer, to earn more money, and to consider themselves happier than those who had not done any work as adolescents. They even had better marriages. This tendency cut across lines of intelligence, ethnic background, and quantity of education. The psychiatrist who directed the study during its later years and who wrote up the results believes that the boys who did not have any work to do failed to acquire the same sense of themselves as worthwhile members of society as those who did. He further associates work with a sense of oneself as a competent person.

More than a hundred years ago, a president of the United States foresaw the problem discussed here, although not, of course, the role that robots would play in it. Speaking in 1880, President Garfield commented that the whole history of the human race divides neatly into two chapters (18): "first the fight to get leisure; and then the second fight of civilization—what shall we do with our leisure when we get it?" As the development of robots accelerates, the need for an answer to that question seems likely to become ever more pressing.

BIBLIOGRAPHY

1. *The New York Times,* Sec. 3, 1 (Jan. 5, 1986).
2. *The New York Times,* Sec. 3, 16 (Mar. 2, 1986).
3. K. Yonemoto, *Robotization in Japan: Socio-economic Impacts, paper,* Japan Industrial Robot Association, Nov. 1985, p. 20.
4. U.S. Office of Technology Assessment, *Exploratory Workshop on the Social Impact of Robotics,* Washington, D.C., 1982, p. 79.
5. J. Deken, *Silico Sapiens: The Fundamentals and Future of Robo*[illegible] Bantam, New York, 1986, pp. 17–35.
6. R. Howard, *Brave New Workplace,* Viking, New York, 19[illegible] 14.
7. N. Wiener, *God and Golem, Inc.,* M.I.T. Press, Cambridge, [illegible] 1964, p. 24.
8. H. Berliner, letter to the author, Dec. 18, 1985.
9. Ref. 5, p. 214.
10. Ref. 4, p. 14.
11. O. B. Hardison, "Education for Utopia," in L. H. La[illegible] *High Technology and Human Freedom,* Smithsonian P[illegible] ington, D.C., 1986, pp. 36, 37.
12. Z. Rovenskii and co-workers, *Machine and Mind* (M[illegible] Office of Technical Services, U.S. Dept. of Commerce, [illegible] D.C. 1961, p. 109.
13. A. J. Cote, *The Search for the Robots,* Basic Books, [illegible] 1967, p. 209.
14. J. Albus, "The Future of Robotics," in V. D. Hunt, ed., *In*[illegible] *Robotics Handbook,* New York, 1983, p. 356.
15. R. Reeves, "The Permissive Dutch," *New York Times Magaz*[illegible] 34 (Oct. 20, 1985).
16. N. H. Cheek and W. R. Burch, *The Social Organization of Leisu*[illegible] *in Human Society,* Harper & Row, New York, 1976, p. 82.
17. G. Vaillant, *The Natural History of Alcoholism,* Harvard University Press, Cambridge, Mass., 1983, pp. 255–258. Dr. Vaillant's view is amplified in *The Reader's Digest,* 95–96 (Jan. 1986).
18. P. F. Douglass and R. W. Crawford, "Implementation of a Comprehensive Plan for the Wise Use of Leisure," in J. C. Charlesworth, ed., *Leisure in America,* American Academy of Political & Social Science, Philadelphia, Pa., 1964, p. 47.